Comprehensive Organic Chemistry Experiments for the Laboratory Classroom

Edited by

Carlos A. M. Afonso
University of Lisbon, Portugal
Email: carlosafonso@ff.ul.pt

Nuno R. Candeias
Tampere University of Technology, Finland
Email: nunorafaelcandeias@tut.fi

Dulce Pereira Simão
University of Lisbon, Portugal
Email: dulcesimao@tecnico.ulisboa.pt

Alexandre F. Trindade
University of Lisbon, Portugal
Email: dulcesimao@tecnico.ulisboa.pt

Jaime A. S. Coelho
University of Lisbon, Portugal
Email: jaimeacoelho@ff.ul.pt

Bin Tan
SUSTC University, Shenzhen, China
Email: tanb@sustc.edu.cn

Robert Franzén
Tampere University of Technology, Finland
Email: robertfranzen@tut.fi

ROYAL SOCIETY OF CHEMISTRY

THE QUEEN'S AWARDS FOR ENTERPRISE: INTERNATIONAL TRADE 2013

Electronic supplementary information (ESI) is available for this book.
Please see http://pubs.rsc.org/en/content/ebook/978-1-84973-963-4 for details.

Print ISBN: 978-1-84973-963-4

A catalogue record for this book is available from the British Library

Translation from the Portuguese language edition:
100 Experiências de Química Orgânica by CARLOS A. M. AFONSO, DULCE PEREIRA SIMÃO, LUÍSA PINTO FERREIRA, MARIA ELISA DA SILVA SERRA, MARIA MANUELA MARQUES RAPOSO

Published by The Royal Society of Chemistry,
Thomas Graham House, Science Park, Milton Road,
Cambridge CB4 0WF, UK

Registered Charity Number 207890

Visit our website at www.rsc.org/books

Printed and bound in the United States of America

Preface

The consolidation of skills in organic chemistry requires learning quite extensive concepts concerning functional groups, molecular structure, properties and reactivity for which laboratory practices are essential. Additionally, laboratory practices frequently lead to many hurdles such as those related with handling of moisture sensitive compounds or purification of mixtures, for which a proper understanding of the reaction intervenients is needed. The manipulation of organic compounds can be a very challenging task, and very skilled organic chemists are often pivotal to performing a needed synthetic modification. Fortunately, this has not prevented researchers from different fields of science from embracing organic chemistry and using their skills in the preparation of new organic compounds, new materials or even in the modification of complex biomolecules such as proteins. In this context, strengthening students from different degrees with synthetic skills can lead to a strong and rewarding asset for their professional prospective. To consolidate skills and knowledge gathered from books, tutors from different courses implement experiments that they consider appropriate toward achieving the desirable learning essentials according to the targeted teaching goals, laboratory availability and costs. This consolidation is typically achieved by selecting, adapting and/or improving experiments selected from books, specific educational journals and other resources, namely scientific journals. *Comprehensive Organic Chemistry Experiments for the Laboratory Classroom* (COCELC) contains 233 experiments covering a range of synthetic transformations from introductory to advanced level. It aims to remedy the lack of experimental diversity of existing sources of organic chemistry experiments for teaching. These experiments were tested in teaching laboratory environments and result from the contribution of authors from universities, colleges and other higher education institutions worldwide.

The optimisation of many laboratory experiments in terms of costs and time have allowed the introduction of organic chemistry laboratory courses in many degrees beyond chemistry majors. The transversal character of organic chemistry, spanning the synthesis of natural products to materials science, is vividly represented by the collection of experiments gathered in this book, focusing on many varied topics such as medicinal chemistry, natural products, green chemistry, polymerisations or asymmetric synthesis.

For each experiment the following details are provided: duration, hazard level, difficulty level, target students, functional groups involved, concepts involved, required chemicals, equipment and experimental techniques, background, complete experimental procedure, safety

Comprehensive Organic Chemistry Experiments for the Laboratory Classroom
Edited by Carlos A. M. Afonso, Nuno R. Candeias, Dulce Pereira Simão, Alexandre F. Trindade, Jaime A. S. Coelho, Bin Tan, and Robert Franzén.

Published by the Royal Society of Chemistry, www.rsc.org

information, interpretation of results, questions for discussion with students and specific comments and advice from the authors on getting the best outcome from the experiment.

Gathering experiments from very different countries and institutions with varied curricula in their courses demanded the creation of a set of rules to ease the normalization of the experiments into different categories regarding hazard, difficulty and study levels, as follows:

Hazard Level

- *Low*: no special measures are needed for the manipulation of the chemicals or the equipment involved in the experiment. Nevertheless, basic safety measures, like the use of a lab coat and safety goggles, should be considered at all times.
- *Moderate*: some of the chemicals used in these experiments are known to be toxic or flammable. The use of fume hoods is advisable. This level can also mean that these experiments may require the handling of some equipment that requires particular attention. For instance, the use of a rotatory evaporator or any vacuum system.
- *High*: such experiments deal with the manipulation of highly toxic chemicals or solvents. They should be executed by students with proper training and who are aware of the risks involved in the handling of such chemicals.

Difficulty Level

- *Low*: the experiment can be easily performed by all kinds of students, since it does not involve the manipulation of hazardous chemicals or any techniques besides elementary ones such as extractions, distillations or crystallizations. The product characterization for these experiments is usually performed by the melting point, infra-red spectrum or some physical aspect of the final product.
- *Medium*: some considerations regarding the procedure should be taken into account to successfully perform the experiment. Some specialist knowledge such as the influence of stirrer speed on the reaction outcome, or the crystallization rate of the final product are examples. This information is given by the author of the experiment to the instructor in the ESI. This level may also relate to the use of equipment for which the student should be previously trained, for instance HPLC or interpretation of NMR spectra.
- *High*: this type of experiment deals with more advanced laboratory techniques. The manipulation of moisture-sensitive chemicals and the need for anhydrous conditions to perform the reactions are examples of what can be found in these experiments. Preparation of compounds in which purification by column chromatography is needed also falls into this category.

Level of study: although it is assumed that students in organic chemistry usually start their studies by having a general overview of the preparation and reactivity of common functional groups, intermediate level experiments can be performed by elementary students, if emphasis on synthetic organic chemistry was given in teaching.

- *Introductory*: recommended for teaching classes of basic organic chemistry, for students that are dealing with organic chemistry for the first time in the laboratory (*e.g.* for early bachelor students). Usually, this represents the manipulation of general functional groups.
- *Intermediate*: recommended for students that have taken previous courses in organic chemistry and feel comfortable with unitary operations, but that are still learning about transformation of functional groups (*e.g.* for final-year bachelor students). Usually, this represents the transformation of more than one functional group, and the preparation of specific molecules. NMR spectral analysis of the products is usually involved.

- *Advanced*: recommended for students that have taken advanced courses in organic chemistry (*e.g.* for masters and PhD students). Experiments regarding the preparation of more demanding molecules and the use of recently developed transformations are focused on.

This book would not have been possible without the commitment and work of the authors of each experiment. They are deeply acknowledged for their contributions. More importantly, students from the different teaching institutions that were anonymously involved in the validation, reproduction, optimisation or creation of these experiments are equally acknowledged.

Carlos A. M. Afonso, Nuno R. Candeias, Dulce Pereira Simão, Alexandre F. Trindade, Jaime A. S. Coelho, Bin Tan and Robert Franzén

Contents

1. **Separation and Purification of Mixtures** 1
1.1. Separation, Purification and Identification of the Components of a Mixture 1
Abel J. S. C. Vieira and Elvira M. S. M. Gaspar
1.2. Isolation of (+)-Limonene from Orange Oil 6
João P. Telo
1.3. Isolation of Plant Pigments from Green and Red Leaves 9
Alice M. Dias and Maria La S. Ferreira
1.4. Extraction of Usnic Acid from Lichen 14
Dietmar K. Kennepohl
1.5. Thin-Layer Chromatography of Plants Pigments 18
Ana Margarida Madureira and Maria-José U. Ferreira
1.6. Isolation of Cinnamaldehyde from Cinnamon 23
Raquel F. M. Frade, Dulce Simão, and Carlos A. M. Afonso
1.7. Isolation and Structural Identification of Piperine, the Major Alkaloid of Black Pepper 26
Andreia Mónico, Angela Paterna, Mariana A. Reis, Ricardo Ferreira, Noélia Duarte, Ana M. Madureira, and Maria-José U. Ferreira
1.8. Caffeine Extraction from Tea and Coffee 30
Ana Margarida Madureira, Abel J. S. C. Vieira, and Maria-José U. Ferreira
1.9. Isolation of the Alkaloid Lupanine from *Lupinus albus* Seeds 35
Ana M. Lourenço and Luísa Maria Ferreira
1.10. Isolation and Purification of Atropine, a Tropane Alkaloid Obtained from *Atropa belladonna* L. (Solanaceae) 38
Gustavo da Silva
1.11. Isolation and Purification of Carnosol from *Salvia officinalis* 43
Filipe Pereira and Patrícia Rijo

Comprehensive Organic Chemistry Experiments for the Laboratory Classroom
Edited by Carlos A. M. Afonso, Nuno R. Candeias, Dulce Pereira Simão, Alexandre F. Trindade, Jaime A. S. Coelho, Bin Tan, and Robert Franzén.

Published by the Royal Society of Chemistry, www.rsc.org

1.12. Analysis of Racemic and (*S*)-Ibuprofen 46
C. Fernandes, H. Cidade, S. Cravo, E. Sousa, and M. Pinto

1.13. Determining Partition Coefficients of Sulfonamides by Reversed-Phase Chromatography 51
E. Sousa, S. Cravo, C. Fernandes, and M. Pinto

2. Modification of sp^3 Carbon **56**

2.1. C-X Bond Formation 56

2.1.1. A S_N1 Reaction: Synthesis of *tert*-Butyl Chloride 56
Emília Valente, Catarina Dias, and Luís Constantino

2.1.2. Optimizing the Reaction Conditions for the Synthesis of *tert*-Pentyl Chloride 61
Jane Brock Greco

2.1.3. Kinetics of a S_N1 Reaction 64
Paulo Coelho and Céu Sousa

2.1.4. Counterion Effects in the Nucleophilic Substitution Reaction of the Acetate Ion with Alkyl Bromides in the Synthesis of Esters 68
Ingrid Montes, M. Elizabeth Valentín, and Waldemar Adam

2.1.5. *N*-Alkylation of Pyrazole: Reaction in an Ionic Liquid 72
Clarissa P. Frizzo, Marcos A. P. Martins, Caroline R. Bender, Paulo R. S. Salbego, Aniele Z. Tier, Geórgia C. Zimmer, Guilherme C. Paveglio, and Helio G. Bonacorso

2.1.6. Conversion of Alcohols into Alkyl Chlorides Using Cyanuric Chloride 76
Eliana Martins, Catarina A. B. Rodrigues, and Carlos A. M. Afonso

2.1.7. Synthesis of Phenacetin 80
Paulo Coelho and Céu Sousa

2.1.8. One-Step Synthesis of 4(3*H*)-Quinazolinones: An Important Heterocyclic Scaffold in Medicinal Chemistry 84
André Dias, Rui Moreira, and Ana S. Ressurreição

2.1.9. Controlled Monoalkylation of the Structurally Rigid Bicyclic System Isomannide 87
M. Manuela A. Pereira

2.1.10. Regioselective *N*-alkylation of Adenine by Nucleophilic Substitution 92
Ana I. Vicente and Carlos A. M. Afonso

2.1.11. Gabriel Synthesis of *n*-Octylamine Under Phase-Transfer Catalysis: The First Step 95
Renata Riva and Luca Banfi

2.1.12. Preparation of Diethyl 2,3-*O*-isopropylidene-L-tartrate 99
Naylil M. R. Capreti and Igor D. Jurberg

2.1.13. Redox-Neutral Synthesis of a Cyclic *N*,*O*-Acetal from Salicylaldehyde and 1,2,3,4-Tetrahydroisoquinoline 103
Claire L. Jarvis and Daniel Seidel

2.2. C-C σ-Bond Formation 107
2.2.1. Selective *C*-acylation of 3-Methyl-1-phenyl-pyrazol-5-one 107
Vanya B. Kurteva, Lubomir A. Lubenov, and Maria A. Petrova

2.2.2. Organocatalytic Asymmetric α-Arylation of Aldehydes 112
Mette Overgaard, Pernille H. Poulsen, and Karl Anker Jørgensen

2.2.3. Matteson Homologation of Pinacol Boronic Ester: An Efficient Method Using the Boronate Complexes 117
Sébastien Balieu, Asmaa Bouyahya, and Jean-Philippe Bouillon

2.2.4. A Practical Organocatalytic Alkylation Reaction with Benzodithiolylium Tetrafluoroborate 120
Pier Giorgio Cozzi, Marco Bandini, Andrea Gualandi, and Luca Mengozzi

2.2.5. Oxazolidinone Mediated Enantioselective Allylation 125
Jonathan D. Sellars, AnnMarie C. O'Donoghue, Ian R. Baxendale, John M. Sanderson, and Elizabeth J. Grayson

3. Substitution at Non-aryl sp^2 Carbon 129
3.1. C-X Bond Formation 129
3.1.1. Synthesis of Paracetamol by Acetylation 129
Catarina Dias, Emília Valente, and Luís Constantino

3.1.2. Synthesis and Characterization of *N,N'*-Dicyclohexyl-*N,N'*-dimethyl-propan-1,3-diamide 133
Ana Paula Paiva and Osvaldo Ortet

3.1.3. Synthesis and Characterisation of an Ester from 4-Nitrobenzoyl Chloride 136
Iain A. Smellie, Matt L. Clarke, Leanne Harris, and Andrew J. Miller

3.1.4. Green Esterification: The Synthesis of Aromas in the Presence of an Acid Resin 139
José E. Castanheiro, António P. S. Teixeira, and António M. D. R. L. Pereira

3.1.5. Acetylation of Cholesterol and Purification by Column Chromatography 142
Emília Valente, Catarina Dias, and Luís Constantino

3.1.6. Synthesis of Hippuric Acid: An Example of Amide Bond Formation 147
Ana S. Ressurreição, Maria de Jesus Perry, Ana Paula Francisco, and Francisca Lopes

3.1.7. Preparation of a Sulfathiazole Prodrug *via* *N*-acylation with Succinic Anhydride 151
Pedro M. P. Gois, Francisca Lopes, and Alexandre F. Trindade

3.1.8. Determination of the Absolute Configuration of Enantioenriched Secondary Alcohols *via* Thin-Layer Chromatography 154
Alexander J. Wagner, Shawn M. Miller, Scott D. Rychnovsky, and Renée D. Link

3.1.9. Anhydride Aminolysis: Synthesis of *N*-arylmaleamic Acids 159
Artur M. S. Silva, Augusto C. Tomé, Diana C. G. A. Pinto, Fernando Domingues, Graça M. Oliveira Rocha, José A. S. Cavaleiro, M. Graça P. M. S. Neves, M. Amparo F. Faustino, and Mário M. Q. Simões

3.1.10. Synthesis of *N*-arylmaleimides 162
Artur M. S. Silva, Augusto C. Tomé, Diana C. G. A. Pinto, Fernando Domingues, Graça M. Oliveira Rocha, José A. S. Cavaleiro, M. Graça P. M. S. Neves, M. Amparo F. Faustino, and Mário M. Q. Simões

3.1.11. Synthesis and Characterization of *N*-Cyclohexyl-*N*-methyloctanamide 165
Ana Paula Paiva and Osvaldo Ortet

3.1.12. Effect of a Catalyst in the Acylation of Alcohols with Acetic Anhydride: Manipulation of a Natural Aroma 169
Mónica S. Estevão and Carlos A. M. Afonso

3.1.13. Synthesis of 4,5-Dichloro-1,2-dicyanobenzene 172
Dulce Simão

3.1.14. Synthesis of *N*-*tert*-Butyloxycarbonyl-3-nitro-L-tyrosine Methyl Ester 175
Susana P. G. Costa, M. Manuela M. Raposo, Cátia I. C. Esteves, and R. Cristina M. Ferreira

3.1.15. Michael Addition Reaction Followed by Elimination Under Solvent-Free Conditions 178
Marcos A. P. Martins, Clarissa P. Frizzo, Aniele Z. Tier, Guilherme C. Paveglio, Caroline R. Bender, Paulo R. S. Salbego, Geórgia C. Zimmer, and Nilo Zanatta

3.1.16. Synthesis of a γ-Keto-amide Derived from Thiophene Using a Carboxyl Ester as Precursor 182
M. Manuela M. Raposo

3.1.17. Cyclic Acetals for Regioselective Protection in Carbohydrate Synthesis: A Comparative Experiment 188
Ana M. Matos, Rafael Nunes, Catarina Dias, and Amélia P. Rauter

3.1.18. Direct Diastereoselective Synthesis of the Tetrahydro-thiazolo[2,3-*b*] isoindole Tricyclic Ring System 194
Maria I. L. Soares, Ana Lúcia Cardoso, Susana M. M. Lopes, and Teresa M. V. D. Pinho e Melo

3.1.19. Asymmetric Cyclocondensation Reaction Induced by Chiral Aminoalcohol 198
Jorge Dourado, Maria Pérez, Rosa Griera, and Maria M. M. Santos

3.1.20. Synthesis and Characterization of Biodiesel Propyl Esters to Determine the Fatty Acid Content of Unknown Plant Oils 202
Heidi R. Vollmer-Snarr, Patrick A. Fracisco, and Naiian Saephanh

3.2. C-C σ-bond Formation 206

3.2.1. Synthesis of 7-Methoxy-4-oxo-*N*-phenyl-4*H*-chromene-2-carboxamide 206
Joana Reis, Alexandra Gaspar, Fernando Cagide, and Fernanda Borges

3.2.2. Synthesis of 2,3-Diphenylindenone 212
João P. Telo

3.2.3. Synthesis of Dimedone 216
Dulce Simão and Ana C. Cerdeira

3.2.4. Acylation Reaction of Enol Ether Using an Ionic Liquid 219
Marcos A. P. Martins, Clarissa P. Frizzo, Geórgia C. Zimmer, Caroline R. Bender, Paulo R. S. Salbego, Aniele Z. Tier, and Nilo Zanatta

4. Addition at Non-aryl sp^2 Carbon 223

4.1. C-X Bond Formation 223

4.1.1. Electrophilic Addition to Alkenes 223

4.1.1.1. Bromination of Cinnamic Acid 223
Paulo Coelho and Céu Sousa

4.1.1.2. Preparation of *Meso*-1,2-Dibromo-1,2-diphenylethane 226
Abel J. S. C. Vieira

4.1.1.3. Bromination of (*E*)-chalcones [(*E*)-1,3-Diarylprop-2-en-1-ones] 229
Artur M. S. Silva, Augusto C. Tomé, Diana C. G. A. Pinto, Fernando Domingues, Graça M. Oliveira Rocha, José A. S. Cavaleiro, M. Graça P. M. S. Neves, M. Amparo F. Faustino, and Mário M. Q. Simões

4.1.1.4. Preparation of *trans*-2-Bromocyclohexanol from Cyclohexanol 231
Matheus P. Freitas

4.1.1.5. Synthesis of *trans*-Cyclohexane-1,2-diol 234
Dulce Simão and Ana C. Cerdeira

4.1.1.6. Synthesis of Isobutylene and Its Use in Esterification Reactions 237
Paula C. Castilho, Pedro Ideia, and Rúben Gonçalves

4.1.1.7. Hydroxyl Group Protection *via* Tetrahydropyranyl Ether Formation 241
Ana I. B. Meirinhos, Filipa Siopa, and Carlos A. M. Afonso

4.1.1.8. Synthesis of (−)-Carvone from (+)-Limonene 244
João P. Telo

4.1.1.9. Glycal Transformation into Surfactant 2-Deoxy Glycosides 247
Catarina Dias and Amélia P. Rauter

4.1.1.10. Preparation of (1*R*,2*R*,3*R*,5*S*)-(−)-Isopinocampheol Through a Hydroboration–Oxidation Reaction 252
Marek P. Krzemiński

4.1.2. 1,2-Addition to Carbonyl Compounds 257

4.1.2.1. Synthesis of Fructone 257
Thomas A. Logothetis

4.1.2.2. Synthesis of Flavone 261
Dulce Simão and Ana C. Cerdeira

4.1.2.3. Synthesis and Characterisation of H_2salen: An Introduction to 1D and 2D NMR Spectroscopy 264
Maria J. Villa de Brito

4.1.2.4. Synthesis of Lophine and Conversion into Dimers 269
Dulce Simão

4.1.2.5. Synthesis of Dibenzalacetone 2,4-Dinitrophenylhydrazone 272
Ana Margarida Madureira, Ana Bela Santana, Emília Valente, and Maria-José U. Ferreira

4.1.2.6. Synthesis and Structural Characterization of an Antitubercular Isoniazid Hydrazone 277
Susana Santos and Filomena Martins

4.1.2.7. Preparation of a Tosylhydrazidyl *N*-Glycosyl Derivative of D-Glucuronic Acid *via* Tosylhydrazone Formation and Intramolecular Ring Closure 282
Nuno M. Xavier

4.1.3. 1,4-Addition to α,β-Unsaturated Carbonyl Compounds 285

4.1.3.1. Green Metrics in a Cyclocondensation Reaction 285
Clarissa P. Frizzo, Marcos A. P. Martins, Paulo R. S. Salbego, Caroline R. Bender, Aniele Z. Tier, and Hélio G. Bonacorso

4.1.3.2. Synthesis of 1*H*-Pyrazoles Using Ball Mill, Grinding and Conventional Thermal Heating 290
Clarissa P. Frizzo, Marcos A. P. Martins, Caroline R. Bender, Paulo R. S. Salbego, Aniele Z. Tier, Guilherme C. Paveglio, and Kelvis Longhi

4.1.3.3. Organocatalytic Enantioselective Michael Addition of Thiophenol to Chalcone 294
Mariola Zielińska-Błajet and Jacek Skarżewski

4.1.3.4. Stereoselective Synthesis of *meso*-1-Allyl-2,6-diphenyl piperidin-4-one 298
Ana-Belén García Delgado, Noelia Quiñones, and María-Paz Cabal

4.1.3.5. Synthesis of a Squarylium Cyanine Dye as Potential Photosensitizer for Photodynamic Therapy (PDT) 301
Marlene L. F. M. Pacheco, Sofia F. P. Friães, Renato E. Boto, Paulo Almeida, Amélia M. Silva, and Lucinda V. Reis

4.2. C-C σ-bond Formation 306

4.2.1. Grignard Addition to Carbonyl Compounds 306

4.2.1.1. The Effects of Stoichiometry and Starting Material on the Product Identity and Yield in Grignard Addition Reactions 306
Jane Brock Greco and Eric Hill

4.2.1.2. Synthesis of Methyl Triphenylmethyl Ether 312
Dulce Simão and Ana C. Cerdeira

4.2.1.3. Grignard-Like Reaction in Water 316
João R. Vale and Carlos A. M. Afonso

4.2.1.4. Cram's Rule – Diastereoselective Grignard Addition to 2-Phenylpropanal 319
Laura M. Hancock, Michael G. Edwards, Graeme R. Jones, and Matthew O'Brien

4.2.1.5. Preparation of (4R,5R)-4,5-Bis(diphenylhydroxymethyl)-2,2-dimethyldioxolane ((−)-TADDOL) 322
Naylil M. R. Capreti and Igor D. Jurberg

4.2.1.6. Synthesis of (S)-Diphenyl(pyrrolidin-2-yl)methanol 327
David Monge, Javier Iglesias-Sigüenza, and Elena Díez

4.2.2. Aldol Reaction and Condensation 331

4.2.2.1. Solvent-Free Aldol Condensation Reactions: Synthesis of Chalcone Derivatives 331
Barbora Morra, Connie Tang, and Jonathon Fossella

4.2.2.2. Synthesis of (E)-Chalcones [(E)-1,3-diarylprop-2-en-1-ones] 335
Artur M. S. Silva, Augusto C. Tomé, Diana C. G. A. Pinto, Fernando Domingues, Graça M. Oliveira Rocha, José A. S. Cavaleiro, M. Graça P. M. S. Neves, M. Amparo F. Faustino, and Mário M. Q. Simões

4.2.2.3. A Solvent-Free Approach for Chalcone Synthesis *via* an Aldol Reaction 339
Sofia F. P. Friães, Renato E. Boto, Paulo Almeida, Amélia M. Silva, and Lucinda V. Reis

4.2.2.4. Preparation of Dibenzylideneacetone 342
Abel J. S. C. Vieira

4.2.2.5. L-Proline Catalyzed Aldol Reaction of 4-Nitrobenzaldehyde with Acetone 345
Fang Fang, Xin-Yuan Liu, and Bin Tan

4.2.2.6. Preparation of a β-Nitrostyrene Derivative by the Henry Reaction: Comparison of a Conventional and a Microwave-Assisted Method 348
Daniel Martins, Joana Reis, Fernanda Borges, and Nuno Milhazes

4.2.2.7. Synthesis of Aurone Derivatives Through Acid-Catalysed Aldol Condensation 352
Ana R. Duarte, Marta P. Carrasco, and Ana S. Ressurreição

4.2.2.8. Synthesis of Pyrazole Heterocycles 355
Thomas A. Logothetis

4.2.2.9. A Green Approach to 3-Carbonylchromones 359
Ana Bornadiego, Jesús Díaz, Ana G. Neo, and Carlos F. Marcos

4.2.2.10. Synthesis of Indigo and Dyeing Process 363
Dulce Simão

4.2.3. 1,2-Addition to Carbonyl Compounds Other than Aldol Reaction 366

4.2.3.1. Knorr Pyrazole Synthesis of Edaravone 366
Ana Paula Francisco, Ana S. Ressurreição, Maria de Jesus Perry, and Francisca Lopes

4.2.3.2. Greener Solvent Substitution in a Verley–Doebner Condensation 370
Jonathon W. Moir and Andrew P. Dicks

4.2.3.3. Knorr Synthesis of Diethyl 3,5-Dimethyl-1*H*-pyrrole-2,4-dicarboxylate 373
M. Soledad Pino-González, Cristina García-Ruiz, Francisco Sarabia, and Gregorio Torres

4.2.3.4. Synthesis of Coumarin-3-carboxylic Acid 376
Paulo Coelho and Céu Sousa

4.2.3.5. Synthesis of Lipophilic Antioxidants Based on Natural Models 380
Sofia Benfeito, Tiago Silva, Diogo Magalhães Silva, E. Manuela Garrido, Fernanda Borges, and Jorge Garrido

4.2.3.6. A Simple and Ecological Preparation of a Chromene-3-carboxamide Derivative 387
Marta Costa and M. Fernanda Proença

4.2.3.7. Synthesis of a Biologically Active Oxazol-5(4*H*)-One *via* Erlenmeyer–Plöchl Reaction 391
Catarina A. B. Rodrigues, José M. G. Martinho, and Carlos A. M. Afonso

4.2.3.8. Synthesis of a Long-Wavelength Absorbing Squaraine Dye 396
Paulo F. Santos and Vânia C. Graça

4.2.3.9. Synthesis of π-Conjugated Systems Using Formylchromone as Building Block 399
Carlos Fernandes, Catarina Oliveira, Fernanda Borges, and Alexandra Gaspar

4.2.3.10. A Multi-Step Synthesis of Imidazolin-5-ones 403
Gurunath Ramanathan and Ashish Singh

4.2.3.11. Baylis–Hillman Reaction Between 4-Nitrobenzaldehyde and Ethyl Acrylate 408
Alexandre F. Trindade, João Ravasco, and Carlos A. M. Afonso

4.2.3.12. Synthesis of Ethyl Mandelate Through a Rhodium-Catalysed Arylation Reaction with Ethyl Glyoxylate and Phenylboronic Acid 411
Carolina S. Marques and Anthony J. Burke

4.2.3.13. Enantioselective Synthesis and Derivatisation of 2-Hydroxy-1,2-diphenylethan-1-one 415
Jonathan D. Sellars, AnnMarie C. O'Donoghue, Ian R. Baxendale, John M. Sanderson, and Elizabeth J. Grayson

4.2.4. 1,4-Addition to α,β-unsaturated Carbonyl Compounds 420

4.2.4.1. Preparation of Chalcone and Its Further Robinson Annulation with Ethyl Acetoacetate 420
Nuno R. Candeias

4.2.4.2. Conjugate Addition of Organocuprates to α,β-Unsaturated Ketones: Synthesis of 3,3-Dimethylcyclohexanone from 3-Methyl-2-cyclohexen-1-one 424
Carlos Gregorio, F. Javier Sardina, and Antonio Mouriño

4.2.5. Multicomponent Reactions 428
4.2.5.1. One-Pot Green Synthesis of Dihydropyran Heterocycles 428
Dennis Russowsky and Camila S. Santos

4.2.5.2. Hantzsch Synthesis of Nifedipine 432
Arno Kraft

4.2.5.3. A Ugi Multicomponent Reaction in the Synthesis of *N*-Cyclohexyl-2-[*N*-(4-methoxybenzyl)acetamide]-2-(thien-2-yl)acetamide 435
Susana P. G. Costa, M. Manuela M. Raposo, and Cátia I. C. Esteves

4.2.5.4. Preparation of Phenylglycine and Hydroxymorpholine Derivatives Through a Petasis Borono–Mannich Reaction 439
Nuno R. Candeias, Roberta Paterna, Pedro M. S. D. Cal, and Pedro M. P. Gois

4.2.5.5. A Multicomponent Reaction of Isocyanides for the Synthesis of 4-Chromanone-2-carboxamides 443
Ana G. Neo, Jesús Díaz, and Carlos F. Marcos

5. Electrophilic Aromatic Substitution **448**
5.1. C-X Bond Formation 448
5.1.1. Synthesis of Methyl Orange 448
Dulce Simão and Ana C. Cerdeira

5.1.2. Halogenation Reactions of Vanillin 451
Iain A. Smellie, Nigel P. Botting, Brian A. Chalmers, and Iain L. J. Patterson

5.1.3. Synthesis of 2-(2,4-Dinitrobenzyl)pyridine 454
Dulce Simão

5.1.4. Regioselectivity in the Nitration of Acylanilines by Electrophilic Aromatic Substitution 456
João P. Telo, Pedro P. Santos, João V. Moreira, Mariana L. Santos, and Ana T. Batista

5.1.5. Synthesis of 2-Nitro-4-methylaniline 462
Dulce Simão and Ana C. Cerdeira

5.1.6. Synthesis of 1-Nitronaphthalene 465
Adam Dzielendziak

5.1.7. Selective Boc-Protection and Bromination of Pyrazoles 468
Thomas A. Logothetis

5.1.8. Preparation of *p*-Bromoaniline 472
Abel J. S. C. Vieira

5.2. C-C Bond Formation 477
5.2.1. Synthesis of 1,4-Di-*t*-butyl-2,5-dimethoxybenzene 477
Dulce Simão and Ana C. Cerdeira

5.2.2. Synthesis of 4,4′-Di-*tert*-butylbiphenyl 480
Dulce Simão

5.2.3. Synthesis of a Macrocycle: *C*-Methyl[4]resorcinarene 482
Christopher Baker, Alexander S. Cragg, Raghuram R. Kothur, Flavia Fucassi, and Peter J. Cragg

5.2.4. Synthesis of 5,10,15,20-Tetrakis(2,6-dichlorophenyl)porphyrin 485
Artur M. S. Silva, Augusto C. Tomé, Diana C. G. A. Pinto, Fernando M. J. Domingues, Graça M. S. R. O. Rocha, José A. S. Cavaleiro, Maria da Graça, P. M. S. Neves, Maria do Amparo F. Faustino, and Mário M. Q. Simões

5.2.5. Solventless Synthesis, Separation and Characterization of Zinc and Free-Base Tetraphenyl Porphyrin 490
Waqar Rizvi, Charles M. Drain, Patrick Moy, and Matthew J. Jurow

5.2.6. Synthesis of 3-Bromo-7-ethylcarbamate-4-methylcoumarin 493
Paula S. Branco and Ana Maria Lourenço

5.2.7. Synthesis of Musk Ketone 496
João Paulo Telo

5.2.8. Synthesis of Methyl 4-oxo-4-(thiophen-2-yl)butanoate 500
M. Manuela M. Raposo

5.2.9. Synthesis of Eosin 504
Dulce Simão

5.2.10. Synthesis and Formylation of 5-Piperidino-2,2′-bithiophene 508
M. Manuela M. Raposo

5.2.11. Synthesis of 3-Bromosalicylaldehyde by *Ortho*-formylation of 2-Bromophenol 514
Trond Vidar Hansen and Lars Skattebøl

5.2.12. Synthesis of Indolo[3,2-*b*]quinolin-11-one by Acid-Catalysed Intramolecular Double Cyclization 517
João Lavrado, Marta Figueiras, David M. Pereira, Sofia A. Santos, Rui Moreira, and Alexandra Paulo

5.2.13. Mild and Fast Friedel–Crafts Acylation Over Zeolites 522
A. F. Brigas, F. Martins, R. Elvas-Leitão, B. S. Santos, A. Martins, and N. Nunes

5.2.14. Reactivity Studies of 1-Propyl-2-(thiophen-2-yl)-1*H*-pyrrole 526
Maria Manuela Marques Raposo, Sara Sofia Marques Fernandes, and Maria Cidália Rodrigues Castro

6. Nucleophilic Aromatic Substitution 532

6.1. Synthesis of 2,3-Quinoxalinedithiol 532
Dulce Simão and Ana C. Cerdeira

6.2. Synthesis of the Antitumoral Drug 2,4,6-Tris(dimethylamino)-1,3,5-triazine *via* Sequential Nucleophilic Substitution 535
Filipa Siopa, Nuno Candeias, and Carlos A. M. Afonso

6.3. Benzotriazole, a Useful Synthetic Auxiliary in Heterocyclizations 539
João Lavrado, Marta Figueiras, Eduardo Ruivo, Susana Lucas, and Alexandra Paulo

6.4. Synthesis of 4,5-Dicyanobenzene-1,2-dithiol 542
Dulce Simão

6.5. Nucleophilic Aromatic Substitution Reactions in 3,6-Bis(3,5-dimethyl-1*H*-pyrazol-1-yl)-1,2,4,5-tetrazine 545
Tiago J. L. Silva, Paulo J. G. Mendes, Ana Isabel Tomaz, and M. Helena Garcia

7. Transition Metal Catalysed Substitution 550
7.1. A Solvent-Free Ullmann Coupling: Synthesis of 2,2′-Dinitrobiphenyl 550
Laurel Goj Habgood and Richard W. Gregor

7.2. Reactivity Studies for the Synthesis of 5-Phenylthiophene-2-carbaldehyde by a Suzuki–Miyaura Coupling 554
Maria Manuela Marques Raposo, Susana Paula Graça da Costa, Rosa Maria Ferreira Batista, and Rosa Cristina Moutinho Ferreira

7.3. Recycling Bromovanillin into Ferulic Acid-Based Antioxidants 559
Tiago Silva, Daniel Chavarria, Lisa Sequeira, and Fernanda Borges

7.4. Sonogashira Coupling Reaction of Aryl Derivatives: A Versatile Method for Acetylide Building Blocks 564
Tiago J. Silva, Paulo J. Mendes, António P. S. Teixeira, M. Paula Robalo, and M. H. Garcia

7.5. Convergent Synthesis of a Suzuki Product 568
Thomas A. Logothetis

7.6. Oxidative Heck Reaction at Room Temperature 573
Rajiv T. Sawant, Ashkan Fardost, and Luke R. Odell

7.7. Synthesis of 2-Methyl-1,1′-binaphthalene *via* Suzuki Cross-Coupling Reaction 577
Javier Iglesias-Sigüenza, David Monge, and Elena Díez

7.8. Green Synthesis of Aromatic Ketones: Decarboxylative Palladium Catalysis Under Microwave Irradiation 581
Jonas Sävmarker, Ashkan Fardost, and Luke R. Odell

7.9. An Efficient Methodology for the Synthesis of the 3-Styryl Coumarin 584
António Manuel D. R. L. Pereira and Sérgio Miguel A. Martins

7.10. Sonogashira Coupling Between a Vinylic Halide and a Terminal Alkyne 588
Carine Maaliki and Jérôme Thibonnet

8. Addition to sp Carbon 592
8.1. Synthesis of Dulcin 592
Dulce Simão

8.2. Microwave-Assisted Solid-Phase Synthesis of Hydantoins 595
Isabelle Parrot and Guilhem Chaubet

9. Preparation of Alkenes 600
9.1. Elimination Reactions 600
9.1.1. Preparation of Cyclohexene 600
Abel J. S. C. Vieira and Elvira M. S. M. Gaspar

9.1.2. A Green Synthesis of 2,3-Dibromo-3-phenylpropionic Acid and the Use of Kinetic Studies to Probe into the Elimination Product When Treated with a Weak Base in Different Solvents 603
Malcolm I. Stewart and Craig D. Campbell

9.1.3. Dehydration of Methylcyclohexanols 608
Cornelis A. Van Walree, Stephan A. Jonker, and Veronica Kaats-Richters

9.1.4. Synthesis of 5-Hydroxymethylfurfural (HMF) from Fructose as a Bioplatform Intermediate 614
Svilen P. Simeonov and Carlos A. M. Afonso

9.2. Wittig Type Reactions 617

9.2.1. Synthesis of *trans*-9-(2-Phenylethenyl)anthracene 617
Dulce Simão

9.2.2. Synthesis of 4-Vinylbenzoic acid 620
Dulce Simão and Ana C. Cerdeira

9.2.3. Preparation of Nitrostilbenes by the Wittig Reaction 623
Francisca Lopes, Maria de Jesus Perry, and Ana Paula Francisco

9.2.4. Building an Alkene Spacer by the Wittig Reaction: Synthesis of 4-[2-(4-Nitrophenyl)ethenyl]benzonitrile 627
António P. S. Teixeira, M. Paula Robalo, and Paulo J. Mendes

9.2.5. Preparation of *trans,trans*-Distyrylbenzene by a Wittig Reaction 631
Arno Kraft

9.2.6. Synthesis and Reactivity of Phosphorus Ylides 635
Ana Lúcia Cardoso, Maria I. L. Soares, Susana M. M. Lopes, and Teresa M. V. D. Pinho e Melo

9.2.7. Synthesis of 4-(Thiophen-2-yl)pyrrolidin-2-one *via* Horner–Wadsworth–Emmons Reaction and Reductive Cyclisation 640
Jonathan D. Sellars, AnnMarie C. O'Donoghue, Ian R. Baxendale, John M. Sanderson, and Elizabeth J. Grayson

9.3. Metathesis 646

9.3.1. A Guided-Inquiry Approach to Ring-Closing Metathesis 646
Hala G. Schepmann and Michelle Mynderse

9.3.2. Sequential Pd-Catalyzed Allylic Alkylation/Ru-Catalyzed Ring-Closing Metathesis 650
Mélanie M. Lorion, Claire Kammerer, Guillaume Prestat, and Giovanni Poli

10. Pericyclic Reactions **656**

10.1. Diels–Alder Reaction of *N*-Phenylmaleimide with *In situ* Generated Buta-1,3-diene 656
Artur M. S. Silva, Augusto C. Tomé, Diana C. G. A. Pinto, Fernando Domingues, Graça M. Oliveira Rocha, José A. S. Cavaleiro, M. Graça P. M. S. Neves, M. Amparo F. Faustino, and Mário M. Q. Simões

10.2. Synthesis of Isomeric Bicyclopropyls from Conjugated Dienes 660
Leiv K. Sydnes and Magne O. Sydnes

10.3. Diels–Alder Reaction Between *p*-Benzoquinone and Cyclopentadiene and Subsequent Photochemical [2π + 2π] Cycloaddition 664
Arno Kraft

10.4. Click Chemistry 668
Susana Dias Lucas, Eduardo F. P. Ruivo, Marta Figueiras, João P. Lavrado, and Rui Moreira

10.5. On-Water Synthesis of a Dipyrromethane *via* Bis-hetero-Diels–Alder Reaction of an Azoalkene with Pyrrole 671
Susana M. M. Lopes, Maria I. L. Soares, Ana Lúcia Cardoso, Américo Lemos, and Teresa M. V. D. Pinho e Melo

10.6. Application of 2,4,6-Trioxo-pyrimidin-5-ylidene Alditol in the Synthesis of Pyrano[2,3-*d*]pyrimidine Containing a Sugar Moiety by Hetero-Diels–Alder Reaction 676
Aleksandra Pałasz and Dariusz Cież

10.7. Synthesis of a Spiroisoxazoline Oxindole by 1,3-Dipolar Cycloaddition 681
Carlos J. A. Ribeiro, Rui Moreira, and Maria M. M. Santos

10.8. Oxonitriles: Four-Step Ozonolysis, Aldol, Conjugate Addition, and Enolate Acylation Sequence 685
Jesus A. Lujan-Montelongo and Fraser F. Fleming

10.9. Flash Vacuum Pyrolysis of *o*-Phenylene Sulfite: Formation and Purification of Cyclopentadienone Dimer 690
R. Alan Aitken and Caroline E. R. Horsburgh

11. Radical Reactions 694

11.1. Synthesis of 4-Methoxymethylbenzoic Acid 694
Dulce Simão and Ana C. Cerdeira

11.2. Synthesis of Benzopinacolone *via* Benzophenone Photoreduction Followed by Pinacol Rearrangement 697
Filipa Siopa and Carlos A. M. Afonso

11.3. Iodosulfonylation–Dehydroiodination of Styrene: Synthesis of (*E*)-β-Tosylstyrene 701
Carmen Nájera, José M. Sansano, and Miguel Yus

12. Oxidations 704

12.1. Ketone and Aldehyde Formation 704

12.1.1. Aerobic Alcohol Oxidation Using a Cu(I)/TEMPO Catalyst System 704
Nicholas J. Hill, Jessica M. Hoover, and Shannon S. Stahl

12.1.2. Chemoselective Oxidation of 1,2-Tetradecanediol 709
Siedlecka Renata and Skarżewski Jacek

12.1.3. Catalyzed Oxidation of Naphthalene to 1,4-Naphthoquinone 713
Jacek Skarżewski

12.1.4. Alcohol Oxidation: Menthone Preparation by Menthol Oxidation Using Pyridinium Chlorochromate Immobilized in Silica Gel 716
Rafael F. A. Gomes and Carlos A. M. Afonso

12.1.5. Heterocyclic Target Synthesis – Three-Step Syntheses from Benzaldehyde 720
Iain A. Smellie, Nigel P. Botting, Brian A. Chalmers, Andrew D. Harper, and Iain L. J. Patterson

12.1.6. Oxidation of Activated Phenols 725
Luísa M. Ferreira and Paula S. Branco

12.2. Oxidative Cyclizations and Dehydrogenation 730

12.2.1. Synthesis of 2-(5-Phenylthien-2′-yl)benzothiazole 730
Maria Manuela Marques Raposo, Susana Paula Graça da Costa, and Rosa Maria Ferreira Batista

12.2.2. Synthesis of *N-tert*-butyloxycarbonyl-[2-(thien-2′-yl)benzoxazol-5-yl]-L-alanine methyl ester 733
Susana P. G. Costa, M. Manuela M. Raposo, Cátia I. C. Esteves and R. Cristina M. Ferreira

12.2.3. Synthesis of Flavones (2-Aryl-4*H*-chromen-4-ones) 737
Artur M. S. Silva, Augusto C. Tomé, Diana C. G. A. Pinto, Fernando Domingues, Graça M. Oliveira Rocha, José A. S. Cavaleiro, M. Graça P. M. S. Neves, M. Amparo F. Faustino, and Mário M. Q. Simões

12.2.4. Hantzsch Synthesis of 3,5-Diethoxycarbonyl-2,6-dimethylpyridine 741
Iain A. Smellie, Nigel P. Botting, Brian A. Chalmers, Iain L. J. Patterson, and Catherine M. Schofield

12.2.5. An "Out-of-the-Box" Example in Heterocyclic Chemistry: Synthesis of 3,6-Bis-(3,5-dimethyl-1*H*-pyrazol-1-yl)-1,2,4,5-tetrazine 745
Tiago J. L. Silva, Paulo J. G. Mendes, Ana Isabel Tomaz, and M. Helena Garcia

12.3. Epoxidation 750

12.3.1. Catalytic Epoxidation of *cis*-Cyclooctene with MTO(VII) and Pyrazole 750
Elisabete P. Carreiro and Anthony J. Burke

12.3.2. Stereoselective Epoxidation of Cholesterol by *m*-Chloroperoxybenzoic Acid 754
Maria de Jesus Perry, Ana Paula Francisco, Ana S. Ressurreição and Francisca Lopes

12.3.3. Green Oxidation of Organic Compounds Using Metalloporphyrins 758
Rose A. Clark, Anne E. Stock, and Edward P. Zovinka

12.3.4. Catalytic Epoxidation of Carbamazepine 762
Artur M. S. Silva, Augusto C. Tomé, Diana C. G. A. Pinto, Fernando M. J. Domingues, Graça M. S. R. O. Rocha, José A. S. Cavaleiro, Maria da Graça P. M. S. Neves, Maria do Amparo F. Faustino, and Mário M. Q. Simões

12.3.5. Regioselective Epoxidation of Geraniol by $VO(acac)_2$ Immobilised in Polystyrene 766
Ana I. Vicente and Carlos A. M. Afonso

12.3.6. Organocatalysed *trans*-Dihydroxylation of Olefins 769
Andreia A. Rosatella and Carlos A. M. Afonso

12.4. Other Oxidations 773
12.4.1. Synthesis of 6-Nitrosaccharin 773
Dulce Simão and Ana C. Cerdeira

13. Reductions 776
13.1. Reduction of a Ketone Using Sodium Borohydride. Control of a Reaction by TLC 776
Luis Constantino, Catarina Dias, and Emília Valente

13.2. Regioselective Catalytic Transfer Hydrogenation of Citral 781
Jaime A. S. Coelho and Carlos A. M. Afonso

13.3. Regioselective 1,2-Reduction of an α,β-Unsaturated Ketone. A Green Experiment 784
M. Manuela A. Pereira

13.4. Reduction of Diphenyl Sulfoxide Catalyzed by the Dioxo-Molybdenum Complex $MoO_2Cl_2(H_2O)_2$ 789
Ana C. Fernandes and Tiago A. Fernandes

13.5. Synthesis of Allylic Esters by Reduction of Fructone Followed by Wittig Olefination 794
Thomas A. Logothetis

13.6. Preparation of a Thia-Tetraaza Macrocyclic Compound Through a Dual-Step Synthesis 798
Judite Costa, João Franco Machado, and M. Fátima Cabral

13.7. Asymmetric Reduction of Acetophenone with Borane Catalyzed by *B*-Methoxy-oxazaborolidine 803
Marek P. Krzemiński

13.8. Synthesis of 4,5-Bis(benzoylthio)-1,3-dithiole-2-thione 807
Dulce Simão and Ana C. Cerdeira

13.9. Organocatalytic Asymmetric Reduction of (*E*)-N,1-Diphenyl-1-propanimine to *N*-(1-Phenylpropyl)aniline with Trichlorosilane 811
Pedro Barrulas and Anthony Burke

14. Rearrangements 815
14.1. Benzilic Acid Rearrangement 815
Paulo Coelho and Céu Sousa

14.2. Preparation of Phenyl Acetate and Its Conversion into 4-Hydroxyacetophenone 818
Eimíle Sheehy and Paul Evans

14.3. Multistep Synthesis of Dilantin 821
Paula C. Castilho and Pedro Ideia

14.4. Preparation of Isoborneol Through the Wagner–Meerwein Rearrangement Reaction of (1*R*)-(+)-Camphene 826
Marek P. Krzemiński

14.5. Umbelliferone: A Natural Scaffold Suitable for the Synthesis of *ortho*-Acetylhydroxycoumarins *via* Fries Rearrangement Reaction 830
Saleta Vazquez-Rodriguez, Maria João Matos, Eugenio Uriarte, Lourdes Santana and Fernanda Borges

14.6. Preparation of 4,5-Functionalized Cyclopentenone *via* Cyclization of a Stenhouse Adduct 834
Jaime A. S. Coelho and Carlos A. M. Afonso

14.7. Synthesis of a Chiral Salen. Examples of the Schmidt Rearrangement and Ultrasound Activation 838
Maria Elisa da Silva Serra and Dina Maria Bairrada Murtinho

15. Biotransformations **841**
15.1. Bioreduction of *N*-Oxide Moiety 841
María Laura Lavaggi, Mercedes González, and Hugo Cerecetto

16. Polymerization Reactions **847**
16.1. Preparation of Nylon 6,6 by Interfacial Polymerization 847
João P. Telo

16.2. Synthesis of Copolymer from Acrylamide (AA), 2-Acrylamido-2-methylpropane-sulfonic Acid (AMPS) and *N,N*-Methyleno-bis-acrylamide (BA) 849
Fátima Coelho and Dulce Simão

16.3. Multi-Step Synthesis of Nylon from Cyclohexene 853
Marisa G. Weaver and Morgan J. Gainer

16.4. Copolymerisation of Styrene and Methyl Methacrylate: An Introduction to Radical Polymerisation and Monomer Reactivity Ratios 859
Valeria Arrighi and Arno Kraft

16.5. Polymerization of ε-Caprolactone Using a Ruthenium(II) Mixed Metallocene Catalyst 864
Andreia Valente, Tiago J. L. Silva, Paulo J. G. Mendes, Ana Isabel Tomaz, and M. Helena Garcia

17. Other Transformations **868**
17.1. The Cannizzaro Reaction: Synthesis of *p*-Chlorobenzyl Alcohol and *p*-Chlorobenzoic Acid 868
Marja Asp-Lehtinen

17.2. The Kemp Elimination in Water: A Laboratory Experiment for Introductory Organic Chemistry 871
Caitlin Williamson, Kenneth E. Maly, and Stephen L. MacNeil

17.3. Synthesis of Veratronitrile 875
Dulce Simão and Ana C. Cerdeira

17.4. Synthesis of Levulinic Acid from Sucrose 878
João P. Telo

17.5. Synthesis of 1-(4-Bromophenyl)-1*H*-pyrrole by the Clauson-Kaas Reaction 881
Maria Manuela Marques Raposo, Maria Cidália Rodrigues Castro, and Sara Sofia Marques Fernandes

17.6. Synthesis of the Manganese(III) Complex of 5,10,15,20-Tetrakis (2,6-dichlorophenyl)porphyrin 885
Artur M. S. Silva, Augusto C. Tomé, Diana C. G. A. Pinto, Fernando M. J. Domingues, Graça M. S. R. O. Rocha, José A. S. Cavaleiro, Maria da Graça P. M. S. Neves, Maria do Amparo F. Faustino, and Mário M. Q. Simões

17.7. Synthesis of 1,3-Dithienylbenzo[*c*]thiophene 888
Tiago J. L. Silva, Paulo J. G. Mendes, Ana Isabel Tomaz, and M. Helena Garcia

17.8. Synthesis of 1-Propyl-2-(thiophen-2-yl)-1*H*-pyrrole 893
M. Manuela M. Raposo

17.9. Rhodium Carbene C–H Insertion in Water and Catalyst Reuse 898
Nuno R. Candeias, Pedro M. P. Gois, and Carlos A. M. Afonso

17.10. Determination of the C–C Bond Strength of Substituted Cyclopropanes and Cyclobutanes using Bomb Calorimetry 903
Steffanie H. Liskey, Alan R. Esker, and J. M. Tanko

18. Chiral Resolutions **908**

18.1. Resolution of a Chiral Amine and Recovery of Unwanted Enantiomer by Racemization: Towards a Greener Industrial Process 908
Pedro P. Santos and Pedro F. Pinheiro

18.2. Synthesis of Racemic Phenylalanine Methyl Ester and its Kinetic Resolution Catalysed by α-Chymotrypsin 914
Luca Banfi and Renata Riva

18.3. Lipase Catalyzed Kinetic Resolution of Racemic 1-Phenylethanol 919
Otto Långvik, Tiina Saloranta, and Reko Leino

18.4. Enzymatic Kinetic Resolution and Separation of *sec*-Alcohols Methodology Based on Fatty Esters 923
Carlos M. Monteiro, Nuno M. T. Lourenço, and Carlos A. M. Afonso

18.5. Enzymatic Kinetic Resolution and Preparative Separation of Secondary Alcohols 927
Nuno M. T. Lourenço, Carlos M. Monteiro, and Carlos A. M. Afonso

18.6. Catalyzed Resolution and Simultaneous Selective Crystallization 931
Jaan Parve, Lauri Vares, Indrek Reile, Tõnis Pehk, Ly Villo, and Omar Parve

Subject Index **935**

1. Separation and Purification of Mixtures

1.1. Separation, Purification and Identification of the Components of a Mixture

Abel J. S. C. Vieira* and Elvira M. S. M. Gaspar

Faculty of Sciences and Technology, Universidade Nova de Lisboa, 2829-516 Caparica, Portugal
*E-mail: ajscv@fct.unl.pt

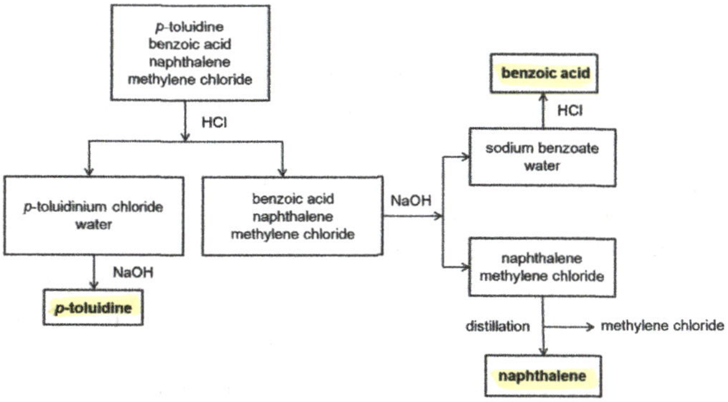

Number of sessions (Duration of each session)	Hazard level	Difficulty level	Level of study
2 or 3 (3 h each)	Low	Low	Introductory

Class names Aromatic compounds, carboxylic acids, amines, hydrocarbons

Concepts involved Acid–base reactions. Fundamental experimental techniques of unitary operations: Extraction, distillation, filtration and thin-layer chromatography. Purification of compounds by recrystallization and characterization through their physical properties and IR spectra

Chemicals needed *p*-Toluidine, benzoic acid, naphthalene, methylene chloride, hydrochloric acid, sodium hydroxide, sodium sulfate, ethanol, petroleum ether (40–60 °C) (also known as light petroleum (40–60 °C))

Equipment and experimental techniques involved Usual laboratory glassware; extraction, filtration, distillation and recrystallization apparatus; TLC plates

Keywords Acid–base reactions, distillation, extraction, filtration, IR spectroscopy, physical properties, thin-layer chromatography

Comprehensive Organic Chemistry Experiments for the Laboratory Classroom
Edited by Carlos A. M. Afonso, Nuno R. Candeias, Dulce Pereira Simão, Alexandre F. Trindade, Jaime A. S. Coelho, Bin Tan, and Robert Franzén.

Published by the Royal Society of Chemistry, www.rsc.org

Background

In this work, typical chemical reactions of organic compounds are not involved, with the exception of acid–base reactions.[1] Its aim is to provide students with the knowledge of fundamental experimental techniques of unitary operations, such as extraction, distillation (under normal and reduced pressure), filtration, recrystallization and thin-layer chromatography.[2,3]

The mixture to be separated is a solution in methylene chloride of three different aromatic compounds: an amine (*p*-toluidine) a carboxylic acid (benzoic acid) and a hydrocarbon (naphthalene). It is a four-component mixture: the (liquid) solvent and the three (solid) solutes. Considering the physical state of these components, one first (easy) attempt of separation could be a simple distillation of methylene chloride, yielding the solvent and a solid residue containing the three solids. However, to isolate each one of the components of the solid residue, complex solid-phase separation techniques would be necessary, which are difficult to perform and, very often, with low efficacy. Therefore, it is much more convenient to proceed to a liquid–liquid extraction.

Although the liquid–liquid extraction is a simple and efficient separation method, in the case of the present mixture, it would be necessary to find a set of solvents (not miscible with methylene chloride) to remove selectively each one of the three solutes, which, obviously, is not easy to obtain. However, taking into account the acid–base properties of two of the solutes, water can be used as the extraction solvent (water is not miscible with methylene chloride and a convenient "green" solvent), by performing very simple acid–base reactions.

Therefore, by treating the starting organic solution (in a separatory funnel) with an acidic aqueous solution, *e.g.* a solution of hydrochloric acid in water, the acid reacts with the *p*-toluidine base, yielding *p*-toluidinium chloride, a water soluble salt. The two phases obtained are easily separated, the aqueous layer containing the ammonium salt of the amine and the organic layer containing the remaining components dissolved in methylene chloride. By treating now this organic phase with a basic aqueous solution, *e.g.* a solution of sodium hydroxide in water, the base reacts with benzoic acid, yielding sodium benzoate, another water-soluble salt. A new separation of the two phases leaves an aqueous layer containing the salt of the acid and an organic layer containing only naphthalene dissolved in methylene chloride.

At this point of the work three isolated solutions are available: one aqueous solution of *p*-toluidinium chloride, one aqueous solution of sodium benzoate and one organic solution of naphthalene in methylene chloride. Naphthalene can be recovered by simply distilling the solvent. The other two components (both insoluble in cold water) can be isolated by "reversing" the pH of the aqueous solutions of the corresponding salts, *i.e.* by adding a base to the *p*-toluidinium chloride solution and an acid to the sodium benzoate solution.

Each one of the three isolated solid compounds can be further purified by recrystallization in the appropriate solvent.

The diagram of the graphical abstract illustrates schematically the sequential separation process.

Additional Safety

The experiment can be performed in a regular laboratory bench, with proper safety equipment, such as protective clothing, gloves and safety goggles. Aromatic compounds are usually toxic and inhalation or contact with eyes or skin must be avoided. The same safety rules apply to methylene chloride, which is quite volatile; good laboratory ventilation is advisable.

p-Toluidine: R:23/24/25-36-40-43-50; S: 1/2-28-36/37-45-61.

Naphthalene: R: 22-40-50/53; S: 2-36/37-46-60-61.

Methylene chloride: R: 40; S: 1/2-23-24/25-36/37.

Hydrochloric acid: R: 23-35; S: 1/2-9-26-36/37/39-45.

Sodium hydroxide: R: 35; S: 1/2-26-37/39-45.
Ethanol: R: 11; S: 2-7-16.

Experimental Procedure

Laboratory Session 1 (3 h)

Separation (ref. 4)

1. In a separatory funnel introduce 150 mL of the mixture to be separated (this volume of solution contains 3 g of *p*-toluidine, 3 g of benzoic acid and 3 g of naphthalene, in methylene chloride).
2. Extract the mixture with 25 mL of an aqueous HCl (3N) solution and shake thoroughly. Allow to stand and separate the aqueous layer. Repeat this procedure twice. Finally, extract the organic layer with 25 mL of water. Collect the aqueous extracts in an Erlenmeyer flask (solution A).
3. Recover the organic phase from the previous extraction and extract it with 25 mL of an aqueous NaOH (3N) solution and shake thoroughly. Allow to stand and separate the aqueous layer. Repeat this procedure twice. Finally, extract the organic layer with 25 mL of water. Collect the aqueous extracts in an Erlenmeyer flask (solution B).
4. Keep the organic phase left after the two previous extractions in an Erlenmeyer flask (solution C) and dry it with anhydrous sodium sulfate.

Recovery

5. To the solution A add a concentrated NaOH solution until no more precipitate formation is observed. Assure that the medium is basic by testing with indicator paper. Cool the mixture in ice, filter with suction at the water vacuum pump and collect the crystals in a Büchner funnel. Wash them with a small amount of ice cold water. Let them dry at the air over filter paper.
6. To the solution B add a concentrated HCl solution until no more precipitate formation is observed. Ensure that the medium is acidic by testing with indicator paper. After this point proceed as described in step 5 for solution A.
7. Filter the solution C through a fluted filter paper and collect the filtrate in a 250 mL round-bottomed flask. Distil the methylene chloride in a rotary evaporator until a small liquid residue of *ca.* 3 mL is left. Transfer this residue to a beaker and cool in ice until the precipitation of the solid is completed. Filter the crystals at the water vacuum pump and let them dry at the air.

Laboratory Session 2 (3 h)

Thin-Layer Chromatography (TLC)

1. Prepare a solution of a small amount of each one of the (well dried) separated compounds in *ca.* 1 mL of methylene chloride.
2. With a capillary glass tube apply a sample of each solution over a silica gel coated aluminium plate with fluorescence indicator. As a reference, apply also a sample of the starting solution.
3. Prepare an elution camera (according to the suggestion of your instructor) and elute with methylene chloride.
4. Upon development completed and drying of the plate identify the spots under a 254 nm UV lamp.

Product Purification

5. Observe the spots on the TLC plate and decide which one of the three recovered compounds is more impure.
6. Recrystallise this compound using the solvent suggested by your instructor or choose the more convenient one after a solvent test.
7. Upon collecting the crystals of the recrystallised compound, dry them at the air over filter paper, keep them in a small flask and dry in a vacuum desiccator.
8. Keep, in addition, the non-recrystallised compounds in appropriate vessels.

(Eventual) Laboratory Session 3 (3 h)

Product Identification

1. After completion of the separation and purification parts of the work, the three recovered products may be quantified and identified (see "Results and product characterisation" subsection).

Waste Disposal

The aqueous solutions resulting from the precipitation steps may be neutralised, diluted and discarded in the sink.

Methylene chloride and solvents used for recrystallization can be collected in a corresponding container to be recycled and reused, or discarded in the appropriate waste collector recipients, for posterior destruction.

Results Interpretation and Additional Questions

Results and Product Characterisation

1. Calculate the R_f value for every spot observed on the TLC plate. Compare the results obtained for the reference (sample of the initial mixture) and for each separated compound. Discuss the purity of the isolated compounds.
2. Weigh the quantity of each isolated compound after separation and calculate the corresponding percentage of recuperation.
3. Measure the melting points of the compounds.
4. For the compound chosen for recrystallization, compare the melting points of a sample of the crude and of the purified compound and calculate the yield of recrystallization. Discuss the differences observed.
5. If suggested by your instructor, record the IR spectra of the isolated compounds. Use dried potassium bromide as support material and dried samples of the compound to prepare transparent disks. Assign the absorption bands characteristic of the functional groups of the compounds.

Additional Questions

1. Write the equations of the acid–base reactions involved in this experiment.
2. Which are the characteristics a solvent must have to be adequate for a liquid–liquid extraction?
3. Why is it convenient to extract with only water after and acidic or basic extraction?

4. In the recovery of the isolated products, why is it necessary to test the pH of the neutralised solutions?
5. Before the recovery of naphthalene by distillation, why is it necessary to dry the methylene chloride solution?

References

1. J. Clayden, N. Greeves, S. Warren and P. Wothers, *Organic Chemistry*, Oxford University Press, 1st edn, 2001, vol. 8, p. 181.
2. R. Adams, J. R. Johnson and C. F. Wilcox, *Laboratory Experiments in Organic Chemistry*, Macmillan, 6th edn, 1970, vol. 6, p. 16.
3. L. M. Harwood, C. J. Moody and J. M. Percy, *Experimental Organic Chemistry*, Blackwell Science, 2nd edn, 1999, vol. 4, p. 204.
4. J. E. Leonard, *J. Chem. Educ.*, 1981, **58**, 1022.

1.2. Isolation of (+)-Limonene from Orange Oil

João P. Telo

Centro de Química Estrutural, Instituto Superior Técnico, Universidade de Lisboa, Av. Rovisco Pais, 1649-003 Lisboa, Portugal. *E-mail: jptelo@tecnico.ulisboa.pt

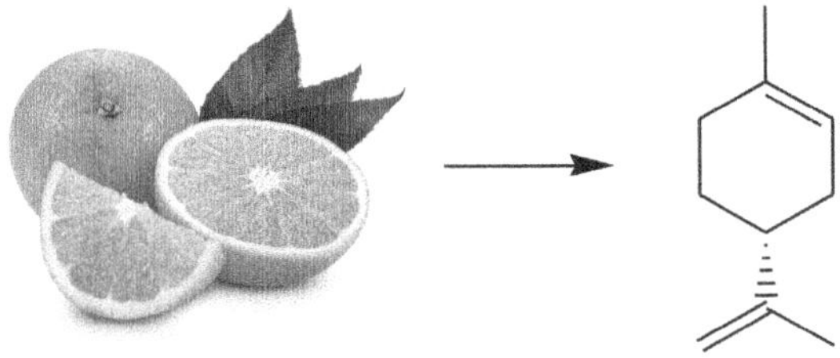

Number of sessions (duration of each session)	Hazard level	Difficulty level	Level of study
1 (3 h)	Low	Low	Introductory
Class names Hydrocarbons (terpenes)			
Concepts involved Extraction and purification of an organic compound from a natural product by steam distillation, liquid–liquid extraction and fractional distillation. The purity of the compound is checked by measuring its refraction index and its absolute rotation			
Chemicals needed Orange oil, petroleum ether (40–60 °C)			
Equipment and experimental techniques involved Steam distillation apparatus, fractional distillation under reduced pressure, liquid–liquid extraction; polarimeter and Abbé refractometer			
Keywords Fractional distillation, natural products, optical rotation, steam distillation, terpenes			

Background

Orange oil, a by-product of the orange juice industry, consists of around 95% (+)-limonene. The oil is contained in small cells inside the peel of the fruit. Minor constituents include myrcene (1–4%), *n*-octanal (1–2%), *n*-decanal and (+)-linalool, among others, in a total of more than 30 compounds. (+)-Limonene is used as a fragrance in the perfume and food industries, as a solvent and as an insect repellent. It is also used in the synthesis of other fragrances, including (−)-carvone (Experiment 4.1.1.8).[1] The isolation of (+)-limonene from orange oil is a useful experiment to exemplify the common techniques of steam distillation, vacuum fractional distillation and liquid–liquid extraction, and is also a good example with which to introduce the study of chirality and optical rotation.

Steam distillation is often used for temperature sensitive materials such as essential oils. Steam is generated outside the system in a boiler and injected in the flask containing the

Comprehensive Organic Chemistry Experiments for the Laboratory Classroom
Edited by Carlos A. M. Afonso, Nuno R. Candeias, Dulce Pereira Simão, Alexandre F. Trindade, Jaime A. S. Coelho, Bin Tan, and Robert Franzén.

Published by the Royal Society of Chemistry, www.rsc.org

mixture. The condensed distillate consists in a two-phase system with an organic layer and an aqueous one, which can be easily separated by partitioning. In the present case, extraction of the aqueous phase with dichloromethane increases the yield of the separation, and the vacuum distillation allows a final purification of (+)-limonene. Vacuum distillation is also used to reduce the boiling point of liquids and decrease the degradation by heating. The use of a fractionation column allows a better separation of liquids with similar boiling points.[2]

The purity of liquid chemicals is normally assessed by comparing their experimental boiling point and index of refraction with literature values. The experimental boiling point (at atmospheric pressure or, in the case of liquids of high boiling point, at a given reduced pressure) is often presented as a range, recording the initial and final distillation temperature of the fraction isolated. The refractive index determines how much light is bent, or refracted, when entering a material and its value is used to characterize liquids. The limonene from orange peels is enantiomerically pure, and this means that only the stereoisomer portrayed in the graphical abstract is isolated. Each enantiomer rotates the plane of polarized light in opposite directions and by the same amount, so the measurement of the optical rotation of a solution of natural limonene (which is dextrorotatory) is also a way to assess its purity by comparing the experimental value with the ones found in the literature.

As a shorter and simpler alternative, (+)-limonene can be isolated by fractional distillation alone of orange oil (method B).

Additional Safety

Orange oil and (+)-limonene are safe natural products, although some allergic skin reactions by prolonged contact with (+)-limonene are reported. Petroleum ether (40–60 °C) is flammable and should be handled in the absence of flames.

Experimental Procedure

Isolation of (+)-Limonene from Orange Oil

Method A

1. Prepare an apparatus for a steam distillation. Steam should be generated in a boiler and injected into the flask containing the orange oil. In the absence of a boiler, produce the water vapour by boiling water in a common flask and conducting the vapour into the other flask containing the orange oil. Collect the distillate in a 500 mL Erlenmeyer flask under the end of the Liebig condenser.
2. Distil 25 mL of orange oil until only water comes out of the condenser. This can be checked by collecting a few millilitres of the condensate in a test tube and checking if there are oily droplets floating in the water.
3. Transfer the distillate into a separatory funnel and separate the lower aqueous layer and the upper organic layer into different Erlenmeyer flasks. Return the aqueous fraction into the separatory funnel and extract it twice with 25 mL of petroleum ether (40–60 °C).
4. Dry the combined organic phases over anhydrous magnesium sulfate and filter it (or simply decant) into a pear-shaped 100 mL flask.
5. Adapt a Liebig condenser, a thermometer and receiver flask. Distil the petroleum ether in a water bath.
6. Adapt a fractionation column (Vigreux column) and distil the (+)-limonene under reduced pressure. Write down the range of temperatures at which the product distils and the pressure used. Pure Limonene should distil at 63 °C at 15 mmHg (174 °C at atmospheric pressure).

7. Weigh the flask with the distilled (+)-limonene to calculate the yield of the extraction (pre-weighting the empty receiver flask before the distillation may be useful).
8. Measure the refraction index with a refractometer and the optical rotation of a solution of 5 g of (+)-limonene in petroleum ether (50 mL of final volume) with a polarimeter.

Method B

Alternatively, the orange oil can be directly distilled at reduced pressure with a fractionation column. Use 25 mL of orange oil and proceed as in Method A, starting at the reduced pressure distillation (step 6).

Waste Disposal

The petroleum ether distilled should be stored for future purification and reutilization.

Results Interpretation and Additional Questions

1. Determine the yield of the extraction.
2. Compare the refraction index and the calculated specific rotation of the product with those of pure limonene found in literature.
3. Draw the structure of the enantiomer of (+)-limonene. The enantiomer should be dextrorotatory, laevorotatory or optically inactive?

References

1. O. S. Rothenberger, S. B. Krasnof and R. B. Rollins, *J. Chem. Educ.*, 1980, **57**, 741.
2. A. I. Vogel, A. R. Tatchell, B. S. Furnis, A. J. Hannaford and P. W. G. Smith, *Vogel's Textbook of Practical Organic Chemistry*, Longman, London, 5th edn, 1996.

1.3. Isolation of Plant Pigments from Green and Red Leaves

Alice M. Dias* and Maria La S. Ferreira

Department of Chemistry, University of Minho, Campus de Gualtar, 4710-057 Braga, Portugal
*E-mail: ad@quimica.uminho.pt

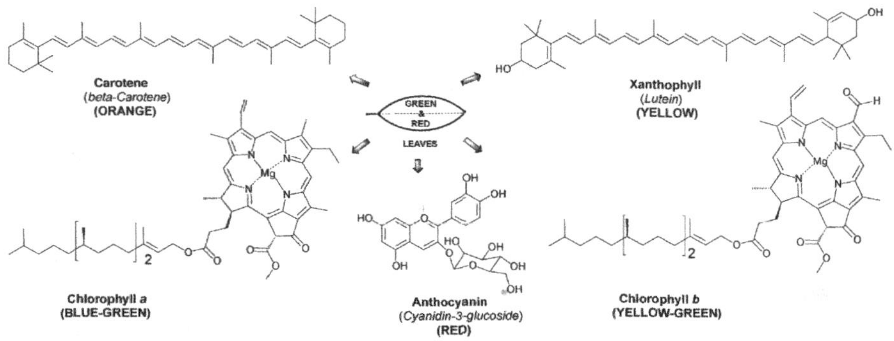

Number of sessions (duration of each session)	Hazard level	Difficulty level	Level of study
1 (4 h)	Moderate	Low	Introductory

Class names Carotenoids, chlorophylls and anthocyanins, alcohols, aldehydes, esters, alkenes and pyrroles

Concepts involved This experiment describes a colourful column chromatography of both lipophilic and hydrophilic plant pigments that agrees with green chemistry principles. A solid extract of green and red plant leaves is prepared by adsorption of crude extract on a household adsorbent. Column chromatography packed with the same adsorbent separates pigment bands with the use of common eluents. Paper chromatography is used to evaluate the eluted fractions

Chemicals needed Green and red leaves of trees, shrubs, indoor and garden plants; potato starch; filter paper; sodium sulfate (or kitchen salt); cotton wool; sand; aluminum foil;
acetone; petroleum ether 40–60 °C; mixtures of petroleum ether (40–60 °C)–acetone, 90 : 10 (v/v) and 75 : 25 (v/v); 1 M aqueous HCl solution (or vinegar)

Equipment and experimental techniques involved 20 mL (2 cm inner diameter × 9 cm external length) disposable syringe and universal support; gravity filtration apparatus; mortar and pestle

Keywords Adsorption, column chromatography, extraction, green chemistry, paper chromatography, plant pigments

Comprehensive Organic Chemistry Experiments for the Laboratory Classroom
Edited by Carlos A. M. Afonso, Nuno R. Candeias, Dulce Pereira Simão, Alexandre F. Trindade, Jaime A. S. Coelho, Bin Tan, and Robert Franzén.

Published by the Royal Society of Chemistry, www.rsc.org

Background

Plant pigments are crucial molecules for chemical and biological sciences. The three main classes of plant pigments are chlorophylls, carotenoids and flavonoids.[1,2] They decorate every spring with various green shades, stained flowers and ripen fruits with a diversity of colours and give brilliant shades of yellow, orange, red, scarlet, magenta, purple to the autumn foliage.[1] However, apart from contributing colour in the vegetal world, they play vital functions in living organisms and find relevant industrial and therapeutic applications.[2] In food, they have organoleptic and nutritive properties.[3] Chlorophylls and carotenoids are photosynthetic pigments, making them essential for life.[4–6] Carotenoids are widely used in food industry due to their colours and broad human benefits, such as vitamin A precursors and antioxidant effects.[2,6] Flavonoids protect plants from oxidative stress and their antioxidant properties are effective in diseases such as cancer or cardiovascular disorders.[2,7] They are key components of wine and can be useful as food preservatives and ecological indicators.

Flavonoids and carotenoids are commonly found in flowers, vegetables and fruits, whereas chlorophylls (*a* and *b*) are the typical pigments of the leaves.[1] Carotenoids, the orange–yellow pigments, are also always present in leaves, but are generally masked by the more abundant green chlorophylls and may only be visible when chlorophylls fade. The most common carotenoids in leaves are β-carotene and lutein. Certain plants also have a significant content of flavonoids in leaves, in particular anthocyanins that give them beautiful red colours.[1] The red cyanidin-3-glycoside is the most abundant anthocyanin in the leaf.[8] The role of the anthocyanins in the leaf has been controversial and the reported hypotheses include antioxidant, anti-UV-B, osmoregulatory and camouflage functions.[1,8,9] These three classes of leaf pigments have distinct physical and chemical properties.[1–7]

Chromatography is a powerful and versatile separation method widely used for analytical, preparative and research purposes.[10] Nowadays, the use of advanced gaseous chromatography (GC) and high-pressure liquid chromatography (HPLC) techniques permits the separation, isolation and possibly identification of closely related components present in complex mixtures, which would be impracticable by other methods. The principles of chromatography are essentially the same for advanced or simpler techniques, like paper chromatography or column chromatography of plant pigments. The use of chromatography dates from 1901, when the Russian botanist Michael Tswett separated plant pigments using a column of calcium carbonate as adsorbent.[10] Since then, column chromatography of chloroplast pigments from green leaves has been used as a historical activity to introduce chromatography in practical classes.[11–14] These techniques usually require liquid–liquid extraction to remove the water present in the vegetal tissues, followed by drying and evaporation of the organic layer. All these manipulations extend the time needed for the activity, waste significant volumes of organic solvents, require expensive/sophisticated equipment and facilitate degradation of pigments. Moreover, many experimental procedures involve the use of expensive and hazardous silica and alumina adsorbents. Often, toxic and/or expensive solvents have been used, including chlorinated solvents, methanol and diethyl ether. Green chemistry philosophy is now being implemented in chemical education and provides more sustainable methods to perform this activity.[14–16]

The objective of this work is the isolation of not only carotenoids and chlorophylls but also anthocyanins from green and red leaves by column chromatography following green chemistry principles. The wide polarity of these three classes of pigments requires the preparation of a solid extract, which is easily obtained by simple and sustainable procedures. Simple maceration of green and red leaves in acetone extracts chlorophylls, carotenoids and flavonoids. Desiccating with kitchen salt or sodium sulfate eliminates the water from vegetal tissues, with no need for further liquid–liquid extraction. Adsorption onto potato starch facilitates rapid evaporation of acetone, avoiding deleterious heating processes or expensive rotary evaporators. Potato starch is a common and safe adsorbent and was also selected for the packing of the column. Common and low toxicity solvents like acetone and petroleum ether were adopted as components for the eluotropic series. A colourful and efficient column is obtained and the composition of eluted fractions is demonstrated by the eco-friendly paper chromatography technique.

Additional Safety

Some species of plants are poisonous and should be avoided or handled with special care. Kitchen salt and potato starch are common household products with no adverse health effects expected from inhalation, ingestion or contact with skin or eyes (in the doses described). Acetone and petroleum ether are skin and lung irritants. These solvents should be manipulated in a laboratory hood, using protective gloves and goggles. They are also flammable liquids, so no open flames should be used in the laboratory when these experiments are being performed. The glass material should be handled with care and safety goggles and gloves should be used.

Experimental Procedure

Part I: Extract Preparation (45 min)

1. Weigh 5 g of green and red leaves. Remove the veins and shop the leaves. Crush the plant tissues in a mortar with 1 g of sand. Add 15 mL of acetone and let macerate for 15 min. The material must be kept covered with aluminium foil, at room temperature.
2. Filter the resulting solution, with the help of a cotton wool plug and funnel, into an Erlenmeyer flask, add two spatulas of sodium sulfate (or salt) and allow to stand for about 10 min.
3. Filter the supernatant liquid through a folded paper filter and retain the filtrate.
4. To prepare the solid extract, weigh 1 g of potato starch in a Petri dish or watch glass, add about 0.5 mL of the filtrate and stir with a glass rod. Allow the mixture to dry until the acetone is completely evaporated (*ca.* 5–10 min).

Part II: Analysis of Extract by Paper Chromatography (10 min)

5. Cut a strip of filter paper with dimensions of 2 cm × 12 cm.
6. In a 250 mL beaker place two round filter papers near the sides of the beaker. Add 10 mL of the mixture petroleum ether/acetone 90 : 10, cover the beaker with a watch glass and allow the solvents to evaporate until equilibrium is reached.
7. With a capillary tube apply samples of the crude extract solution to a line marked at about 1 cm from the bottom of the paper strip. Then apply a sample of the liquid extract obtained at step 3.
8. Locate the filter paper strip in the eluent, ensuring that the pigment spots do not go below the eluent level.
9. Remove paper from eluent once the mobile phase has reached a point about 2/3 of the length of the filter paper. Draw lines to mark the mobile phase front and pigment spots, with a pencil. Identify the pigment spots.

Part III: Separation of Plant Pigments by Column Chromatography (90 min)

10. Place a plug of cotton wool at the end of a 20 mL syringe, approximately 0.5 cm thick, and a 1 cm layer of sand. Set the column in a vertical position. Place a beaker under the column.

11. Prepare a suspension of 12 g of potato starch in 20 mL of petroleum ether in a beaker. Add the potato starch suspension, with the aid of funnel until about three-quarters syringe full (2 cm × 5.5 cm), stirring with a glass rod.
12. When the liquid reaches the surface of the column packing transfer the solid extract to the top of the column.
13. Continue adding petroleum ether to elute and collect the first yellow band (*ca.* 10 min).
14. Use the mixture petroleum ether/acetone (90 : 10) to elute the dark yellow coloured band and collect this band separately from the bluish green band (*ca.* 10 min). Collect the bluish green band (*ca.* 10 min).
15. Pass the mixture petroleum ether/acetone (75 : 25) through the column to collect the yellowish green band (*ca.* 10 min).
16. Pass acetone through the column and collect approximately 4 mL of eluent (*ca.* 10 min).
17. Pass more acetone through the column to collect anthocyanins (*ca.* 4 mL, 10 min). To this last fraction add three drops of 1 M HCl (or vinegar) and observe the colour change.

(Note: a 10 mL syringe can be used, solids and liquids must be reduced to the half, and times are approximately the same.)

Part IV: Analysis of Eluted Extracts by Paper Chromatography (20 min)

18. Cut a strip of filter paper with dimensions of 5 cm × 12 cm. Analyse samples of eluted fractions according to steps 6–9.

Waste Disposal

Acetone and petroleum ether may be reused or collected in a non-chlorinated waste collecting recipient for subsequent destruction. Residues of sodium sulfate and potato starch can be placed directly in the laboratory bin for disposal.

Results Interpretation and Additional Questions

1. Calculate the R_f of different pigments according to the following formula and identify each pigment:

$$R_f = \frac{\text{Migrated distance by pigment}}{\text{Migrated distance by solvent front}}$$

2. Compare the polarity of each class of pigments on the basis of column and paper chromatography results.
3. Rationalise the higher affinity of xanthophylls for the stationary phase, when compared with β-carotene. Apply the same process for molecules of chlorophylls *a* and *b*.
4. Why can anthocyanins band be eluted from the top of the column with acetone?
5. Rationalise the selection of acetone, instead of petroleum ether or ethanol, for the extraction of pigments from the vegetal tissues.
6. Discuss the importance of acetone and water elimination from the crude extract.

Acknowledgement

Reprinted (adapted) with permission from J. Chem. Educ., 2014, 1, 92, 189-192. Copyright (2014) American Chemical Society.

References

1. J. Alkema and S. L. Seager, *J. Chem. Educ.*, 1982, **59**, 183.
2. F. Delgado-Vargas, A. R. Jiménez and O. Paredes-López, *Crit. Rev. Food Sci. Nutr.*, 2000, **40**, 173.
3. H.-D. Belitz and W. Grosch, *Food Chemistry*, Springer-Verlag, Berlin-Heidelberg, 3rd edn, 2004; (a) pp. 232–244; (b) pp. 793–795; (c) pp. 827–831.
4. D. L. Nelson and M. M. Cox, *Lehninger: Principles of Biochemistry*, W. H. Freeman and Company, New York, 5th edn, 2008, pp. 744–761.
5. (a) Z. Liu, H. Yan, K. Wang, T. Kuang, J. Zhang, L. Gui, X. An and W. Chang, *Nature*, 2004, **428**, 287; (b) L. R. Milgrom, *The Colours of Life: An Introduction to the Chemistry of Porphyrins and Related Compounds*, Oxford University Press, New York, 1997.
6. R. Alvarez, B. Vaz, H. Gronemeyer and A. R. de Lera, *Chem. Rev.*, 2014, **114**, 1.
7. A. Castaneda-Ovando, M. L. Pacheco-Hernandez, M. E. Paez-Hernandez, J. A. Rodriguez and C. A. Galan-Vidal, *Food Chem.*, 2009, **113**, 859.
8. Y. Manetas, *Flora*, 2006, **201**, 163.
9. V.-P. Kytridis, P. Karageorgou, E. Levizou and Y. Manetas, *J. Plant Physiol.*, 2008, **165**, 952.
10. C. D. Mickey, *J. Chem. Educ.*, 1981, **58**, 997.
11. H. Strain and J. Sherma, *J. Chem. Educ.*, 1969, **46**, 476.
12. S. Diehl-Jones, *J. Chem. Educ.*, 1984, **61**, 454.
13. D. L. Pavia, G. M. Lampman and G. S. Kriz, *Introduction to Organic Laboratory Techniques: a small scale approach*, Saunders, New York, 1st edn, 1998, pp. 352.
14. A. Johnston, J. Scaggs, C. Mallory, A. Haskett, D. Warner, E. Brown, K. Hammond, M. McCormick and O. McDougal, *J. Chem. Educ.*, 2013, **90**, 796.
15. D. Kimbrough, *J. Chem. Educ.*, 1992, **69**, 987.
16. P. Anastas and N. Eghbali, *Chem. Soc. Rev.*, 2010, **39**, 301.

1.4. Extraction of Usnic Acid from Lichen

Dietmar K. Kennepohl

Athabasca University, 1 University Drive, Athabasca, Alberta, T9S 3A3, Canada
*E-mail: dietmark@athabascau.ca

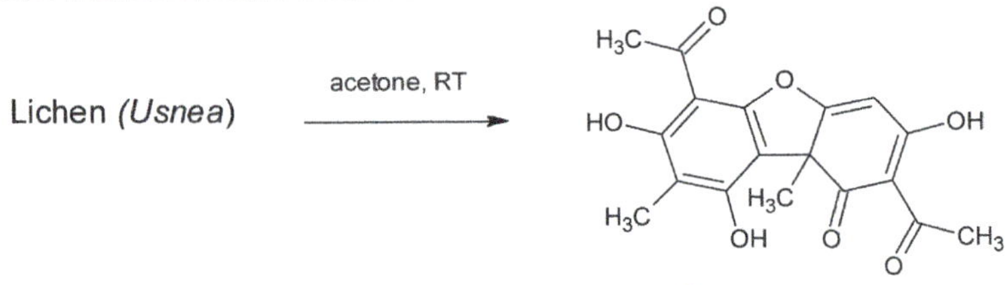

usnic acid (*R* or *S*)

Number of sessions (duration of each session)	Hazard level	Difficulty level	Level of study
1 (3 h)	Low	Medium	Introductory

Functional groups involved Not applicable

Concepts involved The purpose of this experiment is to (1) isolate an enantiomer of usnic acid, (2) learn the technique of liquid–solid extraction and two-solvent recrystallization and (3) determine the specific optical rotation of the optically active product using a polarimeter. Acetone is used to extract usnic acid from the lichen 'Old Man's Beard.' After the usnic acid is extracted and concentrated, the product is recrystallized and weighed. Mixed melting point and infrared spectroscopy (optional) are used to establish product purity and a sample solution is placed in a polarimeter to determine the specific optical rotation

Chemicals needed Lichen sample (*Usnea* sp.) oven dried (40 °C), acetone, ethanol (95%), spectral grade THF, L-(+)-tartaric acid (as the unknown compound)

Equipment and experimental techniques involved Extraction, recrystallization, vacuum filtration, melting point apparatus, rotary evaporator (optional), polarimeter

Keywords Extraction, natural product, optical activity, polarimetry, recrystallization, usnic acid

Background

Organic compounds from natural sources are often referred to as natural products. Their isolation and purification is of great interest to chemists,[1] especially bioactive compounds in medicine and drug development or aromatic essences from spices, herbs, flowers and fruits. In this experiment we will carry out a liquid–solid extraction. Most people have already encountered the extraction process when they make tea. Solid tea leaves are treated with a solvent (boiling water), which draws out desired flavours, colours and even the stimulant caffeine into the water. If you are a coffee drinker you might have tried decaffeinated coffee, which is another example

Comprehensive Organic Chemistry Experiments for the Laboratory Classroom
Edited by Carlos A. M. Afonso, Nuno R. Candeias, Dulce Pereira Simão, Alexandre F. Trindade, Jaime A. S. Coelho, Bin Tan, and Robert Franzén.

Published by the Royal Society of Chemistry, www.rsc.org

of liquid–solid extraction. In this case, the caffeine is specifically removed using supercritical carbon dioxide, leaving the coffee taste without the stimulation in the coffee beans. The aim in this experiment (adapted from Athabasca University teaching laboratories) is to extract one specific natural product (usnic acid) and then purify it by recrystallization.[2] The idea is that the crystals will grow using preferentially the target compound molecules, while leaving behind any minor contaminants in solution.

Usnic acid is a natural antibacterial organic found in lichen called "Old Man's Beard" (*Usnea* sp.) and its history, isolation, characterization and uses have been described in some detail in the literature.[3,4] This compound possesses a chiral centre and so there are two possible enantiomeric forms (*R* and *S*) of the molecule, which are mirror images of each other. They are in fact optical isomers because, despite having identical physical properties, they will rotate plane polarized light in opposite directions. One will be dextrorotary and rotate light to the right, which is denoted by a plus sign (+), and the other is levorotary rotating light to the left, denoted by a minus sign (−). Because of the similar physical properties, a racemic mixture (±) of the enantiomers is difficult to separate, often involving synthesizing diastereomeric derivatives that can be separated and then eventually converted back into the original compound. Fortunately, the lichen we have chosen will produce exclusively either (+) or (−) usnic acid from a particular growth or plant, so no isomer separation is required. Plants and animals preferentially generate only one optical isomer, because the enzymes involved in their creation are also chiral and one-handed in a single organism. The (+)-usnic acid is the most common constituent among *Usnea* species.[5] Both melting point determination and polarimetry will be used to determine the purity of the isolated usnic acid.

In a polarimeter plane-polarized monochromatic light is passed through an optically active sample, which rotates the light, and then continues to a rotatable analyzer that helps to determine the degree of that rotation, called the observed rotation (α).[6] For consistency chemists report polarimetry measurements as a specific rotation, $[\alpha]_D^{20}$, where:

$$[\alpha]_D^{20} = \frac{\alpha}{L \times c}$$

for a solution with a concentration (*c*) using the units g mL^{-1} and a sample cell length (*L*) in decimeters. The superscript 20 denotes the temperature at which the measurement was made and *D* indicates the light source was a sodium lamp (wavelength 589.3 nm).

Additional Safety

Use a fume hood, with proper safety equipment, such as protective clothing, gloves and safety goggles. Usnic acid is harmful if swallowed, inhaled or absorbed through the skin. Acetone vapour is an irritant to the eyes and lungs and harmful to liver and kidneys if swallowed. Should direct contact with the eyes occur with acetone, flush immediately with copious amounts of water. Ethanol (95%) may contain denaturing substances that enhance its toxicity and tetrahydrofuran (THF) is flammable and dissolves latex (use non-latex gloves).

Experimental Procedure

Extraction and Isolation (1.2 h)

1. Place approximately 12–13 g of chopped or crushed lichen (weigh accurately) in a 500 mL Erlenmeyer flask.
2. Add 160 mL acetone and a magnetic stir bar. Stopper flask.
3. Stir the mixture for 1/2 h and immerse any lichen that comes out of the acetone solution to insure maximum extraction.

4. Gravity filter the mixture and collect filtrate in a 250 mL round bottom flask.
5. Use a rotary evaporator in the fume hood to remove the acetone. Allow the last of the acetone to evaporate at room temperature.

Purification and Initial Characterization (1.2 h)

1. Use two-solvent system of acetone and 95% ethanol (10:1) to recrystallize the crude usnic acid. First dissolve the crystals in the minimum amount of hot acetone. Then add the ethanol. (Hint: while crystals are growing, carry out the polarimetry measurement of the unknown.)
2. Isolate resulting yellow crystals by vacuum filtration. Wash with ice cold acetone and allow crystals to dry on a sheet of filter paper.
3. Weigh the dry purified usnic acid.
4. Measure the melting point of the purified product and compare to the literature value. (Optional: confirm identity of usnic acid by mixed melting point technique.)
5. Optional: obtain an IR spectrum of the product and compare it with that of an authentic sample.

Polarimetry of an Unknown Compound (0.3 h)

1. Prepare a 25 mL aqueous solution of your unknown compound (mp 171–174 °C) by dissolving 4–5 g (weigh accurately) in a 25 mL volumetric flask.
2. Rinse and then charge the polarimeter sample cell with distilled water.
3. Following the instructions for your particular polarimeter, take several readings and record them. Your readings will show some variation, but use the mean value in your results. This is a "blank" value (ideally 0.0°) and will have to be subtracted from your "unknown" result.
4. Remove the water from the sample cell. Rinse the cell with a small amount of your unknown solution. Then charge the cell with the unknown solution.
5. Repeat step 3 and take and record several readings.
6. Empty and rinse cell with water.

Polarimetry of Usnic Acid (0.3 h)

You will now use an analogous process to measure the optical rotation of your purified usnic acid, but this time using spectral grade tetrahydrofuran (THF) as the solvent.

1. Weigh 100 mg of your usnic acid and prepare a THF solution (25 mL) in a 25 mL volumetric flask.
2. Rinse and then charge the polarimeter sample cell with THF.
3. Following the instructions for your particular polarimeter; take several readings to determine your blank value.
4. Remove the THF from the sample cell. Rinse the cell with a small amount of your usnic acid solution. Then charge the cell with the unknown solution and take and record several readings.

Waste Disposal

The unknown and its aqueous solutions may be discarded down the sink. Organic solvents and solutions containing usnic acid should be placed in the container provided.

Results Interpretation and Additional Questions

1. Determine the percent usnic acid in the lichen by weight.
2. Report the observed rotation of the unknown compound and the usnic acid. Remember to correct with the reading obtained for the blank [*i.e.* α = α (solution) − α (blank)].
3. Calculate the specific rotation of the unknown compound and the usnic acid.
4. Define the difference between enantiomers and diastereomers using specific examples to illustrate your answer.
5. Draw line–wedge diagrams of the two enantiomers of usnic acid and label the chiral centre with the correct (*R*) or (*S*) configuration.

Acknowledgements

Elaine Birkigt and Robert Carmichael are acknowledged for helpful discussions and practical in-laboratory suggestions.

References

1. *Natural Product Extraction: Principles and Applications*, ed. M. A. Rostagno and J. M. Prado, Royal Society of Chemistry, Cambridge, 2013, no. 21.
2. D. Kennepohl, L. Shaw, D. Law, R. Carmicheal, L. Browne, and A. Last, *Organic Chemistry I Laboratory Manual*, Athabasca University, Edmonton, 2009, pp. 143–154.
3. D. W. Mayo, R. M. Pike and D. C. Forbes, *Microscale Organic Laboratory: With Multistep and Multiscale Syntheses*, John Wiley & Sons, Toronto, 5th edn, 2010, pp. 224–228.
4. M. Cocchietto, N. Skert, P. Nimis and G. Sava, *Naturwissenschaften*, 2002, **89**, 137–146.
5. V. Marcano, V. Rodriguez-Alcocer and A. Morales Méndez, *J. Ethnopharmacol.*, 1999, **66**, 343–346.
6. J. McMurry, *Organic Chemistry*, Cengage Learning, Belmont, CA, 8th edn, 2010, pp. 147–149.

1.5. Thin-Layer Chromatography of Plants Pigments

Ana Margarida Madureira* and Maria-José U. Ferreira

iMed.ULisboa, Faculty of Pharmacy, University of Lisbon, Av. Professor Gama Pinto, 1649-003 Lisboa, Portugal. *E-mail: afernand@ff.ul.pt

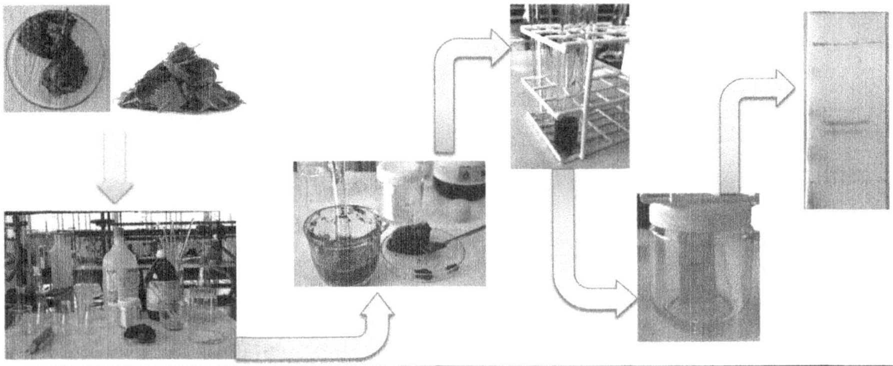

Number of sessions (Duration of each session)	Hazard level	Difficulty level	Level of study
1 (3 h)	Low	Low	Introductory

Concepts involved Separation techniques, thin-layer chromatography, chemical structures *versus* polarity, solid–liquid extraction

Chemicals needed Spinach leaves, TLC plates, acetone, *n*-hexane

Equipment and experimental techniques involved General chemistry laboratory equipment; solid–liquid extraction; chromatographic techniques

Keywords Chlorophylls, chromatography, extraction, plants pigments, TLC (thin-layer chromatography)

Background

Many plant-derived compounds are colored or pigmented. In spring the major color present in plants is green, which appears in a huge variety of hues and shades that change throughout the year. During the autumn, the green color of foliage is replaced by yellows, oranges, reds, magentas and purples and in the winter brown colors predominate. These changes in plant colors result from the production, interaction and breakdown of three major classes of natural products: porphyrins, carotenoids and flavonoids.[1]

Chlorophylls, the green plant pigments, belong to porphyrins and are related with the photosynthetic process. They are able to absorb certain wavelengths of visible light that are converted by chloroplast into chemical energy.[2] Structurally, chlorophylls have a central magnesium ion encased in a large ring consisting of four pyrrole rings that are linked with methine bridges. The four nitrogen atoms, as shown in Figure 1.5.1, surround and bind the magnesium atom.[1] The most important

Comprehensive Organic Chemistry Experiments for the Laboratory Classroom
Edited by Carlos A. M. Afonso, Nuno R. Candeias, Dulce Pereira Simão, Alexandre F. Trindade, Jaime A. S. Coelho, Bin Tan, and Robert Franzén.

Published by the Royal Society of Chemistry, www.rsc.org

porphyrins are undoubtedly chlorophylls *a* and *b*. The main difference between them is the replacement of a methyl group (chlorophyll *a*) by an aldehyde function (chlorophyll *b*).

Carotenoids are colored terpenes, with 40 carbon atoms, that occur in most plants. The extended π-electron system (Figure 1.5.2) confers color to the molecules, which contribute to the yellow, orange, and red pigmentations of plant tissues. The yellow colored β-carotene and xanthophylls (like lutein, violaxanthin and neoxanthin, Figure 1.5.2) are the predominant carotenoids in leaves and other photosynthesizing tissues. The orange color of carrots is related to the presence of β-carotene (Figure 1.5.2). Carotenoids also play an important role in human nutrition and health, providing provitamin A and having anti-cancer activities. β-Carotene and other carotenoids are widely employed as coloring agents for foods, drinks, confectionery and drugs.[3,4]

Anthocyanins, which have the general structure presented in Figure 1.5.3, are a particular group of flavonoids that are mainly responsible for most of the pallet of colors (from orange to blue) found in many flowers, leaves, fruits, seeds and other tissues.[5]

The aim of this work is to extract and separate the major non-polar pigments present in spinach leaves. Spinaches and other kales are rich in pigments that include chlorophylls, carotenoids and flavonoids. Depending on the vegetable sample (fresh or conserved) and the manipulation,

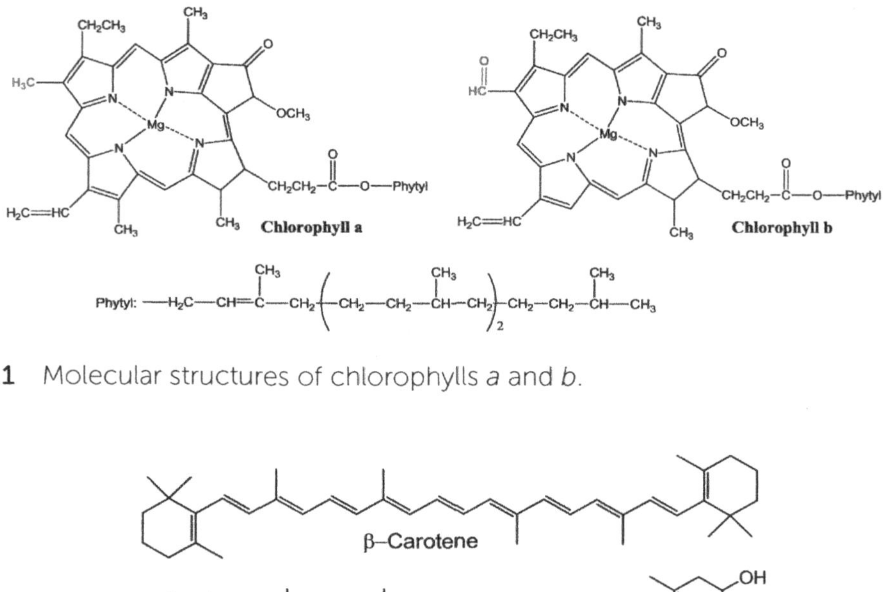

Figure 1.5.1 Molecular structures of chlorophylls *a* and *b*.

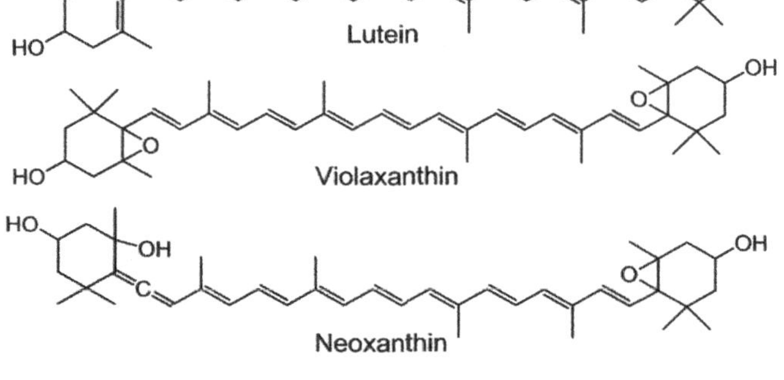

Figure 1.5.2 Molecular structures of some carotenoids.

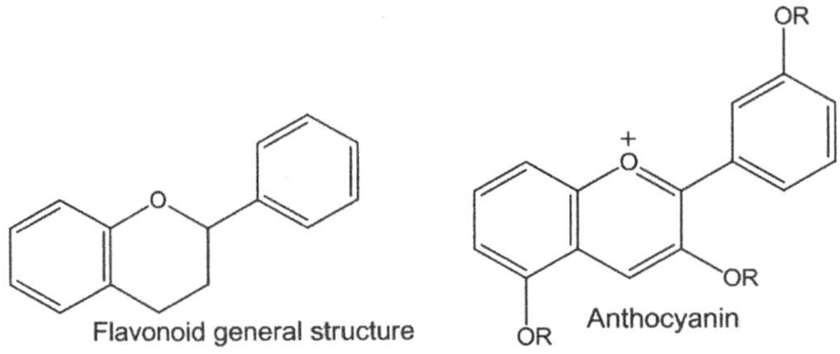

Figure 1.5.3 Flavonoid structures.

some degradation compounds might be present. Frozen spinaches leaves often present variable amounts of pheophytins which result from chlorophylls *a* and *b* degradation (the central Mg^{2+} is replaced by $2H^+$).[5] In this proposed experimental technique only chlorophylls and carotenoids will be extracted and assayed.

The pigments will be separated using the thin-layer chromatography (TLC) technique. TLC is a very versatile chromatography method that allows the rapid separation of the components of a mixture. The ability to quickly and successfully analyze small amounts (10^{-9} g) of complex mixtures with inexpensive equipment makes TLC one of the most used techniques among chemists. In TLC, the separation takes place in a thin layer of adsorbent particles (silica gel) attached to a solid support, as a liquid mobile phase (solvent or eluent) moves along the plate. Several physical processes can be related with the separation of the samples (adsorption, partition, ion-exchange, size-exclusion, *etc.*) but in organic chemistry laboratory most TLC applications are based on adsorption. Usually, the samples are non-volatile liquids or a solution on a volatile solvent that is spotted as a dot or a line with a pipette or a capillary glass tube on the lower edge of the plate. The TLC plate is placed in a chamber with the eluent which rises through capillary action, pulling along the components in the sample. Separation of the mixture occurs because each component has different chemical and physical properties and so interacts with the stationary and mobile phases differently, creating individual bands.[3,6]

Additional Safety

This experiment does not involve any chemicals that are significantly hazardous. However, remember:

1. Always wear goggles in the laboratory.
2. Wear gloves when handling the TLC plates.
3. The use of the glass pipettes and capillaries should be done with extra care to avoid injuries and accidents.
4. *n*-Hexane and acetone are both highly flammable. Avoid the use of flames while working with these solvents.

Experimental Procedure: Laboratory Session: 3 h

Procedures (Adapted from ref. 7)

Sample Preparation

1. Pick around 10 g of fresh/frozen spinach leaves.
2. Tear the leaves into small pieces and place the spinach pieces in a mortar.

3. Add 5 mL of acetone and carefully grind the spinach and acetone together with the pestle. Continue grinding until the acetone becomes dark green.
4. Allow this mixture to stand in contact for a few minutes. If necessary add more acetone.
5. Using a thin stemmed pipette connected with a pipette filler, carefully remove the acetone to a small test tube. Be careful not to extract the small pieces of spinach.
6. Add 3 mL of *n*-hexane to the acetone extract.
7. Mix the tube content.
8. Allow the layers to separate.

Preparation of the TLC Chamber

9. Add developing solvent (70 : 30 mixture of *n*-hexane–acetone) to a TLC chamber (a covered 400 mL glass beaker). The solvent should completely cover the bottom of the chamber to a depth of approximately 0.5 cm. Keep the chamber covered so that evaporation does not change the composition of the solvent.

TLC of the Extract

10. Obtain a TLC plate (a silica gel coated metal sheet) which has been precut and draw a small line (1 cm) in the middle and 1 cm above the short edge of the TLC plate.
11. Fill a capillary tube by placing it in contact with the upper layer of the leaves extract (it will fill by capillary action). Apply the extract on the line by lightly touching with the end of the capillary on the plate, placing small dots next one to each other.
12. Allow to dry (you can gently blow on the strip). Repeat several times to make a concentrated line of extract. Be sure to let the strip to dry between applications. Try to make the line as thin as possible but dark enough to see the color clearly.
13. Place the TLC plate in the chromatography chamber with the sample toward the bottom. Be sure the sample spot is above the level of the solvent. Close the chamber.
14. Allow the TLC plate to develop (separation of pigments) undisturbed until the solvent reaches 1 cm from the top. As the solvent moves up on the TLC plate you should see the different colored pigments separate.
15. Remove the plate from the chamber and *immediately mark the solvent front* using a pencil.
16. Measure and record the pigment distances from the origin (spotting line) as quickly as possible because some pigments (especially the β-carotene) may fade over time. The individual bands might be circled with a pencil.
17. The rate at which a pigment moves up the plate is reported as an R_f value, which is defined as the ratio between the distance moved by the spot and the distance moved by the solvent. Determine the R_f values for each of the pigments, using the formula provided below:

$$R_f = \frac{\text{Distance from origin to component spot}}{\text{Distance from origin to solvent front}}$$

18. Identify each spot based on the data in Table 1.5.1.

Waste Disposal

- Unused spinach leaves should be thrown into the wastebasket.
- The mixture of spinaches and acetone leftover should be disposed into a container labeled "Spinaches – solid waste".

Table 1.5.1 R_f of some spinach pigments when eluted on silica gel precoated plates with *n*-hexane–acetone (70:30).

Pigment	Color	R_f value
Carotene	Yellow–orange	0.80
Pheophytin *a*	Green–grey	0.55
Pheophytin *b*	Grey (may not be visible)	0.47–0.54
Chlorophyll *a*	Blue–green	0.42
Chlorophyll *b*	Green	0.34
Xanthophyll	Yellow	0.30
Xanthophyll	Yellow	0.20
Xanthophyll	Yellow	0.10

- The two-phase mixture of *n*-hexane–acetone should be collected in a container labelled "Spinaches – organic waste".
- All the material used should be washed with water and put in an appropriate container.

Results Interpretation and Additional Questions

1. In the TLC experiment, why should the sample be applied above the level of the developing solvent?
2. Which of the pigment molecules was the least polar?
3. Which of the pigment molecules was the most polar?
4. What do you expect to happen if you increase the percentage of acetone of the eluent?
5. Which of the solvents (A, B or C) would you choose to separate the mixture of compounds **1**, **2** and **3**? Justify your answer:

	Solvent		
Compound	A	B	C
1	0.1	0.8	0.9
2	0.25	0.3	0.33
3	0.4	0.6	0.75

6. Based on their chemical structures, explain why neoxanthin is more polar than carotenes?

References

1. J. Alkema and S. L. Seagerl, *J. Chem. Educ.*, 1982, **59**, 183.
2. L. Taiz and E. Zeiger, *Plant Physiology*, Sinauer Associates Inc, Sunderland, MA, 4th edn, 2006, ch. 1.
3. D. R. Palleros, *Experimental organic chemistry*, John Wiley & Sons, Inc., New York, 2000, experiment 5, p. 172.
4. P. M. Dewick, *Medicinal Natural Products: A Biosynthetic Approach*, John Wiley & Sons Ltd, Chichester, 3rd edn, 2011, ch. 5, p. 298.
5. Y. Tanaka, N. Sasaki and A. Ohmiya, *Plant J.*, 2008, **54**, 733.
6. *Penn State University, Thin-Layer Chromatography*, http://courses.chem.psu.edu/chem36/Experiments/PDF's_for_techniques/TLC.pdf, accessed September 2014.
7. Adapted from: H. T. Quach, R. L. Steeper and G. W. Griffin, *J. Chem. Educ.*, 2004, **81**, 385.

1.6. Isolation of Cinnamaldehyde from Cinnamon

Raquel F. M. Frade[a], Dulce Simão*[b], and Carlos A. M. Afonso*[a]

[a]iMed.ULisboa, Faculdade de Farmácia da Universidade de Lisboa, Av. Prof. Gama Pinto, 1649-003 Lisboa, Portugal; [b]Department of Chemical Engineering, Instituto Superior Técnico, University of Lisbon, Centro de Química Estrutural, Av. Rovisco Pais, 1, 1049-001 Lisbon, Portugal
*E-mail: raquel_frade@hotmail.com, dulcesimao@tecnico.ulisboa.pt, carlosafonso@ff.ulisboa.pt

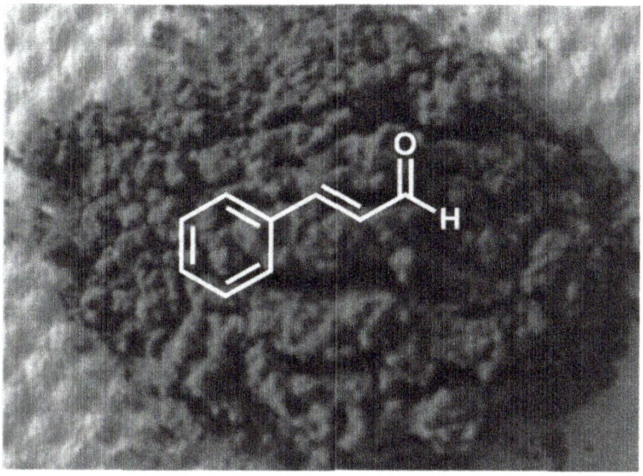

Number of sessions (duration of each session)	Hazard level	Difficulty level	Level of study
2 (4 h each)	Moderate	Low	Intermediate

Class names Aldehydes, α,β-unsaturated carbonyl compounds

Concepts involved This experiment describes two different approaches for the isolation of cinnamaldehyde from cinnamon: steam distillation and extraction with Soxhlet. Isolated product are characterized by infra-red and ^{1}H NMR spectra

Chemicals needed Dichloromethane, anhydrous magnesium sulfate, deuterated chloroform

Equipment and experimental techniques involved Apparatus for the steam distillation and extraction with Soxhlet; heating source (mantle or water bath); separation funnel; rotary evaporator; balance; IR and ^{1}H NMR spectroscopies

Keywords Extraction with Soxhlet, ^{1}H NMR, infra-red, isolation, steam distillation

Background

Nowadays our lifestyle is strongly dependent on a large variety of organic compounds that are used for the production of fuel, food, polymers, solvents, sensors and therapeutic drugs, for instance. The majority of these compounds are generated from the manipulation of natural renewable and non-renewable sources such as crude oil.[1] Natural sources are increasingly important at large-scale (for instance biofuel, carbohydrates, food oils, bio-polymers) and also at small-scale. This

Comprehensive Organic Chemistry Experiments for the Laboratory Classroom
Edited by Carlos A. M. Afonso, Nuno R. Candeias, Dulce Pereira Simão, Alexandre F. Trindade, Jaime A. S. Coelho, Bin Tan, and Robert Franzén.

Published by the Royal Society of Chemistry, www.rsc.org

last is also very important due to their specificity (for instance, flavours, penicillin, taxol, artemisin). Isolation of organic compounds from nature is of crucial importance in industry, namely in the chemical, food and pharmaceutical industries.[1] Cinnamaldehyde is responsible for the flavour and odour of cinnamon, being present in approximately of 2% (w/w) in dried, ground bark from the cinnamon tree. Cinnamaldehyde is used mainly in flavour and perfume industry.[2] However, other biological properties have been identified such as antipyretic, astringent, antimicrobial activity, anti-inflammatory activity, antibacterial[3] and inactivation of *Salmonella* in organic soil.[4] The present experiment achieves the isolation of cinnamaldehyde from cinnamon based on the use of steam distillation and extraction with Soxhlet as key isolation processes.[5]

Additional Safety

Safety equipment such as protective clothing, gloves and safety goggles must always be used. Extraction with Soxhlet should be carried out in a fume hood; in addition, a fume hood should be employed every time dichloromethane is used, because it is a very volatile and toxic solvent. It is harmful if inhaled and in contact with skin, and the vapours can produce irritation in the eyes and respiratory tract. Deuterated chloroform is harmful if inhaled or swallowed and causes respiratory tract, eye and skin irritation; and may cause cancer. It can be flammable in a fire or if heated.

Experimental Procedure (ref. 6)

Laboratory Session 1 (4 h): Isolation by Steam Distillation

Method A

1. Place 15 g of powdered cinnamon, obtained from a local supermarket, in the steam distillation vessel and add 150 mL of distilled water.
2. Perform the steam distillation until a final volume of 100 mL is obtained.
3. Cool the distillate and transfer to a separation funnel.
4. Extract the compound of interest with dichloromethane (2 × 25 mL).
5. Isolate the organic phase and dry with anhydrous magnesium sulfate ($MgSO_4$).
6. Filter the organic phase and evaporate the solvent.
7. Weigh the obtained product.
8. Determine the amount of cinnamaldehyde.
9. Obtain IR and ^{1}H NMR spectra in deuterated chloroform ($CDCl_3$).

Laboratory Session 2 (4 h): Isolation by Extraction with Soxhlet

Method B

1. Weigh 5 g of powdered cinnamon, obtained from a local supermarket, and place it inside the porous thimble and insert it in the inner tube of Soxhlet apparatus.
2. Add 250 mL of dichloromethane to the still pot.
3. Perform continuous extraction with dichloromethane for 1 h.
4. Evaporate the solvent at reduced pressure.
5. Weigh the obtained product.
6. Determine the amount of cinnamaldehyde.
7. Obtain IR and ^{1}H NMR spectra in deuterated chloroform ($CDCl_3$).

Waste Disposal

Dichloromethane should be collected and stored in the chlorinated waste collector recipient for posterior waste processing.

Results Interpretation and Additional Questions

1. Determine the amount of extracted cinnamaldehyde given by the two different methods and identify the method that generates the best product yield.
2. Interpret the acquired IR and ^1H NMR spectra and confirm the obtained structure.
3. Discuss which method gives the most pure sample based on spectral data.
4. Compare the air stability of the cinnamaldehyde and that of the corresponding reduced product (cinnamic alcohol).

References

1. J. O. Metzger, *Angew. Chem., Int. Ed.*, 2006, **45**, 696–698.
2. *The Merck Index*, ed. S. Budavari, A. Smith, M. J. O'Neil, P. E. Heckelman and J. F. Kinneary, Merck & CO., INC, 12th edn, 1996, p. 2356.
3. N. S. Satya, S. Prakash D.V. and V. Meena, *Int. Res. J. Biol. Sci.*, 2012, **1**, 49–51 and references cited therein.
4. N. Yossa, J. Patel, P. Millner and M. Lo, *Foodborne Pathog. Dis.*, 2011, **8**, 311–317.
5. The steam distillation allows the isolation of moderately volatile components by passing steam into a mixture containing the hydrophobic components and water. The extraction with Soxhlet is an efficient method for continuous extraction components present in a solid by a hot solvent. For more detailed explanation see: B. S. Furniss, A. J. Hannaford, P. W. G. Smith and A. R. Tatchell, *Vogel's Textbook of Practical Organic Chemistry*, Longman Group UK Limited, 5th edn, 1989, sections 2.22 and 2.25.
6. This work was adapted from the following book: C. Afonso, D. P. Simão, L. P. Ferreira, M. E. D. S. Serra and M. M. M. Raposo, *100 Experiências de Química Orgânica*, IST Press, 2011.

1.7. Isolation and Structural Identification of Piperine, the Major Alkaloid of Black Pepper

Andreia Mónico, Angela Paterna, Mariana A. Reis, Ricardo Ferreira, Noélia Duarte, Ana M. Madureira, and Maria-José U. Ferreira*

iMed.ULisboa, Faculty of Pharmacy, University of Lisbon, Av. Prof. Gama Pinto, 1649-003 Lisbon, Portugal. *E-mail: mjuferreira@ff.ulisboa.pt

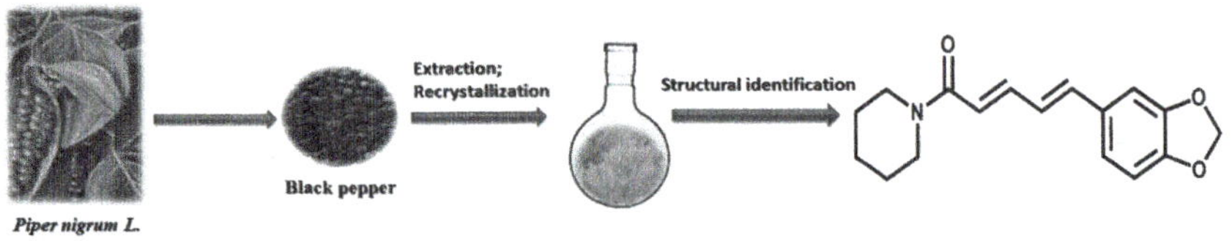

Number of session (duration of each session)	Hazard level	Difficulty level	Level of study
1 (3 h)	Low	Medium	Intermediate

Class names Amides, methylenedioxy, alkenes

Concepts involved This experiment describes a method for isolating and characterizing the alkaloid piperine from black pepper. This experimental procedure is a good way to introduce natural products chemistry, to learn techniques such as solid–liquid extraction, recrystallization and thin-layer chromatography and to practice structural identification, by the analysis of IR, UV, MS, ^{1}H and ^{13}C NMR spectra

Chemicals needed Black pepper, dichloromethane, diethyl ether, acetone, *n*-hexane, ethyl acetate, sulfuric acid, methanol, Dragendorff's reagent

Equipment and experimental techniques involved Solid–liquid extraction; apparatus for reaction at a temperature above room temperature (50 °C); reflux; thin-layer chromatography; melting point; IR; UV; MS; ^{1}H and ^{13}C NMR.

Keywords IR, isolation, NMR, recrystallization, solid–liquid extraction, thin-layer chromatography

Background

Black pepper is the dried unripe fruit of *Piper nigrum* L. (family Piperaceae), which is a native plant to India and other southeastern Asian countries. *Piper nigrum* produces a high content of alkaloids, namely piperine, piperidine and piperettine (Figure 1.7.1), which are responsible for the biological activity of this plant.[1] Alkaloids are a large and structurally diverse group of nitrogen-containing natural products, particularly found in plants, and generally exhibit potent biological activity. They are biosynthesized from amino acids and are classified as alkaloids because of their characteristic basic properties ("alkali-like"). These amines often react with acids to yield water-soluble salts. The nitrogen atoms are mostly present in heterocyclic rings but they are also found as primary amines or as quaternary ammonium groups.[2]

Comprehensive Organic Chemistry Experiments for the Laboratory Classroom
Edited by Carlos A. M. Afonso, Nuno R. Candeias, Dulce Pereira Simão, Alexandre F. Trindade, Jaime A. S. Coelho, Bin Tan, and Robert Franzén.

Published by the Royal Society of Chemistry, www.rsc.org

Black pepper has been used as seasoning and in traditional medicine, due to its carminative, diaphoretic, diuretic and insecticidal properties. Piperine is the major constituent (6–9%) of black pepper and responsible for its hot taste. This compound also possesses biological activities such as antibacterial, anti-inflammatory and anti-oxidant.[1,3]

Piperine concentration in black pepper is variable. It is easily lost as a result of exposing pepper to heat and it also undergoes light-induced isomerization to isopiperine, chavicine and isochavicine (Figure 1.7.2). Consequently, the experimental yield of piperine will depend on the source as well as the method of storing and processing the black pepper.[1]

The aim of this experiment is to isolate and identify piperine, from black pepper. It is a representative experimental procedure to introduce the isolation of natural products to organic chemistry students. It allows students to learn a solid–liquid extraction method and to use other common laboratory techniques such as reflux, recrystallization and thin-layer chromatography. Furthermore, the isolated product gives good data for students to practice structural identification, by the analysis of IR, UV, MS, ^{1}H and ^{13}C NMR spectra of piperine.

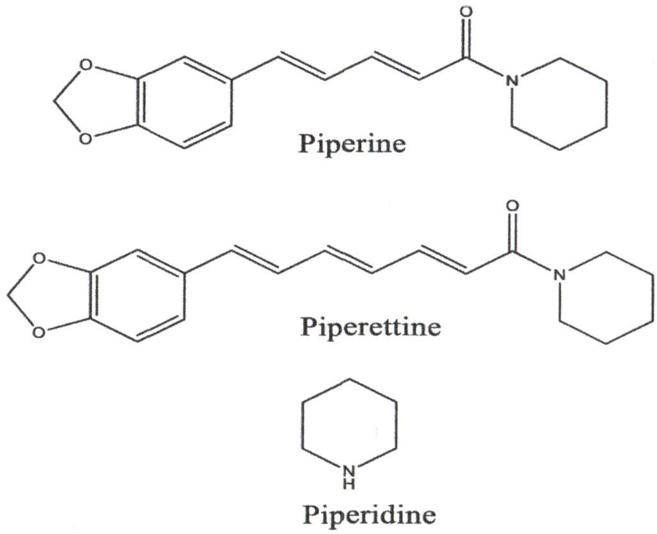

Figure 1.7.1 Alkaloids found in black pepper.

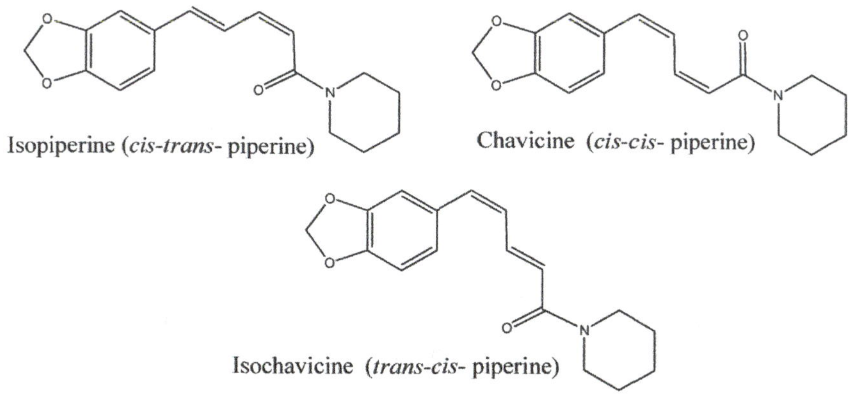

Figure 1.7.2 Piperine isomers.

Additional Safety

The entire experiment must be carried out in a fume hood, using the proper safety equipment such as protective clothing, goggles and safety gloves. Piperine is a lachrymator. Dichloromethane is a very volatile solvent and exposure to high levels can cause skin, ocular and respiratory irritation. Diethyl ether is extremely flammable, hazardous in case of ingestion or inhalation, as well as an irritant in case of skin and eye contact.

Experimental Procedure (ref. 4)

Extraction

1. Add 5.0 g of pure ground pepper and 10 mL of dichloromethane to a 50 mL round-bottom flask, containing a magnetic stir bar.
2. Attach a water condenser to the neck of the flask and turn the water on. Let the mixture reflux for 20 min at 50 °C. Upon completion, cool the flask.
3. Filter the residual solid ground pepper by vacuum filtration with a Büchner funnel and filter paper.
4. Wash the pepper ground twice with 5 mL of dichloromethane and save an aliquot of the filtrate for thin-layer chromatography analysis.

Isolation

5. Transfer the extract to a 50 mL round-bottom flask and evaporate the dichloromethane, under vacuum, until a dark brown and viscous oil is left.
6. Cool the oil in an ice bath and add 3 mL of cold ether, then gently stir for 3–4 min.
7. Remove the solvent, using a rotatory evaporator.
8. Repeat step 6.
9. Allow the flask to sit for 15 min in an ice bath to promote the precipitation of piperine. If this does not occur, scratch the glass with a spatula or repeat the steps 7 and 8.
10. Isolate the yellow crystals of crude piperine, by vacuum filtration. Wash them twice with 2 mL of cold ether and keep a portion of crystals in a capped vial/Eppendorf for thin-layer chromatography analysis.

Recrystallization

11. Place the piperine in a test tube and dissolve it in a minimum amount of hot acetone/*n*-hexane (3 : 2) solution.
12. Let the test tube to sit undisturbed for 15 min at room temperature, then transfer to an ice bath 30 min.
13. Isolate the piperine crystals by using vacuum filtration and wash them with cold ether.
14. When the crystals are dried, obtain their weight.

Piperine Identification

15. Perform thin-layer chromatography (TLC) analysis, using a silica-gel plate and *n*-hexane/ethyl acetate (1 : 1) mixture as eluent. Apply the samples collected during the experiment and the piperine standard to the TLC plate in separate lanes. After development, visualize the TLC plates either using an H_2SO_4/MeOH (1 : 1) solution, followed by heating, or using Dragendorff's reagent that gives orange spots when alkaloids are present.
16. Calculate the retention factor (R_f) of each experimental sample and compare them with the retention factor of the piperine standard.
17. Measure the melting point, and obtain IR, UV, MS, ^{1}H and ^{13}C NMR spectra of piperine.

Waste Disposal

Aqueous solutions may be diluted and discarded in the aqueous waste collector recipient. Diethyl ether may be reused or collected in a non-chlorinated waste container, for proper disposal. Dichloromethane may be reused or collected in a chlorinated waste container, for subsequent proper disposal.

Interpretation of Results and Additional Questions

1. Determine the isolated yield of crude piperine as a weight percentage of black pepper powder used.
2. Explain the importance of thin-layer chromatography to identify compounds.
3. Compare the obtained melting point with that from the literature data.
4. Analyze the IR, UV, MS, ^{1}H and ^{13}C NMR spectra obtained and compare them with those reported in the literature.[5,6]
5. Piperidine and piperettine are two other major alkaloid components of pepper. Explain how you could distinguish between the isolated piperine *versus* piperidine and piperettine, using ^{1}H NMR.

Acknowledgements

Adapted with permission from W. Epstein, D. Netz, J. Seidel, *J. Chem. Educ.* 1993, 70, 598. Copyright (1993) American Chemical Society.

References

1. M. Meghwal and T. Goswami, *Phytother. Res.*, 2013, **27**, 1121.
2. G. Cordell, M. Quinn-Beattie and N. Farnsworth, *Phytother. Res.*, 2001, **15**, 183.
3. A. Thiel, C. Buskens, T. Woehrle, S. Etheve, A. Schoenmakers, M. Fehr and P. Beilsten, *Food Chem. Toxicol.*, 2014, **66**, 350.
4. W. Epstein, D. Netz and J. Seidel, *J. Chem. Educ.*, 1993, **70**, 598.
5. K. Saha, H. Seal and M. Noor, *J. Bangladesh Agric. Univ.*, 2013, **11**, 11.
6. J. Araujo-Junior, E. Cunha, M. Chaves and A. Gray, *Phytochemistry*, 1997, **44**, 559.

1.8. Caffeine Extraction from Tea and Coffee

Ana Margarida Madureira*[a], Abel J. S. C. Vieira[b], and Maria-José U. Ferreira[a]

[a]iMed.ULisboa, Faculty of Pharmacy, University of Lisbon, Av. Professor Gama Pinto, 1649-003 Lisboa, Portugal; [b]Faculty of Sciences and Technology, Universidade Nova de Lisboa, Campus de Caparica, 2829-516 Caparica, Caparica, Portugal. *E-mail: afernand@ff.ul.pt

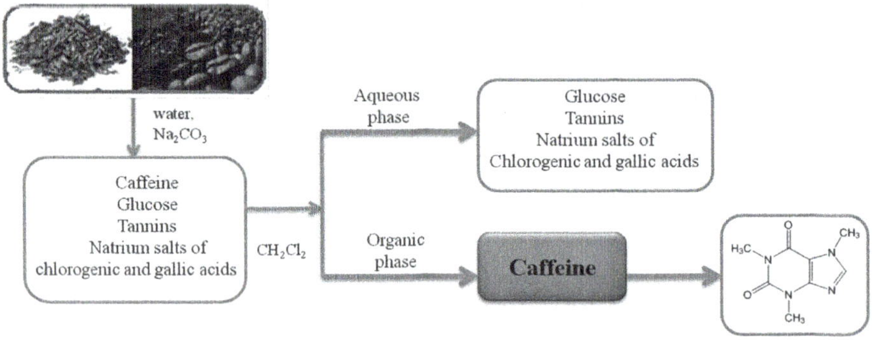

Number of sessions (duration of each session)	Hazard level	Difficulty level	Level of study
2 (3 h each)	Moderate	Medium	Intermediate

Class names Carboxylic acids, amines, amides

Concepts involved This experiment describes a method for isolating and characterizing caffeine from tea leaves and/or coffee. The experimental procedure is a good way to introduce natural products chemistry, to learn techniques such as solid–liquid extraction, recrystallization, sublimation and thin-layer chromatography and to practice structural identification, by the analysis of IR and ^1H and ^{13}C NMR spectra

Chemicals needed Soluble coffee, tea bags, sodium carbonate, dichloromethane, ethanol, potassium bromide, deuterated chloroform

Equipment and experimental techniques involved Separation techniques (liquid–liquid and solid–liquid extraction); recrystallization techniques; melting point; cold finger sublimation apparatus; IR and NMR apparatus

Keywords Extraction, IR, natural products, NMR, recrystallization, sublimation

Background

Tea and coffee are very important in our daily life. Most people begin their day by extracting plant derived-compounds named alkaloids, with boiling water, from tea leaves or roasted coffee beans, which allow the preparation of two of the most common beverages consumed worldwide.

Comprehensive Organic Chemistry Experiments for the Laboratory Classroom
Edited by Carlos A. M. Afonso, Nuno R. Candeias, Dulce Pereira Simão, Alexandre F. Trindade, Jaime A. S. Coelho, Bin Tan, and Robert Franzén.

Published by the Royal Society of Chemistry, www.rsc.org

The impact of the consumption of those two beverages is quite visible in the large quantities of both products production. World black tea production was estimated to be 2.4 million tones and projections for green tea production were around 900 000 tones in 2010. Production of coffee beans was estimated to be around 7.0 million tones (117 million bags) in 2010.[1]

The worldwide consumption of tea and coffee is strongly related to their stimulant properties. The improvements in mental alertness, concentration, fatigue and athletic performance associated with tea and coffee intake are due to their contents of methylxanthines: caffeine (the most active), theobromine and theophylline (Figure 1.8.1).[2] Because they possess a nitrogen atom they are included in the alkaloids family.

Caffeine is also present in several beverages and cocoa derivate foods like carbonated soft drinks, energetic drinks, fruit flavored and water beverages, chocolate bars and pastry. The caffeine content of some of the most commons beverages and foods is presented in Table 1.8.1. Apart from the energy drinks, coffee beverages display the highest caffeine content.[3]

According to the US FDA (Food and Drug Administration), a daily intake up to 400 mg is safe and not associated with adverse health effects.

Caffeine is the most known component of coffee. However, coffee beans (*Coffea* species) are constituted by several other compounds that include cellulose, minerals, sugar, lipids, amino acids, tannins and polyphenols. Additionally, vitamins B3, PP and chlorogenic acid can also be found. The bean growth conditions, industrial treatments (drying processes, roasting), grinding and beverage preparation significantly alter the composition of the final beverage.[6,7] Black and green tea (*Camellia sinensis*) leaves differ from each other by their

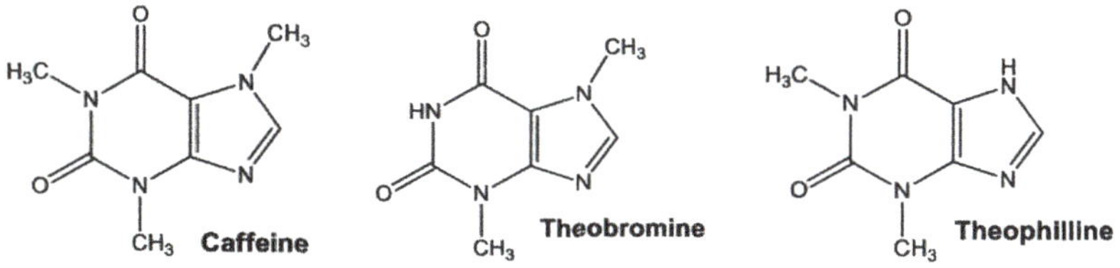

Figure 1.8.1 Methylxanthines from tea and coffee.

Table 1.8.1 Caffeine contents in some food products[2,4,5]

	Per g product[a] (%)	Prepared product
Coffee (beans)	1.2–1.8 (Arabica) 1.8–2.4 (Robusta)	100 mg/50 mL (expresso)
Coffee (soluble)	3.25	30–60 mg/150 mL
Coffee (decaffeinated)	0.0	3 mg/150 mL
Black tea	2.0–5.4	26–40 mg/100 mL
Green tea	1.5–2.9	15–30 mg/100 mL
White tea	3.4–5.7	13–24 mg/100 mL
Coca-Cola	n.a.	10 mg/100 mL
Coca-Cola (decaffeinated)	n.a.	0
Red bull	n.a.	32 mg/100 mL
Monster	n.a.	34 mg/100 mL
Cocoa beans	0.1–0.7	n.a.
Milk chocolate	n.a.	100–145 mg/100 g
Dark chocolate	n.a.	100–145 mg/100 g

[a]n.a. not applicable.

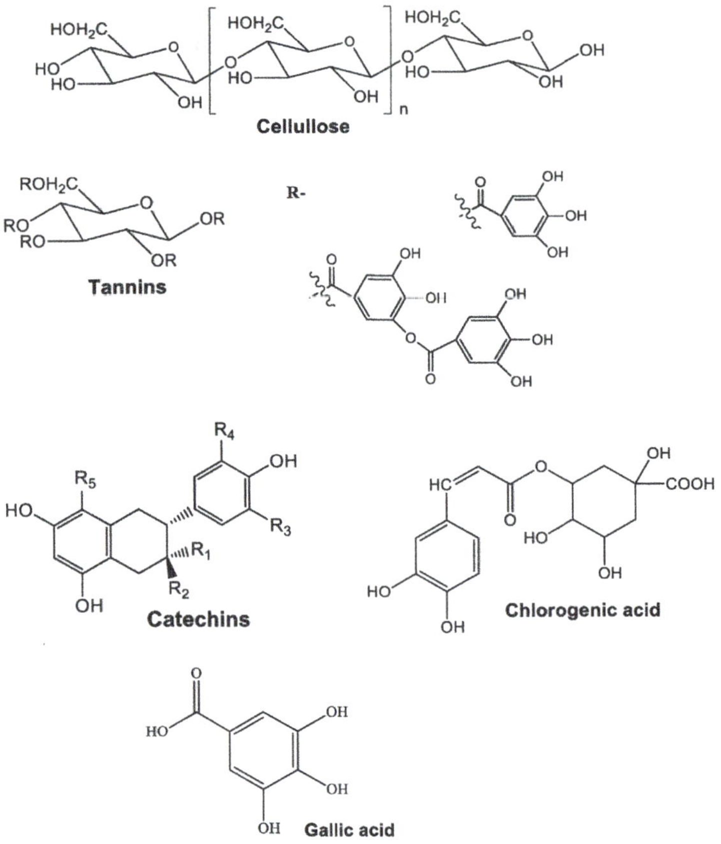

Figure 1.8.2 Some compounds present in coffee beans and tea leaves.

production procedures. Green tea is dried but not wilted, so it does not go through enzymatic oxidation whereas black tea is wilted and fully oxidized.[8] Tea leaves contain, beyond caffeine, theobromine, polyphenols (like tannins that represent 30–35% of the dry tea weight), amino acids, carbohydrates (cellulose), proteins, pigments, vitamins, catechins, minerals and a small amount of saponines.[9,10] Just like coffee, the composition of the final beverages depends on culture, industrial treatments and beverage preparation. Some of those compounds are presented in Figure 1.8.2.

In this experiment, instant coffee or tea leaves will be used. These products have no water insoluble components like proteins, but there still remains several other substances like tannins, glucose, chlorogenic and gallic acids that must be removed to obtain pure caffeine.[11]

Additional Safety

During this laboratory procedure, students will manipulate dichloromethane (methylene chloride). This compound is a low boiling point (40 °C) liquid, immiscible with water and inflammable. Some precautions must be taken:

I. Steps I.5–I.8: the separatory funnel containing water and dichloromethane mixture must be carefully and gently shaken to avoid the formation of emulsions. The emulsions formed can be broken simply by waiting some time (or stirring with a glass rod). If necessary, some saturated salt (sodium chloride) solution can be added.
II. After shorts periods of shaking, open the separatory funnel to make sure that the pressure inside does not rise and no projections occur.
III. Step I.13: the bath should not be too hot otherwise dichloromethane evaporation may occur.
IV. Step II.6–II.7: the sublimation requires some caution, namely the control of the temperature (oil bath) and vacuum applied.

Celite® can cause eye irritation and may be harmful if inhaled since it causes respiratory tract irritation.

Experimental Procedure (ref. 11 and 12)

First Laboratory Session: 3 h

I Isolation of Caffeine from Coffee/Tea

1. Place 10 g of soluble coffee (or 15 g of tea leaves/5–6 bags) and 10 g of powered sodium carbonate in a 250 mL Erlenmeyer flask.
2. Add 175 mL of distilled water and boil the contents for about 10 min.
3. Cool the mixture on an ice-bath.
4. If the experiment is performed with tea leaves proceed to their removal. Then, prepare a mixture of 15 g of Celite® and 100 mL of water and apply it over the disk of filter paper in the Büchner funnel before performing suction with the vacuum. Collect the filtrate.
5. Put the mixture or the filtrate on a 500 mL separatory funnel.
6. Add 35 mL of dichloromethane.
7. Shake the separatory funnel gently for 2–3 min. Allow the two phases to separate.
8. Remove the lower organic phase (dichloromethane).
9. Repeat steps 6–8 three times.
10. Combine all the organic phases obtained.
11. Dry the organic phase with anhydrous Na_2SO_4.
12. Remove the solid by gravity filtration. Rinse the drying Erlenmeyer flask and the funnel three times with dichloromethane and collect all organic phases.
13. Evaporate the dichloromethane on a rotary evaporator. A solid crude extract will be obtained.
14. Determine the weight of the crude product obtained.

Second Laboratory Session: 3 h

II Caffeine Purification

1. Recrystallize the crude solid from 95% ethanol. Use about 5 mL of ethanol per gram of product (as reference value).
2. Small white needles will be formed.
3. Collect the crystals by suction filtration.
4. Air-dry the crystals.
5. Determine the weight of the crystals mass.

Or/and:

6. Place the crude caffeine on a cold-finger apparatus.
7. Heat the oil bath at 180 °C, under reduced pressure (water vacuum pump).
8. Observe the transition from the solid state to gas and the subsequent solidification on the cold finger tube terminus.
9. Determine the yield according to the equations:

$$\%\ \text{yield} = \frac{\text{mass of crude caffeine}}{\text{mass of tea/coffee}} \qquad \%\ \text{recovery} = \frac{\text{mass of purified caffeine}}{\text{mass of crude caffeine}}$$

III Caffeine Identification

From the purified caffeine needles:

1. Determine the melting point and compare the result with the data available in the literature.

2. IR spectrum: prepare a KBr pellet (1 mg caffeine: 100 mg KBr). Obtain the spectrum on the appropriated apparatus. Analyse the absorption bands.
3. ^{1}H and ^{13}C NMR spectra: place a small amount of crystals (5–10 mg) in an NMR tube. Add 600 μL of deuterated chloroform ($CDCl_3$). Trace the proton and carbon spectra on the NMR apparatus. Analyse the signals present and compare with literature data.[13]

Waste Disposal

- Disposable tea bags/leaves can be thrown in the solid residue container.
- After the extraction, the aqueous layer should be disposed into a container labelled "Caf feine – aqueous waste".
- Place the Celite® in the solid residue container.
- The evaporated dichloromethane might be collected for further reutilization in a container labelled "CH_2Cl_2 – evaporated".
- The mother liquor resulting from recrystallization should be disposed into a container labelled "Caffeine – aqueous waste".
- Caffeine crystals should be collected in a container labelled "Caffeine (purified)".

Results Interpretation and Additional Questions

1- Why do you use Na_2CO_3 in this experiment?
2- What are the advantages of the utilization of CH_2Cl_2 (instead of another organic solvent like diethyl ether)?
3- Could you replace dichloromethane with methanol to perform the liquid–liquid extraction?
4- What are the criteria for the selection of recrystallization solvent?
5- Compare the "yields" of crude, purified caffeine by recrystallization and by sublimation and draw the appropriate conclusions.
6- Why is it necessary to remove the stopper from the separatory funnel when the liquid is being drained?
7- Try to explain, using a phase diagram, the conditions at which sublimation may occur.

References

1. Food and Agriculture Organization of the United Nations, *Medium-term prospects for agricultural commodities: Projection to the year 2010*, 2003.
2. D. C. Mitchell, C. A. Knight, J. Hockenberry, R. Teplansky and T. J. Hartman, *Food Chem. Toxicol.*, 2014, **63**, 136.
3. C. J. Stratton, *BYU Stud.*, 1980, **20**, 371.
4. *Caffeine informer*, http://www.caffeineinformer.com/the-caffeine-database, accessed September 2014.
5. J. J. Barone and H. R. Roberts, *Food Chem. Toxicol.*, 1996, **34**, 119.
6. S. I. Mussatto, S. E. M. Machado, S. Martins and J. A. Teixeira, *Food Bioprocess Technol.*, 2011, **4**, 661.
7. R. C. Alves, S. Casal and B. Oliveira, *Quim. Nova*, 2009, **32**, 2169.
8. *Dictionary.com (4 October 2010)*, What's the difference between green and black tea? And who is the the "Earl" in "Earl Grey"?, http://blog.dictionary.com/tea-green-black-earl-grey/, accessed September 2014.
9. *O-cha.net*, Tea components, http://www.o-cha.net/english/cup/pdf/38.pdf, September 2014.
10. D. R. Palleros, *Experimental organic chemistry*, John Wiley & Sons, Inc., New York, 2000, experiment 5, p. 103.
11. K. Williamson and K. Masters, *Macroscale and Microscale Organic Experiments*, Charles Hartford, Brooks/Cole, Cengage Learning, Canada, 6th edn, 2011, unit 6, p. 113.
12. R. Ikan, *Natural products: a laboratory guide*, Academic Press, New York, 2nd edn, 1991.
13. J. Sitkowski, L. Stefaniak, L. Nicol, M. L. Martin, G. J. Martin and G. A. Webb, *Spectrochim. Acta, Part A*, 1995, **51**, 839.

1.9. Isolation of the Alkaloid Lupanine from *Lupinus albus* Seeds

Ana M. Lourenço* and Luísa Maria Ferreira

REQUIMTE, Departamento de Química, Faculdade de Ciências e Tecnologia, Universidade Nova de Lisboa, 2829-516 Caparica, Portugal. *E-mail: ana.lourenco@fct.unl.pt

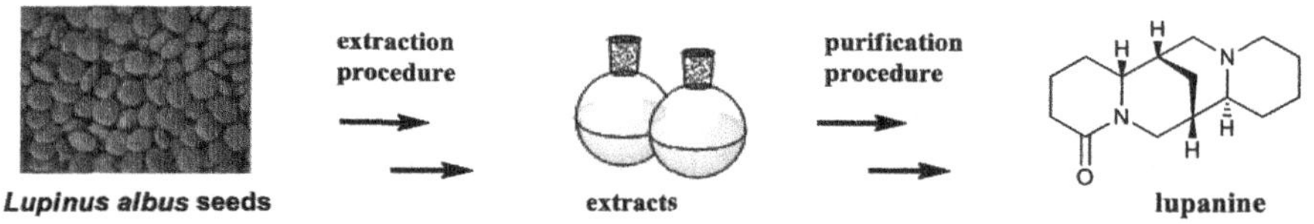

Number of sessions (duration of each session)	Hazard level	Difficulty level	Level of study
One session (4 h)	Moderate	Medium	Intermediate

Class names Tertiary amines, lactams

Concepts involved This experiment consists of the solid–liquid extraction of the alkaloid lupanine from *Lupinus albus* seeds. The process also includes column chromatography purification and NMR analysis for structure characterization

Chemicals needed *Lupinus albus* seeds, potassium hydroxide, Celite®, diethyl ether, dichloromethane, ethyl acetate, ethanol, triethylamine, ninhydrin, anhydrous sodium sulfate, deuterochloroform

Equipment and experimental techniques involved Columns for chromatographic purification, rotatory evaporator, NMR equipment for structural analysis

Keywords Alkaloids, lupanine, NMR, normal-phase chromatography, solid–liquid extraction

Background

The *Lupinus* genus belongs to the Leguminosae family, and is widely used as human food and animal feed due to its high nutrient value, which is comparable to that of soybean. *Lupinus* is used as a source of protein, fiber and for supplementation in food products, like breadmaking and biscuits.[1,2] The genus *Lupinus* typically contains 36–52% protein, 5–20% oil, and 30–40% fiber.[2] Quinolizidine alkaloids (QAs) are also present and have been quantified; in different samples of *L. albus* L. seeds QAs are present from 1.90% to 2.69%.[3] The bitter taste of the seeds is imparted by the QAs that must be separated before consumption. Different procedures to obtain sweet *Lupinus* have been proposed but the process that is widely used for seed debittering is exhaustive leaching in water.

Lupinus albus seeds are rich in QAs, mainly lupanine. This alkaloid has a tetracyclic structure with a quinolizidine nucleus and a lactam subunit. The structure is versatile as starting material for the semi-synthesis of other QAs due to the quinolizidine nucleus. Many attempts have been made to synthesize this subunit, with little success because too many steps are required for the quinolizidine nucleus formation, and also because the overall yield is low. Moreover lupanine is

Comprehensive Organic Chemistry Experiments for the Laboratory Classroom
Edited by Carlos A. M. Afonso, Nuno R. Candeias, Dulce Pereira Simão, Alexandre F. Trindade, Jaime A. S. Coelho, Bin Tan, and Robert Franzén.

Published by the Royal Society of Chemistry, www.rsc.org

also described as reducing high blood pressure and as a hyperglycemia agent.[4] The separation of the alkaloid fraction from *Lupinus albus* seeds is imperative for consumption and the extracted lupanine can be used for different purposes.

The extraction of QAs is currently performed by continuous extraction, or multiple extractions with water at room temperature, which is the way of leaching seeds in plant units. Different methodologies for the isolation of lupanine can be tested. The main goal is to reach a higher amount of extracted QA fraction considering the source used for the extraction.[4–6] The method of extraction assayed in this experiment makes use of ethanol extraction in basic medium (the seeds are milled and macerated with a solution of 25% KOH and Celite® prior contact with the organic solvent). The extract is mainly composed of lupanine in a mixture of other minor QAs. Column chromatography is used for lupanine purification and thin-layer chromatography (TLC) for analysis of the purity of each fraction. At the end of the purification process lupanine is characterized by NMR spectroscopy and comparison with published data.

Additional Safety

It is not essential to work in a fume hood in any step of the experiment. However, if possible in the class laboratory conditions, it is preferred to perform the chromatographic column in a fume hood due to the handling of ethyl ether. Regular laboratory good practices should always be taken into account, especially the use of gloves and safety glasses when handling concentrated hydroxide, organic solvents, separator funnel and vacuum apparatus.

Experimental Procedure

(a) Prior to the Laboratory Session: Extraction of *Lupinus albus* Seeds with Ethanol in Basic Medium

1. Weigh 50 g of *Lupinus albus* seeds and grind them to a fine powder.
2. Add 40 mL of an aqueous solution of 25% KOH. Add 20 g of Celite® to afford a dried powder.
3. Add 200 mL of ethanol to the previous mixture and leave it with gently stirring overnight on a magnetic stirrer (approximately 16 h).

(b) Laboratory Session 1 (4 h): Recovery of the QA Extract and Lupanine Purification

1. Separate the supernatant by filtration through a filter paper and wash the solid material with 100 mL of fresh ethanol.
2. Evaporate the solvent under vacuum to obtain the QA extract.
3. Dissolve the extract in 100 mL of diethyl ether and extract once with 75 mL of water. The organic solvent is dried with anhydrous sodium sulfate, filtered and evaporated under vacuum.
4. Re-dissolve the residue in a small volume of dichloromethane (40–90 mL) and mix it with 15–20 g of silica gel. Let all the mixture dry gently in the rotatory evaporator until a loose powder is formed.
5. Construct a chromatography column with a porous glass (length = 30 cm, diameter = 4 cm). At the bottom place approximately 40 g of silica and then the lupine extract. A small piece of cotton should be placed on top to avoid destruction of the silica gel surface while pouring the eluent.

6. Apply pressure at the top of the column (to elute 27.5 mL min^{-1}) and proceed with the elution successively with diethyl ether (500 mL), ethyl acetate (400 mL), ethyl acetate : *n*-hexane (2 : 1)–5% triethylamine (TEA) (150 mL) and ethyl acetate–5% TEA (400 mL). Diethyl ether removes less polar and yellow compounds and it is poured thorough the silica column until there is a significant decrease of color. Other compounds are extracted with ethyl acetate and the change of this solvent to ethyl acetate : *n*-hexane (2 : 1) occurs when the eluted fraction changes from a light yellow to a colorless fraction.
7. Check by TLC the presence of lupanine in the fraction eluted with ethyl acetate with 5% TEA (silica gel plate/eluent: diethyl ether : dichloromethane (1 : 1), $R_f = 0.25$, stain: ninhydrin). Remove the solvent of this fraction by evaporation under vacuum. Take to dryness the sample with lupanine in the vacuum pump and analyse by nuclear magnetic resonance (NMR).

Waste Disposal

All organic solvents have to be poured into special containers for subsequent disposal in the manner of treatment of chemical waste. The solid residue from the extraction is loaded with potassium hydroxide and should be discharged into a special container for later collection and destruction of chemical waste.

Results Interpretation and Additional Questions

Questions

1. Lupanine may also be extracted from *Lupinus* seeds with water, which is the traditional way of leaching for consumption. Considering the chemical structure of lupanine present a brief discussion about its solubility both in water and in organic solvents.
2. Justify the importance of the use of potassium hydroxide in the extraction process of lupanine from *Lupinus albus* seeds.
3. The TLC silica gel plate to check the purity of lupanine is eluted with diethyl ether : dichloromethane (1 : 1). What change in the R_f value do you expect if TEA at 1% concentration would be added to the eluent?
4. The lupanine structure has some complexity. For better understanding of the space orientation of the quinolizidine structure construct the structure models of both enantiomers of lupanine. Establish the absolute configurations of each asymmetric carbon atom according to CIP (Cahn–Ingold–Prelog) rules.

References

1. G. D. Hill, *Fourth International. Lupin Conference, Geraldton, Western Australia,* 15–22 August 1986, p. 40.
2. A. A. Mohamed and P. Rayasduarte, *Cereal Chem.*, 1995, **72**, 643.
3. M. Muzquiz, C. Cuadrado, G. Ayet, C. Delacuadra, C. Burbano and A. Osagie, *J. Agric. Food Chem.*, 1994, **42**, 1447.
4. P. M. G. Lopez, P. G. de la Mora, W. Wysocka, B. Maiztegui, M. E. Alzugaray, H. Del Zotto and M. I. Borelli, *Eur. J. Pharmacol.*, 2004, **504**, 139.
5. A. K. Przybyl and M. Kubicki, *Tetrahedron*, 2011, **67**, 7787.
6. S. Carmali, V. D. Alvesa, I. M. Coelhoso, L. M. Ferreira and A. M. Lourenço, *Sep. Purif. Technol.*, 2010, **74**, 38.

1.10. Isolation and Purification of Atropine, a Tropane Alkaloid Obtained from *Atropa belladonna* L. (Solanaceae)

Gustavo da Silva

Department of Pharmacological Sciences, Research Institute for Medicines and Pharmaceutical Sciences (iMed.ULisboa), Faculty of Pharmacy – University of Lisbon, Av. Prof. Gama Pinto, 1649-003 Lisbon, Portugal. *E-mail: galfdsilva@ff.ulisboa.pt

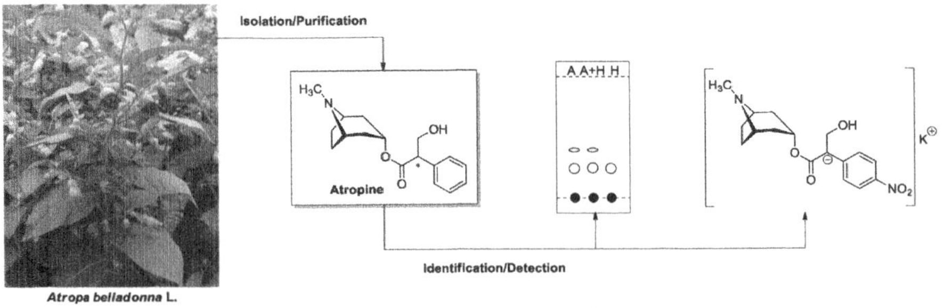

Number of sessions (duration of each session)	Hazard level	Difficulty level	Level of study
1 (3 h)	High	Medium	Intermediate
Class names Tropane alkaloids, tertiary amines, tropic acid esters			
Concepts involved This experiment aims at the extraction of tropane alkaloids from *Atropa belladonna* and includes chromatographic techniques and chromatic reactions for the identification of atropine			
Chemicals needed Acetone, ammonia (concentrated), anhydrous sodium sulfate, diethyl ether, diethylamine, Dragendorff's reagent, ethanol solution of potassium hydroxide (0.1 M), ethyl acetate, fuming nitric acid, methanol, hyoscyamine sulfate, sodium nitrite aqueous solution (10%), sulfuric acid (0.1 N), toluene			
Equipment and experimental techniques involved Ultrasound-assisted extraction, liquid–liquid partition, rotatory evaporator, thin-layer chromatography (TLC), Vitali–Morin reaction			
Keywords Alkaloid-directed extraction, chromatic reaction, TLC analysis, TLC detection, tropane alkaloids			

Background

Atropine is a racemic mixture of pharmaceutical interest, consisting of D-hyoscyamine and L-hyoscyamine.[1] Atropine and its salts are active pharmaceutical ingredients, which can be also used as adjuncts, in many proprietary preparations.[2] The pharmacological interest in the isolation of atropine is reinforced by its inclusion in the WHO Model List of Essential Medicines.[3] The

Comprehensive Organic Chemistry Experiments for the Laboratory Classroom
Edited by Carlos A. M. Afonso, Nuno R. Candeias, Dulce Pereira Simão, Alexandre F. Trindade, Jaime A. S. Coelho, Bin Tan, and Robert Franzén.

Published by the Royal Society of Chemistry, www.rsc.org

wide variety of therapeutic applications of atropine has instigated the production and isolation of this compound. Although atropine can be prepared *via* organic synthesis,[4] it can be easily isolated from plants.[5]

Chemically, atropine is a tropane alkaloid, a 3-tropanol ester of racemic tropic acid.[6] Tropane alkaloids mainly occur in the Solanaceae family but can also be found in other plant families, such as Convolvulaceae, Erytroxylaceae, Proteaceae and Rhizophoraceae.[7] These alkaloids are plant secondary metabolites, which are biosynthetically derived from L-ornithine (Scheme 1.10.1).[8] The deadly nightshade (*Atropa belladonna* L.), native to central and southern Europe, is capable of producing and accumulating high levels of tropane alkaloids, predominantly hyoscyamine.[9]

Atropa belladonna is a perennial herb cultivated in Europe and the United States of America for drug extraction; the tops of three-year-old plants are harvested 2–3 times per year and dried to afford *Belladonnae folium* (*Materia medica*, as atropine, described in monographs of European and American Pharmacopoeias).[10] Roots from plants 3–4 years old have a slightly higher alkaloid content, but are less commonly employed as a source of atropine.[11] The high content in atropine justifies the ethnomedical use of the plant as anesthetic[12] and mydriatic agent.[13]

Atropine is classified as an anticholinergic agent, which competes with acetylcholine for the muscarinic site of the parasympathetic nervous system, thus preventing the passage of nerve impulses. The agonist properties of atropine lead to several useful effects, including antispasmodic action on the gastrointestinal tract, antisecretory effect controlling salivary secretions

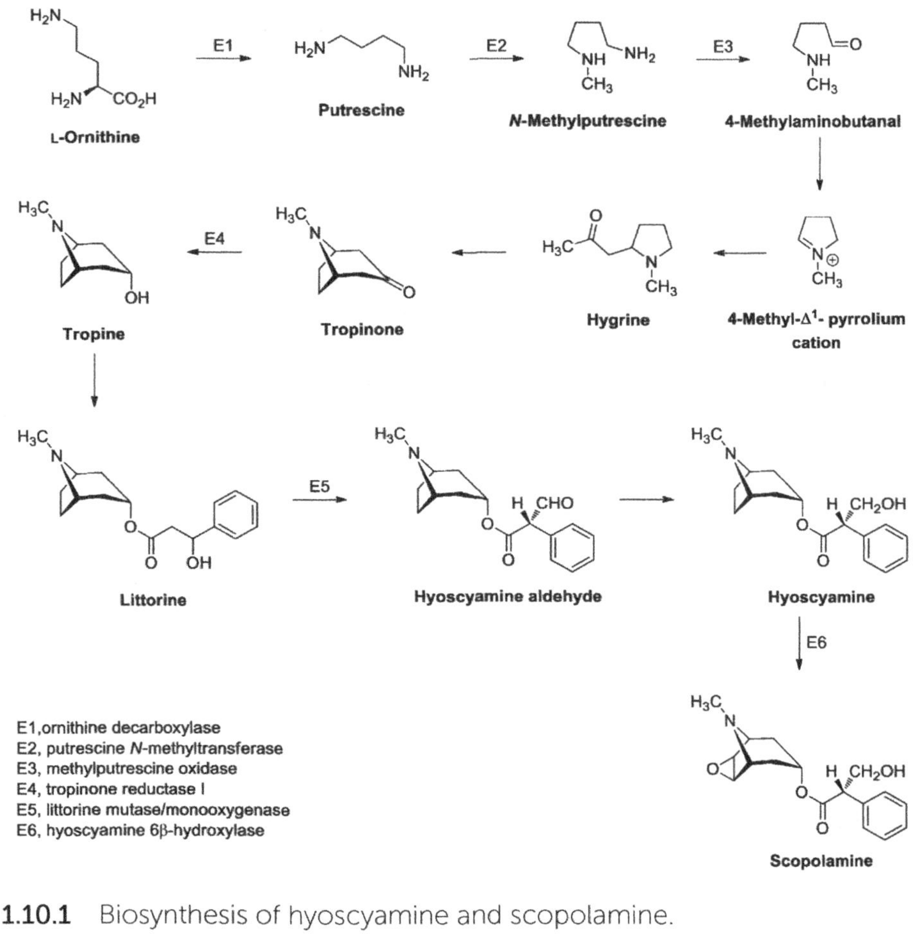

Scheme 1.10.1 Biosynthesis of hyoscyamine and scopolamine.

during surgical operations, and as mydriatics to dilate the pupil of the eye. Atropine is also a useful antidote in cases of poisoning caused by acetylcholinesterase inhibitors (*e.g.* physostigmine and neostigmine) and organophosphate insecticides.[14]

Additional Safety

All three parts of the session should be carried out in a fume hood, with proper safety equipment, such as protective clothing, gloves and safety goggles, to reduce exposure to volatile solvents and toxic chemicals. The use of gloves is essential not only during the manipulation of corrosive chemicals (*e.g.* sulfuric acid), but also because the isolated product (atropine) can be readily absorbed through the skin. Owing to the acute toxicity of atropine and the restricted use of diethylamine (in some countries) it is prudent to check the regulatory schedules. The manipulation of diethyl ether does not incur notable safety precautions, apart from general laboratory good practices. However, keep in mind that diethyl ether can form peroxides (when exposed to atmospheric oxygen, light and/or oxidizing agents), which are explosive and, consequently, during the procedure in the rotary evaporator the use of temperature should be avoided.

Experimental Procedure

This procedure consists of three parts: total alkaloid extraction, product identification and Vitali–Morin reaction.

Part 1: Alkaloid Extraction

1. Accurately weigh 1.0 g of the powdered plant material (*Belladonnae folium*) into a rubber septum sealed tube with 15 mL of sulfuric acid (0.1 N).
2. Sonicate for 5 min and then filter the content of the tube through Grade 1 Whatman paper (or equivalent).
3. Wash the filter with additional sulfuric acid (0.1 N) to a volume of 20 mL of filtrate.
4. Add 1 mL of ammonia (conc.) to the filtrate and transfer the solution into a separatory funnel and perform the extraction with 3 × 20 mL of peroxide-free diethyl ether.
5. After drying over anhydrous sodium sulfate, filter the combined organic layers and remove the solvent *in vacuo* in a rotary evaporator, at room temperature.
6. Label the round bottom flask (which will be used in Part 2) and weigh the residue.
7. Repeat the previous steps, label the round bottom flask (which will be used in Part 3), weigh the residue and determine the mean yield of the total alkaloid extraction.

Part 2: Product Identification

8. Dissolve the total alkaloid extract contained in one of the round-bottom flasks in 0.50 mL methanol.
9. Dissolve 10 mg of hyoscyamine sulfate (reference standard) in 2 mL methanol.
10. Using a micro-capillary pipette, apply 20 μL of the total alkaloid extract (*A*), total alkaloid extract and reference (*A* + *H*), and reference (*H*), in an aluminium TLC plate.
11. In a twin trough chamber (CAMAG or equivalent), previously saturated for 1 h (with filter paper) with 5–10 mL of eluting solvent, develop the plate (as presented in Table 1.10.1) and identify the constituents of the total alkaloid extract.
12. After elution, dry the TLC plate in a stream of cool air.
13. Initially, use a Wood's lamp (UV, 365 nm) for detection purposes and then spray the plate evenly with Dragendorff's reagent.
14. Dry the TLC plate in a cool stream of air and then re-spray the plate evenly with sodium nitrite aqueous solution (10%).

Table 1.10.1 Chromatographic conditions.

Component	Specific setting
Adsorbent	Silica gel 60 F_{254} precoated TLC plates
Solvent system	Toluene – ethyl acetate – diethylamine (70 : 20 : 10) (v/v/v)
Migration	8 cm
Detection	A: UV, 365 nm
	B: Dragendorff followed by sodium nitrite (vis)

15. Using a pencil, mark the major and minor constituents of the total alkaloid extract of *Belladonnae folium* and calculate both R_f values.
16. Based on the intensity and R_f values of both constituents, identify atropine and scopolamine.

Part 3: Vitali–Morin Reaction

17. Transfer the content of the other round bottom flask (obtained in Part 1) to a porcelain evaporating dish and treat the alkaloid extract with 0.5 mL of fuming nitric acid.
18. Using a low flame from a spirit burner, evaporate the solution to dryness and then remove the flame.
19. Dissolve the residue in 5 mL of acetone and then add 1 mL of an ethanol solution of potassium hydroxide (0.1 M) to the dried residue.
20. Observe and register the color alterations.

Waste Disposal

After being neutralized, the aqueous phase from the liquid–liquid extraction should be diluted and discarded in the sink. The recovered diethyl ether may be reused or collected in a non-chlorinated waste collector recipient, along with the remaining chromatographic solvent system, for posterior destruction.

Results Interpretation and Additional Questions

1. Discuss the importance of pH manipulation within the context of extracting atropine from *Belladonnae folium* and discuss alternative procedures for this alkaloid-directed extraction.
2. Compare the total alkaloid extraction yield with the expected values from literature references.
3. Discuss the use of Dragendorff's reagent for the detection of alkaloids and suggest alternative reagents for this propose.
4. Explain the use of sodium nitrite after Dragendorff spraying.
5. Suggest a chemical explanation for the chromatic Vitali–Morin reaction.
6. Justify the use of an ethanol solution of potassium hydroxide during the Vitali–Morin reaction.

References

1. R. Soares, A. K. Singh, E. R. Kedor-Hackmann and M. I. Santoro, *J. AOAC Int.*, 2009, **92**, 1663.
2. A. Brayfield, *Martindale: The Complete Drug Reference*, Pharmaceutical Press, London, 38th edn, 2014.
3. World Health Organization Commission, *WHO Model List of Essential Medicines – 18th List*, WHO Press, Geneva, 18th edn, 2013.

4. A. Ladenburg, *Ann. Chem.*, 1883, **217**, 75; G. Schwenker, W. Prenntzell, U. Gassner and R. Gerber, *Chem. Ber.*, 1966, **99**, 2407; R. Willstätter, *Chem. Ber.*, 1898, **31**, 1537; R. Willstätter, *Ann. Chem.*, 1903, **326**, 23.
5. F. Chemnitius, *J. Prakt. Chem.*, 1927, **116**, 276.
6. W. C. Evans, *Trease and Evans: Pharmacognosy*, Saunders, London, 16th edn, 2009.
7. R. de Simone, L. Margarucci and V. de Feo, *Pharmacologyonline*, 2008, **1**, 70.
8. E. Leete, *Planta Med.*, 1990, **56**, 339; J. Ziegler and P. J. Facchini, *Annu. Rev. Plant Biol.*, 2008, **59**, 735.
9. E. Eich, *Solanaceae and Convolvulaceae: Secondary Metabolites – Biosynthesis, Chemotaxonomy, Biological and Economic Significance (A Handbook)*, Springer, Berlin, 2008.
10. Belladona leaf (Belladonnae folium) Monograph, accessed via European Pharmacopoeia 8th Edition, https://www.edqm.eu/en/european-pharmacopoeia-8th-edition-1563.html.
11. P. M. Dewick, *Medicinal Natural Products: A Biosynthetic Approach*, John Wiley & Sons Ltd, Chichester, 3rd edn, 2009.
12. V. P. Nair and J. M. Hunter, *Contin. Educ. Anaesth. Crit. Care Pain*, 2004, **4**, 164.
13. J. Bruneton, *Toxic Plants – Dangerous to Humans and Animals*, Lavoisier Publishing, Paris, 1999; U. Cikla, S. Turkmen, Y. Karaca, F. A. Ayaz, S. Turedi and A. Gunduz, *Hum. Exp. Toxicol.*, 2011, **30**, 1998; H. Rajput, *Nat. Prod. Chem. Res.*, 2013, **1**, 104.
14. M. Ebadi, *Pharmacodynamic Basis of Herbal Medicine*, CRC Press, Boca Raton, FL, 2nd edn, 2007; J. D. Garven, *J. Pharm. Pharmacol.*, 1956, **8**, 256; D. M. Green, A. W. Muir, J. A. Stratton and T. D. Inch, *J. Pharm. Pharmacol.*, 1977, **29**, 62; J. D. Garven, *J. Pharm. Pharmacol.*, 1956, **8**, 256.

1.11. Isolation and Purification of Carnosol from *Salvia officinalis*†

Filipe Pereira[a] and Patrícia Rijo*[a,b]

[a]Universidade Lusófona Research Center for Biosciences & Health Technologies (CBIOS), Campo Grande, 376, 1749-024 Lisboa, Portugal; [b]Instituto de Investigação do Medicamento (iMed.ULisboa), Faculdade de Farmácia da Universidade de Lisboa, Av. Professor Gama Pinto, 1649-003 Lisboa, Portugal
*E-mail: patricia.rijo@ulusofona.pt

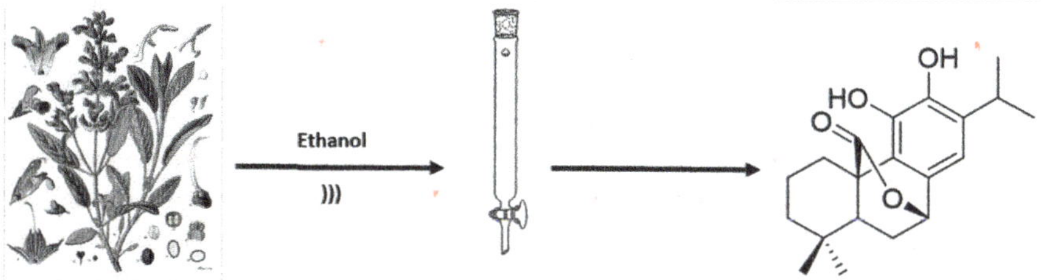

Salvia officinalis L. Column chromatography Carnosol

Number of sessions (Duration of each session)	Hazard level	Difficulty level	Level of study
2 (4 h each)	Moderate	High	Intermediate
Class names Phenolic diterpenes			
Concepts involved The isolation of the diterpene carnosol, from an ethanolic extract of *Salvia officinalis* (Lamiaceae), will be performed using a fast and effective ultrasound method. After precipitation in water the resulting solid will be purified through column chromatography and recrystallization to yield carnosol, which will be characterized by TLC and ^{1}H-NMR spectroscopy			
Chemicals needed Ethanol, ethyl acetate, *n*-hexane, carnosol, silica gel			
Equipment and experimental techniques involved Ultrasound bath, centrifuge, rotatory evaporator, column chromatography, TLC, recrystallization, NMR			
Keywords Column chromatography, evaporation under reduced pressure, NMR, recrystallization, TLC, ultrasound extraction			

Background

The isolation and further purification of carnosol, one of the main diterpenes from *Salvia officinalis* (Lamiaceae), will be carried out. *S. officinalis* is native to the Mediterranean region, but is now widespread all over the world. It has a long history of medicinal and culinary use and more recently as an ornamental garden plant. Carnosic acid and carnosol, the two major phenolic diterpenes present in sage (*S. officinalis*) and rosemary (*Rosmarinus officinalis* L.) extracts, have received attention in food science and biomedicine. In plants, these compounds have been identified as being present in some species of the family Lamiaceae, and the occurrence of these compounds has been observed in leaf extracts of the genus *Salvia*.[1]

†Dedicated to Professor M. Fátima Simões on the occasion of her retirement.

Comprehensive Organic Chemistry Experiments for the Laboratory Classroom
Edited by Carlos A. M. Afonso, Nuno R. Candeias, Dulce Pereira Simão, Alexandre F. Trindade, Jaime A. S. Coelho, Bin Tan, and Robert Franzén.

Published by the Royal Society of Chemistry, www.rsc.org

Carnosol is the product of oxidative degradation of carnosic acid, which is known for its instability. In the presence of methanol for 1 week at room temperature, carnosic acid will oxidize to carnosol. The most common way to obtain a highly purified form of carnosol is through extraction and purification from natural sources such as sage or rosemary.[2] The identification and characterization of antioxidant activity, anti-inflammatory activity and anti-cancer properties of carnosol has attracted significant interest over the years. Recently, the European Union has approved a rosemary extract standardized to carnosol for its application towards food preservation.[3–7]

Liquid extraction constitutes a critical step in the isolation of different compounds from terrestrial plants. The extraction of pharmacologically active compounds from *S. officinalis* by conventional methods (maceration, percolation) is very time consuming, taking between 1 and 2 weeks.[8] The extractions, using polar solvents to extract non-polar organic compounds, can become efficient, if the extraction conditions are optimized. Such efforts could lead to cheap and more eco-friendly processes that could be applied to various extraction processes, providing the compounds to be extracted are stable under the conditions used.[9,10]

Sonication improved the yield of the diterpene from an ethanol extract and shortened the extraction time; it also reduced the solvent effect so that ethanol, which is a poor solvent under conventional conditions, reached a similar level of extraction efficiency as others solvents like ethyl acetate.[11,12]

The objective of this work is the isolation of carnosol from an ethanol extract of the plant material using a quick and effective ultrasound method. After filtration and evaporation the resulting solid will be purified through a silica gel column chromatography and subsequent recrystallization will yield the crystals of carnosol.[13] The structure of the isolated diterpene will be identified by comparison of its spectral properties (^{1}H NMR) with those reported in the literature.[10,14,15]

Additional Safety

The session should be carried out with proper safety equipment, such as protective clothing, gloves and safety goggles. The column chromatography should be performed in a fume hood.

Experimental Procedure

Laboratory Session 1 (4 h): Extraction of Carnosol from *S. officinalis*

1. The dried leaves of *S. officinalis* obtained commercially should be grinded to powder.
2. Place the dried plant material (20 g) in a 250 mL Erlenmeyer flask and add 200 mL of ethanol. All powdered plant should be completely submerged in the ethanol.
3. The flask and its contents should be immersed in an ultrasonic bath and sonicated for 15 min.
4. Filtrate the resultant solution using a filter paper in a 500 mL round-bottom flask to remove insoluble materials, and concentrate to dryness in a rotatory evaporator at 40 °C until constant weight. Determine the residue weight and the extraction yield in w/w (weight of filtrate/weight of plant).
5. Redissolve the residue with ethanol (5 mL). Then, add water (10 mL) and you should observe a precipitate. To separate the solid phase, centrifuge at 2500 rpm for 2.5 min in a conical centrifuge tube.
6. Remove the supernatant with a Pasteur pipette and wash the residue with 5 mL of water.
7. Prepare a TLC plate (silica-gel; *n*-hexane: ethyl acetate (7 : 3)) to identify the carnosol in the precipitate (dissolved in ethyl acetate). Determine the R_f value of carnosol, using a carnosol standard.

Laboratory Session 2 (4 h): Isolation and Purification of Carnosol

1. A column chromatography should be prepared by the slurry method with silica gel (15 g) and *n*-hexane.

2. Dissolve the precipitate in the minimum volume of *n*-hexane (add some ethyl acetate drops) and load the sample on the top of the column.
3. Elute the column with 20 mL of *n*-hexane, 20 mL of a *n*-hexane: ethyl acetate (9 : 1) mixture; 20 mL of *n*-hexane: ethyl acetate (8 : 2); 60 mL of *n*-hexane: ethyl acetate (7 : 3); 20 mL of *n*-hexane: ethyl acetate (5 : 5) and 20 mL of ethyl acetate.
4. Collect small fractions (4 mL) and then use TLC to monitor the column (silica gel; *n*-hexane: ethyl acetate (7 : 3)).
5. Develop the plates to identify the fractions containing carnosol (spot a standard whenever necessary). Combine those fractions in a weighted round-bottom flask.
6. Evaporate all the solvent and evaluate the impure carnosol.
7. Recrystallize carnosol from *n*-hexane (5 mL) to give carnosol crystals. Obtain the weight of pure carnosol.
8. Prepare a solution of carnosol in $CDCl_3$ to obtain a ^1H NMR spectrum. Confirm the sample structure by comparison of the spectral properties with those reported in the literature.[10,14,15]

Waste Disposal

The aqueous solution may be diluted and discarded in the sink. The organic solvents (ethanol, *n*-hexane and ethyl acetate) may be reused or collected in a non-chlorinated waste collector recipient, for posterior destruction.

Results Interpretation and Additional Questions

1. Calculate the extract yield in % (w/w) and the purified carnosol quantity in mg from the raw material.
2. Analyze the TLC plate of the precipitate after centrifugation. Determine the R_f value of carnosol.
3. Discuss the success of carnosol isolation by column chromatography (CC) referring to which fractions contain carnosol and determine the pure and impure carnosol yield.
4. Compare the obtained spectral data (^1H NMR) of carnosol with those described in the literature. (Note: collect them in a table.)

References

1. M. E. Abreu, M. Muller, L. Alegre and S. Munné-Bosch, *J. Sci. Food Agric.*, 2008, **88**, 2648.
2. N. Okamura, Y. Fujimoto, S. Kuwabara and A. Yagi, *J. Chromatogr. A*, 1994, **679**, 381.
3. A. Russo, C. Formisano, D. Rigano, F. Senatore, S. Delfine, V. Cardile, S. Rosselli and M. Bruno, *Food Chem. Toxicol.*, 2013, **55**, 42.
4. V. G. Kontogianni, G. Tomic, I. Nikolic, A. A. Nerantzaki, N. Sayyad, S. Stosic-Grujicic, I. Stojanovic, I. P. Gerothanassis and A. G. Tzakos, *Food Chem.*, 2013, **136**, 120.
5. M. J. Jordán, V. Lax, M. C. Rota, S. Lorán and J. A. Sotomayor, *J. Agric. Food Chem.*, 2012, **60**, 9603.
6. M. T. Huang, C. T. Ho, Z. Y. Wang, T. Ferraro, Y. R. Lou, K. Stauber, W. Ma, C. Georgiadis, J. D. Laskin and A. H. Conney, *Cancer Res.*, 1994, **54**, 701.
7. J. J. Johnson, *Cancer Lett.*, 2011, **305**, 1.
8. M. Sališová, Š. Toma and T. J. Mason, *Ultrason. Sonochem.*, 1997, **4**, 131.
9. F. N. Lugemwa, *Molecules*, 2012, **17**, 9274.
10. F. A. Al-Bayati, *Tikrit J. Pure Sci.*, 2011, **16**, 6.
11. S. Albu, E. Joyce, L. Paniwnyk, J. P. Lorimer and T. J. Mason, *Ultrason. Sonochem.*, 2004, **11**, 261.
12. N. E. Durling, O. J. Catchpole, J. B. Grey, R. F. Webby, K. A. Mitchell, L. Y. Foo and N. B. Perry, *Food Chem.*, 2007, **101**, 1417.
13. M.-E. Cuvelier, C. Berset and H. Richard, *J. Agric. Food Chem.*, 1994, **42**, 665.
14. G. Topçu, M. Öztürk, T. Kuşman, A. A. B. Demirkoz, U. Kolak and A. Ulubelen, *Turk. J. Chem.*, 2013, **37**, 619.
15. A. Pukalskas, T. A. van Beek and P. de Waard, *J. Chromatogr. A*, 2005, **1074**, 81.

1.12. Analysis of Racemic and (*S*)-Ibuprofen

C. Fernandes[a,b], H. Cidade[a,b], S. Cravo[a], E. Sousa*[a,b], and M. Pinto*[a,b]

[a]Laboratório de Química Orgânica e Farmacêutica, Departamento de Ciências Químicas, Faculdade de Farmácia, Universidade do Porto, Portugal; [b]Centro Interdisciplinar de Investigação Marinha e Ambiental (CIIMAR/CIMAR), Universidade do Porto, Portugal. *E-mail: esousa@ff.up.pt, madalena@ff.up.pt

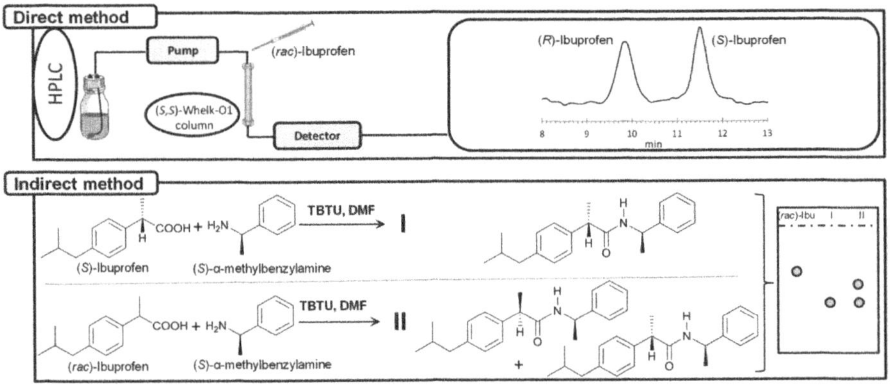

Number of sessions (duration of each session)	Hazard level	Difficulty level	Level of study
2 (3 h + 2 h)	Moderate	Medium	Intermediate

Class names Carboxylic acids, amines, amides

Concepts involved This experiment describes the analysis of a racemic mixture and the (*S*)-enantiomer of ibuprofen using two different methods. By the direct method, through high-performance liquid chromatography (HPLC) with ultraviolet (UV) detection, transient diastereomeric interactions between ibuprofen and a chiral selector lead to the separation of the enantiomers. By the indirect method, the racemic mixture of ibuprofen reacts with an enantiomerically pure amine and the obtained diastereomeric amides are analysed by thin-layer chromatography (TLC). The principle areas of this work are (i) the concept that enantiomers are distinguishable only if they are placed in a chiral environment and (ii) the concepts of analytical chromatography, chiral resolution and synthesis of amides

Chemicals needed Racemic ibuprofen, (*S*)-ibuprofen, (*S*)-(−)-α-methylbenzylamine, *N*,*N*,*N*′,*N*′-tetramethyl-*O*-(benzotriazol-1-yl)uronium tetrafluoroborate (TBTU), dimethylformamide (DMF), sodium hydrogen carbonate (5% wt), anhydrous sodium sulfate, dichloromethane, diethyl ether, petroleum ether (60–80 °C), ethanol (HPLC grade), *n*-hexane (HPLC grade), 2-propanol (HPLC grade), acetic acid (HPLC grade)

Equipment and experimental techniques involved Round-bottomed flask, magnetic stirrer, stirrer bar, separatory funnel, TLC solvent bath, HPLC-UV apparatus, chiral (*S*,*S*)-Whelk-O1 column (250 × 4.6 mm), TLC silica gel HF_{254} plate

Keywords Amidation, chiral high-performance liquid chromatography, diastereoisomers, enantiomers, liquid–liquid extraction, thin-layer chromatography

Comprehensive Organic Chemistry Experiments for the Laboratory Classroom
Edited by Carlos A. M. Afonso, Nuno R. Candeias, Dulce Pereira Simão, Alexandre F. Trindade, Jaime A. S. Coelho, Bin Tan, and Robert Franzén.

Published by the Royal Society of Chemistry, www.rsc.org

Background

Two chemically identical molecular species that differ from each other as non-superimposable mirror images are designated as enantiomers.[1] The spatial relationship between two enantiomers is classically represented by human hands. Enantiomers, in addition to diastereomers and *cis–trans*-isomers, are thus a special case of stereoisomers. The chirality (handedness) of enantiomeric molecules is caused by the presence of one or more chirality elements (chirality axis, chirality plane, or chirality centre, *e.g.*, asymmetric carbon atom) in their structure. Enantiomers present identical physical and chemical properties, when they are in an achiral environment, but it is possible to distinguish them when they interact with chiral systems or by the respective optical activity, that is, their interaction with plane-polarized light: one enantiomer rotates the light to clockwise (dextrorotatory) and the other to counter-clockwise (laevorotatory), referred by the symbols (+) and (−), respectively. The chirality sense of the enantiomers is determined by their absolute configuration, *i.e.*, the spatial arrangement of the atoms in the molecule.[2] The spatial arrangement of the substituents around the stereogenic centre is termed the configuration, which is not necessarily correlated with the signal of the optical activity. Configuration can be assigned by the Cahn–Ingold–Prelog (CIP) system, which expresses the configurations as (*R*) and (*S*) (from the Latin words *rectus* and *sinister* which signify right and left, respectively) by a set of priority rules.[2] Diastereoisomers (or diastereomers) are stereoisomers not related as mirror images. Diastereoisomers are characterized by differences in physical properties, and by some differences in chemical behaviour towards achiral as well as chiral reagents.[1]

Chiral considerations are essential parts of drug research and development, and of the regulatory process taking into account the evolution of chiral drugs in the last few years.[3–6] The different biological/pharmacological activities of enantiomers, in addition to the advances in synthetic and analytical methodologies, have contributed to the considerable increased interest in this field. The gains in potency, efficacy, safety and selectivity obtained by treatment with single enantiomers rather than racemate are generally unquestionable.[4] Accordingly, the distribution of worldwide approved drugs indicates that single enantiomers surpassed achirals while racemic drugs represented the minority,[5] and many of the top-selling drugs are marketed as single enantiomers.[6] In this context, the strategies for resolution and analysis of enantiomers are crucial issues.

The resolution/analysis of enantiomers can be achieved by direct and indirect methods.[7] The indirect method is based on the reaction of the racemic mixture with an optically active reagent to form a pair of diastereomers. They can be separated by taking advantage of the differences of physiochemical properties by conventional procedures such as crystallization, distillation or chromatography. This method has been used extensively in the past;[8,9] however, the remarkable development and application of chiral chromatography has revolutionized the field of enantioseparation, providing a wide range of alternatives to the indirect method. Additionally, enantioselective chromatography is extensively employed in the analysis of the enantiomeric composition (enantiomeric excess, optical purity) of chiral compounds.

The direct resolution of enantiomers can be performed using a chiral selector present in the chromatographic system which must associate preferentially with one of the enantiomers of the mixture, forming transient diastereomeric complexes. The difference in stability between these complexes leads to different retention times; first to be eluted is the enantiomer that forms the less stable complex. Among these methods, direct resolution using chiral stationary phases (CSPs) in high-performance liquid chromatography (HPLC) has been extremely helpful and highly applicable for both analytical[10–12] and preparative purposes.[13,14] (*S*,*S*)-Whelk-O1 is one of the most employed CSPs, with excellent performance,[15] being originally designed in the early 1990s for the separation of enantiomers of non-steroidal anti-inflammatory drugs (NSAIDs).[16]

Ibuprofen (2-(4-isobutylphenyl)propanoic acid), a chiral NSAID widely used for the treatment of inflammatory diseases, is commercialized in most pharmaceuticals as a racemate although the enantiomer with therapeutic interest is the (*S*)-ibuprofen.

The aim of this experimental work is to analyse the racemic mixture and the (*S*)-enantiomer of ibuprofen by two different methods. By the direct method, (*S*,*S*)-Whelk-O1 CSP is chosen to establish transient, diastereomeric interactions with ibuprofen leading to different retention times and, thus, to the separation of the enantiomers, through HPLC with UV detection. By the indirect method, a racemic mixture of ibuprofen reacts with enantiomerically pure (*S*)-(−)-α-methylbenzylamine and the two obtained diastereomeric amides are analysed by TLC. Activation of the carboxyl group of ibuprofen for the formation of the amide bond is achieved *in situ* by the coupling reagent TBTU. Nowadays, TBTU is one of the most popular coupling reagents, proving to be very efficient for the synthesis of different classes of compounds, including amides.[17–20]

Additional Safety

The preparation of the mobile phase in laboratory session 1 (direct method) should be performed in a fume hood. Acetic acid is hazardous in cases of skin and eye contact (irritant), ingestion, and inhalation; ethanol is highly flammable; *n*-hexane is highly flammable, irritating to the skin, harmful (through inhalation and if swallowed), toxic to aquatic organisms and vapours may cause drowsiness and dizziness; 2-propanol is highly flammable, irritating to the eyes, and vapours may cause drowsiness and dizziness. The laboratory session 2 (indirect method) should also be performed in a fume hood. Ibuprofen is harmful if swallowed; (*S*)-(−)-α-methylbenzylamine and DMF are hazardous in cases of skin and eye contact (irritant), ingestion, and inhalation; TBTU has a risk of explosion by shock, friction, fire, or other sources of ignition, and is irritating to eyes, respiratory system, and skin; dichloromethane is irritating to eyes, respiratory system, and skin and vapours may cause drowsiness and dizziness; diethyl ether is extremely flammable, may form explosive peroxides, is harmful if swallowed, may cause skin dryness or cracking, and vapours may cause drowsiness and dizziness; petroleum ether is highly flammable, irritating to the skin, harmful (through inhalation and if swallowed), toxic to aquatic organisms, and vapours may cause drowsiness and dizziness. Proper safety equipment, such as protective clothing, gloves, and safety goggles should be used in all sessions.

Experimental Procedure

Laboratory Session 1 (3 h): Analysis of Racemic and (*S*)-Ibuprofen by the Chiral HPLC

HPLC Conditions

- Column: (*S*,*S*)-Whelk-O1 (250 × 4.6 mm internal diameter, 5 μm particle size, 100 Å pore size);
- mobile phase: *n*-hexane/2-propanol/acetic acid (98 : 2 : 0.05 v/v/v), which is degassed before use;
- flow rate: 0.9 mL min^{-1};
- detection: UV 254 nm;
- temperature: room temperature.

Mobile Phase Preparation

1. Prepare 100 mL of mobile phase: add 2 mL of 2-propanol (graduated pipette) and 0.05 mL of acetic acid (micropipette) to 98 mL of *n*-hexane (graduated cylinder).

Sample Preparation

2. Prepare 5 mL of racemic ibuprofen and (*S*)-ibuprofen solutions (400 μg mL^{-1} in ethanol): weigh out 2 mg of racemic ibuprofen or (*S*)-ibuprofen; place the solid in a 5 mL volumetric flask; add a small volume of ethanol to dissolve; fill the flask to the 5 mL line.

Equilibration of Column

3. Equilibrate the column under the above described conditions at least for 30 min.

Sample Injection and Analysis

4. Inject 20 μL of racemic ibuprofen solution.
5. Determine the retention factor(s).
6. Repeat the same experimental procedure for the (*S*)-ibuprofen.

Laboratory Session 2 (2 h): Analysis of Racemic and (*S*)-Ibuprofen by Diastereomers Synthesis

Amidation Reaction

1. In a fume hood prepare a 10 mL round-bottomed flask with a magnetic stirring bar, and add 50 mg of racemic ibuprofen (0.24 mmol) and 1.5 mL of DMF at room temperature, and stir until all the ibuprofen is dissolved.
2. After complete dissolution, add 1.5 mL of *N*,*N*,*N*′,*N*′-tetramethyl-*O*-(benzotriazol-1-yl)uronium tetrafluoroborate (TBTU) (0.23 g mL^{-1} in DMF, 1.0 mmol), and keep the stirring for 5 min.
3. Subsequently, add 31 μL of (*S*)-(−)-α-methylbenzylamine (0.24 mmol) and stir the reaction mixture at room temperature for 1 h.
4. After 1 h, stop the reaction by adding 3 mL of diethyl ether.
5. Repeat the same experimental procedure for the (*S*)-ibuprofen.

Product Isolation and Analysis

6. Transfer each reaction mixture into a separatory funnel and wash the organic layer with 3 × 2 mL of sodium hydrogen carbonate (5% in water) and then with 3 × 2 mL of water.
7. Dry the organic layer with anhydrous sodium sulfate and filter.
8. Analyse the isolated products as well as dichloromethane solutions of racemic ibuprofen and (*S*)-ibuprofen by TLC using a silica gel HF_{254} plate and a mixture of diethyl ether/petroleum ether (60–80 °C)/acetic acid (5 : 5 : 0.01 v/v/v) as mobile phase, and observe under UV light (254 nm).

Waste Disposal

The organic solvents diethyl ether, petroleum ether, *n*-hexane, and 2-propanol may be reused or collected in a non-chlorinated waste collector recipient, for posterior disposal. DMF and the aqueous solution of sodium hydrogen carbonate (5%) are collected in a non-chlorinated and in a basic waste collector recipient, respectively, for posterior dilution and disposal by a licensed disposal and waste treatment company in accordance with local directives.

Results Interpretation and Additional Questions

1. Interpret and compare the obtained HPLC chromatograms.
2. Calculate the HPLC retention factors (k) using following equation:

$$k = \frac{t_R - t_0}{t_0} \tag{1.12.1}$$

where t_0 is the column dead time (considered to be equal to the peak of the solvent front) and t_R is the retention time of each chromatographic band.

3. Interpret the TLC chromatogram obtained.
4. Calculate the TLC retention factors (R_f) using following equation:

$$R_f = \frac{\text{migration distance of substance}}{\text{migration distance of solvent front}} \tag{1.12.2}$$

5. Compare and discuss the advantages and disadvantages of both direct and indirect methods.

References

1. G. P. Moss, *Pure Appl. Chem.*, 1996, **68**, 2193.
2. S. Allenmark, *Chromatographic Enantioseparation: Methods and Applications*, Ellis Horwood, Chichester, 1991, vol. 2, pp. 24–32.
3. H. Caner, E. Groner, L. Levy and I. Agranat, *Drug Discovery Today*, 2004, **9**, 105.
4. B. Kasprzyk-Hordern, *Chem. Soc. Rev.*, 2010, **39**, 4466.
5. I. Agranat, S. R. Wainschtein and E. Z. Zusman, *Nat. Rev. Drug Discovery*, 2012, **11**, 972.
6. I. Agranat and S. R. Wainschtein, *Drug Discovery Today*, 2010, **15**, 163.
7. B. P. Nagori, M. S. Deora and P. Saraswat, *Int. J. Pharm. Sci. Rev. Res.*, 2011, **6**, 106.
8. H. Cidade and M. Pinto, *IUPAC: Practical Studies for Medicinal Chemistry – An integrating approach for developing countries*, ed. A. Monge and C. R. Ganelli, 2006, vol. I.10, pp. 1–9.
9. S. E. Sen and K. S. Anliker, *J. Chem.*, 1996, **73**, 569.
10. C. Fernandes, M. E. Tiritan, Q. Cass, V. Kairys, M. X. Fernandes and M. Pinto, *J. Chromatogr. A*, 2012, **1241**, 60.
11. C. Fernandes, P. Brandão, A. Santos, M. E. Tiritan, C. Afonso, Q. B. Cass and M. M. Pinto, *J. Chromatogr. A*, 2012, **1269**, 143.
12. E. P. Sousa, M. E. Tiritan, R. V. Oliveira, C. M. M. Afonso, Q. B. Cass and M. M. M. Pinto, *Chirality*, 2004, **16**, 279.
13. M. E. Sousa, M. E. Tiritan, K. R. A. Belaz, M. Pedro, M. S. J. Nascimento, Q. B. Cass and M. M. M. Pinto, *J. Chromatogr. A*, 2006, **1120**, 75.
14. T. C. Lourenço, J. M. Batista Jr, M. Furlan, Y. He, L. A. Nafie, C. C. Santana and Q. B. Cass, *J. Chromatogr. A*, 2012, **1230**, 61.
15. C. Fernandes, A. Palmeira, A. Santos, M. E. Tiritan, C. Afonso and M. M. Pinto, *Chirality*, 2013, **25**, 89.
16. C. Fernandes, M. E. Tiritan and M. Pinto, *Chromatographia*, 2013, **76**, 871.
17. C. Fernandes, K. Masawang, M. E. Tiritan, E. Sousa, V. Lima, C. Afonso, H. Bousbaa, W. Sudprasert, M. Pedro and M. Pinto, *Bioorg. Med. Chem.*, 2014, **22**, 1049.
18. C. Fernandes, L. Oliveira, M. E. Tiritan, L. Leitão, A. Pozzi, J. B. Noronha-Matos, P. Correia-de-Sá and M. M. Pinto, *Eur. J. Med. Chem.*, 2012, **55**, 1.
19. K. Prabakaran, F. N. Khan and J. S. Jin, *Res. Chem. Intermed.*, 2012, **38**, 615.
20. S. Balalaie, M. Mahdidoust and R. Eshaghi-Najafabadi, *J. Iranian Chem., Soc.*, 2007, **4**, 364.

1.13. Determining Partition Coefficients of Sulfonamides by Reversed-Phase Chromatography

E. Sousa[a,b], S. Cravo[a], C. Fernandes*[a,b], and M. Pinto*[a,b]

[a]Laboratório de Química Orgânica e Farmacêutica, Departamento de Ciências Químicas, Faculdade de Farmácia, Universidade do Porto, Portugal; [b]Centro Interdisciplinar de Investigação Marinha e Ambiental (CIIMAR/CIMAR), Universidade do Porto, Portugal. *E-mail: cfernandes@ff.up.pt, madalena@ff.up.pt

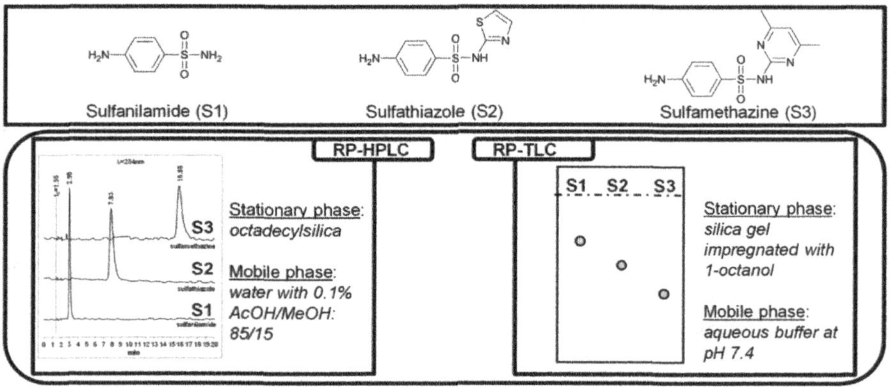

Number of sessions (duration of each session)	Hazard level	Difficulty level	Level of study
2 (1 h each)	Low	Low	Intermediate

Class names Sulfonamides

Concepts involved This experiment describes the determination of the logarithm of partition coefficient (log *P*) of three sulfonamides by two different chromatographic methods: reversed-phase high-performance liquid chromatography (RP-HPLC) and reversed-phase thin-layer chromatography (RP-TLC). By RP-HPLC, the retention factors (expressed as log *k*) are used to estimate the log *P* of each sulfonamide since there is a correlation between these two parameters. By RP-TLC, log *P* values are correlated with the chromatographic parameter R_m. The experimental log *P* values are compared with predicted log *P* values calculated by ChemBioDraw®

Chemicals needed Sulfanilamide, sulfathiazole, sulfamethazine, acetone, 1-octanol (5% diethyl ether), potassium dihydrogen phosphate (0.1 mol L^{-1}), sodium hydroxide (0.1 mol L^{-1}), vanillin (4-hydroxy-3-methoxybenzaldehyde) (10 mg mL^{-1} ethanol), hydrochloric acid (18 wt%), methanol (HPLC grade), Milli-Q water, acetic acid (HPLC grade)

Equipment and experimental techniques Apparatus for RP-TLC and RP-HPLC-UV; reversed-phase octadecyl silica (C_{18}) column (150 × 4.6 mm internal diameter, 5 μm particle size); ChemBioDraw® or other drawing program with log *P* prediction tool

Keywords Lipophilicity, log *P*, partition coefficient, reversed-phase high-performance liquid chromatography, reversed-phase thin-layer chromatography, sulfonamides

Comprehensive Organic Chemistry Experiments for the Laboratory Classroom
Edited by Carlos A. M. Afonso, Nuno R. Candeias, Dulce Pereira Simão, Alexandre F. Trindade, Jaime A. S. Coelho, Bin Tan, and Robert Franzén.

Published by the Royal Society of Chemistry, www.rsc.org

Background

The physicochemical properties of a bioactive compound, such as lipophilicity, are among the most important characteristics[1,2] since it is implicated in both pharmacokinetic and pharmacodynamic events of administration of drugs; because of that the physicochemical properties must be characterized early in drug discovery/development process.[3,4] Lipophilicity is commonly expressed as the partition coefficient (*P*), which is defined as the ratio of the equilibrium concentrations of a dissolved substance in a two-phase system, consisting of two largely immiscible solvents. The logarithm of the partition coefficient (log *P*) is the molecular parameter of choice in numerous (quantitative) structure-activity relationships ((Q)SAR) of diverse classes of compounds and for a wide range of biological effects.[5,6]

Originally, the methods for experimentally determining log *P* did not allow its rapid determination.[7,8] Therefore, other models have been developed such as reversed-phase chromatography, since the retention of a chemical in a reversed-phase chromatographic system is related to its partition coefficient (*P*), which has proved to overcome the difficulties of the traditional "shake-flask" method.[9] Especially, reversed-phase high-performance liquid chromatography (RP-HPLC) and reversed-phase thin-layer chromatography (RP-TLC) have several advantages, *e.g.* they are faster, less sensitive to impurities, and less expensive since they do not require any quantification. Moreover, RP-HPLC is characterized by low sample consumption, good accuracy, excellent reproducibility, and automation possibilities.[9,10]

By RP-HPLC, the analyses are carried out using octadecyl silica as stationary phase and 0.1% aqueous acetic acid/methanol: 85/15 (v/v) as mobile phase. Only retention factors (*k*) need to be recorded, given that there is a correlation between log *k* and log *P*.[10,11]

By RP-TLC, a silica gel plate impregnated with 1-octanol and an aqueous buffer at pH 7.4 as mobile phase are used to perform the analyses. The log *P* values are correlated with the chromatographic parameter R_m, an expression of the lipophilic character of substances.[12,13] The R_m is calculated directly from the R_f using RP-TLC. It was discovered by Hansch that partition coefficients have an additive–constitutive character. By simple summation of the lipophilicity of molecular fragments it is possible to calculate the log *P* of the complete molecule of a substance.[14]

The objective of the practical experiment described here is to evaluate by two different chromatographic methods the log *P* of three sulfonamides – sulfanilamide, sulfathiazole, and sulfamethazine. These compounds have significant historical value as antibacterial drugs, but also nowadays are important as moieties belonging to new orally-available drugs.

Log *P* values are estimated with ChemBioDraw® to support the conclusions drawn from the data generated by these experiments.

Additional Safety

The preparation of the mobile phase in the laboratory session 1 should be performed in a fume hood. Acetic acid is hazardous in case of skin and eye contact (irritant), ingestion, and inhalation; methanol is highly flammable, toxic by inhalation, in contact with skin and if swallowed (danger of very serious irreversible effects); sulfathiazole is irritating to eyes, respiratory system, and skin All of laboratory session 2 (RP-TLC method) should be carried out in a fume hood. Acetone is highly flammable, irritating to the eyes, may cause skin dryness or cracking (repeated exposure) and vapours may cause drowsiness and dizziness; 1-octanol is irritating to eyes and skin; diethyl ether is extremely flammable, may form explosive peroxides, is harmful if swallowed, may cause skin dryness or cracking, and vapours may cause drowsiness and dizziness; sodium hydroxide causes severe burns; vanillin is irritating to the eyes; ethanol is highly flammable; hydrochloric acid causes burns and is irritating to the respiratory system. In both sessions it is necessary to use proper safety equipment, such as protective clothing, gloves, and safety goggles.

Experimental Procedure

Laboratory Session 1 (1 h): Determining the log *P* of Sulfonamides by RP-HPLC

HPLC Conditions

Column: reversed-phase octadecyl silica (C_{18}) column (150 × 4.6 mm internal diameter, 5-μm particle size).
Mobile phase: 0.1% aqueous acetic acid/methanol: 85/15 (v/v) which is degassed before use.
Flow rate: 1 mL min^{-1}
Detection: UV 254 nm
Temperature: room temperature

Sample Preparation

1. Prepare 5 mL of sulfanilamide solution (2 mg mL^{-1} in methanol and diluted 1/1000 in water).
2. Repeat the same experimental procedure for sulfathiazole and sulfamethazine.

Equilibration of Column

3. Connect the column to the solvent delivery system according to the HPLC system requirements and equilibrate under the above described conditions (30 min).

Sample Injection and Analysis

4. Inject separately 20 μL of each sulfonamide solution.
5. Determine the retention factors (k).

Laboratory Session 2 (1 h): Determining the log *P* of Sulfonamides by RP-TLC

Samples

1. Dissolve each sulfonamide (sulfanilamide, sulfathiazole, and sulfamethazine) in acetone (4 mg mL^{-1}).

Stationary Phase

2. Impregnate, by spraying uniformly a precoated silica gel HF_{254} plate (10 × 4 cm) with a solution of 1-octanol (5% in diethyl ether), and let it dry at room temperature.

Mobile Phase

3. Prepare an aqueous buffer at pH 7.4, by the mixture of 25 mL of potassium dihydrogen phosphate solution (0.1 mol L^{-1}) and 19 mL of sodium hydroxide solution (0.1 mol L^{-1}).

Chromatographic Analysis

4. Apply each solution of the sulfonamides as a streak 1 cm from the lower edge of the plate using a glass capillary.

5. Place the plate in the chromatographic chamber until the mobile phase reaches the end of the plate.
6. After removing the plate from the chromatographic chamber dry it with a stream of hot air.
7. Spray the plate with a solution of vanillin hydrochloride (2.5 g vanillin in 250 ml ethanol, containing 15 mL of hydrochloric acid 18 wt%).

Waste Disposal

The RP-HPLC mobile phase (0.1% aqueous acetic acid/methanol: 85/15) may be reused or collected in a non-chlorinated waste collector recipient, for posterior disposal.

The RP-TLC mobile phase (aqueous buffer at pH 7.4, by the mixture of potassium dihydrogen phosphate solution and sodium hydroxide solution) may be reused or collected in an aqueous basic waste collector recipient, for posterior disposal by a licensed disposal and waste treatment company in accordance with the directives of the European Union.

Results Interpretation and Additional Questions

1. Compare the HPLC chromatograms obtained and calculate the logarithms of retention factors (log *k*) using the following equation:

$$\log k = \log\left(\frac{t_R - t_0}{t_0}\right) \tag{1.13.1}$$

   where t_0 is the column dead time (considered to be equal to the peak of the solvent front) and t_R is the retention time of each chromatographic band.

2. Calculate the log *P* using the following equation:

$$y = 0.8304x + 0.6489 \tag{1.13.2}$$

   where *y* is the retention factor values obtained (expressed as log *k*) and *x* the log *P* for the tested sulfonamides.

3. Interpret the TLC chromatogram obtained and calculate the R_f and R_m values of each sulfonamide, using the following equations:

$$R_f = \frac{\text{migration distance of substance}}{\text{migration distance of solvent front}} \tag{1.13.3}$$

$$R_m = \log\left(\frac{1}{R_f} - 1\right) + \log\left(\frac{K_a + [H^+]}{[H^+]}\right) \tag{1.13.4}$$

   where K_a is the dissociation constant of the compound, and $[H^+]$ is the hydrogen ion concentration of the mobile phase.

4. Calculate the lipophilicity substituent or Hansch constant (π) of substituent groups of sulfathiazole and sulfamethazine, using the following equation:

$$\pi_x = R_{m\text{-}RX} - R_{m\text{-}RH} \tag{1.13.5}$$

where RX is the sulfathiazole or sulfamethazine and RH is sulfanilamide (the constant π indicates the change in the logarithm of the partition coefficient resulting from the introduction of a substituent group X).

5. Calculate the log *P* of sulfathiazole and sulfamethazine, considering that the partition coefficient (*P*) of sulfanilamide is 0.15.
6. Compare the experimental log *P* values (for both RP-HPLC and RP-TLC methods) with predicted log *P* values calculated by ChemBioDraw®.

References

1. N. A. Meanwell, *Chem. Res. Toxicol.*, 2011, **24**, 1420.
2. M. P. Gleeson, A. Hersey, D. Montanari and J. Overington, *Nat. Rev. Drug Discovery*, 2011, **10**, 197.
3. P. D. Leeson and B. Springthorpe, *Nat. Rev. Drug Discovery*, 2007, **6**, 88.
4. B. Testa, P. Crivori, M. Reist and P. A. Carrupt, *Perspect Drug Discovery Des.*, 2000, **19**, 179.
5. M. Bajda, S. Boryczka and B. Malawska, *Biomed. Chromatogr.*, 2007, **21**, 123.
6. C. A. Lipinski, F. Lombardo, B. W. Dominy and P. J. Feeney, *Adv. Drug Delivery Rev.*, 1997, **23**, 3.
7. C. Hansch and T. Fujita, *J. Am. Chem. Soc.*, 1964, **86**, 1616.
8. C. Hansch, P. P. Melany and T. Fujita, *Nature*, 1962, **194**, 178.
9. C. V. Eadsforth and P. Moser, *Chemosphere*, 1983, **12**, 1459.
10. L. Ayouni, G. Cazola, D. Chaillou, B. Herbreteau, S. Rudaz, P. Lantéri and P. A. Carrupt, *Chromatographia*, 2005, **62**, 251.
11. S. Carda-Broch and A. Berthod, *Chromatographia*, 2004, **59**, 79.
12. M. A. F. Prado, *J. Chem. Educ.*, 2001, **78**, 533.
13. G. L. Biagi, A. M. Barbaro, M. C Guerra, G. C. Forti and M. E. Fracasso, *J. Med. Chem.*, 1974, **17**, 28.
14. C. Hansch, A. Leo, S. H. Unger, K. H. Kim, D. Nikaitani and E. J. Lien, *J. Med. Chem.*, 1973, **16**, 1207.

2. Modification of sp^3 Carbon

2.1. C-X Bond Formation

2.1.1. A S_N1 Reaction: Synthesis of *tert*-Butyl Chloride

Emília Valente[a], Catarina Dias[a], and Luís Constantino*[a,b]

[a]Faculty of Pharmacy, University of Lisbon, Av. Prof. Gama Pinto, 1649-003 Lisbon, Portugal; [b]Research Institute for Medicines (iMed.ULisboa), Faculty of Pharmacy, Universidade de Lisboa, Av. Professor Gama Pinto, 1649-003 Lisbon, Portugal. *E-mail: constant@ff.ulisboa.pt

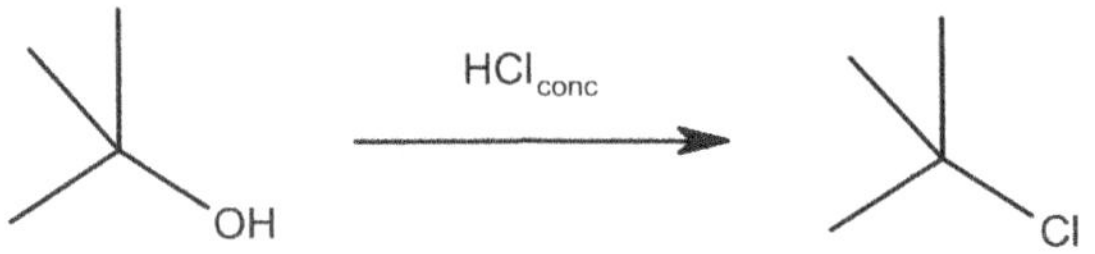

Number of sessions (duration of each session)	Hazard level	Difficulty level	Level of study
1 (3 h)	Moderate	Low	Introductory

Class names Alcohols (tertiary), alkyl halides (tertiary), acids

Concepts involved The purpose of this laboratory experiment is to synthesize an alkyl halide by the corresponding reaction of an alcohol with a hydrogen halide. The tertiary alcohol reacts readily with HCl as the hydrogen halide, and forms the corresponding *tert*-butyl chloride without the need of heat

Chemicals needed *tert*-Butanol, pure hydrogen chloride (36–38%), sodium hydrogen carbonate, anhydrous sodium sulfate

Equipment and experimental techniques involved Heating plate with magnetic stirring, Erlenmeyer flask, separatory funnel, simple distillation equipment, liquid–liquid separation, simple distillation, IR spectroscopy

Keywords Extraction, infrared (IR) spectra, S_N1 reaction, *tert*-butanol, *tert*-butyl chloride

Background

The purpose of this laboratory experiment is to synthesize an alkyl halide by reacting the corresponding alcohol with a hydrogen halide in an easy and inexpensive S_N1 reaction.[1–3] Nucleophilic substitution reactions (S_N) at a sp^3 carbon are central reactions in organic chemistry whose mechanisms are important within the initial knowledge of students.[1–3]

This type of reactions can be schematized as follows (Scheme 2.1.1.1).

In these reactions the attacking species is a nucleophile (Nu), which bears a pair of electrons either as an anion or as a neutral compound. The substrate has a leaving group (X) that is replaced by the nucleophile during the course of the reaction.

Comprehensive Organic Chemistry Experiments for the Laboratory Classroom
Edited by Carlos A. M. Afonso, Nuno R. Candeias, Dulce Pereira Simão, Alexandre F. Trindade, Jaime A. S. Coelho, Bin Tan, and Robert Franzén.

Published by the Royal Society of Chemistry, www.rsc.org

These reactions may be bimolecular nucleophilic substitution (S_N2) or unimolecular nucleophilic substitution (S_N1).

A S_N2 reaction involves a one-step bimolecular mechanism with inversion of configuration (Scheme 2.1.1.2).

In this mechanism the departure of the leaving group occurs simultaneously with the backside attack by the nucleophile. The inversion of configuration can be observed when a stereocentre is formed in the reaction. S_N2 is favoured by an unhindered substrate, an aprotic solvent and a strong nucleophile.

A S_N1 reaction involves a two-step mechanism that is unimolecular in the rate-determining step and has a carbocation as intermediate.

In a S_N1 reaction, a planar carbocation is formed first, which then reacts further with the nucleophile. Since the nucleophile is free to attack from either side, this reaction is associated with racemization when a stereocentre is formed in the reaction (Scheme 2.1.1.3).

Rearrangements may also occur as a carbocation is formed. S_N1 reactions are favoured by a protic solvent and by a stable carbocation intermediate.

As expected, S_N1 and S_N2 are competing mechanisms; however, tertiary substrates will react mainly by the S_N1 mechanism since they form stable carbocations and are too hindered to react

Scheme 2.1.1.1 Nucleophilic substitution reaction (SN) at a sp³ carbon.

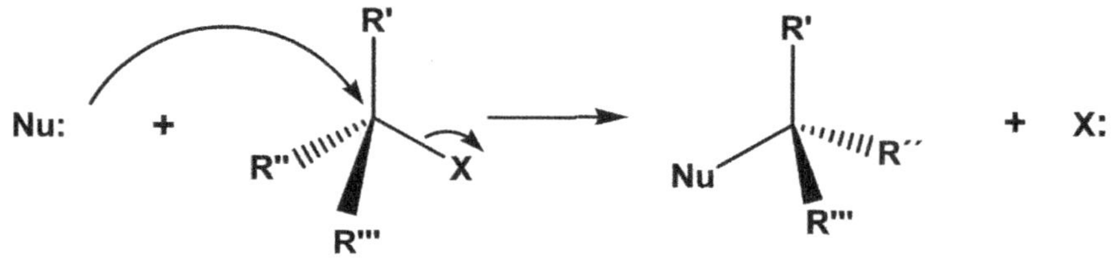

Scheme 2.1.1.2 S_N2 reaction: a one-step bimolecular mechanism with inversion of configuration.

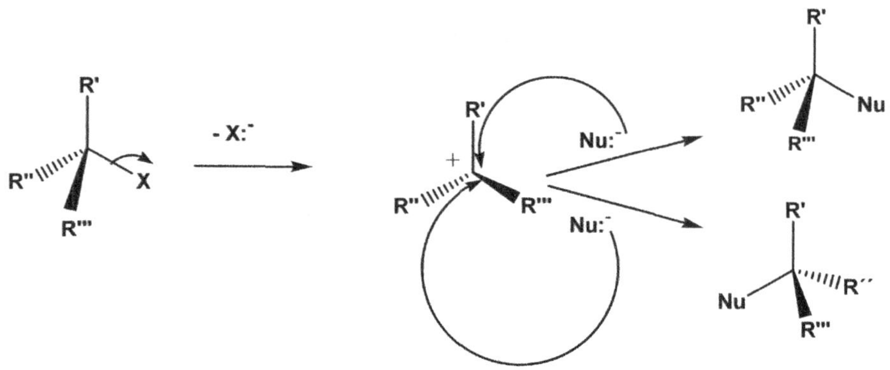

Scheme 2.1.1.3 S_N1 reaction: a two-step unimolecular mechanism that has a carbocation as intermediate. This mechanism is associated with racemization when a stereocentre is formed in the reaction.

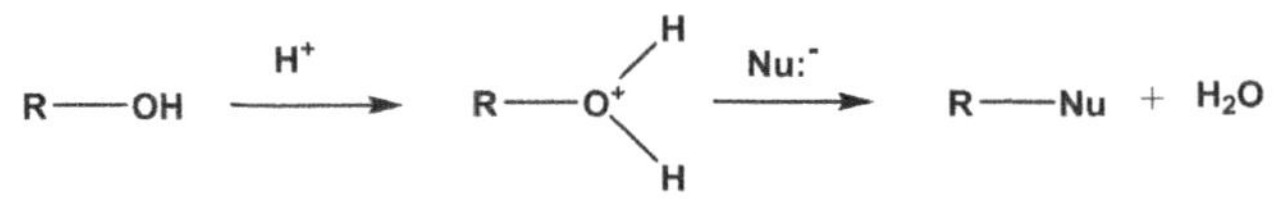

Scheme 2.1.1.4 Activation of a hydroxide group into a good leaving group by protonation.

by a S_N2 mechanism. On the other hand, the steric hindrance in tertiary substrates is too high to allow an effective S_N2 reaction on tertiary substrates.

The main competition between S_N1/S_N2 mechanisms takes place in secondary substrates. A good leaving group, a protic solvent and a good nucleophile facilitate S_N1 reactions while a good nucleophile (in high concentration) and an apolar aprotic solvent facilitate S_N2 reactions.

Hydroxide and alkoxide ions are not good leaving groups; however, they can be activated by means of Lewis or Brønsted acids (Scheme 2.1.1.4).

In the present reaction, the substrate (a tertiary alcohol) reacts with HCl to form a protonated alcohol that has a good leaving group and forms easily a tertiary carbocation. This carbocation then reacts with the chloride ion to form the alkyl halide.

The *tert*-butanol reacts readily with hydrogen halide, and forms the corresponding *tert*-butyl chloride without the need of heat.[3] The reaction is easily and rapidly performed in an Erlenmeyer flask with a stir bar followed by a liquid–liquid separation. It can also be performed in a separatory funnel.[4] After separation of the layers, the organic layer is dried and purified by a simple distillation at low temperature (48–52 °C).

Additional Safety

The experiment involves the use of concentrated hydrogen chloride. Therefore, the use of gloves, laboratory security goggles and long-sleeve laboratory coat during all the operation is mandatory. Manipulation of the acid must be carried out in a fume hood, where the initial reaction should occur. The graduated cylinder needed to measure the acid volume should be kept in the fume cupboard and when not needed should be immersed in an appropriate container half-filled with water in the hood until appropriate material washing.

Experimental Procedure

tert-Butyl Chloride Preparation

1. In a fume hood combine, in a 250 mL Erlenmeyer flask with ground-glass jointware and equipped with a magnetic stirrer, 15 g of *tert*-butanol with 50 mL pure concentrated hydrochloric acid (36–38%).
2. Swirl the solution with caution for 1 min in the uncapped flask. *tert*-Butyl alcohol is miscible with hydrochloric acid and a clear solution results. However, *tert*-butyl chloride is immiscible and the solution rapidly becomes cloudy.
3. Clamp the Erlenmeyer over the magnetic plate, uncap the flask and apply strong stirring for 30 min more.

Reaction Mixture Treatment

4. Place the reaction mixture in a separatory funnel and remove the hydrochloric acid aqueous (bottom) layer.
5. Wash the top layer with 20 mL of saturated sodium bicarbonate. First swirl the uncapped funnel with caution until all gas disappears, and then continue the operation.

6. Discard the alkaline bottom layer.
7. Wash the *tert*-butyl chloride layer with 15 mL of water and discard the aqueous bottom layer.
8. Transfer the crude *tert*-butyl chloride solution to a 50 mL Erlenmeyer and dry with anhydrous sodium sulfate until the solution becomes clear.

tert-Butyl Chloride Isolation and Purification

9. Filter the dry *tert*-butyl chloride to a 50 mL distilling flask.
10. In a fume cupboard add a boiling stone or a magnetic stirrer.
11. In a fume cupboard set up a simple distillation apparatus to purify the alkyl halide (Figure 2.1.1.1).
12. Clamp the neck of the distilling flask in a heating mantle (or in a water bath with heating plate) to a ring stand. To avoid overheating, make sure the flask is not touching the bottom of the mantle or the bath.
13. Add the three-way adapter.
14. Hold the condenser, with the water inlet and the water outlet tubes set up, to the three-way adapter. Clamp the condenser to a second ring stand.
15. Adjust a distillation adapter to the condenser and hold it with stretchy rubber to the water inlet plug.
16. Weigh a 25 mL graduated cylinder with ground-glass joint, and place it in an ice bath.
17. Record the boiling point of the distillate minute by minute and retain the fraction boiling from 48–52 °C in the pre-weighed 25 mL graduated cylinder.
18. Record the volume and the boiling range of the *tert*-butyl chloride. Finally, weigh also the graduated cylinder with the pure received product.

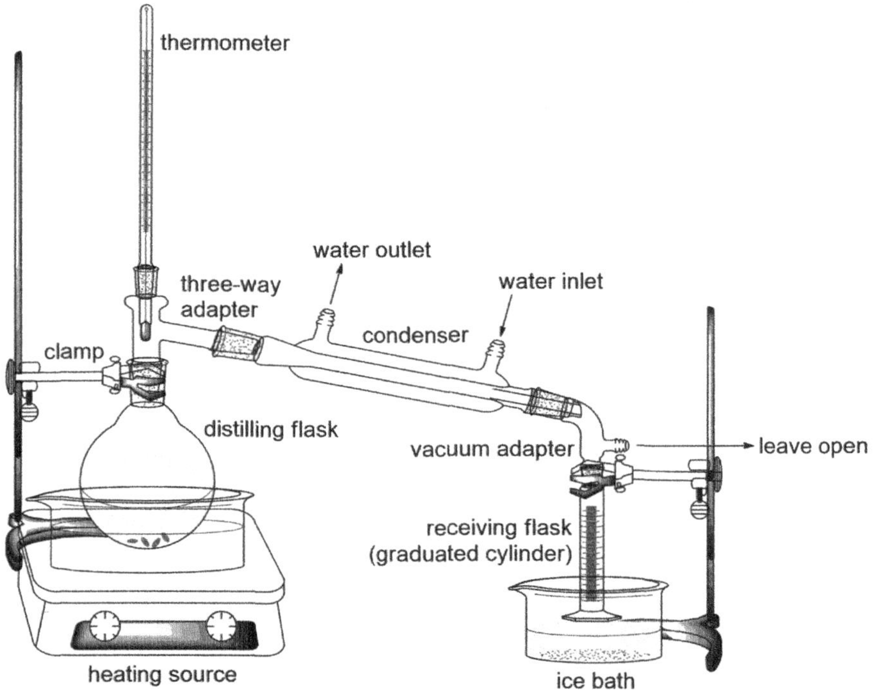

Figure 2.1.1.1 Simple distillation set-up to purify the reaction product *tert*-butyl chloride.

19. If possible, obtain the infrared spectra (IR) of both the reagent and the product.
20. Calculate the reaction yield.

Waste Disposal

The laboratory should have separate, appropriate containers to recover aqueous acid and alkaline solutions where the students will discharge their extraction aqueous phases.

The reaction product should be kept in an appropriate recovery flask.

Results Interpretation and Additional Questions

1. Draw the reaction mechanism.
2. Is a strong nucleophile fundamental for this reaction mechanism? Explain why.
3. What kind of substrates (primary, secondary, or tertiary) gives S_N1 reactions and what types give S_N2 reactions? Explain why.
4. Consider what would happen if you apply the same procedure to *n*-butanol instead of *tert*-butanol. Would you expect to obtain 1-chlorobutane at the end of your work? Explain the reason why.
5. Compare the IR spectra of *tert*-butanol with that of the reaction product and discuss the differences.

References

1. K. P. C. Vollhardt and N. E. Schore, *Organic Chemistry: Structure and Function*, Freeman and Company, New York, 6th edn, 2010.
2. J. Clayden, N. Greeves, S. Warren and P. Wothers, *Organic Chemistry*, Oxford University Press Inc., New York, 1st edn, 2001.
3. F. A. Carey and R. M. Giuliano, *Organic Chemistry*, McGraw-Hill Companies, Inc., New York, 8th edn, 2011.
4. P. M. Ginnings, *J. Chem. Educ.*, 1930, 7(2), 313.

2.1.2. Optimizing the Reaction Conditions for the Synthesis of *tert*-Pentyl Chloride

Jane Brock Greco

Department of Chemistry, Johns Hopkins University, 3400 N. Charles St. Baltimore, MD 21218, USA
*E-mail: jgreco@jhu.edu

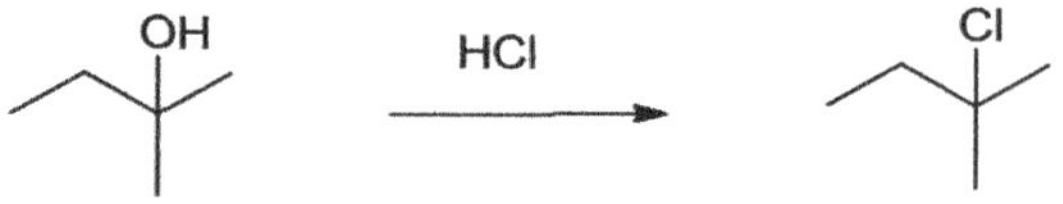

Number of sessions (duration of each session)	Hazard level	Difficulty level	Level of study
1 (4–5 h)	Moderate	Low	Introductory

Class names Alkyl chlorides, alcohols

Concepts involved This study introduces students to the idea of optimizing reaction conditions, and the difference between kinetic and thermodynamic factors in the context of the S_N1 reaction

Chemicals needed 2-Methyl-2-butanol, concentrated (37%) hydrochloric acid, calcium chloride

Equipment and experimental techniques involved Liquid–liquid extraction, distillation, refluxing apparatus (one option)

Keywords Boiling point, distillation, extraction

Background

When organic chemists carry out a reaction for the purpose of synthesizing a product, they often find that the percent yield of the product is lower than ideal. To improve the yield of desired product, a chemist will carry out the reaction under different reaction conditions (time, temperature, solvent, amounts of reagents). This process is referred to as optimizing the reaction. Assuming the chemist carries out the procedure correctly, a low yield of desired product is either due to the formation of side products (through an undesired reaction pathway), or due to the reaction having not gone to completion, meaning that unreacted starting materials are still present.

If a reaction has not gone to completion, it can be due to one of two factors. Either the rate of the reaction is too slow, or the equilibrium constant is not favorable enough. The first case is referred to as a kinetic problem, and the latter case is referred to as a thermodynamic problem. Recall that all chemical reactions are in equilibrium. In some cases, they are not observed to be in equilibrium, because the equilibrium constant is large enough that once the reaction has reached equilibrium, only products are observed. If the equilibrium constant is large and unreacted starting materials are still observed, then the reaction has not yet reached equilibrium. In this case, increasing the temperature or the reaction time will help the reaction reach equilibrium, and result in (nearly) complete conversion of reactants into products.

Comprehensive Organic Chemistry Experiments for the Laboratory Classroom
Edited by Carlos A. M. Afonso, Nuno R. Candeias, Dulce Pereira Simão, Alexandre F. Trindade, Jaime A. S. Coelho, Bin Tan, and Robert Franzén.

Published by the Royal Society of Chemistry, www.rsc.org

On the other hand, if the reaction reaches equilibrium quickly, but the equilibrium constant is not large enough, there will always be unreacted starting materials no matter how long the reaction is allowed to go for, or what temperature the reaction is carried out at. In this case, the equilibrium must be driven towards products, and Le Châtelier's principle dictates that either using an excess of one of the reactants or removing a product will drive the equilibrium to the right. In other words, the strategy that a chemist must employ to optimize a reaction will be different depending on whether the problem is a rate problem or an equilibrium problem.

In this experiment, you will explore these concepts in the context of a familiar reaction: the S_N1 reaction between *tert*-pentanol and chloride ion to produce *tert*-pentyl chloride. The S_N1 designation indicates that this is a substitution reaction with a first order rate law. The rate law for this reaction will be rate = k[*tert*-pentanol]. The rate at which the reaction reaches completion is not dependent on the concentration of the chloride ion. The class will be divided into four groups, and each group will carry out the reaction under a different set of conditions, measuring the percent yield of product obtained in each case. At the end of the experiment, the data will be pooled, and it will be possible to determine whether the suboptimal yield under "standard" conditions is due to a rate problem or an equilibrium problem.

The reaction itself is straightforward: concentrated hydrochloric acid is added to the alcohol and the reaction is stirred vigorously. One of the variables that we look at is the effect of the number of equivalents of HCl on reaction yield: some students perform the reaction with 1.0 equivalent of HCl while others use 3.0 equivalents of HCl. This will examine whether increasing the amount of HCl affects the position of the equilibrium. A third group of students adds $CaCl_2$ to the reaction to determine if increasing the concentration of chloride ions will increase the reaction yield. The fourth group of students will determine if refluxing the reaction has any effect on the product yield. This is to differentiate whether an increase in yield with the addition of more HCl is due to an equilibrium affect, or it is simply increasing the rate of the reaction by a kinetic effect. Refluxing the reaction should increase the rate (if this is important) without changing the position of the equilibrium.

After the reaction is complete the by-products need to be removed and the final product purified. The excess HCl is removed by washing with a solution of sodium bicarbonate. However, it is very important to be careful when venting the separatory funnel. Hydrochloric acid will react with sodium bicarbonate to generate carbon dioxide gas which can lead to the buildup of pressure in the funnel. The organic layer is then washed with a saturated brine solution. It is necessary to carry out the extraction process relatively quickly as water will begin to react with *tert*-pentyl chloride to convert it back into *tert*-pentanol. To lose as little product as possible in the drying step without using additional solvent, $CaCl_2$ is used as the drying agent. The organic solution should be decanted from the $CaCl_2$, and, if necessary, can be filtered through a small piece of glass wool. The product is then purified by distillation.

Additional Safety

Concentrated HCl is extremely corrosive, and should only be handled with appropriate personal protective equipment (gloves, goggles, laboratory coat). The reaction and purification involve volatile organic compounds and should be carried out in a fume hood.

Experimental Procedure

1. In the appropriate sized round bottom flask, combine 6.0 mL of *tert*-pentyl alcohol with:

   Group 1: 1.0 equivalent of concentrated HCl;
   Groups 2 & 4: 3.0 equiv. of concentrated HCl;
   Group 3: 3.0 equiv. of conc. HCl + 1 g $CaCl_2$

2. Stir the reaction for 10 min (Group 4: reflux the reaction for 10 min. Make sure you allow it to cool to room temperature before proceeding.)
3. Pour the mixture into a separatory funnel and use 10 mL of saturated $NaHCO_3$ to wash the rest of the material into the separatory funnel. CAUTION: *This step generated CO_2 gas which can build up in the funnel if it is not properly vented.*
4. Separate the organic and aqueous layers.
5. Wash the organic layer with 10 mL of brine.
6. Drain the organic layer into an Erlenmeyer flask, and dry the organic layer with $CaCl_2$. Decant the organic layer off the $CaCl_2$ into a round bottom flask. You may filter it through a small plug of glass wool if necessary.
7. Set up a simple distillation. Be sure to collect pure (constant boiling) chloride. Record the mass and the boiling point range of your product.

Waste Disposal

The organic materials from the distillation, including the product, should be disposed of as organic waste. In most areas, the sodium bicarbonate/HCl solution can be adjusted to neutral pH and disposed down the drain.

Results Interpretation and Additional Questions

1. What was the percent yield of your reaction?
2. Assuming your yield was not 100%, what do you think were the two most significant reasons for the loss of yield in your reaction?
3. Based on the overall class averages, what conditions led to the overall highest percent yield? Briefly explain why these conditions were the most favorable.
4. Based on the results of these experiments, do you think that the reaction has reached equilibrium after 10 min?

2.1.3. Kinetics of a S_N1 Reaction

Paulo Coelho* and Céu Sousa

Centro de Química – Vila Real, Universidade de Trás-os-Montes e Alto Douro, 5001-801 Vila Real, Portugal
*E-mail: pcoelho@utad.pt

Number of sessions (duration of each session)	Hazard level	Difficulty level	Level of study
1 (2 h)	Low	Low	Introductory
Class names Alkyl halides, alcohols			
Concepts involved The kinetics of *t*-butyl chloride hydrolysis can be determined by a simple acid–base titration. Since this slow reaction produces HCl, the addition of a known amount of base and an acid–base indicator to the reaction mixture allows us to determine the time required for the consumption of the base, which gives us also an indication of the *t*-butyl chloride concentration at that moment. Subsequent addition of equal amounts of base affords more data and permits us to calculate the concentration of the remaining *t*-butyl chloride with time			
Chemicals needed *t*-Butyl chloride, NaOH, acetone, Bromothymol blue			
Equipment and experimental techniques involved Burette, stopwatch, titration			
Keywords Acid–base, *t*-butyl chloride, hydrolysis, kinetics, S_N1, titration			

Background

One of the most studied reactions in organic chemistry is the nucleophilic substitution in which an electron-rich species, called the nucleophile (Nuc), bonds to the positive or partially positive charge of a carbon atom to replace a so-called leaving group (LG). The positive or partially positive atom is referred to as the electrophile. In 1935, Edward D. Hughes and Sir Christopher Ingold proposed that two different mechanisms are at work, both of them competing with each other: S_N1 and S_N1 (S stands for chemical substitution, N for nucleophilic and the number represents the kinetic order of the reaction).[1] The main mechanistic pathway will depend largely on the structure of the starting electrophile:

$$\text{Nuc:} + \text{R-LG} \rightarrow \text{R-Nuc} + \text{LG:} \text{ (nucleophilic substitution reaction)} \quad (2.1.3.1)$$

In the S_N2 mechanism one bond is broken and one bond is formed synchronously (Scheme 2.1.3.1). The nucleophile approaches the substrate from a position 180° away from the leaving group, beginning the formation of the new bond and, at the same time, the breaking of the C–X bond occurs. This is a one-step process with no intermediate. The energy for the breaking of the C–X bond is supplied by the simultaneous formation of the C–Nuc bond.[2] Since two reacting

Comprehensive Organic Chemistry Experiments for the Laboratory Classroom
Edited by Carlos A. M. Afonso, Nuno R. Candeias, Dulce Pereira Simão, Alexandre F. Trindade, Jaime A. S. Coelho, Bin Tan, and Robert Franzén.

Published by the Royal Society of Chemistry, www.rsc.org

species, RX and Nuc, are involved in the slow (rate determining) step, the speed of the reaction depends on the concentrations of both the substrate and the nucleophile.

Alternatively, the alkyl halides may react by a S_N1 mechanism (Scheme 2.1.3.2). This mechanism occurs in two steps: in the first step there is a slow dissociation of the alkyl halide to form a carbocation and a halide anion. Then, the nucleophile attacks quickly the intermediate carbocation to form the final product. According to this mechanism the speed of the reaction depends only on the concentration of the alkyl halide, since this is the only reacting species participating in the slow step. The nature and concentration of the nucleophile does not interfere with the speed of the reaction.

For both mechanisms the order of the reaction of the alkyl halide is 1 but in S_N2 reactions the speed also depends of the nucleophile concentration. The hydrolysis of the *t*-butyl chloride occurs by a S_N1 mechanism and therefore the speed of the reaction depends only on the [*t*-butyl chloride].[3] This is an example of first order kinetics and the corresponding integrated equation, which describes the concentration variation in time, may be written as follows:

$$\text{Rate law } V = -\frac{\Delta[C_4H_9Cl]_t}{\Delta t} = k\,[C_4H_9Cl]_t;\ \ln[C_4H_9Cl]_t = -kt + \ln[C_4H_9Cl]_0 \qquad (2.1.3.2)$$

This equation states that there is a linear variation between $\ln[C_4H_9Cl]$ and the reaction time. Therefore, if the *t*-butyl chloride hydrolysis is a first order reaction, then a representation of $\ln[C_4H_9Cl]$ *versus* time must be a straight line. We can prove that the kinetic order is one by measuring [*t*-butyl chloride] with time. The problem comes down to finding a simple method to follow the *t*-butyl chloride concentration as the reaction proceeds. The *t*-butyl chloride concentration can be followed using conductivity methods, NMR, pH measurements or simple titrations.[4–8]

In this experiment we are going to use successive acid–base titrations to calculate the *t*-butyl chloride concentration: since this reaction affords HCl we can determine the time needed to release a certain amount of HCl just by adding a known amount of NaOH and an acid–base

Scheme 2.1.3.1 Mechanism of a S_N2 reaction

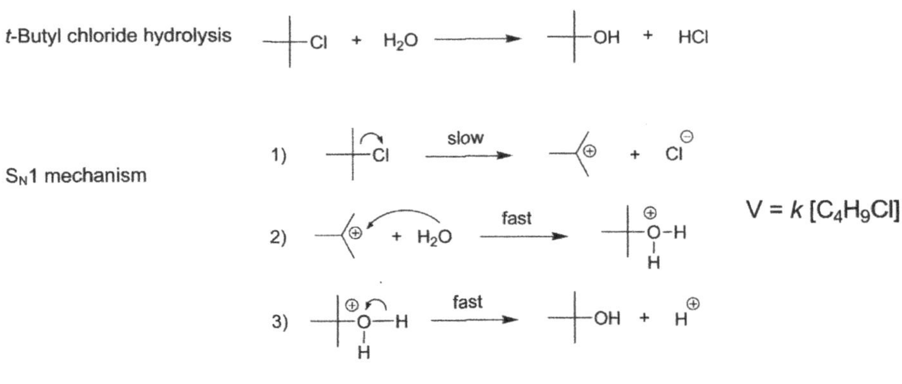

Scheme 2.1.3.2 Mechanism of a S_N1 reaction

indicator to the reaction mixture. In particular we are going to add 5.0×10^{-5} mol of *t*-butyl chloride to a water/acetone solution containing 0.5×10^{-5} mol of NaOH and two drops of Bromothymol blue. The *t*-butyl chloride hydrolysis will start and after a few seconds the colour of the solution will change from blue to yellow indicating that the NaOH has been consumed by the released HCl. At this moment, t_1, we know that 0.5×10^{-5} mol of *t*-butyl chloride has reacted, releasing 0.5×10^{-5} mol of HCl which was neutralized by the NaOH and, therefore, now there is 4.5×10^{-5} mol of *t*-butyl chloride in solution. Adding a second portion of NaOH (0.5×10^{-5} mol) and waiting for the colour change from blue to yellow, t_2, will allow us to determine the time needed to release another 0.5×10^{-5} mol of HCl, indicating that 4.0×10^{-5} mol of *t*-butyl chloride remains in solution at t_2.[9]

Measuring the time needed to change successively the colour of the solution after each NaOH addition will allow us to gather a set of data to plot a graph of $[C_4H_9Cl]$ *versus* time and then $\ln[C_4H_9Cl]$ *versus* time.

It is important to note that in this experiment the hydrolysis reaction never stops and therefore the stopwatch should never be stopped. As the reaction proceeds, the *t*-butyl chloride concentration will decrease along with the speed of the reaction: at the beginning of the hydrolysis the colour changes from blue to yellow after a few seconds of reaction, but after 4 or 5 NaOH additions it will be necessary to wait more than 30 s to see a new colour change.

Additional Safety

This experiment should be carried out in a fume hood. NaOH and HCl are corrosive and should be handled with care.

Experimental Procedure

1. Prepare a 0.050 M solution of freshly distilled *t*-butyl chloride in pure acetone.
2. Prepare 250 mL of an acetone/water (1:9) solution.
3. Place 25 mL of the acetone/water solution in a 100 mL Erlenmeyer and measure the temperature.
4. Fill a burette (25 mL) with a 5.00×10^{-3} M NaOH solution and add 1.00 mL to the Erlenmeyer. Add two drops of Bromothymol blue. The solution becomes blue, indicating a basic solution.
5. Measure exactly 1.00 mL of the 0.050 M *t*-butyl chloride solution using a volumetric pipette, add it quickly to the blue Erlenmeyer solution, stir it manually and start the stopwatch.
6. Record the time for the change of the solution to yellow, add another 1.00 mL of the NaOH solution and stir manually the Erlenmeyer. The solution will change again to blue. Do not stop the stopwatch.
7. Continue to watch the solution and record the time of the second colour change (from blue to yellow).
8. Each time the solution changes to yellow, record the time, add another 1.00 mL of the NaOH solution and stir the solution. Stop after six measurements.
9. Repeat the experiment until you have three consistent assays.

Waste Disposal

The 0.050 M *t*-butyl chloride solution can be stored for some days in a refrigerator. The acetone/water solutions should be discarded in a non-chlorinated waste container.

Results Interpretation and Additional Questions

1. Make a chart with the trial number and the time needed for each colour change (t_{1-6}). Calculate the media of each time and the $[C_4H_9Cl]$ concentration at t_0, t_1, t_2, t_3...t_6.
2. Plot a graph of $[C_4H_9Cl]$ *versus* time and ln $[C_4H_9Cl]$ *versus* time.
3. Calculate the kinetic order of the *t*-butyl chloride hydrolysis reaction.
4. Why should we not continue the experiment for ten NaOH additions and get t_{1-10}? What are the problems with t_{6-10}?
5. The t_{1-6} values are usually quite different from student to student. Why?
6. Why there is no problem if the addition of NaOH is delayed for some seconds after recording the time of each blue–yellow change?

References

1. E. D. Hughes and C. K. Ingold, *J. Chem. Soc.*, 1935, 244–255.
2. J. March, *Advanced Organic Chemistry*, John Wiley & Sons, Inc., New York, 4th edn, 1992, pp. 293.
3. A. R. Olson and R. S. Halford, *J. Am. Chem. Soc.*, 1937, **59**(12), 2644–2647.
4. T. Cyr, J. Prudhomme and M. Zador, *J. Chem. Educ.*, 1973, **50**, 572.
5. T. A. Newton and B. A. Hill, *J. Chem. Educ.*, 2004, **81**, 58.
6. E. A. Moelwyn-Hughes, *J. Chem. Soc.*, 1961, 1517–1522.
7. J. Turner, *J. Chem. Educ.*, 1992, **69**(3), 242.
8. C. F. Most, *Experimental organic chemistry*, John Wiley & Sons, Inc., New York, 1988, p. 460.
9. H. F. Herbrandson, *J. Chem. Educ.*, 1971, **48**(10), 706.

2.1.4. Counterion Effects in the Nucleophilic Substitution Reaction of the Acetate Ion with Alkyl Bromides in the Synthesis of Esters

Ingrid Montes*[a], M. Elizabeth Valentín[a], and Waldemar Adam[a,b]

[a]Department of Chemistry, University of Puerto Rico-Río Piedras Campus, PO Box 23346, San Juan, PR 00931-3346, Puerto Rico, USA; [b]Institute of Organic Chemistry, University of Wuerzburg, Am Hubland, D-97094 Wuerzburg, Germany. *E-mail: ingrid.montes2@upr.edu

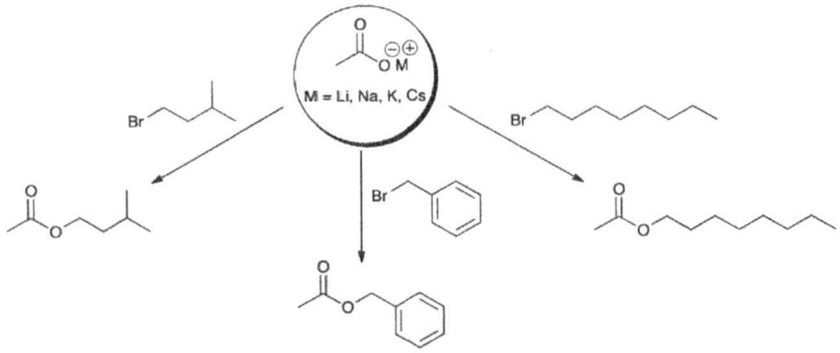

Number of sessions (duration of each session)	Hazard level	Difficulty level	Level of study
2 (4 h each)	Moderate	Medium	Introductory

Class names Alkyl bromides, esters, carboxylic acids, carboxylate salts

Concepts involved This experiment describes a nucleophilic substitution reaction between various alkyl bromides and metal carboxylates. The extent of reaction completion may be rationalized in terms of Pearson's principle of hard and soft acids and bases (HSAB)

Chemicals needed Benzyl bromide, 1-bromooctane, 1-bromo-3-methylbutane, caesium acetate, sodium acetate, potassium acetate, lithium acetate, acetic acid, diethyl ether, ethyl acetate, hexane, sodium bicarbonate, and iodine

Equipment and experimental techniques involved Equipment for refluxing a reaction mixture and magnetic stirring, liquid–liquid extraction apparatus, gas chromatography and thin-layer chromatography

Keywords Esters, gas chromatography, liquid–liquid extraction, nucleophilic substitution, Pearson's HSAB principle, TLC

Background

Esters are commonly employed as flavorings in foods and drinks, since the more volatile ones possess distinctive odors, characteristic of common fruits.[1–4] This laboratory experiment introduces the student to the esterification method, which the student is able to perform in a few hours. In most student's academic experience, Fischer's method is emphasized for esterification in which a carboxylic acid is condensed with the required alcohol.[5–15] Alternatively, the desired

Comprehensive Organic Chemistry Experiments for the Laboratory Classroom
Edited by Carlos A. M. Afonso, Nuno R. Candeias, Dulce Pereira Simão, Alexandre F. Trindade, Jaime A. S. Coelho, Bin Tan, and Robert Franzén.

Published by the Royal Society of Chemistry, www.rsc.org

ester may be synthesized by means of a nucleophilic substitution reaction (S_N2 or S_N1) between an alkyl halide and metal carboxylate.[16–18] Such a method is not commonly used in the organic chemistry teaching laboratory. In particular, the student will learn that the metal counterion plays an important role with regard to the efficiency of the reaction. Moreover, they will learn that the ester yield depends on the type of mechanism, namely, whether it is a S_N2 or S_N1 process, and that the results are well explained in terms of Pearson's principle of hard and soft acid base (HSAB), an empirical concept hardly used in introductory organic chemistry.[19] The HSAB principle classifies acids and bases as hard or soft, according to their polarizability, and states that a hard acid reacts preferably with a hard base, whereas a soft acid favors a soft base. Because the observed trends in the ester yields are interpreted in terms of Pearson's HSAB principle, students become exposed to a successful theory by conducting this experiment.[19–22] The pooling of their results promotes responsibility in obtaining accurate data. The experimental background is focused on the importance of organic synthesis, the use of a specific techniques such as extraction and chromatography, the application of the product prepared, the use of sustainable methods to achieve a given transformation, and the mechanistic rationale of the reaction.

Additional Safety

Students must consult the relevant MSDS before conducting the experiment. The volatile alkyl bromides are eye and skin irritants and flammable liquids, and should be handled with care. Benzyl bromide is a lachrymatory liquid and acetic acid a corrosive substance, neither of which must be inhaled. Ether, ethyl acetate, and hexane are extremely flammable liquids, and harmful if inhaled or if exposed to the skin. Iodine is an eye and skin irritant and harmful if inhaled or swallowed, and should be handled with care. The metal acetates are hygroscopic and cause skin irritation. Students must wear safety goggles and gloves at all times and work in a well-ventilated hood.

Experimental Procedure

First Laboratory Session (4 h)

1. Add 18 mmol of the desired metal acetate, along with 8 mL of acetic acid, to a 50 mL round-bottom flask equipped with a magnetic stirrer (*metal acetates are hygroscopic, work quickly, do not leave any reagents uncovered; acetic acid is a lachrymator substance, use it in a well-ventilated hood*).
2. After partial dissolution of the acetate salt add 12 mmol of alkyl bromide.
3. Quickly connect the reflux condenser (Figure 2.1.4.1), and gently heat under a gentle reflux for 60–90 min.
4. Allow the reaction mixture to cool down, and then slowly add 20 mL of saturated aqueous sodium bicarbonate to neutralize the acetic acid.
5. The reaction mixture is now extracted with diethyl ether (2 × 10 mL) as solvent.
6. Dry the combined organic extracts over $MgSO_4$, remove the drying agent by filtration and evaporate the solvent by mildly heating with a warm water bath (approximately 45 °C).
7. Store your product, tightly sealed, until the next session.

Second Laboratory Session (4 h)

Weigh your product and run a TLC on alumina plates with a 9:1 hexane:ethyl acetate solvent mixture as eluent. Since both the product and the starting material are colorless, they may be visualized in an iodine chamber (for benzyl acetate use an UV lamp).

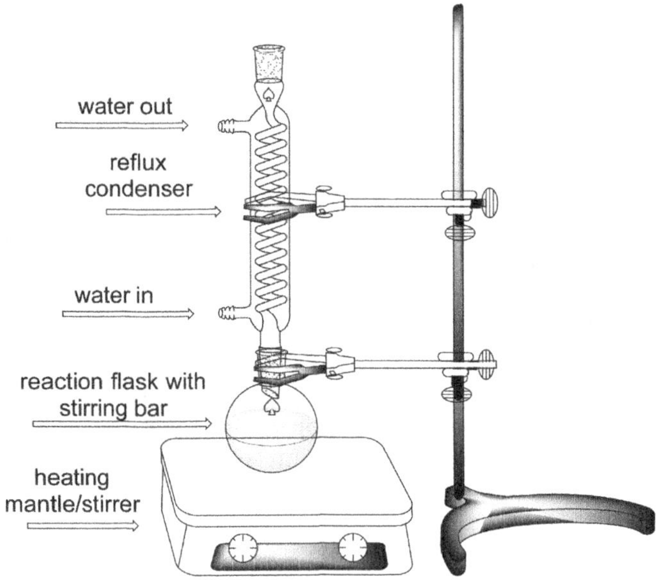

Figure 2.1.4.1 Apparatus for running the reaction under reflux.

1. Prepare your sample for injection into the GC according to the directions provided by the laboratory instructor.
2. Calculate the yield of the crude product, and subsequently the actual product yield by using the purity determined by GC analysis (see GC chromatogram).
3. For result analysis, the students need to exchange data for global results discussion. Pool your results with those of your classmates to assess agreement between the individual data sets; analyze possible trends in reactivity, and discuss the combined results.

Waste Disposal

Neutralized aqueous solutions should be dispensed in the corresponding aqueous waste recipient. Non-halogenated solvents should be dispensed in the corresponding organic waste recipient.

Interpretation of the Results and Additional Questions

1. For each reaction identify the nucleophile, the counterion, the substrate, and the leaving group. Classify each as a hard or soft acid or base.
2. Complete the following table for each of the esters synthesized in the laboratory session.

Reaction equation				
Alkyl halide used				
Acetate counterion	Caesium	Sodium	Potassium	Lithium
Was a precipitate present?				
GC-corrected yield (%)				

3. Is there a relationship between the ionic radius and the hardness of the ion? Briefly explain why.

4. According to your experimental results:

    a. How do your TLC results compare with those of gas chromatography?
    b. Which metal counterion promotes the most efficient reaction?
    c. Which metal counterion promotes the least efficient reaction?

5. In terms of the HSAB principle explain the effect of the metal counterion in an S_N2 reaction.
6. Evaluate and discuss the mechanism that is preferred for each of the alkyl bromides used; justify your answer.

Acknowledgements

Adapted from *Journal of Chemical Education*, 2009, **86** (11), 1315. Copyright by the Division of Chemical Education, Inc. of the American Chemical Society. Used with permission. All rights reserved.

References

1. E. C. Crocker, *J. Chem. Educ.*, 1945, **22**, 567.
2. R. K. Eskew, *J. Chem. Educ.*, 1951, **28**, 326.
3. M. Clarke, A. Brown, D. N. Epp, M. Gallup, J. R. Wilson and J. A. Wuerthele, *J. Chem. Educ.*, 1986, **63**, 1050.
4. A. J. Taylor, *Compr. Rev. Food Sci. Food Saf.*, 2002, **1**, 45–57.
5. W. H. Puterbaugh, C. H. Vanselow, K. Nelson and E. J. Shrawder, *J. Chem. Educ.*, 1963, **40**, 349.
6. M. B. Hocking, *J. Chem. Educ.*, 1980, **57**, 527.
7. K. L. Williamson, *Macroscale and Microscale Organic Experiments*, D. C. Heath Canada, Lexington, 1st edn, 1989, pp. 302–314.
8. D. M. Birney and S. D. Starnes, *J. Chem. Educ.*, 1999, **76**, 1560.
9. C. R. Whitlock and P. A. Bishop, *Chem. Educ.*, 2003, **8**, 352.
10. R. Duarte, J. T. Nielsen and V. Dragojlovic, *J. Chem. Educ.*, 2004, **81**, 1010.
11. S. Lee, *Chem. Educ.*, 2004, **9**, 359–363.
12. I. Montes, D. Sanabria, M. García, J. Castro and J. Fajardo, *J. Chem. Educ.*, 2006, **83**, 628.
13. A. W. C. van den Berg and U. Hanefeld, *J. Chem. Educ.*, 2006, **83**, 292.
14. P. A. Wade, S. A. Rutkowsky and D. B. King, *J. Chem. Educ.*, 2006, **83**, 927.
15. M. K. Reilly, R. P. King, A. J. Wagner and S. M. King, *J. Chem. Educ.*, 2014, **91**, 1706.
16. R. G. Stabile and A. P. Dicks, *J. Chem. Educ.*, 2004, **81**, 1488.
17. M. M. Clennan and E. L. Clennan, *J. Chem. Educ.*, 2005, **82**, 1676.
18. D. P. Brown, H. Durutlic and D. Juste, *J. Chem. Educ.*, 2004, **81**, 1016.
19. R. G. Pearson, *J. Org. Chem.*, 1987, **52**, 2131–2136.
20. T.-L. Ho, *J. Chem. Educ.*, 1978, **55**, 355.
21. R. G. Pearson, *J. Chem. Educ.*, 1987, **64**, 561.
22. R. F. Daley and S. J. Daley, *Organic Chemistry*, http://www.ochem4free.info/, accessed 15 July 2015.

2.1.5. *N*-Alkylation of Pyrazole: Reaction in an Ionic Liquid

Clarissa P. Frizzo*, Marcos A. P. Martins*, Caroline R. Bender, Paulo R. S. Salbego, Aniele Z. Tier, Geórgia C. Zimmer, Guilherme C. Paveglio, and Helio G. Bonacorso

Núcleo de Química de Heterociclos (NUQUIMHE), Department of Chemistry, Federal University of Santa Maria, 97105-900, Santa Maria, RS, Brazil. *E-mail: clarissa.frizzo@gmail.com, mmartins@base.ufsm.br

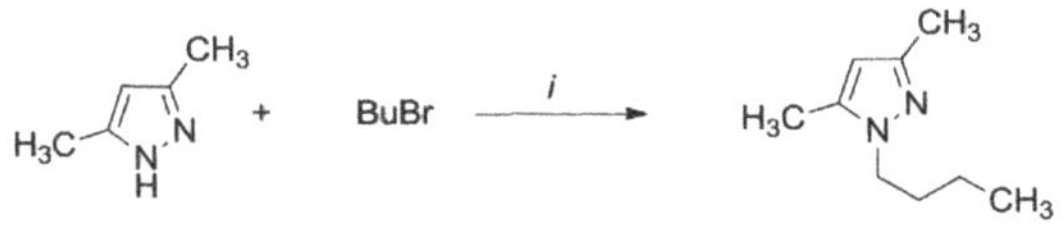

1 2 3

i: [BMIM][BF_4], KOH, 80 °C, 2h.

Number of session (duration of each session)	Hazard level	Difficulty level	Level of study
1 (4 h)	Moderate	Medium	Introductory
Class names Amines, alkyl halides			
Concepts involved This experiment describes a synthesis of *N*-alkylpyrazoles using the ionic liquid (IL) [BMIM][BF_4]. The reaction mechanism involves a nucleophilic substitution with a C–N σ-bond formation. Additionally, the experiment shows the IL effect in the nucleophilic substitution reaction			
Chemicals needed 3,5-Dimethyl-1*H*-pyrazole (pyrazole, **1**), 1-bromobutane (**2**), the IL 1-butyl-3-methylimidazolinium tetrafluoroborate ([BMIM][BF_4]), potassium hydroxide (KOH), diethyl ether, anhydrous sodium sulfate (Na_2SO_4), $CDCl_3$			
Equipment and experimental techniques involved Apparatus for reaction at high temperature (80 °C) and magnetic stirring, rotary evaporator, vacuum pump, NMR			
Keywords Ionic liquids (ILs), liquid–liquid extraction, *N*-alkylation, NMR, pyrazoles, rotary evaporator			

Background of the Experiment

Pyrazoles are important synthetic targets in the pharmaceutical industry because they are part of numerous biologically active compounds,[1] and are used in blockbuster drugs such as Celecoxib[2] and Viagra.[3] Numerous methods have been developed to obtain *N*-alkyl-pyrazoles; for example, the reaction between hydrazines and 1,3-dicarbonyl compounds (or α,β-unsaturated carbonyl compounds).[4–6] Another example is the *N*-alkylation of *NH*-pyrazoles, which is more versatile due to the greater scope of alkyl groups. The *N*-alkylation reactions are useful in the preparation of valuable intermediates, which are used for fungicides and other pharmacologically

Comprehensive Organic Chemistry Experiments for the Laboratory Classroom
Edited by Carlos A. M. Afonso, Nuno R. Candeias, Dulce Pereira Simão, Alexandre F. Trindade, Jaime A. S. Coelho, Bin Tan, and Robert Franzén.

Published by the Royal Society of Chemistry, www.rsc.org

interesting compounds.[7] Due to the broad range of activity and importance of *N*-alkyl-pyrazoles, a simple, green, and high-yielding approach for their synthesis is highly desirable. It is important to mention that the use of nonsymmetrical pyrazoles can result in mixtures of isomers.

In the context of green organic synthesis, ionic liquids (ILs) have attracted great interest.[8] ILs were initially introduced as an alternative green reaction media because of their unique chemical and physical properties; for example, non-volatility, non-flammability, thermal stability, and controlled miscibility. ILs can solubilize both polar and non-polar species and also promote interactions between different components. These structures can be recovered and reused several times, which is particularly interesting in a green context. Furthermore, as a solvent/catalyst they play a significant role in controlling reactions.[9] There are several reports of the *N*-alkylation of N-heterocycles in ILs in the literature.[10–12] Among the N-heterocycles that have been alkylated in ILs are 2,4-thiazolidinones, cyclic imides, pyrroles, benzotriazoles, and phthalimides.

The reaction of nitrogen heterocyclic nucleophiles with alkyl bromides is a great methodology for developing knowledge about nucleophilic substitution and the IL effect in chemical reactions. The performance of IL as a solvent/catalyst can be understood in terms of the activated complex energy of the reaction. Its use results in higher yields in comparison with common aprotic organic solvents. The ability of the IL to accelerate the reaction is probably a result of the decrease in the activation energy of activated complexes, which in turn is most likely due to the effect in the solvation of highly polar or charged intermediates, such as carbocations or carbanions, as well as activated complexes, which could become stabler.[13] Studies on S_N2 reactions have shown that, in general, the use of IL for reactions of highly associating anions, like halides, results in higher reaction rates.[13]

Thus, the main objective of this experiment is to demonstrate the effective use of ILs in the *N*-alkylation reaction, through a nucleophilic substitution mechanism between a 1-bromobutane and *NH*-pyrazole, in the presence of a strong inorganic base.

Additional Safety

All experiments should be carried out in a fume hood, with proper safety equipment such as protective clothing, gloves, and safety goggles. The potassium hydroxide base can irritate the eyes and skin and should be handled with care. Diethyl ether is a flammable solvent and so one must avoid handling it near heat or flames. 1-Bromobutane is also flammable and may cause irritation to the eyes and skin.

Experimental Procedure

Laboratory Session (4 h): *N*-Butyl-3,5-dimethylpyrazole Synthesis

Reaction with Conventional Thermal Heating (Oil Bath) (ref. 14)

1. Weigh 3,5-dimethyl-1*H*-pyrazole (2 mmol, 0.20 g), KOH (2.4 mmol, 0.14 g), and [BMIM] [BF_4] (2.2 mmol, 0.50 g, density of 1.21 g mL^{-1}) all in the same round-bottom flask equipped with a magnetic bar.
2. Using a syringe, add the alkylating agent (1-bromobutane: 2.4 mmol, 0.26 mL, and density of 1.28 g mL^{-1}) to the same round-bottom flask.
3. Place the round-bottom flask, with the reaction mixture, inside an oil bath placed in a magnetic stirring apparatus.
4. Place a reflux condenser with cold running water on the flask.
5. Stir and keep the mixture heated at 80 °C for 2 h.

Product Isolation

6. After completion of the reaction, wait until room temperature is reached.
7. Add diethyl ether (5 mL) and then magnetically stir for approximately 2 min.
8. Remove the diethyl ether by decantation into an Erlenmeyer flask.
9. Repeat the procedure 5 or 6 times.
10. Dry the diethyl ether phase over Na_2SO_4 (~5 g) to remove any traces of moisture.
11. Weigh an empty round-bottom flask (M_1), in which the product will be stored.
12. Filter the mixture into the previously weighed round-bottom flask.
13. Discard the paper filter with Na_2SO_4 salt.
14. Concentrate the remaining organic phase (diethyl ether) using a rotary evaporator, in order to furnish the *N*-butyl-3,5-dimethylpyrazole (152.24 g mol^{-1}) as an orange oil.
15. Keep the product under vacuum (with a vacuum pump) for 30 min, to remove any residual solvent.
16. Perform an NMR analysis in $CDCl_3$ and calculate the reaction yield.

Yield Determination

17. Weigh the round-bottom flask containing the reaction mixture and determine the final weight (M_2).
18. From this value (M_2), subtract the weight of the empty round-bottom flask (M_1), in accordance with eqn (2.1.5.1):

$$M_3 = M_2 - M_1 \qquad (2.1.5.1)$$

19. Calculate the weight of the obtained product (M_3) in mmol.
20. Record the amount of product obtained (in mmol) with the limiting reactant used (in this case the 3,5-dimethyl-1*H*-pyrazole), in order to calculate the reaction yield as per eqn (2.1.5.2):

$$\text{Yield (\%)} = \frac{\text{product (mmol)}}{\text{limiting reactant (mmol)}} \times 100 \qquad (2.1.5.2)$$

Waste Disposal

The diethyl ether may be reused or collected in a non-halogenated waste collector recipient, for later destruction. The Na_2SO_4 salt must be collected in a collector recipient for solid waste. To recover the IL, add dichloromethane to the reaction mixture and remove the KBr salt from the IL phase by successive extractions with water. After this, remove the solvent and place the IL under vacuum (10 mbar) and magnetic stirring, for 8 h. The IL is then recovered and purified.

Interpretation of Results and Additional Questions

1. Interpret the ^1H and ^{13}C NMR spectra of the product obtained and compare your spectra with those reported in the literature (if you wish, the refractive index of the product can also be compared).[14]

2. Indicate the advantages of IL use in chemical reactions in relation to molecular organic solvents.
3. Rationalize a mechanism for *N*-alkyl-pyrazole formation through a nucleophilic substitution, considering the nucleophilicity of the two nitrogen atoms of the pyrazole and the effect of IL in the transition state. Indicate the reaction order.
4. Discuss the use of potassium hydroxide (KOH) in the reaction (remember that the reaction performed without KOH results in a 33% yield).
5. Identify the nucleophile and electrophile and indicate the orbitals (π, σ, n, σ*, and/or π*) involved in this organic reaction. Specify the high energy molecular orbital (HOMO) and low unoccupied molecular orbital (LUMO) involved in the reaction.
6. Analyze the nucleophilicity of the two nitrogen atoms of the pyrazole ring.

Acknowledgement

Experiment adapted from Martins *et al.*, *Catal. Commun.*, 2009, Ionic liquid as catalyst in the synthesis of *N*-alkyl trifluoromethyl pyrazoles, **10**, pp 1153–1156. Copyright (2009) with permission from Elsevier.

References

1. J. Elguero, P. Goya, N. Jagerovic and A. M. S. Silva, *Targets Heterocycl. Syst.*, 2002, **6**, 52.
2. F. R. Souza, V. T. Souza, V. Ratzlaff, L. P. Borges, M. R. Oliveira, H. G. Bonacorso, N. Zanatta, M. A. P. Martins and C. F. Mello, *Eur. J. Pharmacol.*, 2002, **451**, 141.
3. T. D. Penning, J. J. Talley, S. R. Bertenshaw, J. S. Carter, P. W. Collins, S. Docter, M. J. Graneto, L. F. Lee, J. W. Malecha, J. M. Miyashiro, R. S. Rogers, D. J. Rogier, S. S. Yu, G. D. Anderson, E. G. Burton, J. N. Cogburn, S. A. Gregory, C. M. Koboldt, W. E. Perkins, K. Seibert, A. W. Veenhuizen, Y. Y. Zhang and P. C. Isakson, *J. Med. Chem.*, 1997, **40**, 1347.
4. N. K. Terrett, A. S. Bell, D. Brown and P. Ellis, *Bioorg. Med. Chem. Lett.*, 1996, **6**, 1819.
5. T. Norris, R. Colon-Cruz and D. H. B. Ripin, *Org. Biomol. Chem.*, 2005, **3**, 1844.
6. S. Fustero, M. Sánchez-Roselló and A. Simón-Fuentes, *Chem. Rev.*, 2011, **111**, 6984.
7. V. K. Aggarwal, J. Vicente and R. V. J. Bonnert, *J. Org. Chem.*, 2003, **68**, 5381.
8. N. R. Candeias, L. C. Branco, P. M. Góis, C. A. M. Afonso and A. F. Trindade, *Chem. Rev.*, 2009, **109**, 2703.
9. M. A. P. Martins, C. P. Frizzo, A. Z. Tier, D. N. Moreira, N. Zanatta and H. G. Bonacorso, *Chem. Rev.*, 2014, **114**, PR1.
10. M. J. Earle, P. B. Mc Cormac and S. R. Seddon, *Chem. Commun.*, 1998, 2245.
11. Z.-G. Le, Z.-C. Chen, Y. Hu and Q.-G. Zheng, *Synthesis*, 2004, 208.
12. R. Y. Jorapur, J. M. Jeong and D. Y. Chi, *Tetrahedron Lett.*, 2006, **47**, 2435.
13. *March's Advanced Organic Chemistry – Reactions, Mechanisms, and Structure*, ed. M. B. Smith, John Wiley & Sons, Inc., Hoboken, 7th edn, 2013, vol. 9, pp. 347–366.
14. C. P. Frizzo, D. N. Moreira, E. M. Guarda, G. F. Fiss, M. R. B. Marzari, N. Zanatta, H. G. Bonacorso and M. A. P. Martins, *Catal. Commun.*, 2009, **10**, 1153.

2.1.6. Conversion of Alcohols into Alkyl Chlorides Using Cyanuric Chloride

Eliana Martins, Catarina A. B. Rodrigues*, and Carlos A. M. Afonso*

Instituto de Investigação do Medicamento (iMed.ULisboa), Faculty of Pharmacy, University of Lisbon, Av. Prof. Gama Pinto, 1649-003, Lisboa, Portugal
*E-mail: catarinarodrigues@ff.ulisboa.pt, carlosafonso@ff.ulisboa.pt

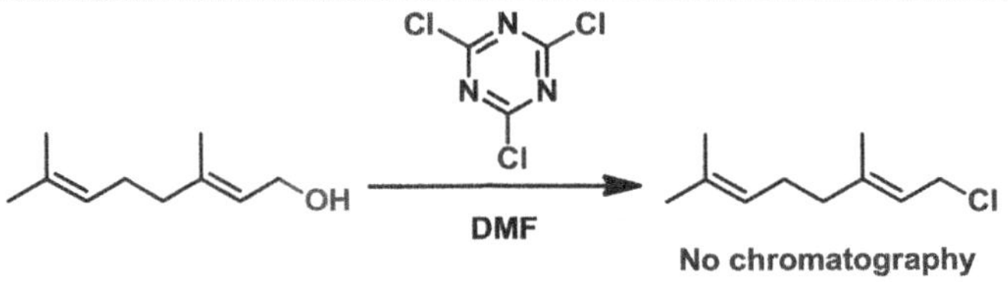

Number of sessions (duration of each session)	Hazard level	Difficulty level	Level of study
1 (3 h)	Moderate	Low	Intermediate

Class names Alcohols, alkyl halides

Concepts involved Mild bimolecular nucleophilic substitution (S_N2), hydroxyl activation

Chemicals needed Geraniol, cyanuric chloride, dry dimethylformamide, dry dichloromethane, calcium chloride, anhydrous magnesium sulfate, saturated solution of sodium carbonate, aqueous solution of 1 M HCl, ethyl acetate, hexane, silica gel, Celite®, deuterated chloroform

Equipment and experimental techniques involved A 50 mL two-necked round bottom flask, magnetic stir bar, stirring plate, dropping funnel, two calcium chloride tubes, water bath apparatus, Buchner funnel, spatulas, Kitasato flask, weighting paper or glass, separatory funnel, thin-layer chromatography plates, 100 mL Erlenmeyer flask, 100 mL round bottom flask, TLC plates (Silica Merck Kieselgel 60 F254 plates), NMR spectrometer (Bruker Avance 400)

Keywords Alcohols, alkyl chlorides, cyanuric chloride, extraction, mild nucleophilic substitution, NMR

Background

Cyanuric chloride is a versatile reagent widely used in the production of antitumor agents, herbicides and dyes.[1–7] The low cost, low toxicity and simple handling conditions of this reagent compared to other alternatives make this compound very attractive. Cyanuric chloride is also very useful in synthetic chemistry for functional groups transformation under mild conditions, such as activation of carboxylic acids,[8] Swern oxidation,[9] Beckmann rearrangement, conversion of formamides into isonitriles and alkyl alcohols into alkyl chlorides, among others (Figure 2.1.6.1).[1,4]

The conversion of alcohols into alkyl chlorides is a very useful transformation for the synthesis of important precursors – alkyl chlorides. Several methods are known to perform this

Comprehensive Organic Chemistry Experiments for the Laboratory Classroom
Edited by Carlos A. M. Afonso, Nuno R. Candeias, Dulce Pereira Simão, Alexandre F. Trindade, Jaime A. S. Coelho, Bin Tan, and Robert Franzén.

Published by the Royal Society of Chemistry, www.rsc.org

transformation: using thionyl chloride, phosphorous halides and phosphines combined with halides sources (such as *N*-chlorosuccinimide or CCl_4[10]), among others. However, the scientific community is continuously seeking for reactions that are green, robust and run under mild conditions to increase the applicability. Another method to convert alcohols into alkyl chlorides that covers these parameters is using cyanuric chloride and DMF under anhydrous conditions (Scheme 2.1.6.1).[11]

Cyanuric chloride forms a reactive intermediate in the presence of DMF within several minutes (Scheme 2.1.6.2, Step A). This intermediate will then react with the alkyl alcohol to give the respective alkyl chloride with inversion of configuration (Scheme 2.1.6.2, Step B).

Due to its attractive properties, this reaction has been widely applied in the synthesis of pharmaceutical relevant molecules.[12,13]

Herein it is intended to explore a mechanistic rich reaction with wide applicability in a teaching laboratory environment. This reaction was used to convert 1-phenylethanol and geraniol (Figure 2.1.6.2) into the respective alkyl chlorides. The first alcohol was used as a standard proof

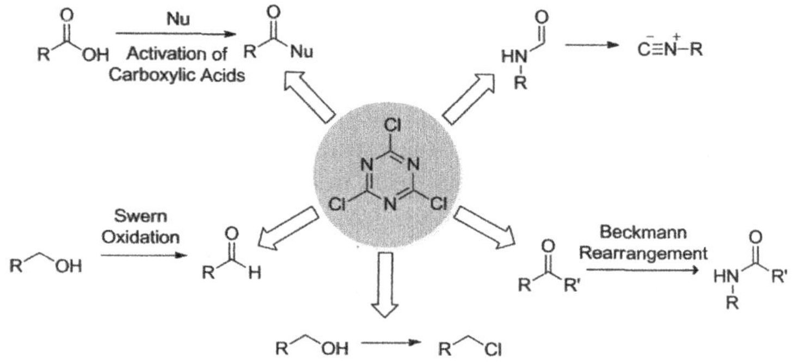

Figure 2.1.6.1 Reported functional group transformations promoted by cyanuric chloride.

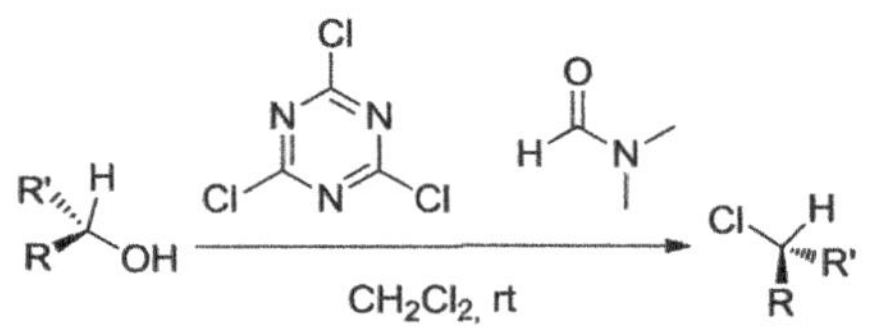

Scheme 2.1.6.1 Conversion of alcohols to alkyl chloride mediated by cyanuric chloride and DMF.

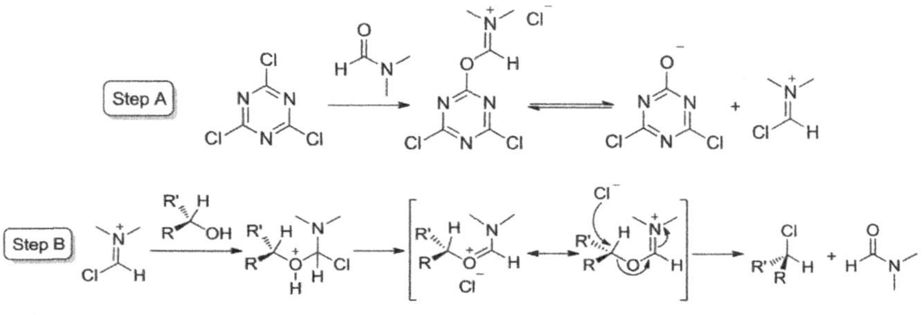

Scheme 2.1.6.2 Mechanism of conversion of alcohols into alkyl chloride mediated by cyanuric chloride and DMF.[11]

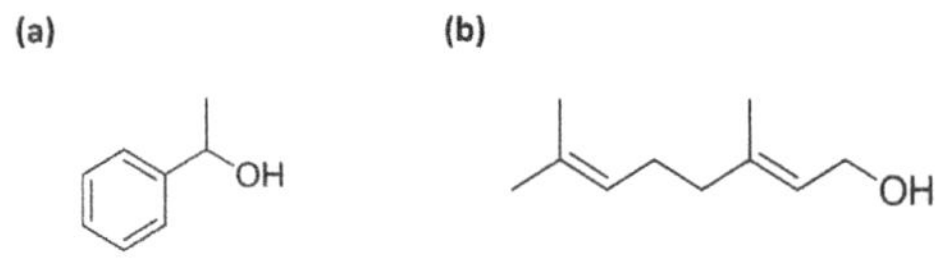

Figure 2.1.6.2 Structures of (a) 1-phenylethanol and (b) geraniol.

of concept and geraniol is an example of application of the described transformation to an industrially relevant molecule. In fact geraniol is widely used as a fragrance in cosmetic (shampoos, soaps, decorative fragrances, *etc.*) and non-cosmetic (detergents, *etc.*) products[14,15] while the resulting product geranyl chloride is a useful intermediate in organic synthesis such as in total synthesis of natural products.[16]

Additional Safety

Gloves should be used throughout the experiment and all the manipulations should be carried out in a fume-hood. Geraniol is irritant in cases of skin and eye contact. Cyanuric chloride has a pungent odor and is harmful if inhaled or ingested and irritant in case of skin and eye contact. Dimethylformamide is an irritant in case of inhalation, ingestion skin and eye contact. Dichloromethane is harmful if swallowed or inhaled and irritating to skin and eyes. Magnesium sulfate is hazardous in cases of ingestion, slightly hazardous in cases of inhalation, of skin or eye contact. Sodium carbonate is hazardous in cases of ingestion, of inhalation, of skin and eye contact. HCl is very hazardous in cases of ingestion, of skin and of eye contact (irritant, corrosive) and slightly hazardous in cases of inhalation (lung sensitizer). Silica gel is slightly hazardous in cases of skin and eye contact (irritant), of ingestion and inhalation – handle this reagent in the fume hood. Celite® is hazardous in cases of inhalation (lung irritant) and slightly hazardous in cases of skin and eye contact (irritant), and of ingestion. Ethyl acetate is hazardous in cases of ingestion, inhalation, skin contact (irritant, permeator) and eye contact (irritant). Hexane is hazardous in cases of skin contact (irritant and permeator), ingestion, inhalation and eye contact (irritant). Deuterated chloroform is hazardous in cases of skin contact (irritant and permeator), eye contact (irritant), ingestion and inhalation.

Experimental Procedure

1. Dry the 50 mL two-necked round bottom flask with a magnetic stir bar in an oven at 100 °C for 1 h.
2. In a fume hood set up the dry conditions reaction apparatus with the previously dried 50 mL two-necked round bottom flask with a magnetic stir bar, a dropping funnel, two calcium chloride tubes, a water bath and a magnetic plate.
3. Place cyanuric chloride (0.5 g, 2.7 mmol) and dimethylformamide (0.55 mL, 6.8 mmol) inside this apparatus.
4. After formation of a white precipitate (15–20 min) add a solution of geraniol (0.4 mL, 2.3 mmol) in dichloromethane (7 mL) through the dropping funnel.
5. Follow the reaction by TLC (eluent hexane–ethyl acetate (9 : 1)), and after 30 min transfer the mixture to a separatory funnel, wash the organic phase with water (15 mL), a saturated solution of sodium carbonate (15 mL) and aqueous HCl (1 M) solution (15 mL).
6. Dry the organic phase with anhydrous magnesium sulfate and filter through a Buchner funnel with Celite® on the bottom (~2 g) and silica gel on the top (~5 g). Wash the silica with more dichloromethane (2 × 5 mL).

7. Remove the solvent under vacuum, weight the obtained residue and determine the yield and purity by spectroscopic techniques (NMR and IR).

Waste Disposal

Residues of hexane and ethyl acetate should be collected in a non-chlorinated waste collector recipient, and residues of dichloromethane should be collected in a chlorinated waste recipient for posterior disposal. The solid residues of Celite® and silica should be placed in the proper solid residues collector recipient.

Results Interpretation and Additional Questions

1. What is the mechanism of the reaction? Does the reaction work with tertiary alcohols?
2. What do you expect to obtain from (−)-menthol?
3. Why should this reaction be carried out under anhydrous conditions?
4. Does the reaction work without or with a catalytic amount of DMF?
5. How do you prepare the bromide derivative from the used substrate?

References

1. G. Blotny, *Tetrahedron*, 2006, **62**, 9507–9522.
2. K. Bretterbauer and C. Schwarzinger, *Curr. Org. Synth.*, 2012, **9**, 342–356.
3. R. Banerjee, D. R. Brown and E. Weerapana, *Synlett*, 2013, **24**, 1599–1605.
4. D. A. Probst, P. R. Hanson and D. A. Barda, *Encyclopedia of Reagents for Organic Synthesis*, John Wiley & Sons, Ltd, Chichester, 2001.
5. B. Barton, S. Gouws, M. C. Schaefer and B. Zeelie, *Org. Process Res. Dev.*, 2003, **7**, 1071–1076.
6. A. Herrera, R. Martínez-Alvarez, P. Ramiro, M. Chioua and R. Chioua, *Synthesis*, 2004, **2004**, 503–505.
7. E. Hollink, E. E. Simanek and D. E. Bergbreiter, *Tetrahedron Lett.*, 2005, **46**, 2005–2008.
8. K. Venkataraman and D. R. Wagle, *Tetrahedron Lett.*, 1979, **20**, 3037–3040.
9. L. De Luca, G. Giacomelli and A. Porcheddu, *J. Org. Chem.*, 2001, **66**, 7907–7909.
10. J. G. Calzada and J. Hooz, *Org. Synth.*, 1988, **50–9**, 634–637.
11. L. De Luca, G. Giacomelli and A. Porcheddu, *Org. Lett.*, 2002, **4**, 553–555.
12. C.-W. Chang, S.-S. Chang, C.-S. Chao and K.-K. T. Mong, *Tetrahedron Lett.*, 2009, **50**, 4536–4540.
13. L. M. Matosziuk, A. S. Harney, K. W. MacRenaris and T. J. Meade, *Eur. J. Inorg. Chem.*, 2012, **2012**, 2099–2107.
14. A. Lapczynski, S. P. Bhatia, R. J. Foxenberg, C. S. Letizia and A. M. Api, *Food Chem. Toxicol.*, 2008, **46**, S160–S170.
15. W. Chen and A. M. Viljoen, *S. Afr. J. Bot.*, 2010, **76**, 643–651.
16. K.-i. Takao, K. Tsunoda, T. Kurisu, A. Sakama, Y. Nishimura, K. Yoshida and K.-i. Tadano, *Org. Lett.*, 2015, **17**, 756–759.

2.1.7. Synthesis of Phenacetin

Paulo Coelho* and Céu Sousa

Centro de Química – Vila Real, Universidade de Trás-os-Montes e Alto Douro, 5001-801 Vila Real, Portugal
*E-mail: pcoelho@utad.pt

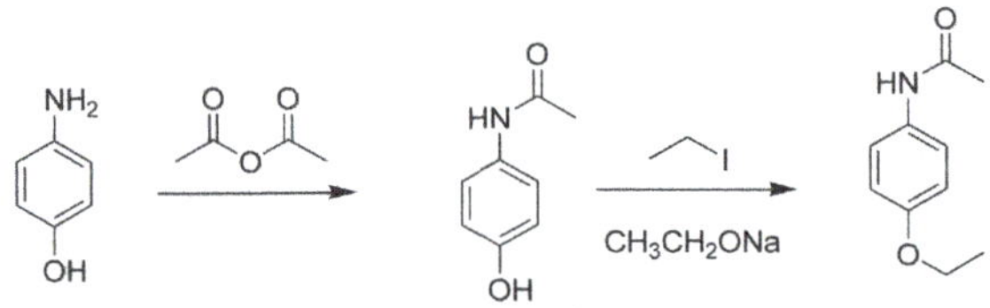

Number of sessions (duration of each session)	Hazard level	Difficulty level	Level of study
2 (2 h each)	Moderate	Low	Intermediate

Class names Phenols, anhydrides, alkyl halides, amines, ethers, amides

Concepts involved The synthesis of phenacetin, an analgesic and antipyretic drug, is accomplished by a two-step procedure involving the acetylation of *p*-aminophenol followed by alkylation with ethyl iodide in basic medium

Chemicals needed *p*-aminophenol, acetic anhydride, sodium, ethyl iodide, ethanol, diethyl ether, petroleum ether (40–60 °C)

Equipment and experimental techniques involved Heating mantle, apparatus for reaction under reflux, vacuum filtration

Keywords Acetylation, ether synthesis, NMR, paracetamol, phenacetin, vacuum filtration

Background

Phenacetin as a drug was introduced in 1887 and widely used as a mild analgesic and antipyretic agent. It was very popular for many decades and often used mixed with aspirin and caffeine as a remedy for fever and pain. However, in 1983, the US Food and Drug Administration (US FDA) ordered the withdrawal of the drugs containing phenacetin, owing to its carcinogenic and kidney-damaging properties. It was then substituted by paracetamol, a similar aniline derivative that is less toxic and suitable for treating mild to moderate pain. It was discovered in 1947 that phenacetin is in fact rapidly metabolised to paracetamol in the human body and the latter molecule is indeed the major metabolite responsible for its analgesic action (Scheme 2.1.7.1).[1] Acetanilide, another aniline derivative with analgesic properties and important toxic effects, is also metabolized to paracetamol.[2] These findings led to the "rediscovery" of paracetamol, an old drug but assumed to be toxic by the end of the nineteenth century. Nowadays, paracetamol is one of the most common analgesics.[3,4]

Comprehensive Organic Chemistry Experiments for the Laboratory Classroom
Edited by Carlos A. M. Afonso, Nuno R. Candeias, Dulce Pereira Simão, Alexandre F. Trindade, Jaime A. S. Coelho, Bin Tan, and Robert Franzén.

Published by the Royal Society of Chemistry, www.rsc.org

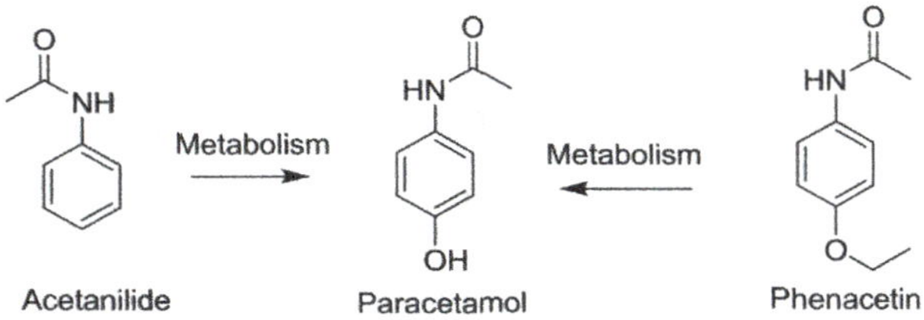

Scheme 2.1.7.1 Formation of paracetamol from acetanilide or phenacetin

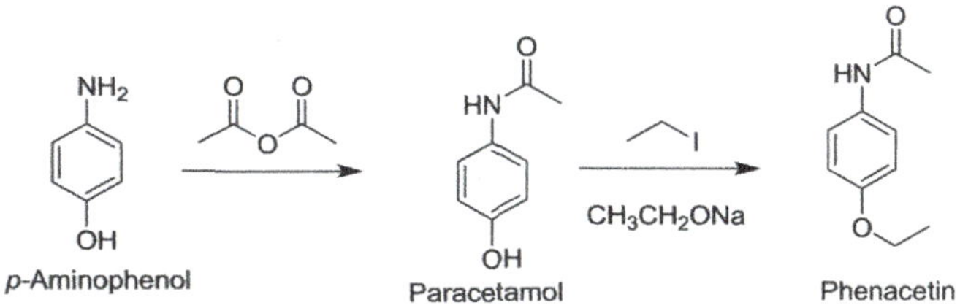

Scheme 2.1.7.2 Synthesis of phenacetin from *p*-aminophenol

The first synthesis of phenacetin was reported in 1878 by H. N. Morse.[5] Phenacetin can be easily prepared in two steps from *p*-aminophenol by acetylation, which initially affords paracetamol, which can undergo subsequent alkylation using ethyl iodide (Scheme 2.1.7.2).

The first step is a classical *N*-acetylation and can be performed by heating a mixture of *p*-aminophenol and acetic anhydride in water without the need of a basic catalysis, although the yields are not very high. Alternatively, a catalyst like pyridine, 4-pyrrolidinopyridine or basic alumina can be used. The *N*-acetyl-*p*-aminophenol (paracetamol) precipitates easily from the reaction mixture.

The second step is an example of the Williamson ether synthesis. Paracetamol is treated with a sodium ethoxide solution (prepared from ethanol and metallic sodium) to afford a phenolate that reacts with ethyl iodide, through a S_N2 mechanism, to give the phenacetin molecule. This method is particularly useful for the preparation of unsymmetrical ethers from alcohols and alkyl halides. The reaction works best with primary halides to avoid side products from competing elimination reactions. Typically it is conducted at 50–100 °C and is takes some hours to be complete. If an unreactive alkylating agent is used (*e.g.* an alkyl chloride) then the rate of reaction can be greatly improved by the addition of a catalytic quantity of a soluble iodide salt (which undergoes halide exchange with the chloride to yield a much more reactive iodide). In extreme cases, silver salts, like silver oxide, may be added. The silver ion coordinates with the halide leaving group to make its departure easier.[6]

Additional Safety

This experiment should be performed in a fume hood. *p*-Aminophenol, acetic anhydride and ethyl iodide are irritant to the skin and eyes – avoid any contact or inhalation. Sodium metal is flammable and reacts violently with water with the formation of highly flammable hydrogen gas, which can cause severe burns. Sodium should be carefully handled with gloves and since it is quickly oxidized it should be manipulated for short periods of time out in the open. In case of fire do not attempt to extinguish with water or a normal fire extinguisher. To extinguish the fire it is ideal to use a specific extinguisher or add large quantities of sand to the fire. Small residues of sodium can be destroyed with ethanol.

Experimental Procedure

Laboratory Session 1 (2 h): Synthesis of Paracetamol

1. Add *p*-aminophenol (2.0 g, 18.32 mmol), acetic anhydride (2.2 mL, 23.28 mmol) and water (10 mL) to an Erlenmeyer (50 mL).
2. Place the Erlenmeyer in a water bath and heat the solution near the boiling point for 20 min. Stir the mixture regularly, using a glass rod, until the complete dissolution of the *p*-aminophenol.
3. Transfer a drop of the reaction mixture to a vial, dilute it with ethyl ether (1 mL) and evaluate the reaction progress by TLC using 80% ethyl ether/petroleum ether as eluent. If the reaction is not finished continue heating for a further 10 min.
4. Cool the flask in an ice bath for 10 min and watch the formation of paracetamol crystals.
5. Assemble a vacuum filtration apparatus using a Büchner funnel. Wet the filter paper with a small amount of cold water and filter the solid. Use small amounts of cold water to help remove any residual product from the flask and to wash the product.
6. Recrystallize the product from water. Place the product in a flask, add water (10 mL) and heat in a heating plate until the complete dissolution of the product. Remove the flask and allow it cool slowly to room temperature. Place the flask in an ice bath for 15 min and collect the crystals by filtration.
7. Dry the product in an oven overnight and measure the weight.

Laboratory Session 2 (2 h): Synthesis of Phenacetin

1. Place ethanol (10 mL) in a round-bottom flask (50 mL), add carefully metallic sodium (0.275 g, 11.96 mmol) and attach a reflux condenser. Connect the water tubing to the condenser ensuring that water goes in the bottom and out the top. Turn on the water so that you have a slow but steady stream of water passing through the condenser.
2. Place a heating mantle below the flask and heat the reaction mixture under reflux for 15 min until the complete consumption of the sodium.
3. Cool the flask to room temperature. Add paracetamol (1.7 g, 11.24 mmol) and ethyl iodide (1.4 mL, 17.41 mmol) drop by drop over 1 min.
4. Heat the reaction mixture under reflux for 1 h.
5. Transfer a drop of the reaction mixture to a vial, dilute it with ethyl ether (1 mL) and evaluate the reaction progress by TLC using 80% ethyl ether/petroleum ether as the eluent.
6. Cool the flask to room temperature, add 10 mL of water and place it in an ice bath. Watch the formation of phenacetin crystals.
7. Assemble a vacuum filtration apparatus using a Büchner funnel. Wet the filter paper with a small amount of cold water and filter the solid. Use small amounts of cold water to help remove any residual product from the flask and to wash the product with cold water.
8. Dry the product in an oven overnight and weigh it.

Waste Disposal

The effluents of the first reaction can be discarded in the sink after dilution. The residue of the second reaction contains a small amount of ethyl iodide and should be collected in an organic waste container.

Results Interpretation and Additional Questions

1. Measure the melting point of both products and compare them with the literature values.
2. Determine the limiting reagent of each reaction and calculate the overall yield.

3. Write a mechanism for the *N*-acetylation of the *p*-aminophenol.
4. What may be the side product of the Williamson synthesis?
5. What are the expected differences in the IR and NMR spectra of paracetamol and phenacetin?
6. Why is the reaction of sodium with water so dangerous?

References

1. B. B. Brodie and J. Axelrold, *J. Pharmacol. Exp. Ther.*, 1949, **97**, 58.
2. D. Lester, L. A. Greenberg and R. P. Carroll, *J. Pharmacol. Exp. Ther.*, 1947, **90**, 68.
3. F. Ellis, *Paracetamol – a curriculum resource,* ed. C. Osborne and M. Pack, 2002, Royal Society of Chemistry, London, http://www.rsc.org/learn-chemistry/content/filerepository/CMP/00/000/047/Paracetamol_web.pdf.
4. S. P. Clissold, *Drugs*, 1986, **32**(Suppl 4), 46.
5. H. N. Morse, *Ber. Dtsch. Chem. Ges.*, 1878, **11**, 232.
6. M. Tanabe and R. H. Peters, *Org. Synth.*, 1981, **60**, 92.

2.1.8. One-Step Synthesis of 4(3*H*)-Quinazolinones: An Important Heterocyclic Scaffold in Medicinal Chemistry

André Dias, Rui Moreira, and Ana S. Ressurreição*

Research Institute for Medicines (iMed.ULisboa), Faculty of Pharmacy, Universidade de Lisboa, Av. Professor Gama Pinto, 1649-003 Lisboa, Portugal. *E-mail: aressurreicao@ff.ulisboa.pt

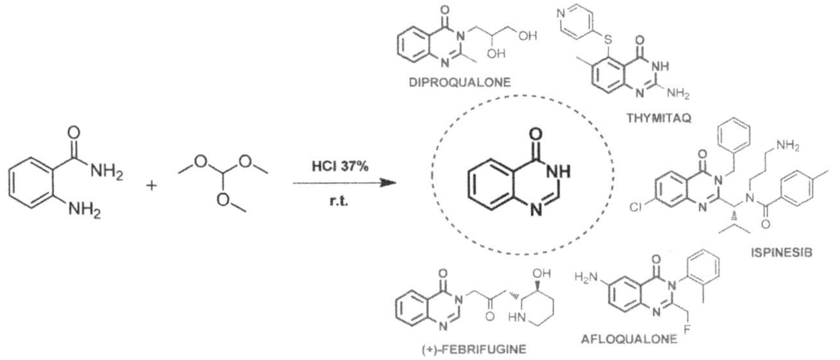

Number of sessions (duration of each session)	Hazard level	Difficulty level	Level of study
1 (3 h)	Moderate	Medium	Intermediate

Class names Amines, amides, orthoesters

Concepts involved This experiment describes a mild method for the synthesis of the 4(3*H*)-quinazolinone scaffold, from 2-aminobenzamide and trimethyl orthoformate, *via* a one-pot cyclisation reaction in acidic medium

Chemicals needed 2-Aminobenzamide, trimethyl orthoformate, hydrochloric acid (37%), sodium hydroxide, sodium chloride, anhydrous sodium sulfate, ethyl acetate

Equipment and experimental techniques involved Liquid–liquid extraction, NMR, magnetic stirring plate, rotatory evaporator, vacuum filtration apparatus, melting point apparatus

Keywords 2-Aminobenzamides, cyclisation reaction, heterocycles, liquid–liquid extraction, NMR, quinazolinones

Background

Heterocycles occupy a central position in organic chemistry, and are of significant importance in the search for new bioactive scaffolds in both the agrochemical and pharmaceutical industries. Indeed, with particular reference to the pharmaceutical industry, heterocyclic motifs are especially prevalent with over 60% of the top retailing drugs containing at least one heterocyclic nucleus as part of the overall topography of the compound.[1]

Comprehensive Organic Chemistry Experiments for the Laboratory Classroom
Edited by Carlos A. M. Afonso, Nuno R. Candeias, Dulce Pereira Simão, Alexandre F. Trindade, Jaime A. S. Coelho, Bin Tan, and Robert Franzén.

Published by the Royal Society of Chemistry, www.rsc.org

From the large family of nitrogen heterocyclic structures, quinazoline (or 1,3-diazanaphthalene, **1**, Figure 2.1.8.1) and its 4-oxo derivative, 4(3*H*)-quinazolinone (**2**, Figure 2.1.8.1), have attracted significant attention due to their wide range of pharmacological properties including, among others, antimalarial,[2] antibacterial,[3] anticonvulsant,[4] anti-inflammatory,[5] and anticancer[6] activity. Furthermore, the quinazolinone scaffold is also a central component of approximately 150 naturally occurring alkaloids isolated from several families of the plant kingdom and microorganisms.[7]

From a synthetic point of view, the first 4(3*H*)-quinazolinone compound (**3**, Figure 2.1.8.2) was obtained as early as 1869 from anthranilic acid and cyanogen.[8] An intense search for a biologically active substrate in this series was stimulated in the early 1950s with the elucidation of an alkaloid, febrifugine (**4**, Figure 2.1.8.2), which is an ingredient of a traditional Chinese herbal medicine effective against malaria.[9] Since then, quinazolinones and their derivatives have become privileged core structures that can be found in numerous marketed drugs. Therefore, unsurprisingly, there are a large number of publications reporting different methods for their synthesis.[10] Among them, the main synthetic routes employ anthranilic acid (2-aminobenzoic acid), anthranilamide (2-aminobenzamide), isatoic anhydride, *N*-arylnitrilium salts, and 3,1-benzoxazinones as appropriate precursors. More recently, several strategies, including metal-catalyzed reactions,[11] multi-component reactions,[12] and microwave heating,[13] have been successfully employed to achieve diversely decorated quinazolinone scaffolds.

The main objective of this work is to illustrate the synthesis of a heterocyclic core structure in medicinal chemistry using a mild and operationally simple methodology, starting from cheap and readily available starting materials. Additionally, the product can be easily isolated from the aqueous reaction medium through liquid–liquid extraction, avoiding the use of expensive purification procedures and equipment.

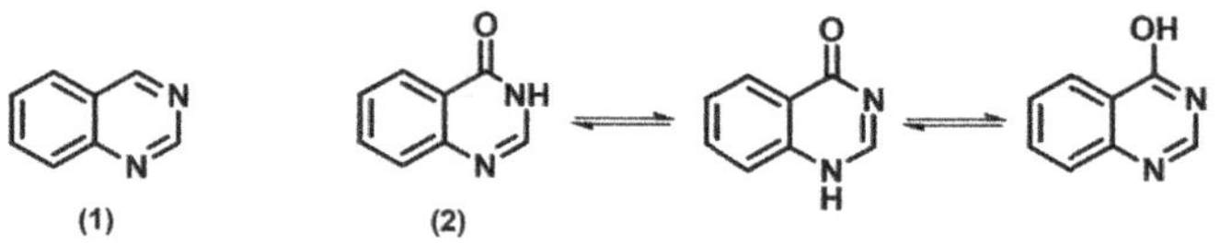

Figure 2.1.8.1 Chemical structures of quinazoline (*1*) and 4(3*H*)-quinazolinone (*2*) and possible tautomers.

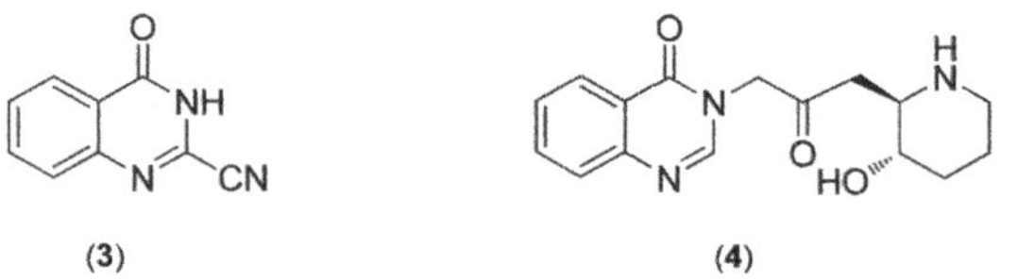

Figure 2.1.8.2 Chemical structures of the first synthetic quinazolinone (2-cyanoquinazolinone, *3*) and febrifugine (*4*).

Additional Safety

The session should be carried out in a fume hood, with proper safety equipment, such as protective clothing, gloves, and safety goggles. Hydrochloric acid (37%) is corrosive and can cause severe skin and eye burns. In addition, hydrochloric acid vapor may cause respiratory irritation. Trimethyl orthoformate and ethyl acetate are toxic and flammable liquids so it is important to keep them distant from ignition sources.

Experimental Procedure

1. Prepare a 10 mL round-bottom flask with a magnetic stir bar and add 2-aminobenzamide (100 mg, 0.74 mmol).
2. In the fume hood use a microsyringe (or a glass pipette) to add trimethyl orthoformate (1.06 mL, 9.68 mmol).

3. Cool the reaction mixture in an ice bath and use a microsyringe to add hydrochloric acid (37%) (92 μL, 2.98 mmol) dropwise at 0 °C, while stirring.
4. Leave the reaction mixture stirring at room temperature for 30 min.

Product Isolation

5. After the reaction reaches completion, dilute the reaction mixture with 10 mL of distilled water and add 5 M NaOH until the pH is 6.
6. Remove the precipitate formed (insoluble impurity) by vacuum filtration and extract the aqueous filtrate with EtOAc (3 × 25 mL).
7. Wash the combined organic phases with brine, dry over anhydrous sodium sulfate, filter and remove the solvent under reduced pressure.
8. Dry the product in a vacuum pump until constant weight.
9. Measure the melting point of the obtained product and prepare the sample for ^{1}H and ^{13}C NMR (approx. 15–20 mg of compound in 500 μL of DMSO-d_6).

Waste Disposal

The neutralized aqueous solution may be poured down the drain after being greatly diluted and swilled down using plenty of running water. Ethyl acetate should be collected in a non-chlorinated waste recipient for posterior destruction.

Results Interpretation and Additional Questions

1. Determine the reaction yield and compare the obtained melting point with the one reported in the literature.[14]
2. Interpret and assign all peaks of the NMR spectra obtained.
3. Discuss the mechanism of the reaction and the importance of the acid catalyst.
4. Discuss the applicability of this reaction starting from substituted 2-aminobenzamides.
5. Based on previous reports in the literature, discuss other synthetic methods towards 4(3*H*)-quinazolinones using 2-aminobenzamides as starting material.

References

1. N. A. McGrath, M. Brichacek and J. T. Njardarson, *J. Chem. Educ.*, 2010, **87**, 1348.
2. H. Kikuchi, S. Horoiwa, R. Kasahara, N. Hariguchi, M. Matsumoto and Y. Oshima, *Eur. J. Med. Chem.*, 2014, **76**, 10.
3. A. M. Alafeefy, M. Ceruso, A. M. Al-Tamimi, S. Del Prete, C. Capasso and C. T. Supuran, *Bioorg. Med. Chem.*, 2014, **22**, 5133.
4. The anticonvulsant activity of quinazolines and derivatives was recently revised in: V. G. Ugale and S. B. Bari, *Eur. J. Med. Chem.*, 2014, **80C**, 447.
5. V. Alagarsamy, V. R. Solomon, R. V. Sheorey and R. Jayakumar, *Chem. Biol. Drug Des.*, 2009, **73**, 471.
6. N. M. A. Gawad, H. H. Georgey, R. M. Youssef and N. A. El-Sayed, *Eur. J. Med. Chem.*, 2010, **45**, 6058.
7. S. B. Mhaske and N. P. Argade, *Tetrahedron*, 2006, **62**, 9787.
8. P. Griess, *J. Prakt. Chem.*, 1869, 369.
9. J. B. Koepfli, J. A. Brockman Jr and J. Moffat, *J. Am. Chem. Soc.*, 1950, **72**, 3323.
10. Recent advances in the synthetic approaches to quinazoline and quinazolinone scaffolds have been reviewed in: (a) I. Khan, A. Ibrar, N. Abbas and A. Saeed, *Eur. J. Med. Chem.*, 2014, **76**, 193; (b) D. J. Connolly, D. Cusack, T. P. O'Sullivan and P. J. Guiry, *Tetrahedron*, 2005, **61**, 10153.
11. X. F. Wu, L. He, H. Neumann and M. Beller, *Chem.–Eur. J.*, 2013, **19**, 12635.
12. M. Dabiri, M. Baghbanzadeh and A. S. Delbari, *J. Comb. Chem.*, 2008, **10**, 700.
13. T. Besson and E. Chosson, *Comb. Chem. High Throughput Screening*, 2007, **10**, 903.
14. S. L. Wang, K. Yang, C. S. Yao and X. S. Wang, *Synth. Commun.*, 2012, **42**, 341.

2.1.9. Controlled Monoalkylation of the Structurally Rigid Bicyclic System Isomannide

M. Manuela A. Pereira

REQUIMTE, Chemistry Department, Faculdade de Ciências e Tecnologia, Universidade Nova de Lisboa, 2827-516 Caparica, Portugal. *E-mail: manuela.pereira@fct.unl.pt

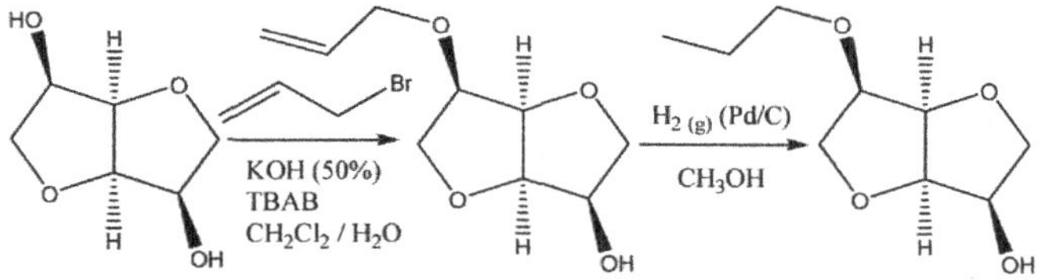

Number of sessions (duration of each session)	Hazard level	Difficulty level	Level of study
3 (3 h each)	High	Medium	Intermediate

Class names Alcohols, allylic ethers, allyl bromides, ammonium salts

Concepts involved Allylic substitution, steric hindrance, phase-transfer catalysis, reduction (hydrogenation over palladium)

Chemicals needed Isomannide, allyl bromide, tetrabutylammonium bromide, palladium on charcoal (10%), hydrogen, potassium hydroxide, sodium hydroxide, phosphorus pentoxide, potassium permanganate, potassium carbonate, sulfuric acid, iodine, ethanol, methanol, hexane, ethyl acetate, dichloromethane, sodium chloride, anhydrous sodium sulfate, Celite®, TLC plates (silica gel 60 F_{254}), silica gel 60 (70–230 mesh ASTM), deuterated chloroform

Equipment and experimental techniques involved Magnetic stirring apparatus, ultraviolet light (254 nm) for chromatography, extraction, catalytic hydrogenation at normal pressure (by using a balloon filled with hydrogen), column chromatography, thin-layer chromatography, rotary evaporator, IR equipment, NMR equipment

Keywords Column chromatography, hydrogenation, IR, NMR, phase-transfer catalyst

Background

The diols isosorbide and isomannide are both obtained from renewable resources like cellulose[1,2] and starch.[3] They represent platform chemicals for the potential replacement of oil-based products.[4] Consequently, they are the focus of interest as starting materials for pharmaceutical applications as well as for organic solvents or fuels and for biopolymers.[4–7]

These compounds have been used as starting materials for the synthesis of biodegradable and high glass transition temperature polymers,[8–10] pharmaceutically important compounds including vasodilators and inhibitors,[11–13] and as chiral ligands or as phase-transfer catalysts in asymmetric synthesis,[14–16] as chiral auxiliaries[17–19] and in the synthesis of chiral ionic liquids.[20]

Comprehensive Organic Chemistry Experiments for the Laboratory Classroom
Edited by Carlos A. M. Afonso, Nuno R. Candeias, Dulce Pereira Simão, Alexandre F. Trindade, Jaime A. S. Coelho, Bin Tan, and Robert Franzén.

Published by the Royal Society of Chemistry, www.rsc.org

Isomannide, also known as (3*R*,3a*R*,6*R*,6a*R*)-hexahydrofuro[3,2-*b*]furan-3,6-diol, is a symmetrical molecule with *C*2-axial symmetry consisting of two *cis*-fused furan rings and two *cis*-oriented hydroxyl groups with both the heterocyclic rings and two hydroxyl groups in same plane; the hydroxy groups at C3 and C6 are endo. Alkylation of hydroxyl substituents gives only two products, the major monosubstituted product and the sterically unfavorable disubstituted as a minor product.[16,21]

The alkylation of each hydroxyl group of isomannide by allyl bromide is a S_N2 substitution.[22] Allyl compounds react rapidly by the S_N2 mechanism because the double bond can stabilize the transition state by conjugation. Since the p orbital in allyl bromide has electrons in it, it shares a pair of electrons with the nucleophile and the leaving group. This process is controlled by steric effects, mainly steric hindrance as a consequence of repulsion between the electrons in all the filled orbitals of the alkyl substituent.

In isomannide the mono allyl ether substituent gives rise to a steric hindrance and avoids the S_N2 replacement of the bromine in allyl bromide by the free hydroxyl group. The effect is superior due to the endo position of both hydroxyl groups, giving rise to the isomannide mono ether as the major compound when isomannide is treated with allyl bromine, even in excess.

Phase-transfer catalysis has long been recognized as a versatile methodology for organic synthesis in both industrial and academic laboratories, featuring its simple experimental operations, mild reaction conditions, inexpensive and environmentally benign reagents and solvents, and possibility to conduct large-scale preparations.[23] In particular, during more than the last two decades, asymmetric phase-transfer catalysis based on the use of structurally well-defined chiral, not racemic, catalysts has become a topic of great scientific interest, and recent enormous efforts have resulted in notable achievements, that make it feasible to perform various bond formation reactions under mild phase-transfer-catalysed conditions.[24–28]

In addition, some isomannide derivatives have been synthesized by phase-transfer-catalysed transformations in the presence of tetrabutylammonium bromide from the corresponding monoalkylated compounds.[21]

In this experiment the commercially available structurally rigid bicyclic system isomannide ((3*R*,3a*R*,6*R*,6a*R*)-hexahydrofuro[3,2-*b*]furan-3,6-diol) is *O*-alkylated under phase-transfer conditions by allyl bromide (allyl-Br) in the presence of tetrabutylammonium bromide (TBAB) and potassium hydroxide. Due to the steric hindrance, the crude product consists mainly of the mono-allylated derivative. It is purified by column chromatography and is hydrogenated over 5% palladium on charcoal to yield the corresponding reduced ether.

Additional Safety

All sessions should be carried out in a fume hood, with proper safety equipment, such as protective clothing, gloves (for splash protection: nitrile rubber with minimum layer thickness of 0.11 mm) and safety glasses.

Catalysts prepared on high surface area supports are highly active and readily cause ignition of hydrogen/air and solvent/air mixtures. Methanol is notable for easy ignition because of its high volatility. Palladium on carbon catalysts containing adsorbed hydrogen are pyrophoric, particularly when dry and at elevated temperatures. Dry catalyst should never be added to an organic solvent in the presence of air.

Palladium on carbon recovered from catalytic hydrogenation reactions by filtration requires careful handling because it is usually saturated with hydrogen and will ignite spontaneously on exposure to air. The filter cake should never be allowed to dry, and the moist material should be added to a large quantity of water and disposed of properly.

Allyl bromide is flammable (Category 2), toxic if swallowed (Category 3), skin corrosive, causes severe skin burns and eye damage (Category 1B), mutagenic (Category 1B), carcinogenic (Category 1B), may cause cancer or genetic defects and is very toxic to aquatic life

(Category 1). Avoid prolonged contact with this material. Overexposure may result in serious illness or death. It is irritating to eyes and skin on contact. Inhalation causes irritation of the lungs and respiratory system. Inflammation of the eye is characterized by redness, watering and itching. Skin inflammation is characterized by itching, scaling, reddening or, occasionally, blistering. Always wear proper protective equipment (protective gloves/protective clothing/eye protection/face protection) when handling this compound: if swallowed immediately call a poison center or doctor; if in the eyes, rinse cautiously with water for several minutes. Remove contact lenses, if present and easy to do. Continue rinsing. (MSDS from Sigma Aldrich.) As allyl bromide is a high hazard compound, special attention should be given to the waste treatment methods. The product or contaminated solvents should be burnt in a chemical incenirator equipped with an afterburner and scrubber, and offer surplus and non-recyclable solutions to a licensed disposal company. The contaminated packaging should be dispose of as unused product.

Ensure adequate ventilation.

Do not breathe dust when using silica gel 60 during chromatographic purification. Use a breathing mask when charging the column.

Experimental Procedure

Laboratory Session 1 (Week 1–3 h): Monoallylation of Isomannide. Synthesis of (3*R*,3a*R*,6*R*,6a*R*)-6-(allyloxy)hexahydrofuro[3,2-*b*]furan-3-ol

1. Prepare a 50 mL round-bottomed flask with a magnetic stir bar and add 0.484 g (3.32 mmol) of D-isomannide, 0.6 mL of KOH (50% solution in water) and finally 0.0334 g (0.104 mmol) of tetrabutylammonium bromide in 8 mL of dichloromethane solution. To this mixture add 0.48 mL (5.5 mmol) of allyl bromide (Caution: allyl bromide is a carcinogenic and a mutagenic compound.)

Leave the two-phase reaction mixture stirring for one week at room temperature. (Considerer weekly laboratory lessons.) At this point, consumption of the starting material is completed.

2. Follow the reaction progress by TLC (6: 4 v/v, ethyl acetate: hexane). The chromatograms are developed by irradiation with an UV lamp (254 nm) or spraying with a potassium permanganate solution (1.5 g of $KMnO_4$, 10 g K_2CO_3 and 1.25 mL 10% NaOH in 200 mL of ethanol) or a sulfuric acid solution (H_2SO_4 and methanol (1: 1 v/v)) followed by heating at 120 °C.

Laboratory Session 2 (Week 2–3 h). Product Isolation, Crude Yield, Product Purification and Reduction of Isomannide Derivative

Product Isolation

1. Add carefully to the reaction mixture 10 mL of water.
2. Transfer carefully the reaction mixture into a separatory funnel and extract the aqueous layer with 4 × 10 mL of dichloromethane. Wash the combined organic layers with 10 mL of brine, dry with anhydrous Na_2SO_4, filter to a round-bottomed flask (tared) and evaporate the solvent in a Rotavapor.
3. Weigh the crude yellow oil and calculate the crude yield.

Product Purification

4. Place the crude yellow oil to be purified by chromatography inside the top of a silica column for chromatography (chromatography by gravity, 10 g of silica gel 60 (0.063–0.200

mm) Merck, ref. 7734 in a column of 1.1 cm diameter). Elute with a hexane/ethyl acetate mixture (6: 4 v/v). Collect 2 mL samples in sample tubes.

5. Analyse the collected samples by TLC and collect together the samples with the desired compound in a flask. Evaporate the solvent to yield the alcohol as a clear oil.
6. Dry in a desiccator under vacuum in the presence of phosphorus pentoxide.
7. Weigh the desired product and calculate the yield. Record the IR, ^{1}H NMR and ^{13}C NMR spectra to confirm the compound structure and the purity.
8. Prepare a 25 mL round-bottomed flask with a magnetic stir bar and a take-off with a ballon filled with hydrogen. Add to the reaction flask 160 mg of the isomannide allyl-ether derivative prepared previously and 7 mL of methanol. After dissolution at room temperature add 30.2 mg of 10% palladium on charcoal and close the flask with the take-off provided with the ballon filled with hydrogen. Turn on the take-off tap into the reaction mixture. The air inside the reaction flask will be enriched in hydrogen and its concentration will be sufficient for alkene reduction (it is not necessary to evacuate the system before hydrogen addition). Stir the heterogeneous reaction mixture.
9. The heterogeneous reaction mixture is stirred for one week at room temperature and under an atmosphere rich in hydrogen until complete consumption of the starting material.
10. Follow the reaction progress by TLC (0.25 mm silica gel 60) (6: 4 v/v, ethyl acetate: hexane). To follow the reaction by TLC, remove the take-off to open the reaction flask and take a sample of the reaction mixture to analyse. If the reaction is not ended, and the presence of alkene is detected, repeat the procedure above to enrich the reaction atmosphere with hydrogen by a balloon filled with hydrogen. The chromatograms are developed by irradiation with an UV lamp (254 nm) or by spraying with a potassium permanganate solution (1.5 g of $KMnO_4$, 10 g K_2CO_3, and 1.25 mL 10% NaOH in 200 mL of ethanol) or a sulfuric acid solution (1: 1, sulfuric acid/methanol) followed by heating at 110 °C or alternatively in iodine vapours.

Laboratory Session 3 (Week 3–3 h) Saturated Ether (3*R*,3a*R*,6*R*,6a*R*)-6-Propoxyhexahydrofuro[3,2-*b*]furan-3-ol) Purification, Yield and Product Characterization

Product Isolation

1. After one week the starting material is completely consumed (TLC analysis).
2. Fill a Hirsch funnel with 2 g of Celite® and filter the solution under vacuum into a round-bottomed flask to remove the catalyst. Wash the Celite® with pure methanol. Evaporate the solvent at reduced pressure by use of a Rotavapor evaporator.
3. Dry the residue in a desiccator under vacuum and over phosphorus pentoxide for 1 h to obtain a clear oil.
4. Weigh the oil and estimate the yield.
5. Record the IR, ^{1}H NMR and ^{13}C NMR spectra to confirm the compound structure and the purity.

Waste Disposal

Attention: The aqueous solutions or organic solvents contaminated with allyl bromide should be discarded and not recycled or reused. Allyl bromide is a carcinogenic and mutagenic compound.

The other aqueous solutions may be diluted and discarded in the sink. Ethyl acetate, hexane, and methanol may be reused or collected in a waste collector recipient.

Celite® with palladium on carbon, and silica gel from column chromatography, may be collected in a waste collector recipient.

Results Interpretation and Additional Questions

1. Identify all the asymmetric carbons in the isomannide structure. Use the Cahn–Ingold–Prelog system to classify them.[29,30]
2. What are the benefits of the phase-transfer catalyst process used in the allylic ether synthesis?
3. Alkene hydrogenation by molecular hydrogen in the presence of palladium is a syn addition of hydrogen to the double bound. Explain the mechanism of this reduction. Why can the syn addition of hydrogen not be followed by this experiment?
4. Interpret the NMR spectra and localize the allylic bond signals both in ^{1}H and ^{13}C NMR of (3*R*,3a*R*,6*R*,6a*R*)-6-(allyloxy)hexahydrofuro[3,2-*b*]furan-3-ol. Verify their absence in ^{1}H and ^{13}C NMR spectra of (3*R*,3a*R*,6*R*,6a*R*)-6-propoxyhexahydrofuro[3,2-*b*]furan-3-ol.

References

1. S. Van De Vyver, J. Geboers, P. A. Jacobs and B. F. Sels, *ChemCatChem*, 2011, **3**, 82–94.
2. P. Gallezot, *Chem. Soc. Rev.*, 2012, **41**, 1538–1558.
3. G. Fleche and M. Huchette, *Starch-Starke*, 1986, **38**, 26–30.
4. M. Rose and R. Palkovits, *ChemSusChem*, 2012, **5**, 167–176.
5. J. Wu, P. Eduard, S. Thiyagarajan, L. Jasinska-Walc, A. Rozanski, C. F. Guerra, B. A. J. Noordover, J. van Haveren, D. S. van Es and C. E. Koning, *Macromolecules*, 2012, **45**, 5069–5080.
6. J. Wu, L. Jasinska-Walc, D. Dudenko, A. Rozanski, M. R. Hansen, D. van Es and C. E. Koning, *Macromolecules*, 2012, **45**, 9333–9346.
7. B. A. J. Noordover, V. G. van Staalduinen, R. Duchateau, C. E. Koning, R. van Benthem, M. Mak, A. Heise, A. E. Frissen and J. van Haveren, *Biomacromolecules*, 2006, **7**, 3406–3416.
8. H. R. Kricheldorf and Z. Gomourachvili, *Macromol. Chem. Phys.*, 1997, **198**, 3149–3160.
9. M. Okada, Y. Okada and K. Aoi, *J. Polym. Sci., Polym. Chem.*, 1995, **33**, 2813–2820.
10. M. Okada, K. Tsunoda, K. Tachikawa and K. Aoi, *J. Appl. Polym. Sci.*, 2000, **77**, 338–346.
11. H. Bidoggia, *Curr. Med. Res. Opin.*, 1987, **10**, 601–611.
12. J. M. Walker, P. V. Curry, A. E. Bailey and S. E. Steare, *Int. J. Cardiol.*, 1996, **53**, 117–126.
13. M. Vogler, U. Koert, K. Harms, D. Dorsch, J. Gleitz and P. Raddatz, *Synthesis*, 2004, 1211–1228.
14. C. Carcedo, A. Dervisi, I. A. Fallis, L. Ooi and K. M. A. Malik, *Chem. Commun.*, 2004, 1236–1237.
15. G. De Coster, K. Vandyck, E. Van der Eycken, J. Van der Eycken, M. Elseviers and H. Roper, *Tetrahedron: Asymmetry*, 2002, **13**, 1673–1679.
16. V. Kumar, C. Pei, C. E. Olsen, S. J. C. Schaffer, V. S. Parmar and S. V. Malhotra, *Tetrahedron: Asymmetry*, 2008, **19**, 664–671.
17. V. Cere, C. Mazzini, C. Paolucci, S. Pollicino and A. Fava, *J. Org. Chem.*, 1993, **58**, 4567–4571.
18. Q. Chao, J. Z. Zhang, L. Pickering, T. S. Jahnke and V. Nair, *Tetrahedron*, 1998, **54**, 3113–3124.
19. A. Loupy and D. A. Monteux, *Tetrahedron*, 2002, **58**, 1541–1549.
20. O. N. Van Buu and G. Vo-Thanh, *Lett. Org. Chem.*, 2007, **4**, 158–167.
21. S. Kumar and U. Ramachandran, *Tetrahedron*, 2005, **61**, 4141–4148.
22. J. Clayden, N. Greeves and S. Warren, *Organic Chemistry*, 2nd edn, 2012.
23. S. S. D. E. V. Dehmlow, *Phase Transfer Catalysis*, 3rd edn, 1993.
24. A. Nelson, *Angew. Chem., Int. Ed.*, 1999, **38**, 1583–1585.
25. M. J. O'Donnell, *Acc. Chem. Res.*, 2004, **37**, 506–517.
26. B. Lygo and B. I. Andrews, *Acc. Chem. Res.*, 2004, **37**, 518–525.
27. K. Maruoka and T. Ooi, *Chem. Rev.*, 2003, **103**, 3013–3028.
28. J. Vachon and J. Lacour, *Chimia*, 2006, **60**, 266–275.
29. E. L. Eliel, S. H. Wilen and L. N. Mander, *Sterochemistry of organic compounds*, John Wiley & Sons, Inc. New York, 1994.
30. O. N. Van Buu, A. Aupoix, N. D. T. Hong and G. Vo-Thanh, *New J. Chem.*, 2009, **33**, 2060–2072.

2.1.10. Regioselective *N*-alkylation of Adenine by Nucleophilic Substitution

Ana I. Vicente* and Carlos A. M. Afonso

iMed.ULisboa, Faculdade de Farmácia da Universidade de Lisboa, Portugal
*E-mail: ana_vicente@ff.ulisboa.pt, carlosafonso@ff.ulisboa.pt

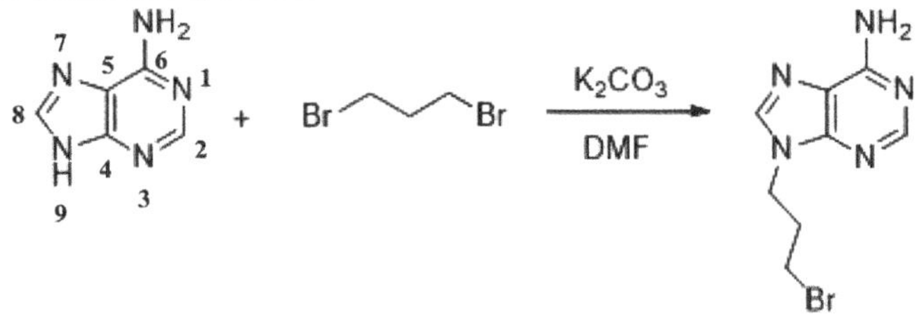

Number of sessions (duration of each session):	Hazard level	Difficulty level	Level of study
2 (0.5 h + 4 h)	Moderate	Medium	Intermediate
Class names Purines, alkyl bromides, amines			
Concepts involved Nucleophilic substitution, regioselectivity, chemical modification of nucleic base, purification by column chromatography and structural analysis by NMR spectroscopy			
Chemicals needed Adenine, potassium carbonate, 1,3-dibromopropane, dimethylformamide, dichloromethane, methanol and silica gel, DMSO-d_6			
Equipment and experimental techniques involved NMR, filtration, stirring hotplate, vacuum drying, TLC plates and column chromatography			
Keywords Adenine, alkylation, bimolecular nucleophilic substitution, column chromatography, NMR, regioselectivity			

Background

Polynucleic acids (DNA and RNA) display remarkable cooperative and selective binding properties that enable molecular recognition or association with complementary guest molecules by hydrogen bonding with the nucleobase moieties.[1] In addition, derivatization of adenine has been an important strategy for the development of selective ligands for the regulation of the activity of pharmacologically important proteins.[2] One of the common synthetic transformations used is *N*-alkylation by a nucleophilic substitution bimolecular S_N2 mechanism in which two reacting species are involved in the slow step, the nucleophile and the substrate.[2] In the Hughes–Ingold mechanism nucleophilic substitution occurs at an aliphatic sp^3 carbon containing an electronegative, stable leaving group attached to it, with simultaneous carbon-nucleophile bond

Comprehensive Organic Chemistry Experiments for the Laboratory Classroom
Edited by Carlos A. M. Afonso, Nuno R. Candeias, Dulce Pereira Simão, Alexandre F. Trindade, Jaime A. S. Coelho, Bin Tan, and Robert Franzén.

Published by the Royal Society of Chemistry, www.rsc.org

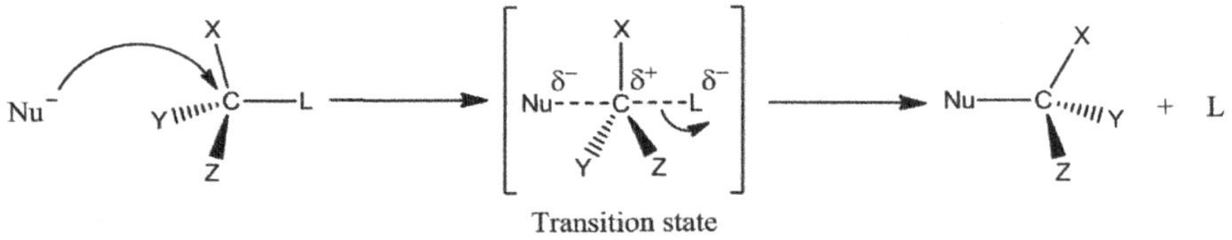

Scheme 2.1.10.1 S_N2 reaction mechanism.

formation and carbon-leaving group bond scission (Scheme 2.1.10.1). This substitution leads to an inversion of stereochemistry if the carbon center under nucleophilic attack is chiral since the nucleophile approaches the carbon bearing the leaving group from the side directly opposite the leaving group. This reaction involves only one step and proceeds through the formation of an unstable arrangement of atoms called a transition state. The S_N2 reaction is favored in the presence of methyl halides relative to more substituted ones due to a steric effect since the spatial arrangement of the atoms attached or near the reacting site of a molecule hinders or retards a reaction.[3] In addition, the solvent affects the S_N2 reaction – a protic solvent (*e.g.*, water, alcohols) solvates the halide ion by forming hydrogen bonds to it and thus it is difficult for the nucleophile to approach the carbon bearing the leaving group. In contrast, polar aprotic solvents (*e.g.*, DMF, DMSO) vastly increase the rates of S_N2 reactions since they are unable to form hydrogen bonds. The nucleophilic attack of adenine to a primary alkyl halide in dimethylformamide (DMF) takes place under optimal experimental conditions to proceed *via* a S_N2 mechanism; it occurs with a primary alkyl bromide in the presence of a good nucleophile and using an aprotic polar solvent. The alkylation of adenine in neutral or slightly basic media gives mostly N3-substituted products while the reaction of adenine anion leads to N9 and N7 derivatives.[4] In polar solvents, N7 substitution is also most favored energetically. In basic pH and nonpolar solvents as well as in the gas phase, N9 substitution is slightly more probable.[5] Mineral hydrides and carbonates are the most frequently used bases for the activation of adenine in alkylation reactions, giving high to moderate regioselectively and with reasonable yields of the N9 derivatives, the most common in biological systems. Adenine can rapidly interconvert in a rapid equilibrium giving different tautomers, molecules with the same molecular formula but different connectivity – constitutional isomers. The main goals of this experiment are the synthesis of an alkylated adenine using a di-halogenated alkyl chain, which allows further linkage of the derivative compound to platforms such as polymers or other structures that confer stability to the recognition system, and its characterization by NMR technique.

Additional Safety

It is necessary to wear protective gloves, eye protection, and face protection to handle the reagents. Adenine is harmful if swallowed and may cause eye, skin, and respiratory tract irritation. Potassium carbonate is hazardous in case of skin contact (irritant), eye contact (irritant), ingestion, inhalation, and is slightly hazardous in case of eye contact (corrosive). 1,3-Dibromopropane forms very sensitive explosive metallic compounds; it is harmful by inhalation and if swallowed and is irritating to eyes and skin; it is flammable and toxic to aquatic organisms causing long-term adverse effects in the aquatic environment. Dimethylformamide is irritating for the eyes, skin, and nose and may cause nausea. Dichloromethane, due to its high volatility, has an acute inhalation hazard and it can also be absorbed through the skin. Methanol causes eye and skin irritation and may be absorbed through intact skin. In handling silica gel it is strictly necessary to use a mask as it is irritating to the respiratory tract and may cause irritation of the digestive tract, and dust from the beads may cause irritation to the skin and eyes.

Experimental Procedure

Laboratory Session 1 (30 min): Alkylation of Sdenine by 1,3-Dibromopropane

1. To a round-bottom flask of 10 mL equipped with a magnetic stirrer add adenine (0.250 g, 1.9 mmol), and K_2CO_3 (0.5 g, 3.6 mmol) in DMF (2 mL) with stirring for approximately 15 min at room temperature.
2. Add slowly 1,3-dibromopropane (1 mL, 7.5 mmol) and leave with stirring for 16 h at room temperature.

Laboratory Session 2 (4 h): Purification of the Alkylated Adenine

1. Confirm by TLC the progress of the reaction using adenine as a reference (CH_2Cl_2/MeOH (10 : 1) as eluent). Dry the TLC before TLC elution until the DMF has evaporated.
2. Evaporate the solvent in a rotary evaporator to dryness. To optimize the evaporation use a piece of paper between the adaptor and the round-bottom flask.
3. Dissolve the obtained solid in 40 mL of a solution of CH_2Cl_2/MeOH (10 : 1) and leave with stirring for approximately 10 min and then filter the mixture and wash the solid with 10 mL of the solution.
4. Evaporate the solvent and purify the product by flash chromatography (silica: approx. 10 g; eluent: first start with CH_2Cl_2 for the five first fractions and then change to CH_2Cl_2/MeOH (10 : 0.5) as eluents and collect fractions of 8 mL).
5. Analyse the obtained reside by TLC using CH_2Cl_2/MeOH (10 : 1) as eluent and by NMR using DMSO-d_6 and calculate the yield of the reaction.

Waste Disposal

Ask the supervisor for the most appropriate way to reuse or to handle the obtained liquid and solid effluents.

Results Interpretation and Additional Questions

1. Verify that you have obtained the desired product and check the purity.
2. Calculate the yield of the reaction.
3. Explain which product we avoided forming with the use of an excess of 1,3-dibromopropane.
4. What evidence provided by ^{1}H NMR confirms the alkylation of the adenine?
5. Draw all the possible tautomers for adenine.
6. Comment on the polarity of the alkylated adenine in comparison with that of the initial adenine.

References

1. S. Sivakova and S. Rowan, *Chem. Soc. Rev.*, 2005, **34**, 9.
2. K. Peter, C. Vollhardt and N. E. Schore, *Organic Chemistry – Structure and Function*, W. H. Freeman, 6th edn, 2010.
3. G. Solomons and C. Fryhle, *Organic Chemistry*, John Wiley & Sons, Inc., New York, 7th edn, 2000.
4. T. Fujii and T. Itaya, *Heterocycles*, 1998, **48**, 1673.
5. A. Stachowicz-Kuśnierz and J. Korchowiec, *Struct. Chem.*, 2016, **27**, 543.

2.1.11. Gabriel Synthesis of *n*-Octylamine Under Phase-Transfer Catalysis: The First Step

Renata Riva* and Luca Banfi

Dipartimento di Chimica e Chimica Industriale, Via Dodecaneso 31, 16146 Genova, Italy
*E-mail: riva@chimica.unige.it

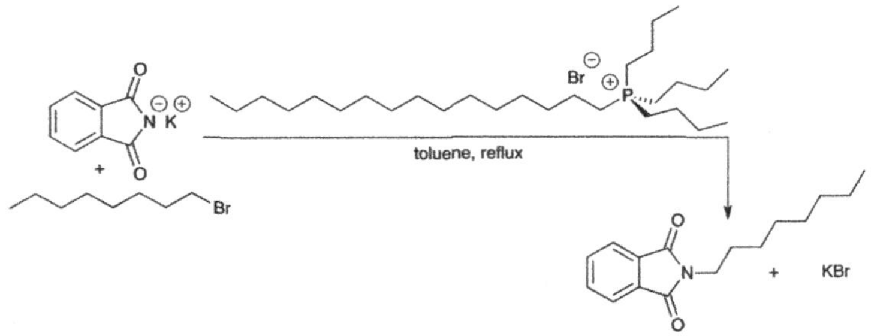

Number of sessions (duration of each session)	Hazard level	Difficulty level	Level of study
1 (5 h)	Moderate	High	Intermediate

Class names Alkyl halides, imides

Concepts involved This example describes an S_N2 reaction performed under phase-transfer catalysis. A reaction at reflux under anhydrous conditions is followed by filtration of solid by-products and purification of the crude mixture by column chromatography

Chemicals needed Potassium phthalimide, *n*-octyl bromide, tributyl(hexadecyl)phosphonium bromide, calcium chloride (anhydrous), toluene (dry), silica gel (particle size: 63–200 μm/70–200 mesh); petroleum ether (40–60 °C), diethyl ether

Equipment and experimental techniques involved Stirring hot plate, reflux condenser, drying tube, silica gel plates for TLC, UV lamp, sintered glass filter, column chromatography, rotary evaporator, vacuum pump, melting point apparatus

Keywords Column chromatography, imide *N*-alkylation, melting point, phase-transfer catalysis, phosphonium salts, S_N2

Background

The synthesis of primary amines may be accomplished by reaction of ammonia with an alkyl halide. Despite its simplicity, this method suffers from an important drawback: even if a strong excess of the nucleophile is used it is nearly impossible to avoid the formation of the corresponding secondary and tertiary amine together with the quaternary ammonium salt.

Comprehensive Organic Chemistry Experiments for the Laboratory Classroom
Edited by Carlos A. M. Afonso, Nuno R. Candeias, Dulce Pereira Simão, Alexandre F. Trindade, Jaime A. S. Coelho, Bin Tan, and Robert Franzén.

Published by the Royal Society of Chemistry, www.rsc.org

The formation of these unwanted by-products can be suppressed by using a nitrogen nucleophile to afford a product that cannot behave as a nucleophile itself. These are indirect methods because the nucleophilic displacement affords a precursor of the primary amine. The desired functional group can be established only through a second reaction. A possible nucleophile is represented by the azide ion: organic azides may be reduced to amines by catalytic hydrogenation[1] or through the Staudinger reaction by means of a phosphane.[2,3]

An alternative route is represented by the classical Gabriel synthesis,[4,5] in which commercially available potassium phthalimide is reacted with an alkyl halide. The product of this reaction is a *N*-alkylphthalimide, which can furnish the primary amine upon treatment with a base, such as sodium hydroxide, or hydrazine (Scheme 2.1.11.1). Both these reactions are examples of double acyl nucleophilic substitution. The hydroxide promoted cleavage is a sequence of two intermolecular reactions, while the hydrazinolysis, using a bifunctional nucleophile, is the result of an intermolecular followed by an intramolecular attack, which, in principle, makes this pathway more favorable.

Only the first step of the Gabriel synthesis is described here. Alcohols could be at first sight appropriate solvents for this reaction, because they can dissolve both reagents. However, solvation of the anion makes it poorly reactive and high temperatures and long reaction times are needed. To overcome this problem an appropriate anion activation is required. An efficient possibility is represented by the use of phase-transfer catalysis, in this case as a solid–liquid procedure.[6] The role of the catalyst, a phosphonium salt, is to transfer the phthalimide anion from the solid into the organic phase. The difficult interaction of the anion with the crowded phosphonium cation leads to a loose ion pair; the apolar solvent (toluene) is unable to efficiently solvate the anion. Therefore, the anion, being "naked", reacts faster with the alkyl halide. An added advantage of this kind of solid–liquid phase-transfer catalysis is that there is no need to perform an extractive work-up as most side-products can be easily removed by filtration. By contrast, because of the low melting point of the product (around 50 °C), the separation of the catalyst cannot be done by crystallization and therefore chromatography on silica gel is the purification method of choice.

Additional Safety

This experiment does not require severe safety measures, besides regular laboratory good practices. In particular, the reaction should be carried out in a fume hood, with proper safety equipment, such as gloves and safety glasses that must be worn all the time. The reaction is performed in refluxing toluene: the hot plate has to be used with care. *n*-Octyl bromide, as with most alkyl halides, is lachrymatory. Care should be taken while picking up the liquid with the pipette. The pipette should be immediately rinsed after use with acetone before taking it out of the hood. Silica gel may be harmful by prolonged and repeated inhalation; take care when weighing, avoiding breathing dust, and possibly wear a paper dust mask.

Additional information on safety issues for all chemicals handled in this experiment can be found in the MSDS (material safety data sheet) provided by the suppliers.

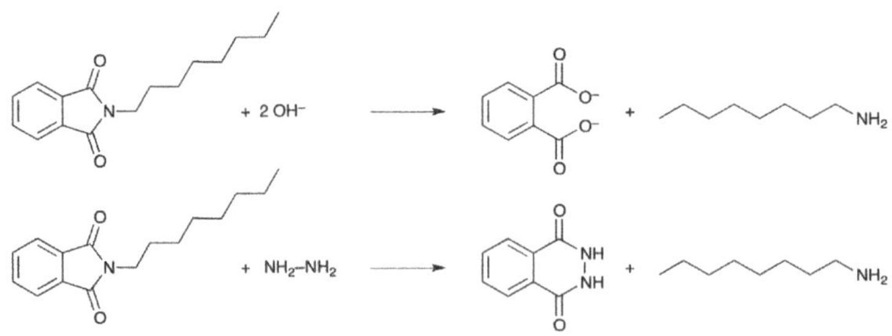

Scheme 2.1.11.1 Transformation of *N*-octylphthalimide into *n*-octylamine.

Experimental Procedure

1. In a one-neck 50 mL flat-bottomed flask (14/19 ground joint) weigh potassium phthalimide (2.41 g, 13.0 mmol). Move the flask to a fume hood.
2. Add a magnetic stir bar and then toluene (10 mL) (if possible use a dry solvent), placing immediately on the flask a bent drying tube filled with dry calcium chloride.
3. In a small test tube weigh tributyl(hexadecyl)phosphonium bromide (0.51 g, 1.00 mmol), also called Landini's catalyst, and pour it into the flask.
4. Finally add *n*-octyl bromide (1.73 mL, 10.0 mmol) to the flask using a 2 mL graduated pipette. Then, to avoid contact with moisture, close, as soon as possible, the flask with a reflux condenser, placing over it the drying tube previously used for the flask. Connect the condenser to the water faucet.
5. Heat at gentle reflux, ensuring that efficient stirring is always maintained. If excessive foam tends to get into the condenser, reduce the temperature of the heating plate.
6. After 1 h of reflux, monitor the reaction by TLC: stop heating and stirring and take 1–2 drops of the solution and dilute them with a few drops of acetone or diethyl ether. While refluxing the reaction again, prepare a silica gel plate (about 4 × 10 cm), placing a small droplet of this solution, a droplet of neutral phthalimide and a superimposition of both analytes. Develop the plate with petroleum ether (40–60 °C)/diethyl ether (7 : 3). This TLC is not useful in establishing the completeness of the reaction, because the limiting reagent (*n*-octyl bromide) cannot be seen by TLC. On the other hand, potassium phthalimide does not run in TLC, because it is too polar and this is the reason why it is not spotted on the TLC. Conversely, its conjugated acid can be seen by TLC; this compound should usually not be present in the reaction. However, in some cases, some amounts of it can be detected in the crude mixture and the student must be aware of that because this unwanted compound renders the following chromatographic separation more difficult. The spots of the TLC can be best pointed out by observing the plate with the help of an UV lamp; an appropriate stain is not necessary.
7. After refluxing for 75 min, cool the flask to room temperature and dilute the content with diethyl ether (10 mL). Then collect the solid by suction in a sintered glass funnel, and wash it at least twice with diethyl ether (about 20 mL overall).
8. Transfer the filtrate into a 100 mL flask and concentrate with a rotary evaporator. It is important to remove as much toluene as possible from the solution; therefore, after evaporation of diethyl ether, the flask must be immersed in a heating bath kept at least at 50 °C.
9. During the refluxing time of the reaction, prepare the column for the chromatography:

   ✓ Use a glass column (3 cm diameter) with a stopcock.
   ✓ Push a small cotton plug into the bottom of the column, using a long glass rod to hold it in the desired position. Fix the column on a stand, using two clamps.
   ✓ Add a small amount of sand in order to have a layer of about 2 cm over the cotton, gently shaking the column to obtain a flat surface.
   ✓ Prepare 350 mL of eluent, mixing petroleum ether (40–60 °C) and diethyl ether in a 6 : 4 ratio, and pour a small amount of it into the column (a layer of about 5 cm).
   ✓ Weigh in a 250 mL Erlenmeyer flask 50 g of silica gel (63–200 µm/70–200 mesh) and add eluent to obtain a fluid slurry that must be carefully poured into the column, without moving the sand layer. Use small amounts of eluent to transfer as much of the silica gel as possible into the column.
   ✓ Open the stopcock and allow the eluent to flow out. The level of the liquid should remain about 2 cm over the silica gel, never going below the top of the layer. Make sure that no bubbles of air remain in the silica gel; otherwise, transfer again the eluent into the column and repeat the procedure until the bubbles have been completely removed.

If necessary, at the end, gently shake the column to obtain a flat surface. Close the stopcock and add about 1 cm of sand. Put a stopper on the top of the column until the sample is ready for chromatography.

10. Dilute the sample with dichloromethane (1–2 mL) to avoid crystallization into the pipette, while transferring it into the column. Then cautiously pipette the sample into the column, avoiding displacement of the sand layer. Rinse the flask 2–3 times with a small amount of eluent (not more than 3–4 mL each) and add the solution into the column. Let the eluent flow out, without going below the sand layer.
11. Add again about 1 cm of sand and pour all the remaining eluent into the column, without perturbing the layers of sand and silica gel.
12. Put a test tube (25 mL capacity) under the column, open the stopcock and start to collect the fractions (about 18 mL each) at a rate of about one test tube every 2 min.
13. Check for the presence of the product by TLC. If the chromatography is not complete, prepare an additional 100 mL of the same eluent and complete the elution.
14. At the end, pour all the fractions containing *N*-octylphthalimide into a tared 100 mL flask, washing the tubes twice with diethyl ether to fully recover the product. Remember not to fill the flask too much; if necessary concentrate first some fractions, then add the remaining ones and complete solvent removal with the rotary evaporator, controlling the temperature of the water bath so that it is lukewarm (around 30 °C).
15. Remove the last traces of the solvent with a single stage oil vacuum pump (at 1–5 mbar) (for at least 15 min), which will help the product to become a solid.
16. Measure the melting point (around 50 °C).

Waste Disposal

The filtered solid can be diluted with water and disposed of in the sink. All solvents should be transferred into the waste bottle supplied for non-halogenated solvents. The content of the column will be disposed of by the technicians.

Results Interpretation and Additional Questions

1. Why does the phase-transfer catalyst improve the reaction rate?
2. What is the composition of the filtered solid?
3. Calculate the overall yield. In addition, using your TLC, calculate the R_f for both *N*-octylphthalimide and phthalimide. Why is the R_f of *N*-octylphthalimide greater than that of phthalimide?
4. Why is KOH a strong enough base for transforming phthalimide into the conjugated base?
5. In your opinion is it possible to use $KHCO_3$ or KNH_2 to obtain potassium phthalimide?
6. Which product do you expect to isolate performing the same reaction on: (a) (*R*)-2-bromopentane; (b) (1*R*,3*R*)-1-bromo-3-isopropylcyclopentane? Consider also the stereochemical issues.

References

1. E. J. Corey and J. O. Link, *J. Am. Chem. Soc.*, 1992, **114**, 1906–1908.
2. H. Staudinger and J. Meyer, *Helv. Chim. Acta*, 1919, **2**, 635–646.
3. W. Q. Tian and Y. A. Wang, *J. Org. Chem.*, 2004, **69**, 4299–4308.
4. S. Gabriel, *Chem. Ber.*, 1897, **20**, 2224–2236.
5. U. Ragnarsson and L. Grehn, *Acc. Chem. Res.*, 1991, **24**, 285–289.
6. D. Landini and F. Rolla, *Synthesis*, 1976, 389–391.

2.1.12. Preparation of Diethyl 2,3-*O*-isopropylidene-L-tartrate

Naylil M. R. Capreti and Igor D. Jurberg*

Institute of Chemistry, State University of Campinas (Unicamp), C.P. 6154, 13083-970, Campinas, SP, Brazil. *E-mail: idjurberg@iqm.unicamp.br

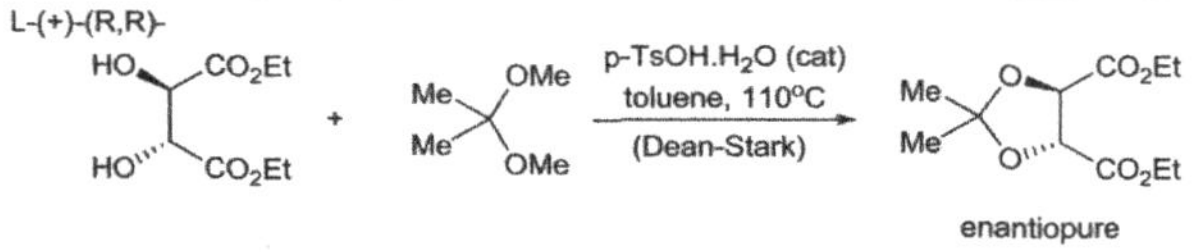

Number of sessions (duration of each session)	Hazard level	Difficulty level	Level of study
1 (5 h)	Moderate	Medium	Intermediate

Class names Alcohols, ketals, esters

Concepts involved Reaction under reflux employing a Dean–Stark apparatus; protection of a 1,2-diol *via* a transketalisation reaction

Chemicals needed L-(+)-Diethyl-tartrate, 2,2-dimethoxypropane, *p*-toluenesulfonic acid monohydrate, sodium bicarbonate, toluene, ethyl acetate

Equipment and experimental techniques involved Liquid–liquid extraction, NMR, heating/stirring plate, rotary evaporator, reflux condenser, Dean–Stark apparatus

Keywords Diethyl 2,3-*O*-isopropylidene-L-tartrate, liquid–liquid extraction, NMR, protection of diols, transketalisation

Background

Protecting Groups

Protection/deprotection sequences are typically a source for loss of efficiency (and elegance!) in the preparation of complex products.[1,2] Nevertheless, targeting a desired transformation on a specific functional group, in the presence of several others, is often a challenging task. Therefore, managing the selective exposure of a functional group *vis-à-vis* others, through strategies of protection/deprotection, can play a pivotal role in a successful synthesis.[3,4]

1,2- and 1,3-Diols

Both 1,2-[5] and 1,3-diols[6] are common moieties found in various polyol natural products, such as polyketides[7] and carbohydrates.[8] They are commonly protected as either cyclic acetals/ketals, cyclic carbonates or silylene ketals.[9] Among these, cyclic acetals and ketals are historically the protecting groups of choice, probably due to their good stability, the ease of installation and removal, and, if desired, for allowing further functional group

Comprehensive Organic Chemistry Experiments for the Laboratory Classroom
Edited by Carlos A. M. Afonso, Nuno R. Candeias, Dulce Pereira Simão, Alexandre F. Trindade, Jaime A. S. Coelho, Bin Tan, and Robert Franzén.

Published by the Royal Society of Chemistry, www.rsc.org

interconversion. For instance, *para*-methoxyphenyl-acetals can be converted into *para*-methoxybenzyl ethers carrying a pendant free alcohol using a combination of Lewis acid and a hydride donor.[10,11]

Introduction and Removal of Cyclic Acetals and Ketals, Selectivity and Stability

Cyclic acetals and ketals, such as dioxolanes and dioxanes, can be prepared by mixing a diol and an aldehyde or ketone, or a synthetic equivalent of those (for instance, one can use acetone or 2-methoxypropene or 2,2-dimethoxypropane) and a catalytic amount of acid (*e.g.* pyridinium *para*-toluenesulfonate, PPTS; camphorsulfonic acid, CSA; *para*-toluenesulfonic acid, PTSA). In this context, it is generally possible to discriminate between 1,2- and 1,3-diols in a 1,2,4-triol system. For instance, acetonides are formed preferentially with 1,2-diols rather than with 1,3-diols,[12] while benzylidene acetals display the opposite preference.[13] Interestingly, in the case of acetonides, their rigid structure also allows for their relative stereochemical determination (1,3-*syn versus* 1,3-*anti*), based on the difference of ^{13}C NMR shifts observed.[14]

Cyclic acetals and ketals are typically stable under basic conditions, but are cleaved with aqueous acid (*e.g.*, $HCl/MeOH/H_2O$, $HCl/THF/H_2O$, or $TFA/MeCN/H_2O$). They are also generally resistant to oxidative (*e.g.* CrO_3/pyridine, OsO_4, $KMnO_4$, *m*-CPBA, Br_2) and reductive (*e.g.* $LiAlH_4$, Na/NH_3, $H_2/Pd/C$) conditions.[9] An important exception is benzylidene acetals, which are more fragile and can be cleaved under numerous circumstances (*e.g.* $H_2/Pd/C$, Na/NH_3, Br_2, *m*-CPBA).[9]

Dioxolane Ring of the TADDOL Backbone

The procedure exhibited here represents the first step of a two-step sequence for the preparation of the enantiopure chiral ligand (−)-TADDOL.[15] This compound has been employed not only as a ligand for several asymmetric metal-catalysed processes,[16] but also as an organocatalyst, by activating substrates *via* hydrogen bonding.[17] Importantly, one should note that the presence of the dioxolane ring in TADDOL serves also as a steric element that interacts with the phenyl rings, thus contributing to their arrangement in space. Indeed, this interaction represents an indirect control of the chiral space available in TADDOL for the substrate to interact with.[18]

Additional Safety

The experiment should be performed inside a fume hood. Safety goggles, gloves and a lab coat must be worn at all times while in the laboratory. L-(+)-Diethyl tartrate is a combustible liquid. 2,2-Dimethoxypropane is a highly flammable liquid, which may cause serious eye irritation and is harmful if swallowed, inhaled or in contact with the skin. *p*-Toluenesulfonic acid is corrosive and may cause severe skin and eye burns.

Experimental Procedure

Laboratory Session (5 h): Preparation of Diethyl 2,3-*O*-Isopropylidene-L-tartrate

1. In a fume hood, at room temperature, charge a 10 mL round-bottomed flask with a magnetic stirring bar, L-(+)-diethyl tartrate (206 mg, 1 mmol, 1 equiv.), 2,2-dimethoxypropane (184 µL, 1.5 mmol, 1.5 equiv.), toluene (2.5 mL, 0.4 M) and *p*-toluenesulfonic acid monohydrate (4 mg, 0.02 mmol, 0.02 equiv.). The resulting solution is colourless.

2. Attach a water reflux condenser, equipped with a Dean–Stark apparatus, to the round-bottomed flask. Heat the mixture to reflux (110 °C) and allow it to stir at this temperature for 3 h. The solution becomes dark yellow upon heating.
3. At the end of the 3 h, you can confirm the complete consumption of the starting material, the L-(+)-diethyl tartrate, by performing thin-layer chromatography (TLC); eluent 9:1 hexanes : AcOEt, R_f (protected diol) = 0.3.
4. Allow the reaction mixture to cool to room temperature. Then, add a saturated solution of $NaHCO_3$ (3 mL).
5. Transfer the contents of the flask into a separating funnel. Separate the organic from the aqueous phase. Extract the aqueous phase with ethyl acetate (2 × 3 mL). Dry the combined organic phases with anhydrous Na_2SO_4.
6. Filter off the solid and collect the solution into a tared round-bottomed flask.
7. Concentrate the solution under reduced pressure.
8. Weigh the flask containing the crude reaction mixture (orange oil), and keep it in a fridge for the next experiment (*cf.* Experiment 4.2.1.5).

Waste Disposal

The aqueous layer recovered from the work-up procedure, as well as the solvent condensed in the rotary evaporator, should be discarded in a recipient for non-halogenated compounds, and sent for destruction.

Results Interpretation and Additional Questions

1. What is the mechanism of the reaction performed?
2. What is the importance of adding a saturated solution of $NaHCO_3$ at the end of the reaction?
3. Interpret the ^{1}H NMR of the crude reaction mixture obtained. What does the singlet at 3.8 ppm suggest? Is the presence of this second product a problem that hampers the use of the crude reaction mixture in the next step, without purification?
4. The same protected diester could be obtained if we had employed acetone instead of 2,2-dimethoxypropane under the very same reaction conditions. Explain why the transketalisation procedure employed here is a better option.

References

1. For an interesting discussion on the development of total syntheses without protecting groups, see: I. S. Young and P. S. Baran, *Nat. Chem.*, 2009, **1**, 193.
2. For selected examples of total synthesis without the use of protecting groups, see: (a) P. S. Baran, T. J. Maimone and J. M. Richter, *Nature*, 2007, **446**, 404; (b) J. A. Enquist Jr and B. A. Stoltz, *Nature*, 2008, **453**, 1228.
3. See for instance: M. Schelhaas and H. Waldmann, *Angew. Chem., Int. Ed.*, 1996, **35**, 2056.
4. For selected examples of total synthesis employing protection/deprotection strategies, see: (a) K. C. Nicolaou, Z. Yang, G.-q. Shi, J. L. Gunzner, K. A. Agrios and P. Gärtner, *Nature*, 1998, **392**, 264; (b) B. M. Trost and G. Dong, *Nature*, 2008, **456**, 485.
5. For reviews on the preparation of members of this class, see: (a) K. C. Bhowmick and N. N. Joshi, *Tetrahedron: Asymmetry*, 2006, **17**, 1901; (b) C. J. R. Bataille and T. J. Donohoe, *Chem. Soc. Rev.*, 2011, **40**, 114; See also: (c) L. Emmanuvel, T. M. A. Shaikh and A. Sudalai, *Org. Lett.*, 2005, **7**, 5071.
6. For reviews on the preparation of members of this class, see: (a) S. E. Bode, M. Wolberg and M. Müller, *Synthesis*, 2006, **4**, 557; (b) P. Gupta, N. Mahajan and S. C. Taneja, *Catal. Sci. Technol.*, 2013, **3**, 2462.
7. C. Hertweck, *Angew. Chem., Int. Ed.*, 2009, **48**, 4688.
8. K. C. Nicolaou and H. J. Mitchell, *Angew. Chem., Int. Ed.*, 2001, **40**, 1576.
9. For a general discussion about these protecting groups and others, see: (a) P. J. Kocienski, *Protecting Groups*, Georg Thieme Verlag, New York, 3rd edn, 2005; (b) T. W. Greene and P. G. M. Wuts, *Protective*

Groups in Organic Synthesis, John Wiley & Sons, Inc., New York, 3rd edn, 1991; (c) K. Jarowicki and P. Kocienski, *J. Chem. Soc., Perkin Trans. 1*, 1998, 4005.

10. (a) M. P. DeNinno, J. B. Etienne and K. C. Duplantier, *Tetrahedron Lett.*, 1995, **36**, 669; (b) D. R. Gauthier Jr, R. H. Szumigala Jr, J. D. Armstrong III and R. P. Volante, *Tetrahedron Lett.*, 2001, **42**, 7011.
11. For other examples of transformations with para-methoxyphenyl acetals, see: Z. Zhang and G. Magnusson, *J. Org. Chem.*, 1996, **61**, 2394.
12. D. R. Williams and S.-Y. Sit, *J. Am. Chem. Soc.*, 1984, **106**, 2949.
13. See for instance: (a) A. Frankowski, D. Deredas, D. Le Nouen, T. Tschamber and J. Streith, *Helv. Chim. Acta*, 1995, **78**, 1837; (b) M. Isobe, H. Takahashi and T. Goto, *Tetrahedron Lett.*, 1990, **31**, 717.
14. (a) S. D. Rychnovsky and D. J. Skalitzky, *Tetrahedron Lett.*, 1990, **31**, 945; (b) D. A. Evans, D. L. Rieger and J. R. Gage, *Tetrahedron Lett.*, 1990, **31**, 7099; (c) S. D. Rychnovsky, B. N. Rogers and T. I. Richardson, *Acc. Chem. Res.*, 1998, **31**, 9.
15. D. Seebach, A. K. Beck and A. Heckel, *Angew. Chem., Int. Ed.*, 2001, **40**, 92.
16. For some examples, see: (a) R. O. Duthaler and A. Hafner, *Chem. Rev.*, 1992, **92**, 807; (b) D. Seebach, G. Jaeschke and Y. M. Wang, *Angew. Chem., Int. Ed.*, 1995, **34**, 2395; (c) G. Jaeschke and D. Seebach, *J. Org. Chem.*, 1998, **63**, 1190; (d) D. Seebach, R. Dahinden, R. E. Marti, A. K. Beck, D. A. Plattner and F. N. M. Kühnle, *J. Org. Chem.*, 1995, **60**, 1788.
17. For some examples, see: (a) A. Cuenca, M. Medio-Simón, G. A. Aguilar, D. Weibel, A. K. Beck and D. Seebach, *Helv. Chim. Acta*, 2000, **83**, 3153; (b) Y. Huang, A. K. Unni, A. N. Thadani and V. H. Rawal, *Nature*, 2003, **424**, 146.
18. For an interesting discussion on this subject, see the consequences of employing an acyclic backbone of the TADDOL ligand in relation to the original structure, in asymmetric gold catalysis: (a) H. Teller, S. Flügge, R. Goddard and A. Fürstner, *Angew. Chem., Int. Ed.*, 2010, **49**, 1949; (b) H. Teller, M. Corbet, L. Mantilli, G. Gopakumar, R. Goddard, W. Thiel and A. Fürstner, *J. Am. Chem. Soc.*, 2012, **134**, 15331.

2.1.13. Redox-Neutral Synthesis of a Cyclic *N,O*-Acetal from Salicylaldehyde and 1,2,3,4-Tetrahydroisoquinoline

Claire L. Jarvis and Daniel Seidel*

Rutgers, the State University of New Jersey, Department of Chemistry and Chemical Biology, Rutgers University, 610 Taylor Road, Piscataway, NJ 08854-8087, USA. *E-mail: seidel@rutchem.rutgers.edu

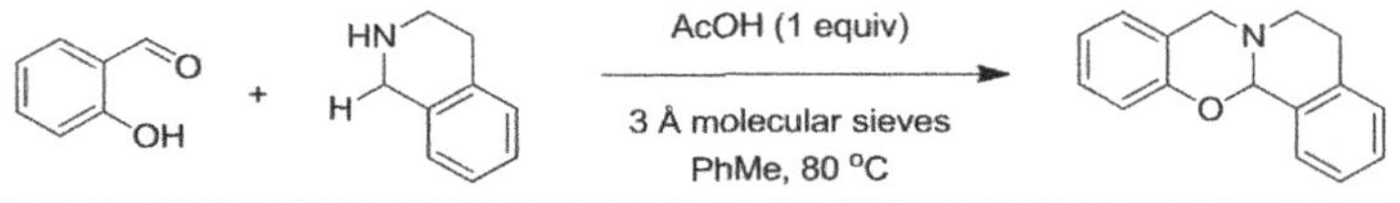

Number of sessions (duration of each session)	Hazard level	Difficulty level	Level of study
1 session (4 h)	Moderate	High	Intermediate

Class names Amines, aldehydes, alcohols

Concepts involved A rapid condensation reaction between salicylaldehyde and 1,2,3,4-tetrahydroisoquinoline in the presence of acetic acid (AcOH) and molecular sieves is followed by an extraction work-up, then purification by silica column chromatography to cleanly give the corresponding *N,O*-acetal product in nearly quantitative yield

Chemicals needed Salicylaldehyde, 1,2,3,4-tetrahydroisoquinoline, acetic acid, toluene, 3 Å molecular sieves (powder), dichloromethane, ethyl acetate, triethylamine, sodium sulfate, hexanes, deuterated chloroform, silica gel for column chromatography, Celite®, sodium bicarbonate

Equipment and experimental techniques involved Apparatus for reaction at high temperature (80 °C) and magnetic stirring, Büchner funnel and flask for vacuum filtration, liquid–liquid extraction apparatus, rotatory evaporator, flash chromatography column, ^{1}H NMR spectroscopy

Keywords Column chromatography, condensation reaction, evaporation under reduced pressure, extraction, NMR, redox chemistry

Background

The *N,O*-acetal moiety can be found in a diverse range of natural products, many of which have been investigated for medicinal applications such as antibacterials,[1] the treatment of HIV[2] and Parkinson's disease.[3] The syntheses of these compounds typically start with a condensation of a prefunctionalized amine (either as an imine or an amino acid[4]) and an alcohol with the use of an external oxidant,[5] often a metal catalyst.[6]

Using simple commercially-available starting materials and acetic acid as the promoter, our research group showed that it is possible to synthesize *N,O*-acetals[7] *via* C–H functionalization of the amine[8] without a net oxidation or reduction (*i.e.*, a redox-neutral reaction). This approach removes the need for an external oxidizing agent which is often required in related amine C–H functionalization reactions.[9] Under our conditions the amine is functionalized *in situ* at the α-C–H position (see Supplementary Information for discussion of mechanism). Molecular sieves are used to absorb water formed in the condensation, driving the reaction forwards. The reaction

Comprehensive Organic Chemistry Experiments for the Laboratory Classroom
Edited by Carlos A. M. Afonso, Nuno R. Candeias, Dulce Pereira Simão, Alexandre F. Trindade, Jaime A. S. Coelho, Bin Tan, and Robert Franzén.

Published by the Royal Society of Chemistry, www.rsc.org

proceeds very quickly in an atom-economical manner (*i.e.*, large excesses of any reagent is not required), furnishing the desired product cleanly without the need for complicated isolation or purification techniques.

Additional Safety

Personal protective equipment including goggles, gloves and appropriate clothing should be worn at all times during this experiment. The reaction, work-up and purification should be conducted in a fume hood. Students should avoid inhaling silica gel or Celite® dust. Care should be taken when handling and disposing of sharp needles and glass pipettes.

Experimental Procedure

Reaction (1 h)

1. In a fume hood heat an oil bath to a steady temperature of 80 °C on a magnetic stirrer plate.
2. Prepare a 25 mL (tared) round-bottom flask with a stir bar. Weigh in 200 mg of 3 Å powdered molecular sieves. Add 0.10 mL of salicylaldehyde (0.122 g, 1 equiv.), 0.17 mL of 1,2,3,4-tetrahydroisoquinoline (0.173 g, 1.3 equiv.), 0.05 mL of glacial acetic acid (0.060 g, 1 equiv.) and 10 mL of anhydrous toluene to the flask.
3. Insert a water condenser into the round-bottom flask and clamp in the oil bath. Stir the reaction for 1 h, maintaining a constant temperature of 80 °C.
4. After 1 h, lift the flask from the oil bath (caution – glassware will be very hot) and allow to cool to room temperature.

Extraction Work-Up (1 h)

5. Once your reaction has cooled to room temperature, transfer all of the solution and solids *via* a glass pipette onto a 1 cm thick plug of Celite® set on top of a Büchner funnel. Use vacuum filtration to collect the liquid in the flask.
6. Wash your reaction flask with 10 mL dichloromethane, then pipette the remaining liquid onto the Celite® plug. Leave the vacuum on for several minutes until all liquid appears to have been collected.
7. Transfer the liquids in your Büchner flask to a separatory funnel. Add 20 mL of saturated aqueous sodium bicarbonate solution. Seal your funnel and shake vigorously, pausing regularly to vent the funnel. Once the two layers have formed, decant off the bottom aqueous layer, then set the organic layer aside in a separate flask. Add the aqueous layer back into the funnel and repeat twice with 20 mL fresh dichloromethane. Pure dichloromethane is denser than water and will now be the bottom layer in the funnel. Combine the organic layers.
8. Add several large spatulas of sodium sulfate drying agent to your organic layers. Shake your flask and leave it standing for a few minutes, then filter off the sodium sulfate on the Büchner funnel.
9. Take one or two drops of the organic solution and transfer into a small capped vial with a few drops of acetone. This will be used in the purification step. Transfer the liquid from the Büchner flask into a 100 mL round-bottom flask (tared) and concentrate on the rotatory evaporator.

Product Purification and Characterization (1.5 h)

10. Prepare a silica column that is 2.5–3 cm wide and 10–15 cm long, using approximately 200 mL of 40 : 59 : 1 ethyl acetate: hexanes: triethylamine solution as the eluent.

11. Weigh the 100 mL round-bottom flask to determine the mass of your crude product (an orange oil), then dissolve it using one pipette full of your eluent. Carefully transfer this solution using a glass pipette onto the top of your silica column. Rinse your round-bottom flask with another pipette of eluent and transfer to the top of the silica column. Drain the column until the liquid level is below the sand and your material is absorbed onto the silica gel, then add the rest of your eluent solution.
12. Collect the liquid from your column in 50–75 mL portions in conical flasks. The desired product is visible as a bright yellow band traveling down the column. Collect the yellow band in a separate beaker then stop the column chromatography. Transfer the solution into a 100 mL round-bottom flask (tared) and concentrate on the rotatory evaporator.
13. Weigh the final product in the round-bottom flask to determine your final yield.
14. Weigh out 15–20 mg of your final product into a small vial and dissolve in 0.5 mL of deuterated chloroform. Transfer the solution to an NMR tube.
15. Acquire the ^{1}H NMR spectrum.

Waste Disposal

All solutions should be disposed of in an appropriate chemical liquid waste container. The dichloromethane should be disposed of in a separate halogenated wastes container. The Celite®, molecular sieves and silica gel should be disposed of in a separate solid wastes container for incineration.

Results Interpretation and Additional Questions

1. Interpret your ^{1}H NMR spectrum, then calculate the percentage purity of your compound. Calculate the percentage yield of your isolated product based on your recorded mass and percentage purity.
2. Label the peaks in your spectrum that correspond to the aromatic protons of your product. Assign all of the remaining alkyl proton peaks.
3. Present a plausible explanation for the color changes observed during the course of the reaction.
4. Based upon the mechanism for this reaction (shown in Scheme 2.1.13.1), how does acetic acid accelerate the reaction?
5. In addition to forming *N,O*-acetals from an alcohol, this reaction can also be adapted to form the corresponding nitrogen analogues (called the aminals) from amines and *N,S*-acetals from thiols (Scheme 2.1.13.2).

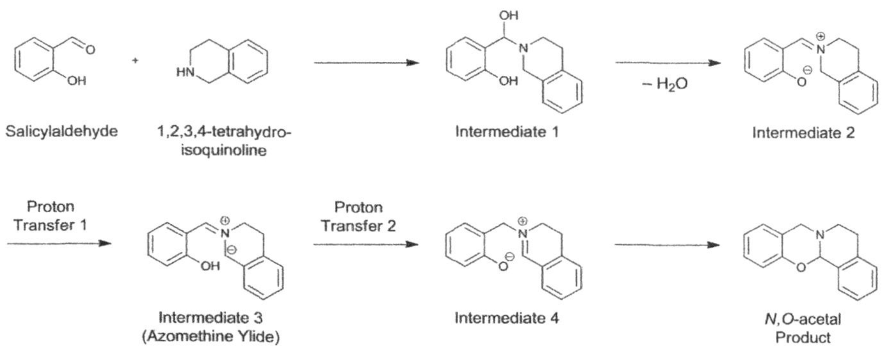

Scheme 2.1.13.1 Mechanism of *N,O*-acetal formation.

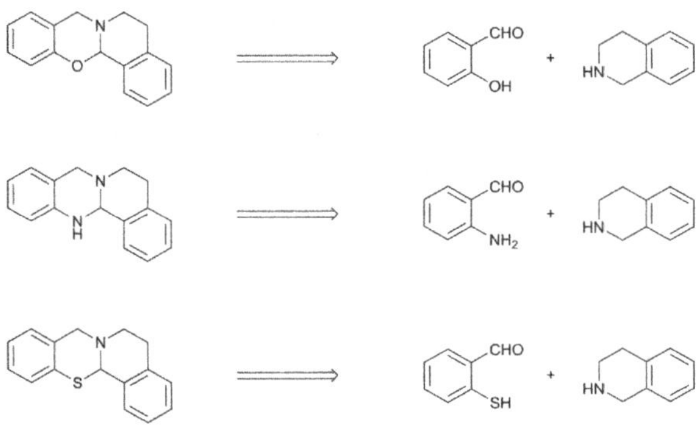

Scheme 2.1.13.2 Synthesis of *N,O*-acetals and the corresponding heteroatom analogues.

Consider (i) the relative electronic properties of oxygen, nitrogen and sulfur and (ii) the proposed reaction mechanism (focusing on Intermediates 2 and 4 in Scheme 2.1.13.1). List the three classes of starting materials in order of increasing reactivity. Explain the cause of this reactivity pattern.

Acknowledgements

Adapted from M. T. Richers *et al. J. Am. Chem. Soc.* 2014, **136**, 6123. Copyright (2014) American Chemical Society.

References

1. B. P Mathew, A. Kumar, S. Sharma, P. K. Shukla and M. Nath, *Eur. J. Med. Chem.*, 2010, **45**, 1502.
2. A. J. Cocuzza, D. R. Chidester, B. C. Cordova, S. Jeffrey, R. L. Parsons, L. T. Bacheler, S. Erickson-Viitanen, G. L. Trainor and S. S. Ko, *Bioorg. Med. Chem. Lett.*, 2001, **11**, 1177.
3. T. M. Bohme, C. E. Angelli-Szafran, H. Hallak, T. Pugsley, K. Serpa and R. D. Schwarz, *J. Med. Chem.*, 2002, **45**, 3094.
4. N. Cohen, J. F. Blount, R. J. Lopresti and D. P. Trullinger, *J. Org. Chem.*, 1979, **44**, 4005.
5. C. L. Mathis, B. M. Gist, C. K. Frederickson, K. M. Midkiff and C. C. Marvin, *Tetrahedron Lett.*, 2013, **54**, 2101.
6. M. L. Deb, S. S. Dey, I. Bento, M. T. Barros and C. D. Maycock, *Angew. Chem., Int. Ed.*, 2013, **52**, 9791.
7. M. T. Richers, M. Breugst, A. Y. Platonova, A. Ullrich, A. Dieckmann, K. N. Houk and D. Seidel, *J. Am. Chem. Soc.*, 2014, **136**, 6123.
8. D. Seidel, *Acc. Chem. Res.*, 2015, **48**, 317.
9. S. A. Girard, T. Knauber and C. J. Li, *Angew. Chem., Int. Ed.*, 2014, **53**, 73.

2.2. C-C σ-Bond Formation

2.2.1. Selective *C*-acylation of 3-Methyl-1-phenyl-pyrazol-5-one

Vanya B. Kurteva*[a], Lubomir A. Lubenov[a], and Maria A. Petrova[b]

[a]Institute of Organic Chemistry with Centre of Phytochemistry, Bulgarian Academy of Sciences, Acad. G. Bonchev str., bl. 9, 1113, Sofia, Bulgaria; [b]University of Chemical Technology and Metallurgy, Department of General and Inorganic Chemistry, 8 Kliment Ohridski blvd., 1756 Sofia, Bulgaria
*E-mail: vkurteva@orgchm.bas.bg

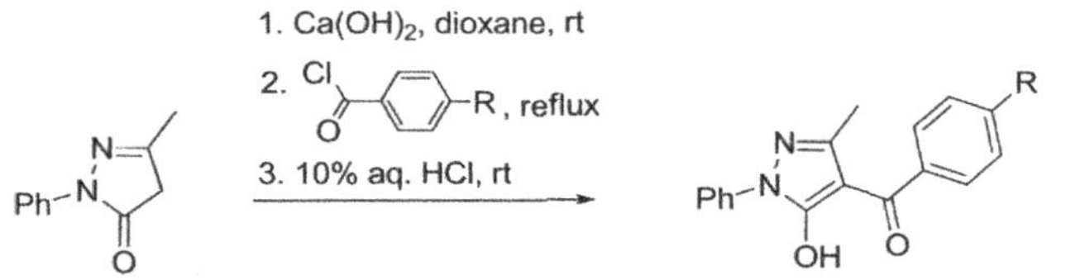

Number of sessions (duration of each session)	Hazard level	Difficulty level	Level of study
3 (5 h + 2 h + 2 h)	Moderate	Medium	Intermediate

Class names Aroyl chlorides, amides, ketones, pyrazolones

Concepts involved The protocol presents the selective *C*-acylation of 3-methyl-1-phenyl-pyrazol-5-one, *i.e.* without contamination of the corresponding *O*-acylated compounds, by one-pot consecutive switching of the pyrazole tautomeric equilibrium towards the enol form and protection of the hydroxyl functionality as a calcium complex, acylation with a series of aroyl chlorides and, finally, decomposition of the complex in acidic media

Chemicals needed 3-Methyl-1-phenyl-pyrazol-5-one, calcium hydroxide, dioxane (dry), aroyl chloride (4-methylbenzoyl chloride, 4-fluorobenzoyl chloride, 4-phenylbenzoyl chloride, 4-trifluoromethylbenzoyl chloride), hydrochloric acid (10%), acetone, methanol, deuterochloroform

Equipment and experimental techniques involved Apparatus for work at reflux with stirring; dropping funnel, calcium chloride tube, mortar with pestle, Büchner funnel with filter paper, TLC plates (basic alumina for Ca-complexes and silica gel for final products) and UV lamp, recrystallization, NMR, melting point determination

Keywords *C*-acylation, chemoselectivity, melting point, NMR, OH-protection, recrystallization

Background

4-Acylpyrazol-5-ones are among the most widely exploited O-donors in coordination chemistry.[1] First studied by Jensen,[2] the chelating behaviour of this class of β-dicarbonyl compounds has received immense attention due to several valuable properties, such as lower pK_a values in comparison with conventional β-diketones, vast extracting ability, great separation power, and intense colour of their complexes. In particular, aroyl pyrazolones have displayed great

Comprehensive Organic Chemistry Experiments for the Laboratory Classroom
Edited by Carlos A. M. Afonso, Nuno R. Candeias, Dulce Pereira Simão, Alexandre F. Trindade, Jaime A. S. Coelho, Bin Tan, and Robert Franzén.

Published by the Royal Society of Chemistry, www.rsc.org

complexation and extraction ability towards metal ions, including 4f- and 5f-elements,[3] which were found to be strongly dependent on the position and nature of the substituent in the aromatic ring of the acyl group, *i.e.* on the electronic, steric, and solubility parameters of the ligand.

Acylpyrazolones have been known since the end of the nineteenth century. The initial synthetic strategies were based on classical Claisen condensation of esters with pyrazolone catalysed by sodium salts of acids or sodium ethoxide or on Fries rearrangement of *O*-acylated pyrazolones to *C*-acylated products by anhydrous zinc chloride.[4] However, both protocols have suffer with such drawbacks as low yields and coloured side-products. An advantageous fast and efficient protocol was developed in 1959 by Jensen,[5] and has practically no alternative and consequently has been intensively exploited until the present day. The method is based on direct acylation of pyrazolones with acyl chlorides or anhydrides in the presence of calcium hydroxide. However, some important synthetic issues in terms of controlling the reaction output have never been discussed in the literature.

The aim of this work is to provide students with the key elements for achieving selective acylation of tautomeric heterocycles.[6] To achieve *C*-acylation at fourth position of pyrazolone, two factors have to be taken into account. First, the staring pyrazolone has to be in enol form (**1a**) to activate the fourth position. Second, the OH group has to be protected to avoid *O*-acylation. In this case, calcium hydroxide operates in both respects by forming a stable complex in basic conditions[7] with hydroxyl functionality (Scheme 2.2.1.1). If acyl chloride is added before complex formation, the corresponding *O*-acylated compound is the only or main reaction product. Additionally, calcium hydroxide is used in two-fold excess to trap the liberated hydrogen chloride, thus keeping the media basic and avoiding decomposition of the complex during the acylation.

The laboratory work also includes product recrystallization, characterization by 1D and 2D NMR spectra, and melting point determination.

Additional Safety

The starting 3-methyl-1-phenyl-2-pyrazolin-5-one possesses acute oral toxicity and irritates skin and eyes. Calcium hydroxide can cause severe skin irritation, chemical burns, blindness, or lung damage. Acyl chlorides are lachrymators and cause skin and eyes corrosion and burns. Dioxane is a carcinogenic flammable liquid, which is irritating to the eyes and respiratory tract. Hydrochloric acid is a toxic compound, which has a corrosive effect on human tissue. The solvents used for TLC and recrystallization – dichloromethane, methanol, and acetone – are flammable

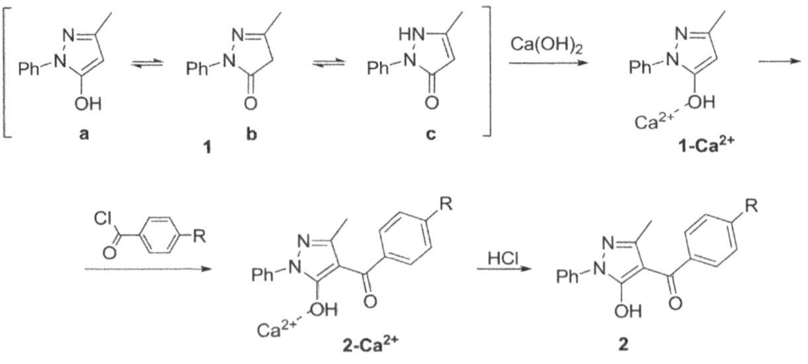

Scheme 2.2.1.1 Three-step one-pot protocol for selective *C*-acylation of 3-methyl-1-phenyl-pyrazol-5-one.

liquids. Dichloromethane causes skin and eye irritation and possesses carcinogenicity. Methanol is acutely toxic if swallowed, in contact with skin, or if inhaled. Acetone causes serious eye irritation and may cause drowsiness or dizziness. Deuterated chloroform is a suspected carcinogen and should be handled with care. Unprotected exposure to these chemicals has to be limited. Avoid breathing solvent vapours and release of reagents to the environment. The hazards of acyl pyrazolones are not known and so the compound have to be handled with care.

Safety glasses, gloves, and laboratory coat are worn during all exercises. Both syntheses and recrystallizations are carried out by using safety heaters with temperature controllers. All exercises, including NMR sample preparations, are performed in a well-ventilated hood.

Experimental Procedure

Session 1 (5 h): Synthesis of 4-Aroyl-3-methyl-1-phenyl-pyrazol-5-ones

1. Set up the necessary apparatus in a well-ventilated hood: a 250 mL two-neck round-bottom flask, condenser with moisture stopper (glass tube with calcium chloride or silica gel with moisture indicator), high turbulence magnetic stir bar, magnetic stirrer, and heating mantle.
2. Grind 3-methyl-1-phenyl-2-pyrazolin-5-one with a pestle in a mortar to achieve easy dissolution.
3. Insert 3-methyl-1-phenyl-2-pyrazolin-5-one (8.7 g, 0.05 mol) and dioxane (60 mL) and heat gently (<70 °C) till full dissolution (15 min).
4. Add $Ca(OH)_2$ (7.4 g, 0.1 mol) at once and reflux for 0.5 h under vigorous stirring to form a Ca-complex. Follow the reaction by TLC on basic alumina till full consumption of pyrazolone; mobile phase DCM.
5. Cool to 0 °C with an ice-bath and add the aroyl chloride (0.05 mol) drop-wise *via* a dropping funnel; *ca.* 2 drops per second.
6. Reflux the reaction mixture for 1.5–2 h depending on the acyl chloride.
7. Cool the reaction mixture to room temperature.
8. Add the reaction mixture to 250 mL 10% aq. HCl under vigorous stirring.
9. After 1.5 h separate the residue formed by filtration and wash with water and then with ethanol.
10. Place the crude product on a filter paper and leave to become air dried in a desiccator.

Remark: The reaction with 4-trifluoromethylbenzoyl chloride (point 6) needs 9 h to be complete, which in combination with the high price of the reagent makes it inappropriate for a student laboratory. If necessary, to obtain the product the main reaction can be carried out overnight.

Session 2 (2 h): Analyses of the Crude Product and Recrystallization

1. Weigh the crude product and determine the yield.
2. Analyse the crude product by TLC on silica gel.
3. Prepare a sample for NMR in $CDCl_3$ and have the spectra recorded.
4. Recrystallize the crude product from methanol–acetone, ethanol, or methanol.

Session 3 (Flexible): Analyses of Pure 4-Aroyl-3-methyl-1-phenyl-pyrazol-5-ones

1. Filter the crystals and dry in air.
2. Weigh the recrystallized compounds and determine the yield.
3. Measure the melting point of the pure compounds.

4. Analyse the signals in the NMR spectra by following the sequence:
    4.1. Analyse the proton spectrum: integral intensities, multiplicity of the signals, characteristic chemical shifts, coupling constants.
    4.2. Analyse the cross peaks in the ^1H-^1H 2D COSY experiment to obtain information about the neighbouring groups.
    4.3. Analyse the cross peaks in the ^1H-^1H 2D NOESY experiment to obtain information about the steric remoteness of the protons.
    4.4. Analyse ^{13}C NMR spectrum and the DEPT experiment.
    4.5. Analyse the cross peaks in the ^1H-^{13}C 2D HSQC experiment to identify the direct proton–carbon connections.
    4.6. Analyse the cross peaks in the ^1H-^{13}C 2D HMBC experiment to assign the signals of quaternary carbons.
    4.7. Organize all spectroscopic data to confirm the structure of the product.

Waste Disposal

The 10% aq. HCl–dioxane solutions have to be collected in a non-chlorinated waste collector recipient for destruction due to the dioxane toxicity. The acetone–methanol mother liquors can be distilled and reused. Deuterochloroform has to be collected in a waste collector for chlorinated solvents.

Results Interpretation and Additional Questions

1. Discuss the influence of the aroyl substituent on the chemical shifts of the NMR signals.
2. Discuss the influence of fluorine on the proton and carbon resonances.
3. Compare the signals in the NMR spectra of acylpyrazolones with those of the corresponding esters.[8]
4. Discuss the role of calcium hydroxide in determining the reaction out-put.
5. Discuss the influence of the type of aroyl substituent on the reaction prolongation and yield.

Acknowledgements

Reprinted (adapted) with permission from V. B. Kurteva, M. A. Petrova, J. Chem. Educ. 2015, 92, 382–384. Copyright (2014) American Chemical Society. The financial support by the EU Social Fund and Ministry of Education and Science, EU operational program "Development of human resources", project BG051PO001-3.3.07-0002 Student Practices, and by the Bulgarian Science Fund, projects UNA-17/2005 and DRNF-02-13/2009, is gratefully acknowledged.

References

1. (a) Y. A. Zolotov and N. M. Kuzmin, *Extraction of Metals by Acylpyrazolones*, Nauka, Moscow, Russia, 1977; (b) C. Pettinari, F. Marchetti and A. Drozdov, in *Comprehensive Coordination Chemistry II*, ed. J. A. McCleverty and T. J. Meyer, Elsevier Ltd., Amsterdam, 2003, vol. 1, ch. 1.6, pp. 97–115; (c) F. Marchetti, C. Pettinari and R. Pettinari, *Coord. Chem. Rev.*, 2005, **249**, 2909–2945; (d) S. Swavey, *J. Chem. Educ.*, 2010, **87**, 727–729.
2. B. S. Jensen, *Acta Chem. Scand.*, 1959, **13**, 1890–1896.
3. (a) R. Meera and M. L. P. Reddy, *Solv. Extr. Ion Exch.*, 2004, **22**, 761–789; (b) K. Binnemans, in *Handbook on the Physics and Chemistry of Rare Earths*, ed. K. A. Gschneider, J.-C. G. Bünzli and V. K. Pecharsky, Elsevier B. V., Burlington, 2005, vol. 35, ch. 225; (c) H. Mukai, S. Miyazaki, S. Umetani, S. Kihara and M. Matsui, *Anal. Chim. Acta*, 1990, **239**, 277–282; (d) M. Atanassova, V. Kurteva and L. Lubenov, *Ind. Eng. Chem. Res.*, 2011, **50**, 12170–12176; (e) M. Atanassova, V. Kurteva, L. Lubenov, S. Varbanov and I. Dukov, *Sep. Purif.*

Technol., 2012, **95**, 58–63; (f) M. Atanassova, V. Kurteva, L. Lubenov and I. Billard, *RSC Adv.*, 2014, **4**, 38820–38829.

4. (a) F. Stoltz, *J. Prakt. Chem.*, 1897, **55**, 145–171; (b) A. Michaelis and F. Engelhardt, *Ber. dtsch. chem. Ges.*, 1908, **41**, 2668–2676; (c) W. Wislicenus, H. Elvert and P. Kurtz, *Ber. dtsch. chem. Ges.*, 1913, **46**, 3395–3407.
5. B. S. Jensen, *Acta Chem. Scand.*, 1959, **13**, 1668–1670.
6. (a) W. Holzer, B. Plagens and K. Lorenz, *Heterocycles*, 1997, **45**, 309–314; (b) W. Holzer, K. Mereiter and B. Plagens, *Heterocycles*, 1999, **50**, 799–818; (c) Y. Akama, M. Shiro, T. Ueda and M. Kajitani, *Acta Crystallogr., Sect. C*, 1995, **51**, 1310–1314; (d) Y. Akama, A. Tong, N. Matsumoto, T. Ikeda and S. Tanaka, *Vib. Spectrosc.*, 1996, **13**, 113–115; (e) W. Holzer, R. M. Claramunt, C. López, I. Alkorta and J. Elguero, *Solid State Nucl. Magn. Reson.*, 2008, **34**, 68–76; (f) P. N. Remya, C. H. Suresh and M. L. P. Reddy, *Polyhedron*, 2007, **26**, 5016–5022.
7. (a) R. N. Jadeja, K. M. Vyas, V. K. Gupta, R. G. Joshi and C. Ratna Prabha, *Polyhedron*, 2012, **31**, 767–778; (b) K. M. Vyas, *Acyl Pyrazolones & their Structural Analogues: Synthesis, Characterization, Crystal Structure and Studies on their Bio-Active Metal Complexes*, Ph.D. Thesis, The Maharaja Sayajirao University of Baroda, Gujarat, India, 2012.
8. V. B. Kurteva and M. A. Petrova, *J. Chem. Educ.*, 2015, **92**, 382–384.

2.2.2. Organocatalytic Asymmetric α-Arylation of Aldehydes

Mette Overgaard, Pernille H. Poulsen, and Karl Anker Jørgensen*

Department of Chemistry, Aarhus University, DK-8000 Aarhus C, Denmark. *E-mail: kaj@chem.au.dk

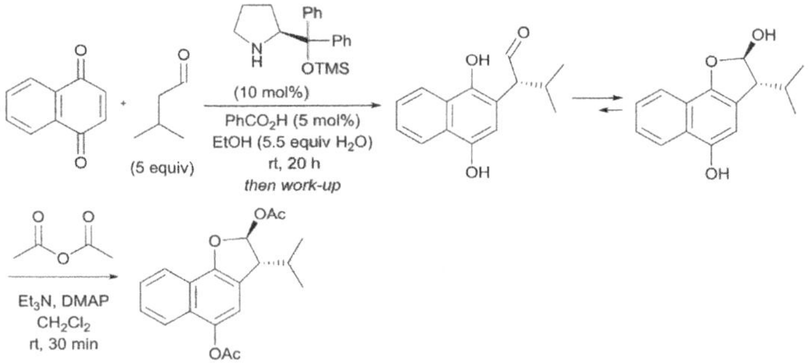

Number of sessions (duration of each session)	Hazard level	Difficulty level	Level of study
3 sessions (2 h + 4 h + 5 h)	Moderate	High	Advanced

Class names Aldehydes, enamines, hemiacetals, alcohols, esters

Concepts involved A sterically shielding organocatalyst is used to activate an aldehyde *via* enamine formation. The aldehyde is enantioselectively α-arylated and the resulting product is acetylated. NMR and chiral HPLC are used to evaluate the stereoselectivity of the reaction

Chemicals needed (*S*)-(−)-α,α-Diphenyl-2-pyrrolidinemethanol trimethylsilyl ether, EtOH (96% vol.), pentane, EtOAc, CH_2Cl_2, benzoic acid, 3-methylbutanal, 1,4-naphthoquinone, 4-(dimethylamino)pyridine (DMAP), Et_3N, Ac_2O, $MgSO_4$, Celite® 545, silica gel, $CDCl_3$, hexane, 2-propanol, sat. aq. $NaHSO_3$, sat. aq. $NaHCO_3$, 1 M HCl

Equipment and experimental techniques involved Magnetic stirrer, rotatory evaporator, separatory funnel, flash column, NMR, chiral HPLC

Keywords Acetylation, chiral HPLC, enamine activation, flash column chromatography, NMR, organocatalysis

Background

A catalyst can have more than one role in a reaction; besides increasing the reaction rate, some catalysts can induce enantioselectivity in the reaction. A chiral catalyst can selectively accelerate the reaction that leads to predominantly one enantiomer. With the use of a chiral catalyst a prochiral starting material can be transformed into either enantiomer of a chiral product.

Asymmetric catalysts have traditionally been either biocatalysts (enzymes)[1] or transition metals with chiral ligands.[2] These two fields suffer with some drawbacks; enzymes can be too

Comprehensive Organic Chemistry Experiments for the Laboratory Classroom
Edited by Carlos A. M. Afonso, Nuno R. Candeias, Dulce Pereira Simão, Alexandre F. Trindade, Jaime A. S. Coelho, Bin Tan, and Robert Franzén.

Published by the Royal Society of Chemistry, www.rsc.org

specific, one enzyme might only catalyze one specific reaction, while transition metals can be expensive, air and moist sensitive and they can leave possibly toxic traces of heavy metals in the products.

Positioned between the two extremes of enzymatic transformations and transition-metal catalysis, organocatalysis emerged in 2000 as a new research field.[3] The catalysts are all-organic systems that are air and water stable, inexpensive and readily available. Within the field of organocatalysis two different mechanistic approaches are generally considered: covalent and non-covalent catalysis. In covalent catalysis, the catalyst and the substrate(s) form a covalent bond, changing the mechanism of the reaction and forming completely new intermediates compared to the uncatalysed reaction. A typical example of a covalent catalyst is secondary amines that are able to catalyze enantioselective functionalizations of carbonyl compounds. Non-covalent catalysis involves mechanisms that rely on non-covalent interactions such as hydrogen bonding or formation of ion pairs, phase-transfer catalysis is a typical example within this category.[4]

Amino acid-derived catalysts and especially chiral secondary amines are some of the most used organocatalysts today in the field of aminocatalysis.[5] Chiral amines can be used to activate carbonyl compounds, thus enabling a large number of enantioselective α-, β- and γ-functionalizations involving enamine, iminium-ion and dienamine intermediates, respectively. Even tetraenamine activation and more remote functionalization strategies involving trienamine intermediates have been described.[6]

The present experiment involves an organocatalytic enantioselective α-functionalization of an aldehyde *via* enamine activation. The α-position of an aldehyde can be considered nucleophilic because of the tautomeric equilibrium between the aldehyde and the corresponding enol. When no further stabilization is present, the equilibrium is almost entirely on the aldehyde form (Scheme 2.2.2.1). To functionalize the aldehyde at the α-position the tautomeric equilibrium must be pushed towards the enol form. This can be achieved by masking the enol as an enamine *via* condensation of the aldehyde with a secondary amine (Scheme 2.2.2.1).

The aldehyde is transformed into the more nucleophilic enamine and the enamine can react with an electrophile at the α-position. The secondary amine works as a catalyst; by hydrolysis the enamine is converted back into an aldehyde and the secondary amine catalyst is regenerated. The catalytic cycle for α-functionalization of aldehydes *via* the enamine intermediate is outlined in Scheme 2.2.2.2.

The chiral secondary amine catalyst used in the present reaction induces stereoselectivity by shielding one face of the enamine with the sterically large diphenyltrimethylsilyl ether group (Scheme 2.2.2.3). This large sterical shield results in the electrophile approaching from the opposite face of the steric group, an attack from below the enamine intermediate.

An important topic in modern organic chemistry is the enantioselective formation of carbon–carbon bonds as it enables rapid construction of complex molecules. Numerous compounds with activities that are relevant to various important research areas such as medicinal chemistry and biology have aromatic moieties as essential substructures.[7] Thus, in the search for methods for arylation, the introduction of an already existing aryl-group into a molecule has been of tremendous importance. Furthermore, the formation of a carbon–carbon bond between an aromatic carbon and the α-carbon of an aldehyde has been a long standing challenge of contemporary organic synthesis. Prior to the current reaction,[8] enantioselective α-arylation of aldehydes mostly relied on transition-metal catalyzed approaches.[9]

Additional Safety

All sessions must be carried out in a fume hood; safety goggles and protective clothing have to be worn throughout the experiment. Gloves should be used when necessary.

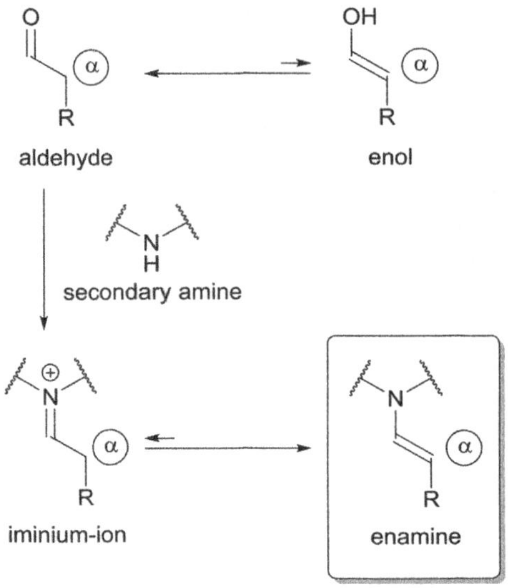

Scheme 2.2.2.1 Activation of aldehydes by shifting the tautomeric equilibrium *via* enamine formation.

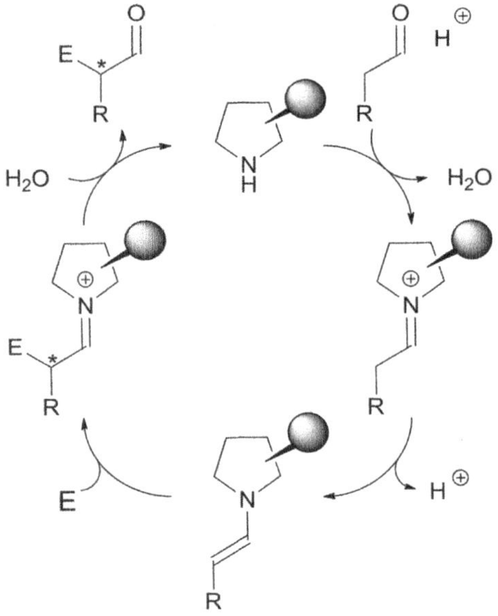

Scheme 2.2.2.2 Catalytic cycle for the aminocatalyzed α-functionalization of aldehydes. E: electrophile.

Experimental Procedure

Laboratory Session 1 (2 h): Organocatalytic Asymmetric α-Arylation of 3-Methylbutanal

1. Weigh out (*S*)-2-(diphenyl-(trimethylsilyloxy)methyl)pyrrolidine (44 mg, 0.14 mmol, 0.10 equiv.) in a glass vial, dissolve it in EtOH (1.4 mL) and transfer the solution to a 10 mL round-bottomed flask equipped with a magnetic stirrer bar.

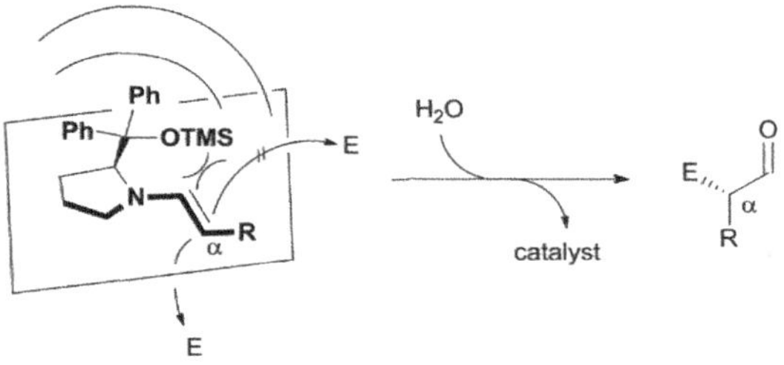

Scheme 2.2.2.3 Sterical shielding of the top face of the enamine intermediate results in a stereoselective α-functionalization. E: electrophile.

2. To the flask add: H_2O (0.135 mL, 0.135 g, 7.77 mmol, 5.5 equiv.), benzoic acid (8.2 mg, 0.068 mmol, 0.050 equiv.), 3-methylbutanal (0.741 mL, 0.581 g, 6.75 mmol, 5.0 equiv.) and 1,4-naphthoquinone (0.214 g, 1.35 mmol, 1.0 equiv.).
3. Stir the reaction mixture for 20 h at room temperature.

Laboratory Session 2 (4 h): Removal of Excess Aldehyde, Acetylation and Work-Up

1. Transfer the reaction mixture to a separatory funnel, dilute with EtOAc (10 mL) and wash with sat. aq. $NaHSO_3$ (4 × 10 mL) and H_2O (1 × 10 mL). Extract the combined aqueous layer with EtOAc (2 × 10 mL). Dry the combined organic layer over $MgSO_4$. Filter off the drying agent by suction and concentrate the filtrate by rotary evaporation.
2. Transfer the crude mixture to a 25 mL round-bottomed flask by dissolving it in EtOAc and evaporate the solvent by rotary evaporation.
3. Equip the flask with a magnetic stirrer bar and add DMAP (17 mg, 0.14 mmol, 0.10 equiv.). Cap the flask with a rubber septum and flush with argon.
4. Using syringes, add to the flask: dry CH_2Cl_2 (6.8 mL), Et_3N (0.471 mL, 0.342 g, 3.38 mmol, 2.5 equiv.) and Ac_2O (0.318 mL, 0.345 g, 3.38 mmol, 2.5 equiv.). Stir the reaction at room temperature for 30 min.
5. Transfer the mixture to a 100 mL separatory funnel, dilute with CH_2Cl_2 (10 mL) and wash with HCl (1 M, 2 × 10 mL), H_2O (10 mL), sat. aq. $NaHCO_3$ (2 × 10 mL) and H_2O (10 mL). Remember to vent frequently as CO_2 gas evolves. Extract the combined aqueous layer with CH_2Cl_2 (2 × 10 mL) and dry the combined organic layer over $MgSO_4$. Filter off the drying agent by suction and concentrate the filtrate by rotary evaporation.
6. Store the crude product in a fridge until the next session.

Laboratory Session 3 (5 h): Flash Column Purification, NMR and Chiral HPLC

1. To dry load the crude on the column make a thick slurry out of the crude, CH_2Cl_2 and Celite® (3 g). Concentrate the solution to a dry powder by careful rotary evaporation.
2. Wet-pack (10% EtOAc/pentane) a glass column (2.3 × 28 cm) with silica. Add the Celite® (with adsorbed product) to the column and top with sand.
3. Elute with 10% EtOAc/pentane and collect in 15 mL fractions. Use TLC (UV visualization) to follow the chromatography.
4. Collect the product fractions and concentrate using rotary evaporation followed by vacuum.
5. Measure the yield and record the NMR spectra of the product ($CDCl_3$). Measure the optical rotation using a polarimeter.
6. Measure the enantiomeric excess (ee) of the product by HPLC using a Chiralpak® IB column and elute with 95% hexane/5% isopropanol and a flow of 1.0 mL min^{-1}.

Waste Disposal

All waste should be collected in appropriate waste containers for posterior destruction. Waste containing CH_2Cl_2 should be collected in containers for chlorinated solvents.

Results Interpretation and Additional Questions

1. Interpret the NMR spectra. What is the diastereomeric ratio (dr)?
2. Calculate the enantiomeric excess (ee). Comment on the ee and the dr, what do the results show? Draw the four possible product configurations, point out which are enantiomers and which are diastereoisomers.
3. Suggest a mechanism for the organocatalytic step. Speculate on the selectivity of the reaction.
4. How is the excess aldehyde removed from the crude mixture by sat. aq. $NaHSO_3$?
5. Show the mechanism of the acetylation step. What is the role of DMAP?

References

1. For a review, see *e.g.*: T. Hudlicky and J. W. Reed, *Chem. Soc. Rev.*, 2009, **38**, 3117.
2. (a) E. N. Jacobsen, A. Pfaltz and H. Yamamoto, *Comprehensive Asymmetric Catalysis*, Springer, Berlin, 1999, vol. 1–3; (b) K. Mikami and M. Lautsen, *New Frontiers in Asymmetric Catalysis*, Wiley-Interscience, Hoboken, New Jersey, 2007.
3. P. Melchiorre, M. Marigo, A. Carlone and G. Bartoli, *Angew. Chem., Int. Ed.*, 2008, **47**, 6138.
4. For a review, see *e.g.*: J.-F. Brière, S. Oudeyer, V. Dalla and V. Levacher, *Chem. Soc. Rev.*, 2012, **41**, 1696.
5. For reviews on aminocatalysis, see: (a) K. L. Jensen, G. Dickmeiss, H. Jiang, Ł. Albrecht and K. A. Jørgensen, *Acc. Chem. Rev.*, 2011, **45**, 248; (b) M. Nielsen, D. Worgull, T. Zweifel, B. Gschwend, S. Bertelsen and K. A. Jørgensen, *Chem. Commun.*, 2011, **47**, 632.
6. For a review on remote functionalizations, see: (a) H. Jiang, Ł. Albrecht and K. A. Jørgensen, *Chem. Sci.*, 2013, **4**, 2287; for an example of tetraenamine activation see: (b) J. Stiller, P. H. Poulsen, D. C. Cruz, J. Dourado, R. L. Davis and K. A. Jørgensen, *Chem. Sci.*, 2014, **5**, 2052.
7. For examples, see: (a) L. M. Salonen, M. Ellermann and F. Diederich, *Angew. Chem., Int. Ed.*, 2011, **50**, 4808; (b) T. J. Ritchie and S. J. F. Macdonald, *Drug Discovery Today*, 2009, **14**, 1011.
8. J. Alemán, S. Cabrera, E. Maerten, J. Overgaard and K. A. Jørgensen, *Angew. Chem., Int. Ed.*, 2007, **46**, 5520.
9. For a representative review, see: (a) F. Bellina and R. Rossi, *Chem. Rev.*, 2010, **110**, 1082; for some examples see: (b) R. Martín and S. L. Buchwald, *Angew. Chem., Int. Ed.*, 2007, **46**, 7236; (c) G. D. Vo and J. F. Hartwig, *Angew. Chem. Int., Ed.*, 2008, **47**, 2127; (d) J. García-Fortanet and S. L. Buchwald, *Angew. Chem., Int. Ed.*, 2008, **47**, 8108.

2.2.3. Matteson Homologation of Pinacol Boronic Ester: An Efficient Method Using the Boronate Complexes

Sébastien Balieu*, Asmaa Bouyahya, and Jean-Philippe Bouillon

Université de Rouen, IRCOF, 1 Rue Tesnières - 76130 Mont Saint Aignan, France
*E-mail: sebastien.balieu@univ-rouen.fr

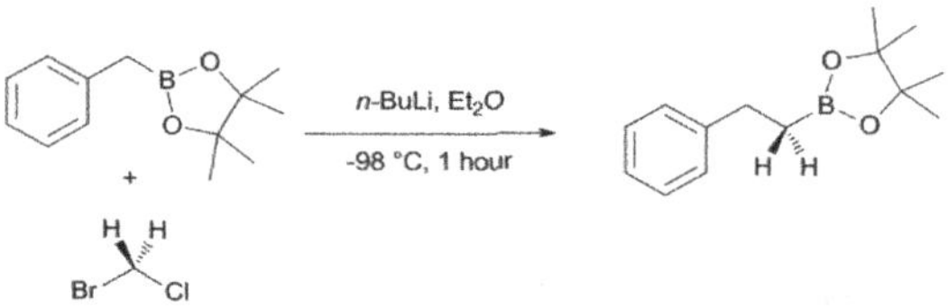

Number of sessions (duration of each session)	Hazard level	Difficulty level	Level of study
1 (4 h)	High	High	Advanced
Class names Alkyl halides and organolithium reagents, boronic esters			
Concepts involved This experiment intends to introduce the students to C–C bond formation using the Matteson homologation and attract their attention to the importance of controlling the temperature of the reaction. This protocol is aimed at advanced level students to set-up the reaction of an alkyl halide with a boronate. The reaction product is obtained pure after filtration over silica gel and the purity can be checked either by ^{1}H NMR or GC-MS (development of analytical skills and characterization of organic compounds)			
Chemicals needed Bromochloromethane, n-butyllithium (1.6 M in hexanes), benzylboronic acid pinacol ester, dry diethyl ether, methanol, liquid nitrogen, silica gel, deuterochloroform			
Equipment and experimental techniques involved Equipment: NMR and/or GC-MS apparatus; techniques: working at low temperature and under inert atmosphere, handling corrosive organic compounds, purification by filtration; materials: dewar, magnetic stirring, syringe pump, Bohlender PTFE tubing, three-way valve			
Keywords Ate-complexes, boronic ester, filtration, GC-MS, homologation, NMR			

Background

The Matteson homologation[1] is an efficient and robust reaction by which to increase the length of a chain including a boronic ester moiety. Indeed, this transformation provides high yields, the migration of the alkyl moiety during the rearrangement of the ate-complex is completely stereoselective, the reaction does not generate by-product (except *n*-butyl bromide and lithium chloride) and is relatively quick to set up. Moreover, the use of boronic ester through the chemistry of boron-ate complexes is experiencing significant development and the number of syntheses involving this function is still growing.[2]

Nevertheless, the key point of the Matteson homologation is the strict control of the temperature of the reaction. Thus, the ate-complex resulting from the addition of the carbenoid on

Comprehensive Organic Chemistry Experiments for the Laboratory Classroom
Edited by Carlos A. M. Afonso, Nuno R. Candeias, Dulce Pereira Simão, Alexandre F. Trindade, Jaime A. S. Coelho, Bin Tan, and Robert Franzén.

Published by the Royal Society of Chemistry, www.rsc.org

the boronic ester rearranges very easily even under low temperature conditions, affording the expected compound. Nevertheless, this compound can subsequently react with a new equivalent of chloromethyllithium to suffer a new homologation. The result will be a mixture of the homologated product and the over-homologated compounds that is difficult to separate. The objective of this work is the synthesis of pure 4,4,5,5-tetramethyl-2-phenethyl-1,3,2-dioxaborolane by controlling the temperature. This means that the *n*-BuLi must be added slowly with a syringe pump through a Bohlender PTFE tubing needle dipping into the reaction mixture.

Additional Safety

The experiment should be carried in a fume hood, with proper safety equipment, such as protective clothing, gloves and safety goggles.

n-Butyllithium (*n*-BuLi) is not considered as toxic (threshold toxicity is not known); however, it is an extremely reactive organometallic compound and should be handled with extreme caution. The solvent, *n*-hexane, is considered to have low toxicity. The principle hazard is the violent exothermic reaction between *n*-BuLi and water in any form, including moist air. Contact with any part of the body can result in immediate caustic and thermal burns and should be strenuously avoided. *n*-BuLi solutions should be used only in a closed system under argon or nitrogen. Containers should be checked regularly for leaks. Any leaking container should be moved to a dry, well-ventilated area and the contents transferred to another approved container. Standard Schlenk techniques will be used. Syringes with locking barrels will be used and the dead-volume will be flushed and quenched.

Liquid nitrogen is an extremely cold liquid (77 K). Because of its low temperature, careless handling of liquid nitrogen may result in cold burns. During its use the liquid nitrogen must be handled carefully with insulator gloves.

Experimental Procedure

Laboratory Session 1 (3 h)

1. In a fume hood prepare a methanol bath in a Dewar at room temperature on a magnetic plate.
2. Prepare a 50 mL round-bottomed flask with a magnetic stir bar, and add 200 mg (0.92 mmol, 1 equiv.) of 2-benzyl-4,4,5,5-tetramethyl-1,3,2-dioxaborolane.
3. Add a three-way valve on the flask and connect it to a manifold and put the flask under vacuum and refill it with nitrogen, repeat this operation two times.
4. Add, with a syringe, through the valve 107 μL (1.65 mmol, 1.8 equiv.) of bromochloromethane and 7 mL of dry Et_2O (dry ether is obtained by storage on 3 Å molecular sieves (20% m/v) for 48 h).
5. Put the flask into the bath of methanol and stir the mixture and then add carefully liquid nitrogen to the bath to reach 175 K.
6. Add with a syringe pump *via* a Bohlender PTFE tubing needle 1.03 mL (1.65 mmol, 1.8 equiv.) of *n*-BuLi (1.6 M in hexane) through the valve during one hour. The temperature must be controlled by regularly addition of liquid nitrogen to stay at 175 K.
7. Once the addition of *n*-BuLi in hexane is finished remove the flask from the bath and continue the stirring until room temperature is reached. A precipitate may appear.
8. Once the reaction mixture has reached room temperature, filtrate the reaction media through a plug of silica gel (4 g of silica/g of crude mixture). Wash the plug twice with Et_2O (10 mL) and evaporate the solvent under reduced pressure (for safety reasons: the absence of peroxides in Et_2O should be checked prior to the evaporation process) to obtain the pure compound.

9. Weigh the desired product and calculate the yield. Record and compare the GC-MS trace of the starting material and the product. Record ^{1}H NMR spectra to confirm the structure of the final compound.

Characterization of the Isolated Product (1 h)

Record ^{1}H NMR and GC-MS spectra. Make the correct attribution of all the protons, and calculate the chemical shift (δ in ppm) and the coupling constants (*J* in Hz). Then, compare ^{1}H NMR and GC-MS spectra with those reported in the supplementary material (ESI).

Waste Disposal

Diethyl ether has to be collected in a non-chlorinated waste collector recipient, for posterior destruction. Silica gel and the solid may be collected in a special waste collector recipient.

Results Interpretation and Additional Questions

1. Propose a mechanism leading to the final compound, then compare with it that reported in the ESI.
2. Explain why the excess of *n*-BuLi and bromochloromethane do not afford over-homologated products.
3. Discuss the importance of the control of the reaction temperature. Propose one possible by-product if the reaction is conducted at higher temperature.

References

1. (a) D. S. Matteson, R. Ray, R. R. Rocks and D. J. S. Tsai, *Organometallics*, 1983, **2**, 1536; (b) D. S. Matteson and D. Majumdar, *Organometallics*, 1983, **2**, 1529; (c) D. S. Matteson and R. Ray, *J. Am. Chem. Soc.*, 1980, **102**, 7590; (d) D. S. Matteson and D. Majumdar, *J. Am. Chem. Soc.*, 1980, **102**, 7588.
2. (a) D. Leonori and V. K. Aggarwal, *Acc. Chem. Res*, 2014, **47**, 3174; (b) M. Burns, S. Essafi, J. R. Bame, S. P. Bull, M. P. Webster, S. Balieu, J. W. Dale, C. P. Butts, J. N. Harvey and V. K. Aggarwal, *Nature*, 2014, **513**, 183; (c) K. Hong, X. Liu and J. P. Morken, *J. Am. Chem. Soc.*, 2014, **136**, 10581; (d) D. S. Matteson, *J. Org. Chem.*, 2013, **78**, 10009; (e) S. N. Mlynarski, A. S. Karns and J. P. Morken, *J. Am. Chem. Soc.*, 2012, **134**, 16449.

2.2.4. A Practical Organocatalytic Alkylation Reaction with Benzodithiolylium Tetrafluoroborate

Pier Giorgio Cozzi*, Marco Bandini, Andrea Gualandi, and Luca Mengozzi

Alma Mater Studiorum – Università di Bologna, Dipartimento di Chimica "G. Ciamician", Via Selmi 2, 40126 Bologna, Italy. *E-mail: piergiorgio.cozzi@unibo.it

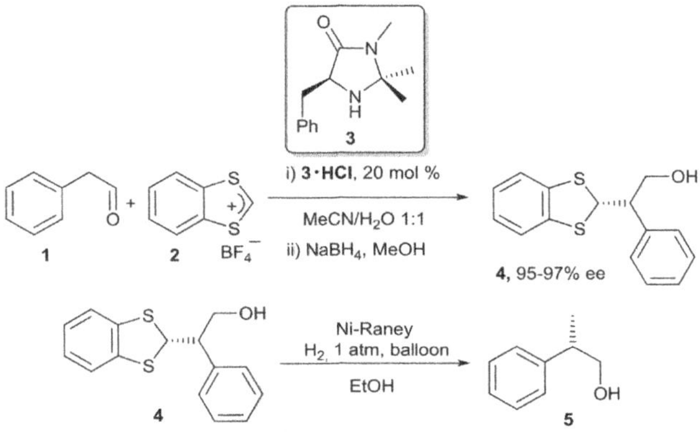

Number of sessions (Duration of each session)	Hazard level	Difficulty level	Level of study
3 (2 h + 4 h + 4 h)	High	High	Advanced

Class names Aldehydes, benzodithioles, alcohols

Concepts involved The experiment is a simple and practical organocatalytic procedure for the enantioselective alkylation of aldehydes using a commercially available and stable carbenium ion. The alkylated adduct **4**, obtained in high enantiomeric excess, is then transformed into the corresponding enantioenriched alcohol **5** by reduction of the benzodithiole group with Raney®-nickel.

Chemicals needed Phenylacetaldehyde, benzodithiolylium tetrafluoroborate, (5*S*)-2,2,3-trimethyl-5-phenylmethyl-4-imidazolidinone monohydrochloride (MacMillan catalyst hydrochloride salt), sodium borohydride, Raney®-nickel, acetonitrile, methanol, ethanol, silica gel, ethyl acetate, cyclohexane, hydrogen

Equipment and experimental techniques involved Vials for organocatalytic reactions (open air), magnetic stirrer, magnetic stirring bar, separating funnel, flask equipped to work with Schlenk techniques, balloon filled with hydrogen, Gooch septum

Keywords Aldehyde alkylation, chiral HPLC, column chromatography, enamine, liquid–liquid extraction, methyl, NMR, organocatalysis

Background

Alkylation using S_N1 or S_N2-type reactions is encountered in the first introductory lessons in courses on organic chemistry. In successive courses, students learn about stereoselective and diastereoselective alkylations of carbonyl compounds.[1] Most α-diastereoselective alkylations of

Comprehensive Organic Chemistry Experiments for the Laboratory Classroom
Edited by Carlos A. M. Afonso, Nuno R. Candeias, Dulce Pereira Simão, Alexandre F. Trindade, Jaime A. S. Coelho, Bin Tan, and Robert Franzén.

Published by the Royal Society of Chemistry, www.rsc.org

carbonyl compounds developed used pre-formed enolates.[2] There are several practical methodologies described in literature for the stereoselective alkylation of carbonyl compounds[3] that rely on the use of chiral auxiliaries.[4] In quite complex and long total synthesis, a quite affordable, standard, and high yielding procedure needs to be employed for the preparation of the starting materials or advanced intermediates.[5] Although a plethora of catalytic methodologies in which metals are employed in catalytic amount have been introduced in recent years,[6] the level of sophistication and skills necessary to reproduce these results is quite high, using expensive ligands, or difficult reaction conditions. Running the reaction at low temperature (−78 °C), under a nitrogen atmosphere, and with carefully dried and deoxygenated solvents are the minimal prerequisites for successful reactions. Quite recently, the use of small organic molecules to catalyse organic transformations has been introduced as a new important field within the domain of stereoselective organic synthesis. This field has received the general name of organocatalysis.[7] The success of organocatalysis is related to the ability to identify specific "activation modes" in which organocatalysts can be used in promoting old or new reactions.[8] One fundamental activation mode is enamine catalysis: the condensation of a primary or secondary chiral amine with a carbonyl (aldehyde or ketone) generates the corresponding enamine, a powerful nucleophile that can react with many different electrophiles, resulting in the functionalization of the substrate at the α-position. Enamine catalysis was reported in 1971,[9] and recently re-discovered in 2000.[10] In this key report enamine catalysis was used to functionalize carbonyl-containing compounds at the α-position in a way that is possible also with many different electrophiles. The organocatalytic reactions are generally insensitive to oxygen and moisture, and therefore there is no need for special reaction vessels, experimental techniques, or for ultra-dry reagents and solvents. Organic catalysts are also non-toxic, and the use of metals is avoided. Organocatalysis in principle can replace the use of chiral auxiliaries. In fact, a direct and easily performed organocatalytic alkylation of aldehydes represents an important key transformation for the preparation of various enantioenriched starting materials. Unfortunately, alkyl halides cannot be employed in these reactions, as they easily react with the chiral secondary or primary amines. Immediately with the discovery of organocatalysis, the α-alkylation of aldehydes was recognized as a difficult transformation, often nominated as the "Holy Grail" in organocatalysis.[11] The purpose of this experiment is to describe a facile and straightforward organocatalytic α-alkylation of aldehydes with a stabilized carbenium ion. We have introduced in the literature the concept of an organocatalytic S_N1 type reaction in which a stabilized carbenium ion is formed by suitable precursors.[12] These conditions are suitable for stereoselective alkylation of enamines formed *in situ* by reaction of secondary amine with aldehyde, while the alkylation of the secondary amine used as catalyst does not take place.[13] We have also shown that the commercially available benzodithiolylium tetrafluoroborate (**2**) can be used in organocatalytic alkylation reactions.[14] The purpose of the experiment is to illustrate the basic principles of aminocatalysis applied to the synthesis of enantioenriched products. Moreover the experiment illustrates the stability of certain carbenium ions, such as commercially available benzodithiolylium tetrafluoroborate, which is handled without any precautions in open air and in the presence of water. In addition, the experiment is a multi-step synthesis in which different reagents are employed that can be used to obtain useful intermediates that are difficult to access through other methodologies. The objective of this work is the preparation of enantioenriched alcohols, by using an organocatalytic reaction as key step. The enamine formed *in situ* by the reaction of the chiral catalyst, a commercially available secondary amine, reacts with the stable carbenium ion **2**, introduced in the reaction mixture. The commercially available MacMillan imidazolidinone catalyst **3**[15] is used as the chiral promoter, while the reaction is performed in a CH_3CN/H_2O mixture. To facilitate separation of the product, the intermediate aldehyde is reduced with $NaBH_4$ to the corresponding alcohol, on the crude reaction mixture. Finally, elimination of the benzodithiole group is quite straightforward and can be performed simply using Raney®-nickel and a balloon filled with hydrogen. This experiment was successfully executed during 3 years by 70 students at the second year of a Master's level course. For 65% of the experiments the isolated yields were over 80%. For all experiments the measured ee was over 90%.

Additional Safety

All sessions should be carried out in a fume hood, with proper safety equipment. Acetonitrile and methanol are toxic if swallowed, in contact with skin or if inhaled and need to be handled using gloves. Hydrogen is a flammable gas that forms an explosive mixture with air. Raney®-nickel is a toxic and pyrophoric material that should be handled under an inert atmosphere using gloves. Raney®-nickel contains significant amounts of hydrogen gas at the end of the reaction and may spontaneously ignite when exposed to air, so it should never be allowed to become a dry powder. Raney®-nickel may cause irritation of the respiratory tract and nasal cavities, and causes pulmonary fibrosis if inhaled. It can also cause eye and skin irritation. Nickel is also rated as being a possible human carcinogen. Sodium borohydride in contact with water releases flammable gases that may ignite spontaneously and it is toxic if swallowed or in contact with skin.

Experimental Procedure

Laboratory Session 1 (2 h): Stereoselective Alkylation of Aldehyde with Benzodithiolylium

1. In a fume hood prepare an ice bath in a Dewar on a magnetic stirrer plate.
2. To a vial equipped with a magnetic stirring bar, add the commercially available MacMillan catalyst hydrochloride salt **3**·HCl (0.04 mmol, 0.2 equiv.), acetonitrile (0.8 mL) and water (0.8 mL), and insert it inside the ice bath with vigorous stirring.
3. After 4 min, add 1,3-benzodithiolylium tetrafluoroborate (**2**) (0.2 mmol, 1 equiv.) and phenyl acetaldehyde (0.3 mmol, 1.5 equiv.). Cover the Dewar with aluminium foil, to maintain the temperature at 0 °C, and stir for 24 h.
4. Prepare the racemic compound (±)-**4** following the identical experimental procedure except by the use of a racemic mixture of the catalyst **3**. Racemic **4** for analytical purposes can be prepared on a 0.1 mmol scale. Take the (*S*)-**3** catalyst (0.01 mmol) and (*R*)-catalyst (0.01 mmol), both commercially available, and mix them together before introduction in the reaction flask.

Laboratory Session 2 (4 h): Reduction of the Product with $NaBH_4$ and Purification

1. Dilute the reaction mixture with ethyl acetate (5 mL) and water (3 mL).
2. Separate the organic layer with a separation funnel. Extract the aqueous phase with ethyl acetate (2 × 5 mL). Collect the organic layers, dry them over Na_2SO_4 and evaporate the solvent under reduced pressure (Rotavapor®).
3. In a fume hood prepare an ice bath on a magnetic stirrer plate.
4. Take the flask containing the residue diluted with methanol (10 mL), introduce a stirring bar and cool at 0 °C inside the ice bath.
5. After 4 min, add carefully $NaBH_4$ (0.8 mmol, 2 equiv.) and stir for 30 min at 0 °C.
6. Add H_2O (2 mL) at 0 °C to the flask and evaporate methanol under reduced pressure (Rotavapor®).
7. Extract the residue with ethyl acetate (3 × 5 mL) and collect the organic layers. Dry these reunited organic layers over Na_2SO_4 and concentrate them under reduced pressure (Rotavapor®).

8. Purify the compound by column flash chromatography on silica gel using a mixture of cyclohexane/ethyl acetate (9:1 to 8:2). TLC control is with basic $KMnO_4$ or ammonium cerium(IV) sulfate/ammonium molybdate stains.
9. Use the same methodology for the racemic mixture, and purify the compound. The racemic compound can be separated by using HPLC equipped with chiral column. For the separation the use of a Daicel Chiralcel IC column is suggested. Conditions for a good separation are: *n*-hexane/*i*-PrOH (96: 4), flow rate 0.50 mL min^{-1}, 30 °C, $\lambda = 232, 254$ nm; t_R(minor) = 35.5 min, t_R(major) = 41.4 min.
10. After separation of the racemic mixture, inject the purified enantioenriched compound in order to determine the enantiomeric excess obtained in the reaction.

Laboratory Session 3 (4 h): Removal of Benzodithiol Group (This Part Can be Omitted)

1. Use a double-necked flask and under a hood connect the flask to a Schlenk line. Insert nitrogen into the flask on a magnetic plate.
2. Add a solution of compound **4** (0.19 mmol) in ethanol (2 mL) into the flask, and add a Raney®-nickel slurry in water (about 0.450 g) using a Pasteur pipette (be careful to avoid exposing traces of dry Raney®-nickel to air; traces of this compound should be washed with water in a flask and disposed of properly).
3. Introduce a stopper into a neck and evacuate the nitrogen using the Schlenk line.
4. Introduce an H_2 atmosphere (1 atm) in the flask using a balloon filled with H_2.
5. Stir the flask for 3 h.
6. Evacuate the H_2 and discharge the balloon under the hood, and fill the flask with nitrogen.
7. Stop the stirrer and with a magnetic stirring bar retrieve Raney®-nickel settled on the bottom of the flask.
8. Remove the solvent with a Pasteur pipette under nitrogen flux, and collect it in a flask under air.
9. Add ethyl acetate (5 mL) to the reaction flask and stir the mixture for 4 min.
10. Repeat steps 7, 8 and 9 twice. The last time, cover the residual Raney®-nickel with H_2O (10 mL) and dispose of properly.
11. Collect the organic solvent and evaporate it under reduced pressure (Rotavapor®).
12. Add ethyl acetate (5 mL) to the residue and separate the organic layer. Extract the aqueous layer with ethyl acetate (2 × 5 mL).
13. Collect the organic layers, wash them with brine (5 mL), dry over Na_2SO_4 and filter over a small pad of silica-gel on a Gooch septum. Evaporate the solvent under reduced pressure (Rotavapor®). Take a sample of compound to acquire the ^{1}H-NMR spectrum and calculate the reaction yield.

Waste Disposal

The neutralized aqueous solution and silica gel are collected in suitable collectors for posterior destruction. Distilled ethyl acetate may be reused or collected in a non-chlorinated waste collector recipient, for posterior destruction. Raney®-nickel needs to be disposed separately, avoiding dry-out.

Results Interpretation and Additional Questions

1. Interpret the ^{1}H NMR spectra obtained, and determine the enantiomeric excess of the product obtained.

2. Rationalize the mechanism of alkylation of aldehydes with benzodithiolylium tetrafluoroborate, proposing a suitable transition state. Write the mechanism for the reaction of formation of enamine between the MacMillan catalyst and phenylacetaldehyde.
3. Discuss the state of the art of alkylation of aldehydes through organocatalysis. Why is the use of CH_3I as alkylating agent in organocatalysis not described?

References

1. *Stereoselective synthesis of drugs and natural products*, ed. V. Andrushko and N.Andrushko, John Wiley & Sons, two volume set, 2013.
2. D. M. Hodgson and A. Charlton, *Tetrahedron*, 2014, **70**, 2207.
3. W. J. Corey and L. Kurti, *Enantioselective chemical synthesis. Methods, logic and practice*, Elsevier, 2013.
4. K. Ross, *Key chiral auxiliary applications*, Elsevier Science & Technology Books, 2014.
5. D. Urabe, T. Asaba and M. Inoue, *Chem. Rev.*, 2015, **115**, 9207–9231.
6. J. Zhou. , *Multicatalyst system in asymmetric catalysis*, John Wiley & Sons, 2014.
7. K. A. Ahrendt, C. J. Borths and D. W. C. MacMillan, *J. Am. Chem. Soc.*, 2000, **122**, 4243.
8. D. W. C. MacMillan, *Nature*, 2008, **455**, 304.
9. (a) U. Eder, G. R. Sauer and R. Wiechert, German patent DE 2014757, 1971; (b) Z. G. Hajos and D. R. Parrish, *J. Org. Chem.*, 1974, **39**, 1615.
10. B. List, R. A. Lerner and C. F. Barbas III, *J. Am. Chem. Soc.*, 2000, **122**, 2395.
11. J. Vesely and R. Rios, *ChemCatChem*, 2012, **4**, 942.
12. A. Gualandi and P. G. Cozzi, *Synlett*, 2013, **24**, 281.
13. B. List, I. Čorić, O. O. Grygorenko, P. S. J. Kaib, I. Komarov, A. Lee, M. Leutzsch, S. C. Pan, A. V. Tymtsunik and M. van Gemmeren, *Angew. Chem., Int. Ed.*, 2014, **53**, 282.
14. (a) A. Gualandi, E. Emer, M. G. Capdevila and P. G. Cozzi, *Angew. Chem., Int. Ed.*, 2011, **50**, 7842; (b) A. Gualandi, D. Petruzziello, E. Emer and P. G. Cozzi, *Chem. Commun.*, 2012, **48**, 3614; (c) A. Gualandi, M. G. Emma, J. Giacoboni, L. Mengozzi and P. G. Cozzi, *Synlett*, 2013, **24**, 449; (d) A. Gualandi, P. Canestrari, E. Emer and P. G. Cozzi, *Adv. Synth. Catal.*, 2014, **356**, 528.
15. G. Lelais and D. W. C. MacMillan, *Aldrichim. Acta*, 2006, **39**, 79.

2.2.5. Oxazolidinone Mediated Enantioselective Allylation

Jonathan D. Sellars*, AnnMarie C. O'Donoghue, Ian R. Baxendale, John M. Sanderson, and Elizabeth J. Grayson

Chemistry Department, University of Durham, South Road, Durham, DH1 3LE, UK
*E-mail: j.d.sellars@durham.ac.uk

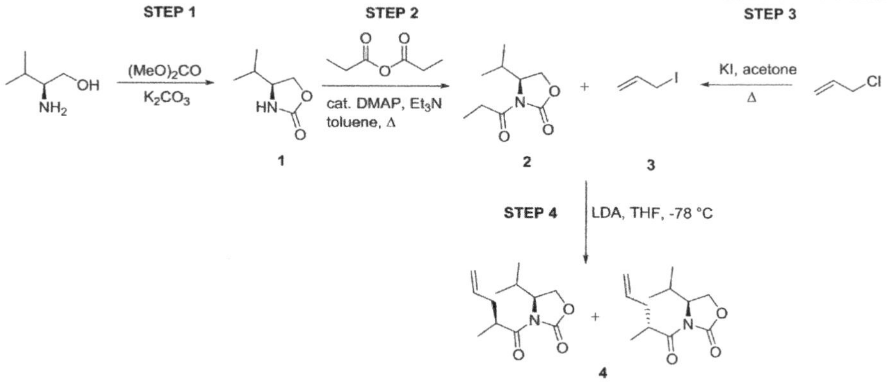

Number of sessions (duration of each session)	Hazard level	Difficulty level	Level of study
12 (7 h each)	High	High	Advanced

Class names Carbamates, anhydrides, alkyl halides enolates, amino alcohols

Concepts involved The purpose of these experiments is to introduce third year undergraduate students to multi-step organic synthesis. In addition, the experiment exposes the student cohort to a wide range of practical (handling pyrophoric reagents, vacuum distillation, recrystallization, utilizing dry solvents, column chromatography) and theoretical (planning experimental quantities, time management) skills, which they will utilize in their fourth-year studies

Chemicals needed Step 1: valinol, dimethyl carbonate, K_2CO_3, $MgSO_4$, dichloromethane, methanol, ammonium hydroxide; step 2: (4*S*)-4-(propan-2-yl)-1,3-oxazolidin-2-one (**1**), *N*,*N*-dimethylaminopyridine, triethylamine, propionic anhydride, 1 M aq. HCl, dry toluene, diethyl ether, *n*-hexane, ethyl acetate, 1 M aq. NaOH, brine, $MgSO_4$; step 3: allyl chloride, KI, diethyl ether, $MgSO_4$, allyl iodide (**3**), acetone; step 4: (4*S*)-4-(propan-2-yl)-3-propanoyl-1,3-oxazolidin-2-one (**2**), allyl iodide (**3**), diisopropylamine, *n*-BuLi (2.5 M in hexanes), dry ice, NH_4Cl (aq.), brine, $MgSO_4$, silica gel, dry THF, acetone, diethyl ether, *n*-hexane, ethyl acetate

Equipment and experimental techniques involved Standard apparatus for low temperature (−78 °C) and room temperature magnetic stirring, liquid–liquid extraction apparatus, flash column chromatography column, distillation apparatus, high vacuum line, NMR spectroscopy, IR spectroscopy

Keywords Asymmetric, chiral auxiliary, flash column chromatography on silica gel, IR, NMR, vacuum distillation

Comprehensive Organic Chemistry Experiments for the Laboratory Classroom
Edited by Carlos A. M. Afonso, Nuno R. Candeias, Dulce Pereira Simão, Alexandre F. Trindade, Jaime A. S. Coelho, Bin Tan, and Robert Franzén.

Published by the Royal Society of Chemistry, www.rsc.org

Background (ref. 1)

"I call any geometrical figure, or group of points, chiral, and say it has chirality, if its image in a plane mirror, ideally realized, cannot be brought to coincide with itself."

This statement by Lord Kelvin in 1904 is universally accepted as the definition of chirality. Introducing chirality to a molecule is a fundamental operation in organic synthesis. Moreover, the ability to assemble complex molecules as single enantiomers is a critical undertaking in modern organic chemistry. For example, single enantiomer drugs are important for toxicological reasons (thalidomide),[2–5] *in vivo* efficacy and metabolism. There are four general approaches to the synthesis of complex single enantiomeric products, involving (i) resolution, (ii) use of the chiral pool, (iii) temporary installation of a chiral auxiliary[6] and (iv) asymmetric catalysis.[7] Utilizing molecules from the chiral pool is by far the most efficient method of introducing chirality; however, limited precursors prevent broad applicability. Likewise, asymmetric catalysis is an elegant approach but is also hampered by substrate generality and predictability. Whilst not so elegant, a chiral auxiliary is frequently employed to introduce chirality to a molecule as it is easily attached and removed, provides predictable chiral outcomes and as diastereoisomers are formed the products allow the utility of classical analytical techniques to measure the ratios and their separation by achiral chromatography. One of the most well-known chiral auxiliaries are the oxazolidinones, derived from amino alcohols. They have been utilized in a plethora of stereoselective reactions, such as aldol reactions,[8] alkylation reactions,[9,10] Diels–Alder reactions[11,12] and Michael additions.[13]

The objective of this experiment is to (a) synthesize a chiral auxiliary; (b) attach this to a simple acyl group; and (c) undertake a stereocontrolled alkylation reaction with subsequent separation of the diastereoisomers.

Additional Safety

All sessions should be undertaken in a fume cupboard. *n*-BuLi (2.5 M in hexanes) is highly flammable and as such should not be exposed to air. Propionic anhydride is corrosive and may cause severe skin burns and eye damage. Diethyl ether is a highly flammable solvent, which may form explosive peroxides, is harmful if swallowed and with repeated exposure may cause skin dryness or cracking. Furthermore, vapors may cause drowsiness and dizziness and so the solvent should be handled in a well-ventilated laboratory.

Experimental Procedure

Step 1: (4*S*)-4-(Propan-2-yl)-1,3-oxazolidin-2-one (1)

1. In a dry 25 mL round-bottom flask under nitrogen, fitted with a septum cap and balloon, place valinol (1 g, 9.69 mmol, 1 equiv.) using dimethyl carbonate (0.7 equiv.) to wash the vial containing valinol.
2. Add the remaining dimethyl carbonate (3.0 equiv.) followed by K_2CO_3 (1.1 equiv.).
3. Reflux the solution for 4 h and monitor by TLC (CH_2Cl_2/MeOH/NH_4OH (4 : 1 : 0.25)).
4. Upon completion, pour the solution into water (20 mL) and extract with CH_2Cl_2 (3 × 10 mL).
5. Dry the organic layer with $MgSO_4$, filter and evaporate.
6. Triturate the solid material with hexane and filter the solid to provide (**1**). Characterize your product (^{1}H NMR, IR, mp, Chiral-LC).

N.B. When submitting a sample for chiral LC, use a concentration of no less than 5 mg mL^{-1}.

Step 2 (4*S*)-4-(Propan-2-yl)-3-propanoyl-1,3-oxazolidin-2-one (2)

1. To a dry round-bottom flask fitted with a reflux condenser under nitrogen add oxazolidinone (**1**) (1 g, 7.74 mmol, 1 equiv.), DMAP (0.1 equiv., cat.) and dry toluene (to produce a 1.1 M solution).
2. Treat the solution with Et_3N (1.2 equiv.) then propionic anhydride (*careful!*) (1.5 equiv.) over 2 min.
3. Heat the solution and once refluxing begins continue to reflux for another 30 min.
4. Add water (10 mL) to the reaction and heat the mixture at reflux for an additional 10 min.
5. Remove the flask from the heat and allow to cool to room temperature.
6. Pour the mixture into a separating funnel containing Et_2O (40 mL), drain the aqueous layer and wash the organic layer with 1 M aq. HCl (30 mL), 1 M aq. NaOH (30 mL) and brine (30 mL).
7. Dry the organic layer with $MgSO_4$, filter and evaporate.
8. Purify the crude material by flash column chromatography (*n*-hexane/EtOAc) to provide (**2**).
9. Characterize your product (2) (^{1}H NMR, IR, mp).

Step 3: 3-Iodoprop-1-ene (3)

1. To a dry round-bottom flask fitted with a reflux condenser under nitrogen add, *via* syringe, allyl chloride (5 mL, 61.4 mmol, 1 equiv.) followed by acetone (to produce a 3 M solution).
2. Add KI (1.25 equiv.) and heat the mixture to reflux for 24 h.
3. Cool the reaction mixture to room temperature and pour into water (30 mL).
4. Add the mixture to a separating funnel and drain the lower organic layer.
5. Extract the remaining aqueous layer with Et_2O (2 × 15 mL) and combine the organic layers.
6. Dry the organic layer with $MgSO_4$, filter and distil the solution at atmospheric pressure to provide (**3**).
7. Characterize your product (^{1}H NMR, IR, bp).

Step 4: (4*S*)-3-[(2*RS*)-2-Methylpent-4-enoyl]-4-(propan-2-yl)-1,3-oxazolidin-2-one (4)

N.B. *Do not* exceed the scale in this procedure! Ensure all equipment is dry when using *n*-BuLi!

1. In a dry round-bottom flask under nitrogen, fitted with a septum cap and balloon, add diisopropylamine (1.2 equiv.) followed by dry THF (5 mL, 0.6 M).
2. Cool the solution to −78 °C with a dry ice/acetone bath and then add dropwise, using a syringe, *n*-BuLi (2.5 M in hexane, 1.2 equiv.).
3. Stir the solution at −78 °C for 5 min, then add *via* a syringe a solution of oxazolidinone (2) (0.5 g, 2.7 mmol, 1 equiv.) dissolved in dry THF (to produce a 0.6 M solution).
4. Stir the solution for 30 min at −78 °C and then add allyl iodide (3) (3 equiv.) with a syringe.
5. Stir the solution for a further 30 min then warm to room temperature.
6. Pour the reaction onto aqueous saturated NH_4Cl (15 mL) and extract with Et_2O (3 × 10 mL).
7. Wash the organic layer with brine (15 mL), dry with $MgSO_4$, filter and evaporate.
8. Purify the crude product by column chromatography, eluting with a *n*-hexane/EtOAc gradient to provide (**4**).
9. Characterize your product (^{1}H NMR, IR).

Waste Disposal

All neutralized aqueous waste produced in each step can be diluted and discarded in the sink. All non-chlorinated waste from each step should be collected together in non-chlorinated waste collectors. Likewise, all halogenated waste should be collected together in the halogenated waste collectors.

Results Interpretation and Additional Questions

You will be assessed by means of a written report. The requested format of your report is designed to provide you with training for writing up experimental data in a publishable form and is the format required for students that undertake a fourth-year research project. As a result we expect to receive precise and clearly laid out write-ups consisting of the following:

- Fully written experimental procedures for *all* the synthetic steps, consisting of fully recorded physical (including comparison to literature data) and spectral data for the products presented in the style specified by the Royal Society of Chemistry.
- Your original spectra (*excluding* expansions used to assign coupling constants), as PDF documents.
- The answers to the following questions:

  1. The Evans oxazolidinone you are synthesizing is a classic chiral auxiliary used in asymmetric synthesis. Give two examples of alternate chiral auxiliaries that could be utilized in this reaction, explaining how asymmetric induction is induced in the reaction.
  2. Based on the NMR spectrum of the final product (**4**), estimate the diastereomeric ratio of the asymmetric alkylation.
  3. Give a mechanistic explanation for the origin of diastereoselectivity in the asymmetric alkylation reaction.
  4. Whilst chiral auxiliaries are extremely effective in asymmetric synthesis, they are costly. Give, by way of two examples, alternative syntheses to asymmetrically introduce an allyl group α- to a carbonyl group.

References

1. T. E. Smith, D. P. Richardson and G. A. Truran, *J. Chem. Educ.*, 2008, **85**, 695.
2. W. G. McBride, *Lancet*, 1961, **2**, 1358.
3. J. H. Kim and A. R. Scialli, *Toxicol. Sci.*, 2011, **122**, 1–6.
4. K. Zwingenberger and S. Wnendt, *J. Inflammation*, 1995, **46**, 177–211.
5. W. Lenz, *Teratology*, 1988, **38**, 203–215.
6. D. J. Ager, I. Prakash and D. R. Schaad, *Chem. Rev.*, 1996, **96**, 835–875.
7. P. I. Dalko and L. Moisan, *Angew. Chem., Int. Ed.*, 2004, **43**, 5138–5175.
8. D. A. Evans, J. Bartroli and T. L. Shih, *J. Am. Chem. Soc.*, 1981, **103**, 2127–2129.
9. D. A. Evans, M. D. Ennis and D. J. Mathre, *J. Am. Chem. Soc.*, 1982, **104**, 1737–1739.
10. T. Tao and R. J. Parry, *Org. Lett.*, 2001, **3**, 3045–3047.
11. D. A. Evans, K. T. Chapman and J. Bisaha, *J. Am. Chem. Soc.*, 1984, **106**, 4261–4263.
12. C. Gaulon, R. Dhal, T. Chapin, V. Maisonneuve and G. Dujardin, *J. Org. Chem.*, 2004, **69**, 4192–4202.
13. J. Dambacher and M. Bergdahl, *Org. Lett.*, 2003, **5**, 3539–3541.

3. Substitution at Non-aryl sp^2 Carbon

3.1. C-X Bond Formation

3.1.1. Synthesis of Paracetamol by Acetylation

Catarina Dias[a], Emília Valente[a], and Luís Constantino*[a,b]

[a]Faculty of Pharmacy, Universidade de Lisboa, Av. Prof. Gama Pinto, 1649-003 Lisboa, Portugal; [b]Research Institute for Medicines (iMed.ULisboa), Faculty of Pharmacy, Universidade de Lisboa, Av. Prof. Gama Pinto, 1649-003 Lisboa, Portugal. *E-mail: constant@ff.ulisboa.pt

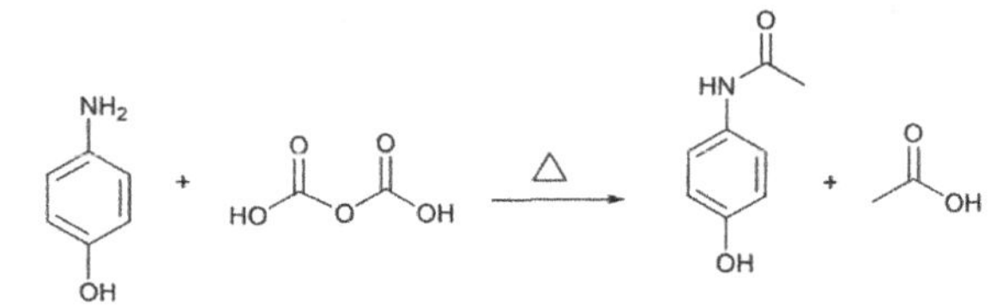

4-Aminophenol Acetic anhydride Paracetamol Acetic acid

Number of sessions (duration of each session)	Hazard level	Difficulty level	Level of study
1 (3–4 h)	Low	Low	Introductory

Class names Amines, phenols, carboxylic anhydrides, amides

Concepts involved This experiment uses the synthesis of paracetamol (acetaminophen; *N*-(4-hydroxyphenyl) ethanamide; *N*-(4-hydroxyphenyl)acetamide) to illustrate an acetylation reaction and an amide synthesis

Chemicals needed Acetic anhydride (ethanoic anhydride), 4-aminophenol, paracetamol (acetaminophen); dichloromethane, acetone, ammonia; KBr for IR spectra and DMSO-d_6 for NMR spectra

Equipment and experimental techniques involved Heating plate with magnetic stirring, water bath, ice cold water bath, vacuum filtration system, heating universal oven, pre-coated TLC silica plates with 254 nm fluorescence indicator, UV lamp for TLC, melting point apparatus, infrared (IR) apparatus, ^{1}H NMR (300 MHz) and ^{13}C NMR (75 MHz) apparatus

Keywords Acetylation, 4-aminophenol, recrystallization, infrared (IR) spectroscopy, paracetamol

Background

Paracetamol (*N*-(4-hydroxyphenyl)ethanamide), also known as acetaminophen in the USA and Japan, is a widely used over-the-counter non-salicylate and non-narcotic analgesic and antipyretic. It is commonly used for the relief of headaches and other minor aches and pains, and it is a major ingredient in numerous cold and flu remedies. It was first used in medicine in 1893 but it only became popular in the second half of the last century. Curiously, paracetamol is the active metabolite of its predecessors (acetanilide and phenacetin) in the same therapeutic class and replaced these compounds in current therapy. It can be administered safely at the

Comprehensive Organic Chemistry Experiments for the Laboratory Classroom
Edited by Carlos A. M. Afonso, Nuno R. Candeias, Dulce Pereira Simão, Alexandre F. Trindade, Jaime A. S. Coelho, Bin Tan, and Robert Franzén.

Published by the Royal Society of Chemistry, www.rsc.org

appropriate therapeutic doses, as it is well tolerated and has a low incidence of gastrointestinal side effects. It is available without a prescription and is used as a common house-hold analgesic. Its low level of side effects make it favourable for child cold and flu and to arthritis patients. Although paracetamol is used to treat inflammatory pain, it is not generally classified as an NSAID because it exhibits only weak inhibition of cyclooxygenase and thus anti-inflammatory activity. The drug is readily metabolised by the liver and toxic metabolites are removed by conjugation with endogenous glutathione. It should be stressed that paracetamol is significantly more toxic than other pain relievers in overdose situations as overuse and acute overdosage can cause severe hepatic and kidney damage that can be fatal if not treated urgently with antidotes that replenish depleted glutathione.[1–4]

The aim of the present experiment is to illustrate the synthesis of this analgesic and antipyretic amide by the acetylation of 4-aminophenol with acetic anhydride.[5] The reaction mechanism is a nucleophilic addition–elimination in which the nucleophilic amino group of the substrate 4-aminophenol attacks the carbonyl carbon of acetic anhydride resulting in formation of the amide product, paracetamol. The substrate has two nucleophile groups (NH_2 and OH) but only the NH_2 group reacts under the conditions of the reaction as the nitrogen lone pair of electrons is more nucleophilic than that of the phenol group.[6]

The reaction is easily performed in a boiling water bath in a fume hood, followed by filtration of the product and single solvent recrystallization from water. The product is then assayed for purity by TLC analysis and melting point determination. The purity of the product may also be evaluated by IR, ^{1}H NMR and ^{13}C NMR spectra, depending on the level of students' understanding of the analysis and interpretation of such spectra. The results are then discussed and compared with literature values.[7,8]

Additional Safety

Besides good laboratory practices, namely wearing a laboratory coat with long sleeves, gloves and eye protection, good ventilation will be necessary. Acetic anhydride is a highly corrosive chemical and reacts with water to form acetic acid. Avoid contact with skin and eyes; breathing can irritate the throat. The addition of the acetic anhydride and the acetylation reaction should be carried out in a fume hood.

Experimental Procedure (3–4 h)

Synthesis of Paracetamol

1. Place 2.75 g (25.2 mmol) of 4-aminophenol and 7.5 mL of distilled water in a 100 mL Erlenmeyer flask containing a magnetic stirring bar. Stir briskly in order to suspend the solid in the water.
2. In a fume hood slowly add 3 mL (31.8 mmol) of acetic anhydride to the suspension. Acetic anhydride is corrosive: handle with care and wipe up any spills.
3. Always in the fume hood, place the reaction mixture in a boiling water bath over a heating plate with magnetic stirrer for 10 min. A precipitate will form.

Product Isolation

4. Cool the reaction mixture and place it in an ice bath.
5. Filter off the solid in a vacuum filtration system. Thoroughly wash the precipitate with ice-cold distilled water.
6. Proceed to the quantitative recovery of the solid to an Erlenmeyer.

Recrystallization by Temperature Gradient

7. Completely dissolve the crude product in the Erlenmeyer using the minimum volume of distilled water at about 80–90 °C. Add 10% more if you need to perform a hot filtration to eliminate insoluble material. Note the volume of solvent that was used.
8. Allow slow cooling until room temperature (pay attention to the appearance of the first crystals); after the flask reaches room temperature, it can be placed on an iced bath.
9. Gently rub the inside surface of the crystallization vessel with a glass rod to induce crystallization if no crystals were formed up to this point.
10. Collect the crystallized product by vacuum filtration, and wash it well with cold water.
11. Dry the crystallized product by pressing it between filter papers and desiccate it in a drying oven and/or in a desiccator until constant weight.
12. Register the weight of the dry crystallized product (note in your laboratory notebook all the weights registered) and calculate the yield.

Physical Characterization/Identification of the Crystalized Paracetamol

Thin-Layer Chromatography Characterization

13. In a test tube dissolve a few crystals of the synthesized paracetamol in acetone (or other organic solvent that dissolves paracetamol).
14. Touch the surface of the solution with the end of a fine capillary tube. The solution will rise up by capillary action.
15. Merely touch the chromatographic sheet with the sample at *ca.* 1 cm from the bottom edge of the plate. Let it dry and if necessary repeat the application to concentrate the compound in a small spot (you can control the application point and the intensity of the spot by seeing the plate under UV_{254} lamp).
16. Following the same procedure, apply the acetone solutions of 4-aminophenol and standard paracetamol.
17. The separation is carried out in a closed chamber by contacting the bottom edge of the layer with the mobile phase dichloromethane–acetone–ammonia (6 : 4 : 0.2).
18. After development of the chromatogram and evaporation of the mobile phase, TLC layers are observed under UV light at 254 nm. The spots are marked and the purity of the synthesized paracetamol is determined by visual evaluation.
19. Determine the R_fs and note the values in a table in your laboratory notebook.
20. Analyse and discuss your TLC results.

Melting Point Determination

21. Measure the melting point (mp) of paracetamol. Compare it with literature values and discuss the purity of the product.

IR and ^{1}H and ^{13}C NMR Spectra

22. Record the IR spectra of paracetamol and 4-aminophenol in a KBr pellet (*ca.* 1 mg product in 100 mg KBr). Analyse and differentiate the absorption bands.
23. If possible, record the ^{1}H and ^{13}C NMR spectra of 4-aminophenol and paracetamol: Place a small amount of crystals (5–10 mg) in a NMR tube, add 600 μL of deuterated dimethyl sulfoxide (DMSO-d_6) and trace the proton and the carbon NMR spectra with the NMR apparatus. Analyse the signals, compare with literature data and discuss your results.

Waste Disposal

The aqueous acid filtrate from the reaction mixture must be put in an acid aqueous solutions container (or carefully neutralized in the fume hood) to be recovered by a waste disposal service.

Paracetamol crystals should be collected to a container labelled "Paracetamol synthesized and purified by students". If not needed for other sessions, this solid must also be dealt through an integrated waste management system.

All the material used must be washed with water and put in an appropriate container to be thoroughly washed.

Results Interpretation and Additional Questions

1. Draw the mechanism of the reaction performed.
2. Yield calculation:
   - Determine the limiting reagent.
   - Determine the expected theoretical mass taking into account the mass and purity of the limiting reagent.
   - Obtain the percent yield of paracetamol in your experiment.
3. Estimate the mass of paracetamol that remained in the cold mother liquor of the crystallization.
4. What is the side product that may be formed along with the main product?
   - Refer to the conditions in which it can be formed.
   - Write the reaction mechanism.
5. When eluting the TLC plate, why must the spots be above the level of the development solvent?
6. Compare and interpret the principal absorption bands in the IR spectra of reagent and product. How can you ensure that no secondary product was formed?
7. Compare and interpret the ^{1}H and ^{13}C NMR signals of reagent and product. What are the main differences?

References

1. Martindale, in *The Complete Drug Reference*, ed. S. C. Sweetman, Pharmaceutical Press, London, 37th edn, 2011, vol. A, p. 112.
2. *Goodman & Gilman's Pharmacological Basis of Therapeutics*, ed. L. L. Brunton, McGraw-Hill, Lisbon, 12th edn, 2011, p. 982.
3. G. G. Graham, M. J. Davies, R. O. Day, A. Mohamudally and K. F. Scott, *Inflammopharmacology*, 2013, **21**, 201.
4. J. A. Hinson, D. W. Roberts and L. P. James, *Handbook of Experimental Pharmacology, Adverse Drug reactions*, Springer, Berlin, 2010, ch. 12, pp. 369–405.
5. T. W. G. Solomons and C. B. Fryhle, *Organic Chemistry*, John Wiley & Sons, Inc., New York, 9th edn, 2008, ch. 18, pp. 778–839.
6. A. I. Vogel, *Vogel's Textbook of Practical Organic Chemistry*, Longman Scientific & Technical, New York, 5th edn, 1989, Experiment 6, vol. 109, p. 985.
7. *Clarke's Analysis of Drugs and Poisons*, ed. A. C. Moffat, M. D. Osselton and B. Widdop, Pharmaceutical Press, London, 4th edn, 2011, volume 2, pp. 1856–1858.
8. The Merck Index, *An Encyclopaedia of Chemicals, Drugs, and Biologicals*, Merck and Co., Whitehouse Station NJ, 14th edn, 2006, p. 9 and p. 78.

3.1.2. Synthesis and Characterization of *N,N′*-Dicyclohexyl-*N,N′*-dimethyl-propan-1,3-diamide

Ana Paula Paiva*[a] and Osvaldo Ortet[a,b]

[a]Centro de Química e Bioquímica, Departamento de Química e Bioquímica, Faculdade de Ciências, Universidade de Lisboa, Rua Ernesto de Vasconcelos, C8, Campo Grande, 1749-016 Lisbon, Portugal; [b]Departamento de Ciência e Tecnologia, Universidade de Cabo Verde, 379C, Praia, Santiago Island, Cape Verde. *E-mail: appaiva@fc.ul.pt

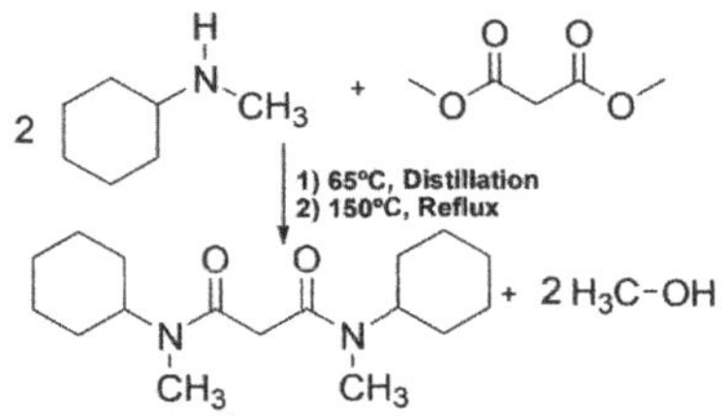

Number of sessions (duration of each session)	Hazard level	Difficulty level	Level of study
1 (4 h)	Moderate	Low	Introductory

Class names Amines, esters, alcohols, amides

Concepts involved This experiment describes a method for preparing a tertiary 1,3-diamide by an addition–elimination reaction of a secondary amine to a diester. The reaction is carried out by reflux, including distillation of an alcohol (which forms as a secondary product)

Chemicals needed Dimethyl malonate, *N*-cyclohexylmethylamine, diethyl ether, ethyl acetate

Equipment and experimental techniques involved Heating-magnetic stirring plate, oil bath, apparatus for simple distillation, apparatus for reflux, glass material for recrystallization

Keywords Addition–elimination reaction, chemistry of carboxylic acid derivatives, crystallization, 1,3-diamide, NMR

Background

Amides are a significant family of organic compounds. Accordingly, several natural materials contain the amide functional group: for instance, amide linkages are the molecular basis of the peptide bonds occurring in proteins and enzymes, and appear frequently in several drugs (*e.g.*, penicillin and LSD).[1,2] The technological relevance of amides should not be forgotten as well, with emphasis in plastic and rubber manufacture, and *nylons* are only an example of an everyday material relying on amide linkages.[2] In the hydrometallurgical area, amide derivatives have been pointed out as efficient extractants to selectively recover basic[3] and precious metals[4] from aqueous chloride media; furthermore, some specific 1,3-diamides (commonly named malonamides) are the basis of the so-called DIAMEX process, aiming at the recovery of lanthanides and actinides from spent radioactive fuels.[5,6]

Comprehensive Organic Chemistry Experiments for the Laboratory Classroom
Edited by Carlos A. M. Afonso, Nuno R. Candeias, Dulce Pereira Simão, Alexandre F. Trindade, Jaime A. S. Coelho, Bin Tan, and Robert Franzén.

Published by the Royal Society of Chemistry, www.rsc.org

In addition to amides, other main families of organic compounds deriving from carboxylic acids are acyl chlorides, organic anhydrides, and esters. When envisaging the mutual transformation of all these functional groups, it should be taken into consideration that acyl chlorides are the most reactive, with amides the lesser ones (anhydrides, acids, and esters are in the middle). Therefore, the forward reaction transforming any of the other compounds in amides is generally favoured.[1] The most convenient synthetic pathway for the production of a given amide depends largely on the nature of the reactants' substituents, but the use of acyl chlorides often provides better yields. The handling of acyl chlorides is a constraint, however, as they are always hazardous and toxic materials.

The present experiment consists of the synthesis of the tertiary *N,N′*-dicyclohexyl-*N,N′*-dimethyl-propan-1,3-diamide from a diester. *N,N′*-Dicyclohexyl-*N,N′*-dimethyl-propan-1,3-diamide can be conveniently synthesized by the addition–elimination reaction of dimethyl malonate with *N*-cyclohexylmethylamine, in an acceptable yield, and through this way the handling of the corresponding acyl chloride is avoided. To help drive the equilibrium to the formation of the diamide in a faster way, the methanol formed is simultaneously distilled; therefore, the total time needed to complete the synthesis is 2 h maximum. The rather different frequency numbers assigned to the stretching vibrations of the ester and amide carbonyl groups in an FTIR spectrum can pertinently be used to conclude about the possible completeness of the reaction in a shorter period of time.[5]

As characterization techniques, determination of the melting point and a plot of the FTIR and ^{1}H NMR spectra of this 1,3-diamide derivative are proposed. The ^{1}H and ^{13}C NMR spectra of *N,N′*-dicyclohexyl-*N,N′*-dimethyl-propan-1,3-diamide are more complex than expected, a phenomenon that typically occurs with some tertiary amides; accordingly, a multiplication of the proton signals is visible, corresponding to all possible conformations the molecule can adopt due to some restrictions of rotation along the C–N bonds.[7–9]

N,N′-Dicyclohexyl-*N,N′*-dimethyl-propan-1,3-diamide shows an efficient and selective behaviour for the recovery of iron(III) from aqueous hydrochloric acid solutions,[3] as well as for lanthanides from spent radioactive fuels.[5]

Additional Safety

The synthesis and purification of the final product should be carried out in a fume hood, with proper safety equipment, such as protective clothing, gloves and safety goggles. Dimethyl malonate is irritating to the eyes, respiratory system and skin; *N*-cyclohexylmethylamine is flammable, toxic and corrosive, causing severe burns; diethyl ether may form explosive peroxides and it is extremely flammable and harmful; ethyl acetate is flammable and irritant. Both diethyl ether and ethyl acetate may cause skin dryness and cracking. A first wash of the glass materials in the fume hood, after contact with the chemicals, is advised.

Experimental Procedure

Laboratory Session (4 h): Synthesis, Recrystallization and Characterization of *N,N′*-Dicyclohexyl-*N,N′*-dimethyl-propan-1,3-diamide

1. In a fume hood, heat an oil bath to maintain a temperature of 75–80 °C on a magnetic stirring hot plate. Mount an apparatus for a simple distillation.
2. Prepare a 50 mL two-neck round-bottomed flask with a magnetic stir bar. Add 9.9 mL (8.5 g, 75 mmol) of *N*-cyclohexylmethylamine and 4.2 mL of dimethyl malonate (4.9 g, 37 mmol). Stir gently. Insert a thermometer (0–200 °C) in one of the necks.
3. Place the round-bottomed flask into the preheated oil bath and include it in the distillation apparatus, with gentle magnetic stirring. A liquid will start to distil, in the range ~50–65 °C.
4. When distillation ceases, remove the distillation apparatus, and insert a reflux condenser. Heat the reaction mixture at a constant and controlled temperature of 150 °C, under stirring, for a maximum of 1 h 30 min. The extent of the reaction can be monitored by FTIR,

if desired, as the stretching vibrations of the carbonyl groups of the ester (reagent) and amide (product) are easily distinguishable, allowing a possible shorter time for the reflux.

5. Pour the yellowish reaction mixture into a beaker, and allow its slow cooling to ambient temperature. When put in an ice bath, crystallization of the crude *N,N′*-dicyclohexyl-*N,N′*-dimethyl-propan-1,3-diamide, as a white solid, can be accelerated by rubbing the walls of the flask with a glass rod.
6. When crystallization is complete, filter the mixture through a Büchner funnel, and wash the solid with diethyl ether. Let the product dry, and register the mass of the synthesized *N,N′*-dicyclohexyl-*N,N′*-dimethyl-propan-1,3-diamide.
7. Recrystallize the crude *N,N′*-dicyclohexyl-*N,N′*-dimethyl-propan-1,3-diamide from ethyl acetate.
8. Record the mass obtained. Determine the melting point of the product.
9. Record the FTIR (KBr pellet) and the ^1H NMR spectrum (in $CDCl_3$) of *N,N′*-dicyclohexyl-*N,N′*-dimethyl-propan-1,3-diamide.

Waste Disposal

Diethyl ether and ethyl acetate should be collected in a non-chlorinated waste collector recipient, for posterior destruction.

Results Interpretation and Additional Questions

1. Bearing in mind the mechanism of formation for *N,N′*-dicyclohexyl-*N,N′*-dimethyl-propan-1,3-diamide, explain the pertinence of introducing a distillation step before the reflux period.
2. Calculate the yield before and after recrystallization, and compare the value of the melting point of the purified product with those reported in literature.
3. Systematize the most important characteristics ethyl acetate should have to be a suitable solvent for the recrystallization of *N,N′*-dicyclohexyl-*N,N′*-dimethyl-propan-1,3-diamide.
4. Interpret and assign the most significant bands appearing in the FTIR spectrum of the product.
5. Interpret and assign the peaks of the ^1H NMR spectrum.

Acknowledgements

Reprinted (adapted) from *J. Chem. Soc., Dalton Trans.*, 1997, pp 649–660 with permission from the Royal Society of Chemistry.

References

1. Any organic chemistry textbook, for instance: T. W. G. Solomons, G. B. Fryhle and S. A. Snyder, Organic Chemistry, John Wiley & Sons, Inc., Hoboken, 11th edn, 2014, ch. 17, pp. 771–820.
2. A. Leggio, E. L. Belsito, M. L. Di Gioia, V. Leotta, E. Romio, C. Siciliano and A. Liguori, *Tetrahedron Lett.*, 2015, **56**, 199.
3. M. C. Costa, I. Pêczek, Z. Sadowski, S. Natu and A. P. Paiva, *Solvent Extr. Ion Exch.*, 2007, **25**, 463.
4. R. A. Grant, European Patent 0 210 004, Matthey Rustenberg Refiners Ltd., 1990.
5. G. Y. S. Chan, M. G. B. Drew, M. J. Hudson, P. B. Iveson, J.-O. Liljenzin, M. Skalberg, L. Spjuth and C. Madic, *J. Chem. Soc., Dalton Trans.*, 1997, 649.
6. A. P. Paiva and P. Malik, *J. Radioanal. Nucl. Chem.*, 2004, **261**, 485.
7. A. H. Lewin, *Tetrahedron Lett.*, 1971, **12**, 3583.
8. J. Á. Bisceglia, M. C. Mollo and L. R. Orelli, *J. Mol. Struct.*, 2012, **1026**, 65.
9. L. Lefrançois, M. Hébrant, C. Tondre, J.-J. Delpuech, C. Berthon and C. Madic, *J. Chem. Soc., Perkin Trans. 2*, 1999, 1149.

3.1.3. Synthesis and Characterisation of an Ester from 4-Nitrobenzoyl Chloride

Iain A. Smellie*, Matt L. Clarke, Leanne Harris, and Andrew. J. Miller

School of Chemistry, University of St Andrews, North Haugh, St Andrews, Fife, KY16 9ST, UK
*E-mail: ias10@st-andrews.ac.uk

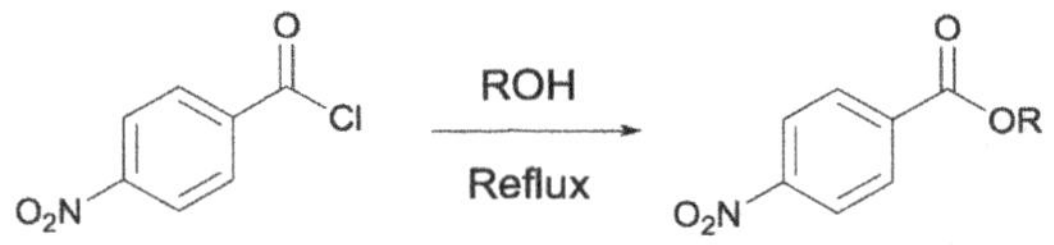

R = Et, Pr, ^{i}Pr

Number of sessions (duration of each session)	Hazard level	Difficulty level	Level of study
2 (3 h each)	Moderate	Low	Introductory

Class names Acyl chlorides, esters

Concepts involved This experiment describes a relatively simple method for aryl ester formation from the corresponding acid chloride and alcohol. In this case the alcohol functions as the nucleophile and the reaction solvent

Chemicals needed Deuterochloroform, diethyl ether, ethanol, magnesium sulfate, 4-nitrobenzoyl chloride, 1-propanol, 2-propanol, sodium hydrogen carbonate

Equipment and experimental techniques involved Heating mantle or hotplate, glassware for reflux assembly, liquid–liquid extraction, vacuum filtration, recrystallization

Keywords Acyl chloride, esterification, liquid–liquid extraction, NMR, recrystallisation

Background

The ester functional group (RCO_2R) is found in many important industrial/pharmaceutical compounds and is also frequently encountered in Nature. Examples include fats and oils[1] and flavour and aroma related compounds[2] (such as benzyl acetate in jasmine). In recent years esters have been the focus of much attention as biofuels;[3] "biodiesels" (such as methyl (*Z*,*Z*)-9,12-octadecadienoate) have been proposed as renewable alternatives to hydrocarbon fuels derived from crude oil.

Esters can be prepared in many ways, perhaps the most familiar being Fischer esterification of acids with alcohols.[4] In many respects addition of alcohols to acid chlorides or anhydrides is more convenient. Indeed, this contributes to the wide use of esters as protecting groups.[5] Since acid chlorides contain a good leaving group (Cl^-), the reaction with alcohols usually proceeds under mild conditions. Although such reactions can be accelerated further by the addition of a tertiary amine base[6] (pyridine derivatives being most common), base additives are not necessary in the experiment reported here.

Comprehensive Organic Chemistry Experiments for the Laboratory Classroom
Edited by Carlos A. M. Afonso, Nuno R. Candeias, Dulce Pereira Simão, Alexandre F. Trindade, Jaime A. S. Coelho, Bin Tan, and Robert Franzén.

Published by the Royal Society of Chemistry, www.rsc.org

Esterification *via* nucleophilic acyl substitution is usually considered as core material within foundation organic chemistry courses. The concepts covered in this experiment are directly relevant to material covered in typical foundation chemistry lecture courses.

The aim of the experiment is to prepare an alkyl 4-nitrobenzoate from 4-nitrobenzoyl chloride using the appropriate alcohol as the reaction solvent. Crude product is obtained after less than 1 h of heating under reflux, the product is obtained after aqueous work-up, and further purification by recrystallization can also be achieved.

Additional Safety

All operations should be performed in a fume hood with appropriate protective clothing (laboratory coat, safety goggles and gloves). 4-Nitrobenzoyl chloride is corrosive and irritating to eyes and skin on contact. Because corrosive HCl gas is formed as the reaction progresses, it is essential to keep the reaction in a fume hood and the acid must be neutralized (as instructed in the main text) with sodium bicarbonate solution. Deuterochloroform is potentially carcinogenic. Ethanol, 1-propanol and 2-propanol are volatile and flammable reagents.

Experimental Procedure

Laboratory Session 1 (3 h): Synthesis of Esters

1. Transfer 1.75 g of 4-nitrobenzoyl chloride to a single-necked 100 mL round-bottomed flask.
2. Add 30 mL of ethanol, 1-propanol or 2-propanol to the acid chloride and two or three anti-bumping granules. Place the flask in a heating mantle, secure it in place with a clamp and equip with a reflux condenser.
3. Heat the solution of acid chloride and alcohol to reflux for 50 min.
4. After the heating period is complete, switch off the heat source and allow the flask to cool to room temperature (It is possible that some of the product may precipitate out at this point.)
5. Add 40 mL of diethyl ether to the reaction flask and transfer the solution to a separating funnel (It may be necessary to rinse the reaction flask with some ether and transfer the washings into the separating funnel.)
6. Add 50 mL of aqueous sodium hydrogen carbonate solution to the separating funnel and shake the funnel gently (Care! hold the stopper firmly and frequently release the pressure build-up in the funnel.) Remove the aqueous layer and discard it.
7. Add 50 mL of water to the separating funnel and wash the organic layer in the same manner as in step 6. Discard the aqueous layer, run off the organic layer into a 100 mL conical flask and dry over anhydrous magnesium sulfate.
8. Filter the ether solution through a small cotton wool plug directly into a clean and dry round-bottomed flask. Wash the cotton wool with ether and press dry with a glass rod.
9. Remove the solvent using a rotary evaporator.

Laboratory Session 2 (3 h): Purification of Product, Yield Determination and Characterization

Purification of ethyl 4-nitro benzoate: Transfer the crude product to a clean, 100 mL conical flask and recrystallize from aqueous ethanol (*ca.* 1 : 1 ethanol : water v/v). When crystallization seems complete, filter off the product under suction using a Büchner funnel and wash with a minimum amount of ice-cold ethanol.

Purification of isopropyl 4-nitro benzoate: Transfer the crude product to a clean, 100 mL conical flask and recrystallize from ethanol. When crystallization seems complete, filter off the product under suction using a Büchner funnel and wash with a minimum amount of ice-cold ethanol.

Purification of propyl 4-nitro benzoate: this product should be sufficiently pure after the solvent is completely removed *in vacuo*.

Weigh the product and determine the % yield. Measure the melting point and record a ^{1}H NMR ($CDCl_3$) spectrum of your product.

Waste Disposal

Aqueous washings from the extraction steps should be diluted with water and disposed of down a sink. All organic solvent waste should be retained and transferred to an appropriate container for disposal.

Results Interpretation and Additional Questions

1. Interpret the ^{1}H NMR spectrum you obtained. Use the NMR spectrum and melting point to confirm the structure of your compound unambiguously and explain why the spectrum fits the proposed structure.
2. Provide a curly arrow mechanism for the reaction you carried out.
3. If your reaction is run in tetrahydrofuran as the solvent in the presence of triethylamine, an insoluble white precipitate appears in addition to the soluble product. This precipitate does not dissolve in most organic solvents, but is soluble in water. Using your mechanistic explanation, what is the identity of this precipitate?

References

1. T. P. Coultate, *Food: The Chemistry of Its Components*, RSC Publishing, Cambridge, 4th edn, 2002, ch. 4, pp. 101–124.
2. A. Fortineau, *J. Chem. Educ.*, 2004, **81**, 45.
3. A. G. King and M. W. Wright, *J. Chem. Educ.*, 2007, **84**, 202.
4. L. F. Fieser and K. L. Williamson, *Organic Experiments*, D. C. Heath and Company, Lexington MA, 7th edn, 1992, ch. 27, pp. 275–282.
5. T. W. Greene and P. G. M. Wuts, *Protective Groups in Organic Synthesis*, John Wiley & Sons, Inc., New York, 3rd edn, 1999, ch. 2, pp. 149–178.
6. R. Murugan and E. F. V. Scriven, *Aldrichim. Acta*, 2003, **36**, 21.

3.1.4. Green Esterification: The Synthesis of Aromas in the Presence of an Acid Resin

José E. Castanheiro*, António P. S. Teixeira*, and António M. D. R. L. Pereira

Departamento de Química, Escola de Ciências e Tecnologia, Centro de Química de Évora, Instituto de Investigação e Formação Avançada, Universidade de Évora, Rua Romão Ramalho, no. 59, 7000-671 Évora, Portugal. *E-mail: jefc@uevora.pt, apsteix@uevora.pt

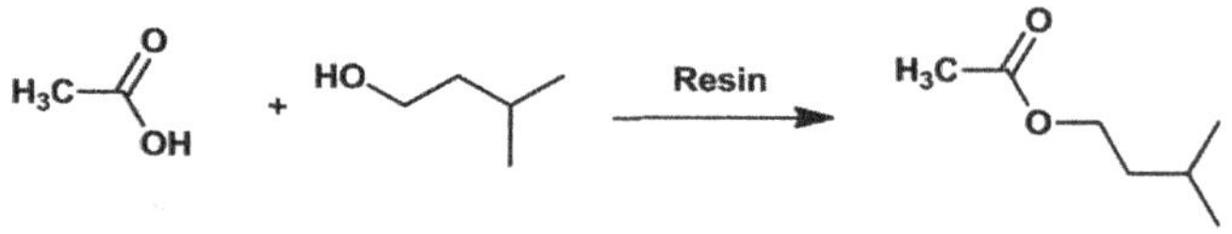

Number of sessions (duration of each session)	Hazard level	Difficulty level	Level of study
3 (3 h each)	Moderate	Low	Introductory

Class names Carboxylic acids, alcohols, esters

Concepts involved Synthesis of isoamyl acetate by heterogeneous catalysis, using a commercial acid resin; Fischer esterification reaction, without a strong inorganic acid; IR and NMR spectroscopy; GC analysis; catalytic activity and stability of resin

Chemicals needed Acetic acid, isoamyl alcohol, diethyl ether, commercial resin (Dowex D50W2), sodium bicarbonate solution (10%), iron(II) sulfate solution (10%), anhydrous magnesium sulfate

Equipment and experimental techniques involved Apparatus for reaction requiring temperature control (80 °C) and magnetic stirring, liquid–liquid extraction apparatus, distillation apparatus, solid–liquid filtration GC equipment, structural analysis (IR and NMR spectroscopy)

Keywords Catalytic activity and stability of resin, Fischer's esterification, heterogeneous catalysis, liquid–liquid extraction, spectroscopic characterization (IR, NMR)

Background

Esterification is a reaction between carboxylic acids and alcohols to generate an ester. When it occurs in the presence of acid catalysis it is named Fischer or Fischer–Speier esterification.[1]

The synthesis of aromas by esterification of a carboxylic acid (acetic or butanoic acid) with an alcohol (*e.g.*, methanol, ethanol, pentanol) has been carried out in the presence of strong homogeneous acid catalysts (*e.g.*, sulfuric acid, hydrochloric acid).[2,3] However, these mineral acids are corrosive and they are usually neutralized after reaction (with consequent salt formation); the use of sulfuric acid usually gives a mixture of products. To overcome the above-mentioned problems, different solid acid catalysts, like zeolites,[4] resins[5,6] heteropolyacids,[7,8] Al-MCM-41[9] and acid clays,[10] have been used in esterification reactions as heterogeneous catalysts. The use of heterogeneous catalysts for these reactions is preferable as they can be easily separated from

Comprehensive Organic Chemistry Experiments for the Laboratory Classroom
Edited by Carlos A. M. Afonso, Nuno R. Candeias, Dulce Pereira Simão, Alexandre F. Trindade, Jaime A. S. Coelho, Bin Tan, and Robert Franzén.

Published by the Royal Society of Chemistry, www.rsc.org

the reactants and products by filtration and can be reused after reaction, becoming thereby a more environmental friendly method.[11] In addition, the use of resin forces the reaction to occur at lower temperature than when an acid is used. This decreases the energy consumption, avoids the use of strong and corrosive inorganic acids and reduces the potentially hazardous nature of the method.

In this context, the synthesis is proposed of an aromas' fruit over a solid commercial catalyst. Their separation and spectroscopic characterization (IR and NMR) are also carried out. Moreover, the experiment is suitable for the determination of catalytic activity and stability of the resin, calculated by GC analysis. This activity can be applied as a *green synthesis* experiment in a laboratory classroom.

Additional Safety

All hazardous materials should be handled using safety standard procedures. The sessions should be carried out in a fume hood, due to the strong smell of carboxylic acids and alcohols. The use of proper safety equipment is also recommended, such as protective clothing, gloves and safety goggles.

Experimental Procedure

Session 1 – Preparation of Isoamyl Acetate

1. In a fume hood heat an oil bath to maintain a temperature of 80 °C on a magnetic hot plate.
2. In a 50 mL two-necked flask with a magnetic stir bar, add 6.5 mL (0.06 mol) of isoamyl alcohol, 13.8 mL (0.24 mol) of acetic acid and 0.2 g of acid resin.
3. Insert a water condenser into the two-necked flask, and put it in the oil bath.
4. Stir the reaction mixture for 2.5 h, at 80 °C.
5. Remove a small sample of reaction mixture for GC analysis.
6. Cool the reaction mixture, immersing the two-necked flask in a cold water bath for 5 min.
7. Separate the resin from the reaction mixture by filtration.
8. Pour the mixture into an Erlenmeyer flask (250 mL) containing about 25 g of crushed ice and stir magnetically for 2 min.
9. Transfer the mixture to a separating funnel and wash the Erlenmeyer with diethyl ether (2 × 10 mL).
10. Add more diethyl ether (25 mL) to the separating funnel and stir carefully.
11. Remove the aqueous phase (lower layer) to an Erlenmeyer, add to a separating funnel 25 mL of a solution of iron(II) sulfate (10%) to wash the organic phase and stir carefully.
12. Remove the aqueous phase (lower layer) to an Erlenmeyer, add to separating funnel 2 × 15 mL of a sodium carbonate solution (10%) to wash the organic phase and stir carefully.
13. Remove the aqueous phase (lower layer) to an Erlenmeyer and transfer the organic phase to another Erlenmeyer.
14. Dry the organic phase with magnesium sulfate for 10 min.

Session 2 – Purification of Isoamyl Acetate

1. Filter the organic phase to remove the drying agent;
2. Distil the diethyl ether solution in a rotary evaporator;
3. Distil the product;
4. Record the IR and NMR spectra.

Session 3 – Reuse of Resin

1. In a fume hood, heat an oil bath to maintain a temperature of 80 °C on a magnetic hot plate.
2. In a 50 mL two-necked flask with a magnetic stir bar, add 6.5 mL (0.06 mol) of isoamyl alcohol, 13.8 mL (0.24 mol) of acetic acid and 0.2 g of acid resin used.
3. Insert a water condenser into the two-necked flask, and put it in the oil bath.
4. Stir the reaction mixture for 2.5 h, at 80 °C.
5. Remove a small sample of reaction mixture for GC analysis.
6. Cool the reaction mixture, immersing the two-necked flask in a cold water bath for 5 min.

Waste Disposal

All chemical waste should be disposed of in accordance with local regulations and laboratory practices. If there are no set of rules in the laboratory, do as follows:

- Neutralize the aqueous phases before placing them in the wastewater.
- Place the drying agent in a container for solids.
- Place the collected organic solvent from the rotary evaporator in a container for non-chlorinated compounds.
- Place the resin in a container for later regeneration.

Results Interpretation and Additional Questions

1. Calculate the yield of the reaction, taking into account the limiting reactant.
2. Analyze the IR and NMR spectra and compare them with data already published.[12]
3. Calculate the catalytic activity of the resin.
4. Calculate the catalytic activity of the resin after the second use.
5. Comment the role of iron(II) sulfate solution.
6. What is the purpose of washing the organic phase with sodium carbonate? What gas is formed during the operation?

References

1. E. Fischer and A. Speier, *Ber. Dtsch. Chem. Ges.*, 1895, **28**, 3252.
2. J. Otera and J. Nishikido, in *Esterification. Methods, Reactions, and Applications*, ed. J. Otera and J. Nishikido, Wiley-VCH Verlag GmbH, Darmstadt, 2nd edn, 2010, ch. 1, vol. 5.
3. Z. Zeng, L. Cui, W. Xue, J. Chen and Y. Che, *Chemical Kinetics*, ed. V. Patel, InTech, Rijeka, Croatia, 1st edn, 2012, ch. 12, vol. 255.
4. A. Corma, H. Garcia, S. Iborra and J. Primo, *J. Catal.*, 1989, **120**, 78.
5. A. Charkrabati and M. M. Sharma, *React. Polym.*, 1993, **20**, 1.
6. M. M. Sharma, *React. Funct. Polym.*, 1995, **26**, 3.
7. P. Dupont, J. C. Védrine, E. Paumard, G. Hecquet and F. Lefebvre, *Appl. Catal., A*, 1995, **129**, 217.
8. J. M. Verhoef, J. Patrica Kooyman, A. J. Peters and H. van Bekkum, *Micropor. Mesopor. Mater.*, 1999, **27**, 365.
9. R. Koster, B. van der Linden, E. Poels and A. Bliek, *J. Catal.*, 2001, **204**, 333.
10. J. T. Kloprogge, *J. Porous Mater.*, 1998, **5**, 5.
11. J. Hagen, *Industrial Catalysis: A Practical Approach*, ed. J. Hagen, Wiley-VCH Verlag GmbH, Darmstadt, 2nd edn, 2006, ch. 1, vol. 1.
12. For instance, at the database: http://sdbs.db.aist.go.jp/sdbs/cgi-bin/cre_index.cgi.

3.1.5. Acetylation of Cholesterol and Purification by Column Chromatography

Emília Valente[a], Catarina Dias[a], and Luís Constantino*[a,b]

[a]Faculty of Pharmacy, Universidade de Lisboa, Av. Prof. Gama Pinto, 1649-003 Lisbon, Portugal; [b]Research Institute for Medicines (iMed.ULisboa), Faculty of Pharmacy, Universidade de Lisboa, Av. Prof. Gama Pinto, 1649-003 Lisbon, Portugal. *E-mail: constant@ff.ulisboa.pt

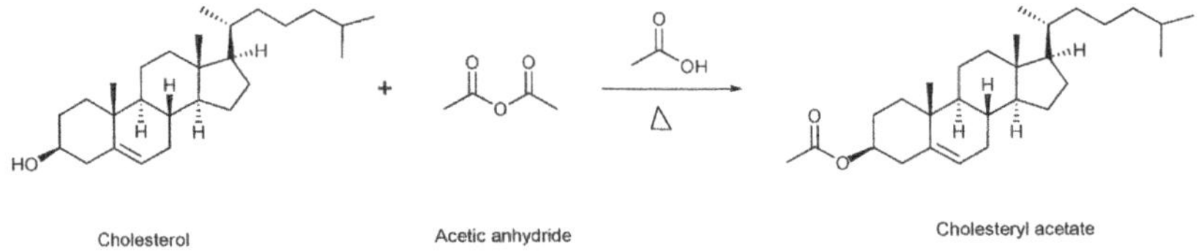

Number of sessions (duration of each session)	Hazard level	Difficulty level	Level of study
1 (4 h)	Moderate	Low	Introductory

Class names Alcohols; esters; anhydrides; carboxylic acids

Concepts involved Ester synthesis under acidic conditions and product purification by column chromatography with thin-layer chromatography control

Chemicals needed Cholesterol, acetic anhydride, glacial acetic acid, sodium hydroxide, methylene chloride, hexane, ethyl acetate, sulfuric acid, ethanol, distilled water

Equipment and experimental techniques involved Erlenmeyer flask, water bath, stirring and heating plates, rotary evaporator system, column for chromatography with silica gel as support phase and mixtures of hexane and ethyl acetate as eluent, thin-layer chromatography (TLC) plates, melting point apparatus

Keywords Acetylation, column chromatography, liquid–liquid extraction, melting point, rotary vacuum evaporation, TLC

Background

This experiment intends to use a simple reaction associated with the purification of the reaction mixture by column chromatography. The aim is to use the acetylation of cholesterol to introduce column chromatography as a potential technique for purification of reaction mixtures.[1,2]

Cholesterol is often treated as a criminal but is in fact a critical component of eukaryotic cell membranes, the precursor to all steroid hormones, a precursor to vitamin D, and the limiting factor that brain cells need to make synapses, making it essential to learning and memory.[3,4] Although cholesterol is clearly essential to life, its deposition in arteries over time causes them to become narrowed and blood flow to the heart to be slowed down or blocked. This deposition is associated with cardiovascular disease and stroke, two leading causes of death in humans.[5] In a healthy organism, an intricate balance is maintained between the biosynthesis, utilization, and transport of cholesterol, keeping its harmful deposition to a minimum.

Comprehensive Organic Chemistry Experiments for the Laboratory Classroom
Edited by Carlos A. M. Afonso, Nuno R. Candeias, Dulce Pereira Simão, Alexandre F. Trindade, Jaime A. S. Coelho, Bin Tan, and Robert Franzén.

Published by the Royal Society of Chemistry, www.rsc.org

Cholesterol occurs in the free form, esterified (cholesterol esters), and in other covalent and non-covalent linkages in animal tissues, including the plasma lipoproteins.[5] In the ester form the hydroxyl of cholesterol is acylated with a long-chain fatty acid.[5,6] Due to the insolubility of the free and the ester forms in the blood, lipoproteins have a fundamental role in their redistribution in the body. Two types of lipoproteins that carry cholesterol (and other compounds) to and from cells are especially important: low-density lipoprotein (LDL) and high-density lipoprotein (HDL).[5]

The so-called "bad" cholesterol, LDL, delivers lipids to the body's cells and is also associated with its deposition at arteries. The so-called "good" cholesterol, HDL cannot deliver cholesterol to cells. Instead, it has the ability to remove excess peripheral cholesterol and return it to the liver for recycling and excretion. This is why instead of focusing only on total cholesterol values in blood the standard cholesterol test also measures LDL cholesterol and HDL cholesterol.

The structure of cholesterol is very important in the fluidity of the animal's cell membrane. The hydroxyl group on cholesterol interacts with the polar head groups of the membrane phospholipids while the bulky tetracyclic ring and the hydrocarbon chain are embedded in the membrane alongside the other lipids non-polar fatty-acid chain.[7]

Depending on the temperature, cholesterol has distinct effects on the fluidity of the eukaryotic cell membrane. At high temperatures, it interferes with the movement of the phospholipid fatty acid chains, decreasing the fluidity of the outer part of the membrane and reduces its permeability to small molecules. At low temperatures, it interferes with interactions between fatty acid chains, and prevents membranes from freezing.[7,8]

Cholesterol acetylation will esterify the cholesterol alcohol function and so a less polar cholesterol ester will be formed. Acetylation of alcohols is an important and routinely utilized transformation in organic chemistry.[9] Acetyl is also one of the most common protecting groups for the hydroxyl group, being stable in acid reaction conditions and easy to remove by mild alkaline hydrolysis. The most commonly used reagent combination for this reaction uses acid anhydride in the presence of acid catalyst.[9,10]

Cholesterol is dissolved in glacial acetic acid and allowed to react with acetic anhydride to form the ester cholesteryl acetate. As the reaction does not go to completion under the conditions proposed, at the end there will be both cholesteryl acetate and unreacted cholesterol.

Separating both compounds by fractional crystallization would be extremely difficult but, as they differ greatly in polarity, they are easily separated by silica gel column chromatography, with the hydroxyl group of cholesterol strongly adsorbed. The extent/yield of the reaction is determined after collecting the easily eluted acetate from the strongly adsorbed alcohol.

The column fractions are controlled for purity by thin-layer chromatography. As both compounds do not absorb visible or UV light, they cannot be detected without first carrying out a chemical reaction. We use an acid solution followed by heating.

Additional Safety

Besides regular laboratory good practices, the students are instructed to work in the fume hood at all times for this experiment.

Experimental Procedure

Acetylation of Cholesterol

1. In a small Erlenmeyer flask cover 0.5 g of cholesterol with 5 mL of glacial acetic acid and swirl. The initial thin slurry soon sets to a stiff paste.
2. Add 1 mL of acetic anhydride.
3. Add a magnetic stirrer and fix the Erlenmeyer to a support so it can be heated in a water bath with magnetic stirring.

4. Heat the mixture in the boiling water bath for *ca.* 30 min. Record the heating period.
5. Cool the reaction mixture.
6. Add 20 mL water and extract with two 25 mL portions of methylene chloride. In all extractions check which are the aqueous layer and the organic layer before going on to the next step.
7. Wash the combined organic extracts twice with 15 mL of distilled water.
8. Wash with 25 mL of 10% sodium hydroxide.
9. Dry the organic extract over anhydrous sodium sulfate in an Erlenmeyer flask.
10. Filter by gravity to a pre-weighed round-bottom flask.
11. Evaporate the solvent under vacuum in a rotary evaporator system.
12. Make a note of the final bulk weight.
13. In a test tube prepare a sample of the product to run later a thin-layer chromatography (TLC) analysis.

Chromatography Column Preparation

While the reaction takes place prepare a 30 cm high and 1.5 cm diameter chromatographic column. Ideally the column will have a porous plaque (or, alternatively, use a bit of wool glass or cotton) and a Teflon stopper at the end (alternatively, use a screw pinch-clamp). Weigh 12.5 g of silica gel as the support phase in a 100 mL beaker. The mobile phase will be a series of solvent mixtures, hexane–ethyl acetate, in different proportions, starting from 9:1 (75 mL), through 1:1 (75 mL) and, if necessary, ethyl acetate by itself (50 mL).

14. Fix the column with a clamp to a support. Close the stopper and fill about 1/4 of the column with the first eluent (hexane–ethyl acetate, 9:1). Make sure a recovery recipient (*e.g.* a beaker), is placed under the column and test if the column stopper is working properly.
15. Make a liquid slurry with the first eluent solvent mixture (hexane–ethyl acetate 9:1). Open the stopper so the solvent starts dropping from the column to the recover recipient and slowly transfer the silica suspension with a glass rod to the column using a wide mouth funnel. Let the solvent drop to the recovery flask, to allow the silica to settle down and the solvent fall to about 4 cm above the silica level. This last operation can be repeated until most of the silica is transferred to the column. Avoid air bubbles during the column preparation.
16. Swirl the silica top to horizontal as much as possible.
17. Cover the silica with sand or cotton to make a protective layer of about 1 cm. Avoid silica dryness.

Sample Preparation and Application

18. Prepare an appropriate test tube support with about 10–12 numbered test tubes (10 mL) and place it under the chromatography column in order to recover the column fractions.
19. Let the solvent drop to *ca.* 1 mm above the protective layer in the column and close stopper.
20. Dissolve your reaction crude product in 3–4 mL of methylene chloride.
21. Apply this sample on top of the column with the aid of a Pasteur pipette.
22. Open the stopper and let the solvent drop until your sample stays at the protective layer level.
23. Close the stopper and apply about 2 mL of the first eluent hexane–ethyl acetate (9:1) with a Pasteur pipette.
24. Let drain until the solvent completely penetrates the protective layer. Repeat this operation twice with 2 mL more of the first solvent mixture.

Column Elution

25. On the top of the chromatography column, place a separating funnel with 75 mL the first solvent mixture (hexane–ethyl acetate, 9 : 1).
26. Slowly let the solvent drop into the column, open the column stopper and allow solvent to drop at the same speed to the recover test tubes. Be careful not to exceed about 8 mL fractions.
27. When the first tube is full simply move the support to the next tube as there is no need to close the column stopper.
28. When the first eluent mixture finishes carry on the elution with the second solvent mixture (75 mL, hexane–ethyl acetate 1 : 1) and finally with 50 mL ethyl acetate if necessary until the two compounds are recovered.
29. Control the composition of the different fractions by thin-layer chromatography (TLC) (see next sub-section).
30. Compare the results with the TLC analysis you made of your reaction crude product and with the TLC analysis for the pure compounds. Select all the fractions with pure cholesteryl acetate.
31. Combine these fractions in a round-bottom flask weighed previously.
32. Evaporate the solvent in a vacuum rotary evaporator until constant weight.
33. Measure your product melting point.
34. Calculate the cholesterol acetylation yield, noting the reaction time and water bath temperature, and compare your result with those of fellow students, noting the different reaction periods.

Thin-Layer Chromatography (TLC)

Chromatographic Development Chamber Preparation

35. Prepare a chromatographic development chamber with the first column eluent as the developing solvent.
36. Place half of a filter paper so that it covers the inside wall of the chamber and allow the solvent to saturate it.
37. In different test tubes prepare standard solutions of cholesterol and cholesteryl acetate in any volatile solvent (*ca.* 5%).
38. Prepare one precoated silica gel plate of a size appropriate to the chamber.
39. Measure 0.5 cm from both ends of the plate in width and 1 cm in height. Softly mark the spots with a pencil.
40. Touch each solution with a different fine capillary tube. The solutions will rise up into the tubing.
41. The chromatographic plate is spotted with the sample (<4 mm) by touching the standard solution-charged tube at the pre-marked spots.
42. The plate must be placed spot end down, upright in the development chamber such as to position the spots above the solvent level.
43. The eluent will climb up the plate until the solvent front is about 1 cm from the upper end of the adsorbent covered region. During this development the system should be enclosed to ensure the atmosphere is saturated with solvent vapour.
44. After development, immediately mark the position to which the solvent front has moved by scratching a line along it.
45. Allow the chromatogram to dry in the fume hood.

46. To visualize the colourless cholesteryl acetate and cholesterol spots it is necessary to spray the chromatogram with a sulfuric acid solution† and place it over a heating plate in the fume hood. Compounds will appear as dark rose spots by oxidation.
47. Measure the migration distance of both compounds from the application point to the middle of the final spot and also the distance of the solvent front from the samples application.
48. Calculate the R_f for both compounds and confirm if the chosen solvent is appropriate. If not, repeat the process with another solvent or solvent mixture.

Chromatography Column Fractions Control

49. Analyse the crude product obtained and the column chromatography fractions by TLC and compare your samples with those of the standards.
50. After identifying which fractions have the pure compounds continue the work described in step 30.

Waste Disposal

Recovery containers must be available for halogenated and non-halogenated organic solvents, and also for aqueous acid and alkaline solutions. The solids (silica gel from the column chromatography and the sodium sulfate drying agent) must be recovered in a special container for special treatment of solids.

Results Interpretation and Additional Questions

1. Draw the mechanism of the acetylation reaction executed.
2. Determine the acetylation reaction yield and compare it with other class results and their reaction temperature and time.
3. Explain the difference between the R_fs of cholesteryl acetate and cholesterol.
4. Explain the reason for a 10% sodium hydroxide solution extraction of the reaction mixture.

Acknowledgements

The authors of this experiment acknowledge Professor Diogo S. Lüdtke and Dr Maria Eduarda Contreira from the Institute of Chemistry, Universidade Federal do Rio Grande do Sul – UFRGS, Brazil, for reproducing and checking this experiment in their own laboratory and for providing data included in the supplementary material.

References

1. K. L. Williamson, *Organic Experiments*, Houghton Mifflin Company, USA, 9th edn, 2004.
2. L. F. Fieser and K. L. Wiliamson, *Organic Experiments*, D. C: Heath and Company Lexington, USA, 6th edn, 1987.
3. I. Björkhem, V. Leoni and S. Meaney, *J. Lipid Res.*, 2010, 512489–512503.
4. F. D. Porter and G. E. Herman, *J. Lipid Res.*, 2011, **52**, 6–34.
5. T. F. Daniels, K. M. Killinger, J. J. Michal, R. W. Wright Jr and Z. Jiang, *Int. J. Biol. Sci.*, 2009, **5**, 474–488.
6. C. Pramfalk, M. Eriksson and P. Paolo Parini, *Eur. J. Lipid Sci. Technol.*, 2012, **114**, 624–633.
7. H. Ohvo-Rekila, B. Ramstedt, P. Leppimaki and J. P. Slotte, *Prog. Lipid Res.*, 2002, **41**, 66–97.
8. F. R. Maxfield and G. van Meer, *Curr. Opin. Cell Biol.*, 2010, **22**, 422–429.
9. G. Sartori, R. Ballini, F. Bigi, G. Bosica, R. Maggi and P. Righi, *Chem. Rev.*, 2004, **104**, 199.
10. A. L. Pearson and W. J. Roush, *Handbook of Reagents for Organic Synthesis: Acetylating Agents and Protecting Groups*, John Wiley & Sons Ltd, Chichester, 1999.

---

†Visualization mixture: sulfuric acid–acetic acid–ethanol (4 : 80 : 16).

3.1.6. Synthesis of Hippuric Acid: An Example of Amide Bond Formation

Ana S. Ressurreição*, Maria de Jesus Perry, Ana Paula Francisco, and Francisca Lopes*

Research Institute for Medicines (iMed.ULisboa), Faculty of Pharmacy, Universidade de Lisboa, Av. Professor Gama Pinto, 1649-003 Lisboa, Portugal. *E-mail: aressurreicao@ff.ulisboa.pt, fclopes@ff.ulisboa.pt

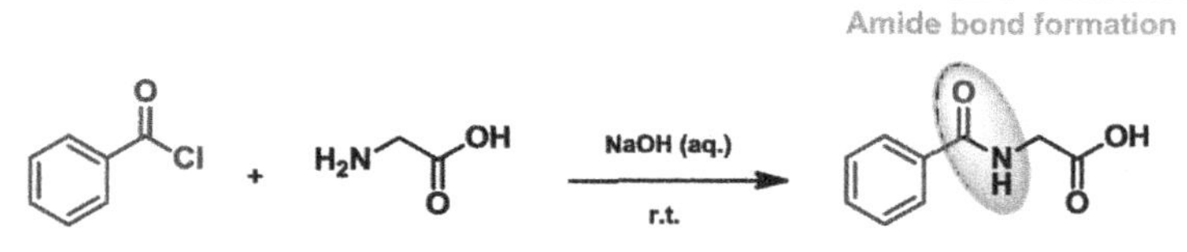

Benzoyl chloride *Glycine* ***Hippuric acid, 1***

Number of sessions (duration of each session)	Hazard level	Difficulty level	Level of study
1 (3 h)	Moderate	Low	Intermediate

Class names Amines, acyl halides, amides

Concepts involved This experiment aims at the preparation of amides from acyl chlorides. This will be illustrated using the synthesis of hippuric acid, starting from glycine and benzoyl chloride in a biphasic basic solvent system (Schotten–Baumann conditions)

Chemicals needed Benzoyl chloride, glycine, sodium hydroxide aqueous solution, hydrochloric acid, diethyl ether, ethanol, dimethyl sulfoxide-d_6

Equipment and experimental techniques involved Recrystallization, NMR, melting point, heating plate, vacuum filtration apparatus

Keywords Amino acid, basic catalysis, benzoylation reaction, hippuric acid, NMR, recrystallization

Background

Amides are an important class of organic chemicals and amide bonds are often found in synthetic and naturally occurring organic molecules. Some familiar examples include nylon (a synthetic polyamide used to make fabrics) and large biomolecules, like proteins. Nevertheless, there are other less known examples where the formation of amide bonds is equally important for a particular biological response, such as the conversion of relatively poorly water-soluble substances into more soluble products, a common metabolic process for the elimination of xenobiotics from our bodies.

Hippuric acid (or *N*-benzoylglycine, **1**), a simple amide, is a major metabolite found in the urine of humans and animals, resulting from an excretory mechanism of benzene-type aromatic compounds from natural and unnatural sources.[1] Our bodies convert benzoic acid, which is present in many colored fruits and vegetables and is a common preservative in food products, into hippuric acid (by reaction with endogenous glycine), in order to detoxify it.[2] Since hippuric acid is more water-soluble than benzoic acid, it will be excreted more efficiently in the urine.

Comprehensive Organic Chemistry Experiments for the Laboratory Classroom
Edited by Carlos A. M. Afonso, Nuno R. Candeias, Dulce Pereira Simão, Alexandre F. Trindade, Jaime A. S. Coelho, Bin Tan, and Robert Franzén.

Published by the Royal Society of Chemistry, www.rsc.org

The work herein described will exemplify amide bond formation through the *in vitro* synthesis of hippuric acid, starting from glycine and the acyl chloride of benzoic acid (benzoyl chloride), which is a more reactive species than the corresponding carboxylic acid due to the nature of the leaving group. The reaction will be carried out in a biphasic basic solvent system, often named as "Schotten–Baumann conditions".[3] The basic medium is necessary because the desired reaction cannot occur if the amino glycine is in the zwitterionic form, meaning that the nitrogen is protonated and does not have a non-bonding electron pair to act as nucleophile. However, treatment of the zwitterionic glycine with an aqueous base (sodium hydroxide) converts the amino acid into a nucleophilic form (Scheme 3.1.6.1a), which rapidly attacks the carbonyl carbon of the benzoyl chloride to initiate amide bond formation (Scheme 3.1.6.1b).

It is important to notice that the acylation of amines with carboxylic acid chlorides leads to the production of one equivalent acid, which will form a salt with unreacted amine and diminish the reaction yield. Therefore, a sacrificial base stronger than glycine's nitrogen must be present in the reaction medium to neutralize this acid. To optimize the conditions and minimize side reactions, the aqueous base is normally added slowly to the reaction mixture.

The main objective of this work is to familiarize students with the concepts of amide bond formation, using an operationally simple methodology involving the *N*-benzoylation of glycine under basic catalysis.

Moreover, it will allow students to master basic chemistry laboratory techniques (*e.g.* filtration, recrystallization) and consolidate structural identification concepts using NMR spectroscopy.

Additional Safety

The experiment should be performed in a fume hood, with proper safety equipment, such as protective clothing, gloves and safety goggles. Benzoyl chloride is a fuming liquid with a pungent odor and very hazardous in case of skin contact (corrosive, irritant) or eye contact (irritant). Aqueous sodium hydroxide solutions are corrosive and can cause skin and eye irritation or burns. Hydrochloric acid (37%) is highly corrosive and can cause severe skin and eye burns. In addition, hydrochloric acid vapour may cause respiratory irritation. Diethyl ether is extremely flammable so it is important to keep it distant from ignition sources.

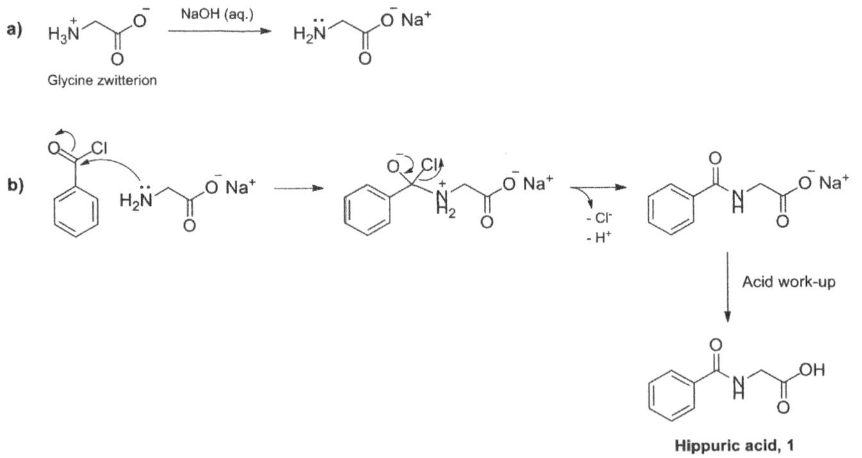

Scheme 3.1.6.1 Mechanism of the *N*-benzoylation of glycine using benzoyl chloride and basic conditions.

Experimental Procedure

1. Place 11 mL of distilled water and 1.5 mL of 6 M NaOH solution in a 100 mL Erlenmeyer flask with ground joint.
2. Add 1.05 g (0.014 mol, 1 equiv.) of glycine to the previous mixture and swirl until a clear solution is obtained.
3. In a fume hood, measure 1.85 mL of benzoyl chloride and quickly add it to the glycine solution. Stopper the Erlenmeyer flask tightly and shake it vigorously until the layers are completely mixed. The reaction is exothermic, so vent it cautiously by removing the stopper occasionally. If the reaction becomes too hot, cool it down in a water bath.
4. Monitor the pH and when the solution becomes acidic, add another 3 mL of 6 M NaOH solution over a 2–3 min period, shaking and ventilating the flask after each addition. If necessary, add an additional 1–2 mL of 6 M NaOH solution to maintain an alkaline pH. The reaction is complete when it forms a clear yellow solution.

Product Isolation

5. Place approximately 7.5 g of crushed ice in a 150 mL beaker and carefully add 7.5 mL of concentrated HCl. Pour the reaction mixture into the acid/ice solution and mix.
6. Collect the precipitate formed by vacuum filtration and discard the aqueous filtrate.

Product Purification

7. Wash the precipitate with 5–10 mL of diethyl ether. Repeat the ether washing two more times.
8. Place the crude product in an Erlenmeyer flask, add 10 mL of ethanol/water solution (1:3, v/v) and heat the mixture until complete solubilization. Allow the solution to cool down to room temperature and, if necessary, place it in an ice-bath to facilitate recrystallization.
9. Collect the crystals (white needles) by vacuum filtration, wash them with cold ether and dry the pure product using a vacuum line or an oven (100 °C).
10. Determine the weight and measure the melting point of the obtained product.
11. Prepare the sample for ^1H and ^{13}C NMR (approx. 20 mg of compound in 500 μL of DMSO-d_6).

Waste Disposal

Neutralize the aqueous filtrate from the crude product and pour it into the container for aqueous inorganic waste. The ether filtrate and the aqueous ethanol filtrate from the recrystallization should be collected in a non-chlorinated waste recipient for posterior destruction.

Results Interpretation and Additional Questions

1. Determine the reaction yield.
2. Compare the melting point of the obtained hippuric acid with the one reported in the literature.[4]
3. Interpret and assign all peaks of the NMR spectra obtained.
4. Why is the hippuric acid soluble in water until hydrochloric acid is added?
5. Why must sodium hydroxide be present for the reaction to take place at a reasonable rate? What would be the disadvantage of adding all the 6 M NaOH solution in one portion at the beginning of the reaction?
6. The present experiment illustrates amide bond formation from acyl chlorides and amines. Discuss other methods for the preparation of amides.

References

1. R. W. Pero, *Curr. Clin. Pharmacol.*, 2010, **5**, 67.
2. J. K. Nicholson and I. D. Wilson, *Nat. Rev. Drug Discovery*, 2003, **2**, 668.
3. For example: (a) D. E. Frantz, D. G. Weaver, J. P. Carey, M. H. Kress and U. H. Dolling, *Org. Lett.*, 2002, **4**, 4717; (b) D. Babinski, I. Soltani and D. E. Frantz, *Org. Lett.*, 2008, **10**, 2901.
4. A. W. Ingersoll and S. H. Babcock, *Org. Synth.*, 1932, **12**, 40.

3.1.7. Preparation of a Sulfathiazole Prodrug *via* *N*-acylation with Succinic Anhydride

Pedro M. P. Gois*, Francisca Lopes, and Alexandre F. Trindade*

Faculty of Pharmacy, University of Lisbon, Av. Prof. Gama Pinto, 1649-003 Lisboa, Portugal
*E-mail: pedrogois@ff.ul.pt, alexandretrindade@ff.ul.pt

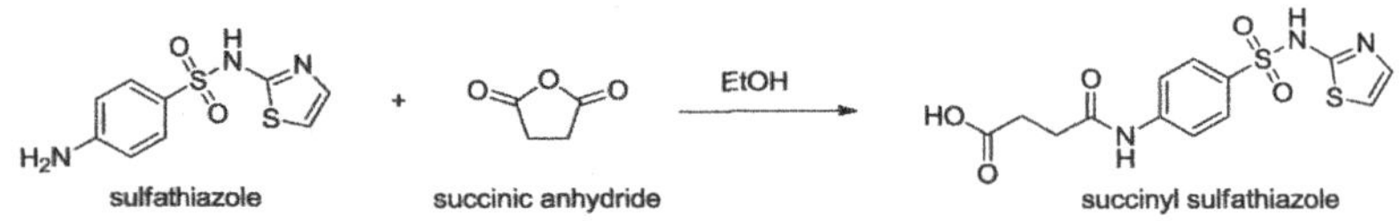

Number of sessions (duration of each session)	Hazard level	Difficulty level	Level of study
1 (3 h)	Low	Low	Intermediate
Class names Anilines, anhydrides, carboxylic acids, amides			
Concepts involved This experiment is based on the synthesis of a prodrug of sulfathiazole through the acylation of its aniline group. Herein is intended for the students to discuss the mechanism of amide synthesis from anhydrides, to improve their ability to isolate organic compounds *via* recrystallization with high boiling point solvents and to elucidate the structure of organic compounds by ^{1}H NMR spectroscopy			
Chemicals needed Sulfathiazole, succinic anhydride, ethanol, water, deuterated dimethyl sulfoxide			
Equipment and experimental techniques involved Magnetic stirring apparatus with heating, Büchner funnel, rotary evaporator, melting point apparatus, NMR spectrometer, distillation under reduced pressure, filtration under vacuum, melting point measurement and NMR spectra analysis			
Keywords Amidation, aniline, cyclic anhydride, NMR, recrystallization, sulfathiazole			

Background

Sulfonamides are a class of antibacterial agents. The first drug of this class ever discovered was sulfamidochrysoidine (Prontosil), which is in fact a prodrug whose main active pharmacophore is a sulfonamide group. This class of antibacterial agents is mainly used to treat intestinal infections and can be the subject of latentiation to obtain prodrugs that are poorly absorbed in the intestine, remaining in this location for a longer time.[1] Typically, the latentiation step involves a reaction of an aromatic amine with anhydrides or activated dicarboxylic acids. In this experiment, a hemisuccinate group (highly hydrophilic) is introduced in the sulfathiazole aniline group to limit the absorption of the prodrug in the gastrointestinal system. The accumulation of this compound in the intestine allows its gradual enzymatic hydrolysis, liberating sulfathiazole that acts locally as an antibiotic.[2]

The reaction of latentiation of sulfonamides is an acylation reaction of an aromatic amine. Mechanistically, the aniline will behave as a nucleophile and the carbonyl group of the anhydride

Comprehensive Organic Chemistry Experiments for the Laboratory Classroom
Edited by Carlos A. M. Afonso, Nuno R. Candeias, Dulce Pereira Simão, Alexandre F. Trindade, Jaime A. S. Coelho, Bin Tan, and Robert Franzén.

Published by the Royal Society of Chemistry, www.rsc.org

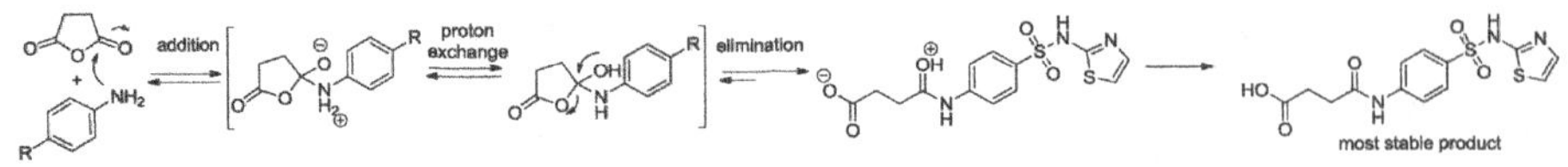

Scheme 3.1.7.1 Mechanism of aniline acylation with cyclic anhydrides.

will act as the electrophile (Scheme 3.1.7.1).[3] The substitution at carboxylic acid derivatives mechanistically involves a formal addition–elimination pathway because the breaking of a carbon–oxygen double bond involves high energy and usually is only achieved by highly reactive nucleophiles, like Grignard reagents or aluminium hydrides. Hence, the non-bonded pair of electrons at the aniline will attack the cyclic anhydride at the carbonyl carbon, producing a tetrahedral intermediate. To regenerate the carbonyl bond the best leaving group will be eliminated in the second step of the mechanism.[4]

Additional Safety

Besides regular laboratory good practice, care should be taken when handling refluxing ethanol, as this organic solvent is highly flammable. Care must be taken on handling sulfathiazole and succinyl sulfathiazole. Succinic anhydride is hazardous in case of skin contact (irritant), eye contact (irritant), ingestion and inhalation.

Experimental Procedure

1. Weigh 0.25 g (0.98 mmol) of sulfathiazole into a 50 mL round-bottomed flask.
2. Add 15 mL of ethanol into the flask, fit a reflux condenser and dissolve the powder under mild heating.
3. Add 0.125 g (1.2 mmol) of succinic anhydride. Heat the mixture under reflux for 45 min.
4. Cool the reaction mixture to room temperature. If a solid precipitates recover it by filtration under vacuum. Evaporate the mother liquors under reduced pressure and combine the two solids.
5. If no precipitate is formed, evaporate the solvent under reduced pressure.
6. Recrystallize the solid residue from hot water/ethanol (3 : 4).
7. Dry the compound, and calculate the isolated yield and measure the melting point.
8. Dissolve about 5 mg of the product in deuterated DMSO for ^{1}H NMR analysis.

Waste Disposal

The residues from the filtrations containing ethanol and water can be disposed in a non-chlorinated organic residues container for later disposal.

Results Interpretation and Additional Questions

1. Comment on the yield and purity determined by NMR and melting point.
2. Identify possible contaminants that can be present in the final product. Comment on how they can be formed.
3. Considering the following three amines indicate the order of nucleophile strength. Justify your answer.

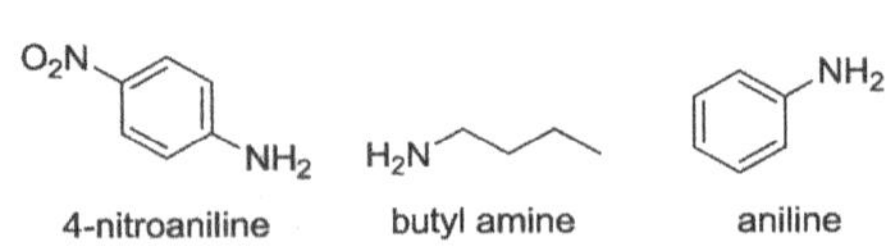

4. In the latentiation reaction of sulfathiazole with succinic anhydride, the carboxylate acts as the leaving group in the addition–elimination mechanism. Why is this true? Consider in your answer that the pK_a of a typical carboxylic acid is 4.5, while the pK_a of the aniline in sulfathiazole is 7.2.

References

1. A. B. Reitz, G. R. Smith and M. H. Parker, *Expert Opin. Ther. Pat.*, 2009, **19**, 1449.
2. G. L. Patrick, *An Introduction to Medicinal Chemistry*, Oxford University Press, 5th edn, 2013.
3. M. L. Moore and C. S. Miller, *J. Am. Chem. Soc.*, 1942, **64**, 1572.
4. J. Clayden, N. Greeves, S. Warren and P. Wothers, *Organic Chemistry*, Oxford University Press, 1st edn, 2001.

3.1.8. Determination of the Absolute Configuration of Enantioenriched Secondary Alcohols *via* Thin-Layer Chromatography

Alexander J. Wagner*, Shawn M. Miller, Scott D. Rychnovsky, and Renée D. Link

Department of Chemistry, University of California–Irvine, 1102 Natural Sciences II, Irvine, CA 92697, USA. *E-mail: ajwagner@uci.edu, shawnmm@illinois.edu, srychnov@uci.edu, rlink@uci.edu

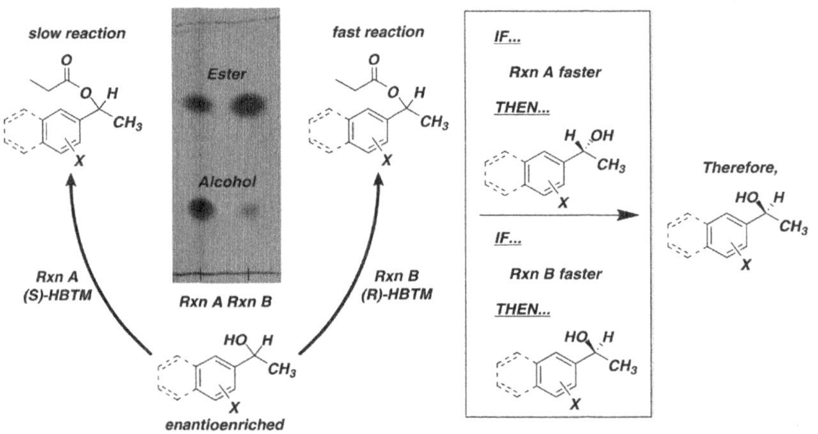

Number of sessions (duration of each session)	Hazard level	Difficulty level	Level of study
1 (3 h)	Moderate	Low	Intermediate

Class names Alcohols, anhydrides, amines, aromatic rings, esters

Concepts involved This experiment describes the use of ^{1}H NMR spectroscopy and the competing enantioselective conversion method to identify the structure and absolute configuration, respectively, of an enantioenriched alcohol

Chemicals needed *For reaction*: (*R*)-1-(*p*-bromophenyl)ethanol, (*S*)-1-(*p*-bromophenyl)ethanol, (*R*)-1-(*m*-methoxyphenyl)ethanol, (*S*)-1-(*m*-methoxyphenyl)ethanol, (*R*)-1-(naphthalen-2-yl)ethanol, (*S*)-1-(naphthalen-2-yl)ethanol, (*S*)-homobenzotetramisole ((*S*)-HBTM), (*R*)-homobenzotetramisole ((*R*)-HBTM), propionic anhydride, triethylamine, toluene, methanol. *For TLC analysis*: hexanes, ethyl acetate. *For stain*: phosphomolybdic acid, ethanol

Equipment and experimental techniques involved TLC materials: plates, chambers, spotters; TLC visualization materials: heat source (oven, hotplate or heat gun); photograph of TLC plate: photo-taking device (smart phone, camera, tablet or computer); ^{1}H NMR spectroscopy: analysis of data, no spectrometer required

Keywords Absolute configuration, catalysis, enantioselective, esterification, NMR analysis, thin-layer chromatography

Comprehensive Organic Chemistry Experiments for the Laboratory Classroom
Edited by Carlos A. M. Afonso, Nuno R. Candeias, Dulce Pereira Simão, Alexandre F. Trindade, Jaime A. S. Coelho, Bin Tan, and Robert Franzén.

Published by the Royal Society of Chemistry, www.rsc.org

Background

Absolute Configuration

With the development of asymmetric transformations and improvements in the isolation of single enantiomers of natural products came a need to be able to determine the absolute configuration of the enantioenriched stereocenter(s).[1–4] Various methods have been developed for the determination of absolute configuration including the advanced Mosher method,[5–7] vibrational circular dichroism,[8] exciton chirality,[9] NMR spectroscopic chiral shift reagents,[10–13] lipase-catalyzed resolutions,[14] and X-ray crystallographic analysis.[15] Several chiral derivatization methods have been transformed into undergraduate experiments, where students convert a single enantiomer into two diastereomers and analyze the diastereomeric products by NMR spectroscopy.[16–18] While effective, the NMR spectroscopic analysis can be complicated for introductory undergraduate organic laboratories.

The Competing Enantioselective Conversion Method

The competing enantioselective conversion (CEC) method is designed to determine the absolute configuration of molecules through the use of enantioselective reactions.[19–22] The CEC method is a modern implementation of the Horeau method.[23] The CEC method utilizes parallel reactions, where the substrate of interest is subjected to reaction conditions that facilitate a chemical transformation (Figure 3.1.8.1a). The transformation is mediated by the use of chiral kinetic resolution reagents. Each reaction contains one enantiomer of the kinetic resolution reagent. The difference in the rate of the transformation between the two parallel reactions results from the difference in energy between the diastereomeric transition states of the enantioenriched substrate with each enantiomer of the kinetic resolution reagent (Figure 3.1.8.1b). The result is a difference in reaction conversion between the two parallel reactions (Figure 3.1.8.1c). The relative "fast" and "slow" reactions can be determined based on reaction conversion. Then, with the use of an empirical mnemonic that was previously developed by testing several substrates with known configurations, the absolute configuration can be assigned.

This experiment modifies the CEC method to be applicable to the undergraduate organic chemistry curriculum with the goal of identifying the structure and absolute configuration of an unknown enantioenriched secondary alcohol. Secondary alcohols were chosen as the reactive functional group for this experiment because of their affordability and importance in chemistry.

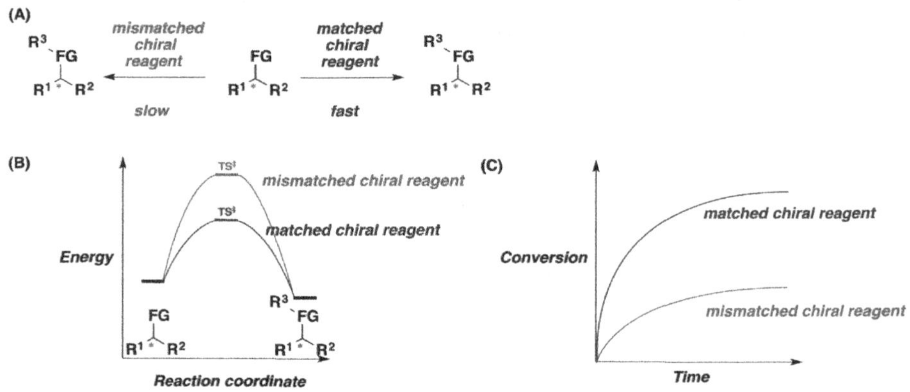

Figure 3.1.8.1 (A) General depiction of the competing enantioselective conversion method; FG = functional group; * = stereogenic center (B) Energy diagram corresponding to the parallel reactions in (A). (C) Kinetic analysis based on the energy diagram measuring the conversion for each reaction. Adapted with permission from *J. Chem. Educ.*, 2014, **91**, pp 716–721. Copyright 2014 American Chemical Society.

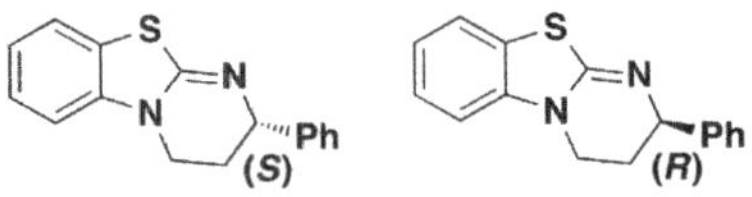

Figure 3.1.8.2 The (*S*)-HBTM (left) and (*R*)-HBTM (right) catalyst.

Additionally, the CEC method for secondary alcohols uses the homobenzotetramisole (HBTM) catalyst[24,25] (Figure 3.1.8.2) at very low molar equivalents. Thin-layer chromatography is used as the characterization technique because it is a cost-effective and scalable technique that is interpreted easily by undergraduate students. A single, unknown enantioenriched alcohol will be reacted with an anhydride in the presence of each enantiomer of the HBTM catalyst in two separate reactions. The progress of each reaction will be determined *via* thin-layer chromatography qualitatively and quantitatively. The structure of the unknown alcohol will be determined *via* ^{1}H NMR spectroscopic analysis and the absolute configuration will be determined by identifying the fast reacting enantiomer of HBTM.

Additional Safety

All work should be carried out in a fume hood, with proper safety equipment, such as protective clothing, gloves, and safety goggles. Organic solvents, such as hexanes, ethyl acetate, toluene, methanol, and ethanol, are volatile and flammable. Phosphomolybdic acid, triethylamine, and the alcohols, anhydrides, and esters involved in this experiment are skin irritants and harmful if swallowed. In addition, triethylamine possesses a pungent odor.

Experimental Procedure

There are two versions of this experiment from which the instructor may choose. Version α is designed to optimize the quality of the qualitative and quantitative analysis of the experimental results. Version β is designed to minimize the amounts of HBTM and enantioenriched alcohols used in the reaction. The procedures for Version α and Version β are identical.

Laboratory Session 1 (3 h): CEC Reaction Set-Up and Product Analysis

1. Acquire the four pre-prepared solutions described in the below table that correspond to the version of the experiment you are performing. Solution 1 contains one of the following enantioenriched alcohols. Obtain the ^{1}H NMR spectrum associated with your Solution 1.

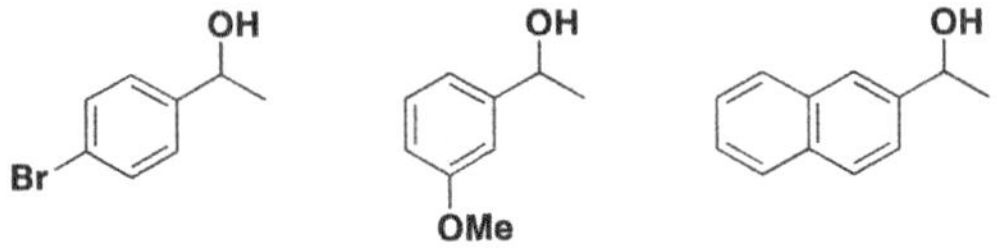

	Stock solution	Compounds in solution
Version α	1	Unknown alcohol (0.30 M) in toluene
	2	(*S*)-HBTM (0.0075 M) and Et_3N (0.60 M) in toluene
	3	(*R*)-HBTM (0.0075 M) and Et_3N (0.60 M) in toluene
	4	Propionic anhydride (0.60 M) in toluene
	Stock solution	Compounds in solution
Version β	1	Unknown alcohol (0.075 M) in toluene
	2	(*S*)-HBTM (0.0009375 M) and Et_3N (6.0 M) in toluene
	3	(*R*)-HBTM (0.0009375 M) and Et_3N (6.0 M) in toluene
	4	Propionic anhydride (6.0 M) in toluene

2. Obtain two glass vials. Label one vial **A** and one vial **B**.
3. Add 0.5 mL of solution **1** to vial **A** and to vial **B**.
4. Add 0.5 mL of solution **2** to vial **A**.
5. Add 0.5 mL of solution **3** to vial **B**.
6. Add 0.5 mL of solution **4** to vial **A**. Swirl the vial and set a timer for 1 h.
7. Add 0.5 mL of solution **4** to vial **B**. Swirl the vial and set a timer for 1 h.
8. After 1 h has passed for reaction **A**, add 0.25 mL of methanol to vial **A**. Swirl the vial.
9. After 1 h has passed for reaction **B**, add 0.25 mL of methanol to vial **B**. Swirl the vial.
10. Prepare a single TLC plate with two lanes. Spot the first lane with solution from reaction **A**, and the second lane with solution from reaction **B**. Attempt to spot the same amount in the first lane as you spot in the second lane.
11. Develop the TLC plate in a TLC chamber with a solvent system of 20% ethyl acetate in hexanes.
12. Stain the TLC plate using phosphomolybdic acid stain and visualize *via* heating.
13. Draw a sketch and take a photograph of the TLC plate.
14. Assign the absolute configuration of your unknown alcohol using both qualitative and quantitative analyses of the results of your CEC experiment with the mnemonic shown below (Figure 3.1.8.3).

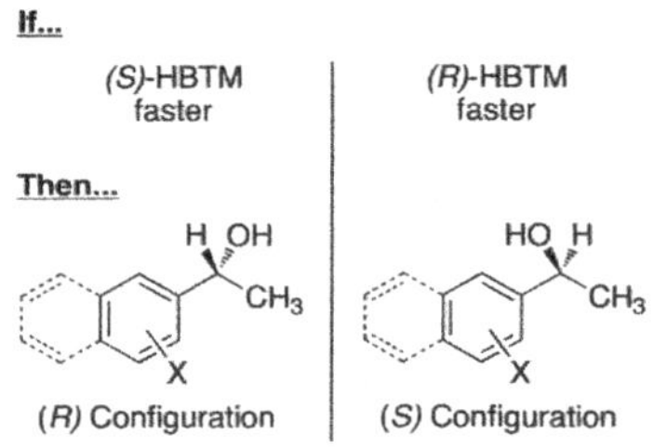

Figure 3.1.8.3 The predictive mnemonic for the CEC method for secondary alcohols. Adapted with permission from *J. Chem. Educ.*, 2014, **91**, pp 716–721. Copyright 2014 American Chemical Society.

Waste Disposal

The reaction solutions and TLC eluent can be collected in a non-chlorinated waste container recipient for proper disposal. The TLC plates and spotters can be collected in the appropriate solid waste containers.

Results Interpretation and Additional Questions

1. Using the ^{1}H NMR spectrum associated with your unknown, identify the structure of your unknown alcohol. Can you establish the absolute configuration of the alcohol using your ^{1}H NMR spectrum? Justify your answer.
2. Perform a qualitative analysis using your TLC sketch and a quantitative analysis *via* the software ImageJ using your photograph. Assign the absolute configuration of your unknown alcohol using the results from your analysis and justify your conclusion.
3. Draw the mechanism of the reaction.
4. Identify the step in your mechanism that is the source of the enantioselectivity of the reaction. Discuss why your unknown enantioenriched alcohol reacted at different rates with each enantiomer of HBTM.

Acknowledgement

Adapted with permission from *J. Chem. Educ.*, 2014, **91**, pp 716–721. Copyright (2014) American Chemical Society.

References

1. E. L. Eliel and S. H. Wilen, *Stereochemistry of Organic Compounds*, Wiley-Interscience, Hoboken, NJ, 1994; pp. 101–147 and 991–1105.
2. J. M. Seco, E. Quiñoá and R. Riguera, *Chem. Rev.*, 2004, **104**, 17–118.
3. T. J. Wenzel and C. D. Chisholm, *Chirality*, 2011, **23**, 190–214.
4. R. S. Cahn, C. Ingold and V. Prelog, *Angew. Chem., Int. Ed. Engl.*, 1966, **5**, 385–415.
5. J. A. Dale, D. L. Dull and H. S. Mosher, *J. Org. Chem.*, 1969, **34**, 2543–2549.
6. I. Ohtani, T. Kusumi, Y. Kashman and H. Kakisawa, *J. Am. Chem. Soc.*, 1991, **113**, 4092–4096.
7. For a review of the advanced Mosher method procedure, see: T. R. Hoye, C. S. Jeffrey and F. Shao, *Nat. Protoc.*, 2007, **2**, 2451–2458.
8. For a review on using vibrational circular dichroism, see: T. B. Freedman, X. Cao, R. K. Dukor and L. A. Nafie, *Chirality*, 2003, **15**, 743–758.
9. For a review on the electronic CD exciton chirality method, see: N. Harada, K. Nakanishi and N. Berova, Electronic CD Exciton Chirality Method: Principles and Applications, in *Comprehensive Chiroptical Spectroscopy, Vol. 2: Applications in Stereochemical Analysis of Synthetic Compounds, Natural Products, and Biomolecules*, ed. N. Berova, P. L. Polavarapu, K. Nakanishi and R. W. Woody, John Wiley & Sons, Inc., Hoboken, NJ, 2012, pp. 115–166.
10. Y. Kobayashi, N. Hayashi and Y. Kishi, *Org. Lett.*, 2002, **4**, 411–414.
11. Y. Kobayashi, N. Hayashi and Y. Kishi, *Tetrahedron Lett.*, 2003, **44**, 7489–7491.
12. I. Ghosh, H. Zeng and Y. Kishi, *Org. Lett.*, 2004, **6**, 4715–4718.
13. I. Ghosh, Y. Kishi, H. Tomoda and S. Omura, *Org. Lett.*, 2004, **6**, 4719–4722.
14. Q. Jing and R. J. Kazlauskas, *Chirality*, 2008, **20**, 724–735.
15. For a review on X-ray crystallographic analysis, see: H. D. Flack and G. Bernardinelli, *Chirality*, 2008, **20**, 681–690.
16. D. A. Allen, A. E. Tomaso, O. P. Priest, D. F. Hindson and J. L. Hurlburt, *J. Chem. Educ.*, 2008, **85**, 698–700.
17. S. Saba, D. D. Clarke, C. Iwanoski and T. Lobasso, *J. Chem. Educ.*, 2010, **87**, 1238–1241.
18. O. S. Fenton and B. R. Sculimbrene, *J. Chem. Educ.*, 2011, **88**, 662–664.
19. For the CEC method with secondary alcohols using ^{1}H NMR spectroscopy, see: A. J. Wagner, J. G. David and S. D. Rychnovsky, *Org. Lett.*, 2011, **13**, 4470–4473.
20. For the CEC method with oxazolidinones, lactams, and thiolactams using ^{1}H NMR spectroscopy, see: M. A. Perry, J. V. Trinidad and S. D. Rychnovsky, *Org. Lett.*, 2013, **15**, 472–475.
21. For the CEC method with primary amines using mass spectrometry, see: S. M. Miller, R. A. Samame and S. D. Rychnovsky, *J. Am. Chem. Soc.*, 2012, **134**, 20318–20321.
22. For the CEC method with secondary alcohols using thin-layer chromatography, see: A. J. Wagner and S. D. Rychnovsky, *J. Org. Chem.*, 2013, **78**, 4594–4598.
23. A. Horeau, Determination of the Configuration of Secondary Alcohols by Partial Resolution, in *Stereochemistry, Fundamentals and Methods*, ed. J. Fiaud, A. Horeau and H. B. Kagan, Georg Thieme, Stuttgart, 1977, Vol. 3, pp. 51–94.
24. V. B. Birman and X. Li, *Org. Lett.*, 2008, **10**, 1115–1118.
25. The HBTM catalyst is commercially available from Sigma-Aldrich (*R*-HBTM: L511730, *S*-HBTM: L511862).

3.1.9. Anhydride Aminolysis: Synthesis of *N*-arylmaleamic Acids

Artur M. S. Silva, Augusto C. Tomé*, Diana C. G. A. Pinto, Fernando Domingues, Graça M. Oliveira Rocha, José A. S. Cavaleiro, M. Graça P. M. S. Neves, M. Amparo F. Faustino, and Mário M. Q. Simões

Department of Chemistry and QOPNA, University of Aveiro, 3810-193 Aveiro, Portugal
*E-mail: actome@ua.pt

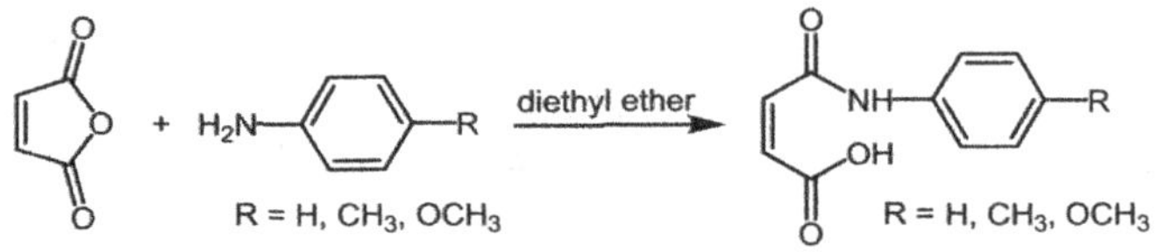

Number of sessions (duration of each session)	Hazard level	Difficulty level	Level of study
1 (4 h)	Moderate	Medium	Intermediate

Class names Anhydrides, amines, amides, carboxylic acids

Concepts involved This experiment illustrates the synthesis of an amide through the reaction of an amine with an anhydride and the characterization of the product by melting point and NMR and IR spectra

Chemicals needed Maleic anhydride, aniline or a substituted aniline, diethyl ether, deuterated dimethyl sulfoxide

Equipment and experimental techniques involved Stirring/heating plate, filtration unit, melting point apparatus, NMR and IR equipment

Keywords Amides, anhydride aminolysis, carboxylic acid derivatives, NMR, vacuum filtration

Background

Carboxylic acid anhydrides are very reactive compounds. They undergo nucleophilic acyl substitution reactions[1] with a range of nucleophiles, typically at room temperature, in the absence of a catalyst. They react with ammonia, amines or alcohols, for instance, to afford amides or esters in high yields.

The structural diversity of anhydrides is enormous: they may be symmetrical ($R^1 = R^2$) or unsymmetrical (or mixed) ($R^1 \neq R^2$), non-cyclic or cyclic. Figure 3.1.9.1 illustrates some of the most frequently used anhydrides in organic synthesis:

When a non-cyclic anhydride reacts with a nucleophile (an amine, for instance) only one acyl group is incorporated in the resulting amide. The remaining acyl group is converted into the corresponding carboxylic acid. However, when a cyclic anhydride reacts with a nucleophile, the resulting product incorporates both acyl units. This experiment illustrates the reaction of a cyclic anhydride (maleic anhydride) with aniline or a substituted aniline. This is a typical nucleophilic

Comprehensive Organic Chemistry Experiments for the Laboratory Classroom
Edited by Carlos A. M. Afonso, Nuno R. Candeias, Dulce Pereira Simão, Alexandre F. Trindade, Jaime A. S. Coelho, Bin Tan, and Robert Franzén.

Published by the Royal Society of Chemistry, www.rsc.org

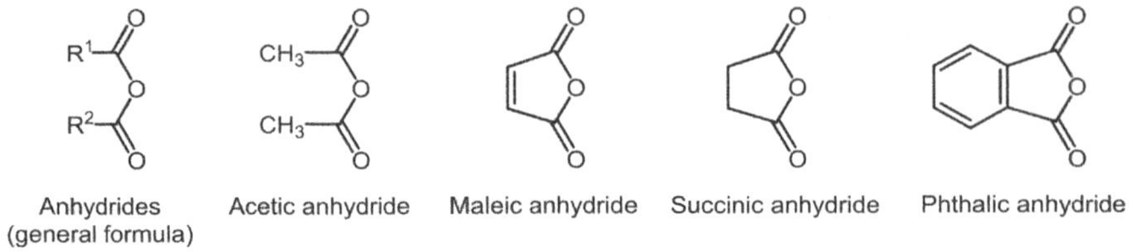

Figure 3.1.9.1 Some frequently used anhydrides in organic synthesis.

acyl substitution reaction and the resulting *N*-substituted maleamic acid incorporates, simultaneously, amide and carboxylic groups.

N-substituted maleamic acids, in particularly the *N*-aryl derivatives, find significant applications in synthetic chemistry. They are precursors of *N*-substituted maleimides, for which important biological properties have been established.[2–9] They are also used in the polymer industry.[10–12] Their potential use as drugs is not well studied although some derivatives were considered as potential hypocholesterolemic agents.[13]

This experiment can be executed safely in organic laboratory courses and the resulting *N*-arylmaleamic acids can be isolated pure just by filtration from the reaction mixture.[14] Analysis of the IR and NMR[15] spectra (^{1}H and ^{13}C) of the obtained *N*-arylmaleamic acids is recommended.

Additional Safety

The students must wear proper personal protective equipment, such as laboratory coat, gloves and safety goggles. The experiment should be carried out in the fume hood. Special precautions should be taken during the experiment as diethyl ether is an easily flammable solvent. Aniline and some aniline derivatives are highly toxic and harmful by contact, if swallowed or by inhalation.

Experimental Procedure

Laboratory Session: 4 h

Synthesis of the N-arylmaleamic Acids

1. In a fume hood, prepare a 250 mL two neck-round bottomed flask with a magnetic stir bar, a condenser and an addition funnel.
2. Place 10 g of maleic anhydride (0.102 mmol) in the flask, add 100 mL of diethyl ether and stir the suspension until complete dissolution of the anhydride.
3. Put the flask in a water bath at room temperature with the condenser and the addition funnel.
4. Prepare a solution of the aniline derivative (0.1 mol) in 25 mL of diethyl ether.
5. Using the addition funnel, add, dropwise, the solution of the aniline derivative to the maleic anhydride solution. Be careful: the reaction is exothermic!
6. Warm up the water bath to 40–45 °C and keep the reaction mixture stirring for about 90 min.

Product Isolation

7. After cooling to room temperature, filter the suspension that was formed using the filtration unit.
8. Transfer the solid to a pre-weighed glass watch and put it in the oven at 75 °C for 20–25 min.
9. Weigh the product obtained.

Structural Characterization

10. Measure the melting point of the dried solid.
11. Place 15 mg of the dry compound in a NMR tube, add 0.5 mL of deuterated dimethyl sulfoxide (DMSO-d_6) [containing tetramethylsilane (TMS) as the internal standard] and shake it until complete dissolution.
12. Confirm the structure of the product using the ^1H and ^{13}C NMR and IR spectra.

Waste Disposal

The diethyl ether from the reaction and from the filtration washings should be collected in a waste recipient for non-chlorinated organic solvents.

Results Interpretation and Additional Questions

1. Determine the reaction yield.
2. Write a plausible mechanism for the reaction.
3. Discuss the insolubility of the *N*-arylmaleamic acid in diethyl ether.
4. Predict the main features of the ^1H NMR spectrum of the synthesized compound. Compare the signals corresponding to the vinyl protons of the *N*-arylmaleamic acid with the signal corresponding to the same protons in maleic anhydride.
5. Interpret the IR and ^1H and ^{13}C NMR spectra.

References

1. F. A. Carey, *Organic Chemistry*, Sixth International Edition, McGraw-Hill, USA, 2006, ch. 20.
2. M. Sortino, F. Garibotto, V. Cechinel Filho, M. Gupta, R. Enriz and S. Zacchino, *Bioorg. Med. Chem.*, 2011, **19**, 2823–2834.
3. J. R. Cashman, M. MacDonald, S. Ghirmai, K. J. Okolotowicz, E. Sergienko, B. Brown, X. Garcia, D. Zhai, R. Dahl and J. C. Reed, *Bioorg. Med. Chem. Lett.*, 2010, **20**, 6560–6564.
4. S. Lopez, M. Castelli, F. Campos, R. Correa, V. Cechinel Filho, R. Yunes, M. Zamora, R. Enriz, J. Ribas, R. Furlan and S. Zacchino, *Arzneim.-Forsch./Drug Res.*, 2005, **55**, 123–132.
5. M. Sortino, V. Cechinel Filho, R. Corrêa and S. Zacchino, *Bioorg. Med. Chem.*, 2008, **16**, 560–568.
6. N. Matuszak, G. G. Muccioli, G. Labar and D. M. Lambert, *J. Med. Chem.*, 2009, **52**, 7410–7420.
7. S. M. Saario, O. M. H. Salo, T. Nevalainen, A. Poso, J. T. Laitinen, T. Järvinen and R. Niemi, *Chem. Biol.*, 2005, **12**, 649–656.
8. S. Watanabe, Y. Igarashi, K. Yagami and R. Imai, *Pestic. Sci.*, 1991, **31**, 45–51.
9. K. Takatori, T. Hasegawa, S. Nakano, J. Kitamura and N. Kato, *Microbiol. Immunol.*, 1985, **29**, 1237–1241.
10. G. Olatunji, Y. Huang and Q. Dai, *Cellulose*, 1997, **4**, 233–238.
11. J. A. Mikroyannidis, *J. Polym. Sci., Part A: Polym. Chem.*, 1992, **30**, 125–132.
12. J. A. Mikroyannidis, *J. Appl. Polym. Sci.*, 1990, **41**, 2625–2636.
13. K.-C. Liu, T.-B. Chen, H.-J. Jan and C.-Y. Shih, *J. Chin. Chem. Soc.*, 1973, **20**, 163–169.
14. M. P. Cava, A. A. Deana, K. Muth and M. J. Mitchell, *Organic Synth., Coll.*, 1973, **5**, 944–946.
15. P. Fellmann, A. Gevertz and C. Cordier, *J. Chem. Educ.*, 2006, **83**, 432–435.

3.1.10. Synthesis of *N*-arylmaleimides

Artur M. S. Silva, Augusto C. Tomé, Diana C. G. A. Pinto, Fernando Domingues*, Graça M. Oliveira Rocha, José A. S. Cavaleiro, M. Graça P. M. S. Neves, M. Amparo F. Faustino, and Mário M. Q. Simões

Department of Chemistry and QOPNA, University of Aveiro, 3810-193 Aveiro, Portugal
*E-mail: fdomingues@ua.pt

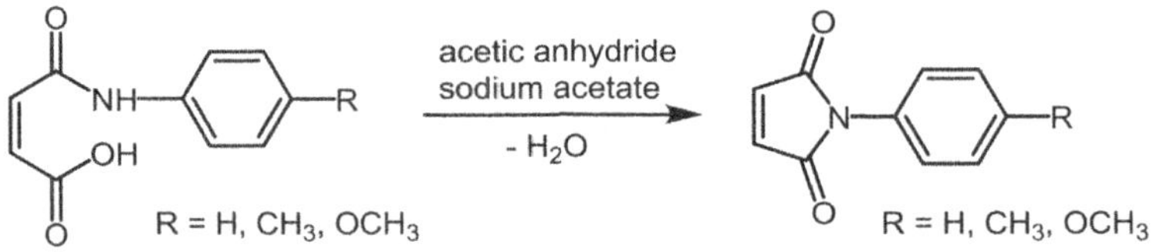

Number of sessions (duration of each session)	Hazard level	Difficulty level	Level of study
1 (4 h)	Moderate	Medium	Intermediate

Class names Carboxylic acids, amides, imides, anhydrides

Concepts involved This experiment illustrates the synthesis of an imide through the intramolecular reaction of an amide group with a carboxylic acid, the purification of the product by column chromatography and the characterization of the purified product by NMR and IR spectroscopy

Chemicals needed *N*-Arylmaleamic acid, acetic anhydride, anhydrous sodium acetate, silica gel, dichloromethane, hexane, deuterated chloroform

Equipment and experimental techniques involved Stirring/heating plate, oil bath, rotatory evaporator, flash column chromatography, melting point apparatus, NMR and IR equipment

Keywords Carboxylic acid derivatives, column chromatography, cyclodehydration reactions, maleimides, NMR

Background

Maleimides are not very common as natural compounds but a few examples illustrate their potential interest as drugs. Pencolide, isolated from *Penicillium multicolor*,[1] showdomycin, isolated from *Streptomyces showdoensis*,[2] and farinomalein, isolated from *Paecilomyces farinosus*,[3] are reported to present antitumor activity or activity against plant pathogenics. *N*-Phenyl- and *N*-phenylalkyl-maleimides display strong antifungal[4,5] and antibacterial[6] activities against a range of microorganisms, and there is evidence that *N*-arylmaleimides can be a good scaffold for developing new anti-cancer therapeutics[7] or for the treatment of CNS diseases.[8]

Comprehensive Organic Chemistry Experiments for the Laboratory Classroom
Edited by Carlos A. M. Afonso, Nuno R. Candeias, Dulce Pereira Simão, Alexandre F. Trindade, Jaime A. S. Coelho, Bin Tan, and Robert Franzén.

Published by the Royal Society of Chemistry, www.rsc.org

Maleimide is also an important building block in organic synthesis. *N*-Arylmaleimide derivatives are easily prepared from the corresponding *N*-arylmaleamic acids that can be obtained from the reaction of maleic anhydride with the appropriate aromatic amines.

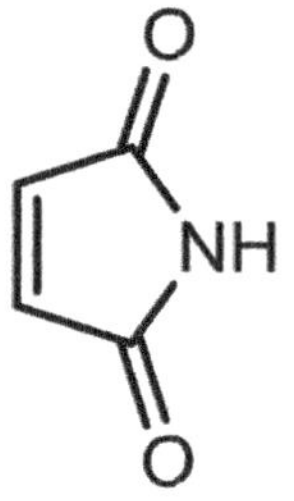

maleimide

From a pedagogical point of view this work illustrates several basic concepts related to carboxylic acid derivatives and the acyl substitution.

This experiment describes the synthesis of *N*-arylmaleimides from *N*-arylmaleamic acids through a cyclodehydration reaction. The *N*-arylmaleimides thus obtained are isolated by filtration and can be further purified by column chromatography and/or recrystallization and characterized through the analysis of their NMR (^{1}H and ^{13}C) and IR spectra.

Additional Safety

The students must wear proper personal protective equipment, such as laboratory coat, gloves and safety goggles. The experiment should be carried out in a fume hood. Special care should be taken with the use of the oil bath. Acetic anhydride, sodium acetate and *N*-arylmaleamic acids are harmful by contact, if swallowed or by inhalation.

Experimental Procedure

Laboratory Session: 4 h

Synthesis of N-Arylmaleimides

1. In a fume hood, prepare a 100 mL round bottomed flask with a magnetic stir bar, a condenser, a drying tube and a heating plate with an oil bath.
2. In the flask place 5 g of the *N*-arylmaleamic acid, 3 g of anhydrous sodium acetate and 30 mL of acetic anhydride.
3. Turn on the heating plate and keep the mixture at 100 °C, for 45 min, with stirring.

Product Isolation

4. After cooling, slowly add the reaction mixture to an ice/water mixture (about 100 mL), in a large beaker, while stirring with a glass rod. A solid should be formed.
5. Filter the solid under vacuum in a Büchner funnel and wash the filtrate several times with small amounts of water. Keep the crude solid under vacuum for more 5 min in order to dry the product and then transfer it to a pre-weighed glass watch and put it in the oven at 75 °C for 25–30 min.
6. Weigh the product obtained and use this value to calculate the crude yield.

Product Purification

7. Dissolve a small portion of the crude product (~1 g) in dichloromethane (2–3 mL) and transfer it to a silica flash column (2.0 cm × 30 cm).
8. Connect the column to an airline and elute with about 200 mL of dichloromethane.
9. Collect the eluate, partially remove the solvent in the rotatory evaporator and add a few drops of hexane. Put the solution aside for crystallization.
10. Collect the crystals in a Büchner funnel and oven dry at 75 °C.

Structural Characterization

11. Measure the melting point.
12. Place 15 mg of the dry compound in the NMR tube, add 0.5 mL of deuterated chloroform ($CDCl_3$) [containing tetramethylsilane (TMS) as the internal standard] and shake it until complete dissolution.
13. Acquire the ^1H and ^{13}C NMR spectra and confirm the structure of the product.

Waste Disposal

Dichloromethane collected in the rotatory evaporator and the recrystallization solvents should be disposed of in an appropriate container for posterior recycling/destruction.

Results Interpretation and Additional Questions

1. Determine the reaction yield.
2. Explain the role of acetic anhydride in this experiment.
3. Propose a mechanism for the reaction.
4. Comment on the solubility differences found between *N*-arylmaleimides and *N*-arylmaleamic acids.
5. Predict the main differences in the ^1H NMR spectrum of the synthesized maleimide derivative when compared with the corresponding maleamic acid.
6. Interpret the ^1H and ^{13}C NMR spectra of the synthesized compound.

References

1. J. H. Birkinshaw, M. G. Kalyanpur and C. E. Stickings, *Biochem. J.*, 1963, **86**, 237–243.
2. S. Roy-Burman, P. Roy-Burman and D. W. Visser, *Cancer Res.*, 1968, **28**, 1106–1610.
3. S. P. Putri, H. Kinoshita, F. Ihara, Y. Igarashi and T. Nihira, *J. Nat. Prod.*, 2009, **72**, 1544–1546.
4. M. Sortino, F. Garibotto, V. Cechinel Filho, M. Gupta, R. Enriz and S. Zacchino, *Bioorg. Med. Chem.*, 2011, **19**, 2823–2834.
5. K. Takatori, T. Hasegawa, S. Nakano, J. Kitamura and N. Kato, *Microbiol. Immunol.*, 1985, **29**, 1237–1241.
6. S. Watanabe, Y. Igarashi, K. Yagami and R. Imai, *Pestic. Sci.*, 1991, **31**, 45–51.
7. J. R. Cashman, M. MacDonald, S. Ghirmai, K. J. Okolotowicz, E. Sergienko, B. Brown, X. Garcia, D. Zhai, R. Dahl and J. C. Reed, *Bioorg. Med. Chem. Lett.*, 2010, **20**, 6560–6564.
8. N. Matuszak, G. G. Muccioli, G. Labar and D. M. Lambert, *J. Med. Chem.*, 2009, **52**, 7410–7420.

3.1.11. Synthesis and Characterization of *N*-Cyclohexyl-*N*-methyloctanamide

Ana Paula Paiva*[a] and Osvaldo Ortet[a,b]

[a]Centro de Química e Bioquímica, Departamento de Química e Bioquímica, Faculdade de Ciências, Universidade de Lisboa, Lisboa, Portugal; [b]Departamento de Ciência e Tecnologia, Universidade de Cabo Verde, Praia, Cape Verde. *E-mail: appaiva@fc.ul.pt

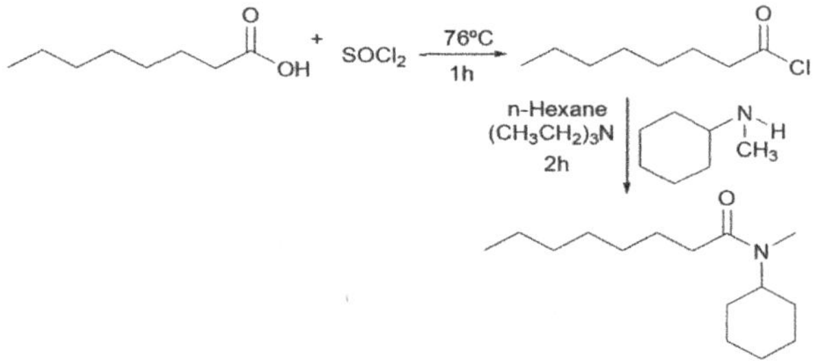

Number of sessions (duration of each session)	Hazard level	Difficulty level	Level of study
2 (4 h + 3 h)	Moderate	Medium	Intermediate

Class names Carboxylic acids, acyl chlorides, amines, amides

Concepts involved This experiment describes a method for preparing a tertiary amide by an addition–elimination reaction of thionyl chloride to a carboxylic acid; the intermediate acyl chloride is similarly converted into a tertiary amide by reaction with a secondary amine. The first reaction is carried out by reflux, the formation of the final amide is accomplished at room temperature, with stirring

Chemicals needed Octanoic acid, thionyl chloride, *N*-cyclohexylmethylamine, petroleum ether (60–80 °C) or *n*-hexane, triethylamine, 6 M hydrochloric acid solution, saturated sodium bicarbonate solution, saturated sodium chloride solution, anhydrous magnesium sulfate

Equipment and experimental techniques involved Heating–magnetic stirring plate, oil bath, apparatus for reflux, rotary evaporator, liquid–liquid extraction, FTIR and NMR spectrometers

Keywords Addition–elimination reactions, chemistry of carboxylic acid derivatives, liquid–liquid extraction, NMR, tertiary amide

Background

Organic amides, a family of compounds derived from carboxylic acids, are characterized by the presence of a nitrogen atom linked to a carbonyl group. Depending on the number of the nitrogen substituents, amides can be classified as primary ($-CO-NH_2$), secondary ($-CO-NHR^1$) and tertiary ($CO-NR^1R^2$) (R^1 and R^2 are alkyl or aryl groups, equal or different).[1] The amide functional group appears in several natural materials, *e.g.*, amide linkages are the molecular basis of the

Comprehensive Organic Chemistry Experiments for the Laboratory Classroom
Edited by Carlos A. M. Afonso, Nuno R. Candeias, Dulce Pereira Simão, Alexandre F. Trindade, Jaime A. S. Coelho, Bin Tan, and Robert Franzén.

Published by the Royal Society of Chemistry, www.rsc.org

peptide bonds characterizing proteins and enzymes, and in several drugs (for instance, penicillin and LSD).[2] The technological importance of amides should not be neglected as well, particularly in the plastic and rubber industries, and several materials rely on amide linkages (*e.g.*, *nylons*).[2] In the hydrometallurgical field, amide derivatives play an important role as extractants to selectively recover basic[3] and platinum-group metals[4] from different sorts of aqueous leaching solutions, as well as lanthanides and actinides from spent radioactive fuels.[5]

The traditional way to produce amides is still through the so-called Schotten–Baumann reaction, first described by German chemists Carl Schotten and Eugen Baumann, which involves the reaction between an acyl chloride and an amine.[6,7] Acyl chlorides are the most reactive compounds of all the carboxylic acid derivatives, with amides being the least reactive (organic anhydrides, acids and esters are in the middle); therefore, the forward reaction transforming acyl chlorides into the corresponding amides is always highly favoured, allowing the manufacture of amides with acceptable purity levels and in high yields.[1]

The present experiment consists of the synthesis from a carboxylic acid of the tertiary amide *N*-cyclohexyl-*N*-methyloctanamide, which shows an efficient and selective behaviour for the recovery of iron(III) from aqueous hydrochloric acid solutions.[3] Tertiary amides can conveniently be produced through the reaction of adequate amines with acyl chlorides, but also with anhydrides or esters; whatever the route chosen, the involved reaction type is always addition–elimination.[1] The particular tertiary amide *N*-cyclohexyl-*N*-methyloctanamide is more conveniently prepared through the acyl chloride route.[3] Hence, octanoic acid is initially transformed into octanoyl chloride by a reflux reaction with thionyl chloride, and then the acyl chloride reacts with *N*-cyclohexylmethylamine, at room temperature, to provide *N*-cyclohexyl-*N*-methyloctanamide in high yield.[3,4] Due to their inherent instability, acyl chlorides are preferably synthesized *in situ* from the correspondent carboxylic acids, although they can be purchased directly. Acyl chlorides are always hazardous and toxic materials; therefore, care should be taken with their handling.

FTIR (Fourier-transform infrared) and ^{1}H NMR spectra are the proposed techniques to characterize *N*-cyclohexyl-*N*-methyloctanamide. The ^{1}H and ^{13}C NMR spectra of this amide compound reveal a typical phenomenon that sometimes occurs with tertiary amides, which is the multiplication of signals.[8–10] For the particular case of *N*-cyclohexyl-*N*-methyloctanamide, the restriction of rotation along the C–N bond leads to the clear observation of the two possible conformations the compound can adopt, which is readily visible through the duplication of some signals. However, if the NMR spectra are plotted at high temperature, the signals will merge completely.[11]

Additional Safety

The chemicals used in this experiment require special care in their handling, particularly thionyl chloride and the intermediate acyl chloride. Accordingly, the synthesis and isolation of the intermediate and final products should always be carried out in a fume hood, with proper safety equipment, such as protective clothing, gloves and safety goggles. Octanoic acid is corrosive and causes burns; thionyl chloride is quite harmful by inhalation, if swallowed, or in contact with the skin; *N*-cyclohexylmethylamine is flammable, toxic and corrosive, causing severe burns; triethylamine is highly flammable, and harmful by inhalation, if swallowed or in contact with the skin; petroleum ether (60–80 °C) and *n*-hexane are highly flammable and toxic, causing serious damage to health by prolonged exposure; 6 M HCl solution is irritant to eyes, respiratory system and skin.

The washing of all the glass material in contact with the chemicals should be carried out in the fume hood. As mentioned above, special care should be taken with thionyl chloride and with the intermediate acyl chloride, as both products react violently with water, liberating toxic gases.

Experimental Procedure

Laboratory Session 1 (4 h): Preparation of *N*-Cyclohexyl-*N*-methyloctanamide

1. In a fume hood, place an oil bath on a magnetic stirring hot plate, and heat it to a temperature of 80–90 °C.
2. Weigh a 50 mL two-neck round-bottomed flask, and insert a stir bar. Add 7.0 mL (6.4 g, 44 mmol) of octanoic acid and 3.8 mL of thionyl chloride (6.2 g, 52 mmol), at room temperature. Stir gently. Place a thermometer (0–200 °C) in one of the necks.
3. Put the round-bottomed flask in the preheated oil bath. Insert a reflux condenser into the other neck. Heat the reaction mixture at a constant temperature of 76 °C, under stirring, for 1 h. Control the temperature by reference to the thermometer inserted in the bulk of the solution.
4. After that time, remove the excess thionyl chloride in a rotary evaporator (location of the equipment in the fume hood is strongly advised). Weigh the flask and register the mass of the synthesized octanoyl chloride.
5. Dissolve the crude product in about 130 mL of petroleum ether (60–80 °C) (or *n*-hexane).
6. In a beaker, prepare a mixture of 6.3 mL of *N*-cyclohexylmethylamine (5.4 g, 48 mmol) and 7.0 mL of triethylamine (5.1 g, 50 mmol). Insert a magnetic stir bar and place the beaker in an ice bath.
7. Under gentle stirring, carefully add the octanoyl chloride solution to the mixture of amines, keeping the beaker inside the ice bath during the addition. When the addition is completed, maintain the reaction mixture at room temperature, under stirring, for about 2 h.

Laboratory Session 2 (3 h): Isolation and Characterization of *N*-Cyclohexyl-*N*-methyloctanamide

1. To isolate *N*-cyclohexyl-*N*-methyloctanamide, wash the reaction mixture with distilled water (2 × 100 mL), 6 M HCl (4 × 50 mL), water (1 × 50 mL), saturated sodium bicarbonate solution (1 × 50 mL), and finally with saturated sodium chloride solution (1 × 50 mL).
2. Dry the organic extract with anhydrous magnesium sulfate, and pass the suspension through a folded filter paper to a previously weighed round-bottomed flask.
3. Evaporate the solvent in the rotary evaporator. Register the mass of product obtained.
4. Record the FTIR (NaCl cells) and the ^{1}H NMR spectra (in $CDCl_3$) of the isolated product.

Waste Disposal

The eventual traces of thionyl chloride, collected in the rotary evaporator, should be placed in a well-stoppered chlorinated waste collector recipient. Petroleum ether or *n*-hexane should be deposited in equivalent containers for non-chlorinated wastes. Remaining portions of 6 M HCl should be neutralized with a basic solution prior to their elimination.

Results Interpretation and Additional Questions

1. Calculate the yield of the final product, and compare it with those reported in the literature.
2. Present the overall mechanism of formation for *N*-cyclohexyl-*N*-methyloctanamide, and rationalize the use of triethylamine in the reaction medium.
3. In the isolation and purification of the product, several washing agents were employed. Explain why they were used, and comment about the order in which they were used.

4. Interpret and assign the most significant bands appearing in the FTIR spectrum of the product.
5. Interpret and assign the peaks of the ^{1}H NMR spectrum. Based on the highly visible duplication of the signals attributed to the C–H proton of the cyclohexyl, and those of the *N*-methyl groups, estimate the ratio of the two conformations evidenced in the NMR spectrum of this amide derivative.

Acknowledgement

Reprinted (adapted) with permission from *European Patent 0 210 004*, 1990. Copyright (1990) Espacenet, European Patent Office.

References

1. Any organic chemistry textbook, for instance: T. W. G. Solomons, G. B. Fryhle, and S. A. Snyder, *Organic Chemistry*, John Wiley & Sons, Inc., Hoboken, 11th edn, 2014, ch. 17, pp. 771–820.
2. A. Leggio, E. L. Belsito, M. L. Di Gioia, V. Leotta, E. Romio, C. Siciliano and A. Liguori, *Tetrahedron Lett.*, 2015, **56**, 199.
3. J. L. Figueira and A. P. Paiva, *Sep. Sci. Technol.*, 2009, **44**, 2928.
4. R. A. Grant, European Patent 0 210 004, Matthey Rustenberg Refiners Ltd., 1990.
5. G. Y. S. Chan, M. G. B. Drew, M. J. Hudson, P. B. Iveson, J.-O. Liljenzin, M. Skalberg, L. Spjuth and C. Madic, *J. Chem. Soc., Dalton Trans.*, 1997, **4**, 649.
6. C. Schotten, *Ber. Dtsch. Chem. Ges.*, 1884, **17**, 2544.
7. E. Baumann, *Ber. Dtsch. Chem. Ges.*, 1886, **19**, 3218.
8. A. H. Lewin, *Tetrahedron Lett.*, 1971, **12**, 3583.
9. J. Á. Bisceglia, M. C. Mollo and L. R. Orelli, *J. Mol. Struct.*, 2012, **1026**, 65.
10. L. Lefrançois, M. Hébrant, C. Tondre, J.-J. Delpuech, C. Berthon and C. Madic, *J. Chem. Soc., Perkin Trans., 2*, 1999, **6**, 1149.
11. R. J. Anderson, D. J. Bendell and P. W. Groundwater, *Organic Spectroscopic Analysis*, Royal Society of Chemistry, Cambridge, 2004, vol. 4, pp. 59–66.

3.1.12. Effect of a Catalyst in the Acylation of Alcohols with Acetic Anhydride: Manipulation of a Natural Aroma

Mónica S. Estevão* and Carlos A. M. Afonso

iMed.UL, Faculdade de Farmácia da Universidade de Lisboa, Av. Prof. Gama Pinto, 1649-003 Lisboa, Portugal
*E-mail: monica.estevao@ff.ulisboa.pt

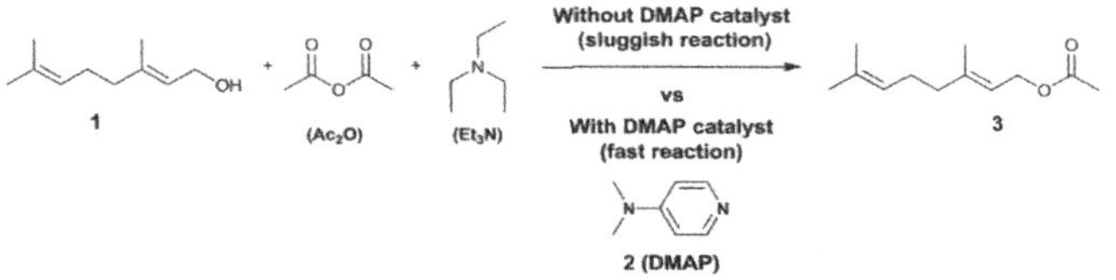

Number of sessions (duration of each session)	Hazard level	Difficulty level	Level of study
1 (4 h)	Moderate	Medium	Intermediate

Class names Alcohols, esters, anhydrides, amines

Concepts involved Demonstration of the mechanism of acylation and the effect of the catalyst

Chemicals needed Geraniol, acetic anhydride, triethylamine, 4-dimethylaminopyridine, dichloromethane, diethyl ether, 3 M HCl, saturated aqueous solution of $NaHCO_3$, anhydrous $MgSO_4$, silica gel, deuterated chloroform

Equipment and experimental techniques involved Standard experimental procedure, use of analytical thin-layer chromatography (TLC) and column chromatography, to monitor the progress of the reaction and for the product purification, respectively. In addition, use of IR and ^1H NMR techniques for product identification

Keywords Acylation, catalysis, chromatography, extraction, NMR

Background

Acetic anhydride (Ac_2O) is frequently used for the introduction of the acetyl group in alcohols (ester formation), thiols (thioester formation) and amines (amide formation).

The reactivity of this anhydride can be considerably increased if used with a catalyst. 4-(Dimethylamino)pyridine (DMAP, **2**) shows unusual properties as a strong activator for the acylation reaction with anhydrides. It is frequently used as catalyst in reactions that require the activation of the acyl group, with anhydride and acyl chloride substrates.[1] More recently, the preparation of compounds by asymmetric acyl transfer using chiral catalysts has received significant attention. One of their major general utilities is the kinetic resolution of secondary alcohols.[2,3] In 1996, Vedejs and Fu independently reported nucleophilic catalysts that could carry out asymmetric acyl transfer, with phosphine and DMAP reactive centres, respectively (Figure 3.1.12.1).[3]

Even though DMAP is frequently used in several reactions, the mechanisms of even the most simple DMAP-catalysed reactions are still an area of great interest.[4] This experiment involves the acetylation of geraniol (**1**) with acetic anhydride using triethylamine as base, in the presence or

Comprehensive Organic Chemistry Experiments for the Laboratory Classroom
Edited by Carlos A. M. Afonso, Nuno R. Candeias, Dulce Pereira Simão, Alexandre F. Trindade, Jaime A. S. Coelho, Bin Tan, and Robert Franzén.

Published by the Royal Society of Chemistry, www.rsc.org

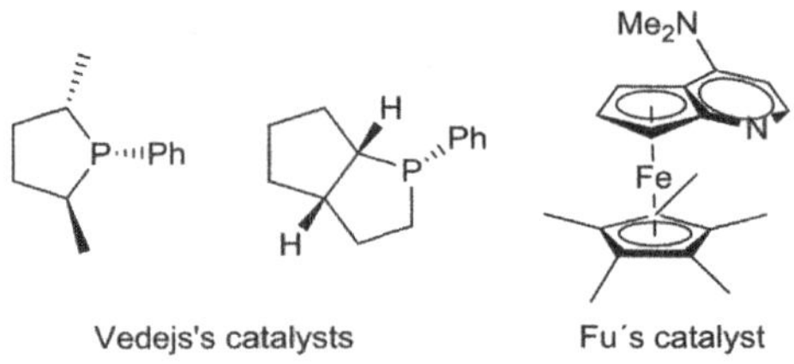

Figure 3.1.12.1 Nucleophilic catalysts for asymmetric acyl transfer.

absence of the catalyst 4-(dimethylamino)pyridine (DMAP, **2**). This experiment aims to illustrate the effect of DMAP as an efficient catalyst for anhydride activation. In addition, geraniol (**1**) is an essential molecule in the flavour and fragrance industries and is a common ingredient in consumer products produced by these industries. It is isolated from citronella oils and from palmarosa oil and its pleasant odour is known to exhibit insecticidal and repellent properties. Geraniol appears as a clear to pale-yellow oil that is insoluble in water, but soluble in most organic solvents. Geraniol has a characteristic rose-like odour and the taste (at 10 ppm) is described as sweet floral rose-like, citrus with fruity and spice-like with a waxy green nuance.[5] Geranyl acetate (**3**) is also widely used as a component of soaps, detergents, creams, lotions and perfumes. As a flavouring ingredient it is found in various fruit, berry, floral and spice flavours.[6]

Additional Safety

The experiment should be carried out in a fume hood, with proper safety equipment, such as protective clothing, gloves and safety goggles. Acetic anhydride is flammable, harmful if swallowed, causes severe skin burns and eye damage and is toxic if inhaled. Triethylamine is a highly flammable liquid and vapour, harmful is swallowed, toxic in contact with skin or if inhaled, causes severe skin burns and eye damage, and may cause respiratory irritation. Geraniol causes skin irritation and serious eye damage. 4-(Dimethylamino)pyridine is toxic if swallowed, fatal in contact with skin, causes skin irritation, serious eye irritation and may cause respiratory irritation. Silica gel is very adsorbent and may have a drying effect on skin and eyes. Use respiratory, eyes and hands protection.

Experimental Procedure

1. In a fume hood prepare a cooling bath with ice and cooking salt (temperature < 0 °C).
2. Prepare a 50 mL round-bottom flask with a magnetic stir bar.
3. Add, successively, 1.0 g (6.5 mmol) of geraniol, dichloromethane (15 mL), triethylamine (1.4 mL, 1.5 equiv.) and acetic anhydride (0.92 mL, 1.5 equiv.).
4. Choose an appropriate eluent and monitor the reaction by TLC for 30 min (detection by UV light or with phosphomolybdic acid).
5. After 30 min add 4-dimethylaminopyridine (0.16 g, 0.2 equiv.) and monitor the reaction by TLC.
6. After reaction completion (≤60 min) dilute the reaction mixture with diethyl ether (50 mL) and transfer it into a separatory funnel.
7. Wash with an aqueous solution of HCl (3 M; 2 × 50 mL) followed by an aqueous saturated solution of sodium hydrogen carbonate (2 × 50 mL).
8. Dry the organic layer with anhydrous magnesium sulfate, filter and remove the solvent on the rotatory evaporator.
9. Weigh the obtained product and perform a TLC analysis.

10. If the obtained product contains starting material, perform a column chromatography with silica using the same eluent chosen for the TLC monitoring (column diameter 30 mm, $weight_{silica} = 30 \times weight_{crude}$).
11. Acquire the IR and ^{1}H NMR spectra of the obtained product and compare with the spectral data of geraniol.

Waste Disposal

Dispose of the dichloromethane in a specified container for halogenated solvents. Dispose of the solid residues in a specified container.

Results Interpretation and Additional Questions

1. Check if you have obtained the desired product and if it is pure.
2. Explain the mechanism of the reaction without DMAP.
3. Mechanistically, how do you explain the role of the catalyst DMAP?
4. Do you think that you could perform the reaction under the same experimental conditions but in the absence of the base triethylamine?
5. Do you think that you could use diethylamine instead of triethylamine?
6. Do you think that the reaction would be faster if you used benzoic anhydride instead of acetic anhydride? Justify your answer.

References

1. (a) G. Hofle, W. Steglich and H. Vorbruggen, *Angew. Chem., Int. Ed.*, 1978, **17**, 569; (b) M. Baidya, S. Kobayashi, F. Brotzel, U. Schmidhammer, E. Riedle and H. Mayr, *Angew. Chem., Int. Ed.*, 2007, **46**, 6176.
2. A. C. Spivey and S. Arseniyadis, *Angew. Chem., Int. Ed.*, 2004, **43**, 5436.
3. A. C. Spivey and S. Arseniyadis, *Top. Curr. Chem.*, 2010, **291**, 233.
4. S. Xu, I. Held, B. Kempf, H. Mayr, W. Steglich and H. Zipse, The DMAP-Catalyzed Acetylation of Alcohols—A Mechanistic Study (DMAP = 4-(Dimethylamino)pyridine), *Chem.–Eur. J.*, 2005, **11**, 4751.
5. (a) S. Budavari, M. O'Neil, A. Smith, P. Heckelman and J. Obenchain, *The Merck Index*, 12th edn, CRC Press, 1996, p. 441; (b) W. Chen and A. M. Viljoen, *S. Afr. J. Bot.*, 2010, **76**, 643.
6. D. L. J. Opdyke, *Food Cosmet. Toxicol.*, 1974, **12**, 885.

3.1.13. Synthesis of 4,5-Dichloro-1,2-dicyanobenzene

Dulce Simão

Department of Chemistry, Instituto Superior Técnico, University of Lisbon, Centro de Química Estrutural, Av. Rovisco Pais, 1, 1049-001 Lisbon, Portugal. *E-mail: dulcesimao@tecnico.ulisboa.pt

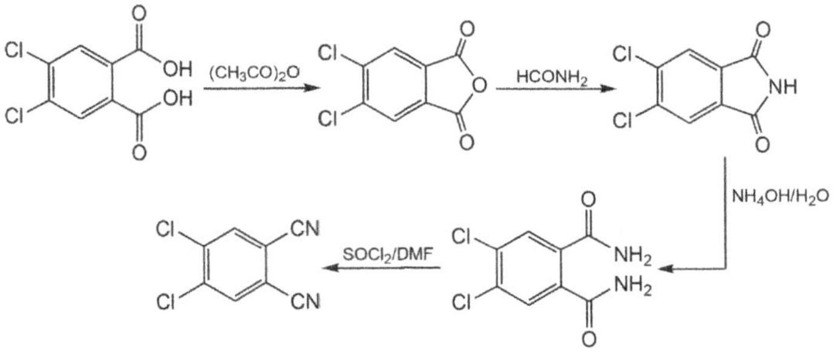

Number of sessions (duration of each session)	Hazard level	Difficulty level	Level of study
5 (5 h each)	Moderate	Medium	Intermediate

Class names Carboxylic acids, anhydrides, imides, amides, nitriles

Concepts involved This experiment introduces undergraduate students to a multi-step organic synthesis involving several functional groups connected to a carbonyl group, under mild conditions and easy work-up

Chemicals needed 4,5-Dichloro-1,2-benzenedicarboxylic acid, acetic anhydride, petroleum ether (40–60 °C), formamide, ammonia (25% and 33%), thionyl chloride, toluene, calcium chloride, anhydrous dimethylformamide

Equipment and experimental techniques involved Magnetic stirring, reflux apparatus, apparatus for anhydrous reaction conditions under inert atmosphere and ice water bath, vacuum filtration, recrystallization, IR, NMR and melting point

Keywords carboxylic acid and derivatives reactions, crystallization, filtration, IR, nitrile formation, NMR

Background

The two main procedures by which alkyl cyanides are prepared, namely, the nucleophilic substitution of alkyl halides by cyanide and the conversion of aldehydes and ketones into cyanohydrins are not suitable for aryl cyanides. The dehydration of amides is one of the most common methods for the synthesis of nitriles. The compound prepared in this experiment, 4,5-dichloro-1,2-dicyanobenzene,[1] is an important precursor in the synthesis of molecular

Comprehensive Organic Chemistry Experiments for the Laboratory Classroom
Edited by Carlos A. M. Afonso, Nuno R. Candeias, Dulce Pereira Simão, Alexandre F. Trindade, Jaime A. S. Coelho, Bin Tan, and Robert Franzén.

Published by the Royal Society of Chemistry, www.rsc.org

organic conductors[2] (Experiment 6.4) and also in obtaining substituted phthalocyanines, which are widely used as pigments and dyes.[3] The synthetic route herein described involves several transformations of carboxylic acids and derivatives that occur through addition and elimination mechanisms together with nucleophilic attack to a carbonyl group. Carboxylic acids have low reactivity due the acid proton and, thus, the first step is its transformation into a more reactive phthalic anhydride. The anhydride can be converted into the imide by reaction with ammonia,[4] although a considerable amount of the corresponding diamide is also formed. Formamide can be used instead to avoid the formation of the diamide side-product and only the desired pure imide is obtained.[3] Further reaction with ammonia solutions at different concentrations leads to diamide. Dehydration accomplished by thionyl chloride produces the corresponding dicyanobenzene.

Additional Safety

This work must be carried out in a fume hood due to corrosive and irritant acetic anhydride and ammonia. Thionyl chloride is a corrosive liquid and may cause burns. It reacts with water to produce hydrogen chloride and sulphur dioxide and since it is used in excess care must be taken when the reaction mixture is added to water.

Experimental Procedure

Preparation of 4,5-Dichlorophthalic Anhydride

Session 1

1. In a 250 mL round-bottom flask, heat 4,5-dichlorophthalic acid (50.0 g, 0.21 mol) and acetic anhydride (45 mL, 0.22 mol) at 130 °C for 3 h.
2. After cooling, collect the product by vacuum filtration and wash well with petroleum ether (40–60 °C).
3. Recrystallize from toluene.
4. Dry the white solid under vacuum.
5. Record the yield, melting point and IR and ^{1}H and ^{13}C NMR spectra.

Preparation of 4,5-Dichlorophthalimide

Session 2

1. In a 250 mL round-bottom flask with magnetic stirring, heat 4,5-dichlorophthalic anhydride (10.0 g, 0.04 mol) and formamide (14 mL) under reflux for 3 h.
2. After cooling, collect the product by vacuum filtration and wash with water (*ca.* 500 mL).
3. Dry under vacuum.
4. Record the yield, melting point and IR and ^{1}H and ^{13}C NMR spectra.

Preparation of 4,5-Dichlorophthalamide

Session 3

In a 250 mL round-bottom flask with magnetic stirring, add 4,5-dichlorophthalimide (8.0 g, 37 mmol) and ammonia solution (25%) (110 mL). Stir for 24 h.

Next day – add ammonia solution (33%) (36 mL) and keep stirring for a further 24 h.

Next day – collect the product by vacuum filtration and wash with water (*ca.* 300 mL). Dry under vacuum. Record the yield, melting point and IR and ^{1}H and ^{13}C NMR spectra.

Preparation of 4,5-Dichloro-1,2-dicyanobenzene

Session 4

1. In a 250 mL four-necked flask with magnetic stirring, add under inert gas, dry *N,N*-dimethylformamide (25 mL).
2. After cooling to 0 °C under stirring add distilled thionyl chloride (18 mL).
3. After 2 h add 4,5-dichlorophthalamide (5.0 g, 0.02 mol) in small portions and stir for 5 h.
4. Remove the cooling bath, change the inert gas inlet by a balloon filled with argon or nitrogen and leave the reaction stirring until next session.

Session 5

1. Pour the reaction mixture over ice/water.
2. Collect the product by vacuum filtration and wash the solid with water (*ca.* 250 mL).
3. Recrystallize from methanol and dry under vacuum.
4. Record the yield, melting point and IR and ^{1}H and ^{13}C NMR spectra.

Waste Disposal

The neutralized aqueous solutions may be diluted and discarded in the sink. Methanol and toluene may be stored for further purification.

Results Interpretation and Additional Questions

1. Determine the reaction yields, and the overall yield.
2. Compare the melting points of the obtained products with the ones reported in the literature.
3. Interpret the IR and ^{1}H and ^{13}C NMR spectra.
4. Write all reaction mechanisms.
5. Explain why all reagents and solvent must be anhydrous in the last step.

References

1. C. A. M. Afonso, D. P. Simão, L. P. Ferreira, M. S. Serra and M. M. M. Raposo, *100 Experiências de Química Orgânica*, IST Press, 2011, p. 313.
2. D. Simão, *et al.*, *Eur. J. Inorg. Chem.*, 2001, **12**, 3119–3126.
3. D. Wöhrle, *et al.*, *Synthesis*, 1993, 194.
4. T. W. G. Solomons, *Organic Chemistry*, John Wiley & Sons Inc., New York, 7th edn, 1998, p. 837.

3.1.14. Synthesis of *N-tert*-Butyloxycarbonyl-3-nitro-L-tyrosine Methyl Ester

Susana P. G. Costa*, M. Manuela M. Raposo, Cátia I. C. Esteves, and R. Cristina M. Ferreira

University of Minho, Department of Chemistry, Campus de Gualtar, 4710-057 Braga, Portugal
*E-mail: spc@quimica.uminho.pt

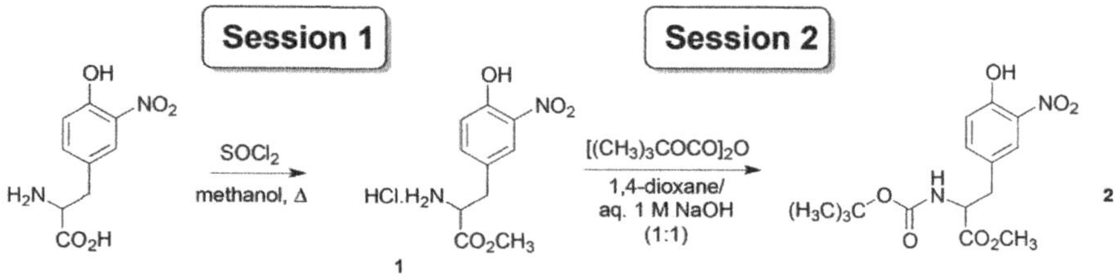

Number of sessions (duration of each sessions)	Hazard level	Difficulty level	Level of study
2 (4 h each)	Moderate	Medium	Intermediate

Class names Amines, alcohols, carboxylic acids, acid chlorides, carbonate esters, carbamates and esters

Concepts involved Synthesis of a tyrosine derivative protected at the amino acid N- and C-terminals with carbamate and ester groups, by acylation and alcoholysis, respectively; rotary evaporation; liquid–liquid extraction; characterization by ^1H NMR and IR spectroscopy

Chemicals needed 3-Nitro-L-tyrosine, thionyl chloride, di-*tert*-butyl dicarbonate, methanol, 1,4-dioxane, sodium hydroxide, magnesium sulfate (anhydrous), ethyl acetate, potassium hydrogen sulfate

Equipment and experimental techniques involved Apparatus for reaction with magnetic stirring at room temperature, heating mantle for heating at reflux, melting point apparatus, rotary evaporation, liquid–liquid extraction funnel, NMR spectrometer, IR spectrophotometer

Keywords α-Amino acid, IR spectroscopy, liquid–liquid extraction, NMR spectroscopy, nucleophilic substitution, protecting group

Background

In synthetic organic chemistry, in order to block the formation of undesired bonds and side reactions, appropriate protecting groups are required so that the reagents undergo transformation into the products as neatly as possible. Temporary protecting groups suppress or diminish the reactivity of functional groups in a molecule, allowing selective formation of bonds between the reactants, and should be removed whenever needed or at the end of the synthetic process. However, the use of protecting groups is usually a laborious necessity, as they must be introduced by a convenient and efficient synthesis, have chemical stability towards different reagents and be selectively removed in the presence of other protecting groups (orthogonality).[1,2]

Comprehensive Organic Chemistry Experiments for the Laboratory Classroom
Edited by Carlos A. M. Afonso, Nuno R. Candeias, Dulce Pereira Simão, Alexandre F. Trindade, Jaime A. S. Coelho, Bin Tan, and Robert Franzén.

Published by the Royal Society of Chemistry, www.rsc.org

For bifunctional molecules such as amino acids, the building blocks in peptide chemistry, the use of protecting groups is essential due to the presence of amino and carboxylic acid functions, the N- and C-terminals, respectively. For the amino group, the commonly employed protecting groups are derived from carbamates and the most usual are the *tert*-butyloxycarbonyl (Boc), benzyloxycarbonyl (Z) and 9-fluorenylmethyloxycarbonyl (Fmoc) groups, which differ in the conditions required for their cleavage (Boc and Z are removed by acidic treatment whereas Fmoc is removed in basic conditions). As for the carboxylic acid group, the most common protecting groups are methyl, ethyl or benzyl esters, which can be cleaved with base.[3,4]

Tyrosine is an aromatic amino acid that is widely present in peptides and in many surface proteins, and is thus exposed and available to undergo reactions such as nitration, resulting in the formation of nitrotyrosine or tyrosine nitrated protein.[5] The presence of a nitrated tyrosine in a protein can dramatically change its structure and conformation, and as a consequence alter its function.[6]

The aim of this work is the synthesis of a protected amino acid derivative, namely, tyrosine, using simple experimental techniques and commercially available reagents. The protection will be accomplished by classical reactions in organic chemistry: preparation of an ester from a carboxylic acid by nucleophilic substitution (conversion of the carboxylic acid into an acid chloride) followed by alcoholysis; preparation of a carbamate from an amine by acylation with a carbonate ester.

Additional information: in case there is interest in conducting Experiment 12.2.2 in this book and if considered adequate by the teacher, the final product can undergo an additional reaction. This additional procedure involves catalytic reduction with hydrogen of the nitro group at the side chain of tyrosine to afford the corresponding amino derivative. Details of the additional procedure are given in the supplementary material.[7,8]

Additional Safety

All sessions should be carried out in a well-ventilated fume hood, with proper safety equipment, such as protective clothing, gloves and safety goggles. 3-Nitro-L-tyrosine may cause skin and eye irritation. Thionyl chloride may cause severe burns and respiratory irritation, being toxic when inhaled. It reacts violently with water, liberating corrosive fumes (SO_2 and HCl). Di-*tert*-butyl dicarbonate may cause skin, eye and respiratory irritation and is a flammable solid. Organic solvents are highly flammable liquids and vapours, 1,4-dioxane may cause eye and respiratory irritation and form explosive peroxides, methanol is toxic in contact with skin, if swallowed and inhaled, and ethyl acetate may cause eye irritation. Solid sodium hydroxide may cause severe skin burns and eye damage.

Experimental Procedure

Laboratory Session 1 (4 h): Synthesis of 3-Nitro-L-tyrosine Methyl Ester Hydrochloride (1)

1. Prepare a 25 mL round-bottomed flask with boiling chips, add 226 mg (1 mmol, 1 equiv.) of 3-nitro-L-tyrosine and methanol (15 mL). Carefully add dropwise 290 μL (4 mmol, 4 equiv.) of thionyl chloride.
2. Place a water condenser with an adapter with calcium chloride filling on the round-bottomed flask and heat at reflux for 3 h in a heating mantle.
3. After cooling down, filter the reactional mixture under vacuum and remove the solvent on the rotary evaporator. Caution: hydrogen chloride and sulfur dioxide may be released, evaporation must be carried out in a fume hood.
4. Wash the obtained solid residue twice with diethyl ether (5 mL), collect the solid and allow it to dry in air in a fume hood. Discard the diethyl ether in the appropriate solvent waste container.
5. Weigh the yellowish solid, acquire a ^{1}H NMR spectrum in DMSO-d_6 and an IR spectrum in a KBr disc. Compare the spectra with literature data (see supplementary material).[8]

Laboratory Session 2 (3 h): Synthesis of *N-tert*-Butyloxycarbonyl-3-nitro-L-tyrosine Methyl Ester 2

1. Prepare a 25 mL round-bottomed flask with a magnetic stirrer and add 277 mg (1 mmol, 1 equiv.) of 3-nitro-L-tyrosine methyl ester hydrochloride (**1**), 1,4-dioxane (2 mL), water (1 mL) and aqueous solution sodium hydroxide (1 M; 1.1 mL, 1.1 mmol, 1.1 equiv.). Stir in an ice bath for about 15 min.
2. Add 240 mg (1.1 mmol, 1.1 equiv.) of di-*tert*-butyl dicarbonate and stir the solution at room temperature for 48 h. Important: this reagent has a low melting point (22–24 °C) and should be weighed rapidly to avoid becoming slurry-like and difficult to weigh.
3. In the rotary evaporator, reduce the solution volume to half (remove 1,4-dioxane). Allow the flask to cool and add ethyl acetate (15 mL). Acidify to pH 3 with 1 M potassium hydrogen sulfate solution (about 10 mL) and transfer the solution to an extraction funnel.
4. Extract with ethyl acetate (3 × 15 mL), separate and combine the organic layers and wash the organic layer with water (3 × 10 mL). Dry the organic layer with anhydrous magnesium sulfate, filter under vaccum and evaporate under reduced pressure.
5. Weigh the yellow solid, measure the melting point, and acquire ^1H NMR spectra in $CDCl_3$ and an IR spectrum in KBr disc. Compare the spectra with literature data (see supplementary material).[8]

Waste Disposal

Solvents can be recovered after evaporation in a rotary evaporator and saved for later distillation and reuse, if possible. If not, methanol, 1,4-dioxane and ethyl acetate should be collected in a non-chlorinated waste collector recipient for subsequent destruction. Aqueous solutions from liquid–liquid extraction should be neutralized and discarded in the sink.

Results Interpretation and Additional Questions

1. Calculate the yield of reaction for each preparation and the overall yield.
2. What is the function of thionyl chloride in the preparation of compound **1**?
3. Write the mechanism for the reactions in sessions 1 and 2 considering the reagents used.
4. Interpret the ^1H NMR spectral data obtained for products **1** and **2**, with the corresponding assignment of signals. Compare with the spectra provided as supplementary material.
5. Interpret the IR spectral data obtained for products **1** and **2**, with the corresponding assignment of bands. Compare with the spectra provided as supplementary material.
6. Relate the wavenumbers of the carbonyl group bands in the IR spectra to the structure of compounds **1** and **2** and the starting reagent (acquire an IR spectra in a KBr disc).

References

1. P. G. M. Wuts and T. W. Greene, *Greene's Protective Groups in Organic Synthesis*, Wiley-Interscience, Hoboken, New Jersey, 4th edn, 2006.
2. J. Robertson, *Protecting Group Chemistry (Oxford Chemistry Primers)*, Oxford University Press, Oxford, 1st edn, 2000.
3. *Amino Acids, Peptides and Proteins in Organic Chemistry, Protection Reactions, Medicinal Chemistry, Combinatorial Synthesis*, ed. A. B. Hughes, Wiley-VCH Verlag GmbH, Weinheim, 1st edn, 2011, vol. 4.
4. A. Isidro-Llobet, M. Álvarez and F. Albericio, *Chem. Rev.*, 2009, **109**, 2455–2504.
5. H. Ischiropoulos, *Biochem. Biophys. Res. Commun.*, 2003, **305**, 776–783.
6. J. M. Souza, G. Peluffo and R. Radi, *Free Radicals Biol. Med.*, 2008, **45**, 357–366.
7. K. Guzow, M. Szabelski, J. Malicka and W. Wiczk, *Helv. Chim. Acta*, 2001, **84**, 1086–1092.
8. S. P. G. Costa, E. Oliveira, C. Lodeiro and M. M. M. Raposo, *Sensors*, 2007, **7**, 2096–2114.

3.1.15. Michael Addition Reaction Followed by Elimination Under Solvent-Free Conditions

Marcos A. P. Martins*, Clarissa P. Frizzo, Aniele Z. Tier, Guilherme C. Paveglio, Caroline R. Bender, Paulo R. S. Salbego, Geórgia C. Zimmer, and Nilo Zanatta

Núcleo de Química de Heterociclos (NUQUIMHE), Department of Chemistry, Federal University of Santa Maria, 97105-900, Santa Maria, RS, Brazil. *E-mail: mmartins@base.ufsm.br, clarissa.frizzo@gmail.com

1 **2a/2b** **3a/3b**

(i) ethanol, 25 °C, 5 min (ii) solvent-free, 25 °C, 5 min

2a,3a = R^1, R^2= H, benzyl **2b,3b** = R^1, R^2= *iso*-Bu

Number of session (duration of each session)	Hazard level	Difficulty level	Level of study
1 (2 h)	Moderate	Medium	Intermediate

Class names α,β-unsaturated ketones and amines

Concepts involved Conjugated addition, elimination, substitution at vinylic carbon, structure of organic compounds elucidated by NMR spectroscopy

Chemicals needed 1,1,1-Trifluoro-4-methoxypent-3-en-2-one (enone **1**), benzylamine (**2a**), *N*,*N*-diisobutylamine (**2b**), ethanol

Equipment and experimental techniques involved Apparatus for reaction, magnetic stirring, vacuum system, separatory funnel

Keywords Amination, liquid–liquid extraction, NMR, solvent-free reaction, vinyl ketones

Background

β-Enaminones are important organic compounds that have multiple applications, due to their adaptable dual behavior as electrophiles and nucleophiles. Their utilization in organic synthesis is currently a subject of great interest.[1] These compounds have attracted much attention because they are important building blocks for the synthesis of many biologically active compounds such as dopamine autoreceptor agonists,[2] acetylcholinesterase inhibitors,[3] and anticonvulsants.[4] They are also useful intermediates for the preparation of several amino acids,[5] aminols,[6] peptides, and alkaloids.[7] Additionally, they have been employed as synthetic building blocks for a wide variety of heterocycles,[8] as well as in the obtainment of other synthetic intermediates.[9]

Comprehensive Organic Chemistry Experiments for the Laboratory Classroom
Edited by Carlos A. M. Afonso, Nuno R. Candeias, Dulce Pereira Simão, Alexandre F. Trindade, Jaime A. S. Coelho, Bin Tan, and Robert Franzén.

Published by the Royal Society of Chemistry, www.rsc.org

These compounds have been synthesized through (i) direct condensation of 1,3-dicarbonyl compounds with amines;[10] (ii) reactions between metal enolates and imine derivatives;[11] (iii) cleavage of heterocycles together with some novel unconventional routes;[12] and (iv) Michael addition reaction followed by elimination. This last reaction involves the initial attack of the amine nitrogen atom in the enone β-carbon, with a charge delocalization to the carbonyl group and subsequent prototropism, making the *O*-alkyl portion a good leaving group. Finally, the last step is the elimination of the alcohol molecule.

One of the principles of green chemistry is to promote the idea of "greener" solvents (non-toxic, benign to environment), which involves replacement in cases that can be substituted with safer alternatives, as well as changes in the methodologies of organic synthesis, when solvents are not needed.[13] In particular, it has been shown that syntheses that are performed under solvent-free conditions are often fast, regio- or chemoselective, occur at high yields, and have environmental and economic advantages. The development of new environmentally favorable routes for the production of chemical intermediates and products is an area of considerable interest to the chemical processing industry and the academic and scientific community. Thus, the aim of this experiment is investigate the synthesis of β-enaminones through the solvent-free and non-toxic solvent reaction of enone **1** with primary and secondary amines.

Additional Safety

All sessions should be carried out in a fume hood, with proper safety equipment such as protective clothing, gloves, and safety goggles. Reagents and solvent should be used carefully because they can be irritating to the eyes and skin. Compounds with a structure similar to **1** are described as flammable and are suspected of causing cancer. The toxicological properties of **1** have not been thoroughly investigated. Benzilamine and diisobutylamine cause skin burn and serious eye damage.

Experimental Procedure

Laboratory Session (2 h): β-Enaminone Synthesis

Method A — Ethanol

1. Prepare a 25/50 mL previously weighed round-bottom flask (M_1) with a magnetic stir bar, and add 0.840 g (5 mmol) of enone **1**, and 2.0 mL of dry ethanol.
2. After this, add 0.536 g (5 mmol) of benzylamine (**2a**) [or 0.646 g (5 mmol) of *N*,*N*-diisobutylamine (**2b**)].
3. Stir the equivalent mixture at 25 °C for 5 min.

Method B — Solvent-Free Reaction

1. Prepare a 25/50 mL previously weighed round-bottom flask (M_1) with a magnetic stir bar, and add 0.840 g (5 mmol) of enone **1**.
2. In a beaker, add 0.536 g (5 mmol) of benzylamine (**2a**) [or 0.646 g (5 mmol) of *N*,*N*-diisobutylamine (**2b**)].
3. With a Pasteur pipette, add the appropriate amine (**2a** or **2b**) to enone **1** at 25 °C.
4. Stir the equivalent mixture at 25 °C for 5 min.

Product Isolation — Reaction with Benzylamine

1. Keep the product under vacuum for 1 h, in order to remove any residual solvent (ethanol is used as the solvent in method A) and/or by-product (methanol is a by-product of the reactions).

Product Isolation — Reaction with N,N-Diisobutylamine

1. Wash the product with distilled water (1 × 5 mL) in a separatory funnel.
2. Keep the product under vacuum (10 mbar) for 1 h, to remove any residual solvent (ethanol is used as the solvent in method A) and/or by-product (methanol is a by-product of the reactions).
3. Perform a NMR analysis and calculate the reaction yield.

Yield Determination

4. After the reaction reaches completion, weigh the round-bottom flask containing the reaction mixture and determine the final weight (M_2).
5. From this value (M_2) and the weight of the empty round-bottom flask (M_1), calculate the mass of the product obtained in the experiment (M_3) using eqn (3.1.15.1):

$$M_3 = M_2 - M_1 \tag{3.1.15.1}$$

6. Report the amount of product obtained (in mmol) with the limiting reactant used (in this case the enone **1**), in order to calculate the reaction yield in accordance with eqn (3.1.15.2):

$$\text{Yield (\%)} = \frac{\text{product (mmol)}}{\text{limiting reactant (mmol)}} \times 100 \tag{3.1.15.2}$$

Waste Disposal

Effluents from this reaction are non-toxic and can be released into the environment.

Interpretation of Results and Additional Questions

1. Interpret the ^{1}H NMR spectra obtained, and compare the obtained melting point with that reported in the Supporting Information.
2. Rationalize the substitution mechanism at the sp^2 carbon involved in the reaction of enone **1** with amine.
3. Indicate the advantages of solvent-free reactions in relation to reactions with molecular solvents.
4. Identify the nucleophile and electrophile and indicate the orbitals involved in this organic reaction. Specify the high energy molecular orbital (HOMO) and low unoccupied molecular orbital (LUMO) involved in the reaction.

Acknowledgement

Experiment adapted with permission from *J. Braz. Chem. Soc.*, 2007, 18 (8), 1486-1491, edited by Sociedade Brasileira de Química (SBQ).

References

1. B. Stanovnik and J. Svete, *Chem. Rev.*, 2004, **104**, 2433.
2. B. W. Caprathe, J. C. Jaen, L. D. Wise, T. G. Heffner, T. A. Pudsley, L. T. Melther and M. Parvez, *J. Med. Chem.*, 1991, **34**, 3726.
3. F. Gatta, M. R. Del Giudice, M. Pomponi and M. Marta, *Heterocycles*, 1992, **34**, 991.

4. K. R. Scott, I. O. Edafiogho, E. C. Richardson, V. A. Farrar, J. A. Moore, E. I. Tietz, C. N. Hinko, H. Chang, A. El-Assadi and J. M. Nicholson, *J. Med. Chem.*, 1993, **36**, 1947.
5. G. Palmieri and C. Cimarelli, *J. Org. Chem.*, 1996, **61**, 5557.
6. (a) G. Bartoli, C. Cimarelli, E. Marcantoni, G. Palmieri and M. Petrini, *J. Org. Chem.*, 1994, **59**, 5328; (b) C. Cimarelli, S. Giuli and G. Palmieri, *Eur. J. Org. Chem.*, 2006, **2006**, 1017; (c) M. I. N. C. Harris and A. C. H. Braga, *J. Braz. Chem. Soc.*, 2004, **15**, 971; (d) N. Zanatta, A. M. C. Squizani, L. Fantinel, F. M. Nachtigall, D. M. Borchhardt, H. G. Bonacorso and M. A. P. Martins, *J. Braz. Chem. Soc.*, 2005, **16**, 1255.
7. L. G. Beholz, P. Benovsky, D. L. Ward, N. S. Bata and J. R. Stille, *J. Org. Chem.*, 1997, **62**, 1033.
8. A. C. Spivey, R. Srikaran, C. M. Diaper and D. J. Turner, *Org. Biomol. Chem.*, 2003, **1**, 1638.
9. (a) F. A. Rosa, P. Machado, M. Rossatto, P. S. Vargas, H. G. Bonacorso, N. Zanatta and M. A. P. Martins, *Synlett*, 2007, 3165; (b) F. A. Rosa, P. Machado, P. S. Vargas, H. G. Bonacorso, N. Zanatta and M. A. P. Martins, *Synlett*, 2008, 1673.
10. (a) J.-Y. Liu, G.-E. Cao, W. Xu, J. Cao and W.-L. Wang, *Appl. Organomet. Chem.*, 2010, **24**, 685; (b) M. K. Samantaray, C. Dash, M. M. Shaikh, K. Pang, R. J. Butcher and P. Ghosh, *Inorg. Chem.*, 2011, **50**, 1840; (c) R. Rezaei and M. Shakeri, *Asian J. Chem.*, 2013, **25**, 7079.
11. (a) T. S. Saleh, M. A. Al-Omar and H. A. Abdel-Aziz, *Lett. Org. Chem.*, 2010, **7**, 483; (b) M. Ghandi and A. H. Jameà, *Tetrahedron Lett.*, 2011, **52**, 4005.
12. (a) D. S. Reddy, W. R. Judd and J. Aubé, *Org. Lett.*, 2003, **5**, 3899; (b) S. Ueno, R. Shimizu and R. Kuwano, *Angew. Chem.*, 2009, **121**, 4613; (c) H. Seki and G. I. Georg, *J. Am. Chem. Soc.*, 2010, **132**, 15512; (d) Y. Wang, X. Bi, W.-Q. Li, D. Li, Q. Zhang, Q. Liu and B. S. Ondon, *Org. Lett.*, 2011, **13**, 1722; (e) T. Miura, Y. Funakoshi, M. Morimoto, T. Biyajima and M. Murakami, *J. Am. Chem. Soc.*, 2012, **134**, 17440; (f) D. Yu, Y. N. Sum, A. C. C. Ean, M. P. Chin and Y. Zhang, *Angew. Chem.*, 2013, **125**, 5229.
13. (a) *Green Chemistry: Designing Chemistry for the Environment*, ed. P. T. Anastas and T. C. Williamson, American Chemical Society, Washington DC, 1996, ACS Symposium Series 626; (b) P. T. Anastas, *Green Chemistry: Theory and Practice*, Oxford University, Press, New York, 1998.

3.1.16. Synthesis of a γ-Keto-amide Derived from Thiophene Using a Carboxyl Ester as Precursor

M. Manuela M. Raposo*

Department of Chemistry, University of Minho, Campus de Gualtar, 4710-057 Braga, Portugal
*E-mail: mfox@quimica.uminho.pt

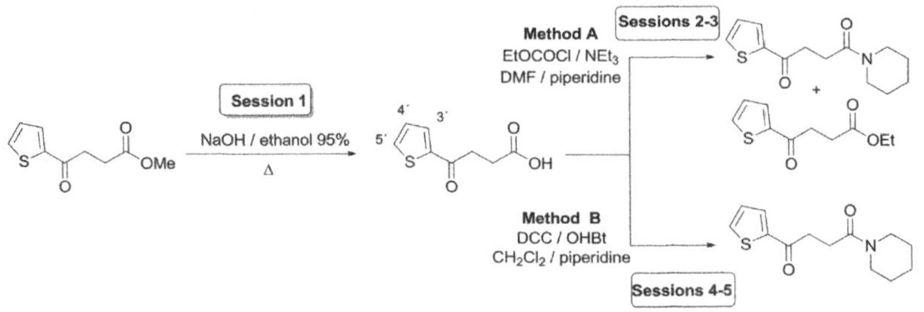

Number of sessions (duration of each session)	Hazard level	Difficulty level	Level of study
5 (4 h each)	Moderate	High	Intermediate

Class names Esters, carboxylic acids, amides

Concepts involved Synthesis of a γ-keto amide derived from thiophene, using two different synthetic methodologies: (i) mixed anhydride or (ii) direct amidation of a carboxylic acid with DCC/OHBt coupling agents; alkaline hydrolysis of a γ-keto ester (saponification) and nucleophilic substitution (conversion of a carboxylic acid into an amide); characterization by ^{1}H NMR and IR spectroscopy

Chemicals needed Methyl 4-oxo-4-(thiophen-2-yl)butanoate (synthesized in experiment 5.2.8), 4-oxo-4-(thiophen-2-yl)butanoic acid commercially available from Acrös [4-oxo-4-(2-thienyl)butanoic acid, 97%, code: 124500250], sodium hydroxide, ethanol (95%), HCl (37%), *N*,*N*-dimethylformamide (dry), triethylamine, ethyl chloroformate, piperidine, dichloromethane, citric acid (5%), anhydrous magnesium sulfate, methanol, ethyl ether, petroleum ether (40–60 °C), 1,3-dicyclohexylcarbodiimide (DCC), 1-hydroxybenzotriazole (HOBt), acetone, ethyl acetate, silica gel for column chromatography (60 Å; 230–400 mesh), thin layer chromatography (TLC) plates with fluorescence indicator

Equipment and experimental techniques involved Apparatus for heating at reflux, apparatus for reaction with magnetic stirring at room temperature and at −5 °C, rotary evaporator, liquid–liquid extraction, gravity and vacuum filtration, thin-layer chromatography (TLC), silica gel column chromatography, melting point

Keywords Amidation of a carboxylic acid, amide, chromatography, ester alkaline hydrolysis, ^{1}H NMR, liquid–liquid extraction

Comprehensive Organic Chemistry Experiments for the Laboratory Classroom
Edited by Carlos A. M. Afonso, Nuno R. Candeias, Dulce Pereira Simão, Alexandre F. Trindade, Jaime A. S. Coelho, Bin Tan, and Robert Franzén.

Published by the Royal Society of Chemistry, www.rsc.org

Background

The amide group plays an essential role in both natural and synthetic molecules. For example, it is present in the structure of proteins which take on essential functions in a wide variety of biological processes, such as enzymatic catalysis (almost all known enzymes are proteins), transport/storage (hemoglobin), immune protection (antibodies) and as mechanical support (collagen). Amides also play an essential role in medicinal chemistry; this functional group exists in more than 25% of known drugs.

These important functions have motivated the development of several synthetic methodologies for the preparation of amides with an aim to increase the efficiency and the selectivity as well as to reduce the quantity of side products and to ease the purification of this class of compounds.

Due to the difficulty of a direct reaction between carboxylic acids and amines, usually it is necessary to activate the carboxyl group by converting it into a more reactive intermediate such as acyl halides, acyl azides, anhydrides, esters, *etc.* to obtain the corresponding amides (Scheme 3.1.16.1). There are several methods of coupling reactive carboxyl derivatives with an amine: (i) an intermediate acylating agent is formed, isolated and then subjected to aminolysis; (ii) a reactive acylating agent is formed from the acid in a separate step(s), followed by immediate treatment with the amine; (iii) the acylating agent is generated *in situ* from the acid in the presence of the amine, by the addition of an activating or coupling agent.

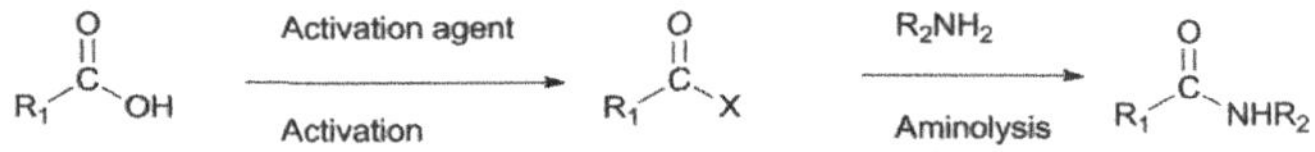

Scheme 3.1.16.1 Activation of the carboxylic acid function followed by aminolysis.

All these methods involve nucleophilic substitution reactions between the ammonia or amine and the acyl carbon. In this experiment the synthesis of an amide will be performed by two methods: mixed anhydride and direct amidation of the carboxylic acid in the presence of coupling agents DCC (dicyclohexyl carbodiimide) and HOBt (1-*hydroxybenzotriazole*).[1–3] The synthesis of amides *via* the mixed anhydride method consists of reacting the carboxylic acid with a chloroformate derivative to give an anhydride, thus activating the carbonyl carbon to make a nucleophilic attack on the amine. Direct amidation with DCC is commonly used in coupling reactions of amino acid units to obtain peptides and also in other reactions of formation of the amide bond using carboxylic acids and primary or secondary amines as precursors. DCC promotes amide bond formation by reaction with the carboxyl acid group, activating it for nucleophilic substitution by a well-known mechanism.[2,4,5] OHBt could also be used in the same context essentially as an additive together with DCC in the case of coupling reactions using sterically hindered compounds or weakly nucleophilic amines.[5] In this case the possible reaction mechanism using DCC/HOBt coupling agents can occur through the formation of the intermediate benzotriazole ester mediated by DCC which subsequently reacts with the amine function yielding the coupling product (amide). The goal of this experiment is the synthesis of a γ-keto amide derived from thiophene,[6] during five sessions and using two methods of synthesis: mixed anhydride and direct amidation in the presence of DCC/HOBt coupling agents. Session 1: synthesis of 4-oxo-4-(thiophen-2-yl)butanoic acid; session 2: synthesis of 1-(piperidin-1-yl)-4-(thiophen-2-yl)butane-1,4-dione (method A: mixed anhydride); session 3: purification of the 1-(piperidin-1-yl)-4-(thiophen-2-yl)butane-1,4-dione using column chromatography with a silica gel; session 4: synthesis of 1-(piperidin-1-yl)-4-(thiophen-2-yl)butane-1,4-dione (method B: direct amidation in the presence of DCC/OHBt coupling agents); session 5:

purification of 1-(piperidin-1-yl)-4-(thiophen-2-yl)butane-1,4-dione. This amide will be used in Experiment 5.2.10 as a precursor in the synthesis of 5-piperidino-2,2′-bithiophene derivative.

Additional Safety

The experimental work in all five sessions should be performed in a well-ventilated fume hood with proper safety equipment such as protective clothing, gloves, and safety goggles. The reaction in session 2 should be performed under anhydrous conditions.

NaOH (session 1) is a corrosive base and may cause severe skin and eye burns. HCl (session 1) is a toxic (oral, dermal, inhalation) and a corrosive acid that may cause severe skin and eye burns. Ethyl chloroformate (session 2), triethylamine, and piperidine are toxic reagents (oral, dermal, inhalation) that may cause severe skin burns and eye damage and are flammable in liquid and vapor form. DMF (session 2) is a toxic reagent (oral, dermal, inhalation) that may cause skin and eye irritation and is flammable in both liquid and vapor form. DCC (session 4) is a strong allergen and sensitizer reagent (oral, dermal, inhalation) that may cause skin rashes and eye damage. OHBt (session 4) is a toxic reagent (oral, dermal, inhalation) that may cause skin and eye irritation. Methanol, ethanol, dichloromethane, diethyl ether, and petroleum ether (40–60 °C) are flammable reagents in both liquid and vapor form and may cause skin, eye, and respiratory tract irritation. Additionally, diethyl ether may form explosive peroxides and the vapor may cause dizziness. The preparation of the silica gel column and the separation of the components of the mixture (session 3) should be carried out in a well-ventilated fume hood to avoid the contact with the silica gel as well as the inhalation of volatile solvents such as dichloromethane, ethyl ether, and petroleum ether. Complete information about the safety, toxicity, and disposal for all reagents can be found, free of charge, at http://www.msds.com/index.asp?open=/protected_public/loginsuccessful.asp.

Experimental Procedure

Note: in laboratory sessions 2 and 4 the reaction mixtures should be stirred constantly at room temperature for 24 h.

Laboratory Session 1 (4 h): Synthesis of 4-Oxo-4-(thiophen-2-yl)butanoic Acid

1. In a fume hood, prepare an apparatus for heating at reflux consisting of a heating mantle, a 500 mL tared round-bottomed flask, a reflux condenser and boiling chips.
2. Add methyl 4-oxo-4-(thiophen-2-yl)butanoate obtained in Experiment 5.2.8 (12 g, 60 mmol) and a solution of NaOH (4.9 g, 122 mmol) in ethanol (95%) to the flask and heat the mixture at reflux for 2 h.
3. Cool the reaction mixture to room temperature and remove the solvent under reduced pressure with a rotary evaporator.
4. Add 75 mL of water to the round-bottomed flask that contains the residue. Transfer the mixture into a separatory funnel and perform the extraction with chloroform (3 × 50 mL) and acidify the aqueous layer with HCl (35%) until reaching pH 1–2.
5. In a fume hood, filter off with the aid of a vacuum the 4-oxo-4-(thiophen-2-yl)butanoic acid that precipitates as a beige solid and wash the filtered solid with cold water.
6. Dry the solid in an oven at 100 °C until constant weight. Weigh the product obtained, determine the melting point and calculate the yield. Prepare a liquid film and record the ^{1}H NMR spectrum in $CDCl_3$ and the IR spectrum. Compare the results with literature data[6] as well as with the spectroscopic data presented in the supplementary information to confirm the compound's purity.

Laboratory Session 2 (4 h): Synthesis of 1-(piperidin-1-yl)-4-(thiophen-2-yl) butane-1,4-dione

Method A: Mixed Anhydride

1. In a fume hood, prepare an apparatus consisting of a previously dried 50 mL two-neck round-bottom flask with a magnetic stirrer, a reflux condenser closed with a calcium chloride tube, a dropping funnel and an ice bath.
2. Add the 4-oxo-4-(thiophen-2-yl)butanoic acid (2 g, 10.8 mmol) and dry DMF (5 mL), and stir the mixture at −5 °C until the 4-oxo-4-(thiophen-2-yl)butanoic acid has completely dissolved.
3. Add dropwise, through the dropping funnel, triethylamine (4.48 mL, 32.4 mmol) and ethyl chloroformate (3.1 mL, 32.4 mmol). After 15 min add piperidine (0.92 g, 10.8 mmol) and keep stirring the mixture at room temperature for 24 h.
4. In the fume hood, filter with the aid of a vacuum the reaction mixture, wash the filtered solid (triethylamine hydrochloride) with DMF (5 mL), and evaporate the solvent to dryness with a vacuum pump. Check the contents of the residue by thin-layer chromatography using a chloroform/methanol (9 : 1) mixture as eluent. Note the presence of two spots on the TLC plate, one corresponding to the amide (a more polar compound with smaller R_f) and another corresponding to the by-product formed, ethyl 4-oxo-4-(thiophen-2-yl)butanoate (a less polar compound with higher R_f).

Laboratory Session 3 (4 h): Purification of 1-(piperidin-1-yl)-4-(thiophen-2-yl)butane-1,4-dione by Column Chromatography on Silica Gel

1. Purify the residue, obtained in session 2, by chromatography through a silica gel column (60 g, 60 Å; 200–300 mesh). Start the elution with 100 mL of a mixture of petroleum ether (40–60 °C)–diethyl ether (3 : 7) and continue with mixtures of petroleum ether (40–60 °C)–ethyl ether of increasing polarity. The first eluted compound will be the by-product ethyl 4-oxo-4-(thiophen-2-yl)butanoate as a pale brown oil [R_f = 0.56; petroleum ether (40–60 °C)–diethyl ether (1 : 1)]. The second eluted product will be the desired amide [(R_f = 0.80, dichloromethane–methanol (9.9: 0.1)].
2. Combine the fractions containing each compound (check by TLC) and evaporate to dryness in a rotary evaporator.
3. Weigh the isolated products and calculate the yield. Record the ^{1}H NMR spectrum in $CDCl_3$ and the IR spectrum of a liquid film. Compare the results with the literature data[6] as well as with the spectroscopic data presented in the supplementary information to confirm the compound's purity.

Laboratory Session 4 (4 h): Synthesis of 1-(piperidin-1-yl)-4-(thiophen-2-yl) butane-1,4-dione

Method B: Direct Amidation with DCC/OHBt

1. In a fume hood, prepare an apparatus consisting of a 500 mL round-bottomed flask with a magnetic stirrer and an ice bath.
2. Add the 4-oxo-4-(thiophen-2-yl)butanoic acid (4 g, 21.6 mmol) and dichloromethane (120 mL), and stir the mixture at 0 °C until the 4-oxo-4-(thiophen-2-yl)butanoic acid is completely dissolved.
3. Add DCC (6 g, 28.4 mmol) and OHBt (4.4 g, 28.4 mmol) to the flask and keep stirring the mixture at room temperature. After 30 min add piperidine (1.84 g, 21.6 mmol) and keep stirring the mixture at room temperature for 24 h.

Laboratory Session 5 (4 h): Purification of 1-(piperidin-1-yl)-4-(thiophen-2-yl)butane-1,4-dione

1. In the fume hood filter with the aid of a vacuum the dicyclohexylurea (DCU) formed as by-product.
2. Transfer the organic solution into a separatory funnel and perform an extraction with 3 × 100 mL of citric acid solution (5%), remove the water from the combined organic layers with the drying agent anhydrous $MgSO_4$, filter and evaporate the solvent under reduced pressure with a rotary evaporator.
3. Dissolve the residue obtained in 50 mL of acetone and leave the solution in a refrigerator at −5 °C until precipitation of the DCU occurs. Filter the dicyclohexylurea from the organic solution and check the purity of piperidine-4-(2′-thienyl)-4-oxobutanamide by TLC using a mixture of dichloromethane–methanol (9.9 : 0.1) as eluent ($R_f = 0.80$). Repeat this operation until the precipitation of DCU stops and check the purity of the amide by TLC.
4. Evaporate the solvent under reduced pressure with a rotary evaporator.
5. Weigh the product obtained and calculate the yield. Record the ^{1}H NMR spectrum in $CDCl_3$ and the IR spectrum of a liquid film. Compare the results with the literature data[6] as well as with the spectroscopic data presented in the supplementary information to confirm the compound's purity.

Waste Disposal

The solvents used in extractions and in column chromatography are recovered after evaporation in a rotary evaporator and should be collected in chlorinated or non-chlorinated waste collector recipients for subsequent disposal. Used silica gel must be stored in a closed container in the fume hood for subsequent disposal. The drying agents should be disposed of in a proper container after filtration from the organic solutions. Special attention should be paid to the treatment and handling of reaction mixtures in sessions 2 and 4 which contain piperidinium hydrochloride and DCU as by-products. In both sessions the by-products should be isolated from the reaction mixtures and stored in a closed container in the fume hood for subsequent disposal.

Results Interpretation and Additional Questions

1. Suggest another synthesis method to accomplish the hydrolysis of methyl 4-oxo-4-(thiophen-2-yl)butanoate (session 1).
2. Explain why the synthesis of the amide using the mixed anhydride method (session 2) should be carried out under anhydrous conditions.
3. Explain the role played by the reagents used for the preparation of the amide in session 2 (EtOCOCl, NEt_3, DMF and piperidine) and in session 3 (DCC and HOBt).
4. Compare the two methods of synthesis used for the preparation of 1-(piperidin-1-yl)-4-(thiophen-2-yl)butane-1,4-dione. Which is the most efficient?
5. Relate the eluent composition to the polarity of the eluted compounds in session 3.
6. Write the reaction mechanism for the synthesis of the amide using the two methods A and B.
7. Interpret the ^{1}H NMR and IR spectra obtained for all compounds,[6] and compare them with those described in the supplementary material.
8. In session 2, in addition to the desired product, the amide, an ester was also obtained as by-product. Describe the major differences in the ^{1}H NMR and IR spectra that permit an unambiguous identification of each compound.

References

1. C. A. G. N. Montalbetti and V. Falque, *Tetrahedron*, 2005, **61**, 10852.
2. F. Albericio, *Curr. Opin. Chem. Biol.*, 2004, **8**, 211–221.
3. (a) M. Bodansky, *Peptide Chemistry: A Practical Textbook*, Springer, New York, 1988; (b) *Handbook of Reagents for Organic Synthesis: Activating Agents and Protecting Groups*, ed. A. J. Pearson and W. R. Roush, John Wiley & Sons Ltd, Chichester, 2000, p. 133.
4. J. G. Moffatt and V. Falque, *J. Org. Chem.*, 1971, **36**, 1909.
5. (a) M. Bodansky, *J. Protein Chem.*, 1989, **8**, 461; (b) *Handbook of Reagents for Organic Synthesis: Activating Agents and Protecting Groups*, ed. A. J. Pearson and W. R. Roush, John Wiley & Sons Ltd, Chichester, 2000, p. 220.
6. M. M. M. Raposo and G. Kirsch, *Heterocycles*, 2001, **55**, 1487.

3.1.17. Cyclic Acetals for Regioselective Protection in Carbohydrate Synthesis: A Comparative Experiment

Ana M. Matos, Rafael Nunes, Catarina Dias, and Amélia P. Rauter*

Centro de Química e Bioquímica, Faculdade de Ciências, Universidade de Lisboa, C8, Piso 5, Campo Grande, 1749-016 Lisboa, Portugal. *E-mail: aprauter@fc.ul.pt

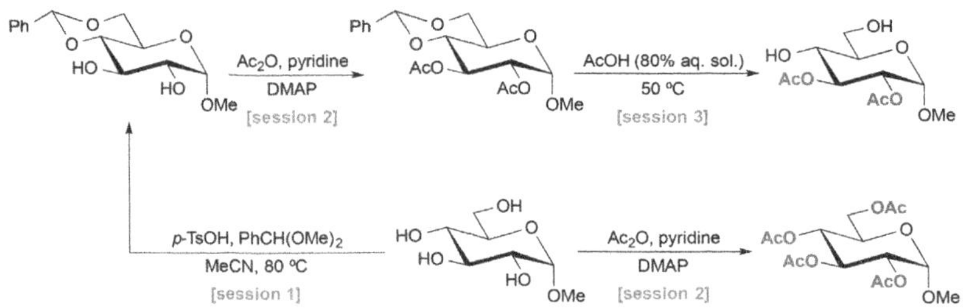

Number of sessions (duration of each session)	Hazard level	Difficulty level	Level of study
3 (4 h each)	Moderate	High	Intermediate

Classes names Alcohols, acetals, esters

Concepts involved This multi-step synthetic route is based on the selective protection of methyl α-D-glucopyranoside at positions 4 and 6 with a benzylidene acetal, acetylation of the remaining free hydroxy groups and selective benzylidene acetal hydrolysis. This protocol illustrates the importance of the benzylidene acetal and of its hydrolysis for sugar regioselective manipulation. Comparison of the target compound's NMR data with those of the peracetylated methyl α-D-glucopyranoside is the starting point for a valuable discussion on concepts such as multifunctionality, regioselectivity and the usefulness of acetal protecting groups in carbohydrate chemistry

Chemicals needed Methyl α-D-glucopyranoside, *p*-toluenesulfonic acid monohydrate, benzaldehyde dimethyl acetal, acetic anhydride, pyridine, 4-(dimethylamino)pyridine, hydrochloric acid (2 M aq. sol.), sulfuric acid, sodium chloride (sat. aq. sol.), sodium hydrogen carbonate (sat. aq. sol.), anhydrous magnesium sulfate, acetic acid (80% aq. sol.), petroleum ether (40–60 °C), dichloromethane, acetonitrile, toluene, methanol, ethyl acetate, deuterated chloroform, silica gel 60 Å (0.040–0.630 mm)

Equipment and experimental techniques involved Rotary evaporator, heating magnetic stirrer with integrated temperature control (alternatively, a heating magnetic stirrer with contact thermometer), separation funnel, reflux condenser, thin-layer chromatography aluminium sheets precoated with silica gel 60 and fluorescent indicator UV_{254}, column chromatography equipment, melting point measurement apparatus, NMR spectrometer

Keywords Carbohydrate chemistry, column chromatography, cyclic acetal, NMR, regioselectivity

Comprehensive Organic Chemistry Experiments for the Laboratory Classroom
Edited by Carlos A. M. Afonso, Nuno R. Candeias, Dulce Pereira Simão, Alexandre F. Trindade, Jaime A. S. Coelho, Bin Tan, and Robert Franzén.

Published by the Royal Society of Chemistry, www.rsc.org

Background

The vast number of structural and functional roles of carbohydrates in living organisms makes them exceptional lead scaffolds for drug discovery. Indeed, it is possible to manipulate their natural interaction with biological targets and that is one of the main reasons why carbohydrates are broadly preferred as synthetic raw materials, and their stereochemistry and multifunctionality are crucial for the conception of selectively functionalized intermediates leading to a wide diversity of organic structures.[1–5]

The use of protecting groups to allow regioselective reactions of hydroxy groups is, therefore, particularly important in synthetic carbohydrate chemistry. While benzyl ethers and esters, such as acetates or benzoates, are commonly used for the indiscriminate protection of primary, secondary or anomeric hydroxy groups, stereochemically hindered trityl or silyl ethers are usually preferred for the selective protection of the primary hydroxy group.[2]

When it comes to regioselective and simultaneous protection of more than one hydroxy group in the sugar, building a cyclic acetal is the correct choice. Its formation results from the acid-catalysed reaction of two suitably oriented hydroxy groups with aldehydes or ketones (Scheme 3.1.17.1). While the former class of compounds preferentially forms six-membered ring acetals involving the primary hydroxy group, ketones form five-membered ring acetals to avoid the unfavourable alkyl group in an axial position.[3]

Isopropylidene and benzylidene groups are commonly used for carbohydrate acetal protection. Whereas the five-membered isopropylidene is formed by reaction of acetone with *cis*-arranged vicinal hydroxy groups (Scheme 3.1.17.2A), the six-membered benzylidene acetal is typically used for regioselective 4,6-*O*-protection of carbohydrates, affording products with free hydroxy groups at positions 1, 2 and 3 (Scheme 3.1.17.2B),[3] which become available for further transformations. Although benzylidene acetal formation employs benzaldehyde as the carbonyl reagent, transacetalation with benzaldehyde dimethyl acetal has proven just as effective and requires much less harsh reaction conditions.[4]

Another major advantage of cyclic acetal protection is the possibility of selective acetal removal. For example, the commercially available methyl α-D-glucopyranoside is itself an acetal and, therefore, labile to acid hydrolysis. However, when protected with benzylidene or isopropylidene groups, the use of mild acidic conditions permits isopropylidene or benzylidene acetal deprotection, keeping the integrity of the anomeric centre, which requires more acidic conditions to be hydrolysed.[5]

This short project is intended to enrich the education of students attending an organic chemistry course by applying synthetic methodologies to monosaccharides as multifunctional and stereochemically complex small molecules. Students will explore and understand the usefulness of cyclic acetals, namely the benzylidene acetal, in the context of a multi-step synthetic route. Throughout the proposed set of experiments, students are challenged to apply different

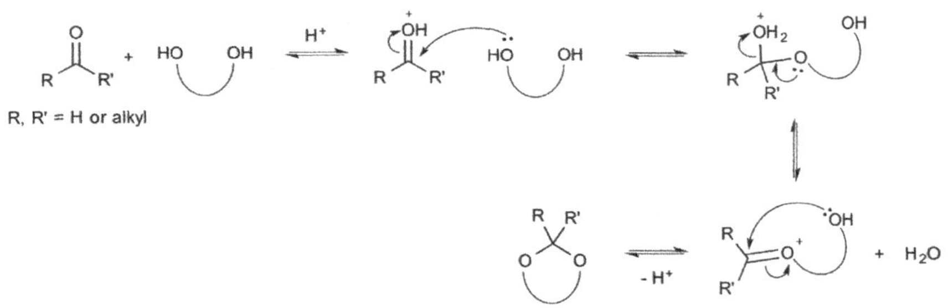

Scheme 3.1.17.1 Mechanism for the acid-catalysed reaction of aldehydes or ketones with diols to give cyclic acetals.

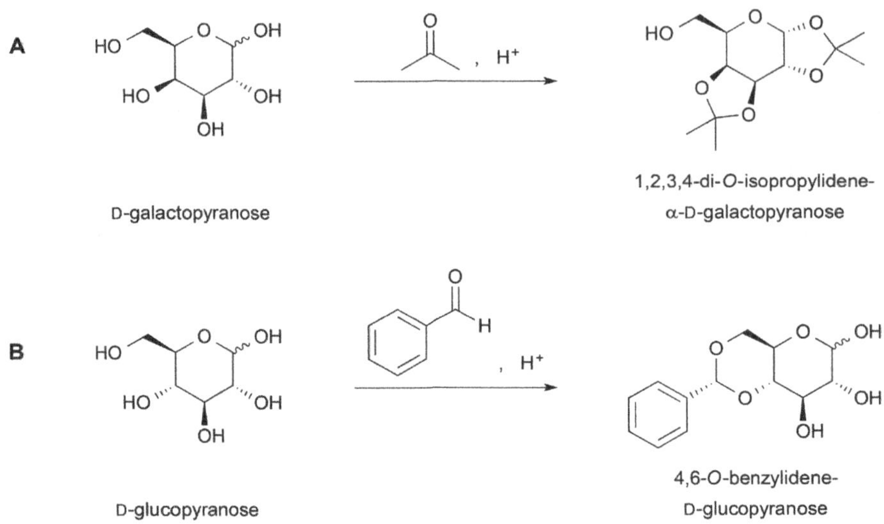

Scheme 3.1.17.2 D-Galactopyranose and D-glucopyranose protection with (A) isopropylidene group and (B) benzylidene group.

purification techniques commonly used in the organic chemistry laboratory. By comparing the ^{1}H NMR spectra of acetylated methyl α-D-glucopyranoside with and without previous benzylidene protection, they will be able to confirm the regioselective manipulation of the starting material and the selective acetal hydrolysis.

Additional Safety

Manipulation of compounds should be carried out in a fume hood, with proper safety equipment, such as protective clothing, gloves and safety goggles. Students are advised to consult GHS hazard and precautionary phrases for all reagents. *p*-Toluenesulfonic acid and hydrochloric acid are corrosive acids and may cause severe skin and eye burns (H314). Long-term exposure to pyridine can cause sterility in males (H360). Benzaldehyde dimethyl acetal and acetic anhydride are harmful in contact with skin and eyes (H312, H318). Chloroform is classified as a carcinogen (H351) and therefore it is not used for extraction purposes, being replaced by dichloromethane, which is much less toxic (H305, H315). Most of the solvents used are flammable, and thus should be kept away from heat, sparks or flames (H226).

Experimental Procedure

Laboratory Session 1 (4 h): Preparation of Methyl 4,6-*O*-Benzylidene-α-D-glucopyranoside

1. In a fume hood, heat an oil bath on a heating magnetic stirrer with integrated temperature control or, alternatively, on a heating magnetic stirrer with contact thermometer, to maintain the reaction temperature at 82 °C.
2. In a 25 mL round-bottomed flask equipped with a magnetic stir bar, prepare a solution of methyl α-D-glucopyranoside (1 g, 5.15 mmol) in acetonitrile (5 mL) and add *p*-toluenesulfonic acid monohydrate (0.02 g, 0.11 mmol, 0.02 equiv.) and benzaldehyde dimethyl acetal (1.55 mL,10.30 mmol, 2 equiv.). Clamp the flask, equipped with a reflux condenser, in the preheated oil bath and let the reaction mixture stir at 82 °C for 2 h.

3. Cool the reaction mixture to room temperature, add dichloromethane (50 mL) and transfer the solution into a separation funnel. Wash the organic phase with an aqueous saturated solution of sodium hydrogen carbonate (50 mL) and then an aqueous saturated solution of sodium chloride (brine, 50 mL).
4. Dry the organic phase with anhydrous magnesium sulfate ($MgSO_4$), filter and evaporate the solvent under reduced pressure.
5. Re-suspend the dry residue in dichloromethane (use the smallest necessary volume for complete dissolution under heat), add petroleum ether (30 mL) and keep the flask on an ice bath.
6. Filter the formed crystals under reduced pressure and transfer them to a tared crystallizing dish with spout and conserve them in a desiccator in vacuum until the next session.

Laboratory Session 2 (4 h): Acetylation Reactions

Beginning of the laboratory session – Weigh the crystallizing dish containing the compound prepared in the previous session, calculate the reaction yield and determine the melting point of the obtained white solid.

Part A – Acetylation of Methyl 4,6-O-Benzylidene-α-D-glucopyranoside

1. Add the earlier synthesized methyl 4,6-*O*-benzylidene-α-D-glucopyranoside (0.9 g, 3.19 mmol, 1 equiv.), 4-(dimethylamino)pyridine (DMAP) (0.03 g, 0.25 mmol, 0.1 equiv.) and pyridine (5 mL) to a 25 mL round-bottomed flask equipped with a magnetic stir bar. Stir at room temperature until complete dissolution (approx. 5 min).
2. Add acetic anhydride (1.21 mL, 12.8 mmol, 4 equiv.) and stir for 30 min at room temperature.
3. Add dichloromethane (10 mL) and transfer the solution to a separation funnel. Neutralize with a 2 M aqueous HCl solution (15 mL) and extract the aqueous phase with dichloromethane (15 mL). Combine the organic phases and repeat the procedure by washing with a 2 M aqueous HCl solution (15 mL) followed by extraction with dichloromethane (15 mL).
4. Combine all the organic phases and wash them with a saturated solution of sodium hydrogen carbonate (50 mL) and then with brine (50 mL).
5. Dry the combined organic phases with anhydrous $MgSO_4$, filter under reduced pressure, pour the filtrate to a previously tared flask and evaporate the solvent with a rotary evaporator until dryness.
6. Keep the flask in a desiccator, under vacuum, until the next session.

Part B – Acetylation of Methyl α-D-Glucopyranoside

1. Add methyl α-D-glucopyranoside (0.5 g, 2.57 mmol, 1 equiv.), DMAP (0.03 g, 0.25 mmol, 0.1 equiv.) and pyridine (5 mL) to a 25 mL round-bottomed flask equipped with a magnetic stir bar. Stir at room temperature until complete dissolution (5 min).
2. Add acetic anhydride (1.95 mL, 20.6 mmol, 8 equiv.) to the solution and stir for 30 min at room temperature.
3. Add dichloromethane (10 mL) and transfer the solution to a separation funnel. Repeat the neutralization process and carry out the extraction as described for part A.
4. Keep the flask in a desiccator, under vacuum, until the next session.

Laboratory Session 3 (4 h): Benzylidene Acetal Hydrolysis and ^1H NMR Data Acquisition

Beginning of the laboratory session – Weigh the flasks containing the compounds prepared in session 2, calculate reaction yields and determine the melting point of the solid compound.

Benzylidene Acetal Hydrolysis

1. In a fume hood, heat an oil bath on a heating magnetic stirrer with integrated temperature control or, alternatively, on a heating magnetic stirrer with contact thermometer to maintain the reaction temperature at 50 °C.
2. Add methyl 2,3-di-*O*-acetyl-4,6-*O*-benzylidene-α-D-glucopyranoside (0.2 g, 0.55 mmol) and an aqueous solution of acetic acid (3 mL, 80% v/v) to a 25 mL round-bottomed flask with a magnetic stir bar. Stir until complete dissolution (*ca.* 5 min). Clamp the flask, equipped with a reflux condenser, in the preheated oil bath and let the reaction mixture stir at 50 °C for 2 h.
3. Meanwhile, fill a chromatography column (3 cm diameter × 10 cm height) with a slurry of silica gel in petroleum ether/ethyl acetate (1:1).
4. After cooling the reaction mixture to room temperature, add toluene (10 mL) and evaporate until dryness.
5. Dissolve the reaction residue in dichloromethane (2 mL), and add one or two drops of methanol to assure complete dissolution.
6. Apply the sample to the chromatography column.
7. Elute with petroleum ether/ethyl acetate (1:1) (150 mL) and collect 20 mL fractions. Control by TLC with the eluent petroleum ether/ethyl acetate (1:1) until all benzaldehyde has come out of the column (see Supplementary material). Staining should be performed by dipping the plate in a 10% methanolic solution of H_2SO_4 followed by heating at 120 °C.
8. Elute with ethyl acetate (100 mL) and discharge the eluted solvent that does not contain benzaldehyde or product as confirmed by TLC.
9. Elute with a mixture of ethyl acetate and methanol (10:1, 200 mL) and collect the fractions, confirming by TLC (same eluent as above) the beginning and the end of methyl 2,3-di-*O*-acetyl-α-D-glucopyranoside elution.
10. Pour the collected fractions containing the product into a 250 mL round-bottomed tared flask and evaporate the solvent with a rotary evaporator until dryness.
11. Determine the mass of the residue, which can be immediately used to run compound ^{1}H NMR spectra. Calculate the crude yield (yield prior to drying the residue in a vacuum desiccator).

Sample Preparation for ^{1}H NMR Spectra Acquisition

1. Dissolve the compound (15 mg) in $CDCl_3$ (500 μL) and transfer the solution to an NMR tube.
2. Dissolve methyl 2,3,4,6-tetra-*O*-acetyl-α-D-glucopyranoside (15 mg), obtained in laboratory session 2, in $CDCl_3$ (500 μL) and transfer the solution to an NMR tube.
3. Record the ^{1}H NMR spectra of the two samples prepared.

Waste Disposal

The neutralized aqueous solutions may be diluted and discarded in the waste collector. Used organic solvents should be collected in appropriate waste collector recipients (chlorinated and non-chlorinated, accordingly).

Results Interpretation and Additional Questions

1. In the process of benzylidene acetal formation, which hydroxy group is expected to react first? Describe the reaction mechanism starting from benzaldehyde dimethyl acetal.
2. Why does benzaldehyde (or its corresponding dimethyl acetal) selectively react with positions 4 and 6 of monosaccharides to form six-membered ring acetals?

3. Why is it advantageous to use DMAP in the acetylation reactions instead of using pyridine alone? Describe the mechanism by which DMAP-catalysed acylation occurs.
4. Analyse the acquired ^1H NMR spectra:
   a. Confirm the presence of two or four acetate groups in the synthesized products.
   b. Compare the chemical shifts of H2, H3, H4 and H6 signals of both derivatives emphasizing their importance for the assignment of the acetyl functionalization pattern.
   c Regarding the spectrum of methyl 2,3-di-*O*-acetyl-α-D-glucopyranoside, assign the signal corresponding to the protons of the methoxy group at the anomeric centre, thus confirming the occurrence of selective acetal hydrolysis at positions 4 and 6.
5. Compare the melting point of methyl 4,6-*O*-benzylidene-α-D-glucopyranoside with that reported in the literature and suggest plausible reasons for any existing differences. Comment on the purity of this synthetic intermediate taking into account the applied purification method.
6. Give the overall yield for the three-step synthesis. Comment on the global efficiency of the synthetic route.

References

1. B. G. Davis and A. J. Fairbanks, *Carbohydrate Chemistry*, Oxford University Press, New York, 2002, ch. 9, pp. 78–89.
2. S. Pétursson, *J. Chem. Educ.*, 1997, **74**(11), 1297.
3. T. K. Lindhorst, *Essentials of Carbohydrate Chemistry and Biochemistry*, Wiley-VCH Verlag GmbH, Weinheim, 2000, ch. 3, pp. 59–67.
4. A. V. Demchenko, P. Pornsuriyasak and C. De Meo, *J. Chem. Educ.*, 2006, **83**(5), 782.
5. B. G. Davis and A. J. Fairbanks, *Carbohydrate Chemistry*, Oxford University Press, New York, 2002, ch. 6, p. 49.

3.1.18. Direct Diastereoselective Synthesis of the Tetrahydro-thiazolo[2,3-*b*]isoindole Tricyclic Ring System

Maria I. L. Soares, Ana Lúcia Cardoso, Susana M. M. Lopes, and Teresa M. V. D. Pinho e Melo*

Department of Chemistry, University of Coimbra, Rua Larga, 3004-535 Coimbra, Portugal
*E-mail: tmelo@ci.uc.pt

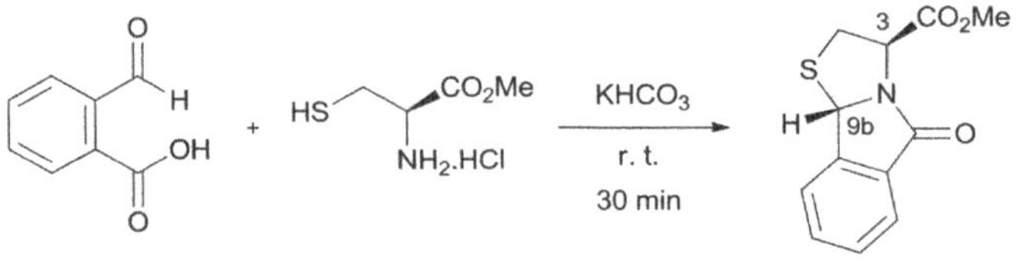

1 2 3

Number of sessions (duration of each session)	Hazard level	Difficulty level	Level of study
1 (4 h)	Moderate	Medium	Advanced

Class names Aldehydes, amines, amides, imines, thiols, carboxylic acids

Concepts involved Diastereoselective synthesis of a tricyclic isoindole derivative using L-cysteine methyl ester and an achiral aldehyde, 2-carboxybenzaldehyde, as starting materials

Chemicals needed L-Cysteine methyl ester hydrochloride, 2-carboxybenzaldehyde, potassium hydrogen carbonate, distilled water, ethanol, dichloromethane, diethyl ether, anhydrous sodium sulfate, chloroform-*d*

Equipment and experimental techniques involved Apparatus to carry out reactions with magnetic stirring, liquid–liquid extraction apparatus, vacuum filtration apparatus, rotary evaporator, NMR, IR, melting point, polarimeter

Keywords Diastereoselectivity, intramolecular acylation, NMR, optical rotation, thiazolidine

Background

Thiazolidines are five-membered heterocycles containing a sulfur and nitrogen atom. 2-Substituted-1,3-thiazolidine-4-carboxylic acid derivatives can be easily obtained from the reaction of aldehydes and L-cysteine or L-cysteine esters. In this process the chiral centre from cysteine (C4 in the thiazolidine) is retained and a new asymmetric centre is created at position 2 of the thiazolidine.[1] Condensation of cysteine with achiral aldehydes is not diastereoselective, giving mixtures of the *cis* and *trans* diastereoisomers **4** (Scheme 3.1.18.1).[2]

Acylation of the diastereoisomeric mixture **4** with acetic anhydride or with acid chlorides can lead to the selective synthesis of *N*-acyl-2-substituted-1,3-thiazolidine-4-carboxylic acid derivatives as pure stereoisomers with (2*R*,4*R*) or (2*S*,4*R*) stereochemistry depending on the reaction conditions (Scheme 3.1.18.2). In fact, 2-substituted-1,3-thiazolidine-4-carboxylic acid derivatives

Comprehensive Organic Chemistry Experiments for the Laboratory Classroom
Edited by Carlos A. M. Afonso, Nuno R. Candeias, Dulce Pereira Simão, Alexandre F. Trindade, Jaime A. S. Coelho, Bin Tan, and Robert Franzén.

Published by the Royal Society of Chemistry, www.rsc.org

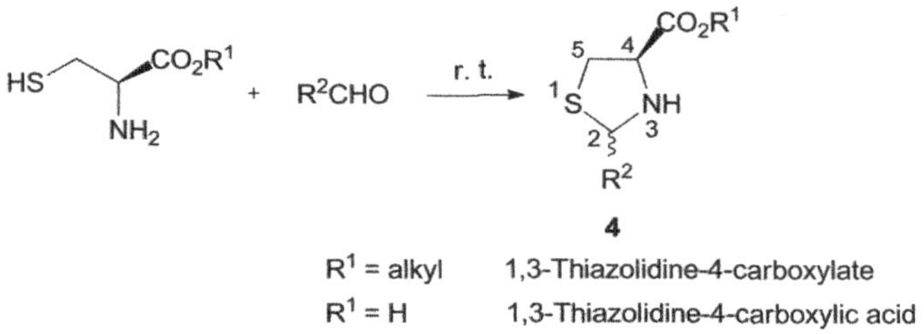

Scheme 3.1.18.1 Synthesis of 2-substituted-1,3-thiazolidine-4-carboxylic acid derivatives (**4**).

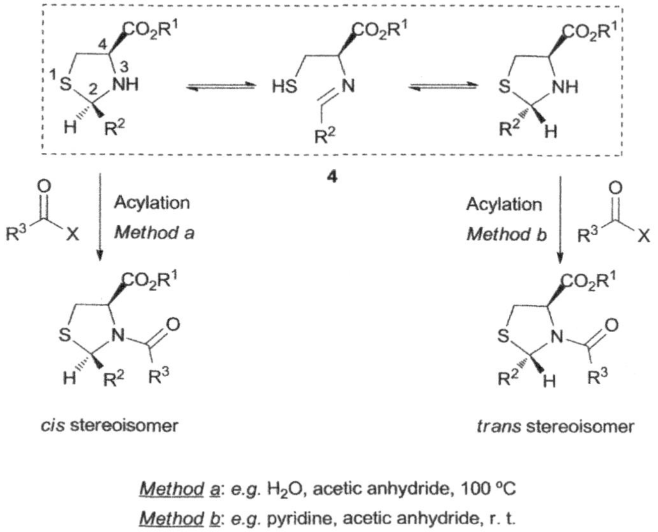

Scheme 3.1.18.2 Diastereoselective *N*-acylation of 1,3-thiazolidine-4-carboxylic acid derivatives.

can undergo selective inversion at C2 through a mechanism involving the opening of the ring, but the protection with the acyl group prevents this epimerization and allows the isolation of pure diastereoisomers.[2–4] Thus, starting from a diastereoisomeric mixture of thiazolidines it is possible to selectively obtain *N*-acylthiazolidines as single diastereoisomers.

The purpose of this work is the diastereoselective synthesis of a tricyclic isoindole derivative, exploring thiazolidines chemistry. The direct synthesis of methyl (3*R*,9b*S*)-5-oxo-2,3,5,9b-tetrahydro-thiazolo[2,3-*b*]isoindole-3-carboxylate (**3**) is performed by reacting 2-carboxybenzaldeyde (**1**) with L-cysteine methyl ester (**2**). The reaction is carried out in a mixture of water/ethanol and the product can be isolated by crystallization.[5,6] The structure of thiazolo[2,3-*b*]isoindole and the stereochemistry of the chiral centres C3 and C9b has been confirmed by X-ray crystallography studies.[6] The stereoselective synthesis of the tricyclic product **3** involves the initial condensation of the aldehyde **1** and cysteine **2** to give a mixture of (2*S*,4*R*) and (2*R*,4*R*) isomers of thiazolidine (**5**). Subsequent *in situ* intramolecular *N*-acylation leads to the selective synthesis of tetrahydro-thiazolo[2,3-*b*]isoindole **3** (Scheme 3.1.18.3).

Additional Safety

The session should be carried out in a fume hood, with proper safety equipment: protective clothing, gloves and safety glasses. L-Cysteine methyl ester hydrochloride, 2-carboxybenzaldehyde and dichloromethane are irritating to eyes, respiratory system and skin. Dichloromethane

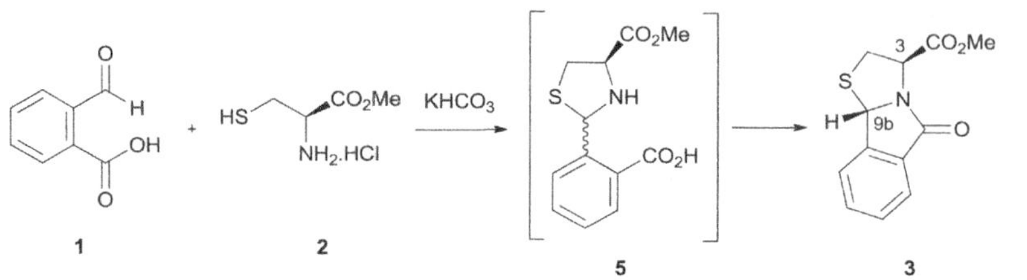

Scheme 3.1.18.3 Synthesis of tetrahydro-thiazolo[2,3-*b*]isoindole **3**.

and diethyl ether vapors may cause drowsiness and dizziness. Diethyl ether is extremely flammable and may form explosive peroxides. It is harmful if swallowed and repeated exposure may cause skin dryness or cracking. Ethanol is highly flammable.

Experimental Procedure

1. To a 50 mL round-bottom flask with a magnetic stirring bar, add L-cysteine methyl ester hydrochloride (1.00 g, 5.83 mmol) and water (5 mL). Stir until all the L-cysteine methyl ester hydrochloride is dissolved.
2. After complete dissolution add potassium hydrogen carbonate (0.58 g; 5.83 mmol). *Caution*: release of carbon dioxide.
3. Add a solution of 2-carboxybenzaldehyde (0.96 g; 6.41 mmol; 1.1 equiv.) in EtOH (7 mL) and leave with stirring at room temperature for 30 min.

Product Isolation

4. Dilute the reaction mixture with water (30 mL) and extract the product with dichloromethane (3 × 30 mL).
5. Dry the combined organic layers over anhydrous Na_2SO_4, filter under vacuum and remove the solvent under reduced pressure. The formation of a white solid is observed.
6. Grind the solid residue with cold diethyl ether, filter off the solid under vacuum and dry.
7. Collect the white solid in a vial, determine the weight and calculate the yield.
8. Determine the melting point of the product and prepare a sample for ^{1}H and ^{13}C NMR in chloroform-*d*. Additional 2D NMR experiments, HMQC and HMBC, can also be performed.
9. Prepare a solution of the product (230 mg in dichloromethane in a 10 mL volumetric flask) and measure the optical rotation.

Waste Disposal

Dichloromethane may be collected in a chlorinated waste collector recipient, for later disposal. Diethyl ether may be collected in a non-chlorinated waste collector recipient, for later disposal. Neutralized aqueous solution may be diluted and discarded in the sink.

Results Interpretation and Additional Questions

1. Compare the obtained melting point with that reported in the literature.
2. Interpret the ^{1}H and ^{13}C NMR spectra of the tetrahydro-thiazolo[2,3-*b*]isoindole **3**. Compare them with the ones reported in the literature.

3. Calculate the specific rotation of the sample using the following formula:

$$[\alpha]_D^t = \frac{\alpha}{cl} \times 100$$

where $[\alpha]_D^t$ is the specific rotation at t °C, α is the observed angle, c is the concentration of the solution (g per 100 mL) and l is the length of the polarimeter sample cell (dm).

4. Rationalize the mechanism involved in the diastereoselective synthesis of tetrahydrothiazolo[2,3-*b*]isoindole **3** and discuss the observed selectivity.

References

1. T. L. Gilchrist, A. M. d. A. R. Gonsalves and T. M. V. D. Pinho e Melo, *Tetrahedron*, 1994, **50**, 13709–13724.
2. L. Szilágyi and Z. Gyorgydeak, *J. Am. Chem. Soc.*, 1979, **101**, 427–432.
3. Z. Györgydeák, M. Kajtár-Peredy, J. Kajtár and M. Kajtár, *Liebigs Ann. Chem.*, 1987, 927–934.
4. F. Benedini, F. Ferrario, A. Sala, L. Sala and P. A. Soresinetti, *J. Heterocycl. Chem.*, 1994, **31**, 1343–1347.
5. T. M. V. D. Pinho e Melo, C. I. A. Santos, A. M. d. A. R. Gonsalves, J. A. Paixao, A. M. Beja and M. R. Silva, *Tetrahedron Lett.*, 2003, **44**, 8285–8287.
6. T. M. V. D. Pinho e Melo, C. I. A. Santos, A. M. d. A. R. Gonsalves, J. A. Paixao and A. M. Beja, *Tetrahedron*, 2004, **60**, 3949–3955.

3.1.19. Asymmetric Cyclocondensation Reaction Induced by Chiral Aminoalcohol

Jorge Dourado[a], Maria Pérez[b], Rosa Griera[b], and Maria M. M. Santos*[a]

[a]Research Institute for Medicines (iMed.ULisboa), Faculty of Pharmacy, Universidade de Lisboa, Av. Prof. Gama Pinto, 1649-003 Lisboa, Portugal; [b]Laboratory of Organic Chemistry, Faculty of Pharmacy and Institute of Biomedicine (IBUB), University of Barcelona, Av. Joan XXIII, s/n, 08028 Barcelona, Spain
*E-mail: mariasantos@ff.ulisboa.pt

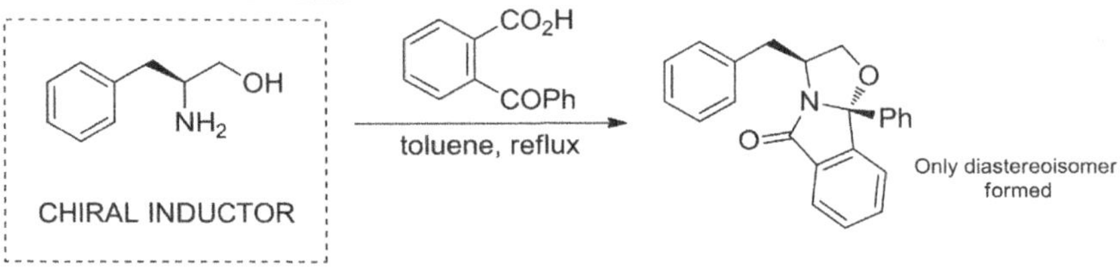

Number of sessions (duration of each session)	Hazard level	Difficulty level	Level of study
2 sessions (1 h + 4 h)	Moderate	Medium	Advanced

Class names Aminoalcohols, keto acids, lactams

Concepts involved This experiment describes the asymmetric synthesis of a tricyclic lactam by the cyclocondensation reaction of (*S*)-phenylalaninol (used as chiral inductor) and 2-benzoylbenzoic acid. The aminoalcohol, besides being the source of chirality, is also used to assemble the polycyclic structure of the product

Chemicals needed (*S*)-Phenylalaninol, 2-benzoylbenzoic acid, toluene, ethyl acetate, *n*-hexane, sodium bicarbonate, silica gel

Equipment and experimental techniques involved Dean–Stark apparatus, condenser, stirring and heating plate, thin-layer chromatography (TLC), rotary evaporator, column chromatography, recrystallization, NMR spectrometer, melting point apparatus

Keywords Asymmetric synthesis, azeotropic mixture, chiral inductor, cyclocondensation reaction, flash chromatography, NMR

Background

Efficient methodologies for the generation of several stereogenic centres with high diastereo- and enantioselectivity in a single synthetic step is one of the most challenging subjects in organic synthesis. Moreover, the synthesis of enantiopure drugs is a highly important issue in the pharmaceutical industry, as different enantiomers are known to have different activities against a target.

There are two main approaches to generate stereogenic centres: (i) reactions using an enantiopure reagent in stoichiometric or catalytic amounts; (ii) reactions in which the starting material is an enantiopure compound that controls the stereochemical course of a diastereoselective reaction.

Comprehensive Organic Chemistry Experiments for the Laboratory Classroom
Edited by Carlos A. M. Afonso, Nuno R. Candeias, Dulce Pereira Simão, Alexandre F. Trindade, Jaime A. S. Coelho, Bin Tan, and Robert Franzén.

Published by the Royal Society of Chemistry, www.rsc.org

The chiral pool (naturally occurring chiral molecules) is often a cheap source by which to easily obtain enantiopure compounds that can be used as chiral auxiliaries to control the stereochemical course of a diastereoselective reaction.[1] In particular, 1,2-aminoalcohols containing a stereogenic centre can be obtained by reduction of the parent natural amino acids without racemization.[2] These compounds are useful for the syntheses of natural products or biological active molecules that contain the aminoalcohol functionality. In many organic syntheses, the chiral auxiliary needs to be removed later in the synthetic sequence.[3] However, there are now several examples of reactions where the chiral inductor not only constitutes the source of chirality but is also used to assemble the final target products.[4,5]

Here we present a simple experiment based on reported literature[6] that allows the students to observe the use of an asymmetric methodology by performing one cyclocondensation reaction using a chiral aminoalcohol as chiral inductor. In this experiment, the chiral inductor, besides being responsible for the stereo-outcome of the final product, is also part of the main skeleton of biologically active molecules.[7] For these reasons, this asymmetric reaction is highly efficient/atom economic.

This experiment is intended for undergraduate or MSc students currently learning advanced organic chemistry. We believe that this is a good experiment for small classes or assigned as a project for a single group of students to further explore the reaction. The students will learn the importance of chiral inductors for the synthesis of enantiopure compounds, the importance of the use of a Dean–Stark apparatus, and also have contact with techniques of purification such as column chromatography and crystallization. In addition, depending on the background of the students, the instructor can use this experiment to (i) rationalize the reaction mechanism, (ii) apply IR, ^{1}H and ^{13}C NMR techniques for structural analysis, and (iii) confirm the formation of only one diastereoisomer by ^{1}H NMR.

This experiment presents the use of (*S*)-phenylalaninol as the chiral aminoalcohol for the synthesis of (3*S*,9b*R*)-3-benzyl-5-oxo-9b-phenyl-2,3,5,9b-tetrahydrooxazolo[2,3-*a*]isoindole (**1**) (Scheme 3.1.19.1). The final product is easily accessible, with very high stereoselectivity, by a cyclocondensation reaction of the chiral aminoalcohol, (*S*)-phenylalaninol, and 2-benzoylbenzoic acid under reflux in toluene using a Dean–Stark apparatus.

Additional Safety

All sessions should be done with proper safety equipment, such as protective clothing, gloves, and safety goggles. Chemicals in the eye should be immediately flushed out with copious amounts of water using the eyewash fountain. To prevent inhalation of organic and inorganic vapours, carry out your experiments in the fume hood or under minihoods on the bench. Do not heat the reaction mixture without attaching the condenser, since the evaporation of all of the solvent causes degradation of the crude reaction residue. Furthermore, some vapours

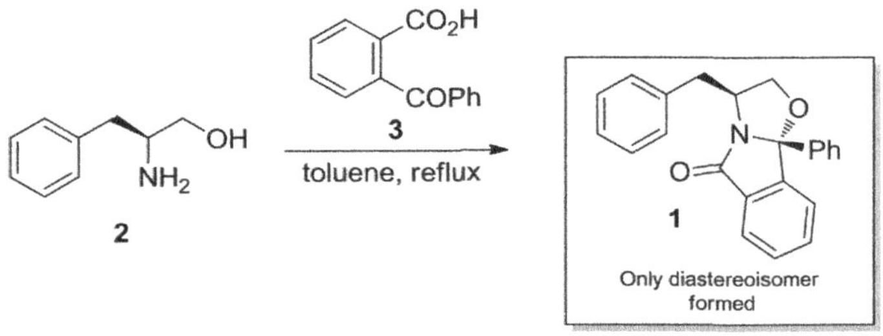

Scheme 3.1.19.1 Synthesis of (3*S*,9b*R*)-3-benzyl-5-oxo-9b-phenyl-2,3,5,9b-tetrahydrooxazolo-[2,3-*a*]isoindole (1).

in contact with the heating plates can be inflamed. In addition, do not heat a closed system because pressure will build up, causing the glass to fail, sending projectiles of glass in all directions.

(*S*)-Phenylalaninol is a corrosive powder that may cause skin burns and eye burn damage. 2-Benzoylbenzoic acid is a powder that causes skin and eye irritation, and may cause respiratory irritation. Toluene is a flammable liquid that causes skin and eye irritation, germ cell mutagenicity and carcinogenicity hazard by aspiration. Ethyl acetate is a flammable liquid that causes eye irritation, narcotic effects by aspiration, and causes skin dryness by repeated contact with the skin. *n*-Hexane is a harmful flammable liquid that causes skin irritation by contact, narcotic effects by aspiration, and is toxic to the aquatic life.

Experimental Procedure

In this experiment, tricyclic lactam **1** will be synthesized in good yield and as single diastereoisomer by heating (*S*)-phenylalaninol (**2**) (commercially available[8] but can also be prepared by the students in a single reaction by reduction of phenylalanine[9]) and 2-benzoylbenzoic acid (**3**) (commercially available[8]) at reflux under Dean–Stark conditions in toluene. The mechanism of the reaction involves the formation of an *N*-acyliminium ion, and the release of water. Purification is achieved by flash-column chromatography and recrystallization.

Session 1

1. Prepare a 25 mL round-bottomed flask with a magnetic stir bar, and add (*S*)-(−)-2-amino-3-phenyl-1-propanol (**2**) (100 mg, 0.66 mmol) and toluene (5 mL).
2. Add 2-benzoylbenzoic acid (**3**) (165 mg, 0.73 mmol, 1.1 equiv.).
3. Equip the reaction flask with a Dean–Stark apparatus and a condenser (Figure 3.1.19.1) and heat to reflux for 24 h.

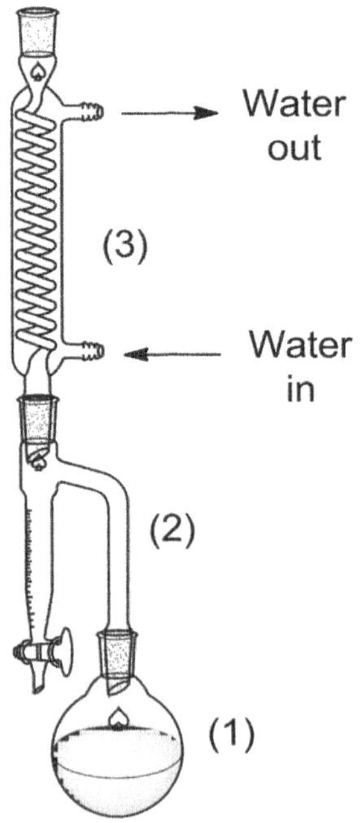

Figure 3.1.19.1 Reaction apparatus: (1) reaction flask, (2) Dean–Stark, and (3) condenser.

Session 2

1. Confirm the reaction is completed (control by TLC, eluent *n*-hexane/EtOAc (1 : 1)).
2. Allow the reaction to cool to room temperature and evaporate the solvent under reduced pressure.
3. Dissolve the residue in EtOAc (10 mL) and wash with a aqueous solution of $NaHCO_3$ (5%; 10 mL), and brine (10 mL).
4. Dry the organic layer with Na_2SO_4, filter and evaporate the solvent using a rotary evaporator.
5. Purify the oil obtained by flash chromatography (300 mL of *n*-hexane/EtOAc (2 : 1)).
6. Evaporate to dryness the column fractions with the desired compound to obtain a colorless thick oil.
7. Dissolve the oil in Et_2O and add *n*-hexane to obtain the pure title compound (**1**) as white crystals.
8. Remove the solvent (Et_2O/*n*-hexane) either by filtration or decantation and dry the product in a vacuum line.
9. Weigh the desired product and calculate the yield. Measure the melting point, and record IR and ^1H and ^{13}C NMR spectra.

Waste Disposal

To minimize damage to the environment chemical wastes must be separated into categories and carefully labelled as to their contents.

Toluene (from the Dean–Stark apparatus and from the distilled solvent of the rotary evaporator apparatus) and *n*-hexane/ethyl acetate (eluent used in the chromatographic purification) should be collected in a non-chlorinated waste collector recipient, for posterior destruction. The aqueous solution of $NaHCO_3$ may be collected in a basic aqueous solutions collector recipient. The Na_2SO_4 and the silica gel can be disposed of in a collector recipient for contaminated solids.

Results Interpretation and Additional Questions

1. Interpret the IR and ^1H and ^{13}C NMR spectra of the isolated compound.
2. Explain the multiplicity of the signals at 2.5 and 3.0 ppm in the ^1H NMR.
3. Compare the obtained melting point with that given in the literature.[6] If a polarimeter is available in the laboratory, determine the specific rotation, [α].
4. Propose the reaction mechanism and discuss the diastereoselectivity of the reaction.
5. Discuss the use and importance of the Dean–Stark apparatus in the overall process.
6. Search the literature for other chiral auxiliaries used in organic synthesis.

References

1. H.-U. Blazer, *Chem. Rev.*, 1992, **92**, 935.
2. D. J. Ager, I. Prakash and D. R. Schaad, *Chem. Rev.*, 1996, **96**, 835.
3. M. D. Groaning and A. I. Meyers, *Tetrahedron*, 2000, **56**, 9843.
4. M. Amat, M. M. M. Santos, O. Bassas, N. Llor, C. Escolano, A. Gómez-Esqué, E. Molins, S. M. Allin, V. McKee and J. Bosch, *J. Org. Chem.*, 2007, **14**, 5193.
5. M. Amat, C. Ramos, M. Pérez, E. Molins, P. Florindo, M. M. M. Santos and J. Bosch, *Chem. Commun.*, 2013, **49**(19), 1954.
6. N. A. L. Pereira, F. X. Sureda, M. Turch, M. Amat, J. Bosch and M. M. M. Santos, *Monatsh. Chem.*, 2013, **144**, 473 (Special issue: 4th Investigator Workshop - EuCheMS Organic Division).
7. J. Soares, N. A. L. Pereira, Â. Monteiro, M. Leão, C. Bessa, D. J. V. A. dos Santos, L. Raimundo, G. Queiroz, A. Bisio, A. Inga, C. Pereira, M. M. M. Santos and L. Saraiva, *Eur. J. Pharm. Sci.*, 2015, **66**, 138.
8. Commercially available from Aldrich and Alfaesar.
9. D. A. Quagliato, P. M. Andrae and E. M. Matelan, *J. Org. Chem.*, 2000, **65**, 5037.

3.1.20. Synthesis and Characterization of Biodiesel Propyl Esters to Determine the Fatty Acid Content of Unknown Plant Oils

Heidi R. Vollmer-Snarr*, Patrick A. Fracisco, and Naiian Saephanh

Department of Chemistry, Stanford University, S.G. Mudd Bldg., 333 Campus Drive, Stanford, CA 94305-4401, USA. *E-mail: hrvsnarr@stanford.edu

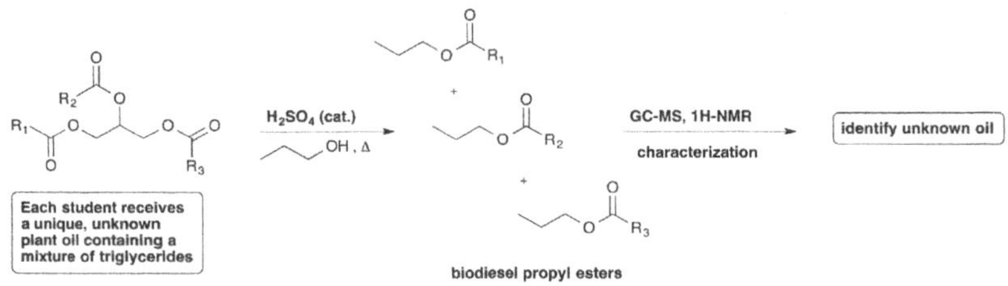

Number of sessions (duration of each session)	Hazard level	Difficulty level	Level of study
1 (4 h)	Moderate	Medium	Intermediate
Class names Esters, alcohols, alkenes			
Concepts involved The student transesterifies triglycerides from one of 15 unknown plant oils with 1-propanol and catalytic sulfuric acid to give a mixture of biodiesel propyl esters. GC-MS and ^{1}H NMR spectroscopy are used to determine the fatty acid derived propyl ester ratio, thereby facilitating the identification of the oil. A viscosity test illustrates that the biodiesel propyl esters have a greater flow rate than the plant oil triglycerides			
Chemicals needed Pure plant oils, 1-propanol, sulfuric acid, sodium chloride, magnesium sulfate, hexane, deuterated chloroform			
Equipment and experimental techniques involved Heating (100 °C) and magnetic stirring apparatus for reaction, centrifuge, GC-MS, ^{1}H NMR spectrometer			
Keywords Biodiesel, extraction, GC-MS, ^{1}H NMR, microscale, transesterification			

Background

We use 934 million tonnes of petrodiesel annually throughout the world.[1,2] Biodiesel, which can be used directly in petrodiesel engines, is a good alternative fuel because it is made from renewable resources. Moreover, the combustion of biodiesel results in less carbon monoxide, sulfur trioxide, particulate matter, and unburned hydrocarbons than the combustion of petrodiesel.[3] Biodiesel is prepared by transesterifying plant or animal-derived triglycerides with a short-chain alcohol in the presence of an acid or base catalyst. Although the transesterification of vegetable oil was patented by Chavanne in 1937,[4,5] the process was not commercialized until 1980.[6,7]

Comprehensive Organic Chemistry Experiments for the Laboratory Classroom
Edited by Carlos A. M. Afonso, Nuno R. Candeias, Dulce Pereira Simão, Alexandre F. Trindade, Jaime A. S. Coelho, Bin Tan, and Robert Franzén.

Published by the Royal Society of Chemistry, www.rsc.org

Due to the pressing need for alternative fuel sources, as well as the relevance of transesterification and lipid chemistry to organic and biochemistry chemistry curricula, several undergraduate-level biodiesel syntheses from new,[8–13] used,[8–10,14] pure,[8–12] and mixed[8–11,13,14] plant oils have been published in the last decade. Some of these syntheses mimic industrial preparations[15] of biodiesel by transesterifying plant oil with methanol and a base catalyst.[8–13] Preparations of biodiesel with acid-catalysts,[12,14] ethanol,[8,11] and 1-propanol[14] have also been described. Characterization and confirmation of the formation of biodiesel esters have been done with proton nuclear magnetic resonance (^{1}H NMR) spectroscopy,[8,11,14] Fourier-transform infrared (FT–IR) spectroscopy,[8–11] viscosity tests,[8,10,14] titration,[9,11,12] gas chromatography-mass spectrometry (GC-MS),[10] and visualization.[13]

Here, we present a new, inquiry-based, microscale approach to the undergraduate biodiesel laboratory where each student in a section starts with a different plant-derived oil, which is unknown to the student. The student performs a sulfuric acid-catalyzed transesterification reaction on the unknown oil with 1-propanol and characterizes the product mixture using GC-MS, ^{1}H NMR spectroscopy, and a viscosity test. GC-MS provides a ratio of the resulting propyl esters, which gives the fatty acid chain ratio of the oil starting material. ^{1}H NMR spectroscopy confirms the propyl ester formation, shows additional trace quantities of products, and illustrates whether the reaction went to completion and formed by-products. The viscosity test is performed to illustrate the advantage of using propyl esters over the bulkier triglycerides as biodiesel fuel in engines. With this information, the student must propose an identity for the unknown plant oil and compare its composition with the oil of another student.

Additional Safety

All work should be done in a fume hood. Protective clothing, including a laboratory coat, gloves, and safety glasses, should be worn at all times. Sulfuric acid is toxic, highly corrosive, and will burn skin on contact.[16] 1-Propanol, plant oils, and biodiesel propyl esters are highly flammable.[17] Deuterated chloroform is highly toxic, harmful if ingested, carcinogenic, and a skin irritant.[18] All of these compounds should be disposed of properly.

Experimental Procedure

Allergen warning: You will be working with one of the following 15 oils. Inform your instructor if you are allergic to any of the following oils.

Almond	Corn	Peanut	Safflower
Avocado	Grapeseed	Pumpkin	Sesame
Camelina	Macadamia nut	Red Palm	Walnut
Coconut	Olive	Rice bran	

Transesterification of Unknown Plant Oils and Product Isolation

1. Use a clamp to secure a 5 mL conical vial containing a spin vane on a magnetic stirrer.
2. Add 0.8 mL of 1-propanol and 0.075 mL of concentrated (>95% or 18 M) sulfuric acid. Stir at medium speed for one minute.
3. Add 2 mL of unknown plant oil. If you have solid oil, warm a small amount in a small beaker, and measure the oil after it liquefies.
4. Place a heating mantle on the magnetic stirrer, and clamp the vial so that it sits in the heating mantle. Insulate the vial with glass wool, and attach a microscale Vigreux column to the vial.
5. The reactants should be refluxed for 1 h at ~97 °C (the boiling point of 1-propanol).

6. Following the reflux, cool the reaction mixture to near-room temperature, and transfer it into a centrifuge tube.
7. Make sure that the centrifuge is balanced with other samples. Centrifuge the mixture for 1 min or until a separation of layers is apparent. The top layer contains biodiesel. The bottom layer contains unreacted propanol, glycerol, and sulfuric acid. Remove the bottom layer from the centrifuge tube.
8. Add 2 mL of 1 M aqueous sodium chloride to the centrifuge tube. Cap the tube, shake the mixture, and remove the cap to release built-up pressure.
9. Centrifuge the tube to separate the layers, and remove the bottom layer. An intermediate white layer may develop, which can be left in the centrifuge tube. The top layer is biodiesel, the middle layer is an emulsion, and the bottom layer is the aqueous layer.
10. Repeat the aqueous wash and centrifugation two more times.
11. Add ~0.25 g of $MgSO_4$ to dry the biodiesel. Cap and shake the centrifuge tube. Centrifuge the mixture to separate the drying agent from the biodiesel.
12. Obtain the mass of a small vial. Prepare a pipette filter by gently packing a Pasteur pipette with cotton wool, and decant the biodiesel supernatant through the pipette filter into the vial.
13. Determine the mass of biodiesel isolated.

GC-MS Sample Preparation

14. Dissolve one drop of biodiesel in 1 mL of hexane.
15. Pipette this solution through a pipette filter into a GC vial.
16. Inject the sample into the GC-MS to obtain the spectra.

^{1}H NMR Sample Preparation

17. Dissolve two drops of biodiesel in 1 mL of $CDCl_3$.
18. Acquire the ^{1}H NMR spectrum.

Viscosity Test

19. The biodiesel and plant oil must be the same temperature. If you are using solid plant oil, melt the oil in a beaker on a hot plate before the test, and heat your biodiesel to the same temperature. With a pipette bulb, draw 1 mL of biodiesel into a pipette and 1 mL of plant oil into another pipette.
20. Clamp the pipettes over a beaker, remove the pipette bulbs, and allow the two pipettes to drain into the beaker simultaneously.
21. Record how long it takes the biodiesel and plant oil to drain.

Waste Disposal

Dispose of the aqueous washes as acidic aqueous waste. Dispose of the biodiesel as organic waste. Rinse all glassware that came into contact with the deuterated chloroform into the chlorinated solvents waste. Rinse all glassware with acetone or ethanol, and dispose of the rinses as organic waste.

Results Interpretation and Additional Questions

1. Which propyl esters did your biodiesel contain? Draw the propyl esters associated with each of the GC peaks and mass spectra, label the molecular ion peaks on each of the mass spectra, and report the relative ratio of propyl esters from the GC results.

2. Draw a generic saturated or unsaturated fatty acid ester on your ^{1}H NMR spectrum, and label the relevant shifts. Has your reaction gone to completion? Do you see any peaks associated with starting material or by-products?
3. Although both vegetable oils and animal fats are composed primarily of triglycerides, they have different physical properties. Use structures to provide an explanation for this.
4. It would be cheaper and easier to burn plant oil without transesterifying it first. Why is this rarely done?
5. Draw a detailed mechanism for the first transesterification reaction (triglyceride → diglyceride) in this experiment. Abbreviate the long alkyl chains as R-groups.

Acknowledgements

We thank the students and teaching assistants of Chem 130, especially Chris Kalnmals for his assistance in putting together the experimental protocol. We are grateful to Chi-Yun Lin for his initial work on a related biodiesel project that led to the development of this project and to Charlie Cox, who assisted us with our statistical analysis. We also thank Allis Chien for her assistance with GC-MS, Stephen Lynch for NMR training, and Zachary Snarr for providing pictures of his reaction. This work was supported by Stanford University.

References

1. M. G. Kulkarni and A. K. Dalai, *Ind. Eng. Chem. Res.*, 2006, **45**, 2901.
2. B. E. Holbein, J. D. Stephen, and D. B. Layzell, *Canadian Biodiesel Initiative*, Final Report, Biocap Canada, Kingston, Ontario, Canada, 2004.
3. E. P. C. Lai, *J. Pet. Environ. Biotechnol.*, 2014, **5**, 122.
4. Historical Perspectives on Vegetable Oil-Based Diesel Fuels, http://lipidlibrary.aocs.org/history/Diesel/file.pdf, accessed April 2015.
5. C. G. Chavanne, Be. Pat., BE422877, 1937.
6. J. J. Bruwer, B. V. D. Boshoff, F. J. C. Hugo, L. M. du Plessis, J. Fuls, C. Hawkins, A. N. van der Walt, and A. Engelbrecht, *Symposium of the South African Institute of Agricultural Engineers*, 1980.
7. J. J. Bruwer, B. V. D. Boshoff, F. J. C. Hugo, J. Fuls, C. Hawkins, A. N. van der Walt, A. Engelbrecht, and L. M. du Plessis, *National Energy Symposium*, American Society of Agricultural Engineers, 1980.
8. N. R. Clarke, J. P. Casey, E. D. Brown, E. Oneyma and K. J. Donaghy, *J. Chem. Educ.*, 2006, **83**, 257.
9. A. P. Ault and R. Pomeroy, *J. Chem. Educ.*, 2012, **89**, 243.
10. M. S. Behnia, D. W. Emerson, S. M. Steinberg, R. M. Alwis, J. A. Duenas and J. O. Serafino, *J. Chem. Educ.*, 2011, **88**, 1290.
11. N. L. B. Pohl, J. M. Streff and S. Brokman, *J. Chem. Educ.*, 2012, **89**, 1053.
12. E. C. Bucholtz, *J. Chem. Educ.*, 2007, **84**, 296.
13. J. Yang, C. Xu, B. Li, G. Ren and L. Wang, *J. Chem. Educ.*, 2013, **90**, 1362.
14. D. Bladt, S. Murray, B. Glitch, H. Trout and C. Liberko, *J. Chem. Educ.*, 2011, **88**, 201.
15. G. Wenzel and P. Schulze Lammers, *J. Agric. Food Chem.*, 1997, **45**, 4748.
16. *Sulfuric Acid*; MSDS No. 339741 [online], Sigma-Aldrich, Saint Louis, MO, October 2014, http://www.sigma-aldrich.com, accessed October 2014.
17. *1-Propanol*; MSDS No. 279544 [online], Sigma-Aldrich, Saint Louis, MO, July 2014, http://www.sigma-aldrich.com, accessed October 2014.
18. *Chloroform-d*; MSDS No. 151858 [online], Sigma-Aldrich, Saint Louis, MO, April 2013, http://www.sigma-aldrich.com, accessed October 2014.

3.2. C-C σ-bond Formation

3.2.1. Synthesis of 7-Methoxy-4-oxo-*N*-phenyl-4*H*-chromene-2-carboxamide

Joana Reis, Alexandra Gaspar, Fernando Cagide, and Fernanda Borges*

Department of Chemistry and Biochemistry, Faculty of Sciences of University of Porto, Rua do Campo Alegre s/n, 4169-007, Porto, Portugal. *E-mail: fborges@fc.up.pt

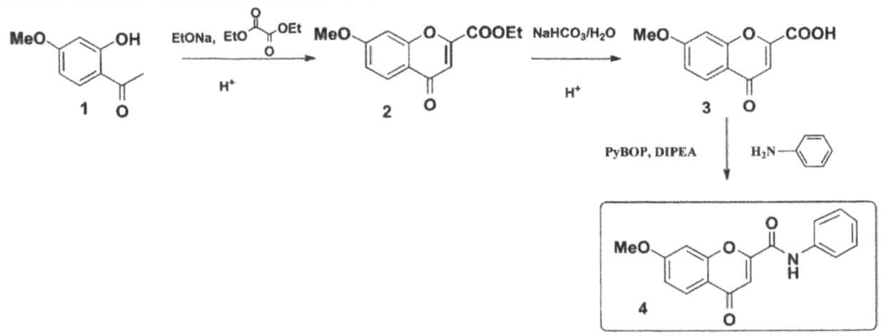

Number of sessions (duration of each session)	Hazard level	Difficulty level	Level of study
3 (4 h each)	Moderate	Medium	Intermediate

Class names Phenols, ketones, esters, carboxylic acids, primary amines, amides

Concepts involved Claisen condensation, ester hydrolysis, amide formation, purification processes, structural analysis

Chemicals needed *Step 1*: 2′-hydroxy-4′-methoxyacetophenone, sodium ethoxide (ethanolic solution), diethyl oxalate, hydrochloric acid, dichloromethane, methanol, sodium sulfate anhydrous. *Step 2*: sodium hydrogen carbonate, hydrochloric acid, dichloromethane, methanol, sodium sulfate anhydrous. *Step 3*: *N*,*N*-diisopropylethylamine, benzotriazol-1-yl-oxytripyrrolidinophosphonium hexafluorophosphate (PyBOP), aniline, dimethylformamide, dichloromethane, methanol, *n*-hexane, silica gel, sodium sulfate anhydrous. Deuterated solvents ($CDCl_3$ and DMSO-d_6) for NMR

Equipment and experimental techniques involved Equipment: balance, stirring and heating plate, UV lamp (254 and 366 nm), rotary evaporator, vacuum oven, NMR spectrometer for ^1H and ^{13}C NMR spectra acquisition. Techniques: filtration under reduced pressure, liquid–liquid extraction, thin-layer chromatography (TLC), flash column chromatography, crystallization, NMR spectroscopy

Keywords Chromone, flash column chromatography, liquid–liquid extraction, multi-step synthesis, NMR

Comprehensive Organic Chemistry Experiments for the Laboratory Classroom
Edited by Carlos A. M. Afonso, Nuno R. Candeias, Dulce Pereira Simão, Alexandre F. Trindade, Jaime A. S. Coelho, Bin Tan, and Robert Franzén.

Published by the Royal Society of Chemistry, www.rsc.org

Background

Chromones are a group of benzopyran compounds that are ubiquitous in Nature, especially in plants and are present in notable amounts in several species. The word chromone is derived from the Greek word *chroma*, meaning "color", and many of its derivatives can exhibit a diversity of colors. Chemically, chromones can be defined as oxygen-containing heterocyclic compounds with a benzoannelated γ-pyrone ring, with chromone (4*H*-chromen-4-one; 4*H*-1-benzopyran-4-one) being the parent compound (Figure 3.2.1.1).[1]

Chromone is a privileged scaffold that can be used in drug discovery processes as several relevant biological properties have been attributed to some of its derivatives, namely, anti-allergic, anti-inflammatory, anticancer, antioxidant, antiviral and antifungal activities.[1,2] The concept of a privileged structure was first introduced by Evans and coworkers[3] in 1988. The original definition has evolved and nowadays privileged structures (scaffolds or motifs) can be defined as molecular frameworks that are able to provide useful ligands for one or more type of receptor or enzyme targets by judicious structural modifications.[4] The prerequisites to label a structural motif as a privileged one still remain unclear. However, two important features needed for a chemical core to be classed as a privileged structure seem to be consensual: the synthetic availability and the simplicity in performing its chemical decoration by diversity-oriented synthesis.[5]

In this context, chemical decoration of the chromone framework has allowed the synthesis of diverse chemical libraries, and the identification of novel lead compounds.[1] As an example, one can highlight chromone-2-carboxamide as an interesting core for the development of adenosine receptor ligands that have potential therapeutic applications in cancer and/or neurodegenerative diseases.[1]

The goal of this experiment is the synthesis of 7-methoxy-4-oxo-*N*-phenyl-4*H*-chromene-2-carboxamide, a relevant chromone derivative with interesting biological properties. The final chromene-2-carboxamide can be obtained following a three-step strategy using 2′-hydroxy-4′-methoxyacetophenone (**1**) as a starting material. The first reaction corresponds to a Claisen condensation (step 1) followed by a subsequent base-catalyzed hydrolysis of the chromone ester derivative (**2**) which gives the corresponding carboxylic acid (step 2), Finally, an amidation reaction using aniline as reagent (step 3) affords 7-methoxy-4-oxo-*N*-phenyl-4*H*-chromene-2-carboxamide (**4**, Scheme 3.2.1.1).

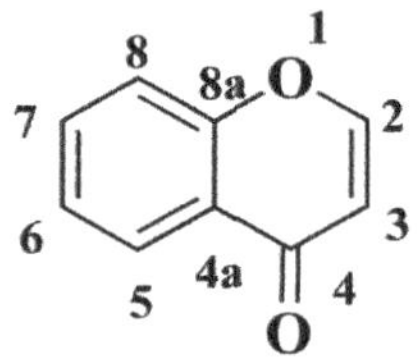

Figure 3.2.1.1 Structure of chromone and numbering according to IUPAC.

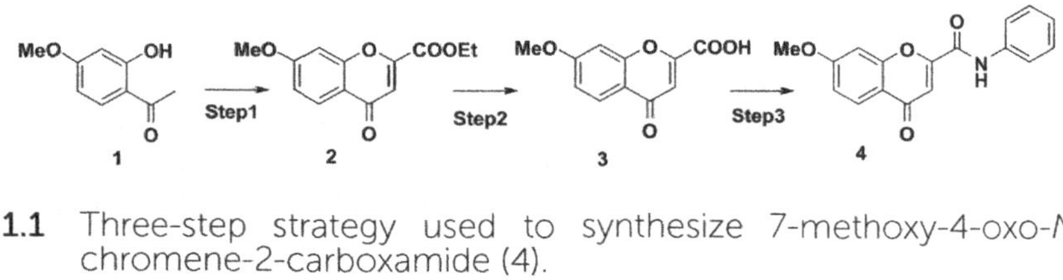

Scheme 3.2.1.1 Three-step strategy used to synthesize 7-methoxy-4-oxo-*N*-phenyl-4*H*-chromene-2-carboxamide (4).

Synthesis of Ethyl 7-Methoxy-4-oxo-4*H*-chromene-2-carboxylate (2) by Claisen Condensation (step 1)

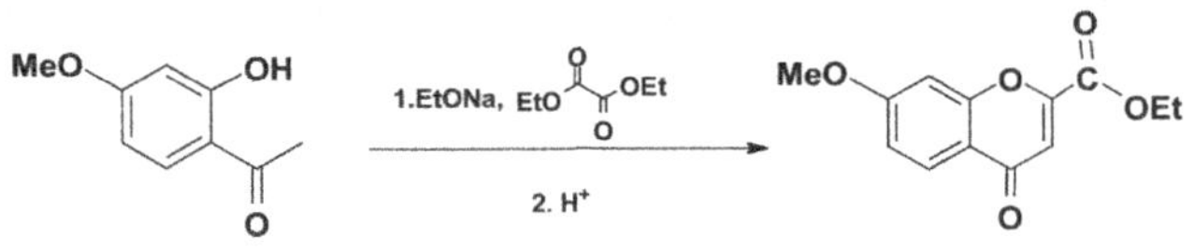

1 2

The synthesis of ethyl 7-methoxy-4-oxo-4*H*-chromene-2-carboxylate (2) involves the reaction of 2′-hydroxy-4′-methoxyacetophenone (starting material) with a strong base, usually sodium ethoxide solution (21 wt% in ethanol), providing an enolate intermediate that reacts with diethyl oxalate. The subsequent cyclisation step, allowing the formation of the chromone, is performed under acidic conditions.[6–8] (Scheme 3.2.1.2).

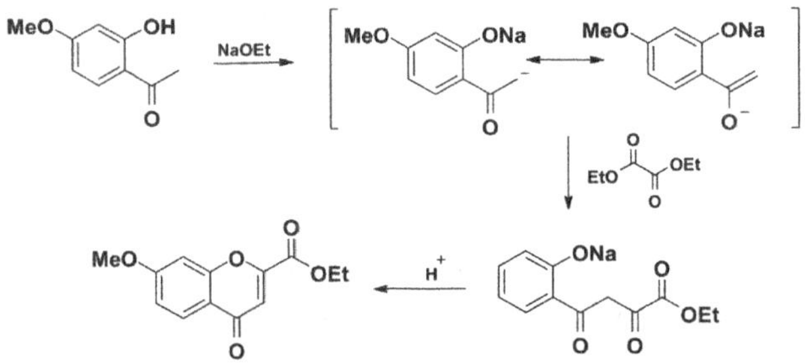

Scheme 3.2.1.2 Synthesis of ethyl 7-methoxy-4-oxo-4*H*-chromene-2-carboxylate.

Synthesis of 7-Methoxy-4-oxo-4*H*-chromene-2-carboxylic Acid (3) by Ester Hydrolysis (step 2)

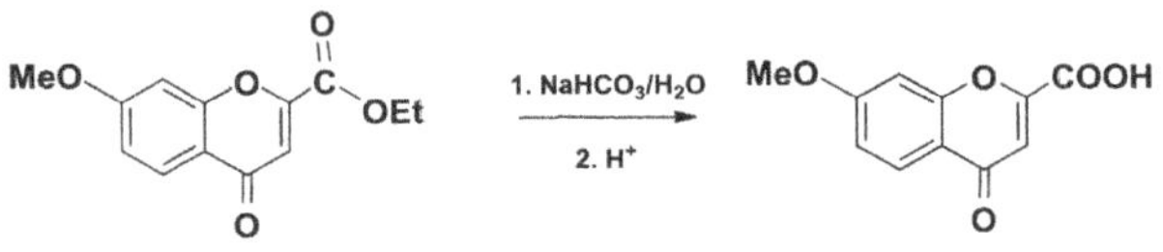

2 3

Esters can be hydrolyzed in the presence of a catalytic amount of acid or a base. In basic medium the reaction involves an addition–elimination mechanism with a direct addition of the nucleophile to the carbonyl group followed by the elimination of the leaving group. In the present case, the nucleophile (HO^-) attacks the carbonyl group of the chromone ester derivative generating a tetrahedral intermediate. The formed intermediate loses the ethoxide group ($CH_3CH_2O^-$), regenerating the carbonyl group. Acidic aqueous (*e.g.* HCl) work-up allows the carboxylic acid to be obtained as a final product[9] (Scheme 3.2.1.3).

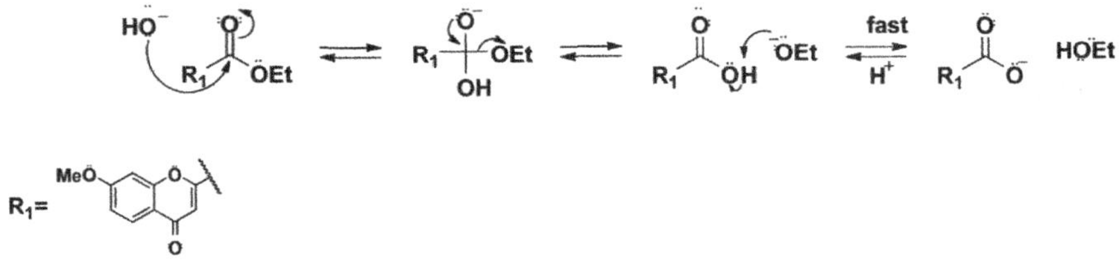

Scheme 3.2.1.3 Mechanism of the hydrolysis of ethyl 7-methoxy-4-oxo-4*H*-chromene-2-carboxylate.

Synthesis of 7-Methoxy-4-oxo-*N*-phenyl-4*H*-chromene-2-carboxamide (4) by Amidation Reaction (step 3)

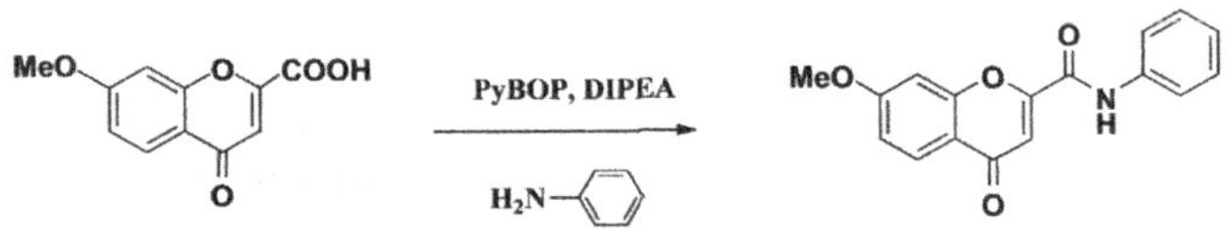

3 4

Despite the amide formation reaction being a key reaction in organic chemistry the direct amide formation between a carboxylic acid and an amine is not a trivial task. Amide bonds are typically formed from amines and pre-activated carboxylic acid derivatives by using coupling reagents.[10] Accordingly, the use of phosphonium-based coupling agents such as BOP (benzotriazol-1-yl-oxy-tris(dimethylamino)phosphonium hexafluorophosphate) or PyBOP (benzotriazol-1-yl-oxy-tris-pyrrolidinophosphonium hexafluorophosphate) for the synthesis of a chromone-2-carboxamide is highlighted.

PyBOP is currently one of the most useful phosphonium coupling agents employed for the formation of an amide bond and is routinely used in peptidic synthesis. This reagent is increasingly employed instead of the former reagent BOP because the by-products are less toxic.[10]

The synthesis of 7-methoxy-4-oxo-*N*-phenyl-4*H*-chromene-2-carboxamide from 7-methoxy-4-oxo-4*H*-chromene-2-carboxylic acid encloses an amidation process and is performed by a one-pot coupling reaction. Thus, the carboxylic function is activated with PyBOP in the presence of DIPEA (*N*,*N*-diisopropylethylamine) to generate an activated ester that undergoes amidation after addition of the chosen amine.

Additional Safety

Overall the experimental procedures bring no additional concerns besides good practice.

Special precaution concerning strong bases, such as sodium ethoxide (EtONa) solution, and strong acids, such as hydrochloric acid (HCl), must be taken. They react violently with water and must be handled with caution. Wear appropriate protective eyeglasses and other personal protective equipment, like protective gloves, respirator and laboratory coat. Throughout the experiments, the operator must stay away from ignition sources. Consultation of the Merck Index is strongly advised.

Experimental Procedure

Laboratory Session 1 (4 h): Synthesis of Ethyl 7-Methoxy-4-oxo-4*H*-chromene-2-carboxylate

Synthesis

1. In a two-neck round-bottom flask, dissolve 2′-hydroxy-4′-methoxyacetophenone (1) (12 mmol) in sodium ethoxide (21% in ethanol, 10 mL) and then add diethyl oxalate (28 mmol).
2. Keep the reaction at reflux for 1 h with continuous stirring.
3. Let the reaction reach room temperature and add concentrated HCl dropwise until acidic reaction (pH ~3). A white precipitate is formed during the last process.
4. Control the reaction progress by TLC (silica gel, dichloromethane/methanol (90 : 10 v/v)).

Work-Up

5. Filter the product (Büchner funnel), and wash it with dichloromethane.
6. Concentrate the filtrate (yellow solution) in a rotary evaporator.
7. Dissolve the slurry in dichloromethane (~50 mL) and extract with water (3 × 25 mL).
8. Dry the organic phase over Na_2SO_4, filter and evaporate the solvent in a rotary evaporator.
9. Recrystallize the light yellow solid from dichloromethane/*n*-hexane.
10. Dry the compound in a vacuum oven (50 °C) and prepare a sample in $CDCl_3$ to acquire ^1H and ^{13}C NMR spectra.

Laboratory Session 2 (4 h): Synthesis of 7-Methoxy-4-oxo-4*H*-chromene-2-carboxylic Acid

Synthesis (Previous Day)

1. In a round-bottom flask, dissolve the chromone ester (2) (10 mmol) in a solution of $NaHCO_3$ (50 mL, 20% in H_2O) and heat at 80 °C for 3 h.
2. Keep the mixture stirred with a magnetic stir bar. Upon completion of the reaction, remove the heat and allow the mixture to cool to room temperature.
3. Control the reaction progress by TLC (silica gel, dichloromethane/methanol (90 : 10 v/v)).

Work-Up

4. After cooling, acidify the solution with concentrated HCl (pH ~3) and extract it with dichloromethane (3 × 25 mL).
5. Wash the organic layer with water (3 × 25 mL) and dry the organic phase over Na_2SO_4.
6. Filter and evaporate it in a rotary evaporator.
7. Wash the solid with cold dichloromethane and diethyl ether, centrifuge and reject the supernatant.
8. Dry the obtained white compound in a vacuum oven and prepare a sample in DMSO-d_6 to acquire ^1H and ^{13}C NMR spectra.

Laboratory Session 3 (4 h): Synthesis of 7-Methoxy-4-oxo-*N*-phenyl-4*H*-chromene-2-carboxamide

Synthesis (Previous Day)

1. In a round-bottom flask dissolve the chromone carboxylic acid (3) (2.63 mmol) in 6 mL of DMF and add 0.37 mL of diisopropylethylamine (DIPEA).

2. Cool the solution at 0 °C in an ice-water bath, and add a solution of PyBOP (2.63 mmol) in dichloromethane (6 mL).
3. Stir the mixture during 30 min and add the aniline (2.63 mmol).
4. Stir the reaction for an additional 4 h at room temperature.
5. Control the reaction progress by TLC (silica gel, dichloromethane/methanol (90:10 v/v)).

Work-Up

6. Add 50 mL of dichloromethane and extract the obtained slurry with water (3 × 15 mL). Then dry the organic phase over Na_2SO_4, filter and finally evaporate in a rotary evaporator.
7. Purify the product by flash-column chromatography (silica gel) using as eluent dichloromethane/methanol (9:1).
8. Recrystallize the final product from dichloromethane/*n*-hexane.
9. Dry the obtained product in a vacuum oven and then prepare a sample in DMSO-d_6 to acquire ^1H and ^{13}C NMR spectra.

Waste Disposal

All organic and aqueous wastes should be placed in an appropriate waste container. After this operation all the glassware must be washed with water. Any glassware stains can be removed with acetone after cleaning with water and soap.

The drying agent and silica gel must be placed in a solid waste container.

Results Interpretation and Additional Questions

1. Determine the reaction yields of each experiment and the reaction overall yield.
2. Interpret the NMR spectra of the synthesized compounds.
3. Propose a reaction mechanism for the formation of 7-methoxy-4-oxo-*N*-phenyl-4*H*-chromene-2-carboxamide.

References

1. A. Gaspar, M. J. Matos, J. Garrido, E. Uriarte and F. Borges, *Chem. Rev.*, 2014, **114**(9), 4960–4992.
2. A. M. Edwards and J. B. Howell, *Clin. Exp. Allergy*, 2000, **30**(6), 756–774.
3. B. E. Evans, K. E. Rittle, M. G. Bock, R. M. DiPardo, R. M. Freidinger, W. L. Whitter, G. F. Lundell, D. F. Veber and P. S. Anderson, *J. Med. Chem.*, 1988, **31**(12), 2235–2246.
4. C. D. Duarte, E. J. Barreiro and C. A. Fraga, *Mini-Rev. Med. Chem.*, 2007, **7**(11), 1108–1119.
5. S. K. Ko, H. J. Jang, E. Kim and S. B. Park, *Chem. Commun.*, 2006, **28**, 2962–2964.
6. G. P. Ellis, Chromone carboxylic acids and their derivatives in *Chromenes*, *Chromanones* and *Chromones*, in *Chemistry of Heterocyclic Compounds*, ed. G. P. Ellis, John Wiley & Sons, Inc., New York, 1977, vol. 31, pp. 943–1042.
7. A. Banerji and N. C. Goomer, *Synthesis*, 1980, **11**, 874–875.
8. I. Hirao, M. Yamaguchi and M. Hamada, *Synthesis*, 1984, **12**, 1076–1078.
9. J. March, *Advanced Organic Chemistry: Reactions, Mechanisms, and Structure*, John Wiley & Sons, Inc., Hoboken, 6th edn, 2006, pp. 378–386.
10. C. A. G. N. Montalbetti and V. Falque, *Tetrahedron*, 2005, **61**, 10827–10852.

3.2.2. Synthesis of 2,3-Diphenylindenone

João P. Telo

Centro de Química Estrutural, Instituto Superior Técnico, Universidade de Lisboa, Av. Rovisco Pais, 1649-003 Lisboa, Portugal. *E-mail: jptelo@tecnico.ulisboa.pt

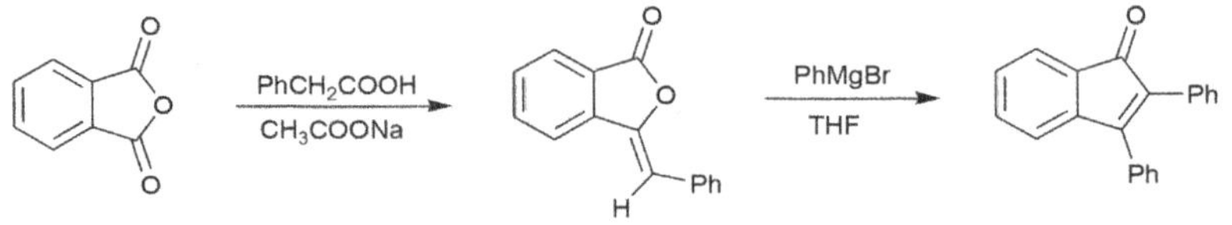

Number of sessions (duration of each session)	Hazard level	Difficulty level	Level of study
3 (4 h each)	Moderate	High	Intermediate

Class names Anhydrides, esters, ketones

Concepts involved 2,3-Diphenylindenone is prepared in two steps involving nucleophilic additions to the carbonyl group. In the first laboratory session benzalphthalide is prepared by a solvent-free condensation of phenylacetic acid and phthalic anhydride. In the second step, the addition of the Grignard reagent phenyl magnesium bromide affords the 2,3-diphenylindenone. The red product can be characterized by visible spectroscopy

Chemicals needed Phthalic anhydride, phenylacetic acid, sodium acetate (freshly molten), ethanol, anhydrous $MgSO_4$, magnesium turnings, bromobenzene, dry THF, toluene, H_2SO_4

Equipment and experimental techniques involved Apparatus for reaction with distillation and sand-bath heating, apparatus for reaction under anhydrous conditions and temperature control, liquid–liquid extraction, rotatory evaporator, vacuum filtration, melting point apparatus, IR spectrometer, UV-Vis spectrophotometer, NMR spectrometer

Keywords Grignard reagents, NMR spectroscopy, reactions under anhydrous conditions, recrystallization, UV-vis spectroscopy

Background

Indanones and indenones are important classes of compounds in organic chemistry. The indenone core-structure can be found in several drugs with pharmaceutical applications.[1,2] Moreover, this class of compounds are synthetic intermediates for the construction of various organic structures.[3,4]

The addition of Grignard reagents to a carbonyl group of an aldehyde, ketone or ester is a fundamental reaction in organic synthesis and an important tool for the formation of carbon–carbon bonds.[5] Grignard reactions and reagents were discovered by and are named after the French chemist François Auguste Victor Grignard, who was awarded the 1912 Nobel Prize in Chemistry for this work. Grignard reagents are prepared by insertion of a magnesium atom in a carbon–halogen bond, transforming the carbon atom into a strong base and strong nucleophile. Grignard reagents are destroyed by week acids like water (and alcohols, terminal alkynes,

Comprehensive Organic Chemistry Experiments for the Laboratory Classroom
Edited by Carlos A. M. Afonso, Nuno R. Candeias, Dulce Pereira Simão, Alexandre F. Trindade, Jaime A. S. Coelho, Bin Tan, and Robert Franzén.

Published by the Royal Society of Chemistry, www.rsc.org

1,3-dicarbonyl compounds, *etc.*), so the reaction has to be carried out under anhydrous conditions. In this work the Grignard reagent phenylmagnesium bromide is added to the ester group of benzalphthalide, which gives 2,3-diphenylindenone after rearrangement of the intermediate adduct followed by dehydration. The benzalphthalide is prepared in the first step by a solvent-free condensation of phenylacetic acid and phthalic anhydride. The final product has an intense red color and can be further characterized by visible-UV spectroscopy. The experimental procedure is an adaptation from that of R. Weiss[6] and T.J. Clark.[7]

Additional Safety

It is advisable to perform these reactions in a fume cupboard. In the first step the temperature reached by the sand-bath heating is high and the necessary precautions should be taken. THF, toluene and the ethanol used in the recrystallization are flammable solvents. The reaction in the second step is highly exothermic and the use of an ice bath for emergency cooling is necessary.

Experimental Procedure

Laboratory Session 1 (4 h): Synthesis of Benzalphthalide

1. Add 10 g of phthalic anhydride (0.067 mol), 11 g of phenylacetic acid (0.080 mol) and 0.26 g of freshly molten sodium acetate in a 250 mL two-neck round-bottomed flask.
2. Put a thermometer in one of the necks and a distillation head, a Liebig condenser and a receiving flask in the other. Make sure the thermometer reaches the reaction mixture.
3. Immerse the reaction flask in a sand-bath and heat the sand bath until the temperature of the solution reaches 230 °C.
4. Raise the temperature gradually to 240 °C, while water and some organic material starts distilling. Keep the reaction at 240 °C until no more water is distilled (around 1 h), and check its conversion by measuring the amount of water. Take a small portion of the brown mass formed out of the reaction flask with a spatula and add it to a test tube with a small volume of ethanol. The reaction is complete when the solid dissolves by heating the test tube in warm water and crystallizes by cooling.
5. Cool the reaction mixture to 90–95 °C and add 40 mL of ethanol to recrystallize the solid.
6. Cool to room temperature and then in an ice bath, filter off the yellow crystals by suction and wash them with cold ethanol (4–5 mL). An analytical sample can be obtained by a second recrystallization from ethanol.

Laboratory Session 2 (4 h): Synthesis of 2,3-Diphenyl-1-indenone (Part 1)

All glassware used in the early stages of this procedure, including the graduated cylinders to measure reactants, should be well dried in an oven.

1. Equip a three-neck 250 mL round-bottomed flask with a reflux condenser protected with a $CaCl_2$ tube (or with another drying agent), a thermometer and a dropping funnel with a pressure equalizer. The apparatus is positioned so that the flask can be easily placed either on a steam bath or on an ice bath.
2. Add to the flask a magnetic stirrer and 1.57 g of dried magnesium turnings.
3. Add 30 mL of dry THF and 6.6 mL of bromobenzene (9.8 g, 0.063 mol) to the dropping funnel and allow a few mL of this mixture to enter into the reaction flask while stirring magnetically. The reaction starts when the solution shows some turbidity. If the reactions fails to start, heat briefly the flask in a water bath (the reaction is very exothermic, do not keep the flask in the warming bath for long).

4. Start the addition of the bromobenzene solution drop-by-drop. After a few minutes the solution turns clear, darker and starts boiling. Use an ice bath to moderate the heat released (for short periods, or otherwise the reaction stops).
5. After the end of the addition heat the flask in a boiling water bath for 15 min. At the end of this period the mixture is dark and shows only residual pieces of metallic magnesium.
6. Take out the water bath (but keep the magnetic stirring), allow the mixture to cool to room temperature and add through the dropping funnel a solution of 5.6 g of benzalphthalide (0.025 mol) in 25 mL of dry THF at a rate of 50–100 drops per minute.
7. After the addition ends, change the glassware attached to the flask for a Liebig condenser with a receiving flask and two stopcocks. Distill most of the solvent in a boiling water bath.
8. Replace the Liebig condenser and one of the stopcocks by the reflux condenser and dropping funnel and add through the dropping funnel a solution of 2.5 mL of concentrated H_2SO_4 in 38 mL of water at a rate that allows the solution to boil gently.
9. After this add rapidly 15 mL of a 6 M H_2SO_4 solution previously heated at 65 °C and stir the mixture vigorously for 5 min with the magnetic stirrer. At this stage the mixture shows an upper dark-red organic layer and a yellow aqueous layer below. Store the mixture in a closed flask.

Laboratory Session 3 (4 h): Synthesis of 2,3-Diphenyl-1-indenone (Part 2) and Characterization of the Product

1. Add 20 mL of toluene to the flask and transfer the mixture to a separatory funnel. Rinse the flask with 10 mL of toluene and add to the separatory funnel.
2. Discharge the lower aqueous layer; wash the organic layer with water
3. Transfer the organic layer to an Erlenmeyer flask and dry it with anhydrous $MgSO_4$.
4. Distill most of the solvent in a rotatory evaporator and add 30 mL of ethanol to the flask. The product precipitates in red crystals.
5. Cool in an ice bath, filter with suction and wash the crystals with a small portion of cold ethanol.
6. Recrystallization can be performed by dissolving the product in warm toluene and adding warm ethanol to the solution. Slow cooling affords beautiful red prisms of 2,3-diphenyl-1-indenone.
7. Prepare a solution of 2,3-diphenyl-1-indenone in toluene in a volumetric flask and take the UV-vis spectrum of the compound. A concentration of 0.15 mM is suitable for a 1 cm optical path. Calculate the λ_{max} and ε_{max} of the band in the visible region.
8. Take the IR spectrum of the compound dispersed in a solid KBr pellet, or in a liquid dispersion in Nujol or CCl_4 between NaCl plates.
9. Take the NMR spectrum (^{1}H and ^{13}C) of 2,3-diphenyl-1-indenone in $CDCl_3$.

Waste Disposal

The solvents and the mother-liquors of the recrystallizations should be disposed of in proper containers for elimination. The acid aqueous solutions should be diluted and discharged in the sink with running water.

Results Interpretation and Additional Questions

1. Draw the reactions mechanisms.
2. Why do we have to use dry THF and oven-dried glassware in the Grignard reaction?

3. Discuss the purity of the products by measuring the melting point and comparing it with the literature values.
4. Analyse the UV-vis, IR and NMR spectra of 2,3-diphenylindenone and compare the λ_{max} and ε_{max} and the IR and NMR peaks with literature values.[8]

References

1. A. Morrell, M. Placzek, S. Parmley, B. Grella, S. Antony, Y. Pommier and M. Cushman, *J. Med. Chem.*, 2007, **50**, 4388.
2. E. M. Coyanis, J. Panayides, M. A. Fernandes, C. B. de Koning and W. A. L. van Otterlo, *J. Organomet. Chem.*, 2006, **691**, 5222.
3. E. F. Ullmm and W. A. Henderson Jr, *J. Am. Chem. Soc.*, 1966, **88**, 4942.
4. G. M. Anstead, J. L. Ensign, C. S. Peterson and J. A. Katzenellenbogen, *J. Org. Chem.*, 1989, **5**, 1485.
5. Grignard Reaction, https://en.wikipedia.org/wiki/Grignard_reaction.
6. R. Weiss, *Org. Synth Coll.*, 1943, **2**, 61.
7. T. J. Clark, *J. Chem. Educ.*, 1971, **48**, 554.
8. AIST, Spectral Database for Organic Compounds, SDBS, http://sdbs.db.aist.go.jp/sdbs/cgi-bin/cre_index.cgi.

3.2.3. Synthesis of Dimedone

Dulce Simão* and Ana C. Cerdeira

Department of Chemistry, Instituto Superior Técnico, University of Lisbon, Centro de Química Estrutural, Av. Rovisco Pais, 1, 1049-001 Lisbon, Portugal. *E-mail: dulcesimao@tecnico.ulisboa.pt

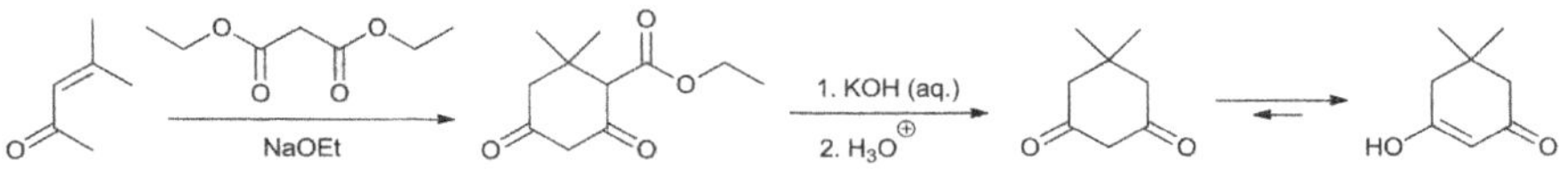

Number of sessions (duration of each session)	Hazard level	Difficulty level	Level of study
2 (3 h each)	High	Medium	Intermediate
Class names Enones, enols, diesters, 1,3-diketones			
Concepts involved Two-step synthesis of dimedone, illustrating a Michael addition followed by an intramolecular Claisen condensation and decarboxylation			
Chemicals needed Sodium metal, absolute ethanol, diethyl malonate (1,3-diethyl propanedioate), mesityl oxide (4-methylpent-3-en-2-one), potassium hydroxide, diethyl ether, hydrogen chloride 37%, acetone, petroleum ether (60–80 °C), anhydrous calcium chloride			
Equipment and experimental techniques involved Liquid–liquid extraction, ^{1}H NMR, vacuum filtration, rotary evaporator, reflux apparatus, IR, melting point apparatus			
Keywords Claisen ester condensation, enol, ^{1}H NMR, IR, liquid–liquid extraction, Michael addition, Robinson annulation			

Background

The Michael addition is a 1,4-addition of an enolate derived from CH-acidic compounds to α,β-unsaturated carbonyl compounds. This reaction is extremely important in the synthesis of 1,5-dicarbonyl compounds, and is commonly used in the synthesis of natural products such as steroids. The Robinson annulation consists of a Michael addition followed by an intramolecular aldol condensation to produce cyclic enones.[1] This experiment[2,3a,4] follows a slightly modified synthetic route, where the aldol condensation is substituted by an intramolecular Claisen ester condensation. Consequently, the first step is the conjugate addition of diethyl malonate to mesityl oxide (4-methylpent-3-en-2-one) followed by cyclization to produce a six-membered ring (Scheme 3.2.3.1).

This cyclic precursor readily undergoes hydrolysis followed by a decarboxylation, to furnish dimedone[5] (5,5-dimethylcyclohexane-1,3-dione), which exists predominantly as its enol tautomer that is formed by transfer of a proton from the CH_2 group of keto form to one of the OH groups.[6,7] Dimedone is generally used to identify aldehydes, because it forms crystalline bisdimedones[3b,8] with aldehydes but not with ketones (Scheme 3.2.3.2). It also has applications in colorimetry, crystallography, luminescence and spectrophotometric analysis.[9]

Comprehensive Organic Chemistry Experiments for the Laboratory Classroom
Edited by Carlos A. M. Afonso, Nuno R. Candeias, Dulce Pereira Simão, Alexandre F. Trindade, Jaime A. S. Coelho, Bin Tan, and Robert Franzén.

Published by the Royal Society of Chemistry, www.rsc.org

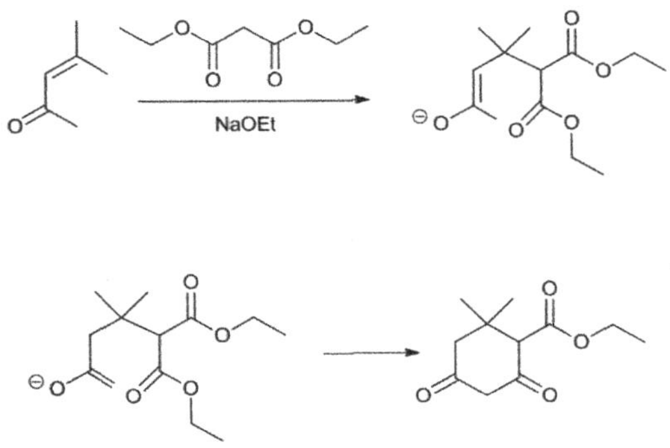

Scheme 3.2.3.1 Conjugate addition and cyclization.

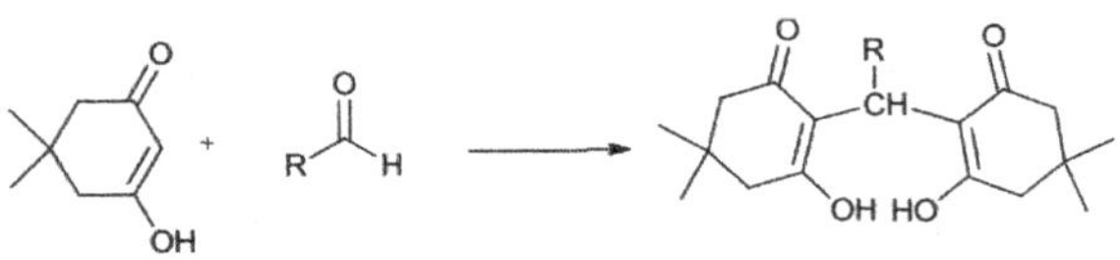

Scheme 3.2.3.2 Bis-dimedone formation.

Additional Safety

This work must be performed under a fume hood since mesityl oxide is a lachrymator. Sodium metal is a corrosive and potentially flammable solid. It reacts violently with water, and in case of fire, use dry chemical powder. Hydrogen chloride is highly corrosive. Flammable solvents, such as ethyl ether and ethanol should be handled carefully.

Experimental Procedure

Session 1

All material and apparatus must be thoroughly dried in a hot oven ($T > 120$ °C) before use.

1. Add absolute ethanol (30 mL) to a 250 mL three-neck round-bottom flask fitted with magnetic stirring, a pressure equalizing dropping funnel, a reflux condenser and a $CaCl_2$ tube.
2. Cut sodium metal (1.15 g, 0.05 mol) into small pieces and add piece by piece, with smooth stirring at such a rate that the mixture boils.
3. Once the sodium is completely dissolved, add diethyl malonate (8.5 mL, 8.0 g, 0.05 mol) dropwise for 5 min.
4. Add absolute ethanol (5 mL).
5. Add mesityl oxide (6.0 mL, 4.9 g, 0.05 mol) also dropwise for 5 min followed by another portion of absolute ethanol (5 mL).
6. Reflux the reaction mixture with stirring for 45 min in a water bath.
7. Dissolve KOH (6.3 g, 0.112 mol) in water (25 mL), add dropwise to the reaction mixture.
8. Reflux for 45 min.
9. Turn the heat off and allow the mixture to cool.
10. Store the reaction mixture until the next session.

Session 2

1. In a rotary evaporator distil off around 35 mL of the ethanol–water mixture.
2. Cool the residue with ice and extract with ethyl ether (25 mL), keeping the aqueous layer.
3. Return the aqueous layer to the reaction flask, acidify to pH 1 with concentrated HCl and reflux for 15 min in an oil bath.
4. Allow the mixture to cool in an ice bath until crystallization is complete.
5. Filter the crude product under vacuum and wash it with water (25 mL) and petroleum ether (60–80 °C) (25 mL), drying the product between washes.
6. Recrystallize from aqueous acetone (1 : 1).

Waste Disposal

Residual sodium must be carefully destroyed with ethanol. The acidic solution resulting from vacuum filtration must be neutralized with a solution of sodium carbonate and discarded in the sink. Ethyl ether used for the liquid–liquid extraction must be placed in a waste collector recipient for non-chlorinated solvents. Petroleum ether from washings may be stored for further purification or be placed in a waste collector recipient for non-chlorinated solvents.

Results Interpretation and Additional Questions

1. Determine the reaction yield.
2. Determine the melting point of the product and evaluate its purity.
3. Record and interpret the IR and ^{1}H NMR spectra.
4. Write all reaction mechanisms.

References

1. K. P. C. Vollhardt and N. E. Schore, *Organic Chemistry*, Freeman & Co., 3rd edn, 1999, p. 807.
2. L. M. Harwood, C. J. Moody and J. M. Percy, *Experimental Organic Chemistry*, Blackwell Science, 2nd edn, 1999, p. 540.
3. A. I. Vogel, *Vogel's Textbook of Practical Organic Chemistry*, Longman Scientific and Technical, 5th edn, 1989(a) p. 1104; (b) p. 1259.
4. C. A. M. Afonso, D. P. Simão, L. P. Ferreira, M. S. Serra and M. M. M. Raposo, *100 Experiências de Química Orgânica*, IST Press, 2011, p. 267.
5. A. Jacobs, *Understanding Organic Reaction Mechanisms*, Cambridge University Press, 1997, p. 242.
6. J. Clayden, N. Greeves, S. Warren and P. Wothers, *Organic Chemistry*, Oxford, 2001, vol. 764, pp. 524–532.
7. J. Jones, Core Carbonyl Chemistry, in *Oxford Chemistry Primers*, Oxford University Press, 1997, p. 76.
8. D. Spencer and T. Henshall, *J. Am. Chem. Soc.*, 1955, 77, 1943.
9. K. Sushma, S. Ghosh and D. Banji, *Int. J. Med. Pharm. Res.*, 2013, **1**, 433.

3.2.4. Acylation Reaction of Enol Ether Using an Ionic Liquid

Marcos A. P. Martins*, Clarissa P. Frizzo*, Geórgia C. Zimmer, Caroline R. Bender, Paulo R. S. Salbego, Aniele Z. Tier, and Nilo Zanatta

Núcleo de Química de Heterociclos (NUQUIMHE), Department of Chemistry, Federal University of Santa Maria, 97105-900, Santa Maria, RS, Brazil. *E-mail: mmartins@base.ufsm.br, clarissa.frizzo@gmail.com

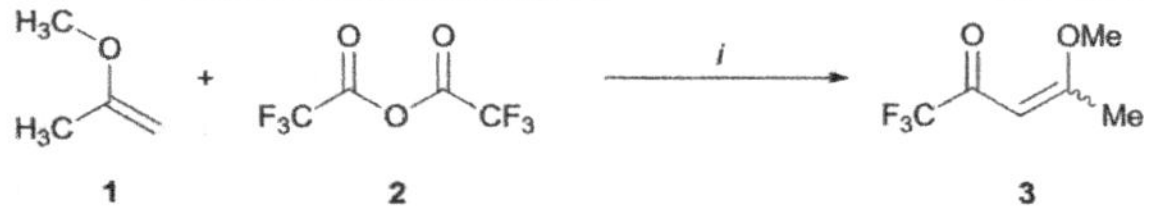

(i) Pyridine, [BMIM][BF_4], 0 °C, then 25 °C, 90 min, 71–85%

Number of sessions (duration of each session)	Hazard level	Difficulty level	Level of study
1 session (4 h)	High	Medium	Advanced

Class names Enol ethers, halogenated anhydrides, enones

Concepts involved *C*-acylation reaction, ionic liquid solvent–catalyst, structure determination of organic compounds by NMR spectroscopy

Chemicals needed 2-Methoxyprop-1-ene (enol ether, **1**), trifluoroacetic anhydride (**2**), pyridine, the IL 1-butyl-3-methyl-imidazolinium tetrafluoroborate ([BMIM][BF_4]), diethyl ether, $LiAlH_4$, anhydrous Na_2SO_4, $CDCl_3$, distilled water

Equipment and experimental techniques involved Apparatus for reaction with controlled temperature (0 °C), magnetic stirring, apparatus for liquid–liquid extraction

Keywords Acylation, ionic liquid, liquid–liquid extraction, NMR, rotary evaporator

Background

The synthesis of acylated enol ethers has been studied by several groups around the world. This synthesis is very important because the acylated enol ethers are used as halogenated building blocks to prepare halogenated heterocycles such as isoxazoles (Scheme 3.2.4.1), pyrazoles, pyrimidines, pyridines, thiazines, and diazepines.[1] There are several basic methods described in the literature for the construction of acylated enol ether compounds: (i) the formation of C2–C3 bonds (*e.g.*, trihaloacylation of acetals and enol ethers); (ii) the formation of C3–C4 bonds (*e.g.*, condensation of carbonyl compounds with 1,1,1-trifluoroacetone derivatives); and (iii) the formation of C1–C2 bonds (*e.g.*, addition of trifluoromethyl-organometallic reactants to acrylic acid esters).[1]

Comprehensive Organic Chemistry Experiments for the Laboratory Classroom
Edited by Carlos A. M. Afonso, Nuno R. Candeias, Dulce Pereira Simão, Alexandre F. Trindade, Jaime A. S. Coelho, Bin Tan, and Robert Franzén.

Published by the Royal Society of Chemistry, www.rsc.org

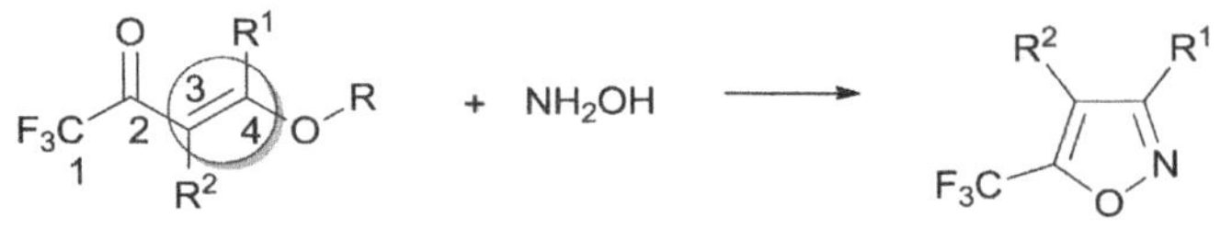

Scheme 3.2.4.1 Reaction of 4-alkoxy-1,1,1-trihalo-3-alken-2-ones with hydroxylamine.

The acylation of enol ethers requires strong activation of the acylating agents. Electronegative substituents, such as the trifluoro- or trichloromethyl groups attached to the carbonyl carbon, increase the reactivity of the acyl component. Trifluoroacetic anhydride (TFAA) or trichloroacetyl chloride (TCAC) are the reactants used for the preparation of trihaloacetylated enol ethers, furnishing β-trihaloacyl-substituted enol ethers in quantitative yields.[2] These reactants are widely used because of their simplicity and adaptability for a wide range of enol ethers.[3] However, it is important to note that TFAA is hydrolyzed in the presence of moisture. Consequently, all reactants should be water-free and any glassware used in the experimental procedure should be dried well. The classical procedure of acylation of enol ethers with trihaloacetyl acylants is performed by using dichloromethane or ethyl ether as the solvent, with a relatively long reaction time (16–48 h).[4a] Methods to improve the synthesis of these compounds are being developed. Ionic liquids (ILs) are attracting much attention, due to being promising "green" solvents and alternatives to the hazardous traditional organic solvents,[5] as well as being able to catalyze important organic transformations without the need to add another catalyst or ligand.[6] Because of this, ILs are being used in organic synthesis and in industry.[7] The objectives of this work are: (i) to perform the acylation of enol ether in the presence of an IL, (ii) to reduce the reaction time, and (iii) to show the C–C σ-bond formation from nucleophilic carbonyl substitution.

Additional Safety

For this experiment it is necessary to use the appropriate safety equipment, such as protective clothing, gloves, and safety goggles. The reactions must be done in a fume hood and the reactants should be used carefully because they can be irritating to the eyes and skin.

Experimental Procedure

Laboratory Session (4 h): Acylation Reaction of Enol Ether Using an IL

1. In a fume hood, prepare the system for an acylation reaction.
2. First, connect an addition funnel to one three-neck round-bottom flask.
3. Place a drying tube above the addition funnel.
4. Seal one mouth of the round-bottom flask with a septum.
5. Use the last mouth of the flask to insert an inert gas flux.
6. Fill the system with an inert gas by controlling the inlet valve for the inert gas.
7. Under inert gas flux, meticulously dry the system using a Bunsen burner or a heat blower.
8. Stop the flow of inert gas and insert a drying tube in this position. The acylation system is shown in Figure 3.2.4.1.
9. In the addition funnel: Add 0.5 mL (5.2 mmol) of enol ether **1**, 0.42 mL (5.2 mmol) of pyridine, and 0.10 mL (0.53 mmol) of [BMIM][BF_4]. It is recommended to use a syringe.
10. Put the round-bottom flask in an ice bath (0 °C) under magnetic stirring.
11. Add 0.75 mL (5.2 mmol) of the trifluoroacetic anhydride **2** to the flask using a syringe.
12. Start the reaction by slowly placing the mixture from the addition funnel (drop by drop) into the round-bottom flask.

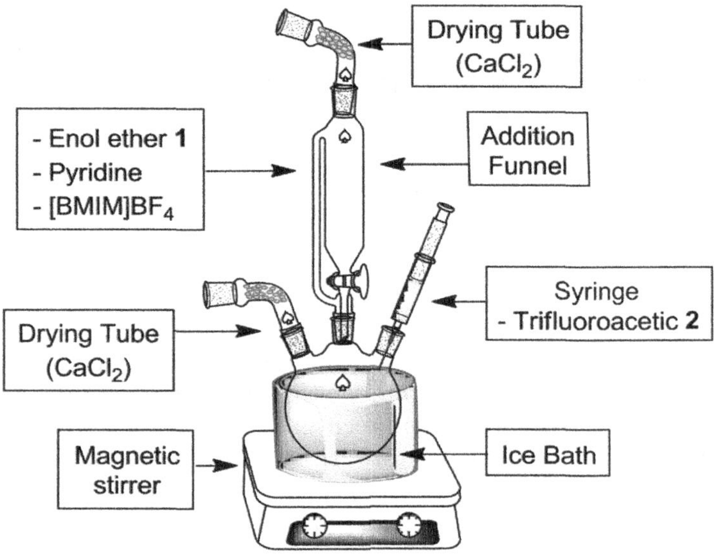

Figure 3.2.4.1 Acylation system.

Note: It is important that the enol ether **1** be distilled from $LiAlH_4$ into a well-sealed flask, in order to prevent the entry of water into the reactant, which could decrease the yield. After distillation, molecular sieves can be used to keep the enol ether dry; however, the use of molecular sieves without previous distillation is not recommended.

[BMIM][BF_4] is highly hygroscopic; therefore, any water should be removed before starting the experimental procedure. The water in the [BMIM][BF_4] can be removed with a vacuum pump (at 10 mbar for 2–3 h).

13. After all the mixture from the addition funnel has been added to the round-bottom flask, remove the ice bath and leave the reaction under magnetic stirring for 90 min.
14. Add diethyl ether (15 mL) and leave under magnetic stirring for 5 min. Then, stop stirring and transfer the diethyl ether phase to an extraction funnel using a pipette.
15. Repeat the previous sequence three times.
16. Wash the diethyl ether phase in the extraction funnel with distilled water (3 × 20 mL).
17. Collect the diethyl ether phase in an Erlenmeyer flask.
18. Add the anhydrous Na_2SO_4 (150 mg) to the diethyl ether phase and shake manually for about 1 min.
19. Filter into a previously weighed flask (M_1).
20. Evaporate the diethyl ether in a rotary evaporator.
21. Perform an NMR analysis using $CDCl_3$, and calculate the reaction yield.

Yield Determination

22. After the solvent has been removed, weigh the round-bottom flask containing the reaction mixture, in order to determine the final weight (M_2).
23. From this value (M_2), and the weight of the empty round-bottom flask (M_1), determine the mass of the product obtained in the experiment (M_3), using eqn (3.2.4.1):

$$M_3 = M_2 - M_1 \tag{3.2.4.1}$$

24. The M_3 value corresponds to the mass of the product obtained in the experiment.
25. Correlate the amount of product obtained (in mmol) with the limiting reactant used (in this case the 2-methoxy-1-propene). Then calculate the reaction yield using eqn (3.2.4.2):

$$\text{Yield}(\%) = \frac{\text{product (mmol)}}{\text{limiting reactant (mmol)}} \times 100 \qquad (3.2.4.2)$$

Waste Disposal

After the reaction is complete, the organic product is separated from the IL using diethyl ether extraction (3 × 15 mL). To remove pyridine salt from the IL phase, add acid solution (HCl, 10%) and evaporate the water and pyridine in a rotary evaporator. After this, the IL is placed in a vacuum pump (10 mbar) for 8 h. The IL is then recovered and purified. The diethyl ether solvent may be reused or collected in a waste collector recipient for later destruction.

Interpretation of Results and Additional Questions

1. Interpret the ^{1}H NMR spectra obtained and compare it with the literature.[3]
2. Calculate the yield and compare it with the literature.[3]
3. Discuss with your partners the importance of using ILs in organic synthesis.
4. Rationalize the mechanism involved in the acylation reaction of enol ether.
5. What are the electrophilic centers of the product obtained? What types of bonds are formed in this reaction?

Acknowledgement

Reprinted (adapted) from M. A. P Martins *et al.*, Synthesis of 1,1,1,-trichloro[fluoro]-3-alken-2-ones using ionic liquids, *J. Mol. Catal. A: Chem.*, 2007, 266, 100–103. Copyright (2007) with permission from Elsevier.

References

1. M. A. P. Martins, W. Cunico, C. M. P. Pereira, A. F. C. Flores, G. Bonacorso and N. Zanatta, *Curr. Org. Synth.*, 2004, 391.
2. A. Colla, M. A. P. Martins, G. Clar, S. Krimmer and P. Fisher, *Synthesis*, 1991, 483.
3. M. A. P. Martins, E. A. Guarda, C. P. Frizzo, E. Scapin, P. Beck, A. C. Costa, N. Zanatta and H. G. Bonacorso, *J. Mol. Catal. A: Chem.*, 2007, **266**, 100.
4. (a) A. Colla, M. A. P. Martins, G. Clar, S. Krimmer and P. Fisher, *Synthesis*, 1991, 483; (b) R. J. Andrew and J. M. Mellor, *Tetrahedron*, 2000, **56**, 7261.
5. (a) L.-W. Xu, Y. Gao, J.-J. Yin, L. Li and X. Chun-Gu, *Tetrahedron Lett.*, 2005, **46**, 5317; (b) T. Welton, *Chem. Rev.*, 1999, **99**, 2071; (c) A. Cosson, M. Johansson and J. E. Backvall, *Chem. Commun.*, 2004, 1494.
6. S. A. Shapi, U. C. Narkhede, S. S. Palimkar, T. Daniel, R. J. Lahoti and K. V. Srinivasan, *Tetrahedron*, 2005, **61**, 3539.
7. (a) N. Jain, A. Kumar, S. Chauhan and S. M. S. Chauhan, *Tetrahedron*, 2005, **61**, 1015; (b) T. Welton, *Coord. Chem. Rev.*, 2004, **248**, 2459.

4. Addition at Non-aryl sp^2 Carbon

4.1. C-X Bond Formation

4.1.1. Electrophilic Addition to Alkenes

4.1.1.1. Bromination of Cinnamic Acid

Paulo Coelho* and Céu Sousa

Centro de, Química – Vila Real, Universidade de Trás-os-Montes e Alto Douro, 5001-801 Vila Real, Portugal
*E-mail: pcoelho@utad.pt

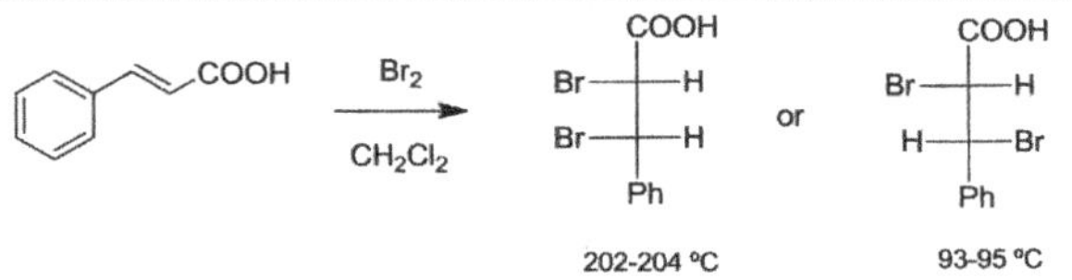

Number of sessions (duration of each session)	Hazard level	Difficulty level	Level of study
1 (2 h)	Moderate	Low	Introductory

Class names Olefins, alkyl halides

Concepts involved The *syn*/*anti* orientation of the bromine addition to cinnamic acid can be easily determined measuring the melting point of the product, 2,3-dibromo-3-phenylpropanoic acid, since the two possible diastereoisomers have very distinctive melting points

Chemicals needed Cinnamic acid, Br_2, CH_2Cl_2, cyclohexene, $Na_2S_2O_3$

Equipment and experimental techniques involved Heating mantle, apparatus for reaction under reflux with addition funnel, filtration

Keywords Bromine addition, cinnamic acid, filtration, mp, olefins, RMN

Background

Alkenes are a class of hydrocarbons with at least one double bond. Double bonds are electron rich and thus they react easily with electrophiles like H_2O, X_2, and HX. This reaction is called an addition since two molecules originate one single molecule. The addition of bromine to an alkene affords a dibromoalkane in which the two bromine atoms are linked to adjacent carbons atoms – vicinal positions. In this addition reaction each bromine atom becomes linked to one carbon atom of the double bond.[1]

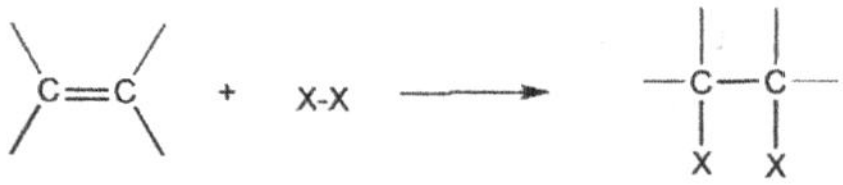

Comprehensive Organic Chemistry Experiments for the Laboratory Classroom
Edited by Carlos A. M. Afonso, Nuno R. Candeias, Dulce Pereira Simão, Alexandre F. Trindade, Jaime A. S. Coelho, Bin Tan, and Robert Franzén.

Published by the Royal Society of Chemistry, www.rsc.org

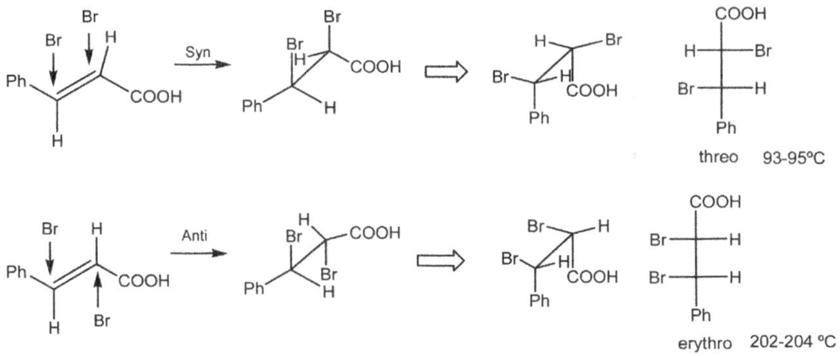

Scheme 4.1.1.1.1 Addition of bromine to a double bond.

This reaction is thermodynamically favorable since the two new covalent C–Br σ bonds that are formed are stronger than the broken C=C π bond. This reaction can also be achieved with other halogens like Cl_2. Fluorine is too reactive and difficult to manipulate, but it can be added to double bonds at low temperature (−78 °C), while iodination is considerable slower than bromination and vicinal diiodo compounds are generally unstable and tend to revert to iodine and the alkene. Mixed halogenations using BrCl, IBr, and ICl can also be performed.[2]

The addition of bromine to alkenes is very fast and can be carried out easily at room temperature.[3] The presence of other reactive groups like aldehydes, ketones, or amines is possible since the addition to the double bond occurs faster. The mechanism is usually electrophilic but when free-radical initiators are present the addition can occur by a free radical mechanism and involve competing substitution reactions, especially if the molecule has allylic hydrogens.[4] Triple bonds are less reactive towards bromine than double bonds; however, two moles of Br_2 can be added to alkynes to give tetrabromo products.

There are two possibilities for the addition of bromine to double bonds: the atoms may enter from the same side of the plane, *syn* addition, or from opposite sides, *anti* addition (Scheme 4.1.1.1.1). Considering cinnamic acid, which possess COOH and Ph substituents in the (*E*)-configuration, the *syn* addition of the two bromine atoms would afford the *threo*-2,3-dibromo-3-phenylpropanoic acid while *anti* addition would give rise to *erythro*-2,3-dibromo-3-phenylpropanoic acid.[2]

Fortunately, these two diastereoisomers exist and have very different melting points. Therefore, measurement of the melting point of the reaction product may elucidate its stereochemistry and consequently the mode of the bromine addition can be determined. This reaction can be easily followed since bromine solutions are red and upon addition to the alkene the coloration disappears progressively. The reaction product, which is racemic, precipitates easily from the reaction mixture.

Additional Safety

This session should be done in a fume hood. Bromine is toxic and corrosive and may cause severe skin and eye burns. Use gloves and safety glasses. Bromine vapor is also very harmful. Notify immediately your supervisor of any bromine spills. Dichloromethane is an irritant. Cyclohexene is a flammable reagent, although not harmful if swallowed or in contact with the skin. It has a bad smell. If necessary, any bromine spills can be neutralized with a sodium thiosulfate solution.

Experimental Procedure

1. Add cinnamic acid (1.0 g, 6.75 mmol) and CH_2Cl_2 (20 mL) to a two-necked 100 mL round bottom flask. Attach a reflux condenser and an addition funnel to the flask. Add also one or two boiling stones.

2. Attach the water tubing to the condenser ensuring that water goes in at the bottom and out at the top. Turn on the water so that you have a slow but steady stream of water passing through the condenser.
3. Place a heating mantle below the flask.
4. Using gloves carefully add Br_2/CH_2Cl_2 (1.0 M, 7.0 mL) into the addition funnel. Seal the top of the addition funnel with one stopper.
5. Heat the flask slowly so as to maintain a gentle reflux of the dichloromethane.
6. Add the bromine solution from the addition funnel about 10–15 drops at a time. Allow the color of the bromine to disappear before adding additional bromine. Continue this process until the bromine color no longer fades. (If you have added all of your bromine and the mixture is colorless, you may need more bromine. Consult your supervisor for instructions.)
7. Continue heating the mixture at reflux for a further 10 min. Remove the heating mantle and turn it off. If the reaction mixture is still red, indicating the presence of bromine, remove the addition funnel and add a few drops of cyclohexene until the mixture is colorless or pale yellow. (Cyclohexene stinks! Do this quickly and keep the bottle of cyclohexene closed!)
8. Remove the condenser and cool the flask in an ice bath for 5–10 min. You should also cool ~5 mL of dichloromethane in a separate container in the ice bath at the same time.
9. While waiting, wash the condenser and the addition funnel with sodium thiosulfate (0.1 M) in the fume hood.
10. Assemble a vacuum filtration apparatus using a Büchner funnel. Wet the filter paper with a small amount of cold CH_2Cl_2 and filter off the solid. Use small amounts of the cold dichloromethane to help remove any residual product from the flask.
11. Collect your product and allow it to air dry for 5–10 min. Weight the product.
12. Measure the melting point of the product and determine the mode of the bromine addition.

Waste Disposal

The washing solution can be diluted and discarded in the sink. The CH_2Cl_2 present in the Büchner funnel should be collected in a chlorinated waste container.

Results Interpretation and Additional Questions

1. Propose a mechanism for the bromine addition to olefins.
2. Determine the limiting reagent and calculate the product yield.
3. Why is the reaction product racemic? How are the two enantiomers formed?
4. This transformation is an example of a stereospecific reaction. What does this mean?
5. Discuss the addition of cyclohexene to remove the bromine excess.
6. Which spectroscopic technique could be used to distinguish between a sample of cinnamic acid and the brominated product you made in this experiment? Justify your answer.

References

1. G. Solomons and C. Fryle, *Organic Chemistry*, John Wiley & Sons, Inc., New York, 7th edn, 2000, p. 322.
2. J. March, *Advanced Organic chemistry*, John Wiley & Sons, Inc., New York, 4th edn, 1992, pp. 735 and 812.
3. S. DeYoung and E. Berliner, *J. Org. Chem.*, 1979, **44**(7), 1088.
4. I. Roberts and G. E. Kimball, *J. Am. Chem. Soc.*, 1937, **59**(5), 947.

4.1.1.2. Preparation of *Meso*-1,2-Dibromo-1,2-diphenylethane

Abel J. S. C. Vieira

Faculty of Sciences and Technology, Universidade Nova de Lisboa, 2829-516 Caparica, Portugal
*E-mail: ajscv@fct.unl.pt

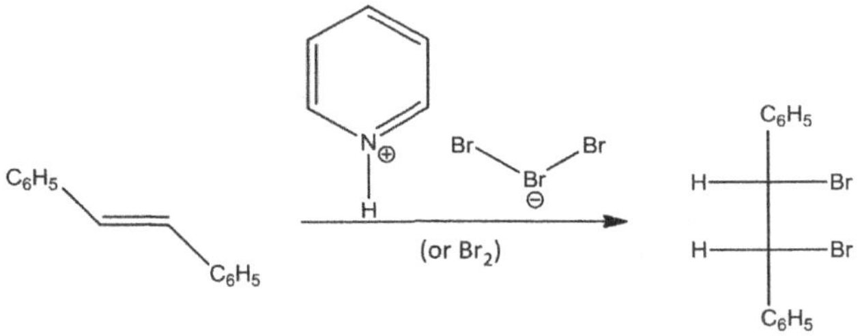

Number of sessions (duration of each session)	Hazard level	Difficulty level	Level of study
1 (3 h)	High	Low	Introductory

Class names Alkenes, vicinal alkyl di-halides

Concepts involved This experiment illustrates a stereospecific addition reaction (addition of bromine to an alkene). Starting from *trans*-stilbene, the *meso* form is the only stereoisomer of the vicinal dibromide obtained

Chemicals needed Pyridine, hydrobromic acid, bromine, acetic acid, *trans*-stilbene, methanol

Equipment and experimental techniques involved Erlenmeyer flask, filtration apparatus, ^1H NMR spectroscopy

Keywords Bromination, crystallization, filtration, NMR, stereospecific addition reaction

Background

Alkenes behave as weak nucleophiles, donating the electron pair of the π bond to strong electrophiles, leading to the formation of addition products (Scheme 4.1.1.2.1).

In the present experiment, the nucleophilic alkene is *trans*-stilbene and the electrophile is the perbromide of pyridinium bromide. The stereospecific addition reaction yields the *meso* form of 1,2-dibromo-1,2-diphenylethane (*meso* stilbene dibromide) as the only stereoisomer formed (Scheme 4.1.1.2.2).[1]

The reason for the stereospecificity of the reaction is the two-step mechanism of the addition. In the first step a three-membered ring bromonium ion is formed as an intermediate. In the second step, the opening of the intermediate by the bromide anion occurs by a S_N2 mechanism, resulting in the formation of the two new C–Br bonds of the product in an *anti*-configuration.

Comprehensive Organic Chemistry Experiments for the Laboratory Classroom
Edited by Carlos A. M. Afonso, Nuno R. Candeias, Dulce Pereira Simão, Alexandre F. Trindade, Jaime A. S. Coelho, Bin Tan, and Robert Franzén.

Published by the Royal Society of Chemistry, www.rsc.org

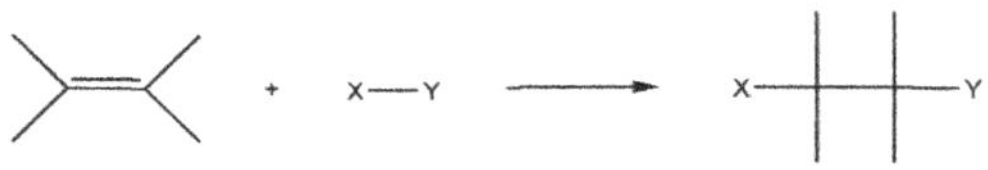

Scheme 4.1.1.2.1 Addition reaction to a C=C double bond.

Scheme 4.1.1.2.2 Addition of bromine to *trans*-stilbene.

The use of perbromide of pyridinium bromide is a good alternative to the use of the more hazardous reactant bromine (Br_2). Although bromine is used to prepare the perbromide, this can be made separately (in a fume hood) and the bromination agent, being a crystalline salt, may be handled safely on the laboratory bench. Therefore, the use of the perbromide salt for bromination is recommended as a procedure of green chemistry.[2]

Additional Safety

The major part of the experiment can be performed on a regular laboratory bench, with proper safety equipment, such as protective clothing, gloves and safety glasses. The step where pyridine and bromine are used must be performed entirely in the fume hood. Bromine is a very dangerous reactant, requiring special care in its manipulation: fume hood and effective protection for eyes and hands. Residues of bromine after reaction must be completely and safely removed.

Pyridine, hydrobromic and acetic acid also require manipulation with caution.
Pyridine: H225; H332; H312; H302; P210; P243; P260; P280
Bromine: H330; H314; H400; P260; P280; P305; P315; P338
Hydrobromic acid: H314; H335; P260; P280; P304; P305
Acetic acid: H226; H314; P260; P280; P303; P305
Methanol: H225; H331; H311; H301; H370; P210; P243; P260; P264; P302

Experimental Procedure

One Laboratory Session (3 h)

Preparation of Perbromide of Pyridinium Bromide ($C_5H_5NH^+Br_3^-$) (ref. 3)

1. In a 100 mL Erlenmeyer flask prepare a mixture of 3 mL of pyridine and 6 mL of a 48% bromidic acid aqueous solution.
2. In a fume hood, cool the mixture with an ice bath and add slowly, under stirring, 5 g of bromine contained in an addition funnel.
3. The product forms as needle-like orange crystals.
4. Cool and collect the crystals in a Büchner funnel, and wash them with cold acetic acid.
5. Recrystallize the solid (without drying it) in 20 mL of acetic acid.
6. Keep the purified perbromide for the next step of the experiment.

Preparation of Meso Stilbene Dibromide

7. In a 100 mL Erlenmeyer flask, dissolve 0.5 g of *trans*-stilbene in 10 mL of acetic acid, heating on a steam bath.
8. Add 1 g of the previously prepared perbromide of pyridinium bromide.

9. Stir and, if necessary, remove the reactant crystals from the flask walls with a small amount of acetic acid, maintaining heating for a further 1–2 min.
10. The dibromide separates almost immediately, in the form of small platelets.
11. Cool the reaction mixture in an ice cold water bath and collect the product in a Büchner funnel.
12. Wash the solid with cold methanol.
13. Dry the product in the air over filter paper and keep it in a vacuum desiccator.

Waste Disposal

The aqueous solutions resulting from the several work-ups may be neutralized, diluted and discarded in the sink. Pyridine residues must be collected in a non-chlorinated waste collector recipient. Special attention is due to handling bromine. After the reaction, the aqueous solutions must be decolorized with sodium bisulfite and diluted before being discarded.

Results Interpretation and Additional Questions

Results and Product Characterization

1. Determine the melting point of the compound (use a well-dried sample).
2. Record the IR spectrum of the compound, using dried potassium bromide as support material. Assign the absorption bands characteristic of the functional groups of the compound.
3. Acquire the ^{1}H NMR spectrum of the compound. Use 30 mg of a well-dried sample of the compound and the deuterated solvent suggested by your instructor. Try to assign the H atoms to the NMR signals.

Additional Questions

1. Discuss the advantage of using perbromide of pyridinium bromide instead of bromine as the electrophile.
2. Write the mechanism of the reaction. Why is the reaction stereospecific but not regioselective?
3. Write, using Fischer projections, all the stereoisomers of 1,2-dibromo-1,2-diphenylethane.
4. What would be the result of the reaction if *cis*-stilbene was used as the starting alkene?

References

1. T. W. G. Solomons, *Organic Chemistry*, John Wiley & Sons Inc., Hoboken, 8th edn, 2003, vol. 8, pp. 334.
2. J. Clayden, N. Greeves, S. Warren and P. Wothers, *Organic Chemistry*, Oxford University Press, 1st edn, 2001, vol. 43, pp. 1155.
3. L. F. Fieser, *Experiments in Organic Chemistry*, D. C. Heath and Company, Boston, 3rd edn, 1957, vol. 10, pp. 64.

4.1.1.3. Bromination of (*E*)-chalcones [(*E*)-1,3-Diarylprop-2-en-1-ones]

Artur M. S. Silva, Augusto C. Tomé, Diana C. G. A. Pinto*, Fernando Domingues, Graça M. Oliveira Rocha, José A. S. Cavaleiro, M. Graça P. M. S. Neves, M. Amparo F. Faustino, and Mário M. Q. Simões

Chemistry Department & QOPNA, University of Aveiro, Campus de Santiago, 3810-193 Aveiro, Portugal
*E-mail: diana@ua.pt

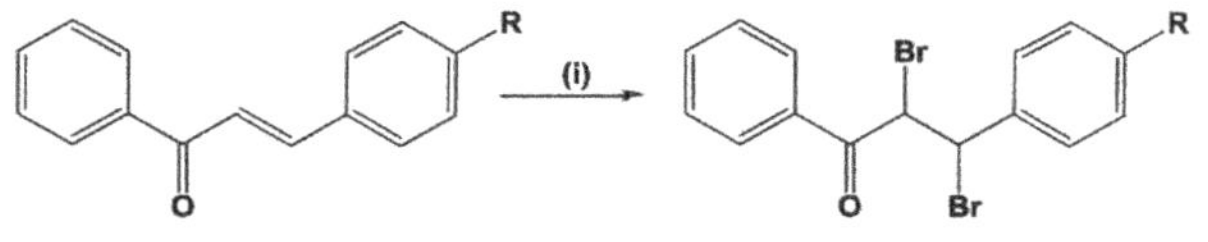

R = H, OCH_3
(i) Br_2 at room temperature

Number of sessions (duration of each session)	Hazard level	Difficulty level	Level of study
1 (4 h)	Moderate	Medium	Intermediate

Class names Alkenes, chalcones, α,β-dibromochalcones

Concepts involved This experiment describes the bromination of chalcones with bromine by an electrophilic addition. The experiment also illustrates the stereochemistry of the bromine addition. Although the 1,3-diaryl-2,3-dibromopropan-1-ones can be obtained directly from the reaction medium by filtration, preparative TLC can be learnt

Chemicals needed (*E*)-1,3-Diarylprop-2-en-1-ones, bromine, dichloromethane, hexane, $CDCl_3$

Equipment and experimental techniques involved Stirring plate, NMR, IR, melting point apparatus

Keywords Chromatography, 1,3-diaryl-2,3-dibromopropan-1-ones, electrophilic additions, filtration, NMR

Background

α,β-Dibromodihydrochalcones, also known as 1,3-diaryl-2,3-dibromopropan-1-ones, are potential intermediates for the synthesis of various heterocyclic compounds such as pyrazoles and hydroxypyrazoline derivatives, compounds known to possess antibacterial and antifungal activity. Consequently, the development of methodologies towards the synthesis of these dibromo derivatives is very important.[1] Some of these procedures, due to their simplicity, can be executed in organic laboratory courses; furthermore, they can be used to exemplify the importance of electrophilic additions[2] in organic synthesis.

This experiment describes the bromination of 1,3-diaryl-2,3-dibromopropan-1-ones with bromine at room temperature. The reaction product can be isolated by filtration and further purified by recrystallization and/or by chromatography. Analysis of the obtained compounds by nuclear magnetic resonance (NMR) is recommended because it will be a good introduction to NMR spectra interpretation. Otherwise, infrared (IR) spectroscopy can be used to illustrate characteristic IR bands.

Comprehensive Organic Chemistry Experiments for the Laboratory Classroom
Edited by Carlos A. M. Afonso, Nuno R. Candeias, Dulce Pereira Simão, Alexandre F. Trindade, Jaime A. S. Coelho, Bin Tan, and Robert Franzén.

Published by the Royal Society of Chemistry, www.rsc.org

Additional Safety

This work requires regular laboratory good practices and personal safety equipment (safety glasses, laboratory coat and nitrile gloves). The synthetic and extraction procedures and chromatographic purifications should be carried out in a fume hood that is working properly and with glass material in perfect condition. It is recommended to avoid direct contact with bromine and pyridinium tribromide because they can be irritating for skin and eye upon contact and also irritant by inhalation. The organic solvent used (dichloromethane) is inflammable when in the presence of ignition sources and is harmful if swallowed, or in contact with the skin.

Experimental Procedure

Laboratory Session: 4 h

1. In a fume hood add to a 100 mL round-bottom flask, equipped with a magnetic stirring bar, the desired chalcone (3.5 mmol) and dichloromethane (40 mL).
2. Put the round-bottom flask on a stirring plate and slowly add bromine (~2 mL) until the reaction mixture becomes red.
3. Maintain the reaction mixture stirring for 30 min.
4. After the stirring period filter the solid formed using a vacuum filtration apparatus.
5. Assemble the vacuum filtration apparatus. Put the filter paper onto the Büchner funnel, wet the filter paper with dichloromethane and open the vacuum valve. Transfer the reaction mixture onto the filter, and wash the product three times with small portions of dichloromethane. Keep the vacuum system connected to dry the filtrated solid.
6. Weigh the product obtained and calculate the reaction yield. Measure the melting point. Check the compound purity by TLC; if it is pure prepare a NMR tube and also obtain the IR spectrum. Otherwise preparative TLC can be used to purify the crude compound (point 7).
7. Dissolve the crude compound in dichloromethane and purify using previously prepared TLC glass plates covered with silica gel 60 F_{254} and elute with a mixture of dichloromethane/hexane (8 : 2).

Waste Disposal

All liquids should be collected in a halogenated waste collector recipient for posterior destruction. Solvents used in chromatographic purifications or extractions should be collected for further distillation and reutilization.

Results Interpretation and Additional Questions

1. Calculate the percent yield of the reaction.
2. Rationalize the reaction mechanism.
3. Draw the obtained product showing the stereochemistry predicted by the halogen addition mechanisms.
4. Using the obtained TLC chromatogram calculate the R_f of the α,β-dibromodihydrochalcone and comment on its purity.
5. Interpret the ^1H and ^{13}C NMR spectra obtained, and compare the obtained melting point with that in the literature.
6. Interpret the IR spectrum.

References

1. D. C. G. A. Pinto, A. M. S. Silva, J. A. S. Cavaleiro and J. Elguero, *Eur. J. Org. Chem.*, 2003, 747–755.
2. F. A. Carey, *Organic Chemistry*, McGraw-Hill, 6th edn, 2006.

4.1.1.4. Preparation of *trans*-2-Bromocyclohexanol from Cyclohexanol

Matheus P. Freitas

Department of Chemistry, Federal University of Lavras, 37200-000, Lavras, MG, Brazil
*E-mail: matheus@dqi.ufla.br

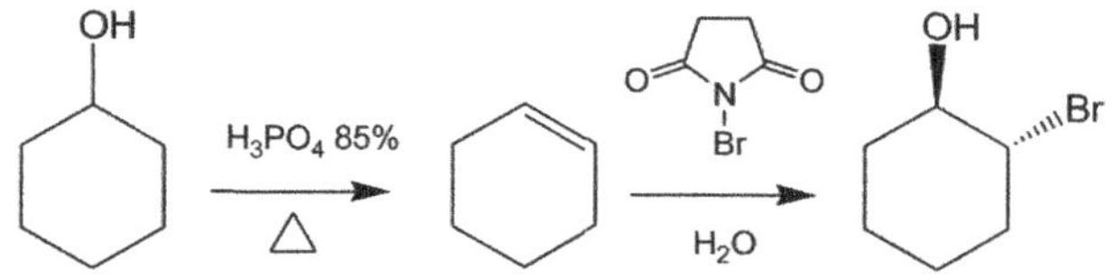

Number of sessions (duration of each session)	Hazard level	Difficulty level	Level of study
2 (4 h each)	Moderate	Low	Introductory

Class names Alcohols, alkenes, halohydrins

Concepts involved Elimination and electrophilic addition reactions, conformational analysis and infrared spectroscopic analysis

Chemicals needed Cyclohexanol, phosphoric acid (85%), *N*-bromosuccinimide, distilled water, sodium chloride, sodium phosphate

Equipment and experimental techniques involved Distillation apparatus (100 mL round-bottom flasks, heater, condenser, thermometer, boiling stones), 250 mL Erlenmeyer flask, mixer and magnetic bar, separation funnel, rotary evaporator, Fourier-transform (FT) infrared spectrometer

Keywords Conformational analysis, distillation, electrophilic addition, elimination, IR spectroscopy

Background

Elimination and electrophilic addition are key reactions in organic chemistry.[1,2] Halohydrins can be obtained from the latter. The end applications of halohydrins, like 2-chloroethanol and enantiomerically pure halohydrins, include dye intermediates, pharmaceuticals, agrochemicals, pesticides, plasticizers, liquid crystals, and others. *trans*-2-Bromocyclohexanol is an halohydrin which undergoes conformational isomerization of its chair-like structure, giving the diaxial and di-equatorial conformers (Figure 4.1.1.4.1). Its conformational equilibrium can be shifted towards either the diaxial or di-equatorial conformation, depending on the medium (solute–solute and/or solute–solvent interactions) and intramolecular forces, such as steric effects, electrostatic interactions, and internal hydrogen bonding between proton acceptors (such as bromine) and donors (such as OH).[3]

The conformational preference of a given compound can affect its physical, chemical, and biological properties. For example, the ligand–enzyme fit responsible for the biological activity of a drug is dependent on its conformation. Infrared spectroscopy is useful for conformational analysis, since it is a technique used to study functional groups and its time scale is faster than

Comprehensive Organic Chemistry Experiments for the Laboratory Classroom
Edited by Carlos A. M. Afonso, Nuno R. Candeias, Dulce Pereira Simão, Alexandre F. Trindade, Jaime A. S. Coelho, Bin Tan, and Robert Franzén.

Published by the Royal Society of Chemistry, www.rsc.org

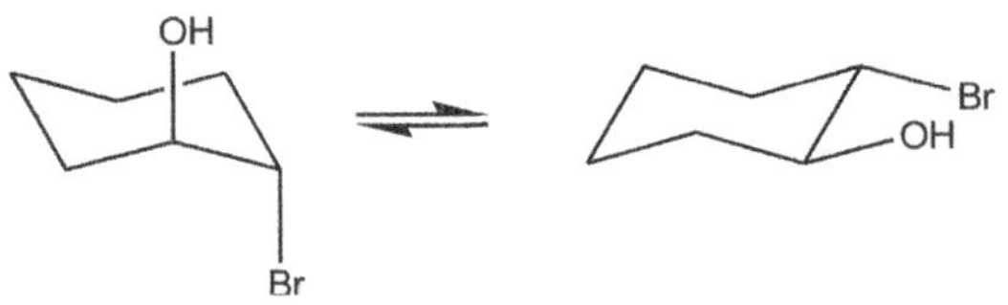

Figure 4.1.1.4.1 Conformational isomerism in *trans*-2-bromocyclohexanol.

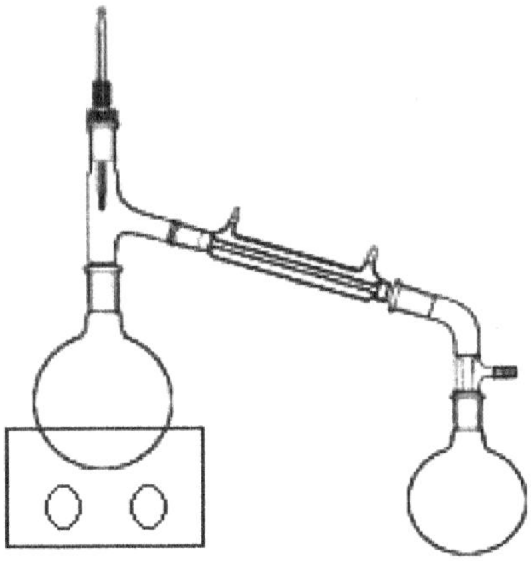

Figure 4.1.1.4.2 Distillation apparatus scheme.

most rotational isomerization processes.[4] In other words, a given functional group of two distinct conformers gives two separate signals, which can be measured and related to the conformers population in the equilibrium using the Beer–Lambert expression ($A = \varepsilon bc$, where A is the intensity of the band, ε is the molar absorptivity (assumed to be similar for the two conformers), b is the optical length, and c is the concentration, or the population, of each conformer).

This experiment is devoted to performing elimination and electrophilic addition reactions, in order to obtain *trans*-2-bromocyclohexanol, as well as to analyzing its conformational equilibrium in terms of inter- and intramolecular interactions using infrared spectroscopy.

Additional Safety

Despite the low risk involved in using most of the chemicals employed in this experiment, phosphoric acid (corrosive) and cyclohexene (toxic effects if inhaled or absorbed) require safety measures. Therefore, a fume hood, protection glasses, and gloves are required.

Experimental Procedure

Session 1: Preparation of Cyclohexene from Cyclohexanol

1. In a 100 mL round-bottom flask, equipped with boiling stones and adapted to a distilling apparatus with thermometer (Figure 4.1.1.4.2), add 20 mL of cyclohexanol and 5 mL of 85% phosphoric acid.
2. Heat the mixture to allow distillation below 100 °C.
3. Stop heating when a small volume is left in the reaction flask.
4. Saturate the distilled content with NaCl and add 5 mL of Na_2CO_3 (10% solution) to basify the solution.

5. Discard the aqueous solution and collect the cyclohexene in a 100 mL Erlenmeyer flask, using a separation funnel.
6. Add 1–2 g of sodium sulfate and mix for 1 min, then filter and proceed to the next step.

The presence of cyclohexene can be identified using either the Br_2/CCl_4 or the Baeyer ($KMnO_4$ solution) tests.

Session 2: Preparation of *trans*-2-Bromocyclohexanol from Cyclohexene

1. In a 250 mL Erlenmeyer flask with a magnetic bar, add 5.0 g (27.8 mmol) of *N*-bromosuccinimide (NBS) and 30 mL of distilled water.
2. Add 5.7 mL of the cyclohexene previously prepared and stir the mixture at room temperature until the solid disappears.
3. Extract the organic portion with diethyl ether (2 × 20 mL) using a separation funnel.
4. Dry with sodium sulfate and evaporate the diethyl ether with a rotary evaporator.
5. Distill the *trans*-2-bromocyclohexanol at reduced pressure (*ca.* 61–62 °C/2 mmHg).[5]

Conformational Analysis using Infrared Spectroscopy

Obtain the infrared spectra of *trans*-2-bromocyclohexanol for the neat liquid (thin film) and solution (0.02 M hexane solution, using a liquid cell with *ca.* 1 mm spacer). Measure the band intensities relative to the OH stretching mode of both diaxial (higher wavenumber) and di-equatorial (lower wavenumber) conformers.[6] Discuss the conformational preferences in terms of inter- (for the neat liquid) and intramolecular (in solution) interactions.[7]

Waste Disposal

Most wastes formed are aqueous solutions and can be discarded, while residual diethyl ether solution used for extraction can be distilled and then reused.

Results Interpretation and Additional Questions

1. Describe both possible mechanisms (E1 and E2) for the reaction sequence.
2. Why is NaCl used in the first step (preparation of cyclohexene)?
3. Describe the mechanism for the synthesis of *trans*-2-bromocyclohexanol.
4. What would be the compound obtained from the reaction between cyclohexene, *N*-bromosuccinimide, and methanol? Discuss the conformational isomerism of this compound compared to *trans*-2-bromocyclohexanol in diluted cyclohexane solution.

References

1. J. E. McMurry, *Organic Chemistry*, Cengage Learning, Belmont, 8th edn, 2011, p. 1376.
2. T. W. G. Solomons, C. B. Fryhle and S. A. Snyder, *Organic Chemistry*, John Wiley & Sons, Inc., Hoboken, 11th edn, 2014, p. 1248.
3. E. L. Eliel and S. H. Wilen, *Stereochemistry of Organic Compounds*, John Wiley & Sons, Inc., New York, 1994, p. 1267.
4. D. L. Pavia, G. M. Lampman, G. S. Kriz and J. A. Vyvyan, *Introduction to Spectroscopy*, Cengage Learning, Belmont, 4th edn, 2008, p. 752.
5. M. P. Freitas, C. F. Tormena, R. Rittner and R. J. Abraham, *J. Phys. Org. Chem.*, 2003, **16**, 27.
6. M. P. Freitas, C. F. Tormena and R. Rittner, *J. Mol. Struct.*, 2001, **570**, 175.
7. C. J. Duarte and M. P. Freitas, *J. Mol. Struct.*, 2009, **930**, 135.

4.1.1.5. Synthesis of *trans*-Cyclohexane-1,2-diol

Dulce Simão* and Ana C. Cerdeira

Department of Chemistry, Instituto Superior Técnico, University of Lisbon, Centro de Química Estrutural, Av. Rovisco Pais, 1, 1049-001 Lisbon, Portugal. *E-mail: dulcesimao@tecnico.ulisboa.pt

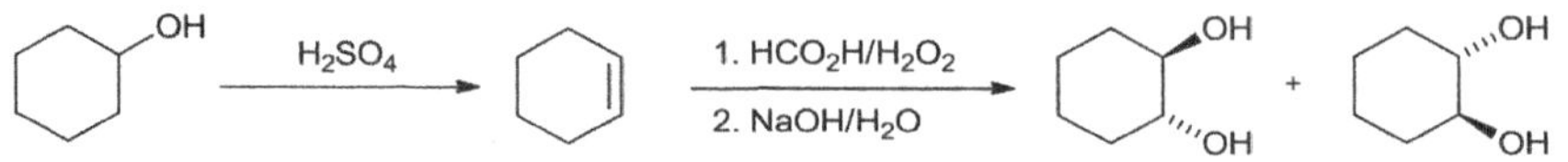

Number of sessions (duration of each session)	Hazard level	Difficulty level	Level of study
3 (4 h each)	Moderate	Medium	Introductory

Class names Alcohols, alkenes, epoxides, diols

Concepts involved Two-step reaction sequence illustrating an E1 elimination (dehydration) of a secondary alcohol, and epoxidation the resulting cyclohexene followed by hydroxylation to give a racemic diol

Chemicals needed Cyclohexanol, concentrated sulfuric acid, sodium chloride, sodium carbonate, anhydrous sodium sulfate or calcium chloride, formic acid (90%), hydrogen peroxide (30%), potassium iodide–starch paper, iron(II) sulfate, sodium hydroxide, ethyl acetate, silicone oil bath

Equipment and experimental techniques involved Fractional and simple distillations, ^1H NMR spectroscopy, liquid–liquid extraction, reflux apparatus with temperature control and magnetic stirring, vacuum filtration, rotary evaporator, IR, refractive index, melting point

Keywords Distillation, E1 elimination, epoxidation, ^1H NMR spectroscopy, S_N2 reaction, *trans*-diol

Background

Cyclohexanol is a secondary alcohol that in the presence of a strong acid (such as concentrated H_2SO_4) undergoes E1 elimination to give cyclohexene.[1a] In the first step, sulfuric acid present as a catalyst is used to protonate the alcohol, turning the hydroxyl group into water, a good leaving group. The elimination results in the second step with loss of a water molecule, so this reaction is also known as dehydration.[2,3a] The carbocation formed is highly reactive so, in the third step, a β-hydrogen atom is abstracted by HSO_4^- to produce cyclohexene. The reaction of formic acid with hydrogen peroxide produces performic acid, a peroxyacid or peracid (HCO_3H) that will react with cyclohexene to give an epoxide, resulting in an addition of oxygen to the double bond in a *cis*-stereoselective mode.[3b,4] Both C–O bonds are formed on the same face of the alkene's π bond, so a *cis*-epoxide is obtained, but in this case it cannot be isolated because formic acid is sufficiently strong to protonate the oxygen atom of the epoxide ring, turning it into a better leaving group and facilitating nucleophilic attack by the formate ion or water[1b] from the side opposite the epoxide group. In this way an inversion of configuration results through an S_N2 reaction.[3c,4,5,6] Both *trans*-hydroxy-formate ester enantiomers undergo base catalyzed hydrolysis to give enantiomeric *trans*-cyclohexane-1,2-diols. Epoxidation followed by hydrolysis is considered a useful method for *anti* hydroxylation of a double bond.

Comprehensive Organic Chemistry Experiments for the Laboratory Classroom
Edited by Carlos A. M. Afonso, Nuno R. Candeias, Dulce Pereira Simão, Alexandre F. Trindade, Jaime A. S. Coelho, Bin Tan, and Robert Franzén.

Published by the Royal Society of Chemistry, www.rsc.org

Additional Safety

Care should be taken while handling formic and concentrated sulfuric acids, which are highly corrosive, and cyclohexene, which is extremely flammable. Avoid water spills into the oil baths, especially when they are hot.

Experimental Procedure

Session 1 – Preparation of Cyclohexene

1. Add cyclohexanol (21.5 mL, 20.64 g, 0.2 mol) and concentrated H_2SO_4 (0.6 mL) to a 250 mL round-bottom flask and stir well.
2. Add two or three boiling stones to the reaction flask, and fit a Vigreux column and a distillation head.
3. Place the receiving flask in an ice/water bath.
4. Heat the reaction mixture flask with a silicone oil bath at 130–140 °C, making sure that the temperature of distillate is never higher than 90 °C. Distill the mixture until only a residue remains in the reaction flask.
5. Transfer the distilled solution to a separatory funnel, saturate it with sodium chloride, add a 5% solution of sodium carbonate (2 mL) and shake.
6. Allow it to stand for a few minutes and then remove the lower aqueous layer.
7. Dry the organic layer with anhydrous sodium sulfate (or calcium chloride), stirring occasionally for 15 min.
8. Filter by gravity to a 100 mL round-bottomed flask and distill the product, collecting the fraction at 81–83 °C. Place the receiving flask in an ice/water bath.
9. Measure the refractive index of cyclohexene using an Abbe refractometer. Record the yield, IR and ^1H NMR spectra.

Preparation of *trans*-Cyclohexane-1,2-diol

Session 2

1. Equip a 250 mL three-neck round-bottom flask with a reflux condenser, thermometer and magnetic stirring.
2. Add formic acid (90%) (75 mL) and H_2O_2 (30%) (17.5 mL), and then slowly add dropwise cyclohexene (13 mL, 10.53 g, 0.13 mol) for 20 min, keeping the temperature of the reaction mixture between 40 and 50 °C. If necessary, cool the mixture with a cold water bath and add cyclohexene at a slower rate.
3. Upon complete addition, place the reaction flask in a water bath at 50–60 °C (reaction mixture at 40 °C) for 1 h.
4. Test the reaction mixture for the presence of peroxides with potassium iodide–starch paper; add $FeSO_4$ (0.5 g) to eliminate them if the test is positive.
5. Remove water and formic acid with a rotary evaporator.
6. Store the residue in a plastic bottle until the next session.

Session 3

1. Pour the residue (diol and formate) in a 250 mL Erlenmeyer flask; add a cold solution of NaOH (10 g) in water (20 mL) in small portions until the residue is completely dissolved. The temperature should not be over 45 °C.
2. Warm this alkaline solution in a water bath to 45 °C for 15 min and immediately transfer it to a separatory funnel.

3. Extract the warmed aqueous layer with ethyl acetate (2 × 50 mL).
4. Combine the organic layers and remove the solvent as much as possible in the rotary evaporator until a precipitate appears.
5. Cool the precipitate in ice an ice bath and collect it by vacuum filtration.
6. Recrystallize the product from ethyl acetate.
7. Record the yield, melting point, and both IR and NMR spectra.

Waste Disposal

Neutralize the aqueous layer from the extractions before discarding in the sink. Ethyl acetate from the extraction and recrystallization can be stored for further purification.

Results Interpretation and Additional Questions

1. Determine the reaction yields.
2. Evaluate the purity of the products by (a) comparing the refractive index of cyclohexene with the one found in the literature and (b) comparing the melting point of *trans*-cyclohexane-1,2-diol with the literature values.
3. Interpret the IR and NMR spectra.
4. Write the reaction mechanisms of the reactions involved.
5. Explain why the synthesis of cyclohexene follows an E1 mechanism.
6. Are there any side products in the E1 reaction? Would there be any side products starting from methylcyclohexanol? Which ones?
7. Indicate a procedure to obtain the *cis* diol instead of the *trans* diol.

References

1. T. W. G. Solomons, *Organic Chemistry*, John Wiley & Sons Inc., Hoboken, 7th edn, 1998(a) p. 294; (b) p. 516.
2. K. P. C. Vollhardt and N. E. Schore, *Organic Chemistry*, Freeman & Co., 3rd edn, 1998, p. 333.
3. J. Clayden, N. Greeves and S. Warren, *Organic Chemistry*, Oxford, 2001(a) p. 478; (b) p. 507; (c) p. 938.
4. G. Procter, *Stereoselectivity in Organic Chemistry*, Oxford Chemistry Primers, 1998, p. 71.
5. C. A. M. Afonso, D. P. Simão, L. P. Ferreira, M. S. Serra and M. M. M. Raposo, *100 Experiências de Química Orgânica*, IST Press, 2011, p. 141.
6. A. I. Vogel, *Vogel's Textbook of Practical Organic Chemistry*, Longman Scientific and Technical, 5th edn, 1989, p. 549.

4.1.1.6. Synthesis of Isobutylene and Its Use in Esterification Reactions

Paula C. Castilho*, Pedro Ideia, and Rúben Gonçalves

Centro de Química da, Madeira, Universidade da Madeira, Campus Universitário da Penteada, Piso 00, 9020-105, Funchal, Portugal. *E-mail: castilho@uma.pt

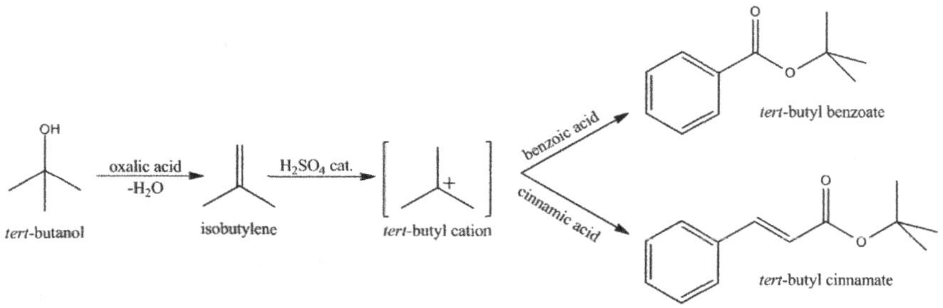

Number of sessions (duration of each session)	Hazard level	Difficulty level	Level of study
2 (4 h each)	Moderate	Medium	Introductory

Class names Alcohols, carboxylic acids, alkenes, esters

Concepts involved Elimination of water from an alcohol requires a Brønsted acid as protonating agent and leads to an alkene, *via* the formation of a carbocation, the stability of which rules the speed and extent of the reaction. *tert*-Butanol is dehydrated to isobutylene, a gas under experimental conditions. This alkene can also be protonated by an acid, generating a tertiary carbocation, a powerful electrophile that reacts easily with the oxygen atom of a carboxylic acid

Chemicals needed *tert*-Butanol, oxalic acid, benzoic (or cinnamic) acid, H_2SO_4, $KMnO_4$, Na_2CO_3 solution, Na_2SO_4, dichloromethane

Equipment and experimental techniques involved Extraction, recrystallization (catalyst recovery), evaporation, heating mantle or oil bath and heating plate, reflux set-up, rotary evaporator; characterization by refractive index, FTIR

Keywords Dehydration, esterification, FTIR, gas phase reagents, liquid–liquid extraction

Background

When esterification reactions are discussed and demonstrated in organic chemistry programmes, the mechanism of Fischer esterification is the one usually presented. This is a reversible reaction that occurs between alcohols and carboxylic acids in the presence of catalytic amounts of mineral acids, which causes the protonation of the carbonyl oxygen of the carboxylic acid. This protonation greatly enhances the affinity of the carbonyl carbon for an electron pair on the oxygen of the alcohol, meaning that the carboxylic acid acts as an electrophile and the alcohol acts as a nucleophile. Fischer esterification reactions are sterically hindered. If either the

Comprehensive Organic Chemistry Experiments for the Laboratory Classroom
Edited by Carlos A. M. Afonso, Nuno R. Candeias, Dulce Pereira Simão, Alexandre F. Trindade, Jaime A. S. Coelho, Bin Tan, and Robert Franzén.

Published by the Royal Society of Chemistry, www.rsc.org

participant acid or the alcohol possesses highly branched groups, the position of equilibrium is less favourable and the rate of esterification is low. In general, the ease of esterification for alcohols, ROH, by the described mechanism is *primary* R > *secondary* R > *tertiary* R with a given carboxylic acid.

Thus, the formation of *tert*-butyl esters follows a different pathway. Tertiary alcohols can undergo protonation in the presence of the acidic catalyst to form the corresponding alkyloxonium and undergo a S_N2 displacement of water from this oxonium ion by the electron pair on the oxygen of the OH group of the carboxylic acid. This means that the alcohol will be the electrophile and the acid will be the nucleophile.[1]

Several methods for the preparation of *tert*-butyl esters from carboxylic acids have been developed. These reactions can be of two types: the first are methods that use irreversible reagents such as alkyl halides, *S*-pyridyl-thiolates in the presence of $CuBr_2$, *tert*-butyl 2,2,2-trichloroacetimidate, *N,N*-dimethylformamide-di-*tert*-butyl acetal or EDCI/DMAP/*tert*-butanol,[2–4] which are expensive reagents. The second method uses isobutylene, preferably generated *in situ*, to minimize the practical issues of handling gaseous isobutylene. For instance, an excess of *tert*-butyl acetoacetate and catalytic amounts of sulfuric acid at room temperature can be used to obtain the desired *tert*-butyl esters, across a range of substrates, in acceptable yields.[5]

From the experimental (and the undergraduate student) point of view, there is no way to tell the difference between the electrophilic and the nucleophilic role of each reagent, since for both mechanisms a carboxylic acid, an alcohol and an acidic catalyst are mixed and warmed up. Only ^{18}O experiments could show the distinction and they are not usually performed at undergraduate level.

A simpler way to demonstrate the nucleophilic attack of the acid over the tertiary reagent is the use of gaseous isobutylene, prepared separately by the dehydration of *tert*-butanol and used immediately over bulky carboxylic acids.

In most programmes of undergraduate organic chemistry, theoretical classes present examples of reactions that occur in gas phase but in laboratory courses the students are mostly confronted with experiments involving liquids and solids. There is no reason to be afraid of gas phase chemistry and this experiment can be a simple illustration of the preparation and use of a gaseous reactant.

1. Dehydration of a tertiary alcohol

Oxalic acid is used as catalyst, instead of the usual sulfuric acid, for the dehydration of *tert*-butanol without any solvent; the organic acid adsorbs the water formed during the reaction and isobutylene is produced as a pure gas that can be identified by the decolouration of a $KMnO_4$ solution. The catalyst can be recovered by a simple recrystallization from water. This step can be performed as an independent experiment, illustrating an elimination reaction. Primary and secondary alcohols can be used under the same experimental conditions to show the differences in the speed of reaction. It is also a simple example of heterogeneous catalysis and solventless reactions.

2. Esterification

Isobutylene is used for the esterification of benzoic acid (or cinnamic acid) by bubbling the gas produced as described in a solution of the chosen acid in dichloromethane containing a small amount of H_2SO_4 (Amberlyst-15 or a zeolite can be used as catalysts but the reaction is slower). Under these conditions, benzoic (or cinnamic) acid acts by nucleophilic addition to the carbocation formed in the protonation of the alkene, as opposed to its electrophilic role in the usual Fischer esterification.

Additional Safety

The whole experimental set-up should be installed in a fume hood, since a gas is produced in the first step. Isobutylene is highly flammable and presents an explosion danger when stored as a compressed, liquefied gas; however, in the present experiment, it is formed in small amounts and immediately consumed, so this risk is minimized. Concentrated sulfuric acid is highly corrosive and should be handled with appropriate gloves. Safety googles should be worn all the time.

Experimental Procedure

Session 1 (4 h)

Generation of Isobutylene

1. In a 100 mL round-bottom flask, mix 15 g of dihydrated oxalic acid and 40 mL of *tert*-butanol. Add a magnetic stir bar and connect a reflux condenser. On top of the condenser, adapt a hose connector with to a piece of tubing about 50 cm long; on the other end of the tubing, insert a Pasteur pipette, as shown in Figure 4.1.1.6.1. Some parafilm or Teflon strip may be needed to ensure there are no gas leaks at this point.
2. Heat up the round-bottom flask with a stirring heating mantle or bath oil (~90 °C) keeping a constant reflux of 1–2 drops per second, with strong magnetic stirring. Keep the tip of the pipette immersed in a test tube containing 2–3 mL of $KMnO_4$ (1% solution), until discoloration is observed.

Esterification Reaction

3. In a 125 mL Erlenmeyer flask, weigh 0.01 mol of benzoic (or cinnamic) acid, add 50 mL of dichloromethane and 0.10 mL of concentrated sulfuric acid and homogenize the mixture.
4. Remove the tip of the pipette from the $KMnO_4$ solution, clean it up (break part of the tip if necessary). Insert the Pasteur's pipette from the isobutylene generation apparatus in a bored cork or rubber stopper so the tip of the pipette will be immersed in the dichloromethane solution when the Erlenmeyer flask is firmly corked (Figure 4.1.1.6.1) and let the gas bubble in the solution for 2 h. Keep the Erlenmeyer flask cooled in an ice bath.
5. At the end of the laboratory period, remove the pipette from the hole in the cork and tightly seal that hole.
6. Leave in a refrigerator for 7 days or until the next laboratory session.

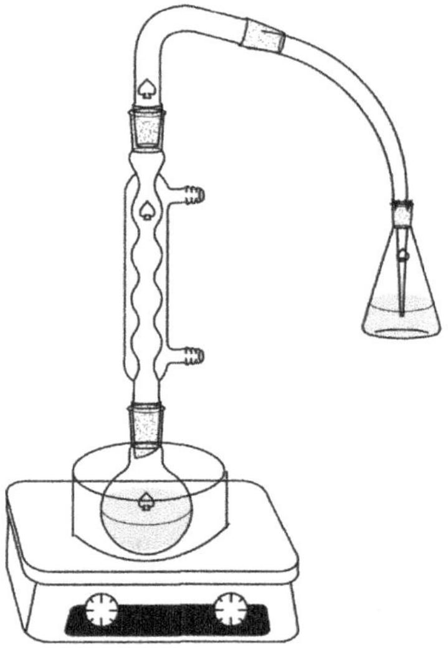

Figure 4.1.1.6.1 Setup for isobutylene generation.

Session 2 (3–4 h)

1. After the resting period, put the Erlenmeyer flask in an ice bath and unseal it.
2. Transfer the solution to a separation funnel and extract twice with 30 mL of 10% Na_2CO_3 solution and twice with 50 mL of distilled water.
3. Combine aqueous phases and save for later use.
4. Dry the organic phase over anhydrous Na_2SO_4, filter and remove the solvent in the rotary evaporator. The ester is obtained as a pure yellow oil.

Characterization

5. Obtain the FTIR spectra of the oil and of the starting acid (benzoic or cinnamic). Obtain NMR spectra (if available) of both reagent and reaction product if available.
6. Weigh the oil and determine the yield of the reaction.
7. Obtain the refractive index of the oil.

Recovery of Unreacted Benzoic (Cinnamic) Acid

8. Take the aqueous phase (point 3) to the fume hood and acidify it with concentrated sulfuric acid. Cork the flask and shake vigorously. Let it rest on a shelf for full crystallization of unreacted acid.

Recovery of Catalyst (Oxalic Acid)

9. Take the round-bottom flask where the isobutylene generation took place (point 1 of session 1), decant any remaining liquid and add just enough hot distilled water to dissolve the solid. Fragment the solid mass with a glass rod and heat up the mixture to guarantee all acid is dissolved.
10. Remove the saturated solution from the heating source and let it cool to room temperature to promote the crystallization of pure oxalic acid, which can be later filtered and stored for reuse.

Waste Disposal

Acidified aqueous phase should be neutralized before pouring down the drain, with the water tap running. Remaining $KMnO_4$ solution should be combined in a clearly labelled flask for later destruction. Dichloromethane should be disposed of in an appropriate container (halogenated waste).

Results Interpretation and Additional Questions

1. Provide the IUPAC name of your reaction product.
2. Interpret the obtained FTIR spectra, comparing the main bands of the reactant and the reaction product.
3. Draw curly arrows to explain the electron flow in all steps of the reaction(s) mechanism(s).
4. Why is it necessary to acidify the aqueous phase to precipitate the unreacted acid?
5. Determine the yield of recovered benzoic (cinnamic) acid and how much of the catalyst you were able to regenerate.

Acknowledgements

Adapted with permission from *Química Nova*, 2003, **26**(3), 425–427. The authors thank the students from the Biochemistry degree of Madeira University, course of Organic Chemistry 2013/14 and 2014/15, who contributed to improve this experiment.

References

1. M. Jones, Jr., *Organic Chemistry*, Norton & Company, New York, 2nd edn, 2000, pp. 896–903.
2. S. Kim and S. I. Lee, *J. Org. Chem.*, 1984, **49**, 1712.
3. M. K. Dhaon, R. K. Olsen and K. Ramasamy, *J. Org. Chem.*, 1982, **47**, 1962.
4. A. Armstrong, I. Brackenridge, R. F. W. Jackson and J. M. Kirk, *Tetrahedron Lett.*, 1988, **29**, 2483.
5. D. F. Taber, D. A. Gerstenhaber and X. Zhao, *Tetrahedron Lett.*, 2006, **47**, 3065.

4.1.1.7. Hydroxyl Group Protection *via* Tetrahydropyranyl Ether Formation

Ana I. B. Meirinhos, Filipa Siopa*, and Carlos A. M. Afonso*

iMed.ULisboa, Av. Professor Gama Pinto, 1649-003 Lisboa, Portugal
*E-mail: filipasiopa@ff.ul.pt, carlosafonso@ff.ul.pt

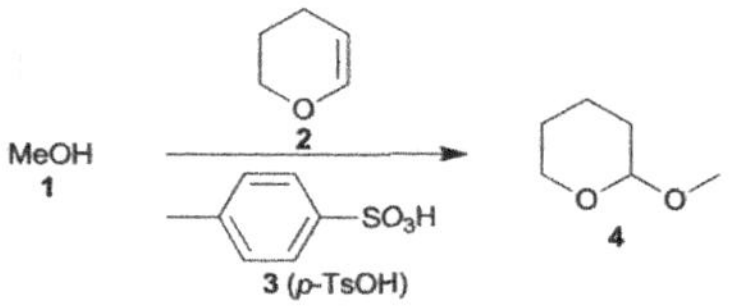

Number of sessions (duration of each session)	Hazard level	Difficulty level	Level of study
1 (3 h)	Moderate	Medium	Intermediate

Class names Alcohols, enol ethers, ketals, acetals

Concepts involved This experiment presents a straightforward method for hydroxyl group protection of methanol, a simple, inexpensive and readily available alcohol, through tetrahydropyranyl ether formation. The product isolation involves a distillation at atmospheric pressure

Chemicals needed Methanol, dichloromethane, dihydro-4*H*-pyran, *p*-toluenesulfonic acid, sodium bicarbonate, magnesium sulfate anhydrous, deuterochloroform

Equipment and experimental techniques involved Apparatus for reaction with magnetic stirring, liquid–liquid extraction apparatus, rotatory evaporator, distillation apparatus, infrared spectrometer and NMR spectrometer

Keywords Addition to enol ethers, distillation, ^{1}H NMR, hydroxyl group protection, IR

Background

The success of organic synthesis is strongly dependent on the possibility to perform multiple and specific functional changes. However, sometimes it is not possible to perform a transformation in a functional group in the presence of a more reactive one. To overcome this problem, the reactivity of the most reactive functional group is temporarily modified through its conversion into a more stable derivative; this synthetic step is called protection. The hydroxyl group is a functional group susceptible to various transformations, so it is frequently necessary to proceed to its protection. The protection of a hydroxyl group by tetrahydropyranyl ether formation is extensively used in organic synthesis, mainly because this group remains stable under various reaction conditions, in particular strongly basic conditions, in the presence of organometallic compounds of lithium and magnesium, reducing and oxidizing conditions and acylation conditions. It is also possible to perform its deprotection in acid and even neutral conditions. Tetrahydropyranylation is generally carried out using dihydro-4*H*-pyran (**2**) in the presence of *p*-toluenesulfonic acid (*p*-TsOH) (**3**), an extremely efficient catalyst in this transformation, but

Comprehensive Organic Chemistry Experiments for the Laboratory Classroom
Edited by Carlos A. M. Afonso, Nuno R. Candeias, Dulce Pereira Simão, Alexandre F. Trindade, Jaime A. S. Coelho, Bin Tan, and Robert Franzén.

Published by the Royal Society of Chemistry, www.rsc.org

having the drawback of catalysing undesirable reactions of **2**, such as its polymerization in a strongly acid environment.[1–3]

Using methanol (**1**) as a representative substrate because it has the advantage of allowing the distillation of the product at atmospheric pressure, the objective of this work is the protection of the hydroxyl group through the formation of 2-methoxytetrahydropyran (**4**).

Additional Safety

This session should be done in a fume hood, with proper safety equipment, such as protective clothing, gloves and safety googles. It is important to take special care in handling the *p*-toluenesulfonic acid, which is a corrosive acid and may cause severe skin and eye burns. Dichloromethane may cause skin, eye and respiratory irritation. Dihydro-4*H*-pyran is a flammable reagent and causes skin, eye and respiratory irritation.

Experimental Procedure

1. In a 250 mL round-bottomed flask equipped with a magnetic stir bar and a calcium chloride tube, prepare a solution of 5.9 mL of methanol (146 mmol), 150 mL of dichloromethane and 13.4 mL of dihydro-4*H*-pyran (1.0 mol equiv.). Place the solution in an ice bath.
2. Add *p*-toluenesulfonic acid (2.8 g, 0.1 mol equiv., *exothermic reaction*) in one portion to the reaction mixture. After a while remove the ice bath and allow the reaction to stir at room temperature.
3. After 1 h transfer the reaction mixture into a separatory funnel and wash with 100 mL of saturated aqueous sodium bicarbonate solution.
4. Remove the water from the organic layer with anhydrous magnesium sulfate, filter and evaporate the solvent in a rotatory evaporator.
5. Distil the reaction at atmospheric pressure and collect the fraction corresponding to the product 2-methoxyhydropyran (**4**) (lit.[4] bp 128–129 °C).
6. Obtain IR and ^{1}H NMR spectra and compare them with those of an authentic sample.

Waste Disposal

The neutralized aqueous solution may be diluted and discarded in the sink. Dichloromethane should be collected in a chlorinated waste collector recipient, for proper disposal.

Results Interpretation and Additional Questions

1. Interpret the ^{1}H NMR and IR spectra obtained and compare them with the spectra of dihydro-4*H*-pyran.
2. Rationalize the mechanism involved in the hydroxyl group protection.
3. Interpret the overall process. Calculate and discuss the atom economy[5] and *E* factor.[6]
4. Explain how the *p*-toluenesulfonic acid can be used as a promoter both in tetrahydropyranyl ether formation and in its deprotection.
5. Another extremely efficient catalyst at room temperature for this transformation is pyridinium *p*-toluenesulfonate.[7] Discuss if the involved reaction mechanism is comparable to that of *p*-toluenesulfonic acid.
6. Discuss if the effected protection can be carried out in the presence of the following compounds: toluene, cyclohexene, 1-phenyl-cyclohexene, acetic acid or acetamide.
7. Suggest a methodology to improve the *E* factor.

References

1. T. W. Greene and P. G. M. Wuts, *Protective Groups in Organic Synthesis*, John Wiley & Sons, 1999, pp. 49–54.
2. A. J. Pearson and W. R. Roush, *Handbook of Reagents for Organic Synthesis*, John Wiley & Sons, 1999, pp. 147–150.
3. J. M. Hornback, *Organic Chemistry*, Cengage Learning, 2005, pp. 1011–1013.
4. G. F. Woods and D. N. Kramer, *J. Am. Chem. Soc.*, 1947, **69**, 2246.
5. B. Trost, *Science*, 1991, **254**, 1471.
6. R. A. Sheldon, *Green Chem.*, 2007, **9**, 1273.
7. M. Miyashita, A. Yoshikoshi and P. A. Grieco, *J. Org. Chem.*, 1977, **42**, 3772.

4.1.1.8. Synthesis of (−)-Carvone from (+)-Limonene

João P. Telo

Centro de Química Estrutural, Instituto Superior Técnico, Universidade de Lisboa, Av. Rovisco Pais, 1649-003 Lisboa, Portugal. *E-mail: jptelo@tecnico.ulisboa.pt

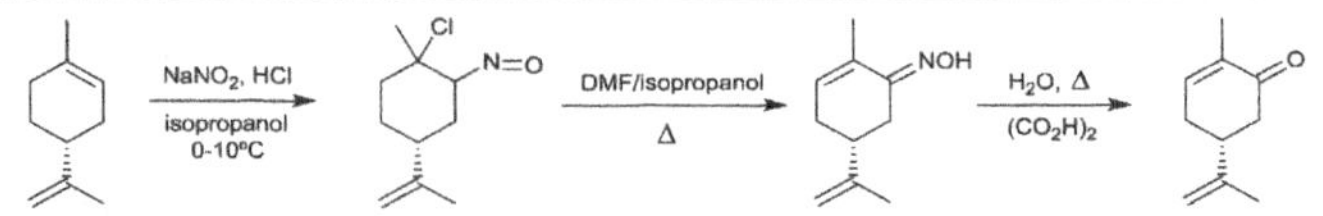

Number of sessions (duration of each session)	Hazard level	Difficulty level	Level of study
3 (4 h each)	Low	Medium	Intermediate

Class names Alkenes, oximes, nitrosos, ketones

Concepts involved Transformation of a natural compound ((+)-limonene) involving electrophilic addition to an alkene, an elimination reaction and hydrolysis of an oxime. The final product ((−)-carvone) can be used to introduce the notions of chirality and optical rotation

Chemicals needed (+)-Limonene, isopropanol, HCl (37%), $NaNO_2$, ethanol, dimethylformamide, oxalic acid, dichloromethane

Equipment and experimental techniques involved Apparatus for reactions under reflux, liquid–liquid extraction, vacuum filtration, steam distillation, vacuum fractional distillation, Abbé refractometer and polarimeter for the characterization of a liquid product

Keywords Alkene electrophilic addition, elimination reactions, fractional distillation, natural products, optical rotation, refraction index, steam distillation, terpenes

Background

(−)-Carvone is a compound of the terpene family and is mostly found in the essential oil from several species of mint, particularly spearmint oil. It is a viscous liquid with an intense spearmint odour which darkens upon exposure to air and light. Interestingly, its enantiomer (+)-carvone smells like caraway seeds, showing that the perception of smell can be different for two enantiomers.[1] (−)-Carvone is mainly used in the flavour and food industry, but also used in oral care and air freshening products. Minor uses as potato sprouting inhibitor, insect repellent and antimicrobial agent have also been reported.[2] It is also employed as starting material for the asymmetric synthesis of several natural products.[2] It can be extracted from spearmint, but most (−)-carvone is prepared by chemical synthesis.

In this procedure (−)-carvone is prepared in three steps from (+)-limonene,[3,4] which is available as a by-product of the orange juice industry. In the first step the regioselective electrophilic addition of *in situ* prepared nitrosyl chloride to (+)-limonene affords limonene nitroso-chloride. NOCl reacts preferentially with the most electron-rich endocyclic (*i.e.*, within the cycle) double bond. Elimination of HCl forms carvoxime (second step), which is hydrolysed to (−)-carvone using oxalic acid as a catalyst in the third step. The use of a weak acid in the final step avoids

Comprehensive Organic Chemistry Experiments for the Laboratory Classroom
Edited by Carlos A. M. Afonso, Nuno R. Candeias, Dulce Pereira Simão, Alexandre F. Trindade, Jaime A. S. Coelho, Bin Tan, and Robert Franzén.

Published by the Royal Society of Chemistry, www.rsc.org

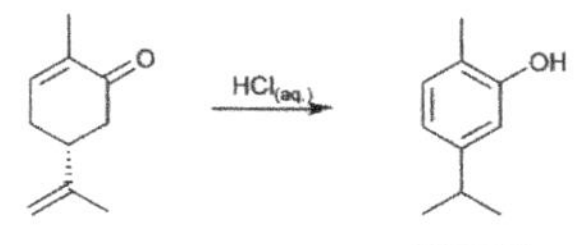

Scheme 4.1.1.8.1 Synthesis of carvacrol from (−)-carvone.

the formation of the aromatic compound carvacrol, which is the main product if aqueous HCl catalysis is used (Scheme 4.1.1.8.1).

Additional Safety

There are no major risks in these procedures. However, strong acids should be handled with care and ethanol or isopropanol are flammable liquids. Dichloromethane is the least toxic of the simple chlorohydrocarbons, but it high volatility may lead to acute inhalation hazards.

Experimental Procedure

Laboratory Session 1 (4 h): Synthesis of Limonene Nitroso-Chloride

1. A four-neck 500 mL round-bottomed flask is charged with 36.5 mL of (+)-limonene and 30 mL of isopropanol and equipped with a reflux condenser, a thermometer and two dropping funnels.
2. The mixture is stirred magnetically and cooled below 10 °C in an ice bath.
3. A solution of 90 mL concentrated hydrochloric acid in 60 mL of isopropanol and a solution of 15.6 g of sodium nitrite in 80 mL of water are dripped simultaneously into the flask through the two dropping funnels, at a rate so that the temperature of the mixture stays below 10 °C.
4. After the addition the mixture is stirred for 30 min with cooling and kept in a freezer for an extra hour (keep the mixture in an ice/salt/water bath if a freezer is not available).
5. The solid is filtered on a Büchner funnel and washed with cold ethanol until a pure-white solid remains on the funnel. Further cooling of the mother-liquors affords an extra crop of limonene nitroso-chloride. Dry the solid in a desiccator under vacuum overnight.

Laboratory Session 2 (4 h): Synthesis of Carvoxime

1. A mixture of 25 g of crude limonene nitroso-chloride, 12 mL of dimethylformamide and 75 mL of isopropanol is heated under reflux for 30 min.
2. The solution is poured into 500 mL of water and ice and stirred vigorously with a glass rod until precipitation of the product.
3. The solid is filtered off in a Buchner funnel and washed with cold water.

Laboratory Session 3 (4 h): Synthesis of (−)-Carvone

1. A mixture of 17 g of carvoxime and 170 mL of a 10% oxalic acid solution is refluxed for 1 h.
2. After this period, an apparatus for steam distillation is adapted to the flask, and the mixture is distilled until only water comes out of the condenser.
3. The upper organic layer is separated and the aqueous fraction is extracted twice with 30 mL of dichloromethane.
4. The solvent is distilled in a rotary evaporator, yielding crude (−)-carvone.

5. Pure (−)-carvone is obtained by fractional distillation under reduced pressure. The fraction that distils between 100 and 115 °C at 15 mmHg is collected.
6. The residue is mostly carvacrol, which can be obtained pure by combining the residues of all the groups and collecting the fraction that distils at 120–130 °C (15 mmHg).
7. The refraction index of (−)-carvone can be measured in the Abbé refractometer and its specific rotation measured in a polarimeter.

Waste Disposal

The dichloromethane should be stored for further distillation.

Results Interpretation and Additional Questions

1. Draw the reactions mechanisms.
2. Why does Cl add to the tertiary carbon atom and NO to the secondary atom of the double bond, and not the other way round?
3. Why does the HCl elimination afford a double bond within the ring, and not an exocyclic double bond?
4. Why is the oxime tautomer of carvoxime more stable than the nitroso one?
5. Why is carvoxime dextrorotatory and limonene laevorotatory if the absolute configuration of the asymmetric carbon atom does not change (R in both)?

References

1. T. J. Leitereg, D. G. Guadagni, J. Harris, T. R. Mon and R. Teranishi, *J. Agric. Food Chem.*, 1971, **19**, 785.
2. C. C. C. R. de Carvalho and M. M. R. da Fonseca, *Food Chem.*, 2006, **95**, 413.
3. O. S. Rothenberger, S. B. Krasnof and R. B. Rollins, *J. Chem. Educ.*, 1980, **57**, 741.
4. R. H. Reitsema, *J. Org. Chem.*, 1958, **23**, 2038.

4.1.1.9. Glycal Transformation into Surfactant 2-Deoxy Glycosides

Catarina Dias and Amélia P. Rauter*

Centro de Química e Bioquímica, Faculdade de Ciências, Universidade de Lisboa, C8, Piso 5, Campo Grande, 1749-016 Lisboa, Portugal. *E-mail: aprauter@fc.ul.pt

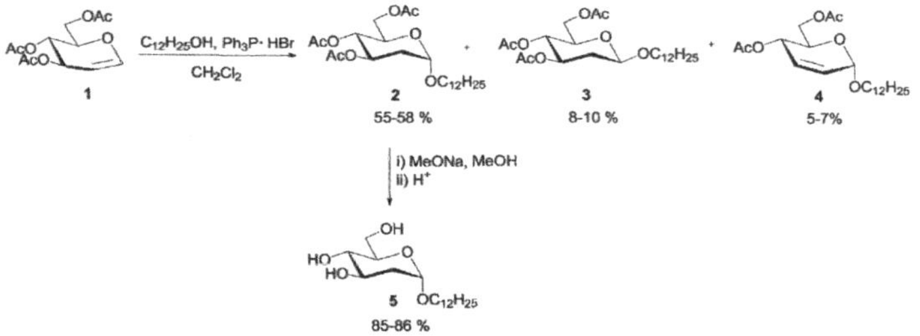

Number of sessions (duration of each session)	Hazard level	Difficulty level	Level of study
3 (4 h + 4 h + 5 h)	Moderate	High	Advanced

Classes names Enol ethers, alcohols, acetals, esters

Concepts involved Acid-catalysed addition of an alcohol to an acetylated glycal (sugar enol ether with a double bond between C1 and C2) for the synthesis of an alkyl 2-deoxyhexopyranoside, purified by column chromatography; Zemplén deprotection by transesterification affords the unprotected target glycoside

Chemicals needed 3,4,6-Tri-*O*-acetyl-1,5-anhydro-2-deoxy-D-*arabino*-hex-1-enitol (**1**), dodecanol, dichloromethane (previously dried over 4 Å molecular sieves), triphenylphosphane hydrobromide, sodium hydrogen carbonate, anhydrous sodium sulfate, cyclohexane (CyHex), *n*-hexane, ethyl acetate, sodium methoxide, methanol (previously dried over 3 Å molecular sieves), Amberlite® IR120 (H^+ form), deuterated methanol, silica gel 60 Å (0.040–0.630 mm), 10% H_2SO_4 solution in MeOH

Equipment and experimental techniques involved Rotary evaporator, heating magnetic stirrer with integrated temperature control (alternatively, a heating magnetic stirrer with contact thermometer), separation funnel, reflux condenser; thin-layer chromatography aluminium sheets pre-coated with silica gel 60 and fluorescent indicator UV_{254}, column chromatography equipment, melting point measurement apparatus, NMR spectrometer

Keywords Anomeric effect, Brønsted catalysis, column chromatography, 2-deoxy-glycoside, Ferrier rearrangement, NMR, oxycarbenium ion, stereoselectivity

Background

Carbohydrates represent a unique family of polyfunctional molecules playing a role in tumour cell metastasis, infection, cell signalling, and fertilization, to name just a few examples of their importance in life processes. They have the potential to be chemically manipulated in a multitude of ways, being converted into compounds of interest for food, clothing, pharmaceutical, and

Comprehensive Organic Chemistry Experiments for the Laboratory Classroom
Edited by Carlos A. M. Afonso, Nuno R. Candeias, Dulce Pereira Simão, Alexandre F. Trindade, Jaime A. S. Coelho, Bin Tan, and Robert Franzén.

Published by the Royal Society of Chemistry, www.rsc.org

agrochemical industries. Sugar-based surfactants are a versatile class of compounds, due to their low toxicity and biodegradation,[1] and their potential to interact with membranes, leading to several biological and medicinal applications.[1,2] The self-organization properties of these molecules depend on the nature of the sugar and the hydrophobic tail length, which therefore determine their application.[3] In particular, surfactant glycosides such as **5** and its analogues have proven to be biologically active, and their activity can be tuned by structural changes in the deoxygenation pattern and the anomeric configuration.[2] This protocol describes the stereoselective synthesis of 2-deoxy glycosides that affords, as major product, the bioactive anomeric configuration.

Glycosides are acetals involving the anomeric centre and their synthesis requires reaction of an activated glycosyl donor with an alcohol. Moreover, the glycosyl donor must be conveniently protected to avoid autoglycosylation. Noteworthy is also the anomeric effect, which occurs in the presence of electronegative substituents (X) linked to the anomeric carbon of a pyranose ring, adopting the C–X bond an axial orientation rather than an equatorial one. This electronic effect occurs because there is a lone pair of electrons of the endocyclic oxygen anti-periplanar to the C–X bond; therefore, a partial donation of this lone pair to the anti-bonding (σ*) orbital of the C–X bond takes place, stabilizing the axial position of X.[4]

Amongst the structural diversity of glycosyl donors, 1,5-anhydro-2-deoxy-1-enitols (Figure 4.1.1.9.1), commonly referred to as glycals, have been recognized as versatile building blocks for the preparation of stereochemically rich molecules, the structural diversity and multifunctionality of which are an important asset for several applications and carbohydrate transformations. Their reactivity can be tuned towards the synthesis of 2-deoxy or 2,3-unsaturated sugars, depending upon the catalyst used. While $Ph_3P{\cdot}HBr$ is known to catalyse sugar 2-deoxygenation,[1–3] other Brønsted and Lewis acids as well as heterogeneous acid zeolite catalysis afford mainly 2,3-dideoxy-2,3-unsaturated glycosides.[5] Moreover, glycals can be converted into 1,2-*trans* glycosides *via* epoxidation followed by treatment with an alcohol, or can react with nucleophiles after direct activation by *N*-iodosuccinimide (NIS) or iodonium di-*sym*-collidine perchlorate (IDCP), *via* nucleophilic opening of the intermediate cyclic iodonium ion. They can also be converted into 2-amino-2-deoxy sugars by reaction with sodium azide in the presence of ceric ammonium nitrate (CAN), followed by reduction (Figure 4.1.1.9.1).[4,6]

In this experiment, the synthesis of surfactant 2-deoxy glycosides *via* reaction of 3,4,6-tri-*O*-acetyl-1,5-anhydro-2-deoxy-D-*arabino*-hex-1-enitol (tri-*O*-acetyl-D-glucal, not recommended name) with dodecanol in the presence of triphenylphosphane hydrobromide (TPHB) is illustrated. The acid-catalysed addition of an alcohol to an acetylated glycal usually occurs with concomitant Ferrier rearrangement, giving the 2,3-unsaturated derivative as major reaction product. However, TPHB, first described by Bolitt *et al.*[7] to prepare 2-deoxy glycosides from glycals, circumvents

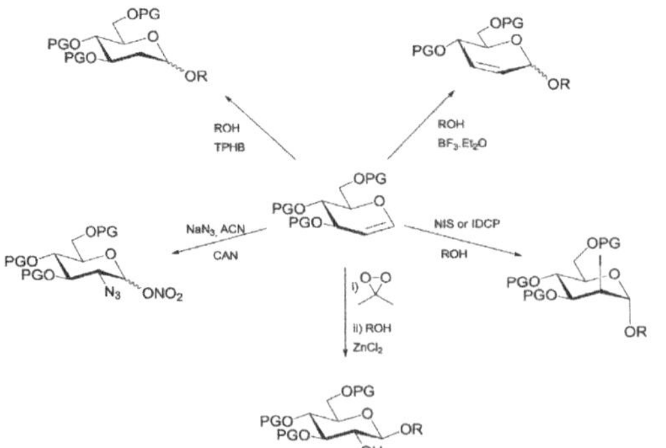

Figure 4.1.1.9.1 Usefulness of 1,5-anhydro-2-deoxy-1-enitols as building blocks in carbohydrate chemistry (PG: protecting group).

> ***Did you know...?***
> *Glycals were given the suffix 'al' by Emil Fischer, who thought he had synthesized an aldehyde. Nowadays, the use of the terms glucal, rhamnal,... is not recommended by IUPAC, since these enol ethers do not exhibit the configuration of glucose, or rhamnose, inferred by the name.*

Ferrier rearrangement. TPHB catalysed glycosylation is carried out under simple experimental conditions, affording sugar-based surfactants type **5** in a stereoselective manner and permitting an easy access to biologically important molecules.

Additional Safety

All experiments should be carried out in a fume hood, with proper safety equipment, such as protective clothing, gloves, and safety goggles. Students are advised to consult GHS hazard and precautionary phrases for all reagents. Dodecanol and TPHB are harmful when in contact with skin and eyes (H314, H315, H319). Sodium methoxide is a flammable solid (H251), reacts violently with water (EUH014) and causes burns by ingestion or inhalation (H314). Therefore, it should be carefully handled.

Experimental Procedure

Laboratory Session 1: Glycosylation Reaction (4 h)

1. In a fume hood, heat an oil bath to 40 °C on a heating magnetic stirrer with integrated temperature control (alternatively, a heating magnetic stirrer with contact thermometer can be used).
2. In a 25 mL round-bottomed flask equipped with a magnetic stir bar, prepare a solution of 3,4,6-tri-*O*-acetyl-1,5-anhydro-2-deoxy-D-*arabino*-hex-1-enitol (**1**, 0.500 g, 1.84 mmol) in dry dichloromethane (2 mL). Add dodecanol (0.43 mL, 1.93 mmol, 1.05 equiv.) and a solution of $Ph_3P{\cdot}HBr$ (0.063 g, 0.184 mmol, 0.1 equiv.) in the same solvent (1 mL). Adapt a reflux condenser and stir the reaction mixture under reflux for 2 h 30 min. Check reaction completion by thin-layer chromatography eluted with CyHex/EtOAc (3 : 1), staining with a 10% H_2SO_4 solution in MeOH, followed by heating at 120 °C (Figure 4.1.1.9.2).
3. Cool the reaction mixture to room temperature and dilute it with dichloromethane (10 mL). Wash the solution with a saturated $NaHCO_3$ solution (2 × 15 mL) and then water (15 mL). After separation of both phases, dry the organic phase with anhydrous $MgSO_4$, remove the desiccant by filtration, and evaporate the solvent under reduced pressure.
4. Keep the residue (yellowish oil) at 0 °C until the next session.

Laboratory Session 2: Product Purification (5 h)

1. To isolate the products from the reaction mixture, assemble a chromatography column:

   - Prepare the eluent CyHex/EtOAc (15 : 1) (1500 mL).
   - Fill a chromatography column (3.5 cm diameter × 12 cm height) with silica gel (30 g) in CyHex/EtOAc (15 : 1).
   - Dissolve the reaction residue in eluent (*ca.* 3 mL), adding one or two drops of dichloromethane to assure complete dissolution, and apply it to the column chromatography.
   - After discharging *ca.* 80 mL of dead volume, collect fractions in a set of 10 mL vials.

2. Follow the course of the column chromatography by eluting TLC plates with EtOAc/CyHex (1 : 3). Stain eluted TLC plates by dipping them in a 10% H_2SO_4 solution in MeOH, followed by heating at 120 °C (Figure 4.1.1.9.3).
3. Combine the fractions containing the pure dodecyl 4,6-di-*O*-acetyl-2,3-dideoxy-α-D-*erythro*-hex-2-enopyranoside [**4**, R_f = 0.71, EtOAc/CyHex (1 : 3)], in a 250 mL labelled and tared round-bottom flask and evaporate the solvent in the rotary evaporator.
4. Combine the fractions containing the pure dodecyl 3,4,6-tri-*O*-acetyl-2-deoxy-α-D-*arabino*-hexopyranoside [**2**, R_f = 0.58, EtOAc/CyHex (1 : 3)] in a 500 mL labelled and tared round-bottom flask and evaporate the solvent in the rotary evaporator.

5. Compound **2** will also be eluted from the column contaminated with the corresponding β-anomer, 3,4,6-tri-*O*-acetyl-2-deoxy-β-D-*arabino*-hexopyranoside (**3**, $R_f = 0.47$; EtOAc/CyHex (1:3)). Combine the fractions with the α/β mixture in a 500 mL labelled and tared round-bottom flask and evaporate the solvent in the rotary evaporator.
6. Weigh the flasks with compounds **2**, **2**/**3**, and **4**.
7. Prepare two NMR samples, one containing the isolated compound **2** (30 mg) and the other containing the mixture of compounds **2** and **3**, by dissolving them (30 mg) in $CDCl_3$ (0.5 mL each). Transfer the solutions to two NMR tubes. Run a ^1H NMR experiment on both samples.

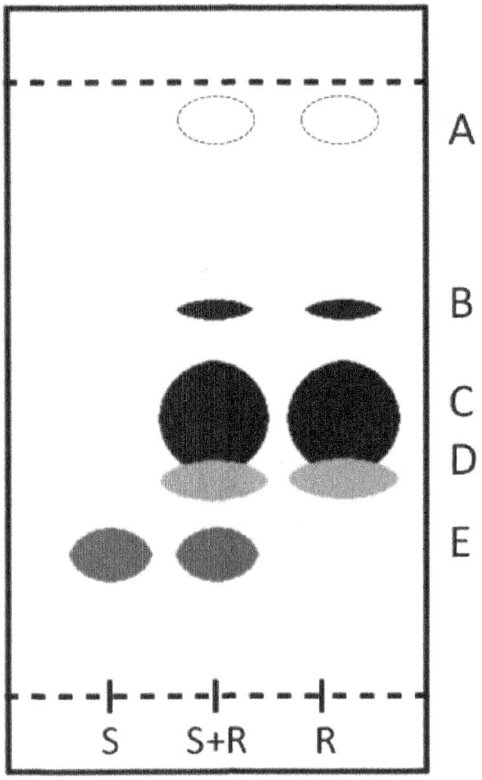

Figure 4.1.1.9.2 Schematic TLC of the reaction (R) and of glycal (starting material, S). A: PPh_3 (only UV-visible); B: compound **4**; C: compound **2**; D: compound **3**; E: glycal **1**.

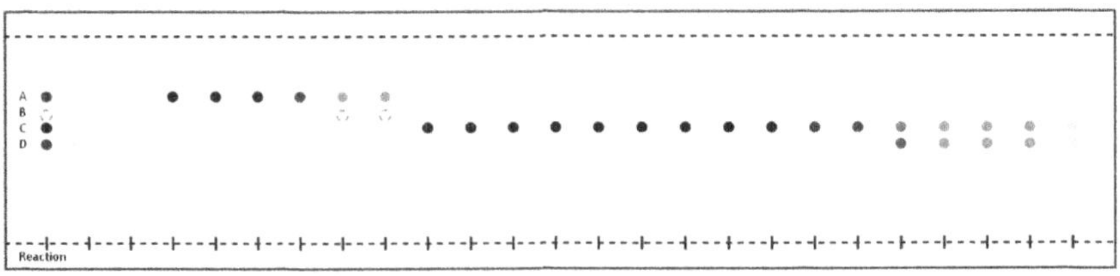

Figure 4.1.1.9.3 Illustrative scheme of the column chromatography followed by TLC. A: dodecyl 4,6-di-*O*-acetyl-2,3-dideoxy-α-D-*erythro*-hex-2-enopyranoside (**4**); B: dodecanol; C: dodecyl 3,4,6-tri-*O*-acetyl-2-deoxy-α-D-*arabino*-hexopyranoside (**2**); D: dodecyl 3,4,6-tri-*O*-acetyl-2-deoxy-β-D-*arabino*-hexopyranoside (**3**).

Laboratory Session 3: Deacetylation Reaction (4 h)

1. Prepare a solution of sodium methoxide in dry methanol (1 M, 1 mL).
2. In a 25 mL round-bottomed flask equipped with a magnetic stir bar, prepare a solution of compound **2** (0.150 g, 0.374 mmol) in methanol (2 mL) and add the previously prepared sodium methoxide solution (0.2 mL). Stir the reaction mixture at room temperature for 1 h 30 min.
3. Neutralize the reaction mixture with Amberlite (IR-120, H^+ form). Filter the resin off and evaporate the solvent.
4. Recrystallize the residue from EtOAc/*n*-hexane (3:1) as follows: dissolve the residue in EtOAc (3 mL) and heat under reflux until complete dissolution. Add hexane (1 mL) and allow the solid to precipitate in an ice bath for 30 min. Filter the solid (dodecyl 2-deoxy-α-D-*arabino*-hexopyranoside, **5**) under reduced pressure and take it to the high vacuum line for 30 min.
5. Measure the melting point (mp 113.9–115.5 °C).
6. Dissolve compound **5** (30 mg) in MeOD (0.5 mL) and transfer the solution to an NMR tube. Run ^{1}H NMR, ^{13}C NMR, COSY, and HMQC experiments.

Waste Disposal

The neutralized aqueous solution may be diluted and discarded in the appropriate waste collectors. Dichloromethane, cyclohexane, *n*-hexane, ethyl acetate, and methanol should be collected in the appropriate waste collector recipients (chlorinated or non-chlorinated organic solvents, accordingly).

Results Interpretation and Additional Questions

1. Calculate the isolated yield for 3,4,6-tri-*O*-acetyl-2-deoxy-α-D-*arabino*-hexopyranoside (**2**) and for dodecyl 4,6-di-*O*-acetyl-2,3-dideoxy-α-D-*erythro*-hex-2-enopyranoside (**4**). Calculate the 2-deoxy glycoside yield obtained in the α/β mixture and give its α/β proportion, as determined by the anomeric signal integration of both anomers in the ^{1}H NMR spectrum of the mixture.
2. Justify mechanistically why the glycosylation reaction affords both α and β anomers and comment on their relative proportion.
3. Compare the melting point obtained for compound **5** with that given in the literature and justify any differences you may find.
4. Using the mono and 2D NMR spectra acquired, assign both ^{1}H and ^{13}C NMR signals for dodecyl 2-deoxy-α-D-*arabino*-hexopyranoside. How do you recognize the anomeric configuration in the ^{1}H NMR spectrum?
5. If the starting material is a benzyl protected glycal, do you expect the Ferrier rearrangement to occur? Justify your answer based on the allylic rearrangement mechanism.

References

1. F. V. M. Silva, M. Goulart, J. Justino, A. Neves, F. Santos, J. Caio, S. Lucas, A. Newton, D. Sacoto, E. Barbosa, M. S. Santos and A. P. Rauter, *Bioorg. Med. Chem.*, 2008, **16**, 4083.
2. A. Martins, M. S. Santos, C. Dias, P. Serra, V. Cachatra, J. Pais, J. Caio, V. H. Teixeira, M. Machuqueiro, M. S. Silva, A. Pelerito, J. Justino, M. Goulart, F. V. Silva and A. P. Rauter, *Eur. J. Org. Chem.*, 2013, **8**, 1458.
3. A. P. Rauter, S. Lucas, T. Almeida, D. Sacoto, V. Ribeiro, J. Justino, A. Neves, F. V. M. Silva, M. C. Oliveira, M. J. Ferreira, M. S. Santos and E. Barbosa, *Carbohydr. Res.*, 2005, **340**, 191.
4. B. G. Davis and A. J. Fairbanks, *Oxford Chemistry Primers #99: Carbohydrate Chemistry and Biochemistry*, Oxford University Press, Oxford, 2000, ch. 7, pp. 59–60.
5. A. P. Rauter, T. Almeida, N. M. Xavier, F. Siopa, A. I. Vicente, S. D. Lucas, J. P. Marques, F. R. Ribeiro, M. Guisnet and M. J. Ferreira, *J. Mol. Catal. A: Chem.*, 2007, **275**, 206.
6. T. K. Lindhorst, *Essentials of Carbohydrate Chemistry and Biochemistry*, Wiley-VCH Verlag GmbH, Weinheim, 2000, ch. 5, pp. 134–139.
7. V. Bolitt, C. Mioskowski, S.-G. Lee and J. R. Falck, *J. Org. Chem.*, 1990, **55**, 5812.

4.1.1.10. Preparation of (1*R*,2*R*,3*R*,5*S*)-(−)-Isopinocampheol Through a Hydroboration–Oxidation Reaction

Marek P. Krzemiński

Department of Organic Chemistry, Faculty of Chemistry, Nicolaus Copernicus University in Toruń, Poland. *E-mail: mkrzem@chem.umk.pl

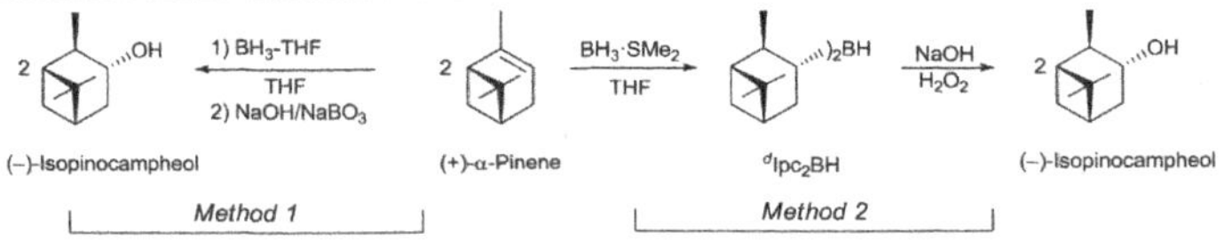

Number of sessions (duration of each session)	Hazard level	Difficulty level	Level of study
2 (4 h each)	Moderate	High	Advanced
Class names Olefins, boranes, alcohols			
Concepts involved This experiment describes a method for a stereoselective transformation of (+)-α-pinene into (−)-isopinocampheol utilizing a hydroboration step with borane followed by oxidation of the resultant organoborane with sodium perborate			
Chemicals needed (+)-α-Pinene, borane–tetrahydrofuran solution (1 M) or 10 M borane–dimethyl-sulfide adduct (BMS), tetrahydrofuran, sodium hydroxide, sodium perborate tetrahydrate or 30% hydrogen peroxide, potassium carbonate, sodium chloride, magnesium sulfate, deuterated chloroform, ethanol			
Equipment and experimental techniques involved Round-bottom flask (100 mL), gas-inlet adapter, oil bubbler, magnetic stirrer, heat gun, plastic syringes (5 and 10 mL), long needles (15 cm), rotary evaporator, polarimeter, melting point equipment, volumetric flask (10 mL), work under an inert atmosphere, sublimation (optional), NMR (optional)			
Keywords Borane, crystallization/sublimation, extraction, hydroboration–oxidation, NMR analysis			

Background

Hydroboration is the addition of a B–H bond to a double or triple carbon–carbon bond to form an organoborane (Scheme 4.1.1.10.1).

Herbert C. Brown with B. C. Subba Rao discovered that diborane in the presence of ethers instantaneously reacts with alkenes to give organoboranes.[1] Terminal and unhindered olefins react with borane to form trialkylboranes without possibility of stopping the reaction at the mono- or dialkylborane stage.[2] Hydroboration of more hindered trisubstituted alkenes can give dialkylboranes. Tetrasubstituted alkenes react rapidly to the monoalkylborane stage.[3] Generally, trialkylboranes are stable compounds as long as they are kept under an inert atmosphere. In contact with air, trialkylboranes, especially those obtained from low molecular weight alkenes, spontaneously ignite with green flames.

Comprehensive Organic Chemistry Experiments for the Laboratory Classroom
Edited by Carlos A. M. Afonso, Nuno R. Candeias, Dulce Pereira Simão, Alexandre F. Trindade, Jaime A. S. Coelho, Bin Tan, and Robert Franzén.

Published by the Royal Society of Chemistry, www.rsc.org

The generally accepted mechanism of hydroboration of an alkene, proposed by Brown, proceeds *via* a four-centred transition state to form an organoborane (Scheme 4.1.1.10.2).[4] The borane adds across the alkene in *syn*-fashion and the boron with its empty p orbital adds to the less substituted end of the alkene.

Several decades of studies on boranes have shown their broad synthetic applications and transformations.[5] The oxidation of organoboranes to alcohols is a very important reaction among possible conversions of carbon–boron bonds. It proceeds with retention of configuration and without any rearrangement of the carbon skeleton. As a consequence, hydroboration–oxidation produces *anti*-Markovnikov *cis*-hydration of the alkene. The standard oxidation conditions utilize 3 M sodium hydroxide with 30% hydrogen peroxide (Scheme 4.1.1.10.3).[6]

Other oxidants, such as sodium perborate,[7] potassium percarbonate,[8] Oxone®[9] and *m*-chloroperbenzoic acid,[10] were also applied for oxidation of organoboranes to alcohols.

In this experiment, the natural, chiral, enantiomerically enriched alkene (+)-α-pinene undergoes hydroboration followed by an oxidation. This reaction is very important because it leads to one of the most important chiral dialkylboranes – (−)-diisopinocampheylborane ($^{d}Ipc_2BH$, after H. C. Brown, the superscript *d* means that Ipc_2BH was obtained from (+)-α-pinene), which is not only a significant hydroborating agent, but also a precursor of many chiral boron reagents.[11] Asymmetric hydroboration–oxidation of prochiral olefins is a valuable method for preparing chiral secondary alcohols. Hydroboration of α-pinene occurs from the less hindered face of the double bond, as shown in Scheme 4.1.1.10.4, confirming all the important features of hydroboration. For clarity, only the first step of hydroboration, the formation of monoisopinocampheylborane, is presented.

Scheme 4.1.1.10.1 Hydroboration of an alkene.

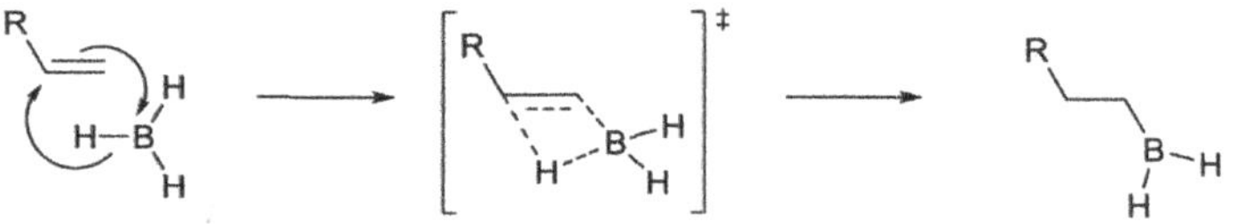

Scheme 4.1.1.10.2 General mechanism of hydroboration.

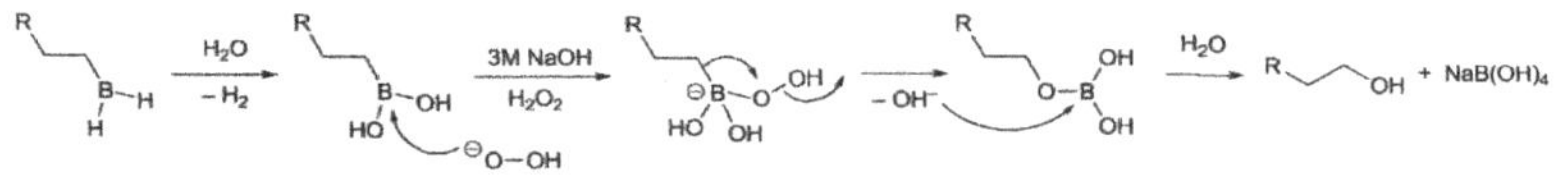

Scheme 4.1.1.10.3 Oxidation of organoboranes.

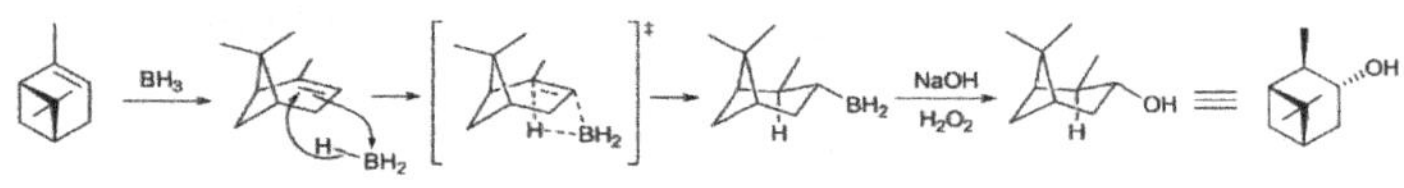

Scheme 4.1.1.10.4 Hydroboration–oxidation of (+)-α-pinene.

Additional Safety

All sessions should be done in a fume hood with protective clothing, gloves, and safety goggles. Borane–tetrahydrofuran solution reacts violently with water and is a flammable liquid and vapour. Hydrogen peroxide is a strong oxidizer and should be handle with great care.

Experimental Procedure

Laboratory Session 1: Hydroboration–Oxidation of (+)-α-Pinene

Method 1 (4 h)

1.1. Prepare a 100 mL round-bottom flask with a magnetic stirring bar. Dry the flask in a flame or use a heat gun, attach gas-inlet adapter capped with a rubber septum and connected with an oil-bubbler, and cool the flask in a stream of nitrogen. Using syringes, add 2.72 g (3.17 mL, 20 mmol, 2 equiv.) of (1*R*)-(+)-α-pinene, 20 mL of dry tetrahydrofuran and 10 mL of 1 M borane–tetrahydrofuran solution (10 mmol, 1 equiv.) at room temperature with stirring.

1.2. Stir the reaction mixture at room temperature for 1 h, cool the flask in an ice-water bath below 10 °C, and carefully add 1 mL of water using a syringe. (*Caution!* Water decomposes all B–H bonds with hydrogen evolution.)

1.3. Add successively 10 mL of water, 10 mL of 3 M sodium hydroxide (30 mmol) and 3.85 g of sodium perborate tetrahydrate (25 mmol). Stir the reaction mixture intensively for 1 h at room temperature to dissolve crystalline sodium perborate.

1.4. Add 10 g of potassium carbonate to the mixture to saturate the aqueous layer. Pour the mixture into the Erlenmeyer flask, close the flask with a glass or rubber stopper and keep the mixture for the next session.

Method 2 (4 h)

2.1. Prepare a 100 mL round-bottom flask with a magnetic stirring bar. Dry the flask in a flame or use a heat gun, attach gas-inlet adapter capped with a rubber septum and connected with an oil-bubbler, and cool the flask in a stream of nitrogen. Using syringes, add 3.40 g (3.96 mL, 25 mmol, 2.5 equiv.) of (1*R*)-(+)-α-pinene and 3 mL of dry tetrahydrofuran. To this solution add slowly 1 mL of 10 M borane–dimethyl-sulfide adduct (BMS) (10 mmol, 1 equiv.) at room temperature with stirring. (*Caution!* A slight exothermic effect can be observed while adding the BMS and also the evolution of hydrogen bubbles if the α-pinene was not completely dry.)

2.2. In the case of solid precipitation within the addition of BMS, the flask should be immersed in an oil bath and heated to 50–55 °C to dissolve the precipitate and to obtain a clear solution.

2.3. Stop the stirrer. Immerse the flask for 1 h in an ice–salt bath, keeping the temperature at −5 to 0 °C with a slow constant flow of nitrogen from the oil-bubbler side, to crystallize $^{d}Ipc_2BH$.

2.4. Using a syringe with a long needle, extending to the bottom of the flask, remove the supernatant liquid. (*Caution!* The supernatant solution contains reactive boranes and must be handled as specified in the handling of the effluents.) Wash the solid in the flask with 2 × 5 mL of dry hexane and add 20 mL of tetrahydrofuran.

2.5. Add carefully 1 mL of water to a suspension of $^{d}Ipc_2BH$ in tetrahydrofuran. (*Caution!* Water decomposes all B–H bonds with hydrogen evolution.) After dissolution of dialkylborane, add 3.7 mL of 3 M sodium hydroxide (11 mmol) and 2.5 mL of 30% hydrogen peroxide (24 mmol). After the addition, stir for 15 min at room temperature and 30 min at 40–50 °C using an oil-bath.

2.6. Cool down the solution to room temperature, add 2.5 g of potassium carbonate to saturate the aqueous layer. Pour the mixture into the Erlenmeyer flask, close the flask with a glass or rubber stopper and keep the mixture for the next session.

Laboratory Session 2: Isolation and Characterization of the Product

Method 1 (4 h)

1.5. Pour the mixture into a separatory funnel and separate the organic layer. Extract the aqueous layer with 3 × 10 mL of diethyl ether. Wash the combined organic layers with 2 × 10 mL of brine. Dry the organic phase with anhydrous $MgSO_4$.

1.6. Filter through a folded filter paper the dried organic solution into a 100 mL round-bottom flask and evaporate the solvents on a rotary evaporator under vacuum.

1.7. Weigh the obtained product and calculate the yield. Measure the melting point and optical rotation, and record ^1H and ^{13}C NMR spectra to confirm the compound purity. Compare with literature data. A 100 mg sample of isopinocampheol can be purified by sublimation under reduced pressure (water aspirator) to measure the melting point and optical rotation.

1.8. Isopinocampheol can be crystallized from pentane to improve its optical purity.

Method 2 (4 h)

2.7. Pour the mixture into a separatory funnel and separate the organic layer. Extract the aqueous layer with 3 × 10 mL of diethyl ether. Wash the combined organic layers with 2 × 10 mL of brine. Dry the organic phase with anhydrous $MgSO_4$.

2.8. Filter the organic solution into a tarred round-bottom flask and evaporate the solvents on a rotary evaporator under vacuum. Weigh the isolated product and calculate the yield. Measure the melting point and record ^1H and ^{13}C NMR spectra to confirm the compound purity. Compare with literature data.

2.9. Using an analytical balance, weigh approximately 200 mg of crude product in a volumetric flask with a capacity of 10 mL and fill the flask to the mark with ethanol. Pour the solution into a polarimetric tube and measure the optical rotation.

Waste Disposal

Organic solvents from the rotary evaporator should be collected in a non-chlorinated waste container. The aqueous solutions may be neutralized, diluted and discarded in the sink. Organic solutions from the synthesis of solid Ipc_2BH (Method 2, point 4) should be hydrolysed with water and collected in a non-chlorinated waste container.

Results Interpretation and Additional Questions

1. Calculate the percent yield of the obtained isopinocampheol.
2. Measure the optical rotation of the starting α-pinene and the product isopinocampheol. Calculate the specific rotations using the following formulas:

$$[\alpha]_D^{\text{temp}} = \frac{\alpha}{l \times (d \,\text{or}\, c)}$$

where: α is the measured angle of rotation of polarized light plane;
temp is the measurement temperature;
D is the wavelength of yellow sodium light line, 589 nm;
l is the length of the polarimetric tube in dm (1 dm is the standard length of polarimetric tube);
d is density in g mL^{-1} if neat α-pinene is measured;
c is the concentration of isopinocampheol in g mL^{-1}.
The specific rotation is written as follows:
For α-pinene:

$$[\alpha]_D^{temp} = + \text{ or } - \ldots \text{ (neat)}$$

For isopinocampheol:

$$[\alpha]_D^{temp} = + \text{ or } - \ldots \ (c \ldots \text{EtOH})$$

Here *c* is the concentration of isopinocampheol in g per 100 mL.

3. Determine the absolute configuration and enantiomeric purity of the starting α-pinene and the product isopinocampheol, comparing the sign and the value of specific rotations with the literature data.

Lit.[12] for (1*R*)-(+)-α-pinene (>99.9% ee): $[\alpha]_D^{25}$ + 51.6 (neat); Lit.[13] for (1*R*,2*R*,3*R*,5*S*)-(−)-isopinocampheol (>99% ee): $[\alpha]_D^{22}$ − 34.0(*c* 20, EtOH).

4. Draw and name the major products obtained in the hydroboration–oxidation reaction of 1-methylcyclopentene, styrene, 1-hexene, and β-pinene with borane.

References

1. H. C. Brown and B. C. Subba Rao, *J. Am. Chem. Soc.*, 1956, **78**, 2582–2588.
2. M. Zaidlewicz and M. P. Krzemiński, Trialkylboranes, in *Science of Synthesis*, ed. D. Kaufmann, D. Matteson, Georg Thieme Verlag, Stuttgart, 2005, vol. 6, Houben-Weil, pp.1097–1215.
3. H. C. Brown, *Organic Syntheses via Boranes*, John Wiley & Sons, Inc., New York, 1975.
4. H. C. Brown and J. Chandrasekharan, *J. Am. Chem. Soc.*, 1984, **106**, 1863–1865.
5. H. C. Brown and M. Zaidlewicz, *Organic Syntheses via Boranes*, Aldrich Chemical Company, Inc., Milwaukee, vol. 2, Recent Developments, 2001.
6. C. F. Lane and J. J. Daniels, *Org. Synth.*, 1972, **52**, 59–62.
7. (a) G. W. Kabalka, T. M. Shoup and N. M. Goudgaon, *Tetrahedron Lett.*, 1989, **30**, 1483–1486; (b) G. W. Kabalka, J. T. Maddox, T. Shoup and K. R. Bowers, *Org. Synth.*, 1996, **73**, 116–119.
8. G. W. Kabalka, P. P. Wadgaonkar and T. M. Shoup, *Organometallics*, 1990, **9**, 1316–1320.
9. D. H. Brown Ripin, W. Cai and S. J. Brenek, *Tetrahedron Lett.*, 2000, **41**, 5817–5819.
10. A. Gavryushin, K. Polborn and P. Knochel, *Tetrahedron: Asymmetry*, 2004, **15**, 2279–2288.
11. (a) H. C. Brown and P. V. Ramachandran, *Acc. Chem. Res.*, 1992, **25**, 16–24; (b) P. V. Ramachandran, *Aldrichimica Acta*, 2002, **35**, 23–35; (c) J. R. Abbott, C. Allais and W. R. Roush, *Org. Synth.*, 2015, **92**, 38–57.
12. W. S. Johnson, B. Frei and A. S. Gopalan, *J. Org. Chem.*, 1981, **46**, 1512–1513.
13. H. C. Brown, J. V. B. Kanth and M. Zaidlewicz, *Tetrahedron*, 1999, **55**, 5991–6000.

4.1.2. 1,2-Addition to Carbonyl Compounds

4.1.2.1. Synthesis of Fructone

Thomas A. Logothetis

University of Southampton, Chemistry, Highfield, Southampton, Hampshire, SO17 1BJ, UK
*E-mail: thomas.logothetis@southampton.ac.uk

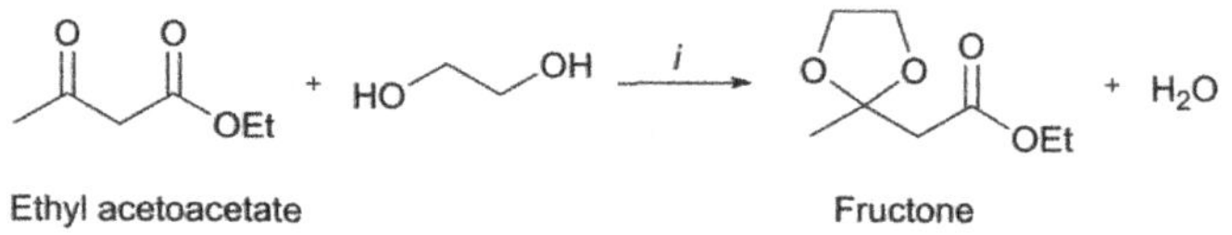

i: *p*-TsOH, $PhCH_3$, ΔT

Number of sessions (duration of each session)	Hazard level	Difficulty level	Level of study
1 (4 h)	Moderate	Medium	Introductory

Class names Ketones, esters, dioxolanes

Concepts involved This experiment describes the synthesis of fructone from ethyl acetoacetate and ethane-1,2-diol using *p*-toluenesulfonic acid as catalyst. The equilibrium is influenced by selectively removing water using a Dean–Stark trap

Chemicals needed Ethyl acetoacetate, ethane-1,2-diol, toluene, *p*-toluenesulfonic acid (pTSA), 2 M sodium hydroxide, brine (saturated), magnesium sulfate (anhydrous)

Equipment and experimental techniques involved Round bottom flasks, Dean–Stark trap (for collection of liquid with higher density), reflux condenser (azeotropic distillation), separating funnel (liquid–liquid extraction), Erlenmeyer flasks, filter funnel, heat source (*e.g.* heating mantle, magnetic stirrer hotplate), rotary evaporator

Keywords Acetal formation, azeotropic distillation, Dean–Stark apparatus, evaporation under reduced pressure, influencing equilibria, IR

Background

Fructone exhibits – as the trivial name indicates – a fruity scent resembling apple aromas with mild woody notes. It is used as a base fragrance in the cosmetics industry. Consequently, the synthesis of fructone is the object of ongoing research, for example to optimise the synthetic methodology[1] and to find more efficient catalysts[2] in an effort to increase yields and to make the process more environmentally friendly.[3]

Comprehensive Organic Chemistry Experiments for the Laboratory Classroom
Edited by Carlos A. M. Afonso, Nuno R. Candeias, Dulce Pereira Simão, Alexandre F. Trindade, Jaime A. S. Coelho, Bin Tan, and Robert Franzén.

Published by the Royal Society of Chemistry, www.rsc.org

Fructone also acts as building block in syntheses of more complex molecules which involve a derivatisation of the ester moiety, *e.g.* selective reduction to an aldehyde,[4] reduction to an alcohol[5] followed by selective oxidation to an aldehyde.[6] These transformations necessitate the protection of the more reactive keto-carbonyl moiety of the readily available ethyl acetoacetate. One possible protection is the acetalisation of the carbonyl, more specifically the formation of a 1,3-dioxolane ring[7] using ethane-1,2-diol, which is another versatile reagent that is typically used as coolant in antifreeze formulations.

The synthesis of fructone from ethyl acetoacetate and ethylene glycol is an equilibrium under thermodynamic control. To shift this to the product side the by-product – water – is constantly removed from the reaction mixture. This is easily achieved with the help of a suitable solvent, one that forms an azeotrope[8] with water, and a Dean–Stark apparatus (Figure 4.1.2.1.1). In the case of a toluene–water mixture the composition of the azeotropic mixture boils at 85 °C (lower than both toluene and water, respectively) and consists of 80% toluene and 20% water. Once this vapour condenses and flows into the Dean–Stark trap the immiscible solvents separate, collecting the water (higher density) at the bottom and allowing only the toluene to overflow and return to the reaction mixture. Thus, as the water is constantly removed, the equilibrium shifts to the product side.

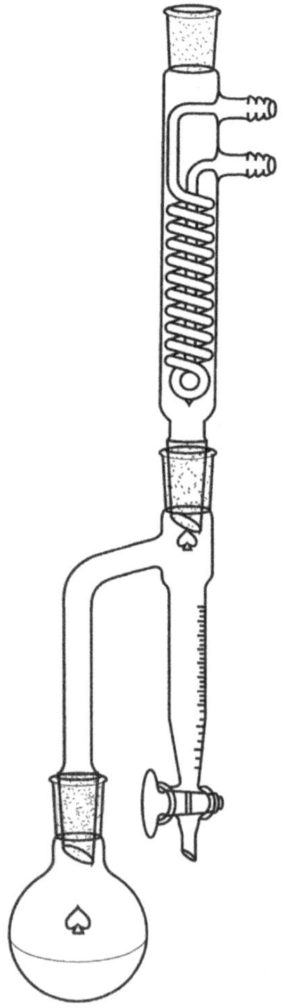

Figure 4.1.2.1.1 Dean–Stark apparatus in an azeotropic distillation.

The azeotropic removal of water using a Dean–Stark apparatus is a common feature in many reactions that are under thermodynamic control in an equilibrium, with water as a by-product and where the reagents and products can withstand elevated temperatures and have higher boiling points than the azeotrope, *e.g.* acetalisations, esterifications. The azeotropic removal of water is a useful technique also to remove residual water from compounds that have been synthesised differently but where traces of water remain in a crude product.

Additional Safety

All procedures should be performed in a well-ventilated fume cupboard, with proper safety equipment such as protective clothing, safety spectacles and gloves. All organic compounds and solvents are flammable and harmful/irritants. pTSA and NaOH (2 M) are corrosive and damaging to the eye.

Experimental Procedure

Reaction

1. In a fume cupboard set up a Dean–Stark apparatus (Figure 4.1.2.1.1) with a water-cooled reflux condenser and a suitable heat source.
2. Prepare a round-bottom flask (100 mL) and add anti-bumping granules or a magnetic stirrer bar. Add toluene (40 mL), ethane-1,2-diol (73.5 mmol, 4.11 mL, 1.05 equiv.), ethyl acetoacetate (70.0 mmol, 8.93 mL) and *p*-toluenesulfonic acid monohydrate (0.26 µmol, 50 mg).
3. Assemble the apparatus and heat the mixture under reflux for 45 min or until no more water is collected in the Dean–Stark trap.
4. Remove the heat source and allow the mixture to cool to room temperature.

Work-Up

5. Transfer the reaction mixture to a separating funnel and discard the contents of the Dean–Stark trap.
6. Wash the organic phase with sodium hydroxide solution (2 M, 20 mL), separate the phases and then discard the aqueous phase.
7. Wash the organic phase with brine (20 mL), separate the phases again and discard the aqueous phase.
8. Dry the organic phase with anhydrous magnesium sulfate, filter off the solids and collect the filtrate in a pre-weighed round-bottom flask (100 mL).
9. Evaporate the solvents in a rotary evaporator (bath temperature 80 °C, 10 mbar).

Analysis

10. Determine the mass of the product and calculate the yield.
11. Record an infrared spectrum of the product and of the starting material (ethyl acetoacetate).

Waste Disposal

The neutralised aqueous solutions – including dissolved $MgSO_4$ – may be diluted and discarded in the sink. Toluene may be recycled or collected as non-chlorinated waste. The product can be used in subsequent experiments (see Experiment 13.5, "Synthesis of allylic esters by reduction of fructone followed by Wittig olefination").

Results Interpretation and Additional Questions

1. Name the product – fructone – using IUPAC nomenclature rules.
2. Rationalise the work-up and isolation procedure and describe how and at which stage the excess reagent, by-product, catalyst and solvent were removed.
3. The equilibrium in this experiment was driven to the product side by selective removal of water, using an azeotropic mixture with toluene.
   a. Which other methods can you envisage in principle to achieve the same outcome?
   b. Assuming heptane and trichloromethane also form azeotropic mixtures with water, can these be used with the same Dean–Stark trap as for toluene? Explain your reasoning!
4. Formulate the mechanism of this reaction and explain the role of the Brønsted acid.
5. Interpret the IR spectra of the product and ethyl acetoacetate and discuss significant differences.

References

1. Y. Y. Wang, R. Wang, L. C. Wu and L. Y. Dai, *Chin. Chem. Lett.*, 2007, **18**, 24.
2. (a) S. Gao, X. Z. Liang, J. G. Yang and M. Y. He, *Sci. China, Ser. B: Chem.*, 2008, **51**, 646; (b) Y. Liu, Y.-T. Wang and D.-J. Tao, *RSC Adv.*, 2014, **4**, 22520.
3. B. Hu, C. Li, S.-X. Zhao, L.-M. Rong, S.-Q. Lv, X. Liang and C. Qi, *Molecules*, 2010, **15**, 5369.
4. D. G. Velazquez and R. Luque, *Tetrahedron Lett.*, 2011, **52**, 7004.
5. S. Raghavan and S. C. Joseph, *Tetrahedron Lett.*, 2003, **44**, 8237.
6. M. Nogawa, S. Sugawara, R. Lizuka, M. Shimojo, H. Ohta, M. Hatanaka and K. Matsumoto, *Tetrahedron*, 2006, **62**, 12071.
7. P. G. M. Wuts and T. W. Greene, *Greene's Protective Groups in Organic Synthesis*, John Wiley & Sons, Inc., Hoboken, 4th edn, 2007, p. 449.
8. J. Gmehling, J. Menke, J. Krafczuyk, K. Fischer, J.-C. Fontaine and H. V. Kehiaian, Azeotropic data for binary mixtures, in *CRC Handbook of Chemistry and Physics*, ed. W. M. Haynes, CRC Press/Taylor and Francis, Boca Raton, FL, 95th edn, 2014.

4.1.2.2. Synthesis of Flavone

Dulce Simão* and Ana C. Cerdeira

Department of Chemistry, Instituto Superior Técnico, University of Lisbon, Centro de Química Estrutural, Av. Rovisco Pais, 1, 1049-001 Lisbon, Portugal. *E-mail: dulcesimao@tecnico.ulisboa.pt

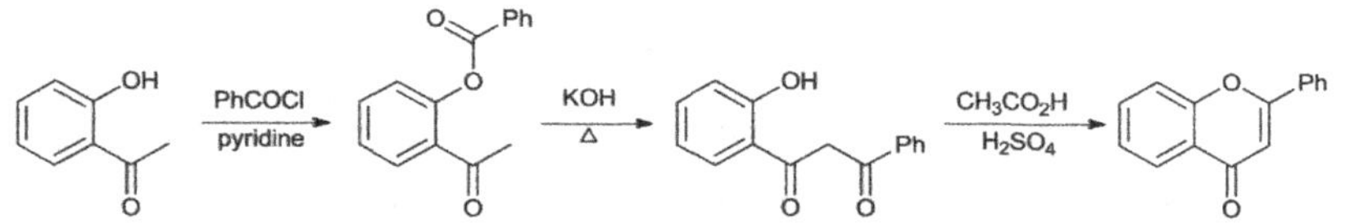

Number of sessions (duration of each session)	Hazard level	Difficulty level	Level of study
3 (3 h each)	Moderate	Medium	Intermediate

Class names Phenols, aryl ketones, acyl halides, esters, 1,3-diketones, flavones

Concepts involved Three-step synthesis of flavone involving nucleophilic substitution on the carbonyl group, base-catalyzed rearrangement and cyclization in acid media

Chemicals needed *o*-Hydroxyacetophenone, benzoyl chloride, dry pyridine, concentrated hydrochloric acid, methanol, anhydrous calcium chloride, potassium hydroxide, acetic acid, glacial acetic acid, concentrated sulfuric acid, petroleum ether (60–80 °C)

Equipment and experimental techniques involved Crystallization, NMR, reaction under anhydrous conditions, mechanical stirring, vacuum filtration, IR

Keywords Baker–Venkataraman rearrangement, cyclization, crystallization, enol, esterification, NMR

Background

Flavone (2-phenylchromen-4-one) belongs to a class of organic compounds known as flavonoids, occurring in nature as plant secondary metabolites. These compounds have attracted a great deal of attention in the past few decades due to their numerous potential health benefits[1–3] and applications in pharmaceutical and nutrition sciences. Although flavone is a colorless compound, the glycoside derivatives are usually colorful and responsible for the pigmentation in many flowers.[4] Most natural yellow dyes are also derivatives from flavone. This multi-step synthesis (known as the Kostanecki–Robinson synthesis of chromones[5]) starts with the reaction an acyl halide with an aromatic alcohol to afford an ester by nucleophilic substitution on the carbonyl group. Acyl chlorides are much more reactive toward nucleophilic addition–elimination than carboxylic acids; for that reason, the reaction occurs rapidly and does not require an acid catalyst. In this experiment[4,6,7] the reaction is performed using benzoyl chloride and *o*-hydroxyacetophenone to give *o*-benzoyloxyacetophenone. Pyridine is added to react as a base with the HCl that is generated, but it may also react as nucleophile with the acyl chloride to form an acylpyridinium ion, an intermediate that is even more reactive towards the nucleophile

Comprehensive Organic Chemistry Experiments for the Laboratory Classroom
Edited by Carlos A. M. Afonso, Nuno R. Candeias, Dulce Pereira Simão, Alexandre F. Trindade, Jaime A. S. Coelho, Bin Tan, and Robert Franzén.

Published by the Royal Society of Chemistry, www.rsc.org

than the acyl chloride is.[8,9] In the second step a base-catalyzed rearrangement occurs *via* enolate formation, also known as Baker–Venkataraman rearrangement, to afford a 1,3-diketone (*o*-hydroxydibenzoylmethane). In the third step a cyclization in acidic medium with subsequent dehydration affords the flavone.[5–10]

Additional Safety

This work should be carried out in a fume hood. *o*-Hydroxyacetophenone is an irritant. Pyridine is irritant to eyes and skin and highly flammable in the presence of open flames and sparks. It is harmful by inhalation, in contact with skin and if swallowed. Benzoyl chloride is corrosive and combustible. Concentrated sulfuric acid is a corrosive and poisonous liquid. Skin contact produce burns. Care should be taken while handling hydrochloric acid, acetic acid and potassium hydroxide solutions, which are also corrosive, and methanol and petroleum ether (60–80 °C) (flammable). The mechanical stirrer should be manipulated with precaution due to its rotational parts.

Experimental Procedure

Session 1 – Preparation of *o*-Benzoyloxyacetophenone

1. To a 250 mL round-bottom flask (fitted with a magnetic stirrer and a reflux condenser with a calcium chloride tube fitted on top) add, with stirring, *o*-hydroxyacetophenone (10 g, 9.0 mL, 0.073 mol), benzoyl chloride (14.4 g, 12 mL, 0.102 mol) and dry, redistilled pyridine (15 mL). The temperature of the reaction mixture will spontaneously rise.
2. Twenty minutes later, pour the mixture into a 1 L beaker containing crushed ice (150 g) and hydrochloric acid 1 M (350 mL) with vigorous stirring.
3. Collect the product by vacuum filtration and wash it first with cool methanol (15 mL) and then with water (15 mL).
4. Recrystallize the crude product from methanol. Record the yield, melting point and IR spectrum.

Session 2 – Preparation of *o*-Hydroxydibenzoylmethane

1. Dissolve *o*-benzoyloxyacetophenone (7 g, 0.029 mol) in dry pyridine (27 mL) in a 250 mL three-neck round-bottom flask, equipped with a mechanical stirrer, a thermometer, and a reflux condenser with a calcium chloride tube fitted on top.
2. Heat this mixture to 50 °C with constant stirring. Pulverize KOH (2.5 g, 0.045 mol) quickly in a mortar pre-heated to 100 °C and add to the reaction mixture.
3. Stir for 15 min until the potassium salt of the product precipitates (yellow).
4. Cool the mixture to room temperature and add 10% acetic acid solution (38 mL) with stirring.
5. Collect the light yellow product by vacuum filtration. No further purification is needed and the product can be used as it is on the next step.
6. Record the yield, melting point and IR spectrum.

Session 3 – Preparation of Flavone

1. In a 100 mL round-bottom flask dissolve *o*-hydroxydibenzoylmethane (4.5 g, 0.019 mol) in glacial acetic acid (25 mL) and add concentrated H_2SO_4 (1 mL) with stirring.
2. Fit the flask with a reflux condenser and heat this mixture in a boiling water bath for 1 h with stirring.

3. Pour the reaction mixture onto crushed ice (130 g) with stirring, and allow the ice to melt completely.
4. Collect the crude product by vacuum filtration and wash it with water until the mother liquor is no longer acidic.
5. Recrystallize the flavone from petroleum ether (60–80 °C).
6. Record the yield, melting point and IR and ^{1}H NMR spectra.

Waste Disposal

The mother liquor resulting from the filtration in the first step contains pyridine and therefore it should be poured into a flask reserved for hazardous acidic aqueous solutions. The acidic mother liquors from the remaining filtrations must be neutralized and diluted, then discarded in the sink. Methanol and petroleum ether used for recrystallizations can be recovered and stored for later purification.

Results Interpretation and Additional Questions

1. Determine the reaction yields.
2. Assess the purity of each compound obtained by comparing the melting points with the ones reported in the literature.
3. Assign the ^{1}H NMR spectrum for flavone and the characteristic absorption bands in the IR spectra for all products.
4. Write the mechanisms for all the reactions.
5. Explain why the first step should be performed free from moisture.

References

1. J.-K. Lin and M.-S. Weng, in *The Science of Flavonoids*, ed. E. Grotewold, Springer Science, New York, 1st edn, 2006, vol. 8, p. 213.
2. S. V. Bhat and B. A. Nagasampagi, *Chemistry of Natural Products*, Springer, 1st edn, 2006.
3. L. H. Yao, Y. M. Jiang, J. Shi, F. A. Tomás-Barberán, N. Datta, R. Singanusong and S. S. Chen, *Plant Foods Hum. Nutr.*, 2004, **59**, 113.
4. L. M. Harwood, C. J. Moody and J. M. Percy, *Experimental Organic Chemistry*, Blackwell Science, 2nd edn, 1999, p. 624.
5. D. T. Davies, *Aromatic Heterocyclic Chemistry*, Oxford Science Publications, 1992, p. 70.
6. A. I. Vogel, *Vogel's Textbook of Practical Organic Chemistry*, Longman Scientific and Technical, 5th edn, 1989, p. 1193.
7. C. A. M. Afonso, D. P. Simão, L. P. Ferreira, M. S. Serra and M. M. M. Raposo, *100 Experiências de Química Orgânica*, IST Press, 2011, p. 185.
8. T. W. G. Solomons, *Organic Chemistry*, John Wiley & Sons, Inc., New York, 7th edn, 1998, p. 830.
9. J. Clayden, N. Greeves and S. Warren, *Organic Chemistry*, Oxford University Press, 2001, p. 281.
10. T. L. Gilchrist, *Heterocyclic Chemistry*, Longman, 3rd edn, 1997, p. 183.

4.1.2.3. Synthesis and Characterisation of H_2salen: An Introduction to 1D and 2D NMR Spectroscopy

Maria J. Villa de Brito

DQB, CQE, Faculdade de Ciências da Universidade de Lisboa, Campo Grande, 1749-001 Lisboa, Portugal
*E-mail: mjbrito@ciencias.ulisboa.pt

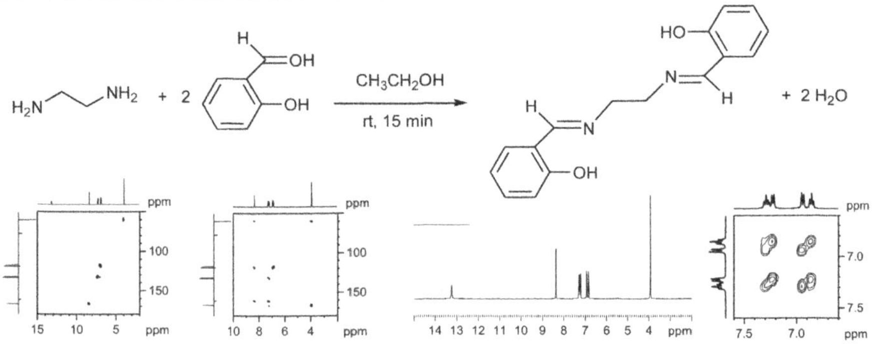

Number of sessions (duration of each session)	Hazard level	Difficulty level	Level of study
2 (2 h each)	Moderate	Low	Intermediate
Class names Amines, aldehydes, phenols, imines			
Concepts involved The main goal of this experiment is the NMR analysis of *N,N'*-bis(salicylidene)ethane-1,2-diamine (H_2salen), synthesised by condensation reactions of salicylaldehyde and ethylenediamine			
Chemicals needed Salicylaldehyde, ethylenediamine, ethanol, diethyl ether, deuterated chloroform, deuterium oxide			
Equipment and experimental techniques involved Magnetic stirring plate, suction filtration apparatus, NMR			
Keywords Condensation reaction, filtration, NMR, Schiff base, tetradentate ligand			

Background

NMR spectroscopy is recognised as one of the most important tools for structural characterisation. With the development of NMR instrumentation, high field FT-NMR with pulsed field gradient probes and gradient pulse sequences have made the use of 2D NMR techniques routine. This has led to the introduction of the basic concepts of NMR theory and the interpretation of ^{1}H and ^{13}C NMR spectra to undergraduate students in the second or third year of chemistry and biochemistry courses in many institutions. Even if students do not have the proper background to fully understand the theory behind the experiments, they can be trained to use 1D and 2D NMR data for solving structural problems.[1–4]

Comprehensive Organic Chemistry Experiments for the Laboratory Classroom
Edited by Carlos A. M. Afonso, Nuno R. Candeias, Dulce Pereira Simão, Alexandre F. Trindade, Jaime A. S. Coelho, Bin Tan, and Robert Franzén.

Published by the Royal Society of Chemistry, www.rsc.org

H_2salen is a widely known Schiff base, often prepared by undergraduate students in a synthesis laboratory course. It is very easily synthesised from inexpensive reagents (salicylaldehyde and ethylenediamine in ethanol) and can be used as a ligand to prepare transition metal complexes with various interesting applications. Several experiments for undergraduate laboratory courses can be found in the literature, involving the synthesis of the ligand, its coordination ability towards different metal centres and studies for different applications. The kinetic study of [Co(salen)] as an oxygen carrier,[5–7] the investigation of a manganese(III) salen complex as mimetic of the superoxide dismutase (SOD) enzymes[8] and the synthesis and use of Jacobsen's catalyst, a chiral salen-based Mn(III) complex,[9,10] in the enantioselective epoxidation of olefins are examples of such experiments that involve further uses of H_2salen or similar compounds as ligand, appropriate for inorganic, bioinorganic, organic or integrated laboratory courses.[11]

This compound gives well-resolved spectra, which can be successfully used for NMR spectroscopic studies. Students can confirm the synthesis of the product by its ^{1}H NMR spectrum, through analysis of the number of resonances, the integrations, the splitting patterns and several effects on the chemical shift, such as hybridisation, electronegativity and H-bonding. However, the complete assignment of all the proton and carbon atoms requires the combination of standard ^{1}H and ^{13}C spectra with 2D NMR techniques.

An NMR spectrum provides a valuable type of information on the basis of the number of resonances observed, which corresponds to the number of chemically distinct types of the nucleus being studied. All the proton or carbon nuclei found in chemically identical environments within a molecule are said to be chemically equivalent and exhibit the same chemical shift in the corresponding NMR spectrum. This most commonly results from molecular symmetry or free rotation about single bonds. If the total number of carbons or protons in a molecule is larger than the number of signals observed in its ^{13}C or ^{1}H NMR spectrum this can suggest that the molecule has some type of symmetry, usually a plane or an axis. Hence, molecular symmetry considerations can often simplify the interpretation of NMR spectra and the structural elucidation.

A routine ^{1}H NMR spectrum provides structure-related information for the protons of the sample, such as the chemical shifts (δ, measured in ppm) of the individual resonances, which reveal the local chemical environment of the corresponding protons, the splitting patterns of the signals and the spin–spin coupling constants (J, measured in Hz) reflecting the interaction through chemical bonds between coupled protons, and the resonance intensities (integrals) which correspond to the relative number of protons giving rise to each resonance. Broad signals can indicate proton exchange or other kinds of dynamic behaviour of the molecule. As signal overlap is common in ^{1}H NMR spectra, it is often difficult to recognise the distinct resonances.

The APT-^{13}C NMR (attached proton test) technique is used to get chemical shift and multiplicity information. It yields a 1D ^{13}C NMR proton-decoupled spectrum in which CH and CH_3 signals are displayed with opposite sign to that of CH_2 and quaternary carbons, hence providing greater information content than a standard ^{13}C NMR experiment. Since couplings to protons are removed, all resonances appear as singlets (except for those carbon atoms coupled to other nuclei such as ^{2}H, ^{19}F or ^{31}P) and peak overlap is rare in ^{13}C spectra. In contrast to proton spectra, integral data are not relevant because, under standard ^{13}C NMR acquisition conditions, peak intensities do not accurately reflect the number of carbon atoms giving rise to each resonance.

Two-dimensional (2D) NMR experiments involve methods that generate two frequency axis: F_2 (x-axis) and F_1 (y-axis). The resulting 2D spectra consist of a two-dimensional contour map plotted as a function of the two frequencies (chemical shifts) in which the contour lines represent the signal intensity (similarly to the way they represent mountain heights in topographical maps). Homonuclear 2D experiments are generally used to correlate spins of high abundance nuclides (most often ^{1}H), while heteronuclear techniques generate 2D correlation spectra between two different nuclei, typically ^{1}H and a less sensitive nucleus such as ^{13}C, ^{15}N, *etc.* In contrast to the homonuclear 2D spectra, heteronuclear shift correlation experiments lack a diagonal, as different nuclides are represented in the two dimensions. Heteronuclear methods currently employ ^{1}H detection for maximum sensitivity, and thus are commonly referred as "inverse experiments". The ^{1}H NMR spectrum (the observed nucleus) lies along the F_2 (x) axis and the ^{13}C NMR spectrum is displayed along the F_1 (y) axis.

^{1}H–^{1}H COSY (correlation spectroscopy) is a 2D experiment that provides a means to identify mutually coupled protons over two or three bonds ($^2J_{HH}$ and $^3J_{HH}$, respectively); it is the most widely used 2D experiment. It yields a 2D NMR plot with the ^{1}H NMR spectrum represented on both axes as external projection. The diagonal peaks (from bottom left to top right of the spectrum) correspond to those observed in the 1D ^{1}H NMR normal spectrum and contain no additional information. The relevant peaks are those off-diagonal, so-called cross-peaks. Each of these is specified by the two frequency coordinates and represents spin–spin or J-coupling between the correlated protons. The COSY spectrum is symmetrical about the diagonal with the cross-peaks on either side of it, mapping the same interaction.

^{1}H–^{13}C HSQC (heteronuclear single quantum coherence) is a 2D NMR method, which correlates the chemical shifts of protons with their directly attached ^{13}C atoms using $^1J_{CH}$ couplings. The 2D spectrum provides a well-defined plot of connectivity peaks, correlating each ^{1}H signal to its directly bonded ^{13}C resonance. Identical correlations (although not as well resolved) can be obtained from the 2D HMQC (heteronuclear multiple quantum coherence) experiment.

^{1}H–^{13}C HMBC (heteronuclear multiple bond correlation) is a long range or multiple bond correlation technique, which was developed to detect ^{1}H–^{13}C couplings across two or three bonds *via* $^2J_{CH}$ and $^3J_{CH}$ (which cannot be differentiated in this experiment). It has the ability to detect correlations to quaternary carbons and across heteroatoms. The 2D HMBC spectrum resembles that of HSQC or HMQC but with a more crowded appearance, due to the larger number of long range correlations associated with each proton and carbon chemical shifts. Despite some drawbacks (sometimes $^1J_{CH}$ couplings are not suppressed, some expected correlations may be missing and, in some cases, $^4J_{CH}$ correlations may be observed) this method is a powerful tool for linking together structural fragments and is typically used to obtain the final answers with respect to a structural problem.

The motivation for the experiment given here was to integrate NMR spectroscopy into the laboratory experience to teach 1D and 2D techniques. This experiment describes the synthesis of H_2salen by slight modifications of published methods[7,8,12] and its characterisation, which will focus on the ^{1}H/^{13}C NMR spectral methods usually first measured on a routine basis for the structural elucidation of organic compounds: ^{1}H, APT-^{13}C, COSY, HSQC and HMBC. A detailed explanation of these topics can be found in spectroscopy textbooks.[13]

Additional Safety

The synthesis session should be carried out in a fume hood. Standard procedures for safe handling of chemicals should be followed. Students should wear a laboratory coat, safety goggles and gloves. Salicylaldehyde, ethylenediamine and ethanol are irritating and harmful if swallowed, inhaled or absorbed through the skin. Ethanol is highly flammable. Ethylenediamine is highly toxic and also flammable.

Experimental Procedure

Laboratory Session 1 (2 h): Preparation of *N*,*N*′-bis(salicylidene)ethane-1,2-diamine, H_2salen

1. In a fume hood place 30 mL of 95% ethanol in a 100 mL Erlenmeyer flask equipped with a magnetic stirring bar. Add 3.2 mL (30 mmol, 2 equiv.) of salicylaldehyde and put the flask on a magnetic stirrer at room temperature. While stirring, add 1.0 mL (15 mmol, 1 equiv.) of ethylenediamine.

2. After formation of a yellow solid (it usually occurs immediately), stir the reaction mixture for an additional 10 min and cool the flask in an ice bath.
3. Collect the bright yellow flaky crystals by suction filtration and wash them twice with cold ethanol (2 × 10 mL) followed by diethyl ether (2 × 10 mL). Dry the crystals under vacuum until they have warmed to room temperature and then in a desiccator for 30 min to 1 h.
4. Weigh the product and calculate the yield. Record the infrared spectrum as KBr pellet and measure the melting point.
5. Record the following set of NMR spectra of your product in $CDCl_3$ solution: ^1H, APT-^{13}C, ^1H–^1H COSY, ^1H–^{13}C HSQC and HMBC. To do this, place approximately 15 mg of your compound and 500 μL of $CDCl_3$ in a 5 mm diameter NMR tube. The solution should fill the lower 4 cm of the NMR tube.
6. After acquisition of the previous set of spectra, add one or two drops of D_2O to the NMR tube, shake the tube well and record a new ^1H NMR spectrum before the next session (optional).

Laboratory Session 2 (2 h): Processing of NMR Data and Analysis of Spectra (Classroom Activity)

1. Process the NMR data and make expansions when necessary (*e.g.* proton resonances for coupling constant analysis or crowded regions on 2D spectra).
2. Analyse the set of NMR spectra and assign all the proton and carbon atoms.

Waste Disposal

Ethanol may be collected in a non-chlorinated waste collector recipient for posterior disposal or distilled from the filtrate diluted solution for reuse.

Results Interpretation and Additional Questions

1. Describe the mechanism of formation of H_2salen, calculate the percent yield and the atom economy[14] for the process.
2. Interpret the infrared spectrum and compare the melting point with that given in the literature.
3. Which proton atoms could you unequivocally assign by analysis of the ^1H NMR spectrum only? Explain your reasoning.
4. How does the ^1H NMR spectrum of H_2salen change upon addition of D_2O? Why are only some of the peaks affected by D_2O?
5. How did you assign the aromatic protons and carbons using 2D spectra?
6. How does H_2salen coordinate to a transition metal ion? Draw the structure of [Co(salen)].

References

1. D. E. Alonso and S. E. Warren, *J. Chem. Educ.*, 2005, **82**, 1385.
2. S. Berger and D. Sicker, *Classics in Spectroscopy*, Wiley-VCH Verlag GmbH, Weinheim, 2009.
3. M. Findeisen and S. Berger, *50 and More Essential NMR Experiments*, Wiley-VCH Verlag GmbH, Weinheim, 2014.
4. K. A. Kalstabakken and A. M. Harned, *J. Chem. Educ.*, 2013, **90**, 943.
5. D. J. Aymes and M. R. Paris, *J. Chem. Educ.*, 1989, **66**, 854.
6. T. G. Appleton, *J. Chem. Educ.*, 1977, **54**, 443.
7. Z. Szafran, R. M. Pike and M. M. Singh, *Microscale Inorganic Chemistry – A Comprehensive Laboratory Experience*, John Wiley & Sons, Inc., New York, 1991, ch. 8, p. 252.
8. G. Vecchio and V. Lanza, *J. Chem. Educ.*, 2009, **86**, 1419.
9. J. Hanson, *J. Chem. Educ.*, 2001, **78**, 1266.

10. J. F. Larrow and E. N. Jacobsen, *Org. Synth.*, 1998, **75**, 1.
11. D. B. Ball and R. Miller, *J. Chem. Educ.*, 2002, **79**, 665.
12. W. A. Herrmann and C. Zybill, in *Synthetic Methods of Organometallic and Inorganic Chemistry*, ed. W. A. Herrmann and A. Salzer, Georg Thieme Verlag, Stuttgart, 1996, ch. 3, vol. 1, p. 81.
13. T. D. W. Claridge, *High Resolution NMR Techniques in Organic Chemistry*, Elsevier Ltd., Oxford, 1st edn, 1999, 2nd imp., 2004; H. Friebolin, *Basic One- and Two-Dimensional NMR Spectroscopy*, Wiley-VCH Verlag GmbH, Weinheim, 4th edn, 2005; D. Pavia, G. Lampman and G. S. Kriz, *Introduction to Spectroscopy*, Brooks/Cole, 3rd edn, 2001.
14. K. M. Doxsee and J. E. Hutchison, *Green Organic Chemistry – Strategies, Tools and Laboratory Experiments*, Thomson Brooks/Cole, USA, 2004, ch. 8, p. 91; B. M. Trost, *Science*, 1991, **254**, 1471.

4.1.2.4. Synthesis of Lophine and Conversion into Dimers

Dulce Simão

Department of Chemistry, Instituto Superior Técnico, University of Lisbon, Centro de Química Estrutural, Av. Rovisco Pais, 1, 1049-001 Lisbon, Portugal. *E-mail: dulcesimao@tecnico.ulisboa.pt

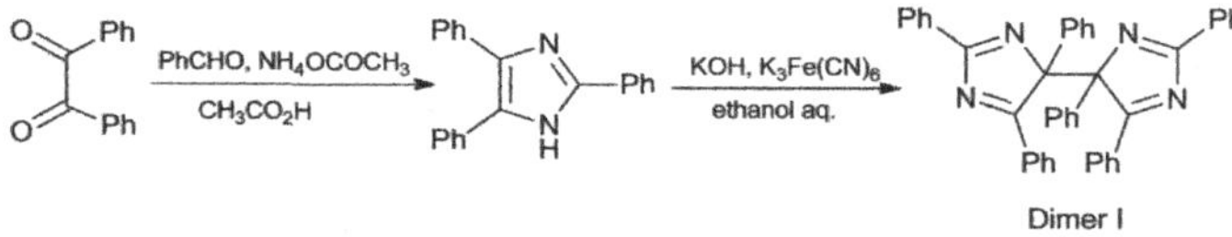

Number of sessions (duration of each session)	Hazard level	Difficulty level	Level of study
2 (3 h each)	Moderate	Low	Intermediate

Class names Aldehydes, 1,2-diketones, and imidazoles

Concepts involved This experiment describes the synthesis of lophine involving a cyclocondensation and tautomerisation. Conversion into dimers showing photochromism, piezochromism and thermochromism with free radical formation

Chemicals needed Benzaldehyde, benzil, ammonium acetate, glacial acetic acid, ethanol, ammonium hydroxide (30% NH_3), potassium ferricyanide, potassium hydroxide, petroleum ether (40–60 °C), toluene, deuterated dimethyl sulfoxide

Equipment and experimental techniques involved Crystallization, IR, NMR, stirring heating plate, vacuum filtration

Keywords Cyclocondensation, free radicals, IR, photochromism, piezochromism, recrystallization, tautomerism, thermochromism

Background

Imidazoles are heterocyclic aromatic compounds. Several derivatives containing the imidazole ring can be found in nature. They have a large industrial application because they are important intermediates in the synthesis of pharmacological and pesticides compounds. One of the best known and useful routes to prepare these compounds is the condensation of 1,2-dicarbonyl compounds with ammonium acetate and an aldehyde.[1] This reaction occurs through a cyclocondensation followed by an irreversible tautomerization.[2] In this experiment, the imidazole lophine is prepared by the previously mentioned method.[3] The three carbon atoms in the imidazole ring are the carbon atoms of the carbonyl groups of benzyl and benzaldehyde. Nitrogen atoms come from ammonia, which is in equilibrium with the ammonium ion from the salt in solution with acetic acid. Further oxidation with potassium ferricyanide generates a lophine dimer (dimer I) that changes color when pressure is applied (piezochromism[4]). In solution it

Comprehensive Organic Chemistry Experiments for the Laboratory Classroom
Edited by Carlos A. M. Afonso, Nuno R. Candeias, Dulce Pereira Simão, Alexandre F. Trindade, Jaime A. S. Coelho, Bin Tan, and Robert Franzén.

Published by the Royal Society of Chemistry, www.rsc.org

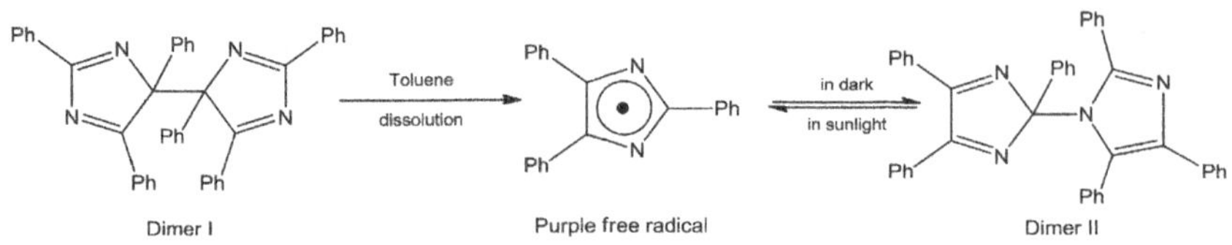

Scheme 4.1.2.4.1 Dimers and free radical formation.

changes color with temperature (thermochromism). This dimer equilibrates with the purple colored free radical (triphenylimidazyl radical) when it is ground in a mortar or when a toluene solution of this dimer is exposed to sunlight or heated. In the dark, radical recombination occurs, producing a photochromic (reversible change of color with light) new colorless dimer (dimer II) (Scheme 4.1.2.4.1). Photolysis of this latter allows a new equilibration with the purple free radical. The photolysis recombination cycle can be run almost infinitely.[5–7]

Additional Safety

This work must be carried out in a fume hood. Benzaldehyde is toxic, benzil and lophine are irritant, ammonium hydroxide and potassium hydroxide are corrosive. Petroleum ether and ethanol, flammables, should be handled with care.

Experimental Procedure

Session 1

Preparation of Lophine

1. In a 250 mL round-bottomed flask with magnetic stirring, dissolve benzil (5.25 g, 25 mmol), benzaldehyde (2.5 mL) and ammonium acetate (10 g, 130 mmol) in acetic acid (100 mL).
2. Heat the mixture to reflux in an oil bath for 1 h with stirring.
3. Cool to room temperature and filter to remove any precipitate that may be present.
4. Add water (300 mL) to the filtrate and collect the solid by vacuum filtration.
5. Neutralize the filtrate with ammonium hydroxide and collect the additional solid again.
6. Recrystallize the product from aqueous ethanol.
7. Record the yield, melting point, and IR and ^{1}H NMR spectra.

Session 2

Preparation of Lophine Dimer I

1. In a 250 mL two-necked flask fitted with a reflux condenser and with magnetic stirring, dissolve potassium hydroxide (12 g), in ethanol (100 mL).
2. Add lophine (1.5 g, 5 mmol) and stir until complete dissolution.
3. Add the mixture to a 1 L beaker containing a large magnetic stirrer bar and cool to 5 °C in an ice salt-water bath with stirring.
4. Prepare a solution of potassium ferricyanide (4.5 g) in water (450 mL) and place it into the addition funnel.
5. Add this solution dropwise to the vigorously stirred ethanolic mixture at such a rate that the temperature remains below 10 °C. An initial violet coloration replaced by a light gray precipitate should be observed.
6. After addition is complete, filter off the precipitate with suction.

7. Wash with water (5 × 50 mL) followed by a rapid wash with cold 50% ethanol (10 mL) and finally with petroleum ether (40–60 °C) (2 × 25 mL).
8. Dry, record the yield and store in the dark.

Piezochromism and Photochromism Observation (Formation of Lophine Dimer II)

9. Place a small quantity of lophine dimer in a mortar, grind it and observe the color change.
10. Dissolve the dimer (20 mg) in toluene (30 mL), seal and store the mixture in the dark.
11. After the solution has become colorless or nearly so, stand the solution in strong sunlight.
12. Return the solution to the dark and observe it after an hour.

Waste Disposal

The neutralized aqueous solutions may be diluted and discarded in the sink. Ethanol and petroleum ether may be stored to further purification.

Results Interpretation and Additional Questions

1. Determine the reaction yields.
2. Compare the melting points of lophine and dimer I with those reported in the literature.
3. Interpret the IR and NMR spectra.
4. Present the overall mechanism of formation as well as of dimerization of lophine.
5. On the basis of the results described, show a qualitative energetic profile (free energy *versus* reaction coordinate) relating the free radical species to dimer I and dimer II.

References

1. D. T. Davies, *Aromatic Heterocyclic Chemistry*, Oxford Science publications, Oxford Chemistry Primers, 1992, pp. 22–23.
2. S. Ege, *Organic Chemistry*, D. C. Heath and Company, 1984, pp. 1021–1022.
3. L. M. Harwood, C. J. Moody and J. M. Percy, *Experimental Organic Chemistry*, Blackwell Science, 2nd edn, 1999, pp. 644–646.
4. D. M. White and J. Sonnenberg, *J. Am. Chem. Soc.*, 1966, 3825–3829.
5. M. Pickering, *J. Chem. Educ.*, 1980, **57**, 833–834.
6. L. A. Cescon, G. R. Coraor, R. Dessauer, E. F. Silversmith and E. J. Urban, Pickering, , *J. Org. Chem.*, 1971, **36**, 2262–2267.
7. S. S. Sathe, D. Ahn and T. F. Scott, *Ind. Chem. Res.*, 2015, **54**, 4203–4212.

4.1.2.5. Synthesis of Dibenzalacetone 2,4-Dinitrophenylhydrazone

Ana Margarida Madureira, Ana Bela Santana, Emília Valente, and Maria-José U. Ferreira*

Faculty of Pharmacy, University of Lisbon, Av. Prof. Gama Pinto, 1649-003 Lisbon, Portugal
*E-mail: mjuferreira@ff.ulisboa.pt

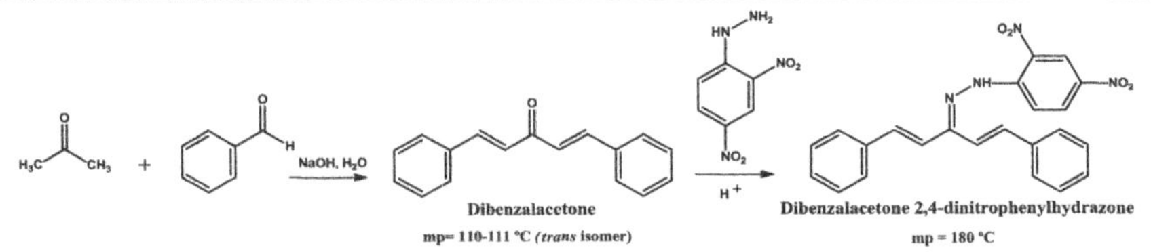

Number of sessions (duration of each session)	Hazard level	Difficulty level	Level of study
2 (4 h each)	High	Medium	Intermediate
Class names Ketones, aldehydes, α,β-unsaturated ketones, hydrazines, hydrazones			
Concepts involved This experiment is an example of a condensation reaction between a ketone (dibenzalacetone) and an amine derivative (2,4-dinitrophenylhydrazine) leading to the formation of an imine (C=N), called a hydrazone. The experiment involves an initial session for synthetizing dibenzalacetone, which results from the reaction of a ketone (acetone) and an aldehyde (benzaldehyde). In this reaction, firstly an α-hydroxyketone is formed that, by a base-catalysed dehydration and a second aldol condensation, is transformed into dibenzalacetone			
Chemicals needed Benzaldehyde, acetone *p.a.*, dibenzalacetone (optional), 2,4-dinitrophenylhydrazine, ethanol (95%), sodium hydroxide solution (10% water solution), glacial acetic acid solution (5% in 95% ethanol), sulfuric acid, deionized water, dichloromethane, deuterated chloroform			
Equipment and experimental techniques involved Thin-layer chromatography, recrystallization, melting point, IR, ^{1}H and ^{13}C NMR			
Keywords Condensation reaction, hydrazone, IR, NMR, precipitation, recrystallization			

Background

The proposed experiment involves a condensation reaction between dibenzalacetone and 2,4-dinitrophenylhydrazine with the formation of an imine – a 2,4-dinitrophenylhydrazone of dibenzalacetone.[1–3]

Hydrazones are a versatile class of nitrogen–nitrogen bond containing compounds for drug design. Often highly crystalline and with sharp melting points, they are characterized by a C–N double bond conjugated with the second nitrogen atom. These structural features are responsible for their physical and chemical properties, namely their higher stability when compared to simple

Comprehensive Organic Chemistry Experiments for the Laboratory Classroom
Edited by Carlos A. M. Afonso, Nuno R. Candeias, Dulce Pereira Simão, Alexandre F. Trindade, Jaime A. S. Coelho, Bin Tan, and Robert Franzén.

Published by the Royal Society of Chemistry, www.rsc.org

imines. Hydrazones have shown relevant biological properties, such as antimicrobial, anticonvulsant, analgesic, anti-inflammatory, antiplatelet, antitubercular and antitumoral activities.[4]

The experiment can be carried out with commercial dibenzalacetone or sequentially to the dibenzalacetone synthesis performed by the same students group in a previous class, and thus represents a suitable way to further confirm the identification of dibenzalacetone based on the melting point of the hydrazone.

The synthesis of dibenzalacetone is also important for learning one of the standard organic reactions – the crossed aldol condensation of aldehydes and ketones, which is extensively used to form C–C bonds and thus to synthesize more complex molecules.[1–3] The experiment involves the application of reactions at both the carbonyl group and the α-carbon of aldehydes and ketones, previously learnt by the students.

In the synthesis of dibenzalacetone, the acetone is used as the enolate source, which reacts with the benzaldehyde. The aldol product readily undergoes a base-catalysed dehydration because of the stability of the resulting product (benzalacetone) that bears an extended conjugated system due to the formation of a new double bond between the carbonyl and to the phenyl group. Benzalacetone, still with α-hydrogens, in the presence of an excess of benzaldehyde undergoes a second aldol condensation to yield the final product, dibenzalacetone.

Synthesis of the dibenzalacetone 2,4-dinitrophenylhydrazone starts by the nucleophilic attack of the terminal amino group from 2,4-dinitrophenylhydrazine to the carbonyl group of the ketone, dibenzalacetone, with formation of the hemiaminal. This intermediate dehydrates quickly and originates a red coloured precipitate insoluble in the alcoholic solvent. The colour is related to the extension of the conjugated system in the resulting hydrazone.

This work has been carried out successfully by second-year undergraduate students on the Integrated Master Degree in Pharmaceutical Sciences at the Faculty of Pharmacy, University of Lisbon, within an intermediate Organic Chemistry course after a basic Organic Chemistry course in the first year in which the basic mechanisms and laboratorial techniques were studied.

The main steps of both reactions are presented in Scheme 4.1.2.5.1.

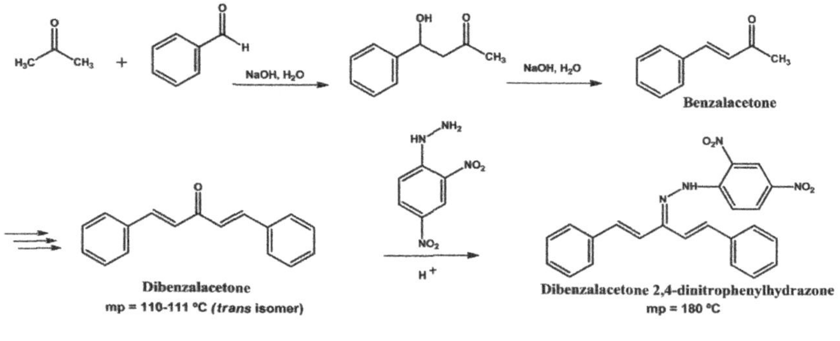

Scheme 4.1.2.5.1 Main steps of dibenzalacetone 2,4-dinitrophenylhydrazone synthesis.

Additional Safety

Session 1: benzaldehyde is reported to be skin and eye irritant. Avoid contact with skin, eyes and clothing. Acetone is volatile, flammable and irritant to respiratory tract and may be harmful if inhaled. Sodium hydroxide is caustic and causes severe burns when in contact with skin and mucous. The addition of the reactants must be done in a fume hood and the use of gloves, laboratory security goggles and long-sleeve laboratory coat during all operations is mandatory.

Session 2: the experiment involves the use of a previously prepared solution of 2,4-dinitrophenylhydrazine in sulfuric acid.† 2,4-Dinitrophenylhydrazine is sensitive to shock and friction so it must be handled with care. Sulfuric acid is a very hazardous chemical product when in contact with skin (corrosive, irritant, permeator), eye (irritant, corrosive), ingestion, or inhalation. It may produce tissue damage particularly on the mucous membranes of eyes, mouth and respiratory tract. Skin contact may produce burns. Therefore, the use of gloves, laboratory security goggles and long-sleeve laboratory coat during all operations is mandatory. The manipulation of 2,4-dinitrophenylhydrazine acid solution must be done in a fume hood. The solution should be placed in a burette, in the fume cupboard, from where the students measure the volume to add to the Erlenmeyer flask with their reaction mixture. Make sure a beaker is placed under the burette, when not in use, to avoid acid solution drops on the fume cupboard.

Experimental Procedure

Session 1 Dibenzalacetone Synthesis and Identification (4 h)

1.A – Synthesis of Dibenzalacetone (ref. 5)

1. In a 125 mL Erlenmeyer flask, with ground-glass jointware, add a magnetic stirring bar and 3.0 mL (3.2 g; 0.030 mol) of freshly distilled benzaldehyde.
2. Add 1.0 mL (0.80 g; 0.014 mol) of acetone plus 25 mL of 95% ethanol.
3. Gently shake the reaction mixture and place the solution under magnetic stirring.
4. Add slowly 30 mL of the sodium hydroxide solution (10%).
5. Keep stirring the reaction mixture in the capped Erlenmeyer for 30 min.
6. Observe your solution carefully, and note that the initial colourless solution turns yellow with a solid precipitate.
7. After 30 min cool the reaction mixture on ice.
8. Collect the yellow crude solid product by vacuum filtration. Clean the Erlenmeyer flask with iced-cold water.
9. Wash the solid precipitate with iced-cold water (three portions of about 40 mL).
10. With the vacuum on, press the solid until no more solvent drops fall from the Büchner funnel.
11. With the vacuum off, break up the solid residue with a spatula.
12. Add 5.5 mL of an iced 5% glacial acetic acid solution in ethanol (95%) to the Büchner funnel, and leave the solution in contact with the residue for 1 min.
13. Turn on the vacuum and remove the solution.
14. Repeat the last two steps (steps 12 and 13).
15. Transfer the yellow solid to a filter paper and press the wet solid to dry the precipitate.

1.B – Dibenzalacetone Purification

16. Transfer the dried crude solid product to a 100 mL Erlenmeyer flask.
17. Crystallize the crude product from ethanol (95%).
18. Observe the formation of bright yellow needles.
19. Collect the crystals by suction filtration.
20. Air-dry the crystals.
21. Determine the weight of the crystals and the product yield.

---

†Preparation of 2,4-dinitrophenylhydrazine reagent: dissolve 8 g of 2,4-dinitrophenylhydrazine in 40 mL of concentrated sulfuric acid. Add this solution, with stirring, to a mixture of 60 mL of water and 200 mL of ethanol.

1.C – Dibenzalacetone Identification

From the purified dibenzalacetone crystals:

22. TLC:
    (a) Prepare an acetone solution of the synthesized dibenzalacetone and of a standard sample.
    (b) Apply the two solutions on a TLC plate with capillary tubes.
    (c) Place the TLC plate on the TLC chamber previously saturated with the mobile phase (dichloromethane).
    (d) After development of the chromatogram and evaporation of the mobile phase, TLC layers are observed under UV light at 254 nm. The spots are marked and the purity of the synthesized dibenzalacetone can be determined by visual evaluation.
23. Determine the *melting point* and compare the result with the data available in the literature.
24. *IR spectrum*: prepare a KBr pellet (1 mg product: 100 mg KBr). Obtain the IR spectrum with the appropriated apparatus. Analyse the absorption bands.
25. *NMR spectra*: place a small amount of crystals (5–10 mg) on a NMR tube. Add 600 μL of deuterated chloroform ($CDCl_3$). Trace the proton and the carbon NMR spectra on the NMR apparatus. Analyse the signals.

Session 2

2.A – Hydrazone Synthesis (ref. 5)

1. In a 50 mL Erlenmeyer flask with ground-glass jointware dissolve 300 mg of dibenzalacetone in 6 mL of ethanol. If necessary warm the mixture in a water bath bellow 60 °C.
2. Add carefully 9 mL of the 2,4-dinitrophenylhydrazine solution (previously prepared†) from a burette placed in the fume hood. Swirl gently your reaction mixture.
3. A red product precipitate will be formed. Allow the caped Erlenmeyer to stand on your bench for 10 min to complete the product precipitation.
4. Collect the solid product by vacuum filtration. Clean the Erlenmeyer flask with iced-cold ethanol.
5. Wash thoroughly the solid product four times: turn the vacuum off, cover the crystals with 9 mL of cold ethanol, allow the contact and turn on the vacuum to eliminate the liquid.
6. Press the solid under vacuum until no more solvent drops from the Büchner funnel.
7. Place the solid crude product in a pre-weighed watch glass and let it to dry in an oven (45 °C) until constant weight (note in your laboratory notebook all the weights registered) and calculate the yield.

2.B – Hydrazone Identification

8. Thin-layer chromatography:
    (a) Prepare an ethyl acetate solution of the synthesized hydrazone and of the reagent (dibenzalacetone).
    (b) Apply the two solutions on a TLC plate with a capillary tube.
    (c) Place the TLC plate in the TLC chamber previously saturated with the mobile phase (dichloromethane).
    (d) After development of the chromatogram and evaporation of the mobile phase, TLC layers are observed under UV light at 254 nm. The spots are marked and the purity of the synthesized hydrazone can be determined by visual evaluation.
9. Determine the melting point of the synthesized hydrazone.

10. IR spectrum: prepare a KBr pellet (1 mg hydrazone: 100 mg KBr). Obtain the spectrum with the appropriated equipment. Analyse the absorption bands.
11. NMR spectra: place a small amount of crystals (5–10 mg) on an NMR tube. Add 600 μL of deuterated chloroform ($CDCl_3$). Trace the proton and the carbon NMR spectra with the NMR equipment. Analyse the signals.

Waste Disposal

Session 1: the aqueous ethanolic alkaline filtrate from the reaction mixture must be collected in an alkaline aqueous solutions container. The aqueous ethanolic acid filtrate from the product washing should be disposed of in an acid aqueous solution container. Dibenzalacetone crystals should be collected in a container labelled "Dibenzalacetone (purified)". All the material used must be washed with water and put in an appropriate container.

Session 2: 2,4-dinitrophenylhydrazine is an irritant and stains the skin. The reagent solution also contains sulfuric acid so the remains of acid hydrazine solution and the aqueous ethanolic acid filtrate from the reaction mixture and washing must be put in an acid aqueous solutions container. All the material used must be washed with water and put in an appropriate container.

Results Interpretation and Additional Questions

1. Propose the mechanisms for the syntheses performed in both sessions.
2. What is the role of the sodium hydroxide solution used in the reaction of session 1?
3. What are the side products that may be formed along with dibenzalacetone? Which conditions favour their synthesis?
4. Why is the product of session 2 so much more colourful than the initial hydrazine reagent?
5. Assign the proton and the carbon NMR spectra of dibenzalacetone and the hydrazone. Compare your IR spectra of both products.
6. Compare the melting points obtained with those described in the literature. Observe your TLC plates. Discuss your results.

References

1. K. P. C. Vollhardt and N. E. Schore, *Organic Chemistry: Structure and Function*, Freeman and Company, New York, USA, 6th edn, 2010, ch. 18, pp. 827–847.
2. J. Clayden, N. Greeves and S. Warren, *Organic Chemistry*, Oxford University Press Inc., New York, USA, 1st edn, 2001, ch. 27, pp. 689–695.
3. T. W. G. Solomons and C. B. Fryhle, *Organic Chemistry*, John Wiley & Sons, Inc., Hoboken, 10th edn, 2009, ch. 19, pp. 869–889.
4. R. Sevim and K. S. Guniz, *Molecules*, 2007, **12**, 1910.
5. K. L. Williamson, *Macroscale and Microscale Organic Experiments*, Houghton Mifflin Company, Boston, USA, 4th edn, 2003, pp. 454–460 and 770.

4.1.2.6. Synthesis and Structural Characterization of an Antitubercular Isoniazid Hydrazone

Susana Santos* and Filomena Martins

Centro de Química e Bioquímica, Departamento de Química e Bioquímica, Faculdade de Ciências, Universidade de Lisboa, Ed. C8, Campo Grande, 1749-016 Lisboa, Portugal
*E-mail: smsantos@ciencias.ulisboa.pt

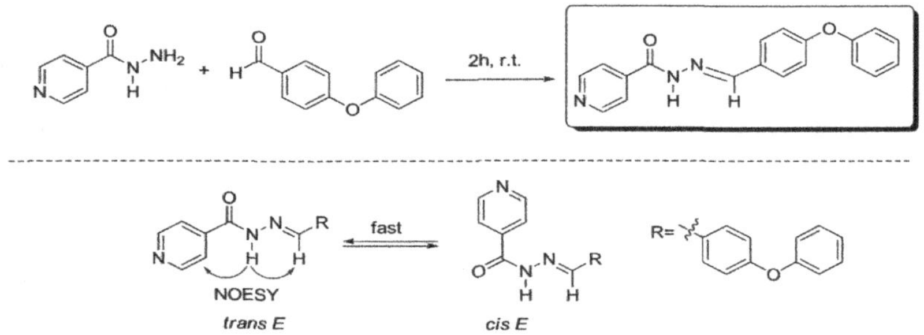

Number of sessions (duration of each session)	Hazard level	Difficulty level	Level of study
2 (3 h each)	Moderate	Medium	Intermediate
Class names Aldehydes, hydrazides, *N*-acylhydrazones			
Concepts involved This experiment illustrates the synthesis of a special kind of imine, an *N*-acylhydrazone, by condensation between a primary amine (isonicotinohydrazide) and 4-phenoxybenzaldehyde. The stereochemical outcome of the reaction is explored using NMR tools			
Chemicals needed Isoniazid, 4-phenoxybenzaldehyde, ethyl acetate, ethanol, acetone, TLC silica gel plates 60 F254 aluminium sheets, silica gel 60 (0.040–0.063 mm) for column chromatography, deuterated DMSO			
Equipment and experimental techniques involved Apparatus for column chromatography and recrystallization, magnetic stirrer, rotary evaporator, UV lamp, pressure source, NMR and melting point apparatus			
Keywords Column chromatography, hydrazones, imine formation, NMR, stereochemistry assignment			

Background

Imines are useful molecules used to synthesize carbon–nitrogen bonds and are key synthetic intermediates in a plethora of important reactions. From a biological point of view, imines are essential components of many systems, such as vision and various enzymatic reactions,[1] and act as pharmacophores in numerous biologically active compounds.[2] For instance, our eyes sense

Comprehensive Organic Chemistry Experiments for the Laboratory Classroom
Edited by Carlos A. M. Afonso, Nuno R. Candeias, Dulce Pereira Simão, Alexandre F. Trindade, Jaime A. S. Coelho, Bin Tan, and Robert Franzén.

Published by the Royal Society of Chemistry, www.rsc.org

light by a photochemical reaction involving an imine, rhodopsin, formed through the reaction between an aldehyde, 11-*cis*-retinal, and a protein, opsin.

Structurally, imines are compounds having the general formula $R^1R^2C{=}NR^3$ and are formed in a reversible acid-catalysed process between ammonia, or a primary amine, and an aldehyde or ketone.

When $R^1{=}H$ the imine is generally named an aldimine, whereas when R^1 and R^2 are both alkyl or aryl groups they are called ketimines. When R^3 is an alkyl or aryl group the imine is named a Schiff base.

The acid-catalysed formation of an imine involves the nucleophilic addition of the amine to the protonated ketone or aldehyde to form an *N*-protonated dipolar tetrahedral intermediate. Intramolecular proton transfer from the nitrogen to the oxygen yields the *O*-protonated hemiaminal. The water group leaves to form the iminium ion, which then transfers the proton to unreacted carbonyl (or water) to produce the imine. When the imine derives from an aldehyde or unsymmetrical ketone, it can exist as a mixture of (*Z*/*E*)-isomers about the N=N bond, which interconvert rapidly at room temperature.

In general, imines are less stable than the starting carbonyl compounds and equilibrium is usually driven to products by water azeotropic elimination. However, in some cases, imine synthesis can be effective in completely aqueous media, without any catalyst and under mild conditions, which is a remarkable putative contradiction. This approach takes advantage of the fact that some solid imines are insoluble in water and precipitate directly from the reaction mixture as they are formed, thereby shifting the equilibrium towards imine formation.[3,4]

Imines are only stable enough to be isolated if either the carbon or the nitrogen of the double bond has an aromatic moiety or an electronegative substituent which reduces the electrophilic character of the imine carbon, as in the case of oximes and hydrazones.[5] Hydrazones, a special type of imine with the general formula $R^1R^2C{=}NNHR^3$, are stable enough to be isolated due to the stabilizing mesomeric effect of the second nitrogen which decreases the electrophilicity of the C=N bond.[5,6] Both (*E*/*Z*)-isomers are sufficiently stable to be detected and separated, although the (*E*)-isomer is usually preferred due to stereochemical reasons. *N*-Acylhydrazones may exist as (*E*/*Z*) geometric isomers about the C=N double bond and, additionally, as *cis*/*trans* amide conformers.[7,8]

Imines and hydrazones exhibit a peculiar chemical behaviour: the electron-rich nitrogen atom may act as a nucleophile, the sp^2 carbon behaves as an electrophile and the C=N double bond can be used as a dienophile or an aza-diene[2] (Figure 4.1.2.6.1). Three of the more important imine

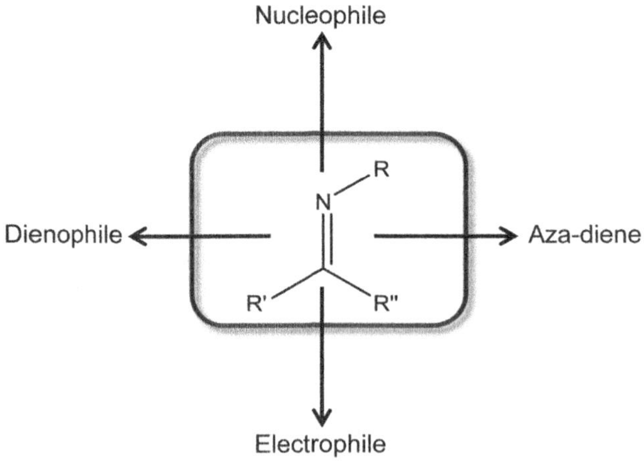

Figure 4.1.2.6.1 Reactive sites of imines.

equilibrium reactions are (a) hydrolysis – the imine reverts back to the precursor amine and carbonyl-containing compound by the addition of water; (b) exchange – the original imine undergoes transamination resulting in the exchange of R groups, and (c) metathesis – after the introduction of a second imine, the exchange of the two R groups occurs.[2] Additionally, imines can be reduced to amines with several reagents, and the use of a multitude of catalysts can provide chiral amines, which have proved to be important in pharmacophore moieties for the design of new drugs.[9]

The objective of this experiment is to synthesize the isoniazid hydrazone *N′*-(*E*)-(4-phenoxybenzylidene)isonicotinohydrazide and to use NMR tools to explore its stereochemistry. This compound is a derivative of isoniazid, one of the two most effective first-line drugs in the market to fight tuberculosis, and shows a good activity against a *Mycobacterium tuberculosis* strain resistant to isoniazid.[10]

Additional Safety

Students are expected to follow standard safety guidelines to work in a chemistry laboratory. Personal protective equipment (laboratory coats, safety goggles and gloves) must be worn at all times. Column chromatography must be performed in the fume hood due to the use of large volumes of volatile solvents, the health risks associated with the use of silica-gel, and also the potential hazards associated with a vacuum pump.

Experimental Procedure

Session 1 (3 h): Synthesis of *N′*-(*E*)-(4-Phenoxybenzylidene) isonicotinohydrazide

1. Assemble a two-neck 50 mL round-bottom flask with an addition funnel and a stir bar.
2. Add 1.00 g (7.29 mmol, 1 equiv.) of isoniazid and 10 mL of water and stir until complete dissolution.
3. Place 1.30 mL (7.29 mmol, 1 equiv.) of 4-phenoxybenzaldehyde in the addition funnel and add it dropwise, over 5 min, to the isoniazid solution.
4. After several minutes a white crystalline solid begins to form.
5. Stir the reaction mixture for an additional period of 2 h, while monitoring the reaction by TLC (ethyl acetate).
6. Collect the resulting solid on a Hirsch funnel using vacuum filtration and wash it with water (25 mL).
7. Allow the solid to air dry on the Hirsch funnel by continuing to suck air through the funnel for 10–15 min.
8. Place the crude solid in a vacuum desiccator and allow it to dry until the next session.

Session 2 (3 h): Purification of *N′*-(*E*)-(4-Phenoxybenzylidene) isonicotinohydrazide

1. Weigh the crude *N′*-(*E*)-(4-phenoxybenzylidene)isonicotinohydrazide.
2. Purify the crude compound using flash column chromatography,[11] which should be performed *in the hood* as follows:

Packing the Column

- Use a ground-glass joint column ($\phi = 4$ cm), clamp it securely and close the stopcock.
- Add a layer of sand using a funnel (approx. 0.5 cm). The sand prevents the silica from running out of the column.

- Pour the ethyl acetate carefully over the sand, up to roughly one third of the height.
- Prepare a slurry of silica-gel and ethyl acetate stirring vigorously to remove all the air from the silica.
- Very carefully, add the slurry to the eluent, a little at a time, while swirling. Make sure that the layer of sand is not too disturbed. Tap gently the column between additions in order to avoid the formation of bubbles.
- Open the stopcock between solvent additions. The top of the silica gel must not be disturbed and should not be allowed to run dry.
- Continue to transfer the slurry to the column until it reaches a depth of 14–15 cm and, afterwards, rinse the inside of the column with a small volume of eluent.
- Fill carefully the column with eluent.
- Connect a flow controller to the top of the column, open the stopcock and use a pressurized air source (pump, nitrogen or air flow) to pack the column. The silica level in the column will contract.
- Adjust the flow rate in order to obtain a fall of the eluent's surface in the column of 5 cm min^{-1}.
- Using the adjusted flow, lower the eluent level until just about 2 cm above the silica-gel level, close the stopcock and switch off the pressure.

Loading the Sample

Because the crude compound is not soluble in ethyl acetate, it is necessary to dry-load it into the column.[12]

- Dissolve 0.60 g of the crude compound in acetone.
- Add dry silica-gel to the compound's solution (approx. two-times the sample's weight).
- Evaporate carefully the solvent in the rotary evaporator until the silica is dry and free-flowing. If it is still an oil, add more silica and repeat the procedure.
- Pour the silica/crude product mixture slowly into the column and allow it to settle. Make sure that the eluent level always remains above the top of the silica and that no bubbles are present.
- Allow the eluent to drain until its level is approximately 1 mm above the top of the silica/crude product mixture.
- Carefully fill the column with the eluent, using a Pasteur pipet, until the level is a few centimetres above the silica/crude product mixture.
- At this point, with the help of as funnel, slowly add the solvent from an Erlenmeyer flask letting it first run down the side of the column. The silica surface must not be disturbed at any time.

Elution of the Column

- Open the stopcock, switch on the pressure and begin "running" the column, collecting 25 mL fractions. Pay attention to the eluent's level and replenish whenever necessary.
- Monitor the content of each fraction by TLC, using ethyl acetate and stop the elution once the compound has all ran out of the column.
- Pool together the fractions containing chromatographically pure *N'*-(*E*)-(4-phenoxybenzylidene)isonicotinohydrazide and concentrate the resulting solution in the rotatory evaporator.

3. If necessary recrystallize the product from ethanol/water (7 : 3).

4. Weight the pure *N′*-(*E*)-(4-phenoxybenzylidene)isonicotinohydrazide and calculate the yield.
5. Determine the melting point and prepare a NMR tube using 10 mg of the compound dissolved in DMSO-d_6 for running 1D and 2D NMR (^{1}H, ^{13}C, HSQC, HMBC and NOESY) experiments.

Waste Disposal

Silica-gel must be disposed as hazardous waste: place the material in a bag or container and dispose of it according to the laboratory rules. Used solvents must be discarded in a non-halogenated waste container. TLC plates and used melting point capillaries can be discarded in a common wastebasket.

Results Interpretation and Additional Questions

1. Draw a detailed mechanism for thc product formation, and explain the reason for the generation of the two geometric isomers.
2. Explain why the (*E*)-isomer is usually the most favoured one.
3. Write the resonance structures of the synthesized hydrazone and try to explain its stability towards hydrolysis.
4. Analyse the ^{1}H NMR and NOESY spectra paying special attention to the signals of the CON<u>H</u> proton and the iminic proton. Which feature is determinant to rule out the existence of the (*Z*)-isomer and which one is important to assign the *trans* amide conformer?
5. The reaction has been monitored by TLC. In this case, was the 2 h reaction time enough to complete the reaction? Explain your answer.
6. In the execution of the column chromatography, once the column is packed, why is it so important to allow the level of the eluent to drop to a 2–3 cm level, before adding your mixture?

References

1. J. Bennett, K. Meldi and C. Kimmell, *J. Chem. Educ.*, 2006, **83**, 1221–1224.
2. L. H. Choudhury and T. Parvin, *Tetrahedron*, 2011, **67**, 8213–8229.
3. V. Saggiomo and U. Lüning, *Tetrahedron Lett.*, 2009, **50**, 4663–4665.
4. A. Simion, C. Simion, T. Kanda, S. Nagashima, Y. Mitoma, T. Yamada, K. Mimura and M. Tashiro, *J. Chem. Soc., Perkin Trans. 1*, 2001, 2071.
5. M. E. Belowich and J. F. Stoddart, *Chem. Soc. Rev.*, 2012, **41**, 2003–2024.
6. J. Kalia and R. T. Raines, *Angew. Chem., Int. Ed. Engl.*, 2008, **47**, 7523–7526.
7. G. Palla, G. Predieri, P. Domiano, C. Vignali and W. Turner, *Tetrahedron*, 1986, **42**, 3649–3654.
8. E. Wyrzykiewicz and A. Blaszczyk, *J. Heterocycl. Chem.*, 2000, **37**, 975–981.
9. T. C. Nugent and M. El-Shazl, *Adv. Synth. Catal.*, 2010, **352**, 753–819.
10. F. Martins, S. Santos, C. Ventura, *et al.*, *Eur. J. Med. Chem.*, 2014, **81**, 119–138.
11. W. C. Still, M. Kahn and A. Mitra, *J. Org. Chem.*, 1978, **43**, 2923–2925.
12. K. Hostettmann, A. Marston and M. Hostettmann, *Preparative Chromatography Techniques: Applications in Natural Product Isolation*, Springer-Verlag, Berlim, 2nd edn, 1998, ch. 5, pp. 63–65.

4.1.2.7. Preparation of a Tosylhydrazidyl *N*-Glycosyl Derivative of D-Glucuronic Acid *via* Tosylhydrazone Formation and Intramolecular Ring Closure

Nuno M. Xavier

Centro de Química e Bioquímica, Faculdade de Ciências, Universidade de Lisboa, 1749-016 Lisboa, Portugal. *E-mail: nmxavier@fc.ul.pt

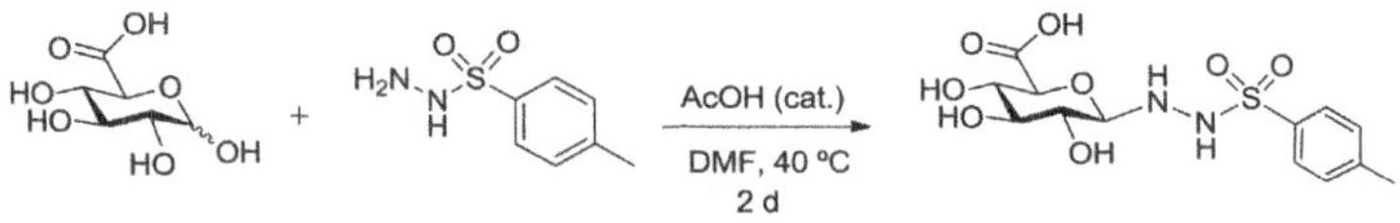

Number of sessions (duration of each session)	Hazard level	Difficulty level	Level of study
2 (1 h + 3 h)	Moderate	Medium	Intermediate

Class names Aldehydes, alcohols, hydrazones, hydrazides

Concepts involved Equilibrium between cyclic and open-chain forms of a reducing sugar, hydrazone formation, intramolecular cyclization

Chemicals needed D-Glucuronic acid, *p*-toluenesulfonyl hydrazide, glacial acetic acid, *N*,*N*-dimethylformamide, diethyl ether, dichloromethane, deuterated methanol

Equipment and experimental techniques involved Evaporation under vacuum, vacuum filtration apparatus, NMR, melting point apparatus, heating plate with temperature control, stirring plate

Keywords Filtration under vacuum, hydrazides, hydrazones, intramolecular ring closure, NMR, reducing sugars

Background

A reducing sugar in solution exists as an equilibrium mixture composed of the cyclic hemiacetal form (lactol) and the open chain aldehyde. The position of this equilibrium highly favors the lactol. Despite the almost negligible concentration of the open chain aldehyde in the mixture, an unprotected reducing sugar may undergo typical chemical reactions of an aldehyde, allowing its selective derivatization at C1 in an acyclic form, through displacement of the equilibrium. Examples of such reactions include those enabling chain extension of the sugar, namely Wittig-type olefinations, addition of Grignard reagents, addition of cyanide, or reduction to produce an open chain alditol.[1] Similarly, with hydroxylamines, hydrazines and acylhydrazines, a reducing sugar effectively reacts in the aldehyde form, giving first the products of condensation, *i.e.* the acyclic aldoximes and hydrazones.[2] These acyclic derivatives evolve, however, to the more stable cyclic isomers, *i.e.* the corresponding glycosylhydroxylamines and glycosylhydrazines/hydrazides, by intramolecular attack of the hydroxyl group at C5 to the sp^2 carbon of the acyclic oximes and hydrazones (Scheme 4.1.2.7.1).

Comprehensive Organic Chemistry Experiments for the Laboratory Classroom
Edited by Carlos A. M. Afonso, Nuno R. Candeias, Dulce Pereira Simão, Alexandre F. Trindade, Jaime A. S. Coelho, Bin Tan, and Robert Franzén.

Published by the Royal Society of Chemistry, www.rsc.org

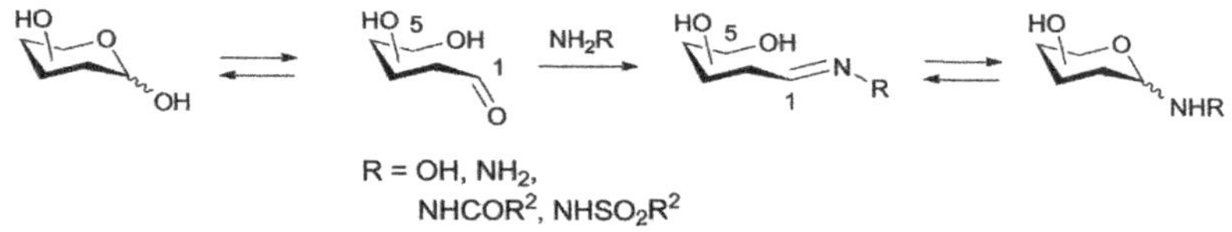

Scheme 4.1.2.7.1 Reaction of aldoses with hydroxylamines and hydrazine derivatives.

In this experiment, D-glucuronic acid, which differs from D-glucose by replacement of the hydroxymethyl group at C5 with a carboxyl group, is treated with tosylhydrazide in the presence of acetic acid. The product obtained is 1-deoxy-1-(2-tosylhydrazin-1-yl)-β-D-glucopyranuronic acid, which most likely arises from ring closure of the intermediate acyclic tosylhydrazone into the pyranose form. The synthesis of *N*-glycosylsulfonohydrazides by this methodology was first reported using partially O-protected monosaccharides with a free anomeric hydroxyl group and more recently the method was applied to unprotected reducing sugars such as glucose, galactose and xylose.[3] In most cases, only *N*-β-glycosyl derivatives were formed. The condensation reaction carried out in this work also furnishes exclusively the β-pyranosyl compound, which simplifies isolation and characterization. Only a small excess of tosylhydrazide and a catalytic amount of acetic acid are necessary to drive the reaction to completion. The product is isolated by precipitation and filtration in high purity, without the need of further purification.

Additional Safety

The experiments should be carried out in a fume hood. The regular precautionary measures are required when handling chemicals and solvents, namely the use of safety glasses, laboratory coat and gloves. Safety glasses are especially mandatory when using the rotatory evaporator due to the risk of implosion when vacuum is applied to the distillation flask.

Although no special precautions are needed for the manipulation of D-glucuronic acid, which is a non-hazardous material, care should be taken when using the other chemicals/solvents in this experiment.

N,N-Dimethylformamide is a flammable liquid. It is harmful in contact with skin, causes eye and respiratory tract irritations and it is toxic if inhaled. *p*-Toluenesulfonyl hydrazide is a flammable and potentially explosive compound. It may cause skin, eye and respiratory tract irritations. It is toxic if swallowed. Glacial acetic acid is a flammable and a corrosive liquid. It causes severe skin, eye and respiratory tract irritations. Both liquid and mist may cause severe burns to all body tissue. It is very hazardous in case of ingestion and inhalation. Diethyl ether is a highly volatile and extremely flammable liquid. It is mildly irritating to the eyes and skin. Inhalation of high concentrations of ether vapour can result in drowsiness, unconsciousness, muscle relaxation, decreased pulse and irregular respiration. Dichloromethane is a highly volatile liquid. It is irritating to the skin, eyes, and respiratory tract. It is toxic by ingestion, inhalation or dermal exposure.

Experimental Procedure

Session 1 (1 h)

1. In a fume hood, turn on a heating plate with set point at 40 °C and heat an oil bath on it.
2. To a pre-dried 25 mL round-bottom flask, add D-glucuronic acid (0.3 g, 1.55 mmol), *N,N*-dimethylformamide (DMF, 2.2 mL), *p*-toluenesulfonyl hydrazide (97%, 0.312 g, 1.63 mmol, 1.05 equiv.) and glacial acetic acid (0.01 mL, 0.18 mmol, 0.11 equiv.).[†]
3. Close the flask with a rubber stopper and put it into the preheated oil bath.
4. Leave the reaction mixture at 40 °C without stirring for 48 h.

[†]To measure the volume of acetic acid, a microsyringe (10, 25 or 50 μL capacity) should be used.

Session 2 (3 h)

1. Take the flask from the oil bath and remove the solvent under reduced pressure (set the water bath temperature of the rotary evaporator at 40 °C and slowly apply the vacuum until an operating pressure of 10–20 mbar is reached).
2. To the resulting yellowish residue, add diethyl ether (15 mL) and dichloromethane (5 mL), place a magnetic stir bar in the flask and leave the mixture under vigorously stirring for 1 h.
3. Filter the mixture under vacuum through a medium-porosity (10–13 μm) filter paper.
4. Wash the solid twice with diethyl ether (15 mL) and twice with dichloromethane (15 mL) and leave it to dry under suction for 10 min to remove the last traces of solvent.
5. Weigh the product, calculate the percentage yield and determine its melting point.
6. Acquire the ^{1}H and ^{13}C NMR spectra of the product in methanol-d_4 as well as two-dimensional homonuclear (COSY) and heteronuclear (HMQC) spectra.

Waste Disposal

Solvent wastes should be collected in the appropriate waste container (labeled as non-halogenated or halogenated solvents).

Results Interpretation and Additional Questions

1. Write the mechanism involved in the formation of the *N*-sulfonohydrazidyl glucuronide consisting of the interconversion between the open-chain (aldehyde and hydrazone) and the pyranose forms.
2. Interpret the NMR spectra obtained for the product and make the assignment of all the signals observed in the ^{1}H and ^{13}C NMR spectra.
3. Draw conclusions about the anomeric configuration of the sugar moiety, based on the coupling constant observed between H1 and H2.
4. Explain the need for mild acidic catalysis in the overall transformation.

References

1. B. G. Davis and A. J. Fairbanks, *Carbohydrate Chemistry: OUP Chemistry Primer No. 99*, Oxford University Press, Oxford, 2002.
2. (a) P. Merino, S. Franco, F. L. Merchan and T. Tejero, *Synth. Commun.*, 1997, **27**, 3529; (b) J. Augé and N. Lubin-Germain, *J. Carbohydr. Chem.*, 2000, **19**, 379.
3. (a) S. E. Mangholz and A. Vasella, *Helv. Chim. Acta*, 1991, **74**, 2100; (b) L. J. G. Edgar, S. Dasgupta and M. Nitz, *Org. Lett.*, 2012, **14**, 4226.

4.1.3. 1,4-Addition to α,β-unsaturated Carbonyl Compounds

4.1.3.1. Green Metrics in a Cyclocondensation Reaction

Clarissa P. Frizzo*, Marcos A. P. Martins*, Paulo R. S. Salbego, Caroline R. Bender, Aniele Z. Tier, and Hélio G. Bonacorso

Núcleo de Química de Heterociclos (NUQUIMHE), Department of Chemistry, Federal University of Santa Maria, 97105-900, Santa Maria, RS, Brazil. *E-mail: clarissa.frizzo@gmail.com, mmartins@base.ufsm.br

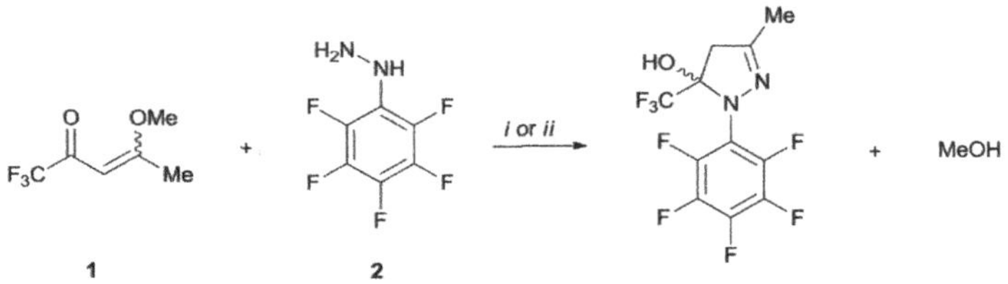

1 2 3

i: [BMIM][BF_4], conventional thermal heating, 80 °C, 1 h; ii: solvent-free, MW, 100 °C, 6 min

Number of sessions (duration of each session)	Hazard level	Difficulty level	Level of study
1 (4 h)	Medium	Medium	Intermediate

Class names α,β-Unsaturated ketones, hydrazines, dihydropyrazoles

Concepts involved This experiment involves a cyclocondensation reaction between 4-methoxy-1,1,1-trifluoro-3-pent-2-one and pentafluorophenyl hydrazine. The reaction is performed under solvent-free conditions under microwave irradiation (MW) or using ionic liquid (IL) under conventional thermal heating. Green metrics such as atom economy (AE), reaction mass efficiency (RME) and environmental factor (E-factor) are used to compare the greenness of each method

Chemicals needed 4-Methoxy-1,1,1-trifluoro-pent-3-en-2-one (enone, **1**), pentafluorophenyl hydrazine (hydrazine, **2**), 1-butyl-3-methyl-imidazolinium tetrafluoroborate ([BMIM][BF_4] – IL), ethyl acetate, dichloromethane, Na_2SO_4, $MgSO_4$, $CDCl_3$, hexane

Equipment and experimental techniques involved Microwave (MW) apparatus, apparatus for reaction with controlled temperature and magnetic stirring, liquid–liquid extraction, NMR, round-bottom flask, condenser, rotary evaporator, vacuum pump

Keywords Atom economy, cyclocondensation reaction, E-factor, liquid–liquid extraction, NMR, reaction mass efficiency

Comprehensive Organic Chemistry Experiments for the Laboratory Classroom
Edited by Carlos A. M. Afonso, Nuno R. Candeias, Dulce Pereira Simão, Alexandre F. Trindade, Jaime A. S. Coelho, Bin Tan, and Robert Franzén.

Published by the Royal Society of Chemistry, www.rsc.org

Background

Green chemistry seeks to reduce chemical related impacts on human health and the environment by the use of alternative, environmentally friendly processes and reaction media. The importance of this concept can be apprehended from the words of the Nobel Laureate Professor Ryoji Noyori, "Green chemistry is not just a mere catch phrase: it is the key to the survival of mankind".[1] In this sense, some tools have been developed to evaluate and achieve less aggressive processes,[2,3] such as solvent selection, by taking into account environmental, health, and safety aspects as well as the life cycle assessment (LCA) and economic criteria.

To evaluate the greenness of a product or process Kralisch *et al.*[4,5] proposed the use of three main criteria: the energy factor (EF), environmental and human health factor (EHF), and finally the cost factor (CF). To support a sustainable path some other metrics can be employed, such as atom economy (AE),[6,7] reaction mass efficiency (RME),[8] environmental factor (E-factor),[9] effective mass yield,[10] mass intensity,[11] and the process profile.[12] The atom efficiency or atom economy concept is a rapid tool for evaluating the amounts of waste that will be generated by alternative processes.[6,7] It is calculated by dividing the molecular weight of the desired product by the sum of molecular weights of all reactants in the stoichiometric equation for the reaction involved. The E-factor is a more broadly employed concept in the chemical industry worldwide since it provides an efficient measurement for waste generation in chemical manufacturing, in order to develop cleaner and more sustainable processes.[13] The E-factor is the real amount of waste produced in the process, defined as everything except the desired product. It takes the chemical yield into account and includes reactants, solvent losses, all process aids and, ideally, even fuel (although this is often difficult to quantify). However, water and ionic liquids (IL) are generally excluded from the E-factor calculation since they are considered to be recovered solvents.

In this context, an organic reaction in a recovered solvent or solvent-free, as well as an experimental procedure for extraction of product that excludes the use of a large amount of solvents (like a simple filtration), is highly desirable from the environmental point of view. Ionic liquids (ILs) in particular have attracted interest as an alternative green reaction solvent due advantages in relation to volatile organic solvents, such as low-volatility, non-flammability, thermal stability, and controlled miscibility. ILs can solubilize both polar and non-polar species, promoting interactions between different components and can be recovered and reused several times, which is of particular interest in a green chemistry context.[14]

The cyclocondensation reaction is an important route for obtaining several substituted pyrazoles.[15,16] Conventional thermal heating (oil bath) is employed for reactions in solution and, generally, presents good performance for cyclocondensation reactions.[14]

However, in the green chemistry context, the thermal heating provided by microwaves (MWs) is a special alternative in relation to conventional thermal heating. The high temperatures reached using MW can avoid the use of catalysts, solvents and other additives in reactions. Additionally, thermal heating provided by MWs results in a more efficient energy transfer to the reaction mixtures, leading to short reaction times and energy economy. In this context, alternatives such as MW and solvent-free methods have been widely used in reactions aimed at waste reduction and atom economy.

Therefore, the main objectives of this experiment are:

(i) perform a cyclocondensation reaction using two different activation modes (conventional thermal heating and MW) and two reaction conditions (IL and solvent-free).

(ii) compare and evaluate the methods using green metrics such as E-factor, atom economy (AE) and reaction mass efficiency (RME), discussing which of these is more useful in evaluating the generated waste in this type of reaction.

Additional Safety

All experiments should be done in a fume hood, with proper safety equipment, such as protective clothing, gloves and safety goggles. The reagents and solvents should be handled with care. Pentafluorophenyl hydrazine has acute toxicity (oral, dermal, inhalation) and can cause skin and eye irritation. Compounds with similar structure to **1** are described as flammable, and are suspected of causing cancer. The toxicological properties of **1** have not been thoroughly investigated. Dichloromethane, ethyl acetate and hexane presents high volatility and are an acute inhalation hazard.

Experimental Procedure

The proposal entails each working group conducting a different method in the laboratory, so that all results can be compared and discussed at the end of the experiment in terms of advantages and disadvantages of each process.

Method A – Reaction with a Conventional Thermal Heating Method (Oil Bath)

1. Prior to the reaction step, it is recommended that the IL be dried under vacuum, with magnetic stirring, at 50 °C to remove any moisture (for around 2 h).
2. Put the mixture of enone **1** (1 mmol, 0.17 g), hydrazine **2** (1.2 mmol, 0.24 g) and [BMIM][BF_4] (1 mmol, 0.23 g, density: 1.21 g mL^{-1}) in a round-bottom flask. The molecular weight of **1**, **2** and the IL is 168.11, 198.09 and 226.02 g mol^{-1}, respectively.
3. The mixture must be stirred and heated at 80 °C for 1 h.

Product Isolation

4. After completion of the reaction, add dichloromethane (10 mL) and wash with water (3 × 10 mL).
5. Dry the organic phase with around 7.40 g Na_2SO_4 (52 mmol) to remove traces of moisture (an exact amount of drying agent is not needed, but the amount used should be weighed and the mass recorded, since it will be important for the green metrics evaluation).
6. Weigh an empty round-bottom flask (M_1), in which the product will be stored.
7. Filter the mixture into the round-bottom flask, previously weighed.
8. Discard the filter paper with the Na_2SO_4 salt.
9. Remove the organic solvent in a rotary evaporator.
10. Subsequently, hold the product under a vacuum pump for 1 h, to remove any residual solvent.

Method B – Reaction Using MW Method

1. Place the mixture of enone **1** (1 mmol, 0.17 g) and hydrazine **2** (1.2 mmol, 0.24 g) in the MW apparatus.
2. The mixture must be set at 100 °C for 6 min.

Product Isolation

3. Add ethyl acetate (20 mL) to the reaction mixture.
4. Place the mixture in a separation funnel and add water (10 mL) and ethyl acetate (20 mL). The product is in the organic phase (ethyl acetate).
5. Repeat the process.
6. Combine the organic phases and then dry with $Mg_2\ SO_4$ (3.25 g, 27 mmol).

7. Weigh an empty round-bottom flask (M_1), in which the product will be stored.
8. Filter the mixture into the round-bottom flask, previously weighed.
9. Discard the filter paper with the Mg_2SO_4 salt.
10. Remove the organic solvent in a rotary evaporator.
11. Subsequently, hold the product under a vacuum pump for 1 h, to remove any residual solvent.
12. Perform a NMR analysis and calculate the reaction yield.

Yield Determination

13. After completion of the reaction, weigh the round-bottom flask containing the reaction mixture and determine the final weight (M_2) (g).
14. From this value (M_2) (g), subtract the weight of the empty round-bottom flask (M_1), following eqn (4.1.3.1.1):

$$M_3 = M_2 - M_1 \tag{4.1.3.1.1}$$

15. M_3 corresponds to the mass (g) of product obtained in the experiment.
16. Calculate the amount of obtained product (M_3) in mmol.
17. Correlate the product obtained with the limiting reactant used (in this case the enone **1**) to calculate the reaction yield following eqn (4.1.3.1.2):

$$\text{Yield}(\%) = \frac{\text{product}(\text{mmol})}{\text{limiting reactant}(\text{mmol})} \times 100 \tag{4.1.3.1.2}$$

Waste Disposal

The aqueous waste must be collected in an aqueous residue recipient. Dichloromethane must be collected in a chlorinated waste collector recipient, for posterior appropriate treatment. Ethyl acetate may be reused or collected in a non-chlorinated waste collector recipient. The drying agents must be collected in a recipient for solid waste. Ionic liquid can be completely recovered and purified under vacuum.

Interpretation of the Results and Additional Questions

1. Interpret the ^{1}H and ^{13}C NMR spectra obtained, and compare the obtained melting point with that given in the literature.
2. Rationalize a mechanism for the synthesis, indicating the nucleophile, electrophile and orbitals involved in this reaction. Indicate the HOMO (high energy molecular orbital) and LUMO (low unoccupied molecular orbital) orbital.
3. Calculate and discuss the AE and RME values obtained.[17]

The AE must be calculated following eqn (4.1.3.1.3):

$$\text{AE} = \left(\frac{mw_{\text{P}}}{mw_{\text{R1}} + mw_{\text{R2}}} \right) \times 100 \tag{4.1.3.1.3}$$

where mw_{P} is the molecular weight of product P, mw_{R1} is the molecular weight of reactant R1, and mw_{R2} is the molecular weight of reactant R2.

The RME must be calculated following eqn (4.1.3.1.4):

$$\text{RME} = \left(\frac{\text{mass of P(g)}}{\text{mass of R1 (g)} + \text{mass of R2 (g)}}\right) \times 100 \qquad (4.1.3.1.4)$$

where "mass of P" is the mass of the isolated product, and mass of R1 and mass of R2 are the mass of reactants R1 and R2, respectively, used to obtain P.

4. Calculate the E-factor and discuss the difference in both methods (conventional and MW). The E-factor must be calculated following eqn (4.1.3.1.5):[17]

$$\text{E-factor} = \left(\frac{m_{\text{total waste}}}{m_{\text{product}}}\right) = \left(\frac{m_{\text{reactants}} - m_{\text{product}}}{m_{\text{product}}}\right) = \left(\frac{m_{\text{input materials}} - m_{\text{product}}}{m_{\text{product}}}\right) \qquad (4.1.3.1.5)$$

where $m_{\text{reactants}}$ is the sum of the reactant masses used and m_{product} is the mass of the final product.

5. Discuss and compare the different metrics used in both methods.

Acknowledgements

Experiment adapted with permission from *Monatsh. Chem.*, 2013, **144**, 1153–1156. With kind permission from Springer Science and Business Media. Copyright (2013).

References

1. R. Noyori, *Tetrahedron*, 2010, **66**, 1028.
2. K. Alfonsi, J. Colberg, P. J. Dunn, T. Fevig, S. Jennings, T. A. Johnson, H. P. Kleine, C. Knight, M. A. Nagy, D. A. Perry and M. Stefaniak, *Green Chem.*, 2008, **10**, 31.
3. C. Capello, U. Fischer and K. Hungerbuhler, *Green Chem.*, 2007, **9**, 927.
4. D. Kralisch, A. Stark, S. Korsten, B. Ondruschka and G. Kreisel, *Green Chem.*, 2005, **7**, 301.
5. D. Kralisch, D. Reinhardt and G. Kreisel, *Green Chem.*, 2007, **9**, 1308.
6. B. M. Trost, *Science*, 1991, **254**, 1471.
7. B. M. Trost, *Acc. Chem. Res.*, 2002, **35**, 695.
8. D. J. C. Constable, A. D. Curzons and V. L. Cunningham, *Green Chem.*, 2002, **4**, 521.
9. P. J. Dunn, S. Galvin and K. Hettenbach, *Green Chem.*, 2004, **6**, 43.
10. T. Hudlicky, D. A. Frey, L. Koroniak, C. D. Claeboe and L. E. Brammer, *Green Chem.*, 1999, **1**, 57.
11. A. D. Curzons, D. J. C. Constable, D. N. Mortimer and V. L. Cunningham, *Green Chem.*, 2001, **3**, 1.
12. C. E. Berkoff, K. Kamholz, D. E. Rivard, G. Wellman and H. Winicov, *CHEMTECH*, 1986, 552.
13. R. A. Sheldon, *Green Chem.*, 2007, **9**, 1273–1283.
14. M. A. P. Martins, C. P. Frizzo, D. N. Moreira and P. Machado, *Chem. Rev.*, 2009, **109**, 4140.
15. C. P. Frizzo, M. R. B. Marzari, L. Buriol, D. N. Moreira, F. A. Rosa, P. S. Vargas, N. Zanatta, H. G. Bonacorso and M. A. P. Martins, *Catal. Commun.*, 2009, **10**, 1967.
16. L. Buriol, C. P. Frizzo, L. D. T. Prola, D. N. Moreira, M. R. B. Marzari, E. Scapin, N. Zanatta, H. G. Bonacorso and M. A. P. Martins, *Catal. Lett.*, 2011, **141**, 1130.
17. M. A. P. Martins., P. H. Beck, L. Buriol, C. P. Frizzo, D. N. Moreira, M. R. B. Marzari, M. Zanatta, P. Machado, N. Zanatta and H. G. Bonacorso, *Monatsh. Chem.*, 2013, **144**, 1043.

4.1.3.2. Synthesis of 1*H*-Pyrazoles Using Ball Mill, Grinding and Conventional Thermal Heating

Clarissa P. Frizzo*, Marcos A. P. Martins*, Caroline R. Bender, Paulo R. S. Salbego, Aniele Z. Tier, Guilherme C. Paveglio, and Kelvis Longhi

Núcleo de Química de Heterociclos (NUQUIMHE), Department of Chemistry, Federal University of Santa Maria, 97105-900, Santa Maria, RS, Brazil. *E-mail: clarissa.frizzo@gmail.com, mmartins@base.ufsm.br

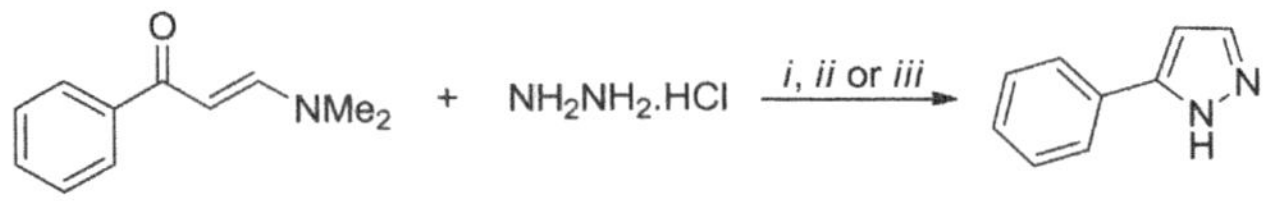

1 2 3

i: TsOH (20 mol%), grinding, 6 min, 25 °C; ii: TsOH (10 mol%), ball mill, 3 min, 25 °C; iii: TsOH (20 mol%), EtOH, conventional thermal heating, 2 h, 80 °C

Number of sessions (duration of each session)	Hazard level	Difficulty level	Level of study
1 (4 h)	High	Medium	Medium

Class names Enaminones, hydrazines, pyrazoles

Concepts involved This experiment describes a synthesis of 1*H*-pyrazoles from enaminone and hydrazine hydrochloride using *p*-toluenesulfonic acid (TsOH) as catalyst. Grinding (i), ball mill (ii) and conventional thermal heating (iii) were used. The first two methods are under solvent-free conditions. The reaction mechanism involves a nucleophilic conjugate addition and a C–N σ-bond formation

Chemicals needed 3-(Dimethylamino)-1-phenylprop-2-en-1-one (enaminone, **1**), hydrazine hydrochloride (hydrazine, **2**), *p*-toluenesulfonic acid, chloroform, deuterated chloroform, ethanol, anhydrous Na_2SO_4, distilled water

Equipment and experimental techniques involved Planetary ball mill, steel milling beaker, five steel milling balls, grinding (mortar and pestle), apparatus for reaction with controlled temperature (80 °C) and magnetic stirring, rotatory evaporator, NMR

Keywords Ball mill, cyclocondensation, grinding, liquid–liquid extraction, NMR, pyrazoles

Background

The development of synthetic methods under solvent-free conditions has attracted interest due to advantages such as mild reaction conditions and reduction in waste and energy costs, making it gainful both for the industry and the environment.[1,2] Furthermore, solvent-free reactions exhibit high atom efficiency and steps such as solvent purification and solvent evaporation are often dispensable.

Mechanochemical techniques, such as ball mill or grinding under solvent-free conditions, are excellent activation methods of "green chemistry" that have been shown to be viable alternatives for saving energy. The efficient mixing and continuous particle refinement allowed by the equipment leads to high reaction rates, higher selectivity and yields when compared to reactions performed in solution.[3–6]

Comprehensive Organic Chemistry Experiments for the Laboratory Classroom
Edited by Carlos A. M. Afonso, Nuno R. Candeias, Dulce Pereira Simão, Alexandre F. Trindade, Jaime A. S. Coelho, Bin Tan, and Robert Franzén.

Published by the Royal Society of Chemistry, www.rsc.org

The grinding method has been reported for Grignard reactions,[7] Reformatsky reactions,[8] Aldol condensations[9] and others.[10] The main advantage of this method is its operational simplicity, requiring only a mortar and pestle. In addition, most of the reactions are performed at room temperature and in absolutely solvent-free medium.

On the other hand, in the ball mill equipment, the possibility of adjusting instrumental parameters such as rotation frequency, time, and oscillation frequency, as well as the number, mass, diameter and material of the milling balls, allows more reproducible results.[11,12] Another advantage of this mechanochemical technique is the capacity for larger scale syntheses.

Enaminones are an important 1,3-dielectrophilic used as three-atom building blocks (CCC) in heterocyclic synthesis. In the presence of hydrazines (NN) as the two-atom fragment, these compounds can generate 1*H*-pyrazoles through a [3+2] cyclocondensation.[13] The nitrogen-containing heterocycles have attracted attention due to their therapeutic and pharmacological applications. In particular, pyrazoles and their derivatives have a wide variety of biological activities, including anti-inflammatory, antipyretic, analgesic, bactericidal, and fungicidal, as well as promising inhibitory activity against monoamine oxidase for the treatment of diseases such as Parkinson's and Alzheimer's.[14] In this context, the goal of this experiment is to synthesize the 1*H*-pyrazole from enaminones and hydrazine in ethanol using conventional thermal heating and under solvent-free conditions using mechanochemical methods such as grinding and ball mill; and to compare the results obtained by different techniques, noting the advantages and limitations of each one.

Additional Safety

All experiments should be performed in a fume hood, with proper safety equipment, such as protective clothing, gloves and safety goggles. *p*-Toluenesulfonic acid and hydrazine hydrochloride are corrosive acids and should be handled with care. Chloroform presents high volatility and is an acute inhalation hazard.

Experimental Procedure

Laboratory Session 1 (4 h): 1*H*-Pyrazole Synthesis

Each class working group can conduct one of the methods in the laboratory, so that all results can be compared at the end of the experiment, discussing the advantages and disadvantages of each process.

Method A – Reactions by Grinding (ref. 15)

1. Place enaminone **1** (1.03 mmol, 0.18 g), hydrazine **2** (1.2 mmol, 0.08 g) and *p*-toluenesulfonic acid (20 mol%, 0.04 g) in a mortar.
2. Grind the mixture with a pestle, at room temperature for 6 min.

Method B – Reactions by Ball Mill (ref. 1)

1. Place the enaminone **1** (3.02 mmol, 0.53 g), hydrazine **2** (3.65 mmol, 0.25 g) and *p*-toluenesulfonic acid (10 mol%, 0.02 g) in a steel milling beaker (25 mL) equipped with five steel milling balls (10 mm diameter). Seal the beaker.
2. Set the parameters in the planetary ball mill: rotation frequency (450 rpm) and milling time (3 min); the milling process can then be initiated.
3. After the milling process, open the beaker and remove the milling balls.

Method C – Reactions by Conventional Thermal Heating Method (Oil Bath) (ref. 15)

1. Add enaminone **1** (1.03 mmol, 0.18 g) and hydrazine **2** (1.2 mmol, 0.08 g) to a round-bottom flask.

2. Dissolve the reactants in ethanol (10 mL).
3. Add *p*-toluenesulfonic acid (20 mol%, 0.04 g).
4. Stir the reaction mixture, under reflux, for 2 h.
5. After completion of the reaction, remove the ethanol under reduced pressure.

Product Isolation for the Three Methods

6. Add water (20 mL) and place the mixture in an extraction funnel. Wash three times with $CHCl_3$ (3 × 20 mL). Due to the toxicity of chloroform, this solvent can be replaced by CH_2Cl_2, providing comparable results.
7. Combine the organic phases of extraction in an Erlenmeyer flask.
8. Dry the organic phase with anhydrous Na_2SO_4 (6 g).
9. Weigh an empty round-bottom flask (M_1), in which the product will be stored.
10. Filter the mixture into the round-bottom flask previously weighed.
11. Discard the paper filter and the Na_2SO_4 salt.
12. Remove the $CHCl_3$ from the round-bottom flask under reduced pressure to give a product with high purity.
13. Subsequently, hold the product under vacuum for 1 h, to remove any residual solvent.

Yield Determination

The yield can be determined in the same way for all three methods.

14. After completion of the reaction, weigh the round-bottom flask containing the reaction mixture and determine the final weight (M_2) (g).
15. From this value (M_2) (g), subtract the weight of the empty round-bottom flask (M_1), following eqn (4.1.3.2.1):

$$M_3 = M_2 - M_1 \qquad (4.1.3.2.1)$$

16. M_3 corresponds to the mass (g) of product obtained in the experiment.
17. Calculate the weight of obtained product (M_3) in mmol.
18. Correlate the product obtained with the limiting reactant used (in this case the enone **1**) to calculate the reaction yield following eqn (4.1.3.2.2):

$$\text{Yield (\%)} = \frac{\text{product (mmol)}}{\text{limiting reactant (mmol)}} \times 100 \qquad (4.1.3.2.2)$$

Waste Disposal

The aqueous waste must be collected in an aqueous residue recipient. Chloroform may be reused or collected in a chlorinated waste collector recipient, for posterior destruction.

Interpretation of the Results and Additional Questions

1. Interpret the ^1H and ^{13}C NMR spectra obtained, and compare with different methods (A, B and C) and with the literature.[1,15]
2. Compare the two mechanochemical methods for obtainment of 1*H*-pyrazole, and indicate the advantages and disadvantages of one over the other.
3. Compare the product yields obtained by the three methods.
4. Considering the nucleophilic and electrophilic centers of enaminone (Figure 4.1.3.2.1), rationalize a mechanism for 1*H*-pyrazole formation, considering the role of *p*-toluenesulfonic acid as catalyst. Discuss the importance of *p*-toluenesulfonic acid in the reaction mechanism.

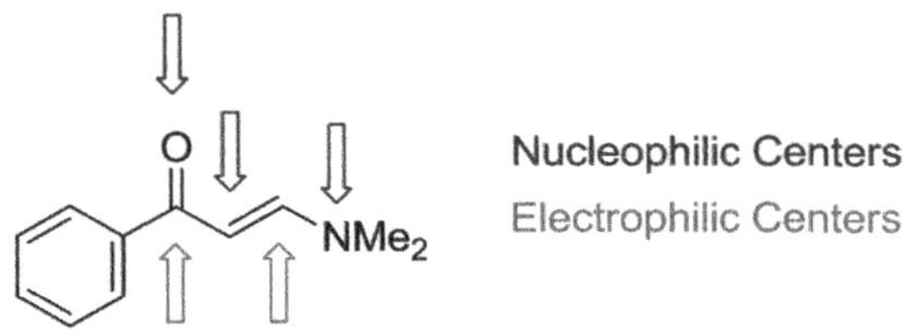

Figure 4.1.3.2.1 Nucleophilic and electrophilic centers of enaminones.

5. Discuss the differences, in terms of kinetics, between the mechanochemical methods without solvent and the reaction performed in molecular solvent.
6. Considering the activation methods used in these experiments, discuss the approach (effective collisions) of the reactants, in this chemical reaction, in the presence or absence of solvent.

Acknowledgements

Experiment adapted with permission from Martins *et al.*, *ACS Sustainable Chem. Eng.*, 2014, **2**, 1895–1901. Copyright (2014) American Chemical Society; and Martins *et al.*, *Tetrahedron Lett.* 51, An efficient solvent-free synthesis of NH-pyrazoles from β-dimethylaminovinylketones and hydrazine on grinding, 3193–3196. Copyright (2010) with permission from Elsevier.

References

1. G. C. Paveglio, K. Longhi, D. N. Moreira, T. S. Muchen, A. Z. Tier, I. M. Gindri, C. R. Bender, C. P. Frizzo, N. Zanatta, H. G. Bonacorso and M. A. P. Martins, *ACS Sustainable Chem. Eng.*, 2014, **2**, 1895.
2. M. A. P. Martins, C. P. Frizzo, D. N. Moreira, L. Buriol and P. Machado, *Chem. Rev.*, 2009, **109**, 4140.
3. T.-X. Métro, J. Bonnamour, T. Reidon, J. Sarpoulet, J. Martinez and F. Lamaty, *Chem. Commun.*, 2012, **48**, 11781.
4. L. Takacs, *Chem. Soc. Rev.*, 2013, **42**, 7649.
5. E. Boldyreva, *Chem. Soc. Rev.*, 2013, **42**, 7719.
6. P. Nun, C. Martin, J. Martinez and F. Lamaty, *Tetrahedron*, 2011, **67**, 8187.
7. F. Toda, H. Takumi and H. Yamaguchi, *Chem. Express*, 1989, **4**, 507.
8. K. Tanaka, S. Kishigami and F. Toda, *J. Org. Chem.*, 1991, **56**, 4333.
9. F. Toda, K. Tanaka and K. J. Hamai, *J. Chem. Soc., Perkin Trans. 1*, 1990, 3207.
10. D. J. Varughese, M. S. Manhas and A. K. Bose, *Tetrahedron Lett.*, 2006, **47**, 6795.
11. R. Schmidt, A. Stolle and B. Ondruschka, *Green Chem.*, 2012, **14**, 1673.
12. T. Szuppa, A. Stolle, B. Ondruschka and W. Hopfe, *Green Chem.*, 2010, **12**, 1288.
13. P. Machado, F. A. Rosa, M. Rossatto, G. S. Sant'Anna, P. D. Sauzem, R. M. S. Silva, M. A. Rubin, J. Ferreira, H. G. Bonacorso, N. Zanatta and M. A. P. Martins, *ARKIVOC*, 2007, **xvi**, 281.
14. (a) A. A. Bekhit and T. Abdel-Aziem, *Bioorg. Med. Chem.*, 2004, **12**, 1935; (b) C. Selvam, S. M. Jachak, R. Thilagavathi and A. K. Chakraborti, *Bioorg. Med. Chem. Lett.*, 2005, **15**, 1793; (c) S. P. Sing, R. Naithani, R. Aggarwal and O. Prakesh, *Indian J. Heterocycl. Chem.*, 2001, **11**, 27; (d) S. W. Djuric, N. Y. Bmaung, A. Basha, H. Liu, J. R. Luly, D. J. Madar, R. J. Sciotti, N. P. Tu, F. L. Wagenaar, P. E. Wiedman, X. Zhou, S. Ballaron, J. Bauch, Y.-W. Chen, X. G. Chiou, T. Fey, D. Gauvin, E. Gubbins, G. C. Hsieh, K. C. Marsch, K. W. Mollison, M. Pong, T. K. Shaughnessy, M. P. Sheets, M. Smith, J. M. Trevillyan, U. Warrior, C. D. Wegner and G. W. Carter, *J. Med. Chem.*, 2000, **43**, 2975; (e) S. S. Woodart, B. C. Hamper, K. Moedritzer, M. D. Rogers, D. A. Mischke, G. A. Dutra and Monsanto Co., *Chem. Abstr.*, 1995, **122**, 1116; (f) F. Chimenti, E. Maccioni, D. Secci, A. Bolasco, P. Chimenti, A. Granese, O. Befani, P. Turini, S. Alcaro, F. Ortuso, R. Cirilli, F. La Torre, M. C. Cardia and D. Simona, *J. Med. Chem.*, 2005, **48**, 7113.
15. K. Longhi, D. N. Moreira, M. R. B. Marzari, V. M. Floss, H. G. Bonacorso, N. Zanatta and M. A. P. Martins, *Tetrahedron Lett.*, 2010, **51**, 3193.

4.1.3.3. Organocatalytic Enantioselective Michael Addition of Thiophenol to Chalcone

Mariola Zielińska-Błajet* and Jacek Skarżewski

Department of Organic Chemistry, Faculty of Chemistry, Wrocław University of Technology, Wyb. Wyspiańskiego 27, 50-370 Wrocław, Poland. *E-mail: mariola.zielinska-blajet@pwr.edu.pl

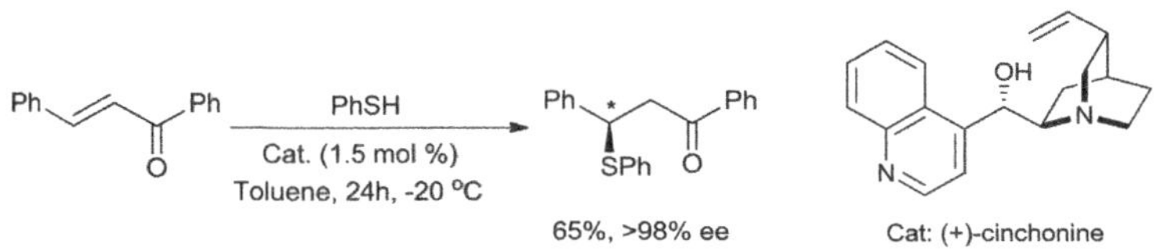

Number of session (duration of each session):	Hazard level	Difficulty level	Level of study
3 (1.5 h + 4 h + 4 h)	Moderate	Medium	Advanced

Class names α,β-Unsaturated ketones (enones), sulfides, ketones, thiols

Concepts involved The presented procedure describes the enantioselective, organocatalytic Michael addition reaction of thiophenol to *trans*-chalcone catalyzed by (+)-cinchonine, and further enantiomeric enrichment by crystallization of the crude product to give optically pure adduct, (+)-(*R*)-3-phenylsulfanyl-1,3-diphenylpropan-1-one

Chemicals needed *trans*-Chalcone, thiophenol, (+)-cinchonine, dry toluene, ethyl acetate, *n*-hexane, dichloromethane, hydrochloric acid (1 N aqueous sol.), sodium hydroxide (10% aq. sol.), anhydrous sodium sulfate, deuterated chloroform

Equipment and experimental techniques involved Magnetic stirring, freezer or cryocooler (−20 °C), access to inert gas (N_2 or Ar), vacuum evaporator, polarimeter, NMR spectrometer, melting point apparatus

Keywords Asymmetric synthesis, catalysis, conjugate addition, crystallization, enantioselectivity, optical rotation

Background

The Michael addition is one of the most important and useful reactions in organic chemistry for the construction of carbon–carbon and carbon-heteroatom bonds. More specifically, the Michael addition reaction is described as a special type of conjugate 1,4-addition, in which the strong nucleophile attacks at the β-carbon atom of an α,β-unsaturated carbonyl compound resulting in a negatively charged enolate intermediate, which after subsequent protonation yields the Michael adduct (Scheme 4.1.3.3.1). This thermodynamically driven reaction generates C–C, C–N, C–S, C–O, C–P and other C–X bonds within the Michael product.[1]

Comprehensive Organic Chemistry Experiments for the Laboratory Classroom
Edited by Carlos A. M. Afonso, Nuno R. Candeias, Dulce Pereira Simão, Alexandre F. Trindade, Jaime A. S. Coelho, Bin Tan, and Robert Franzén.

Published by the Royal Society of Chemistry, www.rsc.org

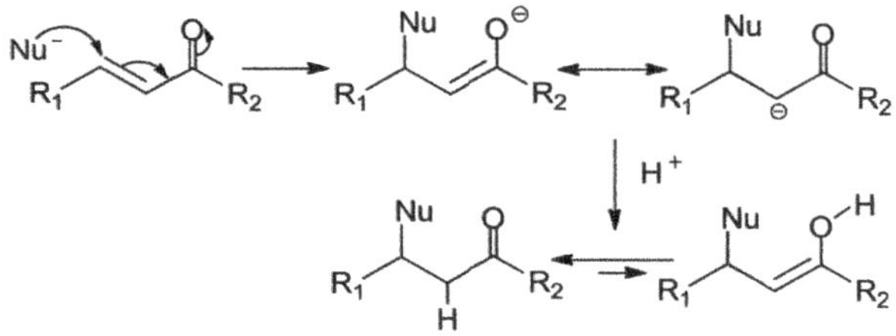

Scheme 4.1.3.3.1 Mechanism for the Michael addition of a nucleophile to an α,β-unsaturated carbonyl compound.

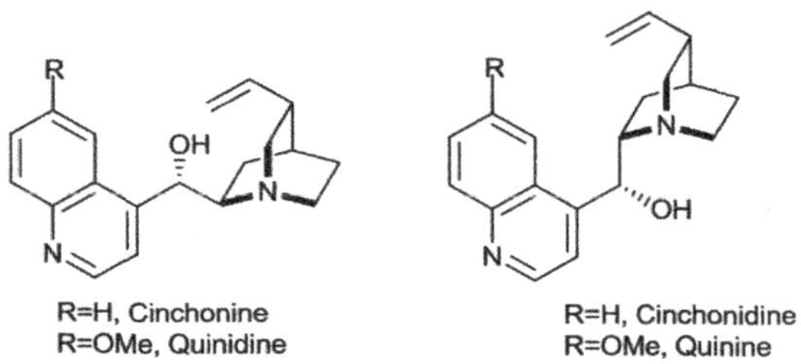

Scheme 4.1.3.3.2 Structure of representative *Cinchona* alkaloids.

Within asymmetric syntheses, considerable attention has been focused on the enantioselective conjugate addition of sulfur nucleophiles to α,β-unsaturated carbonyl compounds, creating a new stereogenic carbon center at the β-position with a new C–S bond formation.[2] The preparation of sulfur-containing molecules is very important, because of the widespread use of these compounds in organic and medicinal chemistry.[3]

Many efficient catalytic systems have been used to produce chiral sulfa-Michael adducts in high optical purity but the application of chiral β-amino alcohols, including *Cinchona* alkaloids, by Wynberg[4] in Michael reactions began the development of organocatalysis (Scheme 4.1.3.3.2).[5]

Organocatalytic enantioselective Michael addition of thiophenol to *trans*-chalcone catalyzed by a chiral bifunctional catalyst such as (+)-cinchonine, followed by crystallization, is a very simple and convenient method to obtain an optically pure adduct, *e.g.* (+)-(*R*)-3-phenylsulfanyl-1,3-diphenylpropan-1-one (over 98% ee, 65% yield). The obtained crude sulfa-Michael adduct with optical purity of about 74% was later enantioenriched by recrystallization from *n*-hexane/CH_2Cl_2. Notably, a single crystallization of the crude adduct gave mostly racemic crystals, while the concentration of mother liquor allows isolation of the enriched (+)-enantiomer. This procedure exemplifies the use of organocatalysis and crystallization as a tool for enantiomeric enrichment. During crystallization of the eutectic mixture two types of crystals are formed: enantiopure product and a racemic one. In our case we are dealing with the formation of racemic compound in the solid state (higher mp) while the mother liquor contains practically pure enantiomer. The composition at the eutectic point is very close to a pure enantiomer. This is not an uncommon case when crystallization is due to enantioenrichment of the lower melting material in the mother liquor.[6]

From a synthetic perspective, the resulting chiral ketone can be readily transformed into different classes of organic compounds under various, mild reaction conditions.[5a,7]

The objective of this experiment is the synthesis of optically pure ketone *via* organocatalytic, enantioselective sulfa-Michael addition reaction in the presence of (+)-cinchonine and familiarization with the technique of enantiomeric enrichment by crystallization.

Additional Safety

Sodium hydroxide, hydrochloric acid, and chloroform are toxic, as well as skin and eye irritants. Deuterated chloroform is also a known carcinogen. Dichloromethane and *n*-hexane are irritant and harmful by inhalation and if swallowed. Ethyl acetate is an extremely flammable liquid and vapor. It causes skin and eye irritation and is harmful if inhaled or swallowed. Toluene is flammable and has toxic effects on the central nervous system by inhalation. Cinchonine is harmful if swallowed. It may cause an allergic skin reaction. Thiophenol is an irritant and very toxic by ingestion, absorption through skin, or inhalation and has an unpleasant, smelly odor. Wear laboratory coats, gloves, and safety goggles at all times. All the experiments should be done in a well-ventilated fume hood to minimize exposure through inhalation.

Experimental Procedure

Laboratory Session 1 (1–1.5 h): Michael Addition of Thiophenol to Chalcone

1. Prepare a 100 mL two-neck round-bottomed flask with a magnetic stir bar, and add 2.08 g (10 mmol, 1 equiv.) of chalcone and 30 mL of dry toluene at room temperature, and stir until all the chalcone is dissolved (5–10 min).
2. Add 0.044 g (1.5% mol) of cinchonine and stir this mixture for 15 min at room temperature.
3. Cool the reaction mixture to *ca.* −20 °C in a cooling bath or freezer for 20 min.
4. Prepare a solution of 1.2 mL (11 mmol, 1.1 equiv.) of thiophenol in 5 mL of dry toluene in a 10 mL conical flask.
5. Add the solution of thiophenol dropwise by syringe to the cooled chalcone solution under a nitrogen atmosphere.
6. Keep the resulting mixture in a freezer (*ca.* −20 °C) for 24 h (until the next session).

Laboratory Session 2 (4 h): Product Isolation and Yield Determination

1. Remove the flask from the freezer and allow it to warm up to room temperature.
2. Add 25 mL of 1 M HCl and transfer the solution into a separatory funnel and perform an extraction with 3 × 20 mL of ethyl acetate.
3. Wash the combined extracts with 20 mL of 10% NaOH, then with brine (20 mL), and dry over anhydrous Na_2SO_4, filter off the solid and evaporate the solvent.
4. Weigh the obtained product and calculate the yield.
5. Measure the melting point, optical rotation and record ^1H and ^{13}C NMR spectra of crude adduct to confirm the purity. Compare the results with the literature data.[5a,8,9]

Laboratory Session 3 (4 h): Enantiomeric Enrichment of Product by Crystallization

1. Place the crude product in a 250 mL round-bottomed flask equipped with a reflux condenser and add 70 mL of *n*-hexane and 20 mL of dichloromethane. Heat until all crude solid dissolves.
2. Allow the flask to cool to room temperature and put the flask in an ice bath for crystallization.
3. Filter off the precipitate under reduced pressure, and concentrate the filtrate (mother liquor) by evaporation. Dry both solids using a vacuum pump or desiccators.
4. Weigh the obtained solids, measure their melting points and optical rotations. Compare the results with the literature data.[5a,8,9]

Waste Disposal

The neutralized solutions of acid and base may be diluted and discarded in the waste collector. Used organic solvents should be collected in appropriate waste containers (halogenated for the mixture of dichloromethane and *n*-hexane after crystallization, and non-halogenated for the mixture of toluene and ethyl acetate after work-up). Sodium sulfate should be collected in a container for solid wastes.

Results Interpretation and Additional Questions

1. Calculate the percent yield for the overall process.
2. Interpret the ^{1}H NMR spectrum of product and compare it with the literature data.[5a]
3. Compare the melting points of the obtained racemic and enantiomerically pure product with the literature data.[5a,8]
4. Calculate the enantiomeric excess (ee, %) of the obtained Michael adducts by comparison with the optical rotation of the pure enantiomer described in the literature data.[5a,9]
5. Draw a mechanism for this reaction and discuss the role of the organocatalyst in the overall process.
6. Discuss the relationship between solubility and melting points of the formed crystals during enantiomeric enrichment by crystallization.

References

1. (a) T. Sasai, T. Arai, Y. Satow, K. N. Houk and M. Shibasaki, *J. Am. Chem. Soc.*, 1995, **117**, 6194; (b) S. Cheng and D. D. Cromer, *Tetrahedron Lett.*, 2001, **42**, 37; (c) K. Jørgensen and D. Perdicchia, *J. Org. Chem.*, 2007, **72**, 3565.
2. (a) M. Mikołajczyk, J. Drabowicz and P. Kiełbasiński, *Stereoselective Synthesis, Methods of Organic Chemistry (Houben-Weyl)*, ed. G. Helmchen, R. W.Hoffmann, J. Mulzer and E. Schaumann, Thieme, Stuttgart, 1995, vol. E21e, p. 5017; (b) K. Tomioka, Y. Nagaoka and M. Yamaguchi, *Comprehensive Asymmetric Catalysis*, ed. E. N. Jacobsen, A. Pfaltz and H. Yamamoto, Springer, Berlin, 1999, p. 1105; (c) N. Krause and A. Hoffmann-Röder, *Synthesis*, 2001, 171; (d) P. Chauhan, S. Mahajan and D. Enders, *Chem. Rev.*, 2014, **114**, 8807.
3. D. Enders, K. Lüttgen and A. A. Narine, *Synthesis*, 2007, **7**, 959.
4. H. Hiemstra and H. Wynberg, *J. Am. Chem. Soc.*, 1981, **103**, 417.
5. (a) J. Skarżewski, M. Zielinska-Błajet and I. Turowska-Tyrk, *Tetrahedron: Asymmetry*, 2001, **12**, 1923; (b) P. McDaid, Y. G. Chen and L. Deng, *Angew. Chem., Int. Ed.*, 2002, **41**, 338; (c) T. C. Wabnitz and J. B. Spencer, *Org. Lett.*, 2003, **5**, 2141; (d) H. Li, J. Wang, L. S. Zu and W. Wang, *Tetrahedron Lett.*, 2006, **47**, 2585; (e) H. Li, L. Zu, J. Wang and W. Wang, *Tetrahedron Lett.*, 2006, **47**, 3145; (f) S. Connon, *J. Chem. Commun.*, 2008, 2499; (g) J. W. Lee, H. B. Jang and C. E. Song, *Cinchona Alkaloids in Synthesis & Catalysis*, ed. C. E. Song, Wiley-VCH Verlag GmbH, Weinheim, 2009, vol. 9, p. 249; (h) R. Kowalczyk, A. E. Nowak and J. Skarżewski, *Tetrahedron: Asymmetry*, 2013, **24**, 505; (i) R. Kowalczyk, A. J. Wierzba, P. J. Boratyński and J. Bąkowicz, *Tetrahedron*, 2014, **70**, 5834.
6. (a) J. Mulzer, *Stereoselective Synthesis, Methods of Organic Chemistry (Houben-Weyl)*, ed. R. W. Hoffmann, J. Mulzer and E. Schaumann, Thieme, Stuttgart, 1995, vol. E21a, p. 77; (b) E. L. Eliel, S. H. Wilen and L. N. Mander, *Stereochemistry of Organic Compounds*, John Wiley & Sons, Inc., New York, 1994, vol. 167, p. 381.
7. (a) J. Skarżewski, R. Siedlecka, E. Wojaczyńska and M. Zielińska-Błajet, *Tetrahedron: Asymmetry*, 2002, **13**, 2105; (b) J. Skarżewski, M. Zielińska-Błajet, S. Roszak and I. Turowska-Tyrk, *Tetrahedron*, 2003, **59**, 3621; (c) M. Zielińska-Błajet, R. Kowalczyk and J. Skarżewski, *Tetrahedron*, 2005, **61**, 5235.
8. A. R. Katritzky, J. Chen and S. A. Belyakov, *Tetrahedron Lett.*, 1996, **37**, 6631.
9. V. V. Krotov, S. M. Staroverov, P. N. Nesterenko and G. V. Lisichkin, *J. Gen. Chem. USSR (Engl. Transl.)*, 1986, **56**, 2177.

4.1.3.4. Stereoselective Synthesis of *meso*-1-Allyl-2,6-diphenylpiperidin-4-one

Ana-Belén García Delgado, Noelia Quiñones, and María-Paz Cabal*

Departamento de Química Orgánica e Inorgánica, Facultad de Químicas, Universidad de Oviedo, Spain
*E-mail: pcabal@uniovi.es

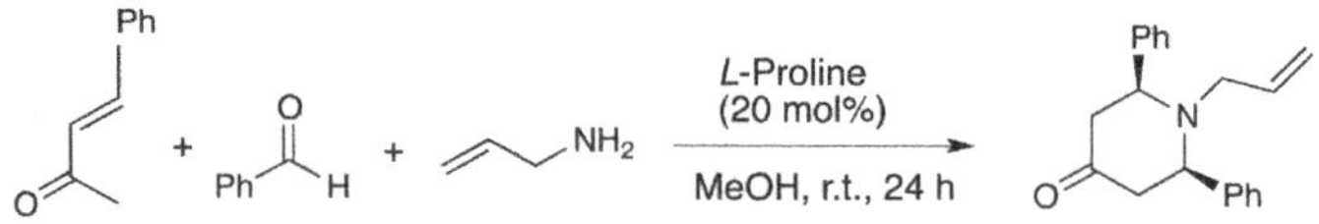

Number of sessions (duration of each session)	Hazard level	Difficulty level	Level of study
1 (4 h)	Moderate	Medium	Advanced

Class names Aldehydes, α,β-unsaturated ketones, amines, amino acids

Concepts involved This experiment involves a multicomponent aza-Diels–Alder reaction catalysed by L-proline. The final product can be isolated by an acid–base extraction work-up

Chemicals needed 4-Phenyl-3-buten-2-one, benzaldehyde, allylamine, L-Proline, methanol, NaOH, HCl, Na_2SO_4, *n*-hexane, ethyl acetate, silica gel 60, deuterated chloroform

Equipment and experimental techniques involved Hot plate with stirring motor, extraction funnel, column chromatography, TLC plates (SiO_2), rotary evaporator, NMR (^1H and ^{13}C)

Keywords Aza-Diels–Alder, column chromatography, 2,6-diarylpiperidin-4-ones, extraction, nitrogen heterocycles, NMR, organocatalysis, L-proline

Background

The piperidine ring is an important framework present in a large variety of natural products.[1] In particular, piperidin-4-ones are versatile building blocks due to the easy manipulation of the carbonyl group for the introduction of different substituents into the six-membered ring.[2] Substitution at the C2 and C6 positions is very frequent and in some cases biological activity studies have shown a relationship between the biological activity and the aromatic ring present in those positions.[3] Even though the synthesis of piperidin-4-ones is still a very important theme of research and several synthetic methods have been developed,[2] in the case of 2,6-diarylpiperidin-4-ones very few methodologies can be found in the literature.[4]

On the other hand, over the past few years organocatalysis has become very popular for a wide range of enantioselective transformations. In particular, L-proline has been reported to catalyse asymmetric aldol, Michael, Mannich, α-amination, Diels–Alder dimerization, and aza-Diels–Alder reactions with cyclic enones.[5] The main goal of this experiment is to use organocatalysis to describe a new synthesis of *meso*-2,6-disubstituted piperidin-4-ones by the aza-Diels–Alder

Comprehensive Organic Chemistry Experiments for the Laboratory Classroom
Edited by Carlos A. M. Afonso, Nuno R. Candeias, Dulce Pereira Simão, Alexandre F. Trindade, Jaime A. S. Coelho, Bin Tan, and Robert Franzén.

Published by the Royal Society of Chemistry, www.rsc.org

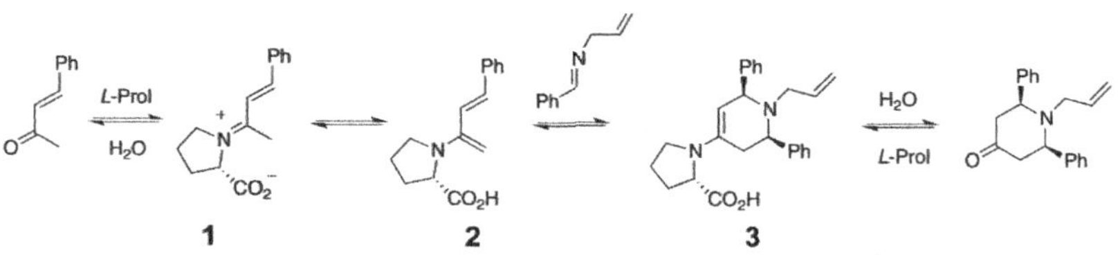

Scheme 4.1.3.4.1

reaction between an acyclic enone and an aldehyde and an amine as a one-pot multi-component reaction, in the presence of L-proline as catalyst.

The following mechanism is proposed for the proline-catalysed imino-Diels–Alder reaction leading to the symmetrical 2,6-diphenyl-4-piperidones (Scheme 4.1.3.4.1). First, the 2-amino-1,3-butadiene (**2**) is generated *in situ* by the reaction of the α,β-unsaturated ketone with L-proline giving the iminium zwitterion (**1**). Equilibration of **1** to 2-amino-1,3-butadiene (**2**), followed by cycloaddition with the imine, provides tetrahydropyridine adduct **3**, which hydrolyzes to the final *N*-allyl-2,6-diphenyl-4-piperidone.

The simplicity and mildness of the reaction protocol make this a very efficient and stereoselective method to obtain *meso*-2,6-disubstituted-4-piperidones. Furthermore, compared to previously reported methods, this L-proline-catalysed cycloaddition methodology has the advantages that neither the use of a metal catalyst nor the restriction of having to isolate the diene intermediate prior to the cycloaddition is required; in addition, the reaction can easily be run on a multi-gram scale.

Additional Safety

All sessions should be done in a fume hood, with proper safety equipment, such as laboratory coat, gloves and safety goggles. 4-Phenyl-3-butenone and benzaldehyde may cause eye, skin and respiratory irritation. Allylamines are lachrymators and skin irritants (LD_{50} oral (rats) = 106 mg kg^{-1}). Methanol causes eye and skin irritation and can be absorbed through the skin. Silica gel 60 is irritating to the respiratory tract, skin and eyes.

Experimental Procedure

Pre-session (30 min and Overnight Reaction)

1. Prepare a 10 mL round-bottom flask with a magnetic stirring bar.
2. Dissolve in MeOH (2 mL) the following reagents in the order given: benzaldehyde (106 mg, 1 mmol, 1 eq.), allylamine (63 mg, 1.1 mmol, 1.1 equiv.) and L-proline (23 mg, 20 mol%).
3. Add 4-phenylbut-3-en-2-one (585 mg, 4 mmol, 4 equiv.) and allow the mixture to stir at room temperature for 24 h.

Laboratory Session (4 h): Isolation and Purification of the Product

1. Evaporate the solvent under reduced pressure and subject the mixture to an acid–base extraction workup to eliminate the excess ketone following the procedure.
2. Add ethyl acetate (10 mL) to the reaction crude and wash the organic layer with 1 M HCl (3 × 5 mL).
3. Immediately collect the aqueous layers from each washing with HCl (1 M) on a cold 1 M NaOH (25 mL) solution and extract the resulting mixture (after the three washings) thoroughly with EtOAc (3 × 20 mL).

4. Dry the combined organic phases over Na_2SO_4, and concentrate them with a rotary evaporator.
5. Use column chromatography – silica gel (25 g), column size 20 mm diameter, eluent: *n*-hexane–EtOAc (10: 1); $R_f = 0.31$ – to obtain the desired pure product (75–80% yield).

Waste Disposal

The neutralized aqueous solution may be diluted and discarded in the sink. Ethyl acetate and methanol may be reused or collected in a non-halogenated waste container for further destruction.

Results Interpretation and Additional Questions

1. In your report the R_f and the yield of the final product should be included.
2. What do you think is the reason for performing an acid–base extraction procedure to isolate the final product?
3. Write the mechanism of the reaction and show the role of the L-proline as a catalyst in the process.
4. Assign the obtained NMR data to confirm the structure of the final product.
5. Calculate the coupling constants for the allylic and olefinic protons.
6. How can you confirm the stereochemistry of the final product by NMR?
7. Choose in the literature another example of an organocatalysis reaction as an important tool for enantioselective organic synthesis.

References

1. (a) P. D. Bailey, P. D. Smith, K. M. Morgan and G. M. Rosair, *Tetrahedron Lett.*, 2002, **43**, 1071; (b) M. Amat, N. Llor, J. Hidalgo, C. Escolano and J. Bosch, *J. Org. Chem.*, 2003, **68**, 1919.
2. (a) P. M. Weintraub, J. S. Sabol, J. M. Kane and D. R. Borcherding, *Tetrahedron*, 2003, **59**, 2953; (b) R. B. C. Jagt, J. G. Vries, B. L. Feringa and A. Minnaard, *Org. Lett.*, 2005, **7**, 2433.
3. (a) N. Rameshkumar, A. Veena, R. Ilavarasan, M. Adiraj, P. Shanmugapandiyan and S. K. Sridhar, *Biol. Pharm. Bull.*, 2003, **26**, 188; (b) C. Ramalingan, S. Balasubramanian, S. Kabilan and M. Vasudevan, *Eur. J. Med. Chem.*, 2004, **34**, 527.
4. (a) D. Ma and N. Ma, *Tetrahedron Lett.*, 2003, **44**, 3963; (b) F. A. Davis, A. Rao and P. J. Carrol, *Org. Lett.*, 2003, **5**, 3855.
5. (a) B. List, *Synlett*, 2001, 1657; (b) W. Notz, F. Tanaka and C. F. Barbas III, *Acc. Chem. Res.*, 2004, **37**, 580; (c) H. Sundén, I. Ibrahem, L. Eriksson and A. Córdova, *Angew. Chem., Int. Ed*, 2005, **44**, 4877.

4.1.3.5. Synthesis of a Squarylium Cyanine Dye as Potential Photosensitizer for Photodynamic Therapy (PDT)

Marlene L. F. M. Pacheco[a], Sofia F. P. Friães[a], Renato E. Boto[b], Paulo Almeida[b], Amélia M. Silva[c], and Lucinda V. Reis*[a]

[a]Department of Chemistry and CQ-VR, UTAD, Quinta de Prados, 5000-801 Vila Real, Portugal; [b]CICS-UBI – Health Sciences Research Centre, University of Beira Interior, Av. Infante D. Henrique, 6200-506 Covilhã, Portugal; [c]Department of Biology and Environment and CITAB, UTAD, Quinta de Prados, 5000-801 Vila Real, Portugal. *E-mail: lucinda.reis@utad.pt

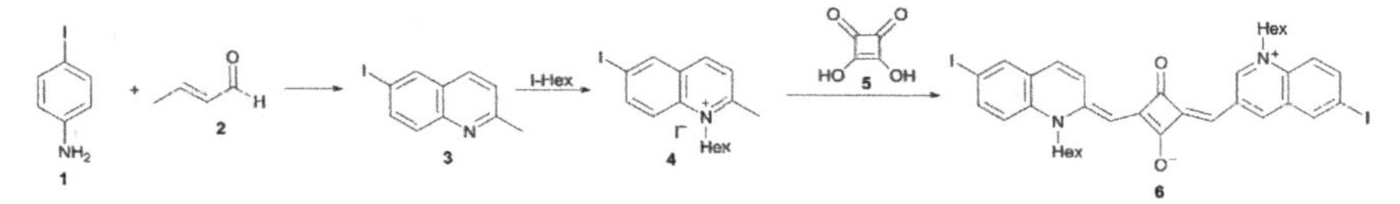

Number of sessions (duration of each session)	Hazard level	Difficulty level	Level of study
3 (4 h each)	Moderate	Medium	Advanced

Class names Aromatic amines, α,β-unsaturated aldehydes, alkyl iodides, ketones

Concepts involved This experiment involves the synthesis of a squarylium cyanine dye in a three-step reaction.6-Iodoquinaldine (**3**), obtained *via* a Doebner–Miller reaction, is alkylated with 1-iodohexane to obtain a quaternary quinaldinium salt **4**, which by condensation with squaric acid (**5**) affords the squarylium cyanine dye **6** with a moderate yield. In addition to this organic chemical knowledge, students have the opportunity to use some isolation, purification and characterization techniques, namely, liquid–liquid extraction, vacuum filtration, recrystallization, melting point determination, IR, VIS, NMR

Chemicals needed 4-Iodoaniline, crotonaldehyde, TBAF (tetrabutylammonium fluoride) solution (1.0 M) in tetrahydrofuran, aqueous solution of 6.0 M HCl, toluene, aqueous solution of 6.0 M NaOH, diethyl ether, anhydrous Na_2SO_4, ethyl acetate, petroleum ether (40–60 °C), 1-iodohexane, acetonitrile, squaric acid, *n*-butanol, pyridine, methanol, dichloromethane

Equipment and experimental techniques involved Apparatus with a 250 mL two-neck round-bottomed flask for reaction with temperature and magnetic stirring, liquid–liquid extraction apparatus, rotary evaporator, vacuum filtration apparatus, recrystallization, NMR, UV/Vis, IR and melting point instruments

Keywords Doebner–Miller reaction, IR, liquid–liquid extraction, NMR, photodynamic therapy (PDT), squarylium cyanine dye

Comprehensive Organic Chemistry Experiments for the Laboratory Classroom
Edited by Carlos A. M. Afonso, Nuno R. Candeias, Dulce Pereira Simão, Alexandre F. Trindade, Jaime A. S. Coelho, Bin Tan, and Robert Franzén.

Published by the Royal Society of Chemistry, www.rsc.org

Background

Photodynamic therapy (PDT) is a treatment that results from a sequence of photochemical and photobiological processes that cause irreversible photodamage to tumor tissues. For it to be effective, three elements are required: visible or near-infrared (NIR) radiation; administration of a photosensitizer and the presence of endogenous molecular oxygen.[1] A desirable sensitizer for PDT, among other parameters, should have strong absorption ($\varepsilon > 10^5$ M^{-1} cm^{-1}) within the longer wavelength region, ideally in 600–900 nm, which is also called the "phototherapeutic window", wherein the tissue penetration by light is higher and light scattering is reduced. Squarylium cyanine dyes, such as compound **6**, are 1,3-disubstituted compounds resulting from the condensation of one equivalent of squaric acid (3,4-dihydroxycyclobut-3-ene-1,2-dione, **5**) with two molar equivalents of electron donating aromatic or heterocyclic methylene bases, and are potential photosensitizers for PDT.[2,3] In early 2000, the synthesis of new cyanine dyes as potential sensitizers for PDT was described.[4,5]

Quite recently, a new squarylium cyanine dye **6** was synthesized bearing two iodine atoms in its aromatic moiety. In this experiment, students have the opportunity to synthesize the squarylium cyanine dye **6**, which absorbs strongly within the so-called "phototherapeutic window" (λ_{max} = 707 nm, log ε = 5.16) in a three-step reaction. The first step consists of the synthesis of 6-iodoquinaldine (**3**) by a Doebner–Miller reaction. This reaction is a well-known condensation between aniline and a α,β-unsaturated carbonyl compound to afford a quinoline. In this experiment the aniline used is the 4-iodoaniline (**1**) and the α,β-unsaturated carbonyl compound is crotonaldehyde (**2**). The reaction is carried out in an aqueous solution of HCl using a two-phase solvent system with toluene, and tetrabutylammonium fluoride as phase-transfer catalyst, following reaction conditions adapted from the literature.[6,7]

In the second step a quaternary quinaldinium salt **4** is synthesized by nitrogen alkylation of the 6-iodoquinaldine (**3**) (6-iodo-2-methylquinoline) with 1-iodohexane, *via* an S_N2 reaction, in acetonitrile under reflux.[8,9] This salt is easily isolated by precipitation with diethyl ether followed by filtration and is used without further purification with regard to satisfactory purity showed by the NMR spectrum. Finally, this quaternary salt (the heterocyclic methylene base), by condensation with 3,4-dihydroxycyclobut-3-ene-1,2-dione (squaric acid, **5**) in a mixture of *n*-butanol/pyridine (10%) under reflux, affords the squarylium cyanine dye **6** in moderate yield.

Additional Safety

All sessions should be done in a fume hood and students should have protective clothing, gloves and safety goggles. NaOH is a caustic base and therefore may cause severe skin burns. HCl is a corrosive acid so students should avoid contact with skin and eyes. Crotonaldehyde is a lachrymator and so it must be handled with care inside of the hood. Prevent skin and eye contact, as well as inhalation.

Experimental Procedure

Laboratory Session 1 (4 h)

Preparation of 6-Iodoquinaldine (3)

1. In a fume hood, heat an oil bath to maintain a temperature of 85 °C on a magnetic hot plate.
2. Prepare a 250 mL two-neck round-bottomed flask with a magnetic stir bar and add 5 g (22.8 mmol, 1 equiv.) of 4-iodoaniline (**1**), 160 mL of an aqueous solution of 6.0 M HCl, 40 mL of toluene and 0.36 mL of TBAF solution (1.0 M) in tetrahydrofuran.

3. Insert a water condenser into a neck of the round-bottomed flask and clamp the flask; then insert it into the 85 °C preheated oil bath.
4. In the other neck of the round-bottomed flask insert a pressure equalizing funnel containing a mixture of crotonaldehyde (2) (3.7 mL, 45.7 mmol, 2 equiv.) and toluene (4.0 mL).
5. Add, dropwise, the mixture of crotonaldehyde to the stirred mixture contained in the round-bottomed flask.
6. Stir the reaction rapidly for 1 h 30 min, maintaining a constant temperature of 85 °C.
7. Remove the flask from the oil bath, cool it down in an ice bath, allowing it to reach room temperature.
8. Transfer the reaction mixture into a separatory funnel, separate the aqueous layer into an Erlenmeyer flask and slowly add an aqueous solution of 6.0 M NaOH (about 150 mL) to neutralize the mixture. Discard the organic layer in a non-chlorinated waste collector recipient for posterior destruction.

Isolation of 6-Iodoquinaldine (3)

9. Transfer the neutralized solution into a separatory funnel and perform the extraction with 3 × 100 mL of diethyl ether. Remove the water from the combined organic layers with anhydrous Na_2SO_4, filter and evaporate the solvent in the rotary evaporator.
10. Recrystallize the so-obtained product by dissolving it in 20 mL of a hot mixture of ethyl acetate–petroleum ether (2 : 1). Let the mixture cool at room temperature for 10–15 min and then put it in an ice bath to complete the precipitation.
11. Filter the crystals in a vacuum filtration apparatus, wash them with cold petroleum ether, dry the obtained pure product (in a vacuum pump) and calculate the yield.
12. Save an aliquot to measure the melting point and record the IR spectrum in the last session.
13. To enhance the yield, concentrate the mother liquors in a total rotary evaporator, cool the round-bottomed flask in an ice bath and add petroleum ether until no turbidity of the solution is observed. Induce crystallization by scraping the bottom of the flask with a metal spatula. Filter the so-obtained crystals in a vacuum apparatus filtration, wash them with cold petroleum ether, dry the obtained product (in a vacuum pump) and calculate the yield.

Preparation of 6-Iodo-N-hexylquinaldinium Iodide (4)

14. In a fume hood, heat an oil bath to maintain a temperature of 100 °C on a magnetic hot plate.
15. Prepare a 50 mL round-bottomed flask with a magnetic stir bar and add 2.0 g (7.04 mmol, 1 equiv.) of 6-iodoquinaldine (3), 30 mL of acetonitrile and 3.3 mL (21.12 mmol, 3 equiv.) of 1-iodohexane.
16. Insert a water condenser into the neck of the round-bottomed flask, clamp the flask and place it in the preheated oil bath.
17. Keep the mixture under reflux during a minimum of 5 days or until next session.

Laboratory Session 2 (4 h)

Isolation of 6-Iodo-N-hexylquinaldinium Iodide (4)

1. Remove the round-bottomed flask from the oil bath and put in on an ice bath to cool to room temperature.
2. Add diethyl ether in order to precipitate the formed quaternary salt **4**.
3. Filter the obtained solid under reduced pressure and then wash it with cold ether.

4. Dry the pure product in a vacuum pump and calculate the yield.
5. Save an aliquot to measure the melting point and record the IR spectrum in the last session.

Synthesis of the Squarylium Cyanine Dye (6)

6. In a fume hood heat an oil bath to maintain a temperature of 130 °C on a magnetic hot plate.
7. Prepare a 250 mL two-neck round-bottomed flask with a magnetic stir bar and add 1.0 g (2.83 mmol, 2 equiv.) of 6-iodo-*N*-hexylquinaldinium iodide (**4**), 0.160 g (1.41 mmol, 1 equiv.) of squaric acid (**5**) and 100 mL of a mixture of *n*-butanol/pyridine (10%).
8. Insert a water condenser into the neck of the round-bottomed flask and clamp the flask, then insert it in the preheated oil bath.
9. Stir the reaction for 2 h under reflux. Over time the mixture becomes an increasingly darker shade of green.
10. Remove the round-bottomed flask from the oil bath and put in a refrigerator until the next session.

Laboratory Session 3 (4 h)

Isolation of the Squarylium Cyanine Dye (6)

1. Filter the formed crystals under reduced pressure and wash them with cold diethyl ether and then with cold distilled water.
2. Transfer mother liquors into a separatory funnel, add 50 mL of CH_2Cl_2 and wash them with 3 × 50 mL of distilled water.
3. Remove the water from the organic layer with anhydrous Na_2SO_4, filter and evaporate the solvent in the rotary evaporator.
4. Combine the remaining golden brown solid with crystals in a round-bottomed flask.

Dye purification and Spectra Recording

5. Recrystallize the obtained golden brown product by dissolving after heat in 30 mL of a hot $MeOH–CH_2Cl_2$ (2 : 1) mixture. Place the round-bottomed flask containing the mixture in an ice bath, scrape the bottom of the flask with a metal spatula to induce crystallization, add a few drops of ethyl ether, and allow stand in an ice bath until precipitation is complete.
6. Filter off the golden brown crystals with a vacuum filtration apparatus, wash them with cold methanol, dry the obtained pure product (in a vacuum pump) and calculate the yield.
7. Measure the melting points of compounds **3**, **4** and **6**, record their IR spectra and also record the Vis spectrum of squarylium cyanine dye **6** in CH_2Cl_2.

Waste Disposal

In session 1 the neutralized aqueous solution should be discarded in an appropriate container. Diethyl ether and dichloromethane may be collected and reused. Acetonitrile and toluene may be collected in a non-chlorinated waste collector recipient for posterior destruction. To the layer aqueous solution of *n*-butanol/pyridine resulting from the dye reaction add an aqueous solution of 1.0 M HCl in order to neutralize the solution and then discard it in an appropriate container.

Results Interpretation and Additional Questions

1. Discuss the importance of TBAF addition and the rapid stirring of the reaction mixture in the 6-iodoquinaldine (3) formation.

2. Compare the melting point of 6-iodoquinaldine (**3**) obtained with the one given in the literature.[10,11]
3. Assign the principal bands in the IR spectra obtained for quaternary salt **4** and dye **6** and compare them with the related ones described in the literature.[8,9]
4. Interpret the NMR spectra of quaternary salt **4** and dye **6** and compare them with the related ones described in the literature.[8,9]
5. Rationalize the mechanisms involved in each reaction step.

References

1. R. Bonnett, *Chemical Aspects of Photodynamic Therapy*, Gordon & Breach Science Publishers, Amsterdam, 1st edn, 2000, vol. 4, pp. 57–87.
2. P. F. Santos, L. V. Reis, I. Duarte, J. P. Serrano, P. Almeida, A. S. Oliveira and L. F. V. Ferreira, *Helv. Chim. Acta*, 2005, **88**, 1135–1143.
3. R. R. Avirah, D. T. Jayaram, N. A. Adarsh and D. Ramaiah, *Org. Biomol. Chem.*, 2012, **10**, 911–920.
4. P. F. Santos, L. V. Reis, P. Almeida, A. S. Oliveira and L. F. V. Ferreira, *J. Photochem. Photobiol., A*, 2003, **160**, 159–161.
5. P. F. Santos, L. V. Reis, P. Almeida, J. P. Serrano, A. S. Oliveira and L. F. V. Ferreira, *J. Photochem. Photobiol., A*, 2004, **163**, 267–269.
6. M. Matsugi, F. Tabuasa and J.-I. Minamikawa, *Tetrahedron Lett.*, 2000, **41**, 8523–8525.
7. K. A. Reynolds, D. J. Young and W. A. Loughlin, *Synthesis*, 2010, **21**, 3645–3648.
8. A. C. Pardal, S. S. Ramos, P. F. Santos, L. V. Reis and P. Almeida, *Molecules*, 2002, **7**, 320–330.
9. L. V. Reis, J. P. Serrano, P. Almeida and P. F. Santos, *Dyes Pigm.*, 2009, **81**, 197–202.
10. K. Jyothish, R. R. Avirah and D. Ramaiah, *ARKIVOC*, 2007, **8**, 296–310.
11. K. Jyothish, K. T. Arun and D. Ramaiah, *Org. Lett.*, 2004, **6**, 3965–3968.

4.2. C-C σ-bond Formation

4.2.1. Grignard Addition to Carbonyl Compounds

4.2.1.1. The Effects of Stoichiometry and Starting Material on the Product Identity and Yield in Grignard Addition Reactions

Jane Brock Greco* and Eric Hill

Department of Chemistry, Johns Hopkins University, 3400 N. Charles St., Baltimore, MD 21218, USA
*E-mail: jgreco@jhu.edu

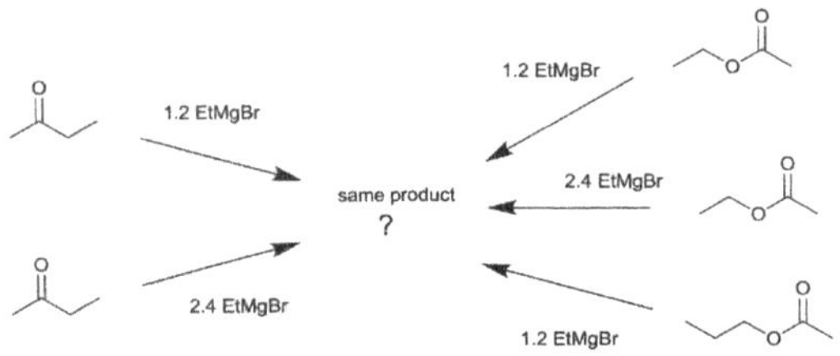

Number of sessions (duration of each session)	Hazard level	Difficulty level	Level of study
1 (5 h)	High	Medium	Introductory
Class names Ketones, esters, alkyl bromides, alcohols, organometallics			
Concepts involved This experiment is designed to illustrate that in a Grignard reaction the relative reactivity of the starting material plays a greater role in product formation than the stoichiometry			
Chemicals needed Ethyl bromide, magnesium, 2-butanone, ethyl acetate, propyl acetate, anhydrous diethyl ether, Drierite, sulfuric acid, magnesium sulfate, sodium bicarbonate, sodium chloride, $CDCl_3$			
Equipment and experimental techniques involved Apparatus for slow addition and reflux, liquid–liquid extraction, distillation, NMR, GC			
Keywords Distillation, extraction, Grignard reaction, NMR, slow addition			

Background

The 1912 Nobel Prize in Chemistry was awarded to Victor Grignard for the carbon–carbon bond forming reaction that now bears his name.[1] The first step of a Grignard reaction is the insertion of Mg into a carbon–halogen bond to generate an organometallic compound, thereby reversing

Comprehensive Organic Chemistry Experiments for the Laboratory Classroom
Edited by Carlos A. M. Afonso, Nuno R. Candeias, Dulce Pereira Simão, Alexandre F. Trindade, Jaime A. S. Coelho, Bin Tan, and Robert Franzén.

Published by the Royal Society of Chemistry, www.rsc.org

the polarity of carbon from partially positive to strongly negative, generating a potent nucleophile. This anionic carbon can then perform nucleophilic addition to a carbonyl, resulting in a new carbon–carbon bond. After the resulting magnesium alkoxide is quenched with acid, an alcohol is produced.

Grignard reactions are frequently performed in undergraduate teaching laboratories, and most organic laboratory textbooks have at least one example of a synthesis using a Grignard reagent. One of the most commonly performed reactions is the synthesis of either benzoic acid or triphenylmethanol from phenylmagnesium bromide. Various inquiry orientated Grignard reactions have been published, including the determination of an unknown by characterization of the product formed by reaction with phenylmagnesium bromide,[2] an investigation of the side products formed in Grignard reactions,[3] and the determination of an unknown ketone after the addition of an alkyl Grignard reagent.[4]

The Grignard reaction serves as an excellent teaching tool, because it is often an early introduction to the concept of side products generated in chemical reactions. While most students are able to recite that ketones are more electrophilic than esters and therefore more susceptible to nucleophilic addition, they fail to predict that a ketone cannot be isolated from a reaction mixture of an ester and one equivalent of a Grignard reagent. This experiment was designed to provide an inquiry investigation into the products formed by the reaction of ethylmagnesium bromide with 2-butanone, ethyl acetate and propyl acetate. The students are given different conditions, some using just over two equivalents of the Grignard reagent and others using just over one equivalent. They then investigate how the number of equivalents used affects both the product formation and the percent yield.

Distillation

Distillation, a millennia old technique, is one of the most common methods of purification used in laboratories, as well as having widespread industrial uses. Distillation takes advantage of the difference in boiling point to isolate desired compounds from an impure mixture, where lower boiling molecules are converted into the gaseous phase at a lower temperature than those that are less volatile.

In our case (see Figure 4.2.1.1.2 below), the heat source raises the temperature of the solution in your round-bottom flask, until diethyl ether is converted into a gas. The warm diethyl ether vapors then recondense on the glassware above the heat source and drain back into the warmed solution. This process repeats itself until the vapors warm the glassware up to the thermometer, raising the observed temperature. Then, by way of the cool water flowing through your reflux condenser, the vapors will be converted back into a liquid, which flows down your reflux condenser and into your collection flask. This process will continue until you have distilled the vast majority of the diethyl ether from your mixture, at which point the temperature on your thermometer will begin to decrease, as insufficient gas vapors remain to maintain the elevated temperature.

With the majority of the diethyl ether now distilled out of your initial mixture, the temperature of your solution will begin to rise above the boiling point of diethyl ether; converting the molecules with higher boiling points into the gaseous phase. These vapors will condense on the glassware and drain back into your solution, until they raise the temperature of the apparatus up to the thermometer and subsequently begin to condense over. You can switch your collection flask as the temperature on your thermometer changes, thereby separating your initially impure mixture into separate fractions, purified by distillation. To avoid your solution splattering, you need to use a round-bottom flask that is greater than twice the volume of the amount of liquid that you want to distill. However, using too large a round-bottom flask can lead to loss of product. In this experiment, you will need to transfer your compound to a smaller round-bottom flask before distilling off your product.

Additional Safety

All sessions should be performed in a fume hood, with proper safety equipment, including protective clothing, gloves and safety goggles. Diethyl ether is extremely volatile and flammable, so all experiments should be performed in the absence of flames or other heat sources. The initiation of a Grignard reagent is autocatalytic, and initiation must be determined before additional alkyl bromide is added. The alkyl bromide must be added slowly. Addition of too much alkyl bromide before initiation has begun, or too quickly thereafter, can lead to a build-up of pressure, and a projectile drying tube. Acidic solutions are corrosive, and the reaction of the magnesium alkoxide with acid is extremely exothermic. Furthermore, the addition of acid to magnesium leads to the production of hydrogen gas, which is flammable and can lead to a build-up of pressure if not properly vented.

Experimental Procedure

1. Make sure that all glassware is dry from the week before. Do not attempt to clean glassware this week.
2. Determine from your instructor which reaction you have been assigned. Calculate the mass and/or volume of the reagents that you will need based on the number of moles given in Table 4.2.1.1.1.
3. Make sure that your drying tubes are filled with fresh Drierite, only a small amount is necessary.
4. Clamp a 250 mL round-bottom flask, fitted with a stir bar, on top of a stir plate and inside a heating mantle that is turned off. Attach the Claisen adaptor to the round-bottom flask. Onto the Claisen adaptor attach both a reflux condenser and separatory funnel, which should both be fitted with drying tubes (Figure 4.2.1.1.1). Attach the water hoses so that the water flows upwards.
5. Lightly grind the magnesium metal in a mortar and pestle. Remember, you are trying to expose a fresh surface, not to create a powder. Transfer the ground turnings into your round-bottom flask.
6. Obtain 40 mL of anhydrous diethyl ether and the proper amount of ethyl bromide, be sure to keep the diethyl ether covered or it will not remain anhydrous for long.
7. Add 10 mL of ether to the magnesium turnings. *Make sure the stopcock on the separatory funnel is closed.* Dissolve the ethyl bromide in 10 mL of ether and transfer this solution to the separatory funnel.

Table 4.2.1.1.1 Number of moles needed.

Scenario	butanone (moles)	Ethyl or propyl acetate (moles)	Ethyl bromide (moles)	Magnesium (moles)
Butanone + 1.2 equiv. ethyl bromide	0.10	—	0.12	0.12
Butanone + 2.4 equiv. ethyl bromide	0.10	—	0.24	0.24
Ethyl acetate + 1.2 equiv. ethyl bromide	—	0.10	0.12	0.12
Ethyl acetate + 2.4 equiv. ethyl bromide	—	0.10	0.24	0.24
Propyl acetate + 2.4 equiv. ethyl bromide	—	0.10	0.24	0.24

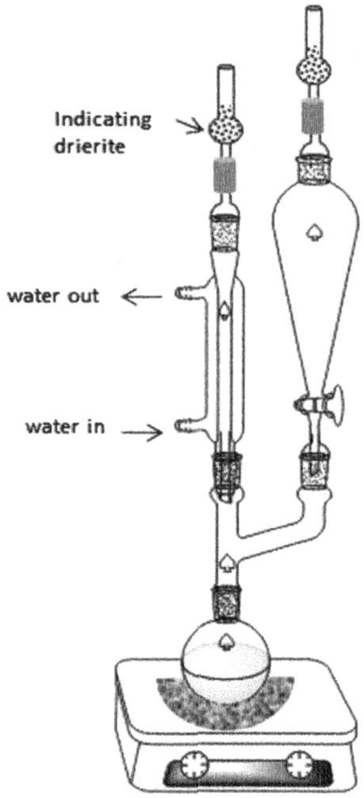

Figure 4.2.1.1.1 Dropwise addition/reflux setup.

8. Turn on the water to the condenser, only a trickle of flowing water is necessary. Using the separatory funnel, add approximately 2–5 mL of the ethyl bromide/ether solution with stirring.
9. At this point, the reaction mixture should spontaneously begin to reflux, which indicates the reaction has initiated. Make sure that you observe bubbles forming on the surface of the magnesium without stirring and that the rate of reaction increases as you add more ethyl bromide. If your reaction does not initiate spontaneously you can add an iodine crystal. However, the reaction with ethyl bromide generally will begin spontaneously and be quite vigorous (students have blown their drying tubes off by adding ethyl bromide too quickly).
10. After the reaction has begun to reflux, quickly remove the drying tube from the reflux condenser, add another 10 mL of diethyl ether then reattach the drying tube.
11. Slowly add the remaining ethyl bromide solution *via* the addition funnel at a rate that maintains a steady reflux. Do not add the solution too quickly.
12. After the addition of ethyl bromide is complete, heat the solution to reflux for 10 min to ensure that the reaction is complete. Remember the boiling point of diethyl ether is quite low, you will not need much heat.
13. Remove the heating attachment and cool the solution in an ice bath. Remember, the Grignard reagent is very moisture sensitive. Do not remove the round-bottom flask from the reaction set up with drying tubes while you are cooling it.
14. Prepare a solution with 0.1 mol of your starting ketone or ester in 10 mL of anhydrous ether, check that the stopcock is closed, then transfer this solution to your separatory funnel.
15. Add the ketone or ester solution dropwise to the Grignard reagent.
16. When addition is complete heat the reaction to reflux for 10 min. Make sure to maintain the reaction volume at a minimum of 50 mL, adding more ether if necessary to maintain this volume of solution.
17. Cool the reaction in an ice-water bath; at this point you may remove the drying tubes.
18. Add ice to the 50 mL mark in a 250 mL Erlenmeyer flask and add 50 mL of 1 M sulfuric acid to the ice.

19. Very slowly pour the cooled reaction mixture into the ice/sulfuric acid slurry.
20. Use an additional 10 mL of 1 M sulfuric acid and 10 mL of diethyl ether to rinse the round-bottom flask.
21. Stir the solution in the Erlenmeyer flask until all of the material dissolves. If the aqueous layer is milky white and/or there is undissolved Mg, you may need to add additional sulfuric acid. When adding additional acid you should use concentrated sulfuric acid, which should be added dropwise. During this process, the diethyl ether will begin to evaporate, add additional ether (not anhydrous) to maintain the ether layer at 50 mL.
22. When the solids ($Mg(OH)_2$ and Mg metal) have dissolved transfer the mixture to the separatory funnel.
23. Remove the aqueous phase.
24. Wash the organic layer with 20 mL of water.
25. Wash the organic layer with 20 mL of saturated sodium bicarbonate solution. Make sure to shake the mixture thoroughly.
26. Wash the organic layer with 20 mL of brine.
27. Dry the organic layer over $MgSO_4$, and remove the salts by gravity filtration into a clean round-bottom flask; if your flask is noticeably more than 50% full, use a larger round-bottom flask. Obtain a GC of the crude organic layer (optional).
28. Place your round-bottom flask into a heating mantle and onto your stir plate and clamp your flask. Attach distilling and outlet adapters as well as a thermometer to your round-bottom flask; the tip of the thermometer should be placed just below the outlet arm of the distilling adapter. To this side arm, attach your reflux condenser (again with the hoses attached so that the water is allowed to flow up), then a vacuum adapter; both of these newly formed connections should be secured with a keck clip (Figure 4.2.1.1.2).
29. Place a tared vial for collection directly underneath your vacuum adapter (Figure 4.2.1.1.2) and begin heating your round-bottom flask. Diethyl ether should distil off at 35 °C. When the temperature increases above 35 °C consider transferring your product to a smaller round-bottom flask to continue the distillation. Carefully record the temperatures at which different products distil over and collect material distilling at constant boiling points into tared (pre-weighed) vials. Obtain a GC and a ^{1}H NMR sample of any fractions with boiling points greater than 40 °C.

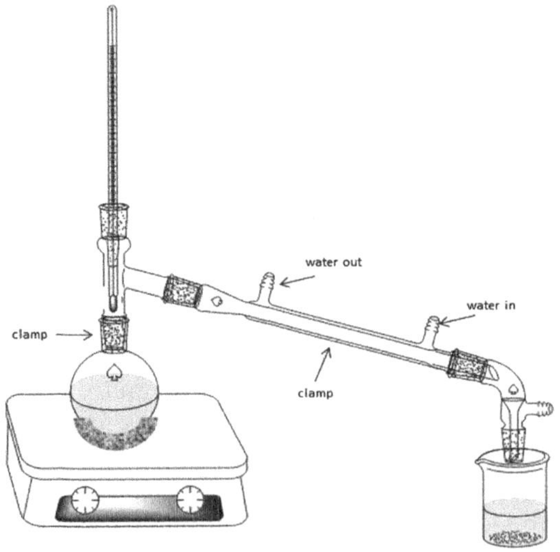

Figure 4.2.1.1.2 Distillation setup.

Waste Disposal

The material from the distillation should be disposed of as organic waste. The acidic aqueous solutions should be neutralized with sodium bicarbonate. The aqueous layers may be either disposed of down the sink or disposed of as chemical waste depending on local environmental regulations.

Results Interpretation and Additional Questions

1. How many equivalents of alkyl bromide would you expect to be necessary to convert your electrophile into 3-methyl-3-pentanol? How does that compare with the number of equivalents that you used?
2. What was the mass and percent yield you obtained?
3. What was the boiling range of the fraction that you collected above 40 °C that had the largest mass? Is this consistent with 3-methyl-3-pentanol or another product?
4. (If obtained) Is the ^{1}H NMR consistent with 3-methyl-3-pentanol or another product? What peaks provide evidence of the alcohol? Are there still impurities present in the spectrum?
5. You and your laboratory partner each set up a reaction. Both of you use 2.4 equivalents of methyl magnesium bromide, but each of you use different electrophiles: you use ethyl acetate, and your laboratory partner uses propyl acetate. Once your reactions are complete, both of you identify two alcohols that have been produced (one of the alcohols you have made, your laboratory partner has also synthesized). Draw and identify the two alcohols produced from your reaction, as well as the two generated in your laboratory partner's reaction.
6. Briefly explain why unreacted ethyl acetate (rather than 2-butanone) is present in the reaction between ethyl acetate and 1.2 equivalents of ethylmagnesium bromide.

References

1. V. Grignard, Nobel, Lecture, 11 December 1912, http://www.nobelprize.org/nobel_prizes/chemistry/laureates/1912/grignard-lecture.pdf.
2. R. E. Rosenberg, *J. Chem. Educ.*, 2007, **84**, 1474.
3. H. M. Weiss, *J. Chem. Educ.*, 1999, **76**, 76.
4. T. S. Everett, *J. Chem. Educ.*, 1998, **75**, 86.

4.2.1.2. Synthesis of Methyl Triphenylmethyl Ether

Dulce Simão* and Ana C. Cerdeira

Department of Chemistry, Instituto Superior Técnico, University of Lisbon, Centro de Química Estrutural, Av. Rovisco Pais, 1, 1049-001 Lisbon, Portugal. *E-mail: dulcesimao@tecnico.ulisboa.pt

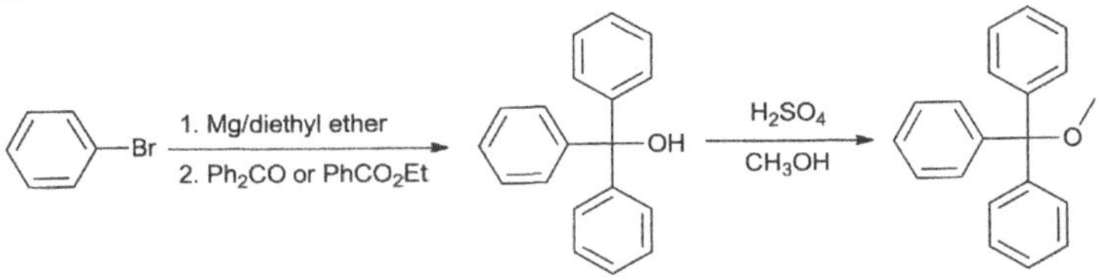

Number of sessions (duration of each session)	Hazard level	Difficulty level	Level of study
3 (4 h each)	Moderate	Medium	Introductory

Class name Aryl halides, alcohols, ketones or esters, ethers

Concepts involved This two-step synthesis illustrates how to obtain a tertiary alcohol *via* Grignard addition, followed by unimolecular nucleophilic substitution (S_N1) to produce an ether

Chemicals needed Magnesium, dry diethyl ether, anhydrous calcium chloride, dry bromobenzene, crystal iodine, benzophenone or ethyl benzoate, dry toluene, concentrated sulfuric acid, sodium hydrogen carbonate, methanol, DMSO-d_6

Equipment and experimental techniques involved Liquid–liquid extraction, NMR, experiment under anhydrous conditions, mechanical stirring, reflux, steam distillation, vacuum filtration, TLC, IR, recrystallization techniques

Keywords Anhydrous conditions, Grignard reagent, liquid–liquid extraction, NMR, nucleophilic unimolecular substitution (S_N1), steam distillation

Background

Grignard reagents (RMgX) are organometallic compounds widely used in syntheses, usually prepared *in situ* by direct reaction of an alkyl or aryl halide with magnesium suspended in diethyl ether. These compounds are easily added to carbonyl groups, which makes them extremely important in the formation of carbon–carbon bonds. In this experiment, the Grignard reagent is phenylmagnesium bromide prepared from bromobenzene and magnesium in diethyl ether. The carbon–magnesium bond is highly polarized towards carbon and can be seen as a carbanion. For that reason Grignard reagents are very strong bases that react with any compound that has hydrogen attached to an electronegative atom. They are also strong nucleophiles and attack the carbon of a carbonyl group.[1a] Consequently, phenylmagnesium bromide reacts with benzophenone (alternatively, ethyl benzoate) followed by acid hydrolysis to give a tertiary alcohol

Comprehensive Organic Chemistry Experiments for the Laboratory Classroom
Edited by Carlos A. M. Afonso, Nuno R. Candeias, Dulce Pereira Simão, Alexandre F. Trindade, Jaime A. S. Coelho, Bin Tan, and Robert Franzén.

Published by the Royal Society of Chemistry, www.rsc.org

(triphenylmethanol).[2] Ethyl benzoate can be used instead benzophenone as esters react with two molar equivalent of the Grignard reagent.[3,4] This alcohol is easily converted into the ether through nucleophilic unimolecular substitution (S_N1),[5] since the carbocation formed (triphenylmethyl or trityl carbocation) is tertiary and the three aromatic rings help to stabilize the positive charge by resonance.[1b] This carbocation can be easily detected when the colourless triphenylmethanol is dissolved in concentrated sulfuric acid, resulting in a red–brown solution that turns colourless when methanol is added.[6]

Additional Safety

This work should be carried out in a fume hood. Care should be taken when refluxing diethyl ether, which is extremely flammable. The same care should be observed when handling methanol as a recrystallization solvent. Concentrated sulfuric acid is corrosive. The mechanical stirrer should be manipulated with precaution due to its rotational parts.

Experimental Procedure

Preparation of Triphenylmethanol

Session 1

All glassware used on this step must be dried in the oven. Remove it and immediately start to assemble the reaction apparatus while the material is still hot.

1. Fit a 250 mL three-neck round-bottom flask with a mechanical stirrer, reflux condenser and a dropping funnel. Use anhydrous calcium chloride tube(s) to protect the reaction mixture from atmosphere moisture. All adjustments needed to guarantee an effective stirring must be performed once the apparatus is completely assembled, without any reactant or solvent inside the flask. Do not measure or weight any chemical before everything is in perfect condition. Make sure the equipment used to measure/weigh is thoroughly dry. Add the reactants to the flask immediately after measured and have an ice bath ready for the addition of bromobenzene.
2. To the flask add magnesium (2.7 g, 0.11 mol) (previously dried in the oven) and diethyl ether (5 mL) that was previously dried over sodium wire.
3. To the dropping funnel add a solution of dry bromobenzene (17.7 g, 12 mL, 0.11 mol) in dry diethyl ether (50 mL).
4. Add 5 mL of this solution to the reaction mixture and verify whether the ether starts to boil gently and the reaction mixture turns to a shade of green, otherwise gently warm until the reaction initiates. If necessary add one crystal of iodine to the mixture (in such a case, the reaction initiates once the iodine colour vanishes).
5. Once the reaction is initiated add the remaining bromobenzene in small portions, at such a rate that the diethyl ether refluxes gently without any heat source. If necessary, cool the reaction flask with an ice bath.
6. Upon complete addition, gently warm the reaction mixture in a water bath until the reaction is complete (when only a small amount of magnesium remains in the flask).
7. To the dropping funnel add a solution of benzophenone (18 g, 0.1 mol) (or ethyl benzoate, 7.5 g, 0.05 mol) in dry toluene (40 mL) and add it dropwise to the Grignard reagent.
8. When this addition is completed, reflux for a further 30 min. Cool and pour over crushed ice (*ca.* 100 g) with stirring, and add concentrated sulfuric acid (12 mL).
9. Add toluene if any of the precipitate remains undissolved. Store the solution until the next session.

Session 2

1. Transfer the solution to a separatory funnel, allow the layers to separate and wash the organic layer with a solution of 5% sodium hydrogen carbonate (25 mL) and water (25 mL).
2. Put the organic layer in a 500 mL round-bottom flask, fit a splash head, a Liebig condenser, a receiver adapter and a 500 mL Erlenmeyer flask.
3. Connect a copper boiler with water previously heated to the splash head in order to introduce the steam into the flask. Collect and remove unreacted starting materials and by-products by steam distillation.
4. When no appreciable amount of water-insoluble material is detected stop the steam distillation. Collect the residue that remain in the flask by vacuum filtration, wash with water and dry. Store a small amount to dry under vacuum and recrystallize the remaining crude product from methanol, collecting a small volume (2–3 mL) of the mother liquor from recrystallization.
5. Dry the product and determine the yield, melting point and record spectral data (IR and ^{1}H NMR).
6. Prepare a TLC plate to determine the R_f of the crude and recrystallized product, benzophenone and the mother liquor. Elute the TLC plate with dichloromethane/petroleum ether (40–60 °C) (1:1)

Session 3 – Preparation of Methyl Triphenylmethyl Ether

1. Place triphenylmethanol (2.0 g, 7.7 mmol) on a 50 mL Erlenmeyer flask, and reduce the crystals to a fine powder with a glass rod.
2. Add concentrated sulfuric acid (20 mL) with stirring to dissolve the alcohol.
3. Slowly pour this mixture into a 250 mL Erlenmeyer flask containing cold methanol (80 mL).
4. Stir and cool in an ice/water bath until the product precipitates. Collect the solid with vacuum filtration, wash with water and recrystallize from methanol.
5. Calculate the yield. Record the melting point and spectral data (IR and ^{1}H NMR).

Waste Disposal

The neutralized aqueous solutions may be diluted and discarded in the sink. The solution obtained from steam distillation can be stored in an aqueous waste collector. Methanol resulting from recrystallization should be placed in a waste collector recipient for non-chlorinated solvents.

Results Interpretation and Additional Questions

1. Evaluate the purity of triphenylmethanol by analysis of the TLC plate and the melting points of both products obtained. Compare the melting points with those reported in literature.
2. Assign the ^{1}H and ^{13}C NMR spectra and the characteristic absorption bands in the IR spectra.
3. Write down all reaction mechanisms.
4. Explain why all the material and reactants must be thoroughly dried in the first step of the reaction. If there were any traces of ethanol in the diethyl ether used, how would that affect the reaction?
5. What are the by-products in this reaction? Draw the chemical equations.
6. What is the purpose of the acidification performed at the end of the synthesis of triphenylmethanol?
7. Unimolecular elimination is often a side reaction of unimolecular nucleophilic substitution. Is it possible to obtain an alkene as a side product in the synthesis of methyl triphenylmethyl ether? Justify your answer.

References

1. J. Clayden, N. Greeves and S. Warren *Organic Chemistry*, Oxford, 2001(a) p. 20; (b) p. 417.
2. A. I. Vogel, *Textbook of Practical Organic Chemistry*, Longman, 5th edn, 1989, p. 540.
3. T. W. G. Solomons, *Organic Chemistry*, John Wiley & Sons Inc., New York, 7th edn, 1998, pp. 549–554.
4. J. Jones, *Core Carbonyl Chemistry, Oxford Chemistry Primers*, Oxford University Press, 1997, vol. 17, p. 45.
5. K. P. C. Vollhardt and N. E. Schore, *Organic Chemistry*, Freeman & Cª, 3th edn, 1999, pp. 350–352.
6. L. F. Fieser and K. L. Williamson, *Organic Experiments*, Houghton Mifflin Company, 8th edn, 1998, pp. 314–316.

4.2.1.3. Grignard-Like Reaction in Water

João R. Vale and Carlos A. M. Afonso*

iMed.UL, Faculdade de Farmácia da Universidade de Lisboa, Av. Prof. Gama Pinto, 1649-003 Lisboa, Portugal
*E-mail: carlosafonso@ff.ulisboa.pt

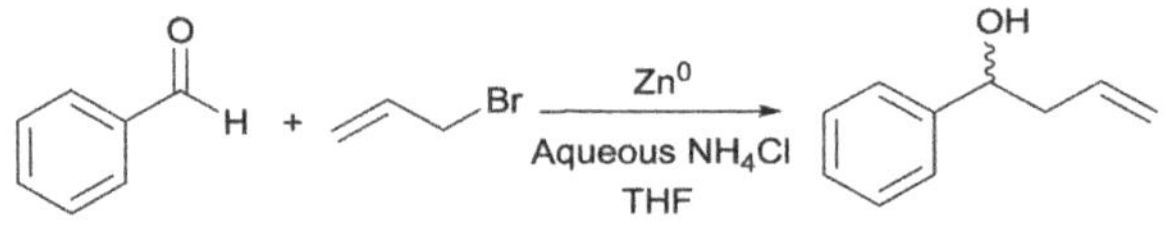

Duration:	Hazard level	Difficulty level	Level of study
1 (2 h)	Medium	Low	Intermediate
Class names Aldehydes, halides, alcohols, allyl zinc			
Concepts involved This experiment involves a Grignard-like reaction where benzaldehyde is allylated with allyl bromide in the presence of zinc metal in aqueous media using THF (tetrahydrofuran) as cosolvent			
Chemicals needed Benzaldehyde, allyl bromide, ammonium chloride, metal zinc, water, THF, diethyl ether, anhydrous magnesium sulfate, deuterated chloroform			
Equipment and experimental techniques involved Apparatus for reaction with magnetic stirring, liquid–liquid extraction and filtration apparatus, NMR			
Keywords Aqueous media, Barbier reaction, Grignard reaction, liquid–liquid extraction, NMR			

Background

Grignard reactions belong to a set of vital transformations in organic chemistry with several synthetic applications since they allow the formation of very important new carbon–carbon bonds. They were discovered by Victor Grignard in 1900,[1] and require the preparation of a Grignard reagent by reacting an organohalide with magnesium metal in an ethereal solvent, producing a reactive organomagnesium halide (Scheme 4.2.1.3.1).

This compound has a carbon bound to a more electropositive atom, magnesium, providing the former with a partial negative charge, making it a strong nucleophile and base. As such, in the presence of a carbonyl group like a ketone or an aldehyde (electrophilic groups) an S_N2-like reaction will take place, after which an acidic workup will provide the alcohol product (Scheme 4.2.1.3.2).[2] An aprotic solvent must be used (usually diethyl ether) because the Grignard reagent is basic and will be consumed in an acid–base reaction when in contact with acidic protons.

Scheme 4.2.1.3.2 shows a classical Grignard reaction, which although very useful has some limitations. Grignard reagents react readily with acidic protons of functional groups such as alcohols and amines, and decompose when in contact with water by the same process, which can make this reaction troublesome to perform in the laboratory, demanding the use of dry solvents, dry glassware and an anhydrous atmosphere. In contrast, the Barbier reaction can overcome this problem by using zinc metal to react with the halide to form a zinc Grignard-like intermediate that has

Comprehensive Organic Chemistry Experiments for the Laboratory Classroom
Edited by Carlos A. M. Afonso, Nuno R. Candeias, Dulce Pereira Simão, Alexandre F. Trindade, Jaime A. S. Coelho, Bin Tan, and Robert Franzén.

Published by the Royal Society of Chemistry, www.rsc.org

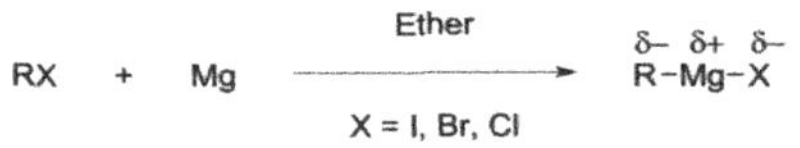

Scheme 4.2.1.3.1 General synthesis of Grignard reagents.

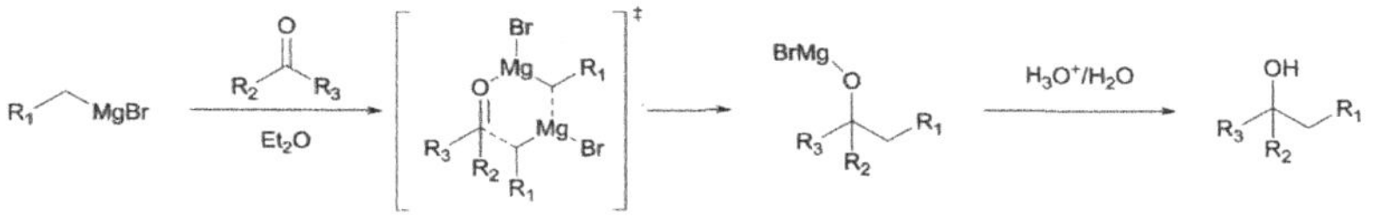

Scheme 4.2.1.3.2 Alkylation of a carbonyl compound with an organomagnesium halide: the Grignard reaction.

a distinct reactivity from the common organomagnesium halide. This intermediate is stable in aqueous media, making it a much more appealing reagent by allowing the use of water as the solvent, which is more ecological compared to commonly used organic solvents, due to its non-toxicity and natural abundance. This reaction is named the "Barbier reaction" after Viktor Grignard's teacher Philippe Barbier and is commonly used in organic chemistry as a useful synthetic tool.[3] The major drawback of this reaction is that alkylations of carbonyl compounds in aqueous media only work with activated alkyl halides.[4] In this experiment, benzaldehyde will be allylated with allyl bromide in the presence of metal zinc in aqueous media using THF as cosolvent.[5]

Another green alternative to perform this kind of transformation is the Prins reaction, which also uses water as solvent and involves an acid-catalyzed addition of alkenes to aldehydes,[6] while also having its limitations as not all reagents can withstand the acidic conditions that it requires.

To evaluate the environmental acceptability of a chemical process R. Sheldon introduced in 1992 two new green chemistry concepts: atom efficiency and the environmental factor (E-factor).[7] The atom efficiency is a tool used to evaluate the waste that will be produced by a chemical process. It can be calculated by dividing the molecular weight of the product by the sum of the molecular weights of all substances formed in the reaction. The E-factor is the actual waste produced in a chemical process, considering waste as everything but the desired product (and any water used). It can be calculated according to eqn (4.2.1.3.1):

$$\text{E-factor} = \frac{\text{Waste (kg)}}{\text{Product (kg)}} \tag{4.2.1.3.1}$$

Additional Safety

This session should be done in a fume hood, with proper safety equipment such as protective laboratory coat, gloves and safety goggles. Special precaution must be taken while handling allyl bromide as it is a flammable, toxic, corrosive and carcinogenic liquid.

Experimental Procedure

1. Prepare a 50 mL round-bottom flask equipped with a magnetic stirrer and suspend it in a water bath at room temperature.
2. Add zinc metal powder (0.78 g, 12.0 mmol), aqueous saturated solution of NH_4Cl (10 mL) and benzaldehyde (1.02 mL, 10.0 mmol) in tetrahydrofuran (THF, 5 mL).
3. With the reaction mixture under vigorous stirring, add dropwise onto the zinc powder allyl bromide (1.04 mL, 12.0 mmol). An immediate reaction should be observed as the zinc metal powder starts to disappear.

4. Let the solution stir for 30–50 min (follow the reaction by TLC, eluent CH_2Cl_2) and then add diethyl ether (10 mL).
5. Filter the reaction mixture and wash the solid with diethyl ether (10 mL).
6. Separate the organic layer, dry it over anhydrous magnesium sulfate (*ca.* 1.5 g) and filter.
7. Evaporate the solvent with a vacuum pump. The compound 1-phenyl-3-buten-1-ol is then obtained as an orange liquid (*ca.* 1.26 g, 85% yield). Proceed to its structural identification by ^{1}H NMR.[5]

Waste Disposal

Ask your instructor for the most appropriate way to reutilize or dispose of the liquid effluents and solids.

Results Interpretation and Additional Questions

1. Confirm whether the desired product was obtained and evaluate its purity.
2. Comment about the atomic efficiency and the environmental-factor (E-factor) of this procedure.
3. Present the mechanism of this transformation.
4. Interpret the experimental approach and the spectral data obtained.
5. Why can't this experiment be performed using magnesium metal instead of zinc metal?

References

1. Y. H. Lai, *Synthesis*, 1981, **8**, 585.
2. J. F. Garst and F. Ungvary, in *Grignard Reagents*, ed. R. S. Richey, John Wiley & Sons, Inc., New York, 1st edn, 2000, ch. 7, pp. 185–275.
3. (a) J. W. Lim, K. H. Kim, B. R. Park and K. N. Kim, *Tetrahedron Lett.*, 2011, **52**, 6545; (b) T. Saloranta, I. Zupkó, J. Rahkila, G. Schneider, J. Wölflin and R. Leino, *Steroids*, 2012, **77**, 110; (c) B. Jiang and H. Tian, *Tetrahedron Lett.*, 2007, **48**, 7942.
4. C. J. Li, *Tetrahedron*, 1996, **52**, 5643.
5. (a) C. Petrier and J. L. Luche, *J. Org. Chem.*, 1985, **50**, 910; (b) G. W. Breton and C. A. Hughey, *J. Chem. Educ.*, 1998, **1**, 85.
6. H. J. Prins, *Chem. Weekbl.*, 1919, **16**, 1072.
7. R. A. Sheldon, *Chem. Ind.*, 1992, **23**, 903.

4.2.1.4. Cram's Rule – Diastereoselective Grignard Addition to 2-Phenylpropanal

Laura M. Hancock, Michael G. Edwards, Graeme R. Jones*, and Matthew O'Brien

Lennard-Jones Laboratories, School of Physical and Geographical Sciences, Keele University, Staffordshire, ST5 5BG, UK. *E-mail: g.r.jones@keele.ac.uk

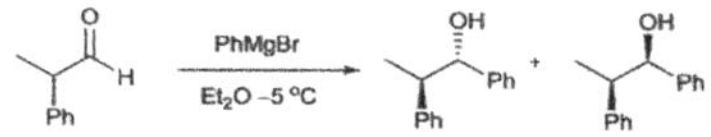

Number of sessions (duration of each session)	Hazard level	Difficulty level	Level of study
1 (3 h)	Moderate	Medium	Intermediate

Class names Aldehydes, Grignard reagents

Concepts involved The experiment repeats the original 1952 experiment reported in the *Journal of the American Chemical Society* by Cram and Ahmed Abd Elhafez, which established Cram's rule; the diastereoselectivity being determined by ^{1}H NMR of a crude reaction mixture and explained using the Felkin–Anh model

Chemicals needed 2-Phenylpropanal, diethyl ether, phenylmagnesium bromide (3 M solution in diethyl ether), saturated ammonium chloride solution, brine, magnesium sulfate, petroleum ether (40–60 °C), deuterated chloroform

Equipment and experimental techniques involved Drying glassware, reaction under inert atmosphere, syringe addition, following a reaction by TLC, liquid–liquid extraction, rotary evaporation, ^{1}H NMR, measuring a diastereomeric excess using ^{1}H NMR, interpreting an NMR spectrum containing a mixture of diastereoisomers

Keywords Cram's rule, diastereoselectivity, Grignard, liquid–liquid extraction, NMR, syringe addition

Background

How does a chiral centre adjacent to an sp^2 carbonyl carbon affect the stereochemical outcome of nucleophilic addition to the carbonyl? Is it possible to predict which diastereoisomer will be preferentially formed or will equal quantities of each diastereoisomer be produced?

This experiment is a favourite of the Year 2 organic chemistry practicals at Keele University where it has run for 20 years. It repeats a pivotal experiment in organic synthesis undertaken in the laboratory of Professor Donald Cram (Nobel Prize in Chemistry, 1987) and which inspired generations of synthetic organic chemists into desperately seeking stereoselectivity.[1] First published in the *Journal of the American Chemical Society* in 1952, Cram's rule has been further refined, first by Felkin[2] and then Anh[3–5] to create the Felkin–Anh model[6] which more correctly describes how the stereoselectivity is controlled and incorporates the Bürgi–Dunitz angle of attack of the nucleophile.[7,8]

The Felkin–Anh model predicts that the two lowest energy conformers will have the large sized group perpendicular to the carbonyl bond as can be seen in Figure 4.2.1.4.1.[9] The nucleophile

Comprehensive Organic Chemistry Experiments for the Laboratory Classroom
Edited by Carlos A. M. Afonso, Nuno R. Candeias, Dulce Pereira Simão, Alexandre F. Trindade, Jaime A. S. Coelho, Bin Tan, and Robert Franzén.

Published by the Royal Society of Chemistry, www.rsc.org

Figure 4.2.1.4.1 The Felkin–Anh model.

can then approach along the Bürgi–Dunitz angle of 107° either eclipsing the small or the medium group. As the approach eclipsing the small group is unhindered then the major diastereoisomer of the alcohol product comes from this attack and the minor diastereoisomer from the more hindered approach eclipsing the medium group.

Additional Safety

Gloves should be worn when using syringes. Instruction should be given to students if this is the first time they have handled syringes and undertake reactions under an inert atmosphere. Appropriate instruction should be given on how to dry glassware using a heat-gun. Phenylmagnesium bromide in diethyl ether is a highly flammable liquid and vapour that causes severe skin burns and eye damage. Contact with water releases flammable gases that may ignite spontaneously.

Experimental Procedure

The apparatus set up is shown in Figure 4.2.1.4.2.

Laboratory Instructions

1. Dry a 100 mL round-bottom flask with a stirrer bead using a heat gun in a stream of nitrogen then stopper with a septum and place under a positive pressure of nitrogen using a balloon.
2. Prepare a solution of 2-phenylpropanal (0.54 g, 4.02 mmol) in 8 mL of anhydrous diethyl ether in a large capped vial.
3. Using a TLC capillary spot two TLC plates with the solution and develop one of the plates in a 1 : 1 mixture of diethyl ether and petroleum ether after the reaction has started.
4. Using a 10 mL syringe with a short needle place the 2-phenylpropanal solution in the dried flask, stir and cool the flask in a salt/ice/water bath.
5. To this stirred solution add a solution of phenylmagnesium bromide in diethyl ether (3 **M**, 1.66 mL, 4.98 mmol) dropwise using a 2 mL syringe with a long needle.

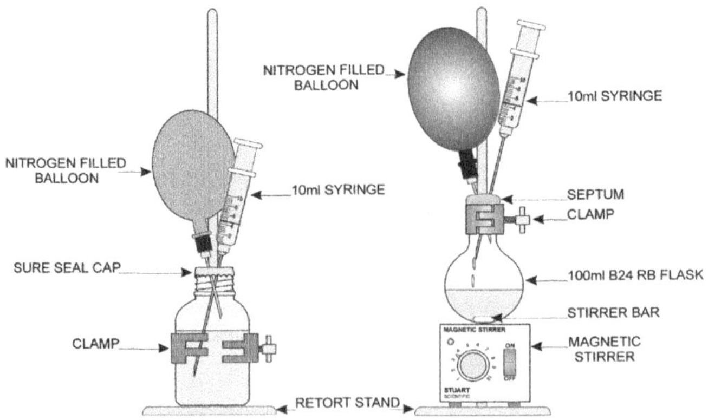

Figure 4.2.1.4.2 Apparatus set up.

6. After 30 min quench the reaction with 10 mL of saturated ammonium chloride solution and pour the mixture into a 100 mL separating funnel.
7. Separate the two phases and extract the aqueous phase three times with 10 mL of diethyl ether.
8. Combine the organic extracts, wash with 10 mL of brine and dry over magnesium sulfate.
9. Spot the reaction mixture onto the second TLC plate (spotted with starting material earlier) and run using the same solvent system as before.
10. Filter the solution and evaporate the solvent on the rotary evaporator to give the diastereomeric mixture of 1,2-diphenylpropan-1-ol.

NMR Analysis

1. Acquire and process (with full expansions, peak picking (in Hz) and integrations) a high-field (250 MHz or above) ^{1}H NMR of the crude product in $CDCl_3$.
2. Predict the chemical shifts and integrations and multiplicities for the equivalent hydrogens within the molecule.
3. As this is a mixture of diastereoisomers, each diastereoisomer will have its own set of signals which will occur at slightly different chemical shifts. Compare your predictions with your NMR spectrum and hence make your assignments.

Waste Disposal

Diethyl ether should be disposed of in the non-halogenated waste. Aqueous solutions should be disposed of down a sink, solid inorganic waste should be disposed of in an appropriate waste bin.

Results Interpretation and Additional Questions

1. Given the coupling constants for the conformers in Figure 4.2.1.4.3 of the two diastereoisomers, work out which is the major diastereoisomer and which is the minor.
2. Use the integrations on your expansions in the ^{1}H NMR to measure the ratio of diastereoisomers and hence calculate the diastereomeric excess.
3. Use the Felkin–Anh model to predict which diastereoisomer is the major and which is the minor one and check that this agrees with your NMR assignment.

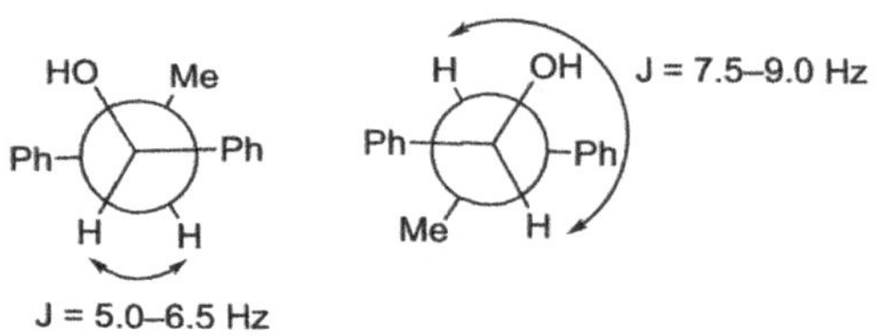

Figure 4.2.1.4.3 Coupling constants of the major and minor diastereoisomers.

References

1. D. J. Cram and F. A. Elhafez, *J. Am. Chem. Soc.*, 1952, **74**, 5828.
2. M. Chérest, H. Felkin and N. Prudent, *Tetrahedron Lett.*, 1968, **9**, 2199.
3. N. T. Anh, O. Eisenstein, J.-M. Lefour and M.-E. Dau, *J. Am. Chem. Soc.*, 1973, **95**, 6146.
4. N. T. Anh and O. Eisenstein, *Tetrahedron Lett.*, 1976, **17**, 155.
5. N. T. Anh and O. Eisenstein, *Nouv. J. Chim.*, 1977, **1**, 61.
6. N. T. Anh, *Top. Curr. Chem.*, 1980, **88**, 146.
7. H. B. Bürgi, J. D. Dunitz and E. Shefter, *J. Am. Chem. Soc.*, 1973, **95**, 5065.
8. H. B. Bürgi, J. D. Dunitz, J. M. Lehn and G. Wipff, *Tetrahedron*, 1974, **30**, 1563.
9. J. Clayden, N. Greeves and S. Warren, *Organic Chemistry*, Oxford University Press, Oxford, 2nd edn, 2012, p. 859.

4.2.1.5. Preparation of (4*R*,5*R*)-4,5-Bis(diphenylhydroxymethyl)-2,2-dimethyldioxolane ((–)-TADDOL)

Naylil M. R. Capreti and Igor D. Jurberg*

Institute of Chemistry, State University of Campinas, C.P. 6154, 13083-970, Campinas, SP, Brazil
*E-mail: idjurberg@iqm.unicamp.br

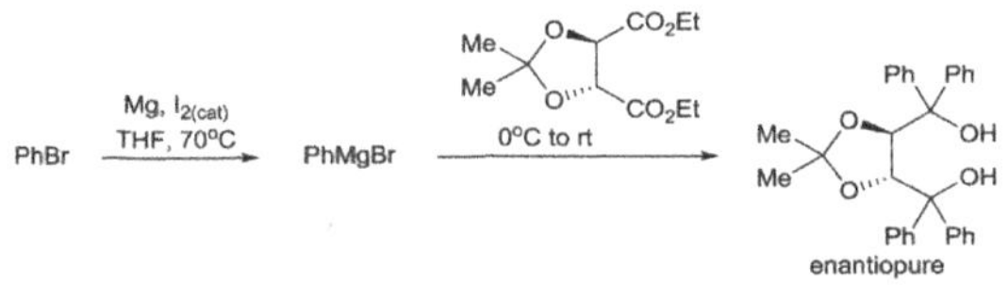

Number of sessions (duration of each session)	Hazard level	Difficulty level	Level of study
3 (2.5 h)	Moderate	Medium	Intermediate
Class names Esters, organomagnesium compounds, alcohols, ketals			
Concepts involved Preparation of a Grignard reagent *via* magnesium insertion into a carbon–halogen bond; Grignard addition to ester groups; electrophilicity of carbonyl compounds			
Chemicals needed Bromobenzene, magnesium, iodine, diethyl 2,3-*O*-isopropylidene-L-tartrate, THF, ethyl acetate			
Equipment and experimental techniques involved Flash column chromatography, NMR, melting point apparatus, polarimeter, heating/stirring plate, rotary evaporator, reaction under reflux and at low temperature, manipulation under anhydrous conditions and inert atmosphere			
Keywords Flash column chromatography, Grignard addition, NMR, phenylmagnesium bromide, TADDOL			

Background

Grignard Reagents: Preparation, Properties and Applications

Arguably, Grignard reagents are among the most popular organometallic reagents in organic chemistry.[1,2] They are named after their inventor, Victor Grignard (University of Nancy, France), who received the Nobel Prize in 1912 for developing the chemistry of these compounds.[3,4] Organomagnesium compounds (Grignard reagents) can be prepared by three main routes: (1) *via* the insertion of magnesium metal into a carbon–halogen bond,[5] (2) *via* the deprotonation of substrates by a magnesiated base (generally an amine)[6] or (3) *via* halogen–magnesium exchange.[7] Importantly, the latter, milder method has greatly expanded the number of functionalized Grignard reagents synthetically available. For instance, nitriles, esters, imines and nitro groups are typically tolerated by this method.

Grignard reagents must be prepared in ethereal solvents (or, less commonly, in Et_3N) because the complexation of the lone pairs of the heteroatom from the solvent stabilizes the organomagnesium compound and allows their formation in good yields. These compounds are classically

Comprehensive Organic Chemistry Experiments for the Laboratory Classroom
Edited by Carlos A. M. Afonso, Nuno R. Candeias, Dulce Pereira Simão, Alexandre F. Trindade, Jaime A. S. Coelho, Bin Tan, and Robert Franzén.

Published by the Royal Society of Chemistry, www.rsc.org

represented in the literature by the chemical formula RMgX. Nevertheless, in ethereal solutions (*e.g.* Et_2O, THF), an organomagnesium compound delivers a mixture of chemicals given by the Schlenk equilibrium:[8]

$$2\ RMgX \rightleftharpoons R_2Mg + MgX_2$$

There are no compounds known that will react only with RMgX and not with R_2Mg and *vice versa*. The reaction with either of them alone will always afford the same product, albeit occasional differences in rates have been noted.[1,2]

Organomagnesium compounds are typically powerful nucleophiles and strong bases that can be employed in a plethora of transformations. They are not only involved in addition reactions,[9] they have also been reported to participate in metal-catalyzed asymmetric transformations[10] and cross-coupling reactions.[11] Furthermore, they also readily undergo transmetalation to afford other organometallic reagents,[12] such as organocopper[13] and organozinc compounds.[14] To determine the precise amount of the Grignard reagent to be employed in such reactions, several titration protocols are available.[15]

As it is historically important, and still often used in the preparation of simple organomagnesium compounds, we will explore in this experimental class the metal insertion strategy aforementioned in (1) to synthesize phenylmagnesium bromide *en route* to (−)-TADDOL. In this context, one should notice that method (1) is in competition with a Wurtz-type coupling,[16] which is a parasitic reaction most commonly found when large R groups are present, when alkyl iodides are used, or especially when allyl or benzyl halides are employed (this secondary reaction can be often minimized or completely suppressed by slow addition of the alkyl halide, the use of a larger excess of magnesium metal or by changing the solvent: Et_2O often performs better than THF).[17] Remarkably, because RMgX is often prepared *via* method (1), which implicates the intermediacy of radicals $R^\bullet$, racemic Grignard reagents are generally obtained, even when starting from enantiomerically pure alkyl halides RX.[18,19]

Electrophilicity of Acid Derivatives R(C=O)X

Acid derivatives display a varied range of electrophilicities depending on the nature of the group X in R(C=O)X. Accordingly, the rate of the addition–elimination reaction observed for these compounds will depend on the ability of X to act as a leaving group and the electronic effects promoted by X towards the carbonyl functional group. The overall sum of these effects will account for the electrophilicity scale found for these compounds: acyl halide > acid anhydride > ester ≈ carboxylic acid > amide > carboxylate ion.[20]

TADDOL

The chiral structures of TADDOL derivatives (α,α,α′,α′-tetraaryl-1,3-dioxolane-4,5-dimethanols),[21] introduced by the group of D. Seebach (ETH Zürich, Switzerland), have been involved in a myriad of applications in organic synthesis. They can be employed as stoichiometric chiral reagents,[22] as ligands in metal-catalyzed asymmetric processes[23] and as organocatalysts.[24]

TADDOL derivatives can be readily accessed in both enantiomeric pure forms starting either from commercially available D-(−)- or L-(+)-diethyl tartrate. After the protection of the 1,2-diol moiety, a Grignard addition is then responsible for inserting the aryl groups in the molecule. One can see from this synthetic route that a huge number of structural changes can be easily introduced in order to deliver a rich library of compounds. Consequently, numerous variations of the original TADDOL structure can be studied in the optimization of reaction protocols.[25] This is probably the main reason why TADDOL derivatives have become so popular in asymmetric synthesis.

Additional Safety

The experiment should be performed inside a fume hood. Safety goggles, gloves and a laboratory coat must be worn at all times while in the laboratory. Diethyl 2,3-*O*-isopropylidene-L-tartrate is not a well-studied compound, and should be handled as toxic. Bromobenzene is flammable, irritating to skin, and toxic to aquatic organisms. Magnesium is a flammable solid. In contact with water, it liberates flammable gases. Iodine is harmful in contact with the skin or if inhaled. It causes severe skin burns and eye damage and can cause allergic reactions.

Experimental Procedure

Laboratory Session 1 (2 h 30 min): Preparation of Phenylmagnesium Bromide

1. In a fume hood, assemble a three-necked flask carrying (1) an addition funnel, (2) a reflux condenser equipped with a nitrogen inlet/outlet, and (3) a rubber septum. Add to the three-necked flask: a stirring bar, magnesium turnings (132 mg, 5.5 mmol, 5.5 equiv.), iodine (1 crystal) and THF (2 mL). Charge the addition funnel with bromobenzene (530 μL, 5 mmol, 5 equiv.) and THF (8 mL).
2. Add approximately one-third of the previously prepared solution of bromobenzene in THF (from the addition funnel) to the mixture of magnesium and iodine in THF (in the three-necked flask).
3. Heat the three-necked flask to 70 °C using an oil bath and allow the reaction mixture to stir at this temperature until the complete consumption of the iodine (*ca.* 30 min). The color of the solution changes from brown to colourless.
4. Slowly add the remaining solution of bromobenzene in THF to the mixture of magnesium in THF at 70 °C. Allow the reaction to stir at this temperature until the magnesium is almost entirely consumed (*ca.* 1 h). The color of the solution changes from colourless to a cloudy grey.
5. Allow the reaction to cool down to room temperature.

Laboratory Session 2 (2 h 30 min): Preparation of (−)-TADDOL

1. In a fume hood, prepare a round-bottomed flask charged with a stirring bar and the diethyl 2,3-*O*-isopropylidene-L-tartrate (246 mg, 1 mmol, 1 equiv.) *prepared in the previous class* (*cf.* Experiment 2.1.12) in THF (5 mL) under a nitrogen atmosphere, at 0 °C.
2. Employing a syringe, transfer the previously prepared solution of phenylmagnesium bromide to the flask prepared in step 1 of this session, at 0 °C.
3. Allow the temperature to rise to the ambient value (*ca.* 23 °C). Allow the solution to stir at this temperature for 1 h 30 min. (Note: It is difficult to follow this reaction by TLC, because R_f (Taddol) = 0.3 ≈ R_f (starting diol), eluent 9 : 1 hexanes: AcOEt.)
4. Quench the reaction mixture with a saturated solution of ammonium chloride (10 mL).
5. Separate the organic phase from the aqueous layer using a separating funnel. Extract the aqueous phase with AcOEt (2 × 10 mL). Dry the combined organic phases with Na_2SO_4, filter off the solid and concentrate the obtained solution under reduced pressure.
6. Remove any remaining trace of solvent in your crude reaction mixture by placing your flask in a high vacuum system. A white/pale brown solid is formed.

Laboratory Session 3 (30 min or 2 h 30 min): Purification of (−)-TADDOL

Two alternatives are possible depending on the time frame. Strategy 1 is faster, but it gives lower yields. Strategy 2 produces better yields, but is slower.

Strategy 1 (30 min): Washing

1. Add hexanes (5 mL) and dichloromethane (500 μL) to the crude reaction mixture.
2. Under vacuum, filter the remaining solid using a sintered funnel and rinse the flask with hexanes. You can use a spatula to help you transfer the solid from the flask to the sintered funnel.
3. Transfer the solid from the sintered funnel to a tared round-bottomed flask.
4. Dry the solid in a high vacuum system, measure its weight, and calculate the yield.

Strategy 2 (2 h 30 min): Flash Column Chromatography (ref. 26)

1. To a beaker, add 20 g of silica, followed by enough hexanes to cover all the silica.
2. Transfer the wet silica to the chromatographic column, pushing it down with compressed air and washing the walls of the column with more hexanes in order to make a well-packed cylinder of silica. It is important to remove all the air inside the silica.
3. Dissolve the crude (−)-TADDOL in a minimum amount of dichloromethane and deposit the solution obtained on the top of silica using a pipette.
4. Use compressed air to push down the solution into the silica. Add sand on the top of the silica in order to have a small cylinder with a height of *ca.* 0.5 cm.
5. Add hexanes (50 mL) to the column. Push down the hexane through the silica employing compressed air until the solvent reaches the level of the sand. You can, but you do not need to, collect fractions at this point.
6. Repeat this process, slowly increasing the polarity of the eluent following the order 95 : 5 hexanes: AcOEt (50 mL) – 9 : 1 hexanes: AcOEt (50 mL) – 8 : 2 hexanes: AcOEt (100 mL). During this process you must collect the fractions in test tubes and monitor the products coming out of the column by TLC analysis.
7. Once your compound has finished eluting from the column (*cf.* TLC analysis), transfer all the clean fractions to a tared round-bottomed flask and concentrate the combined fractions in a rotary evaporator.
8. Weigh the solid obtained, and calculate the yield.

Waste Disposal

The aqueous layer recovered from the work-up procedure should be discarded in a recipient for halogenated chemicals, while the solvent condensed in the rotary evaporator should be discarded in a recipient for non-halogenated chemicals for posterior destruction.

Results Interpretation and Additional Questions

1. What is the mechanism for the formation of the organomagnesium halide starting from R–X and Mg? Describe the main features.
2. What is the role of I_2 in the preparation of PhMgBr? Which other chemical(s) can be employed instead?
3. What would be the major compound of this reaction if only 2 equiv. of PhMgBr were employed? Explain your answer.
4. Interpret the ^1H and ^{13}C NMR spectra obtained for (−)-TADDOL.
5. Compare the melting point and the optical rotation of the obtained (−)-TADDOL with the reported values from the literature.

References

1. For a general overview on organometallic reagents, see: C. Elschenbroich, *Organometallics*, Wiley-VCH Verlag GmbH, Weinheim, 3rd edn, 2006, Grignard reagents are specifically treated in chapter 6, pp. 62–70.
2. For a pedagogical, more detailed presentation of Grignard reagents, see: M. Orchin, *J. Chem. Educ.*, 1989, **66**, 586.

3. (a) V. Grignard, *C. R. Acad. Sci.*, 1900, **130**, 1322; (b) V. Grignard, *Ann. Chim.*, 1901, **7**, 433.
4. The Nobel Prize was shared with P. Sabatier. For an interesting account, see: (a) H. B. Kagan, *Angew. Chem., Int. Ed.*, 2012, **51**, 7376; For Grignard´s Nobel lecture, see: ; (b) http://www.nobelprize.org/nobel_prizes/chemistry/laureates/1912/grignard-lecture.pdf.
5. J. F. Garst and M. P. Soriaga, *Coord. Chem. Rev.*, 2004, **248**, 623.
6. (a) P. E. Eaton, C.-H. Lee and Y. Xiong, *J. Am. Chem. Soc.*, 1989, **111**, 8016; (b) M.-X. Zhang and P. E. Eaton, *Angew. Chem., Int. Ed.*, 2002, **41**, 2169; (c) A. Kraksovskiy, V. Krasovskaya and P. Knoechel, *Angew. Chem., Int. Ed.*, 2006, **45**, 2958.
7. P. Knoechel, W. Dohle, N. Gommermann, F. F. Kneisel, F. Kopp, T. Korn, I. Sapountzis and V. A. Vu, *Angew. Chem., Int. Ed.*, 2003, **42**, 4302.
8. (a) W. Schlenk and W. Schlenk Jr, *Chem. Ber.*, 1929, **62**, 920; (b) W. Schlenk Jr, *Chem. Ber.*, 1931, **64**, 734; See also: ; (c) J. G. Aston and S. A. Bernhard, *Nature*, 1950, **165**, 485.
9. For some examples, see: (a) R. Shitani and G. C. Fu, *Angew. Chem., Int. Ed.*, 2002, **41**, 1057; (b) A. Krasovski and P. Knochel, *Angew. Chem., Int. Ed.*, 2004, **43**, 3333; (c) I. D. Jurberg, B. Peng, E. Wöstefeld, M. Wasserloos and N. Maulide, *Angew. Chem., Int. Ed.*, 2012, **51**, 1950.
10. For some examples, see: (a) A. V. R. Madduri, S. R. Harutyunyan and A. J. Minaard, *Angew. Chem., Int. Ed.*, 2012, **51**, 3164; (b) B. L. Feringa, R. Badorrey, D. Peña, S. R. Harutyunyan and A. J. Minaard, *Proc. Natl. Acad. Sci. U. S. A.*, 2004, **101**, 5834; (c) C.-S. Da, J. R. Wang, X.-G. Yin, X.-Y. Fan, Y. Liu and S.-L. Yu, *Org. Lett.*, 2009, **11**, 5578.
11. For some examples, see: (a) C. E. I. Knappke and A. J. von Wangelin, *Chem. Soc. Rev.*, 2011, **40**, 4948; (b) G. Cahiez, C. Chaboche, C. Duplais and A. Moyeux, *Org. Lett.*, 2009, **11**, 277; (c) G. Cahiez, O. Gager and F. Lecomte, *Org. Lett.*, 2008, **10**, 5255; (d) J. Terao, H. Watanabe, A. Ikumi, H. Kuniyasu and N. Kambe, *J. Am. Chem. Soc.*, 2002, **124**, 4222; (e) R. Martin and A. Fürstner, *Angew. Chem., Int. Ed.*, 2004, **43**, 3955.
12. For some examples, see: (a) F. M. Piller, P. Appukkuttan, A. Gravryushin, M. Helm and P. Knoechel, *Angew. Chem., Int. Ed.*, 2008, **47**, 6802; (b) R. W. Hoffmann and B. Hölzer, *J. Am. Chem. Soc.*, 2002, **124**, 4204; (c) P. Knoechel and B. Betzemeier, *Modern Organocopper Chemistry*, ed. N. Krause, Wiley-VCH Verlag GmbH, Weinheim, 1st edn, 2002, ch. 2, pp. 45–51.
13. E. Nakamura and S. Mori, *Angew. Chem., Int. Ed.*, 2000, **39**, 3750.
14. P. Knoechel and R. D. Singer, *Chem. Rev.*, 1993, **93**, 2117.
15. (a) B. E. Love and E. G. Jones, *J. Org. Chem.*, 1999, **64**, 3755; (b) A. Krazovskiy and P. Knoechel, *Synthesis*, 2006, **5**, 890; (c) Y. Aso, H. Yamashita, T. Otsubo and F. Ogura, *J. Org. Chem.*, 1989, **54**, 5627; (d) J. C. Watson and J. F. Eastham, *J. Organomet. Chem.*, 1967, **9**, 165; (e) H. Kiljunen and T. A. Hase, *J. Org. Chem.*, 1991, **56**, 6950.
16. (a) A. Wurtz, *Ann. Chim. Phys.*, 1855, **44**, 275; (b) A. Wurtz, *Ann.*, 1855, **96**, 364; See also: (c) L. Kürti and B. Czakó, *Strategic Applications of Named Reactions in Organic Synthesis*, Elsevier, California, 1st edn, 2005, pp. 498–499.
17. See for instance: H. Gillman and J. E. Kirby, *J. Am. Chem. Soc.*, 1929, **51**, 1571.
18. For interesting discussions, see: (a) R. W. Hoffmann, *Chem. Soc. Rev.*, 2003, **32**, 225; (b) Z.-N. Chen, G. Fu and X. Xu, *Org. Lett.*, 2011, **13**, 2046.
19. Nevertheless, successful examples do exist. They take advantage of other approaches: (a) R. W. Hoffmann, B. Hölzer, O. Knopff and K. Harms, *Angew. Chem., Int. Ed.*, 2000, **39**, 3072; (b) P. J. Rayner, P. O´Brien and R. A. J. Horan, *J. Am. Chem. Soc.*, 2013, **135**, 8071.
20. For a fundamental discussion on the topic, see: J. Clayden, N. Greeves and S. Warren, *Organic Chemistry*, Oxford University Press Inc., New York, 2nd edn, 2012, ch. 10, pp. 197–221.
21. D. Seebach, A. K. Beck and A. Heckel, *Angew. Chem., Int. Ed.*, 2001, **40**, 92.
22. For selected examples, see: (a) B. Weber and D. Seebach, *Angew. Chem., Int. Ed.*, 1992, **31**, 84; (b) F. Toda, H. Miyamoto and S. Kikuchi, *J. Am. Chem. Soc.*, 1996, **118**, 11315; (c) A. Cuenca, M. Medio-Simón, G. A. Aguilar, D. Weibel, A. K. Beck and D. Seebach, *Helv. Chim. Acta*, 2000, **83**, 3153.
23. For selected examples, see: (a) T. Robert, J. Velder and H. G. Schmalz, *Angew. Chem., Int. Ed.*, 2008, **47**, 7718; (b) J. D. Sieber and J. P. Morken, *J. Am. Chem. Soc.*, 2006, **128**, 74; (c) J.-I. Sakaki, W. B. Schweizer and D. Seebach, *Helv. Chim. Acta*, 1993, **76**, 2654.
24. For selected examples, see: (a) J. D. McGilvra, A. K. Unni, K. Modi and V. H. Rawal, *Angew. Chem., Int. Ed.*, 2006, **45**, 6130; (b) N. Momiyama and H. Yamamoto, *J. Am. Chem. Soc.*, 2005, **127**, 1080; (c) Y. Huang, A. K. Unni, A. N. Thadani and V. H. Rawal, *Nature*, 2003, **424**, 146.
25. See for instance: D. Seebach, A. Pichota, A. K. Beck, A. B. Pinkerton, T. Litz, J. Karjalainen and V. Gramlich, *Org. Lett.*, 1999, **1**, 55.
26. For a detailed discussion on this topic, see: W. C. Still, M. Kahn and A. Mitra, *J. Org. Chem.*, 1978, **43**, 2923.

4.2.1.6. Synthesis of (*S*)-Diphenyl(pyrrolidin-2-yl) methanol

David Monge*, Javier Iglesias-Sigüenza, and Elena Díez*

Departamento de Química Orgánica, Universidad de Sevilla, C/ Prof. García González, no. 1, 41012-Sevilla, Spain. *E-mail: dmonge@us.es, ediez@us.es

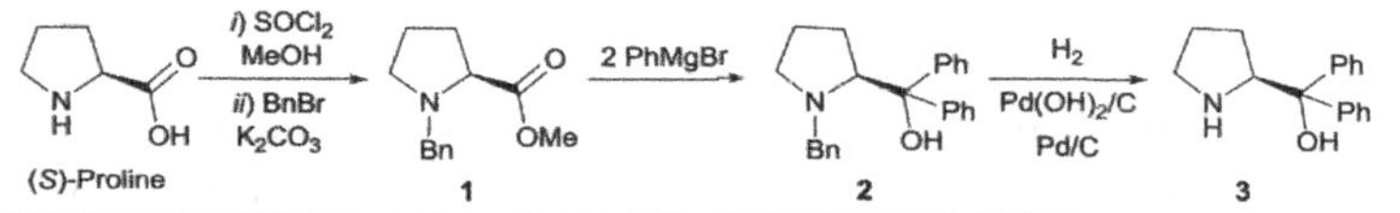

Duration sessions (duration of each session)	Hazard level	Difficulty level	Level of study
7 (3 h + 5 h + 3 h + 3 h + 4 h + 3 h + 3 h)	High	High	Intermediate

Class names Esters, tertiary alcohols, amines

Concepts involved This experiment involves the synthesis of a tertiary alcohol from the amino acid (*S*)-proline using a protecting group strategy. The overall process includes an esterification reaction of an acid *via* acyl chlorides, amine protection by benzylation, Grignard addition to esters to generate tertiary alcohols and hydrogenation reaction. The students will also learn how to work under anhydrous conditions

Chemicals needed (*S*)-Proline, methanol, thionyl chloride, anhydrous potassium carbonate, benzyl bromide, ethyl acetate, sodium chloride, anhydrous sodium sulfate, phenyl magnesium bromide (1 M in THF), anhydrous tetrahydrofuran, dichloromethane, diethyl ether, *n*-hexane, chloroform, ammonium chloride, 10% Pd/C, 10% $Pd(OH)_2/C$, hydrogen cylinder, Celite®, deuterated chloroform

Equipment and experimental techniques involved Equipment for reactions under a nitrogen atmosphere, apparatus for reaction at high temperature and magnetic stirring, vacuum pump and Rotavapor®, liquid–liquid extraction funnel, Büchner vacuum filtration funnel, hydrogen cylinder, NMR spectrometer

Keywords Aminoalcohols, Grignard addition, NMR, (*S*)-proline, protecting group, recrystallization

Background

Enantiopure (*S*)-diphenyl(pyrrolidin-2-yl)methanol (**3**) and derivatives thereof represent an important type of chiral auxiliaries for asymmetric synthesis, including reagents for Corey–Bakshi–Shibata reduction,[1] ligands for several metal catalysed transformations[2] and organocatalysts.[3]

One of the more general synthesis of this amino alcohol is based on a double Grignard addition to an ester moiety using (*S*)-proline as starting material. Herein, we describe an undergraduate-student adapted procedure for the synthesis of **3**, using mild reaction conditions on a medium–large scale, with high yields and reproducibility. In addition, pure products can be obtained by simple recrystallization.

The experiment is a four-step synthesis where the key step is a Grignard addition to an ester to get a tertiary alcohol. The first step consists of an esterification reaction of an acid *via* acyl

Comprehensive Organic Chemistry Experiments for the Laboratory Classroom
Edited by Carlos A. M. Afonso, Nuno R. Candeias, Dulce Pereira Simão, Alexandre F. Trindade, Jaime A. S. Coelho, Bin Tan, and Robert Franzén.

Published by the Royal Society of Chemistry, www.rsc.org

chloride transformation. Next, the experiment employs a general protecting group strategy. Protecting groups are commonly used in laboratory work and they play an important role in multi-step organic synthesis.[4] When a chemical reaction has to be carried out selectively at one reactive site in a multifunctional compound, other reactive sites must be temporarily blocked. Benzyl groups are one way of protecting secondary amines against strong bases (such as Grignard reagents) that might deprotonate them.[5] In this experiment, a benzyl protecting group has been chosen as it can fulfil several requirements. It reacts selectively and in good yield with the amino group and can be easily removed by hydrogenation after Grignard addition.

Additional Safety

All sessions should be carried out in a fume hood, with proper safety equipment, such as protective clothing, gloves and safety glasses. Methanol is highly flammable and toxic. Potassium carbonate is harmful. Benzyl bromide is irritant. $SOCl_2$ is corrosive and toxic and PhMgBr is flammable and corrosive; both react violently with water, so the reactions should be performed under an inert atmosphere (nitrogen or argon). Ethyl acetate and tetrahydrofuran are highly flammable and irritant. Dichloromethane is irritant and breathing its vapours should be avoided. Hydrogenation reactions pose a significant hazard due to the use of flammable solvents (methanol). The presence of hydrogen gas increases the risk of explosion. Chemical fume hood sashes should remain closed at all times during the reaction process. Hood sashes should only be opened during set-up of the reaction, addition of chemical to the reaction vessel as necessary, and during work-up of the reaction mixture. As Celite® may cause respiratory irritation, avoid breathing the dust. Avoid contact with skin and eyes. Avoid formation of dust. Provide appropriate exhaust ventilation at places where dust is formed.

Experimental Procedure

Laboratory Session 1 (3 h): Synthesis of (*S*)-Proline Methyl Ester Hydrochloride

1. Prepare a 1 L two-neck round-bottom flask, equipped with an addition funnel with pressure-equalization arm and a magnetic stir bar.
2. Add 29 g (0.25 mol, 1 equiv.) of (*S*)-proline. The flask should be maintained under a nitrogen atmosphere by three successive vacuum–nitrogen cycles. Then add 125 mL of superdry MeOH *via* a syringe. Keep the system under nitrogen, using a balloon.
3. Stir the resulting solution at 0 °C (ice bath) and add dropwise 20.0 mL (0.3 mol, 1.1 equiv.) of $SOCl_2$ over a period of 1 h (using the addition funnel).
4. Remove the flask from the ice bath to warm to room temperature. Stir the reaction mixture overnight, until the next session.

Laboratory Session 2 (5 h): Synthesis of *N*-Benzyl-(*S*)-proline Methyl Ester (1) (*first part*)

1. Remove MeOH and the excess of $SOCl_2$ under reduced pressure and use the (*S*)-proline methyl ester hydrochloride (light yellow residue) in the next step without further purification.
2. Prepare a 1 L two-neck round-bottom flask, equipped with an addition funnel with pressure-equalization arm and a magnetic stir bar. The flask should be maintained under a nitrogen atmosphere by three successive vacuum–nitrogen cycles.
3. Prepare a solution of 24.8 g (0.15 mol, 1 equiv.) of (*S*)-proline methyl ester hydrochloride in 150 mL of dry MeOH and transfer to the 1 L two-neck round-bottom flask *via* a cannula.

Keep the system under nitrogen, using a balloon. Then, add 62 g (0.5 mol, 3 equiv.) of anhydrous K_2CO_3 and stir this suspension at 0 °C for 1 h.
4. At that point, insert a water condenser into the round-bottom flask and reflux the reaction mixture. Add dropwise 18.3 mL (0.15 mol, 1 equiv.) of benzyl bromide at this temperature over a period of 2 h (using the addition funnel).
5. Reflux the reaction mixture under stirring overnight, until the next session.

Laboratory Session 3 (3 h): Synthesis of *N*-Benzyl-(*S*)-proline Methyl Ester (1) (*second part*)

1. Monitor the reaction by TLC (hexane/Et_2O, 2 : 1).
2. Filter the suspension by suction filtration, washing with MeOH (twice), and evaporate the filtrate under reduced pressure to give crude product **1** as a yellow oil.
3. Dissolve this oil in 300 mL of EtOAc, and pour into a separating funnel. Wash with H_2O (2 × 150 mL) and brine (2 × 150 mL), dry the organic layer by addition of anhydrous Na_2SO_4, filter and remove the solvent under reduced pressure to give *N*-benzyl-(*S*)-proline methyl ester (**1**) as a light yellow oil.

Laboratory Session 4 (3 h): Synthesis of (*S*)-(1-Benzylpyrrolidin-2-yl) diphenylmethanol (2) (*first part*)

1. Prepare a 1 L two-neck round-bottom flask, equipped with an addition funnel with pressure-equalization arm and a magnetic stir bar. The flask should be maintained under a nitrogen atmosphere by three successive vacuum–nitrogen cycles.
2. Add 210 mL (0.21 mol, 3 equiv.) of commercial PhMgBr (1 M in THF) and place the flask in an ice bath. Keep the system under nitrogen, using a balloon.
3. Add dropwise (from the addition funnel) a solution of 15 g (0.07 mol, 1 equiv.) of **1** in 75 mL of anhydrous THF over 45 min and stir the resulting mixture for 1.5 h at 0 °C.
4. Remove the flask from the ice bath, insert a water condenser into the round-bottom flask and reflux (using an oil bath) the reaction mixture under vigorous stirring overnight, until the next session.

Laboratory Session 5 (4 h): Synthesis of (*S*)-(1-Benzylpyrrolidin-2-yl) diphenylmethanol (2) (*second part*)

1. Remove the flask from the oil bath and allow it to cool to room temperature. Monitor the reaction by TLC (hexane/Et_2O, 4 : 1).
2. Add carefully at 0 °C (ice bath) 50 mL of saturated NH_4Cl (100 g of NH_4Cl in 50 mL of H_2O) and pour the mixture into a 1 L separating funnel.
3. Separate the layers (add more water if needed) and extract the aqueous one with additional Et_2O (2 × 100 mL).
4. Dry the combined organic layers by addition of anhydrous Na_2SO_4, filter and evaporate the solvent under reduced pressure to give a light yellow solid.
5. Purify product **2** by recrystallization in Et_2O to yield colourless crystals.

Laboratory Session 6 (3 h): Synthesis of (*S*)-Diphenyl(pyrrolidin-2-yl) methanol (3) (*first part*)

1. Recover crystals of **2** by suction filtration and dry them under vacuum. Weigh the crystals to determine the yield. Keep 30 mg of your product to acquire the ^{1}H NMR spectrum.

2. Prepare a 250 mL round-bottom flask with a magnetic stir bar. Add 4 g (11.7 mmol, 1 equiv.) of alcohol **2**. Evacuate the air and backfill with nitrogen.
3. Dissolve the alcohol in 60 mL of MeOH (partially soluble).
4. Add carefully 1 g of 10% Pd/C and 1 g of 10% $Pd(OH)_2$/C.
5. Attach a balloon of hydrogen to your flask with an adapter that allows the balloon to be closed off from the reaction flask. With the hydrogen balloon closed off, evacuate the flask until the solvent begins to bubble, and then open the balloon to the flask. Repeat the procedure twice more.
6. Stir the reaction mixture at room temperature overnight, until the next session.

Laboratory Session 7 (3 h): Synthesis of (*S*)-Diphenyl(pyrrolidin-2-yl) methanol (3) (*second part*)

1. Monitor the reaction by TLC (hexane/Et_2O, 3 : 1).
2. Remove the catalyst by filtration through a pad of Celite®, washing with MeOH.
3. Evaporate the MeOH under reduced pressure (a white solid should be formed).
4. Purify product **3** by recrystallization (*n*-hexane/$CHCl_3$, 1 : 1) to yield colourless crystals (optional). Weigh the crystals to determine the yield.
5. Dissolve 20 mg of compound **3** in deuterated chloroform (0.6 mL) and transfer to a NMR tube. Repeat the procedure with compound **2**. Acquire both ^1H NMR spectra.

Waste Disposal

The neutralized aqueous solution may be diluted and discarded in the sink. Organic solvents should be collected in a non-chlorinated waste collector recipient. Celite® containing palladium residues should be collected separately and treated as palladium waste.

Results Interpretation and Additional Questions

1. Discuss the importance of the protecting group in the overall process. Propose different protecting groups for this reaction indicating formation and cleavage conditions.
2. Rationalize the mechanism involved in the addition of Grignards to carbonyl or carboxylic derivatives.
3. Calculate the yield of each synthetic step and the overall yield.
4. Analyse the reaction thin-layer chromatography (TLC) results, calculate the R_f of the main products and interpret ^1H NMR spectra obtained for compounds **2** and **3**.

References

1. (a) E. J. Corey, R. K Bakshi and S. Shibata, *J. Am. Chem. Soc.*, 1987, **109**, 5551; (b) E. J. Corey, S. Shibata and R. K. Bakshi, *J. Org. Chem.*, 1988, **53**, 2861.
2. (a) K. Soai, A. Ookawa, T. Kaba and K. Ogawa, *J. Am. Chem. Soc.*, 1987, **109**, 7111; (b) B. M. Trost and C. C. Müller, *J. Am. Chem. Soc.*, 2008, **130**, 2438.
3. (a) C. De Fusco, C. Tedesco and A. Lattanzi, *J. Org. Chem.*, 2011, **76**, 676; (b) M. Marigo, J. Franzén, T. B. Poulsen, W. Zhuang and K. A. Jørgensen, *J. Am. Chem. Soc.*, 2005, **127**, 6964; (c) J. Franzén, M. Marigo, D. Fielenbach, T. C. Wabnitz, A. Kjaersgaard and K. A. Jørgensen, *J. Am. Chem. Soc.*, 2005, **127**, 18296; (d) C. Palomo and A. Mielgo, *Angew. Chem., Int. Ed.*, 2006, **45**, 7876.
4. T. W. Greene and P. G. M. Wuts, *Protective Groups in Organic Synthesis*, Wiley-Interscience, New York, 4th edn, 2006.
5. J. Clayden, N. Greevez and S. Warren, *Organic Chemistry*, Oxford University Press, Oxford, 2nd edn, 2012, ch. 23, p. 528.

4.2.2. Aldol Reaction and Condensation

4.2.2.1. Solvent-Free Aldol Condensation Reactions: Synthesis of Chalcone Derivatives

Barbora Morra*, Connie Tang, and Jonathon Fossella

Department of Chemistry, University of Toronto, Toronto, Ontario, Canada M5S 3H6
*E-mail: bmorra@chem.utoronto.ca

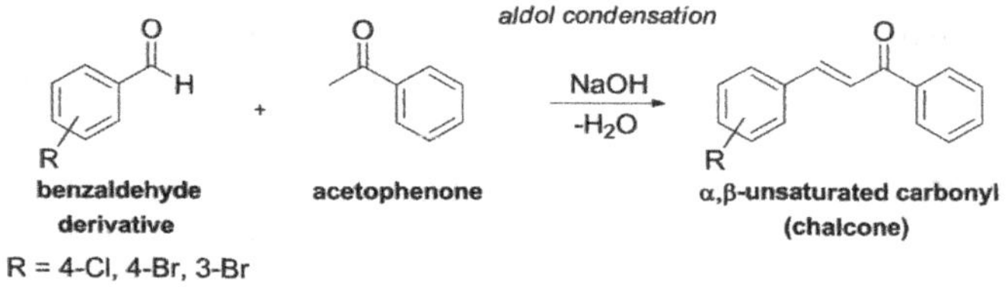

Number of sessions (duration of each session)	Hazard level	Difficulty level	Level of study
1 (3 h)	Moderate	Low	Intermediate

Class names Carbonyls (aldehydes, ketones, α,β-unsaturated ketones)

Concepts involved This experiment involves a solvent-free aldol condensation between acetophenone and various benzaldehyde derivatives in the synthesis of chalcone products

Chemicals needed 4-Chlorobenzaldehyde, 4-bromobenzaldehyde, 3-bromobenzaldehyde, acetophenone, sodium hydroxide (pellets), ethanol (95%)

Equipment and experimental techniques involved Grinding method (mortar and pestle), filtration apparatus (Büchner funnel), recrystallization

Keywords Aldol condensation, chalcones, melting point determination, NMR analysis, recrystallization, solvent-free reaction, α,β-unsaturated ketone

Background

Since its discovery by Wurtz in 1872, the aldol reaction has been one of the most powerful methods by which to form carbon–carbon bonds in organic synthesis.[1] The aldol reaction involves the addition of a nucleophilic enol or enolate of a carbonyl-containing compound to an electrophilic aldehyde or ketone, producing a β-hydroxycarbonyl product. Under certain reaction conditions, the initial aldol addition product can undergo subsequent condensation (loss of H_2O) to give an α,β-unsaturated carbonyl compound. In some cases, the β-hydroxycarbonyl product can

Comprehensive Organic Chemistry Experiments for the Laboratory Classroom
Edited by Carlos A. M. Afonso, Nuno R. Candeias, Dulce Pereira Simão, Alexandre F. Trindade, Jaime A. S. Coelho, Bin Tan, and Robert Franzén.

Published by the Royal Society of Chemistry, www.rsc.org

be isolated since more vigorous conditions such as increased temperature may be required to promote the condensation step. However, in the case of conjugated aldehyde reaction partners, the initial β-hydroxycarbonyl is rarely isolated since it undergoes rapid condensation to give an extended resonance stabilized carbonyl product. In this experiment, aldol condensation reactions will be explored with substituted benzaldehydes and acetophenone to synthesize several chalcone derivatives (Scheme 4.2.2.1.1).

Chalcones (or benzalacetophenones) are an important class of compounds that are biosynthesized in plants to form flavonoids, which are responsible for plant pigmentation, UV filtration and immunity against pathogens and insects. Chalcones also exhibit interesting biological properties including antioxidant, cytotoxic and antimalarial activity, which makes them interesting target molecules for the treatment of viral and cardiovascular disorders as well as parasitic infections.[2]

Although solvents are commonly employed in aldol condensation reactions, this experiment is performed under solvent-free conditions where solid–liquid grinding is used to achieve the chemical transformation. In fact, solvent-free reactivity is just one example of the numerous procedural modifications based on the 12 Principles of Green Chemistry[3] that have been developed for aldol reactions.[4] This experiment provides a convenient method for the synthesis of several chalcone derivatives in high yield and purity while highlighting several sustainable principles including high atom economy,[5] energy efficiency, and waste prevention (solvent-free) while using benign conditions.

This experiment is an inquiry-based activity where an "unknown" benzaldehyde derivative (Figure 4.2.2.1.1) is combined in a mortar and pestle with acetophenone and sodium hydroxide. After grinding for 10 min, the crude chalcone is isolated by suction filtration and recrystallized from 95% ethanol. Once the product is obtained, various techniques, including melting point determination, thin-layer chromatography (TLC) and NMR and IR spectroscopy, are used to determine the structure of the chalcone product.

Additional Safety

This experiment should be performed in a fume hood with proper safety equipment including a laboratory coat, gloves and safety goggles. 4-Chlorobenzaldehyde, 4-bromobenzaldehyde and 3-bromobenzaldehyde are all moderate skin, eye and respiratory irritants and should be kept inside the fume hood to avoid direct contact. Sodium hydroxide is a highly corrosive base and may cause severe skin and eye burns. Both acetophenone and ethanol are flammable liquids and toxic by ingestion.

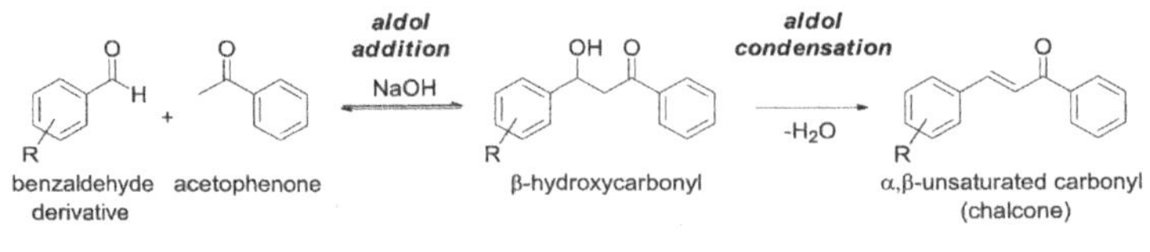

Scheme 4.2.2.1.1 Aldol condensation reaction in the synthesis of chalcone derivatives.

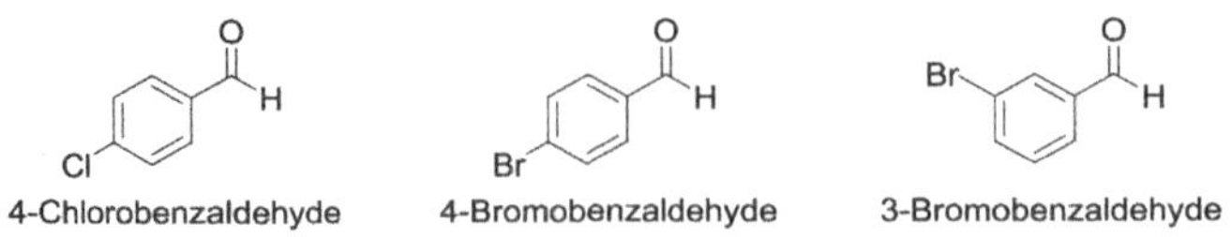

Figure 4.2.2.1.1 Benzaldehyde derivatives.

Experimental Procedure

Laboratory Session: 3 h

Chalcone Synthesis

1. In a fume hood, transfer the pre-weighed unknown benzaldehyde derivative (5 mmol, 1 equiv.) into a mortar and add 0.59 mL (5 mmol, 1 equiv.) of acetophenone and 0.20 g (5 mmol, 1 equiv.) of NaOH pellets (about two pellets).
2. Grind the mixture with a pestle for 10 min while ensuring that the NaOH pellets are crushed completely. Periodically scrape the material from the sides of the mortar. The reaction mixture should resemble a thick, chalky paste.
3. Add 10.0 mL of H_2O, while scraping the sides of the mortar to mix thoroughly.
4. Collect the crude solid by suction filtration on a Büchner funnel. Rinse the mortar and pestle with an additional 5.0 mL of H_2O and add through the funnel.
5. Rinse the crude product with an additional 5.0 mL of H_2O and allow to air dry (5–10 min).
6. Transfer the crude product onto a pre-weighed watch glass and determine the crude yield.

Chalcone Purification

7. For recrystallization, transfer the crude solid into a 50 mL Erlenmeyer flask and dissolve in the *minimum* amount of warm 95% ethanol while heating on a hot plate. Add warm 95% ethanol 1.0 mL at a time, to a maximum of 10.0 mL, until the solution is completely clear.
8. Once the solid has dissolved, remove the flask from the hot plate and leave it to cool to room temperature (5 min) and then cool in an ice bath (5 min).
9. Collect the crystals by suction filtration on a Büchner funnel and wash with approximately 2.0 mL of cold 95% ethanol.
10. Dry the purified product thoroughly under vacuum (5 min) and collect the pure chalcone onto a pre-weighed watch glass.

Chalcone Analysis

11. Weigh the product and calculate the percent yield.
12. Check the purity of the chalcone product by measuring the melting point. Obtain IR and ^1H NMR spectra and perform thin-layer chromatography (TLC) analysis by co-eluting your product against acetophenone using a hexanes/ethyl acetate (3 : 1) solvent system.

Waste Disposal

The water solution may be collected in an aqueous waste container while the ethanol solution may be collected in a non-halogenated flammable waste container for posterior destruction.

Results Interpretation and Additional Questions

1. Compare the appearance and melting point of your product with the literature data to identify the structure of your unknown chalcone.
2. Calculate the R_f values of acetophenone and your chalcone product.
3. Report the percent yield for your chalcone synthesis.
4. Calculating the percent yield of a chemical reaction does not provide a complete view of a transformation as it only considers how much of the desired product is produced. From a green chemistry perspective, the atom economy considers how much of the raw materials (reactants) end up in the final product using the following equation:

$$\text{Atom economy} = \frac{\text{mass of atoms in desired product}}{\text{mass of atoms in reactants}} \times 100\%$$

A chemical reaction can have a high yield but also produce a large amount of waste (low atom economy). When designing a green chemical process, a high yield and atom economy is desirable. Report the atom economy for your chalcone synthesis.

5. Interpret the ^{1}H NMR spectrum of your chalcone product and report the proton assignments. What evidence do you have that your product is the *trans*-isomer?
6. Interpret the IR spectrum of your chalcone product and report the main absorptions.
7. Predict the product of the aldol condensation reaction if acetophenone were replaced with acetone. What product would be expected if excess aldehyde and NaOH were employed?

References

1. (a) A. Wurtz, *Bull. Soc. Chim. Fr.*, 1872, **17**, 436–442; (b) B. M. Trost and C. S. Brindle, *Chem. Soc. Rev.*, 2010, **39**, 1600–1632; (c) *Modern Aldol Reactions*, ed. R. Mahrwald, Wiley-VCH Verlag GmbH, Weinheim, 2004; (d) M. B. Smith and J. March, *March's Advanced Organic Chemistry*, Wiley-Interscience, New Jersey, 6th edn, 2007, pp. 1339–1354; (e) F. A. Carey and R. J. Sundberg, *Advanced Organic Chemistry Part B*, Springer Science, New York, 5th edn, 2007, pp. 64–138.
2. (a) Z. Nowakowska, *Eur. J. Med. Chem.*, 2007, **42**, 125–137; (b) V. J. Ram, A. S. Saxena, S. Srivastava and S. Chandra, *Bioorg. Med. Chem. Lett.*, 2000, **10**, 2159–2161.
3. P. T. Anastas and J. C. Warner, *Green Chemistry: Theory and Practice*, Oxford University Press, Oxford, 1998, pp. 29–56.
4. (a) R. Mestres, *Green Chem.*, 2004, **6**, 583–603; (b) G. W. V. Cave and C. L. Raston, *J. Chem. Educ.*, 2005, **82**, 468–469; (c) D. R. Palleros, *J. Chem. Educ.*, 2004, **81**, 1345–1347; (d) K. M. Doxsee and J. E. Hutchison, *Green Organic Chemistry – Strategies, Tools and Laboratory Experiments*, Brooks-Cole, Pacific Grove, Canada, 2004, pp. 115–119; (e) C. L. Raston and J. L. Scott, *Green Chem.*, 2000, **2**, 49–52; (f) F. Toda, K. Tanaka and K. Hamai, *J. Chem. Soc., Perkin Trans. 1*, 1990, 3207–3209.
5. B. M. Trost, *Science*, 1991, **254**, 1471–1477.

4.2.2.2. Synthesis of (*E*)-Chalcones [(*E*)-1,3-diarylprop-2-en-1-ones]

Artur M. S. Silva*, Augusto C. Tomé, Diana C. G. A. Pinto, Fernando Domingues, Graça M. Oliveira Rocha, José A. S. Cavaleiro, M. Graça P. M. S. Neves, M. Amparo F. Faustino, and Mário M. Q. Simões

Chemistry Department & QOPNA, University of Aveiro, Campus de Santiago, 3810-193 Aveiro, Portugal
*E-mail: artur.silva@ua.pt

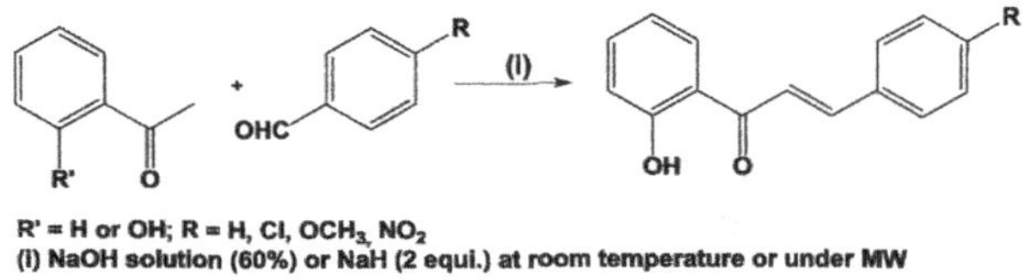

Number of sessions (duration of each session)	Hazard level	Difficulty level	Level of study
2 (3 h each)	Moderate	Medium	Intermediate

Class names Aldehydes, ketones, enones

Concepts involved This experiment involves a mixed aldol reaction of acetophenones with benzaldehydes under basic conditions. The experiment can be performed at room temperature or under MW irradiation conditions. Additionally, several chromatographic techniques (column or TLC) as well as liquid–liquid extraction can be learnt

Chemicals needed Benzaldehyde and acetophenone derivatives, sodium hydroxide and/or sodium hydride, dichloromethane, ethanol, methanol and/or THF, hexane and ethyl acetate for TLC, $CDCl_3$

Equipment and experimental techniques involved Stirring plate or microwave reactor, recrystallization and filtration, NMR, IR and melting point apparatus

Keywords Aldol addition, chromatographic techniques, MW irradiation, NMR, α,β-unsaturated carbonyl compounds

Background

1,3-Diarylprop-2-en-1-ones, commonly named chalcones, are secondary metabolites belonging to the flavonoid family. Chemically they consist of an open-chain flavonoid structure in which the two aromatic rings are joined by a three-carbon α,β-unsaturated carbonyl system. Naturally occurring chalcones present substituents in both aryl rings, namely hydroxy and methoxy groups, and are known to possess many useful biological properties, including anti-inflammatory, antimicrobial, antifungal, antioxidant, cytotoxic, anti-tumour and anticancer activities. Taking into account the prominence of chalcones, the development of efficient methodologies towards their synthesis is very important. Several methodologies for the synthesis of chalcones are based on mixed aldol condensation[1] and can be adapted to an

Comprehensive Organic Chemistry Experiments for the Laboratory Classroom
Edited by Carlos A. M. Afonso, Nuno R. Candeias, Dulce Pereira Simão, Alexandre F. Trindade, Jaime A. S. Coelho, Bin Tan, and Robert Franzén.

Published by the Royal Society of Chemistry, www.rsc.org

intermediate level experiment for an organic chemistry course. The procedure is easy and new developments using MW (microwave) irradiation have also proved to be very efficient and allow the introduction of students to microwave assisted organic synthesis.[2] Recrystallization and filtration techniques can be introduced with this experiment; furthermore, several chromatographic experimental techniques can be illustrated, such as column chromatography and thin-layer chromatography.[3]

The experiment can be used to highlight one of the most important carbonyl compounds reactions, allowing the consolidation of theoretical subjects such as acidity of α-carbonyl hydrogens, aldol and mixed aldol additions, dehydration of the aldol product and α,β-unsaturated carbonyl compounds.[4]

This experiment describes a mixed aldol reaction of benzaldehydes with acetophenones under basic conditions at room temperature; alternatively, this reaction can be performed in a microwave reactor in THF and using sodium hydride. The reaction product can be isolated by filtration and further purified by recrystallization and/or chromatography. The analysis of the obtained compounds by nuclear magnetic resonance (NMR) is recommended because it will be a good introduction to NMR spectra interpretation.

Additional Safety

This work requires regular laboratory good practices and personal safety equipment (safety glasses, laboratory coat and nitrile gloves). The synthetic and extraction procedures and chromatographic purifications should be done in a fume hood that is working properly and with glass material in perfect condition. It is recommended to avoid direct contact with benzaldehyde and acetophenone derivatives and sodium hydroxide solution because they can be irritating to skin. The organic solvents used (methanol and THF) are inflammable when in the presence of ignition sources and are harmful if swallowed, or in contact with the skin. Sodium hydride is a flammable solid that may ignite in moisture and, consequently, should be handled by the class teacher.

Experimental Procedure

Laboratory Session: 3 h

Method A

1. Prepare the aqueous solution of sodium hydroxide by weighing 36 g of sodium hydroxide and then dissolving it in 60 mL of distilled water. Let the solution stand in an ice bath in a fume hood, to reach room temperature.
2. In a fume hood add to a 250 mL round-bottom flask, equipped with a magnetic stirring bar, 1.5 mL of the desired acetophenone and 60 mL of methanol.
3. Put the round-bottom flask in an ice bath placed on a stirring plate and add the sodium hydroxide solution.
4. Add 1.5 equiv. of the desired benzaldehyde and remove the ice bath.
5. Maintain the reaction mixture stirring for 90 min.
6. After the stirring period, pour the reaction mixture into a vessel containing ice (~150 g) and add concentrated hydrochloric acid to adjust the pH to ~2.
7. The solid formed is filtered using a vacuum filtration apparatus.
8. Assemble the vacuum filtration apparatus. Put the filter paper onto the Büchner funnel, wet the filter paper with distilled water and open the vacuum valve. Transfer the reaction mixture onto the filter, and wash the product three times with small portions of iced water. Keep the vacuum system connected to dry the filtrated solid.

9. Alternatively, the reaction mixture can be extracted with dichloromethane (3 × 100 mL) and the solvent evaporated to dryness.
10. The obtained solid can be crystallized in ethanol, analysed by TLC and/or purified by column chromatography (if the experiment is planned for two sessions, see Supplementary information (SI)).
11. For recrystallization dissolve the crude compound in hot ethanol† (5 mL of ethanol per ~500 mg crude compound) and let it crystallize (the recipient can be put in ice for a faster recrystallization).
12. Typical TLC analyses involve the application of the starting acetophenone, the crude or pure compound and co-application (see SI). The TLC can be run using a mixture of hexane/ethyl acetate (7 : 3).
13. The purified compound is carefully transferred to a labelled, pre-weighed watch glass and put in the oven at a temperature of 70 °C for 15 min.
14. Weigh the obtained product and calculate the reaction yield. Measure the melting point, check the compound purity by TLC and prepare the NMR tube.

Method B

1. In a fume hood add to a 250 mL round-bottom flask, equipped with a magnetic stirring bar, 1.5 mL of the desired acetophenone and 60 mL of dried THF.
2. Put the round-bottom flask in an ice bath placed on a stirring plate and slowly add 2 equiv. of sodium hydride.
3. Add 1.5 equiv. of the desired benzaldehyde and remove the ice bath.
4. Maintain the reaction mixture stirring for 60 min (can be performed in a microwave apparatus by irradiation at 400 W for 15 min).
5. After the stirring period, pour the reaction mixture into a vessel containing ice (~150 g) and add hydrochloric acid to adjust pH to ~2.
6. The formed solid is filtered off using a vacuum filtration apparatus.
7. Assemble the vacuum filtration apparatus. Put the filter paper onto the Büchner funnel, wet the filter paper with distilled water and open the vacuum valve. Transfer the reaction mixture onto the filter, and wash the product three times with small portions of iced water. Keep the vacuum system connected to dry the filtrated solid.
8. Alternatively, the reaction mixture can be extracted with dichloromethane (3 × 100 mL) and the solvent evaporated to dryness.
9. The obtained solid can be crystallized in ethanol, analysed by TLC and/or purified by column chromatography (see points 10–12 of the previous method).
10. The purified compound is carefully transferred to a labelled, pre-weighed watch glass and placed in the oven at a temperature of 70 °C for 15 min.
11. Weigh the product obtained and calculate the reaction yield. Measure the melting point, check the compound purity by TLC and prepare the NMR tube.

If the experiment is planned for two sessions purification by column chromatography can be performed in the second session:

1. Check the solvent system by TLC. When the appropriate solvent is chosen, prepare 500 mL of it.
2. Obtain a 20 mm diameter flash chromatography column and use a pen to mark 15 cm. Fill the column with silica gel (in the hood), tapping gently to pack it down. Carefully add 1 cm of sand to the top of the column of silica gel.

---

†For R' = OH it should be boiling ethanol, for R' = H just 30 °C.

3. Clamp the column securely, place a clean 250 mL flask under the column, open the stopcock, and carefully fill the column from the top with your eluent. Attach a piece of tubing equipped with a column adapter to an airline and start a slow flow of air. Hold the column adapter on the top of the column and push the solvent rapidly through the silica gel. When the solvent level is approximately 1 cm from the sand at the head of the column, remove the air. If the column is not completely wetted, refill the column with eluent and repeat the above procedure. Once the column is completely wetted with solvent, lower the eluent level in the column until it is level with the top of the sand. Recombine the solvent you collected during the wetting of the column with any unused eluent.
4. Dissolve the crude product in 2–3 mL of the eluent. Use a Pasteur pipet to add this mixture to the top of the flash column. Then briefly place the air flow back on the column and push this mixture onto the top of the silica gel, taking care not to let the eluent level dip below the level of the silica gel. Rinse the container that contained the crude product with another 2–3 mL of eluent and load this onto the column as you did above. Repeat this procedure several times until there is little or no colour left in the solution.
5. Then carefully fill the column with eluent and use the air to force the eluent through the column at a slow rate, collect 10 mL fractions. (You should have previously obtained about 20 clean test tubes for this purpose. It is not necessary to stop the air flow between fractions. With a little practice the test tubes can be changed without spilling any eluent.) When the eluent level is 1 cm from the top of the sand, remove the air and add more eluent. Continue this process until you have used all your eluent or obtained 20 fractions. At no time should the eluent level be allowed to drop below the level of the silica gel. Do not discard the column until you are sure that you have isolated your product.
6. Check each of the fractions by TLC using whatever eluent you used for your column. Combine the fractions containing your product and remove the solvent by rotary evaporation.

Waste Disposal

All liquids should be collected in a non-halogenated waste collector recipient for posterior destruction. Solvents used in chromatographic purifications or extractions should be collected for further distillation and reutilization.

Results Interpretation and Additional Questions

1. Calculate the percent yield of the reaction.
2. Rationalize the reaction mechanism.
3. Using the obtained TLC chromatogram calculate the chalcone R_f and comment on the purity of the chalcone.
4. Interpret the ^{1}H and ^{13}C NMR spectra obtained, and compare the obtained melting point with the literature value.

References

1. A. M. S. Silva, H. R. Tavares, A. I. N. R. A. Barros and J. A. S. Cavaleiro, *Spectrosc. Lett.*, 1997, **30**, 1655–1667.
2. C. O. Kappe, D. Dallinger and S. S. Murphree, *Practical Microwave Synthesis for Organic Chemists*, Wiley-VCH Verlag GmbH, Weinheim, 2009.
3. B. S. Furniss, A. J. Hannaford, P. W. G. Smith and A. R. Tatchell, *VOGEL's Textbook of Practical Organic Chemistry*, Longman Scientific & Technical, 1989.
4. F. A. Carey, *Organic Chemistry*, 6th edn, McGraw-Hill, 2006.

4.2.2.3. A Solvent-Free Approach for Chalcone Synthesis *via* an Aldol Reaction

Sofia F. P. Friães[a], Renato E. Boto[b], Paulo Almeida[b], Amélia M. Silva[c], and Lucinda V. Reis*[a]

[a]Department of Chemistry and CQ-VR, UTAD, Quinta de Prados, 5000-801 Vila Real, Portugal; [b]CICS-UBI – Health Sciences Research Centre, University of Beira Interior, Av. Infante D. Henrique, 6200-506 Covilhã, Portugal; [c]Department of Biology and Environment and CITAB, UTAD, Quinta de Prados, 5000-801 Vila Real, Portugal. *E-mail: lucinda.reis@utad.pt

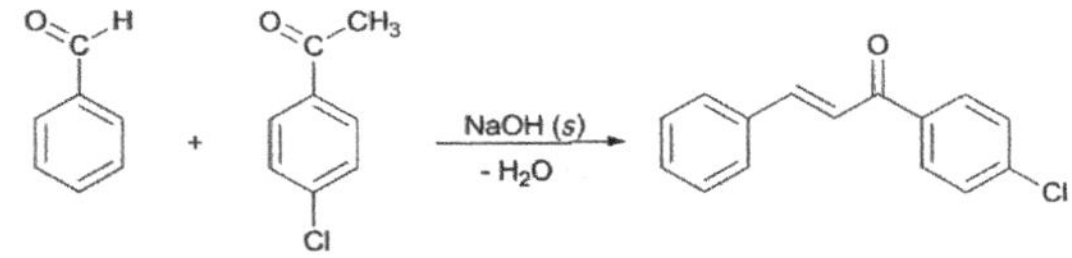

Number of sessions (duration of each session)	Hazard level	Difficulty level	Level of study
1 (3 h)	Low	Low	Intermediate

Class names Aldehydes, ketones, α,β-unsaturated carbonyl compounds

Concepts involved This experiment involves a method for the synthesis of a chalcone by a Claisen–Schmidt aldol reaction between benzaldehyde and 4′-chloroacetophenone in the presence of NaOH (s) under solvent-free conditions

Chemicals needed Benzaldehyde, 4′-chloroacetophenone, sodium hydroxide, ethanol, distilled water

Equipment and experimental techniques involved Porcelain mortar with a pestle, vacuum filtration, recrystallization, oven, IR, NMR, EIMS (evaporative ionization mass spectrometry) and mp determination

Keywords Chalcone, Claisen–Schmidt aldol reaction, EIMS, NMR, recrystallization, mp

Background

Chalcones, also named 1,3-diaryl-2-propen-1-ones, are synthetic or natural compounds belonging to the large flavonoid family. In chemical terms these compounds are open-chain flavonoids with the two aromatic rings joined by a three-carbon α,β-unsaturated carbonyl system (C_6–C_3–C_6). Chalcones exhibit a wide variety of biological activities, depending on the substitution of aromatic rings, highlighting the anti-inflammatory, antifungal, antinociceptive, antimicrobial, anticancer, analgesic, hypolipidemic, antidepressant and antioxidant activities.[1–6]

The main way of obtaining chalcones reported in the literature is to synthesize them by condensation of an aldehyde with a ketone, known as the Claisen–Schmidt reaction.[1,2,4–7]

Several modifications of this procedure can be found in the literature, aimed at more efficient methods, better yields, shorter reaction times, as well as incorporating cleaner technologies.[8] As

Comprehensive Organic Chemistry Experiments for the Laboratory Classroom
Edited by Carlos A. M. Afonso, Nuno R. Candeias, Dulce Pereira Simão, Alexandre F. Trindade, Jaime A. S. Coelho, Bin Tan, and Robert Franzén.

Published by the Royal Society of Chemistry, www.rsc.org

an example, the synthesis of 20 different chalcones by a Claisen–Schmidt aldol reaction, in the absence of solvent, in a few minutes with remarkable yields was reported by Palleros.[8]

The objective of this work is the synthesis of a different chalcone from those reported by Palleros,[8] but also by a Claisen–Schmidt reaction, by mixing benzaldehyde and 4′-chloroacetophenone in the presence of solid NaOH in a mortar with pestle without the use of any organic solvents. Accordingly, after mixing all the reagents in a mortar and grinding them over a period of 20–30 min, chalcone (*E*)-1-(4′-chlorophenyl)-3-phenyl-2-propen-1-one precipitates under the reaction conditions and it is therefore easily isolated by filtration.

As this experiment is carried out in the absence of organic solvents and does not produce products harmful to the environment it is a good example of a green chemistry experiment.

Additional Safety

This experiment should be carried out in a fume hood and students should wear protective clothing, gloves and safety goggles. NaOH is a caustic base and therefore may cause severe skin burns. Benzaldehyde and 4′-chloroacetophenone cause severe eye irritation.

Experimental Procedure

1. In a fume hood put 0.50 mL (5.0 mmol, 1 equiv.) of benzaldehyde into the porcelain mortar and add 0.65 mL (5.0 mmol, 1 equiv.) of 4′-chloroacetophenone.
2. To the mixture add a pellet of NaOH (weighing approximately 0.2 g, 5.0 mmol, 1 equiv.).
3. Grind the mixture with the aid of a pestle for 20–30 min. After a few seconds of grinding the reaction mixture becomes a yellow paste. Finish grinding when the mixture is a solid powder.
4. At the end of grinding add 10 mL of distilled water to the mortar and mix well with the obtained product using the pestle and a spatula to remove the solid from mortar's wall.
5. Collect the resulting solid using vacuum filtration on a Büchner funnel and further wash the mortar and pestle with an additional 10 mL of distilled water in order to remove the remaining solid therein. Add this mixture to the same filter and wash the product on the filter with another 10 mL of distilled water.
6. Regarding the purity of the obtained solid, recrystallize it from ethanol and dry the crystals so obtained in an oven.
7. Once dried, weigh the product, measure the mp and record the IR spectrum.

Waste Disposal

This reaction is carried out in the absence of organic solvents, using just ethanol for recrystallization of the obtained product. The ethanol may be reused after vacuum distillation or collected in a non-chlorinated waste collector recipient for posterior destruction.

Results Interpretation and Additional Questions

1. Determine the reaction yield of (*E*)-1-(4′-chlorophenyl)-3-phenyl-2-propen-1-one so obtained and compare the melting point with the one given in the literature.[2]
2. Assign the principal bands in the IR spectrum and compare them with the ones reported in the literature.[2,9]
3. Interpret the ^{1}H and ^{13}C NMR spectra and compare them with the ones given in the literature.[2,9]
4. Interpret the EIMS and compare it with the one reported in the literature.[9]

5. Rationalize the reaction mechanism.
6. Discuss the importance of using green chemistry stressing the advantages/disadvantages of using this method compared with the classical methods for chalcone synthesis reported in the literature.[1,2,7,8]

References

1. H.-K. Hsieh, T.-H. Lee, J.-P. Wang, J.-J. Wang and C.-N. Lin, *Pharm. Res.*, 1998, **15**, 39–46.
2. R. Corrêa, M. A. S. Pereira, D. Buffon, L. Santos, V. C. Filho, A. R. S. Santos and R. J. Nunes, *Arch. Pharm. Pharm. Med. Chem.*, 2001, **334**, 332–334.
3. M. E. R. Escobedo, L. B. Bermúdez, C. P. Berumen, A. S. Galindo and S. Y. S. Belmares, *Rev. Mex. Cienc. Farm.*, 2012, **43**, 7–14.
4. L. Santos, R. C. Pedrosa, R. Correa, V. C. Filho, R. J. Nunes and R. A. Yunes, *Arch. Pharm. Chem. Life Sci.*, 2006, **339**, 541–546.
5. X. Sui, Y.-C. Quan, Y. Chang, R.-P. Zhang, Y.-F. Xu and L.-P. Guan, *Med. Chem. Res.*, 2012, **21**, 1290–1296.
6. P. M. Sivakumar, P. K. Prabhakar and M. Doble, *Med. Chem. Res.*, 2011, **20**, 482–492.
7. J. M. Guider, T. H. Simpson and D. B. Thomas, *J. Chem. Soc.*, 1955, **30**, 170–173.
8. D. R. Palleros, *J. Chem. Educ.*, 2004, **81**, 1345–1347.
9. SDBS (Spectral Database for Organic Compounds) web site: http://sdbs.db.aist.go.jp/sdbs/cgi-bin/direct_frame_disp.cgi?sdbsno=11606, National Institute of Advanced Industrial Science and Technology, September 2014.

4.2.2.4. Preparation of Dibenzylideneacetone

Abel J. S. C. Vieira

Faculty of Sciences and Technology, Universidade Nova de Lisboa, 2829-516 Caparica, Portugal
*E-mail: ajscv@fct.unl.pt

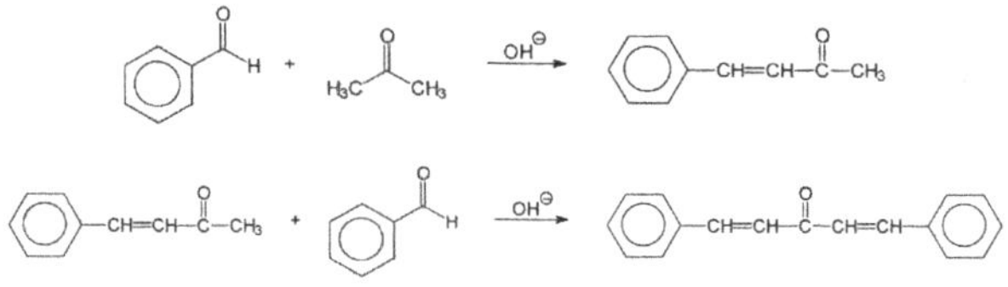

Number of sessions (duration of each session)	Hazard level	Difficulty level	Level of study
1 (3 h)	Moderate	Low	Intermediate

Class names Carbonyls and conjugated carbonyl compounds (enones)

Concepts involved This experiment aims to illustrate a typical reaction of carbonyl compounds. The cross aldol condensation reaction between an aldehyde and a ketone shows the reactivity at both carbonyl and α carbon atoms

Chemicals needed Ethanol, acetone, benzaldehyde, sodium hydroxide

Equipment and experimental techniques involved Apparatus for reaction at room temperature and stirring, recrystallization apparatus, filtration equipment

Keywords Π-Conjugation, cross aldol condensation, mixture stirring, recrystallization, spectra

Background

Carbonyl compounds, aldehydes and ketones, have a chemical reactivity that is characterized by two main types of reaction: at the carbonyl carbon atom and at the α carbon atom.[1,2] The typical reaction at the carbonyl carbon is the nucleophilic addition enabled by the electron deficiency of this atom. Reactions involving the α carbon atom deal with the possibility of the establishment of a keto–enol tautomeric equilibrium, which happens whenever the carbonyl compound possesses at least one hydrogen atom bound to at least one carbon atom adjacent to the carbonyl carbon. The enol or enolate (depending on the acidic or basic catalysis, respectively) behaves as a nucleophile that is able to donate an electron pair from a carbon atom.

In this experiment, two different carbonyl compounds are present together in basic solution: a non-enolizable aldehyde (benzaldehyde) and an enolizable ketone (acetone); therefore, three reactive species are present in the reaction medium: the keto forms of the aldehyde and the ketone and the enolate form of the ketone. In this condition, two aldol addition reactions are possible: the addition of the ketone enolate to the carbonyl of the ketone itself and the addition of the ketone enolate to the carbonyl of the aldehyde. The first reaction is much less favourable

Comprehensive Organic Chemistry Experiments for the Laboratory Classroom
Edited by Carlos A. M. Afonso, Nuno R. Candeias, Dulce Pereira Simão, Alexandre F. Trindade, Jaime A. S. Coelho, Bin Tan, and Robert Franzén.

Published by the Royal Society of Chemistry, www.rsc.org

than the second one, due to the better electrophilicity of the aldehyde and to the greater steric hindrance of the ketone. Therefore, the cross aldol addition of the ketone enolate to the benzaldehyde carbonyl will occur preferentially. The resulting aldol dehydrates spontaneously, even in basic medium and at room temperature, yielding the condensation product (see first reaction scheme). This cross aldol condensation is usually known as the Claisen–Schmidt reaction.

When benzaldehyde is present in excess, as is the case of this experiment, and since the product of the first condensation reaction (benzylideneacetone) is still an enolizable ketone, a new aldol condensation reaction between this latter compound and benzaldehyde can occur, yielding dibenzylideneacetone (see second reaction scheme).

Additional Safety

The experiment can be performed on a regular laboratory bench, with proper safety equipment, such as protective clothing, gloves and safety glasses. Ethanol is a flammable reagent. Benzaldehyde must be handled in a fume hood.

Ethanol: R: 11; S: 2-7-16.
Acetone: R: 11-36-66-67; S: 2-9-16-26.
Benzaldehyde: R: 22; S: 2-24.
Sodium hydroxide: R: 35; S: 1/2-26-37/39-45.

Experimental Procedure

One Laboratory Session (3 h)

Synthesis of Dibenzylideneacetone (ref. 3)

1. Prepare a cold solution of 6.3 g of sodium hydroxide in 63 mL of water and 50 mL of ethanol.
2. Introduce this solution in a 250 mL three-neck round-bottomed flask fitted with a water condenser, an addition funnel and a mechanical stirrer.
3. Place the flask in a cold water bath, keeping the temperature at 20–25 °C.
4. In the addition funnel mix 6.8 g (6.5 mL) of benzaldehyde and 1.9 g (2.3 mL) of acetone.
5. Add about one-half of this mixture to the flask, under vigorous stirring. After a short time a flocky precipitate begins to be formed.
6. Continue stirring for about 15 min and then add the rest of the benzaldehyde/acetone mixture.
7. Keep stirring for a further 30 min at the same temperature (20–25 °C).
8. Cool the contents of the flask in ice cold water, filter off the solid at the water vacuum pump, wash it with ice cold water (in the Büchner funnel) and let it dry in the air on the filter paper.

Product Purification

9. Recrystallize the crude compound in ethanol (use the volume recommended by your instructor).
10. After filtration of the yellow crystals, test for the presence of traces of benzaldehyde (by its characteristic smell).
11. If the test is positive, remove the residual benzaldehyde by washing the crystals with a small amount of ice-cold ethanol.
12. Filter the crystals at the water vacuum pump, dry them in the air on the filter paper and keep the purified product in a vacuum desiccator.

Waste Disposal

The basic mixture of water and ethanol may be neutralized, diluted and discarded in the sink. Ethanol used for recrystallization can be collected in an appropriate container to be recycled and reused, or discarded in a non-chlorinated waste collector recipient, for posterior destruction.

Results Interpretation and Additional Questions

Results and Product Characterization

1. Weigh the dried solid product obtained and calculate the yield of the reaction.
2. Determine the melting point of dibenzylideneacetone. Discuss the level of purity of the synthesized compound by comparing this value with that reported in the literature.
3. Record the IR spectrum of the compound, using dried potassium bromide as support material. Assign the absorption bands characteristic of the functional groups of the compound.
4. Acquire the ^{1}H NMR spectrum of the compound. Use the deuterated solvent suggested by your instructor. Try to assign the H atoms to the NMR signals.

Additional Questions

1. How do you name dibenzylideneacetone according to IUPAC nomenclature rules?
2. Write the base catalysed tautomeric equilibrium of acetone and the mechanism of the addition of the enolate to the carbonyl group of the aldehyde.
3. Rationalize the whole reaction mechanism.
4. Why is the dehydration step spontaneous in this reaction?
5. Suggest how to proceed to obtain the monocondensation product, benzylideneacetone, as the main product.
6. Dibenzylideneacetone is a coloured compound (pale yellow). Colour is due to absorption in the visible region of the electromagnetic spectrum. Try to explain this property in terms of the structure of the compound. (Hint: pay attention to the π system of the structure.)

References

1. T. W. G. Solomons and C. B. Fryhle, *Organic Chemistry*, John Wiley & Sons, Inc. Hoboken, 8th edn, 2003, vol. 17, p. 784.
2. J. Clayden, N. Greeves, S. Warren and P. Wothers, *Organic Chemistry*, Oxford University Press, 1st edn, 2001, vol. 27, p. 689.
3. B. S. Furniss, A. J. Hannaford, P. W. G. Smith and A. R. Tatchell, *VOGEL's Textbook of Practical Organic Chemistry*, Longman Scientific & Technical, 5th edn, 1989, vol. 6, p. 1032.

4.2.2.5. L-Proline Catalyzed Aldol Reaction of 4-Nitrobenzaldehyde with Acetone

Fang Fang*, Xin-Yuan Liu, and Bin Tan

Department of Chemistry, South University of Science and Technology of China, Shenzhen, 518055, People's Republic of China. *E-mail: fang.f@sustc.edu.cn

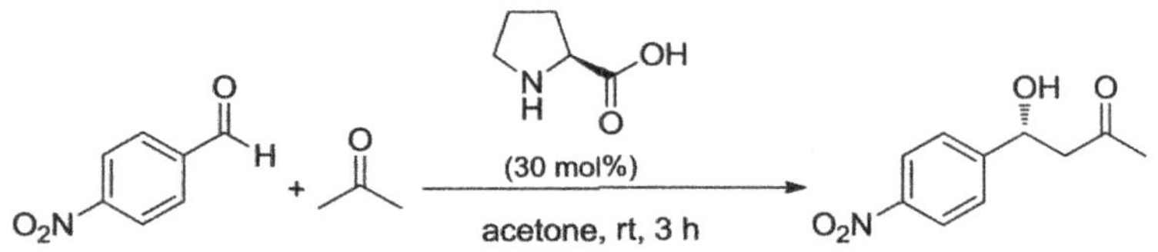

Number of sessions (duration of each session)	Hazard level	Difficulty level	Level of study
2 (4 h each)	Low	Medium	Advanced

Class names Aldehyde, ketone, amino acid

Concepts involved This experiment involves an aldol reaction between acetone and 4-nitrobenzaldehyde catalyzed by L-proline. The experiment can be performed at room temperature over a period of 3 h. The pure product is obtained by column chromatography and the enantiomeric excess of the product determined by chiral HPLC analysis

Chemicals needed Acetone, 4-nitrobenzaldehyde, L-proline, dichloromethane, ethyl acetate, petroleum ether (60–90 °C), isopropanol, hexane, $CDCl_3$, silica for column chromatography

Equipment and experimental techniques involved Stirring plates, column chromatography, rotary evaporator, chiral HPLC

Keywords Aldol reaction, asymmetric synthesis, chiral HPLC, column chromatography, organocatalyst

Background

The catalytic asymmetric aldol reaction is an exceptionally useful fundamental carbon–carbon bond-forming reaction for the stereoselective construction of cyclic and acyclic molecules. In the 1970s, the asymmetric intramolecular aldol reaction was first catalyzed by L-proline with good results.[1,2] In 2000, List and Barbas demonstrated the use of proline as an organocatalyst for the direct intermolecular asymmetric aldol reaction between acetone and various aldehydes.[3] Since L-proline is a nontoxic, inexpensive, and readily available natural amino acid it has been widely used in organocatalytic asymmetric synthesis in recent years.[4]

In this experiment, L-proline as an effective organocatalyst for the direct aldol reaction between unmodified acetone and 4-nitrobenzaldehyde shows both high yield and moderate stereoselectivity.[4] The catalyst loading is 30 mol% in this L-proline catalyzed aldol reaction. The mixture was stirred at room temperature over a period of 3 h. The pure product was obtained by column chromatography on silica gel and the enantiomeric excess of the product was determined by chiral HPLC analysis.

Comprehensive Organic Chemistry Experiments for the Laboratory Classroom
Edited by Carlos A. M. Afonso, Nuno R. Candeias, Dulce Pereira Simão, Alexandre F. Trindade, Jaime A. S. Coelho, Bin Tan, and Robert Franzén.

Published by the Royal Society of Chemistry, www.rsc.org

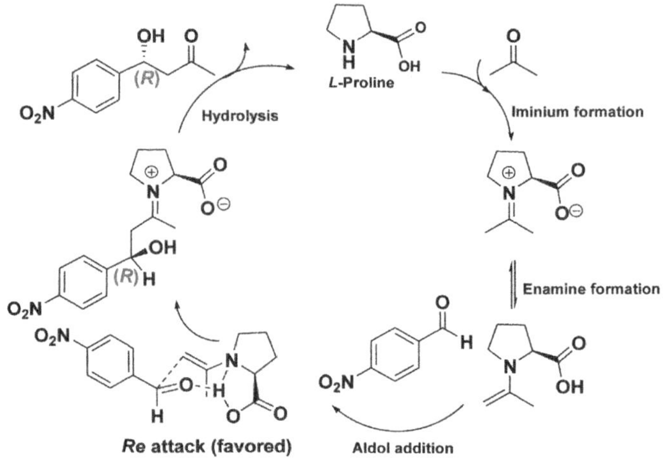

Figure 4.2.2.5.1 Proposed reaction mechanism.

According to previous reports on the proline catalyzed aldol reaction[2,3,5] and other types of transformations,[4] the detailed process of this reaction was proposed as shown in Figure 4.2.2.5.1.

L-Proline is a bifunctional organocatalyst, which provides both the nucleophilic amino group and an acid/base site in the form of the carboxylate. This bifunctional organocatalyst may facilitate each individual step of the mechanism, including the nucleophilic attack of the amino group, the isomerization of iminium to enamine, aldol addition of enamine to the aldehyde's carbonyl group, and hydrolysis of the iminium-aldol intermediate. The configuration of the product was determined by the six-membered chair-like transition state.

Additional Safety

Silica dust is very toxic if inhaled. You should take precautions to avoid breathing it in and you must handle it in a fume cupboard. Large volumes of solvent used for chromatography are also toxic and you should take precautions to avoid breathing in the vapors.

Experimental Procedure

Laboratory Session 1 (4 h): Aldol Reaction

Preparation

1. To a 25 mL round-bottom flask add 4-nitrobenzaldehyde (151.0 mg, 1.0 mmol), L-proline (35.0 mg, 0.3 mmol) and acetone (5.0 mL). Stir the mixture at room temperature for 3 h. The reaction system can be monitored by TLC plate to confirm whether the reaction is complete.
2. First concentrate the reaction mixture and then dilute it with dichloromethane (2.0 mL).

Laboratory Session 2 (4 h): Product Isolation and Characterization

Product Isolation

3. Purify the residue by flash column chromatography on silica gel (petroleum ether: ethyl acetate = 3 : 1) to obtain the final product.
4. Weigh the desired product and calculate the yield.

Characterization

5. Prepare the sample for ^{1}H NMR testing.
6. Determine the enantioselectivity excess of the product by chiral HPLC analysis (Daicel Chiralpak AS-H, isopropanol/hexane = 50/50, 1.0 mL min^{-1}, λ = 254 nm, t_R (major) = 7.4 min, t_R (minor) = 8.7 min).

Waste Disposal

Solvents used in chromatographic purifications should be collected in a non-halogenated waste collector recipient for posterior destruction.

Results Interpretation and Additional Questions

1. Determine the yield of the reaction.
2. Assign the peaks of the NMR spectrum of the product.
3. Determine the absolute configuration of the final product.
4. List one or two catalysts that can catalyze the reaction to obtain the racemic product.
5. Search the literature to find three or four reactions that can be catalyzed by L-proline.

Acknowledgements

Financial support from the South University of Science and Technology of China (SUSTC) is greatly appreciated. We would like to thank Professor Benjamin List for developing the proline-catalyzed asymmetric intermolecular aldol reaction, proposing a reaction mechanism for enantiocontrol. We also thank Dr Dao-Juan Cheng for providing the racemic compound and for HPLC analysis.

References

1. Z. G. Hajos and D. R. Parrish, *J. Org. Chem.*, 1974, **39**, 1615.
2. U. Eder, G. Sauer and R. Wiechert, *Angew. Chem., Int. Ed. Engl.*, 1971, **10**, 496.
3. (a) B. List, R. A. Lerner and C. F. Barbas III, *J. Am. Chem. Soc.*, 2000, **122**, 2395; (b) W. Notz and B. List, *J. Am. Chem. Soc.*, 2000, **122**, 7386; (c) A. B. Northrup and D. W. C. MacMillan, *J. Am. Chem. Soc.*, 2002, **124**, 6798.
4. S. Mukherjee, J. W. Yang, S. Hoffmann and B. List, *Chem. Rev.*, 2007, **107**, 5471.
5. K. Sakthivel, W. Notz, T. Bui and C. F. Barbas III, *J. Am. Chem. Soc.*, 2001, **123**, 5260.

4.2.2.6. Preparation of a β-Nitrostyrene Derivative by the Henry Reaction: Comparison of a Conventional and a Microwave-Assisted Method

Daniel Martins[a], Joana Reis[a], Fernanda Borges[a], and Nuno Milhazes*[a,b]

[a]CIQUP/Department of Chemistry and Biochemistry, Faculty of Sciences of University of Porto, Portugal; [b]CESPU/Superior Institute of Health of Sciences – North, Rua Central de Gandra, 1317, 4585-116, Gandra PRD, Portugal. *E-mail: nuno.milhazes@iscsn.cespu.pt

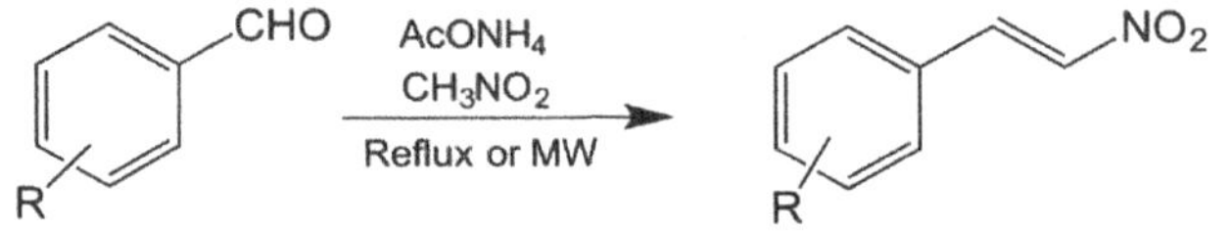

Number of sessions (duration of each session)	Hazard level	Difficulty level	Level of study
2 (3 h each)	Moderate	Low	Intermediate

Class names Nitroalkanes, aldehydes

Concepts involved This experiment involves the synthesis of a β-nitrostyrene derivative by a nitroaldol condensation reaction of the corresponding benzaldehyde with a nitroalkane, by a conventional and a microwave-assisted process

Chemicals needed 4-Hydroxy-3-methoxybenzaldehyde, nitromethane, ammonium acetate, petroleum ether (40–60 °C), diethyl ether, isopropanol, deuterated DMSO

Equipment and experimental techniques involved Apparatus for organic reactions equipped with temperature control and magnetic stirring, UV lamp (254 and 366 nm), rotary evaporator, vacuum oven, microwave reactor, NMR spectrometer for ^{1}H and ^{13}C NMR spectra acquisition, melting point apparatus, separatory funnel, microwave-assisted reactions, filtration under reduced pressure, liquid–liquid extraction, crystallization, NMR spectroscopy, melting point determination

Keywords Benzaldehyde, Henry-type condensation, liquid–liquid extraction, microwave-assisted synthesis, nitrostyrene, NMR

Background

β-Nitrostyrene has been recognized as an highly reactive scaffold that can be readily converted into other useful intermediates, and in medicinal chemistry as scaffolds with noteworthy biological profiles such as anticancer or antimicrobial activities.[1,2] In fact, conjugated nitroolefins are important raw materials with several applications in organic synthesis since they could work as activated alkenes that can be used as starting materials to obtain other nitroalkenes, amines, hydroxylamines, ketoximes, or aldoximes.[3]

Among the diversity of synthetic procedures that could afford nitrostyrenes,[4,5] the most common synthetic route is a nitroaldolic-type condensation, known as Henry reaction, between the

Comprehensive Organic Chemistry Experiments for the Laboratory Classroom
Edited by Carlos A. M. Afonso, Nuno R. Candeias, Dulce Pereira Simão, Alexandre F. Trindade, Jaime A. S. Coelho, Bin Tan, and Robert Franzén.

Published by the Royal Society of Chemistry, www.rsc.org

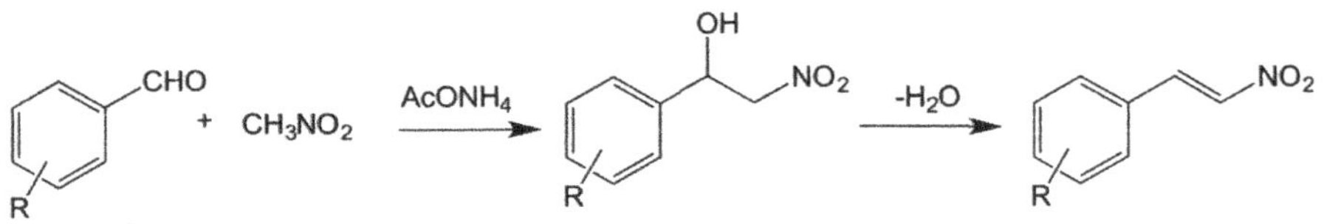

Scheme 4.2.2.6.1 Synthesis of β-nitrostyrene derivatives.

benzaldehyde, with the substitution pattern of the required final compound, and nitroalkane (Scheme 4.2.2.6.1).

In this experiment, the preparation of 4-hydroxy-3-methoxy-β-nitrostyrene is attained by a conventional "heat-and-stir" methodology, using 4-hydroxy-3-methoxybenzaldeyde and nitromethane as starting materials.[2] The nitroalkane is used as the solvent, heated under reflux, and the basic conditions were achieved using ammonium acetate as the catalyst. The preparation of this β-nitrostyrene derivative by a microwave-assisted method is also reported and compared with the traditional method.

The conventional preparation of β-nitrostyrene derivatives requires prolonged heating periods (7–36 h) along with a substantial excess of nitroalkane. The development of fast, green and economical synthetic routes is undeniably an urgent requirement in any modern chemistry laboratory. Microwave-assisted organic synthesis (MAOS) is an effective tool in synthetic organic chemistry. MAOS has been reported to dramatically reduce reaction times, simplify work-up procedures and purification, and increase product yields when compared to conventional heating methods.[6,7] Modern microwave reactors enable accurate monitoring of temperature and pressure under sealed vessels and safe conditions.

In the current experiment, a comparative study between the conventional and microwave-assisted methods to prepare a nitrostyrene derivative, 4-hydroxy-3-methoxy-β-nitrostyrene, its purification and spectroscopic characterization is described. The amount of nitromethane and the overall reaction time are significantly reduced when microwave radiation is used in the synthetic procedure.

Additional Safety

The experimental procedures should be performed in a fume hood (whenever possible) but there are no additional concerns besides good laboratory practice.

Special precaution concerning nitromethane should be taken; it must be handled while wearing safety goggles, protective gloves and clothing to prevent skin exposure. The operator must also stay away from ignition sources since diethyl ether, isopropanol and petroleum ether are flammable reagents.

The obtained β-nitrostyrene also must be handled with extra caution. Wear appropriate eyeglasses or chemical safety goggles, gloves, proper clothing and protective mask to avoid inhalation of the solid product.

Experimental Procedure

Laboratory Session 1 (3 h): Synthesis of 4-Hydroxy-3-methoxy-β-nitrostyrene by a Conventional Method

Previous Day (Performed by the Instructor and/or the Student)

1. In a 100 mL round-bottom flask, dissolve 1.90 g (12.5 mmol) of 4-hydroxy-3-methoxybenzaldehyde in nitromethane (25 mL) and then add 0.23 g (3.0 mmol) of ammonium acetate.

2. Place the round-bottom flask with the reaction solution in an oil bath on a hot plate with magnetic stirring, insert a water condenser and heat the reaction under reflux conditions for 6 h.
3. Remove the flask from the oil bath and allow it to cool to room temperature, keep the reaction at room temperature with stirring until the actual laboratory session.

Laboratory Session 1

4. Control the reaction progress by TLC (petroleum ether (40–60 °C)/diethyl ether, 50:50 (v/v)) and heat the reaction for an additional hour to improve the overall yield of the condensation reaction.
5. Remove the flask from the oil bath and allow it to cool to room temperature.
6. Transfer the reaction to a round-bottom flask and reduce the volume to half (*ca.* 10 mL) in a rotary evaporator apparatus.
7. Dilute the remaining solution with 100 mL of water, transfer the resulting emulsion to a separatory funnel and extract it with 2 × 100 mL of diethyl ether.
8. Dry the combined organic layers under anhydrous magnesium sulfate, filter and evaporate the solvent until an oily residue is obtained.
9. Recrystallize the yellow residue from hot isopropanol.
10. After the compound is dried in a vacuum oven (50 °C), prepare a sample, in DMSO-d_6, to acquire ^{1}H and ^{13}C NMR spectra, and measure the melting point of the final compound to evaluate its identity and purity.

Laboratory Session 2 (3 h): Synthesis of 4-Hydroxy-3-methoxy-β-nitrostyrene by a Microwave-Assisted Method

1. In an appropriate (2–5 mL) microwave vial, dissolve 0.46 g (3.0 mmol) of 4-hydroxy-3-methoxybenzaldehyde and 0.06 g (0.8 mmol) of ammonium acetate in nitromethane (2.5 mL).
2. Place the vial into the microwave reactor and set the temperature to 150 °C for 5 min.
3. Control the reaction mixture by TLC (petroleum ether/diethyl ether, 50 : 50 (v/v)).
4. Transfer the reaction mixture to a round-bottom flask and remove the excess nitromethane in a rotary evaporator apparatus.
5. Dilute the remaining solution with 100 mL of water, transfer the resulting emulsion to a separatory funnel and extract with 2 × 100 mL of diethyl ether.
6. Dry the combined organic layers over anhydrous magnesium sulfate, filter and evaporate the solvent until an oily residue is obtained.
7. Recrystallize the yellow residue from hot isopropanol.
8. After the compound is dried in a vacuum oven (50 °C), prepare a sample, in DMSO-d_6, to acquire ^{1}H and ^{13}C NMR spectra, and measure the melting point of the final compound to evaluate its identity and purity.

Waste Disposal

All organic and aqueous wastes should be placed in appropriate waste containers. Diethyl ether should be recycled or collected in a non-chlorinated waste container. After use, all the glassware must be washed with water. Any glassware stains can be removed with acetone after rinsing with water and appropriate soap. The silica gel residues must be placed in a solid waste container for posterior destruction.

Results Interpretation and Additional Questions

1. Assess compound's purity by ^{1}H and ^{13}C NMR spectra analysis, TLC and melting point evaluation.
2. Determine the reaction yields for each procedure (conventional *versus* microwave-assisted).
3. Explore the advantages of using the microwave-assisted method in comparison with the conventional heating method.
4. Perform a thorough analysis of the NMR spectra of the synthesized compound, determining the coupling constants of protons α and β, the geometry around the double bond and which isomer is formed, (*E*) or (*Z*).
5. Discuss the importance of the ammonium acetate in this reaction and the possibility of using alternative compounds.
6. Suggest variations in the experimental procedure in order to isolate the β-nitroalcohol intermediate.

References

1. J. Reis, C. Oliveira, N. Milhazes, D. Vina and F. Borges, *Lett. Drug Des. Discovery*, 2012, **9**, 958–961.
2. N. Milhazes, R. Calheiros, M. P. M. Marques, J. Garrido, M. N. D. S. Cordeiro, C. Rodrigues, S. Quinteira, C. Novais, L. Peixe and F. Borges, *Bioorg. Med. Chem.*, 2006, **14**, 4078–4088.
3. A. G. M. Barrett and G. G. Graboski, *Chem. Rev.*, 1986, **86**, 751–762.
4. R. S. Varma, K. P. Naicker and P. J. Liesen, *Tetrahedron Lett.*, 1998, **39**, 3977–3980.
5. G. Yan, A. J. Borah and L. Wang, *Org. Biomol. Chem.*, 2014, **12**, 6049–6058.
6. C. O. Kappe, D. Dallinger and S. S. Murphree, *Practical Microwave Synthesis for Organic Chemists*, Wiley-VCH Verlag GmbH, Weinheim, 2009, pp. 291–299.
7. C. D. Wang and S. Wang, *Synth. Commun.*, 2002, **32**, 3481–3486.

4.2.2.7. Synthesis of Aurone Derivatives Through Acid-Catalysed Aldol Condensation

Ana R. Duarte, Marta P. Carrasco, and Ana S. Ressurreição*

Research Institute for Medicines (iMed.ULisboa), Faculty of Pharmacy, Universidade de Lisboa, Av. Professor Gama Pinto, 1649-003 Lisboa, Portugal. *E-mail: aressurreicao@ff.ulisboa.pt

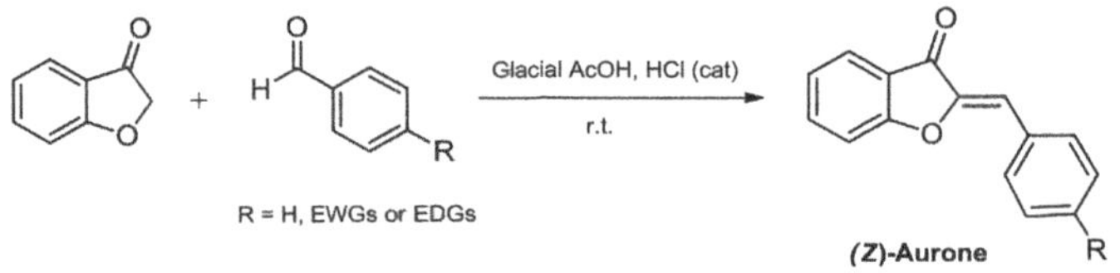

Number of sessions (duration of each session)	Hazard level	Difficulty level	Level of study
1 (3 h)	Moderate	Low	Intermediate

Class names Aldehydes, ketones, α,β-unsaturated ketones

Concepts involved This experiment involves the synthesis of a small library of aurones derivatives through a one-step aldol condensation of benzofuran-3(2*H*)-one and the appropriate benzaldehydes in acidic medium, affording only the (*Z*)-isomer

Chemicals needed Benzofuran-3(2*H*)-one, benzaldehyde and/or 4-(dimethylamino)benzaldehyde and/or 4-nitrobenzaldehyde, glacial acetic acid, hydrochloric acid (37%)

Equipment and experimental techniques involved Magnetic stirring plate, vacuum filtration apparatus, NMR, melting point

Keywords Acid catalysis, aldol condensation, aurones, flavonoids, NMR, precipitation

Background

Flavonoids represent a large class of natural products in the plant kingdom, exhibiting a strong and broad variety of biological activities. Among this large family, aurones (2-benzylidenebenzofuran-3-(2*H*)-ones) are minor tricyclic flavonoids, structural isomers of flavones, that consist of a benzofuranone ring linked through a carbon–carbon double bound to a phenyl moiety. Due to the presence of this exocyclic double bond, these compounds can be found in two geometric isomers, (*E*) or (*Z*), although the latter is generally considered to be thermodynamically more stable and thus more abundant.

In nature, aurones are mostly found in higher plants, playing an important role in the pigmentation of yellow flowers and fruits.[1] Besides their contribution to the pigmentation of flowers, aurones have also shown a wide variety of biological activities (*e.g.* anti-cancer,[2] antimicrobial,[3] anti-parasitic,[4] anti-viral,[5] and anti-inflammatory[3]).

Synthetic aurones are mostly obtained using two different methods:[6] (i) oxidative cyclisation of 2′-hydroxychalcones using various reagents or (ii) aldolic condensation, more specifically a

Comprehensive Organic Chemistry Experiments for the Laboratory Classroom
Edited by Carlos A. M. Afonso, Nuno R. Candeias, Dulce Pereira Simão, Alexandre F. Trindade, Jaime A. S. Coelho, Bin Tan, and Robert Franzén.

Published by the Royal Society of Chemistry, www.rsc.org

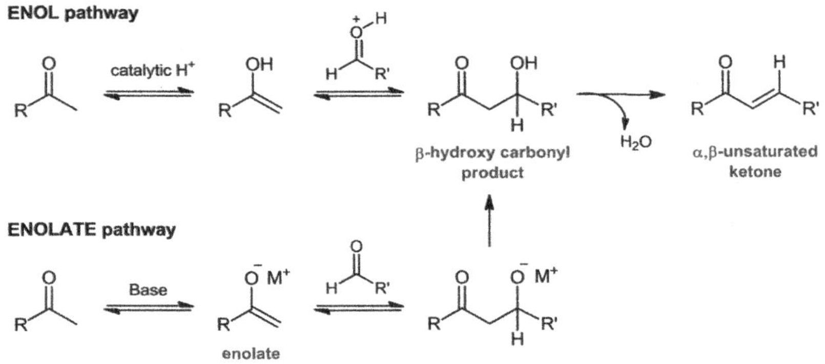

Scheme 4.2.2.7.1 The aldol reaction and subsequent dehydration under acidic and basic conditions.

Claisen–Schmidt condensation, between benzofuran-3(2*H*)-one and benzaldehydes, which can be undertaken under different conditions. In addition to the general good yields, this synthetic pathway is compatible with a wide range of substituents, including the most sensible ones.

Aldol condensations are important reactions in organic synthesis, providing a good way to form carbon–carbon bonds. They are extensively used by chemists to make larger molecules starting from smaller carbonyl compounds. In an aldol condensation an enol (or enolate) reacts with a carbonyl compound to form a β-hydroxyaldehyde or β-hydroxyketone, which readily dehydrates under mild conditions to give a conjugated enone (α,β-unsaturated ketone). The aldol reaction can occur under both acidic and basic conditions (Scheme 4.2.2.7.1).

The reaction in this experiment is a particular type of aldol condensation reaction, namely a Claisen–Schmidt reaction, because it involves the addition of an aldehyde to a ketone to form the corresponding aldol addition product (aurone). The objective of this work is to prepare a small library of aurones, without the use of any organic solvent and metal catalyst, affording only the (*Z*)-isomer in moderate to good yields. The procedure described involves an acid-catalysed Claisen–Schmidt reaction at room temperature; the reaction is compatible with a broad panel of substituted benzaldehydes (either with electron-withdrawing groups (EWGs) or electron-donating groups (EDGs)), however the reaction rate is dependent on the electrophilic character of the substituent. Additionally, the product is quickly precipitated from the aqueous reaction medium and the pure product is easily isolated by filtration, avoiding the use of expensive purification techniques and equipment.

Additional Safety

The session should be done in a fume hood, with proper safety equipment, such as protective clothing, gloves and safety goggles. Glacial acetic acid is flammable so it is important to keep it distant from ignition sources. Both glacial acetic acid and hydrochloric acid (37%) are highly corrosive and can cause severe skin and eye burns. In addition, hydrochloric acid vapour may cause respiratory irritation.

Experimental Procedure

1. Prepare a 25 mL round-bottom flask with a magnetic stir bar and, in a fume hood, add 100 mg (0.75 mmol, 1 equiv.) of benzofuran-3(2*H*)-one and 7.5 mL of glacial acetic acid at room temperature.

2. To the resulting orange solution, add benzaldehyde (0.9 mmol, 1.2 equiv.) and finally the hydrochloric acid (3 drops). As an alternative to benzaldehyde, 4-(dimethylamino)benzaldehyde or 4-nitrobenzaldehyde can also be used.
3. Stir the reaction mixture at room temperature and monitor the progression of the reaction by TLC.

Product Isolation

4. After the reaction reaches completion, put the mixture into cold distilled water (*ca.* 50 mL).
5. Put the resulting suspension in an ice bath (or fridge) for approximately 15 min, to promote complete precipitation of the product.
6. Prepare the vacuum filtration apparatus using a previously tared filter paper.
7. Filter the precipitate and wash it with cold distilled water.
8. Transfer the precipitate (including the filter paper) into a watch glass and put it into a pre-heated oven at 100 °C.
9. After 15 min remove the watch glass from the oven, allow it to cool to room temperature and weigh it. Repeat this step (in 5 min intervals) until a constant weight is obtained.
10. Measure the melting point of the obtained product and prepare the sample for ^{1}H and ^{13}C NMR (approx. 15–20 mg of compound in 500 μL of $CDCl_3$).

Waste Disposal

The aqueous solution obtained in the filtration may contain traces of the benzaldehyde used and should be treated as reaction waste.

Results Interpretation and Additional Questions

1. Determine the reaction yield and compare the obtained melting point with the one reported in the literature.[4,7]
2. Interpret the NMR spectra obtained and based on the information described in the literature[4] determine the stereochemistry of the carbon–carbon double bond.
3. Discuss the mechanism of the reaction and point out the roles of the acid catalyst.
4. Based on previous reports in the literature, indicate other benzaldehydes that are suitable for this reaction.
5. Discuss the influence of the substituents (EWGs and EDGs) on the benzaldehyde reactivity and reaction outcome.

References

1. E. Ono, M. Fukuchi-Mizutani, N. Nakamura, Y. Fukui, K. Yonekura-Sakakibara, M. Yamaguchi, T. Nakayama, T. Tanaka, T. Kusumi and Y. Tanaka, *Proc. Natl. Acad. Sci. U. S. A.*, 2006, **103**, 11075.
2. For example: H. M. Cheng, L. W. Zhang, Y. X. Liu, S. P. Chen, H. Cheng, X. Lu, Z. X. Zheng and G. C. Zhou, *Eur. J. Med. Chem.*, 2010, **45**, 5950.
3. B. P. Bandgar, S. A. Patil, B. L. Korbad, S. C. Biradar, S. N. Nile and C. N. Khobragade, *Eur. J. Med. Chem.*, 2010, **45**, 3223.
4. For example: M. P. Carrasco, A. S. Newton, L. Goncalves, A. Gois, M. Machado, J. Gut, F. Nogueira, T. Hanscheid, R. C. Guedes, D. dos Santos, P. J. Rosenthal and R. Moreira, *Eur. J. Med. Chem.*, 2014, **80**, 523.
5. For example: R. Haudecoeur, A. Ahmed-Belkacem, W. Yi, A. Fortune, R. Brillet, C. Belle, E. Nicolle, C. Pallier, J. M. Pawlotsky and A. Boumendjel, *J. Med. Chem.*, 2011, **54**, 5395.
6. Recent advances in synthetic approaches towards aurones revised in: (a) R. Haudecoeur and A. Boumendjel, *Curr. Med. Chem.*, 2012, **19**, 2861; (b) C. Zwergel, F. Gaascht, S. Valente, M. Diederich, D. Bagrel and G. Kirsch, *Nat. Prod. Commun.*, 2012, **7**, 389.
7. I. Hawkins and S. T. Handy, *Tetrahedron*, 2013, **69**, 9200.

4.2.2.8. Synthesis of Pyrazole Heterocycles

Thomas A. Logothetis*

University of Southampton, Chemistry, Highfield, Southampton, Hampshire, SO17 1BJ, UK
*E-mail: thomas.logothetis@southampton.ac.uk

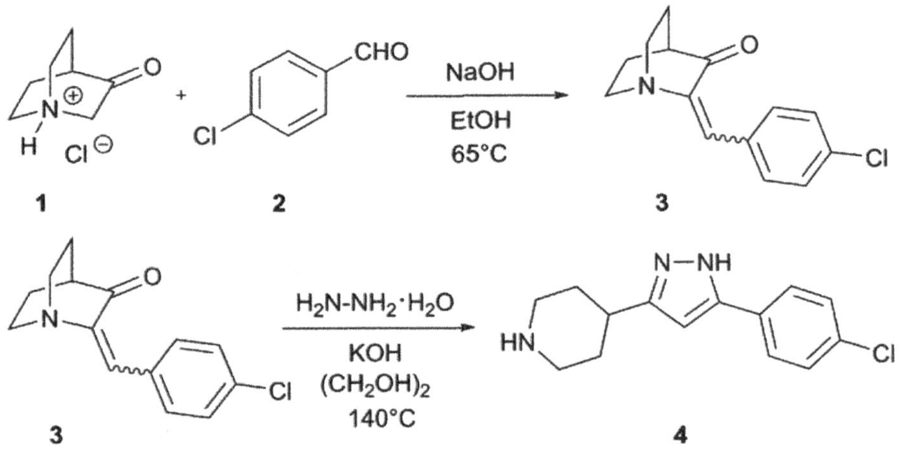

Number of sessions (duration of each session)	Hazard level	Difficulty level	Level of study
2 (4 h each)	Moderate	Medium	Intermediate

Class names Ketone, conjugated ketone, hydrazine, pyrazole, amines

Concepts involved These two experiments feature reactions influenced by the carbonyl moiety of the starting ketone and intermediate, *i.e.* aldol condensation, Michael addition and hydrazone formation. The pathway concludes with an aromatisation that leads to a disubstituted pyrazole. The products are analysed using IR and NMR spectroscopy and electrospray mass spectrometry. Careful analysis of coupling constants and splitting patterns in the ^{1}H NMR of the final product allows the determination of the conformation of the piperidine ring

Chemicals needed 3-Quinuclidinone hydrochloride, 4-chlorobenzaldehyde, sodium and potassium hydroxide, hydrazine hydrate, ethane-1,2-diol, ethanol, deuterochloroform, dichloromethane, methanol, ammonia (aqueous), diethyl ether, acetonitrile, acetone, petroleum ether (40–60 °C), hydrochloric acid (2 M), potassium permanganate

Equipment and experimental techniques involved Erlenmeyer flasks, beaker, vacuum filtration, desiccator, rotary evaporation, magnetic stirrer hotplate, round bottom flask, ice bath, Pasteur pipettes, UV lamp, reflux condenser, silica TLC plates and chamber, thermometer (covering −10 to +200 °C), glass rod, tweezers; stirring at defined temperatures using external cooling and heating, monitoring by TLC, precipitation, filtration, desiccation, recrystallisation; IR, NMR, MS

Keywords Aldol reaction, IR, Michael addition, MS, NMR, precipitation, pyrazole formation

Comprehensive Organic Chemistry Experiments for the Laboratory Classroom
Edited by Carlos A. M. Afonso, Nuno R. Candeias, Dulce Pereira Simão, Alexandre F. Trindade, Jaime A. S. Coelho, Bin Tan, and Robert Franzén.

Published by the Royal Society of Chemistry, www.rsc.org

Background

The synthesis of pyrazole and substituted derivatives thereof is possible *via* different pathways, many of which involve the use of hydrazine to incorporate the nitrogen–nitrogen moiety – rather than building it by forming a bond between independently introduced nitrogen atoms.[1] Other sources for the nitrogen–nitrogen building block include diazo derivatives.[2] The reaction of hydrazine derivatives with 1,3-dicarbonyl compounds[3] – the *Knorr pyrazole synthesis*[4] – is still a major route towards substituted pyrazoles. Another pathway is the conjugate addition of hydrazine derivatives to α,β unsaturated ketones followed by intramolecular hydrazone formation and aromatisation, as used in this experiment.

The precursor for the conjugate addition is synthesised in the first reaction from a substituted benzaldehyde and 3-quinuclidinone. The choice of aldehyde allows for the generation of a library with different substituents and various positions of the aromatic ring. This practical lays the foundation for a scaffold that has five positions that can be manipulated at a later stage (for additional information see experiment 5.1.7 "Selective Boc-protection and bromination of pyrazoles").

The target molecule of this practical – 4-[5-(4-chlorophenyl)-1*H*-pyrazol-3-yl]piperidine – is synthesised in two steps and is of general interest to the pharmaceutical community as it and its derivatives have been associated with possible treatments of schizophrenia,[5] anticoagulant properties[6] and conditions and diseases associated with abnormal metabolism, like obesity, diabetes and related disorders.[7] Unsurprisingly, the industrial laboratories involved in this research area have created several alternative routes to this medicinally very versatile lead, including variations to the scaffold and its substituents.

Noteworthy also is the observation that in case of the dopamine D4 receptor (linked to schizophrenia) the piperidine moiety and its nitrogen substituent are important features while the chlorophenyl moiety is responsible for binding to thrombin, a serine protease linked to the coagulation cascade. This highlights the importance of creating a versatile scaffold that can be tailored with specific modifications to suit the requirements of medicinal chemistry.

Additional Safety

All procedures should be performed in a well-ventilated fume cupboard with proper safety equipment such as protective clothing, safety spectacles and gloves. All organic solvents apart from dichloromethane and deuterochloroform are (highly) flammable solvents. The chlorinated solvents and hydrazine (decontaminate with acetone) are suspected carcinogens while the latter is also a sensitizer and toxic if inhaled; methanol is toxic by all routes; diethyl ether is extremely flammable and may form explosive peroxides; hydrochloric acid and the bases are corrosive. All other reagents are harmful and irritants.

Experimental Procedure

Session 1: Synthesis of 2-(4-Chlorobenzylidene)quinuclidin-3-one (4 h)

1. In a fume cupboard set up a dry round-bottom flask over a magnetic stirrer hotplate with heating device, reflux condenser and temperature control.
2. Charge the flask with ethanol (30 mL) and dissolve 3-quinuclidinone hydrochloride (20 mmol).
3. Under stirring add 4-chlorobenzaldehyde (34 mmol), followed by finely dispersed sodium hydroxide (28 mmol).
4. Continue stirring while heating the reaction mixture to 60–65 °C.

5. Monitor the reaction by TLC (eluent: 40% diethyl ether in light petroleum ether (40–60 °C)) against both starting materials until the limiting reagent has been consumed. Use a sample of the organic phase after a drop of the reaction mixture has been partitioned between diethyl ether and water (each 0.5 mL).
6. When finished, allow to cool down and pour into vigorously stirred water (100 mL) until a precipitate forms, which is collected by vacuum filtration.
7. Recrystallise the crude from ethanol and dry the product by vacuum desiccation.
8. Determine the yield of the product.
9. Measure the melting point.
10. Acquire ^{13}C NMR spectra using $CDCl_3$ as solvent.
11. Acquire an IR (ATR) spectrum.
12. Acquire an ES^+ mass spectrum using acetonitrile as solvent.

Session 2: Synthesis of 4-[5-(4-Chlorophenyl)-1*H*-pyrazol-3-yl]piperidine (4 h)

1. In a well-ventilated fume cupboard set up a dry round-bottom flask over a magnetic stirrer hotplate with heating device and temperature control.
2. Suspend 2-(4-chlorobenzylidene)quinuclidin-3-one (8.0 mmol) in ethane-1,2-diol (15 mL) while stirring before adding potassium hydroxide (16 mmol).
3. Then add hydrazine hydrate (20 mmol). Note: Handle hydrazine hydrate in a well-ventilated fume cupboard only, while protecting yourself with gloves. Decontaminate, clean and/or destroy any unused hydrazine hydrate using acetone (see also "Waste disposal" *vide infra*).
4. Add a reflux condenser to the flask and continue stirring while heating the reaction mixture to 140 °C.
5. Monitor the reaction by TLC against the disappearance of the starting material (eluent: 40% diethyl ether in light petroleum ether (40–60 °C)) and the appearance of the product (eluent: dichloromethane: methanol: ammonia (15 : 4 : 1)). Use a sample of the organic phase after a drop of the reaction mixture has been partitioned between diethyl ether and water (each 0.5 mL).
6. When finished, allow to cool down and slowly pipette the reaction mixture into vigorously stirred, ice-cold water (50 mL) and then collect the precipitate by vacuum filtration.
7. Recrystallise the crude from ethanol and dry the solid product by vacuum desiccation.
8. Determine the yield of the product.
9. Measure the melting point.
10. Acquire ^1H and ^{13}C NMR spectra using $CDCl_3$ as solvent.

Waste Disposal

The neutralised aqueous solutions may be diluted and discarded in the sink. All organic solutions should be recycled or collected as chlorinated waste. Equipment that has been in contact with hydrazine hydrate must be decontaminated with acetone, which must also be used to destroy any accidental spillages or excess reagent. The acetone should then be collected as non-chlorinated waste.

Results Interpretation and Additional Questions

1. Rationalise why acetone is used to clean hydrazine-contaminated glassware and to neutralise spillages.
2. Estimate the pK_a of the protons at the α-position of the carbonyl group and on the nitrogen atom of 3-quinuclidinone hydrochloride.

3. Formulate the mechanism for the formation of 2-(4-chlorobenzylidene)quinuclidin-3-one clearly indicating reversible steps and considering the reaction conditions.
4. Considering the reaction parameters formulate a mechanism for the formation of 4-[5-(4-chlorophenyl)-1*H*-pyrazol-3-yl]piperidine on the basis that a Michael addition and intramolecular hydrazone formation precede the aromatisation in which the pyrazole is created.
5. Interpret the acquired IR, ^{13}C NMR and MS/ES^{+} data for both products and confirm structural features of the expected products.
6. Analyse the ^{1}H NMR spectrum for the second product and determine the conformation and the position (axial or equatorial) of the aromatic substituent on the piperidine ring.

References

1. (a) A. H. Moustafa, C. C. Malakar, N. Aljaar, E. Merisor, J. Conrad and U. Beifuss, *Synlett*, 2013, **24**, 1573; (b) M. R. Kumar, A. Park, N. Park and S. Lee, *Org. Lett.*, 2011, **13**, 3542; (c) B. C. Wray and J. P. Stambuli, *Org. Lett.*, 2010, **12**, 4576.
2. L.-L. Wu, Y.-C. Ge, T. He, L. Zhang, X.-L. Fu, H.-Y. Fu, H. Chen and R.-X. Li, *Synthesis*, 2012, **44**, 1577.
3. S. T. Heller and S. R. Natarajan, *Org. Lett.*, 2006, **8**, 2675.
4. L. Knorr, *Chem. Ber.*, 1883, **16**, 2597.
5. (a) M. Rowley, H. B. Broughton, I. Collins, R. Baker, F. Emms, R. Marwood, S. Patel, S. Patel, C. I. Ragan, S. B. Freedman and P. D. Leeson, *J. Med. Chem.*, 1996, **39**, 1943; (b) M. Rowley, I. Collins, H. B. Broughton, W. D. Davey, R. Baker, F. Emms, R. Marwood, S. Patel, S. Patel, C. I. Ragan, S. B. Freedman, R. Ball and P. D. Leeson, *J. Med. Chem.*, 1997, **40**, 2374; (c) I. Collins, M. Rowley, W. D. Davey, F. Emms, R. Marwood, S. Patel, I. C. Ragan, P. D. Leeson, A. L. Scott and T. Broten, *Bioorg. Med. Chem.*, 1998, **6**, 743.
6. M. J. Hartshorn, C. H. Murray, A. Cleasby, M. Frederickson, I. J. Tickle and H. Jhoti, *J. Med. Chem.*, 2005, **48**, 403.
7. Merck Sharpe & Dohme Corp., J. J. Hale, J. Jiang, D.-M. Shen, Z.-C. Shi, M. Shu, Z. Wu and C. Yang, Novel prolylcarboxypeptidase inhibitors, US Patent WO2011143057 A1, 2011.

4.2.2.9. A Green Approach to 3-Carbonylchromones

Ana Bornadiego, Jesús Díaz, Ana G. Neo, and Carlos F. Marcos*

Departamento de Química e Inorgánica. F. Veterinaria, Universidad de Extremadura, 10071 Cáceres, Spain. *E-mail: cfernan@unex.es

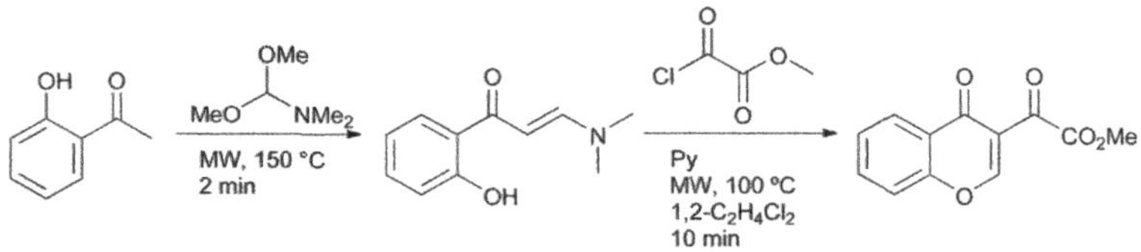

Number of sessions (duration of each session)	Hazard level	Difficulty level	Level of study
1 (4 h)	Moderate	Medium	Intermediate

Class names Carbonyl groups, acetals, enamines, phenols, acid chlorides

Concepts involved This experiment involves the microwave-assisted synthesis of a natural product-like chromone through a sequential procedure involving an aldol condensation, acyl substitution, Baker–Venkataraman rearrangement, addition to an iminium ion and elimination

Chemicals needed *o*-Hydroxyacetophenone, dimethylformamide dimethyl acetal, toluene, hexane, methyl chlorooxoacetate, 1,2-dichloroethane, pyridine, dichloromethane, copper sulfate, sodium chloride, sodium sulfate

Equipment and experimental techniques involved Microwave irradiation is performed in closed vials, using a focused single-mode microwave reactor

Keywords Crystallisation, green chemistry, heterocycles, microwave, NMR

Background

Chromones (benzopyran-4-ones) constitute a large family of plant secondary metabolites, which show important biological activities.[1] They have found a widespread use as pharmaceuticals and synthetic intermediates in the preparation of new heterocyclic systems. Particularly useful are electron-deficient 3-carbonylchromones, which readily react with a range of nucleophiles to afford different types of products, such as pyridines,[2] 2-hydroxybenzophenones,[3] benzocoumarins[4] and xanthones.[5] Various suitable synthetic methods allow organic chemists to obtain diversely substituted chromones. Classic strategies rely on the cyclisation of either β-diketones or chalcones, both of which can be obtained from *o*-hydroxyacetophenones (Scheme 4.2.2.9.1).[6]

In addition, 3-carbonylchromones can be conveniently synthesised by the reaction of 3-(dimethylamino)-1-(2-hydroxyphenyl)prop-2-en-1-one with suitable acylating agents (Scheme 4.2.2.9.2).[7]

The aim of this experiment is the synthesis of 3-methoxalylchromone,[2a] a chromone containing a ketoester group on position 3. Langer and co-workers have described the synthesis of this chromone by the reaction of 3-(dimethylamino)-1-(2-hydroxyphenyl)prop-2-en-1-one with methyl chlorooxoacetate. A modified procedure using microwave activation affords the product in just a few minutes.[5]

Comprehensive Organic Chemistry Experiments for the Laboratory Classroom
Edited by Carlos A. M. Afonso, Nuno R. Candeias, Dulce Pereira Simão, Alexandre F. Trindade, Jaime A. S. Coelho, Bin Tan, and Robert Franzén.

Published by the Royal Society of Chemistry, www.rsc.org

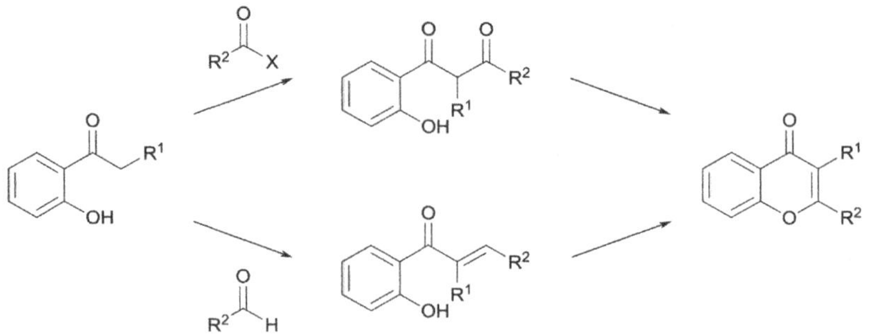

Scheme 4.2.2.9.1 Classical synthetic approaches to chromones.

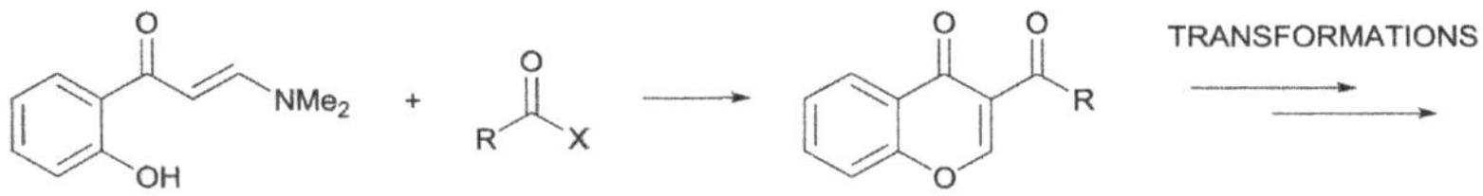

Scheme 4.2.2.9.2 Synthesis of 3-carbonylchromones.

The starting enaminone can be also obtained by a solvent-free, microwave activated aldol condensation of readily available *o*-hydroxyacetophenone and *N*,*N*-dimethylformamide dimethyl acetal,[8] and quickly and efficiently crystallised using microwave heating.

Thus, the target chromone is synthesised through a sequential two-step procedure starting from commercial reagents. Both steps and purification are benefitted by microwave assisted heating, which considerably shortens the reaction times and minimises the required amount of solvents and the generation of residues.

Additional Safety

All the experiments should be carried out in a fume hood. Laboratory coat, disposable gloves and goggles must be worn. High pressures are developed under microwave irradiation, thus the reaction vessels must be allowed to cool down to room temperature before removing them from the microwave reactor.

Hexane and toluene are flammable solvents. Methyl chlorooxoacetate is corrosive and flammable. Pyridine is highly toxic and also flammable. Avoid inhalation and contact with all the chemicals used in this experiment, including the reaction products.

Experimental Procedure

Synthesis of (*E*)-3-(Dimethylamino)-1-(2-hydroxyphenyl)prop-2-en-1-one

1. In a microwave vial provided with a stirring bar add 0.60 mL (5 mmol) of *o*-hydroxyacetophenone and 0.66 mL (5 mmol, 1 equiv.) of *N*,*N*-dimethylformamide dimethyl acetal.
2. Cap the vial and introduce it in the microwave reactor.
3. Program the reactor with the following parameters: power 300 watt, temperature 150 °C, ramp 30 s, hold time 2 min. Run the program.
4. Let the reaction mixture reach room temperature and remove from the microwave reactor.
5. Add 0.5 mL of toluene, recap the vial and place again in the microwave reactor.
6. Reprogram the reactor with the following parameters: power 300 watt, temperature 150 °C, ramp 30 s, hold time 0. Run the program.
7. Allow the vial to reach room temperature, remove it from the microwave reactor and place it in an ice-water bath for 15 min.

8. Filter the resulting crystals under vacuum on a sintered glass filtering plate and wash with 2 mL of cold hexane.
9. Continue passing air through the plate for 10 min, until the crystals are completely dry.
10. Collect and weigh the resulting solid and calculate the yield of the reaction.
11. Keep the product for its use in the following reaction and for further analysis.

Synthesis of Methyl 2-Oxo-2-(4-oxo-4*H*-chromen-3-yl)acetate

1. In a microwave vial provided with a stirring bar prepare a solution of 287 mg (1.5 mmol) of the enaminone obtained in the previous step in 1 mL of 1,2-dichloroethane.
2. Add 267 μL (3.3 mmol, 2.2 equiv.) of recently distilled pyridine and quickly cap the vial.
3. Place the reaction vial in an ice-water bath.
4. In a 10 mL round-bottom flask mix 152 μL (1.65 mmol, 1.1 equiv.) of methyl chlorooxoacetate and 1 mL of 1,2-dichloroethane.
5. With the help of a disposable syringe, add this solution dropwise to the enaminone ice-cooled solution.
6. Cap the vial, dry its outside wall with a paper towel and place it in the microwave reactor.
7. Program the reactor with the following parameters: power 300 watt, temperature 100 °C, ramp 1 min, hold time 9 min. Run the program.
8. Allow the reaction mixture to reach room temperature and then remove it from the microwave reactor.
9. Add 1 mL of dichloromethane and transfer to a small separation funnel, washing the vial with further 2 mL of dichloromethane.
10. Wash with copper sulfate solution (3 × 15 mL) to eliminate the pyridine, and then with brine (3 × 15 mL).
11. Dry the organic phase with Na_2SO_4, filter to a 50 mL round-bottom flask and evaporate the solvent in a rotary evaporator.
12. Collect the resulting solid, transfer it to a filtering plate and wash it with 2 mL of cold hexane.
13. Continue passing air through the plate for 10 min, until the solid is completely dry.
14. Collect and weigh the resulting solid and calculate the yield of the reaction.
15. Keep the product for further analysis.
16. Measure the melting points of the intermediate enaminone and the final product. Record the corresponding IR and ^1H and ^{13}C NMR spectra.

Waste Disposal

Collect all the residues in appropriate waste collector recipients, for posterior destruction. Toluene and hexane residues must be collected in a non-chlorinated waste recipient. Dichloroethane and dichloromethane residues must be collected in the chlorinated waste recipient.

Results Interpretation and Additional Questions

1. Calculate the yield of each of the two reactions and the overall yield of the process.
2. Compare the obtained melting points of the intermediate enaminone and the final products with the literature data. Interpret the IR and NMR spectra of both compounds.
3. Propose a mechanism for the reactions and discuss the role of pyridine.
4. Explain the Baker–Venkataraman rearrangement occurring in the second step of the synthesis.
5. Discuss the process in the context of green chemistry. Which of the 12 principles of green chemistry[9] does it comply with? Calculate the atom economy and E-factor for the overall process.

References

1. A. Gaspar, M. J. Matos, J. Garrido, E. Uriarte and F. Borges, *Chem. Rev.*, 2014, **114**, 4960.
2. (a) S. Mkrtchyan, V. O. Iaroshenko, S. Dudkin, A. Gevorgyan, M. Vilches-Herrera, G. Ghazaryan, D. M. Volochnyuk, D. Ostrovskyi, Z. Ahmed, A. Villinger, V. Y. Sosnovskikh and P. Langer, *Org. Biomol. Chem.*, 2010, **8**, 5280; (b) V. Y. Sosnovskikh, R. A. Irgashev and M. I. Kodess, *Tetrahedron*, 2008, **64**, 2997.
3. B. Appel, S. Rotzoll, R. Kranich, H. Reinke and P. Langer, *Eur. J. Org. Chem.*, 2006, **2006**, 3638.
4. V. O. Iaroshenko, I. Savych, A. Villinger, V. Y. Sosnovskikh and P. Langer, *Org. Biomol. Chem.*, 2012, **10**, 9344.
5. A. Bornadiego, J. Díaz and C. F. Marcos, *Adv. Synth. Catal.*, 2014, **356**, 718.
6. N.-G. Li, Z.-H. Shi, Y.-P. Tang, H.-Y. Ma, J.-P. Yang, B.-Q. Li, Z.-J. Wang, S.-L. Song and J.-A. Duan, *J. Heterocycl. Chem.*, 2010, **47**, 785.
7. I. Yokoe, K. Maruyama, Y. Sugita, T. Harashida and Y. Shirataki, *Chem. Pharm. Bull.*, 1994, **42**, 1697.
8. K. Al-Zaydi and R. Borik, *Molecules*, 2007, **12**, 2061.
9. P. T. Anastas and M. M. Kirchhoff, *Acc. Chem. Res.*, 2002, **35**, 686.

4.2.2.10. Synthesis of Indigo and Dyeing Process

Dulce Simão*

Department of Chemistry, Instituto Superior Técnico, University of Lisbon, Centro de Química Estrutural, Av. Rovisco Pais, 1, 1049-001 Lisbon, Portugal. *E-mail: dulcesimao@tecnico.ulisboa.pt

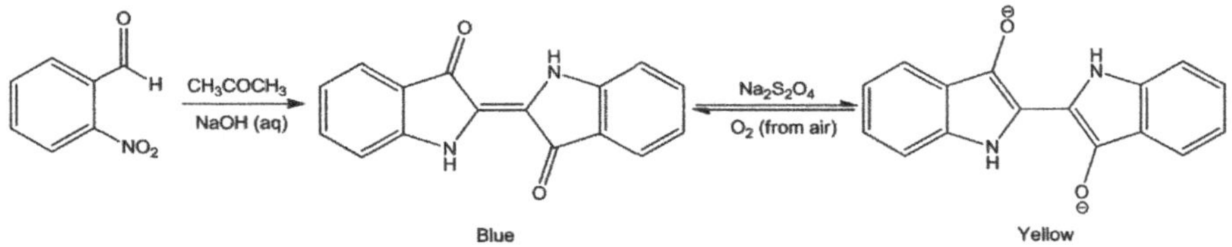

Number of sessions (duration of each session)	Hazard level	Difficulty level	Level of study
2 (3 h each)	Low	Low	Intermediate

Class names Aryl aldehydes, aryl nitros, enols, indoxyls

Concepts involved Indigo syntheses from 2-nitrobenzaldehyde according Baeyer–Drewsen process involving an aldol reaction, followed by several dehydration steps; dyeing a piece of cotton through reduction/oxidation reactions to fix the dye on the fiber

Chemicals needed 2-Nitrobenzaldehyde, acetone, sodium hydroxide, ethanol, sodium hydrosulfite (sodium dithionite), piece of cotton, soap solution

Equipment and experimental techniques involved Magnetic stirring and heating plate, suction filtration, IR, redox dyeing

Keywords Aldol reaction, dyeing, IR, reduction/oxidation, suction filtration

Background

Indigo is the oldest dye known. It was used by the Egyptians, who knew the isolation process from natural sources, and had huge importance in the economy of ancient and medieval times. More recently indigo was used to dye the blue coats supplied by the French to the Americans during the American Revolution. It was used to dye jeans, work clothes at that time, and in recent years to produce blue denim. Although ancient dyes were entirely of natural origin, now dyes are synthetic. Indigo was originally obtained by fermentation of the woad plant (*Isatis tinctoria*) and plants of the *Indigofera* species (*Indigofera tinctoria*).[1] Both plants contain a glucoside that can be hydrolyzed to indoxyl (Figure 4.2.2.10.1), precursor of indigo.

Indigo is considered a vat dye because is not soluble in water. In the indigo dyeing process (vat dyeing), it is converted into a reduced soluble yellow *leuco* form, applied on the fabric and is reoxidized to the insoluble blue dye by air, remaining deposited in the matrix of the cloth fibers.[2] Indigo was first synthesized by Baeyer in 1870. In this work indigo will be prepared using the Baeyer–Drewsen process (known as the Baeyer–Drewsen reaction) involving an aldol reaction of 2-nitrobenzaldehyde with acetone in dilute alkali.[3–6] The $^{-}$OH ion converts acetone

Comprehensive Organic Chemistry Experiments for the Laboratory Classroom
Edited by Carlos A. M. Afonso, Nuno R. Candeias, Dulce Pereira Simão, Alexandre F. Trindade, Jaime A. S. Coelho, Bin Tan, and Robert Franzén.

Published by the Royal Society of Chemistry, www.rsc.org

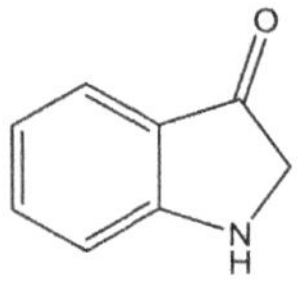

Figure 4.2.2.10.1 Indoxyl structure.

into the corresponding enolate ion that attacks the carbonyl group of 2-nitrobenzaldehyde to form a hydroxy-ketone. The reaction continues in a series of condensations, tautomerizations and dehydrations to give indigo. This dark blue dye will be used to dye a piece of cotton, performing the reduction to the yellow *leuco* form with sodium hydrosulfite and the reoxidation with air.[7–9]

Additional Safety

There is no need to perform the synthesis in a fume hood. The experimental procedure is without additional concerns besides general good practice. Nevertheless care should be taken with sodium hydroxide aqueous solution, which is corrosive, and with acetone, which is highly flammable.

Experimental Procedure

Preparation of Indigo

1. In a 100 mL beaker dissolve 2-nitrobenzaldehyde (1.0 g, 6.6 mmol) in acetone (20 mL).
2. Add water (35 mL) and stir vigorously using a magnetic stirrer.
3. Add sodium hydroxide aqueous solution 2 M (5 mL). The solution turns deep yellow, and quickly become darker with indigo precipitation.
4. Continue to stir for 5 min and collect the purple–blue solid by suction filtration. Wash the product with water until the washings are colorless and then with ethanol (20 mL).
5. Dry the solid with suction for 5 min and then at 100–120 °C for 30 min.
6. Record the yield and IR spectrum.

Dyeing of Cotton

1. Place 200 mg of indigo on a watch glass, add a few drops of ethanol and make a paste by rubbing the mixture with a glass rod.
2. In a 100 mL beaker suspend the paste in water (1 mL), add 2 M sodium hydroxide aqueous solution (3 mL).
3. Add a previously prepared aqueous solution of sodium hydrosulfite (0.3 g/20 mL).
4. Heat the mixture up to 50 °C on a steam bath, and as soon as a clear yellow solution is obtained, add water (40 mL).
5. Immerse the cotton and leave for 1 h at 50 °C.
6. Remove the cotton, squeeze it dry and hang it in the air for 30 min to develop the color. To brighten the color, in a 100 mL beaker, immerse the dyed fabric in a soap solution (0.5%), and heat it on a steam bath (50 °C) for 15 min.
7. Rinse the cotton with water and hang it to dry.

Waste Disposal

The highly dyed washings may be diluted and discarded in the sink.

Results Interpretation and Additional Questions

1. Determine the reaction yield.
2. Interpret the IR spectrum.
3. Write the reaction mechanism for the aldol reaction.

References

1. T. Seilnacht, *Indigo*, http://www.seilnacht.tuttlingen.com/Lexikon/Indigo.htm.
2. D. N. Epp, *J. Chem. Educ.*, 1995, **72**(8), 726.
3. C. A. M. Afonso, D. P. Simão, L. P. Ferreira, M. S. Serra and M. M. M. Raposo, *100 Experiências de Química Orgânica*, IST Press, Instituto Superior Técnico, Lisboa, 2011, p. 207 (translation from the Portuguese language edition).
4. L. M. Harwood; C. J. Moody and J. M. Percy, *Experimental Organic Chemistry*, Blackwell Science, 2nd edn, 1999, p. 622.
5. Kirk-Othmer, *Encyclopedia of Chemical Technology*, John Wiley & Sons, 2nd edn, 1970, vol. 22, pp. 432–437.
6. J. R. Mckee and M. Zanger, *J. Chem. Educ.*, 1991, **68**(10), A242.
7. J. Jack Li, *Name reactions: A collection of detailed reaction mechanisms,* Springer, 2nd edn, 2003, p. 14.
8. A. Mckillop, *Advanced Problems in Organic reaction mechanisms*, Pergamon, 1997, p. 127.
9. V. Bianda, J. A. Constenla, R. Haubrichs and P. L. Zaffalon, *Chem.: Bulgarian J. Sci. Educ.*, 2013, **22**(1), 52.

4.2.3. 1,2-Addition to Carbonyl Compounds Other than Aldol Reaction

4.2.3.1. Knorr Pyrazole Synthesis of Edaravone

Ana Paula Francisco*, Ana S. Ressurreição, Maria de Jesus Perry, and Francisca Lopes*

Instituto de Investigação do Medicamento (iMed.ULisboa), Faculdade de Farmácia da, Universidade de Lisboa, Av. Prof. Gama Pinto, 1649-003 Lisboa, Portugal. *E-mail: afrancis@ff.ulisboa.pt, fclopes@ff.ulisboa.pt

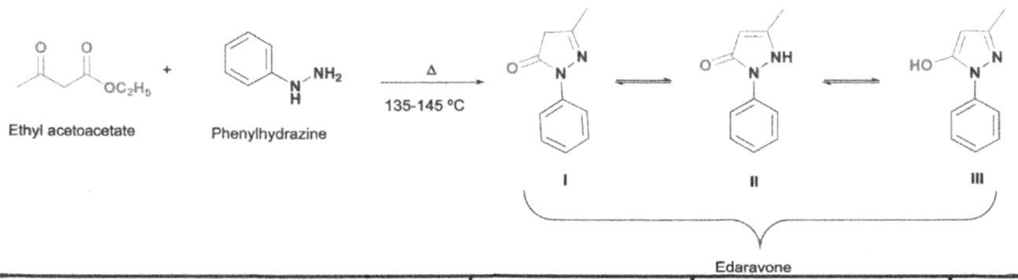

Number of sessions (duration of each session)	Hazard level	Difficulty level	Level of study
1 (3 h)	Moderate	Medium	Intermediate

Class names β-Keto-esters, phenylhydrazines, pyrazolones

Concepts involved Synthesis of edaravone (a pyrazolone heterocycle) by condensation reaction and intramolecular cyclization; the Knorr pyrazole reaction; identification of tautomeric forms of pyrazolones by spectroscopic methods – IR and NMR

Chemicals needed Ethyl acetoacetate, phenylhydrazine, diethyl ether, ethanol, NaOH, NH_4OH, HCl, acetic acid, KBr, DMSO-d_6, $CDCl_3$

Equipment and experimental techniques involved Vacuum filtration, recrystallization, reflux, melting point apparatus, IR, NMR

Keywords Edaravone, heterocycle synthesis, NMR, recrystallization, reflux

Background

The chemistry of pyrazolones started in 1883 when Ludwig Knorr reported the synthesis of antipyrin (Figure 4.2.3.1.1) as an unexpected product of an attempted synthesis of a quinolone derivative using phenylhydrazine as starting material. After its discovery, antipyrin became the most widely used anti-inflammatory drug in the world, until aspirin became more important.

These nitrogen heterocycles are characterized by a five-membered lactam ring, containing two nitrogen atoms and an additional carbonyl group. When the pyrazolones were discovered, they were only known as non-steroidal anti-inflammatory drugs,[1] but recently they have attracted much attention due to their diverse pharmacological properties, such as antioxidant,[2] anticancer,[3]

Comprehensive Organic Chemistry Experiments for the Laboratory Classroom
Edited by Carlos A. M. Afonso, Nuno R. Candeias, Dulce Pereira Simão, Alexandre F. Trindade, Jaime A. S. Coelho, Bin Tan, and Robert Franzén.

Published by the Royal Society of Chemistry, www.rsc.org

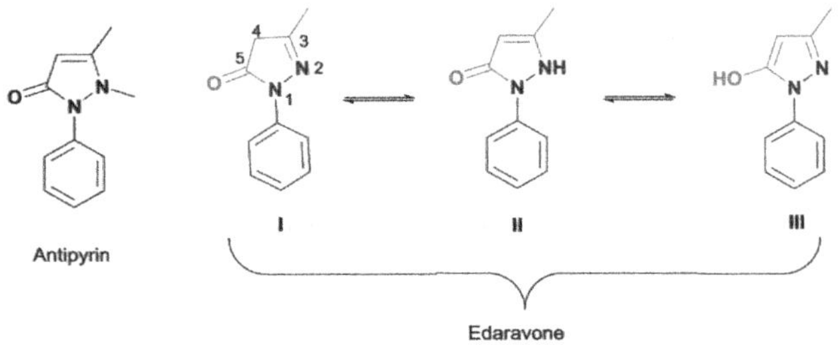

Figure 4.2.3.1.1 Structures of antipyrin and edaravone.

antibacterial,[4] and antiviral.[5] Some pyrazolones (*e.g.* 4-acylpyrazolones) also have complex formation properties. They can coordinate to metal ions due to the several electron-rich donor centres; the carbonyl group from 4-acyl substituent and the 5-OH group from enol form metal chelates, which are highly soluble in most organic solvents, and are useful as solvent extraction reagents.[6,7]

Edaravone (3-methyl-1-phenyl-5-pyrazolone) was first synthesized in 1887. The numbering of the pyrazolone nucleus is based on the pyrazole system (Figure 4.2.3.1.1). Recently, this drug has gained much attention due to its remarkable antioxidant properties and is currently used for the treatment of cerebral thrombosis and cerebral embolism.[2,8,9] Edaravone, has three tautomeric structures: the keto (I), the amine (II), and the enol (III) forms (Figure 4.2.3.1.1). The pK_a value of edaravone is 7.0, implying that it is approximately 50% ionized at physiological pH (pH 7.4). The anionic form may react strongly with reactive oxygen species (ROS) in the brain.[9–11] These chemical properties of edaravone are responsible for its current use in medicine.

Several methods have been developed for the synthesis of pyrazolones,[12] the most common of which involve the condensation of β-keto esters or β-keto aldehydes with substituted or unsubstituted hydrazines.[4] Reactions of hydrazines with 1,3-dicarbonyl compounds were formally named the Knorr pyrazole synthesis after its discovery in the late nineteenth century.

Edaravone is synthesized in this experiment from the reaction of ethyl acetoacetate with phenylhydrazine. When phenylhydrazine is treated with an equimolar quantity of ethyl acetoacetate under reflux and without any solvent, a complete coupling reaction takes place, leading to the formation of 3-methyl-1-phenyl-5-pyrazolone *via* elimination of water and ethanol. The product is isolated from the reaction mixture by addition of diethyl ether and precipitation. The present proposal occurs under the conditions described above, with short reaction times (1 h) and very good yields.

Mechanistically, the reaction proceeds *via* the regioselective attack of the most nucleophilic nitrogen atom of phenylhydrazine on the ketone moiety of ethyl acetoacetate followed by intramolecular cyclization of the intermediate hydrazone (Scheme 4.2.3.1.1).

The chemical and physical properties of edaravone are governed by the tautomerism of the molecule.[13,14] The amphoteric character of edaravone can be assessed by a solubility assay in basic and acid aqueous solutions.

Additional Safety

All the experimental procedures should be done in a fume hood, with proper safety equipment, such as full protective clothes, gloves and safety goggles.

Phenylhydrazine is readily absorbed by the skin and after ingestion and inhalation. It is irritating to the eyes, skin and respiratory tract and presents high acute toxicity and genotoxic potential *in vitro* and *in vivo*.

Diethyl ether is extremely volatile and flammable and precipitation of the compound should be performed far from heat sources. Ethanol is highly flammable. Hydrochloric acid is a toxic

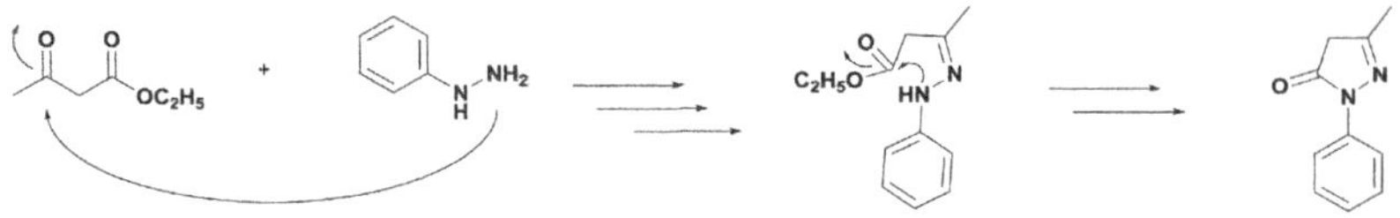

Scheme 4.2.3.1.1 Regioselective attack of phenylhydrazine on the carbonyl compound and intramolecular cyclization of the resultant hydrazone.

compound, and has a corrosive effect on human tissue, with the potential to damage irreversibly respiratory organs, eyes, skin, and intestines.

Deuterochloroform ($CDCl_3$) causes irritation of the skin and respiratory system, may cause chemical burns, and is a possible carcinogen. DMSO-d_6 is toxic by inhalation and causes irritation of the eyes, respiratory system and skin.

Experimental Procedure

Laboratory Session 1 (3 h): Synthesis of 3-Methyl-1-phenyl-5-pyrazolone

1. In a round-bottomed flask, add the ethyl acetoacetate (1.625 mL, 12.5 mmol) and the phenylhydrazine (1.25 mL, 12.5 mmol). This addition is slightly exothermic and must be done slowly and carefully in the fume hood.
2. Assemble a reflux condenser and heat the mixture for 60 min at 135–145 °C.
3. Transfer the heavy syrup formed after heating into a beaker and cool it well in an ice-water bath.
4. Add a portion of 2 mL of diethyl ether and stir the mixture vigorously until the crude powdered pyrazolone is obtained. After initial precipitation, add more three portions of 2 mL of diethyl ether until complete precipitation occurs.
5. Filter the product in a vacuum pump, using a Büchner funnel, and wash the solid thoroughly with diethyl ether.
6. Collect the solid from the Büchner funnel and recrystallize from ethanol 95% (dissolve the product in a minimum of hot ethanol (usually 5–7 mL) in an Erlenmeyer flask while gently and continuously heating the mixture on the hot plate). After dissolution, let the hot solution cool, first to room temperature and then in an ice bath until complete crystallization of the product.
7. Filter off the solid in a Büchner funnel by vacuum filtration and weigh it after drying it in a desiccator.
8. Determine the yield of the reaction and the melting point (125–127 °C) of the pure product.
9. Analyse the sample by ^1H and ^{13}C NMR, prepare two samples, one in DMSO-d_6 and the other in $CDCl_3$ (usually 40 mg of edaravone in 0.5 mL of the deuterated solvents).
10. Analyse the sample by IR.

Solubility Tests

1. Place approximately 10 mg of the solid in a test tube (do not weigh the solid, simply use enough quantity to cover the tip of a small spatula). Then with a Pasteur pipette add the solution of 10% NaOH drop by drop until complete solubilization of the solid while shaking the test tube. Solubility will be indicated by the formation of a homogeneous solution.
2. With another Pasteur pipette add now to the same test tube the necessary amount of acetic acid until a milky suspension appears. Add one or two more drops of the acid until a white precipitate forms in the tube (be careful because if you add excess of the solution the precipitate disappears).

3. Add an excess of the acetic acid until complete disappearance of the precipitate and a homogeneous solution is formed again.

Waste Disposal

Non-halogenated organic solvents must be collected in a non-halogenated collector recipient for further destruction (the mixture of phenylhydrazine/diethyl ether is discarded in this collector). Aqueous solutions of acids and bases should be disposed of in the acids and bases collector recipient.

Results Interpretation and Additional Questions

1. Draw the complete mechanism for the edaravone synthesis reaction.
2. Based on solubility tests, explain the amphoteric character of edaravone using chemical reactions.
3. Interpret the ^1H and ^{13}C NMR spectra obtained in $CDCl_3$.
4. Interpret the ^1H and ^{13}C NMR spectra obtained in DMSO and compare them with those obtained in $CDCl_3$.
5. Interpret the IR spectra obtained.
6. Calculate the percent yield of the reaction.

References

1. F. A. F. Ragab, N. M. Abdel-Gawad, H. H. Georgey and M. F. Said, *Chem. Pharm. Bull.*, 2013, **61**, 834.
2. K. Kikuchi, S. Tancharoen, N. Takeshige, M. Yoshitomi, M. Morioka, Y. Murai and E. Tanaka, *Int. J. Mol. Sci.*, 2013, **14**, 13909.
3. M. H. Norman, L. B. Liu, M. Lee, N. Xi, I. Fellows, N. D. D'Angelo, C. Dominguez, K. Rex, S. F. Bellon, T. S. Kim and I. Dussault, *J. Med. Chem.*, 2012, **55**, 1858.
4. R. V. Ragavan, K. M. Kumar, V. Vijayakumar, S. Sarveswari, S. Ramaiah, A. Anbarasu, S. Karthikeyan, P. Giridharan and N. S. Kumari, *Org. Med. Chem. Lett.*, 2013, **3**, 6.
5. R. Ramajayam, K.-P. Tan, H.-G. Liu and P.-H. Liang, *Bioorg. Med. Chem.*, 2010, **18**, 7849.
6. N. J. Parmar, H. A. Barad, B. R. Pansuriya and R. A. Patel, *J. Coord. Chem.*, 2011, **64**, 688.
7. V. A. Joseph, J. H. Pandya and R. N. Jadeja, *J. Mol. Struct.*, 2015, **1081**, 443.
8. K. Kikuchi, H. Uchikado, N. Miyagi, Y. Morimoto, T. Ito, S. Tancharoen, N. Miura, K. Miyata, R. Sakamoto, C. Kikuchi, N. Iida, N. Shiomi, T. Kuramoto and K. Kawahara, *Int. J. Mol. Med.*, 2011, **28**, 899.
9. H. B. Shang, D. R. Cui, D. H. Yang, S. Liang, W. F. Zhang and W. G. Zhao, *J. Stroke Cerebrovasc. Dis.*, 2015, **24**, 215.
10. K. Hata, A. Urushibara, S. Yamashita, M. Lin, Y. Muroya, N. Shikazono, A. Yokoya, H. Fu and Y. Katsumura, *J. Radiat. Res.*, 2015, **56**, 59.
11. K. Watanabe, M. Taniguchi and M. Shinoda, *Redox Rep.*, 2003, **8**, 157.
12. M. M. Mojtahedi, M. R. Jalali, M. S. Abaee and M. Bolourtchian, *Heterocycl. Commun.*, 2006, **12**, 225.
13. A. N. Queiroz, A. P. S. Mendes, M. S. Leal, A. Neto and R. S. Borges, *J. Comput. Theor. Nanosci.*, 2010, **7**, 153.
14. Q. Wang, Y. Zhang, R. Wang, Y. L. Yang and F. Zhi, *Acta Crystallogr., Sect. E: Struct. Rep. Online*, 2008, **64**, O1924–U3769.

4.2.3.2. Greener Solvent Substitution in a Verley–Doebner Condensation

Jonathon W. Moir and Andrew P. Dicks*

Department of Chemistry, University of Toronto, 80 St. George Street, Toronto, Ontario, Canada M5S 3H6
*E-mail: adicks@chem.utoronto.ca

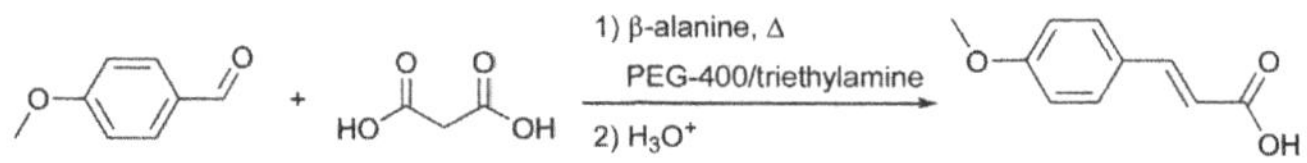

Number of sessions (duration of each session)	Hazard level	Difficulty level	Level of study
1 (4 h)	Moderate	Medium	Intermediate

Class names 1,3-Dicarbonyls, aldehydes, alkenes, carboxylic acids

Concepts involved Organocatalysis and C–C bond formation *via* (i) enolate generation, (ii) nucleophilic attack at an activated carbon and (iii) subsequent decarboxylation/condensation to form an unsaturated carboxylic acid. The experimental discussion includes concepts focused on green chemistry and solvent selection in synthetic organic reactions

Chemicals needed 4-Methoxybenzaldehyde (*p*-anisaldehyde), malonic acid, β-alanine, poly(ethylene glycol)-400 (PEG-400, MW 380–420), triethylamine, concentrated HCl (12 M), absolute ethanol

Equipment and experimental techniques involved Reflux set-up (25 mL round-bottomed flask with stir bar, heating mantle and power source, condenser and water source), vacuum filtration set-up (Büchner funnel, filter paper, filtration flask, vacuum source), recrystallization apparatus (hot plate, 50 mL Erlenmeyer flask), IR measurement equipment (spectrometer, mortar and pestle, Nujol, NaCl plates), NMR measurement equipment (spectrometer, DMSO-d_6 deuterated solvent, NMR tube and cap)

Keywords C–C bond formation, decarboxylation, greener solvents, IR/NMR, vacuum filtration/recrystallization, Verley–Doebner condensation

Background

The Knoevenagel condensation is a fundamental carbon–carbon bond-forming process and a common transformation performed in undergraduate organic chemistry laboratories. This procedurally-simple reaction involves initial formation of an enolate under basic conditions, which attacks a carbonyl group to generate the new carbon–carbon bond.[1] In the Verley–Doebner modification of this reaction, malonic acid is used as the enolate precursor and after nucleophilic attack at an activated carbon atom, decarboxylation occurs to form an α,β-unsaturated carboxylic acid.[2] The reaction is usually stereoselective in that the *trans* isomer of the alkene is exclusively formed. Therefore, the practical Verley–Doebner approach exemplifies many theoretical concepts surrounding enolate generation and reactivity.

Pyridine is frequently used as both solvent and base in the Verley–Doebner reaction, which affords good yields of the unsaturated product. However, the toxicity of pyridine makes it less than ideal for undergraduate practical work, and suitable solvent substitutes are highly desirable.[3,4]

Comprehensive Organic Chemistry Experiments for the Laboratory Classroom
Edited by Carlos A. M. Afonso, Nuno R. Candeias, Dulce Pereira Simão, Alexandre F. Trindade, Jaime A. S. Coelho, Bin Tan, and Robert Franzén.

Published by the Royal Society of Chemistry, www.rsc.org

Indeed, appropriate solvent selection is an important part of modern synthesis, and learning to choose greener solvents for organic reactions is an important skill for new chemists to develop.

In this experiment, use of a poly(ethylene glycol) (PEG-400)/triethylamine mixture as a greener solvent substitute is described for the Verley–Doebner condensation of malonic acid and 4-methoxybenzaldehyde. A procedure was recently incorporated into an undergraduate organic course at the University of Toronto and the University of Pittsburgh where PEG-400 was successfully employed as a solvent for the condensation reaction between 4-nitrobenzaldehyde and dimedone.[5] The non-toxic polyether solvent was found to be easily recovered and reused over several reaction cycles, and many important green chemistry principles involved were incorporated into the student experimental discussion. Triethylamine has previously been utilized as an effective basic solvent in the condensation of malonic acid with aliphatic aldehydes.[6] In addition, the industrial Pfizer solvent selection guide suggests replacing pyridine with triethylamine when the former is used as a basic catalyst in a reaction.[3] As such, a combination of triethylamine and PEG-400 in a 70:30 v/v ratio is a greener replacement for pyridine in a Verley–Doebner reaction which features β-alanine as an organocatalyst. This solvent mixture facilitates student appreciation regarding the importance of medium selection in synthetic processes, whilst maintaining acceptable product yields. The α,β-unsaturated carboxylic acid formed (*trans*-4-methoxycinnamic acid) is also of real-world relevance as an important precursor in the synthesis of UVB light (290–320 nm) absorbers found in sunscreens.

Additional Safety

1. While considered an acceptable substitute for pyridine (which has long-term toxic effects on the liver, kidney and central nervous system), triethylamine is still an acutely toxic substance and is extremely flammable. It should be kept away from all sources of sparks and open flames. Triethylamine is highly caustic and can cause severe skin burns and eye damage. Avoid skin contact and inhalation; use a fume hood and proper protective equipment at all times. This compound is toxic to aquatic environments and should be properly disposed of.
2. Malonic acid is corrosive. It is toxic by inhalation and can cause severe eye damage and skin irritation.
3. Concentrated hydrochloric acid (HCl) is extremely corrosive. It is toxic by inhalation and can cause severe eye damage and skin irritation. Perform all reactions with concentrated HCl in a fume hood.
4. The reflux occurs at a high temperature (~160 °C). Take appropriate precautions when using the heating mantle and reflux apparatus.
5. Appropriate protective equipment, including gloves, eye protection and a laboratory coat should be worn at all times during this experiment.

Experimental Procedure

Laboratory Session (4 h): Synthesis of *trans*-4-Methoxycinnamic Acid

1. To a 25 mL round-bottomed flask equipped with a stir bar, add 4-methoxybenzaldehyde (9.86 mmol, 1.34 g), malonic acid (25.2 mmol, 2.62 g), β-alanine (1.68 mmol, 0.150 g), triethylamine (3.5 mL) and PEG-400 (1.5 mL).
2. Heat the mixture vigorously under reflux for 90 min (~160 °C) using a heating mantle.
3. After this time, remove the flask from the heat and allow the mixture to cool to room temperature. Cool the mixture further in an ice bath for 10 min.
4. Slowly and carefully add concentrated HCl (12 mL, in 1–3 mL portions) to the flask, causing a white solid to precipitate. Collect the precipitate by vacuum filtration and wash the

product with ice-cold water. Allow the crude product to sit on the filter paper under vacuum to assist drying for an additional 5 min.

5. Collect the product and recrystallize it from absolute ethanol. Filter the recrystallized product and wash with a small amount of cold ethanol. Allow the product to dry under vacuum, weigh and calculate the percent yield of the purified *trans*-4-methoxycinnamic acid product.

Waste Disposal

The triethylamine/PEG-400/HCl (aq) filtrate waste should be labelled appropriately and disposed of as organic waste. The absolute ethanol filtrate should be recycled if possible, or labelled appropriately and disposed of as non-halogenated organic waste.

Results Interpretation and Additional Questions

1. After acquiring the ^{1}H NMR spectrum of the final product, assign the observed resonances to each proton in the structure. Does the peak integration match the required number of protons in each case? Be sure to label and identify all impurities if any are visible in the spectrum.
2. Assign the important peaks in the acquired IR spectrum of the product.
3. Write a detailed mechanism for the overall reaction, including use of triethylamine as a base, β-alanine as an organocatalyst, and the decarboxylation step.
4. When the reaction was run in pyridine under reflux, the yield was found to be 93%. When the reaction was run in pure triethylamine under reflux at the same temperature, the yield was found to be 76%. When the reaction was run in pure PEG-400 under reflux at the same temperature, the yield was found to be 28%. Discuss some possible reasons for these differences.
5. Using available MSDS data, compare the toxicities and physical properties of pyridine, triethylamine and PEG-400 (note that this information should have been acquired prior to commencing the experiment). Based on this information, what are some of the advantages and disadvantages of using pyridine, triethylamine and PEG-400 as solvents?
6. Using the Pfizer[3] and the GlaxoSmithKline[4] solvent selection guides, discuss some of the other important considerations when choosing a solvent for an organic reaction. What types of solvents are generally considered to be good candidates for reactions? What types should generally be avoided?

References

1. K. E. Kolb, K. W. Field and P. F. Schatz, *J. Chem. Educ.*, 1990, **67**, A304.
2. R. G. Stabile and A. P. Dicks, *J. Chem. Educ.*, 2004, **81**, 1488, and references therein.
3. K. Alfonsi, J. Colberg, P. J. Dunn, T. Fevig, S. Jennings, T. A. Johnson, H. P. Kleine, C. Knight, M. A. Nagy, D. A. Perry and M. Stefaniak, *Green Chem.*, 2008, **10**, 31.
4. R. K. Henderson, C. Jiménez-González, D. J. C. Constable, S. R. Alston, G. G. A. Inglis, G. Fisher, J. Sherwood, S. P. Binks and A. D. Curzons, *Green Chem.*, 2011, **13**, 854.
5. J. M. Stacey, A. P. Dicks, A. A. Goodwin, B. M. Rush and M. Nigam, *J. Chem. Educ.*, 2013, **90**, 1067.
6. B. V. S. K. Rao, P. Vijayalakshmi and R. Subbarao, *J. Am. Oil Chem. Soc.*, 1993, **70**, 297.

4.2.3.3. Knorr Synthesis of Diethyl 3,5-Dimethyl-1*H*-pyrrole-2,4-dicarboxylate

M. Soledad Pino-González*, Cristina García-Ruiz, Francisco Sarabia, and Gregorio Torres

Organic Chemistry Department, Faculty of Sciences, University of Málaga, 29071 Málaga, Spain
*E-mail: pino@uma.es

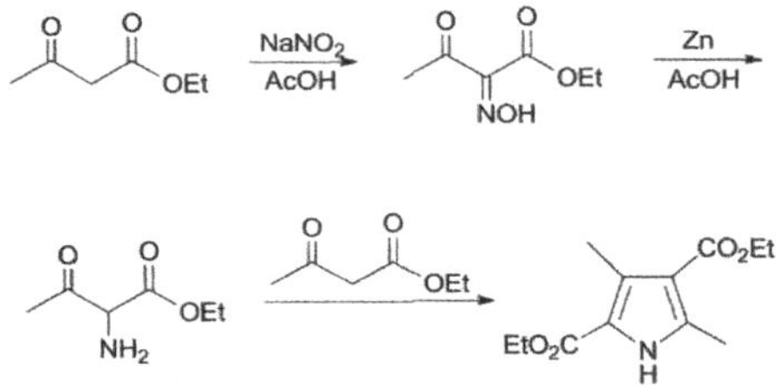

Number of sessions (duration of each session)	Hazard level	Difficulty level	Level of study
3 (1 h + 3 h + 3 h)	Moderate	Medium	Intermediate

Class names Ketones, esters, oximes, amines, pyrroles

Concepts involved This experiment describes the synthesis of a pyrrole derivative with the following steps: carbonyl α-nitrosation, oxime reduction and aromatic ring formation. The student will study concepts of enolization, nucleophilic addition at a carbonyl, formation of heterocyclic ring as well as carbon–nitrogen and carbon–carbon bond formation

Chemicals needed Ethyl acetoacetate, acetic acid, sodium nitrite, zinc dust, ethanol (85% wt)

Equipment and experimental techniques involved One- and two-necked flask equipped with a Liebig-type reflux condenser and with heating–stirring plate, crystallization equipment with filter funnel, Kitasato and Erlenmeyer flasks; heating, stirring, recrystallization

Keywords Knorr synthesis, mp, NMR, recrystallization, reflux

Background

The pyrrole ring[1,2] occurs in many types of compounds with biological activities. Their derivatives can be found in naturally occurring[3] and synthetic molecules with biological relevance.[4,5] Several derivatives have other applications as conducting polymers and in molecular optics or electronics, being used in material science.[6] Among the most common approaches to pyrrole synthesis, the Knorr and Paal–Knorr syntheses have been widely described in review articles.[1,2] In addition, new pyrrole derivatives and improvements for these reactions have been described.[7,8]

The objective of this work is the synthesis of diethyl 3,5-dimethyl-1*H*-pyrrole-2,4-dicarboxylate by the Knorr synthesis.[9,10] Ethyl acetoacetate is the starting material for this experiment. Nitrosation of a half of the initial amount of ethyl acetoacetate with sodium nitrite in acid media produces an oxime that is subsequently reduced by Zn in acetic acid. The obtained 2-amino-3-ketoester attacks the remaining ethyl acetoacetate and builds the aromatic pyrrole ring by double condensation.

Comprehensive Organic Chemistry Experiments for the Laboratory Classroom
Edited by Carlos A. M. Afonso, Nuno R. Candeias, Dulce Pereira Simão, Alexandre F. Trindade, Jaime A. S. Coelho, Bin Tan, and Robert Franzén.

Published by the Royal Society of Chemistry, www.rsc.org

Additional Safety

All sessions should be done in a fume hood, with proper safety equipment, such as protective clothing, gloves and safety goggles. Ethyl acetoacetate is moderately irritating to skin and mucous membranes. Acetic acid is an acid that on ingestion may cause severe corrosions of mouth and G.I. tract. Avoid breathing zinc dust and getting in it the eyes or on skin. Ethanol is a common solvent that may cause nausea, flushing and mental excitement or depression.

Experimental Procedure

Laboratory Session 1 (1 h). Oxime Formation

1. In a fume hood, prepare an ice bath on a magnetic plate.
2. Introduce a 50 mL two-necked round-bottom flask equipped with an internal thermometer and a magnetic stir bar, and add ethyl acetoacetate (3 mL, 23 mmol) and acetic acid (7 mL). Stir vigorously at 0 °C.
3. Over a period of 10 min, add slowly a solution of $NaNO_2$ (0.8 g, 12 mmol) in water (1 mL) with a syringe, maintaining the mixture below 5–10 °C (internal temperature).
4. Stir the mixture for 30 min more and leave for at least 3 h at room temperature.

Laboratory Session 2 (3 h). Oxime Reduction and Pyrrole Formation

1. In a fume hood place an oil bath on the magnetic plate and a Liebig-type reflux condenser in the reaction flask. Fit a calcium chloride guard tube.
2. Add Zn dust (1.5 g) in small portions (10 min period), controlling the temperature (exothermic reaction).
3. Heat for 1 h and maintain an external temperature of 100–110 °C.
4. Remove the flask from the oil bath. Pour the mixture while it is still warm into an Erlenmeyer flask with water and crushed ice (75 mL).
5. Place the flask on an ice bath and stir for 1 h in a cold stirring plate.
6. Collect the precipitate by filtration with a filter funnel, wash it thoroughly with cold water and dry by suction.

Laboratory Session 3 (3 h). Recrystallization, Yield Determination and Product Characterization

1. In a round-flask equipped with a stirrer and reflux condenser place the crystals and add 6–10 mL of EtOH (85%).
2. Heat under reflux till complete dissolution of the crystals, add more EtOH if necessary through the upper side of the condenser.
3. Remove the flask from the oil bath and filter the solution quickly on a conic funnel over an Erlenmeyer flask to remove solid impurities.
4. Let the filtered solution reach room temperature and then place the Erlenmeyer flask in an ice bath for 15 min.
5. Filter the yellow pale crystals through a filter funnel.
6. Collect the crystals in a weighed flask and dried them under vacuum. (A second crop of crystals can be harvested from the mother liquor.)

Yield Determination, mp and NMR Spectra

7. Weigh the round-bottom flask containing the dried crystals and determine the final weight (M_1).

8. Determine the melting point of the obtained product. Compare it with the literature data.[9–11]
9. Prepare an NMR tube with a solution of the obtained product in $CDCl_3$.
10. Acquire the ^1H and ^{13}C NMR spectra to confirm the compound purity. Compare the spectra with literature data.[10,11]
11. Calculate the reaction yield.

Waste Disposal

Collect all solvent wastes in a non-chlorinated waste collector recipient. Solids must be placed in the special solid waste container.

Results Interpretation and Additional Questions

1. Interpret the ^1H and ^{13}C NMR spectra obtained, and compare the experimental melting point with literature data.[10,11]
2. Calculate the percent yield for the overall process.
3. Rationalize the mechanism of nitrosation and oxime formation.
4. Depict the cyclization mechanism that leads to formation of the pyrrole ring.

References

1. V. F. Ferreira, M.-C. B. V. De Souza, A. C. Cunha, L. O. R. Pereira and M. L. G. Ferreira, *Org. Prep. Proc. Int.*, 2001, **33**, 411.
2. D. St. C. Black and G. Maas, in *Science of Synthesis*, Georg Thieme Verlag, Stuttgart, New York, 2001, vol. 9, p. 441.
3. M. Seki and K. Mori, *Eur. J. Org. Chem.*, 2001, 503.
4. M. V. Raimondi, S. Cascioferro, D. Schillaci and S. Petruso, *Eur. J. Med. Chem.*, 2006, **41**, 1439.
5. M. N. Narule, M. K. Gaidhane and P. K. Gaidhane, *J. Pharm. Res.*, 2013, **6**, 626.
6. C.-F. Lee, L.-M. Yang, T.-Y. Hwu, A.-S. Feng, J.-C. Tseng and T.-Y. Luh, *J. Am. Chem. Soc.*, 2000, **122**, 4992.
7. B. Khalili, P. Jajarmi, B. Eftekhari-Sis and M. M. Hashemi, *J. Org. Chem.*, 2008, **73**, 2090.
8. J. V. Cooney, E. J. Beal and R. N. Hazlett, *Org. Prep. Proc. Int.*, 1983, **15**, 292.
9. H. Fischer, *Org. Synth.*, 1935, **15**, 17.
10. E. Fabiano and B. T. Golding, *J. Chem. Soc., Perkin Trans. 1*, 1991, 3371.
11. E. V. Antina, G. B. Guseva, A. E. Loginova, A. S. Semeikin and A. I. V'yugin, *Russ. J. Gen. Chem.*, 2010, **80**, 2374.

4.2.3.4. Synthesis of Coumarin-3-carboxylic Acid

Paulo Coelho* and Céu Sousa

Centro de Química – Vila Real, Universidade de Trás-os-Montes e Alto Douro, 5001-801 Vila Real, Portugal
*E-mail: pcoelho@utad.pt

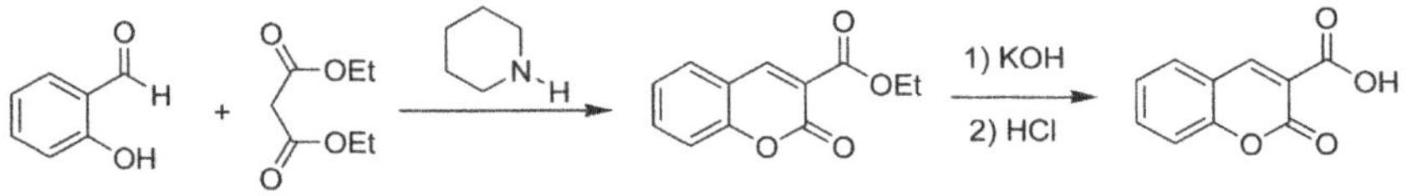

Number of sessions (duration of each session)	Hazard level	Difficulty level	Level of study
2 (2 h each)	Low	Low	Intermediate

Class names Aldehydes, esters, phenols

Concepts involved The synthesis of coumarin-3-carboxylic acid is accomplished by a two-step procedure starting from salicylaldehyde using the Knoevenagel condensation. The reaction of this aldehyde with ethyl malonate, in the presence of a weak base, affords directly the coumarin ester, which is then hydrolysed in strong basic medium to the corresponding coumarin-3-carboxylic acid. Both coumarins are isolated in a pure form by filtration. This experiment has a low level of difficulty, requiring two sessions to prepare these compounds, which belong to a class of well-known natural compounds with a wide range of applications in the pharmaceutical, perfume and cosmetic industries

Chemicals needed Salicylaldehyde, diethyl malonate, piperidine, acetic acid, ethanol, $CaCl_2$, CH_2Cl_2, KOH, HCl

Equipment and experimental techniques involved Heating mantle, apparatus for reaction under reflux, filtration

Keywords Basic hydrolysis, condensation, coumarin, filtration, Knoevenagel, NMR

Background

Coumarins (2*H*-chromen-2-ones) are widespread in nature, being found in many plants of the Umbelliferae, Rutaceae and Compositae families and in several essential oils. Aesculetin (Figure 4.2.3.4.1) can be extracted from horse chestnuts and psoralen from the Indian plant *Psoralea corylifolia*. Psoralen is a mutagen, it intercalates into DNA and, on exposure to ultraviolet (UVA) radiation, induces processes that lead to an increase in the skin pigmentation and inhibition of cell division, being thus associated with the disease psoriasis. Other coumarin derivatives exhibit beneficial properties, such as warfarin, which is an anticoagulant used in the treatment of thrombosis. Coumarins are also used as fragrances in food and cosmetic products.

There are several methods for the synthesis of the coumarin core. Simple coumarins can be easily prepared using the Perkin reaction between *ortho*-hydroxybenzaldehydes and acetic anhydride in the presence of a weak base, like sodium acetate (Scheme 4.2.3.4.1).[1] This reaction starts with an aldol condensation followed by an intramolecular esterification.

Comprehensive Organic Chemistry Experiments for the Laboratory Classroom
Edited by Carlos A. M. Afonso, Nuno R. Candeias, Dulce Pereira Simão, Alexandre F. Trindade, Jaime A. S. Coelho, Bin Tan, and Robert Franzén.

Published by the Royal Society of Chemistry, www.rsc.org

The Pechmann condensation between phenols and β-keto esters is another valuable method for the synthesis of coumarins (Scheme 4.2.3.4.2).[2] The condensation is performed under strong acidic conditions but the yields are generally good. The mechanism involves an aromatic electrophilic substitution of phenol with the protonated carbonyl group followed by lactonization and water elimination to produce the coumarin system. With highly activated phenols such as resorcinol, the reaction can be performed under much milder conditions, providing a useful route to umbelliferone (7-hydroxycoumarin) derivatives. This method has been used for the synthesis of naturally occurring coumarins of biological interest and some industrially important coumarins like 7-hydroxy-4-methylcoumarin and 7-diethylamino-4-methylcoumarin from readily available 1,3-disubstituted compounds.

An alternative route involves the initial addition of *ortho*-metalated phenolic ethers to alkoxymethylene malonic esters followed by an acidic treatment to remove the protective phenolic ether group, followed by lactonization and dehydration (Scheme 4.2.3.4.3).[3]

A different method involves the condensation of *ortho*-hydroxybenzaldehydes with reactive methylene compounds, like malonic ester, cyanoacetic ester or malononitrile, in the presence of a weak base like piperidine. This is known as the Knoevenagel synthesis.[4] This one-pot method is very general, being compatible with different salicylic aldehydes, and uses much milder conditions than the Perkin or the Pechmann reactions, and has thus been used for the synthesis of a large list of coumarins. An interesting variation of this method uses microwave irradiation of a solvent-free mixture of the reagents, affording different coumarins in good yield.[5]

The objective of this work is the preparation of coumarin-3-carboxylic acid, in two steps, through the Knoevenagel condensation of salicylaldehyde with diethyl malonate followed by basic hydrolysis (Scheme 4.2.3.4.4).[6]

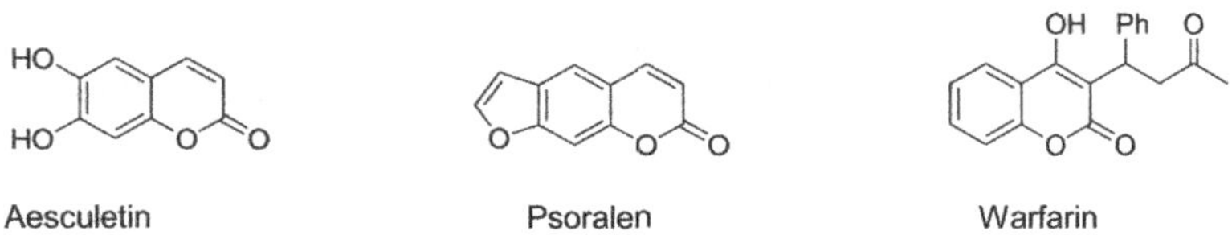

Figure 4.2.3.4.1 Some naturally occurring coumarins.

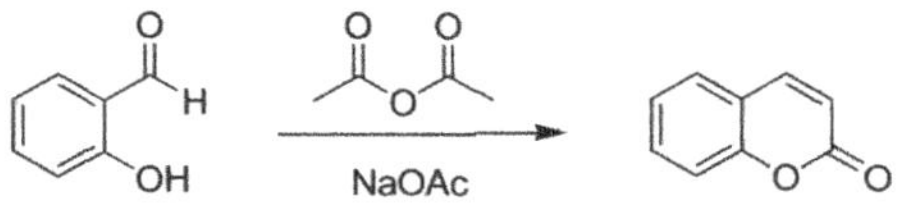

Scheme 4.2.3.4.1 Perkin coumarin synthesis.

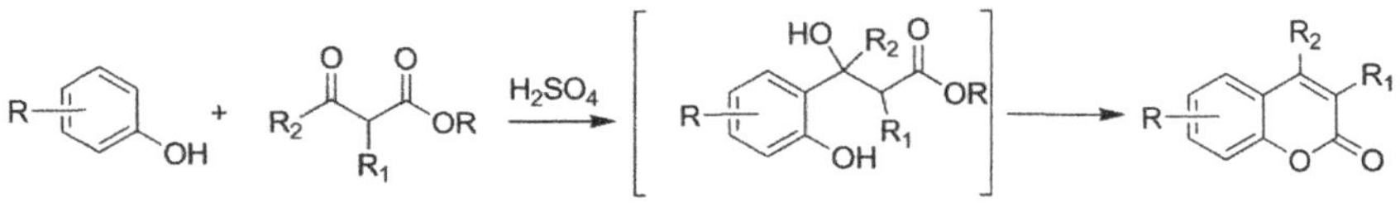

Scheme 4.2.3.4.2 Pechmann condensation.

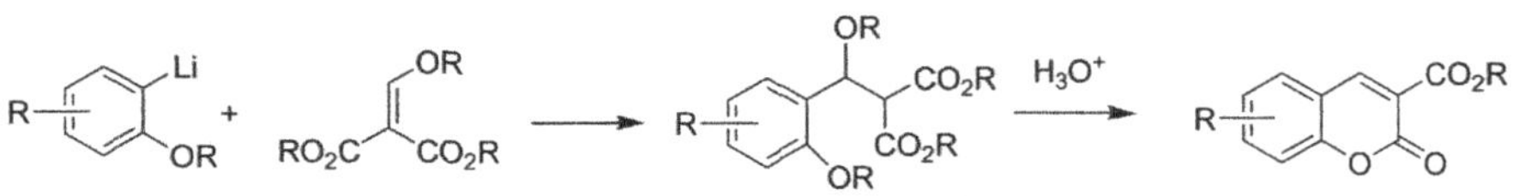

Scheme 4.2.3.4.3 Coumarin synthesis *via* Michael addition.

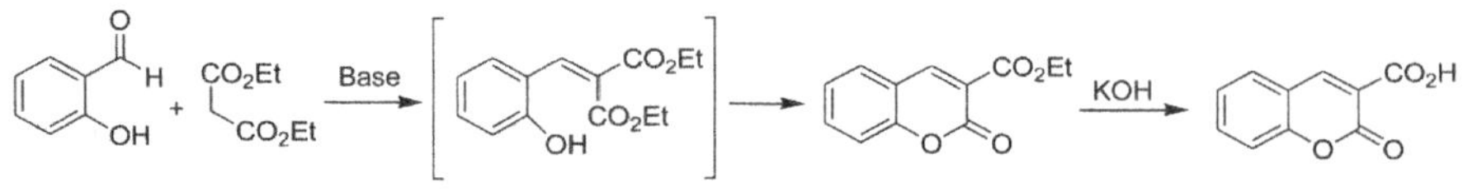

Scheme 4.2.3.4.4 Knoevenagel condensation of salicylaldehyde with diethyl malonate.

Additional Safety

This session should be done in a fume hood. HCl and KOH are very corrosive compounds, handle them with caution. Salicylaldehyde is irritant to the skin, eyes and respiratory tract and has a strong bitter almond odor. Avoid any contact or inhalation. Diethyl malonate can be an irritant. Piperidine is highly flammable and contact with skin or eye can cause burns. The students must handle all compounds in the fume hood and use adequate protection.

Experimental Procedure

Laboratory Session 1 (2 h): Synthesis of the Coumarin

1. Add salicylaldehyde (1.0 g, 8.2 mmol), diethyl malonate (1.5 g, 9.4 mmol), pure ethanol (5 mL), piperidine (0.1 mL) and one drop of acetic acid to a round-bottom flask (25 mL). Add also one or two boiling stones and attach a reflux condenser.
2. Attach the water tubing to the condenser ensuring that water goes in at the bottom and out at the top. Turn on the water so that you have a slow but steady stream of water passing through the condenser.
3. Place a heating mantle below the flask and attach a $CaCl_2$ tube to the top of the condenser. Heat the reaction mixture under reflux for 1 h.
4. Transfer a drop of the reaction mixture to a vial, dilute it with diethyl ether (1 mL) and evaluate the reaction progress by TLC using 60% diethyl ether/petroleum ether (40–60 °C) as eluent.
5. Remove the condenser and allow the flask to cool down, add cold water (10 mL) to the flask and cool it in an ice bath.
6. Assemble a vacuum filtration apparatus using a Büchner funnel. Wet the filter paper with a small amount of cold water, filter the crystals and wash them twice with a cold water/ethanol solution (50/50) (2 × 20 mL).
7. Collect the product and allow it to dry overnight in an oven. Weigh the product.

Laboratory Session 2 (2 h): Hydrolysis of the Coumarin

1. Add the coumarin ester (1.0 g, 4.6 mmol), ethanol (10 mL), water (4 mL) and KOH (1.0 g, 17.8 mmol) to a 50 mL round-bottom flask. Add also one or two boiling stones.
2. Place a heating mantle below the flask and heat the solution to its boiling point until a clear solution is obtained (after around 30 min).
3. Add HCl (2 M, 10 mL) to a 100 mL Erlenmeyer flask, insert a magnetic stirrer and place it in a stirring plate. Add slowly the hot reaction mixture with strong stirring. Observe the formation of the acid and allow the mixture to cool in an ice bath.
4. Wet a filter paper with a small amount of water and filter the solid using a Büchner funnel. Use small amounts of the cold water to help remove any residual product from the flask and to wash the crystals.
5. Transfer a crystal of the product to a vial, dilute it with CH_2Cl_2 (1 mL) and confirm, by TLC, that the reaction is complete using 80% diethyl ether/petroleum ether as eluent.
6. Collect the crystals and dry the product overnight in an oven. Weigh the product.
7. Measure the melting point of both products and calculate the overall yield of this transformation.

Waste Disposal

The washing solutions can be diluted and discarded in the sink. The eluent should be collected in an organic solvent waste container.

Results Interpretation and Additional Questions

1. Propose a mechanism for the Knoevenagel condensation that accounts for the use of the piperidine.
2. Explain the use of $CaCl_2$ at the top of the condenser.
3. Propose a mechanism for the basic hydrolysis of the coumarin ester and justify the use of HCl at the end.
4. Explain the differences between the IR spectrum of the salicylaldehyde and that of the final product.
5. Interpret the NMR data of the coumarin ester and the acid.
6. Calculate the atom economy of this two-step transformation.

References

1. W. H. Perkin, *J. Chem. Soc.*, 1868, **21**, 181; J. R. Johnson, *Org. React.*, 1942, **1**, 210; T. Rosen, *Comput. Org. Syn.*, 1991, 2, 395; T. Eicher and S. Hauptmann, *The Chemistry of Heterocycles*, Thieme, 1995, p. 249.
2. H. V. Pechmann, *Ber. Dtsch. Chem. Ges.*, 1884, **17**, 929; B. P. Mundy, M. G. Ellerd and F. G. Favaloro, *Name reactions and reagents in organic synthesis*, 2nd edn, 2005, John Wiley & Sons, p. 490; E. H. Woodruff, *Org. Synth., Coll.*, 1955, 3, 581.
3. G. A. Kraus and J. O. Pezzanite, *J. Org. Chem.*, 1979, **44**, 2480.
4. F. Bigi, L. Chesini, R. Maggi and G. Sartori, *J. Org. Chem.*, 1999, **64**, 1033; S. B. Phadtare and G. S. Shankarling, *Environ. Chem. Lett.*, 2012, **10**, 363.
5. D. Bogdal, *J. Chem. Res.*, 1998, **8**, 468.
6. E. C. Horning, M. G. Horning and D. A. Dimmig, *Org. Synth.*, 1948, **28**, 24.

4.2.3.5. Synthesis of Lipophilic Antioxidants Based on Natural Models

Sofia Benfeito[a], Tiago Silva[a], Diogo Magalhães Silva[a], E. Manuela Garrido[a,b], Fernanda Borges*[a], and Jorge Garrido*[a,b]

[a]CIQUP/Department of Chemistry and Biochemistry, Faculty of Sciences, University of Porto, Porto, Portugal; [b]Department of Chemical Engineering, School of Engineering (ISEP), Polytechnic of Porto, Porto, Portugal. *E-mail: fborges@fc.up.pt, jjg@isep.ipp.pt

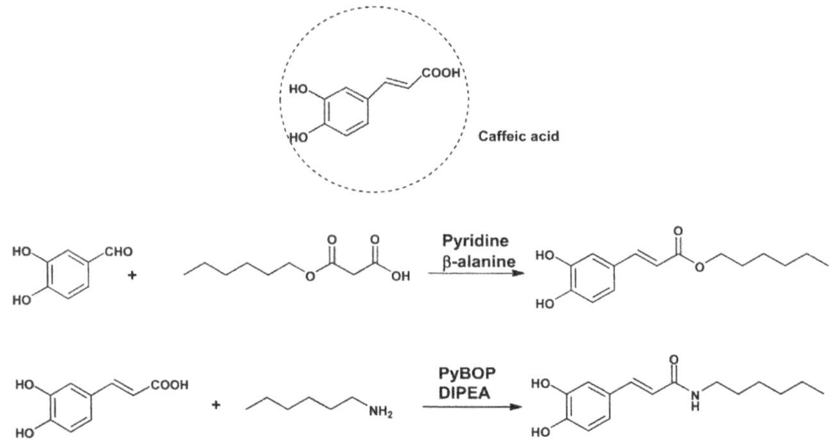

Number of sessions (duration of each session)	Hazard level	Difficulty level	Level of study
4 (6 h + 6 h + 6 h + 1 h)	Moderate	Medium	Intermediate

Class names Aldehydes, alcohols, carboxylic acids, amides

Concepts involved Medicinal and organic chemistry, formation of malonic acid half ester *via* Meldrum's acid, Knoevenagel condensation, nucleophilicity, purification and identification, lipophilicity, antioxidant activity

Chemicals needed Hexan-1-ol, 2,2-dimethyl-1,3-dioxane-4,6-dione (Meldrum's acid), toluene, 3,4-dihydroxybenzaldehyde, 4-hydroxy-3-methoxybenzaldehyde, caffeic acid, pyridine, β-alanine, (benzotriazol-1-yl-oxy) tripyrrolidinophosphonium hexafluorophosphate (PyBOP), *N,N*-diisopropylethylamine (DIPEA), hexylamine, sodium hydrogen carbonate, anhydrous sodium sulfate, cyclohexane, dimethylformamide (DMF), distilled water, diethyl ether, dichloromethane, methanol, ethanol, brine, HCl (1 mol L^{-1}), ethyl acetate, petroleum ether (40–60 °C), silica gel, ethanol, 2,2-diphenyl-1-picrylhydrazyl (DPPH), MeOD and DMSO-d_6

Equipment and experimental techniques involved Stirring and heating plate, UV/vis spectrophotometer, Rotavapor, ultrasound bath, liquid–liquid extraction, filtration, thin-layer and column chromatography, NMR spectroscopy

Keywords Antioxidant assay, flash chromatography, hydroxycinnamic acid derivatives, lipophilicity, NMR

Comprehensive Organic Chemistry Experiments for the Laboratory Classroom
Edited by Carlos A. M. Afonso, Nuno R. Candeias, Dulce Pereira Simão, Alexandre F. Trindade, Jaime A. S. Coelho, Bin Tan, and Robert Franzén.

Published by the Royal Society of Chemistry, www.rsc.org

Background

Phenolic acids, especially hydroxycinnamic acids (HCAs), are a well-known group of natural hydrophilic compounds, which are ubiquitously distributed throughout human diet, namely in fruits, vegetables and spices. HCAs are of particular interest for the development of antioxidants, anti-inflammatory and/or antimicrobial drugs and as UV filters. Currently, they are regarded as potential therapeutic solutions for oxidative-stress related diseases.[1-3] Oxidative stress is best defined in terms of redox deregulation and it is due to an imbalance between the levels of endogenous antioxidants and reactive species, which are inevitably produced as a consequence of aerobic metabolism. In physiological circumstances, reactive species (RS) are essential for cell signaling and immune response, but under pathological conditions they can overcome the toxicity threshold and induce deleterious effects in vital cell structures such as lipids, proteins and nucleic acids.[4] Exogenous antioxidants, such as dietary antioxidants, are substances that can protect cells from the damage caused by an excess RS and therefore could be useful for prevention or treatment of oxidative stress-related diseases. Several results evidence that HCAs antioxidant activity is strongly dependent on their structural features and intrinsically related to the presence of hydroxyl function(s) in the aromatic ring.[5,6] However, the high hydrophilicity of HCAs makes them poorly soluble in aprotic media, which significantly hinders their clinical potential. Most natural antioxidants studied so far have had limited success in human trials, a fact that could be related with their poor bioavailability. In this context, we propose a set of experiments by which the lipophilicity of a naturally occurring HCA (caffeic acid, Figure 4.2.3.5.1) is increased without a significant change of its antioxidant properties. The wide set of feasible chemical modifications of the carboxylic acid function provides an ideal chemical background for the endeavour. An increase in lipophilicity can be easily achieved by the preparation of alkyl esters or amides if the length of the alkyl tether is appropriate for the intended purpose. Previous work from our group[7-10] has shown that the introduction of a six-carbon chain significantly improves lipophilicity, providing derivatives with the desired bioavailability features.

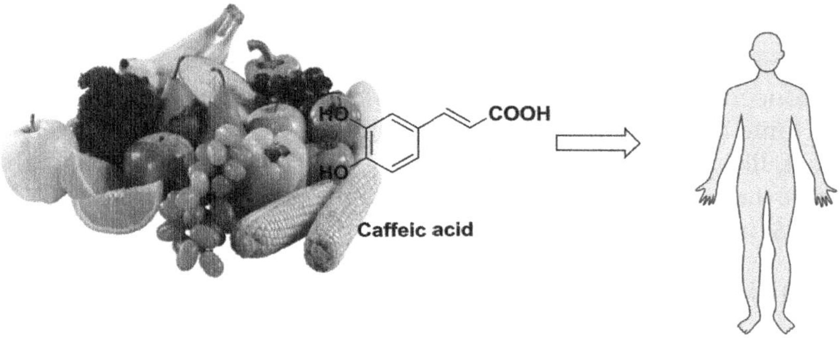

Figure 4.2.3.5.1 Structure of the dietary antioxidant caffeic acid.

The experiments herein described are related to the structural modification of caffeic acid (Figure 4.2.3.5.1), mainly the synthesis of two lipophilic derivatives – hexyl caffeate and the *N*-hexylcarboxamide derivative. The versatility of the experiments allows the instructor to propose the use of a diversity of starting materials in the laboratory classes.

The synthesis of hexyl caffeate is performed in two steps. In step 1 a β-keto acid derivative is obtained using Meldrum's acid (2,2-dimethyl-1,3-dioxane-4,6-dione) and hexan-1-ol (Scheme 4.2.3.5.1). In the second step, the synthesized hexyl malonate reacts with

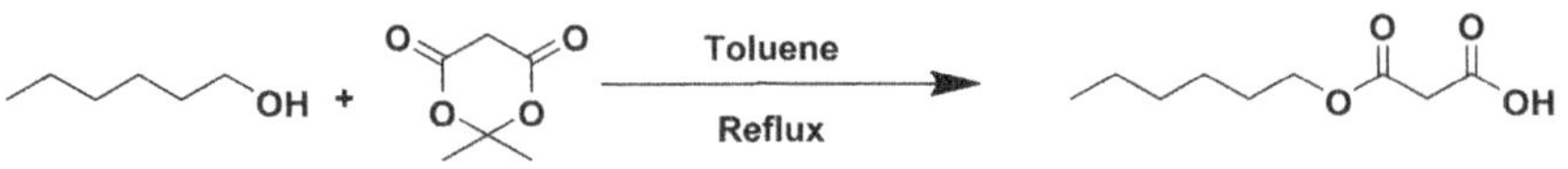

Scheme 4.2.3.5.1 Synthesis of 3-(hexyloxy)-3-oxopropanoic acid.

3,4-dihydroxybenzaldehyde derivatives in the presence of dry pyridine and β-alanine, by a Verley–Doebner modification of the Knoevenagel condensation reaction (Scheme 4.2.3.5.2). The *N*-hexylcarboxamide is synthesized in a one-pot reaction by which caffeic acid is activated with (benzotriazol-1-yloxy)tripyrrolidinophosphonium hexafluorophosphate (PyBOP) followed by addition of hexylamine (Scheme 4.2.3.5.3).

In addition, the partition coefficients and antioxidant properties of the synthesized compounds are assessed and compared with the corresponding acid precursor.

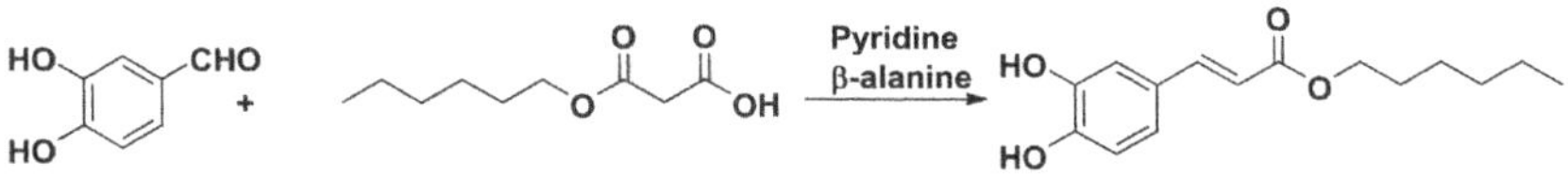

Scheme 4.2.3.5.2 Synthesis of hexyl caffeate.

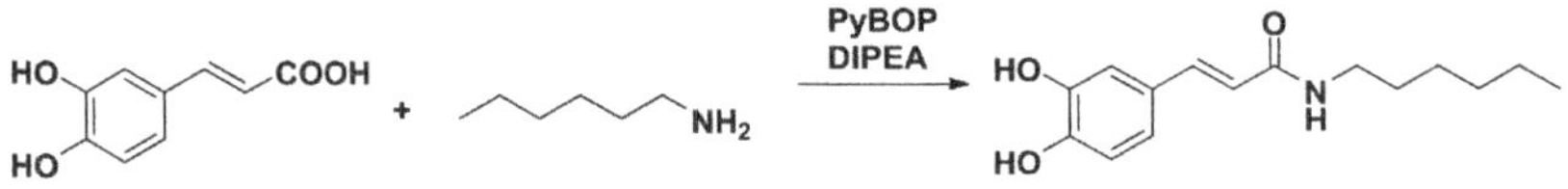

Scheme 4.2.3.5.3 Synthesis of caffeoylhexylamide.

Additional Safety

Overall the experimental procedures bring no additional concerns besides good practice. Special precaution concerning the handling of aromatic amines, such as pyridine, must be taken. Work inside the fume hood and wear appropriate protective eyeglasses and other personal protective equipment, like protective gloves and clothing. Further reading of the toxicological and hazard profiles of the handled chemicals by consulting the Merck Index and Material Safety Data Sheets (MSDS) is strongly advised.

Experimental Procedure

Session 1 (6 h): Synthesis of 3-(Hexyloxy)-3-oxopropanoic acid and of Hexyl (*E*)-3-(3,4-dihydroxyphenyl)acrylate

Synthesis of 3-(Hexyloxy)-3-oxopropanoic Acid

Synthesis

1. In a round-bottom flask add hexan-1-ol (10 mmol), Meldrum's acid (10 mmol), toluene (4 mL) and a magnetic bar.
2. Reflux the mixture for 4 h.
3. Monitor the process by thin-layer chromatography (silica gel and petroleum ether/ethyl acetate (6:4)).

Work-up

1. Cool the reaction mixture to room temperature and then add saturated $NaHCO_3$ solution (10 mL).
2. Neutralize the aqueous layer with HCl (1 mol L^{-1}) to yield the half ester of malonic acid and extract with diethyl ether (3 × 10 mL).
3. Combine the organic phases and wash them with water (3 × 10 mL).
4. Dry the organic solution over anhydrous sodium sulfate, filter and evaporate the solvent. The compound obtained can be used in the following reaction without further purification.

Synthesis of Hexyl (E)-3-(3,4-dihydroxyphenyl)acrylate

Synthesis

1. In a round-bottom flask add the half ester of malonic acid (1.0 mmol), the 3,4-dihydroxy-benzaldehyde (1.0 mmol) and cyclohexane (3 mL).
2. Stir the mixture for 5 min, and then add dry pyridine (1.0 mmol) and β-alanine (1.6 mmol).
3. Reflux the mixture for 6 h.
4. Monitor the process by thin-layer chromatography (silica gel and petroleum ether/ethyl acetate (6 : 4)).

Work-up

1. Remove the flask from the heating plate and cool the reaction mixture on an ice bath.
2. Neutralize with HCl (1 mol L^{-1}), extract with diethyl ether (3 × 10 mL). Combine the organic phases and wash them with water (3 × 10 mL).
3. Dry the organic layer over anhydrous sodium sulfate, filter and evaporate the solvent under reduced pressure.
4. The crude product is purified by flash chromatography using silica gel as stationary phase and petroleum ether/ethyl acetate (6 : 4) as eluent. Collect all fractions with the desired product in a round-bottom flask and evaporate the solvent until dryness. Control the purity of the compound by thin-layer chromatography (silica gel and petroleum ether/ethyl acetate (6 : 4)) and prepare a NMR sample in deuterated methanol (MeOD).

Session 2 (6 h): Purification and Characterization of Hexyl-(E)-3-(3,4-dihydroxyphenyl)acrylate (continuation of session 1) and Synthesis of (E)-3-(3,4-Dihydroxyphenyl)-N-hexylacrylamide

Synthesis

1. In a round-bottom flask add caffeic acid (1 mmol), DIPEA (1 mmol) and DMF (2 mL).
2. Place the flask on ice and add PyBOP (1 mmol) in dichloromethane (2 mL) to the mixture, keeping it stirred at 0 °C for 30 min.
3. Then add hexylamine (1 mmol) and allow the mixture to reach room temperature for an additional period of 6 h.
4. Monitor the reaction progress by TLC (silica gel, dichloromethane/methanol (95: 5)).

Work-up

1. Dilute the mixture with dichloromethane (30 mL) and wash it with water (3 × 10 mL), HCl (1 mol L^{-1}) (3 × 10 mL), $NaHCO_3$ 5% (3 × 10 mL) and brine (2 × 10 mL).

2. Dry the organic layer with anhydrous sodium sulfate, and remove the solvent under reduced pressure.
3. Purify the crude residue by flash-column chromatography (silica gel, dichloromethane/methanol (95 : 5)). Collect the fractions with the desired product and evaporate the solvent to dryness. A light brown solid is obtained. Control the purity of the compound and prepare a NMR sample in d_6-DMSO.

Session 3 (6 h): Purification and Characterization of (E)-3-(3,4-Dihydroxyphenyl)-N-hexylacrylamide (Continuation of Session 2) and Evaluation of Antioxidant Activity by DPPH Radical Assay

One of the most popular and simple spectrophotometric methods to measure the ability of antioxidants to trap free radicals is the DPPH (2,2-diphenyl-1-picrylhydrazyl) radical assay.

Method Fundamentals

The antioxidant activity is often evaluated using the 2,2-diphenyl-2-picrylhydrazyl (DPPH) method, which is often used to determine the radical-scavenging abilities of antioxidant compounds.[11] This method is based on the ability of the antioxidant (AH, Figure 4.2.3.5.2A) to reduce 2,2-diphenyl-2-picrylhydrazyl radical (DPPH$^{\bullet}$) either by electron- or H-donating mechanisms (Figure 4.2.3.5.2A). The radical has a maximum of absorption at 515 nm and its neutralization by an antioxidant can be easily monitored by UV/vis spectroscopy. In the presence of a compound with antioxidant activity, the radical is reduced and a decay in the absorbance at 515 nm is observed. From the data acquired along the assay, the concentration at which 50% of the radical is neutralized (IC_{50}) is easily calculated by plotting the % inhibition *versus* the concentration of the antioxidant. This parameter is then used to compare the antioxidant activity between the synthesized compounds and reference antioxidants, such as vitamin E and trolox (a water-soluble vitamin E analogue) (Figure 4.2.3.5.2B).

Protocol

The radical-scavenging activity is spectrophotometrically evaluated by monitoring the disappearance of DPPH$^{\bullet}$ at 515 nm at room temperature. The following concentrated stock solutions will be used: 1.27 mmol L^{-1} DPPH solution (0.5 mg mL^{-1}) and individual solutions of the

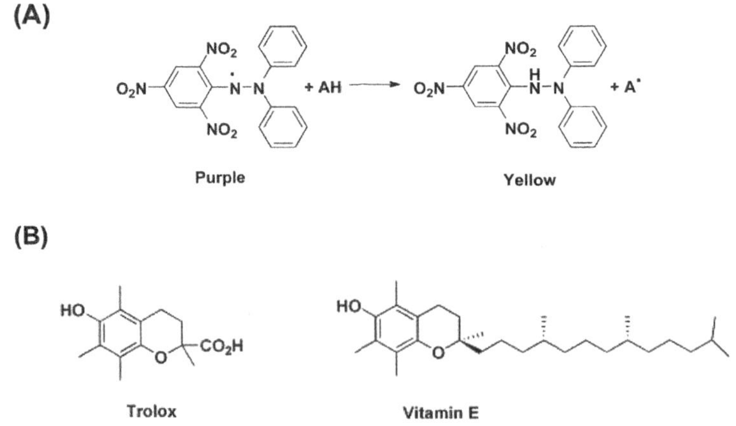

Figure 4.2.3.5.2 (A) Neutralization of the DPPH$^{\bullet}$ radical by an antioxidant (AH). (B) Structure of reference antioxidants.

antioxidants at concentrations ranging from 0.4 to 20 mmol L^{-1} in ethanol. In test tubes conveniently labelled, add 0.1 mL aliquots of ethanolic solutions of the antioxidants to 3.9 mL of the DPPH ethanolic solution. The control solution consists of 0.1 mL of ethanol added to 3.9 mL of the DPPH solution. Gently shake the tubes containing the mixtures and keep them in the dark for 30 min at room temperature. Measure the absorption of DPPH solution (control solution, $A_{control}$) and of each of the prepared samples containing antioxidants (A_{sample}) at 515 nm.

Calculate the % inhibition for each antioxidant and the IC_{50}, as the minimum concentration of each antioxidant required to inhibit 50% of the DPPH radical.

Session 4 (1 h): Theoretical Determination of Partition Coefficients

The lipophilicity of a compound refers to its tendency to partition between lipophilic organic phase (immiscible with water) and polar aqueous phase, and its value is most commonly described as the logarithm of the partition coefficient (*P*) (log *P*) between these two phases. Up to a certain limit, compounds with higher lipophilicity have higher permeation across biological membranes (but lower aqueous solubility). This parameter can be directly modulated *via* chemical structural refinement to improve the odds of obtaining drugs with enhanced bioactivity.

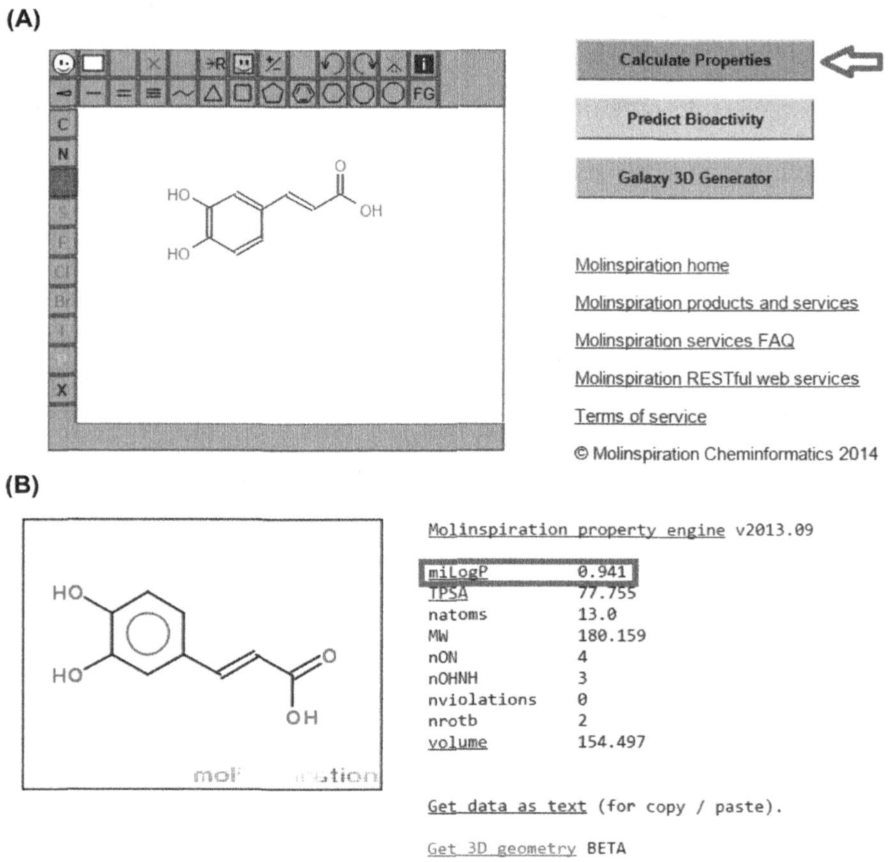

Figure 4.2.3.5.3 (A) Software online platform. Draw the molecule using the available commands in the upper and left-hand sides of the square, then select the option *Calculate Properties* (arrow on top right-hand side). (Molinspiration Cheminformatics Software®.) (B) Drug-like properties of the selected compound. (Molinspiration Cheminformatics Software®.)

The lipophilicity can be evaluated using experimental and theoretical methods. The classic method for measuring this physicochemical parameter is the flask-shaking method,[12,13] which involves the determination of the levels of the test compound in an *n*-octanol/water mixture by HPLC or UV/vis spectroscopy. However, these methods are time consuming and due to time constrains theoretical methods can be used to obtain reliable information. Several freeware online programs like Molinspiration Cheminformatics Software® (http://www.molinspiration.com/cgi-bin/properties, Figure 4.2.3.5.3A) allow a virtual screening of lipophilicity and other important drug-like properties, including total polar surface area (TPSA), number of hydrogen bond donnors and acceptors, and the number of rotatable bonds. Together, these parameters are considered good descriptors of the bioavailability of a drug (Figure 4.2.3.5.3B).

Waste Disposal

The reaction solvents should not be thrown in the sink. Discard the water, organic layers and silica residues in the appropriate containers.

Results Interpretation and Additional Questions

1. Determine the reaction yields.
2. Assign the peaks and interpret the NMR spectra of long-chain alkyl cinnamate and *N*-hexylcarboxamide derivatives.
3. Considering the results obtained with the DPPH assay, which of the compounds is a better antioxidant?
4. Propose another type of benzaldehyde that can be used in the Knoevenagel condensation to increase the antioxidant capacity?
5. Comment on the lipophilicity results obtained for caffeic acid and its derivatives.

References

1. H. M. Cocheme and M. P. Murphy, *Curr. Opin. Invest. Drugs*, 2010, **11**, 426–431.
2. N. Razzaghi-Asl, J. Garrido, H. Khazraei, F. Borges and O. Firuzi, *Curr. Med. Chem.*, 2013, **20**, 4436–4450.
3. P. Fresco, F. Borges, C. Diniz and M. M. Marques, *Med. Res. Rev.*, 2006, **26**, 747–766.
4. D. Pietraforte and W. Malorni, *Antioxid. Redox Signaling*, 2014, **21**, 52–55.
5. M. C. Figueroa-Espinoza and P. Villeneuve, *J. Agric. Food Chem.*, 2005, **53**, 2779–2787.
6. J. S. Menezes, S. P. Kamat, J. S. Cavaleiro, A. Gaspar, J. Garrido and F. Borges, *Eur. J. Med. Chem.*, 2011, **46**, 773–777.
7. F. Mura, T. Silva, C. Castro, F. Borges, M. Zuñiga, J. Morales and C. Olea-Azar, *Free Radical Res.*, 2014, **48**, 1473–1484.
8. J. Teixeira, T. Silva, S. Benfeito, A. Gaspar, E. M. Garrido, J. Garrido and F. Borges, *Eur. J. Med. Chem.*, 2013, **62**, 289–296.
9. F. F. Roleira, C. Siquet, E. Orrù, E. M. Garrido, J. Garrido, N. Milhazes, G. Podda, F. Paiva-Martins, S. Reis, R. Carvalho, E. J. Tavares da Silva and F. Borges, *Bioorg. Med. Chem.*, 2010, **18**, 5816–5825.
10. A. Gaspar, E. M. Garrido, M. Esteves, E. Quezada, N. Milhazes, J. Garrido and F. Borges, *Eur. J. Med. Chem.*, 2009, **44**, 2092–2099.
11. W. Brand-Williams, M. E. Cuvelier and C. Berset, *LWT – Food Sci. Technol.*, 1995, **28**, 25–30.
12. D. N. Brooke, A. J. Dobbs and N. Williams, *Ecotoxicol. Environ. Saf.*, 1986, **11**, 251–260.
13. L. Ropel, L. S. Belvèze, S. K. Aki, M. A. Stadtherr and J. F. Brennecke, *Green Chem.*, 2005, **7**, 83–90.

4.2.3.6. A Simple and Ecological Preparation of a Chromene-3-carboxamide Derivative

Marta Costa[†]* and M. Fernanda Proença

Chemistry Department, University of Minho, Campus de Gualtar, 4710 Braga, Portugal
*E-mail: martafcosta@ecsaude.uminho.pt

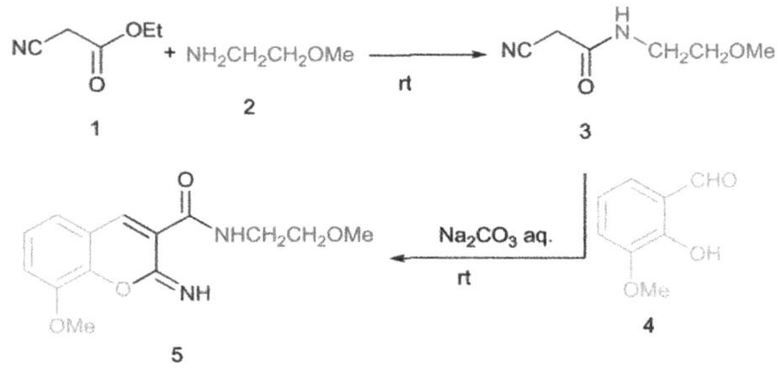

Number of sessions (duration of each session)	Hazard level	Difficulty level	Level of study
2 (2 h)	Moderate	Medium	Intermediate

Class names Esters, amines, amides, alpha-cyano acetamides, aldehydes, phenols, nitriles

Concepts involved This experiment involves: (a) conversion of an ester into an amide for the synthesis of *N*-methoxyethyl cyanoacetamide; (b) Knoevenagel condensation with 3-methoxysalicylaldehyde; (c) intramolecular nucleophilic addition leading to 2-imino-2*H*-chromene-3-carboxamide. The product is characterized by ^{1}H NMR

Chemicals needed Ethyl cyanoacetate, 2-methoxyethylamine, 3-methoxysalicylaldehyde, sodium carbonate, diethyl ether, deuterated dimethyl sulfoxide

Equipment and experimental techniques involved Apparatus for reactions with magnetic stirring, vacuum filtration

Keywords Amide formation, ^{1}H NMR, 2-imino-chromene, Knoevenagel condensation, vacuum filtration

Background

The diverse biological properties and therapeutic applications of chromene derivatives[1] make them valuable leads for the design and synthesis of new pharmacophores in early stage drug discovery. The amide bond present in 2*H*-chromene-3-carboxamides is considered an important asset, common to many bioactive molecules. Previous synthetic approaches to the preparation of these chromene derivatives include (Scheme 4.2.3.6.1):

[†]Current address: Life and Health Sciences Research Institute (ICVS), Health Sciences School, University of Minho, Braga, Portugal and ICVS/3B's, PT Government Associate Laboratory, Braga, Guimarães, Portugal.

Comprehensive Organic Chemistry Experiments for the Laboratory Classroom
Edited by Carlos A. M. Afonso, Nuno R. Candeias, Dulce Pereira Simão, Alexandre F. Trindade, Jaime A. S. Coelho, Bin Tan, and Robert Franzén.

Published by the Royal Society of Chemistry, www.rsc.org

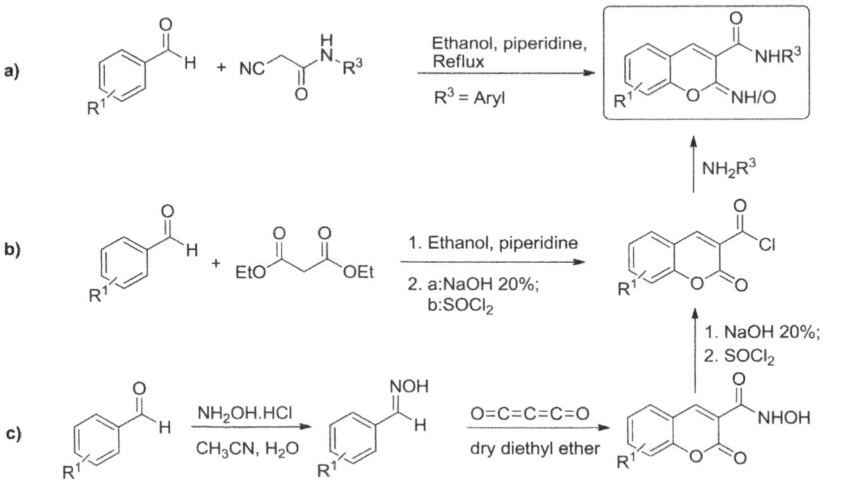

Scheme 4.2.3.6.1 Previous synthetic approaches to 2*H*-chromene-3-carboxamides. Conditions a), b) and c) are described in the text.

(a) the reaction of substituted salicylaldehydes and *N*-aryl cyanoacetamide under reflux in ethanol and piperidine;[2]

(b) the reaction of substituted salicylaldehydes with diethyl malonate, leading to ethyl 2-oxo-2*H*-chromene-3-carboxylate that is hydrolysed to the carboxylic acid, transformed into an acyl chloride and finally reacted with primary amines to generate the amide (diethyl ether, THF or chloroform were used as solvents);[1b,3]

(c) the heterocyclization of benzaldimines and carbon suboxide in diethyl ether.[4]

The search for greener synthetic methods should consider not only the reaction yield, the use of ecological solvents and the toxicity of reagents and products but also the amount of waste produced in the chemical transformation. This is usually quantified by the percent atom economy[5] as the ratio of the molecular mass of the desired product and the molecular mass of all the reagents. Briefly, even if the reaction yield is 100%, an atom economy of 50% means that only half of the mass of the reagent atoms is incorporated in the intended product. The remaining mass corresponds to undesired side-products that can be considered as waste.

This experiment demonstrates the viability of synthesizing a chromene-3-carboxamide derivative, present in diverse bioactive compounds, under simple and eco-friendly conditions. The work is initiated with the solventless synthesis of a cyanoacetamide by combining equimolar amounts of ethyl cyanoacetate and amine. The reaction is fast and quantitative and the product can be directly used for the reaction with salicylaldehyde, in mild aqueous base, at 40 °C and with reduced waste production. The products are isolated by simple filtration in a high purity form.

Additional Safety

All sessions should be done in a fume hood, with proper safety equipment, such as protective clothing, gloves and safety goggles. Ethyl cyanoacetate is harmful by inhalation, in contact with skin and if swallowed. 2-Methoxyethylamine is a highly flammable reagent and may cause burns in contact with the skin. 3-Methoxysalicylaldehyde is harmful if swallowed and is irritating to eyes, respiratory system and skin. Diethyl ether is extremely flammable, may form explosive peroxides and is harmful if swallowed. Vapours may cause drowsiness and dizziness. Ethanol is a highly flammable solvent. Sodium carbonate is irritating to the eyes.

Experimental Procedure

Laboratory Session 1 (2 h): Synthesis of the 2-Cyanoacetamide and Yield Determination

1. In a fume hood, prepare a round-bottomed flask containing a magnetic stirring bar and add 3-aminoethylamine (0.26 g, 300 μL, 3.42 mmol) to ethyl cyanoacetate (0.39 g, 375 μL, 3.44 mmol).
2. Stir the yellow oil at room temperature for 1 h.

Note: the aqueous 0.05 M Na_2CO_3 solution can be prepared while you are waiting for the reaction to occur, as you will need it for the second session.

3. Use an ice bath and your spatula to induce precipitation of the solid product. Leave the precipitate in the ice-bath for approximately 15 min.
4. Add cold diethyl ether (1 mL), filter the yellow solid by vacuum filtration and wash with a few drops of diethyl ether.
5. Collect the solid into an appropriate weighed sample tube.
6. Acquire the ^{1}H NMR spectrum and compare it with the literature data.
7. Calculate the reaction yield and atom economy.

Laboratory Session 2 (2 h): Synthesis of the 2-Imino-2*H*-chromene-3-carboxamide and Yield Determination

1. Add 3-methoxysalicylaldehyde (0.11 g, 0.72 mmol) to a round-bottomed flask equipped with a magnetic stirring bar.
2. Add 3 mL of aqueous 0.05 M Na_2CO_3.
3. Add the previously prepared 2-cyanoacetamide (0.13 g, 0.91 mmol) to the yellow suspension and stir in a water bath at 50 °C for 1 h. Cool the reaction flask in an ice-bath for approximately 5–10 min.
4. Filter off the white solid by vacuum filtration and wash the filtrate with water and a few drops of diethyl ether.
5. Transfer the solid to a weighed sample tube.
6. Acquire the ^{1}H NMR spectrum and compare it with the literature data.
7. Calculate the reaction yield and atom economy.

Waste Disposal

The neutralized aqueous solution may be diluted and discarded in the sink. Diethyl ether should be collected in a non-chlorinated waste collector recipient, for posterior destruction.

Results Interpretation and Additional Questions

1. Compare the yield of the amide and of the final product with literature data.
2. Determine the atom economy for both reactions. Compare the values with the literature data.
3. Analyse the ^{1}H NMR spectra of the amide and of the final product and compare them with literature.
4. Rationalize the mechanism involved in the formation of the 2-cyanoacetamide and in the reaction of this amide with the salicylaldehyde.

Acknowledgements

Reprinted (adapted) from *Green Chem.*, 2008, **10**, 995 with permission from the Royal Society of Chemistry.

References

1. (a) L. Abrunhosa, M. Costa, F. Areias, A. Venâncio and F. Proença, *J. Ind. Microbiol. Biotechnol.*, 2007, **34**, 787; (b) F. Chimenti, B. Bizzarri, A. Bolasco, D. Secci, P. Chimenti, S. Carradori, A. Granese, D. Rivanera, D. Lilli, A. Zicari, M. M. Scaltrito and F. Sisto, *Bioorg. Med. Chem. Lett.*, 2007, **17**, 3065; (c) I. Kostova, *Mini Rev. Med. Chem.*, 2006, **6**, 365; (d) A. Nayyar and R. Jain, *Curr. Med. Chem.*, 2005, **12**, 1873; (e) K. Fylaktakidou, D. Hadjipavlou-Litina, K. Litinas and D. Nicolaides, *Curr. Pharm. Design*, 2004, **10**, 3813; (f) K. Asres, A. Seyoum, C. Veeresham, F. Bucar and S. Gibbons, *Phytother. Res.*, 2005, **19**, 557.
2. C. Huang, F. Wu and Y. Ai, *Bioorg. Med. Chem. Lett.*, 1995, **5**, 2423.
3. S. Robert, C. Bertolla, B. Masereel, J.-M. Dogné and L. Pochet, *J. Med. Chem.*, 2008, **51**, 3077.
4. F. Chimenti, B. Bizzarri, A. Bolasco, D. Secci, P. Chimenti, S. Carradori, A. Granese, D. Rivanera, D. Lilli, M. M. Scaltrito and M. I. Brenciaglia, *Eur. J. Med. Chem.*, 2006, **41**, 208.
5. B. Trost, *Science*, 1991, **254**, 1471.

4.2.3.7. Synthesis of a Biologically Active Oxazol-5(4*H*)-One *via* Erlenmeyer–Plöchl Reaction

Catarina A. B. Rodrigues*[a], José M. G. Martinho[b], and Carlos A. M. Afonso*[a]

[a]Instituto de Investigação do Medicamento (iMed.ULisboa), Faculty of Pharmacy, University of Lisbon, Av. Prof. Gama Pinto, 1649-003, Lisboa, Portugal; [b]Centro de Química-Física Molecular and IN-Institute of Nanoscience and Nanothecnology, Instituto Superior Técnico, Av. Rovisco Pais, 1049-001 Lisboa, Portugal
*E-mail: catarinarodrigues@ff.ulisboa.pt, carlosafonso@ff.ulisboa.pt

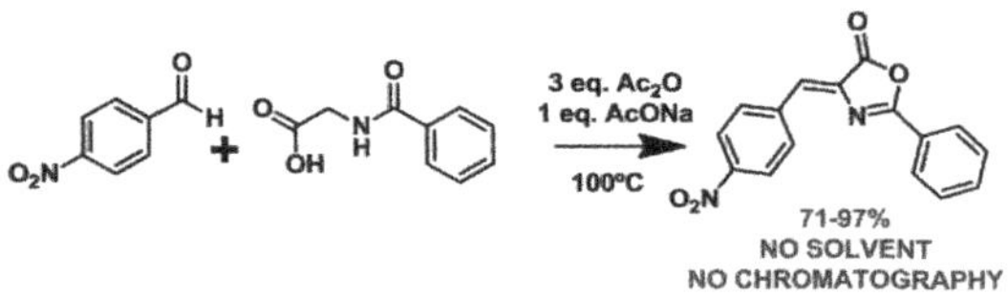

Number of sessions (duration of each session)	Hazard level	Difficulty level	Level of study
1 (3 h)	Medium	Medium	Advanced

Class names Aldehydes, carboxylic acids, amides

Concepts involved Hammett equation, environmentally friendly transformation, carbonyl chemistry, condensation, C–H acidity

Chemicals needed Sodium acetate, hippuric acid, *p*-nitrobenzaldehyde, ethanol, ethyl acetate, hexane, acetic anhydride, deuterated chloroform

Equipment and experimental techniques involved Round-bottom flask (100 or 50 mL), mortar, magnetic stir bar, spatulas, funnel for solids, graduated pipette (1 mL), heating and stirring plate, hot water bath, condenser, Büchner funnel, filter paper, Kitasato flask, weighting paper or glass, TLC plates (silica Merck Kieselgel 60 F254 plates), NMR Spectrometer (Bruker Avance 400)

Keywords Crystallization, E-factor, Erlenmeyer–Plöchl reaction, green chemistry, Hammett equation, NMR, trituration

Background

Oxazol-5(4*H*)-ones (hereafter referred as oxazolones) are small and simple molecules (see general structure presented in Scheme 4.2.3.7.1(a)) with applications in different research fields.[1–4] The presence of several reactive sites (Scheme 4.2.3.7.1(a)) allows the production of a large variety of interesting molecules with applications in chemistry and biology.[1,2] The presence of an exocyclic double bond at position 4 of the oxazolone ring (Scheme 4.2.3.7.1(b)) creates a new reactivity path that allows the construction of various structures[5,6] and creates a highly conjugated system with strong absorbance and tunable fluorescence that enables its application in photochemistry and materials sciences.[4,7,8] The 2-phenyloxazol-5(4*H*)-ones, with such an additional exocyclic double bond, possess a wide range of pharmaceutical properties such as immunomodulatory,[9] antidiabetic,[10] anti-inflammatory,[11] antibacterial[12] and antifungal[13] activity.

Comprehensive Organic Chemistry Experiments for the Laboratory Classroom
Edited by Carlos A. M. Afonso, Nuno R. Candeias, Dulce Pereira Simão, Alexandre F. Trindade, Jaime A. S. Coelho, Bin Tan, and Robert Franzén.

Published by the Royal Society of Chemistry, www.rsc.org

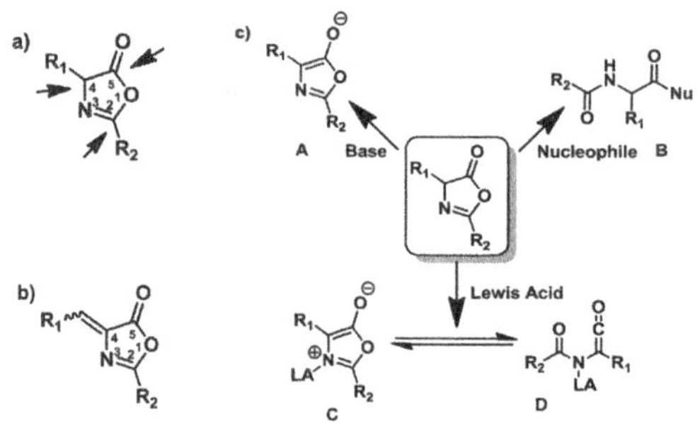

Scheme 4.2.3.7.1 General structure of oxazol-5(4*H*)-ones (a), unsaturated oxazol-5(4*H*)-one (b), and general reactivity (c). (LA = Lewis acid).

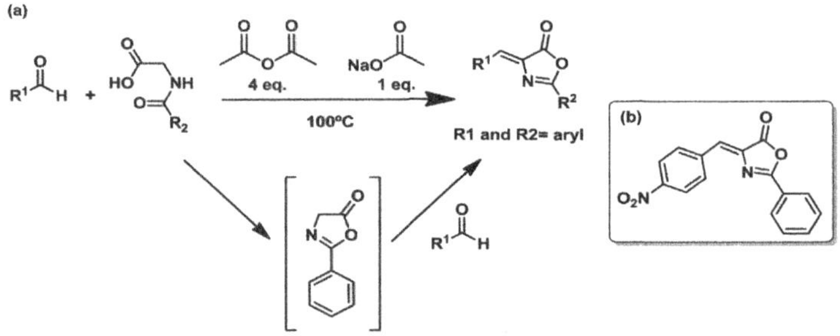

Scheme 4.2.3.7.2 (a) Erlenmeyer–Plöchl method and (b) structure of (*Z*)-4-(4-nitrobenzylidene)-2-phenyloxazol-5(4*H*)-one.

Nowadays the scientific community is seeking to apply green chemistry to their work in order to reduce or eliminate the generation of hazardous substances.[14,15] Hence, the scientific community became interested in the use of reactions in the absence of solvent with the consequent decrease of energy consumption, reaction times, chemical wastes and subsequent reduction of costs.[16,17] One method by which to measure the "environmental acceptability" of a given reaction or process is to calculate the E-factor, which quantifies the amount of waste generated per mass of product.[18]

The synthesis of 2-phenyloxazol-5(4*H*)-ones by the Erlenmeyer–Plöchl method[19] is an example of an environmentally friendly reaction that is feasible in a teaching laboratory. 2-Phenyloxazol-5(4*H*)-ones are easily prepared from hippuric acid and an aldehyde in the presence of sodium acetate and acetic anhydride with no solvent added.[20] The first step of this transformation involves the cyclization of hippuric acid to give 2-phenyl-oxazolone, followed by deprotonation and condensation with an aldehyde to yield the desired oxazolone (Scheme 4.2.3.7.2).

Herein is presented the synthesis of (*Z*)-4-(4-nitrobenzylidene)-2-phenyloxazol-5(4*H*)-one, a potent immunomodulator[9] and tyrosinase inhibitor (IC_{50} = 3.23 μM),[21] by the easy and reproducible Erlenmeyer–Plöchl method. This experiment introduces the students to carbonyl chemistry and the influence of aromatic substituents on the extent of the reaction using the Hammett equation and allows the students to engage with the environmental friendly concept of reactions with no added solvent.

Additional Safety

Gloves should be used during throughout the experiment and the manipulations should be carried out in a fume hood.

Hippuric acid, *p*-nitrobenzaldehyde, sodium acetate, ethanol, ethyl acetate and hexane are hazardous in case of ingestion, inhalation, skin contact and eye contact. Acetic anhydride is extremely hazardous in case of skin contact (irritant, corrosive, permeator), eye contact (irritant), ingestion and inhalation. Deuterated chloroform is hazardous in case of skin contact (irritant and permeator), eye contact (irritant), ingestion and inhalation. (*Z*)-4-(4-Nitrobenzylidene)-2-phenyloxazol-5(4*H*)-one is hazardous in case of skin contact (irritant), eye contact (irritant), ingestion and inhalation (irritant).

Experimental Procedure

1. In a fume hood heat a water bath on a magnetic stirrer with hot plate until it starts to boil.
2. In a mortar mix and grind very well hippuric acid (500 mg, 2.8 mmol), *p*-nitrobenzaldehyde (422 mg, 2.8 mmol) and sodium acetate (230 mg, 2.8 mmol).
3. Transfer the ground mixture to a single-necked round-bottom flask equipped with a magnetic stir bar with the help of a funnel for solids.
4. Add acetic anhydride (1 mL, 11.1 mmol) and stir the mixture until formation of a viscous mixture and homogenization.
5. Attach a condenser to the flask and heat the reaction flask in a boiling water bath for 15 min.
6. Remove the flask from the water bath and leave the reaction mixture to reach room temperature.
7. Add 10 mL of ethanol, detach the yellow solid from the flask wall with a spatula and carefully break up the bigger fragments with a rounded glass rod.
8. Stir the mixture vigorously for 15 min to dissolve the by-products.
9. Filter off the precipitate in a Büchner funnel, wash the precipitate twice with 10 mL of ethanol and dry the precipitate as much as possible on the funnel.
10. When the precipitate is almost dry transfer it to a round-bottom flask and dry it in an oven at 90 °C.
11. When a constant mass is reached determine the yield and the melting point of the product.
12. Confirm the purity of the product by TLC using as eluent a mixture of hexane and ethyl acetate (8 : 2).

Waste Disposal

Residues of ethanol and mixtures of hexane and ethyl acetate should be collected in a non-chlorinated waste collector recipient, for posterior disposal.

Results Interpretation and Additional Questions

1. Propose a mechanism for the Erlenmeyer reaction. Compare your rationalization with reported information.[22,23]
2. Analyse the following statement in the context of the proposed mechanism:

"Other aldehydes instead of *p*-nitrobenzaldehyde were used under the Erlenmeyer conditions. The aldehydes used were benzaldehyde and *p*-(dimethylamine)benzaldehyde. Under these conditions, the respective desired products were not successfully obtained."

Why is the reaction behaviour dependent on the aldehyde substituent [AldX]? Support your answer using the Hammett equation (eqn (4.2.3.7.2)) and the observed Hammett plot for *para*-substituted aldehydes (Figure 4.2.3.7.1) presented below.

Hammett equation

The Hammett equation correlates the structure and reactivity through the quantification of how the reaction rates changes when varying the aromatic substituents. The reaction constant (ρ) is experimentally determined as the slope of a Hammett plot given by eqn (4.2.3.7.1):[24–27]

$$\text{Hammett equation } \rho\sigma_X = \log\frac{k_X}{k_0} \quad (4.2.3.7.1)$$

where σ_X is the substituent constant that quantifies the electronic contribution of substituent X; ρ is a reaction constant that gives information about the mechanistic rate-determining step; k_X and k_0 are the rate constant of the reaction with the ring substituted with X and the unsubstituted ring, respectively.

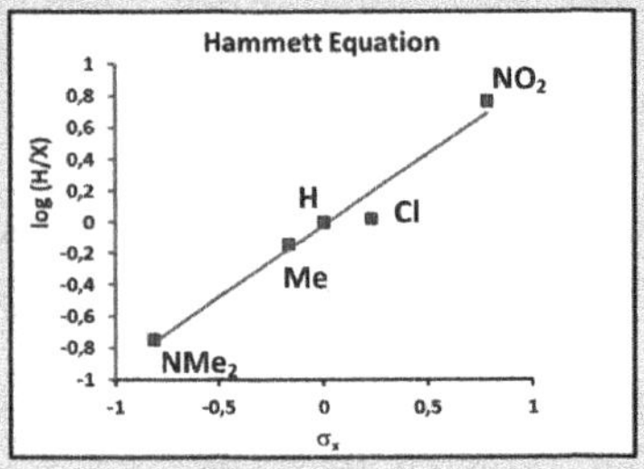

$$\log\frac{[\text{AldH}]}{[\text{AldX}]} = 0.91\,\sigma_X - 0.02 \quad (4.2.3.7.2)$$

Figure 4.2.3.7.1 Hammett plot for *para*-substituted aldehydes.

3. Does the reaction depend on the substituents of the hippuric acid? Use the observed Hammett equation (eqn (4.2.3.7.3)) and plot (Figure 4.2.3.7.2) presented bellow to support your answer.

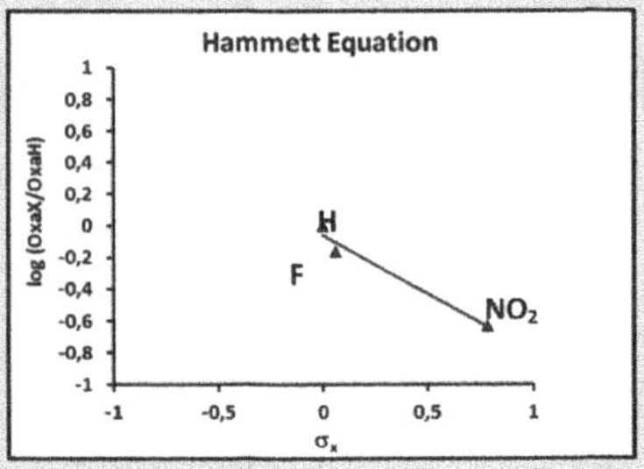

$$\log\frac{[\text{OxaX}]}{[\text{OxaH}]} = -0.74\,\sigma_X - 0.06 \quad (4.2.3.7.3)$$

Figure 4.2.3.7.2 Hammett plot for *para*-substituted hippuric acids.

4. Identify some green chemistry credits of this experiment. Determine the E-factor for this experiment and compare it with reported procedures.

The E-factor is defined as the mass ratio of waste to desired product and is calculated by the following equation:

$$\text{E factor} = \frac{\text{total of waste (g)}}{\text{total of product (g)}}$$

Acknowledgements

We acknowledge the students in the first year of the Pharmaceutical Sciences course at the Faculty of Pharmacy, University of Lisbon, Portugal. Reprinted (adapted) with permission from *J. Chem. Educ.*, 2015, **92**, 1543–1546. Copyright (2015) American Chemical Society.

References

1. J. S. Fisk, R. A. Mosey and J. J. Tepe, *Chem. Soc. Rev.*, 2007, **36**, 1432–1440.
2. R. A. Mosey, J. S. Fisk and J. J. Tepe, *Tetrahedron: Asymmetry*, 2008, **19**, 2755–2762.
3. A. N. R. Alba and R. Rios, *Chem.–Asian J.*, 2011, **6**, 720–734.
4. C. A. B. Rodrigues, I. F. A. Mariz, E. M. S. Macoas, C. A. M. Afonso and J. M. G. Martinho, *Dyes Pigm.*, 2012, **95**, 713–722.
5. L. X. Wu and K. Burgess, *J. Am. Chem. Soc.*, 2008, **130**, 4089–4096.
6. S. L. Wang, J. Ding, B. Jiang, Y. Gao and S. J. Tu, *ACS Comb. Sci.*, 2011, **13**, 572–577.
7. C. A. B. Rodrigues, I. F. A. Mariz, E. M. S. Macoas, C. A. M. Afonso and J. M. G. Martinho, *Dyes Pigm.*, 2013, **99**, 642–652.
8. S. Icli, A. O. Doroshenko, S. Alp, N. A. Abmanova, S. I. Egorova and S. T. Astley, *Spectrosc. Lett.*, 1999, **32**, 553–569.
9. M. A. Mesaik, S. Rahat, K. M. Khan, U. Zia, M. I. Choudhary, S. Murad, Z. Ismail, R. Atta ur and A. Ahmad, *Bioorg. Med. Chem.*, 2004, **12**, 2049–2057.
10. G. Mariappan, B. P. Saha, S. Datta, D. Kumar and P. K. Haldar, *J. Chem. Sci.*, 2011, **123**, 335–341.
11. H. H. Hassanein, M. K. Maha, O. N. El-Samaloty, M. A. El-Rahim, R. A. Taha and M. F. I. Magda, *Arch. Pharm. Res.*, 2008, **31**, 562–568.
12. M. Tandon, D. L. Coffen, P. Gallant, D. Keith and M. A. Ashwell, *Bioorg. Med. Chem. Lett.*, 2004, **14**, 1909–1911.
13. V. Taile, K. Hatzade, P. Gaidhane and V. Ingle, *Turkish J. Chem.*, 2009, **33**, 295–305.
14. P. Anastas and N. Eghbali, *Chem. Soc. Rev.*, 2010, **39**, 301–312.
15. P. T. Anastas and M. M. Kirchhoff, *Acc. Chem. Res.*, 2002, **35**, 686–694.
16. C. E. Ballard, *J. Chem. Educ.*, 2013, **90**, 1368–1372.
17. M. S. Singh and S. Chowdhury, *RSC Adv.*, 2012, **2**, 4547–4592.
18. R. A. Sheldon, *Green Chem.*, 2007, **9**, 1273–1283.
19. J. S. Buck and W. S. Ide, *Org. Synth.*, 1933, **13**, 8–9.
20. L. J. G. Edgar, K. J. Koroluk, M. Golmakani and A. P. Dicks, *J. Chem. Educ.*, 2014, **91**, 1040–1043.
21. K. M. Khan, U. R. Mughal, M. T. H. Khan, U. Zia, S. Perveen and M. I. Choudhary, *Bioorg. Med. Chem.*, 2006, **14**, 6027–6033.
22. F. Chavez, N. Kennedy, T. Rawalpally, R. T. Williamson and T. Cleary, *Org. Process Res. Dev.*, 2010, **14**, 579–584.
23. T. Cleary, T. Rawalpally, N. Kennedy and F. Chavez, *Tetrahedron Lett.*, 2010, **51**, 1533–1536.
24. A. Jacobs, *Understanding Organic Reaction Mechanisms*, Cambridge University Press, 1997.
25. P. S. Marrs, *J. Chem. Educ.*, 2001, **78**, 527.
26. S. L. Keenan, K. P. Peterson, K. Peterson and K. Jacobson, *J. Chem. Educ.*, 2008, **85**, 558.
27. G. K. Ikeda, K. Jang, S. O. C. Mundle and A. P. Dicks, *J. Chem. Educ.*, 2006, **83**, 1341.

4.2.3.8. Synthesis of a Long-Wavelength Absorbing Squaraine Dye

Paulo F. Santos* and Vânia C. Graça

Departamento de Química and Centro de Química – Vila Real, Universidade de Trás-os-Montes e Alto Douro, 5001-801 Vila Real, Portugal. *E-mail: psantos@utad.pt

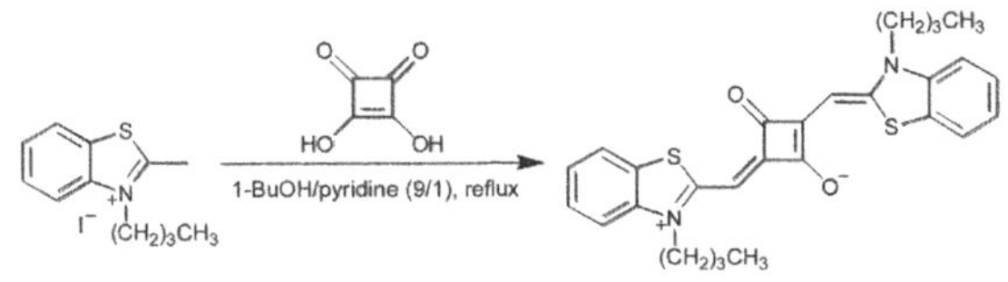

Number of sessions (duration of each session)	Hazard level	Difficulty level	Level of study
2 (4 h each)	Moderate	Medium	Advanced

Class names Enamines, α,β-unsaturated ketones

Concepts involved Nucleophilic addition/elimination reactions

Chemicals needed 3-Butyl-2-methylbenzothiazolium iodide, 3,4-dihydroxycyclobut-3-ene-1,2-dione (squaric acid), 1-butanol, pyridine, diethyl ether, dichloromethane, methanol

Equipment and experimental techniques involved Laboratory reflux apparatus with magnetic stirring, reduced pressure filtration apparatus, vacuum desiccator

Keywords Filtration under reduced pressure, NIR dyes, NMR, squaraine dyes

Background

Squaraines are 1,3-disubstituted derivatives of squaric acid (3,4-dihydroxycyclobut-3-ene-1,2-dione). They form a class of dyes with a resonance-stabilized zwitterionic structure, containing an electron-deficient central four-membered ring connected to two electron-donating groups. Although zwitterionic in nature, because of their polymethine-like structure squaraine dyes have been sometimes classified as cyanines.[1]

Squaraine dyes typically exhibit sharp and intense absorption ($\varepsilon > 10^5$ L mol^{-1} cm^{-1}) in the red and near-infrared regions of the spectrum and good photochemical stability.[2] Because of their special properties squaraine dyes have found wide technological application, mostly in the domain of photonics.[3] Recently there has been increasing interest in this class of dyes for several bio-applications, mainly as sensors, sensitizers and labels for biomolecules.[4]

The objective of this work is the synthesis of symmetrical squaraine dye **2**. Symmetrical squaraine dyes, *i.e.* those bearing identical 1,3-substituent groups, are typically synthesized by a one-pot condensation of one equivalent of squaric acid with two molar equivalents of an electron-rich aromatic or heterocyclic compound in a mixture of 1-butanol/toluene or in 1-butanol/pyridine. Dibutyl squarate (3,4-dibutoxycyclobut-3-ene-1,2-dione) has been proposed as the reactive intermediate.[5,6]

Comprehensive Organic Chemistry Experiments for the Laboratory Classroom
Edited by Carlos A. M. Afonso, Nuno R. Candeias, Dulce Pereira Simão, Alexandre F. Trindade, Jaime A. S. Coelho, Bin Tan, and Robert Franzén.

Published by the Royal Society of Chemistry, www.rsc.org

The reaction illustrates a series of nucleophilic addition/elimination reactions and the use of common structural characterization techniques of organic compounds (IR, UV/vis, NMR).

Additional Safety

All sessions should be carried out in a fume hood, with proper safety equipment, such as laboratory coat, protective mask, gloves and safety goggles. Pyridine is flammable, with a penetrating odour. It is a severe irritant and neurotoxin and is harmful if swallowed, or in contact with the skin. 1-Butanol is irritating to the eyes; repeated contact with the skin can also cause irritation. Diethyl ether is extremely flammable (flash point −45 °C) and may be explosive. When inhaled in high concentrations it has an anaesthetic effect. The high volatility of dichloromethane makes it an acute inhalation hazard. Methanol is a flammable, toxic liquid.

Experimental Procedure

Laboratory Session 1 (4 h): Synthesis of the Squaraine Dye

1. Preheat an oil bath on a magnetic hotplate stirrer.
2. Equip a 250 mL round-bottomed flask (preferably two-necked) with a magnetic stir bar and add 1.00 g (3.00 mmol, 2 equiv.) of 3-butyl-2-methylbenzothiazolium iodide, 0.17 g (1.50 mmol, 1 equiv.) of squaric acid and 100 mL of a 1-butanol/pyridine (9 : 1, v/v) mixture.
3. Insert a water condenser into the vertical ground glass joint of the round-bottomed flask and put the flask in the preheated oil bath, with vigorous stirring.
4. Follow the reaction progress by thin-layer chromatography (TLC) eluting with CH_2Cl_2/MeOH (9 : 1, v/v) (use the side ground glass joint of the round-bottomed flask to remove the aliquots from the reaction mixture). Usually, the reaction takes about 4 h to be completed.
5. Remove the flask from the oil bath and allow it to cool to room temperature. Add diethyl ether (until a slight turbidity appears) and keep the flask refrigerated until the next session.

Laboratory Session 2 (4 h): Product Isolation, Recrystallization and Yield Determination

6. Collect the crystals formed by filtration under reduced pressure in a sintered glass funnel.
7. Wash the crystals, first with water and then with cold diethyl ether.
8. Verify the purity of the crystals by TLC (CH_2Cl_2/MeOH (9 : 1, v/v)).
9. If necessary perform a recrystallization of the product obtained. Dissolve the crystals in the minimum amount of a CH_2Cl_2/MeOH (9 : 1, v/v) mixture, put the solution in ice and add diethyl ether until a slight turbidity appears.
10. Repeat the process of filtration under reduced pressure and wash the crystals with cold diethyl ether.
11. Transfer the crystals into a round-bottomed flask, previously weighed, and dry the crystals in a vacuum desiccator.
12. Weigh the dried product and calculate the reaction yield. Acquire the ^1H NMR, IR and UV/Vis spectra and determine the melting point.

Waste Disposal

Diethyl ether, dichloromethane and methanol used in recrystallization are collected in a chlorinated solvent waste collection vessel for subsequent disposal. The mixture of pyridine/1-butanol/diethyl ether resulting from the filtration of the precipitated compound is collected in a non-chlorinated solvent waste collection vessel for subsequent disposal.

Results Interpretation and Additional Questions

1. Interpret the ^{1}H NMR spectrum obtained. Compare it with the one described in the literature.[7]
2. Compare the melting point obtained with the one reported in the literature.[7]
3. Rationalize the mechanism involved in the formation of the squaraine dye.

References

1. A. Mishra, R. K. Behera, P. K. Behera, B. K. Mishra and G. B. Behera, *Chem. Rev.*, 2000, **100**, 1973.
2. A. Ajayaghosh, *Acc. Chem. Res.*, 2005, **38**, 449.
3. S. Yagi and H. Nakazumi, Heterocyclic Polymethine Dyes: Synthesis, Properties and Applications, in *Topics in Heterocyclic Chemistry*, ed. L. Strekowski, Springer-Verlag, Berlin, 1st edn, 2008, vol. 14, pp. 133–181.
4. L. Beverina and P. Salice, *Eur. J. Org. Chem.*, 2010, 1207.
5. H.-E. Sprenger and W. Ziegenbein, *Angew. Chem., Int. Ed.*, 1968, **7**, 530.
6. K.-Y. Law and F. C. Bailey, *Can. J. Chem.*, 1986, **64**, 2267.
7. M. S. Silva, V. C. Graça, L. V. Reis, P. F. Santos, P. Almeida, J. A. Queiroz and F. Sousa, *Biomed. Chromatogr.*, 2013, **27**, 1671.

4.2.3.9. Synthesis of π-Conjugated Systems Using Formylchromone as Building Block

Carlos Fernandes, Catarina Oliveira, Fernanda Borges, and Alexandra Gaspar*

CIQUP/Department of Chemistry and Biochemistry, Faculty of Sciences of University of Porto, Portugal
*E-mail: alexandra.gaspar@fc.up.pt

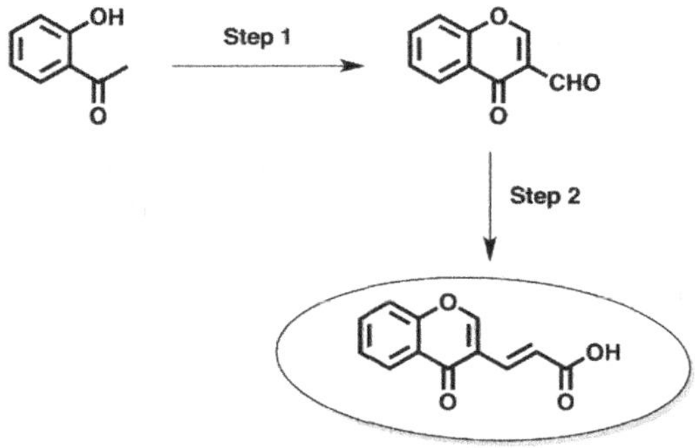

Number of sessions (duration of each session)	Hazard level	Difficulty level	Level of study
2 (4 h each)	Low	Medium	Advanced

Class names Ketones, aldehydes, carboxylic acids, conjugated and vinylic systems

Concepts involved Vilsmeier–Haack reaction, Knoevenagel condensation, ultraviolet spectroscopy, infrared spectroscopy, NMR spectroscopy

Chemicals needed *Step 1*: (a) 2-Hydroxyacetophenone, phosphorus oxychloride, anhydrous *N*,*N*-dimethylformamide; (b) potassium bromide, ethanol, deuterated dimethyl sulfoxide. *Step 2*: (a) malonic acid, pyridine, concentrated hydrogen chloride; (b) potassium bromide, ethanol, deuterated dimethyl sulfoxide

Equipment and experimental techniques involved Balance, UV lamp (254 and 366 nm), rotary evaporator, stirring and heating plate, UV-vis and IR spectrometers; filtration under reduced pressure, liquid–liquid extraction, thin-layer chromatography, crystallization

Keywords Crystallization, flash column chromatography, IR, NMR, UV-vis

Background

Since ancient times natural products, such herbs or extracts, have been used by mankind to treat several diseases, long before the advent of modern medicine. However, the industrial revolution and the expansion of organic chemistry resulted in a predisposition from the pharmaceutical industry for the use of synthetic drugs for therapeutic purposes and with well-studied pharmacological profiles. However, such an approach was not translated into the generation of new and innovative drugs and the rate at which new drugs are introduced onto the market is still falling.[1]

Comprehensive Organic Chemistry Experiments for the Laboratory Classroom
Edited by Carlos A. M. Afonso, Nuno R. Candeias, Dulce Pereira Simão, Alexandre F. Trindade, Jaime A. S. Coelho, Bin Tan, and Robert Franzén.

Published by the Royal Society of Chemistry, www.rsc.org

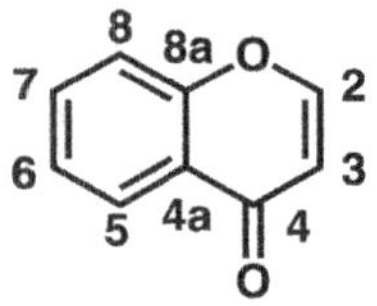

Figure 4.2.3.9.1 Structure and IUPAC numbering of chromone (4*H*-chromen-4-one, 4*H*-1-benzopyran-4-one).

Although natural products have been marginalized by major pharmaceutical companies the landscape in changing in drug discovery programs, favoring the enhancement of nature privileged structures.[2,3] Privileged structures from bioactive natural products, such as indoles, arylpiperazines, biphenyls, benzopyranes, and 1,4-dihydropyridins, can be regarded as starting points to generate drug-like compound libraries and lead compounds. In particular, benzopyran is considered to be a privileged motif for the development of new and potential therapeutic entities, due to its high chemical diversity, biochemical specificity, binding efficiency and propensity to interact with biological targets.[2,3]

Chromones (Figure 4.2.3.9.1) are a class of benzopyran-4-one heterocyclic compounds that are secondary plant metabolites widely distributed in Nature.[4,5] Over the years, chromone was used as a valid scaffold for drug discovery and development programs.[4,5] In fact, several chromone derivatives were reported as compounds possessing compelling biological properties such as anti-allergic, anti-inflammatory, anticancer, antioxidant, antiviral and antifungal activities.[4] The experiments described herein are an example of simple structural modifications of the chromone scaffold, a useful template, to attain libraries of compounds for drug discovery programs.

More recently, chromone carboxamide derivatives were described as new lead compounds for the development of monoamine oxidase-B inhibitors and adenosine receptors ligands, with potential therapeutic application in neurodegenerative and/or cancer diseases.[6,7] In this context, the chemical decoration (introduction of reactive points) of the heterocyclic framework is an important strategy to attain structural diversity and generate chemical libraries to accomplish structure–activity relationships (SARs) studies.

Herein is described a two-step strategy to obtain the chromone-3-(4-oxo-4*H*-chromen-3-yl) acrylic acid. In the first step 4-oxo-4*H*-benzopyran-3-carbaldehyde is synthesized *via* a Vilsmeier–Haack reaction from 2-hydroxyacetophenone[8] and the obtained product is used as the starting material of the second reaction. This second reaction encompasses a Knoevenagel–Doebner condensation[9] to attain the final product: 3-(4-oxo-4*H*-chromen-3-yl)acrylic acid. After the work-up, and purification of the synthesized compounds, each chromone derivative must be identified by infrared (IR) and ultraviolet/visible (UV-vis) spectroscopy. Additionally, the compounds can also be characterized by nuclear magnetic resonance (NMR) techniques. The IR technique can be also used to inspect the success of the reactions, confirming the presence (or absence) of specific functional group(s). Furthermore, analysis of the UV-Vis spectra will allow the introduction of bonding concepts in conjugated systems and lead to discussion of their properties.[10]

Additional Safety

Experimental procedures must be performed within laboratory good practice. Special precaution concerning the handling of some reagents like phosphorus oxychloride ($POCl_3$), concentrated hydrogen chloride (HCl) and aromatic amines such as pyridine must be taken. Work inside the fume hood and wear appropriate protective eyeglasses and other personal protective equipment, like protective gloves and clothing. Throughout all the experiments the operator must stay away from ignition sources. A further reading of the toxicological and hazard profiles of the handled chemicals by consulting Merck Index and Material Safety Data Sheets (MSDS) is strongly advised.

Experimental Procedure

Laboratory Session 1 (4 h): Synthesis of 4-Oxo-4*H*-chromone-3-carbaldehyde *via* Vilsmeier–Haack Reaction

1. *Synthesis*

1.1. In a two-neck round-bottom flask, add dropwise $POCl_3$ (16.6 mmol) to 10 mL of DMF.
1.2 Keep the solution at 50 °C with stirring for 15 min.
1.3. Add 2-hydroxyacetophenone (8.3 mmol) and keep the reaction, with stirring at 50 °C, for an additional 30 min.
1.4. Monitor the reaction progress by thin-layer chromatography (TLC) using silica gel as stationary phase and ethyl acetate/petroleum ether (42–62 °C) (7:3) as eluent. Visualize the TLC spots by UV detection (254 and 366 nm).

2. *Work-Up*

2.1. Pour the reaction mixture into crushed ice and extract with ethyl acetate (3 × 20 mL).
2.1. Combine the organic phases and dry the resultant mixture over anhydrous sodium sulfate.
2.3. After filtration, evapore the solvent under reduced pressure.

3. *Purification*

3.1 Crystallize the crude product from ethyl acetate/*n*-hexane. Dry the compound and check its purity by TLC.

4. *Identification of 4-Oxo-4H-chromone-3-carbaldehyde*

4.1. Prepare an ethanolic solution (10^{-4} M) of the compound and acquire the UV-vis spectrum.
4.2 Prepare a KBr pellet (incorporating approximately 1 mg of the compound into 100 mg of potassium bromide) for acquisition of the IR spectrum.
4.3 Additionally, a sample can be prepared in DMSO-d_6 to acquire the NMR spectra.

Laboratory Session 2 (4 h): Synthesis of 3-(4-Oxo-4*H*-chromen-3-yl)acrylic acid *via* the Knoevenagel–Doebner reaction

1. *Synthesis*

1.1. In a two-neck round-bottom flask, add 4-oxo-4*H*-chromone-3-carbaldehyde (1.0 mmol), malonic acid (2.0 mmol), and pyridine (5 mL) and keep the reaction at reflux for 30–45 min with vigorous stirring;
1.2. Monitor the reaction progress by TLC, using silica gel as stationary phase and dichloromethane/methanol (9:1) as eluent.
1.3. After the reaction is completed, allow the mixture to reach room temperature.

2. *Work-Up*

2.1. Add concentrated HCl (37%) dropwise until pH 1 and keep the reaction mixture under stirring for 15 min.
2.2. Filter the yellow colored solid and wash it with dilute HCl (1 M) and water.
2.3. Dry the attained solid to eliminate solvent residues.

3. Identification of 3-(4-Oxo-4H-chromen-3-yl)acrylic Acid

3.1. Proceed as described previously to acquire the UV-VIS and the IR spectra of the compound.
3.2. Additionally, a sample can be prepared in DMSO-d_6 to acquire the NMR spectra.

Waste Disposal

All organic and aqueous wastes should be placed in an appropriate waste container. The drying agent and Celite® must be placed in a solid waste container.

Results Interpretation and Additional Questions

1. Determine the reaction yields for both reactions.
2. Examine the UV-vis spectrum and determine the λ_{max} for compounds of step 1 and 2. Point out the electronic transitions that occur in the UV-vis spectrum.
3. Analyse the IR spectra for both attained compounds (steps 1 and 2); identify the characteristic IR bands of the compounds.
4. Compare the acquired spectra (steps 1 and 2).
5. Check the key features of each spectrum; correlate the observed changes with the chemical modifications.

References

1. M. S. Kinch, A. Haynesworth, S. L. Kinch and D. Hoyer, *Drug Discovery Today*, 2014, **19**, 1033.
2. K. H. Bleicher, H.-J. Böhm, K. Müller and A. I. Alanine, *Nat. Rev. Drug Discovery*, 2003, **2**, 369.
3. D. D. Baker, M. Chu, U. Oza and V. Rajgarhia, *Nat. Prod. Rep.*, 2007, **24**, 1225.
4. A. Gaspar, M. J. Matos, J. Garrido, E. Uriarte and F. Borges, *Chem. Rev.*, 2014, **114**, 4960.
5. S. K. Ko, H. J. Jang, E. Kim and S. B. Park, *Chem. Commun.*, 2006, **28**, 2962–2964.
6. A. Gaspar, T. Silva, M. Yanez, D. Vina, F. Orallo, F. Ortuso, E. Uriarte, S. Alcaro and F. Borges, *J. Med. Chem.*, 2011, **54**, 5165.
7. A. Gaspar, J. Reis, S. Kachler, S. Paoletta, E. Uriarte, K.-N. Klotz, S. Moro and F. Borges, *Biochem. Pharmacol.*, 2012, **84**, 21.
8. K. C. Rajannaa, F. Solomona, M. M. Alia and P. K. Saiprakashb, *Tetrahedron*, 1996, **52**, 3669.
9. S. Ratnadeep, P. G. Jo Shi, P. V. Mandhane and C. H. G. Badadhe, *Ultrason. Sonochem.*, 2011, **18**, 735.
10. D. L. Pavia, G. M. S. Lampman, G. S. Kriz and J. A. Vyvyan, *Introduction to Spectroscopy*, Brooks Cole, Boston, 4th edn, 2008, pp. 1–752.

4.2.3.10. A Multi-Step Synthesis of Imidazolin-5-ones

Gurunath Ramanathan* and Ashish Singh

Indian Institute of Technology Kanpur, Kanpur 208016, India. *E-mail: gurunath@iitk.ac.in

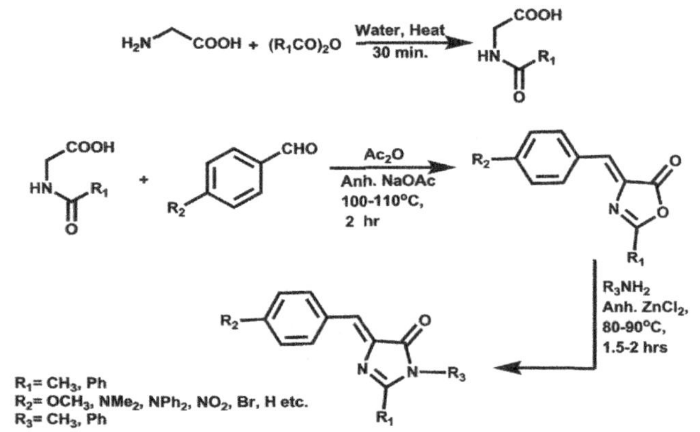

Number of sessions (duration of each session)	Hazard level	Difficulty level	Level of study
4 (3 h each)	Moderate	Medium	Advanced

Class names Aldehydes, amides, oxazolones, imidazolin-5-ones

Concepts involved This exercise trains the student in the multi-step synthesis of a heterocyclic compound. It also introduces column chromatography and NMR techniques. The molecules are fluorescent and thus the student may also study fluorescence properties

Chemicals needed Aromatic aldehyde, glycine, acetic or benzoyl or any anhydride or acid chloride, sodium acetate, methyl amine, aniline, anhydrous zinc chloride, distilled water, neutral alumina, silica gel, ethyl acetate, petroleum ether (60–80 °C), NaOH, dichloromethane, diethyl ether, HCl (conc.)

Equipment and experimental techniques involved Reflux setup, magnetic stirrer, column chromatography, UV-chamber, melting point apparatus, NMR, rotary evaporator

Keywords Column chromatography, cyclization, imidazolin-5-one, Lewis acid, NMR, oxazolone

Background

Imidazolin-5-ones are heterocyclic moieties that are present in fluorescent proteins like the green fluorescent protein (gfp).[1] These fluorescent proteins are often employed as fluorescence probes in molecular and cell biology.[2–4] The synthetically modified imidazolin-5-one also shows medicinal activity[5] and can also be used as an ion sensor.[6,7] There are several methods known for the synthesis of imidazolin-5-ones.[8–10] Tolbert *et al.* reported, in 2010, an easy method for the synthesis of a variety of imidazolinones,[10] but the use of an expensive chemical (ethyl acetimidate hydrochloride) limits this method for a large class of students. Here, the given method is not only economical but also provides an overview of the handling of a multi-step organic

Comprehensive Organic Chemistry Experiments for the Laboratory Classroom
Edited by Carlos A. M. Afonso, Nuno R. Candeias, Dulce Pereira Simão, Alexandre F. Trindade, Jaime A. S. Coelho, Bin Tan, and Robert Franzén.

Published by the Royal Society of Chemistry, www.rsc.org

synthesis. The synthesis of an imidazolin-5-one is accomplished in three steps.[11,12] The first and second steps of the synthesis are discussed in many standard practical organic chemistry text books (*e.g.* ref. 13). The synthesis of acetyl/benzoyl glycine in aqueous medium is the first step. In second step, an oxazolone is prepared by heating the acetyl/benzoyl glycine with an aromatic aldehyde in the presence of NaOAc in acetic anhydride solvent. In the last step, column purified or crystallized oxazolone is heated for 2 h with a primary aromatic or aliphatic amine in the presence of a Lewis acid ($ZnCl_2$) under solvent-free conditions to obtain the desired product (imidazolin-5-one). In brief, the difficulty level ranges from low to high is covered in this three-step synthesis of imidazolin-5-one. By varying the aromatic aldehydes in step 2 and primary amine in step 3, each student can approach a different imidazolin-5-one synthesis following the same procedure. This is useful in the evaluation of students in a large class.

Additional Safety

The student should wear a face shield, goggles, laboratory coat, shoes, and gloves before setting up the reaction to minimize the risks. This experiment needs to be done in a properly ventilated fume hood. Although none of these synthesized compounds has as yet proven themselves as irritants or allergens in any scientific report, caution is still advised.

Experimental Procedure

The synthesis of imidazolin-5-one is completed in the three steps. Each step needs one laboratory session (4 h); the steps are discussed separately below.

Step 1

Acetyl or benzoyl glycine is also commercially available at a reasonable price. These compounds can be purchased and used as such. In the laboratory, this synthesis can be done as follows.

Acetyl Glycine

1. Take 3.75 g of glycine (50 mmol) in a round-bottom flask and dissolve it in water (15 mL). Stir the solution vigorously in the fume hood for 15–20 min.
2. Once a clear aqueous solution of glycine is obtained, add 9.5 mL of acetic anhydride (100 mmol) and boil the solution for 20–30 min. A clear solution eventually forms, which on cooling overnight in a refrigerator yields white crystals or precipitates of acetyl glycine.
3. Filter the crystals or precipitate in a Büchner funnel and wash it with ice-cold water to remove the acetic acid until the filtrate is at the same pH as cold water. Dry the compound in a vacuum desiccator at room temperature. Record the melting point and characterize the pure product by NMR.

Benzoyl Glycine

1. Dissolve 2.5 g of glycine (33 mmol) in 10% NaOH (25 mL) solution in a 100 mL round-bottom flask.
2. Add 4.5 mL of benzoyl chloride† (38 mmol) into the solution of glycine in five portions and vigorously stir. Add ice pieces (50 g) into the reaction mixture and acidify it with conc. HCl, which gives a precipitate of benzoyl glycine.
3. The precipitate of benzoyl glycine contains benzoic acid as an impurity. Wash this precipitate with cold water, to remove HCl, and crystallize it in hot distilled water.

†Caution: strongly lachrymatory and pungent smelling, use only in a fume hood.

4. To get extra pure benzoyl glycine, dissolve these crystals in boiling water with the addition of a small amount of decolorizing charcoal. Filter and cool the solution to obtain benzoyl glycine crystals. Dry the solid in a vacuum desiccator at room temperature. Record the melting point and NMR.

Step 2

This step deals with the synthesis of a five-membered heterocyclic oxazolone. Acetyl or benzoyl glycine obtained in step 1 is used as a starting material.

1. In a two-neck 250 mL round-bottom flask, add substituted aromatic aldehyde (10 mmol), acetyl or benzoyl glycine (13 mmol) and anhydrous sodium acetate (13 mmol) followed by acetic anhydride as solvent (40 mmol).
2. Set up the reflux condenser with water circulation in the fume hood and maintain the reaction temperature between 100 and 120 °C.‡
3. Reflux the reaction mixture with stirring for 2 h. Next cool the solution to around 50 °C and add ice pieces slowly into it and stir vigorously. A precipitation is initiated and is completed within 15–20 min.
4. Filter off the precipitate and wash it with ice-cold water in a Büchner funnel. Wash the compound with 10 mL of diethyl ether to remove any unreacted aldehyde. Dry the compound under vacuum.
5. The crude product must be purified by crystallization or by column chromatography. This is necessary before proceeding to step 3.

(i) In crystallization, put the crude product (500 mg) in boiling tube and dissolve it in 10 mL of ethyl acetate and petroleum ether in the ratio of 1:3. Leave the clear solution overnight to obtain fine crystals of oxazolone (Note: sometimes scratching or seeding might be necessary.)
(ii) In column chromatography: take a column 40 cm long and 2.5 cm in diameter, pack 1/3 of the column with silica gel (100–200 mesh size) by wet process using 5% ethyl acetate: petroleum ether solvent. Unreacted aldehyde elutes out first at less polarity (5–10% ethyl acetate). Check the TLC in 20% ethyl acetate: petroleum ether solvent system. When the aldehyde spot disappears then increase the polarity (10–20% ethyl acetate) and the pure product will elute. Evaporate the solvent and recrystallize as above. Record the melting point of the pure product and characterize it by NMR.

Step 3

Pure oxazolone obtained from step 2, should be used for the synthesis of imidazolin-5-one in step 3. This is a dry reaction (moisture sensitive), done without solvent. The final product (imidazolin-5-one) of this step is a fluorescent compound, having characteristic photophysical properties depending on the amine used.

1. Add the oxazolone (2 mmol) followed by primary amine (2.2 mmol) in a two-neck round-bottom flask. Lastly, add the catalyst $ZnCl_2$ (0.2 mmol) and connect the round-bottom flask to the cold water circulated reflux condenser under a nitrogen atmosphere. This procedure should be performed in a fume hood.
2. Lower the reaction mixture into an oil bath at 80 °C. Increase the temperature of the oil bath in 10 °C intervals, slowly, until the entire reaction mixture just melts. Continue heating at this temperature for 1 h.

‡In some cases, higher temperature may be required.

3. Cool the reaction mixture to room temperature by removing it from the oil bath.
4. Dissolve the reaction mixture in ethyl acetate (>99% pure) and wash it with distilled water 2–3 times to remove $ZnCl_2$. Subsequently, wash the organic layer with a saturated solution of sodium bicarbonate (2–3 times) to remove the inorganic impurities. This operation must be carried out with a separating funnel.
5. Dry the organic layer over anhydrous sodium or magnesium sulfate and evaporate the solvent in a rotary evaporator to obtain the crude product.
6. Purify the crude product by using column chromatography over neutral alumina in a 10% ethyl acetate–petroleum ether solvent system. Increase the polarity on the basis of impurities present in the crude product.
7. The column purified product can be further dissolved in 5 mL of a dichloromethane–petroleum ether mixture (1 : 3 ratio) and kept overnight for crystallization.
8. Record the melting point of the crystals or column purified product. In addition, also record the NMR spectrum to confirm the product formation.

Waste Disposal

Water washes during reaction workup may be discarded in the sink. Recovered ethyl acetate may be reused for the column chromatography. All the organic solvents, used for workup as well as column chromatography, can be collected in a non-chlorinated collector and disposed of suitably.

Results Interpretation and Additional Questions

1. Compare the melting point of the product with that reported in the literature. Interpret the ^1H NMR spectrum and assign the peaks to the corresponding hydrogen (including *J*-values).
2. Calculate the percentage yield of each step and also for the overall process. Compare the experimental yield with the literature yield.
3. Discuss the role of the Lewis acid $ZnCl_2$ in the imidazolinone-5-one synthesis.
4. Discuss the mechanism involved in the oxazolone and imidazolin-5-one synthesis.
5. Discuss the effect of electronic-withdrawing and -donating substitution in the aromatic ring (R_2) in the oxazolone synthesis.
6. Why is purified oxazolone needed for step 3?
7. Why is neutral alumina used for column chromatography instead of silica gel for purifying the imidazolin-5-one?

Acknowledgement

A.S. thanks UGC, India for junior and senior research fellowships (JRF & SRF). This experiment was successfully tried on several groups of advanced undergraduate and post-graduate students at our institute. We thank the Indian Institute of Technology Kanpur for permitting this. An online version of this protocol is available free of charge at http://ictwiki.iitk.ernet.in/wiki/index.php/Multi_step_synthesis_of_imidazolin-5-one.

References

1. F. H. Johnson, O. Shimomura, Y. Saiga, L. C. Gershman, G. T. Reynolds and J. R. Waters, *J. Cell. Comput. Physiol.*, 1962, **60**, 85.
2. R. Y. Tsien, *Ann. Rev. Neurosci.*, 1989, **12**, 227.

3. M. Zimmer, *Chem. Rev.*, 2002, **102**, 759.
4. J. Lippincott-Schwartz, *Science*, 2003, **300**, 87.
5. S. Moorkoth, K. K. Srinivasan, N. G. Kutty, A. Joseph and M. Naseer, *Med. Chem. Res.*, 2013, **22**, 5066.
6. L. Shi, Y. Li, Z. P. Liu, T. D. James and Y. T. Long, *Talanta*, 2012, **100**, 401.
7. X. Y. Liu, L. Shi, Z. Ding and Y. T. Long, *RSC Adv.*, 2014, **4**, 53557.
8. G. Bhattacharjya, G. Savitha and G. Ramanathan, *J. Mol. Struct.*, 2005, **752**, 98.
9. K. Y. Chen, Y. M. Cheng, C. H. Lai, C. C. Hsu, M. L. Ho, G. H. Lee and P. T. Chou, *J. Am. Chem. Soc.*, 2007, **129**, 4534.
10. A. Baldridge, J. Kowalik and L. M. Tolbert, *Synthesis*, 2010, **14**, 2424.
11. B. K. Rajbongshi and G. Ramanathan, *J. Chem. Sci.*, 2009, **121**(6), 973.
12. G. Bhattacharya, Ph.D Thesis "*From Molecular to Material: Design, Synthesis, Characterization and Possible Application of Imidazolin-5-ones*," Indian Institute of Technology, Kanpur, India, 208016, 2006.
13. *Vogel's textbook of practical organic chemistry*, ed. B. S. Furniss, A. J. Hannaford, P. W. G. Smith and A. R. Tatchell, ELBS Longman Singapore Publishers Pvt. Ltd, Singapore, 5th edn, 1996, ch. 8, p. 1155.

4.2.3.11. Baylis–Hillman Reaction Between 4-Nitrobenzaldehyde and Ethyl Acrylate

Alexandre F. Trindade*, João Ravasco, and Carlos A. M. Afonso*

Faculty of Pharmacy, University of Lisbon, Av. Prof. Gama Pinto, 1649-003 Lisboa, Portugal
*E-mail: alexandretrindade@ff.ul.pt, carlosafonso@ff.ul.pt

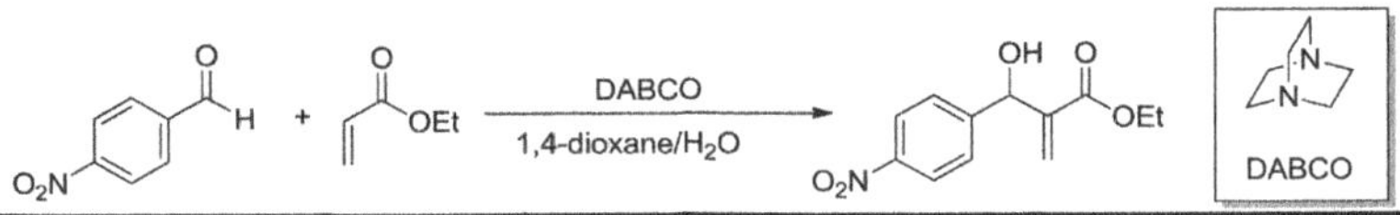

Number of sessions (duration of each session)	Hazard level	Difficulty level	Level of study
2 (4 h + 2 h)	Moderate	Medium	Advanced

Class names Aldehydes, acrylates, tertiary amines, allylic alcohols

Concepts involved The main goal of this experiment is to understand that enolate nucleophiles can be generated under near neutral conditions using a Lewis base/nucleophile and acrylates, and become involved in C–C bond formation leading to allylic alcohols. Students will be encouraged to discuss rationally the reaction mechanism based on up to date literature. They will also be asked to propose experiments of their own to study aldehyde reactivity and mechanistic aspects of the reaction

Chemicals needed 4-Nitrobenzaldehyde, ethyl acrylate, DABCO, dioxane, dichloromethane, water, diethyl ether, deuterated chloroform, anhydrous magnesium sulfate, silica-gel, Celite®

Equipment and experimental techniques involved Magnetic stirring apparatus, extraction funnel, rotary evaporator, IR and NMR spectrometers, filtrations, distillations under reduced pressure, IR and NMR spectra analysis

Keywords Acrylates, Baylis–Hillman reaction, extraction, Lewis bases, NMR, nucleophilic activation

Background

The Baylis–Hillman reaction was described in a German patent in 1972;[1] however, this type of reaction was first described in 1968 by Morita's group, where the reaction was shown to be catalysed by tricyclohexylphosphine.[2] Hence, this reaction is often called the Morita-Baylis–Hillman reaction. This transformation is highly attractive from a synthetic point of view because it affords, in one step, highly complex and functionalized products from commercially available aldehydes or imines and α,β-unsaturated carbonyl compounds with very high atom economy (Scheme 4.2.3.11.1).[3] The Baylis–Hillman or aza-Baylis–Hillman adducts obtained in this fashion will incorporate, respectively, an hydroxyl or amino group in an allylic position of α,β-unsaturated carbonyl compounds. Enones, acrylonitriles or acrylates are typically used as α,β-unsaturated carbonyl compounds.

In the past this reaction had the major inconvenience of being too slow, taking several days to reach good yields. For this reason, tremendous efforts have been made to improve its

Comprehensive Organic Chemistry Experiments for the Laboratory Classroom
Edited by Carlos A. M. Afonso, Nuno R. Candeias, Dulce Pereira Simão, Alexandre F. Trindade, Jaime A. S. Coelho, Bin Tan, and Robert Franzén.

Published by the Royal Society of Chemistry, www.rsc.org

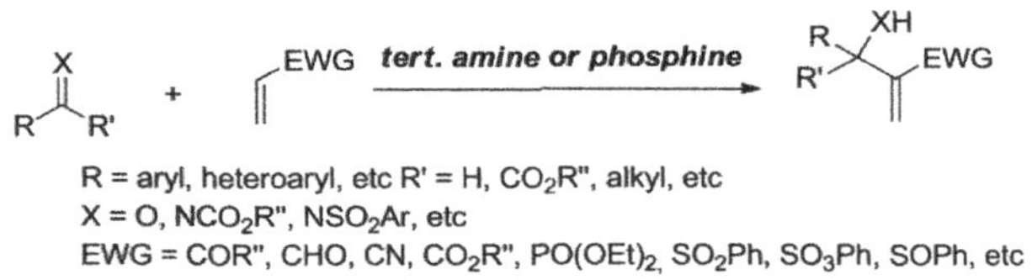

Scheme 4.2.3.11.1

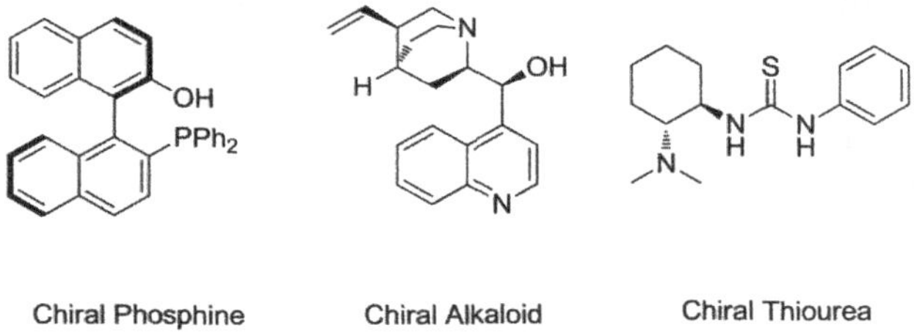

Figure 4.2.3.11.1 Some chiral inductors.

efficiency and structural diversity. This was achieved by the development of efficient catalysts that can not only enhance the reaction kinetics but also can enforce chiral induction into Baylis–Hillman or aza-Baylis–Hillman adducts. Many times it has been observed that the reaction becomes considerably faster when conducted in protic solvents or in the presence of protic additives. This information has inspired the design of the most efficient catalysts for this transformation. Chiral induction can be achieved by using enantioenriched chiral N- or P-based organocatalysts or, alternatively, achiral Lewis bases are used in combination with chiral proton sources (chiral thioureas, phenols, amino acids, *etc.*s Figure 4.2.3.11.1).[4]

Herein, a simple Baylis–Hillman reaction between 4-nitrobenzaldehyde and ethyl acrylate in dioxane/water, using DABCO as promoter, is explored.[5] For the reaction to be completed within the typical time frame of a laboratory class the Lewis base/nucleophile is employed as stoichiometric reagent. The students are encouraged to think about mechanistic details of this powerful carbon–carbon bond forming reaction.

Additional Safety

Besides regular laboratory good practices, care should be taken when handling 1,4-dioxane and diethyl ether, which are quite flammable and can contain peroxides as contaminants. Care must be taken on handling dry Celite® and silica gel, even inside a well-vented hood. All reagents should be handled with care, in particular ethyl acrylate, which should be handled in a well-vented hood. Ethyl acrylate is suspected of being carcinogenic. The Baylis–Hillman adduct isolated at the end of the reaction should also be handled as the most potentially toxic compound, as toxicity data for the adducts is unknown.

Experimental Procedure

Session 1 – Baylis–Hillman Reaction with 4-Nitrobenzaldehyde (4 h)

1. In a 50 mL round-bottomed flask add, in the following order, 4-nitrobenzaldehyde (0.75 g, 5 mmol), ethyl acrylate (1.5 g, 1.6 mL, 15 mmol), and finally, DABCO (0.56 g, 5 mmol) dissolved in 20 mL of 1,4-dioxane/water (1 : 1).
2. After 3 h evaluate the reaction progress by TLC (dichloromethane).
3. Transfer the reaction mixture to an extraction funnel, and extract with 20 mL of diethyl ether.

4 Wash sequentially the organic phase with 20 mL of 1 M HCl solution and distilled water.
5 Dry the organic phase with $MgSO_4$.

Session 2 – Filtration Through a Short Path of Silica Gel/Celite® (2 h)

6 Fill a Büchner funnel (~5 cm diameter) with one layer of silica (~0.5 cm high) followed by a layer of Celite® (~1.5 cm high). Note: please, care must be taken on handling dry Celite® and silica gel, even inside a well-vented hood.
7 Pass through the funnel 10 mL of diethyl ether to wet the silica and Celite®.
8 Filter the organic phase through the silica/Celite® directly into a 100 mL round-bottomed flask. An additional three times, pass 10 mL of diethyl ether to wash the silica gel.
9 Evaporate the solvent under reduced pressure (40 °C @ ~30 mmHg), weigh the residue and verify the purity of the isolated product by IR and ^{1}H NMR (deuterated chloroform) spectroscopy.

Waste Disposal

Diethyl ether can be recovered and reused, or it can be discarded in a non-halogenated solvents container for posterior disposal. The acid water phase from the extractions can be discarded in a container dedicated for aqueous acidic waste.

Results Interpretation and Additional Questions

1. Comment on the yield and purity determined by NMR and IR spectra.
2. Draw the mechanism of the Baylis–Hillman reaction.
3. Discuss, with support from the literature, which step is expected to be the rate-limiting step of the reaction. To answer this question find relevant information in the literature regarding the mechanism of the Morita-Baylis–Hillman reaction using appropriate channels (Scopus, PubMed, Web of Knowledge, SciFinder).
4. Is it predictable that benzaldehyde should react slower than 4-nitrobenzaldehyde? Why? How can you determine the relative reactivity of both aldehydes?
5. How could you evaluate whether the Baylis–Hillman reaction is at equilibrium with the starting substrates?

References

1. A. B. Baylis and M. E. D. Hillman, German Patent 2155113, 1972.
2. (a) K. Morita, Z. Suzuki and H. Hirose, *Bull. Chem. Soc. Jpn.*, 1968, **41**, 2815; (b) K. Morita, Japan Patent, 6803364, 1968.
3. D. Basavaiah, A. J. Rao and T. Satyanarayana, *Chem. Rev.*, 2003, **103**, 811.
4. Y. Wei and M. Shi, *Chem. Rev.*, 2013, **113**, 6659.
5. C. Yu, B. Liu and L. Hu, *J. Org. Chem.*, 2001, **66**, 5413.

4.2.3.12. Synthesis of Ethyl Mandelate Through a Rhodium-Catalysed Arylation Reaction with Ethyl Glyoxylate and Phenylboronic Acid

Carolina S. Marques* and Anthony J. Burke

University of Évora, Chemistry Department, School of Science and Technology, Centro de Química de Évora, Institute for Research and Advanced Training, Rua Romão Ramalho, 59, 7000-671 Évora, Portugal
*E-mail: carolsmarq@uevora.pt

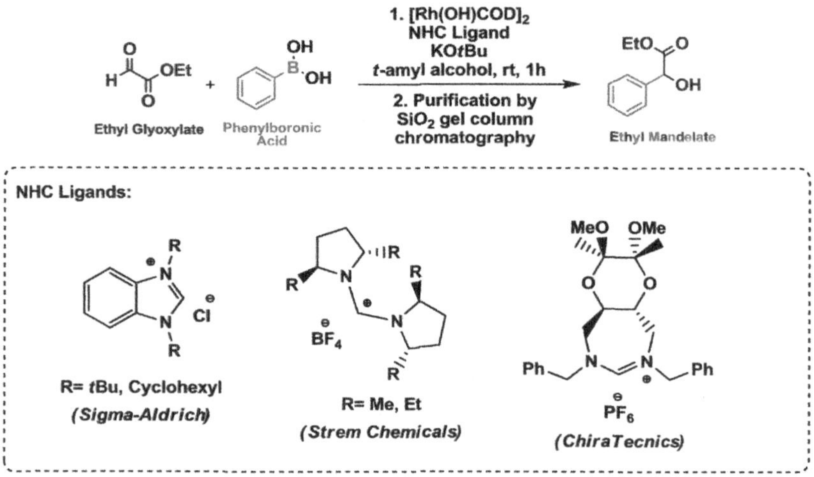

Number of sessions (duration of each session)	Hazard level	Difficulty level	Level of study
2 (2 h each)	Moderate	Medium	Advanced

Class names Aldehydes, α-hydroxyesters, boronic acids

Concepts involved This experiment involves a high yielding method of introducing a phenyl group (using phenylboronic acid as the phenyl transfer reagent) into ethyl glyoxylate, using a metal-based catalyst, giving ethyl mandelate under mild reaction conditions. N-heterocyclic carbene (NHC) ligands, which are phosphine surrogates, are used as the ligands to coordinate and active the metal for catalysis. The product was purified using silica-gel column chromatography

Chemicals needed Ethyl glyoxylate, phenylboronic acid, $[Rh(OH)COD]_2$, NHC ligand, potassium *tert*-butoxide, *tert*-amyl alcohol anhydrous, Celite®, dichloromethane (CH_2Cl_2), silica (SiO_2) gel, hexane, ethyl acetate (AcOEt), TLC plates, phosphomolybdic acid, ethanol

Equipment and experimental techniques involved Reaction vessel (either a two-neck round-bottom flask, or a Schlenk tube or other tube appropriate for catalytic reactions) with magnetic stirring; sintered glass funnel or Buckner funnel with Kitasato flask (filtration under vacuum); glass column for liquid chromatography and test tubes; structural analysis performed by NMR spectroscopy

Comprehensive Organic Chemistry Experiments for the Laboratory Classroom
Edited by Carlos A. M. Afonso, Nuno R. Candeias, Dulce Pereira Simão, Alexandre F. Trindade, Jaime A. S. Coelho, Bin Tan, and Robert Franzén.

Published by the Royal Society of Chemistry, www.rsc.org

Keywords Arylboronic acids, catalysis, ^1H and ^{13}C NMR spectroscopy, organometallic catalyst, phenylation, silica-gel column chromatography

Background

The α-hydroxyester structural function is widespread in a plethora of natural products and is a convenient building block in organic synthesis.[1] Therefore, the availability of efficient methods for accessing enantiopure α-hydroxy carbonyl compounds is of considerable current interest.[2] Several synthetic methods exist in the literature. These include the glyoxylate-ene reaction,[3] the catalytic reduction of α-ketoesters,[4] the esterification of α-hydroxy acids[5] and the benzylic ester rearrangement (BER).[6] In the past, phosphane, phosphinite and phosphate ligands have been used to give palladium based catalysts for the catalytic arylation of α-ketoesters, aldehydes and α,β-unsaturated compounds with phenylboronic acid derivatives.[7] These derivatives of borane (BH_3) are important reagents in organic chemistry enabling many chemical transformations.[8] Boronic acids, boron analogues of carboxylic acids, are not found in nature. They are derived synthetically from primary sources of boron, such as boric acid. Their unique properties of mild Lewis acidity, low reactivity profiles, good stability, wide functional group diversity and ease of handling make them an attractive class of reagents. They have been labelled as green reagents since their degradation product is environmentally friendly boric acid, and they are relatively non-toxic.[8] The use of Rh based catalysts with phosphane ligands[9] and sulfur based ligands[10] has also been successfully investigated. Ruthenium catalysts can also be used.[11]

The arylation of glyoxylate type aldehydes, employing a Pd-phosphane catalytic system for the coupling reaction of arylboronic acids and ethyl glyoxylate, was also reported.[12] Later on, we decided to evaluate this reaction, using NHC (N-heterocyclic carbene) type ligands as well as phosphane and phosphane–phosphite ligands,[13] with Rh pre-catalysts and organoboron reagents.[14] The use of NHC type ligands (cyclic carbenes) became a central topic in coordination chemistry; they form complexes with a wide variety of main transition metals in both high and low oxidation states.[15]

The objective of this experiment is the synthesis of ethyl mandelate *via* the phenylation of ethyl glyoxylate with phenylboronic acid using a Rh–NHC catalyst, under mild reaction conditions. Potassium *tert*-butoxide is used as the inorganic base in order to form the active NHC ligand, deprotonating the commercial NHC-salt, and thus forming the active Rh–NHC active catalyst. The student will learn about the importance of using NHC ligands (as opposed to the more common phosphane ligands) for key catalytic transformations, the utility of boronic acids as versatile synthetic reagents and the application of column silica-gel chromatography for product purification and NMR spectroscopy for product identification.

Additional Safety

This experiment should be performed in a well-ventilated fume hood, using standard safety equipment like protective clothing, gloves and safety glasses. Ethyl glyoxylate (50% solution in toluene), besides being a flammable reagent, can cause serious health problems if swallowed or if it comes in contact with the eyes or the skin. Potassium *tert*-butoxide is a flammable and corrosive base – avoid both skin and eye contact. *tert*-Amyl alcohol is also a flammable reagent and harmful if it comes in contact with eyes and skin. Hexane has been identified by the FDA as a potent neurotoxin and a carcinogen and thus should be treated responsibly. Solid particulate substances like Celite® and silica gel should be managed in a well-ventilated fume hood, as they can cause serious respiratory health problems.

Experimental Procedure

Laboratory Session 1 (2 h): Ethyl Mandelate – Synthesis

1. In a well-ventilated fume hood prepare a 25 mL round-bottomed flask containing a magnetic stir bar on a magnetic stirrer plate.
2. Under a nitrogen atmosphere, add to the flask 3.3 mg of $[Rh(OH)COD]_2$ (1.5 mol%, 7.34 × 10^{-3} mmol), the NHC ligand (see main scheme) (3.3 mol%, 0.015 mmol), 119 mg of phenylboronic acid (2 equiv., 0.98 mmol), 55 mg of KO^tBu (1 eq., 0.49 mmol) and 2 mL of *tert*-amyl alcohol, and then start stirring the mixture.
3. Finally, add 0.1 mL of ethyl glyoxylate (50% in toluene, 0.49 mmol) to the flask with a syringe or micropipette and stir for 1 h. Add 20 mL of CH_2Cl_2 to the flask.
4. Filtration: Method A – using a sintered glass funnel place a layer of Celite® (about 1 cm high) and cover with a silica gel layer (about 3 cm high, depending on the funnel used); then filter the crude reaction mixture under vacuum; wash it with additional 30 mL of CH_2Cl_2; then transfer the filtrate to a suitable round-bottom flask (100 mL) and evaporate the solvent using a rotary evaporator.

Method B – using a Büchner funnel (equipped with filter paper and a layer of Celite® – about 1 cm high – and a layer of silica gel – about 2 cm high, depending on the funnel used) filter the reaction mixture under vacuum; wash it with an additional 30 mL of CH_2Cl_2; then transfer the filtrate to a suitable round-bottom flask (100 mL) and evaporate the solvent using a rotary evaporator.

Laboratory Session 2 (2 h): Ethyl Mandelate – Purification

1. Fill a glass chromatography column with silica gel (stationary phase – about 15 g of silica gel) and hexane/AcOEt (5 : 1) (mobile phase) slurry prepared in a suitable sized Erlenmeyer (conical) flask. When ready (settled), add a solution of the crude impure reaction product (dissolved in the mobile phase)† at the top of the column. The eluent is then passed through the column by gravity.
2. Collect equal sized fractions of 4 mL sequentially (about 15 tubes are needed).
3. TLC (thin-layer-chromatography) analysis (eluent = hexane/AcOEt (1 : 1)) of each fraction collected should be made to identify the ethyl mandelate product. The plates are visualized either by UV light or with phosphomolybdic acid in ethanol.
4. The fractions containing the product should be put in a previously weighed round-bottom flask and the solvents removed by evaporation using a rotary evaporator.
5. Weigh the desired product and calculate the yield. Acquire the ^1H and ^{13}C NMR spectra and compare them with those in the literature.

Waste Disposal

Solid inorganic materials such as Celite® and silica gel should be collected in a proper waste collector recipient. Dichloromethane may be reused or collected in a chlorinated waste collector recipient. The mixtures of hexane/ethyl acetate should be collected in a non-chlorinated waste collector recipient.

Results Interpretation and Additional Questions

1. Interpret the ^1H and ^{13}C NMR spectra obtained.
2. Explain the importance of using liquid chromatography for the product isolation.

†If it does not dissolve well, add a bit more AcOEt and heat using a hot stir gun (with care!) or on a water-bath.

3. Explain how to select the mobile phase in liquid column chromatography based on the compound that you want to purify. Reference should be made to the most common solvents and how to increase the polarity of a solvent during the purification process.
4. Present a reasonable mechanism for this catalytic arylation reaction taking into account some relevant recent literature.[16]
5. In this reaction NHC ligands are used. List the advantages of this class of ligands over the more common classic phosphane ligands.

References

1. (a) H. F. S. G. M. Coppola, *Alpha-Hydroxy Acids in Enantioselective Synthesis*, VCH, Weinheim, 1997; (b) W. Adam, M. Lazarus, C. R. Saha-Moller and P. Schreier, *Acc. Chem. Res.*, 1999, **32**, 837; (c) L. M. West and P. T. Northcote, *J. Org. Chem.*, 2000, **65**, 445.
2. F. A. Davis and B. Chen, *Chem. Rev.*, 1992, **92**, 919.
3. (a) J. Hao, M. Hatano and K. Mikami, *Org. Lett.*, 2000, **2**, 4059; (b) D. A. Evans, C. S. Burgey, N. A. Paras, T. Vojkovsky and S. W. Tregay, *J. Am. Chem. Soc.*, 1998, **120**, 5824; (c) A. Mandoli, S. Orlandi, D. Pini and P. Salvadori, *Tetrahedron: Asymmetry*, 2004, **15**, 3233; (d) K. Mikami, *Pure Appl. Chem.*, 1996, **68**, 639.
4. (a) H.-U. Blaser, H.-P. Jalett, M. Mtiller and M. Studer, *Catal. Today*, 1997, **37**, 441; (b) K. Szöri, M. Sutyinszki, K. Felföldi and M. Bartók, *Appl. Catal., A*, 2002, **237**, 275; (c) M. Jahjah, M. Alame, S. Pellet-Rostaing and M. Lemaire, *Tetrahedron: Asymmetry*, 2007, **18**, 2305; (d) H.-U. Blaser, H.-P. Jalett and F. Spindler, *J. Mol. Catal. A: Chem.*, 1996, **107**, 85; (e) L. Yin, W. Shan, X. Jia, X. Li and A. S. C. Chan, *J. Organomet. Chem.*, 2009, **694**, 2092.
5. (a) T. A. Houston, B. L. Wilkinson and J. T. Blanchfield, *Org. Lett.*, 2004, **6**, 679; (b) T. Maki, K. Ishihara and H. Yamamoto, *Org. Lett.*, 2005, **7**, 5047.
6. (a) C. S. Marques, N. Moura and A. J. Burke, *Tetrahedron Lett.*, 2006, **47**, 6049; (b) A. J. Burke and C. S. Marques, *Mini-Rev. Org. Chem.*, 2007, **4**, 310.
7. (a) P. He, Y. Lu, C.-G. Dong and Q.-S. Hu, *Org. Lett.*, 2007, **9**, 343; (b) P. He, Y. Lu and Q.-S. Hu, *Tetrahedron Lett.*, 2007, **48**, 5283.
8. D. G. Hall, *Boronic Acids, Preparation and Applications in Organic Synthesis and Medicine*, Wiley-VCH Verlag GmbH, Weinheim, 2005.
9. H.-F. Duan, J.-H. Xie, X.-C. Qiao, L.-X. Wang and Q.-L. Zhou, *Angew. Chem., Int. Ed.*, 2008, **47**, 4351.
10. T.-S. Zhu, S.-S. Jin and M.-H. Xu, *Angew. Chem., Int. Ed.*, 2012, **51**, 780.
11. Y. Yamamoto, T. Shirai, M. Watanabe, K. Kurihara and N. Miyaura, *Molecules*, 2011, **16**, 5020.
12. I. N. Francesco, A. Wagner and F. Colobert, *Eur. J. Org. Chem.*, 2008, 5692.
13. C. S. Marques, M. Dindaroglu, H.-G. Schmalz and A. J. Burke, *RSC Adv.*, 2014, **4**, 6035.
14. (a) C. S. Marques and A. J. Burke, *Tetrahedron*, 2012, **68**, 7211; (b) C. S. Marques and A. J. Burke, *Tetrahedron: Asymmetry*, 2013, **24**, 628.
15. (a) R. H. Crabtree, *The Organometallic Chemistry of the Transition Metals*, 4th edn, Wiley-Interscience, 2005; (b) W. A. Herrmann, *Angew. Chem., Int. Ed.*, 2002, 1290.
16. (a) B. J. Truscott, G. C. Fortman, A. M. Z. Slawin and S. P. Nolan, *Org. Biomol. Chem.*, 2011, **9**, 7038; (b) I. Penafiel, I. M. Pastor, M. Yus, M. A. Esteruelas and M. Oliván, *Organometallics*, 2012, **31**, 6154; (c) Y. Yamamoto, K. Kurihara and N. Miyaura, *Angew. Chem., Int. Ed.*, 2009, **48**, 4414; (d) M. J. Burk, M. F. Gross, T. G. P. Harper, C. S. Kalberg, J. R. Lee and J. P. Martinez, *Pure Appl. Chem.*, 1996, **68**, 37; (e) D. Astruc, *Organometallic Chemistry and Catalysis*, Springer Berlin Heidelberg, 2007.

4.2.3.13. Enantioselective Synthesis and Derivatisation of 2-Hydroxy-1,2-diphenylethan-1-one

Jonathan D. Sellars*, AnnMarie C. O'Donoghue, Ian R. Baxendale, John M. Sanderson, and Elizabeth J. Grayson

Chemistry Department, University of Durham, South Road, Durham, DH1 3LE, UK
*E-mail: j.d.sellars@durham.ac.uk

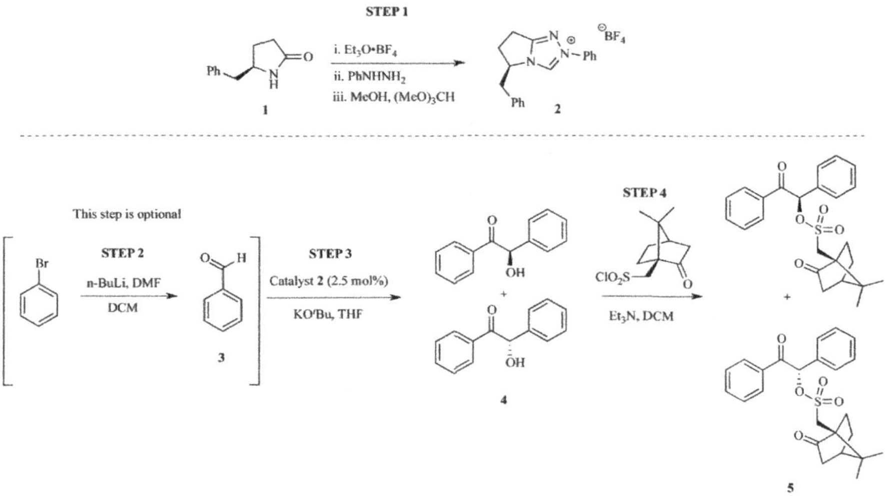

Number of sessions (duration of each session)	Hazard level	Difficulty level	Level of study
8 (3 h + 1 h + 1 h + 7 h + 7 h + 3 h + 7 h + 7 h)	Moderate	High	Advanced

Class names Triazolium salts, alkyl lithium, aldehydes, phenyl halides, lactams, N-heterocyclic carbenes, sulfonyl chlorides, α-hydroxyketones

Concepts involved The purpose of these experiments is to introduce third year undergraduate students to multi-step organic synthesis. In addition, the experiment exposes the student cohort to a wide range of practical (handling pyrophoric reagents, vacuum distillation, recrystallization, utilising dry solvents, column chromatography) and theoretical (planning experimental quantities, time management) skills, which they will utilise in their fourth year studies

Chemicals needed *Step 1*: (5*R*)-5-benzylpyrrolidinone (**1**), triethyloxonium tetrafluoroborate, phenylhydrazine, trimethyl orthoformate, MeOH, dichloromethane (DCM), EtOAc. *Step 2*: bromobenzene, *n*-BuLi (2.5 M in hexanes), *N*,*N*-dimethylformamide, dry ice, NH_4Cl (aq.), $MgSO_4$, THF, acetone, DMF, Et_2O. *Step 3*: benzaldehyde (**3**), (5*R*)-5-benzyl-2-phenyl-6,7-dihydro-5*H*-pyrrolo[2,1-*c*][1,2,4]triazol-2-ium tetrafluoroborate (**2**), KO^{t}Bu, $MgSO_4$, NH_4Cl (aq.), THF, Et_2O, *n*-hexane. *Step 4*: 2-hydroxy-1,2-diphenylethan-1-one (**4**), (1*S*)-(+)-10-camphorsulfonyl chloride, triethylamine, NH_4Cl (aq.), $MgSO_4$, silica gel, DCM, *n*-hexane, EtOAc

Comprehensive Organic Chemistry Experiments for the Laboratory Classroom
Edited by Carlos A. M. Afonso, Nuno R. Candeias, Dulce Pereira Simão, Alexandre F. Trindade, Jaime A. S. Coelho, Bin Tan, and Robert Franzén.

Published by the Royal Society of Chemistry, www.rsc.org

Equipment and experimental techniques involved Standard apparatus for low temperature (−78 °C) and room temperature magnetic stirring, liquid–liquid extraction apparatus, flash column chromatography column, distillation apparatus, high vacuum line, NMR, GC-MS, IR, chiral LC

Keywords Asymmetric reaction, benzoin condensation, chiral LC, derivatisation, flash column chromatography on silica gel, GC-MS, IR, liquid–liquid extraction, NHC carbenes, NMR, umpolung, vacuum distillation

Background

Introducing chirality to a molecule is a fundamental operation in organic synthesis. Moreover, the ability to assemble complex molecules as single enantiomers is a critical undertaking in modern organic chemistry. α-Hydroxy ketones are synthetically important building blocks in the synthesis of natural products.[1] There are a wide range of protocols available for their synthesis: the acyloin condensation involving a reductive coupling of a diester with sodium metal;[2] the oxidation of enolates or silyl enol ethers with oxaziridines or *m*-CPBA (Rubottom oxidation). The former reagent can be chiral, introducing the alcohol functionality asymmetrically;[3–6] and the benzoin condensation, which couples aldehydes together, utilises cyanide in its original guise.[7] Whilst these methods offer viable synthesis routes to α-hydroxy ketones, they are generally achiral and require the use of reactive metals, acidic environments, and toxic reagents and have poor atom economy, all of which can conceivably lead to unwanted side reactions and added expense. Notwithstanding this, new methodology has been developed to overcome these issues through the advent of chiral N-heterocyclic carbene (NHC) catalysts, which are versatile Lewis bases that have received considerable attention for their ability to reverse the polarity of certain functional groups (umpolung reactivity).[8,9] These new organocatalysts have enabled enantioselective condensation reactions to occur readily with excellent yields and resultant ee in the classical benzoin reaction.

Therefore, the objective of this experiment is to (a) synthesise a chiral NHC pre-catalyst; (b) employ this catalyst in a coupling reaction to produce a chiral benzoin product; and (c) derivatise the resultant chiral alcohol to produce diastereoisomers and determine the major enantiomer produced.

Additional Safety

All sessions should be undertaken in a fume cupboard. The Meerwein reagent (triethyloxonium tetrafluoroborate) should be handled with great care. Phenylhydrazine is toxic and may cause cancer. *n*-BuLi (2.5 M in hexanes) is highly flammable and should not be exposed to air.

Experimental Procedure

Step 1 (1 × 3 h, 2 × 1 h Sessions + 2 × Overnight)

(5R)-5-Benzyl-2-phenyl-6,7-dihydro-5H-pyrrolo[2,1-c][1,2,4]triazol-2-ium tetrafluoroborate (2)

1. In a dry 100 mL round-bottom flask under nitrogen, fitted with a septum cap and balloon, place (5*R*)-5-benzylpyrrolidinone (**1**) (0.5 g, 2.86 mmol, 1 equiv.) and add dry dichloromethane with a syringe (to make a 0.14 M solution).
2. Next add triethyloxonium tetrafluoroborate (1.1 equiv.) (*Care*! this should be weighed in a fume cupboard and used straight away) and stir the reaction mixture overnight at room temperature.

3. To this solution add phenylhydrazine (1.1 equiv.) and again stir the solution overnight at room temperature. (Your reaction can be left for longer at this stage.)
4. Remove the solvent *in vacuo* using a rotary evaporator to isolate a crude product.
5. To this crude product add methanol (to produce a 2.86 M solution) and trimethyl orthoformate (the amount to equate to 0.4 M), place a condenser on top of the reaction mixture and heat with stirring to 80 °C overnight.
6. Remove the solvent again using a rotary evaporator and precipitate the product using ethyl acetate to give an off-white/yellow powder.
7. Recrystallize your product from hot methanol to afford **2**.
8. Characterise your product (^{1}H NMR, IR, mp).

Step 2 (2 × 7 h sessions)

Benzaldehyde (3) (Optional)

NB *Do not* Exceed the Scale in this Procedure! Ensure all Equipment is Dry when Using *n*-BuLi!

1. In a dry round-bottom flask under nitrogen, fitted with a septum cap and balloon, add *via* a syringe bromobenzene (3.35 mL, 31.8 mmol, 1 equiv.) and dry THF (to produce a 1.6 M solution).
2. Cool the solution to −78 °C with a dry ice/acetone bath.
3. Add dropwise with a fresh syringe, *n*-BuLi (2.5 M in hexane, 1 equiv.). Once all the *n*-BuLi has been added, stir the solution for 30 min.
4. To the reaction solution, add dropwise with a fresh syringe dry DMF (1.1 equiv.).
5. Stir the solution for 30 min at −78 °C and then allow it to warm to room temperature and stir for a further 1 h.
6. Pour the solution into a conical flask containing saturated aqueous NH_4Cl (25 mL) and separate the organic layer.
7. Extract the aqueous layer with Et_2O (2 × 20 mL) and combine these extracts with the separated organic layer.
8. Wash the single organic phase four times with water (4 × 25 mL), dry the organic layer with $MgSO_4$, filter and evaporate the solvent using a rotary evaporator.
9. Finally, distil the crude oil at atmospheric pressure, collecting using a pig the desired product **3**.
10. Characterise your product (^{1}H NMR, IR, bp).

Step 3 (1 × 3 h session + Overnight)

2-Hydroxy-1,2-diphenylethan-1-one (benzoin) (4)

1. In a dry round bottom flask under nitrogen, fitted with a septum cap and balloon, place *via* a syringe benzaldehyde (**3**) (2.5 mL, 25 mmol, 1 equiv.) and dissolve in dry THF (to produce a 1.25 M solution).
2. Add to this solution solid catalyst (**2**) (0.025 equiv.) followed by solid KO^{t}Bu (0.025 equiv.).
3. Stir the reaction mixture overnight at room temperature then pour the suspension into an aqueous solution of saturated aqueous NH_4Cl (30 mL).
4. Extract the aqueous layer with Et_2O (3 × 30 mL), dry the organic layer over $MgSO_4$, filter and evaporate the Et_2O using a rotary evaporator.
5. Triturate the solid with hexane, filter off the solid and dry in a desiccator to provide **4**.
6. Characterise your product (^{1}H NMR, IR, mp, Chiral-LC).

Step 4 (2 × 7 h Sessions)

2-Oxo-1,2-diphenylethyl[7,7-dimethyl-2-oxobicyclo[2.2.1]heptan-1-yl] methanesulfonate (5)

1. In a dry round-bottom flask under nitrogen, fitted with a septum cap and balloon, weigh out benzoin **4** (0.25 g, 1.2 mmol, 1 equiv.) and dissolve in dry dichloromethane (to produce a 0.24 M solution).
2. Cool the resultant solution to 0 °C and add *via* a syringe Et_3N (1.5 equiv.) followed by solid (1*S*)-(+)-10-camphorsulfonyl chloride (1.1 equiv.).
3. Allow the solution to warm to room temperature and stir for 4 h.
4. Monitor the reaction by TLC and upon complete consumption of starting material add saturated aqueous NH_4Cl (10 mL) and separate the organic layer.
5. Wash the organic layer with saturated aqueous NH_4Cl (2 × 10 mL), dry the organic layer with $MgSO_4$, filter and evaporate using a rotary evaporator.
6. Dry load your sample onto silica gel and perform a chromatographic separation, using a solvent gradient of *n*-hexane/EtOAc to provide **5**.
7. Characterise your product (^{1}H NMR, IR).

Waste Disposal

All neutralised aqueous waste produced for each reaction step can be diluted and discarded in the sink. All non-chlorinated waste from each step should be collected together in the non-chlorinated waste collectors. Likewise, all halogenated waste should be collected together in the halogenated waste collectors.

Results Interpretation and Additional Questions

You will be assessed by means of a written report. The requested format of your report is designed to provide you with training for writing up experimental data in a publishable form and is the format required for students that undertake a 4th year research project. As a result we expect to receive precise and clearly laid out write-ups consisting of the following:

- Fully written experimental procedures for *all* the synthetic steps, consisting of fully recorded physical (include comparison to literature data) and spectral data for the products presented in the style specified by the Royal Society of Chemistry.
- Your original spectra (*excluding* expansions used to assign coupling constants), as PDF documents.
- The answers to the following questions:

  1. Distinguish between nucleophilic and electrophilic singlet carbenes. How are N-heterocyclic carbenes classified?
  2. Discuss the origin of catalysis of the benzoin condensation with reference to the reaction in the absence of catalyst.
  3. Based on the chiral LC of (**4**), estimate the enantiomeric excess of the benzoin condensation step.
  4. Give a mechanistic explanation for the origin of enantioselectivity in the asymmetric benzoin condensation.

References

1. K. C. Nicolaou, G. Vassilikogiannakis, K. B. Simonsen, P. S. Baran, Y.-L. Zhong, V. P. Vidali, E. N. Pitsinos and E. A. Couladouros, *J. Am. Chem. Soc.*, 2000, **122**, 3071–3079.
2. K. T. Finley, *Chem. Rev.*, 1964, **64**, 573–589.
3. F. A. Davis, L. C. Vishwakarma, J. G. Billmers and J. Finn, *J. Org. Chem.*, 1984, **49**, 3241–3243.
4. F. A. Davis, M. S. Haque, T. G. Ulatowski and J. C. Towson, *J. Org. Chem.*, 1986, **51**, 2402–2404.
5. F. A. Davis and B. C. Chen, *Chem. Rev.*, 1992, **92**, 919–934.
6. W. Adam, R. T. Fell, V. R. Stegmann and C. R. Saha-Möller, *J. Am. Chem. Soc.*, 1998, **120**, 708–714.
7. R. Adams and C. S. Marvel, *Org. Synth.*, 1941, **1**, 94.
8. M. S. Kerr, J. Read de Alaniz and T. Rovis, *J. Org. Chem.*, 2005, **70**, 5725–5728.
9. D. Enders and J. Han, *Tetrahedron: Asymmetry*, 2008, **19**, 1367–1371.

4.2.4. 1,4-Addition to α,β-unsaturated Carbonyl Compounds

4.2.4.1. Preparation of Chalcone and Its Further Robinson Annulation with Ethyl Acetoacetate

Nuno R. Candeias

Department of Chemistry and Bioengineering, Tampere University of Technology, Korkeakoulunkatu 8, Tampere, FI-33101 Finland
*E-mail: nuno.rafaelcandeias@tut.fi

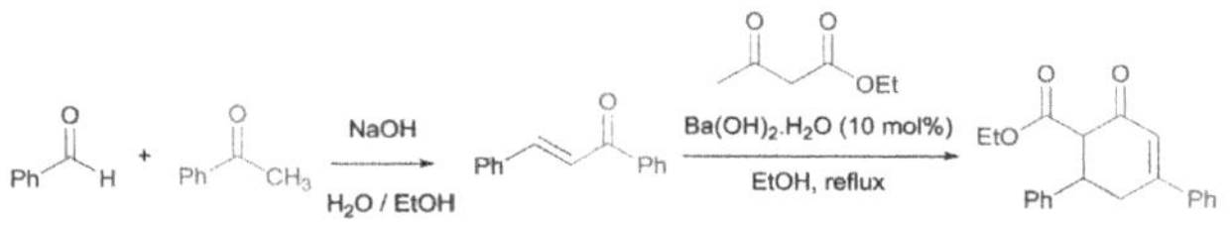

Number of sessions (duration of each session)	Hazard level	Difficulty level	Level of study
3 (4 h each)	Moderate	Medium	Intermediate
Class names Aldehydes, ketones, β-keto esters, α,β-unsaturated ketones			
Concepts involved This experiment aims to demonstrate the different reactivity of carbonyl compounds by performing an aldol condensation followed by a base-catalysed Robinson annulation. The reactions can be followed by TLC and the products purified by recrystallization			
Chemicals needed Benzaldehyde, acetophenone, ethyl acetoacetate, barium hydroxide monohydrate, sodium hydroxide, distilled water, ethanol, hexane, ethyl acetate, deuterated chloroform, deuterated DMSO			
Equipment and experimental techniques involved Stirring and heating plates, reflux apparatus, vacuum filtration, rotatory evaporation			
Keywords Catalysis, condensation, conjugate addition, NMR, recrystallization			

Background

Carbonyl compounds can show different kinds of reactivity depending on the substituents. Aldehydes are more electrophilic than ketones, and so aromatic aldehydes condense with aliphatic or mixed alkyl aryl ketones in the presence of aqueous alkali to form α,β-unsaturated ketones. This reaction is known as the Claisen–Schmidt reaction, a particular case of the aldol condensation, and a very popular method for the preparation of conjugated ketones. For such a reaction to occur successfully, some conditions related with the choice of the reactants must be fulfilled: one partner only must be capable of enolization; the other partner must be incapable of enolization and be more electrophilic than the enolizable partner.[1] The first transformation of this experiment is the preparation of chalcone

Comprehensive Organic Chemistry Experiments for the Laboratory Classroom
Edited by Carlos A. M. Afonso, Nuno R. Candeias, Dulce Pereira Simão, Alexandre F. Trindade, Jaime A. S. Coelho, Bin Tan, and Robert Franzén.

Published by the Royal Society of Chemistry, www.rsc.org

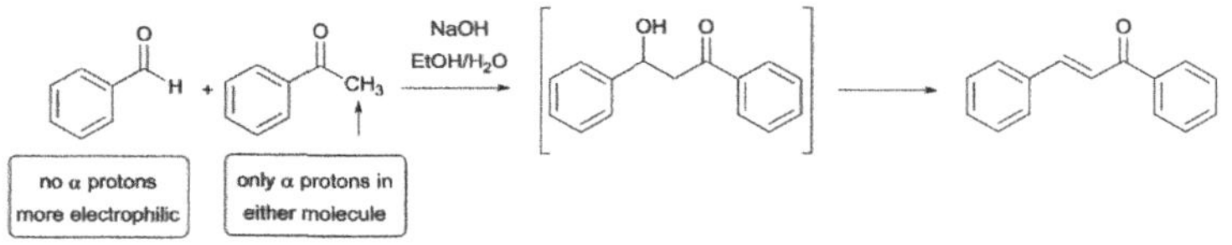

Scheme 4.2.4.1.1 Claisen–Schmidt condensation.

(dibenzylideneacetone) through condensation of benzaldehyde (electrophilic non-enolizable component) and acetophenone (enolizable ketone).[2] The β-hydroxy ketone formed after an aldol reaction quickly dehydrates by E1cB mechanism to afford chalcone (Scheme 4.2.4.1.1) to be used in the next synthetic step.

The second synthetic step of this experiment is the preparation and isolation of a cyclic enone after condensation of ethyl acetoacetate with chalcone under basic catalysis (Robinson annulation).

α,β-Unsaturated ketones have two reactive electrophilic positions; in the presence of nucleophiles enones can react either through direct 1,2-addition or a conjugate 1,4-addition, depending on the conditions of the reaction, the nature of the α,β-unsaturated carbonyl compound and the type of nucleophile.

When placing ethyl acetoacetate in slightly basic medium and in presence of chalcone, this triggers a domino process known as a Robinson annulation reaction. Domino reactions have been defined as reactions involving two or more bond-forming transformations under the same reaction conditions, without further addition of reagents and catalysts, and in which subsequent reaction happens as a result of the functionality formed by bond formation or fragmentation in the previous step. Amongst other advantages, domino reactions allow the formation of several bonds in one sequence without isolating the intermediates, changing the reaction conditions or adding reagents, which ultimately leads to minimization of waste and amount of labour.[3]

The Robinson annulation is a three-step process, consisting of a conjugate addition, an intramolecular aldol reaction and E1cB dehydration (Scheme 4.2.4.1.2). In the first step, a stabilized enolate anion performs a nucleophilic conjugate addition to the α,β-unsaturated ketone to afford the corresponding 1,5-diketone Michael adduct. Such an intermediate, in basic medium, is deprotonated to form an enolate that undergoes an intramolecular aldol reaction to form a six-membered ring keto alcohol. The sequence is finished by the dehydration of the keto alcohol to afford the six-membered ring enone. Notably, the base used is recovered by the end of every stage, and so it can be used in catalytic amounts. Depending on the amount of catalyst, and the reaction conditions (time and temperature), it is possible to isolate the intermediates of this reaction.[4]

Additional Safety

The use of fume hoods is advisable due to the strong odour of benzaldehyde. Chalcone is known to be a skin irritant and so the use of gloves is advised. Handling of corrosive NaOH should be done carefully. Flammable ethanol in the reaction and recrystallization steps should be handled with care and ignition sources should be avoided as much as possible.

Experimental Procedure

This experiment is divided into three sessions of 4 h each.

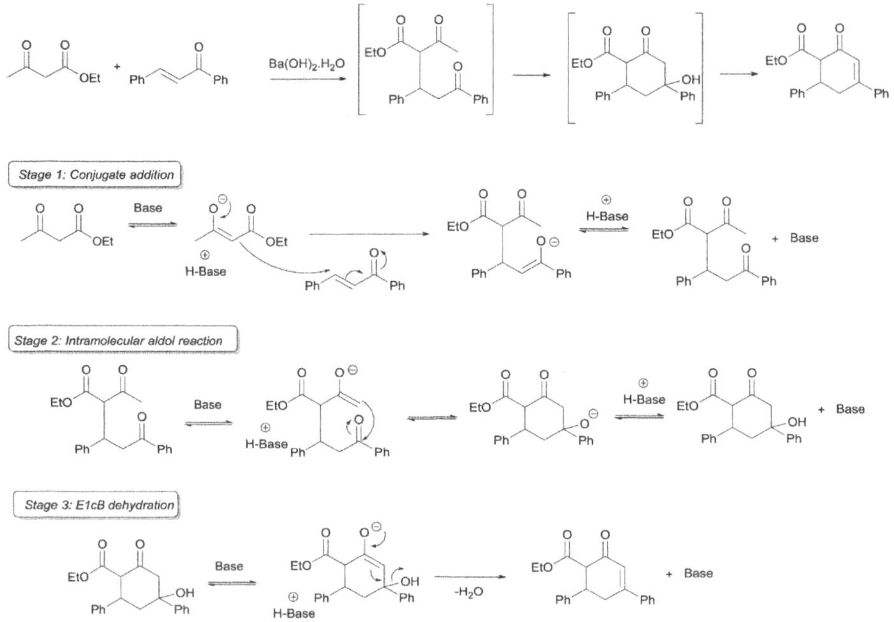

Scheme 4.2.4.1.2 Robinson annulation of ethyl acetoacetate and chalcone.

Session 1 – Preparation of Chalcone

Place a solution of sodium hydroxide (2.2 g, 55 mmol) in water (20 mL) and EtOH (15 mL) in a round-bottomed flask equipped with addition funnel, thermometer and a stirring bar. Immerse the flask in a bath of crushed ice and slowly pour in acetophenone (5.2 g, 43 mmol). Slowly add benzaldehyde (4.6 g, 43 mmol) to the mixture and control the temperature of the reaction to be between 15 and 30 °C, using the ice bath. Stir vigorously until reaction is complete (2–3 h), judged by TLC (SiO_2, AcOEt–hexane (1 : 9)), cool the mixture and filter the product with suction on a Büchner funnel. Wash the product with cold water until the washings are neutral (*ca.* 200 mL), and then with ice-cold EtOH (2 mL). Dry the crude chalcone under suction. Recrystallize from EtOH at 50 °C.

Session 2 – Robinson Annulation

Check the purity of the chalcone by NMR, after preparing the sample in $CDCl_3$.

In a 100 mL round-bottomed flask equipped with a reflux condenser, dissolve the recently prepared chalcone (2.08 g, 10 mmol) in ethanol (25 mL) and add ethyl acetoacetate (1.30 g, 10 mmol). Add barium hydroxide monohydrate (10 mol%, 0.20 g) and heat the mixture at a vigorous reflux. After 1 h at reflux, stop the reflux by cooling the reaction mixture at room temperature, take a TLC of the reaction and analyse the results. Leave the mixture at vigorous reflux overnight.

Session 3 – Purification of Six-Membered Enone

Cool the reaction to room temperature, take a TLC and compare it with the one from the day before. Dilute the resulting suspension with cold water (40 mL) and filter off the pale yellow precipitate with suction. Wash the precipitate with more cold water (20 mL). Recrystallize the crude product in the minimum amount of ethanol by hot filtration–recrystallization and dry it carefully. Evaporate the mother liquors and recrystallize them in ethanol. Prepare an NMR sample of the first crop of crystals in DMSO-d_6.

Waste Disposal

The residual solvent from the first reaction can be neutralized with HCl and discarded into the aqueous container. The washings from the quench of the second reaction can be also discarded into the container for collection of aqueous residues. Ethanol from recrystallizations and remaining mother liquors should be discarded in the organic waste recipient.

Results Interpretation and Additional Questions

1. What would have been the product if benzaldehyde were not added to the reaction vessel in the first reaction? Is the order of addition of the reactants important to obtain a high yield of the chalcone?
2. Why should the recrystallization of chalcone be performed at 50 °C and not in boiling ethanol, as with usual recrystallizations?
3. Considering the NMR of the chalcone, what is the geometry of the double bond? Why is only one stereoisomer formed in this Claisen–Schmidt reaction?
4. Considering the mechanism for the Robinson annulation, what are the other spots in the TLC after 1 h reaction?
5. How would you proceed to stop the Robinson annulation at an intermediary stage so that the 1,5-diketone or the hydroxy ketone could be isolated?
6. Considering the original atom economy definition (molecular mass of desired product)/(sum of molecular masses of stoichiometric reagents),[5] what can you say about the sustainability of the Robinson annulation reaction? Besides the yield of the reaction, determine also the experimental atom economy of the process.

References

1. J. Clayden, N. Greeves and S. G. Warren, *Organic chemistry*, 2nd edn, Oxford University Press, Oxford, 2012, pp. 614–624.
2. A. I. Vogel, A. R. Tatchell, B. S. Furniss, A. J. Hannaford and P. W. G. Smith, *Vogel's Textbook of practical organic chemistry*, 5th edn, Longman Scientific & Technical, London, 1989, pp. 1032 and 1034.
3. L. F. Tietze, *Chem. Rev.*, 1996, **96**, 115–136.
4. L. Delaude, J. Grandjean and A. F. Noels, *J. Chem. Educ.*, 2006, **83**, 1225.
5. B. M. Trost, *Angew. Chem., Int. Ed. Engl.*, 1995, **34**, 259–281.

4.2.4.2. Conjugate Addition of Organocuprates to α,β-Unsaturated Ketones: Synthesis of 3,3-Dimethylcyclohexanone from 3-Methyl-2-cyclohexen-1-one

Carlos Gregorio*, F. Javier Sardina, and Antonio Mouriño

Departamento de Química Orgánica, Facultad de Química, Universidad de Santiago de Compostela, Avda. Ciencias S/N, 15782 Santiago de Compostela, Spain. *E-mail: carlos.gregorio@usc.es

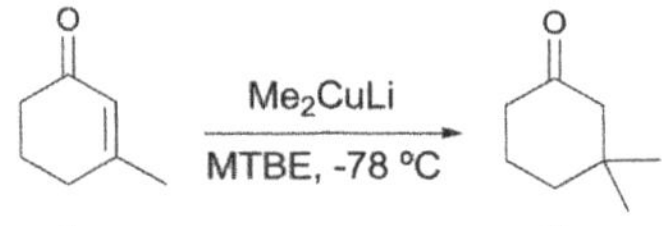

1 2

Number of sessions (duration of each session)	Hazard level	Difficulty level	Level of study
2 (3 h each)	Moderate	Medium	Advanced
Class names α,β-Unsaturated ketones			
Concepts involved (a) Carbon–carbon bond formation by addition of organocuprates to α,β-enones; (b) handling of organometallic reagents under inert atmosphere at low temperature			
Chemicals needed Copper(I) iodide, 3-methyl-2-cyclohexen-1-one, methyllithium (1.6 M in diethyl ether), *t*-butyl methyl ether, silica gel 60 (230–400 mesh), hexanes, ethyl acetate, NH_4Cl, anhydrous Na_2SO_4, *p*-anisaldehyde, Celite® S, deuterated chloroform			
Equipment and experimental techniques involved *Equipment*: round-bottomed flasks, rubber septa, inert gas line, needles, cannulas, balloons, magnetic stirrer, disposable hypodermic polypropylene syringes, dry-ice bath, Büchner flask, beaker, silica TLC plates, UV lamp, chromatography column, separatory funnel, sintered glass funnel, test tubes, rotary evaporator, vacuum pump. *Techniques*: drying glassware, transfer of air–moisture sensitive reagents *via* cannula or syringe, monitoring of air–moisture sensitive reactions by TLC, cooling to −78 °C by dry-ice bath, visualization of starting materials and reaction products by UV or by immersion into a *p*-anisaldehyde stain followed by heating, liquid–liquid separation, filtration through Celite® S, other filtration techniques, flash chromatography, concentration under vacuum, destruction of air–moisture sensitive reagents, ^1H and ^{13}C NMR			
Keywords Air and moisture sensitive reagents, cuprates, flash chromatography, inert atmosphere, NMR, α,β-unsaturated ketones			

Background

Gilman synthesized a monoalkyl copper derivative (MeCu) in 1943 by reaction of methyllithium with 1 equivalent of CuI.[1] Later it was observed that the addition of 1 equivalent of methyllithium (MeLi) to the yellow precipitate of MeCu provides a colorless solution of a dialkyl organocopper reagent (Me_2CuLi), named the Gilman reagent or homocuprate.[2] Organocuprate reagents are usually unstable in ether solutions at room temperature although the stability depends on the nature

Comprehensive Organic Chemistry Experiments for the Laboratory Classroom
Edited by Carlos A. M. Afonso, Nuno R. Candeias, Dulce Pereira Simão, Alexandre F. Trindade, Jaime A. S. Coelho, Bin Tan, and Robert Franzén.

Published by the Royal Society of Chemistry, www.rsc.org

of the alkyl substituent. For example, Me_2CuLi in ether is stable at 0 °C under a nitrogen or argon atmosphere, but secondary or tertiary organocuprates are unstable at temperatures above −20 °C.[3] Applications of organocuprates were further developed by Whitesides,[4] Corey,[5] and Posner.[6] Important reactions include the formation of carbon–carbon bonds using alkyl halides (S_N2), allyl halides or esters (S_N2'), propargyl halides or esters (S_N2'), epoxides and acyl halides. In these reactions the organocuprate transfers only one of the R organic fragments. Lithium reagents add directly to the carbonyl group of α,β-unsaturated ketones to give allylic alcohols. Lithium reagents attack preferentially the harder C=O rather than the softer C=C electrophilic center.[7] The presence of copper makes the lithium reagents undergo conjugate addition. The organocopper reagent is softer than the lithium reagent and attacks in a conjugate fashion the softer C=C double bond.[8]

The formation of 3,3-dimethylcyclohexanone from 3-methyl-2-cyclohexen-1-one was reported by House,[9] Lipshutz,[10] and at a microscale by Newton.[11] A similar experiment was reported by Organ[12] using 2-cyclohexen-1-one and *n*-Bu_2CuLi. The starting 3-methyl-2-cyclohexen-1-one, which is commercially available, is currently prepared in our laboratory according to the procedure of Harwood and Moody.[13] The experiment is described for 1 g of 3-methyl-2-cyclohexen-1-one, but it can be performed using half of the indicated amounts. The reaction is fast and easy to follow by thin-layer chromatography. The quantitative formation of the cuprate is visible by the disappearance of the yellow color of the reaction mixture. The reaction product was purified by flash chromatography, despite the fact that the crude reaction product provides an acceptable ^1H NMR. In summary, the student acquires skills in handling air sensitive organometallic reagents, dry solvents, inert atmosphere, plastic syringes, syringe needless, balloons, flash chromatography, vacuum pump, TLC-monitoring, CO_2-cooling baths, handling air sensitive reagents and reactions, and characterization of reaction products by ^1H and ^{13}C NMR.

Additional Safety

Protective clothing, gloves and safety glasses are required. Methyllithium is flammable in contact with air or moisture and must be handled under an atmosphere of Ar or N_2. Hexanes, ethyl acetate, and *t*-butyl methyl ether (TBME) are flammable liquids and harmful if swallowed. They must be handled in a fume hood. Dry-ice acetone baths must be handled with anti-thermal gloves. Add acetone over CO_2 pellets and not the other way round. After addition of the methyllithium to CuI wash the syringe needle and plastic syringes sequentially with ethyl acetate and water.

Experimental Procedure

First Session (3 h)

1. Oven-dry a one-necked round-bottomed flask (100 mL), equipped with a Teflon® coated magnetic stirring bar. Attach a rubber septum. Insert a short syringe needle through the septum. Then flush the flask with argon gas by means of a long needle connected to an argon-filled balloon through a PVC one way stopcock. This needle should reach the bottom of the flask.
2. Close the stopcock, remove the septum and rapidly charge the flask with CuI† (2.59 g, 13.62 mmol). Reset the septum in place. Remove the short outlet-needle and replace the longer purging-needle with a short one. Open the stopcock and maintain the flask under argon during the reaction.
3. Transfer *t*-butyl methyl ether (TBME, 20 mL)‡ into the reaction flask through the septum by a polypropylene hypodermic syringe equipped with a metallic needle.

---

†CuI (Aldrich Ref: 205540) was used without further purification. A polypropylene powder funnel can be used to transfer the CuI.

‡Dry technical grade *t*-butyl methyl ether (TBME) is used (Scharlau extra pure Ref: ME0550025A or similar). Solvent is dried over anhydrous Na_2SO_4 and stored in a round-bottomed flask.

4. Cool the suspension with an ice bath and stir for 15 min.
5. Add dropwise a solution of methyllithium (MeLi) in diethyl ether (17 mL, 27.2 mmol, 1.6 M)[§] by a 20 mL polypropylene hypodermic syringe equipped with a metallic needle.[¶] During the addition the yellowish suspension turns colorless.
6. Cool the resulting colorless solution of Me_2CuLi to −78 °C using a dry ice–acetone bath. Stir the mixture for 5 min.
7. Prepare a solution of 3-methyl-2-cyclohexen-1-one (1.00 g, 9.08 mmol)[ǁ] in *t*-butyl methyl ether (TBME, 10 mL) under argon (25 mL round bottomed flask) and add it dropwise *via* syringe over at least 10 min.
8. Monitor the reaction is by thin-layer chromatography (TLC).**

Work Up

9. Remove the cooling bath and stir the mixture for 30 min.
10. Remove the rubber septum. Slowly add an aqueous saturated solution of ammonium chloride (30 mL) and stir the mixture for 5 min. Then cap the flask with a plastic stopper until the next laboratory session.

Second Session (3 h)

1. Filter the mixture (blue aqueous phase and organic phase containing suspended solids) through a pad of Celite® S.
2. Transfer the filtrate to a separatory funnel. Extract the blue aqueous phase with TBME (2 × 10 mL). Then wash the combined organic phases with water (2 × 10 mL), dry over anhydrous Na_2SO_4, filter, and concentrate in the rotary evaporator.
3. Purify the oily crude product by flash chromatography (SiO_2, 230–400 mesh, 2 × 10 cm, 5% EtOAc/hexanes).
4. Concentrate the combined chromatography fractions containing the product in a rotary evaporator and remove the residual solvent using a high vacuum pump for about a minute to give a colorless oil. Weigh the oil.
5. Characterize the reaction product **2** by ^1H and ^{13}C NMR using $CDCl_3$ as solvent.

Waste Disposal

Aqueous suspensions of copper should be stored in a labeled container for further processing. TBME, EtOAc and hexanes from the rotary evaporator collecting flask are disposed of in a non-chlorinated waste collector recipient. Flasks containing residual MeLi are sequentially treated with EtOAc and water.

[§]Methyllithium in diethyl ether was purchased from Aldrich (Ref: 197343).

[¶]Needles with Luer hubs were purchased from Aldrich (Ref: Z117102-1EA). The syringes and needles are sequentially washed with EtOAc and water to destroy the remaining MeLi.

[ǁ]3-Methyl-2-cyclohexen-1-one can be purchased from Aldrich (Ref. M39105).

**The reaction was monitored by thin-layer chromatography (TLC) using 2 × 5 cm aluminum-backed silica gel plates (0.2 mm thickness). 30% EtOAc/hexanes was used as the eluent mobile phase in a beaker (100 mL) covered with a watch glass. The glass capillaries were made by heating and stretching Pasteur glass pipettes with a Bunsen lighter. The plates were first examined by ultraviolet illumination at 254 nm and then visualized by immersion in *p*-anisaldehyde stain followed by heating with a heat gun (Bosch, Ref. PHG 500-2). The starting enone can be visualized by UV light (λ = 254 nm). Both the starting enone and the reaction product can be visualized by immersion in *p*-anisaldehyde stain followed by heating with a heat gun. *p*-Anisaldehyde stain is prepared by the addition of *p*-anisaldehyde (2.1 mL) to a solution of H_2SO_4 (6.25 mL) and HOAc (1.8 mL) in ethanol (96%, 170 mL).

Results Interpretation and Additional Questions

1. CuI and MeLi are used to form Me_2CuLi.
   (a) Draw the equipment and glassware required to accomplish this reaction.
   (b) Summarize all the operations carried out to form Me_2CuLi.
2. Thin-layer chromatography (TLC) is used to monitor the formation of the reaction product **2**. The starting material **1** (3-methyl-2-cyclohexen-1-one) can be seen under UV light ($\lambda = 254$ nm), but the reaction product (3,3-dimethylcyclohexanone) is not observed under the same light. Provide an explanation for these observations.
3. The remaining MeLi in the syringe and needles is destroyed by the sequential washing with EtOAc and water. Describe the reactions that take place.
4. If you use MeLi (8 mL, 1.6 M) to prepare **2** and a molar ratio of MeLi : CuI: **1** of 2 : 1 : 1, calculate the amount of **2** that is produced if the overall yield is 70%.
5. Assign the ^1H and ^{13}C NMR peaks corresponding to the reaction product **2**.
6. Propose a mechanism for this transformation.
7. Explain why cuprates prefer to attack the C=C electrophilic center.

References

1. H. Gilman and L. A. Woods, *J. Am. Chem. Soc.*, 1943, **65**, 435.
2. H. Gilman, R. G. Jones and L. A. Woods, *J. Org. Chem.*, 1952, **17**, 1630.
3. G. M. Whitesides, W. F. Fischer Jr, J. San Filippo, R. W. Bashe and H. O. House, *J. Am. Chem. Soc.*, 1969, **91**, 4871.
4. H. O. House, W. L. Respess and G. M. Whitesides, *J. Org. Chem.*, 1966, **31**, 3128.
5. E. J. Corey and G. H. Posner, *J. Am. Chem. Soc.*, 1967, **89**, 3911.
6. G. H. Posner, C. E. Whitten and J. J. Sterling, *J. Am. Chem. Soc.*, 1973, **95**, 7788.
7. V. Pace, L. Castoldi, P. Hoyos, J. V. Sinisterra, M. Pregnolato and J. M. Sánchez-Montero, *Tetrahedron*, 2011, **67**, 2670.
8. (a) G. H. Posner, *An Introduction to Synthesis Using Organocopper Reagents*, John Wiley & Sons, Inc., New York, 1980; (b) B. H. Lipshutz and S. Sengupta, *Organic Reactions*, 2004, vol. 41, Organocopper Reagents: Substitution, Conjugate Addition, Carbo/Metallocupration, and Other Reactions, 2, pp. 135–631.
9. H. O. House and W. F. Fischer, *J. Org. Chem.*, 1968, **33**, 949–956.
10. B. H. Lipshutz, E. L. Ellsworth and T. J. Siahaan, *J. Am. Chem. Soc.*, 1989, **111**, 1351.
11. T. A. Newton, *J. Chem. Ed*, 1991, **68**, A60.
12. M. G. Organ and P. Anderson, *J. Chem. Ed.*, 1996, **73**, 1193.
13. L. M. Harwood and C. J. Moody, *Experimental Organic Chemistry. Principles and Practice*, Blackwell Scientific Pub, Oxford (UK), 2nd edn, 1989, ch. 8, pp. 568–570.

4.2.5. Multicomponent Reactions

4.2.5.1. One-Pot Green Synthesis of Dihydropyran Heterocycles

Dennis Russowsky* and Camila S. Santos

Laboratório de Síntese Organica K-210, Universidade Federal do Rio Grande do Sul, Instituto de Química, Av. Bento Gonçalves, 9500, CEP 91501-970, Porto Alegre, Rio Grande do Sul, Brazil
*E-mail: dennis.russowsky@ufrgs.br

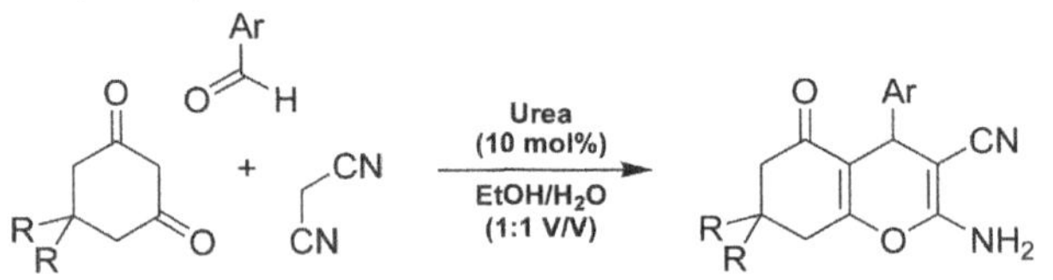

Number of sessions (duration of each session)	Hazard level	Difficulty level	Level of study
2 (3 h each)	Moderate	Low	Intermediate
Class names Aldehydes, ketones, nitriles, ureas, pyrans			
Concepts involved This work describes the synthesis of dihydropyran heterocycle derivatives from aromatic aldehydes, malononitrile and cyclic 1,3-diketones, following a simple and environmentally friendly *one-pot* protocol, which employs urea as organocatalyst			
Chemicals needed Benzaldehyde, malononitrile, dimedone, 1,3-cyclohexanedione, urea, ethanol, deuterated dimethyl sulfoxide (DMSO-d_6), water			
Equipment and experimental techniques involved Round-bottomed flask, magnetic stirring plate, stirring bar, low-pressure filtration apparatus (funnel, Kitasato flask, vacuum pump), desiccator, ^{1}H NMR and ^{13}C NMR spectroscopy			
Keywords Melting point, NMR, vacuum drying, vacuum filtration			

Background

Multicomponent reactions (MCRs) are characterized as a highly convergent process between three or more reagents mixed in the same container (*one pot*), at the same time, to obtain the final product which contains all or almost all atoms of the reagents.[1]

In the field of medicinal chemistry, these reactions play a prominent role in the discovery of new drugs,[2] since they open up the way to obtain different compounds ranging from simple heterocyclic compounds to polycyclic heteroaromatics of greater structural complexity.[3] The control of rate constants of the different possible reactions dictates an organized sequence of these reactions, allowing the formation of a single specific product.[4]

Comprehensive Organic Chemistry Experiments for the Laboratory Classroom
Edited by Carlos A. M. Afonso, Nuno R. Candeias, Dulce Pereira Simão, Alexandre F. Trindade, Jaime A. S. Coelho, Bin Tan, and Robert Franzén.

Published by the Royal Society of Chemistry, www.rsc.org

However, others processes can be performed in the same reaction vessel by adding three or more reagents at different stages of the process. These procedures are called *one-pot* reactions and they should not be confused with the MCRs.[5,6]

These methodologies are practical, efficient and environmentally benign procedures for the synthesis of a plethora of organic compounds, as evidenced by their widespread use in the pharmaceutical industry.[7]

The MCRs or one-pot approaches have been successfully applied as a powerful tool in the synthesis of dihydropyrans.[8] From the biological point of view, 2-amino-3-cyano-4-aryl-4*H*-pyrans, which have a dihydropyran core, are promising compounds in the development of new bioactive compounds, as possible candidates for new drugs. Furthermore, many dihydropyrans have been identified as biologically active compounds with anticancer, antibacterial, antifungal and anti-rheumatic properties.[9]

The simplest method to synthesize the 2-amino-3-cyano-4-aryl-4*H*-pyrans involves a three-component reaction (3-MCR) between 1,3-diketones, aromatic aldehydes and malononitrile promoted by catalysts with a base character.[10] Other methods using other activation processes such as ball milling[11] and ionic liquid media,[12] and also under solvent free conditions, have been reported.[13]

The proposed general mechanistic pathway involves initially the nucleophilic addition of malononitrile (as anionic species) to the aromatic aldehyde, followed by the *in situ* dehydration reaction (Knoevenagel-type step). In the next step, the 1,3-carbonyl compound adds to the adduct formed in the previous step *via* Michael-type addition. Subsequent cyclization with loss of water affords the 2-amino-3-cyano-4-aryl-4*H*-pyrans.[14]

Among the synthetic methods described in the literature, the procedure reported by Brahmachari is of particular interest due to the simple experimental protocol, performing the reaction at room temperature, and also by the use of low toxicity and low cost solvent.[15] Additionally, the use of a urea compound as organocatalyst brings this protocol within the principles of green chemistry.

Organocatalysts are any organic compound that is able to accelerate or promote a given chemical reaction. They do not contain any metals in their structures, are easily storable, stable, non-toxic and of low cost, which arouses interest in this type of catalysis for both small-scale reactions and for industrial processes.[16]

The principal goal of this work is to synthesize different 2-amino-3-cyano-4-aryl-4*H*-pyran derivatives, varying the structures of aromatic aldehydes and 1,3-diketones, through a simple and environmentally friendly methodology.

Additional Safety

It is known that certain aromatic compounds are carcinogenic, and therefore should not be inhaled and skin contact should be avoided (aromatic aldehydes). The manipulation of reagents and solvents (ethanol and DMSO-d_6) should be done in a fume hood. In addition, safety equipment such as protective clothing, gloves and safety glasses should be worn.

Additional Specific Cautions

1. Malononitrile: is toxic in contact with skin; causes eye irritation; is toxic if inhaled.
2. Dimedone: causes eye irritation and is harmful if absorbed through skin. It is harmful if inhaled and causes respiratory tract irritation.
3. DMSO-d_6: used as solvent for NMR spectroscopy is hazardous in case of inhalation or skin and eye contact. DMSO is mutagenic for mammalian somatic cells and is suspected to be a carcinogenic agent.

Experimental Procedure

Laboratory Session 1 (3 h) – (Note 1)†

Synthesis of Dihydropyran 5a

Representative procedure for the synthesis of 2-amino-7,7-dimethyl-5-oxo-4-phenyl-5,6,7,8-tetrahydro-4H-chromene-3-carbonitrile (Note 2).

1. In a fume hood, add to a 100 mL round-bottomed flask 5.0 mmol (0.53 g; 0.51 mL) of benzaldehyde (**1a**),‡ 5.5 mmol (0.37 g) of malononitrile (**2**), 20 mL of a mixture of EtOH/H_2O (1 : 1, v/v) and urea (**4**) 0.5 mmol (0.03 g).
2. Maintain the mixture under rigorous stirring for about 30–40 min, until the appearance of a white precipitate.
3. After this, add 5.0 mmol (0.70 g) of dimedone (**3a**), in small portions.
4. Allow the reaction to stir at room temperature for an additional 4 h (Note 3).

Laboratory Session 2 (3 h)§

Isolation of Product 5a

1. Filter the crude biphasic mixture (composed by a yellowish liquid and a white solid) with the aid of a reduced pressure filtration system and wash carefully with 10 mL cold water (2 × 5 mL) followed by 10 mL of cold ethanol (2 × 5 mL) (Note 4).
2. Leave the white solid thus obtained (the dihydropyran **5a**) to dry in a desiccator, under vacuum, until the next session (Notes 5 and 6).

Determination of the Yield and the Characterization of Product 5a¶

After drying, weigh and calculate the yield of the reaction. Separate a sample and determine the melting point of the solid and compare it with the literature data (Note 7). Typically, the protocol described above is sufficient to obtain a good correlation of the melting points (Note 8). To obtain the NMR spectra, prepare a sample in a NMR tube using DMSO-d_6 as the solvent (Note 9) and record the ^1H and ^{13}C NMR spectra. Compare them with those in the literature (Note 7).

Waste Disposal

The resulting solution of the processes of washing and filtration should be discharged in the appropriate waste collector recipient for aqueous or organic non-chlorinated residues.

Results Interpretation and Additional Questions

1. Why in the synthesis process should you maintain the reaction under rigorous stirring for about 20 min before to add the 1,3-diketone?
2. Suggest a plausible mechanism for the formation of dihydropyrans **5a**.
3. Assign the ^1H and ^{13}C NMR spectra of the product and compare the assignments with the data described in the literature (see Note 7 in Supplementary material).

---

†For notes on Laboratory of Session 1, see ESI.
‡Different aromatic aldehydes can be used, see ESI.
§For notes on Laboratory of Session 2, see ESI.
¶For notes on Laboratory of Session 3, see ESI.

4. Calculate the yield of the one-pot reaction.
5. Discuss the possible advantages of using a one-pot protocol as a synthetic strategy.
6. Discuss the importance of using organocatalysts in organic reactions.

Acknowledgements

The authors would like to acknowledge the Conselho Nacional de Desenvolvimento Científico e Tecnológico (CNPq) and Fundação de Amparo à Pesquisa do Estado do Rio Grande do Sul (FAPERGS) for financial support and fellowship (C.S.S). We are indebted to the Institute of Chemistry of Universidade Federal do Rio Grande do Sul for making available the technical conditions needed to realize this work.

References

1. (a) R. W. Armstrong, A. P. Combs, P. A. Tempest, S. D. Brown and T. A. Keati, *Acc. Chem. Res.*, 1996, **29**, 123; (b) J.-P. Zhu and H. Bienaymé, in *Multicomponent Reactions*, Wiley-VCH Verlag GmbH, Weinheim, 2005.
2. P. Slobbe, E. Ruijter and R. V. A. Orru, *Med. Chem. Commun.*, 2012, **3**, 1189.
3. J. D. Sunderhaus and S. F. Martin, *Chem. Eur. J.*, 2009, **15**, 1300.
4. B. H. Rotstein, S. Zaretsky, V. Rai and A. K. Yudin, *Chem. Rev.*, 2014, **114**, 8323.
5. C. Vaxelaire, P. Winter and M. Christmann, *Angew. Chem., Int. Ed.*, 2011, **50**, 3605.
6. L. Albrecht, H. Jiang and K. A. Jørgensen, *Angew. Chem., Int. Ed.*, 2011, **50**, 8492.
7. W. Zhao and F.-E. Chen, *Curr. Org. Synth.*, 2012, **9**, 873.
8. (a) I. A. Azath, P. Puthiaraj and K. Pitchumani, *ACS Sustainable Chem. Eng.*, 2013, **1**, 174; (b) G. Brahmachari, S. Laskar and B. J. Banerjee, *Heterocycl. Chem.*, 2014, **51**, 303.
9. (a) M. Mahmoodi, A. Aliabadi, S. Emami, S. Maliheh, S. Rajabalian, M. Mohagheghi, A. Khoshzaban, A. Samzadeh-Kermani, M. Lamei, A. Shafiee and A. Foroumadi, *Arch. Pharm. Chem. Life Sci.*, 2010, **343**, 411; (b) C. W. Smith, J. W. Bailey, M. E. J. Billingham, S. Chandrasekhar, C. P. Dell, K. H. Anita, C. A. Hicks, A. E. Kingston and G. N. Wishart, *Bioorg. Med. Chem.*, 1995, **5**, 2783–2788.
10. (a) Z. K. Jaberi and B. Pooladian, *Sci. World J.*, 2012, **2012**, 1; (b) S. Jain, P. K. Paliwal, G. N. Babu and A. J. Bhatewara, *Saudi Chem. Soc.*, 2014, **18**, 535; (c) A. M. Zonouz, D. Moghani and S. Okhravi, *Curr. Chem. Lett.*, 2014, **3**, 71; (d) V. M. Joshi, R. L. Magar, P. B. Throat, S. U. Tekale, B. R. Patil, M. P. Kale and R. P. Pawar, *Chin. Chem. Lett.*, 2014, **25**, 455.
11. M. G. Dekamin and M. Eslami, *Green Chem.*, 2014, **16**, 4914.
12. H. Hu, F. Qiu, A. Ying, J. Yang and H. Meng, *Int. J. Mol. Sci.*, 2014, **15**, 6897.
13. S. Jain, D. Rajguru, B. S. Keshwal and A. D. Acharya, *ISRN Org. Chem.*, 2013, **2013**, 1.
14. S. Banerjee, A. Horn, H. Khatri and G. A. Sereda, *Tetrahedron Lett.*, 2011, **52**, 1878.
15. G. Brahmachari and G. Banerjee, *ACS Sustainable Chem. Eng.*, 2014, **2**, 411.
16. M. Bhanushali and C.-G. Zao, *Synthesis*, 2011, 1815.

4.2.5.2. Hantzsch Synthesis of Nifedipine

Arno Kraft

Heriot-Watt University, Institute of Chemical Sciences, School of Engineering & Physical Sciences, Riccarton, Edinburgh EH14 4AS, UK. *E-mail: A.Kraft@hw.ac.uk

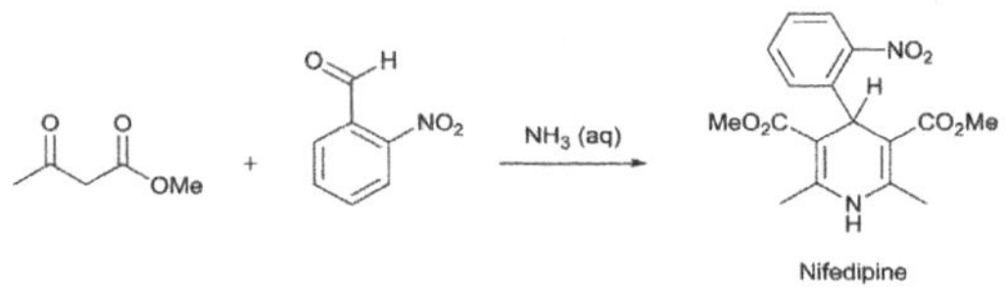

Number of sessions (duration of each session)	Hazard level	Difficulty level	Level of study
2 (4 h each)	Moderate	Medium	Intermediate

Class names Aldehydes, β-ketoesters, enamines

Concepts involved Heterocyclic synthesis, multi-component reaction, Hantzsch dihydropyridine synthesis, pharmaceutical drug

Chemicals needed 2-Nitrobenzaldehyde, methyl acetoacetate, aqueous ammonia (30–35 wt%), methanol, deuterated chloroform

Equipment and experimental techniques involved Apparatus for heating to reflux (90 °C) and magnetic stirring, filtration, recrystallisation, HPLC, NMR, IR and melting point

Keywords 1,4-Dihydropyridine, heterocyclic synthesis, HPLC, multi-component reaction, nifedipine, NMR, pharmaceutical drug, recrystallisation

Background

The German chemist Arthur Rudolph Hantzsch discovered this multi-component synthesis in 1882.[1] The reaction provides easy access to highly substituted symmetrical 1,4-dihydropyridines which are readily oxidised to the corresponding pyridines with oxidants such as cerium(IV) salts, nitric acid or chlorine.[2] The classical Hantzsch synthesis is a one-pot cyclocondensation between an aldehyde, ammonia and two equivalents of a β-ketoester such as ethyl acetoacetate (Scheme 4.2.5.2.1). A discussion of the mechanism of this reaction can be found in the literature.[3,4]

Pyridine is an electron-deficient aromatic heterocycle that is not readily amenable to electrophilic aromatic substitution reactions. Highly substituted pyridines are generally best made by ring-closing reactions from acyclic precursors. The Hantzsch synthesis initially yields a 1,4-dihydropyridine, which has two extra hydrogens at positions 1 and 4 in the ring, with the N atom being defined as position "1". Many 1,4-dihydropyridines are known to exhibit important bioactivity. Lately, the synthetically simple Hantzsch reaction has been used to make nifedipine and related pharmaceutical drugs that act as Ca^{2+} channel blockers for the treatment of high blood pressure.[5]

Comprehensive Organic Chemistry Experiments for the Laboratory Classroom
Edited by Carlos A. M. Afonso, Nuno R. Candeias, Dulce Pereira Simão, Alexandre F. Trindade, Jaime A. S. Coelho, Bin Tan, and Robert Franzén.

Published by the Royal Society of Chemistry, www.rsc.org

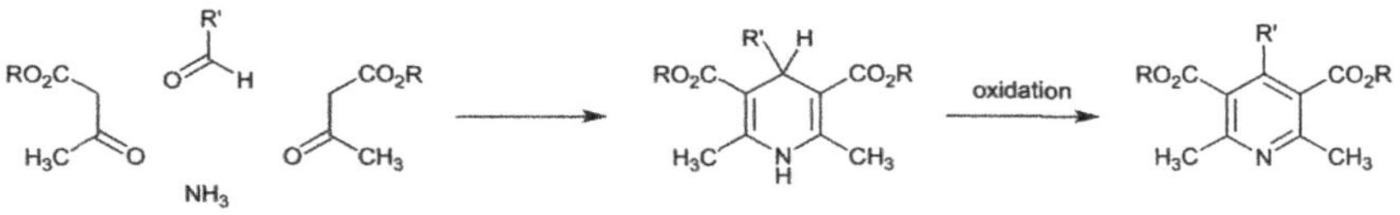

Scheme 4.2.5.2.1 Hantzsch synthesis of 1,4-dihydropyridines and pyridines.

Additional Safety

The preparation should be done in a fume hood, with proper safety equipment, such as protective clothing, gloves and safety goggles. 2-Nitrobenzaldehyde is harmful when swallowed, causes skin and serious eye irritation and potentially respiratory irritation. Methyl acetoacetate causes serious eye irritation. Aqueous ammonia causes severe skin burns and eye damage; it is also very toxic to aquatic life and its release into the environment has to be avoided. Methanol is highly flammable and toxic by inhalation, in contact with skin and if swallowed can cause irreversible damage to the eyes. Nifedipine has to be treated like any other highly toxic compound due to its pharmacological potency and ability to lower blood pressure; a typical daily pharmaceutical dose is only 15–30 mg, and even a small dose can lead to collapse and increased heart rate in people with normal blood pressure.

Experimental Procedure

Laboratory Session 1 (4 h): Nifedipine Synthesis

1. Use a fume cupboard for this experiment.
2. To a 50 mL round-bottom flask add 2.27 g of 2-nitrobenzaldehyde (15.0 mmol), 4.0 mL of methyl acetoacetate (37.1 mmol), 4 mL of methanol and 1.6 mL of concentrated ammonia (35% in water, 35 mmol).
3. Fit with a condenser, then use an isomantle or an oil bath to heat the mixture to reflux for 3.5 h.
4. At the end of the reaction period, allow the reaction mixture to cool. If necessary, use an ice–water bath until a precipitate forms. Keep the stoppered flask at room temperature until the next laboratory session.

Laboratory Session 2 (4 h): Work-Up, Purification and Characterisation

Work Up and Purification

5. Collect the precipitate by suction filtration and wash the filter cake with water (10 mL) and methanol (5 mL).
6. Recrystallise the crude product from methanol, then dry the recrystallised product. If the product turns out to be oily during recrystallisation, cool the solution with an ice–water bath.

Characterisation

7. Weigh the product and determine the yield. Measure the melting point and record an IR spectrum.
8. Record ^{1}H and ^{13}C NMR spectra in deuterated chloroform.
9. Check the purity of the product by HPLC. For this, weigh out 2 mg of the recrystallised sample on an analytical balance, then dissolve it in 2 mL of HPLC-grade methanol. Carry out an HPLC analysis of the product using water/methanol (70 : 30) as the eluent.

Waste Disposal

The neutralised aqueous solution should be collected in an aqueous waste collector recipient, for disposal by an external contractor. Methanol and solutions from the recrystallisation should be collected in a non-halogenated waste collector recipient, for posterior destruction.

Results Interpretation and Additional Questions

1. Interpret the IR and NMR spectra.
2. The mechanism of the Hantzsch dihydropyridine synthesis proceeds *via* many steps and the curly-arrow mechanism is, not surprisingly, rather lengthy. It involves an aldol condensation, a Michael addition, an enamine and a dehydration. Sketch an outline of the main 4–5 intermediates involved in the formation of nifedipine.
3. Comment on the purity of your product.

References

1. A. Hantzsch, *Justus Liebigs Ann. Chem.*, 1882, **215**, 1.
2. J. A. Joule and K. Mills, *Heterocyclic Chemistry*, Blackwell Publishing, Oxford, 4th edn, 2000, pp. 104–105.
3. U. Eisner and J. Kuthan, *Chem. Rev.*, 1972, **72**, 1–42.
4. K. M. Bucholtz, J. D. Hugdahl, A. M. Kiefer, K. M. Kiefer and A. M. S. Maria, *Chem. Educ.*, 2012, **17**, 125–127.
5. F. Bossart, H. Meyer and E. Wehinger, *Angew. Chem., Int. Ed. Engl.*, 1981, **20**, 762.

4.2.5.3. A Ugi Multicomponent Reaction in the Synthesis of *N*-Cyclohexyl-2-[*N*-(4-methoxybenzyl)acetamide]-2-(thien-2-yl)acetamide

Susana P. G. Costa*, M. Manuela M. Raposo, and Cátia I. C. Esteves

University of Minho, Department of Chemistry, Campus de Gualtar, 4710-057 Braga, Portugal
*E-mail: spc@quimica.uminho.pt

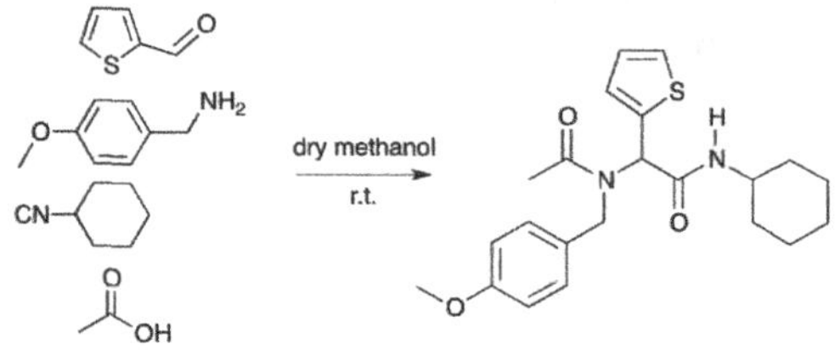

Number of sessions (duration of each session)	Hazard level	Difficulty level	Level of study
2 (4 h each)	Moderate	Low	Intermediate

Class names Aldehydes, amines, carboxylic acids, isocyanides, amides

Concepts involved Preparation of a protected unnatural α-amino acid by an Ugi multicomponent reaction involving an amine, an aldehyde, a carboxylic acid and an isocyanide; filtration under vacuum; rotary evaporation; separation by thin-layer and column chromatography; characterization by ^{1}H and ^{13}C NMR spectroscopy

Chemicals needed Thiophene-2-carboxaldehyde, 4-methoxybenzylamine, acetic acid, cyclohexyl isocyanide, methanol (dry), sodium sulfate (anhydrous), dichloromethane, hexane, silica gel (230–400 mesh), silica gel thin-layer chromatography (TLC) plates with fluorescent indicator

Equipment and experimental techniques involved Apparatus for reaction with magnetic stirring at room temperature, vacuum filtration, thin-layer and column chromatography on silica gel, ^{1}H and ^{13}C nuclear magnetic resonance spectroscopy

Keywords Amino acids, chromatography, multicomponent reaction, NMR, thiophene, Ugi reaction

Background

Multicomponent reactions (MCRs) are processes between three or more reagents that combine according to a sequence of elementary reactions, usually in a one-pot manner, to afford a product that retains most of the atoms in the starting materials. MCRs are carried out with reactants that are compatible with each other and do not undergo alternative irreversible reactions to form other products or by-products. MCRs can be classified by the number of functional groups participating in the reaction (*e.g.*, three reagents is a 3-MCR, while four reagents is a 4-MCR).[1,2]

Comprehensive Organic Chemistry Experiments for the Laboratory Classroom
Edited by Carlos A. M. Afonso, Nuno R. Candeias, Dulce Pereira Simão, Alexandre F. Trindade, Jaime A. S. Coelho, Bin Tan, and Robert Franzén.

Published by the Royal Society of Chemistry, www.rsc.org

MCRs are an alternative approach to chemical synthesis with multiple advantages: minimize time and reagent cost by generating a complex product in a single convergent step; atom economic, since the resulting product is constituted by portions of each individual reactant, usually by loss of a small molecule such as water; generally run at low or ambient temperature, with very simple procedures; and ease of automation. Therefore, whenever possible, it has been the method of choice for the generation of complex polyfunctional molecules using convergent one-pot transformations, for example in combinatorial libraries in pharmaceutical industry and total synthesis of natural products. Common examples of MCRs are Mannich, Petasis, Biginelli, Passerini, Povarov, Bucherer–Bergs, Strecker, Pictet–Spengler, hetero-Diels–Alder cycloaddition and Ugi reactions.[1–4]

The Ugi reaction is an isocyanide-based 4-MCR proposed in 1959 by Ivar Ugi as an alternative to the classical methods of amino acid synthesis, by reacting an amine, a carbonyl compound, an acid and an isocyanide. Isocyanides are very versatile reagents with a functional group of unusual valence structure and reactivity, since it bears a formally divalent carbon that can react with electrophiles and nucleophiles.[5,6]

This reaction can yield a large variety of compounds due to the available components diversity and the possibility of using bifunctional components, which in turn lead to products resulting from intramolecular cyclizations. Generally, the reaction is exothermic and fast at low or ambient temperature, and higher yields can be obtained in concentrated reaction mixtures (0.5–2 M). Following the original application of the Ugi reaction,[7] it can be used for the synthesis of α-amino acids (if an aldehyde is used as the carbonyl component) and α,α-dialkylamino acids (if a ketone is used as the carbonyl component).[8]

Unnatural amino acids of synthetic origin can be regarded as innovative building blocks for the construction of functional peptides, since their incorporation into peptidic chains may result in the appearance of the amino acid properties, such as increased UV/vis absorption or fluorescence, in the resulting peptide. Thiophene and its derivatives exhibit interesting biological activity and important electroluminescent properties with wide application in polymer science, which has led to their application in energy transfer systems in optical and electronic devices, such as sensors and fluorescent markers.[9,10]

The aim of this work is the synthesis of a protected unnatural amino acid using a Ugi multicomponent reaction, namely *N*-cyclohexyl-2-[*N*-(4-methoxybenzyl)acetamido]-2-(thien-2-yl)acetamide. Several experimental techniques will be considered such as thin-layer chromatography (TLC), vacuum filtration, and column chromatography on silica gel. The determination of the structure of the compound synthesized will be performed by NMR spectroscopy.

Additional Safety

All work should be carried out in a well-ventilated fume hood, with proper safety equipment, such as protective clothing, gloves and safety goggles. Cyclohexyl isocyanide is toxic, may cause skin, eye and respiratory tract irritation; it has a foul smell and so it should be handled exclusively in the fume hood, as should all the glassware that comes in contact with it. Thiophene-2-carboxaldehyde may cause skin and eye irritation. 4-Methoxybenzylamine may cause skin burns and eye damage. Acetic acid is a corrosive acid and may cause severe skin and eye burns, and is flammable in liquid and vapor form. The preparation of the silica gel column and the separation of the components of the mixture should be carried out in a well-ventilated fume hood due to the use of volatile solvents such as dichloromethane, methanol and hexane, and silica gel may cause respiratory irritation – do not inhale its dust.

Experimental Procedure

Laboratory Session 1 (4 h): Synthesis of *N*-Cyclohexyl-2-[*N*-(4-methoxybenzyl)acetamide]-2-(thien-2-yl)acetamide

1. Prepare a 25 mL round-bottomed flask with a magnetic stir bar, and add dry methanol (5 mL), 100 μL (1.1 mmol, 1.1 equiv.) of thiophene-2-carboxaldehyde, 130 μL (1 mmol, 1 equiv.) of 4-methoxybenzylamine and anhydrous sodium sulfate (0.5 g). Stir the solution in the stoppered flask for 1 h at room temperature.
2. To the previous solution (containing the imine), add 58 μL (1 mmol) of acetic acid and stir for 15 min at room temperature, keeping the flask stoppered.
3. To the previous reaction mixture add 124 μL (1 mmol, 1 equiv.) of cyclohexyl isocyanide. Stir this mixture in the stoppered flask at room temperature for 24 h (see ESI). *Important*: this isocyanide is a malodorous liquid that should only be handled in a well-ventilated hood and after usage all glassware in contact with it must be placed in a bath of aqueous HCl (1 M) inside the hood.

Laboratory Session 2 (4 h): Product Isolation and Purification

1. In the fume hood, filter with a vacuum the reaction mixture through a sintered glass disc funnel, wash the filtered solid in the funnel with methanol (5 mL), and evaporate the solvent to dryness in a rotary evaporator. Check the purity of the brownish crude residue by thin-layer chromatography (TLC) by comparison with the reagents, aldehyde, amine and isocyanide (eluent: dichloromethane).
2. Purify the residue from evaporation by chromatography through a silica gel column (column height and diameter: 40 and 2.5 cm, respectively; silica weight: 35 g; silica height: ≈ 20 cm; column packing solvent: dichloromethane–hexane, 2 : 1). Dissolve the evaporated residue in the minimum volume of packing solvent and begin eluting with dichloromethane–hexane (2 : 1, ≈50 mL, to collect the unreacted isocyanide), followed by elution with dichloromethane (≈150 mL, to collect the unreacted aldehyde) and then with dichloromethane–methanol (90 : 1, ≈200 mL, to collect the desired product). Check by TLC the contents of each collected fraction (eluent: dichloromethane).
3. Combine the fractions containing the product and evaporate to dryness in a rotary evaporator.
4. Determine the mass of the obtained orange oily product and calculate the yield. Record ^{1}H and ^{13}C NMR spectra in $CDCl_3$ to confirm its purity.

Waste Disposal

The glassware that came into contact with the isocyanide should be placed in a bath of aqueous HCl (1 M) inside the hood, for a pre-cleaning treatment to hydrolyze the isocyanide to the corresponding formamide ($R\text{–}NC + H^{+} \rightarrow R\text{–}NHCHO$). After this treatment, the glassware can be washed with detergent as usual and the dilute acid bath should be further diluted, neutralized and discarded in the sink.

The solvents used in chromatography can be recovered after evaporation in a rotary evaporator and saved for later distillation and reuse, if possible. Fractions containing the unreacted isocyanide should not be recovered by distillation, but collected in an appropriate waste container. Methanol and hexane should be collected in a non-chlorinated waste collector recipient, and dichloromethane should be collected in a chlorinated waste collector for subsequent destruction. Used silica gel must be stored in a closed container in the hood for subsequent disposal. The filtration solid should be disposed of in a proper container.

Results, Interpretation and Additional Questions

1. In session 1, what is the purpose of adding anhydrous sodium sulfate to the reaction mixture?
2. Explain why the synthesis of the imine should be carried out under anhydrous conditions.
3. Write the mechanism of the Ugi multicomponent reaction considering the used reagents and rationalize the several group transformations involved in each elementary step.
4. Although the Ugi multicomponent reaction could be carried out in a truly one-pot manner (all reagents added together at the same time), what is the advantage of using this sequential method, with prior stirring of specific reagents?
5. Interpret the ^{1}H and ^{13}C NMR spectra obtained. Compare with the spectra provided in the ESI.
6. NMR could easily be used to monitor the multicomponent reaction course. What would be the most relevant signal to indicate the level of conversion?
7. Relate the eluent composition to the polarity of the eluted compounds.

References

1. *Multicomponent Reactions*, ed. J. Zhu and H. Bienaymé, Wiley-VCH Verlag GmbH, Weinheim, 1st edn, 2005.
2. *Domino Reactions: Concepts for Efficient Organic Synthesis*, ed. L. F. Tietze, Wiley-VCH Verlag GmbH, Weinheim, 1st edn, 2014.
3. L. Kurti and B. Czako, *Strategic Applications of Named Reactions in Organic Synthesis*, Elsevier Academic Press, London, 1st edn, 2005.
4. D. J. Ramón and M. Yus, *Angew. Chem., Int. Ed.*, 2005, **44**, 1602–1634.
5. A. Dömling and I. Ugi, *Angew. Chem., Int. Ed.*, 2000, **38**, 3168–3210.
6. A. Dömling, *Chem. Rev.*, 2006, **106**, 17–89.
7. G. Gokel, P. Hoffmann, H. Kleimann, H. Klusacek, G. Ludke, D. Marquaerding and I. Ugi, *Isonitrile Chemistry*, ed. I. Ugi, Academic Press, New York and London, 1971, p. 201.
8. S. P. G. Costa, H. L. S. Maia and S. M. M. A. Pereira-Lima, *Org. Biomol. Chem.*, 2003, **1**, 1475–1479.
9. G. Barbarella, M. Melucci and G. Sotgiu, *Adv. Mater.*, 2005, **17**, 1581–1593.
10. (a) S. P. G. Costa, E. Oliveira, C. Lodeiro and M. M. M. Raposo, *Tetrahedron Lett.*, 2008, **49**, 5258–5261; (b) S. P. G. Costa, R. M. F. Batista and M. M. M. Raposo, *Tetrahedron*, 2008, **64**, 9733–9737; (c) C. I. C. Esteves, M. M. M. Raposo and S. P. G. Costa, *Tetrahedron*, 2010, **66**, 7479–7486; (d) E. Oliveira, D. Genovese, R. Juris, N. Zaccheroni, J. L. Capelo, M. M. M. Raposo, S. P. G. Costa, L. Prodi and C. Lodeiro, *Inorg. Chem.*, 2011, **50**, 8834–8849.

4.2.5.4. Preparation of Phenylglycine and Hydroxymorpholine Derivatives Through a Petasis Borono–Mannich Reaction

Nuno R. Candeias*[a], Roberta Paterna[b], Pedro M. S. D. Cal[b], and Pedro M. P. Gois*[b]

[a]Department of Chemistry and Bioengineering, Tampere University of Technology, Korkeakoulunkatu 8, Tampere, FI-33101 Finland; [b]Research Institute for Medicines and Pharmaceutical Sciences (iMed.UL), Faculty of Pharmacy, University of Lisbon, Av. Prof. Gama Pinto, 1649-003 Lisbon, Portugal
*E-mail: nuno.rafaelcandeias@tut.fi, pedrogois@ff.ul.pt

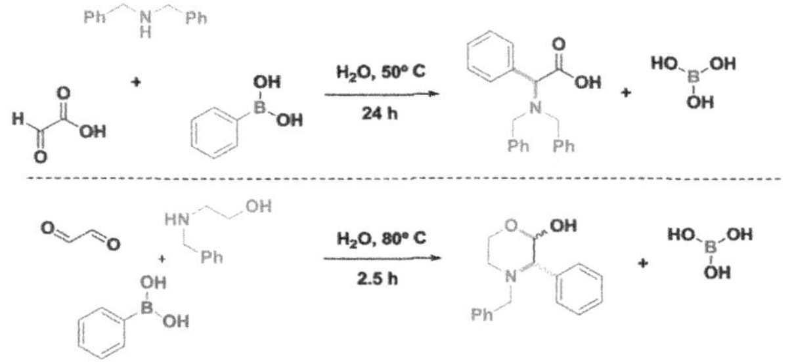

Number of sessions (duration of each session)	Hazard level	Difficulty level	Level of study
1 (4 h)	Low	Medium	Advanced

Class names Aldehydes, carboxylic acids, secondary amines, alcohols, iminium ions, boronic acids

Concepts involved Multicomponent reactions, water as solvent

Chemicals needed Phenylboronic acid, glyoxylic acid, glyoxal aqueous solution, distilled water, 2-(benzylamino)ethanol, *N*,*N*-dibenzylamine

Equipment and experimental techniques involved Filtration over cotton aided by Pasteur pipette, stirring and heating plates, liquid–liquid extraction, rotary evaporator, NMR, melting point

Keywords Diastereomers, filtration, melting point, multicomponent reaction, NMR, water chemistry

Background

Multicomponent reactions (MCRs) are single-step reactions in which a significant part of three or more starting materials are combined in one single product. Considering the formation of several new bonds this makes MCR a remarkable tool for the easy preparation of simple libraries of compounds. In such reactions, the product is formed through consecutive elementary reactions between the several reaction components dependent on delicate chemical equilibria that in the best scenario lead to the irreversible formation of new chemical bonds.[1]

The irreversible formation of a chemical bond shifts the equilibrium towards the product formation, decreasing the reaction time and increasing the selectivity. MCR can be classified as three types, according to the number of irreversible steps involved in the process (Scheme 4.2.5.4.1).[2]

Comprehensive Organic Chemistry Experiments for the Laboratory Classroom
Edited by Carlos A. M. Afonso, Nuno R. Candeias, Dulce Pereira Simão, Alexandre F. Trindade, Jaime A. S. Coelho, Bin Tan, and Robert Franzén.

Published by the Royal Society of Chemistry, www.rsc.org

- Type I: all the intermediates formed along the reaction, including the reaction product, derive from chemical equilibrium between the intervening species. Irreversible steps are absent in this type of reactions.
- Type II: the formation of intermediates happens through chemical equilibria; the last step is an irreversible one resulting in the product formation.
- Type III: all the steps are irreversible. Cascade or domino reactions are one example of this type of reactions.

$$\text{A} + \text{B} \rightleftharpoons \text{C} \rightleftharpoons \ldots \text{O} \rightleftharpoons \text{P} \qquad \text{Type I}$$

$$\text{A} + \text{B} \rightleftharpoons \text{C} \rightleftharpoons \text{D} \ldots \text{O} \longrightarrow \text{P} \qquad \text{Type II}$$

$$\text{A} \longrightarrow \text{B} + \text{C} \longrightarrow \text{D} \longrightarrow \ldots \text{O} \longrightarrow \text{P} \qquad \text{Type III}$$

Scheme 4.2.5.4.1 Types of multicomponent reactions according to number of irreversible steps.

Boronic acids are interesting partners in MCR due to their stability and commercial availability.

The Petasis borono–Mannich reaction is an example of a type II MCR. The reaction product results from condensation of an aldehyde, an amine and a boronic acid. The reaction was named after its similarity with the Mannich reaction and the findings of Nicos A. Petasis that the boronic acid substituent acts as a nucleophile in presence of imines or iminium ions.[1,3]

This reaction is grounded in the fast formation of an iminium ion resultant from the condensation of an aldehyde with an amine, together with the ability of the boronic acid substituent to act as a nucleophile. Such nucleophilic attack of the boronic acid substituent to the iminium ion is irreversible and results in the formation of a new C–C bond (Scheme 4.2.5.4.2). The inertness of the boronic acid towards the aldehyde allows the substituent of the boronic acid to migrate to the iminium ion only after it has been formed. However, for the substituent transfer from the boronic acid to the iminium ion to occur an additional OH-containing functional group vicinal to the iminium ion is needed. The hydroxyl group reacts with the boronic acid to form a tetrahedral boron complex – *ate complex* – that will be responsible for the substituent transfer to the iminium ion.[3]

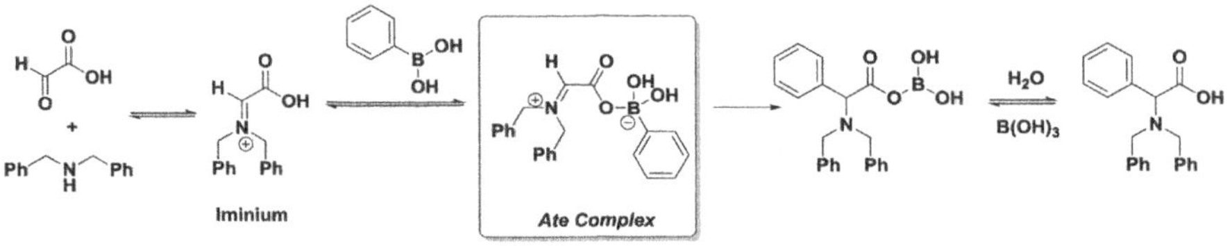

Scheme 4.2.5.4.2 Intermediates in the Petasis borono–Mannich reaction.

The combination of the three components described in the first part of this work will allow the preparation of a non-natural amino acid such as *N*,*N*-dibenzyl-phenylglycine, by using glyoxylic acid as one of the components. Nevertheless, the reaction works with other aldehydes as long as the hydroxyl group is present in the molecule, since this is essential for the *ate complex* formation.[4]

In the second part of the work, a hydroxy-morpholine (4-benzyl-3-phenylmorpholin-2-ol) will be prepared using the same reaction. However, in this case the hydroxyl group needed for the *ate complex* formation will be formed during the reaction (Scheme 4.2.5.4.3).

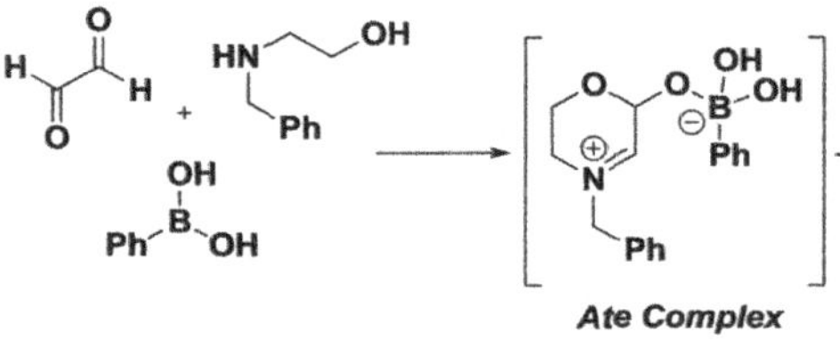

Scheme 4.2.5.4.3 Petasis borono–Mannich reaction between 2-(benzylamino)ethanol, glyoxal and phenylboronic acid.

The use of water as the reaction medium allows the easy isolation of the products after a simple filtration and the use of high temperature to promote the reaction without the liberation of vapours issuing from volatile organic solvents. Nevertheless, the use of this solvent often leads to heterogeneous reaction mixtures, so that proper stirring of the mixture can be difficult to achieve and the reaction completeness is usually achieved due to diffusion processes and mass transfer phenomena.

Additional Safety

This experimental procedure does not require additional measures besides general good practice, laboratory coat, goggles and gloves. The use of a fume hood is advisable only for the extraction with volatile ethyl ether. Vapour inhalation may cause headache, nausea, vomiting, and loss of consciousness. The evaporation of ethyl ether in the rotary evaporator should be done at the lowest possible temperature, depending on the available vacuum. Phenylboronic acid and *N,N*-dibenzylamine are harmful if swallowed, may cause respiratory tract irritation by inhalation, or skin and eyes irritation. *N,N*-Dibenzylamine may be harmful if absorbed through the skin. Glyoxylic acid causes eye and skin irritation and may be harmful if absorbed through the skin or if swallowed. 2-Benzylaminoethanol causes eye and skin irritation.

Experimental Procedure

Preparation, Isolation and Characterization of *N,N*-Dibenzyl-phenylglycine

Previous Day

1. Add a stirrer magnetic bar to a 10 mL round-bottom flask and take note of the weight.
2. Add phenylboronic acid (0.05 g, 0.4 mmol) and water (2.0 mL) to the flask and stir for 5 min, followed by the addition of glyoxylic acid (0.03 g, 0.4 mmol).
3. Heat at 50 °C until complete dissolution (*ca.* 2 min) and add *N,N*-dibenzylamine (0.079 mL, 0.4 mmol),[†] leaving the reaction stirring slowly at that temperature for 24 h.[‡]

Laboratory Session

4. Cool the reaction mixture in an ice bath for 10 min, to promote complete precipitation of the product.
5. Filter the reaction mixture and wash the solid with 3 mL of water.
6. Place the sample in an oven at 70 °C until constant weight.
7. Prepare an NMR sample in DMSO-d_6 and determine the melting point of the obtained solid.

[†]The addition should be done using a gas-tight syringe or a micropipette.

[‡]Due to the heterogeneity of the reaction medium, the stir bar can get stuck during the first hours of the reaction; however, it will become free with time.

Preparation of 4-Benzyl-3-phenylmorpholin-2-ol

1. To a 10 mL round-bottom flask, add phenylboronic acid (0.05 g, 0.4 mmol), water (2.0 mL) and a stirrer magnetic bar.
2. Add glyoxal 40% aqueous solution (0.047 mL, 0.4 mmol), followed by 2-benzylaminoethanol (0.058 mL, 0.4 mmol)[†] to the mixture and leave it stirring at 80 °C for 2.5 h.
3. After cooling the reaction mixture in an ice bath, extract the product with ethyl ether (3 × 6 mL).
4. Dry the combined organic layers over anhydrous sodium sulfate, filter and remove the solvent under reduced pressure.
5. Determine the melting point of the obtained product and prepare the sample for ^{1}H NMR in $CDCl_3$. Determine the diastereoselectivity of the reaction by integration of the doublets at 4.61 ppm (*trans* diastereoisomer, $J = 7.2$ Hz) and 4.87 ppm (diastereoisomer *cis*, $J = 1.7$ Hz).

Waste Disposal

The non-use of organic solvents in this reaction diminishes the need for the effluents' destruction. Although none or few risks are known for the reaction side-products, the reaction solvent should not be disposed of in the sink.

Results Interpretation and Additional Questions

1. Determine the reaction yields obtained from both reactions.
2. Assign the peaks of the NMR spectrum of *N,N*-dibenzyl-phenylglycine.
3. In the formation of 4-benzyl-3-phenylmorpholin-2-ol it is expected that more than one diastereoisomer will be formed. Comment on this statement based on the reactivity of the *ate complex*.
4. Interpret the NMR spectrum of 4-benzyl-3-phenylmorpholin-2-ol. Compare it with the spectrum described in the literature and determine the reaction diastereoselectivity.
5. Compare the melting points of the obtained *N,N*-dibenzyl-phenylglycine and the 4-benzyl-3-phenylmorpholin-2-ol with those reported in the literature.
6. Indicate other aldehydes that are suitable for this reaction, based on previous reports in the literature.

Acknowledgement

Reprinted (adapted) with permission from *J. Chem. Educ.*, 2012, **89** (6), pp. 799–802. Copyright (2012) American Chemical Society.

References

1. (a) *Multicomponent reactions*, URL: http://www.organic-chemistry.org/topics/multicomponent-reactions.shtm, accessed November 2012; (b) A. Doemling, *Org. Chem. Highlights*, 2005, June 5, URL: http://www.organic-chemistry.org/Highlights/2005/05June.shtm.
2. A. Dömling and I. Ugi, *Angew. Chem., Int. Ed.*, 2000, **39**, 3168.
3. N. R. Candeias, F. Montalbano, P. M. S. D. Cal and P. M. P. Gois, *Chem. Rev.*, 2010, **110**, 6169.
4. N. R. Candeias, P. M. S. D. Cal, V. André, M. T. Duarte, L. F. Veiros and P. M. P. Gois, *Tetrahedron*, 2010, **66**, 2736.

4.2.5.5. A Multicomponent Reaction of Isocyanides for the Synthesis of 4-Chromanone-2-carboxamides

Ana G. Neo, Jesús Díaz, and Carlos F. Marcos*

Departamento de Química Orgánica e Inorgánica, F. Veterinaria, Universidad de Extremadura, 10071 Cáceres, Spain. *E-mail: cfernan@unex.es

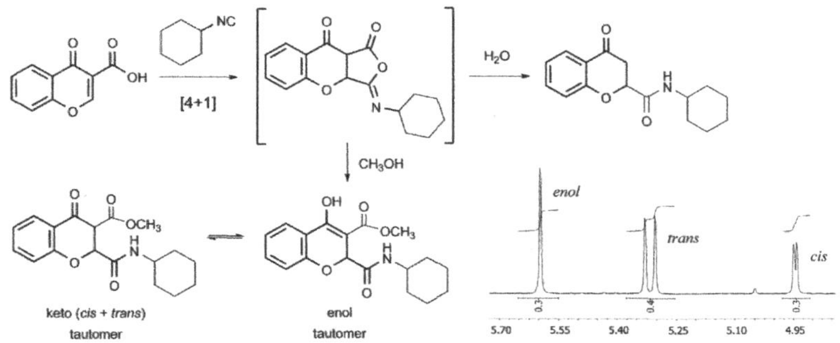

Number of sessions (duration of each session)	Hazard level	Difficulty level	Level of study
2 (2 h each)	Moderate	Medium	Advanced

Class names Carboxylic acids, amides, esters, acid anhydrides, imines, isocyanides, alcohols

Concepts involved This laboratory experiment involves the multicomponent synthesis of chroman-2-carboxamides and chromone amidoesters by a reaction of isocyanides, chromone 3-carboxylic acid and nucleophiles. Chromone amidoesters are obtained as a thermodynamic mixture of one enol and two keto isomers, for which the ratio and interconversion equilibrium constants can be easily calculated from the ^{1}H NMR spectrum

Chemicals needed Chromone-3-carboxylic acid, cyclohexyl isocyanide, methanol, water, ethanol, tetrahydrofuran, scandium trifluoromethanesulfonate, ethyl acetate, hexane, dichloromethane, silica gel, *p*-anisaldehyde, acetic acid, H_2SO_4

Equipment and experimental techniques involved Reactions at room temperature requiring magnetic stirring; microwave irradiation is performed in closed vials, using a focused single-mode microwave reactor (optional); product purification is performed by liquid–liquid extraction and/or by flash column chromatography (optional)

Keywords Decarboxylation, keto–enol tautomerism, liquid–liquid extraction, microwave, multicomponent reactions, NMR

Background

Chromone-2-carboxamides constitute an important family of chromone derivatives with multiple biological activities.[1] In addition, 4-chromanone-2-carboxamides, which have a reduced benzopyran nucleus, are potent antioxidants that inhibit lipid peroxidation more effectively than vitamin E.[2]

Comprehensive Organic Chemistry Experiments for the Laboratory Classroom
Edited by Carlos A. M. Afonso, Nuno R. Candeias, Dulce Pereira Simão, Alexandre F. Trindade, Jaime A. S. Coelho, Bin Tan, and Robert Franzén.

Published by the Royal Society of Chemistry, www.rsc.org

Several syntheses of chromone- and chromanone-2-carboxamides are known. Most of them rely on the coupling of activated 4-chromanone-2-carboxylic acids with different amines. However, an appealing alternative method is the direct conjugate addition of a synthetic equivalent of a carbamoyl anion to chromones. Isocyanides have been used as mild nucleophilic synthetic equivalents of carbamoyl anions.[3] At the same time isocyanides can behave as electrophiles giving α-addition reactions, on which multicomponent reactions of isocyanides are based.[4] Thus, isocyanides have been shown to react with 4-oxo-4*H*-chromene-3-carboxylic acids through a formal [4+1] cycloaddition to give an imino-anhydride intermediate, which can be opened *in situ* with different nucleophiles.[5] When the reaction is performed in the presence alcohols, chromanone derivatives containing an amide on position 2 and an ester on position 3 are obtained (Scheme 4.2.5.5.1). On the other hand, when water is used as nucleophile, the resulting amido carboxylic acid is not isolated, but suffers spontaneous decarboxylation to afford 3-unsubtituted-2-amidochromanones (Scheme 4.2.5.5.2).

This laboratory experiment aims at the preparation of potentially antioxidant chroman-2-carboxamides by a multicomponent reaction of cyclohexyl isocyanide, chromone 3-carboxylic acid and nucleophiles. Two different nucleophiles, water and methanol, are used in parallel experiments. When using methanol, chromone amidoesters are obtained as a mixture of isomers. On the other hand, when water is used, a chromone carboxamide is formed as a single product.

The reaction with methanol provides a mixture of kinetically favored enol and *cis*-keto tautomers. This is equilibrated in solution to a thermodynamic mixture of enol, *cis*-keto and *trans*-keto isomers, the ratio of which can be easily calculated from the ^{1}H NMR spectrum of the reaction product (Scheme 4.2.5.5.3).

Equilibria constants and ΔG_s can be also obtained from these data. Considering a typical enol–*cis*-keto–*trans*-keto ratio of 0.35 : 0.25 : 0.45, $K_{\text{enol-}cis}$ = [*cis*-keto]/[enol] = 0.25/0.35 = 0.71; $K_{\text{enol-}trans}$ = [*trans*-keto]/[enol] = 0.45/0.35 = 1.29.

As $\Delta G = -RT \ln K$, at 25 °C:

$$\Delta G_{\text{enol-}cis} = -8.314\ \text{J K}^{-1}\ \text{mol}^{-1} \times 298\ \text{K} \times \ln 0.71 = 849\ \text{J mol}^{-1} \text{ and } \Delta G_{\text{enol-}trans} = -631\ \text{J mol}^{-1}.$$

The use of microwave irradiation and Lewis acid catalysis allows both chromone-2-carboxamides and chromone-2-carboxamido-3-esters to be obtained in high yields, in just a few minutes. Thus, the simple experimental procedure can be performed in just two laboratory sessions.

Additional Safety

All the experiments should be carried out in a fume hood. Laboratory coat, disposable gloves and goggles must be worn. High pressures are developed under microwave irradiation, thus the reaction vessels must be allowed to cool down to room temperature before removing them from the microwave reactor.

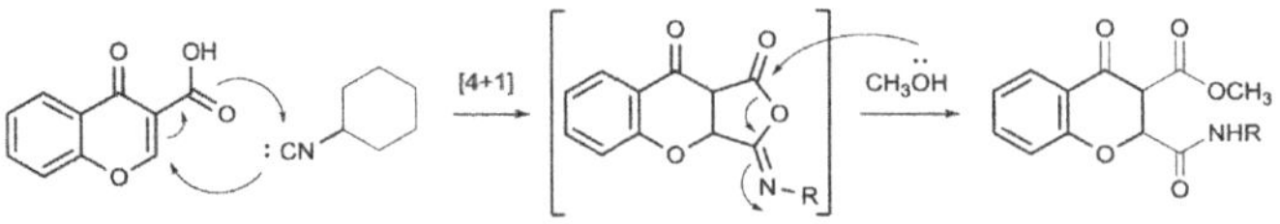

Scheme 4.2.5.5.1 Reaction of cyclohexyl isocyanide, chromone 3-carboxylic acid and methanol.

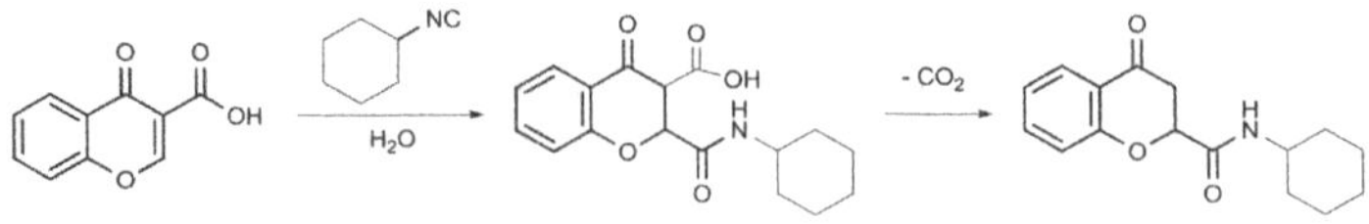

Scheme 4.2.5.5.2 Reaction of cyclohexyl isocyanide, chromone 3-carboxylic acid and water.

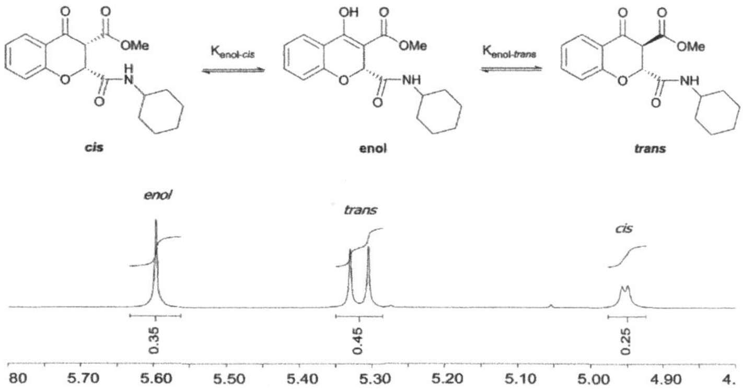

Scheme 4.2.5.5.3 Isomeric equilibrium of the product of chromone 3-carboxylic acid, cyclohexyl isocyanide and methanol and expansion of the NMR of the product (thermodynamic mixture).

Cyclohexyl isocyanide stenches and is toxic. Methanol is toxic. Hexane, tetrahydrofuran and ethyl acetate are flammable solvents. Silica gel can cause respiratory problems if inhaled and should always be handled in a fume hood.

Experimental Procedure

Procedure A: Microwave Assisted Reactions

Laboratory Session 1: Synthesis and Analysis of Methyl 2-(cyclohexylcarbamoyl)-3,4-dihydro-4-oxo-2H-chromene-3-carboxylate

1. In a microwave vial provided with a stirring bar add successively 95 mg (0.5 mmol) of chromone-3-carboxylic acid, 1 mL of tetrahydrofuran, 2.5 mg (0.005 mmol, 0.01 equiv.) of scandium trifloromethanesulfonate, 101 μL (2.5 mmol, 5 equiv.) of methanol and 93 μL (0.75 mmol, 1.5 equiv.) of cyclohexyl isocyanide.
2. Cap the vial and place it in the microwave reactor.
3. Program the reactor with the following parameters: power 150 watt, temperature 100 °C, ramp 1 min, hold time 8 min. Run the program.
4. Let the reaction mixture reach room temperature and remove it from the microwave reactor.
5. Check by thin-layer chromatography (SiO_2, hexane–EtOAc (7 : 3)) if the reaction has gone to completion.
6. Transfer the contents of the vial with a Pasteur pipette to a 100 mL separatory funnel, washing with 4 mL of dichloromethane. Add 15 mL of water, extract with CH_2Cl_2 (3 × 15 mL), dry the combined organic phases with anhydrous Na_2SO_4, filter, and evaporate to dryness in the rotary evaporator.
7. Measure the melting point, and record the IR spectrum.
8. Take about 20 mg of the product and dissolve it in $CDCl_3$. Record the ^1H NMR spectrum.
9. Leave the NMR sample in solution at room temperature and record the ^1H NMR spectrum again after 24 h.

Laboratory Session 2: Synthesis and Analysis of N-Cyclohexyl-3,4-dihydro-4-oxo-2H-chromene-2-carboxamide

1. In a microwave vial provided with a stirring bar add successively 95 mg (0.5 mmol) of chromone-3-carboxylic acid, 1 mL of tetrahydrofuran, 2.5 mg (0.005 mmol, 0.01 equiv.) of

scandium trifloromethanesulfonate, 75 µL (4.2 mmol, 8.3 equiv.) of water and 155 µL (1.25 mmol, 2.5 equiv.) of cyclohexyl isocyanide.
2. Cap the vial and introduce it in the microwave reactor.
3. Program the reactor with the following parameters: power 150 watt, temperature 100 °C, ramp 1 min, hold time 3 min. Run the program.
4. Let the reaction mixture reach room temperature and remove it from the microwave reactor.
5. Check by thin-layer chromatography (SiO_2, hexane–EtOAc (7 : 3)) if the reaction has gone to completion.
6. Transfer the contents of the vial with a Pasteur pipette to a 100 mL separatory funnel, washing with 4 mL of dichloromethane. Add 15 mL of water, extract with CH_2Cl_2 (3 × 15 mL), dry the combined organic phases with anhydrous Na_2SO_4, filter, and evaporate to dryness in the rotary evaporator.
7. Measure the melting point, and record the IR spectrum.
8. Take about 20 mg of the product and dissolve it in $CDCl_3$. Record the ^1H NMR spectrum.

Procedure B: Room Temperature Reactions

Laboratory Session 1: Reaction Setup

Methyl 2-(cyclohexylcarbamoyl)-3,4-dihydro-4-oxo-2*H*-chromene-3-carboxylate

1. In a 10 mL round-bottomed flask provided with a magnetic stir bar add successively 190 mg (1 mmol) of chromone-3-carboxylic acid, 1.5 mL of methanol, 5.0 mg (0.01 mmol, 0.01 equiv.) of scandium trifloromethanesulfonate and 124 µL (1 mmol, 1 equiv.) of cyclohexyl isocyanide.
2. Stir the reaction mixture at room temperature for 24 h, until the next laboratory session.

N-Cyclohexyl-3,4-dihydro-4-oxo-2*H*-chromene-2-carboxamide

3. In a 10 mL round-bottomed flask provided with a magnetic stir bar add successively 190 mg (1 mmol) of chromone-3-carboxylic acid, 1 mL of tetrahydrofuran, 5.0 mg (0.01 mmol, 0.01 equiv.) of scandium trifloromethanesulfonate, 150 µL (4.2 mmol, 8.3 equiv.) of water and 124 µL (1 mmol, 1 equiv.) of cyclohexyl isocyanide.
4. Stir the reaction mixture at room temperature 24 h, until the next laboratory session.

Laboratory Session 2: Workup and Analysis

Methyl 2-(cyclohexylcarbamoyl)-3,4-dihydro-4-oxo-2*H*-chromene-3-carboxylate (Continuation)

1. Check by thin-layer chromatography (SiO_2, hexane–EtOAc (7 : 3)) if the reaction has gone to completion.
2. Add 1 mL of 10% aqueous HCl and stir the mixture for 30 min in order to eliminate traces of the isocyanide.
3. Transfer the contents of the round-bottomed flask with a Pasteur pipette to a 100 mL separatory funnel, washing with 4 mL of dichloromethane. Add 15 mL of water, extract with CH_2Cl_2 (3 × 15 mL), dry the combined organic phases with anhydrous Na_2SO_4, filter, and evaporate to dryness in the rotary evaporator.
4. Perform the analysis as in procedure A.

N-Cyclohexyl-3,4-dihydro-4-oxo-2H-chromene-2-carboxamide (Continuation)

5. Check by thin-layer chromatography (SiO_2, hexane–EtOAc (7 : 3)) if the reaction has gone to completion.
6. Add 1 mL of 10% aqueous HCl and stir the mixture for 1 h in order to eliminate traces of the isocyanide.
7. Transfer the contents of the round-bottomed flask with a Pasteur pipette to a 100 mL separatory funnel, washing with 4 mL of dichloromethane. Add 15 mL of water, extract with CH_2Cl_2 (3 × 15 mL), dry the combined organic phases with anhydrous Na_2SO_4, filter, and evaporate to dryness in the rotary evaporator.
8. Perform the analysis as in procedure A.

Waste Disposal

All the solutions and glassware contaminated with cyclohexyl isocyanide should be treated with a 10% solution of HCl or H_2SO_4 in water–methanol (1 : 1), until the smell disappears. All the residues must be collected in appropriate waste recipients for posterior destruction.

Results Interpretation and Additional Questions

1. Calculate the yield of each of the two reactions and the overall yield of the process.
2. Propose a mechanism for the reactions and discuss why different products are obtained when using methanol and water. Why is a larger excess of isocyanide required in the second case (only protocol A)?
3. Calculate the ratio of the products obtained in the reaction with methanol from the ^{1}H NMR recorded immediately after preparing the solution and after 24 h. Does the ratio change? Why?
4. Calculate the equilibrium constants and ΔG for the interconversion reactions between the enol and each of the two keto tautomers.

References

1. A. Gaspar, M. J. Matos, J. Garrido, E. Uriarte and F. Borges, *Chem. Rev.*, 2014, **114**, 4960.
2. H. Lee, K. Lee, J. K. Jung, J. Cho and E. A. Theodorakis, *Bioorg. Med. Chem. Lett.*, 2005, **15**, 2745.
3. (a) A. Neo, R. Carrillo, J. Delgado, S. Marcaccini and C. Marcos, *Mol. Diversity*, 2011, **15**, 529; (b) R. M. Carrillo, A. G. Neo, L. López-García, S. Marcaccini and C. F. Marcos, *Green Chem.*, 2006, **8**, 787; (c) A. G. Neo, J. Delgado, C. Polo, S. Marcaccini and C. F. Marcos, *Tetrahedron Lett.*, 2005, **46**, 23.
4. (a) A. Dömling, *Chem. Rev.*, 2006, **106**, 17; (b) A. Dömling and I. Ugi, *Angew. Chem., Int. Ed.*, 2000, **39**, 3169.
5. A. G. Neo, J. Díaz, S. Marcaccini and C. F. Marcos, *Org. Biomol. Chem.*, 2012, **10**, 3406.

5. Electrophilic Aromatic Substitution

5.1. C-X Bond Formation

5.1.1. Synthesis of Methyl Orange

Dulce Simão* and Ana C. Cerdeira

Department of Chemistry, Instituto Superior Técnico, University of Lisbon, Centro de Química Estrutural, Av. Rovisco Pais, 1, 1049-001 Lisbon, Portugal. *E-mail: dulcesimao@tecnico.ulisboa.pt

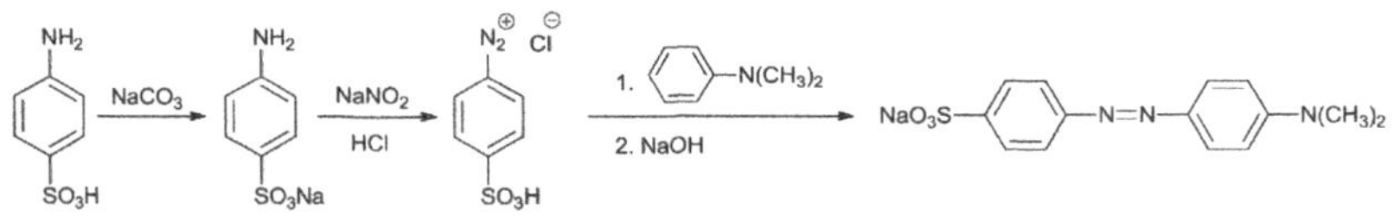

Number of sessions (duration of each session):	Hazard level	Difficulty level	Level of study
1 (4 h)	Moderate	Low	Intermediate

Class names Arylamines, diazonium salts, diazo compounds, aryl sulfonic acids

Concepts involved This experiment illustrates an electrophilic aromatic substitution with arenediazonium salts (diazo coupling). The reaction is performed with temperature control

Chemicals needed Sulfanilic acid, sodium carbonate, sodium nitrite, hydrochloric acid (37%), potassium iodide–starch paper, *N*,*N*-dimethylaniline, glacial acetic acid, sodium hydroxide, sodium chloride, ethanol, deuterated water

Equipment and experimental techniques involved Vacuum filtration, ^1H NMR, reaction with temperature control, magnetic stirring, IR, and UV-vis spectroscopies

Keywords Azo dye, diazo coupling, electrophilic aromatic substitution, ^1H NMR, UV-vis, vacuum filtration

Background

Methyl orange (C.I. Acid orange 52 [1]) is a well-known azo dye with important applications in the textile industry.[2,3] The azo linkage –N=N– between two aromatic rings gives an extended system of delocalized π electrons, responsible for the intense color of this product and allows absorption of light in the visible region. The presence of a sulfonic group makes the product soluble in water and allows adhering to fibers. This dye is also used as an acid–base indicator. In solutions with pH > 4.4 (λ_{max} 460 nm) it is orange–yellow, turning red in an acid solution upon protonation of methyl orange (pH < 3.2, λ_{max} 520 nm) (Scheme 5.1.1.1). These two forms have different chromophores and for that reason absorb at different wavelengths in the range of visible light.[4a]

Comprehensive Organic Chemistry Experiments for the Laboratory Classroom
Edited by Carlos A. M. Afonso, Nuno R. Candeias, Dulce Pereira Simão, Alexandre F. Trindade, Jaime A. S. Coelho, Bin Tan, and Robert Franzén.

Published by the Royal Society of Chemistry, www.rsc.org

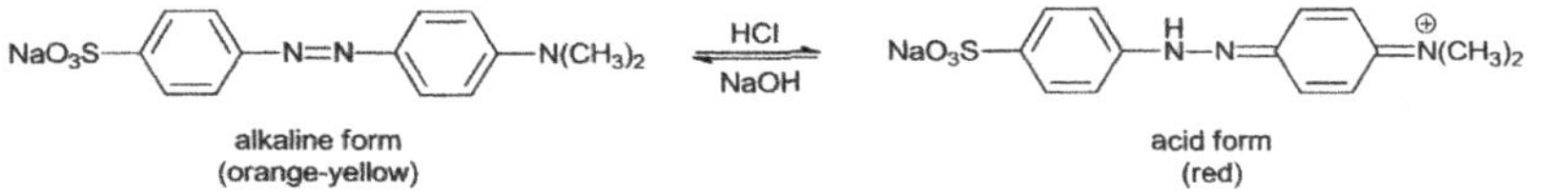

Scheme 5.1.1.1 Alkaline and acidic forms of methyl orange.

The synthesis of methyl orange[5,6] starts with the formation of a diazonium salt. In acid media sodium nitrite produces nitrous acid (HONO) that reacts with the aromatic amine to afford the diazonium ion.[4b] This reaction must be performed at low temperatures otherwise decomposition of the diazonium salt may occur with loss of nitrogen gas, with formation of an aryl cation that is readily attacked by water leading to a phenol.[7a]

The diazonium salt is a aryl cation, so it reacts as electrophile with *N,N*-dimethylaniline through an electrophilic aromatic substitution (diazo coupling), to finally give the azo compound. Diazonium salts are very weak electrophiles and only react with activated aromatic rings such as amines or phenols. In most cases, including the present experiment, the diazo coupling is stereochemically favored to occur at the *para* position. However, if this position is occupied, coupling may occur at the *ortho* position instead.[7b,8-10]

Additional Safety

It is advisable to perform this experiment in a fume hood since *N,N*-dimethylaniline is toxic and a suspected cancer agent. Both sulfanilic acid and sodium carbonate are irritant substances. Care should be taken with corrosive substances (acids and sodium hydroxide) and with the flammable ethanol.

Experimental Procedure

1. To a 250 mL Erlenmeyer flask, add sulfanilic acid (5.5 g, 0.032 mol), sodium carbonate (1.5 g, 0.015 mol) and water (50 mL).
2. Warm the reaction mixture in a water bath (50–60 °C) until the solution is clear.
3. Cool the mixture to about 15 °C in a cold water bath and then add a solution of $NaNO_2$ (2 g, 0.029 mol) in water (5 mL).
4. Cool the mixture to 5 °C with an ice-water bath and add slowly an aqueous solution of HCl (6 M) (11 mL) with stirring, always keeping the temperature under 10 °C.
5. Upon complete addition, wait 5 min and test the reaction mixture for the presence of free nitrous acid with potassium iodide–starch paper. At this stage, the reaction mixture contains the diazonium salt in suspension and must be kept cold in the ice-water bath.
6. Dissolve *N,N*-dimethylaniline (3 mL, 0.024 mol) in glacial acetic acid (1.5 mL) and add this solution to the reaction mixture, with vigorous stirring.
7. Allow the mixture to stand in the cold bath for 10 min, and the acid (red) form of methyl orange will gradually separate.
8. Add slowly, and with stirring, an aqueous solution of 20% NaOH (20 mL). The mixture will turn to a uniform orange color, due to precipitation of the sodium salt of methyl orange as very fine particles; collecting the product by filtration will be difficult as this point.
9. Heat the solution close to the boiling point to partially dissolve the dye, add NaCl (5 g) and continue to heat until the salt has completely dissolved.
10. Allow the mixture to cool for 15 min and then cool in an ice-water bath.
11. Collect the product by vacuum filtration; wash it with cold water first and do the final wash with ethanol.

12. Dry the product thoroughly on the funnel and store under vacuum. Once methyl orange is a salt, the melting point is not well defined.
13. Record the yield, IR, ^{1}H NMR (D_2O) and UV-vis spectrum (conc. 5.2×10^{-5} M in water).

Waste Disposal

The resulting alkaline filtrate, although highly dyed, contains a very small amount of dye in solution, and thus it can be diluted with plenty of water and discarded in the sink.

Results Interpretation and Additional Questions

1. Determine the reaction yield.
2. Write the reaction mechanism.
3. Assign the ^{1}H and ^{13}C NMR spectra and the characteristic absorption bands in the IR spectra.
4. Analyze the UV-vis spectrum and calculate the molar absorptivity coefficient maximum (ε_{max}) for both forms of methyl orange using Lambert–Beer's Law.
5. What precautions can you take to make sure that formation of the side product is minimized?
6. How can you remove the side product once it has formed?

References

1. *Colour Index*, The Society of Dyers and Colourists, 3rd edn, 1971, vol. 1, p. 1091.
2. *Kirk-Othmer Encyclopedia of Chemical Technology*, John Wiley & Sons, 2nd edn, 1967, vol. 2, p. 868.
3. *The Chemistry of Synthetic Dyes*, ed. K. Venkataraman, Academic Press, 1970, vol. 3, p. 30.
4. S. Ege, *Organic Chemistry*, D. C. Heath and Company, 2nd edn 1989(a) p. 955; (b) p. 911.
5. A. R. Tatchell, B. S. Furnis, A. J. Hannaford and P. W. G. Smith, *Vogel's Textbook of Practical Organic Chemistry*, Longman Scientific and Technical, 5th edn, 1989, p. 951.
6. C. A. M. Afonso, D. P. Simão, L. P. Ferreira, M. S. Serra and M. M. M. Raposo, *100 Experiências de Química Orgânica*, IST Press, 2011, p. 389.
7. J. Clayden, N. Greeves, S. Warren and P. Wothers, *Organic Chemistry*, Oxford, 2001(a) p. 572; (b) p. 598.
8. K. P. C. Vollhardt and N. E. Schore, *Organic Chemistry*, Freeman & Co., 3rd edn, 1999, p. 1019.
9. T. W. G. Solomons, *Organic Chemistry*, John Wiley & Sons Inc., 7th edn, 1998, p. 972.
10. M. Sainsbury, *Aromatic Chemistry*, Oxford Chemistry Primers, 1992, p. 57.

5.1.2. Halogenation Reactions of Vanillin

Iain A. Smellie*, Nigel P. Botting, Brian A. Chalmers, and Iain L. J. Patterson

School of Chemistry, University of St Andrews, North Haugh, St Andrews, Fife, KY16 9ST, UK
*E-mail: ias10@st-andrews.ac.uk

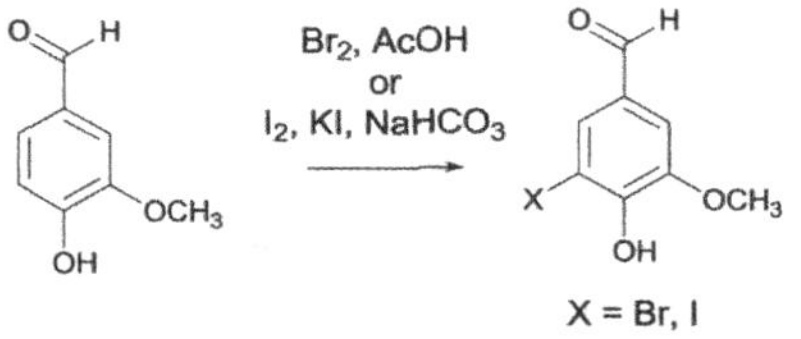

Number of sessions (duration of each session)	Hazard level	Difficulty level	Level of study
4 (3 h each) or 2 (3 h each) if only one halogenation is conducted	Moderate	Low	Introductory

Class names Aryl bromide, aryl iodide, phenoxide ion

Concepts involved This experiment describes methods for bromination and iodination of vanillin by electrophilic aromatic substitution

Chemicals needed Bromine, deuterochloroform, glacial acetic acid, iodine, potassium iodide, sodium hydrogen carbonate, sodium thiosulfate, vanillin

Equipment and experimental techniques involved Magnetic stirring, vacuum filtration, recrystallisation

Keywords Bromination, electrophilic aromatic substitution, iodination, NMR, recrystallisation

Background

Electrophilic aromatic substitution reactions are a regular feature of organic chemistry laboratory courses. Perhaps some of the best known examples are nitration of methyl benzoate,[1] acylation of ferrocene[2] and bromination of acetanilide.[3,4] The aim of this experiment is to prepare halogenated analogues of vanillin (4-hydroxy-3-methoxybenzaldehyde). Vanillin has several advantages for teaching laboratory use:[5] (1) it is relatively electron-rich and is therefore reactive toward electrophiles,[5] (2) it is less toxic than phenol itself[6] and (3) vanillin is cheap and readily sourced due to its availability from paper and wood derived lignin or petrochemical sources.[7–9] Vanillin is reactive enough for bromination to occur without the addition of strong Lewis acids,[5,10] indeed there are many conditions reported in the literature, representative examples include: treatment of vanillin with bromine in water,[5] bromine in acetic acid[10] and sodium bromide[11] and a suitable oxidant. In the experiment described here, a modified version of the procedure of Puri[10] *et al.* has been found to be convenient.

Unlike bromination, iodination of vanillin does not efficiently proceed on treatment with the parent halogen. Many effective protocols use iodide salts or iodine in the presence of an

Comprehensive Organic Chemistry Experiments for the Laboratory Classroom
Edited by Carlos A. M. Afonso, Nuno R. Candeias, Dulce Pereira Simão, Alexandre F. Trindade, Jaime A. S. Coelho, Bin Tan, and Robert Franzén.

Published by the Royal Society of Chemistry, www.rsc.org

oxidant[12–15] although basic conditions can also be employed.[16,17] Iodination in the presence of sodium bicarbonate has been described for the preparation of a fluorescein derivative.[17] In this case, the reaction is likely to be favoured by formation of the corresponding (and more reactive) phenoxide ion.[18] Iodination of vanillin using potassium iodide and sodium bicarbonate is described in the current protocol.

In the reactions described, interpretation of the ^{1}H NMR spectra of the products demonstrate that halogenation is regioselective for the same position in each case.

Additional Safety

All operations should be performed in a fume hood with appropriate protective clothing (laboratory coat, safety goggles and gloves). Glacial acetic acid is flammable and corrosive; exposure can be *via* inhalation, ingestion or skin contact. Bromine and Iodine (vapours in particular) are toxic and can cause severe burns on inhalation or skin contact. Deuterochloroform is potentially carcinogenic. Sodium thiosulfate should be treated as an irritant and is corrosive, especially to the eye. Sodium thiosulfate can release sulfur and sulfur dioxide (corrosive and toxic gas) if exposed to acids.

Experimental Procedure

Laboratory Session 1: Synthesis of 3-Bromo-4-hydroxy-5-methoxybenzaldehyde

1. Transfer 3.0 g of vanillin and 15 mL of glacial acetic acid to a 100 mL conical flask. Add a magnetic stirrer bar and start stirring the solution.
2. From a burette (positioned in a fume hood), slowly add 1.1 mL of bromine to 10 mL of glacial acetic acid in a 50 mL conical flask. Swirl the resulting solution to ensure thorough mixing.
3. Using a Pasteur pipette, transfer small portions of the bromine/acetic acid solution over a period of 10 min to the stirred solution of vanillin.
4. On completion of the bromine addition, the reaction should be allowed to stir for 2 h at room temperature.
5. Pour the reaction mixture onto 50 g of crushed ice and allow the ice to melt.
6. Filter off the resulting white solid under suction and wash the filter cake with distilled water (3 × 15 mL).

Purification of 3-bromo-4-hydroxy-5-methoxybenzaldehyde: Recrystallise the crude product from a minimal volume of ethanol.

Weigh the product and determine the % yield. Measure the melting point and record a ^{1}H NMR ($CDCl_3$) spectrum of your product.

Laboratory Session 2: Synthesis of 4-Hydroxy-3-iodo-5-methoxybenzaldehyde

1. Transfer 100 mL of water, 2.5 g of vanillin, 1.7 g of sodium hydrogen carbonate and 3.3 g of potassium iodide to a 250 mL conical flask. Add a magnetic stirrer bar and start stirring the resulting slurry.
2. Add 8.4 g of iodine (4 × 2.1 g portions) over 30 min. On completion of the iodine addition, the reaction should be allowed to stir for 3 h at room temperature.
3. Seal or stopper the reaction flask and store the reaction flask for 12 h.
4. Filter off the resulting solid under suction and wash the filter cake with dilute (0.05–0.1 M) sodium thiosulfate solution (2 × 15 mL) and water (15 mL).

Purification of 4-hydroxy-3-iodo-5-methoxybenzaldehyde: recrystallise the crude product from a minimal volume of ethanol.

Weigh the product and determine the % yield. Measure the melting point and record a ^1H NMR ($CDCl_3$) spectrum of your product.

Waste Disposal

Waste solutions containing bromine or iodine should be treated with a solution of sodium metabisulfite or sodium thiosulfate. Aqueous washings from the filtration steps should be diluted with water and disposed of down a sink. All organic solvent waste should be retained and transferred to an appropriate container for disposal.

Results Interpretation and Additional Questions

1. Interpret the ^1H NMR spectra you obtained from both reaction steps. Use the NMR spectra and melting points to confirm the structure of your compounds unambiguously.
2. Provide a curly arrow mechanism for the bromination of vanillin.
3. Account for the following observations: the bromination of phenol occurs rapidly in the presence of bromine while the bromination of nitrobenzene requires heating and addition of ferric bromide as a catalyst in addition to bromine in order to obtain the desired product.
4. Account for the following observations: the iodination of vanillin does not efficiently proceed without addition of a base while the bromination of vanillin is favourable without addition of other reagents.

References

1. L. F. Fieser and K. L. Williamson, *Organic Experiments*, D. C. Heath and Company, Lexington MA, 7th edn, 1992, ch. 36, pp. 351–353.
2. C. J. Donahue and E. R. Donahue, *J. Chem. Educ.*, 2013, **90**, 1688.
3. L. M. Harwood and C. J. Moody, *Experimental Organic Chemistry: Principles and Practice*, Blackwell Science, Oxford, 1989, pp. 451–453.
4. P. Cardinal, B. Greer, H. Luong and Y. Tyagunova, *J. Chem. Educ.*, 2012, **89**, 1061.
5. R. G. Fowler, *J. Chem. Educ.*, 1992, **69**, A43.
6. N. I. Sax, *Dangerous Properties of industrial Materials*, Van Nostrand Reinhold, New York, 5th edn, 1979, Section 12.
7. I. A. Pearl, *J. Chem. Educ.*, 1958, **35**, 502.
8. M. B. Hocking, *J. Chem. Educ.*, 1997, **74**, 1055.
9. G. Wu, M. Heitz and E. Chornet, *Ind. Eng. Chem. Res.*, 1994, **33**, 718.
10. S. C. Puri, S. M. Anand and C. K. Atal, *Indian J. Chem.*, 1985, **24B**, 294.
11. D. W. Kang, H. Ryu, J. Lee, K. A. Lang, V. A. Pavlyukovets, L. V. Pearce, T. Ikeda, J. Lazar and P. M. Blumberg, *Bioorg. Med. Chem. Lett.*, 2007, **17**, 214.
12. A. R. Butler, *J. Chem. Educ.*, 1971, **48**, 508.
13. E. Eby and S. T. Deal, *J. Chem. Educ.*, 2008, **85**, 1426.
14. N. C. Ganguly, S. K. Barik and S. Dutta, *Synthesis*, 2010, 1467.
15. R. D. C. Gallo, I. M. Ferreira, G. A. Casagrande, L. Pizzuti, D. Oliveira-Silva and C. Raminelli, *Tetrahedron Lett.*, 2012, **53**, 5372.
16. W. M. Gardner and H. H. Hodgson, *J. Chem. Soc., Trans*, 1909, **95**, 1819.
17. J. V. McCullagh and K. A. Duggat, *J. Chem. Educ.*, 2007, **84**, 1799.
18. E. Grovenstein, N. S. Aprahamian, C. J. Bryan, N. S. Gnanapragasm, D. C. Kilby, J. M. McKelvey and R. J. Sullivan, *J. Am. Chem. Soc.*, 1973, **95**, 4261.

5.1.3. Synthesis of 2-(2,4-Dinitrobenzyl)pyridine

Dulce Simão

Department of Chemistry, Instituto Superior Técnico, University of Lisbon, Centro de Química Estrutural, Av. Rovisco Pais, 1, 1049-001 Lisbon, Portugal. *E-mail: dulcesimao@tecnico.ulisboa.pt

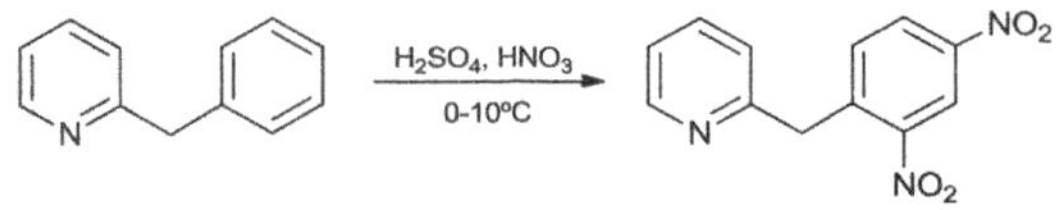

Number of sessions (duration of each session)	Hazard level	Difficulty level	Level of study
1 (4 h)	Moderate	Medium	Introductory

Class names Arenes, pyridines, aryl nitros

Concepts involved Electrophilic aromatic substitution, tautomerism, photochromism, low temperature control reaction

Chemicals needed 2-Benzylpyridine, concentrated sulfuric acid, fuming nitric acid, sodium hydroxide, diethyl ether, ethanol

Equipment and experimental techniques involved Stirring, heating plate, liquid–liquid extraction, rotavapor, recrystallization, NMR, IR, melting point

Keywords Aromatic substitution, electrophilic, IR, liquid–liquid extraction, NMR, photochromism, tautomerism

Background

A photochromic compound is a molecule that reversibly changes color when it absorbs light, normally from uncolored to strongly colored.[1] There are many industry applications of these compounds. For example, lenses of sun glasses, toys, cosmetics and cloths. In this experiment, the product synthesized, 2-(2,4-dinitrobenzyl)pyridine, has this ability as solid and in solution. In the absence of light this compound is tan in color but when exposed to sunlight it turns deep blue in a few minutes. It takes about one day in the dark to lose the coloration and return to the original sandy color. This photochromic behavior is due to the interconversion between two tautomeric forms involving benzyl hydrogen transference to an oxygen atom of the nitro group (Scheme 5.1.3.1).

This compound is prepared by electrophilic aromatic substitution, with 2-benzylpyridine dinitration, under relatively forcing conditions.[2–6] The electrophile is the nitronium ion generated by the interaction of concentrated nitric and sulfuric acid. The nitronium ion reacts preferably with the benzene ring instead the pyridine ring. The presence of a nitrogen atom results in lower reactivity when compared to benzene due to lower stabilization of the intermediate arenium. The two nitro groups are introduced in the benzene ring at the *ortho* and *para* positions in accordance with the substituent directing effect of the benzylpyridine group.

Additional Safety

This work must be carried out in a fume hood due to toxic and irritant, 2-benzylpyridine. Fuming nitric acid and sulfuric acid, which are corrosive, should be handled with care.

Comprehensive Organic Chemistry Experiments for the Laboratory Classroom
Edited by Carlos A. M. Afonso, Nuno R. Candeias, Dulce Pereira Simão, Alexandre F. Trindade, Jaime A. S. Coelho, Bin Tan, and Robert Franzén.

Published by the Royal Society of Chemistry, www.rsc.org

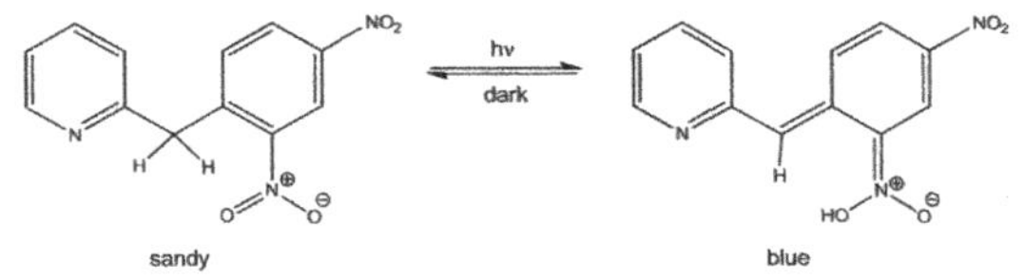

Scheme 5.1.3.1 Tautomerism of 2-(2,4-dinitrobenzyl)pyridine.

Experimental Procedure

1. Cool in an ice bath 2-benzylpiridine (3.2 mL) and fuming nitric acid (4.0 mL) separately.
2. Place sulfuric acid (20 mL) in a 50 mL three-necked flask fitted with a reflux condenser, thermometer, dropping funnel and a stirrer magnetic bar.
3. Stir and cool in an ice-salt bath to below 5 °C. Add cold 2-benzylpyridine dropwise at such a rate that the temperature remains below 10 °C.
4. Add cold nitric acid in the same way. The temperature should remain below 10 °C.
5. Replace the ice bath with a water bath and heat the mixture to 80 °C with stirring. After 20 min at this temperature, pour the contents, carefully, onto crushed ice (*ca.* 200 g) in a 1 L beaker, rinsing the reaction flask with ice-water (*ca.* 40 mL).
6. Place the beaker in an ice bath, replace the bar magnet with one of a larger size, stir gently and add the sodium hydroxide solution (2 M) dropwise until the mixture is strongly alkaline (*ca.* pH 11).
7. To the resulting milky yellow mixture add diethyl ether (300 mL) and keep stirring for a further 15 min to extract the product into the organic phase.
8. Pour into a separating funnel and separate the organic layer, dry it over magnesium sulfate and filter into a 500 mL round-bottomed flask and reduce the volume to 60 mL on the rotary evaporator without external heating.
9. Cool the flask in ice and collect the product by vacuum filtration.
10. Recrystallize from ethanol (use a little decolorizing charcoal if necessary).
11. Record the yield, appearance, melting point and IR and ^{1}H and ^{13}C NMR spectra of the purified product.
12. Observe the effect of sunlight on a few crystals.

Waste Disposal

The neutralized aqueous solutions are to be collected for treatment before disposal according to current, local/national legislation. Ethanol and diethyl ether may be stored for further purification.

Results Interpretation and Additional Questions

1. Determine the reaction yields.
2. Compare the melting point of the obtained product with the one reported in the literature.
3. Interpret the IR and ^{1}H NMR spectra.
4. Write the reaction mechanism and resonance structures of the intermediate carbocation. Explain why the benzylpyridine group is *ortho* and *para* directing.

References

1. M. Irie, *Chem. Rev.*, 2000, **100**(5), 1683–1684.
2. C. A. M. Afonso, D. P. Simão, L. P. Ferreira, M. S. Serra and M. M. M. Raposo, *100 Experiências de Química Orgânica*, IST Press, 2011, p. 361.
3. L. M. Harwood; C. J. Moody and J. M. Percy, *Experimental Organic Chemistry*, Blackwell Science, 2nd edn, 1999, pp. 642–644.
4. E. D. Gilfillan and M. W. Pelter, *J. Chem. Educ.*, 1994, **71**(1), A4.
5. A. Ault, *J. Chem. Educ.*, 2000, **77**, 1386–1387.
6. N. M. Zaczek, W. D. Levy, M. L. Jordan and J. A. Niemyer, *J. Chem. Educ.*, 1982, **59**, 705.

5.1.4. Regioselectivity in the Nitration of Acylanilines by Electrophilic Aromatic Substitution

João P. Telo, Pedro P. Santos*, João V. Moreira, Mariana L. Santos, and Ana T. Batista

CQE, Department of Chemical Engineering, IST, University of Lisbon, Av. Rovisco Pais, 1649-003 Lisbon, Portugal. *E-mail: jptelo@tecnico.ulisboa.pt, pepesantos@tecnico.ulisboa.pt

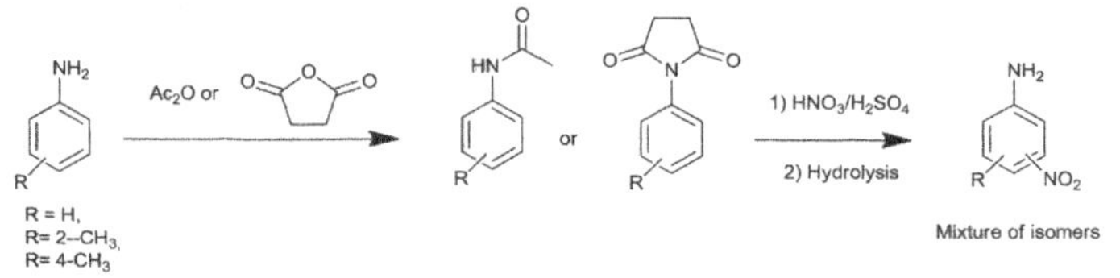

Number of sessions (duration of each session)	Hazard level	Difficulty level	Level of study
3 (4 h each)	Moderate	Medium	Introductory

Class names Aromatic ring, aniline, amide, anhydride, nitro group

Concepts involved This set of experiments highlights the importance of electronic and stereochemical hindrance effects in the nitration orientation of anilines by electrophilic aromatic substitution (EAS)

Chemicals needed Aniline, 4-methylaniline, 2-methylaniline, acetic anhydride, succinic anhydride, glacial acetic acid, ethyl acetate, nitric acid, sulfuric acid, sodium hydroxide, acetonitrile (HPLC grade), distilled water, anhydrous sodium sulfate, methylene chloride, heptane

Equipment and experimental techniques involved Apparatus for reaction with temperature (50–200 °C) and magnetic stirring (hot-plate magnetic-stirrer device), liquid–liquid extraction apparatus, analytical HPLC, RP18 HPLC column, silica TLC plates

Keywords Electrophilic aromatic substitution, HPLC chromatography, nitration, recrystallization, regioselectivity in EAS, TLC chromatography

Background

Electrophilic aromatic substitution[1–3] (EAS) is a fundamental reaction by which to prepare substituted aromatic compounds, both on a laboratory and an industrial scale. Nitration is one of the most useful reactions of this class and its utility is proven by its use in the production of important products for various applications (*e.g.* fragrances, explosives and active pharmaceutical ingredients). The nitro group attached by an EAS reaction can then be transformed into a large array of other functional groups. For example, by reduction of the nitroaromatic to aniline and diazotization, a versatile reactive intermediate (diazonium salt) is formed that can be used for the introduction of several groups in the aromatic ring.

When EAS is performed in a substituted aromatic ring, more than one product is frequently obtained and this may be a drawback for the reaction utility. However, the ease of the experimental nitration procedure by EAS and sometimes easy product separation by physical methods (*e.g.* crystallization, distillation) frequently surpasses this weaknesses. Orientation effects in EAS

Comprehensive Organic Chemistry Experiments for the Laboratory Classroom
Edited by Carlos A. M. Afonso, Nuno R. Candeias, Dulce Pereira Simão, Alexandre F. Trindade, Jaime A. S. Coelho, Bin Tan, and Robert Franzén.

Published by the Royal Society of Chemistry, www.rsc.org

can be anticipated and the product distribution is the result of a kinetically controlled (either by electronic or stereochemical hindrance effects) process. The groups present in the starting material will critically affect the EAS rate and its orientation.

The amine group of aniline is a strongly activating for EAS and more than one substitution may occur.[3] Furthermore, the amine group can be easily oxidized under nitration reaction conditions. To protect the amine group from the oxidizing reaction and to reduce its reactivity in the EAS, an acetylation to the acetanilide derivative is usual. This acetyl-amide derivative is the starting material for the nitration step. After the nitration, the acyl group is removed by acid or basic hydrolysis (Scheme 5.1.4.1).

In this series of experiments, nitration of *N*-acetyl or *N*-succinimidyl derivatives of aniline and 2-methyl- and 4-methylaniline is performed, followed by hydrolysis of the amide or imide group. The regioselectivity of the nitration can be analysed by TLC and/or quantified by HPLC. The objective of the work is to promote discussion on the regioselectivity of the EAS nitration according to the electronic and stereochemical effects of these substituents. Some important notes in terms of understanding the product distribution in the nitration reactions performed in this experiment are:

- Substituents containing an unshared electron pair on the atom connected to the aromatic ring (*e.g.* the NH_2, *N*-acetyl and *N*-succinimidyl (Su) groups in the starting materials used in this experiment) are electron donating by resonance, activating the aromatic ring for EAS. *Ortho*/*para* products are usually observed.
- Alkyl groups are also *ortho*/*para* directing, although less strongly than groups containing an unshared electron pair on the atom connected to the aromatic ring.
- In the sequence NH_2, NHAc and NSu, the electron donating effect of the nitrogen decreases due to the electron withdrawing effect of the carbonyl group.
- Steric hindrance may reduce the reactivity at positions ortho to bulky substituents.
- In the specific case of the succinimidyl derivative of 2-methylaniline, electron donation of the unshared electron pair on the nitrogen to the aromatic ring is impossible, as the molecule cannot assume planarity due to steric hindrance between the methyl group and the succinimide carbonyl groups.

The product distribution obtained for the various nitration reactions can be interpreted by analysis of the electronic and stereoelectronic effects involved. A typical product distribution is shown in Scheme 5.1.4.2.

A rough analysis of the product distribution of the nitration products can be performed using thin-layer chromatography (silica, CH_2Cl_2/pentane), or alternatively quantification can be attained by HPLC (RP18 column, detection at 254 nm).

It is suggested that three different groups of students execute the acylation/nitration/deprotection sequence using one of three anilines as starting materials. Each group can perform the analysis of their product or all the obtained nitroanilines can be analysed together. Analysis of the observed product distribution may be made individually or in a group brainstorming session.

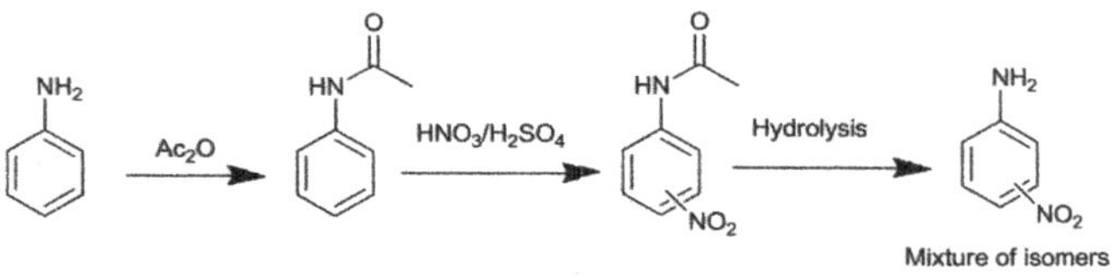

Scheme 5.1.4.1 Nitration of aniline using an acetamide derivative for the EAS step.

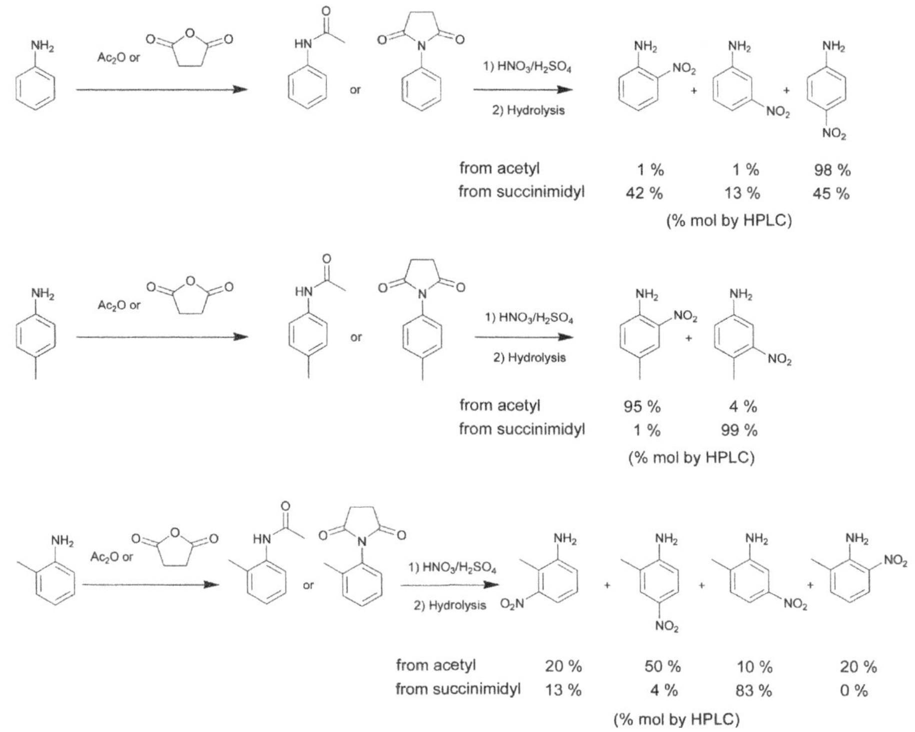

Scheme 5.1.4.2 Typical product distribution obtained in the nitration of *N*-acylaniline and *N*-acylmethylanilines.

Additional Safety

Concentrated nitric and sulfuric acid are strong acids, corrosives and should be handled with care. Gloves, goggles and appropriate laboratory coat must be used during the experimental execution and the experiment should be performed in a well-ventilated fume hood. Anilines and nitroanilines are toxic by inhalation of the vapour, ingestion, or percutaneous absorption. Skin contact should be avoided at all times. Vapours may be liberated during the acidic steps.

Experimental Procedure

Laboratory Session 1 (4 h): Protection of the Aniline

Preparation of the Acetanilide Derivatives

1. Dissolve 0.11 mol of the aniline (10.7 g of aniline or 12.3 g of methylaniline) in 20 mL of glacial acetic acid in a 250 mL round-bottomed three-neck flask equipped with a magnetic stirrer, a thermometer, a reflux condenser and a dropping funnel (Figure 5.1.4.1).
2. Add through the dropping funnel 10.2 g of acetic anhydride at such a rate that the temperature of the mixture does not exceed 80 °C (Caution: exothermic reaction!).
3. After complete addition, allow the mixture to stand under stirring until it reaches room temperature (around 15 min).
4. Pour the solution into a mixture of ice and water (150–200 mL total volume) and stir until full precipitation of the acetanilide.
5. Filter with suction on a Büchner funnel and wash the precipitate with water until the pH of the washings is neutral.
6. Dry the crystals in a desiccator overnight under vacuum.

Preparation of the Succinimidyl Derivatives

1. Add 0.16 mol of the aniline derivative and 16.0 g of succinic anhydride to a 250 mL glass beaker (tall form) and add a thermometer.
2. Place the open beaker directly on a heating plate and heat the mixture slowly in a fume cupboard until the temperature reaches 200 °C (3 h), stirring the mixture occasionally with a glass rod (be careful not to break the thermometer). Water vapour is released during the reaction, but the loss of other materials is negligible (decrease the temperature if some crystals start forming in the top of the vessel by sublimation).
3. Keep the mixture at 200 °C for 1 h. At the end of this period allow the mixture to cool to 60–70 °C and add enough EtOH/H_2O (1 : 1) to solubilize the solids.
4. Allow to cool to room temperature and then to 0 °C by immersing the beaker in a mixture of ice and water.
5. Filter the product under vacuum in a Büchner funnel.

Laboratory Session 2 (4 h): Nitration

Nitration of the Protected Aniline (ref. 4)

1. Add 0.185 mol of the protected aniline derivative and 25 mL of glacial acetic acid to a round-bottomed three-neck flask equipped with a magnetic stirrer, a thermometer, a reflux condenser and a dropping funnel (same apparatus as shown in Figure 5.1.4.1).
2. Add slowly to the well-stirred mixture 50 mL of concentrated sulfuric acid.
3. Cool the flask with a mixture of ice, water and salt and allow the temperature to reach 0–5 °C.
4. Add slowly through the dropping funnel a mixture of 11 mL of concentrated nitric acid and 7 mL of concentrated sulfuric acid while keeping the temperature of the mixture below 10 °C.
5. After the addition, allow the mixture to stand under stirring at room temperature for 1 h (in the case of the succinimidyl anilines warm to 30 °C to complete the nitration).
6. Pour the solution into a mixture of ice and water (around 200 mL), filter the yellow solid off with suction on a Büchner funnel and wash with water. (Note: do not recrystallize the product because this would change the composition of the mixture.)

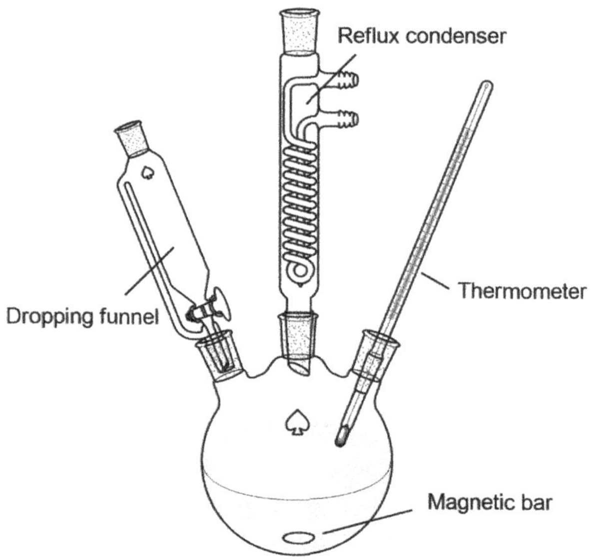

Figure 5.1.4.1 Amide formation apparatus.

Laboratory Session 3 (4 h): Deprotection (hydrolysis) and Nitroaniline Product Distribution Analysis/Quantification

Hydrolysis of the Nitrated Acetyl or Succinimidyl Anilines

1. Heat under reflux a mixture of the nitro acetyl or succinimidyl anilines, 30 mL of concentrated sulfuric acid and 20 mL of water for 45 min.
2. Pour the solution into a mixture of ice and water and precipitate the nitroanilines by neutralizing the mixture with 10% NaOH solution.
3. Extract twice with ethyl acetate (30 mL), dry the organic phase over anhydrous sodium sulfate, filter and evaporate to dryness. The nitroaniline derivatives can be dried in a desiccator.

Quantification of the Nitroaniline Derivative Isomers in the Mixture

Thin-layer chromatography method (TLC): silica 60 G F254. CH_2Cl_2/heptane (1 : 1 or 7 : 3), visualized using a UV 254 nm lamp. Photographs of typical TLCs are included in the ESI.

HPLC method: RP18 column, Inertsil® ODS-3 column (250 mm × 4.6 mm; 5 μm), acetonitrile/water (65/35) at a flow rate of 0.8 mL min^{-1}. Injection solution at 1 mg mL^{-1}, 20 μL loop injector. The UV absorbance was monitored at 254 nm.

(a) For aniline nitration: TLC analysis: R_f (CH_2Cl_2/heptane, 1 : 1, eluted twice): 2-nitroaniline, 0.65; 3-nitroaniline, 0.48; 4-nitroaniline 0.44; HPLC analysis: Approximate retention times (min): 2-nitroaniline, 18.6; 3-nitroaniline, 15.3; 4-nitroaniline 12.5. UV absorption correction factor at 254 nm: 2-nitroaniline, 1.52; 3-nitroaniline, 2.80; 4-nitroaniline 1.00 (dividing the UV areas of each peak on the HPLC chromatogram by the factors above, the relative mass proportion is obtained, allowing a quick product distribution determination).

(b) For 4-methylaniline nitration: TLC analysis: R_f (CH_2Cl_2/heptane, 7 : 3, eluted once): 4-methyl-2-nitroaniline, 0.44; 4-methyl-3-nitroaniline, 0.28; HPLC analysis: approximate retention times (min): 4-methyl-2-nitroaniline, 12.4; 4-methyl-3-nitroaniline, 10.7. UV absorption correction factor at 254 nm: 4-methyl-2-nitroaniline 1.00, 4-methyl-3-nitroaniline 1.30.

(c) For 2-methylaniline nitration: TLC analysis: R_f (CH_2Cl_2/heptane, 1 : 1, eluted twice): 2-methyl-3-nitroaniline, 0.33; 2-methyl-4-nitroaniline, 0.25; 2-methyl-5-nitroaniline, 0.28; 2-methyl-6-nitroaniline, 0.50. HPLC analysis: approximate retention times (min): 2-methyl-3-nitroaniline, 21.5, 2-methyl-4-nitroaniline, 19.8, 2-methyl-5-nitroaniline, 24.7, 2-methyl-6-nitroaniline, 33.0. UV absorption correction factor at 254 nm: 2-methyl-3-nitroaniline, 0.52; 2-methyl-4-nitroaniline, 0.36; 2-methyl-5-nitroaniline, 1.00; 2-methyl-6-nitroaniline, 0.32.

Waste Disposal

Mother liquors from the product filtration must be neutralized with a dilute solution of sodium hydroxide before disposal.

Results Interpretation and Additional Questions

1. Draw the reaction mechanism for the nitration of *N*-acetylaniline.
2. Draw the reaction mechanism for the acyl formation and hydrolysis.
3. Explain the product distribution observed for the nitration of acylaniline derivatives.
4. Explain the product distribution observed for the nitration of 4-methylaniline derivatives.
5. Explain the product distribution observed for the nitration of 2-methylaniline derivatives.

6. Discuss the influence of the position of the methyl groups on the product distribution.
7. In the 2-methylaniline series, which group is the predominant one in directing the EAS? Why?

References

1. T. W. G. Solomons, C. B. Fryhle and S. A. Snyder, *Organic Chemistry, International Student Version*, 11th edn, John Wiley & Sons, Inc., Hoboken, 2014.
2. M. B. Smith, *March's Advanced Organic Chemistry: Reactions, Mechanisms, and Structure*, 7th edn, John Wiley & Sons, 2013.
3. J. Clayden, N. Greeves and S. Warren, *Organic Chemistry*, 2nd edn, Oxford University Press, Oxford, 2012.
4. A. R. Tatchell, B. S. Furnis, A. J. Hannaford and P. W. G. Smith, *Vogel's Text-Book of Practical Organic Chemistry*, 5th edn, Longman, Harlow, 1989.

5.1.5. Synthesis of 2-Nitro-4-methylaniline

Dulce Simão* and Ana C. Cerdeira

Department of Chemistry, Instituto Superior Técnico, University of Lisbon, Centro de Química Estrutural, Av. Rovisco Pais, 1, 1049-001 Lisbon, Portugal. *E-mail: dulcesimao@tecnico.ulisboa.pt

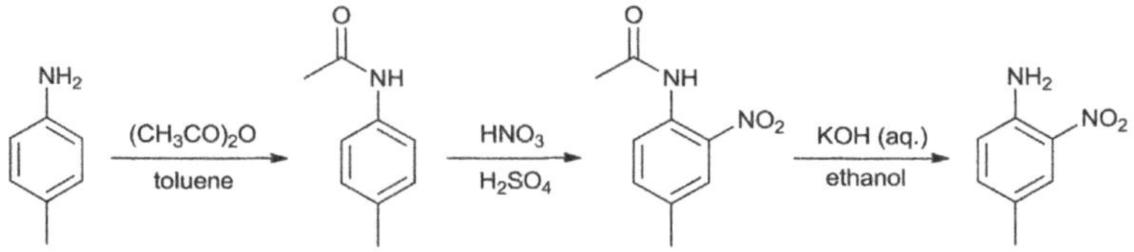

Number of sessions (duration of each session)	Hazard level	Difficulty level	Level of study
3 (4 h each)	Moderate	Medium	Introductory

Class names Aromatic amines, aromatic amides, nitro compounds, anhydrides

Concepts involved Multi-step synthesis illustrating an electrophilic aromatic substitution employing an amine protection/deprotection strategy

Chemicals needed Acetic anhydride, 4-methylaniline, anhydrous calcium chloride, dry toluene, petroleum ether (40–60 and 80–100 °C), ethanol, glacial acetic acid, concentrated sulfuric acid, concentrated nitric acid, potassium hydroxide, DMSO-d_6

Equipment and experimental techniques involved Recrystallization, ^{1}H NMR, reflux apparatus, mechanical stirring, vacuum filtration, heating mantle; reaction with temperature control, IR, melting point

Keywords Amine protection/deprotection, electrophilic aromatic substitution, ^{1}H NMR, nitration, recrystallization, temperature-controlled reactions

Background

2-Nitro-4-methylaniline is an aromatic amine used in the fabrication of dyes, as a diazo component in C.I. Pigment Yellow 1 and C.I. Pigment Red 3.[1] It is prepared from 4-methylaniline, following a three-step synthetic route. Anilines are highly reactive and cause several problems in terms of electrophilic aromatic substitution. In a direct nitration the amine group is easily oxidized with HNO_3 with the formation of several by-products.[2] The strongly acidic nitration conditions (nitric acid and sulfuric acid) causes the protonation of nitrogen to form NH_3^+ which is a strong deactivating group and a *meta* director. The strategy usually employed is to acylate the amine to form an amide, which is much less basic than aniline. Conversion of the amino group into the acetamido group by acylation modifies the interaction of the nitrogen lone pair with the π-electron system of the aromatic ring so that the ring is less activated towards electrophilic attack.[3a–5] Consequently, in this experiment, the first step is the protection of the amine group through acylation to form *N*-acetyl-4-methylamine. The second step is the nitration by electrophilic aromatic substitution where a nitronium ion (NO_2^+) is formed by interaction of sulfuric and nitric acid. The bulky acetamido group preferentially directs the nitration to the *para*

Comprehensive Organic Chemistry Experiments for the Laboratory Classroom
Edited by Carlos A. M. Afonso, Nuno R. Candeias, Dulce Pereira Simão, Alexandre F. Trindade, Jaime A. S. Coelho, Bin Tan, and Robert Franzén.

Published by the Royal Society of Chemistry, www.rsc.org

position rather than the *ortho* position, but in this experiment, due to the presence of methyl group, only *ortho* nitration occurs.[6] The final step is amine deprotection, where the acetyl group is removed with alkaline hydrolysis by an addition–elimination reaction.[3b,4b,7]

Additional Safety

This work should be carried out in a fume hood. 4-Methylamine is toxic, irritant and a possible carcinogenic agent, and it must be handled carefully. Acetic anhydride, concentrated sulfuric and nitric acids are highly corrosive, as is potassium hydroxide. Ethanol and petroleum ether are flammable solvents. The mechanical stirrer should be manipulated with precaution due to its rotational parts.

Experimental Procedure

Session 1 – Preparation of *N*-Acetyl-4-methylamine

1. Fit a 250 mL three-neck round-bottom flask with a thermometer, reflux condenser with a $CaCl_2$ tube fitted on top and a pressure-equalizing dropping funnel.
2. Add 4-methylaniline (10.7 g, 0.1 mol) and dry toluene (20 mL).
3. Stir and slowly add dropwise distilled acetic anhydride (10.2 g, 9.4 mL, 0.1 mol) keeping the temperature below 80 °C.
4. After complete addition, slowly cool the reaction mixture to room temperature until the product precipitates.
5. Collect the product by vacuum filtration and wash it with petroleum ether (40–60 °C) (*ca.* 50 mL). Recrystallize from ethanol/water (4 : 1).
6. Record the melting point and IR and ^1H NMR spectra.

Session 2 – Preparation of *N*-Acetyl-2-Nitro-4-methylaniline

1. Equip a three-neck round-bottom flask with a mechanical stirrer, reflux condenser and thermometer.
2. Add glacial acetic acid (12.5 mL) and *N*-acetyl-4-methylaniline (14 g, 0.094 mol) to the flask and to this mixture slowly add concentrated H_2SO_4 (25 mL).
3. Initiate stirring and cool this mixture in an ice/salt bath.
4. In a 50 mL Erlenmeyer flask prepare a mixture of concentrated HNO_3 (7.5 mL) and concentrated H_2SO_4 (3.5 mL), cool it and place it on a pressure-equalized dropping funnel, which should be fitted on top of the condenser.
5. Once the temperature of the reaction mixture reaches 0–2 °C, initiate the dropwise addition of this acid mixture keeping the temperature below 10 °C.
6. Remove the cooling bath and stir at room temperature for 1 h.
7. Pour the reaction mixture over crushed ice (*ca.* 150 g). Allow it to stand for 15–20 min until the product separates.
8. Filter with vacuum, wash thoroughly with cold water and dry.
9. Recrystallize from ethanol or petroleum ether (80–100 °C).
10. Record the melting point and IR and ^1H NMR spectra.

Session 3 – Preparation of 2-Nitro-4-methylaniline

1. In a 250 mL round-bottom flask equipped with a reflux condenser prepare a solution of KOH (5.7 g, 0.1 mol) in water (8 mL) and ethanol (50 mL).
2. Add *N*-acetyl-2-nitro-4-methylaniline (8.1 g, 0.042 mol) in small portions and reflux the red solution for 1 h in a water bath.

3. Remove the heat and fit a pressure-equalized dropping funnel on top of the condenser containing H_2O (60 mL). Add dropwise the water. The product crystallizes as dark red needles.
4. Cool the mixture in an ice bath, collect the precipitated crystals by vacuum filtration and wash them twice with 20 mL of ethanol/water (1 : 1) and dry.
5. Recrystallize from ethanol. Record the melting point and IR and ^1H NMR spectra.

Waste Disposal

The filtrate from the first step should be disposed of in a waste collector recipient for non-chlorinated solutions. Acid and base filtrates should be neutralized and diluted, then discarded in the sink. Petroleum ether and ethanol from washings and recrystallizations may be stored for further purification or be placed in a waste collector recipient for non-chlorinated solvents.

Results Interpretation and Additional Questions

1. Determine the yields of reaction products.
2. Compare the melting point of the obtained product with the one reported in the literature.
3. Interpret the IR and NMR spectra.
4. Write all reaction mechanisms.
5. Explain why the direct nitration of 4-methylaniline cannot be done.
6. If aniline were used instead 4-methylaniline what would be the product or products obtained in the nitration reaction? Explain why.

References

1. G. Booth, *Ullman's Encyclopedia of Industrial Chemistry*, VCH, Weinheim, 5th edn, 1991, vol. A17, p. 411.
2. J. March, *Advanced Organic Chemistry*, John Wiley & Sons, 4th edn, 1992, p. 523.
3. T. W. G. Solomons, *Organic Chemistry*, John Wiley & Sons Inc., 7th edn, 1998(a) p. 700; (b) p. 841.
4. J. Clayden, N. Greeves, and S. Warren and P. Wothers, *Organic Chemistry*, Oxford University Press, 2001(a) p. 558; (b) p. 293.
5. M. Sainsbury, *Aromatic Chemistry, Oxford Chemistry Primers,* Oxford University Press, 1992, p. 56.
6. L. F. Tietze and T. Eicher, *Reactions and Synthesis in the Organic Chemistry Laboratory*, University Science Books, 1989, p. 141.
7. C. A. M. Afonso, D. P. Simão, L. P. Ferreira, M. S. Serra and M. M. M. Raposo, *100 Experiências de Química Orgânica*, IST Press, 2011, p. 353.

5.1.6. Synthesis of 1-Nitronaphthalene

Adam Dzielendziak

Department of Organic Chemistry, Faculty of Chemistry, Nicolaus Copernicus University in Torun, 87-100 Torun, Poland. *E-mail: adamdz@chem.umk.pl

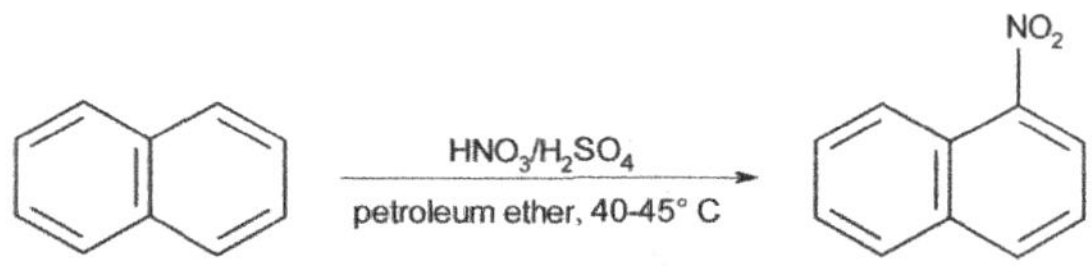

Number of sessions (duration of each session)	Hazard level	Difficulty level	Level of study
2 (4 h + 1.5 h)	Moderate	Low	Intermediate

Class names Polycyclic aromatic hydrocarbons, aromatic nitro compounds

Concepts involved Synthesis – aromatic electrophilic substitution

Chemicals needed Naphthalene, nitric(V) acid (65%), concentrated sulfuric(VI) acid, petroleum ether (40–60 °C), 2-propanol, acetone, toluene, ethyl acetate, acetic acid

Equipment and experimental techniques involved Round-bottom two-neck flask (250 mL), beaker 100 mL, thermometer (0–100 °C), magnetic stirrer with heating, condenser, separatory funnel (250 mL), Erlenmeyer flask (100 mL), round-bottom one-neck flask (250 mL), rotary evaporator, melting point apparatus, TLC plates, TLC chamber, UV lamp, IR spectrometer, ^{1}H NMR spectrometer, Pasteur pipette, capillary tubes; two-phases reaction, washing and separation of crude product, TLC and column chromatography, IR and ^{1}H NMR analysis

Keywords Crystallization, ^{1}H NMR, IR, naphthalene, nitration, TLC and column chromatography, two-phase reaction

Background

The nitration of naphthalene is a typical aromatic electrophilic substitution reaction.[1] The electrophile is the nitronium ion generated by the interaction of concentrated nitric(V) and sulfuric(VI) acids (nitrating mixture). The nitronium ion reacts with naphthalene preferably at position 1 to give the intermediate arenium ion, stabilized by resonance. The intermediate then transfers a proton to the basic sulfate ion, producing 1-nitronaphthalene.

It is very well known that nitration of naphthalene with the nitrating mixture under mild conditions gives only one monoproduct – 1-nitronaphthalene.[1,2] High temperature and prolonged reaction time should be avoided since polynitration is then more likely and oxidative breakdown of the aromatic ring system may occur.[3–6] After the first step of nitration one ring of naphthalene is substituted with a deactivating nitro group. 1,5-Dinitronaphthalene and 1,8-dinitronaphthalene have been found as by-products.

Preparation of 1-nitronaphthalene is described in *Vogel's Practical Organic Chemistry*.[2] The reaction of naphthalene takes place involving concentrated nitric(V) acid and concentrated sulfuric(VI)

Comprehensive Organic Chemistry Experiments for the Laboratory Classroom
Edited by Carlos A. M. Afonso, Nuno R. Candeias, Dulce Pereira Simão, Alexandre F. Trindade, Jaime A. S. Coelho, Bin Tan, and Robert Franzén.

Published by the Royal Society of Chemistry, www.rsc.org

acid at 60 °C. The yield of the final product (1-nitronaphthalene) given in this textbook is 89% and the mp is 61 °C. However, reaction yield and product purity are far from this data.

In this experiment a mixture of nitric(V) and sulfuric(VI) acids reacts with naphthalene dissolved in petroleum ether in a two-phase system.

Additional Safety

Care should be taken in the preparation and handling of the nitrating mixture. Thus, concentrated sulfuric acid should always be mixed with nitric acid very carefully, by pouring into cold nitric acid as a thin stream to prevent acid splashes or spray.

Concentrated sulfuric(VI) acid is very hazardous in case of skin contact (corrosive, irritant), of eye contact (irritant, corrosive), and of ingestion. Skin contact with sulfuric acid may produce burns. In case of contact immediately flush eyes or skin with plenty of water for at least 15 min.

Concentrated nitric(V) acid is also very hazardous in case of skin contact (corrosive, irritant), of eye contact (irritant, corrosive), and of ingestion. In case of contact immediately flush eyes or skin with plenty of water for at least 15 min.

Petroleum ether is harmful if swallowed. It causes eye, skin, and respiratory tract irritation. It is extremely flammable, both as a liquid and vapour.

2-Propanol is flammable and hazardous in case of eye contact (irritant), of ingestion, and of inhalation. It is slightly hazardous in case of skin contact. In case of contact immediately flush eyes or skin with plenty of water for at least 15 min.

1-Nitronaphthalene is toxic if swallowed. It also causes eye and skin irritation, and may cause respiratory irritation, too.

There is no need to perform the synthesis in a fume hood.

Experimental Procedure

Session 1 (4 h)

1. Prepare the nitrating mixture from nitric (1.5 mL) and sulfuric acid (1.5 mL) in a small beaker. Remember to add sulfuric acid to nitric acid, and not *vice versa.*
2. Place 2.56 g of naphthalene, 15 mL of petroleum ether (40–60 °C), and a magnetic stirrer in a 100 mL round-bottom two-neck flask. Attach the thermometer. Put a dropping funnel containing the cold nitrating mixture into the middle neck of the flask.
3. Heat the mixture until the naphthalene dissolves (30–35 °C). Add dropwise the reagent with rapid stirring. The rate of addition must be controlled so that the temperature does not rise above 42 °C. The addition takes 10–15 min.
4. After addition of the nitrating mixture remove the dropping funnel, attach a condenser, and heat the mixture for 0.5 h at 40–45 °C.
5. Then, add 15 mL of petroleum ether and reflux the mixture for 5 min. Remove the condenser and decant immediately the warm petroleum ether solution into a 250 mL Erlenmeyer flask from the mixture of acids.
6. Add 30 mL of petroleum ether to the reaction flask, reflux the mixture for 5 min and repeat decantation.
7. Transfer the solution from the Erlenmeyer flask to the separatory funnel, wash with water (2 × 20 mL), and dry the organic solution with magnesium sulfate(VI).
8. Filter the dry solution into the 250 mL round-bottom flask. Remove the solvent on a rotary evaporator under reduced pressure to dryness.
9. Weight crude product with the flask (weigh the empty flask before or after).
10. Dissolve the product in 2-propanol (12–15 mL) in the same flask, filter the hot solution into a 50 mL crystallization vessel and leave it to cool slowly. Collect crystals by filtration. Leave

the crystals in open air in a fume hood for at least 0.5 h. Taking into consideration slight sublimation of 1-nitronaphthalene keep the product until the next session in a vessel covered with filter paper.

Session 2 (1.5 h)

1. Weigh the product, calculate the yield of crude and pure product, and measure the melting point.
2. Analyse the product by TLC using a plate (2×5 cm^2) of silica gel on aluminium foil with fluorescent indicator (254 nm). Dissolve a small amount of the product in acetone. Dissolve also a small amount of naphthalene in acetone. Apply these two solutions for TLC analysis. Use toluene/ethyl acetate/acetic acid (35:3:1, v:v:v) eluent. Check spots visually and under UV light. Mark all spots with a pencil. Calculate the R_f value of all spots on the TLC plate.
3. Record the IR spectrum of the product in Nujol. Specify all important peaks from the IR spectrum of 1-nitronaphthalene and assign the peaks to appropriate bonds and functional groups.
4. Record the ^{1}H NMR spectrum of the pure product. Note the chemical shifts for all peaks, the amount of splitting, calculate coupling constants, and finally assign the peaks to the appropriate protons of 1-nitronaphthalene.

Waste Disposal

Wash the reaction flask with water, and then with acetone. Acetone waste should be collected in non-halogenated waste container.

The recovered petroleum ether can be used in subsequent experiments without additional purification.

Results Interpretation and Additional Questions

1. Write the equation to show the generation of the electrophile in the nitration reaction.
2. Draw the resonance structures for the carbocation intermediates in the nitration of naphthalene at the 1-position and at the 2-position. Explain why substitution at the 1-position is preferred.
3. Explain why the further substitution of a nitro group in 1-nitronaphthalene at positions 5 and 8 is preferred.
4. Why is naphthalene considerably more reactive than benzene in electrophilic aromatic substitution reactions?
5. Dilute nitric(V) acid is used for the nitration of organic compounds, too. Suggest some simple monosubstituted aromatic compounds.
6. When anilines are nitrated under strong acidic conditions, *meta* orientation is generally observed. If the conditions are less acidic, the orientation is *ortho–para*. Why?

References

1. J. Clayden, N. Greeves and S. Warren, *Organic Chemistry*, Oxford University Press, Oxford, 2nd edn, 2012, p. 476.
2. B. S. Furniss, A. J. Hannaford, P. W. G. Smith and A. R. Tatchell, *Vogel's Textbook of Practical Organic Chemistry*, Longman Scientific & Technical, Harlow, 5th edn, 1989, p. 855.
3. F. Beilstein and A. Kurbatow, *Justus Liebigs Ann. Chem.*, 1880, **202**, 215.
4. H. E. Fierz-David and R. Sponagel, *Helv. Chim. Acta*, 1943, **26**, 98.
5. M. G. Loudon, *Organic Chemistry*, The Benjamin/Cummings Publishing Company, Inc., Redwood City, 3rd edn, 1995, p. 1176.
6. W. Shi, X. Peng, X. Dong, H. Wang and C. Shi, *Asian J. Chem.*, 2012, **24**, 4975.

5.1.7. Selective Boc-Protection and Bromination of Pyrazoles

Thomas A. Logothetis

University of Southampton, Chemistry, Highfield, Southampton, Hampshire, SO17 1BJ, UK
*E-mail: thomas.logothetis@southampton.ac.uk

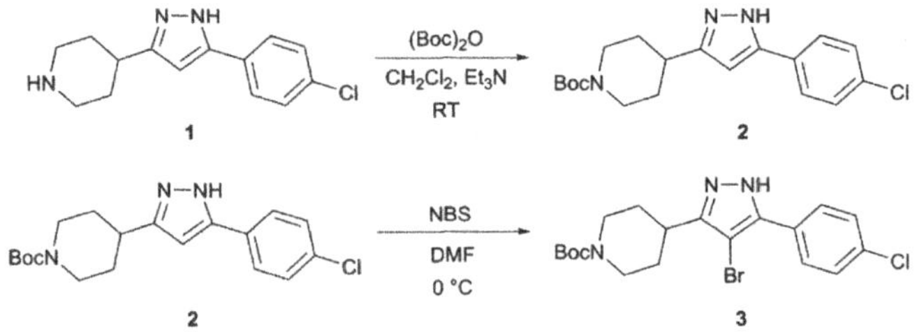

Number of sessions (duration of each session)	Hazard level	Difficulty level	Level of study
2 (3 h each)	Moderate	Medium	Intermediate

Class names Secondary amines, pyrazoles, carbamates, halosuccinimides, pyrocarbonates

Concepts involved This set of two regioselective experiments commences with the selective protection of the secondary piperidine amino group as a carbamate utilising di-*tert*-butyl dicarbonate/*tert*-butyl pyrocarbonate under basic conditions. It proceeds with an electrophilic aromatic substitution on the pyrazole ring consuming *N*-bromosuccinimide (NBS) as a source of "Br^+". The products are analysed by spectroscopic means

Chemicals needed Triethylamine, dimethylformamide, dichloromethane, *tert*-butyl pyrocarbonate/di-*tert*-butyl dicarbonate, magnesium sulfate, light petroleum ether (40–60 °C), hydrochloric acid (2 M), sat. sodium hydrogen carbonate, sat. brine, methanol, ammonia (2 M), *N*-bromosuccinimide (NBS), diethyl ether, ethyl acetate, deuterochloroform, potassium permanganate, 4-[5-(4-chlorophenyl)-1*H*-pyrazol-3-yl]piperidine

Equipment and experimental techniques involved Magnetic stirrer, round bottom flask, ice bath, separating funnel, beakers, Erlenmeyer flasks, TLC analysis, Pasteur pipettes, UV lamp, silica TLC plates and chamber, thermometer (covering −10 to +50 °C), glass rod, tweezers, IR, NMR; stirring at RT and 0 °C using external cooling, monitoring by TLC, slow addition, extractive work-up, rotary evaporation, desiccation, trituration

Keywords Boc-protection, NBS bromination, NMR, pyrazole, trituration

Background

This practical develops the synthesis of pyrazole heterocycles (see Experiment 4.2.2.8. Synthesis of Pyrazole Heterocycles) further by selectively protecting the piperidine amine followed by bromination of the pyrazole ring. Several syntheses have been published for the starting material for this project as it is a valuable intermediate in the pharmaceutical industry.[1] The industry's

Comprehensive Organic Chemistry Experiments for the Laboratory Classroom
Edited by Carlos A. M. Afonso, Nuno R. Candeias, Dulce Pereira Simão, Alexandre F. Trindade, Jaime A. S. Coelho, Bin Tan, and Robert Franzén.

Published by the Royal Society of Chemistry, www.rsc.org

research is also directed to the development of general methodology, though, such as, for example, the functionalization of the nitrogen atoms in the pyrazole.[2]

The creation of a versatile scaffold is important in medicinal chemistry as it allows for the synthesis of a common intermediate in a few steps, an intermediate that has flexibility in-built and opens up the possibility to quickly synthesize a library of many different compounds for screening. The backbone of the starting molecule in this project was synthesized in two steps that already introduced some flexibility (at an early stage, though): choice of substituents (number, position, nature) on the phenyl ring and heteroatoms in the five-membered aromatic ring, *i.e.* a pyrazole; but simply replacing hydrazine with hydroxylamine could produce two isomeric isoxazoles,[1] too (Figure 5.1.7.1). Other possible modifications are the protection or substitution of the pyrazole N–H or the alkylation of the pyrazole N.

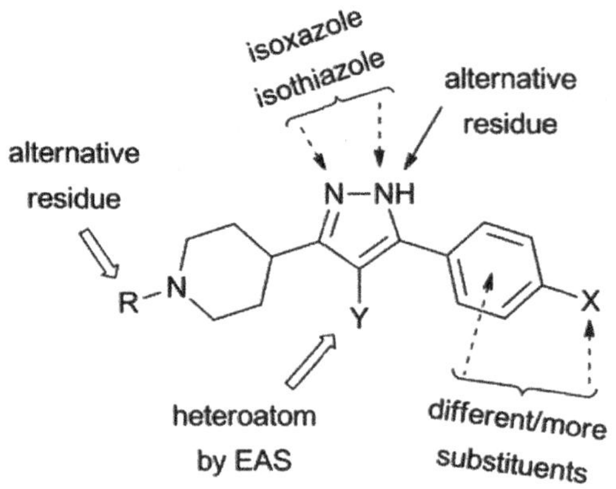

Figure 5.1.7.1 Possible scaffold modifications.

This experiment, however, focusses on selectively protecting the piperidine amine and then introducing a halogen, namely bromine, *via* electrophilic aromatic substitution to the pyrazole, introducing *N*-bromosuccinimide (NBS) as "Br^+" synthon. NBS can supply bromine by releasing the halogen in the presence of a proton source and forming succinimide as a by-product.[3] The additional bromine atom in **3** opens up the possibility to utilize palladium-catalysed cross-coupling reactions[4] for further modification (careful variation of the halogen atoms on both aromatic rings even permits the introduction of two new residues in a regioselective approach).

The protection of the amino group utilizes di-*tert*-butyl dicarbonate – an anhydride $(Boc)_2O$ – under basic conditions to protect the secondary amine as a carbamate, called the *tert*-butyloxycarbonyl group or Boc protection (Figure 5.1.7.2).[5] The anhydride is used because the corresponding chloroformate equivalent is unstable. Boc is a common protecting group for amines in amino acids and peptides. It is stable under basic conditions but cleaved with acid, *e.g.* trifluoroacetic acid or hydrochloric acid (3 M).[6] Boc protection is typically achieved with high yields and the carbamate is easily cleaved using strong acids,[5] releasing carbon dioxide and isobutene in the process.

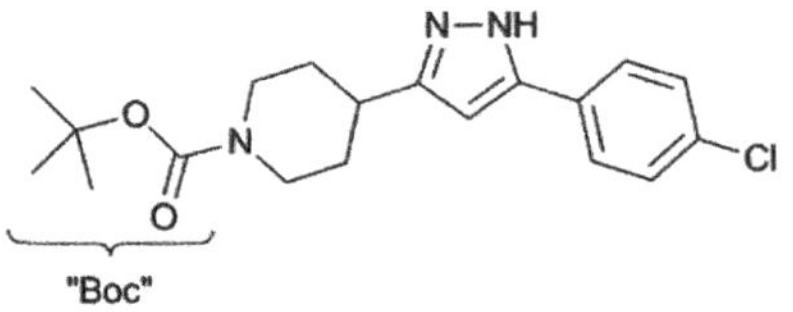

Figure 5.1.7.2 *Tert*-butyloxycarbonyl/Boc carbamate.

Additional Safety

All procedures should be performed in a well-ventilated fume cupboard with proper safety equipment such as protective clothing, safety spectacles and gloves. All organic solvents apart from dichloromethane and deuterochloroform are (highly) flammable solvents. The chlorinated solvents are suspected carcinogens while DMF is a suspected teratogen; methanol is toxic by all routes; diethyl ether is extremely flammable and may form explosive peroxides. Acids and bases are corrosive and harmful in contact with mucous membranes. NBS and Boc_2O are also irritants and the latter a suspect sensitizer.

Experimental Procedure

Session 1: Boc-Protection of Piperidine Moiety (3 h)

1. In a fume cupboard set up a dry round-bottom flask over a magnetic stirrer and suspend the pyrazole heterocycle **1** (4.0 mmol) in dichloromethane (25 mL).
2. Separately, prepare a solution of *tert*-butyl *pyro*carbonate $(Boc)_2O$ (4.2 mmol) in dichloromethane (5.0 mL). Slowly add the solution to the stirred reaction mixture at room temperature using a pipette.
3. Add triethylamine (4.2 mmol) to the reaction mixture and continue stirring at room temperature.
4. Monitor the reaction by TLC and continue stirring at ambient temperature until completion (eluent: CH_2Cl_2:MeOH:NH_3 (15:4:1)). Use a sample of the organic phase after a drop of the reaction mixture has been partitioned between diethyl ether and water (each 0.5 mL).
5. Dilute the reaction mixture with dichloromethane (25 mL) and wash sequentially with hydrochloric acid (2 M, 50 mL), saturated sodium bicarbonate solution (50 mL), water (50 mL) and saturated brine (25 mL).
6. Dry the organic phase with magnesium sulfate before removing the solvent *via* rotary evaporation *in vacuo.*
7. Triturate the oily residue with light petroleum ether (40–60 °C, 10 mL) until a solid forms. Filter off the supernatant and wash the retentate with fresh light petroleum ether (40–60 °C, 10 mL).
8. Dry the white solid by vacuum desiccation.
9. Determine the yield of the product.
10. Measure the melting point.
11. Acquire ^{13}C and DEPT-135 NMR spectra using $CDCl_3$ as solvent.
12. Acquire an IR spectrum.

Session 2: Bromination of Pyrazole (3 h)

1. In a fume cupboard set up a dry round-bottom flask over a magnetic stirrer and dissolve the product of reaction 1 (**2**) (2.5 mmol) in dimethylformamide (10 mL).
2. Stir and cool the solution to 0 °C.
3. Add *N*-bromosuccinimide (NBS) (2.8 mmol) in small portions over a period of 20 min and then continue stirring at 0 °C for a further 30 min.
4. Allow to warm to room temperature and monitor the reaction (eluent: ethyl acetate: petroleum ether (6:4)) until completion. Use a sample of the organic phase after a drop of the reaction mixture has been partitioned between diethyl ether and water (each 0.5 mL).
5. Pour the reaction mixture into water (50 mL) and extract twice with diethyl ether (2 × 40 mL). Then wash the combined organic phases with water (2 × 50 mL) and saturated brine (50 mL).

6. Dry with magnesium sulfate before removing the solvents *via* rotary evaporation *in vacuo*.
7. Triturate the crude product with a 1 : 1 mixture (10 mL) of light petroleum ether (40–60 °C) and diethyl ether until a solid forms.
8. Remove the supernatant by filtration and dry the white solid by vacuum desiccation.
9. Determine the yield of the product.
10. Measure the melting point.
11. Acquire ^{13}C and DEPT-135 NMR spectra using $CDCl_3$ as solvent.
12. Acquire an IR spectrum.

Waste Disposal

The neutralized aqueous solutions – including dissolved $MgSO_4$ – may be diluted and discarded in the sink. All organic solutions should be recycled or collected as chlorinated waste.

Results Interpretation and Additional Questions

1. Assign characteristic stretching vibrations in the IR spectra of both products and confirm the presence of the functional groups.
2. Interpret the carbon NMR spectra of both products and identify key differences between them and the starting material.
3. Formulate the mechanism for the formation of the carbamate and explain the regioselectivity using the pK_a and/or aromaticity of the starting material.
4. Formulate the mechanism of the bromination reaction and rationalize why the pyrazole ring is attacked. How does the resonance stabilization of NBS contribute to the electrophilic character of the bromine atom?

References

1. (a) M. Rowley, H. B. Broughton, I. Collins, R. Baker, F. Emms, R. Marwood, S. Patel, S. Patel, C. I. Ragan, S. B. Freedman and P. D. Leeson, *J. Med. Chem.*, 1996, **39**, 1943; (b) M. Rowley, I. Collins, H. B. Broughton, W. D. Davey, R. Baker, F. Emms, R. Marwood, S. Patel, S. Patel, C. I. Ragan, S. B. Freedman, R. Ball and P. D. Leeson, *J. Med. Chem.*, 1997, **40**, 2374.
2. W. Xu, R. Mohan and M. M. Morrissey, *Bioorg. Med. Chem. Lett.*, 1998, **8**, 1089.
3. J. Clayden, N. Greeves and S. Warren, *Organic Chemistry*, OUP, Oxford, 2nd edn, 2012, ch. 19, p. 441.
4. A. Suzuki, *Angew. Chem., Int. Ed.*, 2011, **50**, 6722.
5. (a) T. W. Greene and P. G. M. Wuts, *Protective Groups in Organic Synthesis*, John Wiley & Sons, Inc., New York, 3rd edn, 1999, ch. 7, pp. 518–525; (b) T. W. Greene and P. G. M. Wuts, *Green's Protective Groups in Organic Synthesis*, John Wiley & Sons, Inc., Hoboken, 4th edn, 2007, ch. 7, pp. 725–735.
6. J. Clayden, N. Greeves and S. Warren, *Organic Chemistry*, Oxford University Press, Oxford, 2nd edn, 2012, ch. 23, pp. 557–559.

5.1.8. Preparation of *p*-Bromoaniline

Abel J. S. C. Vieira

Faculty of Sciences and Technology, Universidade Nova de Lisboa, 2829-516 Caparica, Portugal
*E-mail: ajscv@fct.unl.pt

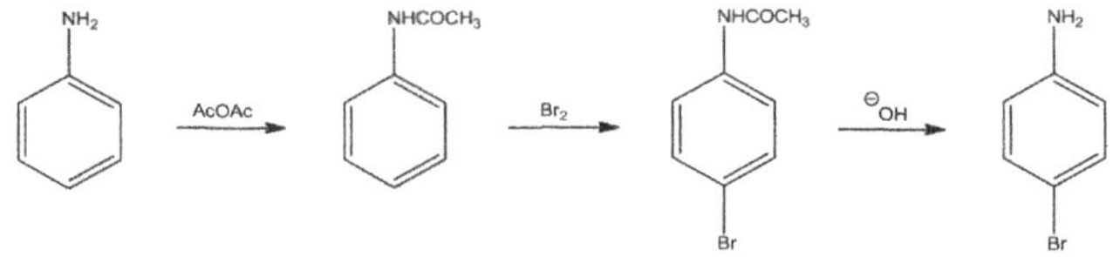

Number of sessions (duration of each session)	Hazard level	Difficulty level	Level of study
2 (3 h each)	High	Medium	Intermediate

Class names Aromatic amines and amides, aromatic halogenated compounds

Concepts involved Electrophilic aromatic substitution reaction. This work consists of a multi-step aromatic synthesis, illustrating the strategy of introduction and removal of protective groups (side-chain reactions)

Chemicals needed Aniline, acetic acid, acetic anhydride, zinc, bromine, sodium bisulfite, potassium hydroxide, ethanol

Equipment and experimental techniques involved Apparatus for reaction with cooling, heating and stirring, recrystallization apparatus, IR and ^{1}H NMR spectroscopies

Keywords Aromatic electrophilic bromination, IR and ^{1}H NMR spectra, multi-step synthesis, protective groups, recrystallization, reflux boiling

Background

The aim of this work is to prepare *p*-bromoaniline starting from aniline.

Given our knowledge of the chemistry of aromatic compounds,[1] the problem seems (at a first sight) to be easy to solve: the only substituent of the benzene ring (the amino group of aniline) is activating towards electrophilic aromatic substitution and *o,p*-directing; therefore, a single bromination reaction should be convenient. However, the amino group is strongly activating (and not very bulky). For this reason, the direct bromination of aniline with Br_2, even in the absence of a catalyst, would directly lead to substitution at all activated positions, yielding the tribrominated product, 2,4,6-tribromoaniline (Scheme 5.1.8.1), which is not the aim.

It is therefore necessary to adopt a strategy of synthesis which, although apparently more complex, can lead to the desired product with a good yield. This can be achieved by using the acetyl group as a protecting group, *i.e.* transforming an amine (aniline) into an amide (acetanilide), by reaction of the amine with acetic anhydride (in the presence of a small amount of Zn powder) (Scheme 5.1.8.2).

Comprehensive Organic Chemistry Experiments for the Laboratory Classroom
Edited by Carlos A. M. Afonso, Nuno R. Candeias, Dulce Pereira Simão, Alexandre F. Trindade, Jaime A. S. Coelho, Bin Tan, and Robert Franzén.

Published by the Royal Society of Chemistry, www.rsc.org

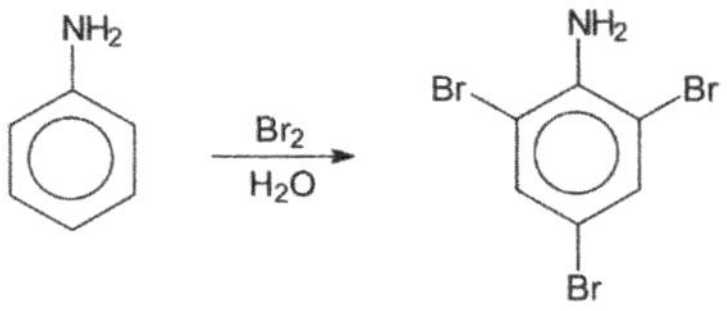

Scheme 5.1.8.1 Direct bromination of aniline.

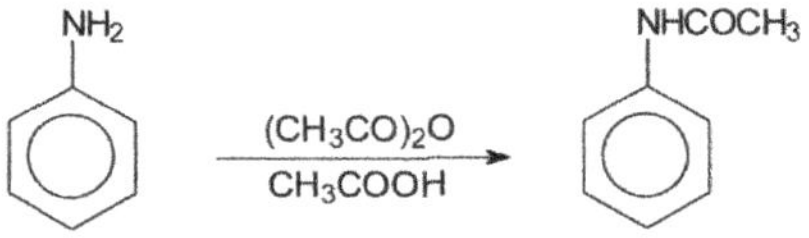

Scheme 5.1.8.2 Introduction of the protective group.

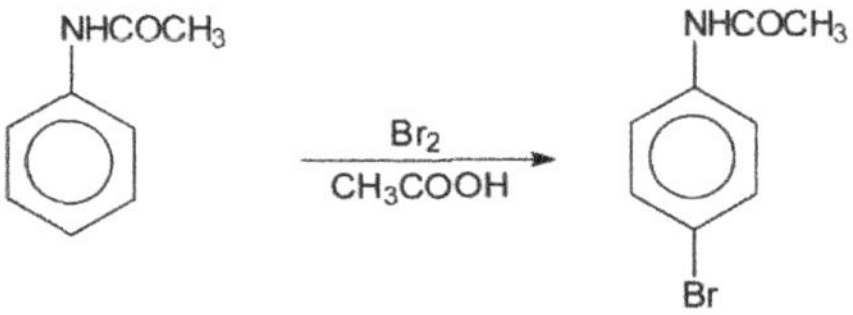

Scheme 5.1.8.3 Bromination of acetanilide.

The acetamido group has several advantages (considering the aim of the experiment) over the amino group. First, although still being activating towards electrophilic aromatic substitution, the acetamido group is much less activating than the amino group, favouring the mono-substitution. Second, as far as regioselectivity is concerned, it is also an *o,p*-directing group, but because it is a bulkier group the substitution at the *ortho* position scarcely occurs, yielding the *para* mono-substituted compound as the major product (Scheme 5.1.8.3).

This compound, *p*-bromoacetanilide, already has the substitution pattern of the benzene ring that is desired for the final product. The only remaining step is the "regeneration" of the amino function, to obtain *p*-bromoaniline. This is achieved by what is usually known as "removal of the protecting group", *e.g.* by a basic (or acidic) hydrolysis reaction of the amide (Scheme 5.1.8.4).

In summary, the synthesis of *p*-bromoaniline from aniline can be achieved by the following sequential reaction steps (Scheme 5.1.8.5).

Additional Safety

The major part of the experiment can be performed at a regular laboratory bench, with proper safety equipment, such as protective clothing, gloves and safety glasses. The step where bromine is used must be performed entirely in the fume hood. The same precaution can be also applied to aniline (an irritating reagent). Bromine is a very dangerous reactant, requiring special care in its manipulation: fume hood and effective eye and hand protection. Residues of bromine after the reaction must be complete and safely removed.

Aniline: H351; H341; H331; H311; H301; H318; H317; H400.
Acetic acid: H226; H314; P260; P280; P303; P305.
Acetic anhydride: H226; H332; H302; H314.
Bromine: H330; H314; H400; P260; P280; P305; P315; P338.
Potassium hydroxide: H314.
Ethanol: H225.

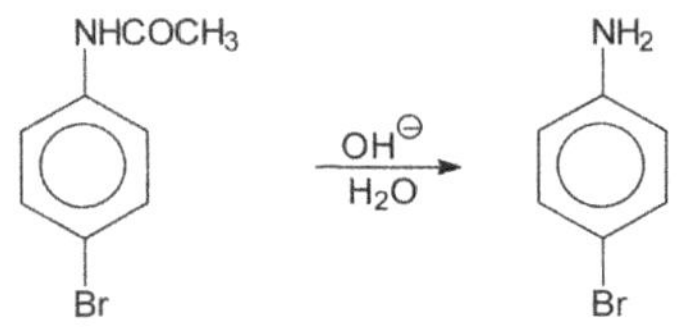

Scheme 5.1.8.4 Removal of the protective group.

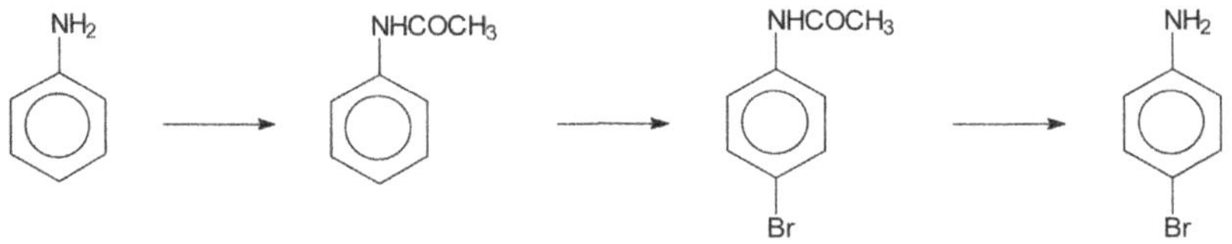

Scheme 5.1.8.5 Overall reaction sequence.

Experimental Procedure

Laboratory Session 1 (3 h)

Preparation of Acetanilide (ref. 2)

1. In a 100 mL round-bottomed flask, fitted with a water condenser, introduce 2.5 mL of glacial acetic acid, 7.5 mL of acetic anhydride and a small amount of zinc powder.
2. Add, while gently shaking under cooling, 5 mL of freshly distilled aniline.
3. Heat the reaction mixture under reflux for 20 min, using a Bunsen burner and a metallic net to protect the flame (alternatively use an electric mantle and an adjustable laboratory jack).
4. Stop heating, let the contents of the flask cool, and pour carefully the reaction mixture over 150 mL of a mixture water/ice (*ca.* 1 : 1). Keep cooling until complete precipitation of the white solid.
5. Filter off this solid at the water vacuum pump, wash it with ice-cooled water and dry in the air over filter paper. To accelerate drying place the product in an oven ($T < 50$ °C) and then in a vacuum desiccator, for several minutes.
6. Use the dried product without further purification for the next step. Keep a small amount for characterization.

Preparation of p-Bromoacetanilide (ref. 3)

7. In the fume hood, place a clamped 100 mL Erlenmeyer flask over a magnetic stirrer plate. Fit a clamped addition funnel, such that its stem slightly enters into the flask.
8. To the Erlenmeyer flask add 5.4 g of acetanilide, dissolved in 40 mL of acetic acid, and start moderate stirring, while cooling the solution in an ice/water bath.
9. In the addition funnel, introduce 20 mL of acetic acid and 2 mL of bromine.
10. Add slowly and dropwise the bromine solution to the solution contained in the Erlenmeyer flask, maintaining stirring and cooling. After the addition is completed, continue stirring for a further 15 min without cooling.
11. Pour the reaction mixture over 200 mL of a 1 : 1 mixture of ice/water.
12. If the mixture is still coloured (light red), add a concentrated aqueous solution of sodium bisulfite until the colour disappears completely. At this point you can leave the fume hood and return to your usual working place in the laboratory.

13. Cool the mixture, filter the precipitate at the water vacuum pump, wash it thoroughly with ice cold water, and dry it in the air over filter paper.
14. Keep the solid in a vacuum desiccator for the next step of the experiment. Reserve a small amount for characterization.

Laboratory Session 2 (3 h)

Preparation of p-Bromoanilne (ref. 3)

15. In a 100 mL round-bottomed flask, fitted with a water condenser, dissolve 6.4 g of *p*-bromoacetanilide in 20 mL of ethanol.
16. Add a solution of 6.0 g of potassium hydroxide in 10 mL of distilled water and boil under reflux for 2 h.
17. Cool the solution and pour it over 60 mL of ice cold water, stirring frequently with a glass rod, while cooling in ice until complete precipitation of the solid. If necessary, add more water.
18. Filter off the solid at the water vacuum pump and wash it with ice cold water.

Product Purification

19. Recrystallize the crude compound in a mixture of ethanol/water (1 : 1).
20. Upon precipitation, filtration and recovery of the crystals, keep them in a vacuum desiccator.

Product Characterization

21. Measure the melting point of each of the prepared compounds.
22. Record the IR spectrum of the final compound (*p*-bromoaniline), using dried potassium bromide as support material.
23. Acquire the ^{1}H NMR spectrum of *p*-bromoaniline. Use the deuterated solvent suggested by your instructor.
24. To evaluate the eventual formation of by-products (*e.g. o*-bromoaniline, tribromoaniline) a simple TLC test can be performed (ask your instructor for help).

Waste Disposal

The aqueous solutions resulting from the several work-ups may be neutralized, diluted and discarded in the sink.

The same applies to the ethanol/water mixture used for recrystallization.

Special attention is required in handling bromine. After the reaction, the aqueous solutions must be decolorized with sodium bisulfite and diluted before being discarded.

Results Interpretation and Additional Questions

1. Compare the melting points obtained with values given in the literature. Note that among the three compounds obtained, only the last one (*p*-bromoaniline) was recrystallized. Comment on the (eventual) observed differences between obtained values and those in the literature.
2. Calculate the yield of each step of the synthesis and also the global yield.
3. Assign the characteristic IR absorption bands of the functional groups of *p*-bromoaniline.
4. Try to assign the H atoms to the NMR signals in the spectrum for *p*-bromoaniline.
5. Write the mechanisms of all the reactions involved in this work.
6. Consider the following hypothesis: even if equimolar quantities of aniline and bromine are used, the direct bromination of aniline is not the better option to prepare *p*-bromoaniline,

as far as either purity or yield of the product is concerned. Comment on this and give your reasons.

7. Why is acetamido a weaker activating group compared to amino (in an aromatic electrophilic substitution reaction)?
8. What is the convenience of using metallic zinc with aniline? (Hint: this amine is very easy to oxidize.)
9. Bromine is a very dangerous reactant; therefore, traces of this compound must be removed after the reaction. Try to rationalize the use of sodium bisulfite for this purpose.

References

1. J. Clayden, N. Greeves, S. Warren and P. Wothers, *Organic Chemistry*, Oxford University Press, 1st edn, 2001, vol. 22, p. 547.
2. B. S. Furniss, A. J. Hannaford, P. W. G. Smith and A. R. Tatchell, *Vogel's Textbook of Practical Organic Chemistry*, Longman Scientific & Technical, Harlow, 5th edn, 1989, vol. 6, p. 916.
3. S. Hünig, G. Märkl and J. Sauer, *Integriertes Organisches Praktikum*, Verlag Chemie GmbH, 1979, vol. 7, p. 620.

5.2. C-C Bond Formation

5.2.1. Synthesis of 1,4-Di-*t*-butyl-2,5-dimethoxybenzene

Dulce Simão* and Ana C. Cerdeira

Department of Chemistry, Instituto Superior Técnico, University of Lisbon, Centro de Química Estrutural, Av. Rovisco Pais, 1, 1049-001 Lisbon, Portugal. *E-mail: dulcesimao@tecnico.ulisboa.pt

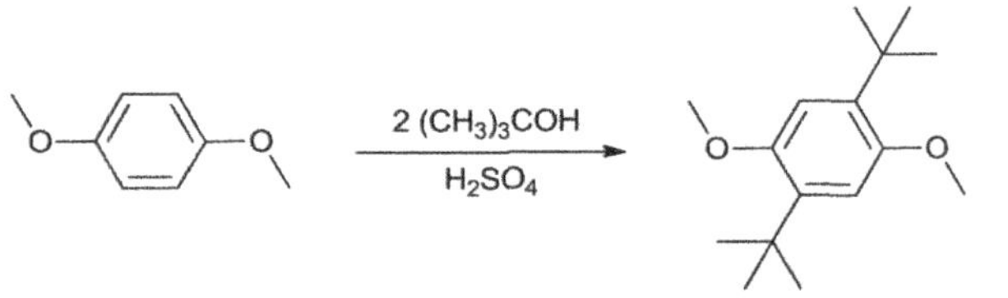

Duration	Hazard level	Difficulty level	Level of study
1 (4 h)	Moderate	Medium	Introductory

Class names Disubstituted benzenes, aromatic ethers, tertiary alcohols

Concepts involved Dialkylation of an activated benzene ring by Friedel–Crafts alkylation; the reaction is performed with temperature control; purification of the final product by recrystallization from mixed solvents

Chemicals needed 1,4-Dimethoxybenzene, *t*-butyl alcohol, acetic acid, concentrated sulfuric acid, fuming sulfuric acid (20% SO_3), methanol, dichloromethane, anhydrous calcium chloride, $CDCl_3$

Equipment and experimental techniques involved Recrystallization, ^{1}H NMR, magnetic stirring, temperature control, rotary evaporator, mixed solvents recrystallization, IR, melting point

Keywords Activated aromatic ring, dialkylation, electrophilic aromatic substitution, Friedel–Crafts alkylation, ^{1}H NMR, recrystallization

Background

In an electrophilic aromatic substitution the hydrogen atom of an aromatic ring is replaced by an electrophile. This mechanism has two steps: attack by an electrophile on the aromatic ring to give a cationic nonaromatic intermediate known as an arenium ion followed by loss of a proton from the cation to restore the aromaticity. A Friedel–Crafts alkylation is an example of electrophilic aromatic substitution[1–4] where the electrophile is a carbocation, which usually is produced by reaction with an alkyl halide in the presence of a Lewis acid (catalyst), with aluminum chloride the most common catalyst.[5] A tertiary alcohol can be used instead of an alkyl halide, in the presence of a strong acid (sulfuric acid).[6] In this experiment the electrophile is a stable tertiary butyl cation generated by treating a tertiary alcohol (*t*-butyl alcohol) with sulfuric acid. The Friedel–Crafts alkylations only occur if the aromatic ring is either unsubstituted or has activating substituents. Another limitation of this reaction is the formation of side products

Comprehensive Organic Chemistry Experiments for the Laboratory Classroom
Edited by Carlos A. M. Afonso, Nuno R. Candeias, Dulce Pereira Simão, Alexandre F. Trindade, Jaime A. S. Coelho, Bin Tan, and Robert Franzén.

Published by the Royal Society of Chemistry, www.rsc.org

through polyaklylation, since the product is more reactive than the starting material, which leads to lower yields.[1–4] In this experiment, the starting material is an aromatic ring activated by two methoxy groups, which will react with the tertiary butyl cation to result in the dialkylated aromatic ether 1,4-di-*t*-butyl-2,5-dimethoxybenzene.

Additional Safety

The use of a fume hood is not mandatory. However, fuming sulfuric acid is highly corrosive and reacts violently with water and it must be handled inside the hood. Wear appropriate protective eyeglasses and gloves to prevent skin exposure. The usual care should be taken with *t*-butyl alcohol and methanol (flammable), and acetic and concentrated sulfuric acids (corrosive).

Experimental Procedure

1. Fit a 250 mL three-neck round-bottom flask with magnetic stirring, reflux condenser, thermometer and a dropping funnel.
2. Add 1,4-dimethoxybenzene (3 g, 0.02 mol), followed by *t*-butyl alcohol (5 mL) and acetic acid (10 mL).
3. Initiate stirring and cool the mixture to 0–3 °C with an ice/water bath.
4. In a 25 mL Erlenmeyer flask prepare a solution of fuming sulfuric acid (20% SO_3) (5 mL) and concentrated sulfuric acid (15 mL), and cool with an ice/water bath. Wipe the outside of the flask dry before pouring the cooled acid solution into the dropping funnel.
5. Add this solution dropwise to the reaction mixture, keeping the temperature between 15 and 20 °C.
6. Stir for a further 5 min at room temperature and then cool in an ice bath.
7. Add crushed ice (*ca.* 100 g) to the reaction mixture to dilute the H_2SO_4 and then add water (50 mL). Cool and collect the solid by vacuum filtration.
8. Wash the product several times with water (*ca.* 300 mL) and dry by pressing down the filter cake with a spatula and allowing it to drain well.
9. To purify the crude product, wash it with methanol (3 × 15 mL), place the solid in a 50 mL Erlenmeyer flask and dissolve it in dichloromethane (8 mL), add anhydrous $CaCl_2$ and stir for 10 min.
10. Filter by gravity the solution to a 100 mL round-bottom flask and add methanol (15 mL).
11. Evaporate the solvent in the rotatory evaporator until the volume is reduced to about 15 mL.
12. Allow the solution to stand and cool for crystallization of the product.
13. Collect the product by vacuum filtration and dry thoroughly.
14. Record the yield and melting point and collect spectroscopic data (IR and ^1H NMR).

Waste Disposal

The neutralized aqueous solutions may be diluted and discarded in the sink. The filtrate resulting from washings with methanol should be placed in a non-chlorinated waste recipient while the mother liquor from recrystallization with dichloromethane/methanol should be stored in the chlorinated waste recipient.

Results Interpretation and Additional Questions

1. Determine the reaction yield.
2. Assign the signals in the ^1H and ^{13}C NMR spectra and the characteristic absorption bands in the IR spectrum.

3. Compare the melting point of the obtained product with the one reported in the literature and evaluate the purity of the product.
4. Write the reaction mechanism.
5. Explain why the reaction does not proceed beyond the dialkylation.
6. Starting from methoxybenzene, what product(s) should be obtained?

References

1. K. P. C. Vollhardt and N. Schore, *Organic Chemistry*, W. H. Freeman & Co., New York, 6th edn, 2011, p. 712.
2. J. Clayden, N. Greeves, S. Warren and P. Wothers, *Organic Chemistry*, Oxford, 2001, p. 553.
3. M. S. Sainsbury, *Aromatic Chemistry*, Oxford University Press, 1992, Oxford Chemistry Primers, p. 14.
4. T. W. G. Solomons, *Organic Chemistry*, John Wiley & Sons Inc., New York, 7th edn, 1998, p. 669.
5. J. March, *Advanced Organic Chemistry*, 4th edn, John Wiley & Sons Inc., New York, 1992, p. 534.
6. L. Fieser and K. Williamson, *Organic Experiments*, Houghton Mifflin, 1998, p. 303.

5.2.2. Synthesis of 4,4′-Di-*tert*-butylbiphenyl

Dulce Simão

Department of Chemistry, Instituto Superior Técnico, University of Lisbon, Centro de Química Estrutural, Av. Rovisco Pais, 1, 1049-001 Lisbon, Portugal. *E-mail: dulcesimao@tecnico.ulisboa.pt

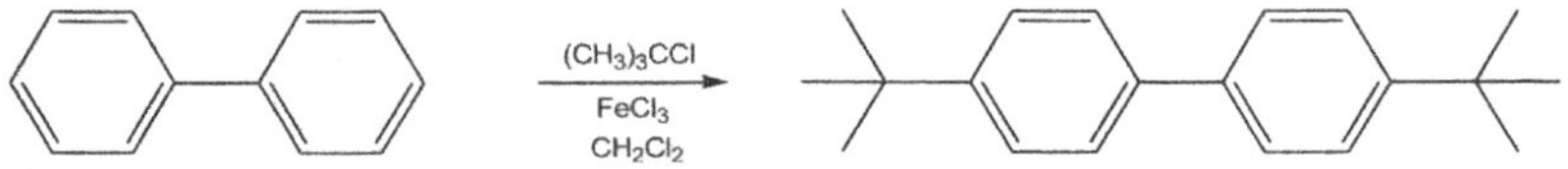

Duration	Hazard level	Difficulty level	Level of study
1 (4 h)	Moderate	Low	Intermediate
Class names Aromatic hydrocarbons, haloalkanes			
Concepts involved Friedel–Crafts alkylation, reactivities and substituent directing effects in the electrophilic aromatic substitution			
Chemicals needed Biphenyl, *tert*-butyl chloride, dry dichloromethane, anhydrous ferric chloride, anhydrous calcium chloride, hydrogen chloride (10%), ethanol, deuterated chloroform			
Equipment and experimental techniques involved Liquid–liquid extraction, NMR, stirring heating plate, use of gas trap to control formed HCl, rotary evaporator, recrystallization, vacuum filtration, melting point, IR			
Keywords Electrophilic aromatic substitution, Friedel–Crafts alkylation, liquid–liquid extraction, NMR			

Background

A Friedel–Crafts alkylation is an example of electrophilic aromatic substitution that leads to carbon–carbon bond formation. The reaction between a haloalkane and benzene in the presence of a Lewis acid catalyst produces an alkylbenzene and hydrogen halide.[1–4] Usually, aluminum chloride is the Lewis acid used but this reaction can be carried out in the presence of other Lewis acid catalysts like ferric chloride.[3] However, this reaction does not stop at the stage of mono-substitution, since the alkylbenzene produced is more reactive than benzene due to the electron-releasing effect of the alkyl group, and so polyaklylation occurs.[3–5] Some mono-alkylbenzenes may be prepared using an excess of benzene.[6] In this experiment the aromatic hydrocarbon is biphenyl instead of benzene, a more hazardous compound. The two bulky *tert*-butyl groups are introduced at the *para* position, according regioselectivity in electrophilic aromatic substitution reactions.[7]

Additional Safety

It is advisable to carry out this reaction in a fume hood to prevent exposure to gaseous products formed during the reaction. Care should be taken with corrosive hydrogen chloride and with the flammable ethanol. Dichloromethane may be harmful if inhaled, causing central nervous system effects, and it is suspected of causing cancer.

Comprehensive Organic Chemistry Experiments for the Laboratory Classroom
Edited by Carlos A. M. Afonso, Nuno R. Candeias, Dulce Pereira Simão, Alexandre F. Trindade, Jaime A. S. Coelho, Bin Tan, and Robert Franzén.

Published by the Royal Society of Chemistry, www.rsc.org

Experimental Procedure

1. In a 250 mL two- or three-necked round-bottomed flask with magnetic stirring, place biphenyl (5 g, 0.03 mol), *tert*-butyl chloride (10 mL, 0.09 mol) and dry dichloromethane (25 mL).
2. Fit a reflux condenser, with a calcium chloride tube connected to a gas absorption trap that must be partially immersed in water.
3. After biphenyl dissolution, add anhydrous ferric chloride (0.2 g, 0.01 mol) and gently stir. HCl is readily observable.
4. Heat the reaction mixture in a water bath and keep refluxing for 1 h.
5. Pour the contents into a separating funnel and wash the organic layer three times with 10% hydrochloric acid (20 mL) and dry over anhydrous calcium chloride.
6. Filter into a 250 mL round-bottomed flask and remove the solvent on a rotary evaporator. The crude product is recrystallized from 95% ethanol.
7. Record the yield and melting point.

Waste Disposal

Acid aqueous solutions from liquid–liquid extraction and the gas trap should be neutralized with hydrogen carbonate, diluted and discarded in the sink. Dichloromethane is to be collected in a chlorinated waste collector recipient. Ethanol may be stored for further purification.

Results Interpretation and Additional Questions

1. Comment on the reaction yield obtained.
2. Compare the melting point of the obtained product with the one reported in the literature and comment on the purity of the obtained compound.
3. Interpret the IR and ^{1}H and ^{13}C NMR spectra.
4. Write the reaction mechanism.
5. Explain the need to work under anhydrous conditions.

References

1. M. Rueping and B. J. Nachtsheim, *Beilstein J. Org. Chem.*, 2010, **6**, 6.
2. M. Bandini, General Aspects and Historical Background, in *Catalytic Asymmetric Friedel–Crafts Alkylations*, Wiley-VCH Verlag GmbH, Weinheim, 2009, ch. 1.
3. T. W. G. Solomons, *Organic Chemistry*, John Wiley & Sons, Inc., New York, 7th edn, 1998, pp. 669–671.
4. K. P. C. Vollhardt and N. E. Schore, *Organic Chemistry*, Freeman & Cª., 3th edn, 1999, pp. 670–673.
5. J. March, *Advanced Organic Chemistry*, 4th edn, John Wiley & Sons, Inc., New York, 1992, p. 536.
6. D. A. Horne, *J. Chem. Educ.*, 1983, **60**(3), 246.
7. A. R. Tatchell, B. S. Furnis, A. J. Hannaford and P. W. G. Smith, *Vogel's Textbook of Practical Organic Chemistry*, Longman Scientific and Technical, 5th edn. 1989, p. 832.

5.2.3. Synthesis of a Macrocycle: *C*-Methyl[4] resorcinarene

Christopher Baker, Alexander S. Cragg, Raghuram R. Kothur, Flavia Fucassi, and Peter J. Cragg*

School of Pharmacy and Biomolecular Sciences, University of Brighton, Huxley Building, Brighton BN2 4GJ, UK. *E-mail: P.J.Cragg@brighton.ac.uk

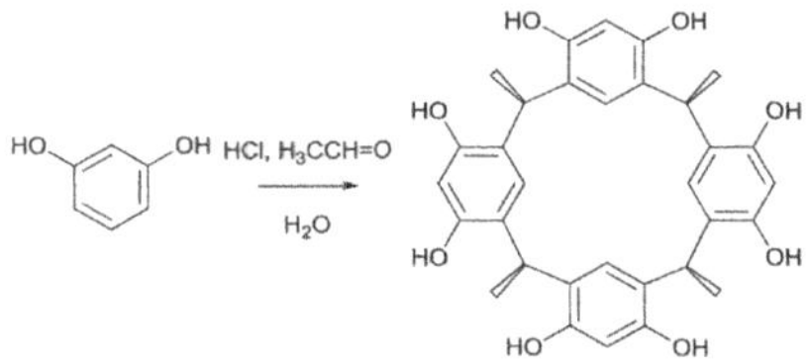

Duration	Hazard level	Difficulty level	Level of study
1 (3 h)	Moderate	Medium	Intermediate

Class names Aldehydes, phenols, alcohols

Concepts involved This experiment demonstrates how a macrocycle can be synthesized in good yield and high purity through an acid-catalysed reaction between resorcinol and acetaldehyde

Chemicals needed Resorcinol, acetaldehyde, concentrated hydrochloric acid (37%), ethanol, distilled water, deuteroacetone (for NMR spectroscopy)

Equipment and experimental techniques involved Conical or round-bottomed flasks (2 × 100 mL), addition funnel, magnetic stirrer bar, magnetic stirrer/hotplate or heating mantle, thermometer, ice bath, Büchner funnel or similar (for filtration); recrystallization, NMR and IR spectroscopy

Keywords Electrophilic aromatic substitution, macrocyclic synthesis, NMR, phenol–formaldehyde condensation, recrystallization

Background

Macrocycles, such as iron-containing porphyrins and the potassium ionophore valinomycin, are found throughout nature and others, like cyclodextrins and phthalocyanines, have great commercial importance. Determining the experimental conditions that preferentially produce macrocycles, rather than polymers, from monomeric precursors requires careful consideration of both the reaction and work up conditions. There are many examples of macrocyclic syntheses available in the scientific literature but, to see the range of compounds that can be prepared, several excellent texts are available that cover theoretical[1,2] and practical[3,4] examples. *C*-Methyl[4] resorcinarene is an excellent example of a synthetic macrocycle which, in addition to having an interesting reaction mechanism, is also an example of a thermodynamic product with C_{4v} symmetry being favoured over the kinetic C_{2v}-symmetric product (Figure 5.2.3.1).

Research on phenol-formaldehyde polymerization in the late nineteenth and early twentieth centuries led to the synthesis of unnatural plastics, the novolacs and resoles, and the preparation

Comprehensive Organic Chemistry Experiments for the Laboratory Classroom
Edited by Carlos A. M. Afonso, Nuno R. Candeias, Dulce Pereira Simão, Alexandre F. Trindade, Jaime A. S. Coelho, Bin Tan, and Robert Franzén.

Published by the Royal Society of Chemistry, www.rsc.org

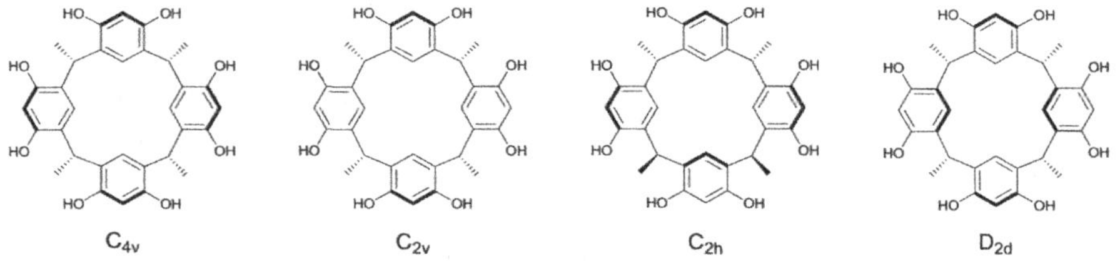

Figure 5.2.3.1 C-Methyl[4]resorcinarene conformers.

of materials such as Bakelite. Pioneering polymer researchers such as Baeyer,[5] Niederl and Vogel[6] also investigated unusually crystalline by-products which were later found to be discrete cyclic molecules rather than linear polymer chains. The [4]resorcinarenes were first described by Baeyer in 1872[1] but it was not until Högberg's work in 1980[7,8] that a set of conditions were described that gave *C*-methyl[4]resorcinarene in reproducibly good yields. Unlike many macrocycles, where the same reaction conditions can be used to prepare a series of structurally similar compounds with different substituents, the conditions required to synthesize [4]resorcinarenes with different bridging groups can be as varied as the groups themselves. Some are acid catalysed in refluxing ethanol, some require lower temperatures, some are base catalysed at 0 °C and yet others require Lewis acids. Fortunately, the synthesis of *C*-methyl[4]resorcinarene requires relatively simple conditions.

Additional Safety

Due to the use of hot corrosive concentrated acid, flammable ethanol and flammable acetaldehyde it is strongly recommended that the reaction is carried out in a fume hood. Filtration, if done relatively speedily, can be carried out in a well-ventilated laboratory. Students should take all normal precautions (correctly worn laboratory coats, safety glasses, suitable footwear, *etc.*). They may additionally wish to wear protective gloves while filtering the crude and recrystallized product to minimize the risk of acid burns from unintended splashes.

Experimental Procedure

1. Dissolve resorcinol (5.5 g, 50 mmol) in water (20 mL) in a 100 mL conical or round-bottomed flask containing a magnetic stirrer bar and place on a stirrer/hotplate or heating mantle (as appropriate).
2. Add hydrochloric acid (5 mL, 37% aqueous solution) dropwise with vigorous stirring.
3. To the stirred solution, add redistilled acetaldehyde (2.6 mL, 2.2 g, 50 mmol) dropwise from the addition funnel.
4. Continue to stir and heat to 75 °C. A precipitate will slowly form over 30–45 min. Continue to heat and stir until a total reaction time of 1 h has been reached then cool in ice.
5. Filter off the beige precipitate and wash with cold distilled water (*ca.* 5 mL).
6. Heat ethanol (5 mL) to boiling in a 100 mL flask and add the crude product. Add more ethanol in 2 mL portions until the product dissolves (*ca.* 35 mL). Once all the sample is in solution remove the flask and cool to room temperature.
7. Isolate the recrystallized product by filtration to give *C*-methyl[4]resorcinarene as a white powder.
8. Obtain a ^{1}H NMR spectrum for your product recorded in acetone-d_6 and also an infrared spectrum.

Waste Disposal

Collect all residues and solutions. Check the solution pH and neutralize with aqueous sodium hydroxide if necessary. Filter off any solids and treat as solid chemical waste. Dispose of neutralized aqueous solutions down the laboratory drain, flushing with copious water.

Results Interpretation and Additional Questions

1. Calculate the percentage yield of your recrystallized product.
2. Many larger *C*-methyl[*n*]resorcinarene macrocycles could be formed in the reaction. Why do you think that *C*-methyl[4]resorcinarene is formed preferentially?
3. *C*-Methyl[4]resorcinarene can exist in four conformers with C_{4v}, C_{2v}, C_{2h} or D_{2d} symmetry. Your product is the C_{4v} form. Why is this conformer isolated preferentially?
4. Interpret the ^{1}H NMR spectrum of your product.
5. Devise a mechanism for the formation of *C*-methyl[4]resorcinarene.

References

1. J. W. Steed and J. L. Atwood, *Supramolecular Chemistry*, Wiley-Blackwell, Chichester, 2nd edn, 2009.
2. F. Davis and S. Higson, *Macrocycles: Construction, Chemistry and Nanotechnology Applications*, John Wiley & Sons Ltd, Chichester, 1st edn, 2011.
3. *Macrocycle Synthesis: A Practical Approach*, ed. D. Parker, Oxford University Press, Oxford, 1st edn, 1996.
4. P. J. Cragg, *A Practical Guide to Supramolecular Chemistry*, John Wiley & Sons Ltd, Chichester, 1st edn, 2005.
5. A. Baeyer, *Ber. Dtsch. Chem. Ges.*, 1872, **5**, 25.
6. J. B. Niederl and H. Vogel, *J. Am. Chem. Soc.*, 1940, **62**, 2512.
7. A. G. S. Högberg, *J. Am. Chem. Soc.*, 1980, **102**, 6046.
8. A. G. S. Högberg, *J. Org. Chem.*, 1980, **45**, 4498.

5.2.4. Synthesis of 5,10,15,20-Tetrakis(2,6-dichlorophenyl)porphyrin

Artur M. S. Silva, Augusto C. Tomé, Diana C. G. A. Pinto, Fernando M. J. Domingues, Graça M. S. R. O. Rocha*, José A. S. Cavaleiro, Maria da Graça, P. M. S. Neves, Maria do Amparo F. Faustino, and Mário M. Q. Simões

Chemistry Department, University of Aveiro, Campus de Santiago, 3810-193 Aveiro, Portugal
*E-mail: grrocha@ua.pt

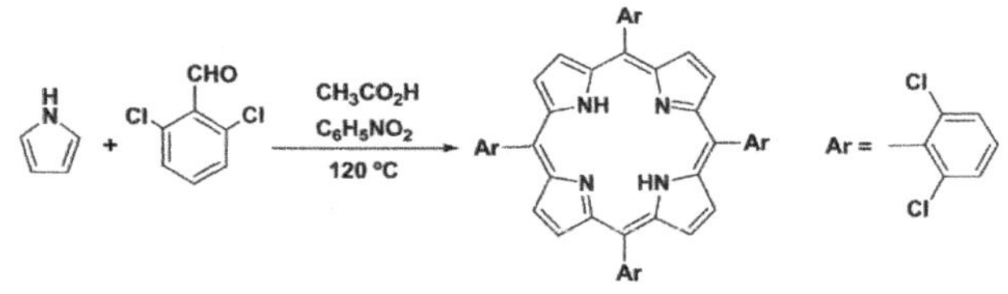

(1)

Number of sessions (duration of each session)	Hazard level	Difficulty level	Level of study
2 (3 h)	Moderate	Medium	Intermediate

Class names Aldehydes, aromatic heterocycles

Concepts involved 5,10,15,20-Tetrakis(2,6-dichlorophenyl)porphyrin is synthesized by the condensation of pyrrole with 2,6-dichlorobenzaldehyde, in a mixture of glacial acetic acid and nitrobenzene at 120 °C

Chemicals needed Glacial acetic acid [64-19-7], nitrobenzene [98-95-3], 2,6-dichlorobenzaldehyde [83-38-5], pyrrole [109-97-7], methanol [67-56-1], dichloromethane [75-09-2], deuterated chloroform [865-49-6]

Equipment and experimental techniques involved Stirring hot plate, oil bath, vacuum filtration apparatus, Rotavapor apparatus, thin-layer chromatography, UV-vis spectrophotometry, ^{1}H NMR spectroscopy

Keywords Condensation, crystallization, ^{1}H NMR spectroscopy, thin-layer chromatography, UV-vis spectrophotometry

Background

Porphyrins are a large class of naturally occurring macrocyclic compounds with many important biological representatives, such as the heme group and chlorophylls, and have applications in a large number of fields, ranging from new materials to medicine. The parent compound, known as porphyrin, consists of four pyrrole type rings linked by methinic bridges, which are usually designed as *meso*-positions (corresponding to the 5, 10, 15, and 20 positions).[1] The structure of the porphyrin nucleus and the IUPAC numbering of the ring positions are represented in Figure 5.2.4.1.

Porphyrins are well known for their intense colours, which are a consequence of the extensive macrocyclic conjugation. The UV-visible absorption spectra of free-base porphyrins (hydrogen atoms linked to two inner nitrogen atoms) consist of two distinct regions due to π–π* transitions.

Comprehensive Organic Chemistry Experiments for the Laboratory Classroom
Edited by Carlos A. M. Afonso, Nuno R. Candeias, Dulce Pereira Simão, Alexandre F. Trindade, Jaime A. S. Coelho, Bin Tan, and Robert Franzén.

Published by the Royal Society of Chemistry, www.rsc.org

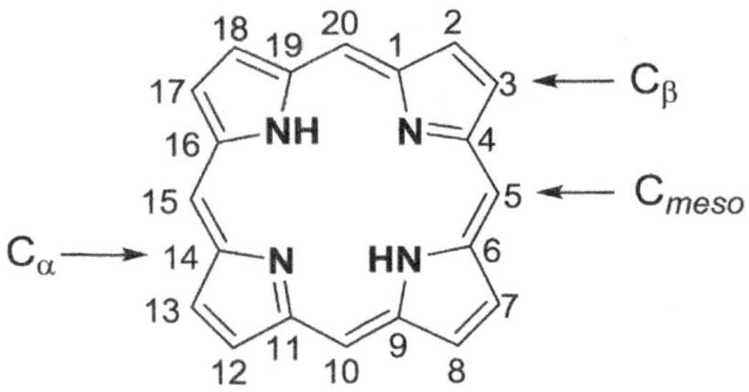

Figure 5.2.4.1 Structure and IUPAC numbering of the porphyrin nucleus.

In the near-UV region (370–430 nm) there is an intense absorption known as the "Soret band", which has typical extinction coefficients of approximately 1×10^5 M^{-1} cm^{-1}. In the visible region (500–700 nm) several absorptions are present, known as the "Q bands", which have extinction coefficients of approximately 1×10^4 M^{-1} cm^{-1}. The free-base porphyrins usually have a four-banded visible absorption spectrum. The relative intensity of these Q bands is dependent on the substituents in the aromatic group.

Porphyrin chemistry has made great progress over the years due to the potential applications of these compounds in various scientific disciplines. The synthetic *meso*-substituted porphyrins are a popular choice due to their simple synthesis and accessibility toward posterior synthetic modifications. The availability of a wide range of aldehydes allows the access to a wide variety of porphyrins. The substituents at the *meso*-position can include alkyl, aryl, heterocyclic or organometallic groups as well as other porphyrins.

The synthesis of *meso*-substituted porphyrins began with the work of Rothemund in 1935.[2] His investigation started with the synthesis of *meso*-tetramethylporphyrins by reaction of acetaldehyde and pyrrole in methanol at various temperatures. Similar studies were performed with other aldehydes, by heating a solution of pyrrole and the aldehyde in methanol under a nitrogen atmosphere in a sealed vessel. The porphyrins were obtained in very poor yields.[3,4] In all these reactions, a second porphyrinic substance, later identified as a chlorin in which one of the pyrrolic units has been reduced, appeared as a contaminant product in 10–20% yield, relative to that of the porphyrin.[4–7] In 1941, Rothemund described in detail the preparative synthesis of 5,10,15,20-tetraphenylporphyrin (TPP). The reaction was performed in pyridine in a sealed vessel. The TPP was obtained with 7.5–9% yield.[8]

In the mid-1960s, Adler, Longo and co-workers made improvements to the Rothemund method. They performed condensations of equimolar quantities of benzaldehyde and pyrrole in propionic acid in an open beaker, yielding 20% of the crystalline porphyrin.[9] The porphyrin synthesized by this method is generally contaminated with 2–10% of chlorin, which is easily removed by treatment with 2,3-dichloro-5,6-dicyano-1,4-benzoquinone (DDQ).[10,11]

In the subsequent years, several modifications have been made to the Adler–Longo method. Lindsey[12,13] and Gonsalves[14] proposed an alternative method, consisting of a two-step one-flask synthesis.

The method of Gonsalves *et al.* involves the treatment of equimolar quantities of pyrrole and the appropriate aliphatic aldehyde in carbon tetrachloride with trifluoroacetic acid. The crude porphyrinogen is isolated and then oxidized, either photochemically in benzene or in acetic acid, or chemically with DDQ or *p*-chloranil in benzene/chloroform. A variety of *meso*-tetraalkylporphyrins were obtained by this method with yields up to 18%.[14]

In the Lindsey approach a solution of pyrrole and the appropriate benzaldehyde (equimolar quantities) in dichloromethane at room temperature is treated with trifluoroacetic acid or boron trifluoride etherate. In the second step, a stoichiometric quantity of DDQ or *p*-chloranil is added, causing the conversion at room temperature of the porphyrinogen into the *meso*-tetraarylporphyrin in yields of 35–40%.[12,13]

In 1985, Gonsalves found that a mixture of nitrobenzene and acetic or propionic acid was extremely efficient in promoting the oxidation of porphyrinogens to porphyrins. A one-step reaction of the aldehyde with pyrrole at 120 °C in acetic or propionic acid containing nitrobenzene afforded the corresponding porphyrin in 45% yield. This methodology allowed the preparation of a wide variety of porphyrins that could not be obtained or gave low yields in previously reported conditions.[15]

The objective of this work is the synthesis of the 5,10,15,20-tetrakis(2,6-dichlorophenyl) porphyrin (**1**, Ar = 2,6-dichlorophenyl) by the condensation of pyrrole with 2,6-dichlorobenzaldehyde, in a mixture of glacial acetic acid and nitrobenzene at 120 °C. The reaction can be monitored by thin-layer chromatography (TLC) and UV-vis. At the end of the reaction the free-base porphyrin is crystallized directly from the reaction medium by the addition of methanol. The structure of **1** is confirmed by ^{1}H NMR spectroscopy.

Additional Safety

All chemicals used in this work should be handled with appropriate care. All can be considered toxic and irritants. Inhalation or any contact with skin and eyes should be avoided. Laboratory coat, safety glasses and adequate gloves are compulsory and all the experimental work should be done in a well-ventilated fume hood.

Nitrobenzene is toxic by inhalation and ingestion, is classified as a carcinogenic group 2B and may be harmful if absorbed by skin. Nitrobenzene should be measured with an organic bottle-top dispenser system, minimizing risks of contact and inhalation. Glacial acetic acid is a flammable liquid, harmful by skin absorption, toxic by inhalation, corrosive, a lachrymator and causes severe skin burns and eye damage. Glacial acetic acid should be measured with an organic bottle-top dispenser system, minimizing risks of contact and inhalation. Pyrrole is a flammable liquid, may cause skin and eye irritation and is toxic by inhalation and ingestion. 2,6-Dichlorobenzaldehyde is corrosive, may be harmful if inhaled and, if absorbed through skin, can cause burns. Dichloromethane is considered to be a possible carcinogen.

Experimental Procedure (Note 1)

First Laboratory Session: 3 h

Synthesis of 5,10,15,20-Tetrakis(2,6-dichlorophenyl)porphyrin (1)

1. In a well-ventilated fume hood, mix 105 mL of glacial acetic acid and 75 mL of nitrobenzene (flasks with bottle-top dispensers should be used for safety reasons), in a two-neck round-bottom flask (1 L), equipped with a reflux condenser, a dropping funnel and a magnetic stir bar.
2. Place the mixture on a heating oil bath at 120 °C, with stirring, and add the 2,6-dichlorobenzaldehyde (6.3 g, 28.8 mmol).
3. After complete dissolution of the aldehyde, slowly add the pyrrole (2.0 mL, 28.8 mmol) through the dropping funnel and leave the reaction under reflux for 45 min.
4. Follow the reaction by TLC (CH_2Cl_2 as eluent) and by UV-vis spectrophotometry (CH_2Cl_2 as solvent).
5. After cooling, the porphyrin can be crystallized directly from the reaction medium by the addition of 200 mL of methanol. Place the reaction flask in a recipient with ice and let it stand in the dark as the product precipitates (Note 2).

Second Laboratory Session: 3 h

Product Isolation

1. Filter the powder through a normal glass funnel with cotton, recovering the mother liquor in a round-bottom flask (1000 mL) (Note 3).
2. Wash the cotton with methanol to remove the black tars and then dissolve the purple porphyrin retained in the cotton with dichloromethane into a clean round bottom flask (250 mL). Evaporate the solvent in a Rotavapor, dissolve the residue in a minimal amount of dichloromethane and add enough methanol (about 50 mL) to crystallize the porphyrin, protected from light.
3. Filter the free-base porphyrin (**1**) crystals, under vacuum, in a Hirsch funnel (small Büchner funnel) with filter paper. Wash the purple crystals 2–3 times with methanol (10 mL).
4. Place the crystals in an oven (60 °C) until dryness. The yield should be around 4–6%.

Structure Determination

5. Confirm the structure of the pure porphyrin by ^1H NMR spectroscopy. Place about 5 mg of the porphyrin in the NMR tube followed by the addition of 0.5 mL of $CDCl_3$ (containing tetramethylsilane as the internal standard).
6. Acquire the ^1H NMR spectrum.

Note 1. At the beginning of the laboratory class, the oil bath should be turned on.

Note 2. Two hours should be long enough to observe the formation of a powder. (a) If the two laboratory sessions are scheduled to take place on the same day, it is recommended that the reaction be performed during the morning period in order to let the porphyrin precipitate during the lunch time break. (b) If the two laboratory sessions are scheduled to take place on different days, the porphyrin will precipitate between the two laboratory sessions.

Note 3. The mother liquor should provide another crop of crystals if left to stand for a couple of hours.

Waste Disposal

The mixture containing the nitrobenzene may be collected in a separate waste bottle, for future destruction/recovery. Dichloromethane may be collected in a waste bottle for chlorinated solvents, for future destruction/recovery. Methanol may be collected in a waste bottle for non-chlorinated solvents, for future destruction/recovery.

Results Interpretation and Additional Questions

1. Interpret the ^1H NMR spectrum obtained for the 5,10,15,20-tetrakis(2,6-dichlorophenyl) porphyrin.
2. Determine the final product yield.
3. Discuss the mechanism involved in the formation of this porphyrin.
4. Discuss the importance of nitrobenzene in the synthetic strategy.

References

1. L. R. Milgrom, *The Colours of Life: an Introduction to the Chemistry of Porphyrins and Related Compounds*, Oxford University Press, Oxford, 1997, ch. 1, p. 1.
2. P. Rothemund, *J. Am. Chem. Soc.*, 1935, **57**, 2010.
3. P. Rothemund, *J. Am. Chem. Soc.*, 1936, **58**, 625.

4. P. Rothemund, *J. Am. Chem. Soc.*, 1939, **61**, 2912.
5. S. Aronoff and M. Calvin, *J. Org. Chem.*, 1942, **8**, 205.
6. M. Calvin, R. H. Ball and S. Aronoff, *J. Am. Chem. Soc.*, 1943, **65**, 2259.
7. R. H. Ball, G. D. Dorough and M. Calvin, *J. Am. Chem. Soc.*, 1946, **68**, 2278.
8. P. Rothemund and A. R. Menotti, *J. Am. Chem. Soc.*, 1941, **63**, 267.
9. A. D. Adler, F. R. Longo, J. D. Finarell, J. Goldmach, J. Assour and L. Korsakof, *J. Org. Chem.*, 1967, **32**, 476.
10. G. H. Barnett, K. M. Smith and M. F. Hudson, *Tetrahedron Lett.*, 1973, **14**, 2887.
11. K. Rousseau and D. Dolphin, *Tetrahedron Lett.*, 1974, **15**, 4251.
12. J. S. Lindsey, H. C. Hsu and I. C. Schreiman, *Tetrahedron Lett.*, 1986, **27**, 4969.
13. J. S. Lindsey, I. C. Schreiman, H. C. Hsu, P. C. Kearney and A. M. Marguerettaz, *J. Org. Chem.*, 1987, **52**, 827.
14. A. M. D. R. Gonsalves and M. M. Pereira, *J. Heterocycl. Chem.*, 1985, **22**, 931.
15. A. M. A. R. Gonsalves, J. M. T. B. Varejão and M. M. Pereira, *J. Heterocycl. Chem.*, 1991, **28**, 635.

5.2.5. Solventless Synthesis, Separation and Characterization of Zinc and Free-Base Tetraphenyl Porphyrin

Waqar Rizvi[a,b], Charles M. Drain[a,b], Patrick Moy[a], and Matthew J. Jurow*[c]

[a]Hunter College of the City University of New York, 695 Park Avenue, New York, New York 10065, USA; [b]The Graduate Center of the City University of New York, 365 Fifth Avenue, New York, New York 10016, USA; [c]The Molecular Foundry, Lawrence Berkeley National Laboratory, Berkeley, California 94720, USA
*E-mail: mjurow@lbl.gov

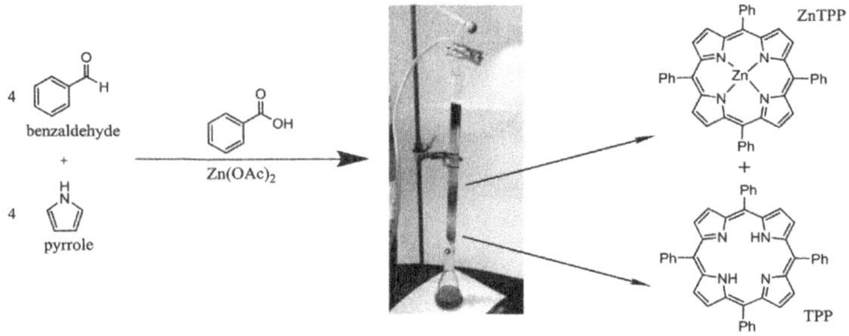

Duration	Hazard level	Difficulty level	Level of study
1 (3 h)	Moderate	Medium	Advanced

Class names Aldehydes, pyrroles

Concepts involved This experiment condenses aldehydes and pyrroles in a solvent-free reaction to rapidly prepare porphyrins, which are chromophores related to natural products; products are separated using column chromatography and characterized by UV-visible absorption spectroscopy and ^{1}H NMR

Chemicals needed Benzoic acid, benzaldehyde, chloroform, pyrrole, zinc acetate, hexanes, CH_2Cl_2, $CDCl_3$, silica gel

Equipment and experimental techniques involved Hot plate w/sand or oil bath, column chromatography, UV-vis spectroscopy, 100 μL syringe, vials, cuvettes

Keywords Column chromatography, UV-vis absorption, NMR, porphyrins, Green Chemistry, symmetry

Background

Porphyrins are deeply colored tetrapyrrolic macrocycles that can chelate almost any metal in the periodic table. The word *porphyrin* is derived from the Greek word *porphura* meaning purple. Porphyrins have distinct UV-visible absorption spectra in two general areas. Porphyrins have a very intense absorption band centered near *ca.* 400 nm called a "Soret" peak, arising from a strongly allowed transition, as well as a series of less intense absorptions at longer wavelengths called "Q-bands". These spectral features provide information about the molecular

Comprehensive Organic Chemistry Experiments for the Laboratory Classroom
Edited by Carlos A. M. Afonso, Nuno R. Candeias, Dulce Pereira Simão, Alexandre F. Trindade, Jaime A. S. Coelho, Bin Tan, and Robert Franzén.

Published by the Royal Society of Chemistry, www.rsc.org

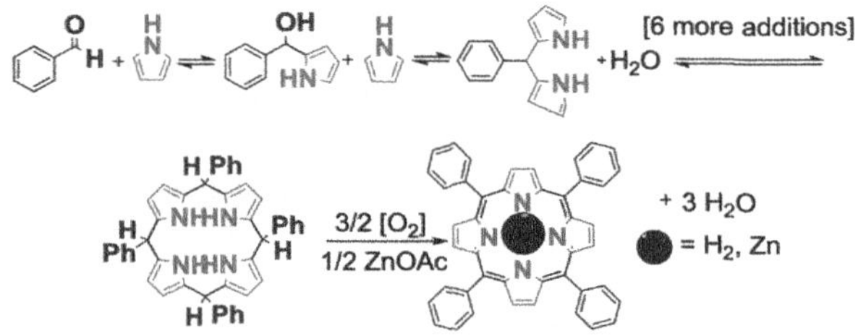

Scheme 5.2.5.1 General synthetic mechanism.

geometry and energies of individual porphyrins. These intense absorption bands and tunable electronic and optical properties make porphyrins important materials in modern technologies like solar cells and photodynamic therapeutics, as well as many of the basic processes of life, like transporting oxygen in red blood cells.[1–4] Porphyrins are also the structural core of chlorophyll A. Tetraarylporphyrins are relatively simple to synthesize from pyrroles and aldehydes containing the appropriate functional groups for the desired application (Scheme 5.2.5.1). Various approaches exist to synthesize porphyrins, including solvent-free methods that are both simple and green.[5,6]

In this experiment, we describe the synthesis of a porphyrin from the condensation of benzaldehyde and pyrrole. The objective of this experiment is to use basic synthetic and purification techniques to synthesize a porphyrin, and to explore the applications of UV-visible and NMR spectroscopy with this important class of compounds.

Tetraphenyl porphyrin (TPP) and zinc tetraphenyl porphyrin (ZnTPP) are easily distinguished by thin-layer chromatography and separated by column chromatography. Each compound elutes as a deeply colored band that is emissive under UV stimulation. Characterization of both products is performed by UV-vis spectroscopy and ^1H NMR.

Additional Safety

This experiment should be carried out in a fume hood with proper ventilation. Safety equipment such as goggles, gloves and protective clothing should be worn at all times. This reaction requires high temperatures (140 °C) in a closed vessel. *Remember*: hot glass looks like cold glass! Be sure to let the bath and reaction vessel cool before handling. Use oven gloves when appropriate. All chemicals should be handled with extreme care. Pyrrole should be added dropwise *via* a syringe.

Experimental Procedure

Laboratory Session 1 (3 h): Synthesis, Separation and Characterization of TPP and ZnTPP

Synthesis

1. In a fume hood heat an oil bath or sand bath to 140 °C on a hot plate.
2. To a 50 mL round-bottomed flask add 60 mg benzoic acid (0.5 mmol), 6 mg zinc acetate (0.03 mmol), and 100 μL of benzaldehyde (1 mmol). Seal the flask with a rubber septum and heat in the oil bath. No stirring is required.
3. Slowly inject 35 μL (0.5 mmol) of pyrrole, dropwise, and let the mixture react until the solution is deeply colored (30–40 min).
4. Remove the flask from the oil bath and allow it to cool to room temperature.

Separation

1. After cooling, open the septum and dissolve the product in a minimum volume of chloroform, making sure to wash the sides of the flask.
2. Pack a short (10 cm × 0.5 cm) column using 20 g silica gel slurry in hexanes. Load the product dissolved in chloroform directly onto the column and elute with a 4:2 (v/v) hexane : CH_2Cl_2 mixture.
3. Collect the two pink bands in separate Erlenmeyer flasks.
4. Transfer the liquid containing each compound to separate, pre-weighed vials and evaporate to dryness. Weigh the dry vials to calculate the yields of each compound.

Characterization of TPP and ZnTPP

1. Dissolve 10–15 mg of each product, ZnTPP and TPP separately after drying well, in 0.75–1 mL deuterated chloroform and pipette into an NMR tube, saving the pipettes and the residue remaining in the product vial. Collect ^1H NMR spectra from 12 to −4 ppm. Label and integrate the peaks.
2. To the few remaining drops of the solution in the vial, add CH_2Cl_2 until the solutions are a pale pink color (1–2 mL).
3. Take the blank UV-visible spectrum with a glass cuvette full of CH_2Cl_2.
4. Add a few drops of the porphyrin solutions to the cuvette, mix, and record the absorption spectra from 370 to 700 nm. Adjust the concentration of the solution until the largest peak has an absorbance value between 1 and 2.5. Repeat for both porphyrins.

Waste Disposal

The waste solutions from the column chromatography should be discarded in the appropriate halogenated organic waste container. The silica gel from the column should be allowed to dry in a fume hood and discarded in a labeled solid waste container.

Results Interpretation and Additional Questions

1. Discuss and explain the similarities and differences in the UV-vis spectra. How did the spectrum change after addition of the metal? How did metalation affect the molecular symmetry?
2. Discuss and explain the similarities and differences in the ^1H NMR spectra. What peaks shifted or changed? What does this tell us about the structure of the molecules?
3. Why do the molecules separate on silica? Which compound elutes first and why?

References

1. D. Dolphin, *The Porphyrins*, ed. D. Dolphin, Academic Press, New York, 1st edn, 1978, vol. 1, ch. 7–10, pp. 290–464.
2. K. Smith, *Porphyrins and Metalloporphyrins*, ed. K. Smith, Elsevier, New York, 1st edn, 1972, vol. 1, ch. 3–5, pp. 106–224.
3. K. M. Kadish, K. Smith and R. Guilard, *The Porphyrin Handbook*, ed. K. Kadish, K. Smith and R. Guilard, Academic Press, San Diego, 1st edn, 2000, vol. 6, ch. 2, pp. 44–67.
4. J. A. S. Cavaleiro, *Handbook of Porphyrin Science*, ed. K. Kadish, K. Smith and R. Guilard, World Scientific, Singapore, 1st edn, 2012, vol. 18, ch. 80, pp. 58–123.
5. J. S. Lindsey, I. C. Schreiman, H. C. Hsu, P. C. Kearney and A. M. Marguerettaz, *J. Org. Chem.*, 1987, **52**, 827.
6. S. Nia, X. Gong, C. M. Drain, M. J. Jurow, W. Rizvi and M. Qureshy, *J. Porphyrins Phthalocyanines*, 2010, **14**, 621.

5.2.6. Synthesis of 3-Bromo-7-ethylcarbamate-4-methylcoumarin

Paula S. Branco* and Ana Maria Lourenço*

Departamento de Química, Faculdade de Ciências e Tecnologia, UNL, 2829-516 Caparica, Portugal.
*E-mail: paula.branco@fct.unl.pt, ana.lourenco@fct.unl.pt

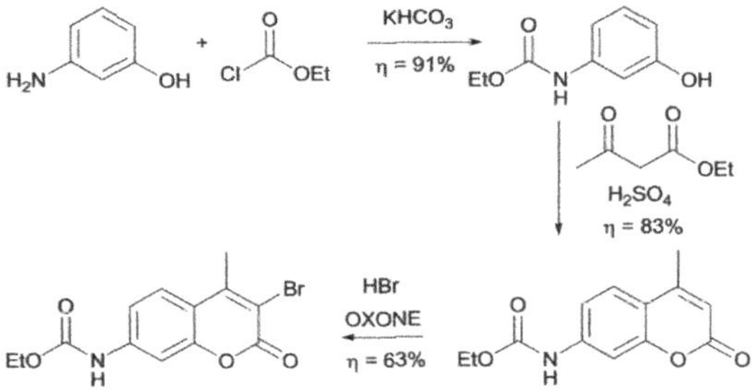

Number of sessions (duration of each session)	Hazard level	Difficulty level	Level of study
3 (4 h each)	Moderate	Medium	Intermediate
Class names Primary amines, phenols, esters, bromines, ketones, carbamates			
Concepts involved Synthesis of heterocycles, acylation reactions, Pechmann condensation, bromination, fluorescent compounds			
Chemicals needed 3-Aminophenol, ethyl chloroformate, ethyl acetoacetate, sulfuric acid, Oxone®, hydrobromic acid, potassium hydrogen carbonate, ethyl acetate, dichloromethane, methanol, spotting solutions for TLC, brine solutions, magnesium sulfate, chloroform			
Equipment and experimental techniques involved Stirring plates, liquid–liquid extraction, rotatory evaporator, NMR, IR, melting point			
Keywords Bromination, extraction, heterocyclic compounds, NMR, Pechmann condensation, synthesis			

Background

The coumarin compounds constitute the largest class of fluorescent dyes widely used as emission layers in organic light-emitting diodes (OLEDs), optical brighteners, nonlinear optical chromophores, fluorescent whiteners, as well as fluorescent labels and probes for physiological measurement.[1,2] Moreover, compounds of this family likewise show extensive and remarkable biological activity.[3] Many articles dealing with their synthesis, reactivity and spectral profile studies have been published.

The beauty of organic chemistry resides in the potential to build fairly complex structures involving multiple reaction in a single synthetic step. One such example is the classical

Comprehensive Organic Chemistry Experiments for the Laboratory Classroom
Edited by Carlos A. M. Afonso, Nuno R. Candeias, Dulce Pereira Simão, Alexandre F. Trindade, Jaime A. S. Coelho, Bin Tan, and Robert Franzén.

Published by the Royal Society of Chemistry, www.rsc.org

Pechmann condensation of coumarins. It is possible to generate the coumarin heterocyclic ring starting from a phenol and a carboxylic acid or ester containing a β-carbonyl group. Esterification/transesterification followed by intramolecular attack on to the activated carbonyl gives the coumarin after dehydration.[4] The condensation is performed under acidic conditions. This synthesis can be applied with success to 3-aminophenol provided that an amine protecting group is used.[5]

Halogenated aromatic compounds are very important and useful synthetic intermediates. Consequently, various methods for the bromination of aromatic compounds have been reported in the literature. The oldest and conventional bromination method typically uses elemental bromine, with its inherent effects as pollutant and as a safety and health hazard. To overcome these difficulties the combination of hydrobromic acid and a suitable oxidant such as Oxone® appears as a useful alternative.[6] The oxidizing agent Oxone® ($KHSO_5$, potassium peroxomonosulfate) is a salt from Caro's acid (H_2SO_5). This fantastic combination affords an *in situ* solution of bromine. The replacement of bromine by a non-toxic stable, simple to handle and more selective reagent is very desirable and represents an important goal in the context of clean synthesis.

Additional Safety

The experimental procedure is without additional risks beyond the care that must be taken in consideration when working with concentrated sulfuric acid. All the experimental procedures should be performed in a fume hood.

Experimental Procedure

Step 1 – Preparation of *m*-(*N*-Ethoxycarbonylamino)phenol

1. Add dropwise 0.5 mL (5.25 mmol) of ethyl chloroformate, in four portions with 10 min intervals with stirring and cooling (10–15 °C), to a suspension of 500 mg (4.58 mmol) of 3-aminophenol and 0.5 g of $KHCO_3$ in 3.5 mL of ethyl acetate and 0.2 mL of water. Then stir the mixture for 1 h. Add water (1.1 mL) and stir the mixture for another 2–3 h.
2. The reaction can be followed by thin-layer chromatography (silicagel $60F_{254}$; methylene chloride/methanol (9.5 : 0.5; iodine vapours or Dragendorff reagent for TLC view and observation under UV at 254 nm)).
3. Separate the aqueous layer, and wash the organic layer successively with water, H_2SO_4 1 M, water, and saturated NaCI solution, dry with $MgSO_4$, and evaporate to dryness. A brown viscous oil (755 mg) of *m*-(*N*-ethoxycarbonylamino)phenol is obtained in around 91% yield.

Step 2 – Preparation of 7-Ethoxycarbonylamino-4-methylcoumarin

1. Add a mixture of 0.5 g (2.76 mmol) of *m*-(*N*-ethoxycarbonylamino)phenol and 0.55 mL of acetoacetic ester in portions with stirring to 1.3 mL of concentrated H_2SO_4, and then stir the mixture for 2 h.
2. The reaction can be followed by thin-layer chromatography (silicagel $60F_{254}$; methylene chloride/methanol (9.5 : 0.5; iodine vapours or Dragendorff reagent for TLC view and observation under UV at 254 nm and 266 nm)).
3. When the reaction has finished, dilute with a mixture of water and ice (6 mL) and stir until crystallization ceases.
4. Remove the precipitate by filtration, wash with water and cold methanol, and dry to give the product 7-ethoxycarbonylamino-4-methylcoumarin (*ca.* 0.56 g (83%), mp 175–177 °C).

Step 3 – Preparation of 3-Bromo-7-ethoxycarbonylamino-4-methylcoumarin

1. To a solution of 7-ethoxycarbonylamino-4-methylcoumarin 440 mg (1.78 mmol) in methylene chloride (6.5 mL) add 690 mg of Oxone followed by 2 mL of aqueous HBr 2 N solution (4.0 mmol).
2. The reaction mixture turns orange immediately; stir the mixture for 2 h 30 min. The reaction can be followed by thin-layer chromatography (silica gel $60F_{254}$; methylene chloride (iodine vapours or TLC view and observation under UV 254 and 366 nm)).
3. After discoloration of the reaction, neutralize it with $NaHCO_3$ saturated solution.
4. To the solution add 10 mL of water. After the methylene chloride has separated successively extract the reaction mixture with chloroform.
5. Dry the organic layer with $MgSO_4$, and evaporate to dryness. A white solid (360 mg) is obtained in *ca.* 63% yield (mp 258–259 °C).

Waste Disposal

Place the desiccant in the container for solid waste. Place the recovered organic solvent from the rotary evaporator in the container for chlorinated solvents.

Results Interpretation and Additional Questions

1. Propose a reaction mechanism for the Pechamnn condensation.
2. Can you provide an explanation for the experimental step of the first session where water was added and the resultant mixture was allowed to react for 2–3 h? What is happening?
3. What is the role of Oxone®? Why does the reaction turn orange after the addition of Oxone?
4. Justify why the bromination occurs on the 3-position and not on the phenyl ring (6 or 8 positions).
5. What is the principal characteristic on the ^1H and ^{13}C NMR spectra of 3-bromo-7-ethoxycarbonylamino-4-methylcoumarin that shows the bromine insertion?
6. What is the mechanism for the bromination reaction? How do you classify this reaction?

References

1. A. Mishra, M. K. R. Fischer and P. Bäuerle., *Angew. Chem., Int. Ed.*, 2009, **48**, 2474.
2. H. S. Jung, P. S. Kwon, J. W. Lee, J. I. Kim, C. S. Hong, J. W. Kim, S. Yan, J. Y. Lee and J. H. Lee, *J. Am. Chem. Soc.*, 2009, **131**, 2008.
3. F. Borges, F. Roleira, N. Milhazes, L. Santana and E. Uriarte, *Curr. Med. Chem.*, 2005, **12**, 887.
4. M. M. Heravi, S. Khaghaninejad and M. Mostofi, *Adv. Heterocycl. Chem.*, 2014, **112**, 1.
5. V. Pozdnev, *Chem. Heterocycl. Compd.*, 1990, **26**, 264.
6. K. M. Kim and I. H. Park, *Synthesis*, 2004, 2641.

5.2.7. Synthesis of Musk Ketone

João Paulo Telo

CQE, Department of Chemical Engineering, IST, University of Lisbon, Av. Rovisco Pais, 1649-003 Lisbon, Portugal. *E-mail: jptelo@tecnico.ulisboa.pt

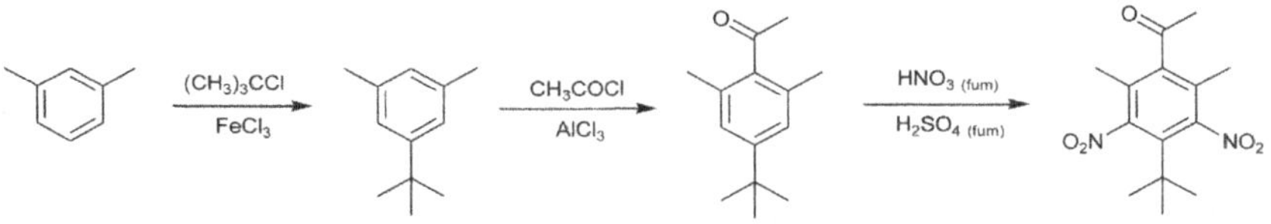

Number of sessions (duration of each session)	Hazard level	Difficulty level	Level of study
3 (4 h each)	High	Medium	Intermediate

Class names Aromatic compounds, ketones, nitroarenes

Concepts involved Synthesis of musk ketone from *m*-xylene in three steps, consisting of three electrophilic aromatic substitutions: Friedel–Crafts alkylation of *m*-xylene to produce 3,5-dimethyl-*tert*-butylbenzene, Friedel–Crafts acetylation and, finally, double nitration to produce the musk ketone; fractional distillation under vacuum of the liquid products and recrystallization (optional) of the solid products; NMR and IR spectroscopies to identify the products

Chemicals needed *m*-Xylene, *t*-butyl chloride, $FeCl_3$, HCl (37%), chloroform, anhydrous $AlCl_3$, acetyl chloride, sodium bicarbonate, methanol, fuming H_2SO_4, fuming HNO_3

Equipment and experimental techniques involved Apparatus for reactions under reflux: three-neck round-bottomed flask, dropping funnels, magnetic stirring; liquid–liquid extraction, vacuum filtration, vacuum fractional distillation, Abbé refractometer for the characterization of a liquid, NMR spectroscopy

Keywords Electrophilic aromatic substitution, fractional distillation, Friedel–Crafts alkylation and acylation, nitration, NMR spectroscopy

Background

The "musk" fragrance used in perfumery was originally obtained from a gland of the male musk deer (*Moschus moschiferus*). The main constituent responsible for the characteristic odor of the deer musk is muscone ((*R*)-3-methylcyclopentadecanone, Figure 5.2.7.1), a 15-carbon atom macrocyclic ketone. Civetone ((*Z*)-cycloheptadec-9-en-1-one), a territorial-marking pheromone of the African civet, is still used in the perfume industry as a musk odorant, although obtained from synthetic routes.[1] It is said that field biologists in Africa used Calvin Klein's perfume "obsession for men", which contains traces of civetone, to attract Jaguars for filming purposes.[2] Musk glands are also found on several other animals, including beetles (*Aromia moschata*), turtles (*Sternotherus odoratus*), crocodiles (*Alligator mississippiensis*) and snakes.

Comprehensive Organic Chemistry Experiments for the Laboratory Classroom
Edited by Carlos A. M. Afonso, Nuno R. Candeias, Dulce Pereira Simão, Alexandre F. Trindade, Jaime A. S. Coelho, Bin Tan, and Robert Franzén.

Published by the Royal Society of Chemistry, www.rsc.org

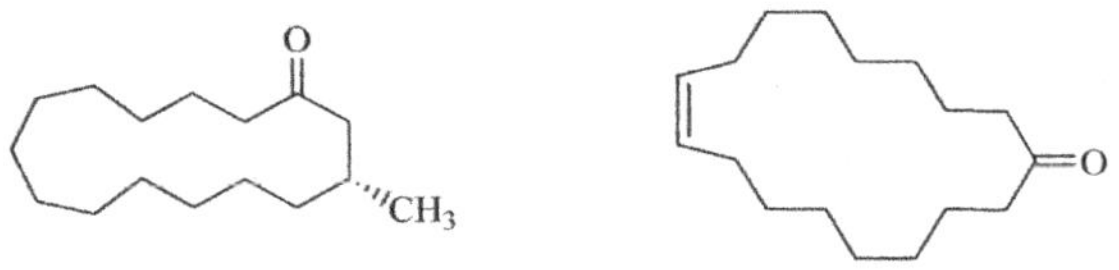

Figure 5.2.7.1 Structures of muscone (left-hand side) and civetone (right-hand side).

Today all musk odorants are obtained synthetically due to obvious ecological reasons. Musk ketone belongs to a series of compounds known as "nitro musks" that, although structurally very different from the natural musks, have similar olfactory properties to the natural musks. The first nitro musk was prepared by the German chemist Albert Baur in 1888. Baur was searching for a more powerful version of TNT (trinitrotoluene) and prepared a series of polynitroaromatics, which, although disappointingly poor explosives, smelled a lot like natural musks. Due to the low-cost production, musk ketone was used widely as a fragrance. Recently the nitro musks have been substituted by other artificial musks due to their toxicity and persistence in the environment.[3] However, significant amounts of nitro musks are still used in Indian incense sticks. Musk ketone is a solid with a reasonably low vapor tension and interestingly a small number of humans seems to be anosmic (*i.e.* cannot detect the smell) to it.

This experiment is based on literature work,[4] but several important changes have been made. The first step is a Friedel–Crafts alkylation of *m*-xylene with *t*-butyl chloride, but the aluminium amalgam has been replaced by the more common Lewis acid catalyst $FeCl_3$. The electophile is the *t*-butyl cation. Friedel–Crafts reactions are known to occur only in aromatic rings that are non-deactivated towards the electrophile aromatic substitution (EAS), which is the case for *m*-xylene, but interestingly the reaction occurs in the *meta* position relative to the activating methyl groups. The second step is a Friedel–Crafts acylation, yielding an aromatic ketone. The electrophile is also a cation (the acylium cation), prepared by reaction of the acetyl chloride with the Lewis acid $AlCl_3$. The third step is a double nitration of the aromatic ring. In this step, a lot of unchanged reactant was obtained when only fuming nitric acid was used. We followed the procedure of another paper that uses sulfuric acid as catalyst, which increased the yield.[5] NMR shows the product to be pure, and the formation of 4-*t*-butyl-2,6-dimethyl-3,5-dinito benzoic acid was not observed. This oxidation was predicted to occur by some reports when sulfuric acid is used at high temperatures.[4]

Additional Safety

The reactions should be done in a well-ventilated fume cupboard due to the release of toxic and corrosive fumes. Anhydrous $AlCl_3$ releases gaseous HCl in contact with (wet) air and should be weighed rapidly in a fume cupboard and introduced into the reaction flask as soon as possible. Fuming nitric and sulfuric acids are corrosive strong acids that release toxic fumes and should be handled with care. Acetyl chloride, chloroform and xylene are toxic by inhalation and by contact with the skin and should be handled with care and in a fume cupboard.

Experimental Procedure

Laboratory Session 1 (4 h): Preparation of *tert*-Butyl-3,5-dimethylbenzene

1. Charge a 250 mL two-neck round-bottomed flask, equipped with a dropping funnel and a reflux condenser with an HCl gas trap, with 40 mL of *m*-xylene and 2 g of anhydrous $FeCl_3$.

2. Add, slowly, 50 mL of *tert*-butyl chloride through the dropping funnel while the contents of the flask are stirred magnetically.
3. Keep stirring after the addition until no more gaseous HCl is released through the reflux condenser.
4. Add 50 mL of water and transfer the mixture to a separatory funnel.
5. Wash the organic phase with diluted HCl, twice with water and dry it with anhydrous sodium sulfate (or anhydrous calcium chloride).
6. Distill the product through a fractionating column under reduced pressure and collect the fraction distilling between 86 and 89 °C (20 mmHg).

Laboratory Session 2 (4 h): Preparation of 4-*tert*-Butyl-2,6-dimethylacetophenone

1. Add to a 250 mL two-neck round-bottomed flask, equipped with a dropping funnel and a reflux condenser with an HCl gas trap, 150 mL of chloroform and 35 g of $AlCl_3$.
2. Stir magnetically and cool the flask in a mixture of ice and salt. Drop, slowly, through the funnel 32.5 g of *tert*-butyl-3,5-dimethylbenzene during around 1 h.
3. After the addition stir the mixture at room temperature for an additional hour.
4. Pour the resultant red mixture into a beaker containing ice (150–200 g) and 30 mL of hydrochloric acid and stir until the red color disappears.
5. Transfer the mixture to a separatory funnel, discharge the aqueous layer and wash the organic phase with a 5% sodium bicarbonate solution and then with water.
6. Separate the organic phase, dry it with anhydrous sodium sulfate and distil the chloroform in a rotatory evaporater.
7. Distill the residue through a fractionating column under reduced pressure and collect the product between 125 and 140 °C (at 20 mmHg, or 107–112 °C at 2.5 mmHg). The pure product distills at 139 °C (20 mmHg) and solidifies within minutes.
8. An analytical sample can be obtained by recrystallization from a small amount of methanol and cooling in the freezer for 1 h before filtration. Partial slow evaporation of the methanol mother-liquor yields an additional crop.

Laboratory Session 3 (4 h): Musk Ketone

1. Charge a 100 mL three-neck round-bottomed flask, equipped with a dropping funnel, a reflux condenser and a thermometer, with 2 g of 4-*tert*-butyl-2,6-dimethylacetophenone and 12 mL of fuming H_2SO_4.
2. Stir magnetically and add slowly through the dropping funnel 8 mL of fuming nitric acid.
3. After the addition heat the mixture in a steam bath for 15 min.
4. Pour the mixture in ice (around 150 g), mix well and let the mixture to settle in order to allow the separation of the product.
5. Filter the solid off with suction and wash it with cold water. The solid can be recrystallized from methanol.

Waste Disposal

The mother-liquors of the recrystallizations should be disposed of in proper containers for elimination. The chloroform distilled should be stored for further purification and reutilization. The acid aqueous solutions should be diluted and discharged in the sink under running water.

Results Interpretation and Additional Questions

Check the purity of the products by measuring the refraction index of the liquids and the melting point of the solids and compare them with the published values. Interprete the NMR and IR spectra of the products.

1. Draw the reaction mechanism for the three reactions.
2. Why does the Friedel–Crafts alkylation occur *meta* to the two methyl groups on *m*-xylene?
3. Quite often polyalkylation is an important side reaction of Friedel–Crafts alkylations. Why is it not observed in this case?
4. Why does the Friedel–Crafts acylation occur between the two methyl groups, and not between a methyl and the *t*-butyl?

References

1. Y.-M. Choo, K.-E. Ooi and I.-H. Ooi, Synthesis of civetone from palm oil products, *J. Am. Oil Chem. Soc.*, 1994, **71**, 911.
2. J. G. Goldma, *Scientific American*, 2013 (10 October), http://blogs.scientificamerican.com/thoughtful-animal/2013/10/10/youll-never-guess-how-biologists-lure-jaguars-to-camera-traps/.
3. H. H. Schmeiser, R. Gminski and V. Mersch-Sundermann, Evaluation of health risks caused by musk ketone, *Int. J. Hyg. Environ. Health*, 2001, **203**, 293–299.
4. E. Nash, E. J. Nienhouse, T. A. Silhavy, D. E. Humbert and M. J. Mish, Aromatic Nitro Musk Synthesis, *J. Chem. Educ.*, 1970, **47**(10), 705.
5. R. C. Fuson, J. Mills, T. G. Klose and M. S. Carpenter, *J. Org. Chem.*, 1947, **12**, 587.

5.2.8. Synthesis of Methyl 4-oxo-4-(thiophen-2-yl) butanoate

M. Manuela M. Raposo

Department of Chemistry, University of Minho, Campus de Gualtar, 4710-057 Braga, Portugal
*E-mail: mfox@quimica.uminho.pt

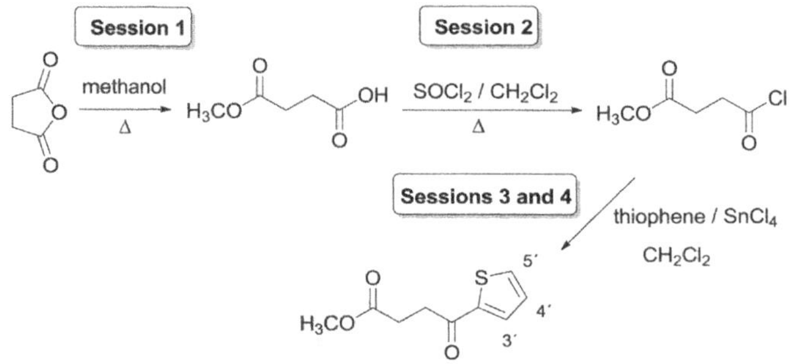

Number of sessions (duration)	Hazard level	Difficulty level	Level of study
4 (4 h + 4 h + 2 h + 2 h)	Moderate	Medium	Intermediate

Class names Anhydrides, carboxylic acids, esters, acid chlorides, thiophenes

Concepts involved Synthesis of a γ-keto ester derived from the thiophene heterocycle through several synthetic transformations; the various steps involved are the preparation of the acylating reagent followed by the synthesis of a 1,4-dicarbonyl compound, alcoholysis, nucleophilic substitution (conversion of a carboxylic acid into an acid chloride) and finally the electrophilic aromatic substitution of the thiophene heterocycle (Friedel–Crafts acylation)

Chemicals needed Succinic anhydride, methanol, thionyl chloride, dichloromethane (dry), tin tetrachloride, thiophene, NaOH (20%), hydrochloric acid (37%), anhydrous sodium sulfate

Equipment and experimental techniques involved Heating mantle, stirring and heating plate, rotary evaporator, vacuum pump, apparatus for heating at reflux under anhydrous conditions, liquid–liquid extraction, thin-layer chromatography (TLC), gravity filtration, ^{1}H NMR, IR

Keywords Alcoholysis, extraction, Friedel–Crafts acylation, ^{1}H NMR, γ-keto ester, thiophene

Background

1,4-Dicarbonyl compounds (*e.g.* γ-keto esters and γ-keto amides) are versatile precursors in organic synthesis, particularly in the preparation of heterocyclic systems, including those containing thiophene, pyrrole and pyridazine amongst others. For the preparation of the thiophene heterocycle the 1,4-dicarbonyl compound will react with a source of sulfur, usually H_2S, Lawesson reagent or P_4S_{10}.[1] On the other hand, pyrrole derivatives are usually prepared by the condensation of 1,4-dicarbonyl compounds with ammonia or with primary amines

Comprehensive Organic Chemistry Experiments for the Laboratory Classroom
Edited by Carlos A. M. Afonso, Nuno R. Candeias, Dulce Pereira Simão, Alexandre F. Trindade, Jaime A. S. Coelho, Bin Tan, and Robert Franzén.

Published by the Royal Society of Chemistry, www.rsc.org

through the Paal–Knorr synthesis.[2] The Friedel–Crafts reaction is one of the most effective methods for introducing an acyl group into an aromatic ring. Thiophene, being an electron-rich heterocycle, reacts easily by electrophilic substitution and preferably at position 2 (α). Nevertheless, substitution at position 3 (β) of the thiophene ring can also occur, especially when the α positions are blocked. Friedel–Crafts acylation of thiophene derivatives is widely used and occurs easily with good yields in the presence of a Lewis acid (*e.g.* $AlCl_3$, $SnCl_4$, *etc.*). The acylation reagents most used are acid chlorides, carboxylic acids, anhydrides, *etc.*[3,4] In this case the acylation takes place exclusively at the α position of the thiophene ring. Thus, the synthesis of 4-(2′-thienyl)-4-oxo-butanoate is possible using cheap, commercial reagents as well a simple synthetic methodology. This compound will be also used in other experimental work as a precursor for the synthesis of the corresponding carboxylic acid obtained by hydrolysis of γ-keto ester. (see Experiment 3.1.16). The objective of this experiment is the synthesis of thiophene α-keto ester derivative[5] in four sessions: session 1: synthesis of the acylating reagent (4-methoxy-4-oxobutanoic acid) by an alcoholysis reaction of succinic anhydride with methanol ; session 2: preparation of methyl 4-chloro-4-oxobutanoate by conversion of the acid precursor into the corresponding acid chloride through a nucleophilic substitution on the carbonyl group using thionyl chloride; session 3: preparation of methyl 4-oxo-4-(thiophen-2-yl)butanoate through Friedel–Crafts acylation of thiophene with methyl 4-chloro-4-oxobutanoate in the presence of $SnCl_4$; session 4: isolation of methyl 4-oxo-4-(thiophen-2-yl)butanoate through extraction followed by evaporation of the organic solvent under reduced pressure.

Additional Safety

Due to the toxicity of some reagents used in this experiment, all sessions should be done in a fume hood, with proper safety equipment, such as protective clothing, gloves and safety goggles. The reactions with thionyl chloride and tin tetrachloride should be performed using dry solvents. Methanol is toxic in contact with skin, if swallowed or inhaled (session 1). Thionyl chloride is a corrosive and toxic reagent (oral, dermal, inhalation), reacts violently with water and other reagents to produce toxic gases and may cause skin burns and eye damage (session 2). Tin tetrachloride is a toxic reagent (oral, dermal, inhalation), fumes on contact with air, giving a stinging odour and may cause skin burns and eye damage (session 3). Sodium hydroxide (session 3) is a corrosive base and may cause severe skin and eye burns. Hydrochloric acid (session 3) is a toxic (oral, dermal, inhalation) and a corrosive acid that may cause severe skin and eye burns. Thiophene is highly flammable in liquid or vapour form, which may cause serious eye irritation and is harmful when inhaled or swallowed. Furthermore it is harmful to aquatic organisms, may cause long-term adverse effects in the aquatic environment (session 3). Dichloromethane is toxic, may cause skin, eye and respiratory tract irritation. The vapours may cause drowsiness and dizziness (session 3). Complete information about the safety, toxicity and disposal of all reagents can be found, free of charge, at http://www.msds.com/index.asp?open=/protected_public/loginsuccessful.asp.

Experimental Procedure

Laboratory Session 1 (4 h): Synthesis of 4-Methoxy-4-oxobutanoic Acid

1. In a fume hood, prepare an apparatus for heating at reflux consisting of a heating mantle, a 250 mL round-bottom flask (with known tare), a reflux condenser and boiling chips.
2. Add succinic anhydride (20 g, 0.2 mol) and methanol (23.7 g, 30 mL, 0.74 mol) to the flask and heat the mixture at reflux for 3 h.

3. Remove the solvent under reduced pressure with a rotary evaporator and then with a vacuum pump.
4. Weigh the product obtained and calculate the yield. Record the ^{1}H NMR and IR spectra and compare with the spectra provided in the ESI to confirm the compound's purity.

Laboratory Session 2 (4 h): Synthesis of Methyl 4-chloro-4-oxobutanoate

1. In a fume hood, prepare an apparatus for heating at reflux in dry conditions, consisting of a heating mantle, a 150 mL round-bottom flask with boiling chips and a reflux condenser closed with a calcium chloride tube.
2. Add 4-methoxy-4-oxobutanoic acid (15 g, 0.114 mol) and 50 mL of dry dichloromethane to the flask and after complete dissolution of the butanoic acid derivative add the previously distilled thionyl chloride (alternatively use thionyl from a freshly open bottle) (12.4 mL, 20.3 g, 0.171 mol).
3. Heat the mixture at reflux for 3 h.
4. Remove the solvent under reduced pressure with a rotary evaporator.
5. Weigh the product obtained and calculate the yield. Record the ^{1}H NMR and IR spectra and compare them with literature data to confirm the compound's purity.

Laboratory Session 3 (2 h): Synthesis of Methyl 4-oxo-4-(thiophen-2-yl) butanoate

1. In a fume hood, prepare an apparatus consisting of a previously dried 250 mL two-neck round-bottom flask with a magnetic stirrer, a reflux condenser closed with a calcium chloride tube, a dropping funnel and an ice bath.
2. Add to the flask dry dichloromethane (65 mL) and $SnCl_4$ (6.7 g, 3.0 mL, 25.7 mmol) at 0 °C.
3. Add dropwise through the dropping funnel, a solution of methyl 4-chloro-4-oxobutanoate (3.5 g, 70 mmol) and thiophene (2.0 g, 2.1 mL, 23.3 mmol) in dry dichloromethane (67 mL) at 0 °C during 1 h.
4. After complete addition stir overnight the reaction mixture at room temperature.

Laboratory Session 4 (2 h): Isolation of Methyl 4-oxo-4-(thiophen-2-yl) butanoate

1. Add 300 mL of cold water to the reaction mixture and acidify with HCl (37%).
2. Transfer the reaction mixture into a separatory funnel and perform an extraction with dichloromethane (3 × 20 mL) and then extract the combined organic layers with a solution of NaOH (20%) (3 × 20 mL) and water (2 × 70 mL). Remove the water from the combined organic layers with anhydrous MgSO4, filter the drying agent and evaporate the solvent under reduced pressure with a rotary evaporator.
3. Determine the mass of the product obtained as a yellow oil and calculate the yield. Record the ^{1}H NMR spectrum in $CDCl_3$ and the IR spectrum in liquid film. Compare the spectra obtained with the literature data[5] and with the spectra provided in the ESI to confirm the compound's purity.

Waste Disposal

The solvents used in extractions are recovered after evaporation in a rotary evaporator and should be collected in chlorinated and non-chlorinated waste collector recipients, for later destruction. Special attention should be paid to the treatment and handling of reaction mixtures in sessions 2 and 3 which contain HCl as a by-product. The neutralized aqueous solution obtained in session 2 may be diluted and discarded in the sink. The neutralized aqueous solution obtained in session 4 should be collected in a waste collector recipient for later destruction since it contains tin.

Results Interpretation and Additional Questions

1. Explain why the reactions in sessions 2 and 3 should be performed under anhydrous conditions.
2. Give examples of other reagents that could be used to perform the conversion of a carboxylic acid group into an acid chloride.
3. Explain the role played by $SnCl_4$ in the Friedel–Crafts reaction performed in session 3 and give examples of other reagents that might be used instead.
4. Write out the reaction mechanism for the reaction in session 3.
5. Explain why the acylation of thiophene occurs at position 2 of the heterocyclic ring.
6. Interpret the ^1H NMR spectra obtained for all compounds, and compare them with those found in the literature.[5]

References

1. T. Ozturk, E. Ertas and O. Mert, *Chem. Rev.*, 2007, **107**, 5210.
2. (a) G. P. Bean, in *Pyrroles The Chemistry of Heterocyclic Compounds*, ed. R. A. Jones, John Wiley & Sons, Inc., New York, 1990, vol. 48, part 1, p. 105; (b) T. L. Gilchrist, *J. Chem. Soc., Perkin Trans. 1*, 1999, 2849; (c) C. Schmuch and D. Rupprecht, *Synthesis*, 2007, 3095.
3. J. A. Joule, K. Mills and G. F. Smith, *Heterocyclic Chemistry*, Chapman & Hall, London, 3th edn, 1995, ch. 2 and 14, p. 22 and pp. 261–262.
4. T. L. Gilchrist, *Heterocyclic Chemistry*, John Wiley & Sons, Inc., New York, 1985.
5. M. M. M. Raposo and G. Kirsch, *Heterocycles*, 2001, **55**, 1487.

5.2.9. Synthesis of Eosin

Dulce Simão

Department of Chemistry, Instituto Superior Técnico, University of Lisbon, Centro de Química Estrutural, Av. Rovisco Pais, 1, 1049-001 Lisbon, Portugal. *E-mail: dulcesimao@tecnico.ulisboa.pt

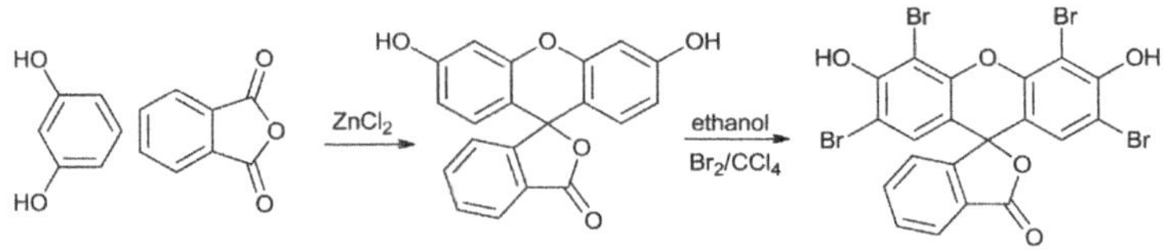

Number of sessions (duration of each session)	Hazard level	Difficulty level	Level of study
2 (4 h each)	High	Medium	Intermediate

Class names Anhydrides, phenols, arenes, bromoarenes, xanthenes

Concepts involved This experiment involves a two-step reaction sequence to obtain eosin, employing a Friedel–Crafts acylation and a bromination by electrophilic aromatic substitution; the ammonium salt of eosin can be prepared and is used to dye a piece of cloth; fluorescence, chemiluminescence, ^{1}H NMR and IR spectroscopy

Chemicals needed Resorcinol, phthalic anhydride, zinc chloride, hydrogen chloride (37%), sodium hydroxide, ethanol, bromine, carbon tetrachloride, ammonia (25%), piece of cloth; bis(2,4,6-trichloro) oxalate, sodium acetate, ethyl acetate, hydrogen peroxide (30%) (for chemiluminescence)

Equipment and experimental techniques involved Apparatus for high temperature control (sand bath), high vacuum pump, vacuum filtration, IR, NMR and melting point

Keywords Chemiluminescence, dye, electrophilic aromatic substitution, fluorescence, Friedel–Crafts acylation, NMR, vacuum filtration

Background

Fluorescein is an acidic xanthene (Figure 5.2.9.1) dye exhibiting fluorescence. It was synthesized by Baeyer in 1871 and involves two successive Friedel–Crafts acylations. The electrophile is an acyl cation generated from phthalic anhydride and zinc chloride, as Lewis acid. The subsequent reaction with two molecules of the activated ring of substituted phenol (resorcinol) and phthalic anhydride produces benzophenone. Fluorescein is formed after a second Friedel–Crafts reaction and intramolecular esterification.[1–4] The disodium and dipotassium salts of fluorescein are known as uranine. Due to the fluorescent property, it is useful for tracing the paths of underground rivers or for tracing sources of contamination of drinking water and leaks. It is also used in medicine as a biological stain and in ophthalmological examinations to visualize scratches on the cornea of the eye, and in forensic chemistry[5] for example.

This compound can exist in two isomeric forms, lactoid structure (yellow) and quinoid form (red)[2,6] as the pH varies. Several studies showed the existence of four possible forms for fluorescein with changing pH[7] (Scheme 5.2.9.1).

Comprehensive Organic Chemistry Experiments for the Laboratory Classroom
Edited by Carlos A. M. Afonso, Nuno R. Candeias, Dulce Pereira Simão, Alexandre F. Trindade, Jaime A. S. Coelho, Bin Tan, and Robert Franzén.

Published by the Royal Society of Chemistry, www.rsc.org

Fluorescein halogenation through electrophilic aromatic substitution affords several important dyes in the chemical industry. The bromination of this compound occurs on the highly activated phenolic aromatic rings to produce eosin, the tetrabromo derivative,[1,3,6] a dye used in lip-stick, nail polish and as a biological stain. This red pigment is extensively used in printing inks.[4,6] The dyeing process can be carried out after conversion into a water-soluble salt like the sodium or ammonium salt. Eosin is also fluorescent when exposed under ultraviolet lamp and can be used as fluorescer or activator (ACT) in a chemiluminescence reaction[8,9] using a peroxyoxalate system with bis(2,4,6-trichlorophenyl oxalate) known as a cyalume.[10] The addition of hydrogen peroxide results in a strong light-orange color emission that lasts for several minutes. The reaction between the oxalic ester and hydrogen peroxide gives an unstable cyclic peroxide. This molecule transmits its energy to the fluorescer (eosin) leading to the electronic excited state of the fluorescer, which decays to the fundamental state emitting the fluorescent orange light (Scheme 5.2.9.2).

A variety of colors can be obtained by changing the fluorescer, usually polycondensed aromatic hydrocarbons. This is the reaction that takes places in the well-known "glow sticks" or

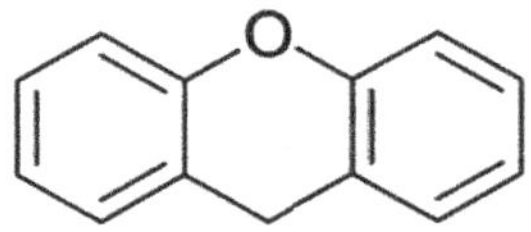

Figure 5.2.9.1 Xanthene structure.

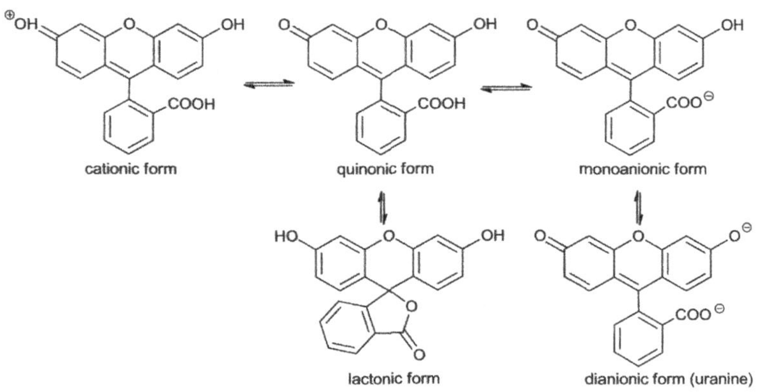

Scheme 5.2.9.1 Possible fluorescein structures with changing pH.

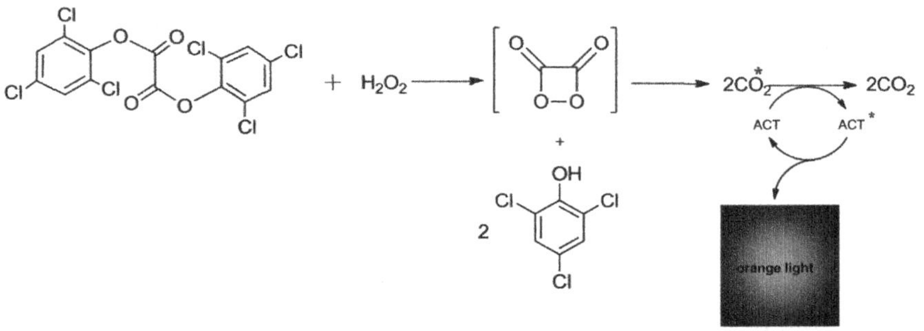

Scheme 5.2.9.2 Chemiluminescence reaction.

"light sticks" used for entertainment events, military operations, training, industrial accidents and natural disasters. When the stick is bent the compartment containing hydrogen peroxide is broken, and the reaction with the ester solution occurs, leading to the glow.[9]

Additional Safety

This work must be carried out in a fume hood due to heating on a sand bath with liberation of moisture and the use of carbon tetrachloride, which is chronically toxic and a human carcinogen. Bromine is a red–brown fuming liquid with an irritating and suffocating odour. It produces a vapor that is very toxic if breathed in. The liquid causes severe burns to the eyes and skin. A sodium thiosulfate solution should be available to deal with spills. Hydrogen peroxide is a corrosive liquid and is a strong oxidant.

Experimental Procedure

Preparation of Fluorescein

1. In a three-necked flask fit a reflux condenser and heat on a sand bath zinc chloride (0.5 g) until no more moisture comes from the condenser.
2. Add resorcinol (1 g) and phthalic anhydride (0.7 g).
3. Stir magnetically and heat at 180 °C. Maintain this temperature for 10 min, at which time the mixture should cease bubbling and become stiff.
4. Cool the mixture to below 100 °C, add water (10 mL) and concentrated hydrogen chloride then stir and grind the solid (trituration) while heating it on a boiling water bath.
5. Remove the liquid with a Pasteur pipette and repeat this process with another similar portion of water and acid.
6. Collect the solid by suction, wash it well with water and allow it to dry.
7. Record the IR spectrum.

Fluorescence

Add fluorescein (0.25 g) to sodium hydroxide (2.5 mL, 3 M aqueous solution) and water (2.5 mL). Dilute the mixture to 500 mL. Examine the solution by transmitted light and then under UV light. Add hydrogen chloride (37%) until pH 2 and examine it again under UV light.

Preparation of Eosin

1. In a three-necked flask place fluorescein (0.5 g) and ethanol (2 mL) and add a stirrer magnetic bar.
2. Stir and add a solution of bromine (1.08 g) in carbon tetrachloride (2.5 mL) dropwise.
3. After addition of half the bromine, a soluble dibromofluorescein solution results. When all the bromine has been added, the tetrabromo compound will precipitate.
4. Stir for a further 15 min, cool the reaction mixture and collect the solid by suction filtration.
5. Wash the solid with ice-cold ethanol and dry it under vacuum.
6. Record the IR and ^{1}H NMR spectra.

Dyeing

Prepare the ammonium salt of eosin by exposing the salt to ammonia vapors as follows. Moisten eosin (0.4 g) with ethanol (few drops) on a small filter paper. Fold up the paper and push it into a reaction tube above ammonia solution 25% (2.5 mL) and cap the tube. After about an hour a portion of the derivative should be completely soluble in the ammonia solution. Remove the filter paper, add a piece of test cloth and heat this solution in a water bath (50–60 °C) for about 10 min. Remove and let it air dry.

Chemiluminescent Reaction

In a test tube place eosin (3 mg), bis(2,4,6-trichloro)oxalate (50 mg) and sodium acetate (50 mg). Add ethyl acetate (10 mL), cap the tube and shake until dissolution. In a darkened room, add hydrogen peroxide 30% (3 mL), cap the tube and shake. Observe the color that is seen, the intensity, and the length of time of the fluorescent reaction.

Waste Disposal

After filtration, the highly dyed washings may be neutralized, diluted and discarded in the sink. Washings from eosin filtration containing carbon tetrachloride should be stored in a chlorinated waste collector and chemiluminescent solution should be stored in a non-chlorinated waste collector for posterior destruction.

Results Interpretation and Additional Questions

1. Determine the reaction yields.
2. Compare the melting points of the obtained products with the ones reported in the literature.
3. Interpret the IR and ^1H NMR spectra.
4. Suggest a plausible mechanism for the formation of fluorescein and eosin.
5. Write out the structures of the fluorescein in acid and basic solution.
6. Write out the structure of the ammonium salt of eosin.
7. Explain the reactivity and regioselectivity observed in the bromination of fluorescein.

Acknowledgement

The author of this experiment acknowledges Professor Diogo S. Lüdtke and Dr Maria Eduarda Contreira from the Institute of Chemistry, Universidade Federal do Rio Grande do Sul – UFRGS, Brazil, for reproducing and checking this experiment in their own laboratory and for providing data included in the ESI.

References

1. C. A. M. Afonso, D. P. Simão, L. P. Ferreira, M. S. Serra and M. M. M. Raposo, *100 Experiências de Química Orgânica*, IST Press, 2011, p. 381.
2. W. R. Orndorff and A. J. Hemmer, *J. Am. Chem. Soc.*, 1927, **49**, 1272–1280.
3. F. Fieser and K. L. Williamson, *Organic Experiments*, Houghton Mifflin Company, 8th edn, 1998, pp. 451–453.
4. J. V. Mc Cullagh and K. A. Dagget, *J. Chem. Educ.*, 2007, **84**(11), 1799.
5. Fluorescein, http://www.absoluteastronomy.com/topics/Fluorescein, accessed May 2015.
6. *Kirk-Othmer Encyclopedia of Chemical Technology*, John Wiley & Sons, 2nd edn, 1970, vol. 22, pp. 432–437.
7. V. Zanker and W. Peter, *Chem. Ber.*, 1958, **91**, 572–580.
8. R. Albertin, M. Arribas, E. Bastos, S. Röpke, P. Sakai, A. Sanches, C. Stevani, I. Umezu, J. Yu and W. Baader, *Quim. Nova*, 1998, **21**(6), 772.
9. J. Alves, A. Boaro, J. S. da Silva, T. L. Ferreira, V. B. Keslarek, C. A. Cabral, R. B. Orfão Jr, L. F. M. L. Ciscato and F. H. Bartoloni, *Photochem. Photobiol. Sci.*, 2015, **14**, 320–328.
10. Cyalume® Technologies, http://www.cyalume.com/, accessed May 2015.

5.2.10. Synthesis and Formylation of 5-Piperidino-2,2′-bithiophene

M. Manuela M. Raposo*

Department of Chemistry, University of Minho, Campus de Gualtar, 4710-057 Braga, Portugal
*E-mail: mfox@quimica.uminho.pt

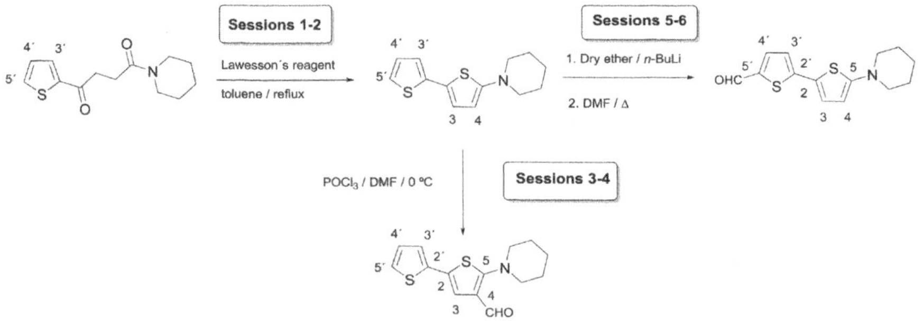

Number of sessions (duration of each session)	Hazard level	Difficulty level	Level of study
6 (4 h each)	High	High	Intermediate

Class names Amides, aldehydes, thiophenes

Concepts involved Synthesis of a 5-*N*,*N*-dialkylamino-2,2′-bithiophene and the corresponding 4- and 5′-formyl derivatives through several synthetic transformations: (i) thionation of carbonyl groups of a tertiary γ-ketoamide derived from thiophene using Lawesson's reagent followed by cyclization of the corresponding thioxoamide through intramolecular nucleophilic attack yielding the 5-*N*,*N*-dialkylamino-2-2′-bithiophene derivative; formylation of the bithiophene derivative using two different synthetic methodologies: (ii) Vilsmeier–Haack reaction or (iii) lithiation followed by electrophilic aromatic substitution; determination of the structure of all compounds through ^{1}H NMR, ^{13}C NMR, IR and UV spectroscopic techniques

Chemicals needed 1-(Piperidin-1-yl)-4-(thiophen-2-yl)butane-1,4-dione (synthesized in Experiment 3.1.16), Lawesson's reagent (LR), toluene, phosphorous(v) oxychloride, DMF (dry), diethyl ether, diethyl ether (dry), *n*-butyllithium (*n*-BuLi) 2.5 M in hexanes, petroleum ether (40–60 °C), chloroform, deuterated chloroform, sodium sulfate (anhydrous), magnesium sulfate (anhydrous), methanol, ethyl acetate, sodium acetate saturated solution, sodium bicarbonate saturated solution, silica gel for column chromatography (60 Å; 230–400 mesh), thin-layer chromatography (TLC) plates with fluorescence indicator

Equipment and experimental techniques involved ^{1}H and ^{13}C nuclear magnetic resonance spectroscopy, UV-visible spectroscopy, IR spectroscopy, apparatus for heating at reflux under anhydrous conditions, apparatus for heating at reflux under anhydrous conditions in an inert atmosphere, rotary evaporator, liquid–liquid extraction, gravity filtration, thin-layer chromatography (TLC), column chromatography, melting point

Keywords Carbonyl thionation, chromatography, ^{1}H NMR, intramolecular cyclisation, thiophene, Vilsmeier–Haack formylation

Comprehensive Organic Chemistry Experiments for the Laboratory Classroom
Edited by Carlos A. M. Afonso, Nuno R. Candeias, Dulce Pereira Simão, Alexandre F. Trindade, Jaime A. S. Coelho, Bin Tan, and Robert Franzén.

Published by the Royal Society of Chemistry, www.rsc.org

Background

Oligothiophenes are heterocyclic compounds that have attracted widespread interest for many years due to their electronic/optical properties, high thermal and chemical stability and their relative ease of synthesis.[1] They have been used for the fabrication of (opto)electronic devices such as organic light emitting diodes (OLEDs),[2] photovoltaic and dye sensitized solar cells (OPVs and DSSCs).[3] Applications in other scientific areas include nonlinear optics (NLO),[4] chemosensors,[5] photochromic[6] and biology.[7] In particular, the past decade has been marked by a growing interest in the design of NLO chromophores for two-photon absorption (TPA) applications. These efforts have been driven by the many applications of TPA phenomenon in various fields including material science, biology and medicine.[8]

The oligothiophene derivatives may be synthesized by several methods such as cross-coupling reactions (Stille, Suzuki or others) and by procedures involving thiophene ring formations.[1b] A common approach to the synthesis of thiophene heterocycle involves the use of 1,4-dicarbonyl compounds which react with a source of sulfur, usually H_2S and HCl, phosphorus(v) sulfide or Lawesson's reagent (LR) [2,4-bis(4-methoxyphenyl)-1,3,2,4-dithiadiphosphetane-2,4-disulfide] which was first used as a thionation reagent by Lawesson and collaborators in 1978.[9a] (See the structure and mechanism of dissociation of Lawesson reagent in the ESI for Experiment 17.8.) This reagent is mainly used in the conversion of carbonylic compounds into thiocarbonylic derivatives (*e.g.* thioesters, tioacetones, tiopeptides, *etc.*) and is currently considered one of the most efficient thionation reagents, having several advantages over other thionation reagents.[9]

Formyl (oligo)thiophene derivatives are versatile "crossroads" intermediates that can be used in the synthesis of push–pull substituted π-conjugated systems for several applications (*e.g.* medicinal, supramolecular, optoelectronic materiais, *etc.*).[4b,5,8] Not surprisingly, a large number of methods have been developed to obtain these functionalized heterocycles. Vilsmeier–Haack formylation and metalation followed by quenching with DMF are the most important routes for the preparation of formyl-substituted (oligo)thiophenes. The choice of the most appropriate method of synthesis depends on the desired selectivity of the formylation reaction. The Vilsmeier formylation is an aromatic electrophilic substitution reaction in which the electrophile (formylating agent) is obtained by reacting a substituted amide, in this case *N,N*-dimethylformamide (DMF), with phosphorus oxychloride ($POCl_3$) to produce the reactive species (the Vilsmeier reagent: $ClCH = NMe_2^+$), which in turn reacts with electron rich (hetero)aromatic compounds, yielding the corresponding aldehydes. This is a classic method of introducing formyl groups in (hetero)aromatic compounds. On the other hand, heterocyclic organolithiums prepared by direct deprotonation of acidic hydrogens using, for example, *n*-BuLi, *sec*-BuLi, or *tert*-BuLi as strong bases are also able to react with DMF to afford the corresponding formyl derivatives.[10]

The aim of this experiment is the synthesis of a 5-piperidino-2,2′- bithiophene[11] as well as the 4- and 5′-formyl derivatives[12] through several synthetic transformations, namely: (i) thionation of carbonyl groups of a tertiary γ-ketoamide derived from thiophene using Lawesson's reagent[9] followed by cyclization of the corresponding thioxoamide through intramolecular nucleophilic attack[13] yielding the 5-piperidino-2,2′- bithiophene; formylation of the bithiophene derivative using two different synthetic methodologies: (ii) Vilsmeier–Haack reaction or (iii) lithiation followed by electrophilic aromatic substitution. The formylation of 5-piperidino-2,2′- bithiophene *via* two different methods of synthesis affords selectively two different products (4- or 5'-formyl derivatives).

Additional Safety

All sessions should be done in a fume hood, with proper safety equipment, such as protective clothing, gloves and safety goggles. Lawesson's reagent is a flammable solid with a strong odor that in contact with water liberates toxic gas (session 1). $POCl_3$ is a corrosive and toxic reagent (oral, dermal, inhalation), reacts violently with water and other reagents to produce toxic gases

(H_2S) and may cause skin burns and eye damage (session 3). *n*-Butyllithium (session 5) is a toxic and corrosive reagent (oral, dermal, inhalation) that may cause skin burns, eye damage, or respiratory tract irritation. It reacts violently with water and is hazardous to the aquatic environment; it is also flammable in solid, liquid and vapour forms. Therefore, particular attention should be taken when manipulating it, especially when concentrated solutions (2.5 M or higher) are exposed to air due to its flammability. Organic solvents (diethyl ether, petroleum ether, methanol, chloroform, ethyl acetate) are highly flammable liquids and vapours: methanol is toxic in contact with skin, if swallowed and inhaled, ethyl acetate may cause severe skin burns and eye damage, chloroform and diethyl ether are toxic reagents (oral, dermal, inhalation) that may cause skin and eye irritation, petroleum ether is a toxic reagent (oral, dermal, inhalation) that may cause skin and eye irritation and is hazardous to the aquatic environment. Complete information about the safety, toxicity and disposal for all reagents can be found, free of charge at: http://www.msds.com/index.asp?open=/protected_public/loginsuccessful.asp.

Experimental Procedure

Note: The synthesis of the precursor 1-(piperidin-1-yl)-4-(thiophen-2-yl)butane-1,4-dione is described in Experiment 3.1.16.

Laboratory Session 1 (4 h): Synthesis of 5-Piperidino-2,2′-bithiophene

1. In a fume hood, prepare an apparatus for heating at reflux consisting of a heating mantle, a 250 mL round-bottomed flask, a reflux condenser and boiling chips.
2. Add 1-(piperidin-1-yl)-4-(thiophen-2-yl)butane-1,4-dione (2.5 g, 10 mmol), Lawesson's reagent (4.8 g, 12 mmol) and 50 mL of dry toluene to the flask and heat the mixture at reflux for 35 min. Check the contents of the reaction mixture by thin-layer chromatography (TLC) through a comparison with the amide precursor using a mixture of chloroform/methanol (9: 1) as eluent.
3. Evaporate the solvent under reduced pressure in a rotary evaporator.

Laboratory session 2 (4 h): Purification of 5-Piperidino-2,2′-bithiophene

1. Purify the residue resulting from evaporation by chromatography through a silica gel column (60 Å; 200–300 mesh).
2. Combine the fractions containing the product (R_f = 0.61) and evaporate to dryness in a rotary evaporator.
3. Weigh the mass of the obtained pale yellow solid and calculate the yield. Record the ^{1}H and ^{13}C NMR spectra in $CDCl_3$, IR spectrum in liquid film, and UV-vis spectrum in ethanol and determine the melting point. Compare the results with the literature data[11] as well as with the spectra provided in the ESI to confirm the compound's purity.

Laboratory Session 3 (4 h): Synthesis of 4-Formyl-5-piperidino-2,2′-bithiophene

1. Prepare a 10 mL round-bottomed flask with a magnetic stirrer and add DMF (dry) (0.351 g, 0.37 mL, 4.8 mmol) and $POCl_3$ (0.74 g, 0.45 mL, 4.8 mmol) at 0 °C. Stir in an ice bath at 0 °C for about 15 min.
2. To the previous reaction mixture add dropwise the 5-piperidino-2,2′-bithiophene (1 g, 4.0 mmol) dissolved in 2 mL of DMF. Insert a water condenser with an adapter with calcium chloride filling into the round-bottomed flask and clamp the flask in a pre-heated oil bath. Stir the reaction for 2 h at the constant temperature of 60 °C.

3. Check the evolution of the transformation using TLC by comparing the contents of the reaction mixture with the precursor bithiophene [eluent: petroleum ether (40–60 °C)/diethyl ether (1 : 1)].
4. Cool the reaction mixture to room temperature, add it slowly to 75 mL of a saturated solution of sodium acetate and stir the mixture for 30 min at room temperature.
5. Transfer the obtained mixture into a separatory funnel and dilute the organic layer with 50 mL of diethyl ether. Separate and wash the organic layer with 100 mL of a saturated solution of $NaHCO_3$. Dry the organic layer with anhydrous Na_2SO_4, filter the drying agent and evaporate the solvent under reduced pressure with a rotary evaporator.

Laboratory Session 4 (4 h): Purification of 4-Formyl-5-piperidino-2,2′-bithiophene

1. Purify the residue obtained from evaporation by chromatography using a silica gel column (60 Å; 200–300 mesh).
2. Combine the fractions containing the product (R_f = 0.65) and evaporate to dryness in a rotary evaporator.
3. Determine the mass of the obtained pale beige solid and calculate the yield. Record the ^1H and ^{13}C NMR spectra in $CDCl_3$, IR spectrum in liquid film, UV-vis spectrum in ethanol and determine the melting point. Compare the results with the literature data[12] as well as with the spectra provided in the ESI to confirm the compound's purity.

Laboratory Session 5 (4 h): Synthesis of 5′-Formyl-5-piperidino-2,2′-bithiophene

1. Prepare an apparatus for stirring under anhydrous conditions with an inert atmosphere consisting of a 10 mL three-neck round-bottomed flask with a magnetic stirrer, a water condenser with an adapter with calcium chloride filling into the round-bottomed flask, and two rubber septa.
2. Fill the round-bottomed flask with an inert gas such argon or nitrogen, add through one of the septa a solution of 5-piperidino-2,2′-bithiophene (0.5 g, 2.0 mmol) in 2 mL of dry diethyl ether at 0 °C.
3. Add dropwise at 0 °C a solution of 2.5 M of *n*-BuLi in hexanes (0.88 mL, 2.2 mmol) to the flask and stir for about 1 h. Remove the flask from the ice bath, allow it to warm up to room temperature and then stir for 15 min.
4. Add dropwise, at room temperature, DMF (0.10 mL, 2.0 mmol) dissolved in 4 mL of dry diethyl ether and heat at reflux for 1 h.
5. Cool the reaction mixture to room temperature and add 40 mL of water.
6. Transfer the obtained mixture into a separatory funnel and extract with ethyl acetate (3 × 50 mL). Separate and wash the combined organic layers with 100 mL of water. Dry the organic layer with anhydrous $MgSO_4$. Filter the drying agent and evaporate the solvent under reduced pressure with a rotary evaporator.

Laboratory Session 6 (4 h): Purification 5′-Formyl-5-piperidino-2,2′-bithiophene

1. Purify the residue from evaporation by chromatography using a silica gel column (60 Å; 200–300 mesh). Dissolve the residue in the minimum amount of chloroform.
2. Combine the fractions containing the product (R_f = 0.58) and evaporate to dryness in a rotary evaporator.

3. Determine the mass of the obtained orange solid and calculate the yield. Record the ^{1}H and ^{13}C NMR spectra in $CDCl_3$, IR spectrum in liquid film, and UV-vis spectrum in ethanol and determine the melting point. Compare with the literature data[12] as well as with the spectra provided in the ESI to confirm the compound's purity.

Waste Disposal

The solvents used in extractions and in column chromatography are recovered after evaporation in a rotary evaporator and should be collected in chlorinated or non-chlorinated waste collector recipients, as appropriate, for subsequent destruction. The aqueous solutions from liquid–liquid extraction should be neutralized and discarded in the sink. After filtration from the organic solutions, the drying agents should be disposed of in a proper container, as well as used silica gel. Special attention should be paid to the treatment and handling of reaction mixtures in sessions 1, 3 and 5, which contain H_2S, HCl and *n*-BuLi, respectively. Hydrogen sulfide is a toxic colourless gas that can be recognized by its unpleasant "rotten eggs" smell. This gas could be trapped by bubbling it through a solution that reacts with the SH group. Solutions of NaOH, chlorine bleach or potassium permanganate in water can be used for this propose. The excess of *n*-BuLi can be destroyed by dilution with hydrocarbon solvent to a concentration of approximately 5 wt%, followed by gradual addition to water with vigorous stirring under an inert atmosphere.

Results Interpretation and Additional Questions

1. Explain why some of the reactions should be carried out under anhydrous conditions (sessions 3 and 5) and with an inert atmosphere (session 5).
2. What is the function of $POCl_3$ and DMF in session 3 and *n*-BuLi and DMF in session 5?
3. Interpret the ^{1}H and ^{13}C NMR, IR and UV-vis spectra obtained with the corresponding assignments of signals (bands). Compare your interpretation with the spectra provided in the ESI.
4. ^{1}H NMR and IR spectra can be used to confirm the conversion of 1-(piperidin-1-yl)-4-(thiophen-2-yl)butane-1,4-dione into 5-piperidino-2,2′-bithiophene (session 1). What are the principal differences observed in the ^{1}H NMR and IR spectra of 1-(piperidin-1-yl)-4-(thiophen-2-yl)butane-1,4-dione (ESI for Experiment 3.1.16) and 5-piperidino-2,2′-bithiophene which indicate that the conversion has occurred?
5. ^{1}H NMR can easily be used to monitor the formylation reactions performed in sessions 3 and 5 through the two different synthetic procedures. Explain how you can distinguish between the two formylated derivatives (5′-formyl-5-piperidino-2,2′-bithiophene and 4-formyl-5-piperidino-2,2′-bithiophene) by ^{1}H NMR.
6. Write out the mechanism for the Vilsmeier formylation of 2,2′-bithiophene and suggest an explanation for the fact that the formylation of 2,2′-bithiophene *via* metalation with *n*-BuLi followed by addition of DMF (session 5) yielded 5′-formyl-5-piperidino-2,2′-bithiophene whereas Vilsmeier–Haack formylation of the same precursor (section 3) yields 4-formyl-5-piperidino-2,2′-bithiophene.

References

1. (a) A. Mishra, C.-Q. Ma and P. Baeuerle, *Chem. Rev.*, 2009, **109**, 1141; (b) P. Bäuerle, in *Handbook of Oligo- and Polythiophenes*, ed. D. Fichou, Wiley-VCH Verlag GmbH, Weinheim, 1999, ch. 3, The Synthesis of Oligothiophenes, pp. 89–173.
2. X. H. Zhu, J. B. Peng, Y. Cao and J. Roncali, *Chem. Soc. Rev.*, 2011, **40**, 3509.
3. A. Misha, M. K. R. Fischer and P. Bauerle, *Angew. Chem., Int. Ed.*, 2009, **48**, 2479.

4. (a) L. R. Dalton, P. A. Sullivan and D. H. Bale, *Chem. Rev.*, 2010, **110**, 25; (b) A. Wojciechowski, M. M. M. Raposo, M. C. R. Castro, W. Kuznik, I. Fuks-Janczrek, F. Bureš and I. V. Kityk, *J. Mater. Sci.: Mater. Electron.*, 2014, **25**, 1745.
5. M. M. M. Raposo, B. García-Acosta, T. Ábalos, P. Calero, R. Martínez-Mãnez, J. V. Ros-Lis and J. Soto, *J. Org. Chem.*, 2010, **75**, 2922.
6. P. Coelho, M. C. R. Castro, A. M. C. Fonseca, S. S. M. Fernandes and M. M. M. Raposo, *Tetrahedron Lett.*, 2012, **53**, 4502.
7. M. L. Capobianco, G. Barbarella and A. Manetto, *Molecules*, 2012, **17**, 910.
8. E. Genin, V. Hugues, G. Clermont, C. Herbivo, A. Comel, M. C. R. Castro, M. M. M. Raposo and M. Blanchard-Desce, *Photochem. Photobiol. Sci.*, 2012, **11**, 1756.
9. (a) B. S. Pedersen, S. Scheibye, N. H. Nilsson and S.-O. Lawesson, *Bull. Soc. Chim. Belg.*, 1978, **87**, 223; (b) M. P. Cava and M. I. Levinson, *Tetrahedron*, 1985, **41**, 5061; (c) R. A. Cherkasov, G. A. Kutyrev and A. N. Pudovick, *Tetrahedron*, 1985, **41**, 2567; (d) T. Ozturk, E. Ertas and O. Mert, *Chem Rev.*, 2007, **107**, 5210.
10. (a) *Comprehensive Heterocyclic Chemistry II*, ed. C. W. Bird, Pergamon, Oxford, 1984, vol. 4, pp. 751–754 and p. 771; (b) G. Jones and S. P. Stanforth, The Vilsmeier Reaction of Fully Conjugated Carbocycles and Heterocycles, in *Organic Reactions*, ed. L. A. Paquete, John Wiley & Sons, Inc., New York, 1997, vol. 49, pp. 15–17; (c) J. March, *Advanced organic chemistry: reactions, mechanisms and structure*, John Wiley & Sons, Inc., New York, 4th edn, 1992, pp. 542–543; (d) L. Kurti and B. Czakó, *Strategic applications in named reactions in organic synthesis*, Elsevier, San Diego, USA, 2005, p. 468; (e) M. Schlosser, *Angew. Chem., Int. Ed.*, 2005, **44**, 376; (f) R. Chinchilla, C. Nájera and M. Yus, *Chem. Rev.*, 2004, **104**, 2667.
11. M. M. M. Raposo and G. Kirsch, *Heterocycles*, 2001, **55**, 1487.
12. M. M. M. Raposo and G. Kirsch, *Tetrahedron*, 2003, **59**, 4891.
13. T. Nisho, *Helv. Chim. Acta*, 1998, **81**, 1207.

5.2.11. Synthesis of 3-Bromosalicylaldehyde by *Ortho*-formylation of 2-Bromophenol

Trond Vidar Hansen*[a] and Lars Skattebøl[b]

[a]Department of Pharmaceutical Chemistry, School of Pharmacy, University of Oslo, PO Box 1068 Blindern, N-0316 Oslo, Norway; [b]Department of Chemistry, University of Oslo, PO Box 1033 Blindern, N-0315 Oslo, Norway. *E-mail: t.v.hansen@farmasi.uio.no

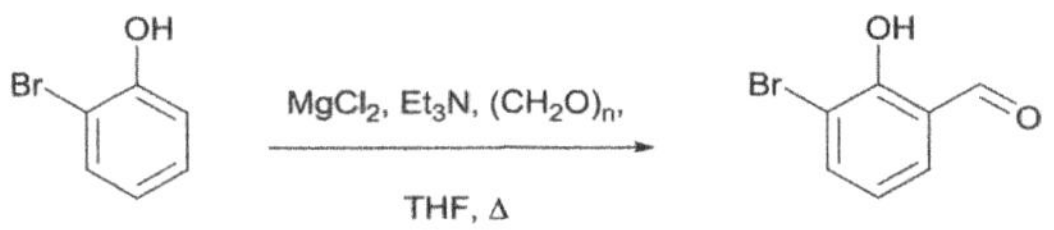

Number of sessions (duration of each session)	Hazard level	Difficulty level	Level of study
2 (4 h + 2 h)	Low	Low	Advanced

Class names Phenols, aldehydes

Concepts involved This experiment involves a method for the preparation of 3-bromosalicylaldehyde from 2-bromophenol using the *ortho*-formylation reaction. The formyl group is regioselectively introduced in the presence of magnesium chloride and triethylamine in refluxing tetrahydrofuran (THF) with paraformaldehyde as the formylation reagent

Chemicals needed 2-Bromophenol, paraformaldehyde, magnesium chloride ($MgCl_2$, anhydrous), triethylamine (anhydrous), diethyl ether, hydrochloric acid (HCl, aqueous 1 N), magnesium sulfate ($MgSO_4$, anhydrous), *n*-hexane

Equipment and experimental techniques involved Apparatus for reaction with temperature controller and magnetic stirring, liquid–liquid extraction apparatus

Keywords Electrophilic aromatic substitution, formylation, FTIR, liquid–liquid extraction, melting point, NMR, regioselectivity

Background

The formyl group is one of the most useful functional groups in organic chemistry. Although it is available by oxidation of the corresponding primary alcohol, work was initiated at the end of the nineteenth century for the direct introduction of the formyl group onto aromatic compounds.[1] This was not an easy task since chloroformaldehyde, an obvious choice for the formyl carbocation, is not available for practical use. However, the use of other chemical equivalents resulted in several formylation reactions that are classified as electrophilic aromatic substitutions. Among these are the established methods of Gattermann–Koch, Vilsmeier–Haack, and Duff.[2] These reactions may provide good yields of aldehydes, but they suffer from several drawbacks: (a) poor regioselectivity as both *ortho*- and *para*-substitution takes place; (b) the use of harsh reaction conditions; (c) noxious reagents, such as carbon monoxide, zinc cyanide or phosphoryl chloride, are involved. Moreover, the regioselectivity and yield depend very much on the

Comprehensive Organic Chemistry Experiments for the Laboratory Classroom
Edited by Carlos A. M. Afonso, Nuno R. Candeias, Dulce Pereira Simão, Alexandre F. Trindade, Jaime A. S. Coelho, Bin Tan, and Robert Franzén.

Published by the Royal Society of Chemistry, www.rsc.org

substituent(s) on the aromatic ring. In addition, with regard to (b) and (c), not all substituents will tolerate the reagents and conditions. Although the hydroxyl group, being an activating substituent, favors the reactions, the disadvantages (a)–(c) apply to formylations of phenols as well. Hence, the application of synthetic protocols with respect to (a)–(c) is desirable.

Some years ago Casiraghi and co-workers reported that the reaction of paraformaldehyde and magnesium phenoxides, the latter formed from the respective phenol and ethylmagnesium bromide in benzene, resulted in mono-formylation exclusively at the *ortho* position of the hydroxyl.[3] Stoichiometric amounts of hexamethylphosphoramide (HMPA), a potential carcinogen, were necessary to achieve high regioselectivity, and also proved necessary for obtaining reasonable yields. The magnesium ion plays an essential role in this reaction. A milder protocol was developed in which the Grignard reagent, working as a base in the Casiraghi reaction, was replaced with the magnesium chloride–triethylamine base system.[4] This combination afforded much better yields, fewer by-products and renders the use of the metal complexing co-solvent HMPA superfluous. Excellent regioselectivity of the salicylaldehyde was observed. Compared to the other methods mentioned above, this protocol is far more environmentally benign and much simpler to execute in the laboratory.

The objective of this work is the synthesis of 3-bromosalicylaldehyde. We will use the *ortho*-formylation reaction with 2-bromophenol as the substrate in the presence of paraformaldehyde, anhydrous magnesium chloride ($MgCl_2$) and triethylamine in refluxing THF. This reaction occurs only when all ingredients are present. The 3-bromosalicylaldehyde product should be obtained in 70–90% isolated yield after extractive work-up and one recrystallization. Most often the product is sufficiently pure for further applications, rendering purification by chromatography unnecessary.

Additional Safety

All sessions should be done in a fume hood, with proper safety equipment, such as protective clothing, gloves and safety goggles. Hydrochloric acid (HCl) is a corrosive acid and may cause severe skin and eye burns. 2-Bromophenol is a flammable liquid that causes eye, skin, and respiratory tract irritation, and is harmful if swallowed.

Experimental Procedure

Laboratory Session 1 (4 h): Synthesis of 3-Bromosalicylaldehyde

1. Prepare a dry 50 mL, three-necked round-bottomed flask, equipped with a stirring bar, a reflux condenser, a thermometer and a rubber cap.
2. Add anhydrous magnesium chloride (952 mg, 10 mmol), paraformaldehyde (450 mg, 15 mmol), triethylamine (1.01 g, 10 mmol) and tetrahydrofuran (25 mL) to the flask.
3. Then add, dropwise, 2-bromophenol (865 mg, 5 mmol) to the reaction mixture using a syringe or an addition funnel.
4. Stir the reaction mixture magnetically while heating under reflux in an argon atmosphere for 3 h.
5. Cool the reaction mixture to room temperature using an ice/water bath.
6. When cooled, add 10 mL of diethyl ether, and transfer the reaction mixture to a separatory funnel.
7. Wash the organic phase successively with 1 N HCl (3 × 10 mL) (Caution! gas evolves) and then with water (3 × 10 mL).
8. Dry the combined organic phase over anhydrous magnesium sulfate ($MgSO_4$) and filter into a tared round-bottom flask.
9. Remove the solvent by rotatory evaporation, to leave a pale yellow solid consisting mainly of 3-bromosalicylaldehyde.

Laboratory Session 2 (2 h): Determination of Yield and Chemical Purity, Product Recrystallization and ^{1}H NMR

1. Weigh the tared round-bottom flask containing the product and determine the amount of product. From this, calculate the yield of the reaction using the following equation

$$\text{percent yield} = \frac{\text{actual yield}}{\text{theoretical yield}} \times 100$$

The unit of measure for both practical and theoretical yield must be the same.

2. Determine the melting point of the product.
3. Purify the product obtained by recrystallization from hexane (a minimum amount, usually 3–5 mL) to give the aldehyde as colorless needles.
4. Determine the chemical purity of the product by thin-layer chromatography (TLC).
5. Record and assign the ^{1}H NMR spectrum.

Waste Disposal

Diethyl ether may be reused or collected in a non-chlorinated waste collector recipient, for posterior destruction. The aqueous phases should be neutralized, diluted and discarded.

Results Interpretation and Additional Questions

1. Interpret the ^{1}H NMR spectrum of the product and compare the observed melting point with the literature value[5] (mp 52 °C). Calculate the percentage yield.
2. Discuss the advantages of using tetrahydrofuran as the solvent. In addition, discuss why diethyl ether, dichloromethane and acetonitrile are all suited for this reaction, but the reaction is sluggish when using benzene as solvent.
3. Discuss the importance of the magnesium cation. This reaction also works well with magnesium bromide or magnesium iodide, but when any of the magnesium salts are replaced with the corresponding calcium salts, the reaction does not work. Explain this observation.
4. Explain why 2-halogen substituted phenols all work, but when using either 2- or 4-cyanophenol as the starting material the reaction is sluggish, yielding the corresponding 3- or 5-substitured cyanosalicylaldehydes in less than 25% yield.
5. Discuss and explain why 2-methoxyphenol (*ortho*-anisole) affords a much poorer yield than 4-methoxyphenol (*para*-anisole). When employing the latter, yields of >90% after 3 h reaction time, are most often observed.

References

1. (a) G. A. Olah, L. Ohannasian and M. Arvanaghi, *Chem. Rev.*, 1987, **87**, 671; (b) T. Laird, in *Comprehensive Organic Chemistry*, ed. J. F. Stoddart, Pergamon, Oxford 1979, vol. 1; p. 1105; (c) W. Kantlehner, *Eur. J. Org. Chem.*, 2003, 2530.
2. A. Hassner and C. Stumer, *Organic Syntheses based on name reactions and unnamed reactions*, Pergamon 1994.
3. (a) G. Casnati, M. Crisafulli and A. Ricca, *Tetrahedron Lett.*, 1965, **6**, 243; (b) G. Casiraghi, G. Casnati, M. Cornia, A. Pochini, G. Puglia, G. Sartori and R. Ungaro, *J. Chem. Soc. Perkin Trans. 1*, 1978, 318; (c) G. Casiraghi, G. Casnati, G. Puglia, G. Sartori and G. Terenghi, *J. Chem. Soc. Perkin Trans. 1*, 1980, 1862.
4. (a) N. U. Hofsløkken and L. Skattebøl, *Acta Chem. Scand.*, 1999, **53**, 252; (b) T. V. Hansen and L. Skattebøl, *Org. Synth.*, 2005, **82**, 64; (c) T. V. Hansen and L. Skattebøl, *Discuss. Addendum: Org. Synth.*, 2012, **89**, 220.
5. J. Müller, *Ber.*, 1909, **42**, 1166.

5.2.12. Synthesis of Indolo[3,2-*b*]quinolin-11-one by Acid-Catalysed Intramolecular Double Cyclization

João Lavrado[a], Marta Figueiras[a], David M. Pereira[a], Sofia A. Santos[a], Rui Moreira[a,b], and Alexandra Paulo*[a,b]

[a]Research Institute for Medicines (iMed.ULisboa), Faculty of Pharmacy, Universidade de Lisboa, Av. Prof. Gama Pinto, 1649 003 Lisboa, Portugal; [b]Pharmaceutical and Medicinal Chemistry Department, Faculty of Pharmacy, Universidade de Lisboa, Av. Prof. Gama Pinto, 1649 003 Lisboa, Portugal
*E-mail: mapaulo@ff.ulisboa.pt

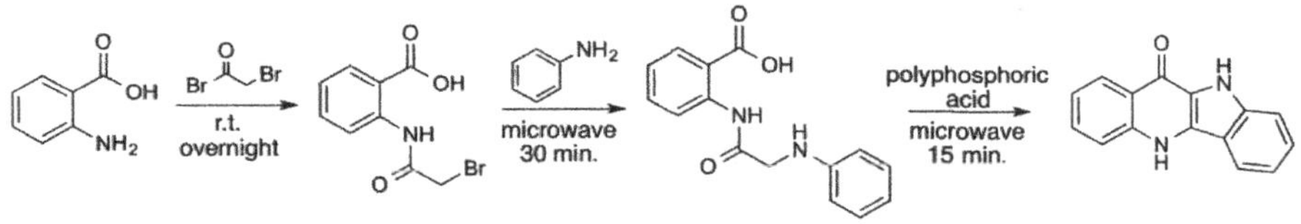

Number of sessions (duration of each session)	Hazard level	Difficulty level	Level of study
3 (1 h + 4 h + 4 h)	Moderate	Medium	Advanced

Class names Aromatic amines, alkanoyl halides, haloalkanes, enols, carboxylic acids, heterocycles

Concepts involved This experiment describes a three-step method for a microwave assisted synthesis of a tetracyclic heteroaromatic compound by an intramolecular double cyclization reaction catalysed by polyphosphoric acid, followed by liquid–liquid extraction and recrystallization procedures. The reaction involves C–C bond formation *via* a Claisen-type condensation, aromatic electrophilic substitution and formation of aromatic or conjugated systems by dehydration

Chemicals needed Polyphosphoric acid, anthranilic acid, bromoacetyl bromide, aniline, potassium hydroxide (10% and saturated solutions), hydrogen bromide (5%), ethyl acetate, anhydrous sodium sulfate, sodium chloride (saturated solution), dimethylformamide, 1,4-dioxane, diethyl ether, hexane, deuterated dimethyl sulfoxide

Equipment and experimental techniques involved Microwave apparatus, plate with magnetic stirring, rotary evaporator, vacuum apparatus, NMR, liquid–liquid extraction, recrystallization by the pair of solvents technique

Keywords Condensation, crystallization, 1D and 2D NMR, dehydration, heterocycles, liquid–liquid extraction

Background

Nitrogen-containing heterocycles are present in many natural molecules with important biological functions, such as for example purine and pyrimidine bases in nucleic acids of DNA and the neurotransmitter serotonin, which structure contains an indole ring. For this reason, heterocycles are important chemical fragments of many drugs and, therefore, the study of their methods of synthesis is of major importance in any organic or medicinal chemistry course. Indoloquinolines

Comprehensive Organic Chemistry Experiments for the Laboratory Classroom
Edited by Carlos A. M. Afonso, Nuno R. Candeias, Dulce Pereira Simão, Alexandre F. Trindade, Jaime A. S. Coelho, Bin Tan, and Robert Franzén.

Published by the Royal Society of Chemistry, www.rsc.org

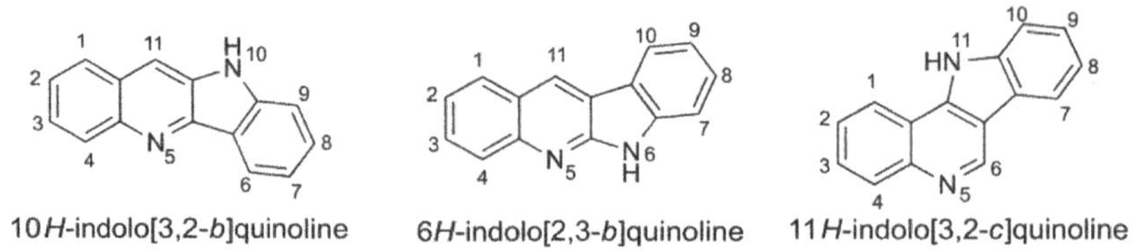

Figure 5.2.12.1 Structures of natural indoloquinoline skeletons.

(Figure 5.2.12.1) are natural alkaloids found in an African traditional herbal drug. Due to their broad spectrum of biological activities the indoloquinoline tetracyclic aromatic skeleton has been used as scaffold for the development of new drug leads.[1] For example, the indolo[3,2-*b*]quinoline skeleton has been extensively explored for malaria[2–5] and cancer[6,7] therapies. The natural 10*H*-indolo[3,2-*b*]quinoline has been synthesized by ten different methods.[8] One common synthetic approach uses the 5*H*-indolo[3,2-*b*]quinolin-11(10*H*)-one as key intermediate mainly due to the chemical diversity of the library that can be obtained from it, but also to the good yields usually obtained, the facility of the three-step synthetic method and the commercial availability of starting building blocks.

The synthesis of 5*H*-indolo[3,2-*b*]quinolin-11(10*H*)-one by polyphosphoric acid catalysed intramolecular cyclization was first purposed by Görlitzer and Weber in 1981[9] and later optimized to be achievable starting from 2-aminobenzoic acids, bromoacetyl bromide and anilines.[10] Now, we have adapted this method to a much faster procedure by using microwave irradiation in two of the three steps. Microwave irradiation is nowadays a common tool used in organic synthesis to speed-up reactions and improve yields.[11] Polyphosphoric acid is a moderately strong mineral acid that has been widely used to perform intramolecular and intermolecular acylations, heterocyclic synthesis, and acid-catalysed rearrangements.[12–15] It has powerful dehydration properties and is able to catalyse both nucleophilic and electrophilic reactions. The mechanism proposed for this intramolecular double cyclization involves C–C bonds formation by a Claisen-type reaction and an aromatic electrophilic substitution, as well the formation of aromatic systems by dehydration.

Additional Safety

All steps should be done in a fume hood, with proper safety equipment, such as protective clothing, gloves and safety goggles. Bromoacetyl bromide is a very strong acid, causing severe skin burns and liberating toxic gas, and thus must be handled with extreme caution. Aniline may be harmful if inhaled and cause cyanosis. In addition, skin permeation can occur and cause allergic skin reactions. Polyphosphoric acid, a strong acid, is highly corrosive and may cause skin burns. Dilution of the acid reaction mixture with water and the neutralization steps must be done very slowly to avoid spills.

Experimental Procedure

Laboratory Section 1 (1 h)

Synthesis of 2-(2-Bromoacetamido)benzoic Acid

1. In a 10 mL round-bottomed flask put 1 g of 2-aminobenzoic acid (anthranilic acid) and dissolve the powder in a mixture of 2 mL of DMF and 2 mL of 1,4-dioxane.
2. Cool the solution to 0 °C in an ice bath and add dropwise (very slowly) 0.8 mL of bromoacetyl bromide while maintaining the solution in the ice bath. The temperature of the solution must be kept between 0 and 5 °C.
3. After addition, remove the ice bath and stir the reaction mixture overnight at room temperature.

Laboratory Section 2 (4 h)

1. Isolation of 2-(2-Bromoacetamido)benzoic Acid

1. Transfer the reaction mixture to a flask containing 30 mL of ice water.
2. Collect the precipitate that is formed by vacuum filtration and wash it with water until neutral pH. Five portions of 50 mL each may be required.
3. Dry the solid as much as you can by pressing the solid onto the filter paper and continue the suction for about 5 min.
4. Then spread the solid out on a watch glass and place it in an oven at 80–85 °C until the product is completely dried.
5. Weigh the beige solid obtained (1.3–1.4 g) and determine the melting point of the product (mp 162–165 °C).
6. Record the ^{1}H NMR spectrum in d_6-DMSO.

2. Synthesis of 2-[2-(Phenylamino)acetamido]benzoic Acid

1. In the appropriate microwave vessel dissolve 0.6 g (2.33 mmol) of the previously synthesized 2-(2-bromoacetamido)benzoic acid in 3 mL of DMF and add 0.68 mL of aniline and a magnetic stirrer.
2. Heat the reaction mixture under microwave irradiation at 130 °C (200 W) for 30 min.
3. Cool the reaction mixture to a safe handling temperature and transfer it to a beaker containing 60 mL of ice-water.
4. Adjust the pH of the solution to 10 with 10% KOH solution (approx. 5 mL).
5. Transfer the solution to a separatory funnel and wash the aqueous solution with 25 mL of dichloromethane. Discard the organic phase.
6. Acidify to pH 2 the aqueous layer with a 5% HBr solution (approx. 9 mL).
7. Collect the white solid formed by vacuum filtration, wash it with water (3 × 10 mL) and then let it dry under vacuum until the next session.

Laboratory Session 3 (4 h)

Synthesis of 5H-Indolo[3,2-b]quinolin-11(10H)-one

1. Weigh the 2-[2-(phenylamino)acetamido]benzoic acid previously synthesized (325–410 mg) and determine the melting point (mp 194-–197 °C).
2. Put a 100 mL Erlenmeyer flask (or another appropriate vessel for your microwave apparatus) on a balance and add 0.250 g (0.9 mmol) of 2-[2-(phenylamino)acetamido]benzoic acid (powder) followed by 15 g of polyphosphoric acid (very dense liquid) that must cover and dissolve the powder reagent.
3. Add a medium size magnetic stir bar, insert a condenser into the Erlenmeyer flask and put it in the microwave apparatus.
4. Stir vigorously the mixture for 15 min at 130 °C, 200 W.
5. At the end of the reaction time, remove the vessel from the apparatus and allow the reaction mixture to cool to a warm temperature, so it can be safely handled but the viscosity of the mixture has not yet increased too much.
6. Put a 200 mL beaker containing 20 mL of cold water in an ice bath, in the fume hood and add slowly the reaction mixture to the water. A glass rod can be used to pour the reaction mixture along. Use another 10 mL of cold water to rinse the reaction vessel if necessary.
7. Neutralize the reaction mixture with saturated KOH solution (150–200 mL may be required).
8. Transfer the solution into a separatory funnel and then extract it with 2 × 100 mL of ethyl acetate. To avoid emulsions a gentle shaking of the two phases for 1–2 min is advisable.

Wash the combined organic layers with 100 mL of water. Then dry the organic solution with first 100 mL of brine (NaCl saturated solution) and then anhydrous Na_2SO_4. Filter and evaporate completely the solvent using a rotary evaporator.

9. Weigh the remaining reaction product.
10. Recrystallize the dark solid obtained by the mixed solvent technique, using ethyl acetate as good solvent and a mixture of diethyl ether: hexane (9 : 1) as the poor solvent.
11. Recover the formed yellow crystals by suction filtration. Wash the crystals with the poor solvent and dry the crystals. Weigh the dried crystals and record ^1H and two-dimensional ^1H–^1H COSY NMR spectra in d_6-DMSO.
12. Record also the ^1H NMR spectrum of the starting material (2-[2-(phenylamino)acetamido] benzoic acid) in d_6-DMSO.

Waste Disposal

The neutralized aqueous solutions may be diluted and discarded in the sink. Acid and basic solutions must be put in an appropriate waste collector for posterior destruction. Dichloromethane solution must be put in the chlorinated waste collector. Ethyl acetate may be reused or collected in a non-chlorinated waste collector recipient for posterior destruction. In addition, the mixture of organic solvents from the recrystallization procedure must be put in the non-chlorinated waste collector recipient.

Results Interpretation and Additional Questions

1. Calculate the yield of each step of the experiment and the overall yield.
2. Interpret the ^1H NMR spectra of 2-(2-bromoacetamido)benzoic acid and 2-[2-(phenylamino)acetamido]benzoic acid. Attempt to assign each signal based on chemical shift and multiplicity.
3. Interpret the ^1H and ^1H–^1H COSY NMR spectra of 5*H*-indolo[3,2-*b*]quinolin-11(10*H*)-one and compare with data published in the literature.[16] Two four-proton aromatic systems must be clearly observed.
4. Based on ^1H NMR spectra and melting points comment on the purity of the compounds and discuss appropriate techniques by which to evaluate compounds' purity.
5. Propose a possible mechanism of double cyclization considering that in the overall process 2 moles of water are lost and the mechanism involves a Claisen-type condensation and an aromatic electrophilic substitution.

References

1. J. Lavrado, R. Moreira and A. Paulo, *Curr. Med. Chem.*, 2010, **17**, 2348–2370.
2. J. Lavrado, A. Paulo, J. Gut, P. J. Rosenthal and R. Moreira, *Bioorg. Med. Chem. Lett.*, 2008, **18**, 1378–1381.
3. J. Lavrado, K. Gaani, P. A. Nobre, S. A. Santos, P. Figueiredo, D. Lopes, V. do Rosário, J. Gut, P. J. Rosenthal, R. Moreira and A. Paulo, *Bioorg. Med. Chem. Lett.*, 2010, **20**, 5634–5637.
4. J. Lavrado, G. G. Cabal, M. Prudêncio, M. M. Mota, J. Gut, P. J. Rosenthal, C. Diaz, R. C. Guedes, D. J. V. A. dos Santos, E. Bichenkova, K. T. Douglas, R. Moreira and A. Paulo, *J. Med. Chem.*, 2011, **54**, 734–750.
5. A. Paulo, M. Figueiras, M. Machado, C. Charneira, J. Lavrado, S. A. Santos, D. Lopes, J. Gut, P. J. Rosenthal, F. Nogueira and R. Moreira, *J. Med. Chem.*, 2014, **57**, 3295–3313.
6. J. Lavrado, A. P. Reszka, R. Moreira, S. Neidle and A. Paulo, *Bioorg. Med. Chem. Lett.*, 2010, **20**, 7042–7045.
7. J. Lavrado, P. M. Borralho, S. A. Ohnmacht, R. E. Castro, C. M. P. Rodrigues, R. Moreira, D. J. V. A. dos Santos, S. Neidle and A. Paulo, *ChemMedChem*, 2013, **8**, 1648–1661.
8. J. Lavrado, R. Moreira and A. Paulo, *Heterocyclic Targets in Advanced Organic Synthesis*, ed. M. C. Carreiras and J. Marco-Contelles, Research Signpost, Kerala, India, 2011, ch. 5, pp. 109–124.
9. K. Görlitzer and J. Weber, *Arch. Der Pharm.*, 1981, **314**, 850.

10. D. E. Bierer, L. G. Dubenko, P. Zhang, Q. Lu, P. A. Imbach, A. W. Garofalo, P. W. Phuan, D. M. Fort, J. Litvak, R. E. Gerber, B. Sloan, J. Luo, R. Cooper and G. M. Reaven, *J. Med. Chem.*, 1998, **41**, 2754–2764.
11. M. A. Surati, S. Jauhari and K. R. Desai, *Arch. Appl. Sci. Res.*, 2012, **4**, 645–661.
12. H. G. Bonacorso, R. Andrighetto, N. Zanatta and M. A. P. Martins, *Tetrahedron Lett.*, 2010, **51**, 3752–3755.
13. X. Liu, Q. Zhang, D. Zhang, X. Xin, R. Zhang, F. Zhou and D. Dong, *Org. Lett.*, 2013, **15**, 776–779.
14. M.-Y. Chang, T.-W. Lee and M.-H. Wu, *Org. Lett.*, 2012, **14**, 2198–2201.
15. J. H. Dodd, Polyphosphoric Acid, in *Encyclopedia of Reagents for Organic Synthesis*, John Wiley & Sons, Ltd, 2001.
16. J. Lavrado, PhD thesis, *Faculty of Pharmacy*, University of Lisbon, Lisbon, 2010, p. 233, http://repositorio.ul.pt/handle/10451/11987.

5.2.13. Mild and Fast Friedel–Crafts Acylation Over Zeolites

A. F. Brigas*[a,b], F. Martins[b], R. Elvas-Leitão[b,c], B. S. Santos[d], A. Martins[b,c], and N. Nunes[b,c]

[a]Universidade do Algarve, Departamento de Química e Farmácia, Centro de Investigação em Química do Algarve (CIQA), Campus de Gambelas, 8005-139 Faro, Portugal; [b]Universidade de Lisboa, Faculdade de Ciências, Centro de Química e Bioquímica (CQB), C8, Campo Grande, 1749-016 Lisboa, Portugal; [c]Instituto Politécnico de Lisboa, Instituto Superior de Engenharia de Lisboa (ISEL), Área Departamental de Engenharia Química, Rua Conselheiro Emídio Navarro, 1, 1959-007 Lisboa, Portugal; [d]Universidade de Coimbra, Departamento de Química, 3004-535 Coimbra, Portugal. *E-mail: abrigas@ualg.pt

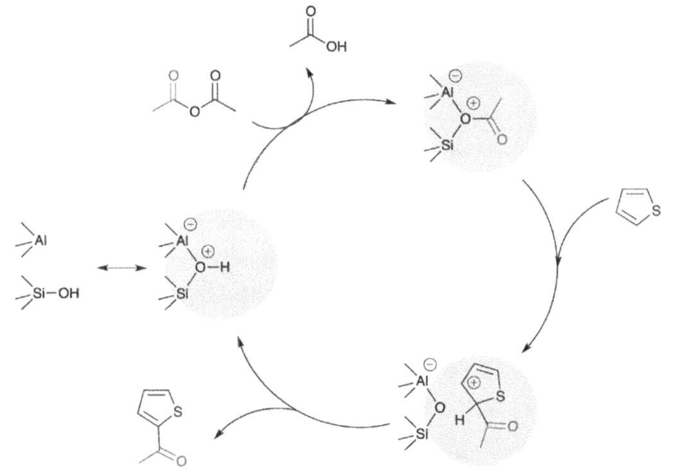

Number of sessions (duration of each session)	Hazard level	Difficulty level	Level of study
1 (3 h)	Moderate	Medium	Advanced

Class names Thiophenes, aromatic ketones, alkyl anhydrides

Concepts involved This experiment involves an alternative method to classical Friedel–Crafts acylation of poorly reactive aromatic compounds under both mild (60 °C) and fast (<1 h) experimental conditions. Acidic commercial zeolites are used as heterogeneous catalysts. Reaction progress is tracked by GC analysis, allowing the kinetic treatment of results and also the establishment of correlations between zeolite structures and conversion, selectivity and product yields

Chemicals needed Acidic commercial zeolite structures such as BEA, FAU, MFI, MOR; thiophene, acetic anhydride

Equipment and experimental techniques involved Heating plate with temperature control and magnetic stirring, gas chromatograph, HP-5-MS capillary GC column or equivalent, syringe, nanofilter membrane

Keywords Environmentally friendly processes, Friedel–Crafts acylation reactions, GC, heterogeneous catalysis, nanofilter membrane, zeolites

Comprehensive Organic Chemistry Experiments for the Laboratory Classroom
Edited by Carlos A. M. Afonso, Nuno R. Candeias, Dulce Pereira Simão, Alexandre F. Trindade, Jaime A. S. Coelho, Bin Tan, and Robert Franzén.

Published by the Royal Society of Chemistry, www.rsc.org

Background

The synthesis of highly functionalized molecules from simple heteroatomic compounds continues to be a keystone in organic synthesis. Several important organic compounds possess in their structure pyrrole, thiophene and furan rings, making the functionalization of these compounds through different experimental procedures an ongoing challenge.[1,2] Multiple approaches have been proposed for their modification. One of the most used reactions is the Friedel–Crafts acylation, which consists basically in the production of an aromatic ketone by the reaction of an aromatic substrate with an acyl compound in the presence of a catalyst.

However, these reactions usually take place in the presence of environmentally damaging, harmful, and difficult to recover homogenous Lewis acids such as $AlCl_3$, BF_3 or strong protonic acids (H_2SO_4 and HF).[1–3] Besides the corrosive waste produced, the procedures involved usually also require an amount of homogeneous catalyst above reaction stoichiometry. This student research project aims to provide greener and more efficient alternatives to common Friedel–Crafts acylation procedures. Environmental and economic issues are the driving forces for the search of alternative synthetic methods, which need to be optimized to improve product selectivity and to promote, simultaneously, environmentally friendlier reaction conditions. Accordingly, the heterogenization of reactions that typically occur in homogeneous media has been a field of intense research in the last few years and the use of zeolites as catalyst or catalyst support has grown considerably.[4,5] Zeolites are crystalline microporous aluminosilicates known since 1756. Nowadays, there are more than 250 synthetic zeolite structures and about 50 natural ones.[6] The unique properties of these materials, namely high porous volume, ion exchange capacity, mechanical and thermal stability, shape selectivity and intrinsic acidity, make them suitable for vast applications such as catalysis, separation and purification processes.[7] The use of zeolites in Friedel–Crafts acylation of aromatic compounds was reported by Derouane *et al.*[8,9] and several other research groups.[10,11] The mechanism of these reactions under heterogeneous catalysis is thought to be similar to acid catalysis occurring in homogeneous media, as illustrated in the graphical abstract scheme adapted from ref. 11, where zeolite surface active sites interact strongly with adsorbed reagents and, desirably, weakly with products.

By establishing appropriate reaction conditions in these heterogeneous media, it is possible to enhance reaction rate and selectivity on the basis of the intrinsic properties of zeolites, namely their acidity and microporosity that can be customized by performing several simple treatments described in the literature. For example, zeolite acidity can be modified through ion exchange procedures.[12] On the other hand, porosity can be tuned by performing desilication treatments.[13] Additionally, the thermal and mechanical stability of these materials allow the recycling and reuse of the catalyst.

The purpose of this work is to demonstrate the use of zeolites in the Friedel–Crafts acylation reaction using, as an example, the acetylation reaction of thiophene under mild and fast liquid-phase conditions, acetic anhydride as acylating agent, and a commercial zeolite, BEA, as the heterogeneous catalyst.[†].

Additional Safety

Students are expected to follow standard safety rules to work in a chemistry laboratory. Experimental work should take place in a well-ventilated place, preferably in a fume hood. Thiophene and acetic anhydride are flammable, corrosive and toxic substances. Therefore, individual safety equipment such as protective clothing, gloves and goggles are mandatory.

[†]From the authors own experience, the best results are obtained with BEA or FAO zeolites but any commercial zeolite can be used.

Experimental Procedure

Part A: Friedel–Crafts Acylation

1. Using a magnetic hot plate, heat in a fume hood an oil bath so that a temperature of 60 °C is reached.
2. Prepare a 50 mL two-neck round-bottom flask (tared) with a magnetic stirrer bar. The second neck of the round-bottom flask should be fitted with a rubber septum cap. Weigh accurately 200 mg of zeolite. Add 0.88 g (10.5 mmol) of thiophene and 5.40 g (52.9 mmol) of acetic anhydride.
3. Insert a water condenser into the round-bottom flask and clamp the flask in the preheated oil bath. Stir vigorously the reaction for about 1 h.
4. During the course of the reaction, withdraw small aliquots (about 0.5 mL) of the reaction mixture using a syringe equipped with a nanofilter, to separate the zeolite powder from the reactants and products. Six aliquots should be taken, at least, at the following reaction times: 3, 6, 10, 20, 30 and 60 min. The aliquots should be placed in stoppered flasks and, preferably, stored in the refrigerator.

Part B: Analysis of the Reaction Mixture by Gas Chromatography (GC)

1. Connect the GC equipment and programme injector (θ = 250 °C) and detector (θ = 300 °C) temperatures. In addition, programme the GC oven as follows: 60 °C for 2 min; ramp of 15 °C min^{-1}; then 200 °C for 3 min.
2. Inject about 0.02–0.04 µL of each aliquot in the GC using a micro-syringe and run the analysis. The total analysis takes about 10 min, plus 5 min cooling to initiate the next run.
3. If a more advanced approach is intended, an internal standard should be used in order to account for differences in the detector response and, thus, to correctly quantify reactants and reaction products. However, in a more basic approach the response factors for all substances (reactants and products) can be considered to be approximately the same.

Waste Disposal

All wastes should be stored in appropriate and well identified recipients for further disposal by a certified company, in accordance with SDS regulations.[14]

Results Interpretation and Additional Questions

1. Interpret the chromatograms by comparing the obtained retention times for reactants and reaction products with those on reference chromatograms (GC or GC-MS).
2. Analyse the reaction progress of products and reactants over time:[15]

    2.1. Plot the % conversion of thiophene *versus* time.
    2.2. Show that the reaction proceeds *via* first order kinetics. Explain why this reaction is truly a pseudo-first order reaction.
    2.3. Calculate the rate constant of thiophene conversion.
    2.4. Calculate the half-life of this reaction.

3. Calculate product yields.
4. Compare the results with those found in the literature.[9] If several zeolite structures are available, students may use different structures to rationalize differences in terms of structure–reactivity relationships and selectivity. In this case, students should make a critical analysis focusing on the effect of the zeolite structure on conversion and product yields.

References

1. G. Frank and J. W. Stadelhofer, *Industrial Aromatic Chemistry*, Springer-Verlag, Berlin, 1988.
2. T. B. Poulsen and K. A. Jørgensen, *Chem. Rev.*, 2008, **108**, 2903–2915; S. You, Q. Cai and M. Zeng, *Chem. Soc. Rev.*, 2009, **38**, 2190–2201.
3. G. A. Olah, *Friedel–Crafts Chemistry*, John Wiley & Sons, New York, 1973.
4. A. Breda, M. Signoretto, E. Ghedini, F. Pinna and G. Cruciani, *Appl. Catal., A*, 2006, **308**, 216–222.
5. P. Marion, R. Jacquot, S. Ratton and M. Guisnet, in *Zeolites for Cleaner Technologies*, ed. M. Guisnet and J. P. Gilson, Imperial College Press, London, 2002, p. 281.
6. C. Baerlocher and L. B. McCusker, *Database of Zeolite Structures*, http://www.iza-structure.org/databases, accessed September 2014.
7. S. C. Larsen, *J. Phys. Chem. C*, 2007, **111**, 18464–18474.
8. E. G. Derouane, C. J. Dillon, D. Bethell and S. D. B. Hamid, *J. Catal.*, 1999, **187**, 209–218.
9. V. F. D. Álvaro, A. F. Brigas, E. G. Derouane, J. P. Lourenço and B. S. Santos, *J. Mol. Catal. A: Chem.*, 2009, **305**, 100–103.
10. G. Sartori and R. Maggi, *Chem Rev.*, 2006, **106**, 1077–1104.
11. K. Smith and G. A. El-Hiti, *Curr. Org. Chem.*, 2006, **10**, 1603–1625.
12. M. F. Ribeiro and A. Fernandes, in *Catalysis from Theory to Application*, ed. J. L. Figueiredo, M. M. Pereira and J. Faria, Coimbra University Press, Coimbra, 2008, p. 107.
13. D. P. Serrano, J. M. Escola and P. Pizarro, *Chem. Soc. Rev.*, 2013, **42**, 4004–4035.
14. European Agency for Safety and Health at Work, Checklist for the prevention of accidents in laboratories, E-Facts 20, https://osha.europa.eu/en/publications/e-facts/efact20, accessed in March 2015.
15. P. W. Atkins and J. de Palma, *Physical Chemistry*, Oxford University Press, Oxford, 9th edn, 2010.

5.2.14. Reactivity Studies of 1-Propyl-2-(thiophen-2-yl)-1*H*-pyrrole

Maria Manuela Marques Raposo*, Sara Sofia Marques Fernandes, and Maria Cidália Rodrigues Castro

University of Minho, Department of Chemistry, Campus de Gualtar, 4710-057 Braga, Portugal
*E-mail: mfox@quimica.uminho.pt

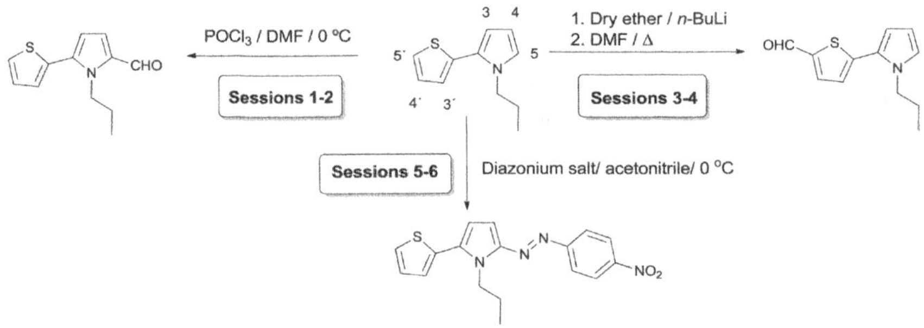

Number of sessions (duration)	Hazard level	Difficulty level	Level of study
6 (4 h each)	High	High	Advanced

Class names Aldehydes, diazenes, pyrroles, thiophenes

Concepts involved Study of the reactivity of a thienylpyrrole heterocyclic system focusing on the selective functionalization of the pyrrole ring through electrophilic aromatic substitution (Vilsmeier–Haack formylation and azo coupling) or the selective functionalization of the thiophene heterocycle through metalation followed by electrophilic aromatic substitution; structure determination of the synthesized products through ^{1}H NMR, ^{13}C NMR, IR and UV spectroscopic techniques

Chemicals needed DMF (dry), phosphoryl chloride, *n*-BuLi (2.5 M in hexanes), chloroform, deuterated chloroform, diethyl ether (dry), saturated aqueous solution of sodium bicarbonate, saturated aqueous solution of sodium acetate, magnesium sulfate (anhydrous), sodium sulfate (anhydrous), 4-nitroaniline, HCl (37%), sodium nitrite, acetonitrile, glacial acetic acid, ethyl acetate, diethyl ether, petroleum ether (40–60 °C), silica gel for column chromatography (60 Å; 230–400 mesh) and silica gel thin-layer chromatography (TLC) plates with fluorescent indicator

Equipment and experimental techniques involved Reflux apparatus, reflux apparatus under anhydrous conditions and/or inert atmosphere, liquid–liquid extraction, gravity and vacuum filtration, evaporation under vacuum using a rotary evaporator, melting point and column chromatography

Keywords Aldehyde, azo dye, chromatography, diazotization, NMR, Vilsmeier–Haack formylation

Comprehensive Organic Chemistry Experiments for the Laboratory Classroom
Edited by Carlos A. M. Afonso, Nuno R. Candeias, Dulce Pereira Simão, Alexandre F. Trindade, Jaime A. S. Coelho, Bin Tan, and Robert Franzén.

Published by the Royal Society of Chemistry, www.rsc.org

Background

Several methods of synthesis can be used to synthesize functionalized pyrrole derivatives (*e.g.* Knorr, Paal–Knorr, Hantzsch, Van Leusen, Barton–Zard cross-coupling reactions, *etc.*). However, the yields are usually low and several by-products, such as undesired regioisomers, are obtained. Additionally, pyrroles are susceptible to chemical degradation as they are rather easily oxidized; this further hampers their synthesis and especially their isolation and purification.[1,2] Therefore, the development of efficient methods for the synthesis of pyrrole derivatives remains an important synthetic challenge, in particular due to their application as key building blocks for the preparation of more complex heterocyclic systems with a wide range applications in several areas such as medicinal, supramolecular chemistry and materials for optoelectronics.[3]

The reactivity of pyrrole, thiophene, furan and their derivatives is ruled largely by their aromaticity and is based on the chemistry of benzene. As expected for aromatic systems, the 1-hetero-2,4-cyclopentadienes (*e.g.* pyrrole and thiophene) undergo electrophilic substitution. There are two possible sites of possible attack, at C2 and at C3. Both modes benefit from the presence of the resonance-contributing heteroatom, but an attack at C2 leads to an intermediate with the additional resonance form, making this position the preferred site of substitution. However, because C3 is also activated to electrophilic attack, mixtures of products can form, depending on conditions, substrates and electrophiles.[1–4] The relative nucleophilic reactivity of benzene and the three heterocycles increases in the order of benzene << thiophene < furan < pyrrole. The reactivity of pyrrole is comparable to that of an electron-rich benzene derivative such as aniline (aminobenzene) or phenol (hydroxybenzene). In a typical electrophilic aromatic substitution such as bromination, pyrrole is about 5.9×10^8 times more reactive than thiophene. As a consequence, electrophilic substitution reactions of thienylpyrroles (*e.g.* Vilsmeier–Haack formylation and diazo coupling) were found to be very selective with the electrophile substituted selectively on the pyrrole ring.[3a,b,d]

The Vilsmeier–Haack formylation and metalation followed by reaction with DMF are the most relevant methods for the preparation of thiophene and pyrrole formylated derivatives.[1,3,4] The Vilsmeier formylation is an aromatic electrophilic substitution reaction in which the electrophile (formylating agent) is obtained by reacting a substituted amide, in this case *N,N*-dimethylformamide (DMF) with phosphorus oxychloride ($POCl_3$) to produce the reactive species (the Vielsmeier reagent: $ClCH = NMe_2^+$) which in turns reacts with electron rich (hetero)aromatic compounds, yielding the corresponding aldehydes.[1,3,4]

On the other hand, heterocyclic organolithium prepared by direct deprotonation of acidic hydrogens using, for example, *n*-BuLi, *sec*-BuLi, *tert*-BuLi, strong bases are also able to react with a wide variety of electrophilic compounds such as DMF to afford the corresponding formyl derivatives.[5]

The diazo-coupling as an aromatic electrophilic reaction yields also the azo derivative by selective functionalization in the pyrrole ring of the thienylpyrrole system.[3a,b,e]

In this experiment, reactivity studies of 1-(*n*-propyl)-2-(2′-thienyl)pyrrole will be carried out through recently described methods, with the objective of selectively functionalizing the pyrrole or the thiophene ring of the thienylpyrrole system.[3] The selective functionalization of the pyrrole ring *via* electrophilic aromatic substitution reactions (Vilsmeier–Haack formylation and azo coupling reactions) will yield the formylated derivative[3d] and the diazene.[3b] On the other hand, formylation in the thiophene ring[3d] of the thienylpyrrole derivative occurs *via* metalation (lithiation) followed by aromatic electrophilic substitution.

Additional Safety

All sessions should be carried out in a fume hood, with proper safety equipment, such as protective clothing, gloves and safety goggles.

$POCl_3$ is a corrosive and toxic reagent (oral, dermal, inhalation), reacts violently with water and other reagents to produce toxic gases (HCl) and may cause skin burns and eye damage (session 1). *n*-Butyllithium (session 3) is a toxic and corrosive reagent (oral, dermal, inhalation) that may cause skin burns, eye damage, respiratory tract irritation, reacts violently with water, is flammable in solid, liquid and vapor forms and is hazardous to the aquatic environment. 4-Nitroaniline and $NaNO_2$ are very toxic reagents (oral, dermal, inhalation) that may cause skin and eye irritation. Additionally, $NaNO_2$ is an oxidizing reagent and is hazardous to the aquatic environment.

Organic solvents (diethyl ether, petroleum ether, chloroform, ethyl acetate, *N*,*N*-dimethylformamide) are highly flammable liquids and vapors, ethyl acetate may cause severe skin burns and eye damage, chloroform, ethyl ether and *N*,*N*-dimethylformamide are toxic reagents (oral, dermal, inhalation) that may cause skin and eye irritation, petroleum ether is a toxic reagent (oral, dermal, inhalation) that may cause skin and eye irritation and is hazardous to the aquatic environment. Glacial acetic acid and HCl are corrosive acids and may cause severe skin and eye burns, and are flammable in liquid and vapor form.

Complete information about the safety, toxicity and disposal for all reagents can be found, free of charge, at: http://www.msds.com/index.asp?open=/protected_public/loginsuccessful.asp.

Experimental Procedure

Laboratory Session 1 (4 h): Synthesis of 1-Propyl-5-(thiophen-2-yl)-1*H*-pyrrole-2-carbaldehyde

1. Prepare a 10 mL round-bottomed flask with a magnetic stirrer and add DMF (dry) (0.18 g, 0.2 mL, 2.5 mmol) and $POCl_3$ (0.38 g, 0.2 mL, 2.5 mmol) at 0 °C. Stir in an ice bath at 0 °C for about 15 min.
2. To the previous reaction mixture add dropwise the 1-(*n*-propyl)-2-(2′-thienyl)pyrrole (0.4 g, 2.1 mmol) dissolved in DMF (2 mL). Insert a water condenser with an adapter with calcium chloride filling into the round-bottomed flask and clamp the flask in a pre-heated oil bath. Stir the reaction for 2 h at a constant temperature of 60 °C.
3. Check the evolution of the reaction using TLC by a comparison of the contents of the reaction mixture with the precursor thienylpyrrole (eluent: petroleum ether (40–60 °C)/diethyl ether (7 : 3)).
4. Cool down the reaction mixture at room temperature, add it slowly to 30 mL of a saturated aqueous solution of sodium acetate and stir the obtained mixture for 30 min.
5. Transfer the obtained mixture into a separatory funnel and dilute the organic layer with 30 mL of diethyl ether. Separate and wash the organic layer with 50 mL of a saturated solution of $NaHCO_3$. Dry the organic layer with anhydrous Na_2SO_4, filter the drying agent and evaporate the solvent under reduced pressure with a rotary evaporator.
6. Check the contents of the residue using TLC by a comparison with the precursor thienylpyrrole (eluent: petroleum ether (40–60 °C)/diethyl ether (7 : 3)).

Laboratory Session 2 (4 h): Purification of 1-Propyl-5-(thiophen-2-yl)-1*H*-pyrrole-2-carbaldehyde

1. Purify the residue obtained from evaporation using chromatography with a silica gel column (60 Å; 200–300 mesh). Dissolve the residue in the minimum amount of chloroform and start the elution with petroleum ether (40–60 °C)/diethyl ether (9 : 1) and continue the elution with mixtures of petroleum ether (40–60 °C)/ethyl ether of increased polarity. Check by TLC the contents of each collected fraction (eluent: petroleum ether (40–60 °C)/diethyl ether (7 : 3).
2. Combine the fractions containing the product (R_f = 0.62) and evaporate to dryness in a rotary evaporator.

3. Weigh the mass of the obtained product (green oil) and calculate the yield. Record the ^{1}H and ^{13}C NMR spectra in $CDCl_3$, IR spectrum in liquid film and UV-vis spectrum in ethanol. Compare them with the literature data[3d] as well as with the spectra provided in the ESI to confirm the compound's purity.

Laboratory Session 3 (4 h): Synthesis of 5-(1-Propyl-1*H*-pyrrol-2-yl) thiophene-2-carbaldehyde

1. Prepare an apparatus for stirring under an inert atmosphere in anhydrous conditions consisting of a 25 mL three-neck round-bottomed flask with a magnetic stirrer, a water condenser with an adapter with calcium chloride filling into the round-bottomed flask, and two rubber septa.
2. Fill the round-bottomed flask with an inert gas such argon or nitrogen, through one of the septa add a solution of thienylpyrrole derivative (0.4 g, 2.1 mmol) in 2 mL of dry diethyl ether at 0 °C.
3. Add dropwise at 0 °C a solution of 2.5 M of *n*-BuLi in hexanes (1 mL, 2.5 mmol) to the flask and stir for about 1 h. Remove the flask from the ice bath, allowing it to warm up to room temperature and then stir for 15 min.
4. Add dropwise, at room temperature, DMF (0.20 mL, 2.5 mmol) dissolved in 4 mL of dry diethyl ether and heat at reflux for 1 h.
5. Check the evolution of the reaction using TLC by a comparison of the contents of the reaction mixture with the precursor thienylpyrrole (eluent: petroleum ether (40–60 °C)/diethyl ether (7 : 3)).
6. Cool down the reaction mixture at room temperature and add 40 mL of water.
7. Transfer the obtained mixture into a separatory funnel and extract with ethyl acetate (3 × 50 mL). Separate and wash the combined organic layers with 100 mL of water. Dry the organic layer with anhydrous $MgSO_4$, filter the drying agent and evaporate the solvent under reduced pressure with a rotary evaporator.
8. Check the contents of the residue using TLC by comparison with the precursor thienylpyrrole (eluent: petroleum ether (40–60 °C)/diethyl ether (7 : 3)).

Laboratory Session 4 (4 h): Purification of 5-(1-Propyl-1*H*-pyrrol-2-yl) thiophene-2-carbaldehyde

1. Purify the residue obtained from evaporation by chromatography using a silica gel column (60 Å; 200–300 mesh). Dissolve the residue in the minimum amount of chloroform and start the elution with petroleum ether (40–60 °C)/diethyl ether (9 : 1) and continue the elution with mixtures of petroleum ether (40–60 °C)/diethyl ether of increased polarity. Check by TLC the contents of each collected fraction (eluent: petroleum ether (40–60 °C)/diethyl ether (7 : 3)).
2. Combine the fractions containing the product (R_f = 0.49) and evaporate to dryness in a rotary evaporator.
3. Determine the mass of the obtained product as an orange oil and calculate the yield. Record the ^{1}H and ^{13}C NMR spectra in $CDCl_3$, IR spectrum in liquid film and UV-vis spectrum in ethanol. Compare with the literature data[3d] and the spectra provided in the ESI to confirm the compound's purity.

Laboratory Session 5 (4 h): Synthesis of (*E*)-2-(4-Nitrophenyl)-1-(1-propyl-5-(thiophen-2-yl)-1*H*-pyrrol-2-yl)diazene

(i) Diazotization of 4-Nitroaniline

1. Add to a beaker 4-nitroaniline (0.29 g, 2.09 mmol), $NaNO_2$ (0.114 g, 2.09 mmol) and 5 mL of water. Stir until a homogeneous solution is formed.

2. Prepare an apparatus for stirring at 0–5 °C consisting of a 25 mL round-bottomed flask with a magnetic stirrer, and an ice bath and add to the flask 1.5 g of ice and 1.5 mL of HCl (37%), $d = 1.18$.
3. Add to the flask the aqueous solution of aniline and $NaNO_2$ contained in the beaker. Stir for 30 min at 0–5 °C.

(ii) Coupling of the diazonium salt with 1-(*n*-propyl)-2-(2′-thienyl)pyrrole

1. Prepare an apparatus for stirring at 0–5 °C consisting of a 50 mL round-bottomed flask with a magnetic stirrer, and an ice bath.
2. Prepare a solution of the pyrrole derivative (0.4 g, 2.08 mmol) in 20 mL of acetonitrile and add to this solution a few drops of acetic acid in the 50 mL round-bottomed flask.
3. Add dropwise the diazonium salt previously prepared to the solution of thienylpyrrole in acetonitrile. Stir for 2 h at 0 °C.
4. Allow the mixture to reach room temperature and add 10 mL of petroleum ether (40–60 °C) and 20 mL of water.
5. Isolate by vacuum filtration the solid that precipitates.
6. Transfer the filtrate solution obtained into a separatory funnel, and extract with 20 mL of chloroform.
7. Separate and extract the organic layer with water (2 × 30 mL), separate and dry the combined organic layers with anhydrous $MgSO_4$, filter the drying agent and evaporate the solvent under reduced pressure with a rotary evaporator.
8. Check the contents of the residue by TLC through a comparison with the precursor thienylpyrrole (eluent: petroleum ether (40–60 °C)/diethyl ether (7 : 3)).

Laboratory Session 6 (4 h): Purification of (*E*)-2-(4-Nitrophenyl)-1-(1-propyl-5-(thiophen-2-yl)-1*H*-pyrrol-2-yl)diazene

1. Purify the residue obtained from evaporation by chromatography using a silica gel column (60 Å; 200–300 mesh). Dissolve the residue in the minimum amount of chloroform and start the elution with petroleum ether (40–60 °C)/diethyl ether (9 : 1) and continue the elution with mixtures of petroleum ether (40–60 °C)/diethyl ether of increased polarity until (1 : 1). Check by TLC the contents of each collected fraction (eluent: petroleum ether (40–60 °C)-diethyl ether (7 : 3)).
2. Combine the fractions containing the product ($R_f = 0.64$) and evaporate to dryness in a rotary evaporator.
3. Determine the mass of the obtained product as a green solid with metallic luster and calculate the yield. Record the ^1H and ^{13}C NMR spectra in DMSO-d_6, IR spectrum in liquid film, and UV-vis spectrum in ethanol. Compare them with the literature data[3b] and the spectra provided in the ESI to confirm the compound's purity.

Waste Disposal

The solvents used in extractions and in column chromatography are recovered after evaporation in a rotary evaporator and should be collected in chlorinated or non-chlorinated waste collector recipients, for subsequent destruction. The aqueous solutions from liquid–liquid extraction should be neutralized and discarded in the sink. The drying agents should be disposed of in a proper container after filtration from the organic solutions. Used silica gel must be stored in a closed container in the fume hood for subsequent disposal. Special attention should be paid to the treatment and handling of reaction mixtures in sessions 1 and 5 that contain HCl as by-product and in session 3 due to the use of *n*-BuLi as a metalation reagent. The excess *n*-BuLi can be destroyed by dilution with hydrocarbon solvent to a concentration of approximately 5 wt%, followed by gradual addition to water with vigorous stirring under an inert atmosphere.

Results Interpretation and Additional Questions

1. What is the function of $POCl_3$ and DMF in session 1 and *n*-BuLi and DMF in session 3?
2. Write down all the resonance structures obtained as a consequence of electrophilic attack at C2 and C3 in the pyrrole heterocycle.
3. Write down the two mechanisms for the formylation reactions of 1-(*n*-propyl)-2-(2′-thienyl) pyrrole and suggest an explanation for the fact that the formylation of this compound *via* metallation with *n*-BuLi followed by addition of DMF (session 3) yielded 5-(1-propyl-1*H*-pyrrol-2-yl)thiophene-2-carbaldehyde whereas Vilsmeier–Haack formylation of the same compound (Section 1) gave 1-propyl-5-(thiophen-2-yl)-1*H*-pyrrole-2-carbaldehyde.
4. Write down the mechanism for the diazo coupling reaction (session 5).
5. Interpret the ^1H and ^{13}C NMR, IR and UV-vis spectra obtained with the corresponding assignments of signals (bands). Compare them with the spectra provided in the ESI.
6. ^1H NMR can easily be used to monitor the formylation reactions performed in sessions 1 and 3 through two different synthetic procedures. Explain the principal differences observed in the ^1H NMR spectra of the two formylated derivatives 5-(1-propyl-1*H*-pyrrol-2-yl)thiophene-2-carbaldehyde and 1-propyl-5-(thiophen-2-yl)-1*H*-pyrrole-2-carbaldehyde.
7. Explain the spectroscopic differences (IR and ^1H NMR) observed between 1-(*n*-propyl)-2-(2′-thienyl)pyrrole (ESI for Experiment 17.8) and 5-(1-propyl-1*H*-pyrrol-2-yl) thiophene-2-carbaldehyde.
8. Explain the reasons that lead to the higher value for the absorption maxima (λ_{max}) of the 2-(4-nitrophenyl)-1-(1-propyl-5-(thiophen-2-yl)-1*H*-pyrrol-2-yl)diazene compared to the precursor 1-(*n*-propyl)-2-(2′-thienyl)pyrrole.

References

1. A. H. Jackson, M. Artico, H. J. Anderson, C. E. Loader, A. Gossauer, P. Nesvadba and N. Dennis, *The Chemistry of Heterocyclic Compounds*, ed. R. A. Jones, John Wiley & Sons, Inc., New York, 1990, vol. 48, part 1, p. 295.
2. (a) R. E. Niziursky-Mann and M. P. Cava, *Heterocycles*, 1992, **34**, 2003; (b) L. I. Belen'kii, *Heterocycles*, 1994, **37**, 2029.
3. (a) M. M. M. Raposo, Recent developments in the chemistry of 2-thienylpyrroles: synthesis, reactivity and applications, *Targets in Heterocyclic Systems: Chemistry and Properties*, ed. O. A. Attanasi and D. Spinelli, Italian Society of Chemistry, 2007, vol. 11, pp. 122–154; (b) M. M. M. Raposo, A. M. R. C. Sousa, A. M. C. Fonseca and G. Kirsch, *Tetrahedron*, 2005, **61**, 8249; (c) M. M. M. Raposo, A. M. R. C. Sousa, G. Kirsch, F. Ferreira, M. Belsley, E. Matos Gomes and A. M. C. Fonseca, *Tetrahedron*, 2005, **61**, 11991; (d) M. M. M. Raposo, A. M. R. C. Sousa, A. M. C. Fonseca and G. Kirsch, *Tetrahedron*, 2006, **62**, 3493; (e) M. M. M. Raposo, M. C. R. Castro, A. M. C. Fonseca, M. F. S. Cardoso, M. Belsley, L. M. Carvalho and P. Coelho, *Dyes Pigm.*, 2011, **91**, 62; (f) P. J. Coelho, M. C. R. Castro and M. M. M. Raposo, *J. Photochem. Photobiol. Chem., A*, 2013, **259**, 59; (g) J. Garcia-Amorós, M. Reig, M. C. R. Castro, A. Cuadrado, M. M. M. Raposo and D. Velasco, *Chem. Commun.*, 2014, **50**, 6704.
4. (a) *Comprehensive Heterocyclic Chemistry II*, ed. C. W. Bird, Pergamon, Oxford, 1984, vol. 4, pp. 751–754 and pp. 771–773; (b) G. Jones and S. P. Stanforth, The Vilsmeier Reaction of Fully Conjugated Carbocycles and Heterocycles, *Organic Reactions*, ed. L. A. Paquete, John Wiley & Sons, Inc., New York, 1997, vol. 49, pp. 15–17; (c) J. March, *Advanced organic chemistry: reactions, mechanisms and structure*, John Wiley and Sons, Inc., New York, 4th edn, 1992, pp. 542–543; (d) L. Kurti and B. Czakó, *Strategic applications in named reactions in organic synthesis*, Elsevier, San Diego, 2005, p. 468; (e) K. Peter, C. Vollhardt and N. E. Schore, *Organic Chemistry: structure and function*, W. H. Freeman, New York, 5th edn, 2007, pp. 979–981 and p. 1041.
5. (a) M. Schlosser, *Angew. Chem., Int. Ed.*, 2005, **44**, 376; (b) R. Chinchilla, C. Nájera and M. Yus, *Chem. Rev.*, 2004, **104**, 2667.

6. Nucleophilic Aromatic Substitution

6.1. Synthesis of 2,3-Quinoxalinedithiol

Dulce Simão* and Ana C. Cerdeira

Department of Chemistry, Instituto Superior Técnico, University of Lisbon, Centro de Química Estrutural, Av. Rovisco Pais, 1, 1049-001 Lisbon, Portugal. *E-mail: dulcesimao@tecnico.ulisboa.pt

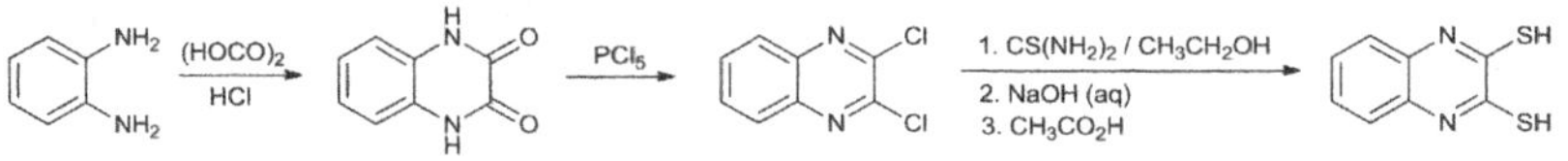

Number of sessions (duration of each session)	Hazard level	Difficulty level	Level of study
3 (4 h each)	Moderate	Low	Intermediate

Class names *o*-Phenylenediamines, α-dicarboxylic acids, quinoxalines, *o*-dichloroquinoxalines, *o*-dithiolquinoxalines

Concepts involved Three-step synthesis of an aromatic dithiol involving a cyclic condensation reaction, nucleophilic aromatic substitution with thiourea as nucleophile and dithiol purification based in acid/base reactions

Chemicals needed *o*-Phenylenediamine, oxalic acid dihydrate, concentrated hydrochloric acid, anhydrous phosphorus pentachloride, anhydrous calcium chloride, thiourea, ethanol, sodium hydroxide, glacial acetic acid

Equipment and experimental techniques involved Crystallization, IR, heating and magnetic stirring plate and heating mantle, reflux, vacuum filtration, NMR, melting point

Keywords Aromatic nucleophilic substitution, cyclic condensation reaction, IR, NMR, vacuum filtration

Background

In this experiment, 2,3-quinoxalinedithiol an aromatic *ortho*-arene dithiol will be prepared. This compound is used as ligand due to its ability to coordinate with transition metals to produce metallic complexes, which are key building blocks for conducting and magnetic molecular materials.[1–4] The most common route to prepare quinoxalines is the condensation of 1,2-diaminobenzene with a 1,2-dicarbonyl compound.[5–7] In this work *o*-phenylenediamine reacts with oxalic acid to form an amide like structure, the 1,2,3,4-tetrahydroquinoxaline-2,3-dione, which can tautomerize[8,9] as shown in Scheme 6.1.1.

This compound is then chlorinated with phosphorus pentachloride, to give 2,3-dichloroquinoxaline. This intermediate is also used for the synthesis of condensed heterocycles with pharmacological interest.[10] The aromatic nucleophilic substitution of the chlorine with thiourea produces 2,3-quinoxalinedithiol after acid/base reactions.[11,12] The thiourea sulfur atom is nucleophilic and readily takes part in the reaction. The two nitrogen atoms of the pyrazine ring make

Comprehensive Organic Chemistry Experiments for the Laboratory Classroom
Edited by Carlos A. M. Afonso, Nuno R. Candeias, Dulce Pereira Simão, Alexandre F. Trindade, Jaime A. S. Coelho, Bin Tan, and Robert Franzén.

Published by the Royal Society of Chemistry, www.rsc.org

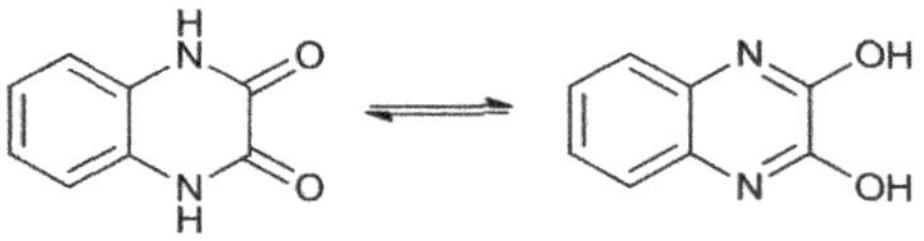

Scheme 6.1.1 Tautomerism of 1,2,3,4-tetrahydroquinoxaline-2,3-dione.

it more reactive towards nucleophilic substitution at the 2-position once the intermediate anion is stabilized by electronegative nitrogen and by delocalization around the ring.[8,9]

Additional Safety

This work should be carried out in a fume hood. Oxalic acid is corrosive and may cause severe eye and skin irritation with possible burns. Care should be taken while handling all acid and base solutions, which are also corrosive as well phosphorus pentachloride, a corrosive and poisonous solid that can react with water or steam to form phosphoric acid, and hydrogen chloride. Thiourea and *o*-phenylenediamine cause eye and skin irritation and there is evidence of a carcinogenic effect. It is advisable weigh all chemicals under the fume hood.

Experimental Procedure

Session 1 – Preparation of 1,2,3,4-Tetrahydroquinoxaline-2,3-dione

1. In a 250 mL round bottomed flask dissolve *o*-phenylenediamine (8 g, 74 mmol) and oxalic acid dihydrate (9.3 g, 74 mmol) in an aqueous solution of 4 N HCl (44 mL).
2. Heat the mixture to reflux in a heating mantle for 15 min and then cool until a precipitate appears.
3. Filter under vacuum and wash the precipitate with cold water.
4. Dry the product under vacuum.
5. Record the yield and melting point of the product and obtain its IR spectrum.

Session 2 – Preparation of 2,3-Dichloroquinoxaline

1. With a pestle and mortar quickly grind together 1,2,3,4-tetrahydroquinoxaline-2,3-dione (5 g, 30.8 mmol) and PCl_5 (12.7 g, 60.9 mmol).
2. Place this mixture in a 250 mL round-bottom flask fitted with a water cooled condenser with a $CaCl_2$ tube (anhydrous) on top.
3. Heat this mixture in an oil bath to 160 °C. At this temperature the reactants start to melt and the reaction begins.
4. Two hours later turn off the heat and allow the mixture to cool until room temperature. Pour it over crushed ice (*ca.* 200 g) to remove excess PCl_5.
5. Filter the product under vacuum, wash it well with water and dry under vacuum.
6. Record the yield and melting point of the product and obtain the IR spectrum.

Session 3 – Synthesis of 2,3-Quinoxalinedithiol

1. In a 500 mL round-bottomed flask prepare a solution of 2,3-dichloroquinoxaline (5 g, 24.8 mmol) and thiourea (10 g, 132 mmol) in ethanol (125 mL), and reflux in a water bath for 2 h.
2. Remove the heat source, cool with an ice bath and add an aqueous solution of NaOH 1.6 M (250 mL) dropwise.

3. Reflux for another hour, and then gravity filter with a pleated paper filter.
4. To the filtered solution slowly add glacial acetic acid until neutral pH.
5. Cool, filter under vacuum and wash the brownish-orange precipitate with water.
6. Record the yield and melting point of the product and obtain the IR spectrum.

Waste Disposal

The acidic mother liquor must be neutralized, diluted and discarded in the sink.

Results Interpretation and Additional Questions

1. Determine the reaction yields.
2. Assess the purity of the obtained products by comparing the melting points with the ones reported in the literature.
3. Assign the ^{1}H and ^{13}C NMR spectra and the characteristic absorption bands in the IR spectra.
4. Write out all the reaction mechanisms.
5. Explain why the dithiol can be purified based on acid/base reactions.

References

1. R. M. Metzger, *et al.*, *Low-Dimensional Systems and Molecular Electronics*, Plenum Press, New York, 1991, vol. 248, NATO ASI Series B, p. 1.
2. V. Z. Kresin and W. A. Little, *Organic Superconductivity*, Plenum Press, 1990, p. 295.
3. J. S. Miller, *Extended Linear Chain Compounds*, Plenum Press, 1983, vol. 3, p. 319.
4. D. Simão, *et al.*, *Synth. Met.*, 1999, **102**, 1613–1614.
5. T. L. Gilchrist, *Heterocyclic Chemistry*, Longman, 3rd edn, 1997, p. 286.
6. N. Paudler, *Contemporary Heterocyclic Chemistry*, John Wiley & Sons, Inc., 1982, p. 221.
7. A. R. Katritzky, *Handbook of Heterocyclic Chemistry*, Pergamon Press, 1985, p. 473.
8. J. Clayden, N. Greeves, S. Warren and P. Wothers, *Organic Chemistry*, Oxford University Press, 2001, p. 1152.
9. D. T. Davies, *Aromatic Heterocyclic Chemistry*, Oxford University Press, 1992, Oxford Chemistry Primers, p. 39.
10. Y. A. Ammar, A. G. Al-Sehemi, A. M. Sh. El-Sharief and M. S. A. El-Gaby, *Phosphorus, Sulfur Silicon Relat. Elem.*, 2009, **184**, 660.
11. L. J. Theriot, K. K. Ganguli, S. Kavarnos and I. Bernal, *J. Inorg. Nucl. Chem.*, 1969, **31**, 3133.
12. C. A. M. Afonso, D. P. Simão, L. P. Ferreira, M. S. Serra and M. M. M. Raposo, *100 Experiências de Química Orgânica*, IST Press, 2011, p. 185.

6.2. Synthesis of the Antitumoral Drug 2,4,6-Tris(dimethylamino)-1,3,5-triazine *via* Sequential Nucleophilic Substitution

Filipa Siopa*[a], Nuno Candeias*[b], and Carlos A. M. Afonso*[a]

[a]Instituto de Investigação, do Medicamento (iMed.ULisboa), Faculdade de Farmácia, Universidade de Lisboa, Av. Prof. Gama Pinto, 1649-003 Lisboa, Portugal; [b]Department of Chemistry and Bioengineering, Tampere University of Technology, Korkeakoulunkatu 8, Tampere, FI-33101, Finland
*E-mail: filipasiopa@ff.ul.pt, nuno.rafaelcandeias@tut.fi, carlosafonso@ff.ul.pt

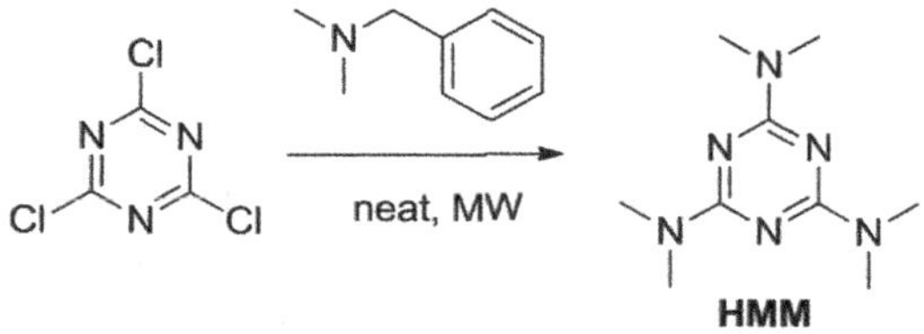

Number of sessions (duration of each session)	Hazard level	Difficulty level	Level of study
1 (3 h)	High	Medium	Intermediate

Class names Amines, heteroaryl halides, triazines

Concepts involved This procedure involves the preparation of a triamine derivative *via* a formal nucleophilic substitution; a tandem addition–elimination and bimolecular displacement performed under solvent-free and microwave heating

Chemicals needed Cyanuric chloride, *N*,*N*-dimethylbenzylamine, dichloromethane, ethanol, hexane, silica

Equipment and experimental techniques involved Vial, syringe, spatula, round bottom flask, Büchner funnel, filter paper, domestic kitchen-type microwave instrument, melting point apparatus, NMR spectrometer

Keywords Hexamethylmelamine (HMM), crystallization/filtration, melting point, ^{1}H and ^{13}C NMR spectroscopy, nucleophilic substitution, microwaves

Background

1,3,5-Triazines or *s*-triazines derivatives are an important class of organic compounds that have applications in different fields such as in textile,[1] plastic[1,2] and rubber industries[1,2] and are used as pesticides,[1] dyestuffs[1] and for the production of herbicides.[3,4] In addition, these compounds also have broad applications in pharmaceutical science due to their important anti-metastatic,[3–5] anti-viral[6] and anti-malarial properties[7] 2,4,6-Tris(dimethylamino)-1,3,5-triazine or hexamethylmelamine (HMM) is an important representative of this class of compounds due to its anti-tumor activity against ovarian adenocarcinoma, lymphoma, and breast cancer. In fact, Altretamine (formerly HMM; Hexalen capsules, MGI PHARMA, Inc., Bloomington, MN) is used for the treatment of ovarian cancer administrated orally as tablets or capsules.[8–10]

Comprehensive Organic Chemistry Experiments for the Laboratory Classroom
Edited by Carlos A. M. Afonso, Nuno R. Candeias, Dulce Pereira Simão, Alexandre F. Trindade, Jaime A. S. Coelho, Bin Tan, and Robert Franzén.

Published by the Royal Society of Chemistry, www.rsc.org

HMM was synthesized by reaction of cyanuric chloride with *N,N*-dimethylbenzylamine. The proposed experiment is based on the formation of a relatively stable electron withdrawing quaternary ammonium substituent in the triazine ring, after the first aromatic nucleophilic reaction (Scheme 6.2.1). The accumulation of the positive charge in the triazine ring, after the first aromatic nucleophilic reaction, increased the reactivity of the ring to the next aromatic nucleophilic substitution. This process favors the formation of trisubstituted products and decreases the formation of partially substituted products, hence reducing product contamination by reaction intermediates (Scheme 6.2.1).[11]

The use of microwave irradiation as the energy source in organic reactions presents some advantages to the thermal method.[12,13] For instance, reactions can be performed under solvent-free conditions or using water as solvent (decreasing the environmental impact of the reaction); and products can be obtained in higher yields and shorter times due to acceleration effects caused by the selective absorption of microwave by polar molecules.[5,14]

Additional Safety

Place the domestic kitchen-type microwave instrument in the fume hood. Cyanuric chloride and *N,N*-dimethylbenzylamine are fatal if inhaled and harmful if swallowed. Benzyl chloride that is formed during the reaction is corrosive, toxic and can cause severe skin burns and eye damage. Ethanol is highly flammable. Hexane is highly flammable, may cause skin irritation, may be fatal if swallowed and enters airways, and may cause damage to organs through prolonged or repeated exposure. Students should wear protective clothing, gloves, and safety goggles.

Experimental Procedure

Laboratory Session 1 (3 h): Synthesis of Hexamethylmelamine (HMM)

Procedure

1. Add cyanuric chloride (500 mg, 2.7 mmol) to a 5 mL vial (ϕ = 13 mm).
2. In the fume hood add *N,N*-dimethylbenzylamine (1.6 mL, 10.8 mmol) to the vial containing cyanuric chloride.

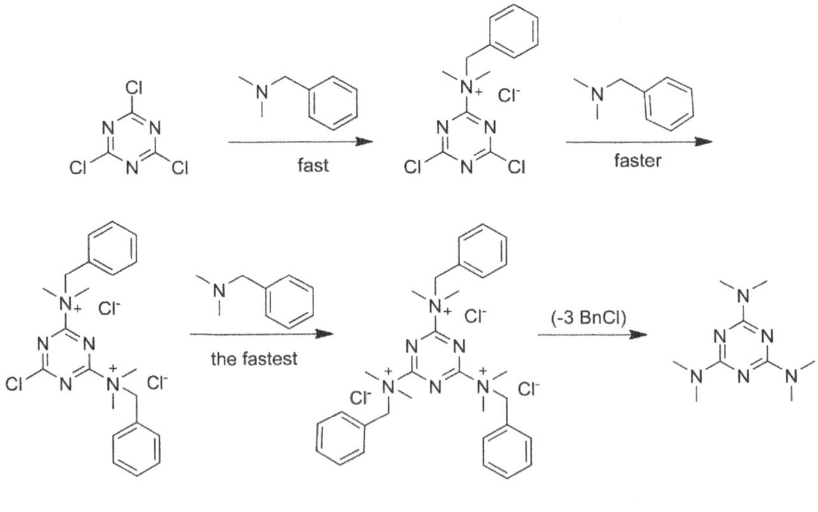

Scheme 6.2.1 Mechanism of HMM formation by reaction of cyanuric chloride with *N,N*-dimethylbenzylamine.

3. Place the reaction mixture in the domestic kitchen-type microwave instrument. The intensity of the microwave is adjusted to medium-high (*ca.* 800 W) and the reaction time is 3 min (*Note 1*).
4. Cool the reaction mixture to room temperature and remove the vial (*Note 2*).
5. Transfer the reaction mixture to a round-bottom flask by using ethanol (50 mL) and gently use a spatula to facilitate the dissolution (*Note 3*).
6. Add silica gel (2.5 g) to the round-bottom flask containing the reaction mixture and remove the volatiles by using a Rotovapor® (*Note 4*).
7. Add 20 mL of hexane and stir for 30 min. Filter by suction and remove the solvent in the rotatory evaporator to obtain a powder white solid.
8. Add 10 mL of hot hexane and put the flask in an ice bath.
9. The HMM should crystallize after 5 min approximately. Collect the crystals by vacuum filtration. Determine the product yield, measure the melting point and, optionally, acquire the ^{1}H and ^{13}C NMR spectra.

Note 1 A domestic kitchen-Crown Japan microwave instrument was used and the intensity was adjusted to medium high. Before the microwave irradiation, the reaction mixture is a white solid and a colorless liquid and after the microwave irradiation the reaction mixture is a yellow oil. The experiment was also reproduced in a CEM-equipment; however, we deliberately studied this transformation in a domestic microwave oven so that the experiment can more easily be implemented in a teaching laboratory.

Note 2 At the end of the reaction let the vial cool down for *ca.* 3 min inside the microwave. The temperature of the reaction, after the microwave irradiation, was measured and varies between 150 and 200 °C.

Note 3 Ensure that the entire sample is removed from the vial to the round-bottom flask; if necessary use more ethanol. To facilitate the compound dissolution in the transference process, add hot ethanol and use a spatula to gently stir.

Note 4 All the solvent must be removed during the evaporation process.

Waste Disposal

The hexane and ethanol waste should be collected in a non-chlorinated recipient for posterior disposal.

Results Interpretation and Additional Questions

1. Explain the reaction mechanism for the synthesis of HMM using *N,N*-dimethylbenzylamine, a tertiary amine.[11] Compare the reaction mechanism with a tertiary amine to the reaction with secondary or primary amines.
2. Discuss the competition between the benzyl and the methyl group in the nucleophilic substitution of the ammonium salt.
3. Considering the use of *N,N*-dimethylbenzylamine, 4-nitro-*N,N*-dimethylbenzylamine and 4-methoxy-*N,N*-dimethylbenzylamine in the synthesis HMM, analyze the ^{1}H NMR data for each reaction obtained after 30 min reaction at 150 °C of cyanuric chloride with 7 mol equivalents of the corresponding amine. Discuss the influence of the substituent at the *para* position of the amine aromatic ring for the preparation of HMM.
4. Explain why microwave irradiation is more advantageous in some reactions when compared to conventional heating.

References

1. G. Blotny, *Tetrahedron*, 2006, **62**, 9507–9522.
2. K. Bretterbauer and C. Schwarzinger, *Curr. Org. Synth.*, 2012, **9**, 342–356.
3. C. A. M. Afonso, N. M. T. Lourenco and A. D. Rosatella, *Molecules*, 2006, **11**, 81–102.
4. R. Banerjee, D. R. Brown and E. Weerapana, *Synlett*, 2013, **24**, 1599.
5. V. B. Kurteva and C. A. M. Afonso, *Green Chem.*, 2004, **6**, 183–187.
6. M. Venkatraj, K. K. Ariën, J. Heeres, J. Joossens, J. Messagie, J. Michiels, P. Van der Veken, G. Vanham, P. J. Lewi and K. Augustyns, *Bioorg. Med. Chem. Lett.*, 2012, **22**, 7174–7178.
7. M. Sharma, K. Chauhan, S. S. Chauhan, A. Kumar, S. V. Singh, J. K. Saxena, P. Agarwal, K. Srivastava, S. Raja Kumar, S. K. Puri, P. Shah, M. I. Siddiqi and P. M. S. Chauhan, *MedChemComm*, 2012, **3**, 71–79.
8. J. K. Chan, V. Loizzi, A. Manetta and M. L. Berman, *Gynecol. Oncol.*, 2004, **92**, 368–371.
9. J. K. Sun, Y. G. Deng, S. L. Wang, J. N. Cao, X. F. Gao and X. D. Dong, *Drug Delivery*, 2010, **17**, 164–170.
10. P. Dornan, C. N. Rowley, J. Priem, S. T. Barry, T. J. Burchell, T. K. Woo and D. S. Richeson, *Chem. Commun.*, 2008, 3645–3647.
11. B. Kolesinska and Z. J. Kaminski, *Tetrahedron*, 2009, **65**, 3573–3576.
12. P. Lidström, J. Tierney, B. Wathey and J. Westman, *Tetrahedron*, 2001, **57**, 9225–9283.
13. F. Bassyouni, S. Abu-Bakr and M. Rehim, *Res. Chem. Intermed.*, 2012, **38**, 283–322.
14. Á. Díaz-Ortiz, J. Elguero, A. de la Hoz, A. Jiménez, A. Moreno, S. Moreno and A. Sánchez-Migallón, *QSAR Comb. Sci.*, 2005, **24**, 649–659.

6.3. Benzotriazole, a Useful Synthetic Auxiliary in Heterocyclizations

João Lavrado*[a], Marta Figueiras[a], Eduardo Ruivo[a], Susana Lucas[a], and Alexandra Paulo[a,b]

[a]Research Institute for Medicines (iMed.ULisboa), Faculty of Pharmacy, Universidade de Lisboa, Av. Prof. Gama Pinto, 1649-003 Lisboa, Portugal; [b]Pharmaceutical and Medicinal Chemistry Department, Faculty of Pharmacy, Universidade de Lisboa, Av. Prof. Gama Pinto, 1649-003 Lisboa, Portugal
*E-mail: jplavrado@ff.ulisboa.pt, mapaulo@ff.ulisboa.pt

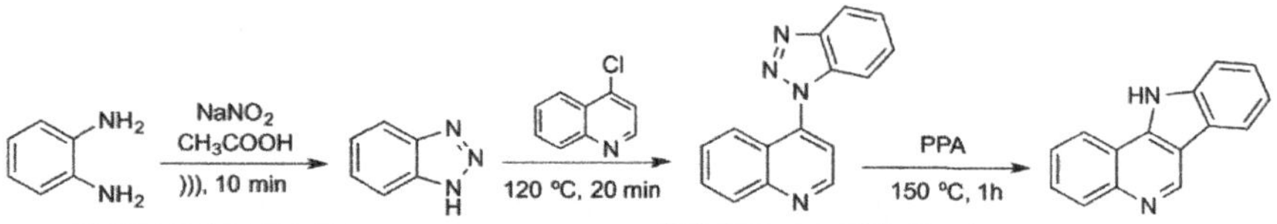

Number of sessions (duration of each session)	Hazard level	Difficulty level	Level of study
1 (4 h)	Moderate	Medium	Intermediate
Class names Amines, diazonium salts, 1,2,3-triazoles			
Concepts involved This experiment illustrates the importance of benzotriazole as a useful synthetic auxiliary with a large range of applications, including the synthesis of heterocyclic compounds with biological relevance			
Chemicals needed 1,2-Diaminobenzene, sodium nitrite, acetic acid, 4-chloroquinoline, polyphosphoric acid, dichloromethane, brine (sodium chloride saturated solution), anhydrous sodium sulfate			
Equipment and experimental techniques involved Ultrasonic bath (can be substituted by a heating apparatus), heating apparatus with temperature control (150 °C) and magnetic stirring, liquid–liquid extraction and rotary evaporator apparatus, vacuum apparatus, NMR			
Keywords Diazonium salt, Graebe–Ullmann reaction, liquid–liquid extraction, NMR, 1,2,3-triazole			

Background

Heterocycles, in particular nitrogen heterocycles, are a main class of pharmacologically active compounds exhibiting diverse biological and pharmacological activities. The development of novel, economically viable and efficient protocols for the synthesis of heterocyclic scaffolds has always been full of challenges and it is perhaps the ultimate goal of synthetic organic chemists in the search for novel bioactive lead structures.

The benzotriazole methodology has been drawn attention during recent decades, being explored in organic synthesis as a synthetic auxiliary, mainly because it is inexpensive, non-toxic, easy to introduce in scaffolds through a variety of reactions, stable during the course of the reactions, and, finally, can be removed easily at the end of the synthetic sequence.[1–4] Being a good leaving group, benzotriazole can react with nucleophiles and electrophiles, being prominent in

Comprehensive Organic Chemistry Experiments for the Laboratory Classroom
Edited by Carlos A. M. Afonso, Nuno R. Candeias, Dulce Pereira Simão, Alexandre F. Trindade, Jaime A. S. Coelho, Bin Tan, and Robert Franzén.

Published by the Royal Society of Chemistry, www.rsc.org

many useful types of reactions like aryl-alkylations and heteroalkylations,[1] benzannulations,[5] conjugate additions,[6,7] acylations[8] and heterocyclizations.[9–11]

The objective of this work is to synthetize the natural indolo[3,2-*c*]quinoline by using a benzotriazole (BTA) auxiliary group, which has been shown to be the best synthetic approach.[12] From commercially available reagents, the heterocyclization mediated BTA demonstrates the potential of the synthetic auxiliary group. The experiment encompasses the synthesis of the BTA intermediate by an ultrasound-assisted methodology, which is known to reduce reaction times and improve yields.[13] However, in the absence of an ultrasound bath, the reaction can be done with thermal heating (70–80 °C for approximately 1 hour). The nucleophile character of the BTA moiety is also demonstrated by the formation of the C–N bond with a heterocyclic quinoline system in a solvent-free reaction and, finally, taking advantage of its thermal instability, the formation of the indolo[3,2-*c*]quinoline heterocyclic product is favoured.[14–16]

Additional Safety

All steps should be carried out in a fume hood, with proper safety equipment, such as protective clothing, gloves and safety goggles. Acetic acid is a strong eye, skin, and mucous membrane irritant. Prolonged skin contact with acetic acid may result in tissue destruction. Polyphosphoric acid is toxic by ingestion, a strong irritant and may cause burns. Handling must be done carefully.

Experimental Procedure

Step 1. Synthesis of 1*H*-Benzo[*d*][1,2,3]triazole

1. In a 50 mL round-bottom flask containing 1,2-diaminobenzene (100 mg, 0.92 mmol) in acetic acid (5 mL) at 5 °C, add, gently, sodium nitrite (191 mg, 2.76 mmol).
2. Sonicate the mixture in the ultrasonic bath for 10 min at 5–10 °C in ice water. After this period, add the mixture to 20 mL of cold water in a 100 mL beaker and finally transfer the solution into a separator funnel, extracting the reaction product with dichloromethane (3 × 25 mL).
3. Wash the combined organic extracts with water (3 × 25 mL), dry with brine (sodium chloride saturated solution) and with anhydrous Na_2SO_4.
4. Transfer the extracts into a 100 mL round-bottom flask (previously weighed) and remove the solvent in a rotary evaporator apparatus. After complete evaporation of the solvent, a white solid (88–92% yield) should be obtained.
5. Measure the melting point and acquire the ^1H NMR spectrum in acetone-d_6.

Step 2. Synthesis of 4-(1*H*-Benzo[*d*][1,2,3]triazol-1-yl)-quinoline

1. To a 10 mL round-bottom flask add a thoroughly homogenized mixture of 4-chloroquinoline (50 mg, 0.31 mmol) and 1*H*-benzo[*d*][1,2,3]triazole (36.4 mg, 0.31 mmol).
2. Heat the mixture in a pre-heated oil bath at 110 °C for approximately 20 min with magnetic stirring.
3. Cool the resulting solid to room temperature (95–98% yield).
4. Measure the melting point and acquire the ^1H NMR spectrum in DMSO-d_6.

Step 3. Synthesis of the Indolo[3,2-*c*]quinoline Heterocycle

1. In a 50 mL round-bottom flask containing 4-(1*H*-benzo[*d*][1,2,3]triazol-1-yl)-quinoline (50 mg, 0.20 mmol), add carefully polyphosphoric acid (10 g).

2. Heat the mixture in a pre-heated oil bath at 150 °C until the release of N_2 ceases (approximately 1 h and verified by the end of the formation of small bubbles in the reaction medium).
3. Add carefully the slightly cooled reaction mixture to cold water (50 mL), wash the reaction round-bottom flask with cold water (25 mL) and combine both solutions.
4. Collect the resulting precipitate by filtration, wash it with cold water and place it in an oven at 70–80 °C until the product is completely dried. A white solid (78–82% yield) should be obtained.
5. Measure the melting point and acquire the ^{1}H NMR spectrum in DMSO-d_6.

Waste Disposal

Acid aqueous solutions may be collected in an acid waste collector, for posterior destruction. Neutralized aqueous solutions may be diluted and discarded in the sink. Organic solvents must be collected in a chlorinated waste collector recipient, for posterior destruction, as well as any solvents resulting from the optional recrystallization of 4-(1*H*-benzo[*d*][1,2,3]triazol-1-yl)-quinoline.

Results Interpretation and Additional Questions

1. Calculate the yield of each step of the experiment and the overall yield of the indolo[3,2-*c*]quinoline synthesis by the BTA methodology.
2. Interpret the ^{1}H NMR spectra considering chemical shifts, intensity and multiplicity of signals, and comment on the spectral changes observed compared with the ^{1}H NMR spectra of the starting materials.
3. Comment on the purity of the final indolo[3,2-*c*]quinoline and of intermediate compounds based on melting points and ^{1}H NMR spectra.
4. Comment on the importance of the benzotriazole auxiliary group in heterocyclizations and on the potential advantage of its use in the development of substituted indolo[3,2-*c*]quinolines.
5. Propose a mechanism for each of the three reactions involved.

References

1. A. R. Katritzky and X. F. Lan, *Chem. Soc. Rev.*, 1994, **23**, 363.
2. A. R. Katritzky and S. A. Belyakov, *Aldrichimica Acta*, 1998, **31**, 35.
3. A. R. Katritzky, X. Lan, J. Z. Yang and O. V. Denisko, *Chem. Rev.*, 1998, **98**, 409.
4. R. R. Kale, V. Prasad, P. P. Mohapatra and V. K. Tiwari, *Monatsh. Chem.*, 2010, **141**, 1159.
5. A. R. Katritzky, J. Q. Li and L. H. Xie, *Tetrahedron*, 1999, **55**, 8263.
6. A. R. Katritzky and M. Qi, *Collect. Czech. Chem. Commun.*, 1998, **63**, 599.
7. W. Liu, D. Z. Zhang, S. C. Zheng, Y. S. Yue, D. G. Liu and X. M. Zhao, *Eur. J. Org. Chem.*, 2011, 6288.
8. A. R. Katritzky, K. Suzuki and Z. Q. Wang, *Synlett*, 2005, 1656.
9. A. R. Katritzky, S. A. Henderson and B. Z. Yang, *J. Heterocycl. Chem.*, 1998, **35**, 1123.
10. A. R. Katritzky, *J. Heterocycl. Chem.*, 1999, **36**, 1501.
11. R. Abonia, E. Rengifo, J. Quiroga, B. Insuasty, J. Cobo and M. Nogueras, *Tetrahedron*, 2004, **60**, 8839.
12. P. T. Parvatkar, P. S. Parameswaran and S. G. Tilve, *Curr. Org. Chem.*, 2011, **15**, 1036.
13. C. M. P. Pereira, H. A. Stefani, K. P. Guzen and A. T. G. Orfao, *Lett. Org. Chem.*, 2007, **4**, 43.
14. C. Graebe and F. Ullmann, *Liebigs Ann. Chem.*, 1896, **291**, 16.
15. P. Nantkana and J. Kalinows, *Acta Pol. Pharm.*, 1971, **28**, 219.
16. L. R. Whittell, K. T. Batty, R. P. Wong, E. M. Bolitho, S. A. Fox, T. M. Davis and P. E. Murray, *Bioorg. Med. Chem.*, 2011, **19**, 7519.

6.4. Synthesis of 4,5-Dicyanobenzene-1,2-dithiol

Dulce Simão

Department of Chemistry, Instituto Superior Técnico, University of Lisbon, Centro de Química Estrutural, Av. Rovisco Pais, 1, 1049-001 Lisbon, Portugal. *E-mail: dulcesimao@tecnico.ulisboa.pt

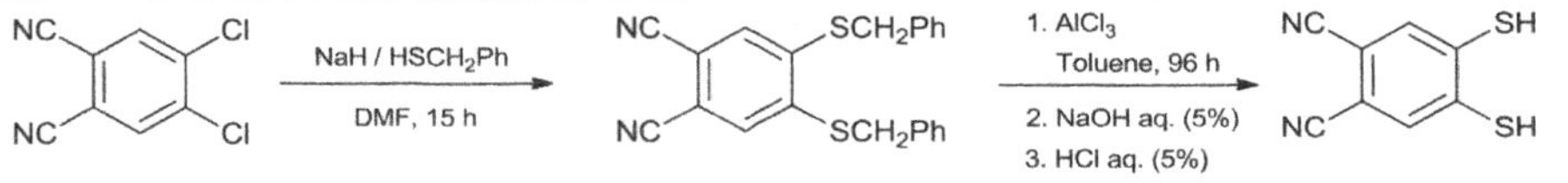

Number of sessions (duration of each session)	Hazard level	Difficulty level	Level of study
4 (4 h each)	High	High	Intermediate
Class names Aromatic halides, aromatic thioethers, aromatic thiols			
Concepts involved Aromatic dithioether preparation involving a nucleophilic aromatic substitution followed by thioether cleavage to obtain the aromatic *ortho*-arene dithiol; this experiment is conducted under inert gas atmosphere and temperature control			
Chemicals needed Benzyl mercaptan, sodium hydride (60%, dispersion in mineral oil), dry *N*,*N*-dimethylformamide, 4,5-dichloro-1,2-dicyanobenzene, chloroform, methanol, dry toluene, dry aluminium chloride, 5% (w/v) aq. sodium hydroxide, 5% (v/v) aq. hydrogen chloride (37%), KBr, CD_2Cl_2, $(CD_3)_2CO$			
Equipment and experimental techniques involved Magnetic stirring, apparatus for dry conditions reaction with inert atmosphere and temperature control, liquid–liquid extraction, vacuum filtration, recrystallization, melting point, IR and NMR spectroscopy			
Keywords Dithiols, liquid–liquid extraction, NMR, nucleophilic aromatic substitution, thioether cleavage			

Background

In this experiment, 4,5-dicyanobenzenedithiol an aromatic *ortho*-arene dithiol will be prepared. This compound is used as ligand due to its ability to coordinate with transition metals to produce metallic complexes, which are key building blocks for conducting and magnetic molecular materials.[1–3] Benzyl mercaptan is a thiol and is considered to be an organic acid. In the presence of sodium hydride this thiol forms, by acid–base reaction, the thiolate anion, the nucleophile in the nucleophilic aromatic substitution to prepare the aromatic dibenzyl sulfide.[4] The mechanism involves addition of the nucleophile followed by elimination of the leaving group – the addition–elimination mechanism. This reaction requires the presence of one or more strongly electron-withdrawing groups located *ortho* or *para* to the leaving group of benzene ring. In this reaction two cyano groups are in the *para* position to chlorines atoms. The attack by the nucleophile produces an anion known as a Meisenheimer complex. Electron-withdrawing groups like cyano groups stabilize this through delocalization of the negative charge with the cyano groups. The forward reaction eliminates chloride ion and reforms the aromatic ring.[5a] Nevertheless, problems were found when this reaction was tried for the first time with the usual nucleophiles like NaSH, when only one of chlorine atom was displaced.[6] The best method found to achieve the aromatic dithiol is the one described in this experiment with the aromatic dithioether preparation followed by thioether

Comprehensive Organic Chemistry Experiments for the Laboratory Classroom
Edited by Carlos A. M. Afonso, Nuno R. Candeias, Dulce Pereira Simão, Alexandre F. Trindade, Jaime A. S. Coelho, Bin Tan, and Robert Franzén.

Published by the Royal Society of Chemistry, www.rsc.org

cleavage to obtain the aromatic *ortho*-arene dithiol. Thioethers are in general cleaved by reduction with sodium/ammonia, by acid-catalysed hydrolysis, or by reaction with a heavy metal ion such as silver(I) or mercury(II), followed by hydrogen sulfide.[7,8] In the present work a Lewis acid like aluminium chloride is used to accomplish the cleavage.[3,6,9,10]

The acid–base properties of thiols allow them to be isolated from the reaction mixture. Extraction of the organic reaction mixture with an aqueous sodium hydroxide solution converts the thiol into the water-soluble salt, which dissolves into the aqueous layer. Acidification after separation of the aqueous layer affords the thiol again, which is insoluble in water and precipitates.[5b]

Additional Safety

This work must be carried out in a fume hood due to high toxicity, flammability and extremely unpleasant smell of benzyl mercaptan. A bleach trap can be used to oxidize the thiol to less malodorous compounds before venting to the environment. Sodium hydride (60%, dispersion in mineral oil) reacts with water and may release a flammable and/or toxic gas. Aluminium chloride, anhydrous is a poisonous, hygroscopic and corrosive solid. It reacts violently with water to produce hydrogen chloride. Both sodium hydroxide and hydrogen chloride are corrosive.

Experimental Procedure

Preparation of 1,2-Bis(*S*-benzylthio)-4,5-dicyanobenzene

Session 1

1. Fit a 250 mL four-necked flask with a reflux condenser, a thermometer, a pressure-equalizing dropping funnel, a magnetic stirrer bar and an inert gas (nitrogen or argon) inlet system.
2. Under an inert atmosphere, add sodium hydride (60%) (1.0 g) through a solid addition funnel, followed by dry DMF (40 mL) washing down any solid that may have adhered to the funnel.
3. Stir the resulting suspension and add dropwise benzyl mercaptan (3 mL, 26 mmol) keeping the temperature between 10 and 15 °C with an ice/water bath.
4. When the solution remains clear (all solid must have disappeared), add in small portions (*ca.* 0.3 g every 2 min) 4,5-dichloro-1,2-dicyanobenzene (2.5 g, 13 mmol) at room temperature until all has been added.
5. Change the inert gas inlet by a balloon filled with argon or nitrogen and leave the reaction mixture overnight with stirring.

Session 2

1. Collect the resultant yellow powder by vacuum filtration and recrystallize from chloroform/methanol, dissolving in the minimum amount of hot chloroform and further addition of the same amount of methanol.
2. Cool it in an ice bath and filter under vacuum the pale yellow needles. Dry under vacuum.
3. Record the yield, melting point and IR and, ^{1}H and ^{13}C NMR spectra.

Preparation of 4,5-Dicyanobenzenedithiol

Session 3

1. Fit a 250 mL four-necked flask with a reflux condenser, a pressure-equalizing dropping funnel, a magnetic stirrer bar and an inert gas (nitrogen or argon) inlet system.
2. Weigh quickly the finely powdered anhydrous aluminium chloride (2.2 g, 1.61 mmol) in the fume hood and add to the flask under inert atmosphere.

3. Add dry toluene (25 mL) and stir the resulting suspension. Add dropwise a solution of 1,2-bis(*S*-benzylthio)-4,5-dicyanobenzene (1.5 g, 4.03 mmol) in dry toluene (50 mL). This addition should take 1 h approximately.
4. Change the inert gas inlet by a balloon filled with argon or nitrogen and leave the reaction stirring at room temperature for 96 h.

Session 4

1. Add water with crushed ice (60 mL) with stirring to the reaction mixture and remove the pink precipitate by gravity filtration.
2. Transfer the filtrate to a 500 mL separatory funnel and separate the organic layer and wash it with water (3 × 45 mL).
3. Extract the organic layer with a 5% (w/v) aqueous sodium hydroxide (3 × 40 mL).
4. Combine the three alkaline layers, cool in an ice bath and acidify with 5% (v/v) aqueous hydrogen chloride (*ca.* 80 mL) until a yellowish solid precipitates (pH 2).
5. Collect the product by vacuum filtration. Wash well with water. Dry under vacuum.
6. Record the yield, melting point and IR and ^{1}H and ^{13}C NMR spectra.

Waste Disposal

All glassware and other laboratory ware that came into contact with benzyl mercaptan should be submerged in a bleach solution in the fume hood for 24 h. Disposable items such as gloves and paper towels should be enclosed in a zip-lock bag, placed in a wide-mouth plastic jar and disposed of as hazardous waste. Aqueous solutions are diluted and collected in a waste collector recipient for aqueous solutions, chloroform in a chlorinated waste collector recipient. Methanol and DMF are collected in a non-chlorinated waste collector. Toluene may be stored for further purification.

Results Interpretation and Additional Questions

1. Determine the reaction yields.
2. Assess the purity of 1,2-bis(*S*-benzylthio)-4,5-dicyanobenzene and the 4,5-dicyanobenzenedithiol, comparing the melting points with the ones reported in the literature.
3. Assign the ^{1}H and ^{13}C NMR spectra and the characteristic absorption bands in the IR spectra.
4. Write out all reaction mechanisms.
5. What is the purpose of extracting the organic phase with an aqueous sodium hydroxide solution and further acidification?
6. Discuss the effect of cyano groups in nucleophilic aromatic substitution.

References

1. J. M. Williams, *Organic Conductors, Progress in Inorganic Chemistry*, 1985.
2. J. S. Miller, *Extended Linear Chain Compounds*, Plenum Press, 1983, vol. 3, p. 319.
3. D. Simão, *et al.*, , *Eur. J. Inorg. Chem.*, 2001, **12**, 3119–3126.
4. A. M. Richter, V. Engels, N. Beye and E. Fanghanel, *Z. Chem.*, 1989, **29**, 444.
5. J. Clayden, N. Greeves, S. Warren and P. Wothers, *Organic Chemistry*, Oxford University Press, 2001,(a) p. 1249; (b) p. 186.
6. D. Simão, *Síntese de sais de transferência de carga baseados em ditiolatos aromáticos como aceitadores electrónicos*, PhD thesis, 1998.
7. T. W. Greene, *Protective Groups in Organic Synthesis*, John Wiley & Sons, 1981.
8. G. H. Whitham, *Organosulfur Chemistry*, Oxford University Press, Oxford Chemistry Primers, 1995, p. 15.
9. C. A. M. Afonso, D. P. Simão, L. P. Ferreira, M. S. Serra and M. M. M. Raposo, *100 Experiências de Química Orgânica*, IST Press, 2011, p. 181.
10. D. E. L. Carrington, K. Clark and R. M. Scrowston, *J. Chem. Soc. C*, 1971, **23**, 3262.

6.5. Nucleophilic Aromatic Substitution Reactions in 3,6-Bis(3,5-dimethyl-1*H*-pyrazol-1-yl)-1,2,4,5-tetrazine

Tiago J. L. Silva*[a,b], Paulo J. G. Mendes[b], Ana Isabel Tomaz[a], and M. Helena Garcia[a]

[a]Centro de Química Estrutural, Faculdade de Ciências da Universidade de Lisboa, Campo Grande, 1049-016 Campo Grande, Lisboa, Portugal; [b]Departamento de Química, Escola de Ciências e Tecnologia, Centro de Química de Évora, Instituto de Investigação e Formação Avançada, Universidade de Évora, Rua Romão Ramalho 59, 7000-671 Évora, Portugal. *E-mail: tjsilva@ciencias.ul.pt

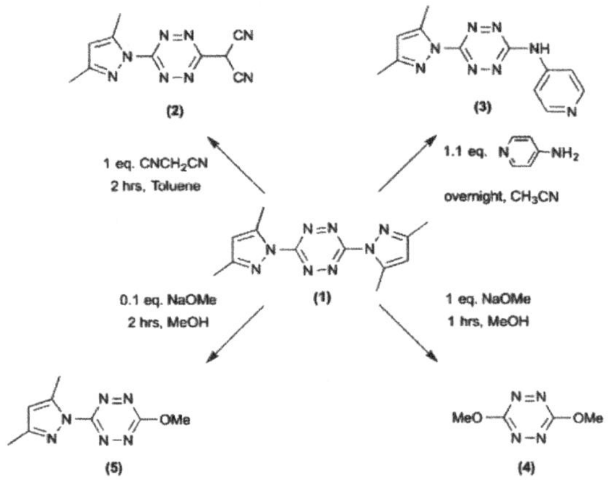

Number of sessions (duration of each session)	Hazard level	Difficulty level	Level of study
3 (4 h each)	Moderate	Medium	Advanced

Class names Tetrazines, pyrazoles, nitriles, pyridines

Concepts involved This experiment involves four examples of nucleophilic aromatic substitutions in tetrazine rings. The N-, C- and O-nucleophiles chosen have a completely different reactivity, providing the students an easy way to understand the nucleophilicity of the different compounds

Chemicals needed 3,6-Bis(3,5-dimethyl-1*H*-pyrazol-1-yl)-1,2,4,5-tetrazine, 4-aminopyridine, malononitrile, sodium methoxide, acetonitrile, triethylamine, toluene, ethanol, methanol, anhydrous magnesium sulfate, neutral aluminium oxide for chromatography, deuterated chloroform, deuterated acetone, deuterated dimethyl sulfoxide

Equipment and experimental techniques involved Vacuum filtration, liquid–liquid extraction, column chromatography, NMR spectroscopy, laboratory weighing equipment, heating plate with magnetic stirring, rotatory evaporation apparatus, preparative thin-layer chromatography

Keywords Column chromatography, NMR spectroscopy, nucleophilic aromatic substitution, preparative thin-layer chromatography, tetrazines

Comprehensive Organic Chemistry Experiments for the Laboratory Classroom
Edited by Carlos A. M. Afonso, Nuno R. Candeias, Dulce Pereira Simão, Alexandre F. Trindade, Jaime A. S. Coelho, Bin Tan, and Robert Franzén.

Published by the Royal Society of Chemistry, www.rsc.org

Background

Tetrazines are six-membered aromatic rings with four nitrogen atoms, whose isomeric structures are depicted in Figure 6.5.1.

1,2,4,5-Tetrazines, also known as *s*-tetrazines or *sym*-tetrazines (symmetric tetrazines), are the most studied isomers in this family of molecules.[1] These compounds exhibit a panel of rare and useful characteristics that make them attractive for several applications in different domains, ranging from materials science,[1] functional molecular systems for the development of sensors and optical switches,[2] electroactive polymers[3] or as building blocks in solar cells,[4] to coordination chemistry[1] and biochemical sciences.[5]

Tetrazine reactivity is quite different from other heterocyclic aromatic systems due to the unique electronic properties of these molecules.[6] For example, they are known for reacting readily and without a catalyst with various ethynyl-based compounds in a Diels–Alder reaction (Scheme 6.5.1), affording pyridazines that are important in drug development.[5,7]

Thus, the demand for various tetrazine derivatives that can be used as precursor compounds for the synthesis of novel bioactive compounds is high. These derivatives can be prepared by aromatic nucleophilic substitution (S_NAr) at the 3- and 6-positions of an adequate tetrazine precursor, particularly if the initial 3 and 6 substituents are good leaving groups such as halogens, thiolates, sulfonates or azolyl groups (Scheme 6.5.2). A nucleophilic attack at these positions by C-, N-, O- or S-nucleophiles leads to the formation of new symmetrical or unsymmetrical 1,2,4,5-tetrazines, depending on the experimental conditions used. The starting 1,2,4,5-tetrazines used in the S_NAr reactions are usually the 3,6-dichloro-*s*-tetrazine, the 3,6-bis(methylthio)-*s*-tetrazine or the 3,6-bis(3,5-dimethylpyrazolyl)-*s*-tetrazine.

3,6-Bis(3,5-dimethyl-1*H*-pyrazol-1-yl)-1,2,4,5-tetrazine is a very versatile reagent for which the replacement of the 3,5-dimethylpyrazolyl groups by N-, O- and C-based nucleophiles *via* S_NAr reaction is very well documented.[8] The reactivity of this 1,2,4,5-tetrazine derivative towards different nucleophiles is strongly dependent on the experimental conditions. Hence, a careful choice of the experimental conditions employed, for example room temperature reactions in apolar solvents *versus* high temperature reactions in polar protic or aprotic solvents, leads to control of the reaction product obtained.

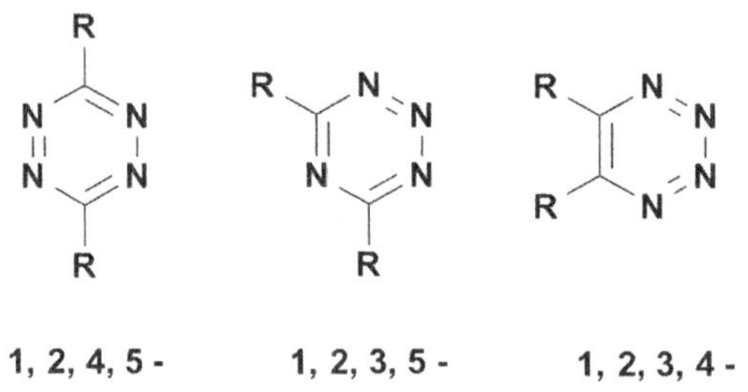

Figure 6.5.1 Generic structures of the three possible tetrazine isomers.

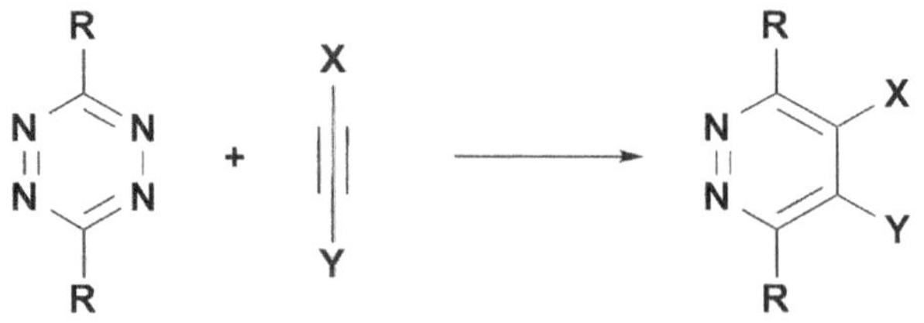

Scheme 6.5.1 Reaction of 1,2,4,5-tetrazine compounds with alkynyl compounds, affording pyridazines.

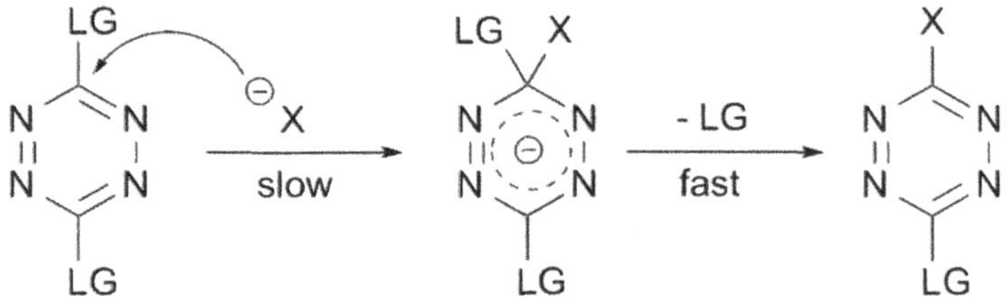

Scheme 6.5.2 Nucleophilic aromatic substitution (S_NAr) on 1,2,4,5-tetrazine rings.

Hence, the goal of this work is to illustrate this feature through the synthesis of a series of different asymmetric and symmetric tetrazines, namely 2-[6-(3,5-dimethyl-1*H*-pyrazol-1-yl)-1,2,4,5-tetrazin-3-yl]malononitrile (**2**), 6-(3,5-dimethyl-1*H*-pyrazol-1-yl)-*N*-(pyridin-4-yl)-1,2,4,5-tetrazine-3-amine (**3**), 3,6-dimethoxy-1,2,4,5-tetrazine (**4**) and 3-(3,5-dimethyl-1*H*-pyrazol-1-yl)-6-methoxy-1,2,4,5-tetrazine (**5**). These compounds are synthesized from the same precursor – 3,6-bis(3,5-dimethyl-1*H*-pyrazol-1-yl)-1,2,4,5-tetrazine (**1**) – under different experimental conditions that illustrate the marked difference in the reactivity of *s*-tetrazine rings towards different types of nucleophiles.

Additional Safety

In all laboratory sessions the use of basic safety protection equipment (protective clothing, gloves and safety goggles) is mandatory. All laboratory sessions should be done in a fume hood. 4-Aminobenzonitrile, malononitrile and sodium methoxide are toxic, and toluene, dichloromethane and acetonitrile are flammable solvents, harmful if swallowed, inhaled, or in contact with the skin and eyes. Care should be taken when handling all these substances.

Experimental Procedure

Session A (4 h) – Reaction with C- and N-Nucleophiles

Reaction 1 (3 h) – Synthesis of 2-[6-(3,5-Dimethyl-1H-pyrazol-1-yl)-1,2,4,5-tetrazin-3-yl]malononitrile (2)

1. Prepare a 50 mL round-bottomed flask with a magnetic stir bar and add 0.2 g (0.74 mmol, 1 equiv.) of solid 3,6-bis(3,5-dimethyl-1*H*-pyrazol-1-yl)-1,2,4,5-tetrazine (**1**). Add 25 mL of toluene.
2. To the red suspension, add 0.076 g (1.1 mmol, 1.5 equiv.) of malononitrile followed by 140 μL of triethylamine (1.0 mmol, 1 equiv.). Stir the reaction for 2 h at room temperature. During this time, the colour of the reaction mixture will turn to dark purple.
3. Evaporate the solvent in a rotatory evaporator, and wash the dark purple solid with 25 mL of a 10% hydrochloric acid solution.
4. Filter off the solid in a sintered glass filter, and wash it with 100 mL of water (divided into several smaller volumes).
5. Transfer the solid into a beaker and add 100 mL of acetone. Dry the purple solution with anhydrous sodium sulfate, filter the mixture and evaporate the solvent in the filtrate until dryness using a rotatory evaporator.

Reaction 2 – Synthesis of 6-(3,5-Dimethyl-1H-pyrazol-1-yl)-N-(pyridin-4-yl)-1,2,4,5-tetrazine-3-amine (3) (Part 1)

1. Prepare a 100 mL round-bottomed flask with a magnetic stir bar and add 0.5 g (1.8 mmol) of 3,6-bis(3,5-dimethyl-1*H*-pyrazol-1-yl)-1,2,4,5-tetrazine (**1**). Add 40 mL of acetonitrile.

2. Add 0.435 g of 4-aminopyridine (3.6 mmol, 2 equiv.) followed by 3 mL of triethylamine (22 mmol, 12 equiv.). Place a cap on the flask, and heat it in an oil bath up to 40 °C. Allow it to stir overnight. (*The remaining steps should be performed in the next session.*)

Session B (4 h) – Reaction with O-Nucleophiles

Reaction 2 – Synthesis of 6-(3,5-Dimethyl-1H-pyrazol-1-yl)-N-(pyridin-4-yl)-1,2,4,5-tetrazine-3-amine (3) (Part 2)

1. Remove the reaction flask from the oil bath, and allow it to cool to room temperature. Evaporate the solvent in a rotatory evaporator, after removing the magnetic stirrer bar.
2. Fill in a chromatography column (3 cm diameter × 20 cm height) with a slurry of neutral aluminium oxide in dichloromethane. To the dried flask containing the product, add 1 mL of dichloromethane and carefully load the chromatography column with this solution.
3. Chromatograph the mixture by using 100 mL of a mixture of dichloromethane and ethanol (9 : 1). Collect the first orange coloured band (R_f = 0.75) into a 100 mL flask (tared) and evaporate the solvent in the rotatory evaporator.

Reaction 3 (1.5 h) – Synthesis of 3,6-Dimethoxy-1,2,4,5-tetrazine (4)

1. Prepare a 50 mL round-bottomed flask with a magnetic stir bar and add 0.2 g (0.74 mmol, 1 equiv.) of 3,6-bis(3,5-dimethyl-1*H*-pyrazol-1-yl)-1,2,4,5-tetrazine (**1**). Add 25 mL of methanol and stir until all the red solid is dissolved.
2. Add 0.2 g of anhydrous magnesium sulfate (1.66 mmol, 2.2 equiv.) to the reaction mixture followed by 0.070 g of sodium methoxide (0.9 mmol, 1.2 equiv.).
3. Follow the reaction by TLC using hexane/ethyl acetate (3 : 2) as the eluent, until no starting material is present in the TLC plate (which takes about 1 h).
4. Evaporate the solvent in the rotatory evaporator making sure that *no heating is used* during this process.
5. Fill a chromatography column (3 cm diameter × 20 cm height) with a slurry of flash silica gel in dichloromethane. To the dried flask containing the product add 1 mL of dichloromethane and transfer all of the solution to the chromatography column.
6. Chromatograph the mixture by using 200 mL of dichloromethane. Collect the coloured first band (R_f = 0.95) to a 100 mL flask (tared) and evaporate the solvent in the rotatory evaporator making sure again that no heating is used.

Session C – Reaction with O-Nucleophiles – Part 2 (4 h)

Reaction 4 (4 h) – Synthesis of 3-(3,5-Dimethyl-1H-pyrazol-1-yl)-6-methoxy-1,2,4,5-tetrazine (5)

1. Prepare a 50 mL round-bottomed flask with a magnetic stir bar and add 0.2 g (0.74 mmol, 1 equiv.) of 3,6-bis(3,5-dimethyl-1*H*-pyrazol-1-yl)-1,2,4,5-tetrazine (**1**). Add 25 mL of methanol and stir it until the entire red solid is dissolved.
2. To the previous solution, add 0.2 g of anhydrous magnesium sulfate (1.66 mmol, 2.2 equiv.) followed by 0.010 g of sodium methoxide (0.18 mmol, 0.2 equiv.).
3. Follow the reaction by TLC using hexane/ethyl acetate (3 : 2) until no starting material is present in the TLC plate. Then evaporate the solvent in the rotatory evaporator making sure that *no heating is used in this process.*
4. Prepare a preparative TLC plate (20 cm × 20 cm) and, with a pencil carefully draw a thin line 2 cm above the bottom (the application line). Dissolve the solid obtained in the minimum amount of dichloromethane in the previous step, and apply it to the chromatographic plate.

5. Chromatograph the mixture by using 100 mL of a mixture of hexane/ethyl acetate (3 : 2). Scratch out the red band (R_f = 0.80), and collect the coloured silica in a beaker.
6. Dissolve the product in the minimum amount of dichloromethane (20–30 mL) and filter the solution to remove the silica.
7. Evaporate the solvent in the rotatory evaporator.

NMR Spectra

Prepare four NMR samples by dissolving 20 mg of each final product in 0.5 mL of an adequate solvent in separate NMR tubes. For compound **2** use DMSO-d_6, for compound **3** use $(CD_3)_2CO$, and for compounds **4** and **5** use $CDCl_3$.

Waste Disposal

Hydrochloric acid solution should be neutralized, diluted and discarded in the sink. Non-chlorinated and chlorinated organic solvents should be collected in a non-chlorinated waste collector recipient or a waste recipient for chlorinated compounds, respectively. Silica gel and drying agents (magnesium sulfate) should be collected in an appropriate solid waste collector recipient for posterior destruction.

Results Interpretation and Additional Questions

1. Calculate the yields of the reactions.
2. Draw the mechanism of the nucleophilic aromatic substitution occurring in *Reaction 4*.
3. Considering the experiment procedures, namely solvent, temperature and reaction time, discuss the reactivity of the N-, C- and O-nucleophiles towards S_NAr in 1,2,4,5-tetrazine derivatives.
4. Interpret the ^1H and ^{13}C NMR spectra.

References

1. G. Clavier and P. Audebert, *Chem. Rev.*, 2010, **110**, 3299.
2. K. Parimal, E. H. Witlicki and A. H. Flood, *Angew. Chem., Int. Ed.*, 2010, **49**, 4628.
3. P. Audebert, F. Miomandre, G. Clavier, M. C. Vernières, S. Badré and R. Méallet-Renault, *Chem.–Eur. J.*, 2005, **11**, 5667.
4. Z. Li, J. Ding, N. Song, J. Lu and Y. Tao, *J. Am. Chem. Soc.*, 2010, **132**, 13160.
5. N. K. Devaraj and R. Weissleder, *Acc. Chem. Res.*, 2011, **44**, 816.
6. P. Audebert, F. Miomandre, G. Clavier, M. C. Vernières, S. Badré and R. Méallet-Renault, *Chem.–Eur. J.*, 2005, **11**, 5667.
7. (a) N. K. Devaraj, R. Weissleder and S. A. Hilderbrand, *Bioconjugate Chem.*, 2008, **19**, 2297; (b) N. K. Devaraj, R. Upadhyay, J. B. Hatin, S. A. Hilderbrand and R. Weissleder, *Angew. Chem., Int. Ed.*, 2009, **48**, 7013.
8. R. I. Ishmetova, N. I. Latosh, I. N. Ganebnykh, N. K. Ignatenko, S. G. Tolshchina and G. L. Rusinov, *Russ. J. Org. Chem.*, 2008, **45**, 1102.

7. Transition Metal Catalysed Substitution

7.1. A Solvent-Free Ullmann Coupling: Synthesis of 2,2′-Dinitrobiphenyl

Laurel Goj Habgood* and Richard W. Gregor

Rollins College, 1000 Holt Avenue, Box 2743, Winter Park, FL 32789, USA. *E-mail: lhabgood@rollins.edu

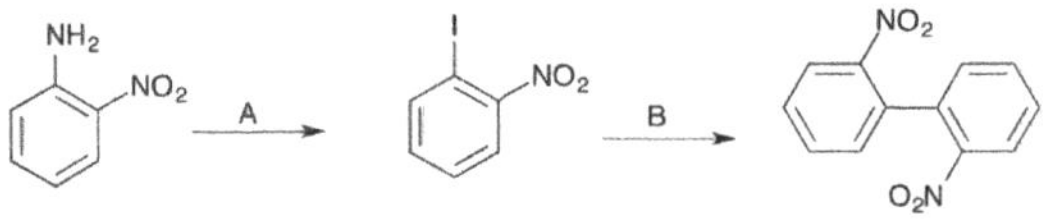

A: 1. H_2O, p-TsOH, $NaNO_2$ 2. KI

B: Cu, heat

Number of sessions (duration of each session)	Hazard level	Difficulty level	Level of study
2 (3 h each)	Moderate	Medium	Intermediate
Class names Substituted arenes, aryl halides			
Concepts involved This experiment involves the two-step synthesis of 2,2′-dinitrobiphenyl; an electrophilic aromatic substitution converts 2-nitroaniline into 1-iodo-2-nitrobenzene, which subsequently undergoes a heterogeneous copper-catalysed Ullmann coupling reaction to form the product			
Chemicals needed 2-Nitroaniline, *p*-toluenesulfonic acid monohydrate, sodium nitrite, potassium iodide, sodium sulfite (10% aqueous solution by weight), β-naphthol, copper powder, sodium hydroxide, dichloromethane, hexane, ethyl acetate, silica (60 Å), deuterated chloroform			
Equipment and experimental techniques involved Mortar and pestle for grinding, test-tube for heterogeneous catalytic reaction, column chromatography, thin-layer chromatography, melting point, NMR spectroscopy			
Keywords C–C bond formation, column chromatography, electrophilic aromatic substitution, NMR, solvent-free synthesis, Ullman coupling			

Background

In the past decade undergraduate organic chemistry textbooks have begun to incorporate transition metal-catalyzed carbon–carbon bond formation (coupling) reactions in the main body of the work.[1–3] In the laboratory, students are most familiar with the use of acetylide anions, Grignards, and Diels–Alder reactions, but the transition metal-based approaches are accessible. The most common methods utilize palladium catalysts, which bring issues of expense, air and/or moisture sensitivity, and lack of commercial availability when translating to the setting of a teaching laboratory. One of the classic coupling reactions is the copper based Ullmann reaction. Copper offers the advantages of low cost, air stability, biological friendliness, and student familiarity with the use of organocuprates as a carbon–carbon bond-forming reagent (*e.g.* Gilman reagent).

Comprehensive Organic Chemistry Experiments for the Laboratory Classroom
Edited by Carlos A. M. Afonso, Nuno R. Candeias, Dulce Pereira Simão, Alexandre F. Trindade, Jaime A. S. Coelho, Bin Tan, and Robert Franzén.

Published by the Royal Society of Chemistry, www.rsc.org

The copper-catalyzed synthesis of 2,2′-dinitrobiphenyl was designed with several objectives in mind.[4] Students perform reactions either identical (electrophilic aromatic substitution (EAS)) or analogous (use of copper rather than palladium for carbon–carbon bond formation) to those found in textbooks. Two different topics, EAS and C–C bond formation, are linked to assist students in appreciating how diverse reactions can be sequenced together. The class is introduced to a two-week laboratory sequence in which the product of the first week is the substrate in the second week. Confirmation of product synthesis can be performed by both melting point and ^{1}H NMR methods depending on time and facilities. As many green chemistry principles as feasible are incorporated with the recognition that it is not always possible to create a fully green process.[5]

In week one, the synthesis of 1-iodo-2-nitrobenzene is achieved by subsequent additions/grindings of reagents and purified with a small silica column. The procedure avoids the diazotization of the aromatic amine which often uses solutions of caustic hydrochloric acid and sodium carbonate or involves discrete formation of the complexed dye rather than the reactive diazonium intermediate.[6,7] Students with low yields either repeat the experiment, time permitting, or augment with commercially available 1-iodo-2-nitrobenzene. In week two, a test tube is charged with the reagents, heated, and the product isolated using a small silica column. The identity and purity of the 2,2′-dinitrobiphenyl is confirmed with melting point and ^{1}H NMR spectroscopy. As an alternative ending to the experiment (no column purification), a ^{1}H NMR spectrum of the crude reaction mixture can be analyzed to illustrate the conversion of 1-iodo-2-nitrobenzene into 2,2′-dinitrobiphenyl and a percent conversion calculated.

The synthesis of 2,2′-dinitrobiphenyl introduces students to the techniques of solvent-free reactions by crushing and heating while reinforcing essential skills such as chromatography and spectroscopy. Students experience one of the issues synthetic chemists face, low yields or poor purity impacting future experiments. The experiment lends itself to a discussion of what is green about the reactions (microscale method leading to reduced waste, use of catalyst) and where improvements can still be made (finding substitutes for halogenated reagents/solvents). Given the direct correlation to classroom material, reaction methodologies, and low cost per student, we believe the synthesis of 2,2′-dinitrobiphenyl to be an interesting experiment for students and instructors alike.

Additional Safety

Students must wear safety glasses and gloves while performing the experiments under a fume hood. All compounds are to be handled in a manner consistent with general chemical safety principles and recommendations from the Material Safety Data Sheets (MSDS) provided by the manufacturer. Hexane, dichloromethane, and ethyl acetate are flammable and irritants. Dichloromethane is a probable carcinogen. *p*-Toluenesulfonic acid monohydrate and sodium hydroxide are corrosive and may cause burns. 2-Nitroaniline and sodium nitrite are listed as toxic if inhaled/swallowed. 2-Nitroaniline, sodium nitrite, potassium iodide, sodium sulfite, β-naphthol, hexane, 1-iodo-2-nitrobenzene, copper, dichloromethane, ethyl acetate, and silica are all irritants (skin, eye, and respiratory).

Experimental Procedure

Laboratory Session 1 (3 h): Synthesis of 1-Iodo-2-nitrobenzene

1. Crush 2-nitroaniline (0.276 g, 2 mmol, 1 equiv.), using a mortar and pestle, until it is a fine powder. Add water (0.2 mL, 4 drops), performing the entire reaction with mortar and pestle, to form a paste.
2. Add *p*-toluenesulfonic acid monohydrate (1.141 g, 6 mmol, 3 equiv.) and blend the resulting paste for about 2 min.
3. Introduce sodium nitrite (0.345 g, 5 mmol, 2.5 equiv.) to the mixture and periodically crush/stir over a 10 min interval.

4. At this point the diazonium tosylate is formed and may be confirmed using the β-naphthol test. Place a small amount (<2 mg) of the paste on a watch glass and cover with several drops of the β-naphthol solution. A positive result is indicated by a dark brown color characteristic of the azo dye.
5. To the bright yellow paste add potassium iodide (0.830 g, 5 mmol, 2.5 equiv.); a darkening of the paste to a light brown is observed.
6. Over the course of 20 min, grind the mixture with the pestle continuously, perhaps using a spatula to keep the material from "climbing" the walls of the mortar.
7. At the end of the time period scrape the product onto a glass filter frit, wash thoroughly with 15 mL of 10% aqueous sodium sulfite solution, and dry the solid using vacuum filtration. Dispose of the filtrate as instructed.
8. Repeat the β-naphthol test on the product to ensure that the diazonium tosylate is no longer present. If a positive result is noted, rinse the solid with additional sodium sulfite solution.
9. A silica TLC plate should be spotted with ~3 mg samples of reagent 2-nitroaniline and the crude product then developed using dichloromethane : hexane (30 : 70) as the eluent.
10. Prepare a chromatography column (1.5 cm diameter requires ~3 g of dry silica gel).
11. Add the dry crude product to the top of the column and carefully elute with dichloromethane : hexane (30 : 70). The bright yellow 1-iodo-2-nitrobenzene product should be collected as the first fraction from the column in about 30 min (drip rate of about one drop every 2–3 s).
12. The solvent should be removed from the fraction and the mass, melting point, and ^1H NMR spectrum (in $CDCl_3$) of the solid taken.

Laboratory Session 2 (3 h): Synthesis of 2,2′-Dinitrobiphenyl

1. In a 15 cm test tube mix 1-iodo-2-nitrobenzene (0.150 g ideally, 0.6–1.0 mmol), copper powder (0.200 g, 3 mmol), and sand (0.200 g). Save ~3 mg of 1-iodo-2-nitrobenzene to use for a TLC control later.
2. In an area well removed from any flammable material prepare a heating bath by adding sand until the well of the heating mantle is 2/3 full. Adjust the power supply and use a suspended thermometer to heat the sand bath to a temperature of about 350 °C at a depth of about 1–2 cm. The flame from a Bunsen burner is an acceptable alternative heat source.
3. Be careful to point the mouth of the test tube away from anyone. Carefully submerge the test tube (until the reagents are covered) in the sand for *between 20 and 30 seconds*. You will see the 1-iodo-2-nitrobenzene melt and then boil as the reaction proceeds. Repeat the heating procedure 2–3 more times. There may be some yellow condensate part way up the test tube.
4. Once the test tube is *cooled to room temperature*, mix the solid material in the bottom of the tube.
5. Remove two samples (2–3 mg each) for TLC and ^1H NMR. Small amounts of the TLC and NMR solvents may be used to extract the organic components from the crude mixture.
6. Perform TLC (dichloromethane : hexane (30 : 70)) of the crude product and the reserved 1-iodo-2-nitrobenzene to identify the components in the crude mixture.
7. Obtain a ^1H NMR spectrum with integrated peaks to identify unreacted 1-iodo-2-nitrobenzene and the 2,2′-dinitrobiphenyl product.
8. Prepare a small chromatography column identical to the column used earlier. Gently add the dry crude product on the top of the silica gel and use dichloromethane : hexane (30 : 70) until the first colored fraction (1-iodo-2-nitrobenzene) is collected (about 30 min). Switch to dichloromethane : ethyl acetate (90 : 10) and collect the second colored fraction (2,2′-dinitrobiphenyl).
9. Evaporate the solvent from the second fraction to determine the yield of the product. Confirm the identity and purity of the newly synthesized 2,2′-dinitrobiphenyl by determining the melting point.

Waste Disposal

The organic liquid waste was collected in a single glass bottle, as per instructions by our waste management provider. Solid waste was collected in a separate plastic container for appropriate waste disposal.

Results Interpretation and Additional Questions

1. Draw the structure of the diazonium intermediate.
2. Provide the chemical equation for the coupling reaction between the diazonium salt of 2-nitroaniline and β-naphthol. How does this explain the intense color of the product?
3. Propose a mechanism for the potassium iodide substitution.
4. Is the copper catalyst acting as an electrophile or a nucleophile? Is this consistent with the behavior of a typical transition metal catalyst?
5. Provide assignments (including peak pattern and *J* values) for the ^{1}H NMR spectra of 1-iodo-2-nitrobenzene and 2,2′-dinitrobiphenyl.
6. Suppose you took a ^{1}H NMR spectrum of only the crude product. How could you use integration of the peaks to calculate the "extent of reaction" or "conversion ratio", that is, the percentage of reagent converted into product?

Acknowledgement

Reprinted (adapted) with permission from *J. Chem. Educ.*, 2011, **88** (3), pp 331–333. Copyright (2011) American Chemical Society.

References

1. P. Y. Bruice, *Organic Chemistry*, Pearson Prentice Hall, Upper Saddle River, NJ, 7th edn, 2014, ch. 12, pp. 541–548.
2. J. G. Smith, *Organic Chemistry*, McGraw Hill, New York, 4th edn, 2013, ch. 26, pp. 1046–1052.
3. L. G. Wade Jr, *Organic Chemistry*, Pearson, Upper Saddle River, NJ, 8th edn, 2013, ch. 17, pp. 792–796.
4. R. A. Schambach, *J. Chem. Educ.*, 1976, **53**, 735.
5. T. E. Goodwin, *J. Chem. Educ.*, 2004, **81**, 1187.
6. J. W. Lehman, *Multiscale Operational Organic Chemistry*, Pearson Prentice Hall, Upper Saddle River, NJ, 2nd edn, 2009, parts II and III, pp. 380–388, 415–423 and 561–562.
7. A. M. Schoffstall, B. A. Gaddis and M. L. Druelinger, *Microscale and Miniscale Organic, Chemistry Laboratory Experiments*, McGraw Hill, New York, 2nd edn, 2004, ch. 22, pp. 441–441 and 451–452.

7.2. Reactivity Studies for the Synthesis of 5-Phenylthiophene-2-carbaldehyde by a Suzuki–Miyaura Coupling

Maria Manuela Marques Raposo*, Susana Paula Graça da Costa, Rosa Maria Ferreira Batista, and Rosa Cristina Moutinho Ferreira

University of Minho, Department of Chemistry, Campus de Gualtar, 4710-057 Braga, Portugal
*E-mail: mfox@quimica.uminho.pt

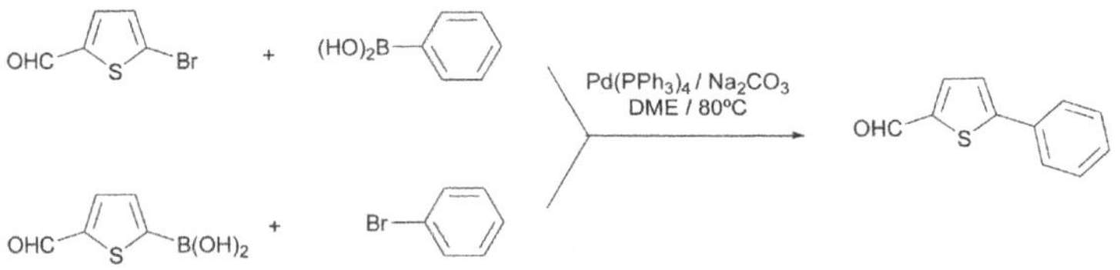

Number of sessions (duration of each session)	Hazard level	Difficulty level	Level of study
3 (2 h + 2 h + 4 h)	Moderate	Medium	Intermediate

Class names Aldehydes, (hetero)arylbromides, boronic acids

Concepts involved Preparation of a formyl-arylthiophene derivative by Suzuki–Miyaura coupling between hetero(aryl)bromides and hetero(aryl)boronic acids

Chemicals needed Tetrakis(triphenylphosphine)palladium(0), 5-formyl-2-thiopheneboronic acid, phenylboronic acid, 5-bromo-2-thiophenecarboxaldehyde, bromobenzene, sodium carbonate, 1,2-dimethoxyethane (DME), ethyl acetate, deuterated chloroform, ethanol, saturated aqueous NaCl solution, aqueous solution of 10% NaOH, magnesium sulfate (anhydrous), diethyl ether, petroleum ether (40–60 °C), silica gel (230–400 mesh), silica gel thin-layer chromatography (TLC) plates with fluorescent indicator

Equipment and experimental techniques involved ^1H and ^{13}C nuclear magnetic resonance spectroscopy, assembly for heating at reflux (oil bath) with stirring under anhydrous conditions and inert atmosphere, liquid–liquid extraction, gravity and vacuum filtration, rotary evaporator, thin-layer chromatography (TLC), column chromatography on silica gel, melting point apparatus

Keywords Aldehyde, boronic acid, chromatography, NMR, Suzuki–Miyaura cross-coupling, thiophene

Background

In 2010 the Nobel Prize in Chemistry was awarded to Richard F. Heck, Ei-ichi Negishi and Akira Suzuki for the development of palladium-catalyzed cross coupling reactions. Few reactions have contributed to enhancing the efficiency of organic synthesis as much as the palladium-catalyzed cross-coupling reactions. This chemical tool has vastly improved the possibilities for chemists to create sophisticated chemicals, for example carbon-based molecules as complex as those created by nature itself. To create these complex molecules, chemists

Comprehensive Organic Chemistry Experiments for the Laboratory Classroom
Edited by Carlos A. M. Afonso, Nuno R. Candeias, Dulce Pereira Simão, Alexandre F. Trindade, Jaime A. S. Coelho, Bin Tan, and Robert Franzén.

Published by the Royal Society of Chemistry, www.rsc.org

need to be able to join carbon atoms together. However, carbon is stable and carbon atoms do not easily react with one another. The first methods used by chemists to bind carbon atoms together were therefore based upon various techniques for rendering carbon more reactive. Such methods worked when creating simple molecules, but when synthesizing more complex molecules chemists ended up with too many unwanted by-products in their reaction mixtures. Palladium-catalyzed cross coupling solved that problem and provided chemists with a more precise and efficient tool to work with. In the Heck reaction, Negishi reaction and Suzuki reaction, carbon atoms meet on a palladium atom, whereupon their proximity to one another kick-starts the chemical reaction. Palladium-catalyzed cross coupling is used in research worldwide, as well as in the commercial production of, for example, pharmaceuticals and molecules used in the electronics industry.

The palladium-catalyzed cross-coupling reaction between organoboron compounds and organic halides or triflates provides a powerful and general methodology for the formation of carbon–carbon bonds. Recently, this reaction has been called the Suzuki coupling, Suzuki reaction, or Suzuki–Miyaura (SM) coupling (Scheme 7.2.1). The availability of the reagents and the mild reaction conditions contribute to the versatility of this reaction. The coupling reaction offers several additional advantages, such as being largely unaffected by the presence of water, tolerating a broad range of functional groups, and proceeding generally regio- and stereoselectively. Moreover, the inorganic by-product of the reaction is non-toxic and easily removed from the reaction mixture thereby making the Suzuki coupling suitable not only for laboratories but also for industrial processes.[1–3]

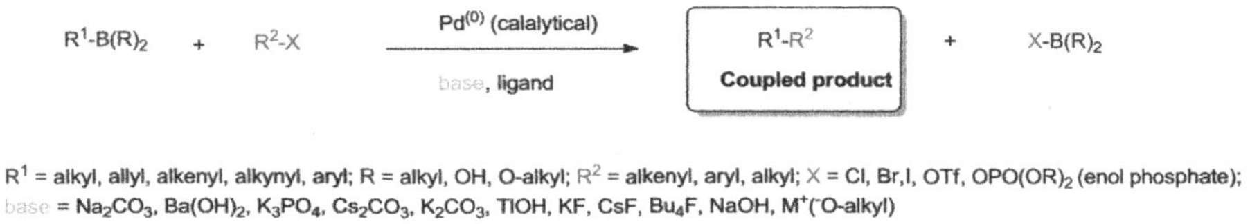

Scheme 7.2.1 Suzuki–Miyaura cross-coupling. Reprinted from Strategic Applications of Named Reactions in Organic Synthesis, L. Kurti and B. Czako, p. 448. Copyright © 2005, Elsevier Inc. All rights reserved.[3]

The mechanism of the Suzuki–Miyaura cross-coupling is analogous to the catalytic cycle for the other cross-coupling reactions. The oxidation reaction occurs with formally electrophilic organic groups, whereby palladium becomes oxidized through its donation of electrons to form the new Pd–C bond. However, transmetalation occurs with formally nucleophilic organic groups, which are transferred from boron to palladium (see more complete information about the mechanism in the ESI).[3]

Recently, a large variety of arylthiophene-2-carboxaldehydes have been synthesized in high yield using the Suzuki–Miyaura coupling.[4] Thiophene derivatives have attracted great interest due to their wide range of device applications such as light emitting devices, dye sensitized organic solar cells, chemosensors, and nonlinear optical materials.[5] In view of these attractive characteristics, it is highly desirable to synthesize thiophene derivatives with different (hetero) aryl groups in order to tailor the thermal and photophysical properties.

The aim of this work is the synthesis of a formyl-arylthiophene derivative, using simple experimental techniques and commercially available reagents. The synthesis of this compound involves a palladium catalysed Suzuki–Miyaura cross-coupling, between a (hetero)arylboronic acid derivative and an (hetero)aryl-brominated compound using DME as solvent and Na_2CO_3 as base. Different coupling components will be used in order to determine the influence of the structure of the boronic acid as well as of the brominated compound on the yield of the Suzuki coupling reaction. The determination of the structure of the compound synthesized will be also performed through ^{1}H and ^{13}C NMR spectroscopy.

Additional Safety

The experimental work should be performed in a well-ventilated fume hood with proper safety equipment such as protective clothing, gloves, and safety goggles.

1,2-Dimethoxyethane, petroleum ether (40–60 °C) and diethyl ether are highly flammable and may form explosive peroxides. The vapours may cause dizziness. Ethyl acetate may cause eye irritation. 5-Bromo-2-thiophenecarboxaldehyde, 5-formyl-2-thiopheneboronic acid and phenylboronic acid may cause skin, eye and respiratory tract irritation. Bromobenzene is flammable and irritating to skin. Silica gel may cause respiratory irritation, do not breathe the dust. Complete information about the safety, toxicity and disposal for all reagents can be found, free of charge, at: http://www.msds.com/index.asp?open=/protected_public/loginsuccessful.asp.

Experimental Procedure

Laboratory Session 1 (2 h): Synthesis of 5-Phenylthiophene-2-carbaldehyde

The procedure is the same for both pairs of reagents.

Half of the students will do the experiment by route A and the other half will perform the experiment by route B.

Synthetic route A: 5-formyl-2-thiopheneboronic acid (1.2 mmol, 1.2 equiv., 187 mg) and bromobenzene (1.0 mmol, 1.0 eq., 0.11 mL).

Synthetic route B: phenylboronic acid (1.2 mmol, 1.2 equiv., 146 mg) and 5-bromo-2-thiophenecarboxaldehyde (1.0 mmol, 1.0 equiv., 0.12 mL)

1. Prepare an apparatus for stirring under an inert atmosphere consisting of a 100 mL three-neck round-bottomed flask with a magnetic stirrer, a water condenser with an adapter with calcium chloride filling and two rubber septa.
2. Add to the flask the brominated compound (1 mmol, 1 equiv.), 20 mL of DME, and $Pd(PPh_3)_4$ (58 mg, 0.05 mmol, 0.05 equiv.). Fill the round-bottomed flask with an inert gas such as argon or nitrogen and stir the solution under inert atmosphere for 10 min at room temperature.
3. Add through one of the septa a solution of the boronic acid (1.2 mmol, 1.2 equiv.) dissolved in the minimum volume of ethanol, and aqueous 2 N Na_2CO_3 (2 mL). Heat the reaction mixture at 80 °C for 24 h with stirring in an oil bath. Allow the flask to cool to room temperature and save for the next session with the flask stoppered.

Laboratory Session 2 (2 h): Isolation of 5-Phenylthiophene-2-carbaldehyde

1. Filter the reaction mixture obtained in the previous session under vacuum through a sintered glass disc funnel.
2. Add 15 mL of saturated aqueous NaCl to the reaction mixture and extract with ethyl acetate (3 × 10 mL).
3. Combine the organic extracts and wash separately with water (15 mL) and 15 mL of aqueous solution of 10% NaOH.
4. Dry the organic layer over anhydrous $MgSO_4$, filter and evaporate the solvent under reduced pressure with a rotary evaporator.

5. Check the crude composition by TLC in comparison with the reagents (eluent: petroleum ether (40–60 °C)/diethyl ether, 1:1).
6. Keep the flask with the crude in a vacuum excicator till the next session.

Laboratory Session 3 (4 h): Purification of 5-Phenylthiophene-2-carbaldehyde

1. Purify the previously obtained crude by column chromatography on silica gel (column height and diameter: 40 and 2.5 cm, respectively; silica weight: 35 g; silica height: ≈20 cm; column packing solvent: petroleum ether (40–60 °C)). Begin eluting with a mixture of petroleum ether (40–60 °C)/diethyl ether (9:1) (≈100 mL). Continue the elution with mixtures of petroleum ether (40–60 °C)/diethyl ether of increasing polarity until 1:1 (≈200 mL). Collect numbered fractions of about 25 mL each.
2. Check the various fractions collected by TLC using as eluent a mixture of dichloromethane and petroleum ether (40–60 °C) (1:1) and combine the fractions containing the product.
3. Evaporate the solvent to dryness under reduced pressure with a rotary evaporator.
4. Weigh the yellow solid and check the purity by ^{1}H and ^{13}C NMR spectroscopy and melting point.
5. Preparation of the NMR sample: take 15 mg of the product, dissolve it in 0.7 mL of $CDCl_3$ and transfer to a NMR tube. Acquire the ^{1}H and ^{13}C NMR spectra.
6. Compare the results with the literature data[4b] as well as with the spectra presented in Supporting material.

Waste Disposal

Petroleum ether, diethyl ether, 1,2-dimethoxyethane and ethyl acetate should be collected in a non-chlorinated waste collector recipient. Used silica gel must be stored in a closed container in the hood for subsequent disposal. The filtration solid containing palladium should be collected in a palladium waste container. The aqueous solutions should be neutralized and disposed of in the sink.

Results Interpretation and Additional Questions

1. Interpret the ^{1}H and ^{13}C NMR spectra obtained, and compare the obtained melting point with the literature data.[4b]
2. Why should the Suzuki–Miyaura reaction be carried out in an inert atmosphere?
3. Explain briefly the function of $Pd(PPh_3)_4$ and Na_2CO_3 used in the Suzuki cross-coupling reaction.
4. (a) Suggest other bases and solvents that could be used in the Suzuki–Miyaura reaction.
   (b) In which cases could aryl chlorides be used as the electrophilic coupling components in Suzuki–Miyaura cross coupling reactions?
5. Write the Suzuki cross-coupling reaction mechanism as well as the structure of the by-products during transmetalation and reductive elimination.
6. In which of the synthetic routes was the best yield obtained? Explain why based on the different reactivity of reagents.

References

1. (a) N. Miyaura and A. Suzuki, *Chem. Rev.*, 1995, **95**, 2457; (b) A. Suzuki, *J. Organomet. Chem.*, 1999, **576**, 147; (c) F. Bellina, A. Carpita and R. Rossi, *Synthesis*, 2004, 2419; (d) H. Doucet, *Eur. J. Org. Chem.*, 2008, 2013.
2. A. J. J. Lennox and G. C. Lloyd-Jones, *Chem. Soc. Rev.*, 2014, **43**, 412.

3. L. Kurti and B. Czako, *Strategic Applications of Named Reactions in Organic Synthesis*, Elsevier, San Diego, USA, 2005, p. 448.
4. (a) S. Ali, N. Rasool, A. Ullah, F.-U.-H. Nasim, A. Yaqoob, M. Zubair, U. Rashid and M. Riaz, *Molecules*, 2013, **18**, 14711; (b) S. P. G Costa, R. M. F. Batista, P. Cardoso, M. Belsley and M. M. M. Raposo, *Eur. J. Org. Chem.*, 2006, **17**, 3938.
5. (a) C. Marin-Hernandez, L. E. Santos-Figueroa, M. E. Moragues, M. M. M. Raposo, R. M. F. Batista, S. P. G. Costa, T. Pardo, R. Martínez-Máñez and F. Sancenón, *J. Org. Chem.*, 2014, **75**, 2922; (b) R. M. F. Baptista, D. Isakov, M. M. M. Raposo, M. Belsley, I. Bdikin, A. L. Kholkin, S. P. G. Costa and E. de Matos Gomes, *J. Nanopart. Res.*, 2014, **16**, 2502; (c) R. M. F. Batista, E. Oliveira, S. P. G. Costa, C. Lodeiro and M. M. M. Raposo, *Tetrahedron*, 2011, **67**, 7106; (d) J. Pina, J. Seixas de Melo, R. M. F. Batista, S. P. G. Costa and M. M. M. Raposo, *J. Phys. Chem. B*, 2010, **114**, 4964; (e) M. M. M. Raposo, B. García-Acosta, T. Ábalos, P. Calero, R. Martínez-Manez, J. V. Ros-Lis and J. Soto, *J. Org. Chem.*, 2010, **75**, 2922; (f) R. M. F. Batista, S. P. G. Costa, M. Belsley and M. M. M. Raposo, *Dyes Pigm.*, 2009, **80**, 329; (g) J. Pina, J. Seixas de Melo, H. D. Burrows, R. M. F. Batista, S. P. G. Costa and M. M. M. Raposo, *J. Phys. Chem. A*, 2007, **111**, 8574.

7.3. Recycling Bromovanillin into Ferulic Acid-Based Antioxidants

Tiago Silva, Daniel Chavarria, Lisa Sequeira, and Fernanda Borges*

CIQUP/Department of Chemistry and Biochemistry, Faculty of Sciences, University of Porto, Rua do Campo Alegre s/n, 4169-007, Porto, Portugal. *E-mail: fborges@fc.up.pt

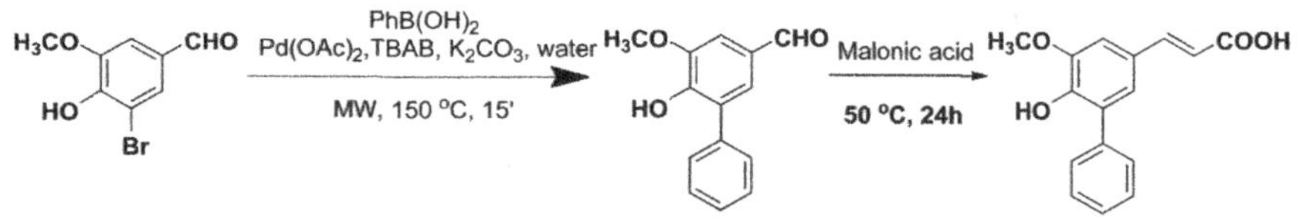

Duration	Hazard level	Difficulty level	Level of study
3 (4 h each)	Moderate	Medium	Intermediate

Class names Aldehydes, carboxylic acids, aryl bromides, boronic acids

Concepts involved Microwave-assisted synthesis, Suzuki–Miyaura C–C coupling, palladium catalysis, green chemistry, aldol condensation, purification, lipophilicity, structural analysis

Chemicals needed 5-Bromovanillin, phenylboronic acid, tetrabutylammonium bromide, potassium carbonate, palladium(II) acetate, malonic acid, piperidine, pyridine, hydrochloric acid, brine, anhydrous sodium sulfate, sodium bicarbonate

Equipment and experimental techniques involved Scale, stirring and heating plate, microwave reactor, UV lamp (254 and 366 nm), rotary evaporator, NMR spectrometer; filtration under reduced pressure, liquid–liquid extraction, thin-layer chromatography (TLC), flash column chromatography, crystallization, NMR spectroscopy

Keywords Aldol condensation, antioxidants, flash chromatography, hydroxycinnamic acids, NMR, Suzuki–Miyaura cross coupling

Background

The cascade of oxidative damage is behind several processes with economic and social impact, from the deterioration of foods to the development of complex human diseases.[1,2] Hence, the development of novel antioxidants based on natural compounds is a valid strategy for the discovery of operative additives with potential application in food, cosmetic and pharmaceutical industries.[3–5] An antioxidant is a substance that, when in presence of an oxidizable substrate, is able to prevent or delay the oxidation of the latter. Vanillin (4-hydroxy-3-methoxybenzaldehyde) is an aromatic aldehyde, responsible for the characteristic odor of vanilla and is the main component of the extract of the vanilla bean.[6] Vanillin is a valuable industrial aromatic compound involved in flavoring, preservatives, pharmaceuticals, and fragrances.

Although vanillin itself has no noticeable antioxidant activity, it can be used as a precursor for the synthesis of other antioxidants such as hydroxycinnamic acids, a class of naturally occurring effective antioxidants. In fact, vanillin is a valuable building block for the design, and obtention of novel

Comprehensive Organic Chemistry Experiments for the Laboratory Classroom
Edited by Carlos A. M. Afonso, Nuno R. Candeias, Dulce Pereira Simão, Alexandre F. Trindade, Jaime A. S. Coelho, Bin Tan, and Robert Franzén.

Published by the Royal Society of Chemistry, www.rsc.org

antioxidants with enhanced physicochemical and activity properties, which may ultimately lead to the development of antioxidants of commercial interest.[6]

In this particular case, the synthetic strategy uses a vanillin derivative and was specifically designed to obtain ferulic acid analogues. Ferulic acid (*trans*-4-hydroxy-3-methoxycinnamic acid) is ubiquitous in plants and is well-known due to its antioxidant properties and sun-protective benefits to the skin.[7] However, ferulic acid has only limited solubility in hydrophobic media, which reduces its application as antioxidant.[7,8] The introduction of an aromatic ring at C5 is a suitable chemical modification in order to increase its lipophilicity.[8] To achieve this goal, we propose a two-step synthetic strategy that enfolds key aspects of modern and classic organic synthesis. The first involves a microwave-assisted organic synthesis (MAOS), an ever-expanding field within organic synthesis. When compared to classic heating methods, MAOS has been shown to dramatically reduce reaction times, increase ease of reaction work-up, improve product yields and reduce unwanted side reactions.[9] Modern microwave reactors enable accurate monitoring of temperature and pressure under sealed vessel and safe conditions. Microwave chemistry is based on the efficient heating of materials by microwave dielectric heating effects and depends on the ability of these materials to absorb microwave energy and convert it into heat. Major heating mechanisms induced by microwave irradiation are dipolar polarization and ionic conduction, which promote molecular friction, collisions and motion of the free molecules within the reaction vessel.[9] In accordance, a microwave irradiation technique is used in the first step of the proposed synthetic strategy (Scheme 7.3.1), which consists of a Suzuki–Miyaura cross coupling in aqueous media between a halogenated vanillin derivative (5-bromovanillin) and a boronic acid with the appropriate substitution pattern (phenylboronic acid). To simplify the synthetic pathway our experiments were designed starting with commercially available 5-bromovanillin instead of the naturally occurring vanillin. However, this substrate can be easily prepared by bromination of vanillin in glacial acetic acid.[10–12]

The reaction product results from the formation of a C–C bond between the halogenated position of 5-bromovanillin and the phenyl group of the boronic acid (Scheme 7.3.2). The new C–C

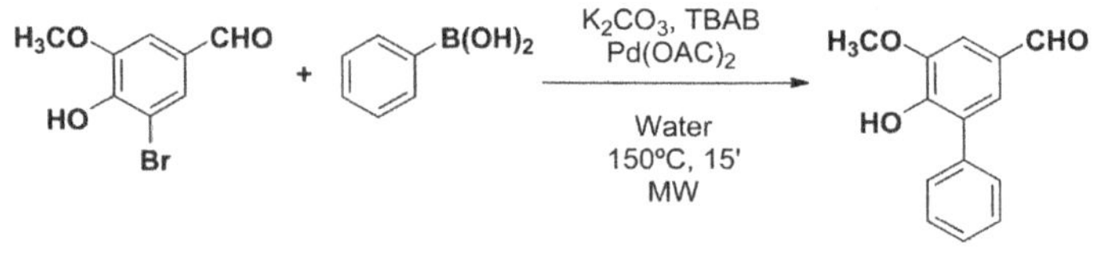

Scheme 7.3.1 Synthesis of 5-phenylvallin from 5-bromovanillin.

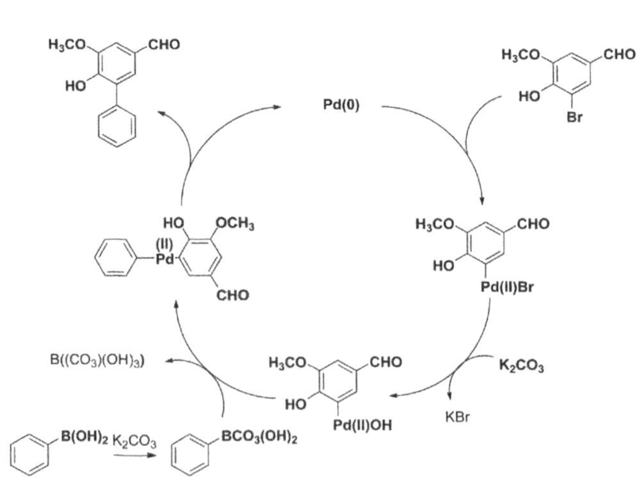

Scheme 7.3.2 Mechanism of Suzuki–Miyaura cross coupling reaction.

bond is formed in a Pd-catalyzed cycle involving the oxidative addition of the aldehyde to the Pd complex, followed by a transmetalation with the boronic acid and a reductive elimination of Pd.[13] The reaction is conducted in water, a cheap and non-toxic solvent. Tetrabutylammonium bromide (TBAB) was added to the reaction, acting as a phase-transfer catalyst. In this particular case, the use of TBAB facilitates solvation of the organic substrates in the solvent medium and enhances the rate of the coupling reaction by activating the boronic acid by formation of $[PhB(OH)_3]^{-}[R_4N]^{+}$.[14] The reaction can be performed with diverse arylboronic acids containing different electron donating and withdrawing groups, allowing an additional laboratory class with the study of their influence in the physicochemical/antioxidant properties. The versatility of the systems will certainly create a dynamic laboratory class.

The following step involves the transformation of the aromatic aldehyde into an α,β-unsaturated carboxylic acid, which can be realized *via* a Knoevenagel–Döbner condensation.[8,13] Therefore, the nucleophilic addition of malonic acid to the carbonyl group of 5-phenylvanillin was carried out in the presence of pyridine/piperidine yielding the corresponding ferulic acid derivative (Scheme 7.3.3).

This reaction occurs *via* a classic aldol mechanism that involves the initial formation of an enol intermediate in the presence of base followed by the attack of the enol onto the carbonyl. Lastly, the Döbner rearrangement occurs provoking the decarboxylation of the reaction intermediate (Scheme 7.3.4).

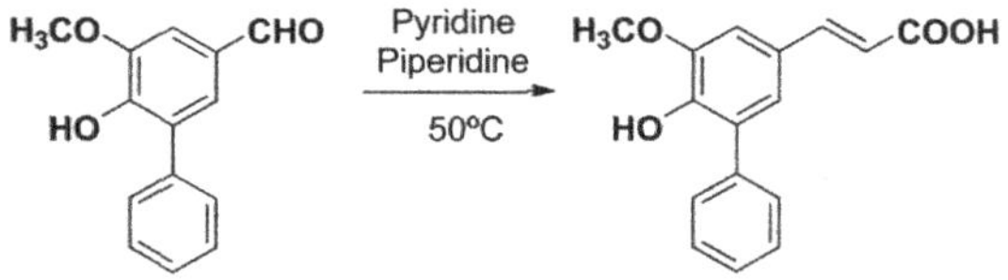

Scheme 7.3.3 Synthesis of 5-phenylferulic acid from 5-phenylvanillin.

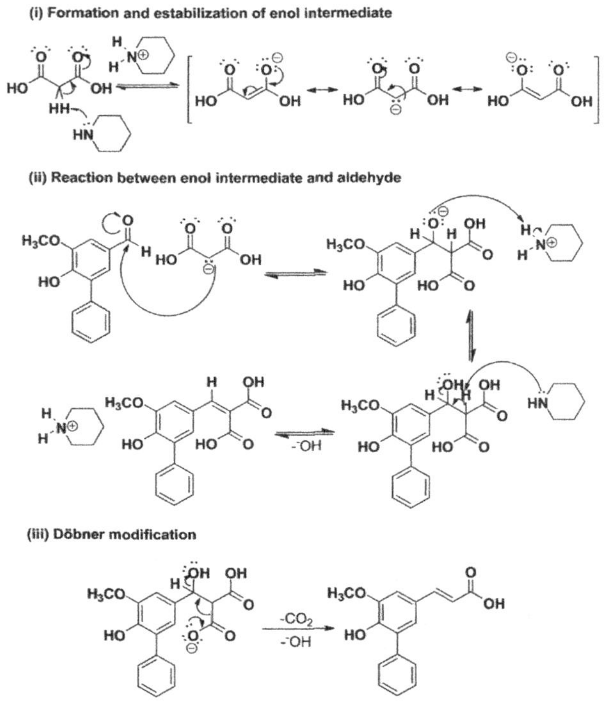

Scheme 7.3.4 Mechanism of Knoevenagel–Döbner condensation.

Additional Safety

Special precaution concerning the handling of aromatic amines, such as pyridine and piperidine, must be taken. Work inside the fume hood and wear appropriate protective eyeglasses and other personal protective equipment, like protective gloves and clothing. Throughout the experiments the operator must stay away from ignition sources. Further reading of the toxicological and hazard profiles of the handled chemicals by consulting the Merck Index and Material Safety Data Sheets (MSDS) is strongly advised.

Experimental Procedure

Session 1 (4 h): Synthesis, Isolation and Characterization of 5-Phenylvanillin

Synthesis

1. To an appropriate microwave vial add 5-bromovanillin (0.50 g, 2.16 mmol), phenylboronic acid (0.26 g, 2.16 mmol), TBAB (0.70 g, 2.16 mmol), K_2CO_3 (0.90 g, 6.49 mmol), palladium acetate (0.002 g, 8.901×10^{-3} mmol), water (15 mL) and a magnetic bar.
2. Place the vial in the microwave reactor and set temperature to 150 °C for 15 min.
3. Control the reaction progress by TLC (dichloromethane/methanol, 95 : 5 (v/v)).

Work-Up

4. Add Celite® to a filtering plate and filter the reaction mixture to discard the palladium residues.
5. Rinse with water (3 × 10 mL) and ethyl acetate (3 × 10 mL). Discard the aqueous layer and wash the organic layer with 1 M HCl (3 × 10 mL), water (3 × 10 mL) and brine (10 mL).
6. Dry the organic layer over anhydrous sodium sulfate, filter to a 100 mL round-bottom flask and evaporate.
7. Recrystallize the product from ethyl acetate/*n*-hexane.
8. Prepare a NMR sample in $CDCl_3$ or DMSO-d_6.

Session 2 (4 h): Purification and Characterization of 5-Phenylvanillin (Continuation of Session 1, If Needed) and Synthesis of 5-Phenylferulic Acid

Synthesis

1. In a round-bottom flask add 5-phenylvanillin (0.25 g, 1.09 mmol), malonic acid (0.14 g, 1.31 mmol), pyridine (5 mL, 60.82 mmol), piperidine (4 drops) and a magnetic bar.
2. Heat the mixture at 50 °C for 24 h.
3. Control the reaction progress by TLC (dichloromethane/methanol/formic acid, 90 : 10 : 0.1 (v/v)).

Session 3 (4 h): Isolation and Characterization of 5-Phenylferulic Acid

Work-Up

1. Remove the flask from the heating plate and cool it on an ice bath.
2. Add water (20 mL), neutralize with 1 M HCl and keep the mixture on ice with agitation for 10 min.
3. Extract with ethyl acetate (3 × 10 mL) and discard the aqueous layer. Extract with 1 M $NaHCO_3$ (3 × 10 mL) and discard the organic layer.
4. Treat the combined aqueous layers with 1 M HCl until acid pH and extract with ethyl acetate (3 × 10 mL). Wash the organic layer with water (10 mL) and brine (2 × 10 mL). Discard the aqueous layer, gather and dry the organic layers with sodium anhydrous sulfate.

5. Filter to a round-bottom flask, remove the solvents with reduced pressure and recrystallize the product in dichloromethane/*n*-hexane.
6. Prepare an NMR sample in DMSO-d_6.

Waste Disposal

All organic and aqueous wastes should be placed in the appropriate waste container. The drying agent and Celite must be placed in a solid waste container.

Results Interpretation and Additional Questions

1. Determine the reaction yields.
2. Analyze the NMR spectra and assign the peaks of 5-phenylvanilin and 5-phenylferulic acid.
3. Interpret the NMR spectrum of 5-phenylferulic acid. Analyzing the coupling constants of protons α and β, determine which of the isomers is formed: (*E*)-5-phenylferulic acid or (*Z*)-5-phenylferulic acid.
4. Use the appropriate software (*e.g.* ChemDraw®) to estimate the partition coefficients of ferulic acid and of the final product. Considering the goal of the experiment, compare the values obtained and discuss the final results.

References

1. N. A. Simonian and J. T. Coyle, Oxidative stress in neurodegenerative diseases, *Annu. Rev. Pharmacol. Toxicol.*, 1996, **36**, 83–106.
2. S. Reuter, S. C. Gupta, M. M. Chaturvedi and B. B. Aggarwal, Oxidative stress, inflammation, and cancer: how are they linked?, *Free Radical Biol. Med.*, 2010, **49**(11), 1603–1616.
3. J. Garrido, E. M. Garrido and F. Borges, Studies on the Food Additive Propyl Gallate: Synthesis, Structural Characterization, and Evaluation of the Antioxidant Activity, *J. Chem. Educ.*, 2012, **89**, 130–133.
4. T. Silva, C. Oliveira and F. Borges, Caffeic acid derivatives, analogs and applications: a patent review (2009-2013), *Expert Opin. Ther. Pat.*, 2014, **24**(11), 1257–1270.
5. S. Benfeito, C. Oliveira, P. Soares, C. Fernandes, T. Silva, J. Teixeira and F. Borges, Antioxidant therapy: still in search of the 'magic bullet', *Mitochondrion*, 2013, **13**(5), 427–435.
6. K. Anuradha, B. N. Shyamala and M. M. Naidu, Vanilla-its science of cultivation, curing, chemistry, and nutraceutical properties, *Crit. Rev. Food Sci. Nutr*, 2013, **53**(12), 1250–1276.
7. E. Graf, Antioxidant potential of ferulic acid, *Free Radical Biol. Med.*, 1992, **13**(4), 435–448.
8. T. Silva, J. Bravo, T. Summavielle, F. Remião, C. Pérez, C. Gil, A. Martínez and F. Borges, Biology-oriented development of novel lipophilic antioxidants with neuroprotective activity, *RSC Adv.*, 2015, **5**, 15800–15811.
9. C. O. Kappe, D. Dallinger and S. S. Murphree, *Practical Microwave Synthesis for Organic Chemists. Strategies, Instruments and Protocols*, 2009, ch. 1 and 2, John Wiley & Sons.
10. H. V. Ryswyk, A general chemistry experiment incorporating synthesis and structural determination, *J. Chem. Educ.*, 1997, **74**(7), 842–844.
11. S. I. McGowens and E. F. Silversmith, Organic Reactions Involving Bromine, *J. Chem. Educ.*, 1998, **75**(19), 1293–1294.
12. D. F. Taber, S. Patel, T. M. Hambleton and E. E. Winkel, Vanillin Synthesis from 4-Hydroxybenzaldehyde, *J. Chem. Educ.*, 2007, **84**(7), 1158.
13. R. K. Parashar, *Reaction mechanisms in organic synthesis*, 2009, ch. 5, John Wiley & Sons.
14. N. E. Leadbeater and M. Marco, Rapid and amenable Suzuki cross coupling reaction in water using microwave and conventional heating, *J. Org. Chem.*, 2003, **68**(3), 888–892.

7.4. Sonogashira Coupling Reaction of Aryl Derivatives: A Versatile Method for Acetylide Building Blocks

Tiago J. Silva*[a], Paulo J. Mendes[b], António P. S. Teixeira[b], M. Paula Robalo[c,d], and M. H. Garcia[a]

[a]Centro de Ciências Moleculares de Materiais, Faculdade de Ciências da Universidade de Lisboa, Portugal; [b]Departamento de Química, Escola de Ciências e Tecnologia, Centro de Química de Évora, Instituto de Investigação e Formação Avançada, Universidade de Évora, Évora, Portugal; [c]Área Departamental de Engenharia Química, Instituto Superior de Engenharia de Lisboa, Portugal; [d]Centro de Química Estrutural, Complexo I, Instituto Superior Técnico, Universidade de Lisboa, Portugal. *E-mail: tjsilva@fc.ul.pt

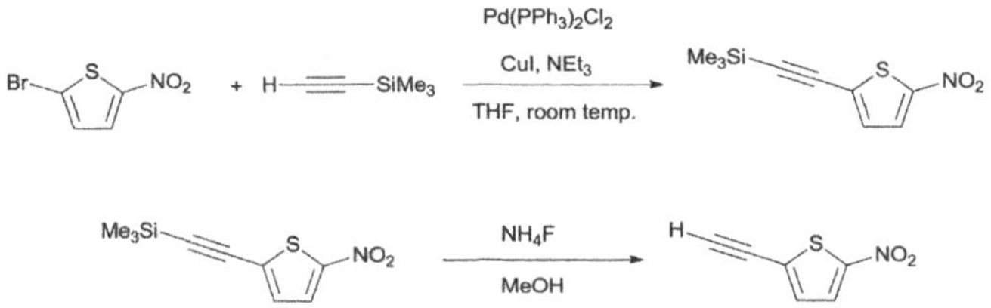

Number of sessions (duration of each session)	Hazard level	Difficulty level	Level of study
2 (5 h + 3 h)	Moderate	High	Intermediate

Class names Aryl halides, terminal alkynes, alkylsilanes

Concepts involved Palladium catalysis, cross coupling reaction, deprotection of alkynylsilanes

Chemicals needed 5-Bromo-2-nitrothiophene, triethylamine, copper(I) iodide, ammonium fluoride, bis(triphenylphosphine)palladium(II) dichloride, ethynyl(trimethyl)silane, tetrahydrofuran, *n*-hexane, diethyl ether, methanol, silica gel, anhydrous sodium sulfate, saturated sodium bicarbonate solution, 5% hydrochloric acid solution

Equipment and experimental techniques involved Liquid–liquid extraction, column chromatography, NMR, heating plate with magnetic stirring, vacuum/nitrogen line, rotatory evaporation apparatus

Keywords Chromatography, cross coupling, heterocyclic chemistry, NMR, palladium catalysis, terminal alkynes

Background

The coupling of aryl or vinyl halides with terminal alkynes catalysed by palladium, commonly known as the Sonogashira[1] cross-coupling reaction, is one of the most important and widely used sp^2–sp carbon–carbon bond formation reactions in organic synthesis. The obtained products have frequently found applicability in several areas of chemistry, such as natural products, biologically active molecules, dyes, molecular electronics, dendrimers and conjugated polymers. The reaction name arises from the discovery in 1975 by Sonogashira *et al.* that this process could be performed easily at room temperature using a palladium catalyst, combined with a co-catalytic amount of CuI in an amine as solvent (Scheme 7.4.1).[1]

Comprehensive Organic Chemistry Experiments for the Laboratory Classroom
Edited by Carlos A. M. Afonso, Nuno R. Candeias, Dulce Pereira Simão, Alexandre F. Trindade, Jaime A. S. Coelho, Bin Tan, and Robert Franzén.

Published by the Royal Society of Chemistry, www.rsc.org

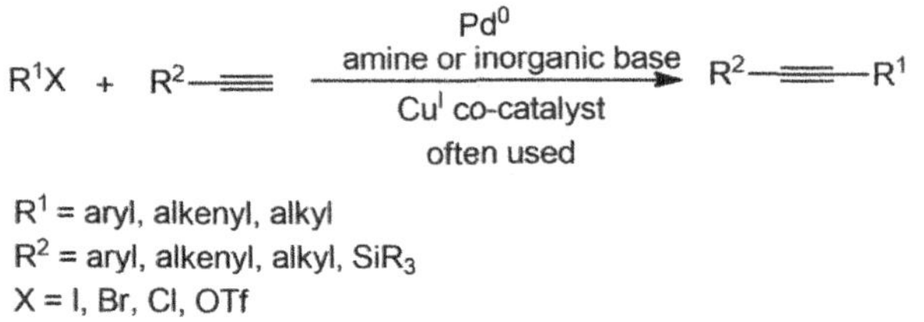

Scheme 7.4.1 Sonogashira coupling reaction.

Since the original work of Sonogashira, this palladium-catalysed reaction has been extensively investigated to improve the reaction conditions and a wide variety of modifications have been reported thus far. This includes the use of a variety of additives, solvents, copper-free and biphasic versions, phase transfer catalysts and polymer/solid-supported Pd complexes. In addition to the copper salts, the use of other metal co-catalysts such as zinc, tin, boron and aluminium has also been reported as well as the use of microwave-mediated transition metal free coupling. These several aspects, as well as mechanism insights, have been extensively reviewed.[2–6]

The mechanism of the Sonogashira reaction is not understood in detail, mainly due to the difficulties of analysing the combined action of the two metal catalysts present. Mechanistic variations have been proposed depending on the cross-coupling methodology used, for instance when dealing with different catalysts, reaction conditions and procedures, in particular for copper-co-catalysed and copper-free versions. The first catalytic cycle was proposed by the Sonogashira group in its original paper.[1] Later representations of the Sonogashira mechanism (copper-co-catalysed version) included two independent cycles for catalysis: one for the production of the cuprous acetylides (the so-called "copper-cycle") and the other for their coupling with the desired aryl species (the "palladium-cycle").[3] The palladium catalysis (Pd-cycle) is typical for C–C cross-coupling formations[7] and follows an oxidative addition–transmetalation–*trans*/*cis* isomerization–reductive elimination mechanism. The copper catalysis (Cu-cycle) is still poorly understood. It is supposed that the base (*e.g.*, an amine) abstracts the acetylenic proton of the terminal alkyne, thus forming a copper acetylide in the presence of the copper(I) salt. However, the generally employed amines are usually not basic enough to deprotonate the alkyne in order to generate the anionic nucleophile that should form the copper acetylide. Therefore, a π-alkyne Cu complex could be involved in the cycle, thus making the alkyne proton more acidic for easier abstraction.[8]

The catalytically active species, Pd^0L_2, are generated *in situ* from the initial palladium complex and can be of colloidal nature and/or stabilised by other ligands present, including the base used and/or solvent molecules. These Pd^0L_2 complexes can be formed from Pd^0 catalysts such as $Pd(PPh_3)_4$, or can be generated from the reduction of more soluble and air stable Pd^{II} complexes, namely $Pd(OAc)_2$, $Pd(PPh_3)_2Cl_2$, $Pd(k^2\text{-dppe})Cl_2$, $Pd(k^2\text{-dppp})Cl_2$ or $Pd(k^2\text{-dppf})Cl_2$ through a σ-complexation–dehydropalladation–reductive elimination mechanism.[9,10]

Organosilicon reagents have several applications in organic chemistry, namely as protecting agents for amines, carboxylic acids, amides, alcohols and thiols and alkynes. For instance, the trimethylsilyl group is widely used for protection of alcohols and alkynes. Protection/deprotection conditions can be fine-tuned by the presence of different substituents at the silicon atom. Deprotection of trimethylsilylacetylene, used in this experiment, is easily made by fluorinated compounds like ammonium fluoride, tetrabutylammonium fluoride (TBAF) or potassium fluoride in methanol.[11]

The objective of this work is to obtain trimethyl[(5-nitrothiophen-2-yl)ethynyl]silane as an example of Sonogashira coupling reaction between a terminal alkyne (ethynyl(trimethyl)silane) and an aryl compound (2-bromo-5-nitrothiophene) in a first step. The protected alkyne is then deprotected by the use of ammonium fluoride to afford 2-ethynyl-5-nitrothiophene in excellent yield. The experiment was adapted from Tiago J. L. Silva *et al.*[12] and can be extended to several other easily commercially accessible bromo and iodo thienyls such as 2-iodo-5-nitrothiophene, 5-bromo (or iodo) thiophene-2-carbaldehyde, 5-bromo (or iodo) thiophene-2-carbonitrile or the corresponding 1,4-benzene derivatives. This experiment constitutes a good introduction to basic organometallic reactions involved in Pd catalysed organic synthesis, namely oxidative addition, reductive elimination and transmetalation, as well as to the use of dry solvents and inert atmosphere conditions.

Additional Safety

All hazardous materials should be handled using safety standard procedures. The sessions should be done in a fume hood, with proper safety equipment, such as protective clothing, gloves and safety goggles.

Experimental Procedure

To execute this experiment a vacuum line and previous know-how on working under nitrogen atmosphere using standard *Schlenk* techniques are needed. THF, dichloromethane and triethylamine should be previously dried using standard methods.[13]

Reaction 1 (5 h) – Synthesis of Trimethyl[(5-nitrothiophen-2-yl)ethynyl]silane

1. Prepare a 100 mL three-necked round-bottomed flask with a magnetic stir bar and add 1.0 g (5 mmol) of 5-bromo-2-nitrothiophene followed by 70 mg of $Pd(PPh_3)_2Cl_2$ (0.1 mmol, 0.02 equiv.) and 19 mg of copper(I) iodide (0.1 mol, 0.02 equiv.).
2. Degas the flask and add 10 mL of freshly distilled and degassed THF followed by 1.65 mL of dried triethylamine (8 mmol, 1.6 equiv.). Stir the mixture for 15 min and place the flask in a cold water bath (15–20 °C) that can be prepared by adding a small amount of crushed ice into a water recipient.
3. Prepare an addition funnel charged with a solution of 0.55 g of ethynyl(trimethyl)silane dissolved in 10 mL of freshly dried and degassed THF.
4. Adapt the addition funnel into the flask and add the ethynyl(trimethyl)silane solution dropwise (1 drop every 2–3 s). Stir the solution for 2 h.
5. Filter the resulting suspension and wash the solid with 10 mL of dichloromethane. Transfer the combined THF and dichloromethane into a flask and evaporate in a rotatory evaporator.
6. Dissolve the resulting dark solid in 20 mL of dichloromethane and transfer the solution into a separatory funnel. Wash the organic phase twice, in the following order: a 5% hydrochloric acid solution, a saturated sodium bicarbonate solution, and finally water.
7. Dry the organic phase with anhydrous sodium sulfate, and filter to a flask. Evaporate the solvent in the rotatory evaporator.
8. Fill a chromatography column (3 cm diameter × 20 cm height) with a slurry of silica gel in hexane. Chromatograph the obtained solid using a hexane/diethyl ether (9 : 1) mixture. In test tubes, collect fractions of the eluent (approx. 10 mL each) and analyze them by TLC.
9. Evaporate the solvent from the combined collected fractions that present the first product (corresponding to the first yellow band). Weigh the flask containing the brownish-yellow powder.

Reaction 2 (3 h) – Synthesis of 2-Ethynyl-5-nitrothiophene

1. Prepare a 25 mL round-bottomed flask with a magnetic stir bar and add 0.34 g (1.5 mmol) of trimethyl[(5-nitrothiophen-2-yl)ethynyl]silane. Add 10 mL of methanol and stir the solution for 5 min.

2. Add 61 mg (1.7 mmol, 1.13 equiv.) of ammonium fluoride and stir the mixture at room temperature for 2 h.
3. Evaporate the solvent in the rotatory evaporator.
4. Extract the product by adding portions of 15 mL of *n*-hexane, filter to a flask and evaporate the solvent in the rotatory evaporator. Weigh the flask containing the yellow solid.

Waste Disposal

All chemical waste should be disposed of in accordance with local regulations and laboratory practices. If there is no set of rules in the laboratory, put the non-halogenated, halogenated and metal waste solutions in appropriate container recipients for further waste disposal treatment.

Results Interpretation and Additional Questions

1. Calculate the yield of each reaction and the global yield.
2. Acquire and interpret the NMR spectra (^{1}H, ^{13}C NMR) of the final product by inserting 20 mg of the compound in a vial. Add the solvent (500 μL of $CDCl_3$) and transfer the sample to an NMR tube. Record the ^{1}H and ^{13}C spectra. Make the correct assignments for all the protons and carbons of both products and compare their differences.
3. Acquire the IR spectra for both protected and deprotected product. Assign the principal vibrational modes.
4. The use of iodine derivatives should promote the cross-coupling reaction. Explain why.

References

1. K. Sonogashira, Y. Tohda and N. Hagihara, *Tetrahedron Lett.*, 1975, **16**, 4467.
2. H. Doucet and J.-C. Hierso, *Angew. Chem., Int. Ed.*, 2007, **46**, 834.
3. R. Chinchilla and C. Nájera, *Chem. Rev.*, 2007, **107**, 874.
4. R. Chinchilla and C. Najera, *Chem. Soc. Rev.*, 2011, **40**, 5084.
5. M. Schilz and H. Plenio, *J. Org. Chem.*, 2012, **77**, 2798.
6. M. Bakherad, *Appl. Organomet. Chem.*, 2013, **27**, 125.
7. L. Xue and Z. Lin, *Chem. Soc. Rev.*, 2010, **39**, 1692.
8. P. Bertus, F. Fecourt, C. Bauder and P. Pale, *New J. Chem.*, 2004, **28**, 12.
9. E. Negishi and A. de Meijere, *Handbook of Organopalladium Chemistry for Organic Synthesis: Volume 1*, John Wiley & Sons, 2003.
10. L. S. Hegedus and B. C. G. Söderberg, *Transition Metals in the Synthesis of Complex Organic Molecules*. University Science Books: 2010.
11. T. W. Greene and P. G. M. Wuts, *Greene's Protective Groups in Organic Chemistry*, John Wiley & Sons, 2006, ch. 8, pp. 927–933.
12. T. J. L. Silva, P. J. Mendes, A. M. Santos, M. H. Garcia, M. P. Robalo, J. P. P. Ramalho, A. J. P. Carvalho, M. Büchert, C. Wittenburg and J. Heck, *Organometallics*, 2014, **33**, 4655.
13. D. D. Perrin, W. L. F. Amarego and D. R. Perrin, *Purification of Laboratory Chemicals*, 2nd edn, Pergamon, New York, 1980.

7.5. Convergent Synthesis of a Suzuki Product

Thomas A. Logothetis

University of Southampton, Chemistry, Highfield, Southampton, Hampshire, SO17 1BJ, UK
*E-mail: thomas.logothetis@southampton.ac.uk

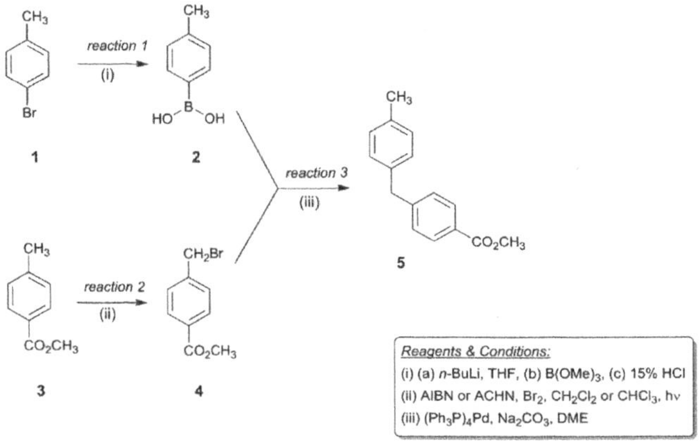

Number of sessions (duration of each session)	Hazard level	Difficulty level	Level of study
3 (6 h each)	High	High	Intermediate

Class names Benzyl bromides, bromobenzenes, lithiated benzenes, boronic acids

Concepts involved The main goal of this practical is to synthesise a diarylmethane product in three convergent steps. The final step is a Suzuki reaction – a palladium-catalysed cross-coupling reaction between an aryl boronic acid and a benzyl bromide. The latter is accessible *via* light-induced radical bromination of a toluene derivative whereas the former is made from bromotoluene in a sequence involving metal–halogen exchange, transmetalation and hydrolysis. Experimental concepts include column chromatography, safe handling and decontamination of toxic, corrosive, lachrymatory and pyrophoric reagents

Chemicals needed 4-Bromotoluene, methyl 4-toluate, *n*-butyllithium, trimethylborate, tetrahydrofuran, dichloromethane, ACHN/Vazo-88® or AIBN, bromine, 1,2-dimethoxyethane, sodium hydroxide, sodium carbonate, tetrakis(triphenylphosphine)palladium, pentane, heptane, diethyl ether, light petroleum ether (40–60 °C), toluene, hydrochloric acid, ethanol, acetone, dry ice, silica for flash chromatography (40–63 μm), magnesium sulfate, sodium hydrogen carbonate, sodium thiosulfate

Equipment and experimental techniques involved Two three-neck round-bottom flasks (100 mL, 250 mL), round-bottom flask (50 mL), septum, nitrogen inlet/bubbler, reflux condenser, heating bath, Dewar or cooling bath, pressure equalised dropping funnel, TLC chamber and silica TLC plates, UV lamp, permanganate dip, tweezers, beakers, Erlenmeyer flasks, separating funnel, halogen lamp, glass column (~50 cm, Ø 3 cm), bellows, Pasteur pipettes, magnetic stirrer hotplate, Büchner and Hirsch flasks and filters, Drechsel bottles, disposable needles and syringes (1–10 mL), test tubes in rack, glass rod, evaporation flask. Performing reactions in anhydrous solvents under inert gas atmosphere, using a cardice bath for reactions at low temperatures, safe handling of pyrophoric, toxic and corrosive reagents, light-induced reactions, recrystallisation, flash column chromatography, rotary evaporation, desiccation, vacuum filtration, monitoring by TLC, spectroscopic analysis

Comprehensive Organic Chemistry Experiments for the Laboratory Classroom
Edited by Carlos A. M. Afonso, Nuno R. Candeias, Dulce Pereira Simão, Alexandre F. Trindade, Jaime A. S. Coelho, Bin Tan, and Robert Franzén.

Published by the Royal Society of Chemistry, www.rsc.org

Keywords Boronic acid, flash column chromatography, metal–halogen exchange, NMR, palladium-catalysed cross-coupling, radical bromination, Suzuki reaction

Background

Most syntheses of complex molecules require many steps until the target is reached. The overall yield depends on the individual yield of each synthetic step but is also influenced by the pathway. In a linear arrangement of all steps the overall yield is much lower than in a partially or fully convergent route. Figure 7.5.1 displays both situations assuming that each individual reaction produces a yield of 80%. The overall yield of the linear synthesis is then only 21% whereas the partial convergent pathway affords a total yield of 51%. Therefore, the task of a skilled chemist is foremost to create a synthetic plan that consists of as few steps as absolutely necessary and to devise as convergent a route as possible.

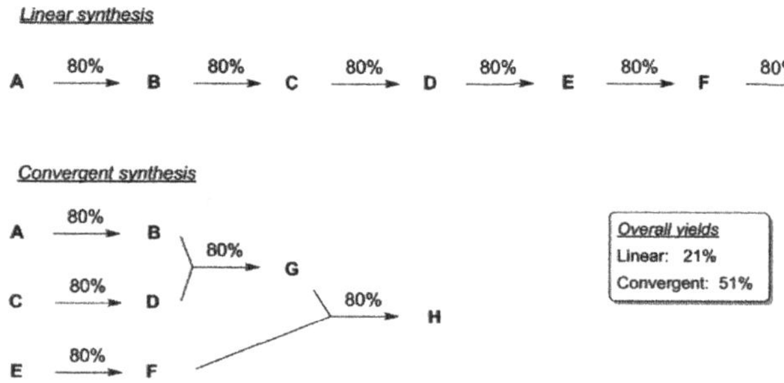

Figure 7.5.1 Yields in arbitrary convergent and linear pathways.

In this practical three steps are arranged in a convergent route that separately synthesises a boronic acid and a benzyl bromide derivative and then combines these two building blocks to the final product in the last convergent step.

The synthesis of *p*-tolylboronic acid (**2**) starts from 4-bromotoluene (**1**) with a halogen–metal exchange utilising *n*-butyllithium. The lithiated arene[1] is then subjected to a transmetalation reaction when treated with trimethylborate. This affords a boronic ester that can be hydrolysed to the corresponding boronic acid – a widely employed sequence to produce the starting materials for many different palladium-catalysed cross-coupling reactions.[2] The boronic acid can reversibly dehydrate and form an anhydride that easily cyclises to form a stable boroxin derivative[3] that also undergoes Suzuki coupling.

The synthesis of methyl 4-(bromomethyl)benzoate (**4**) starts with methyl 4-methylbenzoate (**3**). In a radical bromination[4] reaction the mono-brominated product is selectively produced when an excess of starting material is treated slowly with a diluted solution of an equimolar amount of bromine. The initiation can be achieved either by using a radical initiator and heat or simply with a light-induced homolysis of bromine. In the spirit of green chemistry more environmentally friendly variations in water have been developed that also omit the toxic and corrosive bromine altogether.[5]

The products **2** and **4** are then combined in a palladium-catalysed cross-coupling reaction, namely the Suzuki reaction,[6–8] where a new carbon–carbon bond forms in the target compound **5**. Palladium-catalysed reactions[9] are very versatile and have found widespread application in the synthesis of complex molecules as they typically combine two precursors that themselves can be relatively complex. It is therefore a type of reaction that lends itself to a convergent strategy!

Common to this type of reaction is an interaction of the palladium with a π-system (olefin or aromatic) of a halogen-bearing moiety into which the metal inserts before it reacts with a nucleophile (borate or otherwise) and finally eliminates the product with the new carbon–carbon bond. The ideals of green chemistry influence also this reaction and newer variations can be conducted solely in water using ligands based on chiral amino acids rather than phosphines.[10]

Additional Safety

All procedures should be performed in a well-ventilated fume cupboard, with proper safety equipment such as protective clothing, safety spectacles and gloves. *n*-Butyllithium is a pyrophoric reagent and its solutions in hexane are highly flammable. It must be handled under an inert gas atmosphere. Syringes and needles should be rinsed with ethanol followed by water before disposal. Sand should be kept ready for dealing with spillages and as extinguishing medium. All organic compounds and solvents bar dichloromethane are flammable and harmful/irritants. Ethers are extremely flammable and may form explosive peroxides. Dichloromethane is a suspected carcinogen. Methanol is toxic by inhalation, in contact with skin and when ingested. Silica should be weighed in a fume cupboard and inhalation of dust must be avoided. Dry ice poses a risk of frost burn. Gloves should be used when handling it or the cooling bath. Bromine is a toxic, corrosive and very volatile reagent and must be handled with thick gloves only in a well-ventilated fume cupboard. Spillages and equipment must be decontaminated with sodium thiosulfate solution. All other reagents are flammable, harmful and irritant. Benzyl bromide derivatives are sensitisers and potent lachrymators; they must be handled in a well-ventilated fume cupboard and spillages decontaminated with ethanol. Acids and bases are corrosive and harmful in contact with mucous membranes.

Experimental Procedure

All reactions are carried out in dry glassware under a protective inert gas atmosphere. All solvents used in reactions are anhydrous, unless stated otherwise and degassed where indicated.

Session 1: Synthesis of 4-Tolylboronic acid (2) (6 h)

1. In a fume cupboard assemble a three-neck round-bottom flask (100 mL) with a gas inlet, gas bubbler and stopper in the three necks and a stirrer bar over a magnetic stirrer. Always ensure there is a positive pressure of inert gas in the apparatus, particularly when cooling the contents.
2. Add 4-bromotoluene (15.0 mmol, 1.85 mL) and replace the stopper with a rubber septum.
3. Add tetrahydrofuran (25 mL) and then cool the solution with an external cardice–acetone bath to −78 °C.
4. Slowly add with a syringe *n*-butyllithium (2.5 M in hexanes, 7.00 mL, 17.5 mmol) and continue stirring at −78 °C for a further 10 min.
5. Under rapid stirring add quickly with a syringe trimethylborate (22.5 mmol, 2.5 mL). Then remove the cooling bath and allow to warm to ambient temperature.
6. Pour the reaction mixture into hydrochloric acid (6 M, 35 mL) and vigorously stir the biphasic mixture for 10 min.
7. Separate the phases and then extract the aqueous phase with dichloromethane (2 × 25 mL).
8. Wash the combined organic phases with water (2 × 35 mL) followed by brine (35 mL).
9. Dry the organic phases with $MgSO_4$ and after filtration concentrate *in vacuo*.
10. Dissolve the crude residue in toluene (25 mL) and concentrate *in vacuo*; then repeat the process with diethyl ether (25 mL).
11. Triturate the crude with pentane (30 mL), separate the solid by filtration and desiccate the resulting solid until its weight remains constant.

Session 2: Synthesis of Methyl 4-(Bromomethyl)benzoate (4) (6 h)

1. In a fume cupboard assemble a three-neck round-bottom flask (250 mL) with a stopper, a reflux condenser with gas outlet and a pressure-equalised dropping funnel with stopper in the three necks, and a stirrer bar over a magnetic stirrer.
2. Connect the gas outlet to two Drechsel bottles and fill the second gas wash bottle with sodium hydroxide solution (2 M, 40 mL).
3. Under a gentle stream of inert gas add methyl 4-methylbenzoate (20 mmol, 3.00 g) and 1,1′-azobis(cyclohexanecarbonitrile) (ACHN) (0.005 equiv., 0.1 mmol, 24 mg) to the round-bottom flask and then replace the stopper with a septum.
4. Dissolve the contents in degassed dichloromethane (100 mL).
5. Add degassed dichloromethane (50 mL) to the dropping funnel followed by bromine (20 mmol, 1.03 mL). *Care*: use thick gloves when handling bromine. For convenience, this mixture can be pre-mixed in a small conical flask and then transferred to the dropping funnel.
6. Heat the stirred reaction mixture until refluxing and irradiate it with a halogen lamp placed next to the round-bottom flask.
7. Slowly drop the bromine solution to the stirred reaction mixture over a period of ~30 min.
8. Monitor the reaction by TLC. *Care*: the product is a lachrymator and sensitiser.
9. When finished concentrate the reaction mixture *in vacuo* and cool the residue on ice.
10. Recrystallise the crude product from heptane.

Session 3: Synthesis of Methyl 4-(4-Methylbenzyl)benzoate (5) (6 h)

1. In a fume cupboard set up a round-bottom flask (50 mL) with stirrer bar over a magnetic stirrer hotplate.
2. Add degassed 1,2-dimethoxyethane (15 mL) and dissolve equimolar amounts (3.0 mmol) of 4-tolylboronic (**2**) (408 mg) acid and methyl 4-(bromomethyl)benzoate (**4**) (687 mg) in the stirred solvent.
3. Then add a degassed aqueous solution of sodium carbonate (2 M, 4.5 mL) followed by tetrakis(triphenylphosphine)palladium(0) (0.01 equiv., 0.03 mmol, 35 mg).
4. Attach a reflux condenser and purge the apparatus with an inert gas.
5. Reflux the reaction mixture and monitor the reaction by TLC until completion.
6. After cooling, dilute with dichloromethane (30 mL) and water (30 mL) and separate the phases.
7. Extract the aqueous phase with dichloromethane (30 mL) and then wash the combined organic phases with saturated sodium bicarbonate (2 × 50 mL), water (50 mL) and saturated brine (3 × 50 mL).
8. Dry with magnesium sulfate, and after filtration, concentrate *in vacuo.*
9. Purify the crude product by flash-column chromatography using silica and as eluent a mixture of 10% diethyl ether in light petroleum ether.

Waste Disposal

The neutralised aqueous solutions – including dissolved $MgSO_4$ – may be diluted and discarded in the sink. Silica waste should be collected separately. Equipment that has been in contact with bromine should be rinsed and decontaminated with sodium thiosulfate solution. Equipment that has been in contact with benzyl bromide derivatives must be decontaminated with ethanol. All organic solutions should be recycled or collected as chlorinated waste.

Results Interpretation and Additional Questions

Interpret all acquired spectroscopic and spectrometric data and formulate mechanisms for the reaction, thus becoming aware of possible by-products and actions of all reagents. Consult relevant literature (from database searches) to find additional resources that support the analysis of your reaction products and form a basis for the decision process.

1. What is the purpose of the Drechsel bottles and how do they work chemically?
2. Analyse the ^{1}H NMR spectrum of the product of the bromination reaction taking into account possible impurities, *i.e.* dibromination and/or unreacted starting material. Decide how to proceed depending on the outcome of the analysis.
3. A possible alternative product of the synthesis of the boronic acid is the corresponding boroxin, a trimeric cyclic boronic anhydride. Choose an appropriate analytical technique and find out which product has formed in your reaction. In what way is the outcome relevant?
4. The palladium-catalysed cross-coupling reactions are versatile C–C bond formations. Find out about these type of reactions from the cited literature and then formulate the mechanism for the Suzuki reaction you performed.

References

1. J. Clayden, Regioselective Synthesis of Organolithiums by X-Li Exchange, in *Organolithiums: Selectivity for Synthesis*, ed. J. E. Baldwin and R. M. Williams, Elsevier Science, Oxford, 2002, vol. 23, ch. 3, Tetrahedron Organic Chemistry Series, pp. 111–126.
2. C. C. C. Johansson Seechurn, M. O. Kitching, T. C. Colacot and V. Snieckus, *Angew. Chem., Int. Ed.*, 2012, **51**, 5062.
3. (a) R. M. Washburn, E. Levens, C. F. Albright and F. A. Billig, *Organic Syntheses*, 1963, vol. 4, coll, p. 68; (b) 1959, vol. 39, p. 3.
4. J. Clayden, N. Greeves and S. Warren, *Organic Chemistry*, Oxford University Press, Oxford, 2nd edn, 2012, pp. 984–989.
5. A. Podgoršek, S. Stavber, M. Zupana and J. Iskra, *Tetrahedron Lett.*, 2006, **47**, 7245.
6. J. Clayden, N. Greeves and S. Warren, *Organic Chemistry*, Oxford University Press, Oxford, 2nd edn, 2012, pp. 1082–1090.
7. (a) A. R. Martin and Y. Yang, *Acta Chem. Scand.*, 1993, **47**, 221; (b) S. P. Stanford, *Tetrahedron*, 1998, **54**, 263.
8. A. Suzuki, *Angew. Chem., Int. Ed.*, 2011, **50**, 6722.
9. B. M. Trost and T. R. Verhoven, Organopalladium complexes in Organic Synthesis and Catalysis, in *Comprehensive Organometallic Chemistry*, ed. G. Wilkinson, F. G. A. Stone and E. Abel, Pergamon, Oxford, 1982, ch. 8, p. 799.
10. G. Zhang, Y. Luan, X. Han, Y. Wang, X. Wen and C. Ding, *Appl. Organomet. Chem.*, 2014, **28**, 332.

7.6. Oxidative Heck Reaction at Room Temperature

Rajiv T. Sawant, Ashkan Fardost, and Luke R. Odell*

Organic Pharmaceutical Chemistry, Department of Medicinal Chemistry, Uppsala University Biomedical Center, Box 574, SE-751 23, Uppsala, Sweden. *E-mail: luke.odell@orgfarm.uu.se

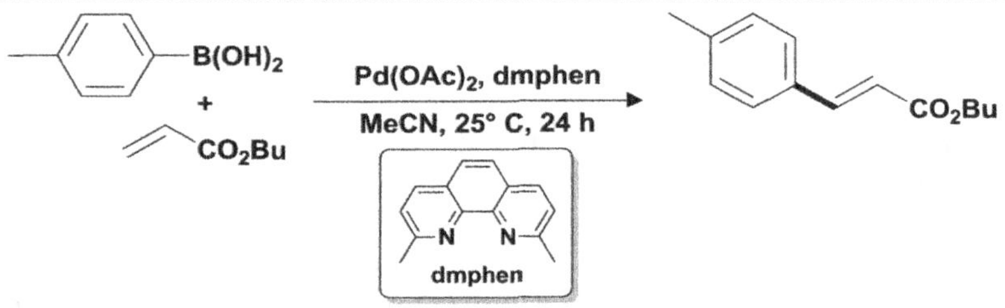

Number of sessions (duration of each session)	Hazard level	Difficulty level	Level of study
2 (2 h + 4 h)	Moderate	Medium	Advanced

Class names Boronic acids, olefins

Concepts involved Oxidative Heck reaction: this experiment involves a method for the palladium-catalysed arylation of *n*-butyl acrylate using *p*-tolylboronic acid

Chemicals needed *p*-Tolylboronic acid, *n*-butyl acrylate, palladium acetate, 2,9-dimethyl-1,10-phenanthroline, acetonitrile, ethyl acetate, isohexane, dimethyl acetamide, brine solution, sodium sulfate, deuterated chloroform

Equipment and experimental techniques involved Liquid–liquid extraction, magnetic stirring, rotary evaporation, thin-layer chromatography, NMR spectroscopy

Keywords Boronic acid, Heck reaction, liquid–liquid extraction, NMR spectroscopy, olefin, palladium(II) catalysis

Background

The formation of carbon–carbon bonds is one of the most important and fundamental processes in modern organic chemistry. Amongst the myriad of methods available those that are catalysed by transition metals are particularly attractive due to their high selectivity, ease of use and environmental friendliness. The importance of these transformations has recently been acknowledged by The Royal Swedish Academy of Sciences who awarded the 2010 Nobel Prize in Chemistry to Richard F. Heck, Ei-Ichi Negishi and Akira Suzuki for "palladium catalyzed cross-couplings in organic synthesis".

The Mizoroki–Heck reaction, generally defined as the substitution of a vinylic hydrogen with an aryl, vinyl or benzyl group, was independently discovered by Heck[1] and Mizoroki[2] in the early 1970s and is today regarded as one of the premier synthetic tools for the construction of new C–C bonds.[3,4] The palladium(II)-catalysed oxidative Heck reaction, which differs from the palladium(0)-catalysed Mizoroki–Heck reaction in the generation of the key arylpalladium species *via* non-redox palladium(II) process, has emerged as a useful alternative over the past decade.[5] At present, efficient

Comprehensive Organic Chemistry Experiments for the Laboratory Classroom
Edited by Carlos A. M. Afonso, Nuno R. Candeias, Dulce Pereira Simão, Alexandre F. Trindade, Jaime A. S. Coelho, Bin Tan, and Robert Franzén.

Published by the Royal Society of Chemistry, www.rsc.org

procedures are available to allow the use of boronic acids,[6–8] trifluoroborates,[9] stannanes,[10] phosphonates,[11] carboxylic acids,[12] sulfinates[13] as well as arenes and heteroarenes[14] as substrates in the oxidative Heck reaction. This use of boronic acids is particularly attractive as they are widely commercially available, tolerate a broad variety of functional groups and are easy to handle.

In general, highly electrophilic palladium(II) salts such as $Pd(OAc)_2$ or $Pd(TFA)_2$ are employed as catalysts in the oxidative Heck reaction and reactivity and selectivity can be further modulated through the addition of various bidentate nitrogen ligands (*e.g.* dmphen or bipyridine). The oxidative Heck reaction begins with the formation of the arylpalladium intermediate **B** *via* mechanisms that vary for each class of starting materials (Scheme 7.6.1). In its most common incarnation, the oxidative Heck reaction with aryl boronic acids, the catalytic cycle begins with the formation of **B** *via* transmetalation. In step 2, an olefin coordinates to the arylpalladium intermediate to form the cationic π-complex, $\mathbf{C}_\pi$. Migratory insertion (step 3) generates the σ-complex **C**σ, and internal rotation followed by syn β-hydride elimination (step 4) generates complex **D**. The arylated olefin dissociates (step 5) from the intermediate **D** to form the palladium hydride species **E**. Finally, the starting palladium(II) catalyst **A** is regenerated from the palladium hydride intermediate **E**, by oxidation (step 6).

Palladium(II)-catalysed coupling reactions normally require a dedicated reoxidant to regenerate the active catalytic species and avoid the formation of non-productive palladium black. Commonly used reoxidants are metal salts, such as copper and silver salts, *p*-benzoquinone (*p*-BQ) and other quinones, and various peroxides. The use of copper or silver salts is costly and *p*-BQ is toxic and relatively expensive, and, in addition, all of these reoxidants produce stoichiometric amounts of waste, which may complicate purification procedures and have a negative environmental impact. Hydrogen peroxide, on the other hand, is cheap and produces water as by-product, but it is a strong oxidant that is incompatible with many functionalities. The ultimate reoxidant, however, is molecular oxygen, preferably as ubiquitous air (or even hypoxic air), as it is virtually cost-free and environmentally benign.

In this experiment, *p*-tolylboronic acid, will be used to substitute the vinylic hydrogen of butyl acrylate, in an open vessel reaction setup, under palladium(II) catalysis. The reaction introduces students to environmentally friendly palladium(II)-catalysed oxidative Heck processes and serves as an entry point into the discussion of regioselectivity in Heck reactions.

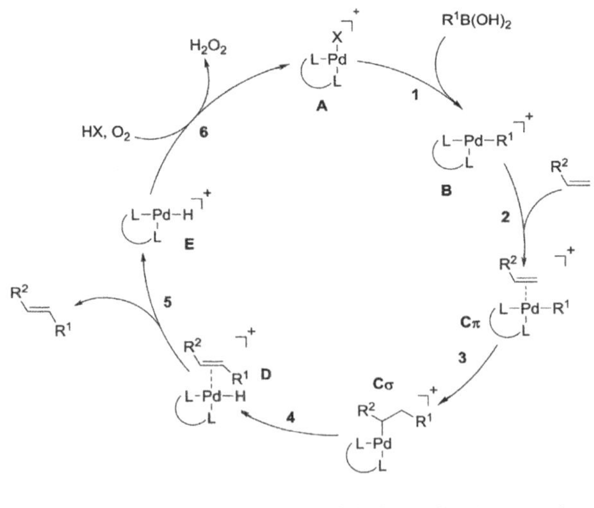

Scheme 7.6.1 General catalytic cycle for the oxidative Heck reaction. Adapted from Figure 1 in *Comprehensive Organic Synthesis II*, 2nd edn, 2014, vol. 7, L. R. Odell, J. Sävmarker, J. Lindh, P. Nilsson, M. Larhed, 7.18 Addition reactions with formation of carbon–carbon bonds: (v) the oxidative Heck reaction, pp. 492–537, with permission from Elsevier.

Additional Safety

All sessions should be conducted in a fume hood and all participants should be equipped with protective clothing, gloves and safety glasses. Palladium acetate is corrosive and exposure may cause skin, eye and respiratory tract burns. 2,9-Dimethyl-1,10-phenanthroline is an irritant and can cause skin, eye and respiratory tract irritation. Acetonitrile is a flammable solvent, harmful if swallowed or in contact with skin and an eye irritant. Ethyl acetate is a highly flammable liquid and vapour, causes serious eye irritation and may cause drowsiness or dizziness. Isohexane is a highly flammable liquid, which may be fatal if it enters the airways or if swallowed, and may cause drowsiness or dizziness, while also being toxic to aquatic life with long lasting effects. Dimethylacetamide (DMA) is harmful in contact with skin or if inhaled, causes serious eye irritation and may damage the unborn child. It is recommended that only small quantities of DMA be available to the students during this experiment.

Experimental Procedure

Experimental Procedure

Laboratory Session 1 (2 h): Oxidative Heck Reaction

1. Prepare a 50 mL round-bottom flask equipped with a magnetic stir bar.
2. Add 136 mg of *n*-butyl acrylate (1.0 mmol), 135 mg of *p*-tolylboronic acid (1.05 mmol), 4.5 mg of palladium(II) acetate (0.02 mmol), 5 mg of 2,9-dimethyl-1,10-phenanthroline (dmphen) (0.024 mmol), and 5.0 mL of acetonitrile.
3. Stir the open flask vigorously at room temperature until the next laboratory session.
4. Determine a suitable thin-layer chromatography (TLC) eluent system for monitoring the reaction. The goal here is to find an ethyl acetate/isohexane mixture that gives good separation between product and starting materials. To do this you will need to obtain an authentic sample of the product from your instructor. Dissolve the sample in a small amount of acetone and spot it onto a TLC plate alongside your starting materials. Elute your TLC plate with a 3:1 mixture of isohexane/ethyl acetate. Visualize your plate under a UV lamp and calculate the R_f values for the product and starting materials. Adjust the polarity of your eluent system to give a product R_f between 0.4 and 0.8.

Laboratory Session 2 (4 h): Product Isolation and Yield Determination

1. Check that your reaction has reached completion (*i.e.* full consumption of *p*-tolylboronic acid) using your TLC eluent system from step 4.
2. Remove the magnetic stir bar and concentrate the reaction mixture *via* rotary evaporation.
3. To this crude product, add slowly 10 mL of saturated aqueous sodium bicarbonate and 10 mL of ethyl acetate.
4. Transfer the solution into a separating funnel and extract with 2 × 20 mL of ethyl acetate. Wash the combined organic layer with brine and dry over anhydrous Na_2SO_4, filter and concentrate *via* rotary evaporation.
5. Weigh the round-bottom flask containing the product and calculate the crude product mass (CM) and yield. Compare this value with the literature data in ref. 6.

Yield Determination

The reaction yield will be determined using ^{1}H NMR.

1. Add 40 mg of the crude product (CP) to a 2 mL sample vial followed by 10 mg of the internal standard (IS) dimethyl acetamide. It is important to know the exact amount of crude product and internal standard added.

2. Add 0.65 mL of $CDCl_3$ and mix the solution by drawing it up in a Pasteur pipette a few times.
3. Transfer the mixture to a NMR tube, run a ^1H experiment and complete the following steps:
   - Calculate the number of moles of IS added (*A*).
   - Identify and integrate the CH_3 peak belonging to your product and the acetyl (CH_3CO) peak from the internal standard (compare with literature values).
   - Calculate the ratio CH_3/CH_3CO (*B*).
   - Calculate the number of moles (*C*) of product in your mixture ($A \times B$).
   - Calculate the mass of product (*D*) in your NMR mixture ($C \times$ molar mass of product).
   - Calculate the mass of product formed in your reaction ($D \times$ CM/CP).
   - Calculate the NMR yield.

Waste Disposal

The aqueous solution may contain traces of palladium and should be treated as reaction waste. Ethyl acetate and isohexane may be reused or collected in a non-chlorinated waste collector recipient, for disposal.

Results Interpretation and Additional Questions

- Interpret the ^1H NMR spectrum and assign all the signals.
- What possible isomers can be formed during this reaction? Which isomer has formed in your reaction? What evidence do you have for its formation?
- Describe the fundamental steps in the catalytic cycle.
- Which step(s) in the cycle determine the reactions regioselectivity and stereoselectivity?
- An enterprising young student had just attended a lecture on microwave-assisted organic synthesis and figured that using microwave heating would allow the reaction time to be reduced from hours to minutes. The student then proceeded to set up the reaction in a standard 5 mL microwave vial, closed the lid and heated the reaction in a microwave reactor at 120 °C for 20 min. Unfortunately, the crude yield after the reaction was less than 5%. Discuss the possible reasons for the poor conversion observed in this reaction.
- A GC-MS analysis of the crude reaction mixture shows the presence of two additional by-products with *m*/*z* values of 92 and 182? Suggest a structure for these two by-products and by what general process have they been formed.

References

1. R. F. Heck and J. P. Nolley, *J. Org. Chem.*, 1972, **37**, 2320.
2. T. Mizoroki, K. Mori and A. Ozaki, *Bull. Chem. Soc. Jpn.*, 1971, **44**, 581.
3. *The Mizoroki–Heck Reaction*, ed. M Oesterich, John Wiley & Sons, Ltd, Chichester, 2009.
4. *Science of Synthesis; Cross-Coupling and Heck Reactions Volume 3: Metal-Catalyzed Heck-Type Reactions and C—H Couplings*, ed. M. Larhed, Georg Thieme Verlag KG, Stuttgart, 2012.
5. L. R. Odell, J. Sävmarker, J. Lindh, P. Nilsson and M. Larhed, *Compr. Org. Synth. (2nd Ed.)*, 2014, **7**, 492.
6. J. Lindh, P.-A. Enquist, Å. Pilotti, P. Nilsson and M. Larhed, *J. Org. Chem.*, 2007, **72**, 7957.
7. K. S. Yoo, C. H. Yoon and K. W. Jung, *J. Am. Chem. Soc.*, 2006, **128**, 16384.
8. L. R. Odell, J. Lindh, T. Gustafsson and M. Larhed, *Eur. J. Org. Chem.*, 2010, **2010**, 2270.
9. J. Sävmarker, J. Lindh, P. Nilsson, P. J. R. Sjöberg and M. Larhed, *ChemistryOpen*, 2012, **1**, 140.
10. J. P. Parrish, Y. C. Jung, S. I. Shin and K. W. Jung, *J. Org. Chem.*, 2002, **67**, 7127.
11. A. Inoue, H. Shinokubo and K. Oshima, *J. Am. Chem. Soc.*, 2003, **125**, 1484.
12. A. G. Myers, D. Tanaka and M. R. Mannion, *J. Am. Chem. Soc.*, 2002, **124**, 11250.
13. X. Zhou, J. Luo, J. Liu, S. Peng and G.-J. Deng, *Org. Lett.*, 2011, **13**, 1432.
14. E. M. Beck, N. P. Grimster, R. Hatley and M. J. Gaunt, *J. Am. Chem. Soc.*, 2006, **128**, 2528.

7.7. Synthesis of 2-Methyl-1,1′-binaphthalene *via* Suzuki Cross-Coupling Reaction

Javier Iglesias-Sigüenza*, David Monge, and Elena Díez

Departamento de Química Orgánica, Universidad de Sevilla, C/ Profesor García González, nº 1, 41012-Sevilla, Spain. *E-mail: jiglesias@us.es

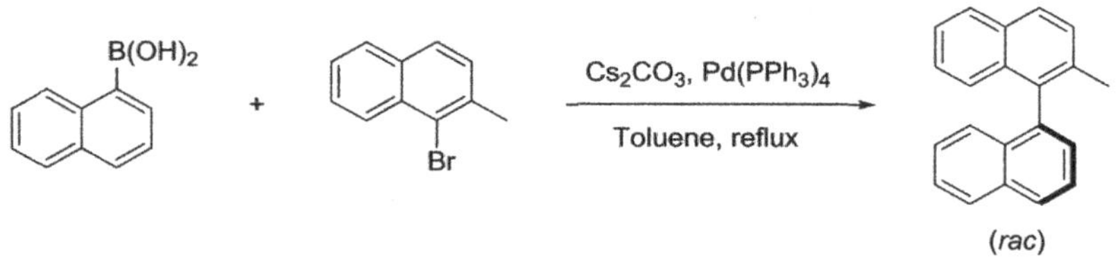

(*rac*)

Number of sessions (duration of each session)	Hazard level	Difficulty level	Level of study
2 (3 h + 4 h)	Moderate	Medium	Advanced

Class names Boronic acids, aryl halides

Concepts involved This experiment shows the importance of the use of organometallic compounds in the synthesis of useful complex organic molecules. In this way, this experiment introduces the concept of metal–ligand interaction and allows the study of typical organometallic mechanisms, such as oxidative addition, transmetalation or reductive elimination. In addition, the model system proposed allows the introduction of the concept of axial chirality

Chemicals needed 1-Bromo-2-methylnaphthalene, naphthalene-1-boronic acid, tetrakis(triphenylphosphine)palladium(0), caesium carbonate, toluene, Celite®, diethyl ether, silica gel, pentane, ammonium molybdate tetrahydrate, cerium(IV) sulfate, sulfuric acid, deuterated chloroform

Equipment and experimental techniques involved Equipment for reactions under nitrogen/argon atmosphere, hot plate stirrer, vacuum pump and Rotavapor, column chromatography

Keywords Column chromatography, cross-coupling reactions, NMR, palladium catalysis, Suzuki reaction

Background

The Suzuki coupling of a boronic acid or ester with a vinyl, aryl halide or triflate is probably the most commonly used of all cross-coupling reactions.[1] In fact, the Nobel Prize in Chemistry 2010 was awarded jointly to Richard F. Heck, Ei-Ichi Negishi and Akira Suzuki "for palladium-catalyzed cross couplings in organic synthesis".[2] The wider application of this reaction is the preparation of biaryl compounds and analogues,[3] which are very important for the development of biologically active molecules or materials with technological applications.

Comprehensive Organic Chemistry Experiments for the Laboratory Classroom
Edited by Carlos A. M. Afonso, Nuno R. Candeias, Dulce Pereira Simão, Alexandre F. Trindade, Jaime A. S. Coelho, Bin Tan, and Robert Franzén.

Published by the Royal Society of Chemistry, www.rsc.org

Despite the reaction versatility, it is interesting to know different aspects about the reaction conditions:

- Firstly, all kinds of carbon–boron bonds including $C_{sp}{}^3$–B, $C_{sp}{}^2$–B or C_{sp}–B bonds can be used as cross-coupling partners in these coupling reactions.
- Any organoboron compound can be employed in this reaction except alkynylboranes because they are stronger Lewis acids and easily hydrolyzed in the presence of bases. In addition, boron esters can be used instead of boronic acids.
- On the other hand, the reaction works with halides and triflates. The relative reactivity is R-I > R-OTf > R-Br ≫ R-Cl, so the use of chlorides should be avoided. These reagents substituted with electron-withdrawing groups are suitable substrates for this cross-coupling reaction.
- The most commonly used bases in these coupling are Cs_2CO_3 or Na_2CO_3, but these bases are often useless with sterically demanding substrates. In these cases, K_3PO_4 or hydroxides are used. Finally, fluoride salts such as CsF or Bu_4NF are useful for base-sensitive substrates.
- The most frequently used catalysts are $Pd(PPh_3)_4$ or $Pd_2(dba)_3$ as a Pd(0) source. But Pd(II) catalysts, like $Pd(OAc)_2$ or $PdCl_2$, can also be used.
- By-products such as homocoupling products and deboronation products are often obtained.

The mechanism of this coupling is described in Scheme 7.7.1. The first step is the oxidative addition of the halide **2** or triflate to the initial palladium(0) complex to form the organopalladium(II) species **3**. Reaction with the base gives intermediate **4**, which *via* transmetalation with the boron-ate complex **6** forms the organopalladium species **8**. The new palladium(II) complex with two organic ligands undergoes reductive elimination to give the coupled product **9** and the original palladium(0) catalyst **1**, ready for another catalytic cycle.

The important difference between the Suzuki reaction and other coupling reactions, such as the Stille coupling, is the transmetalation step, which explains the need for an additional base in the Suzuki coupling. The base accelerates the transmetalation step, leading to the borate directly, presumably *via* a more nucleophilic "ate" complex.

This reaction tolerates a wide range of functional groups, it is highly chemoselective, it can be carried out under mild conditions and finally it uses boron compounds, which show superior features such as easy manipulation, ready availability and low toxicity.

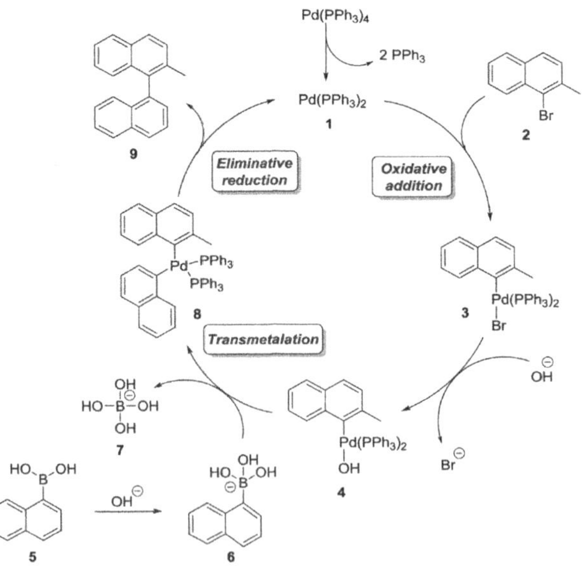

Scheme 7.7.1 Suzuki cross-coupling mechanism.

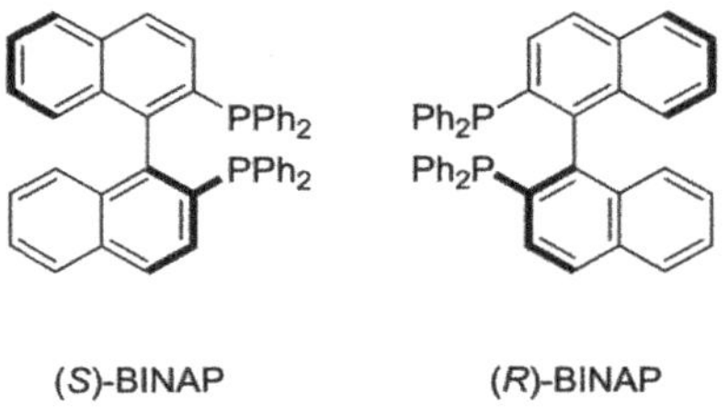

Figure 7.7.1 Chiral biaryl systems.

For these reasons, this cross-coupling is widely used and it has found great popularity in the pharmaceutical industry. For example, on the order of about 100 metric tons per year of 2-cyano-4′-methylbiphenyl is manufactured *via* Suzuki coupling as a pharmaceutical intermediate product.[4]

In this experiment, we describe the synthesis of 2-methyl-1,1′-binaphthalene from commercial starting materials *via* Suzuki reaction. Purification using flash chromatography leads to the corresponding biaryl compound in high and reproducible yield. The axially chiral biaryl compound is obtained as a racemic mixture.

Some biaryl compounds, such as the important bisphosphine in Figure 7.7.1, known as BINAP, exist as two separate enantiomers because rotation around the aryl–aryl bond is restricted, due to steric hindrance. The enantiomers of axially chiral compounds are usually given the stereochemical labels R_a and S_a. The designations are based on the same Cahn–Ingold–Prelog priority rules used for tetrahedral stereocenters.[5]

Additional Safety

1-Bromo-2-methylnaphthalene, naphthalene-1-boronic acid, caesium carbonate and cerium(IV) sulfate could produce dermal and eye irritation (H315 and H319) and, besides, they could irritate respiratory tracts (H335); sulfuric acid causes severe skin burns and eye damage (H314). For this reason, it is completely necessary avoid the contact with them. The students have to wear proper personal protective equipment, such as laboratory coat, gloves and goggles. On the other hand, pentane is a flammable compound (H225) and it could be fatal if swallowed (H304). Celite® and silica gel may cause respiratory irritation – avoid breathing the dust. Avoid contact with skin and eyes. Avoid formation of dust. Provide appropriate exhaust ventilation at places where dust is formed. Toluene is a highly flammable liquid and vapor (H225) and can causes skin irritation (H315), diethyl ether is an extremely flammable liquid and vapor (H224), is harmful if swallowed (H302) and may form explosive peroxides (EUH019).

At the end of the first laboratory session, the reaction mixture is refluxed for 20 h using an oil bath on a magnetic hot plate. Hot oil is dangerous and can cause severe burns. It also has a bad odor and makes flasks dangerously slippery. Accidental addition of water to a hot oil bath can cause serious splattering.

All sessions should be carried out in a fume hood.

Experimental Procedure

Laboratory Session 1 (3 h): Suzuki Reaction

1. Toluene should be deoxygenated prior to use. This operation involves bubbling an inert gas, like argon, through the solvent for 30 min. The inert gas can be introduced to the solvent by a long needle attached to the gas source. A small needle could be used as oxygen exit from the vessel that contains the solvent. Care should be taken to ensure the flow rate of gas is not too high, because it could involve evaporation of the solvent.

2. Prepare a Schlenk or a round-bottom flask with a magnetic stir bar, add 190 mg (1.5 mmol, 1.5 equiv.) of naphthalene-1-boronic acid, 456 mg (2 mmol, 2 equiv.) of caesium carbonate and 81 mg (0.1 mmol, 10 mol%) of tetrakis(triphenylphosphine)palladium(0) as catalyst.
3. The Schlenk should be maintained under an argon atmosphere by three successive vacuum–argon cycles. To do this, seal the Schlenk with a greased stopper, evacuate the air, and then fill with inert gas from the vacuum line. Repeat this process three times. Keep the system under an argon atmosphere.
4. Add 3 mL of deoxygenated toluene and 109 μL (1 mmol, 1 equiv.) of the aryl bromide. Then, reflux the reaction mixture for 20 h using an oil bath on a magnetic hot plate, until the next session.

Laboratory Session 2 (4 h): Purification of the Product

1. To confirm that the reaction is complete, monitor it by TLC, using pentane as solvent. Then, filter the reaction mixture through a pad of Celite® washing with diethyl ether. Remove the solvent under reduced pressure.
2. Purify the resulting residue by flash chromatography using pentane as eluent.
3. Weigh the round-bottom flask containing the final product and determine the final weight. Calculate the yield.
4. Take 15 mg of the product, dissolve it in 0.6 mL of $CDCl_3$ and transfer to a NMR tube. Acquire a ^1H NMR spectrum to confirm the compound purity. Compare with literature data.

Waste Disposal

Manipulations of dry silica should be performed only in the fume hood. The solvents may be collected in a non-chlorinated waste collector recipient. Silica waste should be collected in an appropriate collector recipient.

Results Interpretation and Additional Questions

1. Calculate the R_f of the product from the TLC analysis and interpret the ^1H NMR spectrum obtained.
2. Schematize the mechanism involved in the Suzuki reaction with these reagents. Discuss the role of the base in the mechanism.
3. Discuss the similarities and differences between the Suzuki reaction and other palladium catalyzed cross-coupling reactions.
4. Indicate what kind of chirality exists in the synthesized compound. What requirement is necessary for this type of chirality? Assign the configuration of both enantiomers.

References

1. (a) D. Steinborn, *Fundamentals of Organometallic Catalysis*, Wiley-VCH Verlag GmbH, New York, 2012; (b) S. Kotha, K. Lahiri and D. Kashinath, *Tetrahedron*, 2002, **58**, 9633; (c) N. Miyaura and A. Suzuki, *Chem. Rev.*, 1995, **95**, 2457.
2. *The Nobel Prize in Chemistry*, 2010, http://www.nobelprize.org/nobel_prizes/chemistry/laureates/2010/.
3. O. Baudoin, *Eur. J. Org. Chem.*, 2005, 4223.
4. *Organometallics as Catalysts in the Fine Chemical Industry*, ed. M. Beller and H.-U. Blaser, Springer-Verlag Berlin and Heidelberg GmbH & Co. KG, 2012, Topics in Organometallic Chemistry Series, Palladium-Catalysed Coupling Reactions, J. G. Vries, ch. 1, pp. 1–34.
5. *IUPAC. Compendium of Chemical Terminology. (the "Gold Book")*, ed. A. D. McNaught and A. Wilkinson. Blackwell Scientific Publications, Oxford, 2nd edn, 1997.

7.8. Green Synthesis of Aromatic Ketones: Decarboxylative Palladium Catalysis Under Microwave Irradiation

Jonas Sävmarker, Ashkan Fardost, and Luke R. Odell*

Organic Pharmaceutical Chemistry, Department of Medicinal Chemistry, Uppsala University Biomedical Center, Box 574, SE-751 23, Uppsala, Sweden. *E-mail: luke.odell@orgfarm.uu.se

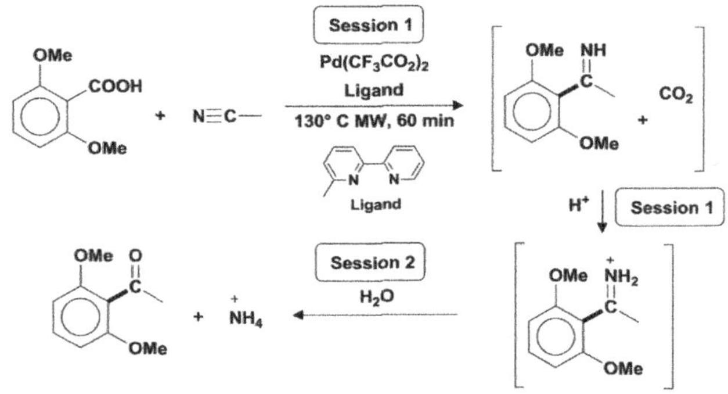

Number of sessions (duration of each session)	Hazard level	Difficulty level	Level of study
2 (2 h + 3 h)	Moderate	Medium	Advanced

Class names Carboxylic acids, nitriles, ketones, imines

Concepts involved Palladium(II) catalysed decarboxylation and ketone formation: this experiment gives a method for the transformation of a carboxylic acid into a ketone *via* a ketimine intermediate using microwave irradiation. The reaction yield is determined by ^{1}H NMR using the internal standard method

Chemicals needed 2,6-Dimethoxybenzoic acid, palladium trifluoroacetate, 6-methyl-2,2′-dipyridyl, acetonitrile, formic acid, ethyl acetate, isohexane, dimethylacetamide, sodium bicarbonate, brine solution, sodium sulfate, deuterated chloroform

Equipment and experimental techniques involved Liquid–liquid extraction, rotary evaporation, thin-layer chromatography, microwave reactor, NMR spectroscopy

Keywords Carboxylic acid, decarboxylation, ketone, liquid–liquid extraction, nitrile, NMR spectroscopy, palladium(II) catalysis

Background

Microwave heating has, over the past 15 years, become an indispensable weapon in the modern organic chemists arsenal.[1–3] This is mainly because modern instrumentation allows the "super-heating" of reaction mixtures in a controlled and safe manner. This, in turn, equips chemists with the ability to conduct high temperature reactions to both reduce reaction rates (*i.e.* increase productivity and lower energy costs) and develop new chemical transformations.

The formation of carbon–carbon bonds is one of the most important and fundamental processes in modern organic chemistry. Amongst the myriad of methods available those that are

Comprehensive Organic Chemistry Experiments for the Laboratory Classroom
Edited by Carlos A. M. Afonso, Nuno R. Candeias, Dulce Pereira Simão, Alexandre F. Trindade, Jaime A. S. Coelho, Bin Tan, and Robert Franzén.

Published by the Royal Society of Chemistry, www.rsc.org

catalysed by transition metals are particularly attractive due to their high selectivity, ease of use and environmental friendliness. Indeed, the importance of these transformations has recently been acknowledged by The Royal Swedish Academy of Sciences who awarded the 2010 Nobel Prize in Chemistry to Richard F. Heck, Ei-Ichi Negishi and Akira Suzuki for "palladium catalysed cross-couplings in organic synthesis".

Aromatic ketones are ubiquitous in organic molecules, and can be found in everything from pharmaceuticals and natural products to plastics and organic electronics. While there are myriads of ways to form them, most processes require harsh oxidative conditions with limited selectivity, or toxic, expensive and unstable substrates (Grignard reagents, aryl halides and arylboronic acids).

Since the seminal decarboxylative Heck-type reaction reported by Myers and co-workers in 2002,[4] a multitude of palladium(II)-catalysed decarboxylative reactions have been discovered.[5–9] In this experiment, decarboxylative palladium catalysis is employed as an environmentally friendly one-step route to aromatic ketones by reacting aromatic benzoic acids with nitriles under microwave irradiation.[10,11]

Additional Safety

All sessions should be conducted in a fume hood and all participants should be equipped with protective clothing, gloves and safety glasses. Palladium trifluoroacetate and 6-methyl-2,2′-dipyridyl are irritants and can cause skin, eye and respiratory tract irritation. Acetonitrile is a flammable solvent, harmful if swallowed, harmful in contact with skin and an eye irritant. Ethyl acetate is a highly flammable liquid and vapour, causes serious eye irritation and may cause drowsiness or dizziness. Formic acid is a flammable liquid and vapour and it can cause severe skin burns and eye damage.

Dimethylacetamide (DMA) is harmful in contact with skin or if inhaled, causes serious eye irritation and may damage the unborn child. It is recommended that only small quantities of DMA be available to the students during this experiment.

The reaction is conducted at a temperature above the boiling point of the solvent. This should only be carried out using a dedicated microwave instrument.

Care should also be taken to release any pressure built up during the reaction by piercing the septa before opening the reaction vial.

Experimental Procedure

Laboratory Session 1 (2 h): Reaction Set-Up and Acid Treatment

1. Prepare a 2–5 mL microwave transparent process vial equipped with a magnetic stirring bar.
2. Add 13.3 mg $Pd(CF_3CO_2)_2$ (0.04 mmol), 8.2 mg 6-methyl-2,2′-bipyridine (0.048 mmol), 91 mg 2,6-dimethoxybenzoic acid (0.5 mmol), 200 mg (μL) H_2O and dissolve in 2 mL MeCN.
3. Seal the tube with an appropriate cap and expose to microwave heating for 60 min at 130 °C.
4. Determine a suitable thin-layer chromatography (TLC) eluent system for monitoring the reaction.
5. Allow enough time for the tube to reach room temperature and pierce the septa with a needle before opening the vial.
6. Check that your reaction has reached completion (*i.e.* full consumption of 2,6-dimethoxybenzoic acid) using an appropriate TLC eluent system from step 4.
7. Add 1 mL of formic acid and stir the reaction at room temperature until the next laboratory session.

Laboratory Session 2 (3 h): Product Isolation and Yield Determination

1. Transfer the reaction mixture to a round-bottomed flask, rinse with EtOAc and concentrate the crude material *via* rotary evaporation.

2. To this crude product, add slowly 10 mL of saturated aqueous sodium bicarbonate and 10 mL of ethyl acetate.
3. Transfer the solution into a separating funnel, separate the organic phase and extract the aqueous phase with 2 × 20 mL of ethyl acetate. Wash the combined organic layer with brine and dry over anhydrous Na_2SO_4, filter and concentrate *via* rotary evaporation. Calculate the crude mass (CM) and yield. The reaction yield will be determined by ^{1}H NMR.
4. Add 30 mg of the crude product (CP) to a 2 mL sample vial followed by 10 mg of the internal standard (IS) dimethyl acetamide. It is important to know the exact amount of crude product and internal standard added.
5. Add 0.65 mL of $CDCl_3$ and mix the solution by drawing it up in a Pasteur pipette a few times.
6. Transfer the mixture to a NMR tube, run a ^{1}H experiment and complete the following steps:

   - Calculate the number of moles of IS added (*A*).
   - Identify and integrate a CH_3 peak belonging to your product and the acetyl (CH_3CO) peak from the internal standard (compare with literature values).
   - Calculate the ratio CH_3/CH_3CO (*B*).
   - Calculate the number of moles (*C*) of product in your mixture ($A \times B$).
   - Calculate the mass of product (*D*) in your NMR mixture ($C \times$ molar mass of product).
   - Calculate the mass of product formed in your reaction ($D \times$ CM/CP).
   - Calculate the NMR yield.

Waste Disposal

The aqueous solution may contain traces of palladium and should be treated as reaction waste. Ethyl acetate may be reused or collected in a non-chlorinated waste collector recipient, for disposal.

Results Interpretation and Additional Questions

1. Interpret and assign the ^{1}H and ^{13}C NMR peaks for the product.
2. Discuss the advantages and disadvantages of carboxylic acids as coupling partners in cross-coupling chemistry.
3. What other carboxylic acids could be used as suitable substrates in this reaction?
4. Suggest an alternative synthetic route to the product obtained in this experiment. The route should include at least one C–C bond forming step. How does your route compare with the one in the experiment? Think about issues such as selectivity, cost, ease of synthesis and greenness.

References

1. F. Russo, L. R. Odell, K. Olofsson, P. Nilsson and M. Larhed, *Microwaves in Organic Synthesis*, ed. A. de la Hoz and A. Loupy, Wiley-VCH Verlag GmbH, Weinheim, 3rd edn, 2012, vol. 2, ch. 15, pp. 607–672.
2. J. Gising, L. R. Odell and M. Larhed, *Org. Biomol. Chem.*, 2012, **10**, 2713.
3. C. O. Kappe, *Angew. Chem., Int. Ed.*, 2004, **43**, 6250.
4. A. G. Myers, D. Tanaka and M. R. Mannion, *J. Am. Chem. Soc.*, 2002, **124**, 11250.
5. N. Rodríguez and L. J. Goossen, *Chem. Soc. Rev.*, 2011, **40**, 5030.
6. L. J. Goosen, N. Rodríguez and K. Goosen, *Angew. Chem., Int. Ed.*, 2008, **47**, 3100.
7. J. Rydfjord, F. Svensson, A. Trejos, P. J. R. Sjöberg, C. Sköld, J. Sävmarker, L. R. Odell and M. Larhed, *Chem.–Eur. J.*, 2013, **19**, 13803.
8. W. Jia and N. Jiao, *Org. Lett.*, 2010, **12**, 2000.
9. L. J. Goossen, N. Rodríguez, B. Melzer, C. Linder, G. J. Deng and L. M. Levy, *J. Am. Chem. Soc.*, 2007, **129**, 4824.
10. A. Axelsson, J. B. Veron, J. Sävmarker, J. Lindh, L. R. Odell and M. Larhed, *Tetrahedron Lett.*, 2014, **55**, 2379.
11. J. Lindh, P. J. R. Sjöberg and M. Larhed, *Angew. Chem., Int. Ed.*, 2010, **49**, 7733.

7.9. An Efficient Methodology for the Synthesis of the 3-Styryl Coumarin

António Manuel D. R. L. Pereira*[a,b] and Sérgio Miguel A. Martins[b]

[a]Departamento de Química, Escola de Ciências e Tecnologia, Universidade de Évora, Colégio Luís António Verney, Rua Romão Ramalho, N°, 59 7000-671, Portugal; [b]Laboratório HERCULES, Universidade de Évora, Largo Marquês de Marialva, N°, 8 7000-809 Évora, Portugal. *E-mail: amlp@uevora.pt

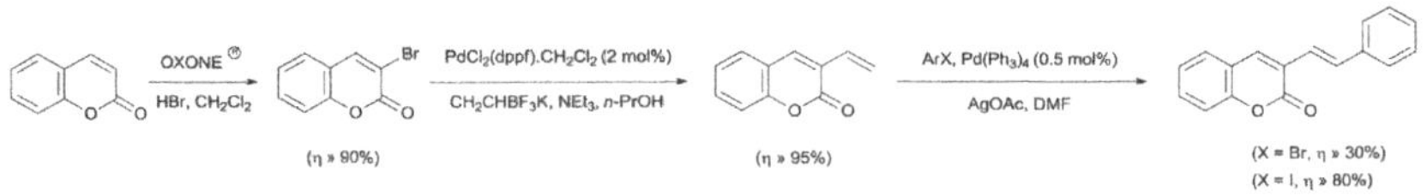

Number of sessions (duration of each session)	Hazard level	Difficulty level	Level of study
3 (4 h each)	Moderate	Medium	Advanced
Class names α,β-Unsaturated carbonyls, aryl halides, styryls			
Concepts involved This experiment gives a useful method for the synthesis of 3-styrylcoumarin, using an effective α-unsaturated carbonyl halogenation and high yield palladium coupling reactions			
Chemicals needed Coumarin (2*H*-chromen-2-one), Oxone® (potassium peroxymonosulfate), hydrobromic acid, triethylamine, dichloromethane, dimethylformamide, hexane, *n*-propanol, anhydrous sodium sulfate, calcium chloride, silica gel (230–400 mesh), potassium vinyltrifluoroborate, [1,1′-bis(diphenylphosphino)ferrocene]dichloropalladium(II)·CH_2Cl_2, bromobenzene or/and iodobenzene, tetrakis(triphenylphosphine)palladium(0), silver acetate			
Equipment and experimental techniques involved Apparatus for reaction with temperature and magnetic stirring, liquid–liquid extraction apparatus, flash column chromatography, NMR spectrometer. UV-vis spectrophotometer			
Keywords Heck reaction, liquid–liquid extraction, NMR, Suzuki reaction, α,β-unsaturated carbonyl halogenation, UV			

Background

Coumarins (or benzopyranones), whether naturally occurring or synthetic, have attracted the interest of the scientific community due to their broad pharmacological activities,[1,2] such as antiprotozoal,[3] anticancer,[4] antibacterial,[5] among others. Depending on the nature and substitution pattern, coumarins show exceptional optical properties,[6] as they constitute the largest class of fluorescent dyes,[7] widely used as emission layers in organic light-emitting diodes (OLED),[8] optical brighteners,[9] nonlinear optical chromophores,[10] fluorescent whiteners,[11] fluorescent labels and probes for physiological measurement[12] and, more recently, in labelling[13] and caging.[14] Developments from the last decade show that the introduction of appropriate substituents into the coumarin ring contribute to structures with improved photophysical and spectroscopic proprieties.[15] Many articles dealing with their synthesis, reactivity and spectral profile

Comprehensive Organic Chemistry Experiments for the Laboratory Classroom
Edited by Carlos A. M. Afonso, Nuno R. Candeias, Dulce Pereira Simão, Alexandre F. Trindade, Jaime A. S. Coelho, Bin Tan, and Robert Franzén.

Published by the Royal Society of Chemistry, www.rsc.org

have been published. In particular, it seems that the presence of an aryl or heteroaryl moiety on the 3-position of the coumarinic system induces specific activities.[16]

The objective of this work is the synthesis of 3-styrylcoumarin using the regioselective and highly efficient Heck coupling reaction.[17] To carry out the synthesis of 3-styrylcoumarin, the synthesis of 3-bromocoumarin was essential, which was accomplished with Oxone®[18] and HBr on reaction with coumarin. Subsequent Suzuki cross-coupling reaction[19] of 3-bromocoumarin with potassium vinyltrifluoroborate[20] allowed 3-vinylcoumarin, the scaffold used to obtained 3-styrylcoumarin. The increase in conjugation is reflected in the absorbance of the synthesized compound revealing a pronounced bathochromic shift and hyperchromic effect.

Additional Safety

All sessions should be done in a fume hood, with proper safety equipment, such as protective clothing, gloves and safety goggles. Hydrobromic acid is a corrosive acid and may cause severe skin and eye burns, and can irritate the respiratory system. Triethylamine is a flammable reagent, is harmful if swallowed, or in contact with the skin, and may cause burns. Bromobenzene is a flammable reagent, is harmful if swallowed, or in contact with skin, and can irritate the eyes and the respiratory system. Iodobenzene is harmful if swallowed, or in contact with skin, and can irritate the eyes.

Experimental Procedure

Laboratory Session 1 (4 h): Synthesis of 3-Bromocoumarin

1. Prepare a 25 mL round-bottomed flask with a magnetic stir bar, and add 438 mg (3.0 mmol, 1.0 equiv.) of coumarin, 1.29 g (4.2 mmol) of Oxone® (potassium peroxymonosulfate) and 12 mL of CH_2Cl_2.
2. Add 3.3 mL (6.6 mmol) of 2 M HBr, in one portion, to the solution prepared in 1. The final solution exhibits dark red coloration.
3. After stirring 2 h at room temperature add, cautiously, 1.25 mL of NEt_3 (9.0 mmol). The solution color changes immediately to yellow.
4. After stirring for a further 1 h at room temperature, transfer the solution into a separatory funnel, already containing 20 mL of distilled water, and perform an extraction with 3 × 25 mL of CH_2Cl_2. Remove the water from the combined organic layers with anhydrous Na_2SO_4, filter and evaporate the solvent.
5. Weigh the desired product and calculate the yield. Record UV and ^1H and ^{13}C NMR spectra.

Note: the next step of the reaction can be conducted with the obtained crude product, with a small reduction in the expected reaction yield (5–10%). Optionally, the reaction product can be purified by flash-column chromatography on silica gel (230–400 mesh; hexane/CH_2Cl_2 (5 : 5)).

Laboratory Session 2 (4 h): Synthesis of 3-Vinyl Coumarin

1. Prepare a 50 mL round-bottomed flask with a magnetic stir bar, and add 252 mg (1.120 mmol, 1.0 equiv.) of 3-bromocoumarin, 180 mg (1.344 mmol) of potassium vinyltrifluoroborate, 18 mg (0.021 mmol) of $PdCl_2(dppf){\cdot}CH_2Cl_2$, 0.156 mL (1.120 mmol) of NEt_3 and 18 mL of *n*-PrOH.
2. Insert a water condenser into the round-bottomed flask, equipped with a calcium chloride drying tube, and clamp the flask in the oil bath.
3. Reflux the mixture while stirring magnetically for 15 min.

4. After cooling to room temperature, transfer the solution into a separatory funnel already containing 20 mL of distilled water, and perform an extraction with 3 × 25 mL of CH_2Cl_2. Remove the water from the combined organic layers with anhydrous Na_2SO_4, filter and evaporate the solvent.
5. Weigh the desired product and calculate the yield. Record UV and ^1H and ^{13}C NMR spectra.

Note: the next step of the reaction can be conducted with the obtained crude product, with a small reduction in the expected reaction yield (5–10%). Optionally, the reaction product can be purified by flash-column chromatography on silica gel (230–400 mesh; hexane/CH_2Cl_2 (5 : 5)).

Pre-Laboratory Session 3 (10–15 min): Synthesis of 3-Styryl Coumarin

Prepare a 50 mL round-bottomed flask with a magnetic stir bar, and add 104 mg (0.600 mmol) of 3-vinylcoumarin, 0.062 mL (0.545 mmol, 1.0 equiv.) of iodobenzene (or 0.057 mL (0.545 mmol, 1.0 equiv.) of bromobenzene), 31 mg (0.027 mmol) of $Pd(PPh_3)_4$, 100 mg (0.600 mmol) of CH_3CO_2Ag and 1.5 mL of DMF. Insert a water condenser into the round-bottomed flask and clamp the flask in the oil bath, and stir the mixture under a nitrogen atmosphere, maintaining a constant temperature of 80 °C for 72 h.

Laboratory Session 3 (4 h): Synthesis of 3-Styryl Coumarin

1. After cooling the reaction mixture to room temperature, transfer the solution into a separatory funnel already containing 20 mL of distilled water, and perform an extraction with 3 × 25 mL of CH_2Cl_2. Remove the water from the combined organic layers with anhydrous Na_2SO_4, filter and evaporate the solvent.
2. Purify the reaction product by flash-column chromatography on silica gel (230–400 mesh; hexane/CH_2Cl_2 (2 : 8)).
3. Weigh the desired product and calculate the yield. Record UV and ^1H and ^{13}C NMR spectra.

Waste Disposal

The neutralized aqueous solution may be diluted and discarded in the sink. CH_2Cl_2 may be reused or collected in a chlorinated waste collector recipient, for posterior destruction.

Results Interpretation and Additional Questions

1. Calculate the product yield for each step.
2. Interpret the ^1H and ^{13}C NMR spectra.
3. Interpret and compare UV spectra.
4. Present all reactions mechanisms.

References

1. M. E. Riveiro, N. De Kimpe, A. Moglioni, R. Vazquez, F. Monczor, C. Shayo and C. Davio, *Curr. Med. Chem.*, 2010, **14**, 1325–1338.
2. J. R. S. Hoult and M. Paya, *Gen. Pharmacol.*, 1996, **27**, 713–722.
3. A. J. Pierson, T. Dumetre, S. Hutter, F. Delmas, M. Laget, J. P. Finet, N. Azas and S. Combes, *Eur. J. Med. Chem.*, 2010, **45**, 864–869.
4. S. Combes, P. Barbier, S. Douillard, A. M-Florin, V. B-Rey, J. T. Pierson, A. Y. Fedorov, J. P. Finet, J. Boutonnat and V. Peyrot, *J. Med. Chem.*, 2011, **54**, 3153–3162.
5. Y. P. Chin, W. J. Huang, F. L. Hsu, Y. L. Lin and M. H. Lin, *Arch. Pharm.*, 2011, **344**, 386–393.
6. N. A. Kuznetsova and O. L. Kaliya, *Russ. Chem. Rev.*, 1992, **61**, 1243–1267.
7. A. R. Jagtap, V. S. Satam, R. N. Rajule and V. R. Kanetkar, *Dyes Pigm.*, 2009, **82**, 84–89.

8. M. T. Lee, C. K. Yen, W. P. Yang, H. H. Chen, C. H. Liao, C. H. Tsai and C. H. Chen, *Org. Lett.*, 2004, **6**, 1241–1244.
9. J. Kido and Y. Iizumi, *Appl. Phys. Lett.*, 1998, **73**, 2721–2723.
10. S. J. Benight, L. E. Johnson, R. Barnes, B. C. Olbricht, D. H. Bale, P. J. Reid, B. E. Eichinger, L. R. Dalton, P. A. Sullivan and B. H. Robinson, *J. Phys. Chem. B*, 2010, **114**, 11949–11956.
11. K. A. Wilze and A. K. Johnson, *Handbook of Detergents, Chemistry, Production, and Application of Fluorescent Whitening Agents*, CRC Press, Taylor & Francis, Boca Raton, 2007, Part F.
12. H. S. Jung, P. S. Kwon, J. W. Lee, J. I. Kim, C. S. Hong, J. W. Kim, S. H. Yan, J. Y. Lee, J. H. Lee, T. Joo and J. S. Kim, *J. Am. Chem. Soc.*, 2009, **131**, 2008–2012.
13. S. Heiner, H. Detert, A. Kuhn and H. Kunz, *Bioorg. Med. Chem.*, 2006, **14**, 6149–6155.
14. H. Yu, J. Li, D. Wu, Z. Qiu and Y. Zhang, *Chem. Soc. Rev.*, 2010, **39**, 464–473.
15. S. Mizukami, S. Okada, S. Kimura and K. Kikuchi, *Inorg. Chem.*, 2009, **48**, 7630–7648.
16. M. J. Matos, D. Vina, C. Picciau, F. Orallo, L. Santana and E. Uriarte, *Bioorg. Med. Chem. Lett.*, 2009, **19**, 5053–5055.
17. I. Beletskaya and A. Cheprakov, *Chem. Rev.*, 2000, **100**, 3009–3066.
18. K. M. Kim and I. H. Park, *Synthesis*, 2004, **16**, 2641–2644.
19. S. R. Chemler, D. Trauner and S. J. Danishefsky, *Angew. Chem., Int. Ed.*, 2001, **40**, 4544–4568.
20. G. A. Molander and R. Figueroa, *Aldrichimica Acta*, 2005, **38**, 49–56.

7.10. Sonogashira Coupling Between a Vinylic Halide and a Terminal Alkyne

Carine Maaliki and Jérôme Thibonnet*

UFR Sciences et Techniques de Tours, Parc Grandmont, Bat Yves Chauvin, 32 Av. Monge 37200 Tours, France. *E-mail: jerome-thibonnet@univ-tours.fr

Number of sessions (duration of each session)	Hazard level	Difficulty level	Level of study
1 (4 h)	Moderate	High	Advanced

Class names Vinylic halides, terminal alkynes

Concepts involved Sonogashira cross-coupling of a terminal alkynes with a vinyl halide under palladium catalysis under anhydrous conditions using Schlenk techniques

Chemicals needed (3*E*)-Iodoprop-2-enoic acid, phenylacetylene, *N*,*N*-dimethylformamide, diisopropylamine, copper iodine, triphenylphosphine, diisopropylamine, palladium(II) acetate, ammonium chloride, diethyl ether, hydrochloric acid, anhydrous magnesium sulfate, sodium hydroxide, deuterium chloroform

Equipment and experiment techniques involved Magnetic stirring, Schlenk, vacuum line, separatory funnel, acid–base extraction, crystallization, ^{1}H NMR, ^{13}C NMR

Keywords Acid–base washing, C(sp)–C(sp) bond formation, NMR, Sonogashira coupling

Background

Over the past few decades, transition-metal-catalyzed cross-coupling reactions have emerged as one of the most general and reliable methods for forming sp^2–sp^2 or sp–sp^2 carbon–carbon bonds in the field of organic synthesis. Among them, the reaction between alkenyl or aryl halides or triflates and terminal alkynes catalyzed by a Pd(II)/Cu(I) system is known as the Sonogashira coupling, and is one of the most powerful methods for the straightforward construction of sp^2–sp carbon–carbon bonds to prepare aryl-alkynes and conjugated enynes (Scheme 7.10.1).[1]

This methodology has been widely applied to prepare biologically active molecules, natural products, conducting polymers/engineering materials and macrocycles with acetylene links.[2]

The copper-co-catalyzed Sonogashira reaction is believed to take place through two independent catalytic cycles (Scheme 7.10.2).

The reaction's initial step results in the reduction of the Pd(II) complex under addition of triphenylphosphine to obtain a Pd(0) complex (Pd(0)L_2).

Comprehensive Organic Chemistry Experiments for the Laboratory Classroom
Edited by Carlos A. M. Afonso, Nuno R. Candeias, Dulce Pereira Simão, Alexandre F. Trindade, Jaime A. S. Coelho, Bin Tan, and Robert Franzén.

Published by the Royal Society of Chemistry, www.rsc.org

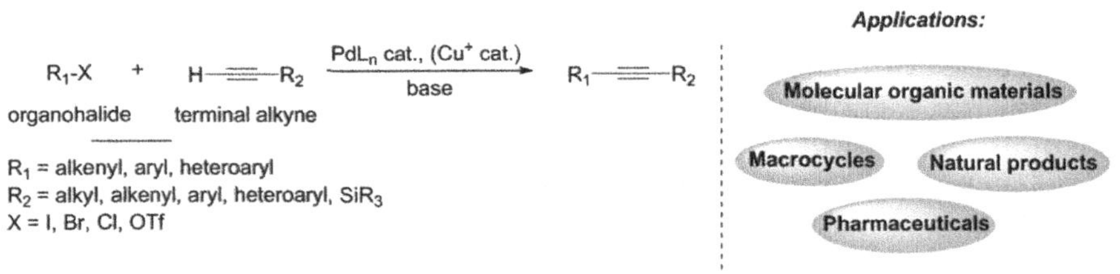

Scheme 7.10.1 General procedure for Sonogashira coupling and applications.

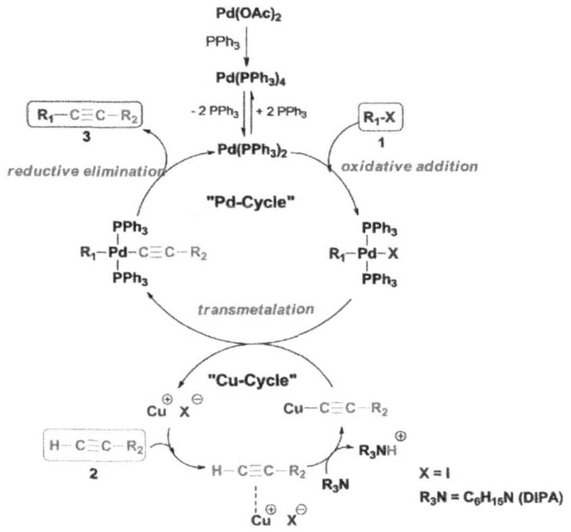

Scheme 7.10.2 Proposed catalytic cycles in the copper-co-catalyzed Sonogashira reaction.

The palladium catalyzed mechanism begins with the fast oxidative addition of the organohalide to the Pd(0)L_2 to form a Pd(II) complex. In the oxidative addition step, the characteristics of the R_1–X substrate are crucial, with this step being facilitated if X = I or OTf and if the electronic density is reduced on the C–X bond by the presence of electron-withdrawing groups.[3]

In parallel, the terminal alkyne and the copper co-catalyst forms the organocopper reagent. Transmetalation occurs between the Pd(II) complex (R_1PdL_2X) and the organocopper reagent and forms the Pd(II) alkyne complex ($R_1Pd(C{\equiv}CR_2)L_2$), and gives the copper halide. Reductive elimination gives the final coupled alkyne after *trans*/*cis* isomerization. The catalyst is then regenerated. The catalytic cycle can begin again.

The Cu-cycle is still less understood. In presence of the copper(I) salt, the base (generally an amine) is supposed to abstract the acetylenic proton of the terminal alkyne, thus forming a copper acetylide ($Cu(C{\equiv}CR_2)$) which is the organocopper reagent.

This experiment aims at the synthesis of enynoic acids under Sonogashira conditions using palladium(II) acetate, triphenylphosphine, copper iodide, diisopropylamine and DMF as solvent at room temperature.

Additional Safety

Phenylacetylene, diisopropylamine and DMF are inflammable; they must be kept away from sources of ignition (GHS02). Diisopropylamine is corrosive to metal (GHS05). All the reagents used in the procedure are irritating to the eyes, skin and respiratory tract (GHS07 and GHS08).

Vinyl halides are potentially carcinogenic. Aqueous solutions of sodium hydroxide and hydrochloric acid can cause eye and skin irritation.

A glass flaw on the vacuum line and Schlenk lines can constitute an implosion hazard. Implosions can be violent and can produce high velocity glass shrapnel. Therefore, Schlenk lines must be visually inspected, especially during and immediately after installation. Cracks of any sort should be repaired by a competent glass technician. All repairs should be annealed or changed.

Using a vacuum pump means paying attention of joints, stopcocks and connectors. They must be cleaned/greased often, simply to prevent annoying leaks. Leaking manifolds allow air to enter a vacuum system. The liquid nitrogen trap can condense the oxygen into a liquid state (blue liquid) and create an explosion in the trap.

If you remove your nitrogen dewar and notice the presence of liquid, you must stop the vacuum pump, release the vacuum and open the ports to increase the pressure.

Experimental Procedure

The difficulty with this experiment is to maintain the reaction mixture under inert conditions. Between each addition of reagents, the reaction mixture is purged with argon.

The vessel used must be very dry otherwise the reaction will not work. The Schlenk tube must be heated with a paint burner for 3–4 min under vacuum. Under argon, the Schlenk tube cools down to room temperature.

Protocol

1. Under argon, charge a dry Schlenk tube with 0.5 g (2.02 mmol) of (3*E*)-iodoprop-2-enoic acid and 10 mL of DMF (dimethylformamide). Stir the solution rapidly.
2. Put in a cooling bath.
3. Add in the following reagents in the order given: 0.44 mL (0.41 g, 4.04 mmol) phenylacetylene, 0.7 mL diisopropylamine (5 mmol), 96.5 mg copper iodine (20 mol%) (solution becomes green), 132 mg of triphenylphosphine (20 mol%) and 28 mg $Pd(OAc)_2$ (5 mol%) (reaction mixture turns red). After 20 min, the reaction mixture turns brown and a white solid appears. If the stirring is not working well, DMF can be added.
4. Stir the reaction mixture for 60 min.
5. Hydrolyse the reaction mixture by addition of 40 mL diethyl ether, 10 mL saturated solution of ammonium chloride and 5 mL of 2 M aqueous solution of hydrochloric acid.
6. Transfer the biphasic solution to a separatory funnel. The acidity of the aqueous phase is monitored with pH paper.
7. Then extract the aqueous phase three times with diethyl ether (3 × 20 mL). The combined organic phases are washed with an aqueous solution of ammonium chloride (3 × 10 mL) and with distilled water (1 × 10 mL).
8. Organic phases are then dried over anhydrous magnesium sulfate. After filtration on cotton, the solvent is removed under reduced pressure with a rotary evaporator.
9. Weight the round-bottomed flask containing the crude product in order to determine the final weight (w_1).

Product Isolation

Acid–Base Washing

The crude product is purified by acid–base washing. Dissolve the crude product with 2 equivalents of NaOH (1 M), and wash the aqueous phase three times with diethyl ether to remove impurities. (The expected product is in the aqueous phase, the organic phase is left out.) Add

50 mL diethyl ether and 10 mL HCl 1 M to transfer the product from the aqueous phase to the organic phase. (Observe here that the aqueous phase loses it color and that the organic phase becomes beige. The acidity of the aqueous phase must be controlled by pH paper.) Finally, extract the aqueous phase twice with diethyl ether and dry the combined organic phases over anhydrous magnesium sulfate. After filtration, remove the solvent with reduced pressure until the flask has a constant weight.

Crystallisation

Dissolve the product in a minimum quantity of chloroform. By evaporation at room temperature, formation of crystals of the products occurs.

1. Weigh the round-bottomed flask containing the purified product and determine the final weight (w_2).
2. Place 10 mg of the isolated product in an NMR tube and add 0.6 mL $CDCl_3$.
3. Acquire the ^1H and ^{13}C NMR spectrum.
4. With 5 mg of the product, acquire the FT-IR spectrum.

Waste Disposal

The copper aqueous phase may be diluted and discarded in the metallic inorganic waste collector recipient. The pH of the aqueous phase must be controlled and the solution must be placed in the acid or basic container for subsequent destruction. $MgSO_4$ and cotton may be placed in the bin. Diethyl ether must be collected in a non-chlorinated waste collector recipient.

Results Interpretation and Additional Questions

1. Compare w_1 and w_2 and discuss the result.
2. Calculate the yield of the reaction.
3. Determine the melting point.
4. Analyse the NMR spectra.
5. Assign the signals of the FT-IR spectrum.
6. Explain what reaction occurred during washings, firstly with the basic aqueous solution and then with the aqueous acid solution. How is this useful?
7. Propose a catalytic cycle for the reaction, using the reagents.

References

1. (a) K. Sonogashira, Y. Tohda and N. Hagihara, *Tetrahedron Lett.*, 1975, **16**, 4467; (b) J. S. Moore, *Acc. Chem. Res.*, 1997, **30**, 402; (c) A. Meijere and F. Diederich, *Metal-Catalyzed Cross-Coupling Reactions, second edition*, Wiley-VCH Verlag GmbH, Weinheim, Germany, 2002, p. 478; (d) L. Brandsma, S. F. Vasilevsky and H. D. Verkruijsse, *Application of Transition Metal Catalysts in Organic Synthesis*, Springer, Berlin, Germany, 1999, p. 179; (e) K. Sonogashira, *Handbook of Organopalladium Chemistry for Organic Synthesis*, Wiley-VCH Verlag GmbH, New York, 2002, p. 493; (f) K. Sonogashira, *J. Organomet. Chem.*, 2002, **653**, 46; (g) E. Negishi and L. Anastasia, *Chem. Rev.*, 2003, **103**, 1979.
2. (a) B. M. Trost, T. N. Nanninga and D. M. T. Chan, *Organometallics*, 1982, **1**, 1543; (b) F. D. Lewis, D. K. Howard, S. V. Barancyh and J. D. Oxman, *J. Am. Chem. Soc.*, 1986, **108**, 3016; (c) B. Samuelson, *Angew. Chem., Int. Ed. Engl.*, 1983, **22**, 805; (d) I. E. Ronning and H. A. Frank, *J. Food Prot.*, 1988, **51**, 1643.
3. R. Chinchilla and C. Najera, *Chem. Rev.*, 2007, **107**, 874.

8. Addition to sp Carbon

8.1. Synthesis of Dulcin

Dulce Simão

Department of Chemistry, Instituto Superior Técnico, University of Lisbon, Centro de Química Estrutural, Av. Rovisco Pais, 1, 1049-001 Lisbon, Portugal. *E-mail: dulcesimao@tecnico.ulisboa.pt

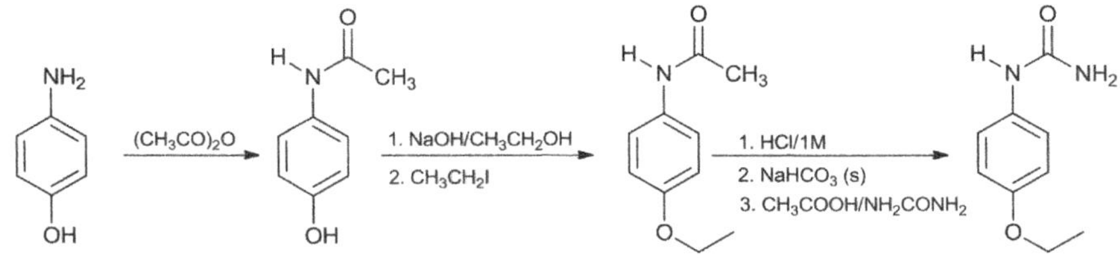

p-aminophenol paracetamol phenacetin dulcin

Duration	Hazard level	Difficulty level	Level of study
3 (4 h each)	Moderate	Low	Intermediate
Class names Aromatic amines, aromatic amides, aromatic alcohols, aromatic ethers, ureas			
Concepts involved Three-step reaction sequence of an analgesic and its and its transformation into a sweetener, illustrating amino protection, Williamson synthesis, amide hydrolysis and nucleophilic addition of an aryl amine with an isocyanate			
Chemicals needed *p*-Aminophenol, acetic anhydride, sodium hydroxide, ethanol, ethyl iodide, decolorizing charcoal, hydrogen chloride (37%), sodium hydrogen carbonate, glacial acetic acid, urea			
Equipment and experimental techniques involved Recrystallization, IR, NMR, stirring heating plate, vacuum filtration			
Keywords Amide hydrolysis, amine protection, crystallization, IR, isocyanate, NMR, Williamson synthesis			

Background

Paracetamol (*p*-hydroxyacetanilide or acetaminophen) is an analgesic that acts slowly in comparison with aspirin and is also a strong antipyretic. It is the active constituent of Tylenol® and Ben-u-ron®. It is easily prepared by acetylation of *p*-amino phenol,[1] involving a nucleophilic addition–elimination reaction.[2] This acetylation is essential to protect the amino group in the next step.[3,4] Phenacetin is not as strong as antipyretic and several studies revealed that its administration in large quantities could lead to kidney and liver damages. Animal tests showed that phenacetin could induce cancer when taken in tablets and was banned from the market in 1981.[5,6] Phenacetin is obtained in good yield from paracetamol through a Williamson synthesis, an important route by which to obtain unsymmetrical ethers.[7] The reaction of paracetamol

Comprehensive Organic Chemistry Experiments for the Laboratory Classroom
Edited by Carlos A. M. Afonso, Nuno R. Candeias, Dulce Pereira Simão, Alexandre F. Trindade, Jaime A. S. Coelho, Bin Tan, and Robert Franzén.

Published by the Royal Society of Chemistry, www.rsc.org

with sodium hydroxide produces a phenoxide ion that reacts with ethyl iodide in a bimolecular nucleophilic substitution (S_N2) affording an aromatic ether.[8–10] This compound is used to obtain dulcin, one of the first artificial sweeteners to reach the market. In 1951, toxicity studies led to its removal from the market.[11] This compound is synthesized by acid hydrolysis of the amide function of phenacetin[12] followed by nucleophilic addition of the amine to an isocyanate[13] which is produced *in situ* by reaction of urea with acetic acid. This experiment can be performed on a micro scale, without the use of halogenated solvents. In addition, dulcin can be prepared directly from a paracetamol tablet, minimizing costs and effluents,[8] where all reaction products can be recrystallized from water, thus contributing to green chemistry.

Additional Safety

Acetic anhydride and acetic acid are corrosive to the skin, eyes and mucous membranes; *p*-aminophenol is irritant; sodium hydroxide and hydrogen chloride are corrosive; ethyl iodide is irritant and a flammable liquid, as is ethanol, and they should be handled with care.

Experimental Procedure

Session 1 – Preparation of Paracetamol (*p*-Hydroxyacetanilide or Acetaminophen)

1. In a 250 mL round-bottom flask place *p*-aminophenol (11 g, 100 mmol) and water (30 mL).
2. Add a magnetic bar and acetic anhydride (12 mL). Fit a reflux condenser. Start vigorous stirring and heat in a water bath (40–50 °C). Observe the solid dissolution.
3. After 10 min, cool the flask in an ice bath and collect the precipitate by vacuum filtration.
4. Wash the precipitate with water and recrystallize it from water.
5. Record the yield, mp and acquire IR spectrum.

Session 2 – Preparation of Phenacetin

1. In a 50 mL round-bottom flask, place *p*-hydroxyacetanilide (2.1 g, 13.9 mmol) and a solution of NaOH in ethanol (15.5 mL, 1 M).
2. Add a magnetic stir bar, fit a reflux condenser. Stir and reflux the mixture for 15 min with a water bath.
3. Remove the heating source and add ethyl iodide (2.1 mL, 26.1 mmol) through the top of the condenser with a 50 mL pressure-equalizing dropping funnel.
4. Reflux again for a further 15 min.
5. Filter under vacuum the hot reaction mixture into a filter flask with ice and water (*ca.* 40 mL). Phenacetin precipitates on contact with the cold water.
6. Shake the mixture until the ice melts completely. Filter the mixture by vacuum to collect the product, which is recrystallized from water, and use decolorizing charcoal to remove colored impurities.
7. Record the yield and mp and acquire IR spectrum.

Session 3 – Preparation of Dulcin

1. In a three-necked 50 mL round-bottom flask place phenacetin (1.8 g, 10 mmol) and aqueous HCl solution (8.3 mL, 50 mmol, 6 M). Add a magnetic bar and fit a reflux condenser. Stir and reflux for 15 min using a silicone oil bath. Observe the phenacetin dissolution.
2. Remove the heating source but continue stirring. When reflux stops, add, in small portions, $NaHCO_3$ until the pH is in the range 6–6.5 (use pH indicator strips).

3. Add urea (2.4 g, 15.9 mmol) and glacial acetic acid (3 drops).
4. Reflux the mixture for 1 h. Dulcin starts to precipitate as white crystals.
5. Cool the mixture with an ice water bath and collect the product by vacuum filtration.
6. Recrystallize the compound from water.
7. Record the yield and mp.

Waste Disposal

The neutralized aqueous solutions may be diluted and discarded in the sink.

Results Interpretation and Additional Questions

1. Determine the yields of reaction products.
2. Compare the melting point of the obtained product with the one reported in the literature.
3. Interpret the IR and NMR spectra.
4. Write out all the reaction mechanisms.
5. Explain the importance of the first step in dulcin synthesis.
6. Explain why a stronger base than NaOH is not needed to remove a proton from *p*-hydroxyacetanilide.

References

1. A. R. Tatchell, B. S. Furnis, A. J. Hannaford and P. W. G. Smith, *Vogel's Textbook of Practical Organic Chemistry*, Longman Scientific and Technical, 5th edn, 1989, p. 985.
2. T. W. G. Solomons, *Organic Chemistry*, John Wiley & Sons Inc., New York, 7th edn, 1998, p. 837.
3. R. S. Ward, *Bifunctional Compounds*, Oxford University Press, 1st edn, 1994, vol. 17, Oxford Chemistry Primers.
4. T. W. Greene, *Protective groups in Organic Synthesis*, John Wiley & Sons, 1st edn, 1981, p. 249.
5. U. C. Dubach, B. Rosner and T. Stürmer, *N. Engl. J. Med.*, 1991, **324**, 155–160.
6. *Consolidated List of Products Whose Consumption And/or Sale Have Been Banned, Withdrawn, Severely Restricted Or Not Approved by Governments*, United Nations Publications, 1983, p. 172.
7. T. W. G. Solomons, *Organic Chemistry*, John Wiley & Sons, Inc., New York, 7th edn, 1998, p. 1026.
8. B. D. Williams, B. Williams and L. Rodino, *J. Chem. Educ.*, 2000, **77**, 357–359.
9. E. J. Volker, E. Pride and C. Hough, *J. Chem. Educ.*, 1979, **56**, 831.
10. K. Ditts and M. Durand, *J. Chem. Educ.*, 1990, **67**, 74.
11. R. H. Goldsmith, *J. Chem. Educ.*, 1987, **64**, 954–955.
12. T. W. G. Solomons, *Organic Chemistry*, John Wiley& Sons, Inc., New York, 7th edn, 1998, p. 840.
13. J. March, *Advanced Organic Chemistry*, John Wiley and Sons, 4th edn, 1992, p. 903.

8.2. Microwave-Assisted Solid-Phase Synthesis of Hydantoins

Isabelle Parrot* and Guilhem Chaubet

IBMM UMR 5247 CNRS-Université Montpellier 1 et 2, Faculté des Sciences, Pl. Eugène Bataillon, 34 095 Montpellier Cedex 5, France. *E-mail: isabelle.parrot@um2.fr

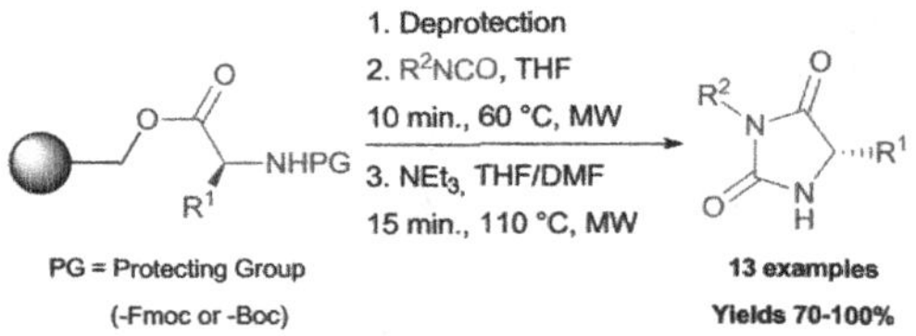

Duration	Hazard level	Difficulty level	Level of study
2 (3 h each)	High	Medium	Advanced

Class names Esters, carbamates, amines, isocyanates, ureas, hydantoins

Concepts involved This experiment aims to initiate undergraduate students into the use of both solid-phase synthesis and microwave irradiation *via* the preparation of diversely substituted hydantoins. Starting from commercially available resins preloaded with protected amino acids, this three-step procedure offers rapid access to the desired heterocycles by a cyclative cleavage strategy, with excellent yields and without the need of purification. This gives the opportunity to the students to learn another synthesis tool – rather than the classical solution phase synthesis/work-up/column chromatography purification – and shows them that tedious work-up and purifications can be avoided

Chemicals needed Commercially available Wang and/or Merrifield resins preloaded with *N*-Fmoc and/or *N*-Boc protected amino acids, piperidine and/or TFA, triethylamine, *N*-alkyl and/or *N*-aryl isocyanates, DMF, DCM, THF, MeOH, solutions for the Kaiser test prepared in advance by the instructor in charge to avoid harmful problems (phenol, EtOH, KCN, pyridine, ninhydrin)

Equipment and experimental techniques involved Microwave especially designed for chemistry, orbital shaker (not essential), vacuum pump for suction filtration, rotary evaporator; ^{1}H and ^{13}C NMR, HPLC, low resolution mass spectrometry for characterizations

Keywords Cyclative-cleavage, evaporation under reduced pressure, hydantoins, NMR, solid-phase synthesis

Background

Through this laboratory activity it is intend to familiarize undergraduate students with microwave-assisted organic synthesis. Microwave technology has been used for organic chemistry purposes since the mid-1980s and displays numerous advantages compared to conventional heating methods.[1] Among them, the most important benefit is the acceleration of the heating process due to microwave irradiations.[2] In a conventional heating process, heat is typically transferred from an external source – typically an oil bath or a heating mantle – to the walls

Comprehensive Organic Chemistry Experiments for the Laboratory Classroom
Edited by Carlos A. M. Afonso, Nuno R. Candeias, Dulce Pereira Simão, Alexandre F. Trindade, Jaime A. S. Coelho, Bin Tan, and Robert Franzén.

Published by the Royal Society of Chemistry, www.rsc.org

of the flask first before it reaches the reaction mixture. Overall, this process is wasteful and costly in terms of energy, and does not represent an efficient way to conduct energy to the reaction mixture. In contrast, microwave irradiations are directly transmitted from the source to the polar molecules of the reaction mixture, due to the use of microwave-transparent materials for the reaction vessels, resulting in almost no energy loss. Very high temperatures can then be reached within a minute, leading to shortened reaction times and enhanced reaction rates compared to conventional heating, the so-called "microwave effect".[3] The objective of this work here is to combine this strategy with the use of solid-phase synthesis for a rapid, cheap and efficient synthesis of hydantoins. Hydantoins are heterocyclic five-membered rings found in many bioactive compounds with a broad range of activities including fungicidal, antibacterial, herbicidal, antiviral, anti-inflammatory, analgesic, neuroprotective, antiarrhythmic, diuretic or antihypertensive properties.[4–12] Since substituent variations can dramatically influence the biological activities of these heterocycles, methods are needed to synthesize rapidly and efficiently libraries of diversely substituted hydantoins.[13–15] Using a solid-phase strategy allows us to bypass tedious work-up and purification steps by simply filtrating and washing the resin. A cyclative-cleavage strategy is performed in a one-pot process, the heterocyclic moiety being built with a concomitant release from the resin and collected with a great purity. Thus, microwave irradiation combined with the solid-phase approach helps to prepare a small library of hydantoins with reaction times in minutes instead of hours.

Additional Safety

List of dangerous chemicals involved in this experiment: trifluoroacetic acid (TFA), piperidine, phenyl isocyanate, triethylamine, methanol (MeOH), dichloromethane (DCM), tetrahydrofuran (THF) and dimethylformamide (DMF).

Many of the chemicals used in this activity are flammable and poisoning may occur from ingestion, inhalation, or percutaneous absorption. In addition, phenyl isocyanate and triethylamine are also lachrymators, while triethylamine, piperidine and TFA are corrosive and may damage the skin. Thus, besides regular laboratory good practices (*e.g.* wearing laboratory coat, safety spectacles and protective gloves), students must work in a fume hood. Kaiser solutions must be prepared beforehand by the instructor. Risks can be minimized by carefully reading in advance the hazard and precautionary statements of each chemical.

Even though high temperatures may be used during this activity, every microwave device is normally equipped with a cooling system, dramatically reducing the risks of burning for the students or the instructor.

Experimental Procedure

Two laboratory sessions are required. The first will concern the deprotection of the *N*-protected amino acids anchored on preloaded Wang and/or Merrifield resins, involving two different procedures relying on the nature of protecting group (*i.e.* either a resin preloaded with *N*-Fmoc or *N*-Boc protected amino acid, Schemes 8.2.1 and 8.2.2, respectively). During the second session, free amino functional groups of the resin will react with an isocyanate to build the corresponding intermediate urea on a solid-support (Scheme 8.2.3). The desired hydantoin is then released by a basic treatment of beads (Scheme 8.2.4).

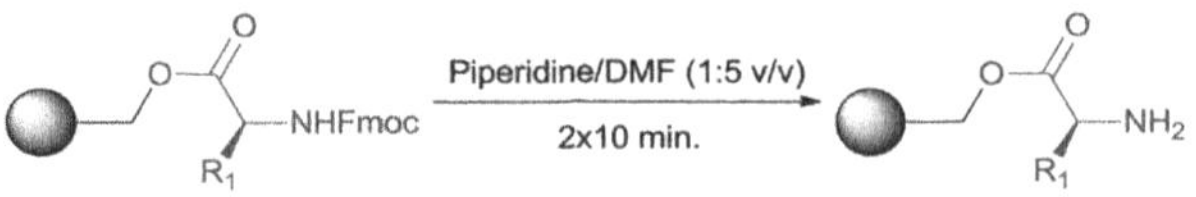

Scheme 8.2.1 *N*-Fmoc deprotection.

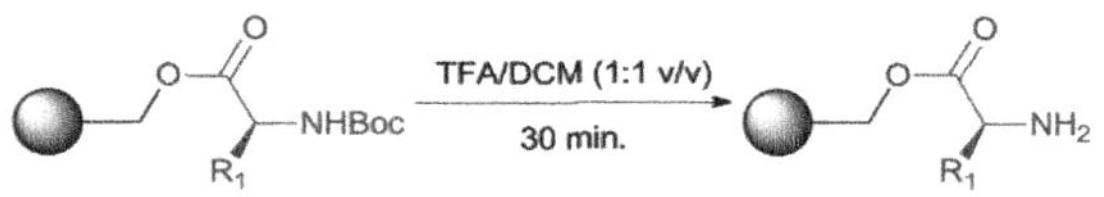

Scheme 8.2.2 *N*-Boc deprotection.

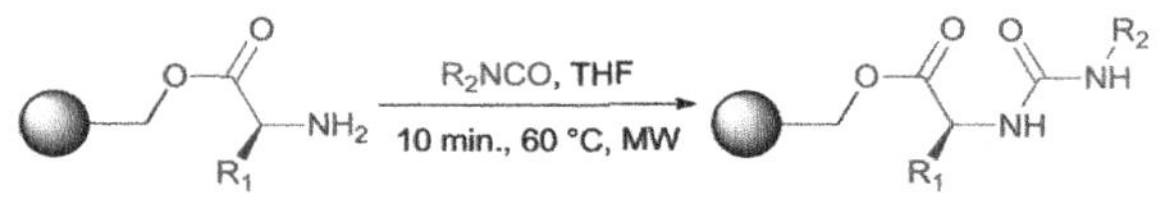

Scheme 8.2.3 Urea bond formation.

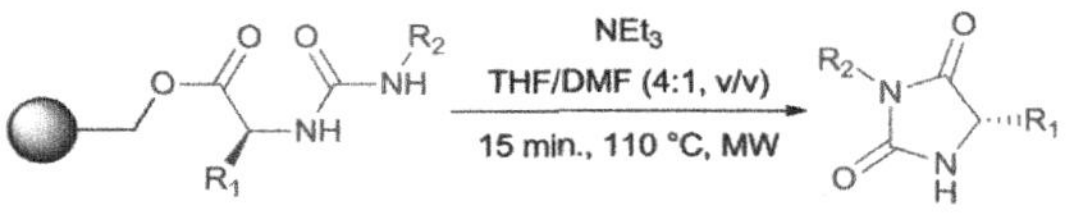

Scheme 8.2.4 Hydantoin formation by cyclative-cleavage.

Laboratory Session 1 (3 h)

Put 1–5 mg of the starting resin aside as a negative reference for Kaiser tests.

N-*Carbamate Deprotection*

Method A (*N*-Fmoc Deprotection).

1. In a fume hood, place the resin (0.76 mmol) in a reactor with a sufficient amount of a 1 : 5 v/v solution of piperidine/DMF (10 mL typically) to cover and allow the swelling of the resin.
2. Stir the reaction for 10 min using, preferentially, an orbital shaker. Otherwise, a very gentle magnetic stirring could be applied to the reaction, placing a magnetic stir bar inside the reactor.
3. The resin is then swollen and washed carefully with 30 mL DCM, then 30 mL MeOH, and finally with 30 mL DCM. Repeat this washing cycle three times. With regard to the swelling and constriction properties of resin, use a suction filtration process. The washing process should take approximately 30 min.
4. Perform a Kaiser test using 1–5 mg of resin. A positive test indicates completion of the deprotection, a negative test requires a repetition of the deprotection procedure from the beginning. Usually, two deprotection cycles are necessary.

Method B (*N*-Boc Deprotection).

1. In a fume hood, place the resin (0.76 mmol) in a reactor with a sufficient amount of a 1 : 1 v/v solution of TFA/DCM (10 mL typically) to cover and allow the swelling of the resin.
2. Stir the reaction for 30 min using, preferentially, an orbital shaker. Otherwise, a very gentle magnetic stirring could be applied to the reaction, placing a magnetic stir bar inside the reactor.
3. The resin is then swollen and washed carefully with 30 mL DCM, then 30 mL MeOH, and 30 mL of a 1 : 1 v/v solution of triethylamine/DCM and finally with 30 mL DCM. Repeat this washing cycle three times. With regard to the swelling and constriction properties of resin, use a suction filtration process. The washing process should take approximately 40 min.

4. Perform a Kaiser test using 1–5 mg of resin. A positive test indicates completion of the deprotection, a negative test means a repetition of the deprotection procedure from the beginning is required. Usually, one deprotection cycle is sufficient.

Colorimetric Kaiser Test

To monitor the completion of the deprotection and to confirm the presence of free amines, a colorimetric Kaiser test is essential. Use the solution previously prepared by the instructor. Two solutions are required. Solution A: a solution of phenol at 80% in EtOH in which 2 mL of a solution of a 0.001 M aqueous solution of KCN in 98 mL of pyridine was added. Solution B: a solution of ninhydrin in 5% in EtOH.

1. To 1–5 mg of resin placed in a test tube, add 3 drops of each Kaiser solution.
2. Cautiously stir manually while heating the beads at 120 °C for 1–3 min.
3. The same test should be done on the starting material as a negative test. Blue-stained resin beads indicate the presence of free amines, while a negative test is indicated by yellow beads. The Kaiser test should take approximately 15 min.

Laboratory Session 2 (3 h)

Isocyanate Coupling

Formation of the urea bond is realised by heating the previously deprotected primary amine anchored on the polystyrene matrix with an isocyanate source.

1. Place the resin obtained in session 1 (0.76 mmol, 1 equiv.) in a clean, dry microwave reactor and add sufficient THF to just cover and swell the beads.
2. Add a solution of isocyanate source (3.04 mmol, 4 equiv.) dissolved in a minimum of THF. Seal the reactor and heat the reaction mixture under microwave irradiation for 10 min at 60 °C.
3. After the cooling process, drain the resin by simple suction filtration, swell and wash successively with 10 mL THF, 10 mL DMF, 10 mL MeOH, and 10 mL DCM. Repeat the washing cycle two times.
4. Perform a Kaiser test using 1–5 mg of resin, a negative test indicating the completion of the coupling step. The washing process and Kaiser test should take approximately 50 min.

Cyclative-Cleavage and Product Isolation

The cyclative-cleavage is performed, resulting in the isolation of the hydantoin.

1. Place the resin obtained previously (0.76 mmol, 1 equiv.) in a clean, dry microwave reactor and add a solution of triethylamine (0.43 mL, 3.04 mmol, 4 equiv.) in a 4 : 1 v/v mixture of THF/DMF to cover and allow the beads to swell.
2. Seal the reactor and heat under microwave irradiation for 15 min at 110 °C.
3. After the cooling process, drain the resin by simple suction filtration. *Do not waste the filtrates as they contain the final product*! Swell and wash the resin successively with 10 mL THF, and 10 mL DCM. Repeat the washing cycle two times. The washing process should take approximately 20 min.
4. Combine the organic filtrates in a tared vacuum round-bottomed flask, dried carefully *in vacuo* to remove all residual DMF, and weigh to determine the yield.
5. Prepare LC (LC/MS) sample(s) to evaluate the purity before preparing NMR tubes and record ^{1}H and ^{13}C spectra to confirm the compound structure.

Waste Disposal

All the filtrates generated in the two sessions are to be disposed in adequate solvent wastes, regarding the policy of the university where this activity is conducted. The remaining resin beads obtained are disposed in solid wastes, as well as the different beads used for Kaiser tests. Test tubes are disposed in a glass waste container.

Results Interpretation and Additional Questions

1. Explain why a washing process with a solution of triethylamine in DCM is used for the *N*-Boc deprotection step in session 1.
2. Provide a detailed mechanism for the Kaiser test and explain why a blue coloration is observed.
3. Interpret the LC (LC-MS) data recorded and the NMR spectra obtained. Calculate the percent yield and compare your results with those of other students.
4. Compare results (yields/purity) obtained with various polystyrene matrix, linkers, and/or amino acids. Give proposals to explain the influence of the resin and/or the amino acid side chain on the cyclative-cleavage step.
5. Propose a common method for synthesizing hydantoins. Describe the advantages and disadvantages of such a method compared to the one used in this experimental procedure.
6. Propose another family of heterocycles that could be synthesized on the basis of the strategy developed here.

Acknowledgement

The authors are thankful to Dr Thibault Coursindel and Professor Jean Martinez for their help during the preparation of this manuscript.

References

1. P. Lidström, J. Tierney, B. Wathey and J. Westman, *Tetrahedron*, 2001, **57**, 9225.
2. B. L. Hayes, *Aldrichimica Acta*, 2004, **37**, 66.
3. C. O. Kappe, *Angew. Chem., Int. Ed.*, 2004, **43**, 6250.
4. W. J. Brouillette, V. P. Jestkov, M. L. Brown, M. S. Akhtar, T. M. DeLorey and G. B. Brown, *J. Med. Chem.*, 1994, **37**, 3289.
5. R. M. Schelkun, P. Yuen, K. Serpa, L. T. Meltzer, L. D. Wise, E. R. Whittemore and R. M. Woodward, *J. Med. Chem.*, 2000, **43**, 1892.
6. H. U. Stilz, W. Guba, B. Jablonka, M. Just, O. Klingler, W. König, V. Wehner and G. Zoller, *J. Med. Chem.*, 2001, **44**, 1158.
7. E. Ősz, L. Somsák, L. Szilágyi, L. Kovács, T. Docsa, B. Tóth and P. Gergely, *Bioorg. Med. Chem. Lett.*, 1999, **9**, 1385.
8. J. J. Sciciński, M. D. Barker, P. J. Murray and E. M. Jarvie, *Bioorg. Med. Chem. Lett.*, 1998, **8**, 3609.
9. R. Kuang, J. B. Epp, S. Ruan, L. S. Chong, R. Venkataraman, J. Tu, S. He, T. M. Truong and W. C. Groutas, *Bioorg. Med. Chem.*, 2000, **8**, 1005.
10. R. Kuang, J. B. Epp, S. Ruan, H. Yu, P. Huang, S. He, J. Tu, N. M. Schechter, J. Turbov, C. J. Froelich and W. C. Groutas, *J. Am. Chem. Soc.*, 1999, **121**, 8128.
11. T. Matsugi, M. Kageyama, K. Nishimura, H. Giles and E. Shirasawa, *Eur. J. Pharmacol.*, 1995, **275**, 245.
12. H. Ohta, T. Jikihara, K. Wakabayashi and T. Fujita, *Pestic. Biochem. Physiol.*, 1980, **14**, 153.
13. E. Colacino, F. Lamaty, J. Martinez and I. Parrot, *Tetrahedron Lett.*, 2007, **48**, 5317.
14. T. Coursindel, J. Martinez and I. Parrot, *J. Chem. Educ.*, 2010, **87**, 640.
15. G. Chaubet, G. Cazals, A. Lebrun, J. Martinez and I. Parrot, *Synlett*, 2014, **25**, 574.

9. Preparation of Alkenes

9.1. Elimination Reactions

9.1.1. Preparation of Cyclohexene

Abel J. S. C. Vieira* and Elvira M. S. M. Gaspar

Faculty of Sciences and Technology, Universidade Nova de Lisboa, 2829-516 Caparica, Portugal
*E-mail: ajscv@fct.unl.pt

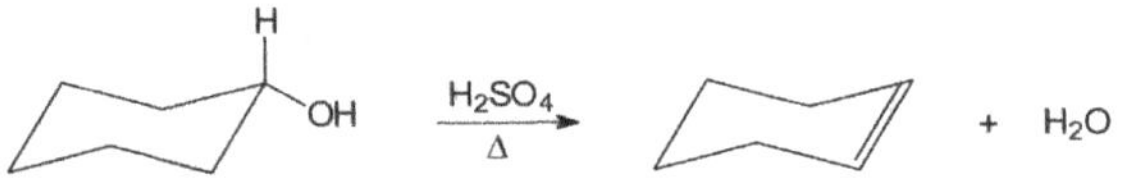

Number of sessions (duration of each session)	Hazard level	Difficulty level	Level of study
1 (3 h)	Moderate	Low	Introductory
Class names Alcohols, alkenes			
Concepts involved This experiment aims to illustrate an elimination reaction (dehydration of a secondary alcohol in an acidic medium); example of hydroxyl as a poor leaving group (need of acidic medium to perform the reaction)			
Chemicals needed Cyclohexanol, sulfuric acid, sodium carbonate			
Equipment and experimental techniques involved Apparatus for simple and fractional distillation, IR and NMR spectroscopy			
Keywords Fractional and simple distillation, IR, NMR, unimolecular elimination			

Background

Alcohols are usually bad substrates for nucleophilic substitution or elimination reactions. This is because the OH functional group is a very poor leaving group (hydroxide ion is a strong base). However, in acidic medium, usually in the presence of strong inorganic acids, the hydroxyl group of the alcohol exits in its protonated form, becoming a very good leaving group, which leaves the substrate in the form of a water molecule (a weak base) (Scheme 9.1.1.1).

In the case of primary alcohols, this reaction can be used to prepare symmetric ethers (*e.g.* diethyl ether). With secondary and tertiary alcohols, the main reaction is elimination (dehydration), especially at high temperatures; the product being an alkene or a mixture of alkenes.[1,2]

As far as dehydration of cyclohexanol (a symmetric secondary alcohol) is concerned, there is no possibility of regioselectivity in the elimination reaction; therefore, cyclohexene is obtained as the unique dehydration product (Scheme 9.1.1.2).

Comprehensive Organic Chemistry Experiments for the Laboratory Classroom
Edited by Carlos A. M. Afonso, Nuno R. Candeias, Dulce Pereira Simão, Alexandre F. Trindade, Jaime A. S. Coelho, Bin Tan, and Robert Franzén.

Published by the Royal Society of Chemistry, www.rsc.org

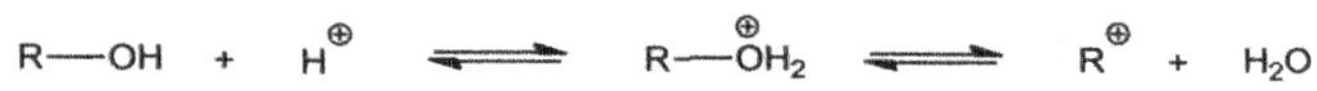

Scheme 9.1.1.1 Protonation of an alcohol.

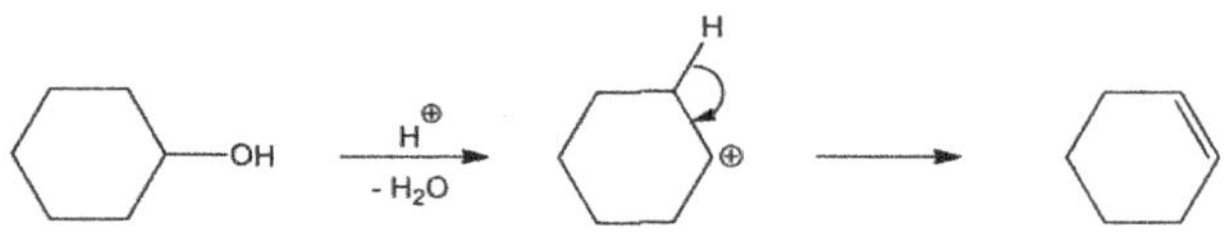

Scheme 9.1.1.2 Dehydration of cyclohexanol.

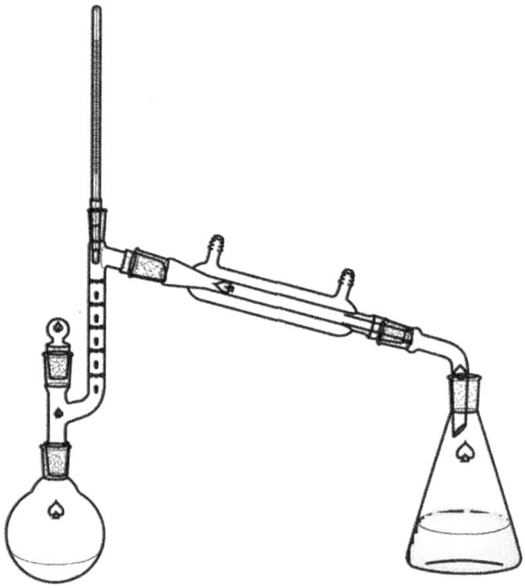

Figure 9.1.1.1 Apparatus used in the synthesis of cyclohexene.

Additional Safety

The experiment can be performed on a regular laboratory bench, with proper safety equipment, such as protective clothing, gloves and safety glasses. Cyclohexene is flammable, toxic by inhalation and has a characteristic (unpleasant) smell. It must be kept in a well-closed recipient. Sulfuric acid is a corrosive liquid that can cause burns and must be handled with care.

Cyclohexanol: R: 20/22-37/38; S: 2-24/25.

Sulfuric acid: R: 35; S: 1/2-26-30-45.

Experimental Procedure

One Laboratory Session (3 h) (Figure 9.1.1.1)

Preparation of Cyclohexene (ref. 3)

1. In a 250 mL round-bottomed flask introduce carefully, with gentle stirring, 20 g of cyclohexanol, 0.6 mL of concentrated sulfuric acid and some anti-bumping granules.
2. Fit the flask with a Vigreux fractionation column, a still head with a thermometer, a Liebig condenser, a receiving adapter and a receiving flask immerged in an ice cold water bath.
3. Clamp the whole apparatus and do not forget the adjustable laboratory jack under it.

4. Heat the reaction mixture with an electric mantle, keeping a regular boiling, such as the temperature of the distillate does not exceed 90 °C.
5. Stop the distillation when only a small dark residue remains in the distillation flask (or when the temperature of the distillate is higher than 90 °C).
6. Keep the distillate collected in the receiving flask well cooled.

Product Recovery (Work-Up) and Purification

7. Transfer the distillate to a separatory funnel, add 2 mL of a 5% sodium carbonate aqueous solution and saturate the aqueous layer with sodium chloride, while efficiently shaking.
8. Let the mixture rest until a clear separation of two layers is observed.
9. Reject the (lower) aqueous phase (test it with a drop of water to be sure that it is the aqueous layer).
10. Collect the (upper) turbid organic phase and dry it with anhydrous magnesium sulfate added in small portions, shaking occasionally, until a clear liquid is obtained.
11. Decant this liquid to a well-dried round-bottomed 100 mL flask.
12. Proceed to a simple distillation, heating with an electric mantle. Collect the fraction distilling between 81 and 85 °C (at atmospheric pressure).

Product Characterisation

13. Record the boiling point of cyclohexene. Note that this experimental result is the temperature interval at which the distillate was collected and the atmospheric pressure measured in the laboratory at the same time.
14. Measure the refractive index of cyclohexene. Record also the temperature at which this measurement was made.
15. Calculate the yield of the reaction.
16. Record the IR spectrum of the compound, using sodium chloride discs as support material.
17. Acquire the ^{1}H and ^{13}C NMR spectra of the compound. Use the deuterated solvent suggested by your instructor.

Waste Disposal

The aqueous phase resulting from separation and neutralisation of traces of acid may be diluted and discarded in the sink.

Results Interpretation and Additional Questions

1. Assign the IR absorption bands characteristic of the functional groups of the compound.
2. Try to assign the H and C atoms to the NMR signals.
3. Write the mechanism of the reaction. Is this reaction stereospecific? Explain why.
4. Give an example of an alcohol that, upon dehydration, can yield more than one alkene.
5. Try to rationalise the reason for the following experimental details: use of fractional distillation during the reaction and the limit of 90 °C in collecting the distillate.
6. What are sodium carbonate and sodium chloride used for in the work-up?
7. Suggest a simple visual test to distinguish cyclohexene from cyclohexane.

References

1. A. Streitwieser, C. H. Heathcock and E. M. Kosower, *Introduction to Organic Chemistry*, MacMillan Pub. Co., 4th edn, 1992, vol. 11, p. 268.
2. T. W. G. Solomons and C. B. Fryhle, *Organic Chemistry*, John Wiley & Sons, 8th edn, 2003, vol. 7, p. 298.
3. B. S. Furniss, A. J. Hannaford, P. W. G. Smith and A. R. Tatchell, *Vogel's Textbook of Practical Organic Chemistry*, Longman Scientific & Technical, 5th edn, 1989, vol. 5, p. 491.

9.1.2. A Green Synthesis of 2,3-Dibromo-3-phenylpropionic Acid and the Use of Kinetic Studies to Probe into the Elimination Product When Treated with a Weak Base in Different Solvents

Malcolm I. Stewart* and Craig D. Campbell

CRL, University of Oxford, 12 Mansfield Road, Oxford, OX1 3TA, UK
*E-mail: Malcolm.stewart@chem.ox.ac.uk

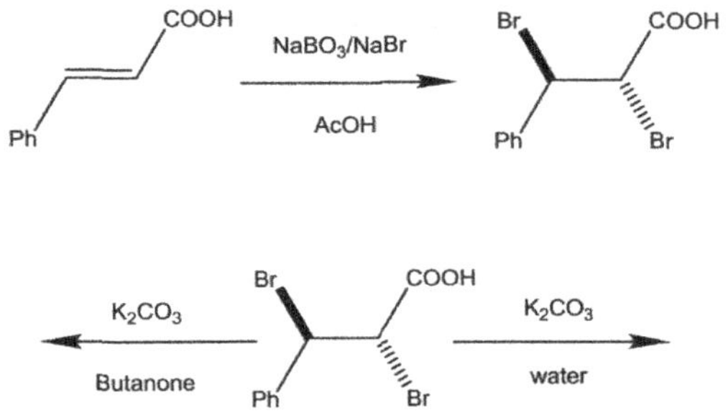

Number of sessions (duration of each session)	Hazard level	Difficulty level	Level of study
2 (4 h + 6 h)	Moderate	Medium	Introductory

Class names Alkene, carboxylic acid, alkyl halides

Concepts involved Stereoselective addition to double bonds, solvent involvement in the elimination pathway, physical organic chemistry

Chemicals needed Sodium bromide, cinnamic acid, sodium perborate, acetic acid, water, diethyl ether, magnesium sulfate, chloroform, potassium carbonate and butan-2-one (MEK)

Equipment and experimental techniques involved Round bottom flasks, measuring cylinders, hotplate-stirrer, separatory funnels, funnels, sinter funnels, filter paper, rotary evaporator, UV spectrometer, IR spectrometer, melting point apparatus, reflux apparatus, thermometer, rubber tubing, computer

Keywords Evaporation under reduced pressure, green chemistry, group work, infrared spectroscopy, mechanism elucidation, physical organic chemistry

Background

The carbon to carbon double-bond, commonly referred to as an alkene or olefin, is an important functional group in organic chemistry. This unsaturated system can undergo reaction with a range of reagents, including the halogens, to produce addition products. This is an important strategy to introduce a desired functional group into a molecule. Important considerations for such reactions are the regioselectivity and stereoselectivity of the processes, and there are often different products possible depending on the choice of reagent and conditions. For example,

Comprehensive Organic Chemistry Experiments for the Laboratory Classroom
Edited by Carlos A. M. Afonso, Nuno R. Candeias, Dulce Pereira Simão, Alexandre F. Trindade, Jaime A. S. Coelho, Bin Tan, and Robert Franzén.

Published by the Royal Society of Chemistry, www.rsc.org

Markovnikov's rule is used to predict the regioselectivity of such additions, but in some cases this rule is not enough to explain the obtained products.

In the first part of the following experiment, bromine will be added across the double bond functional group of the naturally occurring compound cinnamic acid. Bromine is a toxic, corrosive liquid that has led to many accidents in laboratories around the world. As a result, a new, safer, method developed by Kabalka *et al.*[1] will be used to generate the bromine *in situ*. Sodium perborate is used as an oxidant to allow this generation of bromine to happen.

The second step is a group project to investigate the consequences of differences in mechanism on the elimination reactions in a polar protic and polar aprotic solvent.[2]

When it comes to the mechanism of the elimination, there are two common pathways available, defined as "E1" and "E2". In these experiments, both mechanisms are possible. Physical chemistry tells us that both of these possibilities have different variables which affect the rate of the reaction. In the E_1 mechanism, or unimolecular elimination, *only* the *concentration of the substrate* affects the rate of the reaction. This is because the rate determining step is the breaking of the carbon–leaving group bond. Following the E_2 mechanism, or bimolecular elimination, *both* the *concentration of the substrate* and the *concentration of the base* affect the rate of reaction. This time the breaking of the carbon –leaving group bond *and* the carbon–hydrogen bond happen simultaneously and so both are involved in the rate determining step. The mechanism of these pathways can be seen in Figure 9.1.2.1.

If a compound has a choice of different groups to undergo elimination, the reaction conditions, such as temperature, base and solvent, may influence which elimination occurs.

Each student (pair) will be told the mass of the dibromide synthesised in the previous step and instructed how much carbonate base (K_2CO_3) and which solvent to perform the reaction with, either water (a polar protic solvent) or butan-2-one (a polar aprotic solvent).

Sophisticated machinery can be used to follow reactions, but a far easier way to follow the progress of this particular reaction is to measure the amount of carbon dioxide being released. As gas is evolved as a part of the reaction, the rate of gas production is proportional to the rate of the reaction. Combining all data from each group, the data can be represented by four distinct graphs:

1. Water is the solvent and only the base concentration is varied.
2. Water is the solvent and only the dibromide concentration is varied.
3. Butan-2-one is the solvent and only the base concentration is varied.
4. Butan-2-one is the solvent and only the dibromide concentration is varied.

Each graph will clearly show how the variable alters the rate of the reaction, enabling a hypothesis for the mechanism in each case to be determined.

Additional Safety

As with all experiments, proper personal protective equipment should be worn at all times. Since acetic acid is used in this procedure and bromine is generated *in situ*, the addition reaction must be carried out in a fume hood. Chloroform is a suspected carcinogen so care must be taken when

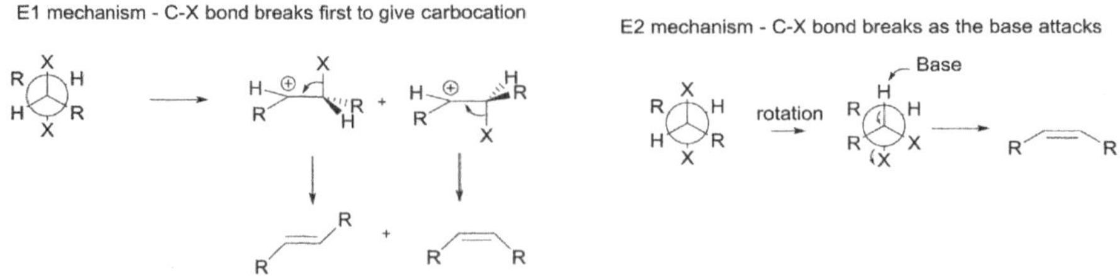

Figure 9.1.2.1 Possible mechanisms of the elimination.

using it and it must be disposed of in your halogenated waste stream. If possible, the second step should also be carried out in a fume hood; however, the products are relatively safe and it is possible to carry out the work on the bench.

Experimental Procedure

Session 1 – Green Synthesis of the Dibromide Compound (4 h)

This experiment (Scheme 9.1.2.1) must be done in a fume cupboard!

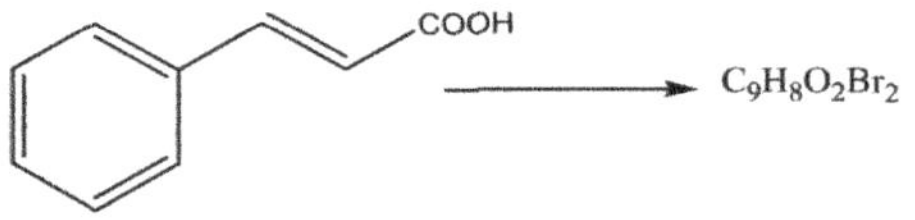

Scheme 9.1.2.1 The synthesis of the dibromide.

1. Add sodium bromide (3.26 g, 30.0 mmol) to a round-bottomed flask (100 mL) containing a mixture of cinnamic acid (2.0 g, 13.5 mmol) and sodium perborate (2.29 g, 15.0 mmol) in glacial acetic acid (25 mL).
2. Stir the mixture at room temperature for 2 h before diluting with 2 M HCl (aqueous, 50 mL).
3. Extract the product using diethyl ether (2 × 30 mL), combine the ether extracts and wash with water (3 × 50 mL). Dry the ether layer over anhydrous magnesium sulfate. Note the weight of a clean, dry round bottomed flask (250 mL), and filter the dried solution into this.
4. Evaporate the solvent on a rotary evaporator and obtain a crude yield.
5. Place the resulting solid in a sintered funnel and wash with a small amount of ice-cold chloroform using suction filtration.
6. Calculate the yield and obtain a melting point and UV and IR spectra of the sample. Compare your product to the starting material and comment on this in your results. Retain the pure sample in a pre-weighed, clearly labelled sample bottle.

Session 2 – Elimination Step (6 h)

You will be instructed on how much dibromocinnamic acid and potassium carbonate to use, and whether to use water or butan-2-one as a solvent.

1. Add the dibromocinnamic acid and potassium carbonate to a round-bottomed flask (100 mL) containing a stirrer bar, add 20 mL of [solvent], and set up the equipment on a hot-plate stirrer.
2. Insert a pipette into a thermometer adapter, making sure it is a tight fit and then attach a reflux condenser to your flask before putting the thermometer adapter onto the top of it. Attach a length of rubber tubing to the pipette. Invert a 100 mL measuring cylinder filled with water in an ice bucket (also containing water), taking care to minimise the air that gets into the measuring cylinder. Insert the other end of the rubber tubing into the measuring cylinder, but keep the end below the 100 mL mark and clamp the cylinder in place. Check that the quick-fit joints are all tight (gas will escape from the smallest gap) and that the condenser has water running through it.
3. Turn on the heating and stirring, and note the start time, *i.e.* time = 0. In the next 5 min, bubbles of gas will start to enter the measuring cylinder. Note the time that the gas reaches a level of 20 mL and then when it reaches 100 mL. In addition, note the time at which the reaction reaches reflux. When 100 mL of gas has been collected, you can remove the rubber tubing from the water and the thermometer adaptor on the top of the reflux condenser.

4. Heat the reaction at reflux until it has been reacting for a total of 1 h 20 min. Follow the appropriate instructions below.

Elimination in Water

5. After the reaction has been heated at reflux for 1 h 20 min, remove it from the heat, and let it cool to room temperature.
6. Extract the product using diethyl ether (2 × 20 mL).
7. Dry over anhydrous magnesium sulfate and weigh a round-bottomed flask (250 mL), then filter the mixture into the flask under gravity. Evaporate the solvent on a rotary evaporator to leave a yellowish oil.
8. Calculate the yield and obtain UV and IR spectra of the sample.

Elimination in Butanone

5. Leave the reaction to reflux for 1 h 20 min before cooling the mixture and diluting with water (20 mL).
6. Extract the product using diethyl ether (2 × 20 mL) and dry over anhydrous magnesium sulfate. Filter under gravity, and remove the solvent on a rotary evaporator to leave a yellowish oil.
7. Calculate the yield and obtain UV and IR spectra of the sample.

Calculating the Rate of Reaction and Determining the Order of Reaction

To calculate the rate of gas production, which is proportional to the rate of elimination, divide the volume of gas by the time taken to collect that volume of gas. Your rate will now have units of $cm^3\ s{-}^1$.

$$\text{rate} \propto \frac{dV}{dt} \approx \frac{\text{volume}}{\text{time}}$$

Everyone in the class will submit their results (remember to note the number of moles of substrate and base you used, and whether you carried out the reaction in butanone or water) and by plotting a graph of *rate* against the *amount used* you will be able to determine the order of reaction. You should have four graphs (water & base varied, water & substrate varied, butanone & base varied and butanone & substrate varied).

Waste Disposal

There is no special waste generated so the organic waste from the procedures can go out with any other organic waste and any aqueous waste can be poured down the sink.

Results Interpretation and Additional Questions

Bromination Reaction

Given that the sodium perborate is used to generate BrOH, discuss the chemistry involved in the *in situ* generation of the bromine within the reaction.

1. The addition of bromine to the starting material can result in the formation of four different stereoisomers. Draw all of them and assign (*R*) or (*S*) to each of the new stereogenic centres.
2. Given that the (*RS*) and (*SR*) dibromides have a melting point of 204 °C and the (*RR*) and (*SS*) derivatives both melt at 91–93 °C, what is the stereochemistry of the bromine addition

in your reaction? How else could the stereochemistry of the product be discovered? Discuss the mechanism of this addition reaction to explain your answer to the rest of this question.

Elimination Reaction

1. What is the order of reaction with respect to the substrate concentration in each of the solvents? What is the order of reaction with respect to the base concentration in each of the solvents?
2. Postulate a mechanism for elimination in each solvent which corresponds to the orders of reaction you have observed.
3. Where is the CO_2 evolved from, and how might the volume of CO_2 be affected by any acetic acid remaining from synthesis of the dibromide? What are the main sources of error in this experiment and how would you improve it?

References

1. G. W. Kabalka, K. Yang, N. K. Reddy and C. Narayana, *Synth. Commun.*, 1998, **28**, 925.
2. (a) H. Mestdagh and A. Puechberty, *J. Chem. Educ.*, 1991, **68**, 515; (b) L. Corvari, J. R. McKee and M. Zanger, *J. Chem. Educ.*, 1991, **68**, 161.

9.1.3. Dehydration of Methylcyclohexanols

Cornelis A. Van Walree*[a,b], Stephan A. Jonker[a], and Veronica Kaats-Richters[a]

[a]Department of Chemistry, Utrecht University, Budapestlaan 4b, 3584 CD, Utrecht, The Netherlands; [b]School of Chemical and Physical Sciences, Flinders University, GPO Box 2100, Adelaide 5001, Australia
*E-mail: c.a.vanwalree@uu.nl

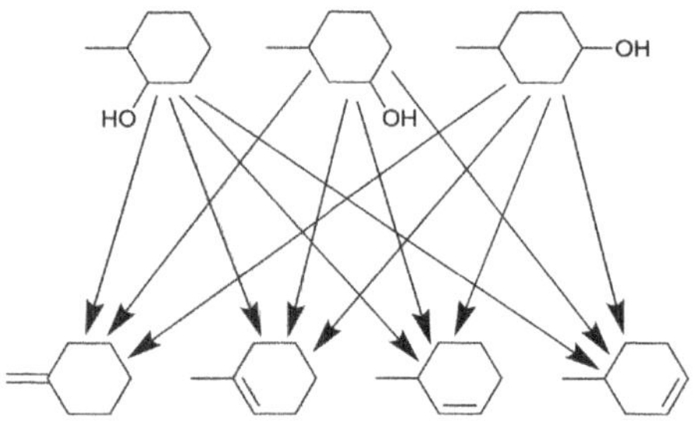

Number of sessions (duration of each session)	Hazard level	Difficulty level	Level of study
2 (4 h each)	Moderate	Medium	Intermediate

Class names Alcohols, olefins

Concepts involved In this experiment students perform an acid-catalysed dehydration of 2-, 3- or 4-methylcyclohexanol and study the product distribution with GC-MS and ^{1}H NMR; they will apply knowledge about E1 and E2 mechanisms, Zaitsev's rule and the stability of carbocations

Chemicals needed 2-Methylcyclohexanol, 3-methylcyclohexanol, 4-methylcyclohexanol (all mixtures of *cis*- and *trans*-isomers), 85% phosphoric acid; for reference to instructors: 1-methylcyclohexene, 3-methylcyclohexene, 4-methylcyclohexene, methylidenecyclohexane

Equipment and experimental techniques involved Set-up for distillation, hotplate–magnetic stirrer, liquid–liquid separation, NMR spectroscopy, GC-MS, infrared spectroscopy

Keywords Carbocations, dehydration, distillation, GC-MS, IR, NMR

Background

Eliminations are one of the most important types of reactions within organic chemistry. In these reactions, a single molecule splits into two molecules. Typically, one of these molecules is a small (diatomic or triatomic) compound, while in the other fragment a π-bond is formed. The general form is shown in Scheme 9.1.3.1, in which X represents a leaving group. A hydrogen atom from the carbon atom adjacent to the leaving group is abstracted.

Elimination reactions often lead to the formation of multiple compounds, as hydrogen atoms can be abstracted from chemically different carbons atoms. The product composition is determined by the stability of the resulting alkenes, the mechanism involved and the conformational dependence of the elimination process. For instance, dehydrohalogenation of menthyl chloride

Comprehensive Organic Chemistry Experiments for the Laboratory Classroom
Edited by Carlos A. M. Afonso, Nuno R. Candeias, Dulce Pereira Simão, Alexandre F. Trindade, Jaime A. S. Coelho, Bin Tan, and Robert Franzén.

Published by the Royal Society of Chemistry, www.rsc.org

Scheme 9.1.3.1 General elimination reaction.

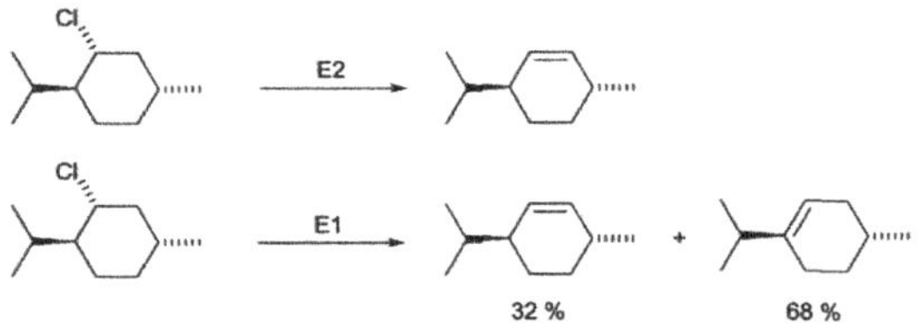

Scheme 9.1.3.2 Possible dehydrohalogenation pathways of menthyl chloride.

under E2 conditions gives 100% 2-menthene, while under E1 conditions 32% 2-menthene and 68% 3-menthene are obtained (Scheme 9.1.3.2).[1]

In this experiment, you will perform an acid-catalysed dehydration of a methylcyclohexanol. The dehydration agent is concentrated (85%) phosphoric acid. You will do the reaction with one of three possible starting compounds: 2-methylcyclohexanol, 3-methylcyclohexanol, or 4-methylcyclohexanol. Other members of your group will start with another methylcyclohexanol. Note that mixtures of the *cis*- and *trans*-isomers of the methylcyclohexanols will be used. You will determine the product distribution by the use of GC-MS and ^{1}H NMR spectroscopy, while infrared spectroscopy also will be used. By comparing the results obtained for different starting alcohols, together you will gain surprising insights into the mechanism of the dehydration reactions.

Before You Start

In your organic chemistry textbook, study the theory of elimination reactions in general, and dehydration of alcohols in particular. Read about E1 and E2 mechanisms, the stability of carbocations, rearrangements and Zaitsev's rule.

Question: What mechanism is the acid-catalysed dehydration of the methylcyclohexanols likely to follow?

Additional Safety

Because of the odour of the products, all manipulations should be done in a fume hood. Students should wear a laboratory coat and safety goggles, and be aware of the flammability of the organic substances. 85% Phosphoric acid is corrosive and can lead to skin and eye damage upon contact.

Experimental Procedure

Laboratory Session 1 (4 h): Synthesis of the Alkenes

1. In a fume hood, put a 100 mL three-necked flask on a stand. Place an appropriately sized magnetic stirring bar in the flask.
2. Equip the side arms of the flask with a pressure-equalizing dropping funnel and a thermometer. Connect a distillation head to the middle neck, and attach a thermometer and a suitable condenser to it. Take care that connections are closed by using clamps where required. Use a 100 mL conical flask as receiver.
3. Place appropriately sized heating-stirring equipment under the flask.
4. Charge the flask with 10 mL of 85% phosphoric acid, and put 0.25 mol of the selected methylcyclohexanol in the dropping funnel.

5. Heat the phosphoric acid to boiling. Then start with the addition of the alcohol (caution: the first drops may react quite vigorously). The addition should occur at about the same rate as the product is distilled off; preferably, the contents of the flask should not increase in volume. Take care that the temperature does not become too high, in order to prevent distillation of the phosphoric acid.
6. The volatile distillate might be cooled in an ice bath.
7. When the addition of the alcohol is complete, heating should be continued until organic distillate is no longer collected.
8. Transfer the biphasic distillate to a small separatory funnel and remove the water layer. Dry the organic layer over anhydrous magnesium sulfate. Remove the drying agent by filtration. The filtrate is the product.
9. Determine the yield, and calculate the percentage yield.
10. Record infrared, ^{1}H NMR and GC-MS spectra of the product (or have them recorded).

Session 2 (4 h): Analysis of the Product(s)

For convenience, the expected products are labelled **A** to **E** (Figure 9.1.3.1).

Infrared Spectroscopy

The infrared spectra of compounds **A–D** are shown in Fig. 9.1.3.2; that of **E** was not available. You should notice that the spectra are all rather similar, and that it is not easy to obtain detailed information about the composition of the product from its infrared spectrum. Nevertheless, some important conclusions can be drawn from the infrared spectrum by answering the following questions:

- How can you tell whether any starting compound is still present?
- Which signals indicate the presence of alkene product(s)?

GC-MS

From the GC-MS chromatograms and mass spectra:

- Use the mass spectra to assign the peaks in the gas chromatogram to compounds. Note that the differences between the MS spectra are small, which makes it difficult to assign the spectra. Nevertheless, it can be done with the use of the reference spectra provided.
- Give the relative amounts of the different alkene products.

^{1}H NMR

In ^{1}H NMR, the number of signals, the chemical shifts, signal multiplicities and signal intensities are characteristic for a given product. For the dehydration products of the methylcyclohexanols, interpretation of the spectra is complicated by the overlap of the signals of the different

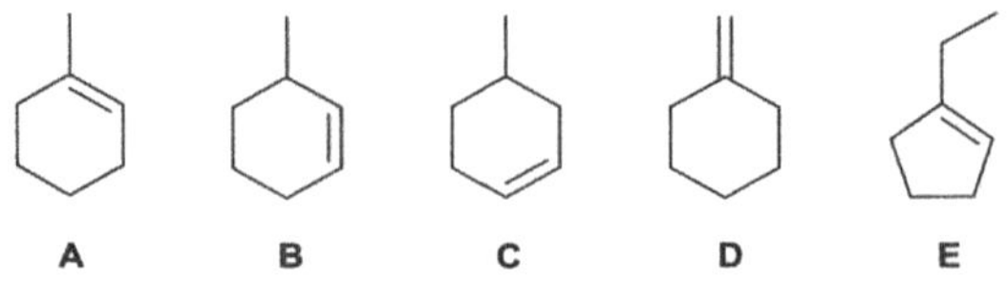

Figure 9.1.3.1 Expected products.

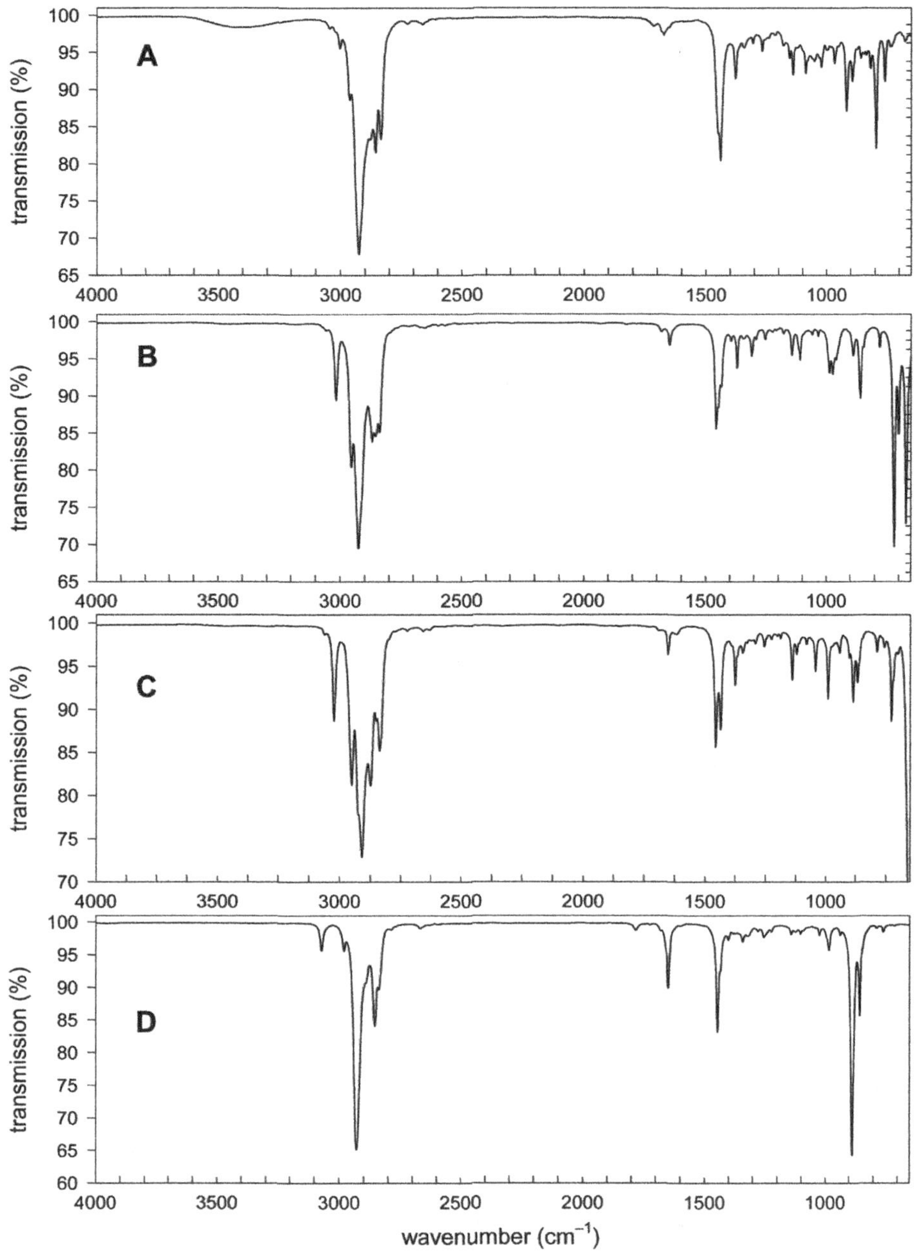

Figure 9.1.3.2 Infrared spectra of possible elimination products **A–D**.

compounds, in particular in the region 2.5–0.8 ppm. Fortunately, information about the product distribution can be obtained with the use of the spectra of the pure compounds, in particular the vinylic region shown in Figure 9.1.3.3. Use the integrals of the characteristic peaks of products **A–D** to calculate their relative concentrations. To this end, it is convenient to reproduce Table 9.1.3.1 below in your laboratory notebook, and fill in the data. The relative amount of **C** can be calculated by first taking the integral of the overlapping signals of two H atoms of **C** and one H atom of **B** ($2_C + 1_B$). Subtraction of the known contribution of one H atom of **B** (1_B) then affords the integral belonging to the H atoms of **C** only.

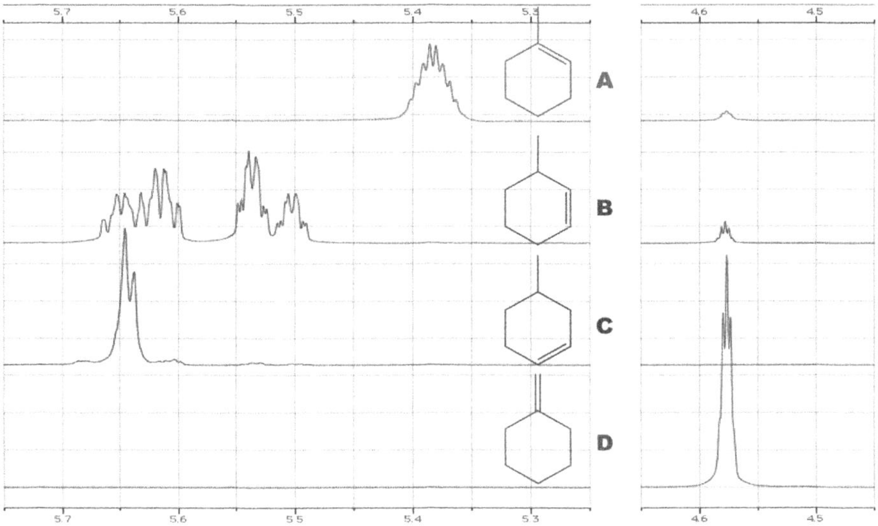

Figure 9.1.3.3 Magnification of the vinylic region of the ^{1}H NMR spectra of possible elimination products. The signals in the spectra of **A** and **B** near 4.6 ppm are due to contamination with **D**.

Table 9.1.3.1 ^{1}H NMR integral ratios and GC-MS ratios of **A–D**.

	Number of H atoms	Integral	Integral/H	Ratio NMR	Ratio GC-MS
A	1_A				
B	1_B				
C+B	$2_C + 1_B$				
C	2_C				
D	2_D				
Total				100%	100%

Waste Disposal

All waste should be thrown in the organic waste container, halogen free if available.

Results Interpretation and Additional Questions

1. Are the results obtained from ^{1}H NMR in line with the results from GC-MS? Which analysis technique do you prefer? Why?
2. Are your results comparable with those of other students who used the same starting alcohol? Explain possible differences.
3. To what extent do your results differ from those of group members who used another alcohol?
4. Go back to the question you answered before the start of the experiment, about the mechanism. Is the formation of the products found consistent with the proposed mechanism?
5. If product formation is not consistent with the proposed mechanism, what conclusions must be drawn? Or is another mechanism consistent with the results?

To obtain more insight in these reactions, the literature[2–6] may be consulted. It will show you that the reactions are complicated, and that the contents of textbooks may give a simplified picture of the actual course of affairs. Moreover, it is very important to realize that science is dynamic. Ideas and concepts that seem to be settled may become subject to reinterpretation when new experimental evidence is found that is not in line with them.

References

1. J. E. McMurry, *Organic Chemistry*, Brooks/Cole, 8th edn, 2012, p. 406 and p. 1163.
2. J. W. Lehman, *Multiscale Operational Organic Chemistry. A Problem-Solving Approach to the Laboratory Course*, Prentice Hall, Upper Saddle River, New Jersey, 2002, Experiment 21.
3. D. Todd, *J. Chem. Educ.*, 1994, **71**, 440.
4. J. J. Cawley and P. E. Lindner, *J. Chem. Educ.*, 1997, **74**, 102.
5. M. M. Clennan and E. L. Clennan, *J. Chem. Educ.*, 2011, **88**, 646.
6. J. B. Friesen and R. Schretzman, *J. Chem. Educ.*, 2011, **88**, 1141.

9.1.4. Synthesis of 5-Hydroxymethylfurfural (HMF) from Fructose as a Bioplatform Intermediate

Svilen P. Simeonov*[a,b] and Carlos A. M. Afonso*[a]

[a]Research Institute for Medicines and Pharmaceuticals Sciences, Faculty of Pharmacy, University of Lisbon, Av. Prof. Gama Pinto, 1649-003, Lisbon, Portugal; [b]Institute of Organic Chemistry with Centre of Phytochemistry, Bulgarian Academy of Sciences, Acad. G. Bonchev str., bl.9, 1113 Sofia, Bulgaria
*E-mail: svilen.p.simeonov@gmail.com, carlosafonso@ff.ulisboa.pt

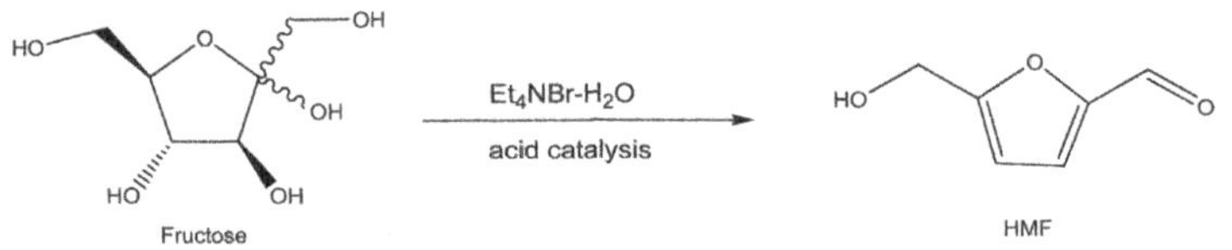

Number of sessions (duration of each session)	Hazard level	Difficulty level	Level of study
1 (3 h)	Low	Low	Intermediate
Class names Hydroxyl, aldehyde, furan			
Concepts involved This experiment gives a method for dehydration of fructose to 5-hydroxymethylfurfural (HMF), employing crystallization of the reaction media as a tool for the isolation of the final product			
Chemicals needed Tetraethylammonium bromide, fructose, Amberlyst 15, silica gel, EtOAc, EtOH, $CDCl_3$			
Equipment and experimental techniques involved Magnetic stirrer with heating, water bath, standard glassware			
Keywords Biorefinery, carbohydrates, crystallization, dehydration, ^{1}H NMR, 5-hydroxymethylfurfural (HMF)			

Background

Diminishing natural resources and their growing prices has opened up a new space for research on the synthesis of new molecules derived from biorenewable resources that can serve as energy sources[1] or chemical intermediates.[2] Among the already reported large number of chemical building blocks and intermediates accessible from biorenewable resources[3] derived from carbohydrates, 5-hydroxymethylfurfural (HMF) has been described as one of the most promising for industry[4] and a number of its possible applications and transformations, such as production of biofuel 2,5-dimethylfuran, as well as chemical building blocks for the polymer industry, such as furan-2,5-dicarboxylic acid and furan-2,5-dicarbaldehyde, have been reported[5] (Scheme 9.1.4.1).[6]

HMF is synthesized mainly by the dehydration of monosaccharaides, requiring the loss of three water molecules. The process is usually acid catalyzed and fructose is observed to be the best starting material.[7] Nevertheless, glucose is more desirable as a starting material because it is the most abundant monosaccharide in nature. Unfortunately, glucose is much less reactive since it is in the form of a more stable six-membered ring compared to the five-membered ring for fructose, and normally the formation of HMF from glucose requires its initial isomerization to fructose.[8]

Comprehensive Organic Chemistry Experiments for the Laboratory Classroom
Edited by Carlos A. M. Afonso, Nuno R. Candeias, Dulce Pereira Simão, Alexandre F. Trindade, Jaime A. S. Coelho, Bin Tan, and Robert Franzén.

Published by the Royal Society of Chemistry, www.rsc.org

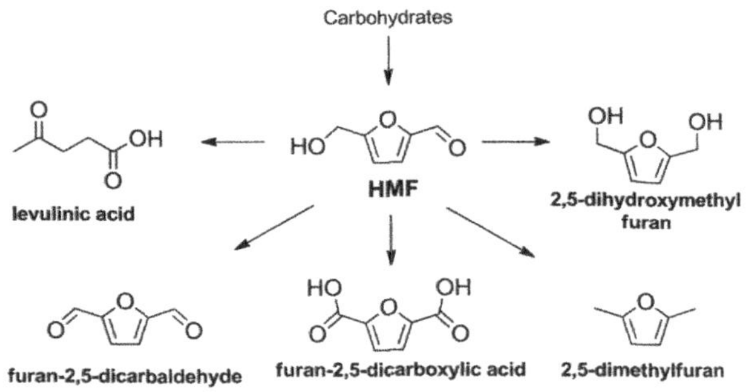

Scheme 9.1.4.1 Possible chemical building blocks derived from HMF.

Several problems limit the production of HMF at industrial scales. Probably the major one is the isolation of HMF.[6] Because of carbohydrates low solubility in organic solvents, water or biphasic systems water/organic solvents are widely studied as reaction media but HMF is difficult to extract because the distribution coefficient between organic and aqueous phases is not favorable.[4] This problem has been partially overcome by using more efficient extracting solvents, such as methyl isobutyl ketone (MIBK).[9,10] Continuous extraction procedures have also been applied.[11] Even though improvements in this field have been achieved in the last few years, more effective and industrially-applicable isolation techniques still need to be developed.

A method for the synthesis of HMF from fructose by using acid catalysis and precipitation of the reaction media has been developed as a student laboratory experiment as the main goal to illustrate a very simple and industrially useful separation technology that may open up a new space for higher scale production of HMF.[12] In addition, this experiment meets other student learning goals such as more environmentally friendly processes, involved reaction mechanism, spectroscopic structural elucidation (^{1}H NMR) high purity evaluation (HPLC) and consolidation of general organic chemistry preparative practices.

Additional Safety

All sessions should be done in a fume hood, with proper safety equipment, such as protective clothing, gloves and safety goggles.

Ethyl acetate is flammable and may cause CNS depression on inhalation.

Tetraethylammonium bromide is slightly hazardous in case of skin contact (irritant), of eye contact (irritant), and of ingestion.

HMF can cause eye and skin irritation. It may produce yellow stains on the skin which are considered harmless.

$CDCl_3$ is hazardous in case of eye contact (irritant), of ingestion, and of inhalation. It is slightly hazardous in case of skin contact (irritant).

Experimental Procedure

Session (3 h)

1. In a fume hood heat a water bath on a magnetic hot plate until it starts to boil.
2. Prepare a single-necked 250 mL round-bottom flask equipped with a magnetic stir bar and charge it with 9.1 g of TEAB (1% water content w/w), 1 g of fructose, 0.1 g Amberlyst 15 and 0.9 mL of water.
3. Place the reaction mixture in the boiling water bath and stir it for 15 min.
4. Remove the flask and cool the reaction mixture to room temperature and evaporate the water using a rotary evaporator.

5. Wash the solid with 50 mL of EtOAc, decant and collect the solvent, then add 5 mL of EtOH to the solid and heat it until it is fully dissolved.
6. Add 300 mL of EtOAc to the hot EtOH solution under vigorous stirring and filter out the resulting precipitate and the catalyst.
7. Prepare a filter with a path of silica gel (10 g) and filter through it the combined solutions from step 5 and 6.
8. Evaporate the solvent on a rotary evaporator and determine the HMF yield.

Waste Disposal

Tetraethylammonium bromide can be reused for further cycles or disposed as halogenated waste. The used solvents should be collected in a non-chlorinated waste collector recipient, for posterior waste disposal.

Results Interpretation and Additional Questions

1. Interpret your ^{1}H NMR spectra. Calculate the percentage yield of your isolated product based on your recorded mass and the theoretical yield.
2. Propose a reaction mechanism. The students can compare their rationalization with reported information.
3. Identify some green chemistry credits of this experiment, including the E-factor.
4. Discuss the possible industrial applications of 2,5-dihydroxymethyl furan, furan-2,5-dicarbaldehyde, furan-2,5-dicarboxylic acid, as well as the current chemical building blocks that they can replace.

Acknowledgement

Reprinted (adapted) with permission from *J. Chem. Educ.*, 2013, **90** (7), pp 1373–1375. Copyright (2013) American Chemical Society.

References

1. J. L. Blatti and M. D. Burkart, *J. Chem. Educ.*, 2011, **89**(2), 239–242.
2. (a) B. Kamm, *Angew. Chem., Int. Ed. Engl.*, 2007, **46**(27), 5056–5068; (b) G. B. Kauffman, *J. Chem. Educ.*, 2011, **89**(2), 190–191.
3. A. Corma, S. Iborra and A. Velty, *Chem. Rev.*, 2007, **107**(6), 2411–2502.
4. J. Lewkowski, *ARKIVOC*, 2001, **2**, 17–54.
5. (a) S. Subbiah, S. P. Simeonov, J. M. S. S. Esperanca, L. P. N. Rebelo and C. A. M. Afonso, *Green Chem.*, 2013, **15**(10), 2849–2853; (b) R. F. M. Frade, J. A. S. Coelho, S. P. Simeonov and C. A. M. Afonso, *Toxicol. Res.*, 2014, **3**(5), 311–314.
6. A. A. Rosatella, S. P. Simeonov, R. F. M. Frade and C. A. M. Afonso, *Green Chem.*, 2011, **13**(4), 754–793.
7. B. F. M. Kuster, *Starch-Starke*, 1990, **42**(8), 314–321.
8. S. P. Simeonov, J. A. S. Coelho and C. A. M. Afonso, *ChemSusChem*, 2013, **6**(6), 997–1000.
9. S. Lima, P. Neves, M. M. Antunes, M. Pillinger, N. Ignatyev and A. A. Valente, *Appl. Catal., A*, 2009, **363**(1–2), 93–99.
10. C. Moreau, R. Durand, S. Razigade, J. Duhamet, P. Faugeras, P. Rivalier, P. Ros and G. Avignon, *Appl. Catal., A*, 1996, **145**(1–2), 211–224.
11. S. Hu, Z. Zhang, Y. Zhou, B. Han, H. Fan, W. Li, J. Song and Y. Xie, *Green Chem.*, 2008, **10**(12), 1280–1283.
12. S. P. Simeonov, J. A. S. Coelho and C. A. M. Afonso, *ChemSusChem*, 2012, **5**(8), 1388–1391.

9.2. Wittig Type Reactions

9.2.1. Synthesis of *trans*-9-(2-Phenylethenyl)anthracene

Dulce Simão

Department of Chemistry, Instituto Superior Técnico, University of Lisbon, Centro de Química Estrutural, Av. Rovisco Pais, 1, 1049-001 Lisbon, Portugal. *E-mail: dulcesimao@tecnico.ulisboa.pt

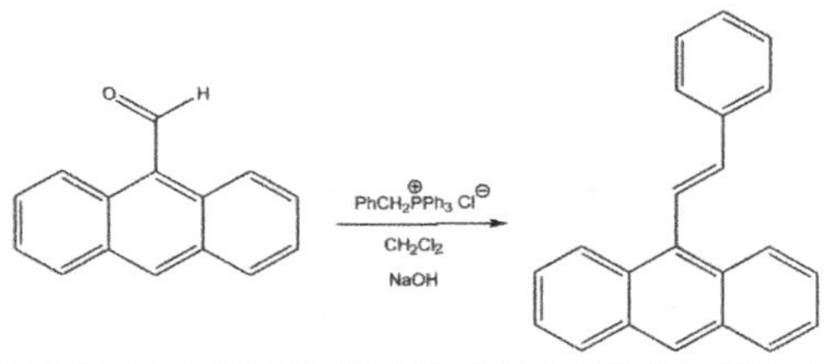

Number of sessions (duration of each session)	Hazard level	Difficulty level	Level of study
1 (4 h)	Low	Low	Introductory

Class names Aldehydes, alkenes, phosphonium salts, peroxyoxalates, oxalate diesters

Concepts involved This experiment involves a Wittig reaction to obtain an alkene that is fluorescent, and acts as activator in a chemiluminescent reaction

Chemicals needed Benzyltriphenylphosphonium chloride, 9-anthraldehyde, dichloromethane, sodium hydroxide, anhydrous calcium chloride, isopropanol, ethanol, hexane, bis(2,4,6-trichlorophenyl) oxalate, sodium acetate, ethyl acetate, hydrogen peroxide (30%)

Equipment and experimental techniques involved Stirring plate, liquid–liquid extraction, Rotavapor, recrystallization, TLC, IR, melting point

Keywords Chemiluminescence, fluorescence, IR, liquid–liquid extraction, Wittig reaction

Background

Phosphorus ylide addition to aldehyde or ketone gives an alkene and is known as the Wittig reaction. If the phosphorus ylide is unstable, it is generated *in situ* by treatment of a phosphonium salt with a base. The phosphonium salts are usually prepared from the phosphine and an alkyl halide.[1,2] Commercially available benzyltriphenylphosphonium chloride is used as the starting material, instead of preparing the phosphonium salt by reaction of triphenylphosphine with the alkyl halide. Typically, alkyllithium or phenyllithium are needed to prepare the ylide; however, in this experiment a 50% sodium hydroxide solution is basic enough to remove the proton at the α-position to phosphorus. The ylide reacts easily with the aldehyde

Comprehensive Organic Chemistry Experiments for the Laboratory Classroom
Edited by Carlos A. M. Afonso, Nuno R. Candeias, Dulce Pereira Simão, Alexandre F. Trindade, Jaime A. S. Coelho, Bin Tan, and Robert Franzén.

Published by the Royal Society of Chemistry, www.rsc.org

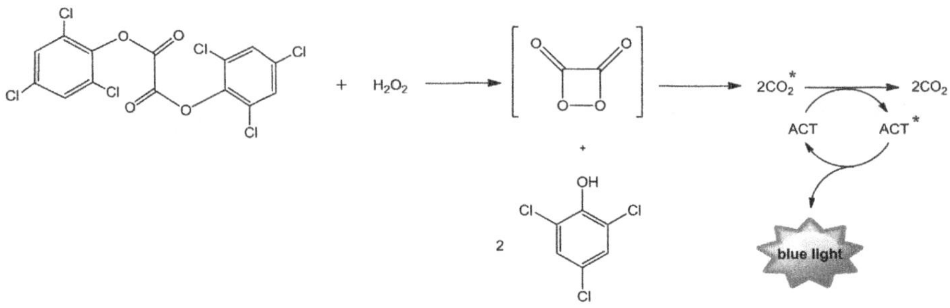

Scheme 9.2.1.1 Chemiluminescence reaction.

to produce the alkene and triphenylphosphine oxide. The reaction mechanism involves the formation of oxaphosphetane, a four-membered ring intermediate. Wittig reactions can give either the (*E*)- or (*Z*)-isomer of the alkene depending on the nature of the phosphonium reagent but in this experiment (*E*)-isomer is the only one to be obtained.[3–6] The beautiful yellow color of the alkene is related to the highly conjugated unsaturated system that absorbs light in the visible region. This compound is fluorescent when exposed under ultraviolet lamp and can be used as fluorescer or activator in a chemiluminescence reaction[7–9] using a peroxyoxalate system with bis(2,4,6-trichlorophenyl) oxalate known as a cyalume.[10] The addition of hydrogen peroxide results in a light-blue color emission that lasts for several minutes. The reaction between the oxalic ester and hydrogen peroxide affords an unstable cyclic peroxide. This molecule transmits its energy to the fluorescer (*trans*-9-(2-phenylethenyl)anthracene) leading to the electronic excited state of the fluorescer, which decays to the fundamental state emitting the fluorescent blue light (Scheme 9.2.1.1).

Various colors can be obtained on changing the fluorescer, which is usually a polycondensed aromatic hydrocarbon. This is the reaction that takes place in well-known "glow sticks" or "light sticks" used at entertainment events, military operations, training, industrial accidents and natural disasters.[8] When the stick is bent the compartment containing hydrogen peroxide is broken, the reaction with the ester solution occurs, leading to the glow.[10]

Additional Safety

Benzyltriphenylphosphonium chloride and 9-anthraldehyde are irritants. The reaction is performed in a round-bottom flask with a pressure-equalizing dropping funnel to prevent dichloromethane inhalation; dichloromethane is an irritant and toxic. For the same reason, liquid–liquid extraction should be carried out in a fume hood. Care should be taken with corrosive sodium hydroxide and with the flammables ethanol and isopropanol. Hydrogen peroxide is a corrosive liquid and is a strong oxidant.

Experimental Procedure

1. In a 50 mL round-bottom flask place benzyltriphenylphosphonium chloride (0.97 g) and 9-anthraldehyde (0.57 g), add dichloromethane (3 mL) and a magnetic stir bar.
2. Fit a pressure-equalizing dropping funnel. Vigorously stir the mixture and slowly add dropwise 50% aqueous sodium hydroxide (1.3 mL). The addition takes 4–6 min. Keep stirring for a further 30 min.
3. Transfer the reaction mixture to a 150 mL separatory funnel using water (10 mL) and dichloromethane (10 mL) to complete the transfer. Shake the mixture, remove the organic layer, and then extract the aqueous layer additionally with dichloromethane (5 mL).

4. Dry the combined organic layers with anhydrous calcium chloride, filter into a 50 mL round-bottom flask, evaporate it to dryness on a Rotavapor.
5. Recrystallize the crude product from isopropanol. The product is obtained as thin yellow plates.
6. Record the mp and IR (KBr) spectrum. Examine a dilute ethanol solution of the product (5 mg per 50 mL) under an ultraviolet lamp.

TLC: In a 4 cm × 9 cm silica gel TLC plate with fluorescent indicator, apply a 9-anthraldehyde solution and product solution. Eluent: hexane/ethyl acetate (5 : 1). After development examine the chromatogram under an ultraviolet lamp.

Chemiluminescent Reaction

7. In a test tube place *trans*-9-(2-phenylethenyl)anthracene (3 mg), bis(2,4,6-trichlorophenyl) oxalate (50 mg) and sodium acetate (50 mg). Add ethyl acetate (10 mL), cap the tube and shake until dissolution. In a dark room, add hydrogen peroxide 30% (3 mL), cap the tube and shake. Observe the color that is seen, the intensity, and the duration of the fluorescent reaction.

Waste Disposal

The neutralized aqueous solutions may be diluted and discarded in the sink. Dichloromethane and isopropanol may be stored for further purification. TLC eluent and chemiluminescent solution should be stored in a non-chlorinated waste collector for posterior destruction.

Results Interpretation and Additional Questions

1. Determine the reaction yield.
2. Compare the melting point of the obtained product with the one reported in the literature.
3. Interpret the IR and ^{1}H NMR and spectra.
4. Write out all reaction mechanisms.
5. Calculate R_f for each spot on the chromatogram.
6. Contrarily to other phosphonium salts, why is sodium hydroxide solution strong enough to deprotonate the benzyltriphenylphosphonium chloride?

References

1. K. P. C. Vollhardt and N. E. Schore, *Organic Chemistry*, Freeman & Cª., 3ª edn, 1999, p. 757.
2. W. G. Solomons, *Organic Chemistry*, John Wiley & Sons Inc., New York, 7th edn, 1998, pp. 745–749.
3. J. Clayden, N. Greeves, S. Warren and P. Wothers, *Organic Chemistry*, Oxford University Press, 2001, p. 814.
4. C. Jaworek and S. Iacobucci, *J. Chem. Educ.*, 2002, **79**, 111.
5. E. F. Silversmith, *J. Chem. Educ.*, 1986, **63**, 645.
6. H. D. Becker and K. Andersson, *J. Org. Chem.*, 1983, **48**, 4542.
7. L. F. Fieser and K. L. Williamson, *Organic Experiments*, Houghton Mifflin Company, 8th edn, 1998, p. 527.
8. R. Albertin, M. Arribas, E. Bastos, S. Röpke, P. Sakai, A. Sanches, C. Stevani, I. Umezu, J. Yu and W. Baader, *Quím. Nova*, 1998, **21**(6), 772.
9. J. Alves, A. Boaro, J. S. da Silva, T. L. Ferreira, V. B. Keslarek, C. A. Cabral, R. B. Orfão Jr., L. F. M. L. Ciscato and F. H. Bartoloni, *Photochem. Photobiol. Sci.*, 2015, **14**, 320–328.
10. Cyalume® Technologies, http://www.cyalume.com/, accessed May 2015.

9.2.2. Synthesis of 4-Vinylbenzoic acid

Dulce Simão* and Ana C. Cerdeira

Department of Chemistry, Instituto Superior Técnico, University of Lisbon, Centro de Química Estrutural, Av. Rovisco Pais, 1, 1049-001 Lisbon, Portugal. *E-mail: dulcesimao@tecnico.ulisboa.pt

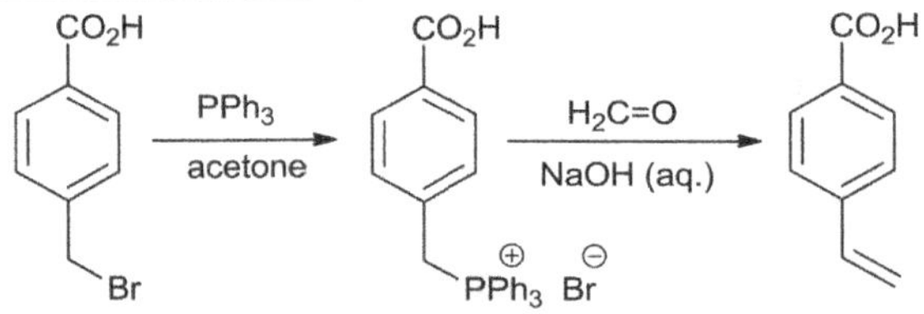

Number of sessions (duration of each session)	Hazard level	Difficulty level	Level of study
2 (3 h each)	Moderate	Low	Introductory
Class names Benzyl halides, phosphines, phosphonium salts, ylides, aldehydes, alkenes		**Concepts involved** This two-step synthesis illustrates a phosphorus ylide formation followed by a Wittig reaction to obtain an aromatic alkene	
Chemicals needed 4-(Bromomethyl)benzoic acid, triphenylphosphine, acetone, diethyl ether, formaldehyde (37% aqueous solution), sodium hydroxide pellets, ethanol, concentrated hydrogen chloride			
Equipment and experimental techniques involved Vacuum filtration, IR, magnetic stirring, heating mantle, reflux apparatus, melting point			
Keywords Alkene, IR, phosphorus ylide, vacuum filtration, Wittig reaction			

Background

The Wittig reaction is frequently used in the selective synthesis of alkenes, through the addition of phosphorus ylides to carbonyl compounds. The phosphorus ylide (or phosphorane) is usually prepared in two steps: the first is a S_N2 reaction between a phosphine and an alkyl halide or an aryl halide to furnish a phosphonium salt.[1–3] In this experiment triphenylphosphine, which is an excellent nucleophile, reacts readily with 4-(bromomethyl)benzoic acid to furnish the triphenylphosphonium salt. In the second step, deprotonation by a relatively strong base gives the triphenylphosphonium ylide, which is not isolated. Bases such as phenyllithium, butyllithium or alkoxides are commonly used in this step; however, in this work sodium hydroxide aqueous solution is basic enough to remove the proton at the α-position to phosphorus because the ylide formed is stabilized through resonance with the aromatic ring and carboxyl group.[4,5] The ylide reacts easily with formaldehyde to produce the alkene and triphenylphosphine oxide. The reaction mechanism involves the formation of an oxaphosphetane, a four-membered ring intermediate. Wittig reactions can give either the (*E*)- or (*Z*)-isomer of the alkene depending on the nature of the ylide and carbonyl compound, but in this experiment with formaldehyde only one product is obtained.[1–3]

Comprehensive Organic Chemistry Experiments for the Laboratory Classroom
Edited by Carlos A. M. Afonso, Nuno R. Candeias, Dulce Pereira Simão, Alexandre F. Trindade, Jaime A. S. Coelho, Bin Tan, and Robert Franzén.

Published by the Royal Society of Chemistry, www.rsc.org

Additional Safety

The use of a fume hood is mandatory for this experiment. 4-(Bromomethyl)benzoic acid, triphenylphosphine and 4-carboxybenzyltriphenylphosphonium bromide are irritants. Formaldehyde is toxic, a suspected cancer agent, and should be handled carefully. Care also should be taken while handling acetone, diethyl ether and ethanol (flammable solvents), sodium hydroxide and hydrogen chloride (both corrosive).

Experimental Procedure

Session 1 – Preparation of 4-Carboxybenzyltriphenylphosphonium Bromide

1. To a 100 mL round-bottom flask equipped with a reflux condenser, add 4-(bromomethyl) benzoic acid (4.3 g, 20 mmol) and triphenylphosphine (5.2 g, 20 mmol).
2. Dissolve with acetone (60 mL).
3. Reflux for 45 min in a heated water bath.
4. Cool the reaction mixture and collect the product by vacuum filtration.
5. Wash the solid with diethyl ether (2 × 20 mL) and dry on the funnel with vacuum. No further purification is needed and the product can be used as it is for the next step.
6. Record the yield and melting point. Obtain an IR spectrum.

Session 2 – Preparation of 4-Vinylbenzoic acid

1. Add 4-carboxybenzyltriphenylphosphonium bromide (3.76 g, 8 mmol), aqueous formaldehyde (37%) (32 mL) and water (15 mL) to a 250 mL Erlenmeyer flask equipped with a magnetic stirrer.
2. Clamp the flask and stir vigorously.
3. Add a solution of NaOH (2.5 g, 62.5 mmol) in water (15 mL) over a 10 min period and stir for an additional 45 min.
4. Filter off the precipitate with a vacuum and wash it with water (*ca.* 100 mL).
5. Acidify the filtrate and washings by the slow addition of concentrated hydrochloric acid until pH ~1 (use pH paper strips).
6. Filter off the resulting precipitate with vacuum.
7. Recrystallize the crude product from ethanol/H_2O (7 : 3).
8. Record the yield and melting point of the product and obtain an IR spectrum.

Waste Disposal

The filtrate resulting from the first step contains acetone and diethyl ether, and should be stored in a non-chlorinated waste collector for posterior destruction. Triphenylphosphine oxide, obtained as side product and isolated by vacuum filtration, can be stored for further purification. The acidic filtrate from the second step should be neutralized and stored in a waste collector recipient for aqueous solutions. Ethanol used for recrystallization can be stored for further purification.

Results Interpretation and Additional Questions

1. Compare the melting points with those reported in the literature. What does this tell us about the purity of the obtained products?
2. Identify the characteristic absorption bands in the IR spectra and interpret the ^1H NMR data from the literature.[6]

3. Write out all the reaction mechanisms.
4. If benzaldehyde were used instead formaldehyde, what would be the product or products obtained? Explain why.

References

1. P. Vollhardt and N. E. Schore, *Organic Chemistry*, Freeman & Co., 3rd edn, 1999, p. 757.
2. W. G. Solomons, *Organic Chemistry*, John Wiley & Sons, Inc., New York, 7th edn, 1998, p. 745.
3. J. Clayden, N. Greeves, S. Warren and P. Wothers, *Organic Chemistry*, Oxford University Press, Oxford, 2001, p. 814.
4. L. M. Harwood, C. J. Moody and J. M. Percy, *Experimental Organic Chemistry*, Blackwell Science, 2nd edn, 1999, pp. 563–566.
5. L. F. Tietze and T. H. Eicher, *Reactions and Synthesis in Organic Chemistry Laboratory*, University Science Books, 1989, p. 179.
6. AIST, Spectral Database for Organic Compounds (SDBS), http://sdbs.db.aist.go.jp/sdbs/cgi-bin/cre_index.cgi?lang=eng, accessed September 2015.

9.2.3. Preparation of Nitrostilbenes by the Wittig Reaction

Francisca Lopes*, Maria de Jesus Perry, and Ana Paula Francisco

Research Institute for Medicines (iMed. ULisboa), Faculdade de Farmácia da, Universidade de Lisboa, Av. Prof. Gama Pinto, 1649-003 Lisboa, Portugal. *E-mail: fclopes@ff.ulisboa.pt

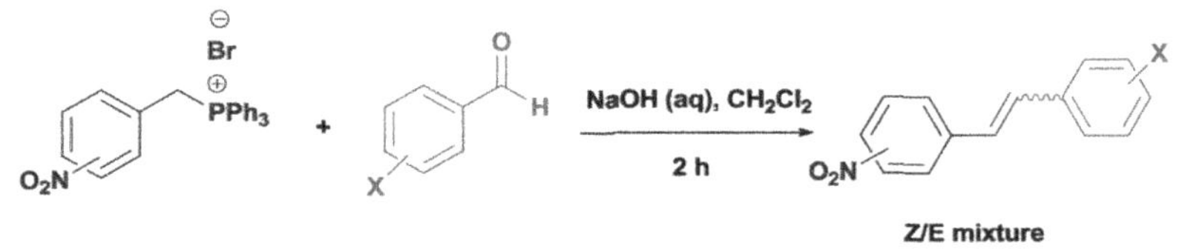

Number of sessions (duration of each session)	Hazard level	Difficulty level	Level of study
2 (4 h)	Moderate	Medium	Intermediate
Class names Aldehydes, alkenes, phosphonium ylides			
Concepts involved Carbon–carbon double bond formation, stereoselective reactions, phase-transfer catalysis (PTC)			
Chemicals needed 4-Chlorobenzaldehyde, 4-methoxybenzaldehyde, 3-nitrobenzyltriphenylphosphonium bromide, sodium hydroxide, dichloromethane, silica gel (60 Å; mesh 230–400)			
Equipment and experimental techniques involved Stirring plate, rotatory evaporator, column chromatography, thin-layer chromatography (TLC), liquid–liquid extraction, NMR			
Keywords Alkene synthesis, extraction, flash chromatography, NMR, stereoselectivity, Wittig reaction			

Background

Nitrostilbenes are important building blocks of several compounds with potential therapeutic value such as 4-pyridonimines,[1] anthranilic acids[2] and phthalimides[3,4] (Scheme 9.2.3.1). Unsymmetrical nitrostilbenes are easily synthesized by the Wittig reaction.

The Wittig reaction, discovered in 1954, is a convenient and reliable method for the regioselective formation of carbon–carbon double bonds and is widely used in the preparation of alkenes. It is one of the easiest and least expensive ways to synthesize alkenes from aldehydes and ketones. The reactive species in this reaction is a unique intermediate known as an ylide. Ylides are overall neutral species although they have adjacent atoms with opposite charges. In terms of reactivity, the ylide acts as a carbanion and thus it is nucleophilic.

Mechanistically (Scheme 9.2.3.2) the Wittig reaction was traditionally described as deprotonation of the phosphonium salt to generate the phosphorus ylide, which attacks the π^* orbital of the carbonyl to generate a betaine intermediate. Cyclization of the betaine then gives the

Comprehensive Organic Chemistry Experiments for the Laboratory Classroom
Edited by Carlos A. M. Afonso, Nuno R. Candeias, Dulce Pereira Simão, Alexandre F. Trindade, Jaime A. S. Coelho, Bin Tan, and Robert Franzén.

Published by the Royal Society of Chemistry, www.rsc.org

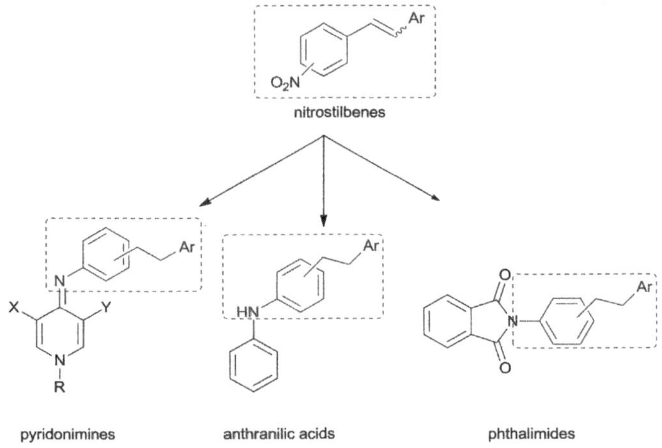

Scheme 9.2.3.1 Compounds with therapeutic interest prepared from nitrostilbene building blocks.

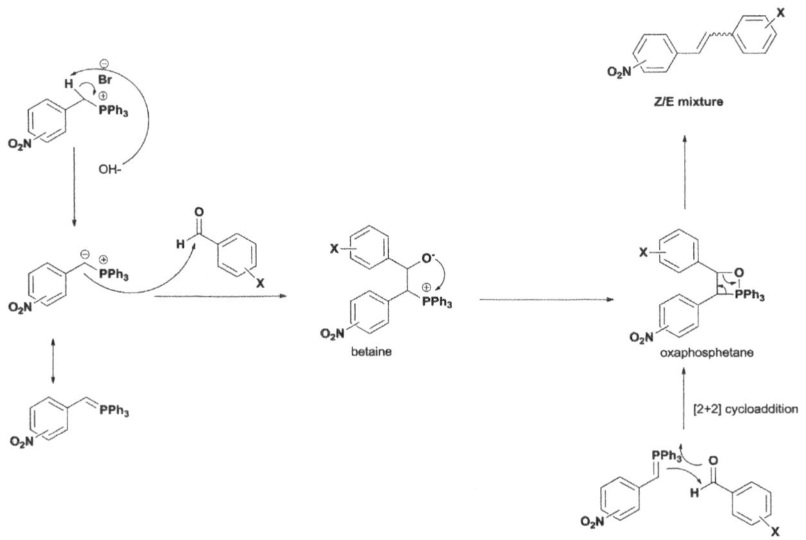

Scheme 9.2.3.2 Proposed pathways for the mechanism of the Wittig reaction.

oxaphosphetane, a four-membered ring containing both oxygen and phosphorous.[5] However, the exact mechanism of this reaction is still under study. More recently, an alternative pathway has been proposed involving a concerted [2 + 2] cycloaddition between the ylide and the carbonyl group to give the oxaphosphetane.[5,6] After this both pathways followed the same process of cycloreversion of the oxaphosphetane, which results in the formation of the alkene and the triphenylphosphine oxide (Scheme 9.2.3.2). Due to this cycloreversion step, the reaction can evolve under thermodynamic or kinetic control. Stereoselectivity of the reaction is mostly dependent on steric and electronic effects. While unstabilized ylides (with electron donating groups) give almost exclusively the (*E*)-alkene (kinetic control), stabilized ylides (with electron withdrawing groups) give preferentially the (*Z*)-alkene (thermodynamic control). The ylides can be prepared from alkyl or benzyl halides by substitution reaction with a phosphine (usually

PPh_3). The resulting phosphonium salt is further deprotonated to give the ylide. The phosphonium salt can also be purchased as it is commercially available.

The objective of this experiment is the synthesis of nitrostilbenes under phase-transfer catalysis (PTC) conditions. The method is a modified process of the original Wittig reaction and provides the synthesis of nitrostilbenes in excellent yields, at room temperature, in very reasonable time and in a stereoselective way.[7] The solvent system will be a mixture of water and dichloromethane, which are not co-soluble and will give two separate layers. Thus the ionic hydroxide and the phosphonium salt can go into the water, and the aldehyde and the product can go into the dichlorometane. The reaction will proceed in the heterogeneous phase with the phosphonium salt acting as the phase-transfer catalyst.[8,9] For most of aldehydes it is possible to observe qualitatively the formation of both diastereoisomers by TLC as they have different R_f values. Complementarily, the ^{1}H NMR spectral analysis allows the assignment of the *E*/*Z* ratios.

Additional Safety

The experimental procedure should be done in a fume hood, with proper safety equipment, such as protective clothes, gloves and safety goggles. All starting aldehydes are considered to be eye and skin irritants. Nitrobenzyltriphenylphosphonium chloride is highly toxic by ingestion. Sodium hydroxide (0.1 N) is extremely damaging to the eyes and skin. Handle the dry silica gel carefully in a fume hood. The solid silica gel forms an aerosol that is damaging to lungs. The room must be well ventilated. Deuterochloroform ($CDCl_3$) causes irritation of the skin and respiratory system, may cause chemical burns, and is a possible carcinogen.

Experimental Procedure

Laboratory Session 1 (4 h): Synthesis of Nitrostilbenes

1. Each pair of students in the laboratory select one aldehyde (*e.g.* benzaldehyde, 4-methoxybenzaldehyde, 4-chlorobenzaldehyde, 4-trifluoromethylbenzaldehyde).
2. In a round-bottomed flask equipped with a magnetic stirrer add the phosphonium salt (0.84 mmol) and the aldehyde (0.84 mmol).
3. Suspended the mixture in 3 mL of dichloromethane.
4. Add an aqueous solution of 0.1 N NaOH (1 mmol). It is possible to observe two phases. Analyse the composition of the reaction mixture by TLC (eluent: hexane: diethyl ether (4:1)). As the reaction mixture is biphasic, introduce the capillary or the Pasteur pipette (covered on the top with your finger) into the lower layer (dichloromethane) then remove your finger from the top of the capillary and let the solvent ascend by capillarity.
5. Stir the reaction mixture vigorously for 2 h. The end of the reaction can be confirmed by TLC.
6. Dilute the reaction mixture with 3 mL of dichloromethane and pour the mixture into the separatory funnel.
7. Shake the mixture vigorously, and separate the organic layer into a 50 mL Erlenmeyer flask.
8. Add an additional 5 mL of dichloromethane to the separatory funnel, and shake vigorously again.
9. Pour the organic layer into the same Erlenmeyer flask that has the first dichloromethane phase.
10. Dry the combined organic layers with sodium sulfate.
11. Filter the organic solution into a round-bottomed flask and remove the solvent in a rotatory evaporator. Keep the flask in the refrigerator until the next session.

Laboratory Session 2 (4 h): Product Isolation, Characterization and Yield Determination

The crude product is purified by flash chromatography.

1. Weigh 10 g of dry silica gel (230–400 mesh) and fill the column.
2. The column can be packed by inserting the tube of a water pump at the bleed of the column with the valve open and gently applying vaccum for 10 min. Once the column is packed add a protective layer (1 cm thick) of sand on top of the silica.
3. Using a Pasteur pipette, load the sample dissolved in a small amount of dichloromethane (about 1–1.5 mL).
4. After loading the sample, elute the chromatography column with dichloromethane. Monitor the solvent level and make sure it does not go below the top of the silica. Add solvent as necessary.
5. Apply air pressure on the top of the column.
6. Collect fractions of 5 mL sequentially in test tubes correctly labelled for later TLC analysis.
7. Assemble the tubes containing the mixture of (*E*/*Z*)-isomers in a weighed balloon and evaporate the solvent in the rotary evaporator.
8. Determine the weight of the residue and calculate the yield of the reaction.
9. Analyse the mixture by ^{1}H NMR.

Waste Disposal

The neutralized aqueous NaOH solution may be diluted and collected in a waste collector recipient for further destruction. Non-halogenated organic solvents must be collected in a non-halogeneted collector recipient for further destruction. The silica gel used must be placed in a container and disposed of during the regular hazardous waste pickups.

Results Interpretation and Additional Questions

1. What is an ylide? How does it behave? Discuss the importance of NaOH in the process.
2. What side products are generated during this reaction? During your analysis of the filtrate, how many of these will you observe on the TLC plate?
3. Interpret the ^{1}H NMR spectra obtained. Assign the identity of the isomers based on the magnitude of the coupling constants. Determine the *E*/*Z* ratio.
4. Calculate the percent yield of the reaction.

References

1. T. Rodrigues, R. C. Guedes, D. J. V. A. dos Santos, M. Carrasco, J. Gut, P. J. Rosenthal, R. Moreira and F. Lopes, *Bioorg. Med. Chem. Lett.*, 2009, **19**, 3476.
2. L. J. Simons, B. W. Caprathe, M. Callahan, J. M. Graham, T. Kimura, Y. Lai, H. LeVine, III, W. Lipinski, A. T. Sakkab, Y. Tasaki, L. C. Walker, T. Yasunaga, Y. Ye, N. Zhuang and C. E. Augelli-Szafran, *Bioorg. Med. Chem. Lett.*, 2009, **19**, 654.
3. K. Motoshima, M. Ishikawa, K. Sugita and Y. Hashimoto, *Biol. Pharm. Bull.*, 2009, **32**, 1618.
4. K. Motoshima, T. Noguchi-Yachide, K. Sugita, Y. Hashimoto and M. Ishikawa, *Bioorg. Med. Chem.*, 2009, **17**, 5001.
5. P. A. Byrne and D. G. Gilheany, *Chem. Soc. Rev.*, 2013, **42**, 6670.
6. B. Albrecht, *J. Chem. Educ.*, 2014, **91**, 2182.
7. T. Rodrigues, F. Lopes and R. Moreira, *Synth. Commun.*, 2012, **42**, 747.
8. J. C. Warner, P. T. Anastas and J. P. Anselme, *J. Chem. Educ.*, 1985, **62**, 346.
9. J. Gillois, G. Guillerm, M. Savignac, E. Stephan and L. Voquang, *J. Chem. Educ.*, 1980, **57**, 161.

9.2.4. Building an Alkene Spacer by the Wittig Reaction: Synthesis of 4-[2-(4-Nitrophenyl)ethenyl]benzonitrile

António P. S. Teixeira*[a], M. Paula Robalo[b,c], and Paulo J. Mendes[a]

[a]Departamento de Química, Escola de Ciências e Tecnologia, Centro de Química de Évora, Instituto de Investigação e Formação Avançada, Universidade de Évora, Rua Romão Ramalho 59, 7000-671 Évora, Portugal; [b]Área Departamental de Engenharia Química, Instituto Superior de Engenharia de Lisboa, Rua Conselheiro Emídio Navarro, 1, 1959-007 Lisboa, Portugal; [c]Centro de Química Estrutural, Complexo I, Instituto Superior Técnico, Universidade de Lisboa, Av. Rovisco Pais, 1049-001 Lisboa, Portugal
*E-mail: apsteix@uevora.pt

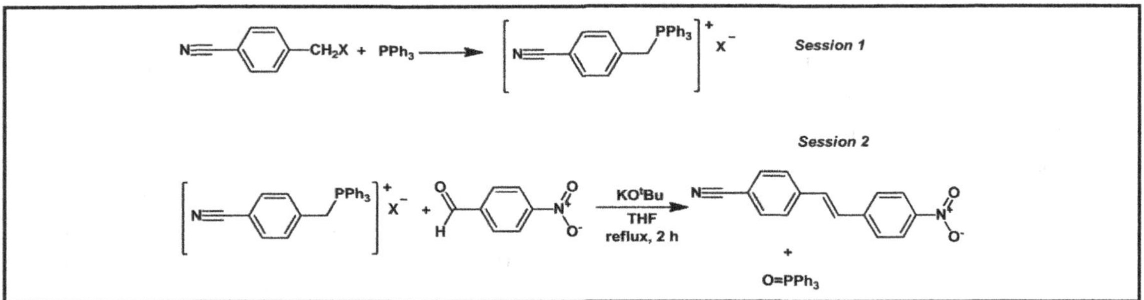

Number of sessions (duration of each session)	Hazard level	Difficulty level	Level of study
2 (3 h)	Moderate	Medium	Intermediate

Class names Benzyl halides, phosphorus ylides, aldehydes, alkenes

Concepts involved Wittig reaction, double bond formation, spectroscopic characterization, *E*/*Z* isomers, *E*/*Z* isomeric ratio

Chemicals needed 4-Nitrobenzaldehyde, 4-(bromomethyl)benzonitrile, triphenylphosphine, potassium *tert*-butoxide, toluene, THF, diethyl ether, dichloromethane, methanol, anhydrous magnesium sulfate

Equipment and experimental techniques involved Heating mantle or oil bath with magnetic stirring, Büchner funnel, reflux apparatus, rotary evaporator, structural analysis (NMR, IR, MS)

Keywords Aldehydes, alkenes, phosphorus ylides, precipitation, spectroscopic characterization, Wittig reaction

Background

The Wittig reaction was discovered by George Wittig (Nobel Prize in Chemistry in 1979) in 1954.[1] The Wittig reaction is one of the most common and valuable methods for the formation of alkenes, starting from a carbonyl compound and a primary or secondary alkyl halide. The reaction involves the formation of an alkene from an aldehyde or a ketone using a phosphorous ylide derivative, $R_2C^-\text{-}P^+(C_6H_5)_3$. This latter species is obtained from a nucleophilic substitution reaction of an alkyl halide by triphenylphosphine, followed by deprotonation of the

Comprehensive Organic Chemistry Experiments for the Laboratory Classroom
Edited by Carlos A. M. Afonso, Nuno R. Candeias, Dulce Pereira Simão, Alexandre F. Trindade, Jaime A. S. Coelho, Bin Tan, and Robert Franzén.

Published by the Royal Society of Chemistry, www.rsc.org

corresponding phosphonium salt by a strong base. This reaction has wide applicability, using different aldehydes and ketones and alkyl halides, despite its relatively slow rate reaction with ketones. Knowledge of the exact position of the double bond is also an advantage of the method.

Nevertheless, one limitation of the method is the formation of a mixture of the geometric (*E*)- and (*Z*)-isomers. The control of double bond geometry remains a challenge in organic synthesis and total stereoisomeric purity is hard to achieve, but the (*E*/*Z*) isomeric ratio can be modified by control of the reaction conditions (temperature, polarity of solvent, stabilizing salts and base). In addition, the use of a stabilized ylide substituted with an electron-withdrawing group (*e.g.*, -CO_2R, -CN, -SO_2R) ensures a high selectivity for the more thermodynamically stable (*E*)-isomer.[2–5] The Schlosser modification of the Wittig reaction, using phenyllithium, allows the stabilization of the *erythro* betaine and interconversion into the corresponding *threo* intermediate, leading to the formation of (*E*)-isomer from non-stabilized ylides.[6]

Despite the great number of Wittig reactions already done, the mechanism of this reaction remains controversial and there is no unified mechanism to explain all results and a continuum of possible mechanistic pathways could be proposed, due to the diversity of structural and stereoelectronic variations among the reagents involved. While the formation of oxaphosphetanes is easy to identify during the reaction, betaines can still have a transient existence and a few have been identified.[7] According to the classic mechanism, the Wittig reaction begins with the nucleophilic addition of the ylide carbanion to the carbonyl atom to afford the betaine. The formation of a P–O bond gives the four-membered cycle oxaphosphetane, which is unstable and readily decomposes to form the phosphine oxide and the alkene. The formation of the kinetic *erythro*- or the thermodynamic *threo*-betaines leads to the formation of the corresponding oxaphosphetanes, which decompose to afford the (*Z*)- or the (*E*)-isomers, respectively.[8]

The aim of this work is the synthesis of the 4-[2-(4-nitrophenyl)ethenyl]benzonitrile and the isolation of the (*E*)-isomer. This example illustrates the use of the Wittig reaction to join two reacting species by the formation of a new carbon–carbon double bond, with the consequent increase in size and complexity of the molecule. The new double bond links two phenyl rings, extending the carbon chain and the electronic delocalization through the molecule, allowing electronic transfer processes between the substituents of each ring. The presence of a nitrile group allows the coordination to transition metal moieties or the transformation into other functional groups. Isolation of the (*E*)-isomer is possible by selective precipitation.

Additional Safety

All hazardous materials should be handled using standard safety procedures. The sessions should be done in a fume hood, with proper safety equipment, such as protective clothing, gloves and safety goggles.

Triphenylphosphine and dichloromethane are toxic and irritant. Diethyl ether is flammable and THF and toluene are flammable and irritant and should be handled with care.

Since the toxicity of (4-cyanophenyl)methyltriphenylphosphonium bromide was not determined, it should be handled with care, as similar salts have been described as possibly irritant and hygroscopic.

Experimental Procedure (ref. 9)

Session 1 (3 h)

Synthesis of (4-Cyanophenyl)methyltriphenylphosphonium Bromide

1. In a round-bottomed flask containing a stirring bar, place 40 mL of toluene.
2. Weigh 2.00 g (10.2 mmol) of 4-(bromomethyl)benzonitrile and 2.67 g (10.2 mmol, 1.0 equiv.) of triphenylphosphine and transfer to the flask.

3. Insert a water condenser into the round-bottomed flask and turn on the cooling water in order to circulate and to condense escaping vapours.
4. Reflux the mixture for 2 h using a heating mantle or an oil bath with magnetic stirring.
5. Remove the flask from the oil bath or the heating mantle and allow the mixture to cool to room temperature. The white solid produced in the reaction should begin to precipitate out of solution during the cooling.
6. Separate the white solid from the solution by filtration through a Büchner funnel.
7. Wash the solid with several portions of toluene and diethyl ether.
8. Dry the solid.
9. Weigh the white solid.
10. Record the ^{1}H and ^{31}P NMR spectra of the compound.

Session 2 (3 h)

Synthesis of 4-[2-(4-Nitrophenyl)ethenyl]benzonitrile

1. In a round-bottomed flask containing a stirring bar, place 20 mL of THF.
2. Weigh 2.35 g (5.12 mmol) of (4-cyanobenzyl)methyltriphenylphosphonium bromide and transfer to the flask, followed by 0.72 g (6.14 mmol, 1.2 equiv.) of potassium *tert*-butoxide.
3. After gentle stirring for 10 min, add, slowly, a THF solution (2 mL) with 0.77 g (5.12 mmol, 1.0 eq.) of 4-nitrobenzaldehyde.
4. Insert a water condenser into the round-bottom flask and turn on the cooling water.
5. Reflux for 2 h in an oil bath or a heating mantle with magnetic stirring.
6. Remove the flask from the oil bath and allow the mixture to cool to room temperature.
7. Filter the dark-brownish solution from the solid residue.
8. Add water to the filtrate to precipitate the yellow solid.
9. Collect the yellow solid by vacuum filtration.
10. Dissolve the crude solid in dichloromethane and dry the solution with anhydrous magnesium sulfate.
11. Filter the solution and evaporate the solvent.
12. Weigh the solid to calculate the yield of the crude Wittig reaction isomers.
13. Record the ^{1}H NMR spectrum to determine the compound purity and the (*E*/*Z*) isomeric ratio.

Isolation of (E)-4-[2-(4-Nitrophenyl)ethenyl]benzonitrile

14. Dissolve the solid in methanol in a round-bottomed flask.
15. Concentrate the solution in a rotary evaporator until a small portion of solid is precipitated.
16. Decant the supernatant and wash the solid with a small amount of diethyl ether.
17. Repeat the procedure until 1/3 of the solid obtained in the previous reaction is collected.
18. Record the ^{1}H and ^{13}C NMR spectra, to determine the isomeric purity, and other spectra (IR and MS).
19. Determine the melting point.

Waste Disposal

All chemical waste should be disposed of in accordance with local regulations and laboratory practices. If there are no set of rules in the laboratory, the solutions with only toluene, diethyl ether, THF or mixtures of these solvents should be collected in a non-chlorinated waste collector recipient and solutions containing dichloromethane should collected in a recipient for chlorinated compounds.

Results Interpretation and Additional Questions

1. Calculate the yields of the compounds (4-cyanophenyl)methyltriphenylphosphonium bromide (session 1) and 4-[2-(4-nitrophenyl)ethenyl]benzonitrile (session 2).
2. Propose a mechanism for the formation of both isomers.
3. Interpret the spectra obtained.
4. Determine the (*E*/*Z*) isomeric ratio of the Wittig reaction.
5. Justify the formation of the major isomer ((*E*)-isomer) according to the substituent of the benzyl halide.

References

1. G. Wittig and U. Schöllkopf, *Chem. Ber.*, 1954, **87**, 1318.
2. B. E. Maryanoff and A. B. Reitz, *Chem. Rev.*, 1989, **89**, 863.
3. E. Vedejs and C. F. Marth, *J. Am. Chem. Soc.*, 1990, **112**, 3905.
4. B. E. Maryanoff, A. B. Reitz, M. S. Mutter, R. R. Inners, H. R. Almond, Jr., R. R. Whittle and R. A. Olofson, *J. Am. Chem. Soc.*, 1986, **108**, 7664.
5. A. B. Reitz, S. O. Nortey, A. D. Jordan, M. S. Mutter and B. E. Maryanoff, *J. Org. Chem.*, 1986, **51**, 3302.
6. M. Schlosser and K. F. Christmann, *Angew. Chem., Int. Ed.*, 1966, **5**, 126.
7. R. A. Neumann and S. Berger, *Eur. J. Org. Chem.*, 1998, 1085.
8. P. Galatsis, *Name Reactions for Homologations – Part I*, ed. J. J. Li, John Wiley & Son, Inc., Hoboken, NJ, 2009, ch. 2.13, pp. 588–612.
9. Adapted from: A. P. S. Teixeira, Síntese de compostos derivados de ferro(II) com potenciais propriedades em óptica não linear, PhD Thesis, Universidade de Évora, 2004, ch 6, pp. 253–254.

9.2.5. Preparation of *trans,trans*-Distyrylbenzene by a Wittig Reaction

Arno Kraft

Heriot-Watt University, Institute of Chemical Sciences, School of Engineering & Physical Sciences, Riccarton, Edinburgh EH14 4AS, United Kingdom. *E-mail: A.Kraft@hw.ac.uk

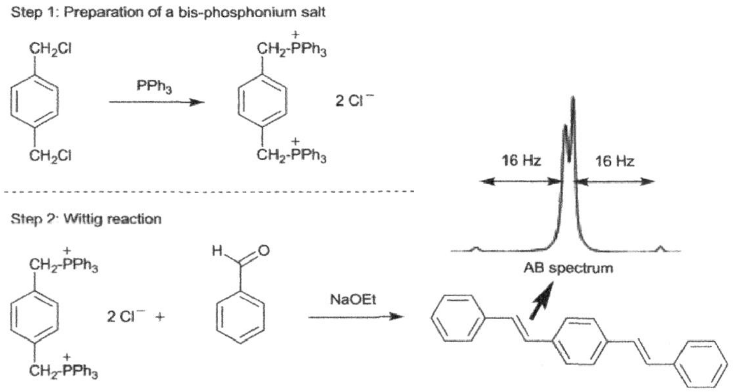

Number of sessions (duration of each session)	Hazard level	Difficulty level	Level of study
3 (3 h)	Moderate	Medium	Intermediate

Class names Benzyl chlorides, phosphonium salts, aldehydes, alkenes

Concepts involved Wittig reaction, NMR analysis of *cis/trans* alkene mixture, AB spectrum

Chemicals needed Triphenylphosphine, 1,4-bis(chloromethyl)benzene, dimethylformamide, diethyl ether, benzaldehyde, sodium ethoxide, ethanol, toluene, iodine, petroleum ether (60–80 °C), ethyl acetate, deuterated dichloromethane

Equipment and experimental techniques involved Apparatus for heating to reflux (150 °C) and magnetic stirring, filtration, recrystallisation, NMR

Keywords Alkene, *cis/trans* isomers, FTIR, NMR, phosphonium salt, recrystallisation, Wittig reaction

Background

The Wittig reaction is named after its discoverer, Georg Wittig, co-winner of the 1979 Nobel Prize for Chemistry.[1] It is an effective way of making an alkene carbon–carbon double bond from an aldehyde or ketone with the help of an organophosphorus intermediate.

Triphenylphosphine (PPh_3) is a good nucleophile and can easily undergo an S_N2 reaction with a primary or secondary alkyl halide to give an alkyltriphenylphosphonium salt (**1**) in high yield (Scheme 9.2.5.1). Deprotonation of the phosphonium salt **1** with a suitable base results in the reactive intermediate **2**, which is also called an ylide or a phosphorane. A phosphorane can be

Comprehensive Organic Chemistry Experiments for the Laboratory Classroom
Edited by Carlos A. M. Afonso, Nuno R. Candeias, Dulce Pereira Simão, Alexandre F. Trindade, Jaime A. S. Coelho, Bin Tan, and Robert Franzén.

Published by the Royal Society of Chemistry, www.rsc.org

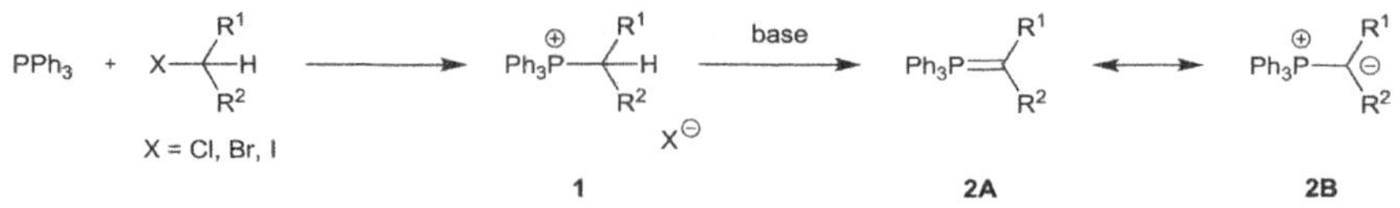

Scheme 9.2.5.1 Preparation of a phosphonium salt and ylide.

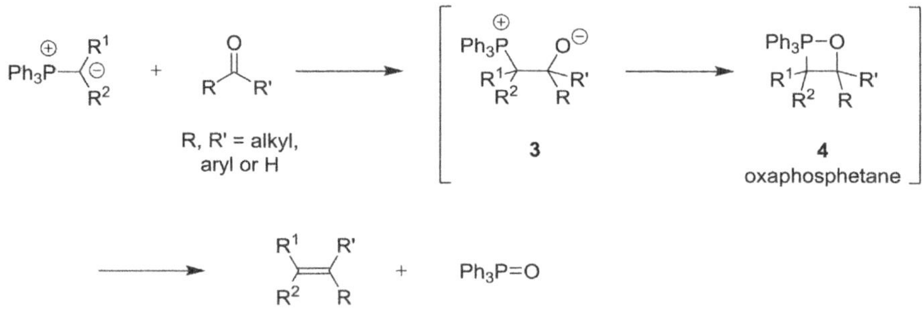

Scheme 9.2.5.2 Proposed intermediates in a Wittig reaction.

regarded as an organophosphorus compound with a formal double bond between phosphorus and carbon (**2A**); its reactivity is, however, best described by the alternative resonance structure of ylide **2B** with a negative charge on carbon, which is stabilised by electron-donation into the d-orbitals of the phosphorus.[2]

Reaction of **2** with an aldehyde or a ketone leads to the formation of an alkene (Scheme 9.2.5.2). The alkene C=C bond will always end up exactly where the C=O group was before, and the oxygen atom of the carbonyl compound is formally replaced by the = CR^1R^2 part of the phosphorane.

The Wittig reaction is a versatile method for making mono-, di- and trisubstituted alkenes; tetrasubstituted alkenes cannot be obtained due to steric hindrance. The mechanism of the Wittig reaction is still debated.[3,4] The reaction proceeds *via* a four-membered cyclic intermediate called an oxaphosphetane (**4**) that is formed by a [2 + 2] cycloaddition, but may also proceed *via* a betaine-type transition state **3** (Scheme 9.2.5.2). Triphenylphosphine oxide is a by-product of the reaction, and the high strength of the P=O double bond is the thermodynamic driving force behind the Wittig reaction.

A stabilised ylide results when one of the substituents next to the negatively charged ylide carbon is an electron-withdrawing group. The fact that such an ylide can be isolated and an NMR spectrum is easily obtained indicates its stability. Unstabilised ylides (where R^1 and R^2 are alkyl groups or H) normally require strong bases such as $NaNH_2$ or BuLi, which are not only more expensive but also demand special care and attention when working with them. The ylide in this experiment can be made with the milder, and easier to handle, base sodium ethoxide.

Stabilised ylides yield predominantly *trans* alkenes, whereas unstabilised ylides have a preference for *cis* alkenes. In this laboratory experiment students will investigate whether there is a preference for the bis-phosphonium salt used to form a *cis* or *trans* ((*Z*) or (*E*)) alkene or a mixture of the two isomers.

Additional Safety

The preparation of the phosphonium salt should be carried out in a fume hood, with proper safety equipment, such as protective clothing, gloves and safety goggles. Triphenylphosphine is harmful if swallowed and may cause an allergic skin reaction; it may cause long lasting harmful effects to aquatic life. 1,4-Bis(chloromethyl)benzene is harmful if swallowed, causes skin and serious eye irritation; it is very toxic to aquatic life. Dimethylformamide (DMF) is flammable, harmful in contact with skin and if inhaled; it causes serious eye irritation and may damage the fertility of the unborn child. Skin contact with dimethylformamide should be avoided. Diethyl ether is extremely flammable – make sure there are no naked flames or hot surfaces nearby as diethyl ether ignites readily even when coming into contact with a heated hotplate; diethyl ether can form explosive peroxides, is harmful if swallowed, and vapours may cause drowsiness and dizziness. Phosphonium salts are irritating to eyes, respiratory system and skin. Benzaldehyde is harmful if swallowed. Sodium ethoxide can cause severe skin burns and eye damage; in case of contact with eyes, rinse immediately with plenty of water and seek medical advice. Ethanol (methylated spirits) is flammable. Toluene is flammable, harmful when inhaled, irritating to the skin, may cause lung damage if swallowed, and is suspected of damaging fertility or the unborn child. Iodine is harmful in contact with skin and if inhaled, and is very toxic to aquatic life. Petroleum ether is highly flammable, may cause lung damage if swallowed, and is toxic to aquatic organisms; its vapours may cause drowsiness and dizziness and it is suspected of damaging fertility or the unborn child. Ethyl acetate is flammable, irritating to the eyes and vapours may cause drowsiness and dizziness.

Experimental Procedure

Laboratory Session 1 (3 h): Bis-phosphonium Salt Preparation

Preparation of the Bis-Phosphonium Salt

1. Reflux a mixture of 2.850 g of triphenylphosphine (11.0 mmol, 2.2 equiv.), 0.875 g of 1,4-bis(chloromethyl)benzene (5.00 mmol) and 10 mL of DMF for 2 h using an isomantle. A precipitate should start to form after about half an hour.
2. Allow the mixture to cool to room temperature, then cool with an ice–water bath, filter the solid, wash with a small amount of DMF followed by diethyl ether. Dispose of the filtrate, then apply suction to the filtered *p*-xylylenebis(triphenylphosphonium chloride) until the product is dry.
3. Determine the yield of the bis-phosphonium salt.

Laboratory Session 2 (3 h): Wittig Reaction and Work-Up and Purification

Wittig Reaction

1. In a 100 mL round-bottom flask dissolve benzaldehyde (6.5 mmol, 2.6 equiv.) and *p*-xylylene-bis(triphenylphosphonium chloride) (1.75 g, 2.5 mmol) in dry ethanol (15 mL).
2. Add a solution of sodium ethoxide (6.25 mmol, 2.5 equiv.) in dry ethanol (30 mL), then attach a drying tube filled with calcium chloride, and stir the reaction mixture at room temperature for 2 h.

Work Up

3. Add water (*ca.* 25 mL) and collect the precipitate by suction filtration. Wash with 60% ethanol (or 60% methylated spirits) and air-dry the product for at least 2 days.

Laboratory Session 3 (3 h): Purification and Characterisation

Purification

1. Dissolve the crude product in boiling toluene (20 mL) in the presence of a pinch of iodine and reflux for 15 min, then allow to crystallise. Collect the precipitate by suction filtration and dry overnight.

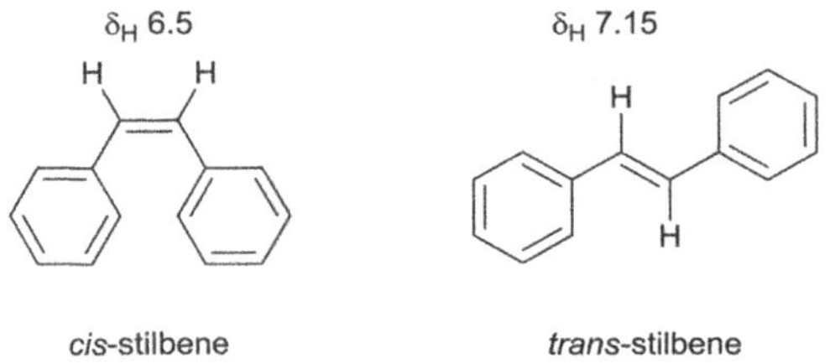

Figure 9.2.5.1 Characteristic alkene ^{1}H NMR chemical shifts for *cis*- and *trans*-stilbene.

Characterisation

2. Determine the yield. Measure the melting point and record an IR spectrum.
3. Carry out a thin-layer chromatography analysis (eluent: petroleum ether/ethyl acetate (5 : 1)) of both your crude and recrystallised product.

Waste Disposal

Residues from filtration and recrystallisation should be collected in the appropriate non-halogenated/halogenated waste container, for disposal by an external contractor. NMR samples in $CDCl_3$ or CD_2Cl_2 should be collected in a chlorinated waste collector recipient, for posterior destruction.

Results Interpretation and Additional Questions

1. Analyse the ^{1}H NMR spectrum of the recrystallised product obtained in this Wittig reaction. Suggest a structure for the product of your Wittig reaction, including its stereochemistry.
2. Compare this with the ^{1}H NMR spectrum of the crude product of your Wittig reaction. *Hint*: A *cis*-alkene and a *trans*-alkene have characteristic coupling constants; *cis*-stilbene and *trans*-stilbene also show clear differences in their alkene proton chemical shifts (Figure 9.2.5.1).
3. What conclusions can you draw regarding the *cis*/*trans* selectivity in this example of a Wittig reaction?
4. What is the purpose of the iodine during the recrystallisation step?

References

1. The Nobel Prize in Chemistry 1979, http://www.nobelprize.org/nobel_prizes/chemistry/laureates/1979/, accessed October 2014.
2. J. Clayden, N. Greeves and S. Warren, *Organic Chemistry*, Oxford University Press, 2nd edn, Oxford, 2012, pp. 689–693.
3. Z. Chen, Y. Nieves-Quinones, J. R. Waas and D. A. Singleton, *J. Am. Chem. Soc.*, 2014, **136**, 13122.
4. B. Albrecht, *J. Chem. Educ.*, 2014, **91**, 2182.

9.2.6. Synthesis and Reactivity of Phosphorus Ylides

Ana Lúcia Cardoso, Maria I. L. Soares, Susana M. M. Lopes, and Teresa M. V. D. Pinho e Melo*

Department of Chemistry, University of Coimbra, Rua Larga, 3004-535 Coimbra, Portugal
*E-mail: tmelo@ci.uc.pt

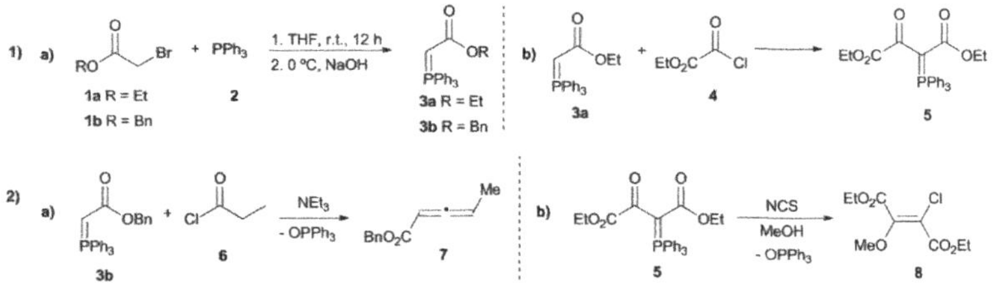

Number of sessions (duration of each session)	Hazard level	Difficulty level	Level of study
4 (3 h)	Moderate	Medium	Advanced

Class names Alkyl halides, phosphorus ylides, ketenes, ketones, halogens, alcohols, alkenes

Concepts involved In this experiment the reactivity of phosphorus ylides will be explored for the construction of C–C double bonds leading to the synthesis of allenes (Wittig reaction) and tetrasubstituted alkenes (non-classical Wittig reaction)

Chemicals needed Triphenylphosphine, ethyl bromoacetate, benzyl bromoacetate, sodium hydroxide, ethyl oxalyl chloride, toluene, triethylamine, tetrahydrofuran, propionyl chloride, *N*-chlorosuccinimide, sodium bisulfite, silica gel, anhydrous sodium sulfate, ethyl acetate, *n*-hexane, diethyl ether, chloroform, chloroform-*d*, methanol, dichloromethane

Equipment and experimental techniques involved Apparatus to carry out reactions with magnetic stirring and under inert atmosphere, liquid–liquid extraction apparatus, flash column chromatography, Rotavapor, NMR, IR, melting point

Keywords Alkenes, column chromatography, ketenes, NMR, phosphorus ylides, Wittig reaction

Background

Phosphorus ylides (also called phosphoranes) are versatile intermediates in synthetic organic chemistry.[1] These species are neutral dipolar molecules possessing a formal negatively charged carbon directly attached to a formal positively charged phosphorus atom, with both atoms having full octets of electrons. When alkyl halides are treated with phosphines, they undergo direct displacement (S_N2) reactions to form phosphonium salts, which upon treatment with bases originate phosphorus ylides (Scheme 9.2.6.1).

Phosphorus ylides (often called Wittig reagents) can react with an aldehyde or ketone to give substituted alkenes and triphenylphosphine oxide in a transformation called the Wittig reaction. This reaction was developed by the German chemist Georg Wittig in 1954, and as full recognition for developing one of the most synthetically useful reactions in organic chemistry,

Comprehensive Organic Chemistry Experiments for the Laboratory Classroom
Edited by Carlos A. M. Afonso, Nuno R. Candeias, Dulce Pereira Simão, Alexandre F. Trindade, Jaime A. S. Coelho, Bin Tan, and Robert Franzén.

Published by the Royal Society of Chemistry, www.rsc.org

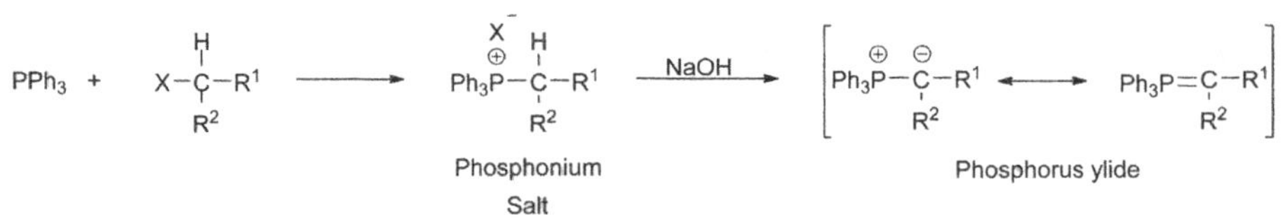

Scheme 9.2.6.1 Synthesis of phosphorus ylides.

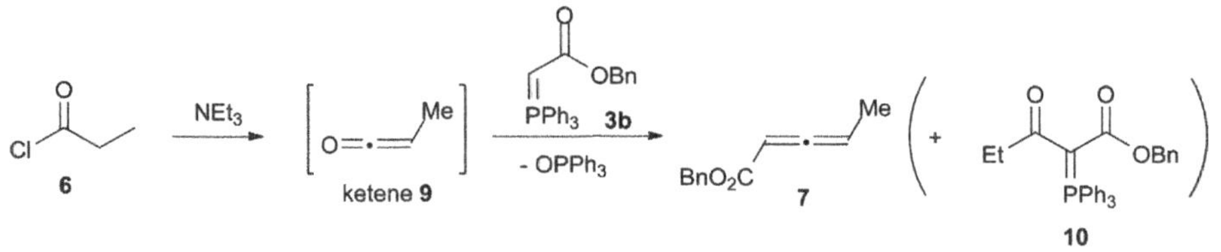

Scheme 9.2.6.2 The Wittig reaction.

Wittig was awarded the Nobel Prize in Chemistry in 1979. The mechanism of the Wittig reaction is complex, still under debate and the subject of ongoing research; therefore we will include only some details. The first step is the attack of the carbanion part of the phosphorus ylide on the carbonyl group to form a charged intermediate called a betaine, which can close to form a four-membered ring (oxaphosphetane). This ring then opens to form the alkene product and a phosphine oxide (Scheme 9.2.6.2). The stereochemical outcome of this reaction is strongly influenced by the nature of the phosphorus ylide and of the carbonyl substrate.

In the first part of this work two different phosphorus ylides (**3a** and **3b**) will be prepared by the reaction of triphenylphosphine (**2**) with alkyl halides (**1**) followed by treatment of the corresponding phosphonium salts with base.[2] Subsequently, the stabilized ylide diethyl 2-oxo-3-triphenylphosphoranylidinebutanedioate (**5**) will be prepared by reaction of phosphorus ylide **3a** with ethyl oxalyl chloride (**4**). The reactivity of these ylides will be explored for the construction of C=C double bonds leading to the synthesis of allenes and tetrasubstituted alkenes.

Allenes are important building blocks in organic chemistry.[3,4] They have two cumulated C=C double bonds, a structure feature that can be obtained *via* a Wittig reaction.[5] In the second part of the work, an allene will be prepared by the Wittig reaction of the *in situ* generated ketene **9** (from propionyl chloride (**6**) and triethylamine) with phosphorus ylide **3b**. In this reaction, the formation of disubstituted ylide **10**, resulting from the reaction of phosphorus ylide **3b** with propionyl chloride (**6**), is also observed as by-product (Scheme 9.2.6.3).[6]

The reaction of ylide **5** with *N*-chlorosuccinimide (NCS) (electrophile) in the presence of methanol (nucleophile) will be carried out to provide the tetrasubstituted alkene **8** (Scheme 9.2.6.4).[7] In this non-classical Wittig reaction, the electrophile binds exclusively to the carbon of the phosphoranylidene group of the starting ylide with concomitant elimination of the oxygen from the keto carbonyl group.

Additional Safety

All sessions should be carried out in a fume hood, with proper safety equipment such as protective clothing, gloves and safety glasses. Triphenylphosphine is harmful if swallowed or in contact with skin. Propionyl chloride is an easily flammable reagent and is corrosive. Ethyl bromoacetate and benzyl bromoacetate are very toxic reagents, irritants and may be harmful if in contact with the skin or eyes. Ethyl oxalyl chloride is a flammable reagent and may cause severe skin burns and serious damage to eyes. Sodium hydroxide is a corrosive reagent and may cause severe skin burns and serious damages to eyes. Triethylamine is an easily flammable and

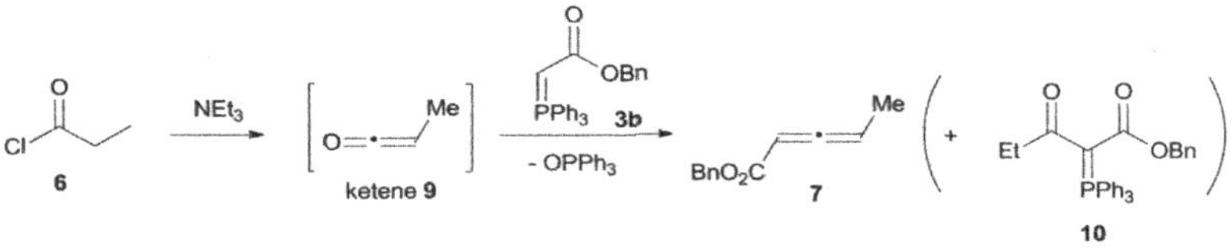

Scheme 9.2.6.3 Wittig reaction of phosphorus ylide **3b** and ketene **9**.

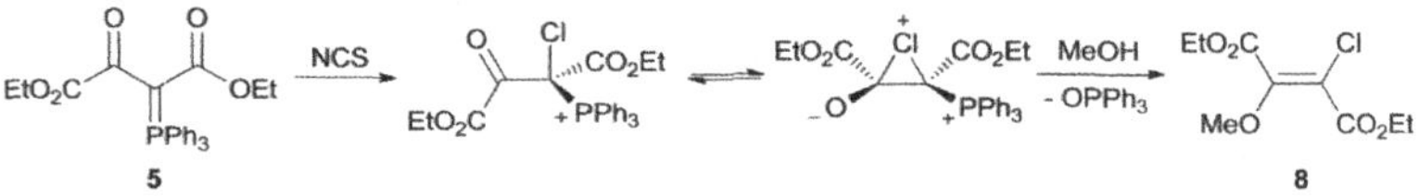

Scheme 9.2.6.4 Non-classical Wittig reaction of phosphorus ylide **5** and NCS in the presence of methanol.

corrosive reagent and may be harmful if in contact with the skin or eyes. Silica gel may be harmful if inhaled or swallowed, and may cause irritation in contact with the skin or eyes.

Experimental Procedure

Laboratory Session 1 (3 h): Synthesis of Ethyl 2-(triphenylphosphoranylidene)acetate (3a) and Benzyl 2-(triphenylphosphoranylidene)acetate (3b)

Previous Day

1. In a fume hood, prepare a 100 mL round-bottom flask with a magnetic stirring bar.
2. Add toluene (30 mL) and triphenylphosphine (6.5 g, 0.025 mol) to the flask and stir for 5 min, followed by the slow addition of ethyl bromoacetate (2.7 mL, 0.024 mol).
3. Leave the reaction stirring at room temperature for 12 h.
4. Repeat this procedure for the synthesis of phosphorus ylide **3b** using benzyl bromoacetate (3.9 mL, 0.024 mol) instead of ethyl bromoacetate.

Product Isolation

5. Filter off the phosphonium salt ($PPh_3CHCO_2Et^+$ Br– or $PPh_3CHCO_2Bn^+$ Br–) from the reaction mixture and wash it with 3 mL of toluene.
6. Transfer the salt to a 2 L round-bottom flask with a magnetic stirring bar, add water (500 mL) and cool the solution to 0 °C in an ice bath.
7. Prepare an aqueous sodium hydroxide solution (16.5 g, 0.41 mol, 500 mL) and add it, slowly, to the previous solution while maintaining vigorous stirring. Observe the precipitation of the ylide.
8. Filter off the solid, wash it with 5 mL of cold diethyl ether and let it dry.
9. Transfer each product to a vial, determine the weight and calculate the yield. Determine the melting points of the two compounds.

Laboratory Session 2 (3 h): Synthesis of Diethyl 2-Oxo-3-triphenylphosphor anylidenebutanedioate (5)

1. In a fume hood, prepare a 100 mL round-bottom flask with a magnetic stirring bar, and add toluene (38 mL) and phosphorus ylide **3b** (5.2 g, 15 mmol) prepared in the previous session.

2. After cooling the reaction mixture to 5–10 °C, add ethyl oxalyl chloride (2.1 g, 15 mmol) dropwise.
3. Stir the mixture at 5–10 °C for 5 min and then at room temperature for 30 min.
4. Remove the solvent under reduced pressure, add water to the residue and extract with $CHCl_3$.
5. Dry the combined organic layers over anhydrous sodium sulfate, filter and remove the solvent under reduced pressure.
6. After grinding of the residue with cold diethyl ether, filter the solid and transfer it to a vial, determine the weight and calculate the yield.
7. Determine the melting point of the obtained product and prepare the sample for ^1H, ^{13}C and ^{31}P NMR in $CDCl_3$.

Laboratory Session 3 (3 h): Synthesis of Benzyl Penta-2,3-dienoate (7)

Previous Day

1. In a fume hood, prepare a 100 mL round-bottom flask with a magnetic stirring bar.
2. Add THF (50 mL) and phosphorus ylide **3a** (2.0 g, 4.87 mmol) prepared in the previous session to the flask and stir under a nitrogen atmosphere.
3. Slowly add triethylamine (0.7 mL, 4.87 mmol), followed by the addition of a solution of propionyl chloride (0.42 mL, 4.87 mmol) in THF (1 mL). Leave the reaction stirring at room temperature for 12 h.

Product Isolation

4. Filter off the solid (triethylamine salt) from the reaction mixture and remove the solvent under reduced pressure.
5. Dissolve the residue in dichloromethane, wash with water and dry over anhydrous sodium sulfate.
6. Filter and remove the solvent under reduced pressure.
7. Purify the compound by column chromatography using ethyl acetate/*n*-hexane (1:2) as eluent (the allene is the first product to be eluted).
8. Determine the yield and prepare a sample for ^1H and ^{13}C NMR in $CDCl_3$.

Laboratory Session 4 (3 h): Synthesis of Diethyl 2-chloro-3-methoxybutenedioate (8)

1. In a fume hood, prepare a 50 mL round-bottom flask with a magnetic stirring bar, and add $CHCl_3$ (5 mL) and phosphorus ylide 5 (0.5 g, 1.1 mmol) prepared in the previous session.
2. Add a solution of *N*-chlorosuccinimide (0.15 g, 1.1 mmol) in MeOH (8 mL) to the stirred solution.
3. Stir the reaction for 10 min at room temperature.
4. Wash the reaction mixture with a solution of $NaHSO_3$, dry the organic layer over anhydrous sodium sulfate, filter and remove the solvent under reduced pressure.
5. Purify the compound by column chromatography using ethyl acetate/*n*-hexane (2:1) as eluent (the alkene is the first product to be eluted).
6. Determine the yield and prepare a sample for ^1H and ^{13}C NMR in $CDCl_3$.

Waste Disposal

Diethyl ether, toluene, tetrahydrofuran and *n*-hexane may be collected in a non-chlorinated waste collector recipient, for disposal. Dichloromethane and chloroform may be collected in a chlorinated waste collector recipient, for disposal. Neutralized aqueous solution may be diluted and discarded in the sink.

Results Interpretation and Additional Questions

1. Explain why the phosphorus ylides **3** and **5** are known as stabilised phosphorus ylides.
2. Interpret the ^{1}H and ^{13}C NMR spectra of the allene **7** and the alkene **8**. Compare them with those reported in the literature.
3. Interpret the ^{31}P NMR spectrum of phosphorus ylide **5**.
4. Rationalize the mechanism involved in the formation of allene **7** and alkene **8**.
5. Outline a synthetic strategy for preparing each of the following compounds based on Wittig reactions:

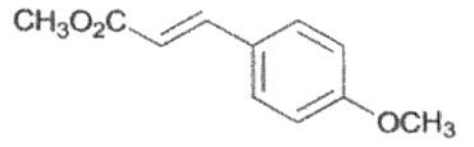

Methyl (*E*)-4-methoxycinnamate

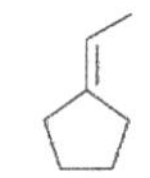

Ethylidenecyclopentane

References

1. O. I. Kolodiazhnyi, *Russ. Chem. Rev.*, 1997, **66**, 225–254.
2. M. P. Cook Jr and D. L. Burman, *J. Org. Chem.*, 1982, **47**, 4955–4963.
3. S. Yu and S. Ma, *Angew. Chem., Int. Ed.*, 2012, **51**, 3074–3112.
4. B. Alcaide and P. Almendros, *Chem. Soc. Rev.*, 2014, **43**, 2886–2887.
5. I. Tömösközi and H. J. Bestmann, *Tetrahedron Lett.*, 1964, 1293–1295.
6. T. M. V. D. Pinho e Melo, A. L. Cardoso, A. M. d'A. Rocha Gonsalves, J. C. Pessoa, J. A. Paixão and A. M. Beja, *Eur. J. Org. Chem.*, 2004, 4830–4839.
7. A. M. d'A. Rocha Gonsalves, A. M. D. P. V. Cabral, T. M. V. D. Pinho e Melo and T. L. Gilchrist, *Synthesis*, 1997, 673–676.

9.2.7. Synthesis of 4-(Thiophen-2-yl)pyrrolidin-2-one *via* Horner–Wadsworth–Emmons Reaction and Reductive Cyclisation

Jonathan D. Sellars*, AnnMarie C. O'Donoghue, Ian R. Baxendale, John M. Sanderson, and Elizabeth J. Grayson

Chemistry Department, University of Durham, South Road, Durham, DH1 3LE, UK
*E-mail: j.d.sellars@durham.ac.uk

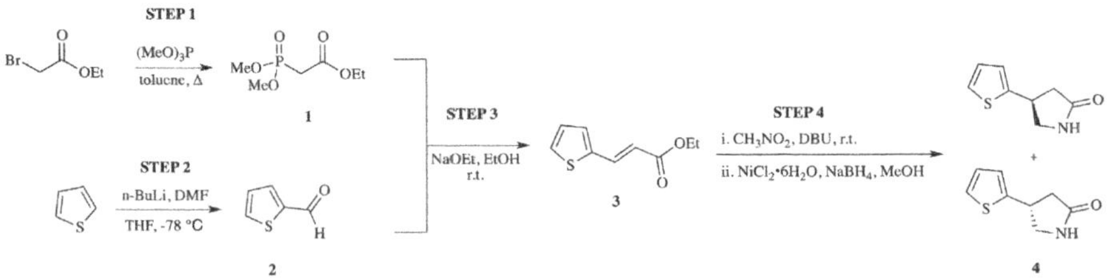

Number of sessions (duration of each session)	Hazard level	Difficulty level	Level of study
12 (7 h)	High	High	Advanced

Class names Ylides, aldehydes, α,β-unsaturated esters, thiophenes, pyrrolidin-2-ones

Concepts involved The purpose of these experiments is to introduce third year undergraduate students to multi-step organic synthesis. In addition, the experiment exposes the student cohort to a wide range of practical (handling pyrophoric reagents, vacuum distillation, recrystallization, utilising dry solvents, column chromatography) and theoretical (planning experimental quantities, time management) skills, which they will utilise in their fourth year studies

Chemicals needed *Step 1*: ethyl bromoacetate, trimethyl phosphite, toluene. *Step 2*: Thiophene, dry Ice, *n*-BuLi (2.5 M in hexanes), dry THF, dry DMF (dimethylformamide), DCM (dichloromethane), NH_4Cl (aq.), $MgSO_4$, silica gel. *Step 3*: sodium metal, ethyl 2-(dimethoxyphosphoryl)acetate (**1**), thiophene-2-carbaldehyde (**2**), dry ethanol, dry THF, *n*-hexane, ethyl acetate, 10% (v/v) aq. HCl, $NaHCO_3$ (aq.), $MgSO_4$, silica gel. *Step 4*: ethyl (2*E*)-3-(thiophen-2-yl)prop-2-enoate (**3**), DBU, nitromethane, *n*-hexane, ethyl acetate, diethyl ether, methanol, $NiCl_2{\cdot}6H_2O$, $NaBH_4$ (s), PMA, $Ce(SO_4)_2$, K_2CO_3, ethyl acetate, brine, silica gel

Equipment and experimental techniques involved Standard apparatus for low temperature (−78° C) and room temperature magnetic stirring, liquid–liquid extraction apparatus, flash chromatography column, distillation apparatus, high vacuum line

Keywords Conjugate addition, flash column chromatography on silica gel, GC-MS, Horner–Wadsworth–Emmons, IR, *in situ* reduction-cyclisation, liquid–liquid extraction, NMR, *ortho*-lithiation, vacuum distillation, ylide

Comprehensive Organic Chemistry Experiments for the Laboratory Classroom
Edited by Carlos A. M. Afonso, Nuno R. Candeias, Dulce Pereira Simão, Alexandre F. Trindade, Jaime A. S. Coelho, Bin Tan, and Robert Franzén.

Published by the Royal Society of Chemistry, www.rsc.org

Background

Alkenes are present in a vast array of chemical molecules and are important building blocks in organic chemistry.[1] There are a wide range of protocols available for their synthesis: by elimination of water; from the insertion of carbenes generated from chloroalkanes; Julia olefination of aldehydes or ketones with a phenylsulfonyl carbanion;[2–4] McMurry reaction;[5,6] Peterson reaction of aldehydes or ketones with α-silylcarbanions;[7,8] reduction of alkynes;[9–12] and the Wittig reaction.[13–17] The experiment most synonymous with alkene formation is the Wittig reaction, which was named after its discoverer, the Nobel Prize winner Georg Wittig. In this reaction a phosphonium ylide, generated by removal of an α-proton from an alkyltriphenylphosphonium salt followed by addition of an aldehyde, gives rise to an alkene possessing predominantly (*Z*) geometry. To access the (*E*) geometry, stabilised ylides such as those derived from a phosphonate have to be utilised, with one such modification known as the Horner–Wadsworth–Emmons reaction.[18–22] Another important chemical building block is the 2-pyrrolidinones, which appear in a vast array of natural products and drugs.[23–30] Whilst substitution of 2-pyrrolidinones is most often undertaken at either the nitrogen, C3 [31] or C5 centres,[32–34] 4-substituted 2-pyrrolidinones remain a synthetic challenge. Nevertheless, their synthesis has been achieved using a variety of methods, utilising a phosphoramidite gold(I)-catalysed diastereo- and enantioselective reaction,[35] C–H insertion using dirhodium complexes[36,37] and reductive amination and cyclisation.[38]

Therefore, the objective of this experiment is to synthesise this challenging substrate through a series of experiments involving (a) synthesising a stabilised ylide; (b) employing this in the synthesis of an (*E*)-alkene; (c) utilising the alkene in subsequent chemistry to access a 4-substituted pyrrolidinone as a racemic mixture.

Additional Safety

All sessions should be undertaken in a fume cupboard. *n*-BuLi (2.5 M in hexanes) is highly flammable and as such should not be exposed to air. Ethyl bromoacetate is very toxic by inhalation, in contact with skin and if swallowed and so should always be handled in a fume cupboard. Sodium metal causes burns and reacts violently with water liberating extremely flammable gases and as such should be handled with extreme care. Diethyl ether is a highly flammable solvent, which may form explosive peroxides, is harmful if swallowed and with repeated exposure may cause skin dryness or cracking. Furthermore, vapours may cause drowsiness and dizziness and so the solvent should be handled in a well-ventilated laboratory.

Experimental Procedure

Step 1

Ethyl 2-(Dimethoxyphosphoryl)acetate (1)

1. Place ethyl bromoacetate (10.8 mL, 64 mmol, 1 equiv.) (Care! this reagent is lachrymatory and should be handled in a fume cupboard) in a round-bottom flask and add toluene (to produce a 2 M solution), and add to this solution trimethyl phosphite (1.25 equiv.).
2. Reflux the mixture for 2 h, and then allow it to cool.
3. Remove a sample to submit for GC-MS. If the reaction has gone to completion, remove volatile materials on the rotary evaporator and then place the sample on the high vacuum line for 15 min. If the reaction has not gone to completion, then reflux for a further 2 h.
4. The product (**1**) does not require further purification.
5. Characterise your product (^{1}H NMR, IR, GC-MS).

Step 2

Thiophene-2-Carbaldehyde (2)

NB Do not exceed the scale in this procedure! Ensure all equipment is dry when using *n*-BuLi!

1. In a dry round-bottom flask under nitrogen, fitted with a septum cap and balloon, place thiophene (5 mL, 62.5 mmol, 1 equiv.) and dry THF (to produce a 1.8 M solution) using syringes.
2. Cool the flask to −78 °C using an acetone/dry ice bath.
3. Add *n*-BuLi (2.5 M in hexane, 1.2 equiv.) dropwise using a dry syringe.
4. Stir for 30 min at −78 °C then add dry DMF (1.25 equiv.) dropwise using a syringe.
5. After a further 15 min remove the cooling bath and stir for 4 h at room temperature (The reaction mixture can also be left overnight.)
6. Pour the reaction mixture into a saturated aqueous NH_4Cl solution (20 mL) and stir well.
7. Decant the organic layer into a round-bottom flask and remove most of the THF and hexane on the rotary evaporator (Care! Do not evaporate to dryness.)
8. Combine the residue and the aqueous phase and extract with DCM (3 × 20 mL).
9. Wash the combined DCM extracts with water (4 × 10 mL), dry the organic layer with $MgSO_4$, filter and evaporate.
10. Then, using dry apparatus, vacuum distil your product (2) from molecular sieves onto a small amount of molecular sieves at approximately 16 mbar.
11. Characterise your product (^{1}H NMR, bp, IR).

Step 3

Ethyl (2E)-3-(Thiophen-2-yl)prop-2-enoate (3)

1. Weigh sodium (4 equiv.) in light petroleum (40–60 °C) and cut into small pieces.
2. Add the sodium a piece at a time to dry ethanol (to produce a 1.5 M solution) in a dry round-bottom flask.
3. Place the reaction mixture under a nitrogen atmosphere by fitting an adaptor with a septum cap and balloon.
4. In a second dry flask weigh crude **1** (2 equiv.), and place under nitrogen, fit the flask with a septum cap and balloon and add dry THF using a syringe (to produce a 3 M solution).
5. Transfer this solution to the sodium ethoxide (first solution) using cannula transfer techniques and stir the mixture for 1 h at room temperature.
6. Weigh thiophene-2-carbaldehyde (2) (0.83 g, 7.4 mmol, 1 equiv.) in a third dry flask, place under a nitrogen atmosphere using a septum cap and balloon and add THF (to produce a 0.25 M solution).
7. Add the solution of aldehyde (2) *via* cannula to the previous reaction solution.
8. Stir for 2 h, and then run a TLC of the reaction mixture against the thiophene carboxaldehyde (hexane/ethyl acetate (7.5 : 2.5)) (visualise under UV). (The reaction can also be left to stir overnight.)
9. If the reaction has gone to completion, remove the stirrer bar and then remove volatile materials on the rotary evaporator.
10. To the residue add DCM (40 mL) and 10% (v/v) aqueous HCl (10 mL) and transfer to a separating funnel.
11. Separate the DCM phase and wash it with 10% (v/v) aqueous HCl (*ca.* 2 mL), until just acidic, then saturated aqueous sodium bicarbonate (until slightly basic), and then water.
12. Dry the DCM solution over $MgSO_4$, filter and evaporate.

13. Purify the crude product by column chromatography with *n*-hexane/EtOAc to provide **3**.
14. Characterise your product (^{1}H NMR, IR).

N.B. The column fractions containing the product should be evaporated on the rotary evaporator using a cold bath, and do not put the product on the high vacuum line as it is slightly volatile!

Step 4

4-(Thiophen-2-yl)pyrrolidin-2-one (4)

Stage One

1. Weigh the ester 3 (0.6 g, 4.19 mmol, 1 equiv.) into a round-bottom flask and add nitromethane (to produce a 4 M solution) to ensure that the mixture can be stirred with a magnetic stirrer bar.
2. Add a solution of DBU (1 equiv., 3 M) in nitromethane dropwise *via* a syringe.
3. Continue stirring at room temperature and monitor the reaction by TLC (*n*-hexane/EtOAc) after 3 h.
4. When the reaction has reached completion, dilute the reaction mixture with Et_2O (20 mL) and add water (20 mL), separate the organic layer and extract the aqueous layer with Et_2O (2 × 20 mL).
5. Dry the combined organic layers with $MgSO_4$, filter and evaporate.
6. Purify the crude product by column chromatography using *n*-hexane/EtOAc to give the nitro ester intermediate.
7. Characterise your product (^{1}H NMR, IR).

N.B. The column fractions containing the product should be evaporated on the rotary evaporator using a cold bath, and do not put the product on the high vacuum line as it is volatile!

Stage 2

1. Dissolve the intermediate nitro ester (0.4 g, 1.6 mmol, 1 equiv.) into a round-bottom flask and add methanol (to produce a 0.1 M solution) followed by $NiCl_2 \cdot 6H_2O$ (1 equiv.).
2. Cool in an ice/water bath, then add $NaBH_4$ in small portions (5 equiv.) to the stirred solution and leave to stir for 1 h.
3. Run a TLC (elute with ethyl acetate and visualise under UV to look for the disappearance of starting material, then dip in PMA/$Ce(SO_4)_2$ to look for products).
4. If all the starting material has reacted add K_2CO_3 (4 equiv.) and stir at room temperature for 2 h. Run another TLC using only the dip to visualise products.
5. Once complete, extract the reaction mixture with ethyl acetate (3 × 60 mL).
6. Wash the combined organic extracts with brine (20 mL), dry over $MgSO_4$, filter and evaporate.
7. Purify the crude product by column chromatography with EtOAc to provide **4**.
8. Characterise your product (^{1}H NMR, IR, mp).

Waste Disposal

All neutralised aqueous waste produced in each step can be diluted and discarded in the sink. All non-chlorinated waste from each step should be collected together in the non-chlorinated waste collectors. Likewise, all halogenated waste should be collected together in the halogenated waste collectors.

Results Interpretation and Additional Questions

You will be assessed by means of a written report. The requested format of your report is designed to provide you with training for writing up experimental data in a publishable form and is the format required for students that undertake a 4th year research project. As a result we expect to receive precise and clearly laid out write-ups consisting of the following:

- Fully written experimental procedures for *all* the synthetic steps, with fully recorded physical (including comparison to literature data) and spectral data for the products presented in the style specified by the Royal Society of Chemistry.
- Your original spectra (*excluding* expansions used to assign coupling constants), as PDF documents.
- The answers to the following questions:

  1. Discuss the origin of selectivity in the Horner–Wadsworth–Emmons reaction. Explain, with reference to other ylides, how you arrive at product **3** in this experiment.
  2. Based on the NMR spectrum of **3**, how do you know you have the *trans* isomer?
  3. The NMR spectrum of **4** produces distinctive multiplicities and low aromatic integrals. Account for these observations.
  4. Pyrrolidinones are important chemical building blocks. The final step of this experiment involves a two-stage reaction to produce **4**. Give a mechanistic argument for the reaction, explaining the selectivity of addition of nitromethane. In addition, provide a retrosynthetic analysis and describe alternative synthetic routes to **4**.

References

1. J. Clayden, N. Greeves, S. Warren and P. Wothers, *Organic Chemistry*, 1st edn, Oxford University Press, Oxford, 2001.
2. J. Pospíšil and I. E. Markó, *Org. Lett.*, 2006, **8**, 5983–5986.
3. J. Pospíšil, T. Pospíšil and I. E. Markó, *Org. Lett.*, 2005, **7**, 2373–2376.
4. G. E. Keck, K. A. Savin and M. A. Weglarz, *J. Org. Chem.*, 1995, **60**, 3194–3204.
5. D. X. Zeng and Y. Chen, *Synlett*, 2006, **2006**, 0490–0492.
6. X.-F. Duan, J. Zeng, J.-W. Lü and Z.-B. Zhang, *J. Org. Chem.*, 2006, **71**, 9873–9876.
7. M. Iguchi and K. Tomioka, *Org. Lett.*, 2002, **4**, 4329–4331.
8. W. Adam and C. M. Ortega-Schulte, *Synlett*, 2003, **2003**, 0414–0416.
9. R. Shen, T. Chen, Y. Zhao, R. Qiu, Y. Zhou, S. Yin, X. Wang, M. Goto and L.-B. Han, *J. Am. Chem. Soc.*, 2011, **133**, 17037–17044.
10. J. Li, R. Hua and T. Liu, *J. Org. Chem.*, 2010, **75**, 2966–2970.
11. A. M. Whittaker and G. Lalic, *Org. Lett.*, 2013, **15**, 1112–1115.
12. T. Schabel, C. Belger and B. Plietker, *Org. Lett.*, 2013, **15**, 2858–2861.
13. R. Zhou, C. Wang, H. Song and Z. He, *Org. Lett.*, 2010, **12**, 976–979.
14. D. F. Taber and R. B. Sheth, *J. Org. Chem.*, 2008, **73**, 8030–8032.
15. A. El-Batta, C. Jiang, W. Zhao, R. Anness, A. L. Cooksy and M. Bergdahl, *J. Org. Chem.*, 2007, **72**, 5244–5259.
16. D.-J. Dong, H.-H. Li and S.-K. Tian, *J. Am. Chem. Soc.*, 2010, **132**, 5018–5020.
17. C. J. O'Brien, J. L. Tellez, Z. S. Nixon, L. J. Kang, A. L. Carter, S. R. Kunkel, K. C. Przeworski and G. A. Chass, *Angew. Chem., Int. Ed.*, 2009, **48**, 6836–6839.
18. A. Teichert, K. Jantos, K. Harms and A. Studer, *Org. Lett.*, 2004, **6**, 3477–3480.
19. T. D. W. Claridge, S. G. Davies, J. A. Lee, R. L. Nicholson, P. M. Roberts, A. J. Russell, A. D. Smith and S. M. Toms, *Org. Lett.*, 2008, **10**, 5437–5440.
20. T. Umezawa, T. Seino and F. Matsuda, *Org. Lett.*, 2012, **14**, 4206–4209.
21. E. Quesada and R. J. K. Taylor, *Synthesis*, 2005, **2005**, 3193–3195.
22. A. Ianni and S. R. Waldvogel, *Synthesis*, 2006, **2006**, 2103–2112.
23. C. R. Ganellin and D. J. Triggle, *Dictionary of Pharmacological Agents*, Taylor & Francis, 1996.

24. B. Winblad, *CNS Drug Rev.*, 2005, **11**, 169–182.
25. J. Alves, *Org. Chem. Int.*, 2011, 2011.
26. A. K. Ghosh, S. Leshchenko-Yashchuk, D. D. Anderson, A. Baldridge, M. Noetzel, H. B. Miller, Y. Tie, Y.-F. Wang, Y. Koh, I. T. Weber and H. Mitsuya, *J. Med. Chem.*, 2009, **52**, 3902–3914.
27. F. Haaf, A. Sanner and F. Straub, *Polym. J.*, 1985, **17**, 143–152.
28. D. T. Davies, N. Kapur and A. F. Parsons, *Tetrahedron Lett.*, 1999, **40**, 8615–8618.
29. F. A. Amer, M. Hammouda, A. A. S. El-Ahl and B. F. Abdel-Wahab, *Synth. Commun.*, 2009, **39**, 416–425.
30. P. Moutevelis-Minakakis, E. Papavassilopoulou and T. Mavromoustakos, *Molecules*, 2012, **18**, 50–73.
31. P. A. Reddy, B. C. H. Hsiang, T. N. Latifi, M. W. Hill, K. E. Woodward, S. M. Rothman, J. A. Ferrendelli and D. F. Covey, *J. Med. Chem.*, 1996, **39**, 1898–1906.
32. L. E. Overman and T. P. Remarchuk, *J. Am. Chem. Soc.*, 2001, **124**, 12–13.
33. V. Singh, a. Rashmi Saxena and S. Batra, *J. Org. Chem.*, 2004, **70**, 353–356.
34. G. Blay, V. Hernández-Olmos and J. R. Pedro, *Org. Lett.*, 2010, **12**, 3058–3061.
35. A. Z. González, D. Benitez, E. Tkatchouk, W. A. Goddard III and F. D. Toste, *J. Am. Chem. Soc.*, 2011, **133**, 5500–5507.
36. M. Anada and S. Hashimoto, *Tetrahedron Lett.*, 1998, **39**, 79–82.
37. A. G. H. Wee, A. Baosheng Liu and D. D. McLeod, *J. Org. Chem.*, 1998, **63**, 4218–4227.
38. B. M. Kenda, A. C. Matagne, P. E. Talaga, P. M. Pasau, E. Differding, B. I. Lallemand, A. M. Frycia, F. G. Moureau, H. V. Klitgaard, M. R. Gillard, A. Bruno Fuks and P. Michel, *J. Med. Chem.*, 2003, **47**, 530–549.

9.3. Metathesis

9.3.1. A Guided-Inquiry Approach to Ring-Closing Metathesis

Hala G. Schepmann*[a] and Michelle Mynderse[b]

[a]Southern Oregon University, 1250 Siskiyou Boulevard, Ashland, Oregon, 97520, USA; [b]2371 Henley Avenue, Berkley, Michigan, 48072, USA. *E-mail: schepmah@sou.edu

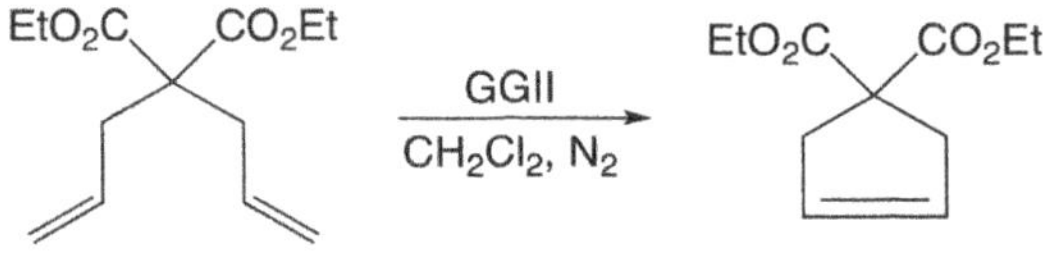

Number of sessions (duration of each session)	Hazard level	Difficulty level	Level of study
3 (3 h each)	Moderate	High	Intermediate

Class names Alkenes

Concepts involved The design and implementation of a guided-inquiry experiment for the organic laboratory is described. A Grubbs second-generation catalyst is used to effect the ring-closing metathesis of diethyl diallylmalonate. The reaction is carried out under an inert atmosphere at room temperature and monitored by argentic TLC. The crude reaction is purified with activated carbon, and the product is analyzed using GC-MS and ^{1}H, COSY, ^{13}C, and DEPT-135 FT-NMR. Additionally, a guided-inquiry approach is employed wherein students design a procedure to determine the influence of various parameters on the progress of the reaction and the purity and yield of the product

Chemicals needed Diethyl diallylmalonate, Grubbs second-generation catalyst, dichloromethane, nitrogen gas, dimethyl sulfoxide, potassium permanganate, sodium carbonate, sodium hydroxide (5%), silver nitrate impregnated TLC plates, activated carbon, silica gel, deuterochloroform

Equipment and experimental techniques involved Inert atmosphere apparatus, vacuum filtration apparatus, rotary evaporator, argentic TLC, GC, GC-MS, NMR

Keywords Evaporation, filtration, inquiry-based, NMR characterization, olefin metathesis, organometallics

Background

Many organic chemistry students have difficulty using higher-order thinking to solve experimental problems. To improve students' understanding of experimental procedures and provide them an opportunity to experience the satisfaction of designing an experiment and the excitement of making a new discovery, we developed an advanced guided-inquiry experiment for the organic laboratory sequence.[1] This laboratory sequence guides students through the process of scientific investigation while allowing for independent decision-making. We hoped that such an experience would build

Comprehensive Organic Chemistry Experiments for the Laboratory Classroom
Edited by Carlos A. M. Afonso, Nuno R. Candeias, Dulce Pereira Simão, Alexandre F. Trindade, Jaime A. S. Coelho, Bin Tan, and Robert Franzén.

Published by the Royal Society of Chemistry, www.rsc.org

upon students' acquired laboratory skills, provide opportunities for analysis, creation, and synthesis of information, and bolster their enthusiasm for and appreciation of scientific research.

A ring-closing metathesis (RCM) experiment was selected to expose students to an important, widely used, and environmentally-friendly catalytic reaction. In 2005, the Nobel Prize in Chemistry was awarded to Yves Chauvin, Robert H. Grubbs, and Richard R. Schrock for the development of organometallic metathesis.[2] RCM is a simple, rapid, effective, and low-cost method to form alkenes in rings, most notably in medium-sized rings containing 8–13 ring atoms. This reaction has revolutionized organic synthesis and is frequently employed in the synthesis of pharmaceutical compounds. RCM reactions use a metal alkylidene catalyst to convert acyclic dienes into cyclic alkenes. In addition, ethene gas (or other simple alkenes) is produced and drives the reaction to completion. The mechanism, proposed by Yves Chauvin and Jean-Louis Hérisson in 1971,[3] involves the repetitive formation and cleavage of metallacyclobutanes.

This experiment involves the cyclization of diethyl diallylmalonate using Grubbs second-generation (GGII) ruthenium catalyst.[4] Use of GGII allows RCM reactions to be conducted in the presence of a diverse range of functional groups under mild conditions and provides high product yields. Since the catalyst is *slightly* air and moisture sensitive, the reaction was carried out in distilled dichloromethane under a blanket of nitrogen. The progress of the reaction was monitored by argentic TLC. Upon complete conversion of the starting diene into the cyclic olefin product (typically between 90 and 120 min), activated carbon was added to remove the ruthenium catalyst. Filtration of the solution followed by evaporation of the solvent resulted in a relatively pure product in 60–90% yield. After carrying out the initial experiment, a guided-inquiry approach[5] was used to determine the influence of various parameters on reaction time, reaction progress analysis, product purity, and product yield.

Additional Safety

Dichloromethane is a possible cancer agent; diethyl diallylmalonate is an irritant; dimethyl sulfoxide is an eye, lung, and skin irritant; silica gel and activated carbon are inhalation hazards; silver nitrate (impregnated in the silica gel coated TLC plates) is a skin and eye irritant; and Grubbs second-generation catalyst is a flammable solid. Toxicity data for Grubbs second-generation catalyst is not available; therefore, the catalyst should be handled with care.

Experimental Procedure

Laboratory Session 1 (3 h): Ring-Closing Metathesis of Diethyl DiallylMalonate (DDM)

1. Remove a 50 mL round-bottomed flask and a magnetic stir bar from the oven and cool open to air to room temperature. Note: If the experiment is carried out in a humid environment, it is best to seal the flask with a threaded septum and cool under N_2. If Schlenk lines are not available, a nitrogen-filled balloon apparatus may be used.
2. In the 50 mL round-bottomed flask containing a magnetic stir bar, add DDM (75 μL) using a 100 μL glass microsyringe.
3. Add 20 mL distilled, purified dichloromethane to the round-bottomed flask containing the DDM.
4. Add GGII catalyst (0.012 g) to the DDM/solvent solution in the round-bottomed flask.
5. Seal the round-bottomed flask with a threaded septum.
6. Place and maintain the reaction solution under N_2, degas briefly (3×).
7. Stir the reaction solution.

8. Monitor the reaction by argentic TLC using a degassing needle and a long TLC spotter (3 spots *maximum*). Use dichloromethane to develop your plate. Visualize spots with $KMnO_4$ stain.
9. After the reaction is complete (as determined by argentic TLC), remove the septum. Add activated carbon (20 weight equivalents relative to GGII) to the reaction mixture, replace the septum, place under N_2 and degas briefly (3×). Continue to stir the reaction under N_2 for a minimum of 48 h.

Laboratory Sessions 2 and 3 (3 hours each): Product Isolation and Design, Implementation, and Optimization of Modified Experiment

Product Isolation (Laboratory Session 2)

1. Using vacuum filtration, filter the crude product solution through a filter paper disk (1 cm) in a Hirsch funnel into an evacuated 50 mL side arm Erlenmeyer flask. Optional: filter the purified solution using a fritted glass filter to remove residual activated carbon.
2. Transfer the filtrate to a 50 mL round-bottomed flask.
3. Concentrate the filtrate using rotary evaporation. Obtain your product weight and calculate the reaction yield.
4. Acquire GC-MS and ^{1}H, COSY, ^{13}C, and DEPT-135 NMR spectra.

Design, implementation, and optimization of modified experiment (Laboratory sessions 2 and 3)

5. Select an experimental question from Table 9.3.1.1
6. Design and conduct a modified procedure to answer the selected experimental question.
7. Assess the experiment outcomes.
8. Optimize the experimental design and technique and repeat experiment, as necessary.

Waste Disposal

Dichloromethane and dimethyl sulfoxide solvent waste was collected and sent out for disposal. Ruthenium-contaminated silica gel and plates and activated carbon solid waste was collected and sent out for disposal.

Table 9.3.1.1 Description of experimental questions.

Experimental question	Analysis	Number of trials assigned
Degree of catalyst's sensitivity to water	Reaction time, product purity and yield	3
Minimum amount of catalyst required	Reaction time, product yield	3
Minimum amount of solvent required	Reaction time, product yield	3
Optimal amount of activated carbon required to purify crude reaction[6]	Product purity and yield	4
Optimal amount of DMSO required to purify crude reaction[7]	Product purity and yield	4
Optimal amount of silica gel required to purify reaction product[6]	Product purity and yield	4
Optimal GC method to analyze progress of reaction	Retention times, peak shapes, and run times	5

Results Interpretation and Additional Questions

1. Standard Protocol

a. Determine the millimolar ratio of the DDM starting material to the GGII catalyst used.
b. Calculate the R_f values of the starting material and the product of the reaction.
c. Analyze and interpret the GC-MS data to confirm product formation and determine purity.
d. Analyze and interpret the ^1H, COSY, ^{13}C, and DEPT-135 NMR spectra of the product of your reaction to confirm product formation and determine purity.
e. Taking into consideration the ^1H NMR data regarding the purity of your product, calculate the percent yield of your product.

2. Modified Protocol

a. Analyze and interpret the data acquired.
b. Based on your data, describe what additional experiments you would like to conduct and what alterations you would recommend.

References

1. Reprinted (adapted) with permission from H. G. Schepmann and M. Mynderse, *J. Chem. Educ.*, 2010, **87**, 721. Copyright (2010) American Chemical Society.
2. (a) The Nobel Foundation Nobel Prizes Page, http://nobelprize.org/nobel_prizes/chemistry/laureates/2005/, accessed October 2014; (b) C. P. Casey, *J. Chem. Educ.*, 2006, **83**, 192; (c) W. F. Coleman, *J. Chem. Educ.*, 2006, **83**, 236.
3. J.-L. Hérisson and Y. Chauvin, *Makromol. Chem.*, 1971, **141**, 161.
4. (a) G. C. Fu, S. T. Nguyen and R. H. Grubbs, *J. Am. Chem. Soc.*, 1993, **115**, 9856; (b) S. T. Nguyen, R. H. Grubbs and J. W. Ziller, *J. Am. Chem. Soc.*, 1993, **115**, 9858; (c) M. Scholl, S. Ding, C. W. Lee and R. H. Grubbs, *Org. Lett.*, 1999, **1**, 953; (d) T. M. Trnka and R. H. Grubbs, *Acc. Chem. Res.*, 2001, **34**, 18; (e) Y. Schrodi and R. L. Pederson, *Aldrichimica Acta*, 2007, **40**, 45.
5. (a) B. A. Gaddis and A. M. Schoffstall, *J. Chem. Educ.*, 2007, **84**, 848; (b) K. Monteyne and M. S. Cracolice, *J. Chem. Educ.*, 2004, **81**, 1559; (c) J. R. Mohrig, *J. Chem. Educ.*, 2004, **81**, 1083.
6. J. H. Cho and B. M. Kim, *Org. Lett.*, 2003, **5**, 531.
7. Y. M. Ahn, K. Yang and G. I. Georg, *Org. Lett.*, 2001, **3**, 1411.

9.3.2. Sequential Pd-Catalyzed Allylic Alkylation/ Ru-Catalyzed Ring-Closing Metathesis

Mélanie M. Lorion[a,b], Claire Kammerer[†], Guillaume Prestat[‡,*], and Giovanni Poli*[a,b]

[a]UPMC Univ Paris 06, IPCM UMR CNRS 8232, FR2769, Institut Parisien de Chimie Moléculaire, Paris, France; [b]CNRS, UMR 8232, F-75005, Paris, France
*E-mail: guillaume.prestat@parisdescartes.fr, giovanni.poli@upmc.fr

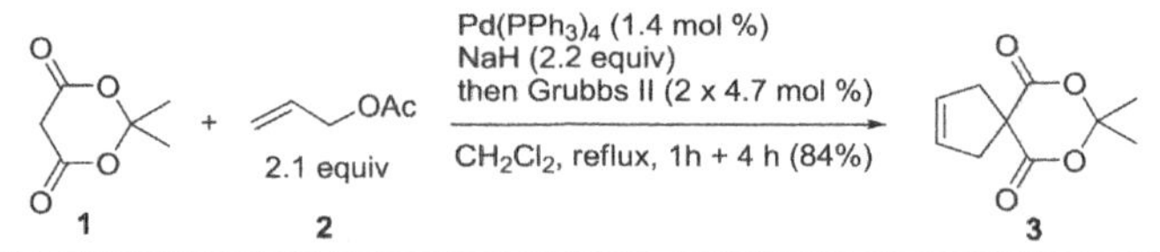

Number of sessions (duration of each session)	Hazard level	Difficulty level	Level of study
2 (6 h + 2 h)[1]	Moderate	High	Advanced

Class names Ester (malonate), allylic acetate, olefin

Concepts involved The protocol deals with a sequence of two transition-metal catalyzed reactions accomplished in the same pot. Meldrum's acid (**1**) is first bis-allylated with two equivalents of allyl acetate (**2**) under palladium catalysis. Then, a ruthenium catalyst triggers ring-closing metathesis, affording a spirocyclic cyclopentene product

Chemicals needed 2,2-Dimethyl-1,3-dioxane-4,6-dione (Meldrum's acid), allyl acetate, sodium hydride (60% in mineral oil), tetrakis(triphenylphosphine)palladium(0), [1,3-Bis(2,4,6-trimethylphenyl)-2-imidazolidinylidene]dichloro(phenylmethylene)(tricyclohexylphosphine)ruthenium (Grubbs II catalyst), methylene chloride (dried for the reaction and normal for the work up), NH_4Cl, NaCl, $MgSO_4$, TLC plates, silica gel, cyclohexane, ethyl acetate for chromatography

Equipment and experimental techniques involved Equipment: a magnetic stirrer with heating plate, an oil bath, two 25 mL round-bottomed flasks, two magnetic stir bars, a cannula (double ended needle), a 10 mL syringe fitted with a long needle, a 250 µL syringe fitted with a needle, a reflux condenser, Schlenk line (or oven + inert gas line), 100 mL separatory funnel, funnel, Erlenmeyer flasks, 250 mL round-bottomed flask, TLC plates, column chromatography (20 mm, 30 cm), test tubes and rack for column chromatography. Techniques: addition *via* cannula, TLC monitoring, liquid–liquid extraction, column chromatography

Keywords Allylic alkylation, column chromatography, NMR, palladium, ring-closing metathesis (RCM), ruthenium

Background

The Mechanisms

Transition metal (TM) catalyzed domino reactions are valuable tools in modern organic synthesis. Such processes allow both atom- and step-economy, thereby limiting the production of waste, energy, and time. Most of the TM-catalyzed reactions are based on the use of a

[†]Present address: Université de Toulouse, CEMES, UPR 8011, Toulouse, France.
[‡]Present address: Université Paris Descartes, LCBPT, UMR 8601 CNRS, Paris, France.

Comprehensive Organic Chemistry Experiments for the Laboratory Classroom
Edited by Carlos A. M. Afonso, Nuno R. Candeias, Dulce Pereira Simão, Alexandre F. Trindade, Jaime A. S. Coelho, Bin Tan, and Robert Franzén.

Published by the Royal Society of Chemistry, www.rsc.org

single TM-catalyst. To increase the molecular diversity, the range of accessible reactions may be enhanced by the use of two different catalysts in a single step. This experiment demonstrates the interest in this emerging area by an original one-pot sequence using a Pd-catalyzed allylation and a Ru-catalyzed olefin metathesis, which are two particularly powerful and widely used tools for the synthesis of complex molecules. The most classical version of the former reaction is catalyzed by a palladium(0) complex and enables the allylation of several carbon- as well as heteroatom-based soft nucleophiles (Scheme 9.3.2.1).[2]

The mechanism of this reaction involves the reversible coordination of the olefin to the palladium(0) complex, followed by oxidative addition of the palladium to the allylic system through departure of the allylic leaving group. The resulting π-allylpalladium(II) complex, in turn generated passing from the fugitive σ-allyl complex, undergoes a (usually irreversible) nucleophilic attack on one of the two distal carbon atoms, generating the olefinic product coordinated to palladium(0). Eventually, product decoordination closes the catalytic cycle (Scheme 9.3.2.2).

Interest in the olefin metathesis reaction, which involves the mutual exchange of two alkylidene groups, has grown considerably over recent decades.[3,4] In particular, the development of ruthenium carbenes has overcome the instability problems due to extreme sensitivity to oxygen and water encountered with previous catalysts such as molybdenum carbenes,[5] while combining high activity and good functional tolerance. Among the many catalysts available today, the complexes described by Grubbs *et al.* are the most popular ones. The *first-generation* catalyst *Grubbs I*, having two chlorides and two tricyclohexylphosphines as ancillary ligands, has a limited activity with tri-, tetra and electronically-depleted olefins.[6] However, the replacement of a phosphine with an *N*-heterocyclic carbene (NHC) such as saturated 1,3-dimesitylimidazolin-2-ylidene (SIMes) has generated a catalyst known as *second-generation Grubbs II*, with superior activity and a wider field of application (Figure 9.3.2.1).[7]

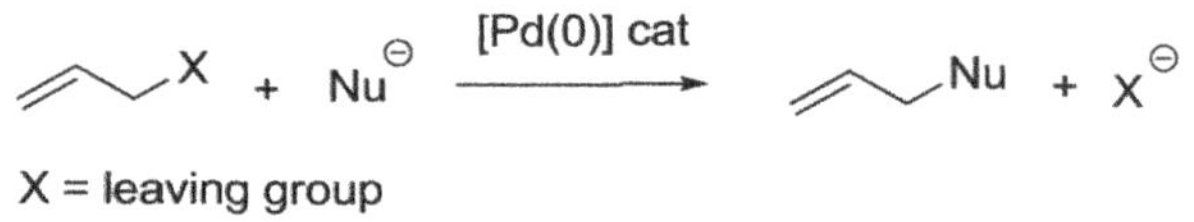

Scheme 9.3.2.1 Palladium-catalyzed allylation of nucleophiles.

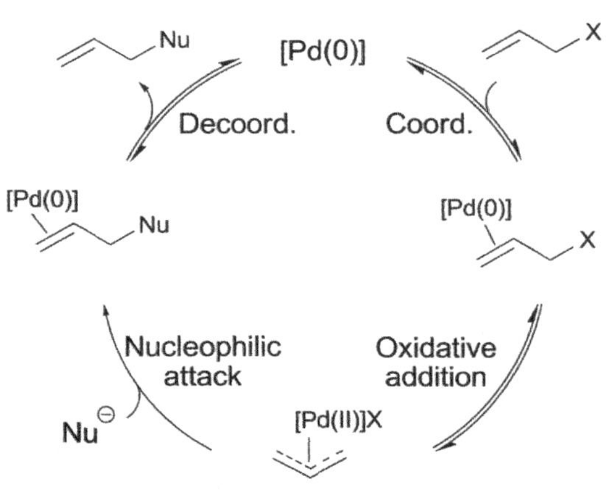

Scheme 9.3.2.2 Mechanism of the Pd(0)-catalyzed allylation of nucleophiles.

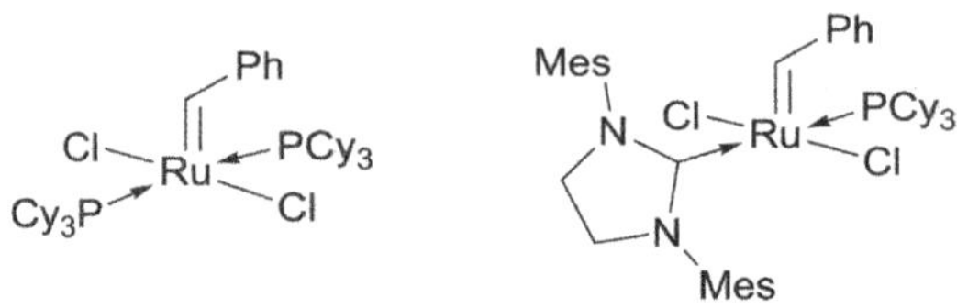

Figure 9.3.2.1 Ruthenium-based pre-catalysts for the metathesis reaction.

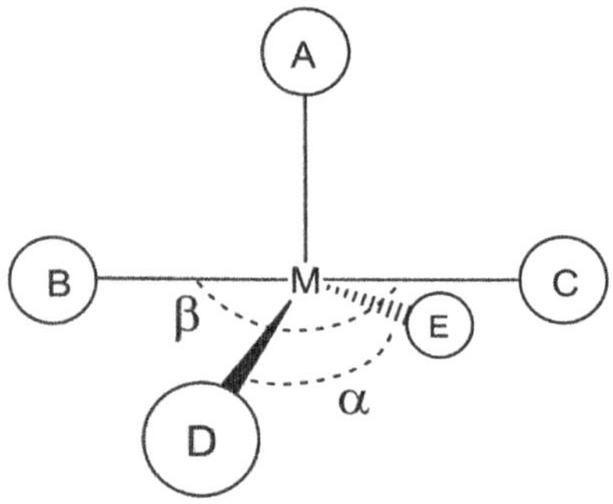

Figure 9.3.2.2 Generic pentacoordinated complex.

Pentacoordinated metal complexes can adopt different geometries, spanning from pure square-pyramidal to pure trigonal-bipyramidal extremes. It is possible to compute the Addison–Reedijk degree of trigonality (τ) of such complexes by applying the formula:

$$\tau = \frac{\beta - \alpha}{60}$$

where β is the largest basal angle (B$\hat{M}$C), and α is the second largest one (Figure 9.3.2.2).[8,9]

From a mechanistic viewpoint, the catalytically active 14 valence electron (VE) species is formed from the pre-catalysts through tricyclohexylphosphine decoordination.[10] The thus generated vacant site is then occupied by an olefin, which coordinates to ruthenium before undergoing a [2 + 2] cycloaddition to yield a ruthenacyclobutane intermediate, as proposed by Chauvin in 1971.[11] Subsequent retrocycloaddition generates the metathesis product, coordinated to a new ruthenium carbene. Alkene decoordination releases the product and regenerates the catalytically active species. All steps in the catalytic cycle being reversible, the obtained distribution of products reflects their relative stability. However, the equilibrium can be shifted in favor of the desired product by increasing the concentration of one reactant or by taking advantage of the volatility of one of the products (Scheme 9.3.2.3).

The present experiment exploits the two aforementioned catalytic processes in a sequential one-pot procedure.[12] This type of transformation is considered of high interest in modern times, as it allows complexity in molecules to be built up in a single synthetic operation, starting from simple substrates. According to the timing of the reagents added as well as the nature of the catalytic processes involved, these sequences have been named and classified under different names over recent years.[13,14] In the present case, we have two mechanistically independent catalytic processes in sequence, driven by two different metals. Consequently, the product of the first catalytic cycle becomes the substrate for the second (Scheme 9.3.2.4).

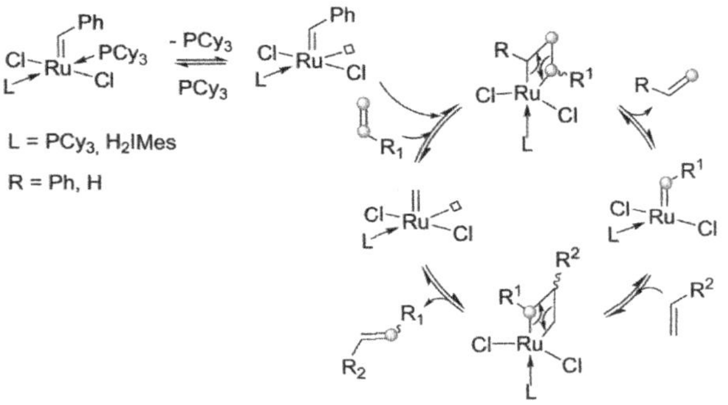

Scheme 9.3.2.3 Mechanism of the Ru-(Grubbs I and II)-catalyzed olefin metathesis.

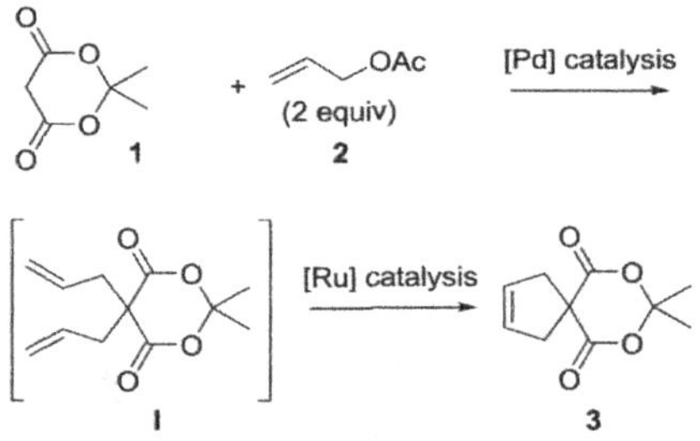

Scheme 9.3.2.4 Global path in the transformation of **1** + **2** to **3**.

The Nucleophile

The nucleophile used in the first step, Meldrum's acid,[15,16] deserves some comments. Although in this experiment it is deprotonated with a strong base such as NaH, its pK_a is abnormally low (4.8–4.9 in H_2O; 7.2 in DMSO), *i.e.* eight orders of magnitude lower than its open chain analogue, diethyl malonate (13 in H_2O; 15.9 in DMSO). This feature has intrigued chemists for a long time, and the search for its origin is a matter of study even to the present day.[17]

Additional Safety

All sessions should be done in a fume hood, with proper safety equipment, such as protective clothing, gloves and safety goggles. Silica gel is harmful if powders are inhaled. Meldrum's acid is *irritant*, allyl acetate is *toxic* and *highly flammable*, sodium hydride is *highly flammable*, cyclohexane and ethyl acetate are *irritant* and *highly flammable*.

Experimental Procedure

First Session (6 h)

1. Prepare a dried 25 mL round-bottomed flask (n° 1) with a magnetic stir bar and add 100 mg (0.7 mmol, 1 equiv.) of Meldrum's acid, 12 mg (0.010 mmol, 1.4 mol%) of $Pd(PPh_3)_4$ and 3 mL of dried methylene chloride under an argon atmosphere through a rubber septum. Warning! $Pd(PPh_3)_4$ is oxygen sensitive and should be handled under inert atmosphere.
2. Prepare a second dried 25 mL round-bottomed flask (n° 2) with a magnetic stir bar and add 61 mg (60% in mineral oil, 1.5 mmol, 2.2 equiv.) of NaH and 6 mL of dried methylene chloride. The resulting suspension is stirred at room temperature until NaH is finely dispersed.
3. After disappearance of NaH, transfer Meldrum's acid and $Pd(PPh_3)_4$ from flask n° 1 to flask n° 2 *via* a cannula[18] (hydrogen gas evolution). Add 3 mL of dried methylene chloride to flask n° 1, and transfer it to flask n° 2 *via* a cannula.

4. Add 157 μL (1.5 mmol, 2.1 equiv.) of allyl acetate to flask n°2.
5. Install an oil bath and a reflux condenser, flushing with an inert gas during this operation so as to avoid entrance of adventitious moisture or oxygen, and stir the solution at reflux.
6. After complete consumption of Meldrum's acid, monitored by TLC (cyclohexane/EtOAc (80 : 20), $KMnO_4$ detection, R_f (intermediate **I**) = 0.5) (1 h), add 28 mg (0.033 mmol, 4.7 mol%) of Grubbs II catalyst and stir the solution at reflux. After 1 h stirring, let the solution cool to room temperature and carry out a TLC analysis (cyclohexane/EtOAc (95 : 5), $KMnO_4$ detection) to identify intermediate **I** (R_f = 0.33) and product **3** (R_f = 0.13). If the reaction is incomplete, add 28 mg (0.033 mmol, 4.7 mol%) of Grubbs II catalyst and stir the solution at reflux.
7. After complete consumption of intermediate **I** monitored by TLC (cyclohexane/EtOAc (95 : 5)) (3 h), let the solution cool down to room temperature.[19]

Product Isolation and Yield Determination

8. Quench the reaction with a saturated aqueous solution of NH_4Cl (10 mL).
9. Transfer the resulting biphasic mixture into a separatory funnel and separate the organic layer.
10. Perform an extraction of the aqueous layer with 2 × 20 mL of methylene chloride.
11. Wash the combined organic layers with a saturated aqueous solution of NaCl (15 mL) and dry them over anhydrous $MgSO_4$.
12. Filter the resulting suspension over filter paper or cotton wool into a 250 mL round-bottomed flask and remove the solvent with a rotary evaporator.
13. Weigh the crude product.

Second Session (2 h)

1. Purify the crude product by column chromatography (SiO_2, $m(SiO_2)$ = 30 × weight of the crude product, cyclohexane/AcOEt (80 : 20) as eluent) and collect the fractions containing the product in a round-bottomed flask.
2. Remove the solvents at the rotary evaporator, and, possibly, at the high-vacuum line.
3. Weigh the desired product and calculate the yield.
4. Determine the melting point of the product and record IR and NMR spectra.

Waste Disposal

The neutralized aqueous solution may be diluted and discarded. Methylene chloride may be collected in a chlorinated waste collector recipient. Silica must be carefully collected into a silica waste collector recipient using a disposable facemask when handling it.

Results Interpretation and Additional Questions

1. On the basis of the general mechanisms discussed in the Background section, propose the mechanisms of the two specific steps, with appropriate curly arrows.
2. Knowing that H_2 ($H_2 + H_2O \leftrightarrows H^- + H_3O^+$) has a p$K_a$ of 35, justify the fact that the deprotonation of Meldrum's acid with NaH is quantitative.
3. (a) Predict the multiplicity as well as the number of groups of signals expected in the ^{1}H NMR spectrum of product **3**. (b) Show the enantiotopic hydrogens in product **3**.
4. Predict what you would expect to happen if the second step of the transformation is carried out in an ethylene atmosphere.
5. Knowing that Pd(0) has a d^{10} electron configuration, assign the total valence electron (VE) count to all the intermediates involved in the first catalytic cycle.
6. Knowing that in the crystal structure of Grubbs II catalyst the Cl–Ru–Cl angle is 167.7° and the C–Ru–P angle is 164.3°, obtain the degree of trigonality (τ) of this complex and deduce if its geometry is square-pyramidal or trigonal-bipyramidal.

7. The Grubbs II catalyst can be regarded either as a d^6 Ru(II) complex or as a d^4 Ru(IV) complex, depending on the electron counting formalism used. Compute the total VE count of this complex using each formalism.
8. Suggest a plausible synthesis of Meldrum's acid from malonic acid. Then, check the literature for the actual preparation[20] and propose a possible mechanism for its generation.

Acknowledgements

The authors warmly thank Professor Deryn E. Fogg (University of Ottawa) for fruitful discussions.

References

1. The indicated number of sessions and duration is the optimal one, allowing complete formation and conversion of intermediate I. However, if for technical reasons a shorter duration time is needed, the ring-closing metathesis step can be prematurely stopped at any time. This action will leave some unreacted intermediate I together with product **3**, which can be chromatographically separated with ease during the reaction protocol. For example, halting the duration of the second step after 1 h reflux (instead of 3 h) will afford a ~60% conversion of I into **3**.
2. (a) J. Tsuji, in *Handbook of Organopalladium Chemistry for Organic Synthesis*, ed. E. Negishi, John Wiley & Sons, Inc., New York, 2002, vol. 2, pp. 1669–1687; (b) C. Kammerer-Pentier, F. Liron, G. Prestat and G. Poli, Transition Metal Catalyzed Enantioselective Allylic Substitution in Organic Synthesis, in *Topics in Organometallic Chemistry*, ed. U. Kazmaier, Springer-Verlag, Berlin Heidelberg, 2012, vol. 38, pp. 1–64.
3. (a) G. Vougioukalakis and R. H. Grubbs, *Chem. Rev.*, 2009, **110**, 1746; (b) C. Samojłowicz, M. Bieniek and K. Grela, *Chem. Rev.*, 2009, **109**, 3708; (c) *Handbook of Metathesis*, ed. R. H. Grubbs, Wiley-VCH Verlag GmbH, Weinheim, 2003, vol. 1; (d) B. J. van Lierop, J. A. M. Lummiss and D. E. Fogg, in *Olefin Metathesis-Theory and Practice*, ed. K. Grela, Wiley-VCH Verlag GmbH, Weinheim, 2014, ch. 3, pp. 85–152; (e) K. Zukowska and K. Grela, in *Olefin Metathesis-Theory and Practice*, ed. K. Grela, Wiley-VCH Verlag GmbH, Weinheim, 2014, ch. 2, pp. 39–84; (f) *Metathesis in Natural Product Synthesis: Strategies, Substrates and Catalysts*, ed. J. Cossy, S. Arseniyadis and C. Meyer, Wiley-VCH Verlag GmbH, Weinheim, 2010.
4. The Nobel Prize was awarded in 2005 to Y. Chauvin, R. H. Grubbs and R. R. Schrock for the "development of the metathesis method in organic synthesis".
5. R. R. Schrock, J. S. Murdzek, G. C. Bazan, J. Robbins, M. DiMare and M. O'Regan, *J. Am. Chem. Soc.*, 1990, **112**, 3875.
6. P. Schwab, R. H. Grubbs and J. W. Ziller, *J. Am. Chem. Soc.*, 1996, **118**, 100.
7. (a) J. Huang, E. D. Stevens, S. P. Nolan and J. L. Petersen, *J. Am. Chem. Soc.*, 1999, **121**, 2674; (b) M. Scholl, T. M. Trnka, J. P. Morgan and R. H. Grubbs, *Tetrahedron Lett.*, 1999, **40**, 2247; (c) M. Scholl, S. Ding, C. W. Lee and R. H. Grubbs, *Org. Lett.*, 1999, **1**, 953.
8. A. W. Addison, T. N. Rao, J. Reedijk, J. van Rijn and G. C. Verschoor, *J. Chem. Soc., Dalton Trans.*, 1984, 1349.
9. Thus, for a perfectly square-pyramidal geometry, $\beta = \alpha = 180°$, and $\tau = 0$, while for a perfect trigonal bipyramid geometry, $\beta = 180°$, $\alpha = 120°$, and $\tau = 1$.
10. (a) E. L. Dias, S. T. Nguyen and R. H. Grubbs, *J. Am. Chem. Soc.*, 1997, **119**, 3887; (b) M. S. Sanford, M. Ulman and R. H. Grubbs, *J. Am. Chem. Soc.*, 2001, **123**, 749.
11. J.-L. Hérisson and Y. Chauvin, *Makromol. Chem.*, 1971, **141**, 161.
12. C. Kammerer, G. Prestat, T. Gaillard, D. Madec and G. Poli, *Org. Lett.*, 2008, **10**, 405.
13. (a) G. Prestat and G. Poli, *Chemtracts: Org. Chem.*, 2004, **17**, 97; (b) G. Poli and G. Giambastiani, *J. Org. Chem.*, 2002, **67**, 9456.
14. For reviews see: (a) A. Ajamian and J. L. Gleason, *Angew. Chem., Int. Ed.*, 2004, **43**, 3754; (b) D. E. Fogg and E. N. dos Santos, *Coord. Chem. Rev.*, 2004, **248**, 2365; (c) J. M. Lee, Y. Na, H. Han and S. Chang, *Chem. Soc. Rev.*, 2004, **33**, 302; (d) J.-C. Walsike, J. O. Obrey, R. T. Baker and G. C. Bazan, *Chem. Rev.*, 2005, **105**, 1001; (e) J. Park and S. Hong, *Chem. Soc. Rev.*, 2012, **41**, 6931; (f) C. Robert and C. M. Thomas, *Chem. Soc. Rev.*, 2013, **42**, 9392; (g) J. A. Mata, F. E. Hah and E. Peris, *Chem. Sci.*, 2014, **5**, 1723.
15. V. D. B. Bonifácio, *Synlett*, 2004, 1649.
16. CSID:15418, http://www.chemspider.com/Chemical-Structure.15418.html, accessed 6 November 2014.
17. See for example: D. Chopra, V. V. Zhurov, E. A. Zhurova and A. A. Pinkerton, *J. Org. Chem.*, 2009, **74**, 2389 and papers quoted therein.
18. The movie of the cannulation is provided as ESI. It can be downloaded and run on a standard media player.
19. If judged more practical, the reaction can be left running overnight at room temperature at this stage, until the following day.
20. See for example: T.-S. Jin, R.-Q. Zhao, M. Li, Y. Zhao and T.-S. Li, *ARKIVOC*, 2006, (xiv), 53.

10. Pericyclic Reactions

10.1. Diels–Alder Reaction of *N*-Phenylmaleimide with *In situ* Generated Buta-1,3-diene

Artur M. S. Silva, Augusto C. Tomé, Diana C. G. A. Pinto, Fernando Domingues, Graça M. Oliveira Rocha, José A. S. Cavaleiro, M. Graça P. M. S. Neves, M. Amparo F. Faustino*, and Mário M. Q. Simões

Department of Chemistry and QOPNA, University of Aveiro, 3810-193 Aveiro, Portugal
*E-mail: faustino@ua.pt

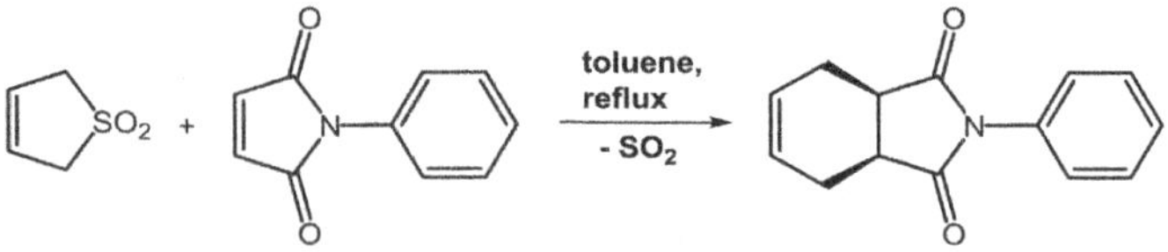

Number of sessions (duration of each session)	Hazard level	Difficulty level	Level of study
2 (3 h each)	Moderate	Medium	Intermediate
Class names C=C double bonds (alkenes and conjugated dienes)			
Concepts involved This experiment involves the reaction of *N*-phenylmaleimide with 3-sulfolene (a synthetic equivalent to buta-1,3-diene). The process involves the thermal cheletropic extrusion of SO_2 from 3-sulfolene to generate the conjugated diene, followed by a cycloaddition reaction with *N*-phenylmaleimide. The resulting cycloadduct is purified by preparative thin-layer chromatography (TLC) and characterized by nuclear magnetic resonance spectroscopy			
Chemicals needed *N*-Phenylmaleimide, 3-sulfolene, toluene, dichloromethane, deuterated chloroform			
Equipment and experimental techniques involved Controlled temperature heating/stirring plate, round bottomed flasks (50 mL and 100 mL), magnetic stir bar, condenser, oil bath, thermometer, analytical TLC plates with fluorescent indicator, small chromatographic chamber, vacuum manifold, rotary evaporator, beaker, preparative thin-layer chromatography plates (20 × 20 cm) with fluorescent indicator and chromatographic chamber with lid, UV-lamp, sintered glass funnel with side-arm for vacuum and with ground-glass joint, and NMR tube			
Keywords Cheletropic reaction, Diels–Alder reaction, diene, dienophile, NMR, preparative TLC			

Background

The Diels–Alder reaction is a [4π + 2π] cycloaddition reaction. It is one of the most representative examples of pericyclic reactions and is considered one of the most versatile reactions in organic chemistry. The Diels–Alder reaction occurs between a conjugated diene and an alkene (or alkyne) and leads to the formation of a six-membered ring.[1] The alkene (or alkyne) that adds to the diene (the 4π component) is called the dienophile (the 2π component). The product of the reaction is

Comprehensive Organic Chemistry Experiments for the Laboratory Classroom
Edited by Carlos A. M. Afonso, Nuno R. Candeias, Dulce Pereira Simão, Alexandre F. Trindade, Jaime A. S. Coelho, Bin Tan, and Robert Franzén.

Published by the Royal Society of Chemistry, www.rsc.org

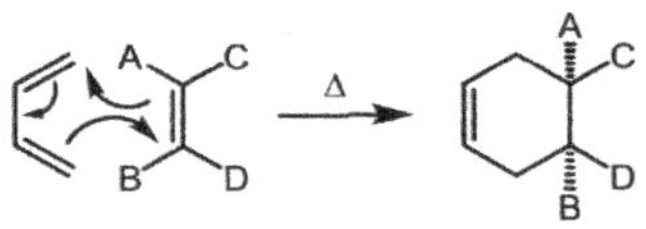

Scheme 10.1.1 Mechanism of the Diels–Alder reaction.

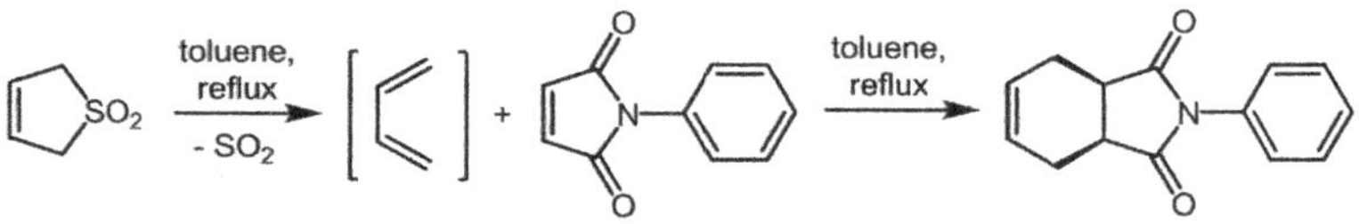

Scheme 10.1.2 Diels–Alder reaction between *in situ* generated 1,3-butadiene and *N*-phenylmaleimide.

called the adduct or cycloadduct. The Diels–Alder reaction occurs in a concerted way, through a cyclic transition state, and the new six-membered ring results from the formation of two new carbon–carbon single bonds at both ends of the diene system (Scheme 10.1.1). The presence of electron withdrawing substituents directly linked to the dienophile reduces dramatically the activation energy barrier and the reaction proceeds in high yields at moderate temperatures.[2,3]

The main goal of this work is to demonstrate the versatility of the Diels–Alder reaction for the preparation of cyclohexene derivatives. The cycloaddition reaction will occur between buta-1,3-diene and *N*-phenylmaleimide (an electron-poor dienophile). However, since buta-1,3-diene is a toxic and extremely flammable gas, requiring particular precautions for its use, in this work it will be safely generated *in situ* by thermal extrusion of SO_2 from 3-sulfolene (Scheme 10.1.2).[4] Thus, 3-sulfolene is a stable and convenient synthetic equivalent of buta-1,3-diene.

This work allows the development of several experimental skills such as reflux heating, monitoring a reaction by TLC, the purification of a compound by preparative TLC and its analysis by NMR (^{1}H and ^{13}C spectra).

Additional Safety

This work requires regular laboratory good practices and personal safety equipment (safety glasses, laboratory coat and nitrile gloves). It is highly recommended to avoid direct contact with 3-sulfolene and *N*-phenylmaleimide as they can be irritating to skin and eyes. The organic solvents used (toluene and dichloromethane) are inflammable when in the presence of ignition sources and are harmful if swallowed, or in contact with the skin. Most of the experiment should be done in the fume hood.

Experimental Procedure

Laboratory Sessions

Session 1: (3 h)

Synthesis of the Cycloadduct

1. In the fume hood, prepare a 50 mL round-bottomed flask with a magnetic stir bar and a condenser.
2. Place about 2 g (17 mmol) of 3-sulfolene and 1.2 equiv. (3.5 g) of *N*-phenylmaleimide in the flask. Then add 15 mL of toluene.

3. Put the flask in an oil bath, turn on the heating plate and keep the reaction mixture stirring for 90 min at 130 °C (oil bath temperature).

Product Isolation

4. After cooling the reaction mixture to room temperature, evaporate the toluene in the rotary evaporator with the water bath at 70 °C. The crude reaction product is obtained as an oil.
5. On an analytical TLC plate (~12 cm × 2.5 cm), at about 1 cm from the bottom, spot the reaction product and the starting *N*-phenylmaleimide, for comparison, both dissolved in dichloromethane.
6. Develop the TLC plate using dichloromethane as the eluent in a small chromatographic chamber. Under a UV-lamp, identify the main product of the reaction and the unreacted *N*-phenylmaleimide.
7. Label conveniently your round-bottomed flask, close it with a stopper and keep it in a safe place until the next session.

Session 2: (3 h)

Product Purification

1. Weigh approximately 100 mg of the crude product in a small beaker and dissolve it in 1 mL of dichloromethane.
2. Using a Pasteur pipet, apply the diluted sample on a preparative TLC plate (20 cm × 20 cm) as a continuous line parallel to the bottom (at about 2 cm from the bottom).
3. Add 15 mL of dichloromethane to the TLC chamber and, carefully, place the TLC plate inside the TLC chamber.
4. Develop the chromatogram until the eluent reaches 1 cm below the top of the plate. Remove the TLC plate from the chamber and leave the solvent to evaporate from the plate surface.
5. Under UV light visualize the bands corresponding to the different components of the mixture and outline the main bands with a spatula. By comparing with the previous TLC, identify the band corresponding to the cycloadduct.
6. Using a spatula, scrape the silica corresponding to the band onto a piece of aluminium foil, transfer it to a sintered glass funnel connected to a 100 mL round-bottomed flask and wash it with dichloromethane (~20 mL), under reduced pressure (use a water jet vacuum pump).
7. Evaporate the solvent in the rotary evaporator at a temperature below 40 °C. The cycloadduct is obtained as a white solid.
8. Dry the cycloadduct under a vacuum pump for at least 30 min.

Structural Characterization

9. Place ~10 mg of the dry compound in an NMR tube, add 0.5 mL of deuterated chloroform ($CDCl_3$) (containing tetramethylsilane (TMS) as the internal standard) and shake it until complete dissolution.
10. Acquire the ^1H and ^{13}C NMR spectra and confirm the structure of the product.

Waste Disposal

The recovered toluene and dichloromethane should be collected in waste recipients for non-chlorinated or chlorinated organic solvents, respectively. The silica gel should be disposed of in an appropriate container.

Results Interpretation and Additional Questions

1. Determine the retention factor of the cycloadduct in the TLC plate using dichloromethane as the eluent.
2. Predict the main differences in the ^{1}H NMR spectrum of the cycloadduct when compared with the spectrum of the *N*-phenylmaleimide.
3. Interpret the ^{1}H and ^{13}C NMR spectra.
4. Propose a mechanism for the reaction.

References

1. F. A. Carey, *Organic Chemistry*, McGraw-Hill, USA, 6th International edn, 2006, ch. 10.
2. J. Sauer and R. Sustmann, *Angew. Chem., Int. Ed. Engl.*, 1980, **19**, 779.
3. D. Craig, *Chem. Soc. Rev.*, 1987, **16**, 187.
4. 3-Sulfolene derivatives are frequently used as synthetic equivalents of conjugated dienes. See, for instance: C. M. A. Alonso, M. G. P. M. S. Neves, A. C. Tomé, A. M. S. Silva and J. A. S. Cavaleiro, *Tetrahedron Lett.*, 2000, **41**, 5679.

10.2. Synthesis of Isomeric Bicyclopropyls from Conjugated Dienes

Leiv K. Sydnes*[a] and Magne O. Sydnes[b]

[a]Department of Chemistry, Allégt. 41, 5007 Bergen, Norway; [b]Faculty of Science and Technology, Department of Mathematics and Natural Science, 4036 Stavanger, Norway. *E-mail: leiv.sydnes@uib.no

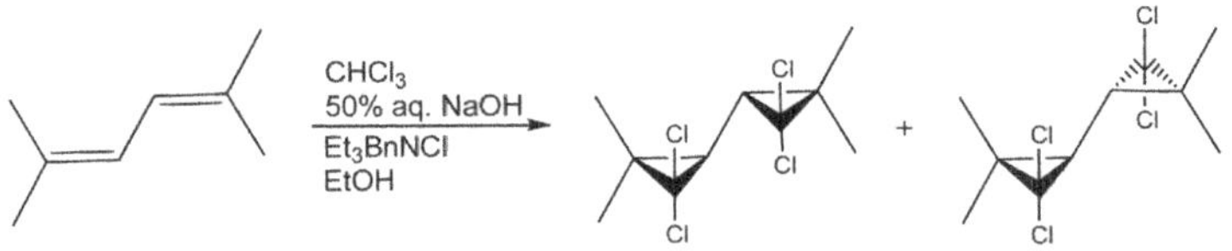

Number of sessions (duration of each session)	Hazard level	Difficulty level	Level of study
3 (1 h + 4 h + 1 h)	Moderate	Medium	Advanced

Class names Alkenes, halogenated cyclopropanes

Concepts involved (a) Phase-transfer catalysis under aqueous conditions; (b) generation of a short-lived intermediate which reacts in a concerted fashion with an alkene; (c) formation of diastereoisomeric mixtures

Chemicals needed 2,5-Dimethyl-2,4-hexadiene, chloroform, sodium hydroxide, water, triethylbenzylammonium chloride, diethyl ether, anhydrous $MgSO_4$, hexanes

Equipment and experimental techniques involved *Equipment required*: three-necked round-bottom flask, dropping funnel, condenser, stirring rod with shaft and plastic blade, stirring engine, separatory funnel, Erlenmeyer flasks, beakers, Pasteur pipette, Büchner funnel, rotary evaporator, NMR spectrometer, gas chromatograph (optional). *Experimental techniques*: addition operations, stirring, liquid–liquid extraction, recrystallization, IR and NMR spectroscopy, gas chromatography (optional)

Keywords Crystallization, cyclopropanation, dienes, extraction, NMR, phase-transfer catalysis

Background

Cyclopropanes have become both valuable intermediates and interesting target molecules in organic synthesis. Although such compounds have been known for more than 100 years[1] they remained an obscure class of compounds until the 1950s when *gem*-dihalocyclopropanes became easily available due to the pioneering work of Hine[2–4] and Doering.[5] Their findings initiated a tremendous research activity that is still ongoing, as reflected in the chemical literature. If the classical *Houben-Weyl* book series is available in the library it should be consulted to learn about methods of preparation and the chemical reactivity of cyclopropanes.[6]

An interesting development in recent years has been the isolation and structure elucidation of an increasing number of natural products with one or several more or less functionalized

Comprehensive Organic Chemistry Experiments for the Laboratory Classroom
Edited by Carlos A. M. Afonso, Nuno R. Candeias, Dulce Pereira Simão, Alexandre F. Trindade, Jaime A. S. Coelho, Bin Tan, and Robert Franzén.

Published by the Royal Society of Chemistry, www.rsc.org

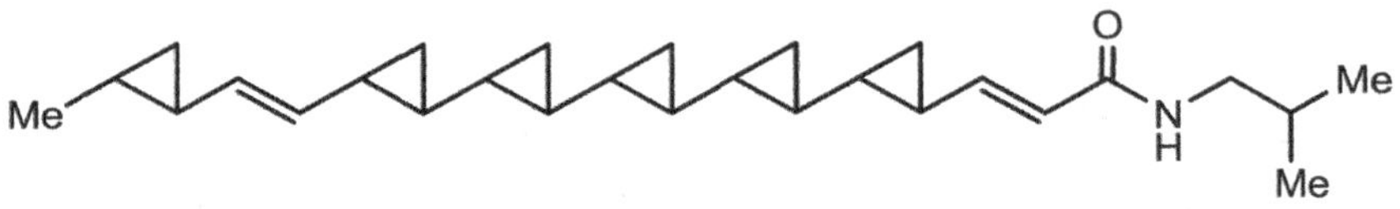

Figure 10.2.1 Structure of U-106305 isolated from the fermentation broth of *Streptomyces* sp. UC-11136.

cyclopropane moieties incorporated in the molecular backbone.[7,8] Among these compounds are some fascinating oligo-cyclopropanes with contiguous cyclopropanes,[7,9,10] among which a compound denoted U-106305 is particularly intriguing due to its very unusual structure (Figure 10.2.1). The presence of six cyclopropane rings, of which five are next to each other, made the structure elucidation by Kuo *et al.* a formidable task,[10] but by a stepwise approach that also included the synthesis of various bicyclopropyl derivatives both the structure elucidation and synthesis were achieved.

Bicyclopropyl compounds can be prepared by addition of carbenes to buta-1,3-dienes, and a variety of such syntheses has been reported in the literature.[11–25] The complexity of a synthesis depends on the nature of the carbene involved and the precursor used, and quite a few of the preparations are somewhat demanding to carry out because of the chemicals involved, the reaction conditions, and by-product formation.[26] Perhaps the safest and simplest carbene sources to use are chloroform and bromoform, which can be converted into the corresponding dihalocarbenes by several methods.[27] Some of these are tedious, for instance the pioneering Doering–Hoffmann method, which requires anhydrous conditions and low temperature,[5] whereas others are quite straightforward. The latter is first and foremost the case with Makosza's method, which is performed under aqueous conditions and at room temperature,[28,29] and this is the method that will be applied in this synthesis.

Makosza's method is based on phase-transfer catalysis with an ammonium salt as the catalytic agent. The ammonium salt functions as a catalyst by converting a fairly water-soluble sodium salt ($NaCX_3$) into an ammonium salt, which dissolves in the organic phase. Thus, the ammonium cation facilitates the transfer of CX_3^- from the aqueous phase to the organic phase (hence the term phase-transfer catalysis) where the anion subsequently decomposes to: CX_2 (dihalocarbene) and X^-. The dihalocarbene then reacts with the C=C bonds (instead of with water, which was what happened in Hine's experiments) and forms the corresponding cyclopropanes. The cyclopropanation is a so-called cheletropic reaction, which is a class of pericyclic reactions that have been described mechanistically on the basis of orbital symmetry considerations.[30]

The synthesis is a good example of a reaction that has minimal environmental impact if performed properly. The by-products formed are sodium chloride and water (and some formate from reaction of dichlorocarbene with water and/or hydroxide); unreacted chloroform can be recovered by distillation, and the solvents used for extraction and recrystallization by evaporation. The synthesis is easy to control, the reaction time is not crucial, and the product mixture is stable when kept at ambient temperature and below, so it can easily be carried out in parallel to or in between other preparations.

Additional Safety

To use a fume hood is not mandatory, but is recommended. Safety goggles are required. Besides regular good laboratory practices, caution should be exercised when the concentrated sodium hydroxide solution is handled. As an additional safety measure, the extraction may be carried out with dichloromethane (DCM) instead of diethyl ether.

Experimental Procedure

Laboratory Session 1 (1 h): Starting the Reaction

1. Connect a 250 mL round-bottom flask – equipped with a condenser, a dropping funnel, and a stirring rod with a Teflon blade and a shaft – to a stirring motor, immerse it in a bath with cold water (or ice/water), and fasten it to a rod in a fume hood.
2. Transfer triethylbenzylammonium chloride (TEBA) (0.2 g), 2,5-dimethylhexa-2,4-diene (4.4 g, 40 mmol), ethanol (3 drops), and chloroform (23.9 g, 200 mmol) to the flask.
3. Transfer a 50% aqueous NaOH solution (11 mL from a stock solution or 16.5 g made from 8.25 g of NaOH and 8.25 g of H_2O; 0.21 mol) to the dropping funnel.
4. Stir rapidly (300 rpm is a good speed) and add the NaOH solution, dropwise, over a period of 8–10 min. Leave the resulting mixture stirring overnight (14 h) in the water bath, which reaches room temperature.

Laboratory Session 2 (4 h): Quenching, Extraction, Work-Up, and Recrystallization

1. Replace the water in the bath with cold water, immerse the flask in the bath, and continue stirring during addition of 3 M hydrochloric acid (15–20 mL).
2. Transfer the resulting hydrolysate to a separatory funnel and extract with diethyl ether (3 × 30 mL). Combine the organic phases and dry ($MgSO_4$).
3. After having dried for some 1.5 h, remove the drying agent by filtration and concentrate the filtrate under vacuum, to leave behind a yellowish solid.
4. Recrystallize the crude product from a mixture of hexanes by first dissolving the product in boiling hexanes and then letting the solution cool slowly. Store the solution cold to facilitate crystallization.

Laboratory Session 3 (1 h): Isolation of the Product

1. Filter the precipitate by suction filtration. When all the hexanes seemed to have been removed, apply suction for a while to dry the solid on the filter.
2. Determine the melting point.

With a pure sample of 2,2,2′,2′-tetrachloro-3,3,3′,3′-tetramethylbicyclopropyl in hand, the ^1H and ^{13}C NMR spectra may be recorded and compared with reported data. The proton spectrum of the mixture may then be used to determine the isomeric composition of the product, and this result may be compared with that obtained by GLC analysis.

Waste Disposal

If the aqueous phase is not slightly acidic after extraction (check by using pH paper), it should be neutralized before it is discarded in the sink. The organic solvent used for extraction, either diethyl ether or methylene chloride, may be reused; if not, it has to be transferred to the appropriate waste-solvent bottle or can for subsequent destruction. As for the recovered chloroform, this may be collected and stored for later use; if not it has to be discarded in the waste bottle dedicated for storage of halogenated solvents.

Results Interpretation and Additional Questions

1. Usually there is a slight variation in the isomer composition of the product. Why may this result in different melting points from that reported in the literature?[16,25]

2. The isomeric composition can conveniently be determined by ^1H NMR spectroscopy. When the NMR spectrum is recorded at high field (>200 MHz) the mixture gives six singlets in total, three from each isomer. When the peaks are numbered 1–6 from low to high field and the relative area of peak i is A_i (i = 1–6), what is the expression for the percentage of the more abundant isomer?
3. The product consists of a *meso* form and a pair of enantiomers that exhibit rather different polarity due to different dipole moments. Study this observation by building molecular models of the two forms and explore how the total dipole moment varies with the rotation about the inter-ring C–C bond.[25] Which isomer has the highest dipole moment?
4. Why is it important that the dihalocarbene intermediate, which is very reactive, is generated in the absence of water?
5. Why are the alkyl groups in the ammonium cation important for the phase transfer?

Acknowledgements

Thanks are due to Chris S. Stavaas, University of Bergen for recording the NMR spectra incorporated in the ESI.

References

1. G. Gustavson, *J. Prakt. Chem.*, 1890, **42**, 495.
2. J. Hine, *J. Am. Chem. Soc.*, 1950, **72**, 2438.
3. J. Hine and A. M. Dowell Jr, *J. Am. Chem. Soc.*, 1954, **76**, 2688.
4. J. Hine, A. M. Dowell Jr and J. E. Singley Jr., *J. Am. Chem. Soc.*, 1956, **78**, 479.
5. W. v. E. Doering and A. K. Hoffmann, *J. Am. Chem. Soc.*, 1954, **76**, 6162.
6. *Carbocyclic Three-membered Ring Compounds; Methods of Organic Chemistry (Houben-Weyl)*, ed. A. de Meijere, Thieme Verlag, Stuttgart Germany, 4th edn, 1997, vol. E17a and E17b.
7. J. Pietruszka, *Chem. Rev.*, 2003, **103**, 1051.
8. L. A. Wessjohann, W. Brandt and T. Thiemann, *Chem. Rev.*, 2003, **103**, 1625.
9. M. Yoshida, M. Ezaki, M. Haqshimoto, M. Yamashita, N. Shigematsu, M. Okuhara, M. Kohsaka and K. Horikoshi, *J. Antibiot.*, 1990, **43**, 748.
10. M. S. Kuo, R. J. Zielinski, J. I. Cialdella, C. K. Marschke, M. J. Dupuis, G. P. Li, D. A. Kloostermann, C. H. Spilman and V. P. Marshall, *J. Am. Chem. Soc.*, 1995, **117**, 10629.
11. C. G. Overberger, R. E. Zangaro, R. E. K. Winter and J.-P. Anselme, *J. Org. Chem.*, 1971, **36**, 975.
12. E. Müller and O. Roser, *J. Prakt. Chem.*, 1932, **133**, 291.
13. (a) I. E. Dolgii, E. A. Shapiro and O. M. Nefedov, *Izv. Akad. Nauk SSSR, Ser. Khim.*, 1979, **28**, 1282; (b) *Chem. Abstr.*, 1979, **91**, 157317k.
14. T. Armaroli, C. Dell'Erba, A. Gabellini, F. Gasparrini, A. Mugnoli, M. Novi, G. Petrillo and C. Tavani, *Eur. J. Org. Chem.*, 2002, 1284.
15. J. Martelli and R. Carrié, *Tetrahedron*, 1978, **34**, 1163.
16. H. Komrsova and J. Farkas, *Collect. Czech. Chem. Commun.*, 1960, **25**, 1977.
17. U. Schöllkopf and J. Paust, *Chem. Ber.*, 1965, **98**, 2221.
18. L. Skattebøl, *J. Org. Chem.*, 1964, **29**, 2951.
19. D. Seyferth, J. M. Burlitch, R. J. Minsasz, J. Y.-P. Mui, H. D. Simmons Jr., A. J. H. Treiber and S. R. Dowd, *J. Am. Chem. Soc.*, 1965, **87**, 4259.
20. P. Weyerstahl, D. Klamann, M. Fligge, C. Finger, F. Nerdel and J. Buddrus, *Liebigs Ann. Chem.*, 1967, **710**, 17.
21. G. Schrumpf and W. Lüttke, *Liebigs Ann. Chem.*, 1969, **730**, 100.
22. E. V. Dehmlow and J. Schönefeld, *Liebigs Ann. Chem.*, 1971, **744**, 42.
23. E. V. Dehmlow, *Tetrahedron*, 1972, **28**, 175.
24. K. Kleveland and L. Skattebøl, *Acta Chem. Scand., Ser. B*, 1975, **29**, 191.
25. C. Rømming and L. K. Sydnes, *Acta Chem. Scand.*, 1976, **B 30**, 963.
26. *Carbene(oide) Carbine; Methoden der Organischen Chemie (Houben-Weyl)*, ed. M. Regitz, Thieme Verlag, Stuttgart, Germany, 4th edn, 1989, vol. E19b.
27. L. K. Sydnes, *Preparation of gem-dihalocyclopropanes; a short summary of some classical work*, 2014, available within the ESI.
28. M. Makosza and M. Wawrzyniewicz, *Tetrahedron Lett.*, 1969, 4659.
29. M. Makosza and M. Fedorynski, *Synth. Commun.*, 1973, **3**, 305.
30. R. B. Woodward and R. Hoffmann, *Angew. Chem., Int. Ed.*, 1969, **8**, 781.

10.3. Diels–Alder Reaction Between *p*-Benzoquinone and Cyclopentadiene and Subsequent Photochemical [2π + 2π] Cycloaddition

Arno Kraft

Heriot-Watt University, Institute of Chemical Sciences, School of Engineering & Physical Sciences, Riccarton, Edinburgh EH14 4AS, United Kingdom. *E-mail: A.Kraft@hw.ac.uk

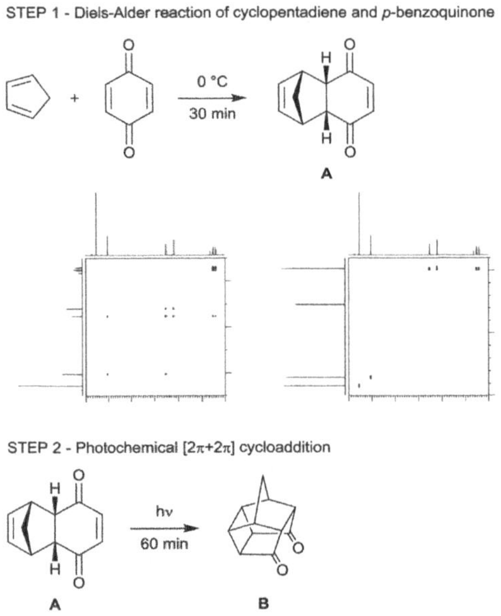

Number of sessions (duration of each session)	Hazard level	Difficulty level	Level of study
2 (3 h each)	Moderate	Medium	Advanced

Class names Alkenes, dienes, α,β-unsaturated ketones

Concepts involved Thermal cycloaddition, Diels–Alder reaction, photochemical cycloaddition, photochemistry, 2D NMR

Chemicals needed Benzoquinone, cyclopentadiene, petroleum ether (60–80 °C), ethyl acetate, deuterated chloroform

Equipment and experimental techniques involved Apparatus for heating to reflux (90 °C) and magnetic stirring, filtration, recrystallisation, preparation of NMR sample, photochemical apparatus (consisting of a black-light and cardboard box), NMR, IR, melting point

Keywords Cycloaddition, Diels–Alder reaction, diene, molecular orbital theory, NMR, 2D NMR, photochemical cycloaddition, recrystallisation

Comprehensive Organic Chemistry Experiments for the Laboratory Classroom
Edited by Carlos A. M. Afonso, Nuno R. Candeias, Dulce Pereira Simão, Alexandre F. Trindade, Jaime A. S. Coelho, Bin Tan, and Robert Franzén.

Published by the Royal Society of Chemistry, www.rsc.org

Background

The reaction of conjugated dienes with alkenes leads to the formation of six-membered rings and is known as the Diels–Alder reaction, named after Otto Diels and Kurt Alder who received the Nobel Prize for Chemistry in 1950 for discovering this highly useful reaction.[1] It is probably the best known cycloaddition reaction and is a powerful way of making six-membered rings in high yield and with high stereospecificity.

Not surprisingly, the Diels–Alder reaction has been used extensively in the synthesis of complex natural products. The reaction mechanism of a Diels–Alder reaction involves no charged or radical intermediates but proceeds in a single concerted step as indicated by the three "simultaneous" curly arrows in Scheme 10.3.1.[2] The Diels–Alder reaction is fast when the alkene is electron-deficient, the diene is electron-rich and/or has a fixed cisoid conformation, like cyclopentadiene.

In this experiment, students use cyclopentadiene, which is prepared by the thermal cracking of dicyclopentadiene (Scheme 10.3.2). A bottle of cyclopentadiene can be kept in an ice bath during laboratory hours but needs to be stored in the freezer between laboratory sessions since cyclopentadiene dimerises (by a Diels–Alder reaction) within a few hours at room temperature.

One of the first cycloadditions performed by Diels and Alder was the reaction of cyclopentadiene with *p*-benzoquinone.[3,4] The Diels–Alder reaction is the classical example of a thermal [4π + 2π] cycloaddition between a diene (a 4π component) and an alkene (a 2π component), also called the dienophile. Any cycloadditions where the total number of π electrons involved amounts to $4n + 2$ (with $n = 1, 2 \ldots$) are "allowed" under thermal conditions.

Cycloadditions involving $4n$ π electrons are forbidden under thermal conditions but they are "photochemically allowed".[5] Consequently, a [2π + 2π] cycloaddition between two alkenes will only occur when one alkene is in its excited state. Since alkenes do not absorb visible light, this usually means irradiating the reaction mixture with UV light. In this experiment, students will use a black-light, an inexpensive compact fluorescent tube with an emission maximum at 352 nm, for converting cycloadduct **A** into cycloadduct **B**, which contains an extra cyclobutane ring.

Additional Safety

The preparation should be carried out in a fume hood, with proper safety equipment, such as protective clothing, gloves and safety goggles. *p*-Benzoquinone is toxic if inhaled or swallowed, causes skin and serious eye irritation and possible respiratory irritation; its release into the environment should be avoided as it is very toxic to aquatic life. Cyclopentadiene and dicyclopentadiene are highly flammable, harmful by inhalation and if swallowed; they are irritating to eyes, respiratory system and skin, as well as toxic to aquatic life. Cyclopentadiene is quite volatile (its boiling point is 40–42 °C) and, due to contamination with dicyclopentadiene, tends to have a

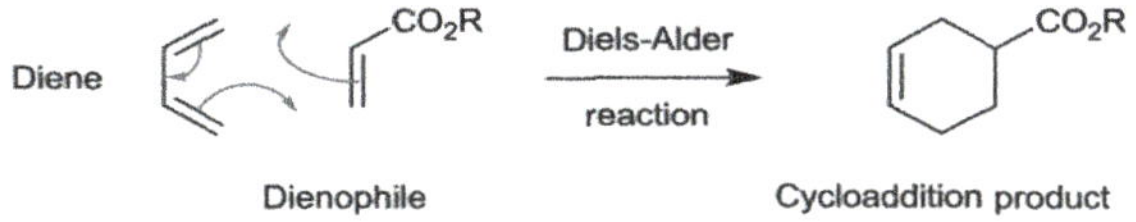

Scheme 10.3.1 Example of a Diels–Alder reaction.

Scheme 10.3.2 Thermal cracking of dicyclopentadiene.

disagreeable smell; it should always be handled in a fume cupboard. Petroleum ether is highly flammable, harmful, suspected of damaging fertility or the unborn child, and toxic to aquatic organisms; inhalation should be avoided since vapours may cause drowsiness and dizziness. Ethyl acetate is highly flammable and vapours can cause eye irritation, drowsiness or dizziness. Methylated spirits are highly flammable. Chloroform causes skin irritation, is harmful if swallowed and suspected of causing cancer, and may also cause damage to organs through prolonged or repeated exposure. UV light causes damage to the eyes, so immediately after switching on the black-light, the UV source should be covered with a cardboard box to minimise exposure to UV light.

Experimental Procedure

Laboratory Session 1 (3 h): Diels–Alder Reaction

Thermal Cycloaddition

1. Use a fume cupboard for this experiment.
2. In a conical flask cool a solution of 2.91 g of *p*-benzoquinone (27.0 mmol) in 12 mL of petroleum ether/ethyl acetate (5 : 1) with an ice–salt bath.
3. To the stirred solution add 2.5 mL of cyclopentadiene (25 mmol, measured with a graduated pipette). Try to keep the reaction mixture at 0 °C or below. Continue stirring the reaction mixture for 30 min and for a further 30 min at room temperature.

Work Up and Purification

4. Cool again with an ice–salt bath and collect the solid by filtration.
5. Recrystallise the crude product from petroleum ether, then dry the recrystallised product. This gives product **A**.

Laboratory Session 1 (3 h): Photochemical Cycloaddition

Characterisation

1. Weigh the product and determine the yield. Measure the melting point and record an IR spectrum.

Photochemical Cycloaddition

2. Prepare a solution of 20–30 mg of **A** in 0.7 mL of $CDCl_3$ and pour it into an NMR tube. Place the NMR tube in the rack underneath a black-light, cover with the cardboard box and irradiate for 60 min. This will produce product **B**.

Characterisation

3. Record a ^1H NMR spectrum of your sample of **B**.

Waste Disposal

Residues from recrystallisation and acetone washing contaminated with cyclopentadiene should be collected in the non-halogenated waste container, for disposal by an external contractor. The NMR sample in $CDCl_3$ should be disposed of in a chlorinated waste collector recipient, for posterior destruction.

Results Interpretation and Additional Questions

1. The ^{1}H NMR spectra of **A** and **B** are more complicated than most that you will have seen before, since strained norbornene and other polycyclic systems show a lot of small vicinal and long-range couplings, which complicate the interpretation of the ^{1}H NMR spectrum, and also because the spin systems in **A** and **B** are partly higher order. Use the set of 2D NMR spectra that has been provided to assign all the ^{1}H and ^{13}C NMR signals of **A**. Note that the two protons on the CH_2 group are diastereotopic and have therefore different chemical shifts. The NOESY evidence should allow you to assign them without ambiguity.
2. Confirm the stereochemistry of **A** using the evidence from the NOESY.
3. **A** and **B** have the same molecular formula. With the help of the IR and NMR evidence, identify which functional groups are no longer present in **B**. Suggest what has happened.

References

1. *The Nobel Prize in Chemistry 1950*, http://www.nobelprize.org/nobel_prizes/chemistry/laureates/1950/, accessed October 2014.
2. J. Clayden, N. Greeves and S. Warren, *Organic Chemistry*, Oxford University Press, Oxford, 2nd edn, 2012, pp. 877–885.
3. O. Diels and K. Alder, *Liebigs Ann. Chem.*, 1928, **460**, 98.
4. C. C. Nawrat and C. J. Moody, *Angew. Chem., Int. Ed.*, 2014, **53**, 2056.
5. J. Clayden, N. Greeves and S. Warren, *Organic Chemistry*, Oxford University Press, Oxford, 2nd edn, 2012, pp. 896–898.

10.4. Click Chemistry

Susana Dias Lucas*, Eduardo F. P. Ruivo, Marta Figueiras, João P. Lavrado, and Rui Moreira

Research Institute for Medicines (iMed.ULisboa), Faculty of Pharmacy, Universidade de Lisboa, Av. Prof. Gama Pinto, 1649-003 Lisboa, Portugal. *E-mail: sdlucas@ff.ulisboa.pt

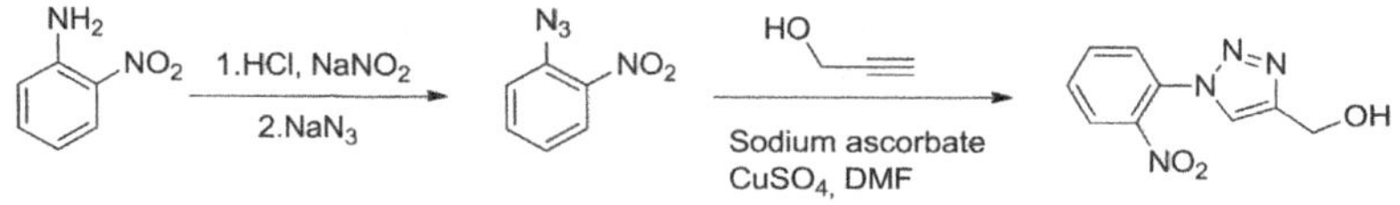

Number of sessions (duration of each session)	Hazard level	Difficulty level	Level of study
1 (3 h)	Moderate	Medium	Advanced
Class names Amines, diazoniums, azides, alkynes			
Concepts involved This experiment involves a method for the synthesis of an aromatic azide from nitroaniline *via* a diazonium salt. This will be the starting material for reaction with an alkyne to afford the desired 1,2,3-triazole, using a copper-catalyzed Huisgen cycloaddition			
Chemicals needed 2-Nitroaniline, hydrochloric acid (37%), sodium nitrite, propargyl alcohol, dimethylformamide, sodium ascorbate, copper sulfate pentahydrate, sodium sulfate			
Equipment and experimental techniques involved Apparatus for magnetic stirring, vacuum filtration apparatus, vacuum pump (optional), thin-layer chromatography, liquid–liquid extraction apparatus, Rotavapor, NMR, melting point			
Keywords Click chemistry, diazonium salt, Huisgen cycloaddition, liquid–liquid extraction, NMR, precipitation			

Background

The “click chemistry” approach was pioneered by Sharpless group in order to develop an expanding set of powerful, selective, and modular “blocks” that work reliably in both small- and large-scale applications. Click reactions are defined as high yielding chemical transformations, wide in scope, that create only by-products that can be removed without chromatography, stereospecific, simple to perform, and can be conducted in easily removable or benign solvents. Click transformations achieve their required characteristics by having a high thermodynamic driving force, usually greater than 20 kcal mol^{-1} and tend to be highly selective for a single product. Examples of potential click reactions are cycloadditions of unsaturated species, nucleophilic substitutions, carbonyl chemistry or additions to carbon–carbon multiple bonds.[1]

Huisgen cycloaddition is a 1,3-dipolar cycloaddition of azides and alkynes that has become one of the most important click transformations, especially given the ~10^6-fold acceleration of the copper(I)-catalyzed 1,4-disubstituted 1,2,3-triazole formation in a regioselective manner.[2]

Comprehensive Organic Chemistry Experiments for the Laboratory Classroom
Edited by Carlos A. M. Afonso, Nuno R. Candeias, Dulce Pereira Simão, Alexandre F. Trindade, Jaime A. S. Coelho, Bin Tan, and Robert Franzén.

Published by the Royal Society of Chemistry, www.rsc.org

The triazole products are more than just passive linkers; they readily associate with biological targets, through hydrogen bonding and dipole interactions. Their applications are increasingly found in all aspects of drug discovery, ranging from lead finding through combinatorial chemistry and target-templated *in situ* chemistry, to proteomics and DNA research, using bioconjugation reactions.[3]

This experiment was developed to demonstrate to students of advanced organic chemistry level the feasibility of cycloadditions towards the straightforward synthesis of important chemical scaffolds. The proposed procedure includes the synthesis of the starting material, an aromatic azide from nitroaniline *via* diazonium salt.[4] Aryl diazonium salts react directly with azide ions without catalysts to give the corresponding aryl azides; this reaction does not take place with cleavage of the C–heteroatom bond but occurs with attack of the azide on the diazonium ion with formation of aryl pentazoles and subsequent products.[5]

Huisgen cycloaddition is then performed with the addition of the aryl azide to an acetylene in the presence of a Cu(I) catalyst generated *in situ*, leading selectively to the 1,4-disubstituted triazole.

Additional Safety

All sessions should be done in a fume hood, with proper safety equipment, such as protective clothing, gloves and safety goggles. HCl has a corrosive effect on human tissue. Sodium azide is toxic (LD_{50} oral (rats) = 27 mg kg^{-1}) and may be fatal in contact with skin or if swallowed; it decomposes explosively upon heating above 275 °C. Do not use metal spatulas for weighing and transferring azides. The preparation and addition of sodium azide aqueous solution should be done carefully in the fume hood with the sash positioned as low as possible to avoid direct contact with any trace of hydrazoic acid and its toxic vapors that may be formed. Sodium nitrite is harmful if swallowed, causes eye, skin, and respiratory tract irritation. If for any reason storage of the intermediate 2-nitrophenylazide is necessary this should be done below room temperature and away from sources of heat, light, pressure, and shock. Propargyl alcohol is a flammable liquid, toxic by inhalation and corrosive.

Experimental Procedure

Step 1. Synthesis of 2-Nitrophenyl Azide

1. Prepare a 100 mL round-bottom flask with a magnetic stirring bar with an ice bath for cooling.
2. Dissolve 2-nitroaniline (500 mg, 3.6 mmol) in a 5 M HCl aqueous solution (31.5 mL).
3. Add a previously prepared solution of $NaNO_2$ (263 mg, 3.8 mmol, 1.05 equiv.) in 2.5 mL of water, dropwise, and let the mixture react for 1 h, cooling throughout with the ice bath. The solution color will change from orange to colorless.
4. Add a previously prepared solution of NaN_3 (236 mg, 3.6 mmol, 1 equiv.) in 2.5 mL of water, cooling throughout with the ice bath. The product will start to precipitate immediately as a white solid.
5. Promote full precipitation of the desired product by keeping the mixture at 0–5 °C for 10 min. Collect the precipitate by filtration and wash the solid with cooled water. Dry the product under reduced pressure.
6. Run a TLC of the obtained product against the starting material using EtOAc/*n*-hexane (1 : 2) as eluent.
7. Weigh the desired product and calculate the yield. Measure the melting point and record ^{1}H and ^{13}C NMR spectra to attest to the compound purity.

Step 2. Synthesis of (1-(2-Nitrophenyl)-1*H*-1,2,3-triazol-4-yl)methanol

8. Prepare a 25 mL round-bottom flask with a magnetic stirring bar.
9. Dissolve the product that you obtained in the first step (2-nitrophenyl azide, 100 mg, 0.61 mmol) in DMF (5.5 mL).
10. Add propargyl alcohol (36 μL, 0.61 mmol, 1 equiv.) and freshly prepared solutions of sodium ascorbate (1 mL, 0.1 M) and $CuSO_4$ (1 mL, 0.1 M).
11. Follow the reaction by TLC using EtOAc/*n*-hexane (1:2) as eluent. The reaction should take 15 min to complete.
12. Pour the reaction mixture into 20 mL of water and extract with 3 × 20 mL of ethyl acetate using a separatory funnel. Combine the organic layers and dry them with anhydrous Na_2SO_4. Filter the Na_2SO_4 and concentrate the liquid with a rotatory evaporator, using a previously weighted round-bottom flask. The product is isolated as a pale yellow solid with no need of further purification.
13. Weigh the desired product and calculate the yield. Measure the melting point and record ^1H and ^{13}C NMR spectra to attest to the compound purity.

Waste Disposal

The neutralized aqueous solution may be diluted and discarded in the sink. Ethyl acetate may be reused or collected in a non-halogenated waste container, for further destruction.

Results Interpretation and Additional Questions

1. Calculate the R_f for both products obtained in this experiment.
2. Assign the obtained NMR data and compare the obtained melting point with the literature.[4]
3. Calculate the yields for both reaction steps and the overall yield.
4. Give the mechanism for the reaction that affords the desired azide in step one, *via* diazonium salt.
5. Give the catalytic cycle involved in copper-catalyzed Huisgen cycloaddition to afford the 1,2,3-triazole from the reaction of an azide with an acetylene. Discuss the regioselectivity of the reaction.
6. Choose a literature example in which the triazole click chemistry is enhanced as a powerful tool for the study of biomolecular systems. Discuss the utility of this approach in dissecting biological events.

References

1. H. C. Kolb, M. G. Finn and K. B. Sharpless, *Angew. Chem., Int. Ed.*, 2001, **40**, 2004.
2. V. V. Rostovtsev, L. G. Green, V. V. Fokin and K. B. Sharpless, *Angew. Chem., Int. Ed.*, 2001, **41**, 2596.
3. (a) P. Thirumurugan, D. Matosiuk and K. Jozwiak, *Chem. Rev.*, 2013, **113**, 4905; (b) J. Martell and E. Weerapana, *Molecules*, 2014, **19**, 1378.
4. P. A. S. Smith and J. H. Boyer, *Org. Synth.*, 1957, **37**, 1.
5. S. Brase, C. Gil, K. Knepper and V. Zimmermann, *Angew. Chem., Int. Ed.*, 2005, **44**, 5188.

10.5. On-Water Synthesis of a Dipyrromethane *via* Bis-hetero-Diels–Alder Reaction of an Azoalkene with Pyrrole

Susana M. M. Lopes[a], Maria I. L. Soares[a], Ana Lúcia Cardoso[a], Américo Lemos[b], and Teresa M. V. D. Pinho e Melo*[a]

[a]Department of Chemistry, University of Coimbra, Rua Larga, 3004-535 Coimbra, Portugal; [b]CIQA, FCT, University of Algarve, Campus de Gambelas, 8005-139 Faro, Portugal. *E-mail: tmelo@ci.uc.pt

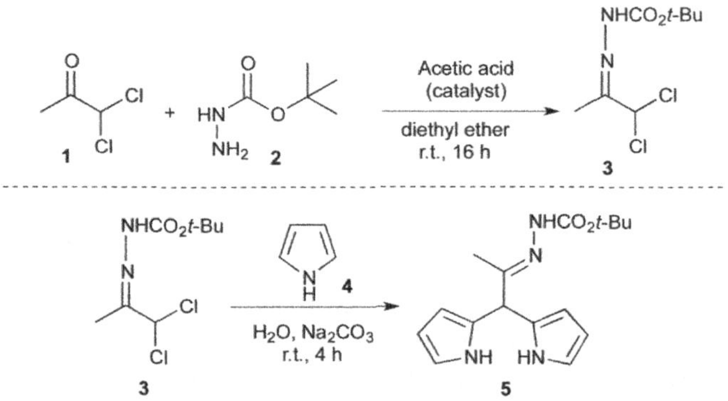

Number of sessions (duration of each session)	Hazard level	Difficulty level	Level of study
2 (2 h + 6 h)	Moderate	Medium	Advanced

Class names Hydrazines, ketones, hydrazones, azoalkenes

Concepts involved This experiment describes a method for the synthesis of a dipyrromethane *via* two consecutive Diels–Alder reactions using water as solvent

Chemicals needed 1,1-Dichloro-2-propanone, *tert*-butyl carbazate, diethyl ether, glacial acetic acid, sodium carbonate, pyrrole, dichloromethane, *n*-hexane, anhydrous sodium sulfate, chloroform-*d*

Equipment and experimental techniques involved Apparatus to carry out reactions with magnetic stirring, liquid–liquid extraction apparatus, rotary evaporator, NMR, melting point

Keywords Azoalkene, dipyrromethane, hetero-Diels–Alder reaction, liquid–liquid extraction, NMR, recrystallization

Background

Porphyrins are macrocyclic compounds featuring a highly conjugated system consisting of four pyrrole rings connected together by four methine bridges (=CH–). The base structure of porphyrins contains 22 π-electrons, 18 of which form an aromatic conjugated system following Hückel's rule ($4n + 2$). Tetrapyrrolic macrocycles belong to a class of structures that play important roles in nature such as *heme*, the pigment in red blood cells responsible for the oxygen transport, and *chlorophylls*, a green pigment critical in photosynthesis (Figure 10.5.1).[1] An important application of porphyrins is as photosensitizers in the photodynamic therapy of cancer.[2]

Comprehensive Organic Chemistry Experiments for the Laboratory Classroom
Edited by Carlos A. M. Afonso, Nuno R. Candeias, Dulce Pereira Simão, Alexandre F. Trindade, Jaime A. S. Coelho, Bin Tan, and Robert Franzén.

Published by the Royal Society of Chemistry, www.rsc.org

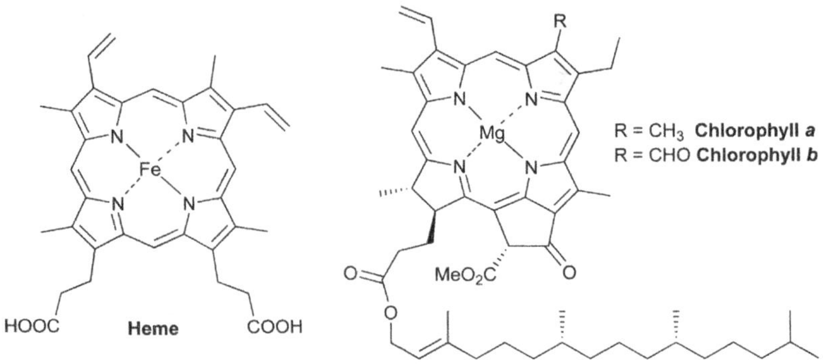

Figure 10.5.1 Structure of heme and chlorophylls.

This experiment describes a method to synthesize dipyrromethanes, which are valuable building blocks for the synthesis of porphyrins.[3] The classical method for the preparation of dipyrromethanes is the acid-catalysed condensation of aldehydes (or ketones) with pyrrole. The drawback of this approach is the competitive formation of tripyrromethanes and higher oligopyrromethanes, as well as N-confused dipyrromethane derivatives. Therefore, over the years, some new synthetic approaches have been developed to overcome these disadvantages.[4]

The Diels–Alder reaction is a [4 + 2] cycloaddition reaction of a conjugated diene and a dienophile (alkene or alkyne) and is one of the most versatile methods for the construction of six-membered rings. A variant of this reaction is the hetero-Diels–Alder reaction, in which either the diene or the dienophile contains a heteroatom. This reaction is a powerful tool for the synthesis of six-membered ring heterocycles. In this work, a 5-substituted dipyrromethane (**5**) will be prepared *via* hetero-Diels–Alder reaction of an azoalkene with pyrrole.

Azoalkenes, unsubstituted in the 4-position, are generally prepared *in situ* from the corresponding α-halohydrazones by treatment with base and are intercepted in the Diels–Alder reaction with electron-rich heterocycles or nucleophilic olefins to give 1,4,5,6-tetrahydropyridazines or open chain hydrazones. In many cases, evidence for the *in situ* generation of azoalkenes comes from their characteristic yellow, orange, red or amaranth colours, a consequence of their conjugation. On the other hand, changes in the colour can be used to check the progress of the reaction. Azoalkenes with substituents in the 4-position are usually stable enough to be isolated.[5]

An example of the hetero-Diels–Alder reaction of azoalkene **7** with the heterocyclic dienophiles dihydrofuran (**8**) and pyrrole (**4**) is illustrated in Scheme 10.5.1. Azoalkene **7**, generated *in situ* from the α-bromohydrazone **6** by treatment with sodium carbonate, reacts with dihydrofuran and pyrrole to give 1,4,5,6-tetrahydropyridazine **9** and the open chain hydrazone **11**, respectively. The formation of hydrazone **11** can be rationalized by considering the rearomatization of the initially formed cycloadduct, the bicyclic 1,4,5,6-tetrahydropyridazine **10**.[6]

On the other hand, the base-mediated dehydrohalogenation of α,α-dihalohydrazones in the presence of pyrrole leads to two consecutive hetero-Diels–Alder reactions affording dipyrromethanes in a one-pot procedure. These hetero-Diels–Alder reactions are accelerated and give higher yields using water as solvent, allowing simpler purification procedures than the ones for the reactions carried out in dichloromethane or in the absence of solvent.[7] The "on water" reactions occur at the oil–water interface of oil (reactants) droplets. The structure of water in this interface allows free OH groups to protrude into the organic phase, playing a crucial role in the catalytic process *via* hydrogen bonds. Thus, rate acceleration of water reactions is expected when stronger hydrogen bonds are formed between these OH groups and the transition state than with reactants.[8]

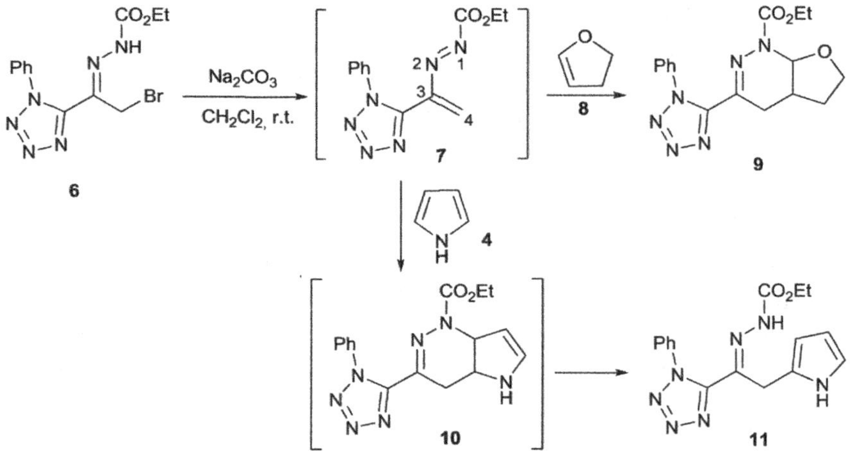

Scheme 10.5.1 Hetero-Diels–Alder of azoalkene **7** with dihydrofuran and pyrrole.

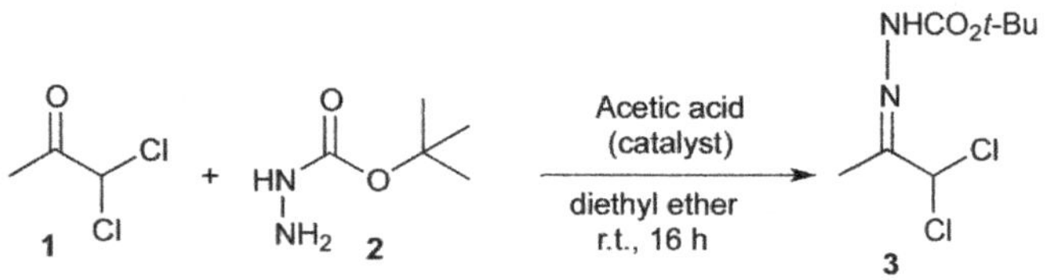

Scheme 10.5.2 Synthesis of α,α-dihalohydrazone **3**.

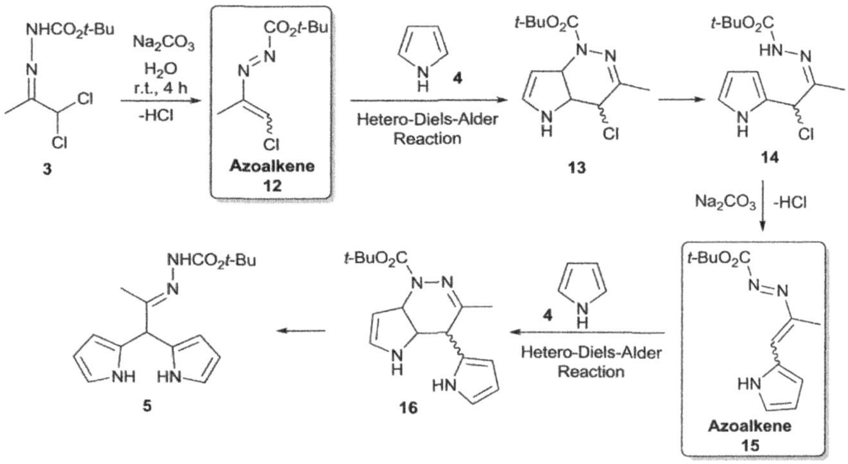

Scheme 10.5.3 Mechanism proposal for the synthesis of dipyrromethane **5**.

The dihalohydrazone **3** was obtained by reacting 1,1-dichloro-2-propanone (**1**) with *tert*-butyl carbazate (**2**) in diethyl ether in the presence of a catalytic amounts of acetic acid (Scheme 10.5.2).

The synthesis of dipyrromethane **5** can be rationalized as outlined in Scheme 10.5.3. The transient azoalkene **12**, generated *in situ* by treatment of hydrazone **3** with sodium carbonate, is intercepted in the first Diels–Alder reaction with pyrrole to give the bicyclic

1,4,5,6-tetrahydropyridazine **13** followed by rearomatization to the open chain hydrazone **14**. Dehydrohalogenation of hydrazone **14** promoted by the base, generates the new azoalkene **15**, which is intercepted in the second Diels–Alder reaction with another molecule of pyrrole to give bicyclic 1,4,5,6-tetrahydropyridazine **16**. Rearomatization of the pyrrolic unit of **16** gives the target dipyrromethane **5**.

Additional Safety

All sessions should be done in a fume hood, with proper safety equipment, such as protective clothing, gloves and safety goggles. 1,1-Dichloro-2-propanone is a flammable reagent, harmful if swallowed. Acetic acid is a flammable, corrosive acid and may cause severe burns. Pyrrole is an organic reagent that is flammable, harmful by inhalation, acutely toxic and may cause serious damage to eyes. *tert*-Butyl carbazate is a highly flammable organic reagent, with danger of serious damage to health by prolonged exposure, irritating to eyes, respiratory system and skin, toxic in contact with skin and if swallowed. Dichloromethane is a flammable organic solvent irritating to eyes, respiratory system and skin, with limited evidence of a carcinogenic effect and whose vapours may cause drowsiness and dizziness. *n*-Hexane is a highly flammable organic solvent, irritating to the skin, it poses danger of serious damage to health by prolonged exposure through inhalation, is toxic to aquatic organisms, poses a possible risk of impaired fertility, may cause lung damage if swallowed and its vapours may cause drowsiness and dizziness.

Experimental Procedure

Laboratory Session 1 (2 h): Synthesis of 1,1-Dichloroacetone *t*-Butoxycarbonyl Hydrazone (3)

Previous Day

1. In a fume hood, prepare a 50 mL round-bottom flask with a magnetic stirring bar, and add 1 mL of 1,1-dichloro-2-propanone (10.40 mmol, 1.00 equiv.), 25 mL of dry diethyl ether and 1.58 g of *tert*-butyl carbazate (11.96 mmol, 1.15 equiv.). Add two drops of glacial acetic acid to the reaction mixture using a Pasteur pipette. Leave the reaction stirring at room temperature for 16 h.

Product Isolation

2. Evaporate the solvent and add 3 mL of diethyl ether at 0 °C. Filter off the solid in suspension, wash with 2 mL of diethyl ether at 0 °C and dry the solid under vacuum.
3. Transfer the white solid to a vial, determine the weight and calculate the yield. Make a TLC analysis to confirm the compound purity.
4. Determine the melting point of the obtained product and prepare a sample for ^{1}H and ^{13}C NMR in chloroform-*d*.
5. Store the solid in a closed vial in a refrigerator until the next session.

Laboratory Session 2 (6 h): Synthesis of Dipyrromethane 5

1. In a fume hood, prepare a 100 mL round-bottom flask with a magnetic stirring bar, and add 4.40 g of Na_2CO_3 (42.00 mmol, 10.00 equiv.), 55 mL of distilled water, 5.76 mL of pyrrole (83.00 mmol, 20.00 equiv.) and 1.00 g of 1,1-dichloroacetone *t*-butoxycarbonyl hydrazone (**3**) (4.2 mmol, 1.00 equiv.), prepared in the previous session. Leave the reaction stirring at room temperature for 4 h.

Product Isolation

2. Transfer the mixture into a separatory funnel, extract with dichloromethane (3 × 40 mL) and dry over anhydrous Na_2SO_4. After filtration, remove the solvent and the excess of pyrrole (bp 131 °C) under reduced pressure.
3. Recrystallize the product with 20 mL diethyl ether/*n*-hexane (1 : 20), filter and dry the solid under vacuum.
4. Collect the grey solid in a vial, determine the weight and calculate the yield. Make a TLC analysis to confirm the compound purity.
5. Determine the melting point of the obtained product and prepare a sample for ^1H and ^{13}C NMR in chloroform-*d*.

Waste Disposal

Diethyl ether and *n*-hexane may be collected in a non-chlorinated waste collector recipient, for later disposal. Dichloromethane may be collected in a chlorinated waste collector recipient, for later disposal. Neutralized aqueous solution may be diluted and discarded in the sink.

Results Interpretation and Additional Questions

1. Interpret the ^1H and ^{13}C NMR spectra of compound **3** and dipyrromethane **5** and compare the melting point of these compounds with the ones reported in the literature.[6]
2. Calculate the yields of the synthesis of compounds **3** and **5**.
3. Explain the different outcome observed in the reaction of azoalkene with dihydrofuran and with pyrrole (Scheme 10.5.1).
4. Analyse the mechanism of the synthesis of dipyrromethane **5** outlined in Scheme 10.5.3.
5. The reaction of dihalohydrazone **3** with pyrrole (17 equiv.) in dichloromethane at room temperature for 66 h leads to compound **5** in only 18% yield. Make a comparison between these observations and the results obtained performing this reaction using water as solvent.

References

1. L. R. Milgrom, *The Colours of Life An Introduction to the Chemistry of Porphyrins and Related Compounds*, Oxford University Press, Oxford, 1997.
2. J. F. Lovell, T. W. B. Liu, J. Chen and G. Zheng, *Chem. Rev.*, 2010, **110**, 2839–2857.
3. D. T. Gryko, D. Gryko and C.-H. Lee, *Chem. Soc. Rev.*, 2012, **41**, 3780–3789.
4. N. A. M. Pereira and T. M. V. D. Pinho e Melo, *Org. Prep. Proced. Int.*, 2014, **46**, 183–213.
5. O. A. Attanasi, L. D. Crescentini, G. Favi, P. Filippone, F. Mantellini, F. R. Perrulli and S. Santeusanio, *Eur. J. Org. Chem.*, 2009, 3109–3127.
6. S. M. M. Lopes, A. F. Brigas, F. Palacios, A. Lemos and T. M. V. D. Pinho e Melo, *Eur. J. Org. Chem.*, 2012, 2152–2160.
7. N. A. M. Pereira, S. M. M. Lopes, A. Lemos and T. M. V. D. Pinho e Melo, *Synlett*, 2014, **25**, 423–427.
8. R. N. Butler and A. G. Coyne, *Chem. Rev.*, 2010, **110**, 6302–6337.

10.6. Application of 2,4,6-Trioxo-pyrimidin-5-ylidene Alditol in the Synthesis of Pyrano[2,3-*d*]pyrimidine Containing a Sugar Moiety by Hetero-Diels–Alder Reaction

Aleksandra Pałasz* and Dariusz Cież

Department of Organic Chemistry, Faculty of Chemistry, Jagiellonian University, Ingardena 3 St, 30-060 Kraków, Poland. *E-mail: palasz@chemia.uj.edu.pl

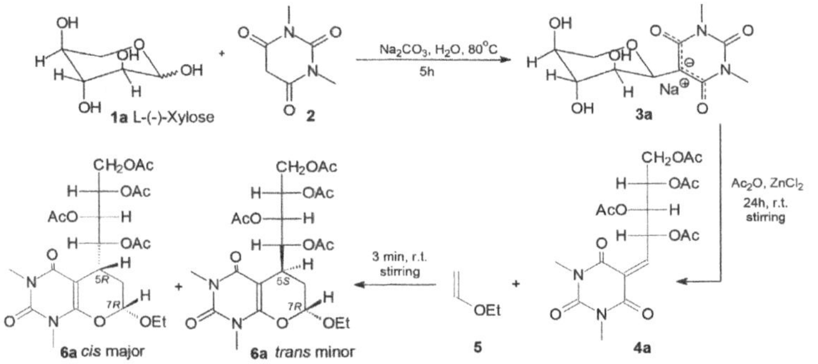

Number of sessions (duration of each session)	Hazard level	Difficulty level	Level of study
4 (7 h + 2 h + 2 h + 6 h)	High	Medium	Advanced

Class names Carbonyls, hydroxyls, carbamoyl groups

Concepts involved This experiment describes a method for the synthesis of enantiomerically pure *cis* and *trans* diastereoisomers of pyrano[2,3-*d*]pyrimidine (fused uracil) with a sugar moiety by solvent-free hetero-Diels–Alder cycloaddition of *O*-acetylated pyrimidin-5-ylidene alditol with vinyl ether

Chemicals needed L-(−)-Xylose, 1,3-dimethylbarbituric acid, sodium carbonate, acetic anhydride, anhydrous zinc chloride, ethyl vinyl ether, anhydrous magnesium sulfate, dichloromethane, ethanol, methanol, *tert*-butyl methyl ether, petroleum ether (40–60 °C)

Equipment and experimental techniques involved Apparatus for reaction with temperature (80 °C) and magnetic stirring, column chromatography, liquid–liquid extraction

Keywords Column chromatography, hetero-Diels–Alder cycloaddition, Knoevenagel condensation, NMR, pyrano[2,3-*d*]pyrimidine, solvent free reaction

Background

Uracil derivatives are one of the most researched areas in medicinal chemistry. Uracil (Scheme 10.6.1) is a common and naturally occurring pyrimidine derivative and one of the four nucleobases in the nucleic acid of RNA. In RNA, uracil binds to adenine *via* two hydrogen bonds. In DNA, the uracil nucleobase is replaced by thymine. Uracils are considered as privileged structures in

Comprehensive Organic Chemistry Experiments for the Laboratory Classroom
Edited by Carlos A. M. Afonso, Nuno R. Candeias, Dulce Pereira Simão, Alexandre F. Trindade, Jaime A. S. Coelho, Bin Tan, and Robert Franzén.

Published by the Royal Society of Chemistry, www.rsc.org

drug discovery, having a wide array of biological activities. Antiviral and anti-tumor are the two most widely reported activities of uracil analogs;[1,2] however, they also possess herbicidal, insecticidal and bactericidal activities.[3–5]

Pyran derivatives are common structural subunits in various important natural products, including carbohydrates, alkaloids, polyether antibiotics, pheromones, and iridoids.[6,7] The preparation of the compounds containing a pyran and an uracil ring poses a significant synthetic challenge. Recently, it was shown that fused uracils such as pyrano[2,3-*d*]pyrimidines with an aryl substituent at carbon C5 in the ring system can be efficiently synthesized by hetero-Diels–Alder reactions of 5-arylidene derivatives of barbituric acids with the vinyl ethers (Scheme 10.6.2).[8]

The Diels–Alder reaction formally describes the addition of 1,3-diene and dienophile to give a six-membered ring system with one or two double bonds depending on the type of dienophile used. This synthetic procedure meets the requirements of a modern synthetic method by showing an excellent chemo- and regioselectivity as well as a high simple and induced diastereoselectivity in many cases. The scope of the Diels–Alder reaction is very high – not only allowing the synthesis of cyclohexenes, but also giving access to a multitude of different heterocycles.[7] According to an *endo-* or *exo*-orientation of the dienophile and an (*E*)- or (*Z*)-configuration of the heterodiene, four different transitions structures can be discussed.[7] The hetero-Diels–Alder reaction of α,β-unsaturated carbonyl compounds representing an 1-oxa-1,3-butadiene system with electron rich alkenes such as enol ethers gives excellent access to 2-substituted 3,4-dihydro-2*H*-pyrans and belongs to the Diels–Alder reactions with inverse electron demand.

Introducing a sugar moiety instead of an aryl group at the C5 position of pyrano[2,3-*d*]pyrimidine can increase the potential pharmacological activity of the fused uracil. In this experiment, first 5-ylidene barbituric acids bearing the carbohydrate substituent are constructed. The Knoevenagel condensation[9–11] is a common synthetic method for formation of arylidene derivatives of active methylene compounds by reaction with various aromatic and heteroaromatic aldehydes. The Knoevenagel condensation under aqueous conditions is a very convenient method for the preparation of pure *C*-glycosidic ketones in one step directly from unprotected sugar.[12–14] The use of water-soluble unprotected carbohydrates allows the synthesis of compounds without a protection–deprotection protocol. Knoevenagel condensation followed by hetero-Diels–Alder (HDA) reaction of α,β-unsaturated carbonyl compounds representing an 1-oxa-1,3-butadiene system with enol ethers is an efficient method for the synthesis of 3,4-dihydro-2*H*-pyrans.[15–23]

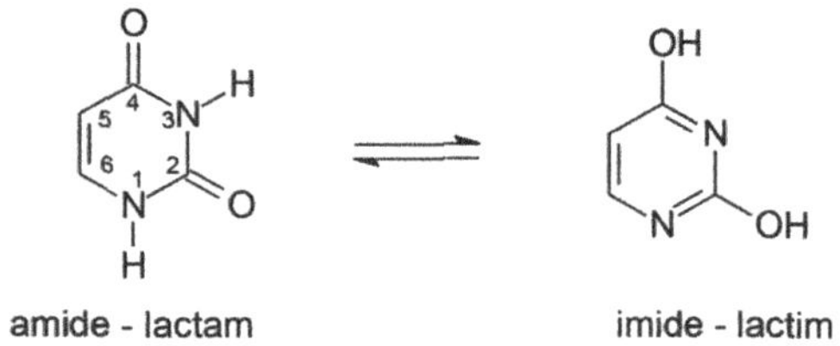

Scheme 10.6.1 Structure of uracil with numbering system. Uracil tautomers.

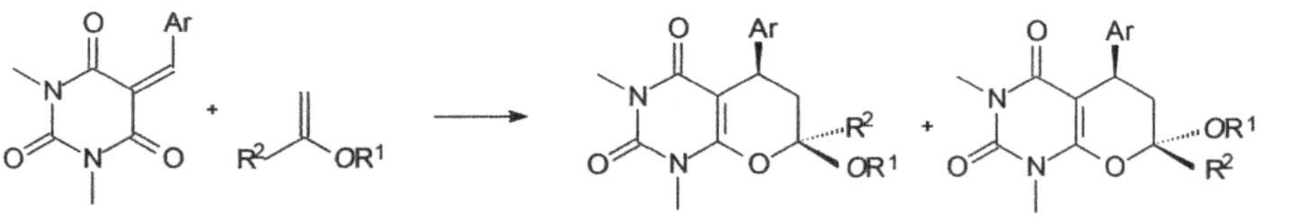

Scheme 10.6.2 Hetero-Diels–Alder reactions of 5-arylidene derivatives of barbituric acids with the vinyl ethers.

The objective of this experiment is the synthesis of fused uracils – pyrano[2,3-*d*]pyrimidines containing a sugar moiety by inverse-electron demand hetero-Diels–Alder reactions of the 5-ylidene derivatives of an *N*,*N*′-dimethylbarbituric acid derivative and ethyl vinyl ether.

Additional Safety

All sessions should be done in a fume hood, with proper safety equipment, such as protective clothing, gloves and safety goggles. 1,3-Dimethylbarbituric acid is harmful if swallowed and causes serious eye damage. Acetic anhydride is a flammable reagent, harmful if swallowed or if inhaled and causes severe skin burns and eye damage. Zinc chloride is harmful if swallowed, causes severe skin burns and eye damage and is very toxic to aquatic life with long lasting effect. Ethyl vinyl ether is a highly flammable reagent, causes skin irritation, serious eye irritation and may cause respiratory irritation. Ethanol is highly flammable solvent. *tert*-Butyl methyl ether is highly flammable solvent and causes skin irritation. Petroleum ether is a highly flammable reagent, may be fatal if swallowed and enters airways, may cause genetic defects and may cause cancer. Dichloromethane is harmful if swallowed or if inhaled, causes skin irritation, causes serious eye irritation, is suspected of causing cancer and may cause damage to organs through prolonged or repeated exposure. Methanol is highly flammable solvent, toxic if swallowed or if inhaled and toxic in contact with skin, and causes damage to organs.

Experimental Procedure

Laboratory Session 1 (7 h): Knoevenagel Condensations of L-(−)-Xylose (1a) with 1,3-Dimethylbarbituric Acid (2) in Water Solution in the Presence of Sodium Carbonate (Ref. 12)

1. In a fume hood heat an oil bath to maintain a temperature of 80 °C on a magnetic hot plate.
2. Prepare a 100 mL round-bottomed flask with a magnetic stir bar, and add 1.56 g (10 mmol) of 1,3-dimethylbarbituric acid (2), 30 mL of water, 1.1 g (10 mmol, 1 equiv.) of sodium carbonate and 1.5 g (10 mmol, 1 equiv.) of L-(−)-xylose (**1a**).
3. Adapt a water condenser to the round-bottomed flask and clamp the flask in the preheated oil bath. Stir the reaction for 5 h, maintaining a constant temperature of 80 °C.
4. Remove the flask from the oil bath and cool it to room temperature.
5. Concentrate the reaction mixture under reduced pressure in a rotary evaporator to 5 mL and add 15 mL of ethanol.
6. Collect precipitated product **3a** by filtration under reduced pressure and dry the compound at 80 °C under vacuum until next session.

Laboratory Session 2 (2 h): Determination of Yield of Sodium 5-Glycopyranosyl-1,3-dimethylbarbiturate (3a) and Acetylation of Knoevenagel Condensation Product (Ref. 12)

1. Weigh the dried product **3a** and calculate the yield.
2. In a fume hood prepare a 100 mL round-bottomed flask with a magnetic stir bar, and add 50 mL of acetic anhydride, 10.0 g of anhydrous zinc chloride (73.0 mmol, 18.25 equiv.), and 1.2 g (4 mmol) of sodium 5-glycopyranosyl-1,3-dimethylbarbiturate (**3a**). Close the flask with ground-in stopper.
3. In a fume hood stir the reaction mixture for 1 h on a magnetic stirrer and leave it standing for 24 h at room temperature.

Laboratory Session 3 (2 h): Isolation of Acetylation Product 4a

1. In a fume hood pour the reaction mixture into a beaker containing 100 mL of ice-water.
2. Extract the mixture with 3 × 20 mL of dichloromethane.
3. Dry the combined organic layers over anhydrous $MgSO_4$ and evaporate the solvent under reduced pressure.
4. Purify the oily residue by crystallization from 15 mL of ethanol. Cool the solution in ice or in a freezer.
5. Collect precipitated product **4a** by filtration under reduced pressure and leave it drying under vacuum at room temperature until the next session.

Laboratory Session 4 (6 h): Yield Determination of Acetylation Product 4a and Synthesis of Pyrano[2,3-*d*]pyrimidine 6a by Solvent-Free Hetero-Diels–Alder Reaction (Ref. 23)

1. In a fume hood prepare a 50 mL round-bottomed flask with a magnetic stir bar, and add 0.91 g (2 mmol) of alditol (**4a**) and 2.88 g, 3.8 mL (40 mmol, 20 equiv.) of ethyl vinyl ether (**5**) and close the flask with a ground-in stopper.
2. Stir the reaction mixture without solvent at room temperature for 3 min. Solid alditol (**4a**) dissolves in excess of ether **5** while stirring and the new product **6a** precipitates.
3. Control the progress of the reaction by TLC on a silica gel plate using dichloromethane/methanol (20 : 1, v/v).
4. Evaporate of the excess of ether **5** under reduced pressure in a rotary evaporator.
5. Record the ^1H NMR spectrum of the crude mixture of the *cis*/*trans* diastereoisomers of the pyrano[2,3-*d*]pyrimidines **6a**.
6. Isolate the major diastereoisomer (*cis*) of pyrano[2,3-*d*]pyrimidine **6a** from the crude reaction mixture by crystallization from *tert*-butyl methyl ether/petroleum ether (3 : 1, v/v).
7. Collect precipitated major product **6a** by filtration under reduced pressure and dry the sample under vacuum at room temperature in a vacuum drier.
8. To isolate the minor diastereoisomer (*trans*) of pyrano[2,3-*d*]pyrimidine **6a**, take the filtrate obtained after filtration of the major *cis* diastereoisomer of **6a** and evaporate the solvent under reduced pressure in a rotary evaporator.
9. Separate and purify the mixture of *cis*/*trans* diastereoisomers **6a** by column chromatography on silica gel using dichloromethane/methanol mixture as eluent (20 : 1, v/v), column length 60 cm. Collect fractions in test-tubes and control the composition of each fraction by TLC (CH_2Cl_2/CH_3OH (20 : 1)). Collect the fractions with the same composition.
10. Weigh the desired major, *cis* diastereoisomer purified by crystallization and calculate the yield. Measure the melting points, and record ^1H and ^{13}C NMR spectra of pure *cis* and *trans* diastereoisomers of **6a** to confirm the compound purity.

Waste Disposal

The aqueous solutions may be diluted and discarded in the sink. The aqueous solution with $ZnCl_2$ must be collected in a zinc compounds waste collector recipient, for posterior destruction. Organic solvents may be reused or collected in chlorinated or non-chlorinated waste collector recipients, for posterior destruction.

Results Interpretation and Additional Questions

1. Calculate the yield of each reaction.
2. Interpret the ^1H and ^{13}C NMR spectra of the pure *cis* and *trans* diastereoisomers of the pyrano[2,3-*d*]pyrimidine.

3. Determine the ratios of the *cis*/*trans* diastereoisomers of the pyrano[2,3-*d*]pyrimidines on the basis of ^{1}H NMR spectra of crude product.
4. Propose and present the mechanism of formation of *C*-glycosylbarbiturate (**3a**).
5. Compare the melting points of the prepared *cis*/*trans* diastereoisomers of the pyrano[2,3-*d*] pyrimidines with the ones reported in the literature.[23]

References

1. W. B. Parker, *Chem. Rev.*, 2009, **109**, 2880.
2. T. Pathak, *Chem. Rev.*, 2002, **102**, 1623.
3. R. Soladino, C. Crestini, A. T. Palamara, M. C. Danti, F. Manetti, F. Corelli, E. Garaci and M. Botta, *J. Med. Chem.*, 2001, **44**, 4554.
4. C. Zhi, Z. Long, A. Manikowski, N. C. Bron, P. M. Tarantino, K. Holm, E. J. Dix, G. E. Wright, K. A. Foster, M. M. Butler, W. A. LaMarr, D. J. Skow, I. Motorina, S. Lamothe and R. Storer, *J. Med. Chem.*, 2005, **48**, 7063.
5. S. Fustero, J. Piera, J. F. Sanz-Cervera, S. Catalan and A. C. Ramirez, *Org. Lett.*, 2004, **6**, 1417.
6. D. L. Boger and S. N. Weinreb, *Hetero Diels–Alder Methodology in Organic Synthesis*, Academic Press, San Diego, 1987.
7. L. F. Tietze and G. Kettschau, *Top. Curr. Chem.*, 1997, **189**, 1.
8. A. Pałasz, *Monatsh. Chem.*, 2008, **139**, 1397.
9. E. Knoevenagel, *Chem. Ber.*, 1896, **29**, 172.
10. E. Knoevenagel, *Chem. Ber.*, 1898, **31**, 730.
11. L. F. Tietze and U. Beifuss, in *Comprehnsive Organic Synthesis*, ed. B. M. Trost, I. Fleming and C. H. Heathcock, Pergamon, Oxford, 1991, vol. 2, ch. 1.11, p.341.
12. M. A. Gonzalez, J. L. J. Requejo and J. C. P. Albarran, *Carbohydr. Res.*, 1986, **158**, 53.
13. F. Rodrigues, Y. Canac and A. Libineau, *Chem. Commun.*, 2000, 2049.
14. J. Wang, Q. Li, Z. Ge and R. Li, *Tetrahedron*, 2012, **68**, 1315.
15. L. F. Tietze, U. Hartfiel, T. Hubsch, E. Voss, K. Bogdanowicz-Szwed and J. Wichmann, *Liebigs Ann. Chem.*, 1991, 275.
16. L. F. Tietze, *J. Heterocycl. Chem.*, 1990, **27**, 47.
17. K. Bogdanowicz-Szwed and A. Pałasz, *Monatsh. Chem.*, 2001, **132**, 393.
18. A. Pałasz, *Org. Biomol. Chem.*, 2005, **3**, 3207.
19. A. Pałasz, K. Jelska, M. Ożóg and P. Serda, *Monatsh. Chem.*, 2007, **138**, 481.
20. A. Pałasz and K. Bogdanowicz-Szwed, *Monatsh. Chem.*, 2008, **139**, 647.
21. A. Pałasz, *Synthesis*, 2010, **23**, 4021.
22. A. Pałasz and T. Pałasz, *Tetrahedron*, 2011, **67**, 1422.
23. A. Pałasz, J. Kalinowska-Tłuścik and M. Jabłoński, *Tetrahedron*, 2013, **69**, 8216.

10.7. Synthesis of a Spiroisoxazoline Oxindole by 1,3-Dipolar Cycloaddition

Carlos J. A. Ribeiro, Rui Moreira, and Maria M. M. Santos*

Research Institute for Medicines (iMed.ULisboa), Faculty of Pharmacy, Universidade de Lisboa, Av. Professor Gama Pinto, 1649-003 Lisbon, Portugal. *E-mail: mariasantos@ff.ulisboa.pt

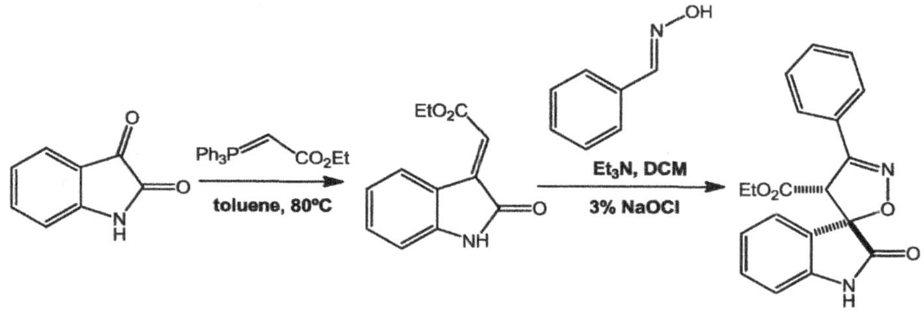

Number of sessions (duration of each session)	Hazard level	Difficulty level	Level of study
2 (8 h + 4 h)	Moderate	High	Advanced

Class names Ketones, oximes, alkenes, nitrile oxides

Concepts involved Wittig reaction, pericyclic reaction, 1,3-dipolar cycloaddition, dipolarophile, 1,3-dipole, HOMO, LUMO

Chemicals needed Isatin, (carbethoxymethylene)triphenylphosphorane, (*E*)-benzaldehyde oxime, sodium hypochlorite solution ~3%, triethylamine, toluene, ethyl acetate, *n*-hexane, dichloromethane, sodium chloride, diethyl ether, silica gel, deuterated chloroform

Equipment and experimental techniques involved Stirring and heating plate, condenser, rotary evaporator, liquid–liquid extraction, column chromatography, NMR spectrometer, melting point apparatus

Keywords 1,3-Dipolar cycloaddition, flash chromatography, nitrile oxide, NMR, pericyclic reaction, spirooxindole

Background

A pericyclic reaction is a chemical reaction that proceeds through a cyclic transition state without formation of any intermediates, progressing in concerted fashion.[1] Three major classes of pericyclic reactions can be described: cycloaddition reactions, sigmatropic rearrangements, and electrocyclic reactions.

This experiments focus on one particular case of cycloadditions termed 1,3-dipolar cycloadditions. This class of reactions has been used extensively in organic synthesis, offering a very useful method for the preparation of five-membered ring heterocycles.[2] These pericyclic reactions, described electronically as an allowed [4 + 2] cycloaddition, involve two reagents: the 1,3-dipole

Comprehensive Organic Chemistry Experiments for the Laboratory Classroom
Edited by Carlos A. M. Afonso, Nuno R. Candeias, Dulce Pereira Simão, Alexandre F. Trindade, Jaime A. S. Coelho, Bin Tan, and Robert Franzén.

Published by the Royal Society of Chemistry, www.rsc.org

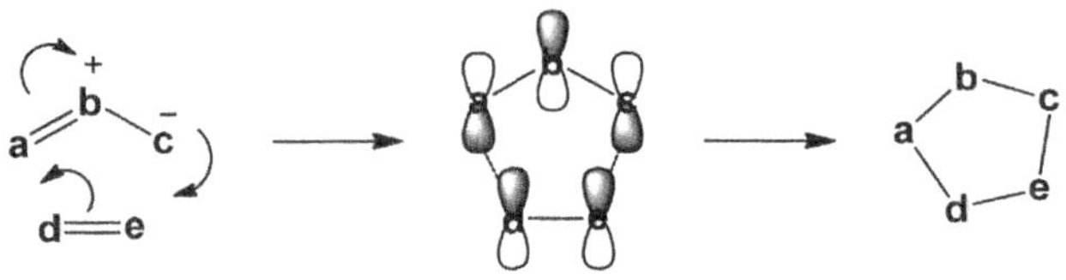

Scheme 10.7.1 Mechanism of the 1,3-dipolar cycloaddition reaction.

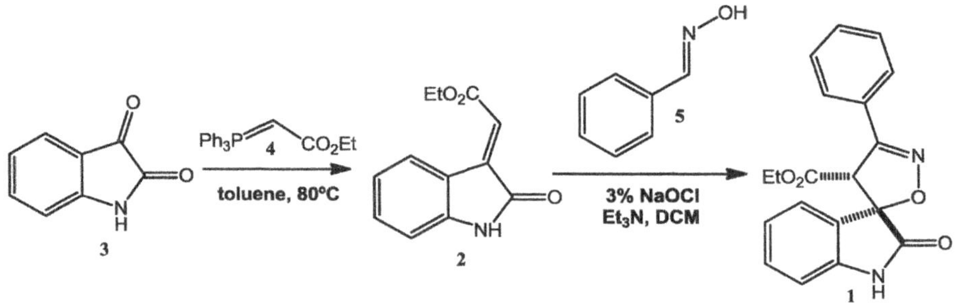

Scheme 10.7.2 Synthesis of spiroisoxazoline oxindole **1**.

and the dipolarophile. The latter (*e.g.* alkenes, alkynes, carbonyls and nitriles) contributes with a pair of π electrons, and the 1,3-dipole with four electrons, in the bond-making process. Mechanistically, the concerted cycloadditions can be represented as going through a transition state in which the four-electron component of the dipole interacts with the 2π-electron component of the dipolarophile, with the formation of two new sigma bonds at the same time (Scheme 10.7.1).[3,4]

The cycloaddition outcome (reactivity, regioselectivity and stereoselectivity) is controlled by the frontier orbitals HOMO (highest occupied molecular orbital) and LUMO (lowest unoccupied molecular orbital) of the two reactants. Furthermore, although two possible HOMO/LUMO pairs can be envisioned, the reaction will take place between the HOMO and LUMO that has the smaller energy gap and that allows a better overlap between them.[1]

In this experiment a nitrile oxide will be used as 1,3-dipole and an alkene (3-methylene indolin-2-one) as dipolarophile, with consequent formation of a new isoxazoline ring. The isoxazolines are useful not only as potential biologically active substances[5,6] (in particular spiroisoxazoline oxindole derivatives,[7,8] such as compound **1**), but also as precursors of stereoselective compounds, including β-hydroxyketones and γ-amino alcohols, due to the easy cleavage of the oxygen–nitrogen bond in the ring.

As nitrile oxides are extremely reactive, and easily dimerize to form the corresponding furoxan, they are normally generated *in situ* in the presence of the dipolarophile.[3]

In the first session dipolarophile **2** will be synthesized by a Wittig reaction from commercially available starting materials **3** and **4**. The 1,3-dipole, benzonitrile oxide, will be generated *in situ* from the corresponding oxime **5** in the presence of triethylamine and aqueous NaOCl in dichloromethane, followed by the cycloaddition reaction in the presence of the dipolarophile **2** (Scheme 10.7.2).

Additional Safety

Both sessions should be done in a fume hood, with proper safety equipment, such as protective clothing, gloves and safety goggles. Isatin, (carbethoxymethylene)triphenylphosphorane, and sodium chloride are not hazardous substances according to the Regulation (EC) No. 1272/2008, and EC directives 67/548/EEC or 1999/45/EC. (*E*)-Benzaldehyde oxime causes skin irritation, serious eye irritation and may cause respiratory irritation. Triethylamine is a highly flammable liquid causing acute oral, dermal, and inhalation toxicity. It may cause skin corrosion, and it is harmful if swallowed. Sodium hypochlorite solution 3–8% (household bleach) can cause skin irritation, and serious eye damage. Toluene and *n*-hexane are highly flammable liquids that cause skin irritation and reproductive toxicity. These two solvents may be fatal if swallowed and enter airways. They may cause damage to organs through prolonged or repeated exposure. Ethyl acetate is also a highly flammable liquid that causes serious eye irritation. Dichloromethane causes skin, respiratory and serious eye irritation. It is suspected of causing cancer, and may cause damage to organs through prolonged or repeated exposure. Diethyl ether is an extremely flammable liquid and it is harmful if swallowed. All five organic solvents presented may cause drowsiness or dizziness. Silica gel is not a hazardous substance according to Regulation (EC) No. 1272/2008 nor dangerous according to Directive 67/548/EEC. However, the use of a mask while handling it is recommended. Chloroform-d_6 is harmful if swallowed and toxic if inhaled. It causes skin and serious eye irritation. It is suspected of causing cancer and damage to fertility or to the unborn child and also may cause damage to organs through prolonged or repeated exposure.

Experimental Procedure

Session 1 (8 h)

Synthesis of (E)-Ethyl 2-(2-Oxoindolin-3-ylidene)acetate

1. In a fume hood set up a stirring and heating plate to a temperature of 80 °C.
2. Prepare a 25 mL round-bottomed flask (tared) with a magnetic stir bar, and add 100 mg (0.68 mmol, 1 equiv.) of isatin, 1 mL of toluene, and 237 mg (0.68 mmol, 1 equiv.) of (ethoxycarbonylmethylene)-triphenylphosphorane.
3. Insert a water condenser into the round-bottomed flask and clamp the flask in the pre-heated plate. Stir the reaction for 30 min, maintaining a constant temperature of 80 °C (verify completion of the reaction by TLC, using as eluent the mixture *n*-Hex : EtOAc (3 : 1)).
4. Remove the flask from the plate and allow it to cool to room temperature.
5. Evaporate the solvent using a rotary evaporator.
6. Add 25 mL of ethyl acetate and wash with 25 mL of brine (2×).
7. Dry the organic phase over anhydrous sodium sulfate.
8. Filter the mixture.
9. Evaporate the solvent using a rotary evaporator.
10. Purify the residue by flash chromatography on silica gel (d = 2 cm; h = 15 cm), using as eluent: *n*-Hex : EtOAc (90 : 10) 100 mL; *n*-Hex : EtOAc (85 : 15) 100 mL; *n*-Hex : EtOAc (80 : 20) 300 mL; *n*-Hex : EtOAc (75 : 25) 50 mL.
11. Evaporate to dryness the column fraction with the desired compound to obtain an orange solid.
12. Before proceeding to the characterization of the compound obtained, perform steps 1 and 2 of the next reaction.
13. Weigh the desired product and calculate the yield. Measure the melting point, and record ^{1}H and ^{13}C NMR spectra to confirm the compound purity. To prepare the NMR sample, add 20 mg of the final compound, and 600 µL of $CDCl_3$ to an NMR tube.

Synthesis of Ethyl 2-Oxo-3′-phenyl-spiro[indoline-3,5′-isoxazoline]-4′-carboxylate

1. Prepare a 25 mL round-bottomed flask (tared) with a magnetic stir bar, and add 50 mg (0.23 mmol, 1 equiv.) of methylene indolin-2-one derivative prepared in the previous step, 1 mL of dichloromethane, 10 µL triethylamine (0.07 mmol, 0.3 equiv.), and 3 mL NaOCl (3–8%; household bleach).

2. On a magnetic stirrer, chill the reaction mixture in an ice bath, and add dropwise a solution of aldoxime (75 μL, 0.69 mmol, 3 equiv.) in 1 mL dichloromethane, prepared previously in an Eppendorf.
3. Stir vigorously the mixture at room temperature for 24 h.

Session 2 (4 h)

4. Verify completion of the reaction by TLC, using as eluent the mixture *n*-Hex:EtOAc (3:1).
5. Add 20 mL of dichloromethane and wash with 20 mL of brine (2×).
6. Dry the organic phase over anhydrous sodium sulfate.
7. Filter the mixture.
8. Evaporate the solvent using a rotary evaporator.
9. Purify the residue by flash chromatography on silica gel (d = 2 cm; h = 15 cm), using as eluent: *n*-Hex:EtOAc (90:10) 50 mL; *n*-Hex:EtOAc (85:15) 50 mL; *n*-Hex:EtOAc (80:20) 200 mL; *n*-Hex:EtOAc (75:25) 200 mL; *n*-Hex:EtOAc (70:20) 50 mL.
10. Evaporate to dryness the column fraction with the desired compound.
11. Wash the residue with diethyl ether to eliminate the minor diastereoisomer.
12. Weigh the desired product (white powder) and calculate the yield. Measure the melting point, and record ^{1}H and ^{13}C NMR spectra to confirm the compound purity. To prepare the NMR sample, add 20 mg of the final compound and 600 μL of $CDCl_3$ to an NMR tube.

Waste Disposal

Hexane, toluene, ethyl acetate, and diethyl ether have to be collected in a non-halogenated waste collector recipient, and dichloromethane in a halogenated waste collector recipient for posterior destruction. The aqueous phases resulting from liquid–liquid extractions may be discarded in the sink.

Results Interpretation and Additional Questions

1. Interpret the NMR spectra of the pure compounds, and compare the obtained melting points with the literature.[7,9]
2. Propose the reaction mechanism leading to compound **2**.
3. Propose the reaction mechanism leading to the final compound **1**, starting from compounds **2** and **5**.
4. Determine the NMR ratio between the two spirooxindole isomers formed.
5. Based on the NMR spectra of product **1**, discuss the regioselectivity and stereoselectivity of the 1,3-dipolar cycloaddition.

References

1. I. Fleming, *Molecular Orbitals and Organic Chemical Reactions*, Student edn, John Wiley & Sons, Ltd, Chichester, 2009.
2. V. Jager and P. A. Colinas, *The Chemistry of Heterocyclic Compound: Synthetic applications of 1,3-dipolar cycloaddition chemistry towards heterocycles and natural products*, ed. A. Padwa and W. H. Pearson, John Wiley & Sons, Inc., 2002, vol. 59, ch. 6, pp. 361–472.
3. J. Clayden, N. Greeves and S. Warren, *Organic chemistry*, Oxford University Press, 2nd edn, 2012, ch. 34, pp. 877–908.
4. M. Kissane and A. R. Maguire, *Chem. Soc. Rev.*, 2010, **39**, 845.
5. D. Rane and M. Sibi, *Curr. Org. Synth.*, 2011, **8**, 616.
6. G. P. Savage, *Curr. Org. Chem.*, 2010, **14**, 1478.
7. C. J. A. Ribeiro, S. P. Kumar, R. Moreira and M. M. M. Santos, *Tetrahedron Lett.*, 2012, **53**, 281.
8. C. J. A. Ribeiro, J. D. Amaral, C. M. P. Rodrigues, R. Moreira and M. M. M. Santos, *Bioorg. Med. Chem.*, 2014, **22**, 577.
9. H. B. Rasmussen and J. K. MacLeod, *J. Nat. Prod.*, 1997, **60**, 1152.

10.8. Oxonitriles: Four-Step Ozonolysis, Aldol, Conjugate Addition, and Enolate Acylation Sequence

Jesus A. Lujan-Montelongo[a] and Fraser F. Fleming*[b]

[a]Departamento de Química, Centro de Investigación y de Estudios Avanzados (Cinvestav), Avenida Instituto Politécnico Nacional 2508, San Pedro Zacatenco 07360, México D. F., Mexico; [b]Department of Chemistry, Drexel University, 32 South 32nd St., Philadelphia, PA 19104, USA. *E-mail: flemingf@duq.edu

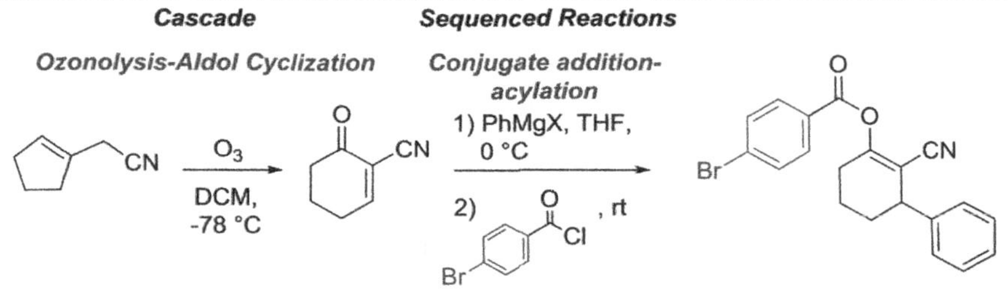

Number of sessions (duration of each session)	Hazard level	Difficulty level	Level of study
3 (3 h each)	High	High	Advanced

Class names Nitriles, carbonyls, enolates, acyl halides, esters, aryl halides

Concepts involved The experiment illustrates an ozonolysis–aldol cyclization cascade and a sequenced conjugate addition–alkylation to prepare an enol ester. Four reactions are involved: ozonolysis, aldol cyclization, conjugate addition, and enolate acylation

Chemicals needed 2-(1-Cyclopentenyl)acetonitrile, oxygen, acetone, dry ice, dimethyl sulfide, phenylmagnesium chloride or bromide solutions, *p*-bromobenzoyl chloride, dichloromethane, tetrahydrofuran, ethyl acetate, sodium chloride, sodium bicarbonate, ammonium chloride, sodium sulfate, silica gel, chloroform-*d*

Equipment and experimental techniques involved Ozonolysis, conjugated addition and enolate trapping, ozonator, magnetic stirrers, manifold (optional), rotary evaporator, separatory funnel, column chromatography (15 mm × 30 cm), 10 mL round bottom flasks, plastic syringes, plastic bin or short Dewar flask for cold bath, syringe needles, test tubes. NMR and IR analysis: NMR tubes, Pasteur pipettes, Pasteur pipette rubber bulb

Keywords Aldol, column chromatography, conjugate addition, enolate acylation, infrared spectroscopy, liquid–liquid extraction, nuclear magnetic resonance, ozonolysis

Background

Ozonolysis is a fundamental reaction infrequently performed in the sophomore organic chemistry laboratory.[1] The two-fold challenge lies in safely and efficiently performing ozonolyses with many students and employing the resulting carbonyl compound in a meaningful reaction. Over several years an ozonolysis protocol has been developed that is safe, rapid, and suitable for a

Comprehensive Organic Chemistry Experiments for the Laboratory Classroom
Edited by Carlos A. M. Afonso, Nuno R. Candeias, Dulce Pereira Simão, Alexandre F. Trindade, Jaime A. S. Coelho, Bin Tan, and Robert Franzén.

Published by the Royal Society of Chemistry, www.rsc.org

group of 20 students. The experiment features the ozonolysis of 2-(1-cyclopentenyl)acetonitrile (**1**) in a robust, multi-step laboratory sequence that illustrates ozonolysis, aldol condensation, conjugate addition, and enolate acylation reactions.

Conceptually, the experiment is divided into two parts: a cascade reaction and a sequenced conjugate addition–enolate acylation. In the first part, the ozonolysis[2] of 2-(1-cyclopentenyl)acetonitrile (**1**) affords an intermediate ozonide **2** that is readily reduced to the ketonitrile **3** upon addition of Me_2S (Scheme 10.8.1). The ketonitrile **3** spontaneously suffers an aldol cyclization–dehydration without the need for added reagents, to afford 6-oxocyclohex-1-ene-1-carbonitrile (**4**).[3] The synthesis of **4** proceeds with yields in excess of 90%, illustrating the power of cascade reactions.[4]

Oxonitrile **4** is a highly reactive dienophile[5] and Michael acceptor.[6] Particularly attractive are the conjugate additions of **4** with Grignard reagents,[7] which proceed without the need for catalysts. The resulting enolates **5** can be alkylated or acylated to afford various six-membered nitriles that are valuable precursors for synthesizing *cis*- and *trans*-decalins (Scheme 10.8.2).[7g]

The ozonolysis–aldol route to 6-oxocyclohex-1-ene-1-carbonitrile (**4**) is virtually quantitative, requiring only a solvent exchange for the subsequent conjugate addition. Adding phenylmagnesium bromide or chloride to the crude oxonitrile **4** followed by *p*-bromobenzoyl chloride[8] affords a crystalline material that is readily isolated in good yield. The sequence is robust, readily halted at several stages, uses inexpensive reagents, and develops practical skills in organic synthesis. Students develop a proficiency with organometallics in the context of a multi-step synthesis, purification and isolation techniques, and the interpretation of analytical data such as IR and NMR spectra. The experiment is readily accomplished in three, 3 hour sessions and is robust and efficient having been tested over four years with cohorts of 20 students.

Additional Safety

Eye protection, laboratory coat and nitrile gloves must be worn while performing this experiment. Mishandling of the dry ice/acetone bath can result in serious burns. Ozone forms explosive peroxides with alkenes and many other organic molecules, thus ozone should be handled in solution at low temperature (perform the ozonolysis in a dry ice/acetone cold bath and use always small scale). Ozone itself is highly explosive. Ozone must be generated in a well-ventilated

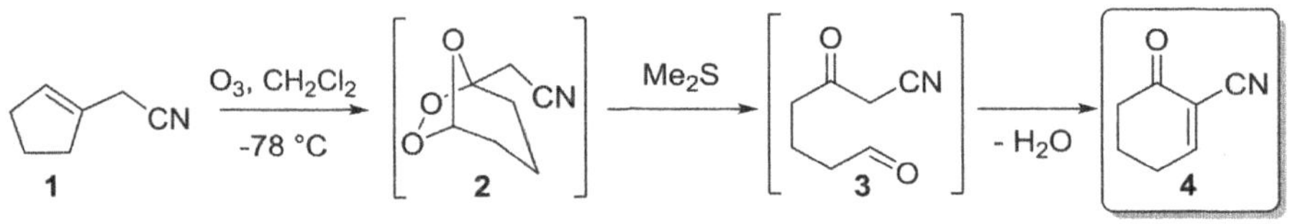

Scheme 10.8.1 Synthesis of 6-oxocyclohex-1-ene-1-carbonitrile.

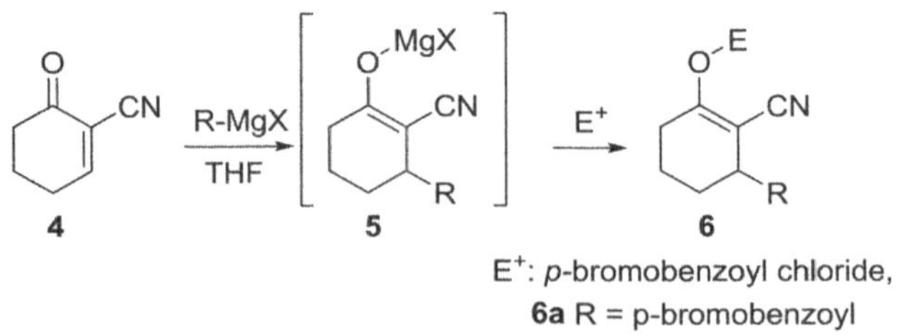

Scheme 10.8.2 1,4-Grignard addition and enolate trapping of oxonitrile **4**.

fume hood because ozone is a strong irritant to the eyes, upper respiratory tract and lungs. Keep the fume hood sash at the lowest recommended level all the time. Reagents should be dispensed in a hood to avoid inhalation of vapors. 2-(1-Cyclopentenyl)acetonitrile causes irritation of the skin, is harmful if inhaled, and if swallowed. Dichloromethane vapors are narcotic in high concentrations and can cause nausea and irritation if inhaled. Dimethyl sulfide has an unpleasant odor even at low concentrations, is an eye and skin irritant, and is harmful if swallowed. Tetrahydrofuran is highly flammable and harmful if inhaled or upon contact with the body. Phenylmagnesium chloride and phenylmagnesium bromide are very reactive and corrosive; they react violently with water and cause bodily harm upon contact or inhalation. *p*-Bromobenzoyl chloride is very corrosive and harmful by contact. Sodium bicarbonate and ammonium chloride can cause serious eye irritation.

Experimental Procedure

Session 1 (3 h): Synthesis of 6-Oxocyclohex-1-ene-1-carbonitrile (4)

1. Weigh out neat 2-(1-cyclopentenyl)acetonitrile (0.2 g, 1.86 mmol) into a 10 mL round-bottomed flask containing a magnetic stirring bar and then add 5 mL of dichloromethane.
2. Attach the ozone delivery device (Figure 1, ESI) and cool the solution to −78 °C by immersing the flask into an acetone/dry ice bath.
3. Carefully, open the manifold port to allow a gentle stream of dry ozone to pass through the solution with the following flow rate: 5 psi O_2 feed, 2 standard liters per minute (SLPM) of ozone enriched oxygen. This will give an approximate flow rate of 0.5 SLPM of O_3 enriched oxygen per port.
4. Continue the ozonolysis until the distinctive blue color of excess ozone is first observed (usually 5–10 min) and then close the ozone delivery port (Figure 2, ESI). Leave the set up attached until all four reaction flasks attached to the manifold turn blue (usually within 10 min).
5. Remove the excess ozone by lowering the ozonator voltage to zero and purging the −78 °C solution with a stream of oxygen (typically about 5–10 min).
6. After the solution becomes colorless, remove the cooling bath and replace the ozonolysis adapter with a rubber septum.
7. Add neat dimethyl sulfide (0.2 mL, 2.73 mmol) *via* syringe to the flask (this addition must be performed in a fume hood!).
8. Allow the mixture to stir at room temperature for 5 min and then store the solution in the dark until the next laboratory period (at least 36 h is required for the aldol cyclization–dehydration).

Session 2 (3 h): Synthesis of 2-Cyano-1,4,5,6-tetrahydro-[1,1′-biphenyl]-3-yl 4-bromobenzoate (6a)

1. Concentrate the resulting solution under reduced pressure using a rotary evaporator keeping the water bath temperature at or slightly below 40 °C. Dilute the resulting thick syrup with 4 mL of ethyl acetate.
2. Wash the resulting solution sequentially with saturated sodium bicarbonate (1 mL), water (1 mL), and saturated sodium chloride solution (1 mL) in a small separatory funnel or test tube. Separate and collect the organic phase.
3. Dry the organic phase by adding solid sodium sulfate, filter the reaction mixture through a filter paper or cotton plug, and then concentrate the solution in a 10 mL round-bottom flask using a rotary evaporator.
4. Place a magnetic stirring bar into the flask containing the dry, crude oxonitrile, and then cap the flask with a septum.

5. Purge the flask with a stream of nitrogen from a tank, a nitrogen filled balloon, or using Schlenk techniques with a vacuum manifold and, if necessary, recap the flask with the rubber septum.
6. Add 2 mL of dry tetrahydrofuran *via* syringe, connect a nitrogen-filled balloon fitted with a syringe by forcing the needle through the septum on the reaction flask, and then cool the flask to 0 °C by immersion in an ice/water bath.
7. Add the PhMgCl or PhMgBr solution by syringe (1.1 equiv. based on the crude weight of oxonitrile).
8. After 5 min, remove the cooling bath, remove the septum and add solid *p*-bromobenzoyl chloride (1.2 equiv.) in one portion.
9. Replace the septum and then stir the mixture. The reaction mixture changes from an orange suspension to a translucent solution.
10. After 10 min, add saturated ammonium chloride solution (5 mL) and stir the reaction mixture for 1 min. Add ethyl acetate (3 mL), separate the two phases, and extract the aqueous phase with ethyl acetate (2 × 3 mL) using a small separatory funnel or a test tube.
11. Combine the organic extracts and then sequentially wash the combined organic phase with saturated sodium bicarbonate solution, water, and saturated sodium chloride solution (3 mL each).
12. Dry the organic phase over anhydrous sodium sulfate and then concentrate the solution. Tightly cap the material until the next session.

Session 3 (3 h): Purification of 2-Cyano-1,4,5,6-tetrahydro-[1,1′-biphenyl]-3-yl 4-Bromobenzoate (6a)

1. Load a 15 mm × 30 cm chromatography column with a hexanes suspension of silica gel (10 g).
2. Add sufficient silica gel to the crude enol ester to form a homogeneous mixture. Load the absorbed enol ester silica gel onto the top of the column and perform the separation by using hexanes : dichloromethane : diethyl ether (90 : 5 : 5) as eluent. Collect 3 mL fractions and identify the test tubes containing the enol ester by TLC analysis (use hexanes : ethyl acetate (85 : 15) as the developing solvent).
3. Combine the test tubes containing the pure enol ester nitrile and then concentrate the combined material. Weigh the pure enol ester nitrile for a yield.

Waste Disposal

All organic waste has to be collected in a non-chlorinated waste collector recipient, except dichloromethane which has to be disposed of in a chlorinated waste collector recipient. Treat all the aqueous waste with excess bleach and dispose in an aqueous waste collector recipient. Additional bleach is useful to destroy traces of dimethyl sulfide from glassware. Solid waste such as used sodium sulfate, used silica gel, and cotton should be disposed of in a solid waste collector recipient. All the sharps should be disposed of in a sharp waste container.

Results Interpretation and Additional Questions

1. Calculate the product yield. Obtain a melting point and compare it with the reported value (133–133.5 °C).
2. Interpret the ^{1}H NMR and IR spectra, identifying all the characteristic functional group signals.
3. Suggest why the oxoketonitrile **3** evolves spontaneously to the enol **5**.
4. Oxonitrile **4** has at least three electrophilic centers. Why does the Grignard attack the olefin and not the ketone?

5. Suggest a reason why the enolate acylation occurs on oxygen to give *O*-acylation instead of *C*-acylation.
6. Rationalize the mechanisms involved in the reaction sequence.

Acknowledgements

Financial support from the National Science Foundation (USA, CHE 1111406 and IRD) is gratefully acknowledged. J. A. Lujan-Montelongo wishes to thank CONACYT (Mexico). Mitch Fedak, Xun Yang, and Dinesh Nath are thanked for their input into experimental optimization of the procedure in earlier undergraduate laboratory sessions. Any opinions, findings and conclusions or recommendations expressed in this material are those of the authors and do not necessarily reflect the views of the National Science Foundation.

References

1. Although not common, examples of ozonolysis experiments have been reported: (a) b. M. Branan, J. T. Butcher and L. R. Olsen, *J. Chem. Ed.*, 2007, **84**, 1979; (b) D. J. Sponholtz, M. A. Walters, J. Tung and J. J. BelBruno, *J. Chem. Ed.*, 1999, **76**, 1712; (c) C. M. Rhoads, G. R. Farquar and W. F. Wood, *J. Chem. Ed.*, 1997, **74**, 1220.
2. (a) R. Criegee, *Angew. Chem., Int. Ed.*, 1975, **14**, 745; (b) C. Geketbeky and S. Berger, *Eur. J. Org. Chem.*, 1998, 1625.
3. (a) F. F. Fleming, A. Huang, V. A. Sharief and Y. Pu, *J. Org. Chem.*, 1997, **62**, 3036.
4. K. C. Nicolaou, D. J. Edmonds and P. G. Bulger, *Angew. Chem., Int. Ed.*, 2006, **12**, 7134.
5. (a) B. Witte, L. Meyer and P. Margaretha, *Helv. Chim. Acta*, 2000, **83**, 554; (b) J.-L. Zhu, K.-S. Shia and H.-J. Liu, *Chem. Commun.*, 2000, 1599 and references included within.
6. F. F. Fleming and B. C. Shook, *Org. Synth.*, 2002, **78**, 254.
7. (a) M.-T. Hsieh, K.-S. Shia, H.-J. Liu and S.-C. Kuo, *Org. Biomol. Chem.*, 2012, **10**, 4609; (b) Y.-C. Wong, M.-T. Hsieh, P. K. Amancha, C.-L. Chin, C.-F. Liao, C.-W. Kuo and K.-S. Shia, *Org. Lett.*, 2011, **13**, 896; (c) C.-L. Chin, C.-F. Liao, H.-J. Liu, Y.-C. Wong, M.-T. Hsieh, P. K. Amancha, C.-P. Chang and K.-S. Shia, *Org. Biomol. Chem.*, 2011, **9**, 4778; (d) M. J. Campbell, P. D. Pohlhaus, G. Ming, K. Ohmatsu and J. S. Johnson, *J. Am. Chem. Soc.*, 2008, **130**, 9180; (e) F. F. Fleming, B. C. Shook, T. Jiang and O. W. Steward, *Tetrahedron*, 2003, **59**, 737; (f) L.-R. Kung, C.-H. Tu, K.-S. Shia and H.-J. Liu, *Chem. Commun.*, 2003, 2490; (g) F. F. Fleming, B. C. Shook, T. Jiang and O. W. Steward, *Org. Lett.*, 1999, **1**, 1547; (h) F. F. Fleming, Y. Pu and F. Terceck, *J. Org. Chem.*, 1997, **62**, 4883.
8. H. O. House, R. A. Auerbach, M. Gall and N. P. Peet, *J. Org. Chem.*, 1973, **38**(3), 514.

10.9. Flash Vacuum Pyrolysis of *o*-Phenylene Sulfite: Formation and Purification of Cyclopentadienone Dimer

R. Alan Aitken* and Caroline E. R. Horsburgh

EaStCHEM School of Chemistry, University of St Andrews, North Haugh, St Andrews, Fife KY16 9ST, UK
*E-mail: raa@st-and.ac.uk

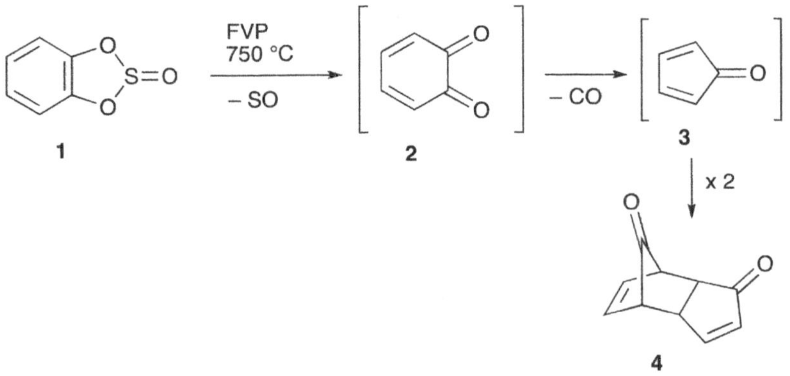

Number of sessions (duration of each session)	Hazard level	Difficulty level	Level of study
2 (3 h each)	High	High	Advanced

Class names Cyclic sulfites, 1,3-dienes

Concepts involved This experiment illustrates a sequence of cycloreversion and thermal extrusion in the gas phase generating a reactive 1,3-diene which undergoes Diels–Alder dimerisation upon condensation in the cold trap. Purification by preparative TLC gives the pure crystalline dimer, which is analysed by mp, IR and NMR

Chemicals needed *o*-Phenylene sulfite (1,3,2-benzodioxathiole 2-oxide)

Equipment and experimental techniques involved Flash vacuum pyrolysis equipment, preparative TLC equipment

Keywords Diels–Alder cycloaddition, flash vacuum pyrolysis, gas-phase extrusion, NMR, preparative TLC

Background

Flash vacuum pyrolysis (FVP) is a valuable but little-known synthetic technique allowing preparation of many unusual products difficult to access by more conventional methods.[1–6] Particularly for processes involving the thermal generation and intramolecular reaction of "reactive intermediates" such as carbenes, nitrenes, arynes and radicals, it avoids the side-reactions commonly occurring when such processes are attempted in solution. This means that the products may be obtained directly in pure form. A range of different experimental set-ups have been described,[1,7] but the one used here is robust and easily constructed from standard laboratory equipment with the addition of a tube furnace and a simple vacuum line with pump. Despite the high temperatures involved

Comprehensive Organic Chemistry Experiments for the Laboratory Classroom
Edited by Carlos A. M. Afonso, Nuno R. Candeias, Dulce Pereira Simão, Alexandre F. Trindade, Jaime A. S. Coelho, Bin Tan, and Robert Franzén.

Published by the Royal Society of Chemistry, www.rsc.org

(500–1000 °C), the high vacuum flow conditions mean that each molecule only spends a few milliseconds in the hot zone so the technique is actually quite gentle and sensitive products including chiral compounds can be obtained in good yield without significant decomposition or racemisation. The experiment involves pyrolysis of the simple heterocyclic compound *o*-phenylene sulfite (**1**) which is readily prepared by heating catechol (benzene-1,2-diol) with thionyl chloride. When it is subjected to FVP at 750 °C and 10^{-1}–10^{-2} Torr, it first loses sulfur monoxide to give *o*-benzoquinone (**2**) and this then loses carbon monoxide. The resulting cyclopentadienone (**3**) is a reactive 1,3-diene, which readily undergoes Diels–Alder cycloaddition with itself to form a [4 + 2] dimer.[8,9] The SO fragment eliminated in the first step is an unusual and unstable oxide of sulfur, which disproportionates by a complex mechanism to give ultimately SO_2 and elemental sulfur. The cyclopentadienone dimer (**4**) will be purified by preparative thin-layer chromatography. This technique is ideal for separation of complex mixtures on a scale of 0.1–1 g and involves running the mixture, applied as a linear band, up a large (20 × 20 cm) TLC plate of silica or alumina. After visualisation under UV light the component bands can be scraped off individually and the products extracted from the support material.

Additional Safety

The use of a furnace heated to 750 °C involves a significant fire risk in the presence of flammable materials such as organic solvents and the pyrolysis process also produces toxic gases. The pyrolysis equipment must be situated in a dedicated fume hood and no flammable solvents should be used nearby. Heat-resistant gloves should be used when assembling and manipulating the pyrolysis equipment. Liquid nitrogen-cooled traps should never be left open to the air at any time. The starting compound *o*-phenylene sulfite (**1**) is a mild lachrymator and should be weighed out in the fume hood. Care should be taken when using the UV lamp to visualise the TLC plate to avoid eye and skin exposure.

Experimental Procedure

Laboratory Session 1 (3 h): FVP of *o*-Phenylene Sulfite (1)

1. Make sure that the furnace has been switched on and has stabilised at 750 °C.
2. Close taps A and B (Figure 10.9.1) and fill the Dewar flask on the trap nearest the pump with liquid nitrogen. When it has cooled down, switch on the pump.
3. While the system is pumping down, weigh out *o*-phenylene sulfite (0.50 g) (kept in the refrigerator) into the inlet tube.
4. Now close tap C and open taps B and D and check the vacuum using the vacuum gauge. Keep tap D closed except when actually taking a pressure reading.

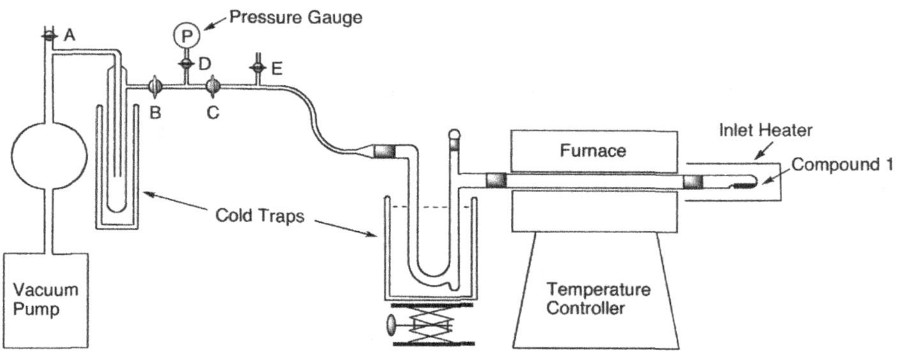

Figure 10.9.1 Schematic diagram of FVP apparatus.

5. Carefully apply silicone grease to the three joints of the product trap and to the inlet tube and assemble these as shown in Figure 10.9.1. The system must be strain free: make sure the furnace tube is straight and sitting lightly in the furnace. Use heat resistant gloves to avoid burning yourself and remember to have the small ridge of the inlet tube at the bottom (this prevents the liquid running into the furnace).
6. Fill the wide Dewar flask with liquid nitrogen and, after checking that taps D and E are closed, slowly open tap C and immediately raise the Dewar by means of a lab-jack around the product collection cold trap. When the Dewar is fully raised give each joint a slight turn to ensure a good seal and then open tap D and record the vacuum. After closing tap D, the pyrolysis is now ready to be carried out.
7. Carefully slide the inlet heater over the inlet until it is against the furnace and switch on the heater to the required setting.
8. As the pyrolysis proceeds the products will be seen collecting in bands according to volatility in the cold-trap. When the substrate has all disappeared from the inlet (20–30 min) the pyrolysis is complete.
9. Close tap C, open tap A and immediately switch off the pump and remove the liquid nitrogen Dewar from the trap by the pump (*in that order*).
10. For sensitive products the system can now be filled with nitrogen gas through tap E; however, in this case the product is stable, so slowly open tap E to the air. Immediately lower the Dewar flask from the product trap. Slide the inlet heater off and using heat resistant gloves disconnect the product trap from the furnace tube and the vacuum line and allow it to warm up *in the fume hood*. (If the trap is stuck to the furnace tube consult a demonstrator – *do not use force*.)
11. Switch off the inlet heater and furnace. When the product has warmed up, wash it out using dichloromethane (*ca.* 5 cm^3) into a small rotary evaporator flask and evaporate off the solvent. Clean out the inlet and trap. The product can be kept at this stage in the stoppered flask at room temperature until the next laboratory session.

Laboratory Session 2 (3 h): Purification of Product 4 by Preparative TLC

1. Redissolve the crude product in dichloromethane (1–2 cm^3) and using the template provided and a dropper apply the solution very carefully in as narrow a line as possible about 1 cm up from the end of the silica on a preparative TLC plate (Figure 10.9.2). Avoid digging a trench in the silica with the end of the dropper and allow each application to dry before making the next to maintain a narrow band.
2. When all the product has been applied place the plate carefully in the developing tank containing just enough of a 3 : 1 mixture of diethyl ether and petroleum ether (40–60 °C) to come up to the silica all the way along the plate. Allow the plate to develop until the solvent front comes within 1 cm of the top, which should take about 30–40 min.
3. After allowing the solvent to dry off in a fume hood, observe the plate under the UV lamp (254 nm). (*Danger*: do not look directly at the UV light or allow it to shine on your skin.) Refer to Figure 10.9.3 and check with a demonstrator that you can identify which band corresponds to the product and mark it off using the corner of a spatula.
4. Carefully scrape off the band into a conical flask using a powder funnel in a fume hood. Grind the silica to a fine powder using a spatula then add dichloromethane (30–40 cm^3).

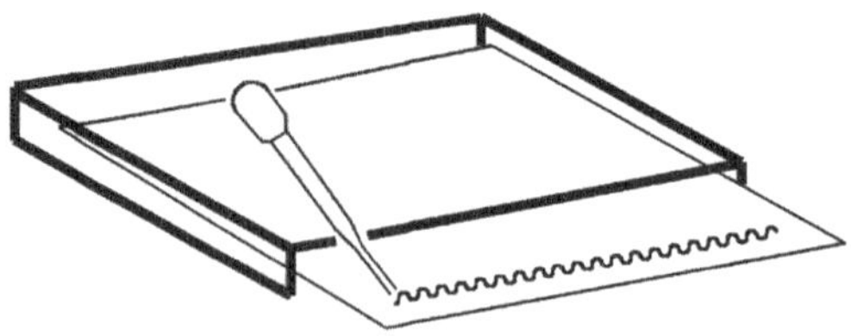

Figure 10.9.2 Use of plastic template to achieve even application of sample to plate.

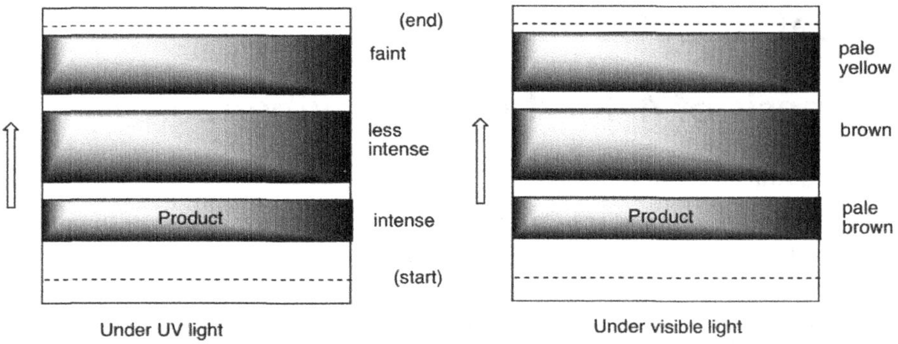

Figure 10.9.3 Appearance of preparative TLC plate after development.

5. After the flask has stood for 20 min, filter off the silica with suction using a sintered glass funnel, wash with CH_2Cl_2 (5 cm^3) and evaporate the combined solution.
6. Record the yield, mp and IR and ^1H NMR spectra of the product.

Waste Disposal

Waste organic solvents should be disposed of in the waste solvent container for incineration. Used silica should be sent for solid waste disposal.

Results Interpretation and Additional Questions

1. Give arrow-pushing mechanisms both for formation of intermediate **2** from **1** and for formation of the product **4** from diene **3**. What other diene forms a dimer of similar structure?
2. When the product trap was warming up you might have seen a yellow solid in the coldest region of the trap. What is this and why did it not collect in the region corresponding to its boiling point?
3. Interpret the IR and ^1H NMR spectra of the product **4** and, in particular, explain why there are two separate IR stretches in the C=O region and why the NMR signals in both the alkene ($\delta > 6.0$) and alkane (δ 2.0–3.5) regions occur in a 3 : 1 ratio.
4. Write an equation for the formation of the substrate **1** from benzene-1,2-diol (catechol) and thionyl chloride. How might the corresponding cyclic sulfate **5** be prepared and how might it behave upon FVP?

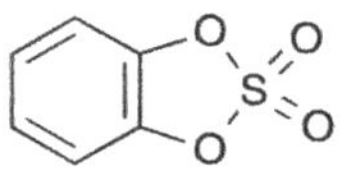

5

References

1. R. F. C. Brown, *Pyrolytic Methods in Organic Chemistry*, Academic Press, New York, 1980.
2. C. Wentrup, *Adv. Heterocycl. Chem.*, 1981, **28**, 231–361.
3. U. E. Wiersum, *Recl. Trav. Chim. Pays-Bas*, 1982, **101**, 317–332, 365–381.
4. M. Karpf, *Angew. Chem., Int. Ed. Engl.*, 1986, **25**, 414–430.
5. R. A. Aitken and A. W. Thomas, in *Chemistry of the Functional Groups*, ed. S. Patai, John Wiley & Sons, Inc., New York, 1997, supplement A3, pp. 473–536.
6. H. McNab, *Aldrichim. Acta*, 2004, **37**, 19–26.
7. C. Wentrup, *Aust. J. Chem.*, 2014, **67**, 1150–1165.
8. D. C. DeJongh, R. Y. Van Fossen and C. F. Bourgeois, *Tetrahedron Lett.*, 1967, **8**, 271–276.
9. D. C. DeJongh and R. Y. Van Fossen, *J. Org. Chem.*, 1972, **37**, 1129–1135.

11. Radical Reactions

11.1. Synthesis of 4-Methoxymethylbenzoic Acid

Dulce Simão* and Ana C. Cerdeira

Department of Chemistry, Instituto Superior Técnico, University of Lisbon, Centro de Química Estrutural, Av. Rovisco Pais, 1, 1049-001 Lisbon, Portugal. *E-mail: dulcesimao@tecnico.ulisboa.pt

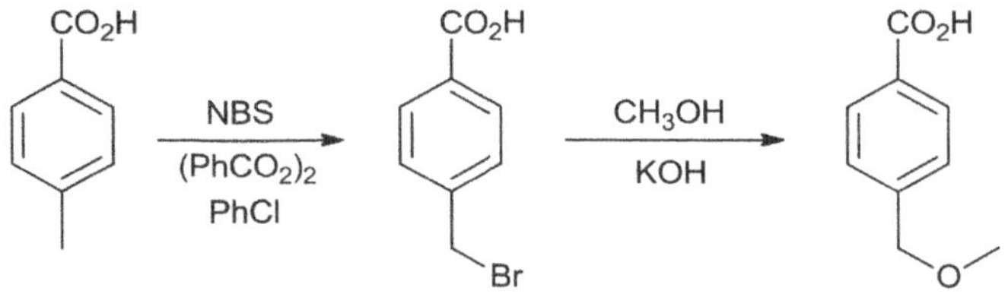

Number of sessions (duration of each session)	Hazard level	Difficulty level	Level of study
2 (3 h each)	Moderate	Low	Intermediate
Class names Alkanes, haloalkanes, *N*-halosuccinimides, peroxides, alcohols, ethers		**Concepts involved** Radical bromination using *N*-bromosuccinimide and nucleophilic substitution on a saturated carbon atom	
Chemicals needed 4-Methylbenzoic acid, *N*-bromosuccinimide (NBS), dibenzoyl peroxide, chlorobenzene, petroleum ether (40–60 °C), methanol, potassium hydroxide, hydrogen chloride			
Equipment and experimental techniques involved Crystallization, IR, stirrer and oil bath, reflux apparatus, rotary evaporation, vacuum filtration, ^{1}H NMR, melting point			
Keywords Crystallization, IR, NBS, nucleophilic substitution, radical bromination			

Background

Bromoalkanes are usually formed *via* a radical reaction of an alkane with bromine in the presence of light or heat. In this work, *N*-bromosuccinimide (NBS) is used as the bromine source and benzoyl peroxide as the initiator that undergoes homolytic cleavage of the O–O bond simply on heating. The radicals formed are unstable and lead to formation of CO_2 and phenyl radicals by a fragmentation process.[1a,2] Under these conditions a small concentration of bromine radicals is generated from NBS, triggering a chain reaction where molecular bromine is released during the reaction. The HBr produced in the substitution reaction reacts with the NBS to produce succinimide and bromine in low concentration[1b,3a,4a,5] (Scheme 11.1.1).

The hydrogen atoms of the methyl group of 4-methylbenzoic acid are bonded to a benzylic carbon atom, adjacent to an aromatic ring, being easily substituted in free radical reactions once the radical formed is stabilized by resonance.[3b,6] The second step of this synthesis illustrates

Comprehensive Organic Chemistry Experiments for the Laboratory Classroom
Edited by Carlos A. M. Afonso, Nuno R. Candeias, Dulce Pereira Simão, Alexandre F. Trindade, Jaime A. S. Coelho, Bin Tan, and Robert Franzén.

Published by the Royal Society of Chemistry, www.rsc.org

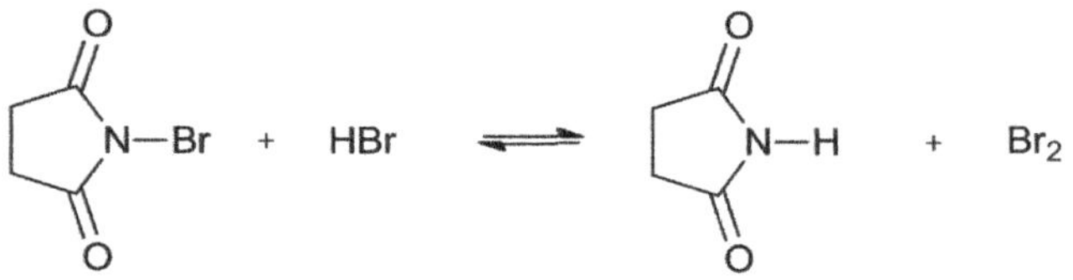

Scheme 11.1.1 Bromine formation with NBS.

how benzyl halides undergo ready nucleophilic substitution. A benzylic compound can react either by S_N1 or S_N2 mechanisms.[4b] The presence of electron-donating groups in the *ortho* or *meta* positions in relation to the leaving group will favor the S_N1 mechanism stabilizing the carbocation intermediate. On the other hand, strong electron-withdrawing groups will favor the S_N2 mechanism stabilizing the transition state.[1c] In this experiment 4-bromomethylbenzoic acid contains the electron-withdrawing carboxyl group (COOH), which is readily converted into 4-methoxymethylbenzoic acid using methanol in basic medium.[5,7–9]

Additional Safety

The use of a fume hood is not mandatory for this experiment. However, care should be taken when handling benzoyl peroxide since it is an oxidizing agent and potentially explosive if heated or ground as the dry solid. Chlorobenzene and petroleum ether are flammable solvents and should be handled carefully as well potassium hydroxide that is corrosive. Methanol is toxic, so precautions should be taken during evaporation in a rotary evaporator in order to avoid vapor inhalation.

Experimental Procedure

Session 1 – Preparation of 4-Bromomethylbenzoic Acid

1. To a 100 mL round-bottom flask add 4-methylbenzoic acid (2.72 g, 20 mmol) and *N*-bromosuccinimide (3.60 g, 20 mmol).
2. Add benzoyl peroxide (0.20 g, 0.83 mmol) through a solid addition funnel, followed by chlorobenzene (25 mL) washing down any solid that may have adhered to the sides of the flask.
3. Reflux for 1 h, and then cool in an ice bath for 10 min.
4. Filter off the precipitate with vacuum and wash it with petroleum ether (40–60 °C) (3 × 10 mL).
5. Transfer the solid to a beaker, add water (50 mL) and stir vigorously to dissolve any succinimide left.
6. Filter with vacuum and wash the precipitate with water (2 × 10 mL) and petroleum ether (2 × 10 mL).
7. Allow the precipitate to dry on the funnel as thoroughly as possible.
8. Record the yield and the melting point of the crude material, and obtain the IR (KBr) spectrum. The product is pure enough to be used for the next step, but it can be recrystallized from ethyl acetate.

Session 2– Preparation of 4-Methoxymethylbenzoic acid

1. Add KOH (1.1 g, 20 mmol), methanol (25 mL, 19.8 g, 0.62 mol) and 4-bromomethylbenzoic acid (1.1 g, 5 mmol) to a 100 mL round-bottom flask.
2. Fit the flask with a reflux condenser, a $CaCl_2$ tube on top and gently reflux for 45 min in a water bath.

3. Remove the methanol on a rotary evaporator with gentle warming.
4. Dissolve the alkaline residue in water (30 mL) and acidify with an aqueous solution of HCl (10%) (verify with pH paper).
5. Filter the precipitate with vacuum, wash it with petroleum ether (2 × 15 mL) and dry the product thoroughly in the funnel.
6. Recrystallize the crude product from water and wash it with petroleum ether.
7. Record the yield, melting point, and IR (KBr) spectrum.

Waste Disposal

The filtrate from the first step containing chlorobenzene should be placed in a waste collector recipient for chlorinated solutions. Petroleum ether resulting from washings and recrystallization should be stored for posterior purification and recovery. The acidic filtrate from the second step should be neutralized, diluted and discarded in the sink.

Results Interpretation and Additional Questions

1. Determine the yields of reaction products.
2. Compare the melting points of the obtained products with the ones reported in the literature and evaluate their purity.
3. Interpret the IR spectra obtained and the NMR spectra available in literature.
4. Write out the mechanisms of the radical bromination reaction.
5. Considering the experimental conditions, determine whether the nucleophilic substitution mechanism is unimolecular (S_N1) or bimolecular (S_N2). Write both possible mechanisms.

References

1. J. Clayden, N. Greeves, S. Warren and P. Wothers, *Organic Chemistry*, Oxford, 2001, (a) p. 1023; (b) p. 1039; (c) p. 426.
2. S. H. Pine, *Organic Chemistry*, McGraw-Hill, 5th edn, 1987, p. 922.
3. S. Ege, *Organic Chemistry*, D. C. Heath and Company, 2nd edn, 1989, (a) p. 828; (b) p. 832.
4. J. March, *Advanced Organic Chemistry*, John Wiley & Sons, Inc., New York, 4th edn, 1992, (a) p. 694; (b) p. 306.
5. D. L. Tuleen and B. A. Hess, *J. Chem. Educ.*, 1971, **48**, 476.
6. M. Hornby and J. Peach, *Foundations of Organic Chemistry, Oxford Chemistry Primers*, Oxford University Press, 1993, p. 75.
7. L. M. Harwood, C. J. Moody and J. M. Percy, *Experimental Organic Chemistry*, Blackwell Science, 2nd edn, 1999, p. 468.
8. L. F. Tietze and T. H. Eicher, *Reactions and Synthesis in Organic Chemistry Laboratory*, University Science Books, 1989, p. 48.
9. C. A. M. Afonso, D. P. Simão, L. P. Ferreira, M. S. Serra and M. M. M. Raposo, *100 Experiências de Química Orgânica*, IST Press, 2011, p. 67.

11.2. Synthesis of Benzopinacolone *via* Benzophenone Photoreduction Followed by Pinacol Rearrangement

Filipa Siopa* and Carlos A. M. Afonso*

iMed.ULisboa, Av. Professor Gama Pinto, 1649-003 Lisboa, Portugal
*E-mail: filipasiopa@ff.ul.pt, carlosafonso@ff.ul.pt

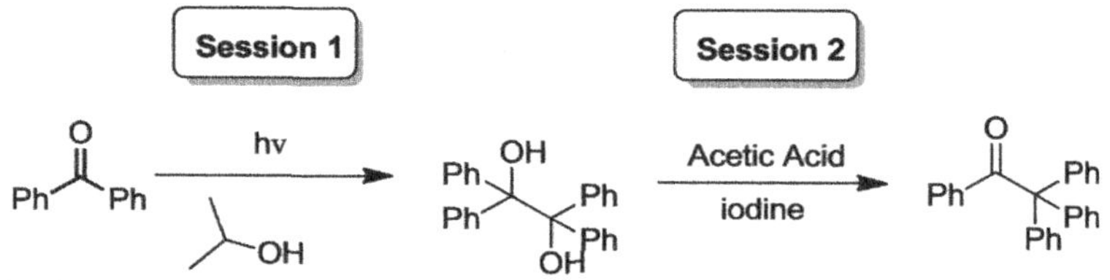

Number of sessions (duration of each session)	Hazard level	Difficulty level	Level of study
2 (1 h + 3 h)	Moderate	Medium	Intermediate

Class names Ketone, alcohol

Concepts involved Demonstration of the use of sunlight in benzophenone reduction to benzopinacol, and the radical mechanism involved; the pinacol–pinacolone rearrangement experiment allows the student to explore the carbocation stability and the 1,2-rearrangement

Chemicals needed Benzophenone, isopropanol, acetic acid, iodine, ethanol, TLC eluent hexane/ethyl acetate (7 : 3)

Equipment and experimental techniques involved Round bottom flask, spatula, graduated cylinder, Pasteur pipette, rubber septa, Büchner funnel, filter paper, melting point apparatus, hotplate stirrer, slow crystallization, infrared spectrometer and NMR spectrometer (optional)

Keywords ^{1}H and ^{13}C NMR spectroscopy, IR spectroscopy, photoreaction, pinacol rearrangement, radical mechanism, slow crystallization

Background

The electromagnetic radiation in the UV/Vis region is used in a large number of organic reactions.[1] The UV/Vis radiation promotes the substrate activation and formation of radical intermediates by homolytic cleavage of the labile bond or by the formation of unstable species resulting from electronic transitions.[2] The photoexcited benzophenone abstracts a hydrogen from the 2-propanol and dimerizes, yielding benzopinacol and acetone.[3] Benzopinacol can be easily isolated by spontaneous crystallization from the reaction medium and characterized by analytical techniques.

Due to the structural specificity of benzophenone it is possible to perform the photochemical transformation with sunlight. A photon of light with 355 nm wavelength (UV region) is absorbed

Comprehensive Organic Chemistry Experiments for the Laboratory Classroom
Edited by Carlos A. M. Afonso, Nuno R. Candeias, Dulce Pereira Simão, Alexandre F. Trindade, Jaime A. S. Coelho, Bin Tan, and Robert Franzén.

Published by the Royal Society of Chemistry, www.rsc.org

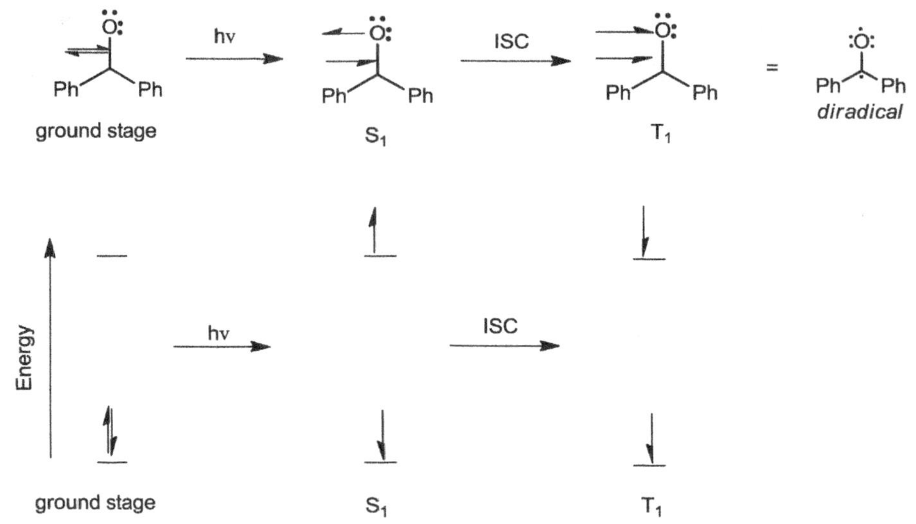

Scheme 11.2.1 Formation of benzophenone diradical by sunlight *absorption*.

by benzophenone, since the energy of the photon corresponds to the *energy* difference of orbitals, *i.e.* electronic levels (Scheme 11.2.1). Thus an electron is promoted to a higher energy orbital and the molecule is in the excited singlet state (S_1), with spin opposite to the electron that is in the lower energy orbital (spin-paired electrons). Intersystem crossing (ISC) can then occur and the excited electron change the direction of its spin, such that it is now *unpaired* with the electron that is in the lower energy orbital. This excited stage is called the triplet state (T_1). Benzophenone is in the T_1 state when it reacts with the isopropanol (Scheme 11.2.1).

The quantum yield is the fraction of absorbed light that produces a particular result and measures the efficiency of a photochemical reaction. There are several types of quantum yields: the primary quantum yield is related to the number of species that disappears directly as a result of the molecular absorption in a single step process; however, this is very difficult to measure. Instead, the product quantum yield (usually designated Φ) is normally used, and is calculated by dividing the number of molecules (moles) of product formed by the number of photons absorbed.[4]

The pinacol rearrangement,[5] normally addressed in organic chemistry courses, consists of the transformation of a vicinal diol into a ketone *via* carbon group migration under acidic conditions. One of the hydroxyl groups is protonated and loses water to form a relatively stable tertiary carbocation. The 1,2-migration of the aryl group from the adjacent carbon forms an even more stable carbocation, since the positive charge is delocalized by heteroatom resonance, yielding the benzopinacolone (Scheme 11.2.2).

In this work we attempt to study two synthetic transformations: the reduction of benzophenone to benzopinacol with concomitant oxidation of isopropanol to acetone promoted by the sunlight radiation and the pinacol rearrangement.

Additional Safety

All procedures should be performed with safety equipment, such as, gloves, safety goggles and laboratory coat. The pinacol rearrangement should be done in the fume hood. Benzophenone is toxic and hazardous. Isopropanol and ethanol are flammable and should be handled with good ventilation, avoiding all ignition sources. Acetic acid is a flammable reagent and causes severe burns.

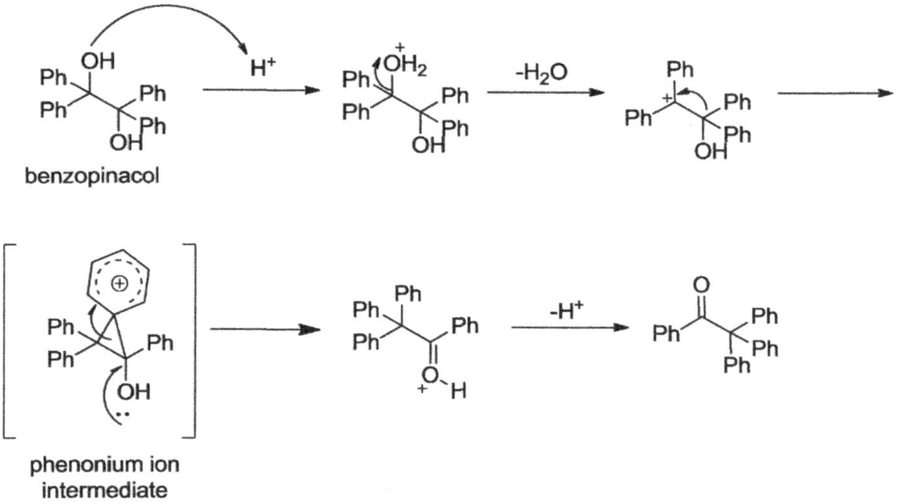

Scheme 11.2.2 Mechanism of the benzopinacol–benzophenone transformation.

Experimental Procedure

Laboratory Session 1 (1 h): Synthesis of Benzopinacol

Procedure – Synthesis of Benzopinacol

1. In a 100 mL round-bottom flask prepare a solution of benzophenone (5 g, 27.5 mmol) and isopropanol (50 mL), and add 1 drop of glacial acetic acid. Place a septum or stopper and seal the round-bottom flask, and label the flask with your name or group reference.
2. Place the reaction in a safe sunny location (next to a window), until the benzopinacol stops precipitating, normally one sunny week is enough.

Laboratory Session 2 (3 h): Isolation of Benzopinacol and Preparation of Benzopinacolone

Procedure – Isolation of Benzopinacol

1. Recover the benzopinacol by filtration of the reaction mixture in a Büchner funnel and wash it with cold isopropanol (10 mL). Allow the solid to air dry.
2. Weigh the solid and perform a TLC with the benzophenone and an authentic sample of benzopinacol.
3. Measure the melting point and obtain an infrared spectrum and ^{1}H NMR spectrum (optional) of your sample.
4. Determine the yield of the reaction.

Procedure – Preparation of Benzopinacolone

1. In a 100 mL round-bottom flask with a magnetic stir bar, add the benzopinacol (4 g, 10.9 mmol) previously prepared, acetic acid (20 mL) and iodine (two or three crystals).
2. Insert a water condenser into the round-bottom flask and heat the reaction mixture under reflux during 5 min; observe the dissolution of all starting materials.
3. Allow the solution to cool and observe the formation of a yellow precipitate; the benzopinacolone will separate from the solution.

4. Isolate the product by vacuum filtration using a Büchner funnel and wash the solid with ethanol.
5. Measure the melting point and obtain an infrared spectrum and ^{1}H NMR spectrum (optional) of your sample.

Waste Disposal

The isopropanol and ethanol waste should be collected to a non-chlorinated waste collector recipient, for posterior destruction.

Results Interpretation and Additional Questions

1. For the products synthesized analyse the analytical data (IR, melting point and ^{1}H NMR), verify the product purity and compare the melting point with the literature value.
2. Calculate the yield, atom economy, and E-factor.
3. Explain mechanistically both transformations performed.
4. For the benzopinacol synthesis describe the function of isopropanol, and the use of a catalytic amount of acetic acid.
5. For the benzopinacolone procedure, comment on whether the rearrangement can be performed in the presence of a catalytic amount of acid.

References

1. (a) N. Hoffmann, *Chem. Rev.*, 2008, **108**, 1052; (b) M. A. Tasdelen and Y. Yagci, *Angew. Chem., Int. Ed.*, 2013, **52**, 5930.
2. J. D. Coyle, *Photochemistry in Organic Synthesis*, Royal Society of Chemistry, 1986, pp. 63–64.
3. A. R. Tatchell, B. S. Furnis, A. J. Hannaford and P. W. G. Smith, *Vogel's Textbook of Practical Organic Chemistry*, Longman, Harlow, 5th edn, 1989, p. 528.
4. (a) M. Smith, *March's Advanced Organic Chemistry: Reactions, Mechanisms, and Structure*, John Wiley & Sons, Inc., Hoboken, 7th edn, 2013, pp. 306–307; (b) R. K. Bansal, *Organic Reaction Mechanisms*, Tata McGraw-Hill, 3th edn, 1998, p. 464.
5. (a) T. Laue and A. Plagens, *Named Organic Reactions*, John Wiley & Sons, 2nd edn, 2005, p. 229; (b) M. Lezaeta, W. Sattar, P. Svoronos, S. Karimi and G. Subramaniam, *Tetrahedron Lett.*, 2002, **43**, 9307–9309.

11.3. Iodosulfonylation–Dehydroiodination of Styrene: Synthesis of (*E*)-β-Tosylstyrene

Carmen Nájera*, José M. Sansano, and Miguel Yus

Departamento de Química Orgánica and Centro de Innovación en Química Avanzada (ORFEO-CINQA). Universidad de Alicante, Ctra. San Vicente s/n, E-03690-San Vicente del Raspeig (Alicante), Spain
*E-mail: cnajera@ua.es

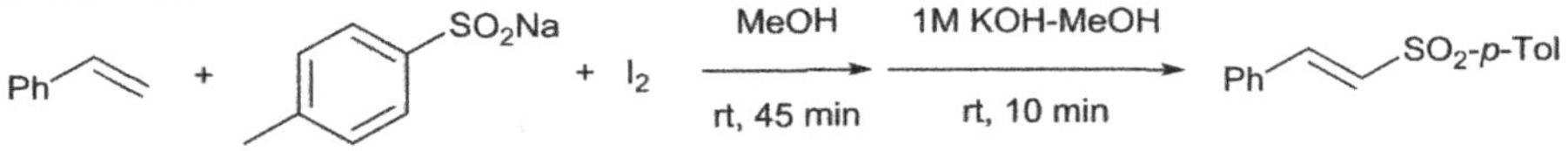

Number of sessions (duration of each session)	Hazard level	Difficulty level	Level of study
1 (3 h)	Moderate	Medium	Advanced

Class names Sulfones, sulfinates, sulfonyl iodides, iodoalkanes, alkenes

Concepts involved This experiment involves the regioselective addition reaction of *in situ* generated *p*-toluenesulfonyl iodide onto styrene through a free radical mechanism, followed by a stereoselective *anti*-β-elimination

Chemicals needed Sodium *p*-toluenesulfinate, iodine, styrene, methanol, potassium hydroxide, sodium hydrogen sulfite, ethyl acetate, 95% ethanol, hexane, deuterated chloroform

Equipment and experimental techniques involved Simple transparent glassware material (sunlight) and powerful magnetic stirring, liquid–liquid extraction, separating funnel, distillation–evaporation, Rotavapor, crystallization

Keywords Alkenes, elimination, IR, mp, NMR, radical addition, recrystallization, sulfones

Background

Sulfones constitute a very interesting family of compounds, which are very easily obtained and purified.[1] The most attractive chemical features of the sulfonyl group reside in the stabilization of α-anions, in its electron-withdrawing group nature allowing, for example, α,β-unsaturated sulfones to act as good Michael-type acceptors, and in its ability to act as good leaving group. So, in a few words, a sulfonyl group is introduced in the organic skeleton, immediately directs the desired reactivity and, finally, it is removed under mild conditions.[2–4] The synthesis of sulfones can be accomplished using several methods depending on the final structure. In this experiment, a general synthesis of vinyl sulfones from alkenes, employing the iodosulfonylation process[5] followed by a β-elimination of hydrogen iodide, is described.[6] The reaction operates under a radical pathway,[7] which is another important aspect for the student from an academic point of view. This free-radical transformation occurs at room temperature (preferably with sunlight) and so, in this respect, no special safety precautions are recommended. Products and side materials are not toxic and can be separated easily from the reaction mixture.

Comprehensive Organic Chemistry Experiments for the Laboratory Classroom
Edited by Carlos A. M. Afonso, Nuno R. Candeias, Dulce Pereira Simão, Alexandre F. Trindade, Jaime A. S. Coelho, Bin Tan, and Robert Franzén.

Published by the Royal Society of Chemistry, www.rsc.org

The objective of this work is the synthesis of β-tosylstyrene[8] in a one-pot process from styrene employing 1.5 equivalents of sodium *p*-toluenesulfinate and stoichiometric amounts iodine in methanol. The whole procedure is designed to be benign with the environment (methanol is finally distilled from the reaction mixture). Iodine is the last reagent to be added (slowly) to the stirred white suspension, in such a way that it is possible to see its consumption (from yellow to white suspension). After the generation of the intermediate β-iodosulfone, a 1 M methanolic solution of potassium hydroxide is added, favoring the β-elimination of hydrogen iodide to yield the final product, which will be purified by liquid–liquid extraction and further recrystallization.

Additional Safety

All sessions should be done in a fume hood, with proper safety equipment, such as protective clothing, gloves and safety goggles. Styrene is flammable and toxic, avoid contact with eyes and skin. Methanol is toxic and flammable, and is harmful if swallowed. Potassium hydroxide is corrosive, avoid contact with the skin.

Experimental Procedure

Laboratory Session 1 (1 h): Iodosulfonylation of Styrene

1. Place in a fume hood a 50 mL one-necked round bottomed flask with a magnetic stir bar onto a magnetic stirrer.
2. Add sodium *p*-toluenesulfinate tetrahydrate (375 mg, 1.5 mmol), methanol (5 mL) and styrene (115 µL, 1.0 mmol) and stir vigorously the reaction mixture.
3. Add slowly (5 min) iodine (255 mg, 1.0 mmol) at room temperature and continue stirring for 45 min.

Laboratory Session 1 (0.5 h): β-Elimination of Hydrogen Iodide

4. Add slowly 1 M methanolic potassium hydroxide solution (1 mL, 1 mmol) stirring the new suspension for an additional 10 min.
5. Add 1 M aqueous solution of sodium hydrogen sulfite (10 mL) and methanol that has been distilled under reduced pressure (~15 mmHg).

Laboratory Session 1 (1.5 h): Isolation and Purification of β-Tosylstyrene

6. Extract the aqueous layer with ethyl acetate in a separating funnel (2 × 5 mL) and dry the resulting organic phase (anhydrous magnesium sulfate) and then evaporate under reduced pressure (~15 mmHg).
7. Recrystallize the residue from warm 95% ethanol (8 mL). After cooling the solution at 0 °C, filter and dry the resultant colorless crystals (205–215 mg, 78–83% yield).

Waste Disposal

Methanol is distilled after the reaction is complete, as well as ethyl acetate to isolate the final product. The neutralized aqueous solution containing iodide salts, bisulfite and tetrathionate salts may be diluted and discarded in the sink.

Results Interpretation and Additional Questions

1. Determine the chemical yield of β-tosylstyrene.
2. Compare the obtained melting point with the one published in the literature.

3. Interpret the ^{1}H and ^{13}C NMR and FTIR spectra (previously obtained and provided by the instructor).
4. Rationalize the involved radical mechanism. Describe the initiation, propagation and termination steps. Justify the regioselection of the process.
5. Justify the double bond configuration of the final product using a conformational analysis.
6. What is the role of the aqueous sodium hydrogen sulfite solution?

Acknowledgement

Reprinted (adapted) with permission from *J. Chem. Educ.*, 1995, 72 (7), pp. 664–665. Copyright (1995) American Chemical Society.

References

1. (a) *Syntheses of Sulphones, Sulphoxides and Cyclic Sulphides*, ed. S. Patai, and Z. Rappoport, John Wiley & Sons, Inc., New York, 1st edn, 1995; (b) K.-M. Roy, Sulfones and sulfoxides, in *Ullmann's Encyclopedia of Industrial Chemistry*, Wiley-VCH Verlag GmbH, Weinheim, 1st edn, 2002, vol. 34, p. 705.
2. Diaryl sulfones have been used in the treatment of tuberculosis and leprosy. Other sulfones are antibiotics, and some of them are used for the treatment of diabetes and its complications. X. Chen, S. Hussain, S. Parveen, S. Zhang, Y. Yang and C. Zhu, *Curr. Med. Chem.*, 2012, **19**, 3578.
3. For the application of sulfones in total synthesis of natural products, see: E. N. Prilezhaeva, *Russ. Chem. Rev.*, 2001, **70**, 897.
4. For the application of sulfones in synthesis of biologically active compounds, see: E. N. Prilezhaeva, *Russ. Chem. Rev.*, 2000, **69**, 367.
5. (a) C. Nájera, B. Baldó and M. Yus, *J. Chem. Soc., Perkin Trans. I*, 1988, 1029; (b) C. Nájera and M. Yus, *J. Chem. Soc., Perkin Trans. I*, 1989, 1387.
6. C. Nájera, J. M. Sansano and M. Yus, *J. Chem. Educ.*, 1995, **72**, 664.
7. S. Z. Zard, *Radical Reactions in Organic Synthesis*, Oxford University Press, Oxford, 1st edn, 2003.
8. (a) β-Tosylstyrene has been employed in the synthesis of azides: D. Sahu, S. Dey, T. Pathak and B. Ganguly, *Org. Lett.*, 2014, **16**, 2100; (b) In Michael-type additions: T. Nishimura, Y. Takiguchi and T. Hayashi, *J. Am. Chem. Soc.*, 2012, **134**, 9086; (c) In electrocyclizations: N. A. Magomedov, P. L. Ruggiero and Y. Tang, *J. Am. Chem. Soc.*, 2004, **126**, 1624; (d) In 1,3-dipolar cycloadditions: V. Padmavathi, R. A. V. Bhaskar, R. P. Sumathi, A. Padmaja and R. D. Bhaskar, *Indian J. Chem., Sect. B: Org. Chem. Incl. Med. Chem.*, 1998, **37B**, 1286.

12. Oxidations

12.1. Ketone and Aldehyde Formation

12.1.1. Aerobic Alcohol Oxidation Using a Cu(I)/TEMPO Catalyst System

Nicholas J. Hill, Jessica M. Hoover, and Shannon S. Stahl*

Department of Chemistry, University of Wisconsin-Madison, 1101 University Avenue, Madison, WI, 53706, USA. *E-mail: stahl@chem.wisc.edu

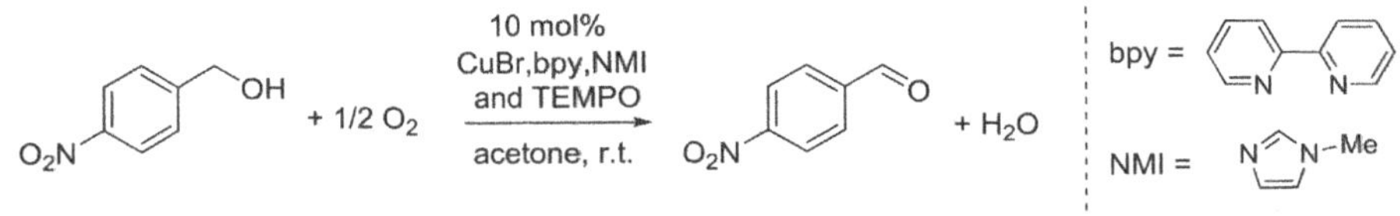

Number of sessions (duration of each session)	Hazard level	Difficulty level	Level of study
1 (3 h)	Moderate	Low	Intermediate
Class names Benzyl alcohols, aldehydes			
Concepts involved This experiment gives a "green" oxidation method using a CuBr/TEMPO catalyst system and ambient air as the terminal oxidant to convert 4-nitrobenzyl alcohol into the corresponding aldehyde			
Chemicals needed 4-Nitrobenzyl alcohol, copper(I) bromide, TEMPO (2,2,6,6-tetramethylpiperidine-1-oxyl), *N*-methylimidazole (NMI), 2,2′-bipyridine, acetone, anhydrous magnesium sulfate, pentane, deuterochloroform			
Equipment and experimental techniques involved Apparatus for magnetic stirring, liquid–liquid extraction apparatus, hot plate or rotary evaporator			
Keywords Extraction, green chemistry, NMR, oxidation, transition metal catalysis			

Background

Conversions of alcohols into aldehydes and ketones are among the most important and widely used oxidation reactions in organic chemistry. Modern undergraduate organic chemistry textbooks typically present several methods to effect these reactions, and among the most commonly featured oxidants are Cr^{IV} and Mn^{IV} reagents, such as pyridinium chlorochromate, pyridinium dichromate, CrO_3/H_2SO_4 (Jones reagent), $KMnO_4$, and MnO_2.[1] These reagents have found widespread utility in organic chemistry, but they are toxic, difficult to dispose, and used in stoichiometric quantities. The latter considerations explain why such oxidants are rarely used

Comprehensive Organic Chemistry Experiments for the Laboratory Classroom
Edited by Carlos A. M. Afonso, Nuno R. Candeias, Dulce Pereira Simão, Alexandre F. Trindade, Jaime A. S. Coelho, Bin Tan, and Robert Franzén.

Published by the Royal Society of Chemistry, www.rsc.org

in large-scale industrial synthesis of organic chemicals and are often avoided even in laboratory-scale organic synthesis. Furthermore, the use of such oxidizing agents in undergraduate instructional laboratories is increasingly discouraged.

Over the past decade, extensive efforts have been directed toward the development of catalytic methods for alcohol oxidation that are capable of using atmospheric oxygen as the stoichiometric oxidant.[2] Molecular oxygen is perhaps the ideal stoichiometric oxidant from a "green chemistry" perspective because it is abundant, non-toxic, renewable, and does not generate hazardous waste products. Such aerobic oxidation methods are attractive from practical, environmental, and pedagogical viewpoints; however, catalysts composed of expensive noble metals, together with the common use of halogenated solvents and/or pure O_2 as the oxidant have hindered the implementation of aerobic oxidation methods into the instructional laboratory. Homogeneous Cu/TEMPO catalyst systems avoid most of these limitations, and a recent (bpy)CuI/TEMPO catalyst (bpy = 2,2′-bipyridine, TEMPO = 2,2,6,6-tetramethyl-1-piperidinyloxyl) is particularly versatile and effective.[3] Here, we describe an inexpensive and operationally simple implementation of this catalyst system for the aerobic oxidation of 4-nitrobenzyl alcohol. The protocol provides undergraduate students with practical experience of a "green" alcohol oxidation method while introducing the important concepts of metal-catalyzed reactions.

Catalytic Cycle

The catalyst system used in this experiment is a modification of that recently reported by the Stahl group at the University of Wisconsin-Madison.[3] A simplified catalytic cycle of this reaction is shown in Scheme 12.1.1.1.

The overall reaction can be divided into two stages: a catalyst oxidation stage (Stage 1) and a substrate oxidation stage (Stage 2). CuBr enters the catalytic cycle bonded to bpy and NMI, (*N*-methylimidazole) producing the red-brown Cu^I coordination complex [Cu(bpy)(NMI)]

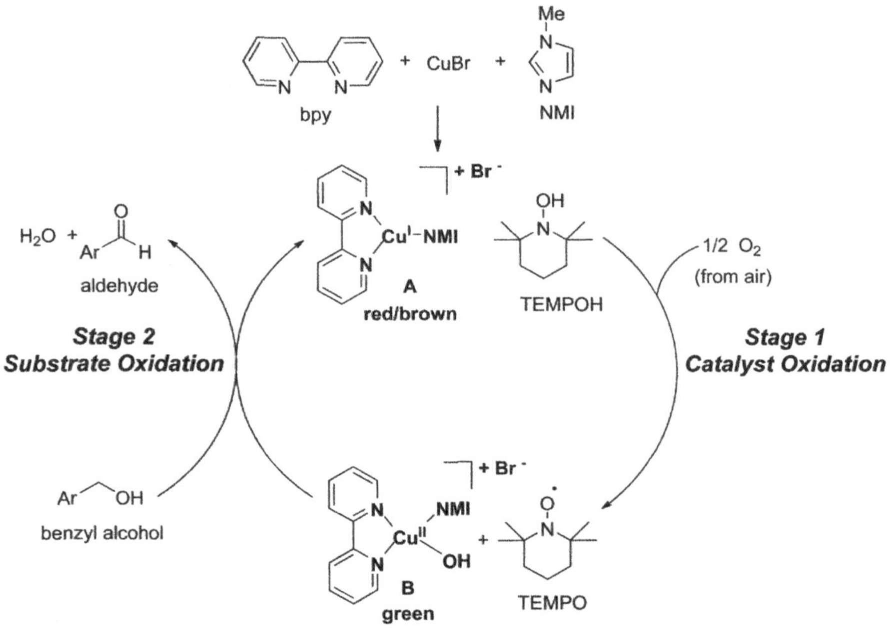

Scheme 12.1.1.1 Simplified catalytic cycle for Cu/TEMPO oxidation of a benzyl alcohol.

Br (**A**). In Stage 1 of the catalytic cycle, **A** is oxidized to the green Cu^{II}-hydroxide species [Cu(bpy)(OH)(NMI)]Br (**B**) by atmospheric O_2, together with the oxidation of TEMPOH to the TEMPO radical. In Stage 2 of the cycle, the benzyl alcohol is oxidized to the aldehyde by the Cu^{II}-hydroxide species **B** and TEMPO. The Cu^{I} and TEMPOH species formed in Stage 2 continue the catalytic cycle until the benzyl alcohol substrate is fully consumed. The copper species remaining after complete conversion of the substrate is likely to be a Cu^{II}-OH complex similar to **B**.

A subtle mechanistic point is worth clarifying for students. The stable form of TEMPO is the oxidized free radical species and thus TEMPOH is not actually present when the reaction begins (*i.e.* during the first catalytic oxidation stage). The Cu^{II}-OH species **B** is initially generated by the direct oxidation of **A** by atmospheric oxygen.

$$2\,[(\text{bpy})\text{Cu}^{I}(\text{NMI})]^{+}\text{Br}^{-} + H_2O + 1/2\,O_2 \rightarrow 2\,[(\text{bpy})\text{Cu}(\text{OH})(\text{NMI})]^{+}\text{Br}^{-}$$

This reaction is slower than oxidation of the Cu^{I} species **A** in the presence of TEMPOH but is still sufficiently rapid to begin the catalytic cycle, and explains how the reaction is initiated in the absence of TEMPOH.

The substrate oxidation sequence (Stage 2 of the catalytic cycle) can be presented as a pair of simplified electron-pushing mechanisms. The benzyl alcohol substrate $ArCH_2OH$ reacts with the Cu^{II}-OH complex **B** to form the Cu^{II}-alkoxide intermediate **C** with concomitant release of water (Scheme 12.1.1.2). This reaction, which may involve initial coordination of the alcohol to the Cu^{II} center, corresponds to a ligand-substitution reaction.

Another ligand substitution reaction results in replacement of the NMI ligand in **C** by TEMPO to form the Cu-TEMPO adduct **D** (Scheme 12.1.1.3). The TEMPO group in complex **D** then removes a proton from the alkoxide, and formation of the aldehyde results in the flow of electrons into the Cu center to form the reduced species **E**. NMI then reacts with Cu complex **E** to regenerate **A** by substituting TEMPOH (which undergoes isomerization to place the proton on

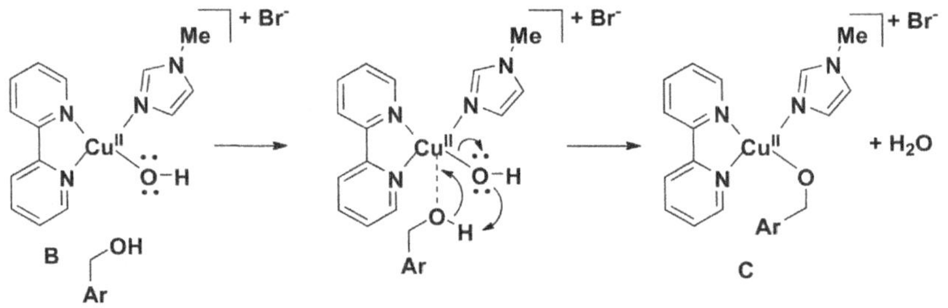

Scheme 12.1.1.2 Formation of Cu^{II}-alkoxide intermediate **C** from complex **B**.

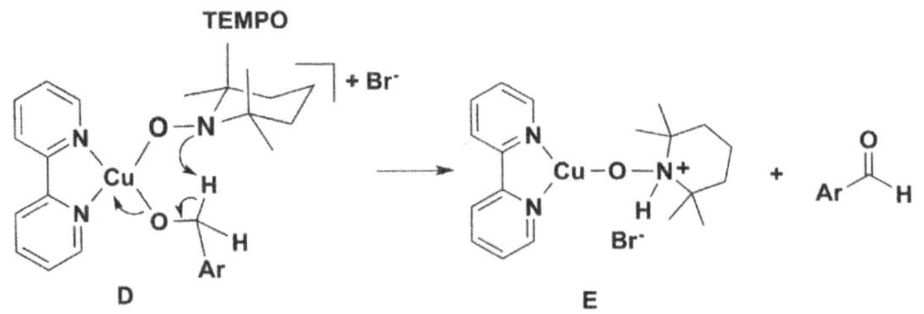

Scheme 12.1.1.3 Formation of aldehyde product and generation of complex **E**.

the oxygen atom). Complex **A** and TEMPOH then continue in the catalytic cycle until all of the benzyl alcohol substrate has been consumed.

Additional Safety

Appropriate personal protective equipment (gloves, goggles, laboratory coat) should be worn at all times, and the experiment is ideally performed in a fume hood. The aerobic oxidation reactions presented herein employ organic solvents in the presence of air; therefore, precautions appropriate for the handling of flammable solvents (acetone, pentane) should be employed.[4]

Experimental Procedure

1. Add 4-nitrobenzyl alcohol (0.40 g, 2.5 mmol) and acetone (15 mL) to a 50 mL Erlenmeyer flask and stir the mixture until the solid is fully dissolved.
2. Weigh CuBr (35 mg, 0.24 mmol), bpy (40 mg, 0.26 mmol), and TEMPO (40 mg, 0.26 mmol) into separate glass sample vials and add each reagent sequentially to the rapidly stirred acetone solution of the benzyl alcohol. Rinse each sample vial with ~2 mL acetone following each addition of solid to the solution and add these washings to the reaction flask (for a total solvent volume of ~20 mL).
3. Add 4 drops of NMI (*N*-methylimidazole) (approx. 0.08 mL, 1 mmol) and stir the mixture until the reaction mixture changes from red-brown to green.
4. Following the color change, stir the mixture for an additional 5 min and transfer to a 250 mL separatory funnel. Add water (30 mL) and pentane (25 mL) to the separatory funnel (add ~10 mL acetone if a precipitate appears in the separatory funnel). Extract the reaction mixture into the pentane layer with shaking and allow the organic and aqueous layers to separate. Isolate the organic layer.
5. Extract the aqueous layer with an additional portion of pentane (10 mL), combine the organic layers, and stir over anhydrous $MgSO_4$. Remove the solid by gravity filtration and transfer the filtrate into a weighed beaker. Evaporate the solvent from the filtrate in a fume hood using a hot plate on a low-heat setting[5] or use a rotary evaporator. Allow the product to cool to room temperature. Add ice-cold pentane (5 mL) to the flask, stir briefly, and isolate the solid by vacuum filtration. Rinse the solid with ice-cold pentane (2 × 2 mL) and allow to air-dry. Obtain the mass of the dry solid and calculate the %yield. Characterize the product by melting range, ^{1}H NMR ($CDCl_3$) and IR spectroscopy.

Waste Disposal

Solid and liquid waste should be disposed into appropriately labeled containers. Flush all aqueous solutions down the drain. Spread the spent magnesium sulfate on a tray or filter paper in the fume hood to evaporate any solvent, and then place it in the non-hazardous solid waste container. Transfer the acetone and pentane recovered from the reaction mixture to the container for non-halogenated organic solvent.

Results Interpretation and Additional Questions

1. What evidence would appear in the ^{1}H NMR and IR spectra of the isolated 4-nitrobenzaldehyde indicating (a) residual 4-nitrobenzyl alcohol; (b) partial over-oxidation to the corresponding benzoic acid? Is such evidence present in your data?

2. Analyze your experimental ^{1}H NMR and IR spectroscopic data and comment upon the purity of the isolated 4-nitrobenzaldehyde. If appropriate, use the ^{1}H NMR data to calculate the amount of residual 4-nitrobenzyl alcohol.
3. Briefly explain the following experimental observations; (a) the color of the acetone solution of 4-nitrobenzyl alcohol and CuBr changes from pale green to red upon addition of bpy; (b) the color of the reaction mixture changes from red/brown to green upon completion of the oxidation reaction; (c) upon addition of water and pentane to the reaction mixture the aqueous layer becomes pale blue.
4. Acetone was used as the solvent for the aerobic oxidation reaction. Briefly explain why pentane was added to the reaction mixture during the work-up process.
5. What was the purpose of rinsing the isolated 4-nitrobenzaldehyde with ice-cold pentane?

Acknowledgement

Adapted with permission from N. J. Hill, J. M. Hoover and S. S. Stahl *J. Chem. Educ.*, 2013, **90**, 102. Copyright (2013) American Chemical Society.

References

1. For a summary of undergraduate organic chemistry textbook coverage of oxidation and reduction reactions, see the following: I. A. Shibley Jr, K. E. Amaral, D. J. Aurentz and R. J. McCaully, *J. Chem. Educ.*, 2010, **87**, 1351.
2. For reviews, see: (a) B. L. Ryland and S. S. Stahl, *Angew. Chem., Int. Ed.*, 2014, **53**, 8824; (b) C. Parmeggiani and F. Cardona, *Green Chem.*, 2012, **14**, 547; (c) M. J. Schultz and M. S. Sigman, *Tetrahedron*, 2006, **62**, 8227; (d) T. Mallat and A. Baiker, *Chem. Rev.*, 2004, **104**, 3037; (e) I. E. Markó, P. R. Giles, M. Tsukazaki, I. Chellé-Regnaut, A. Gautier, R. Dumeunier, F. Philippart, K. Doda, J.-L. Mutonkole, S. M. Brown and C. J. Urch, *Adv. Inorg. Chem.*, 2004, **56**, 211.
3. (a) B. L. Ryland, S. D. McCann, T. C. Brunold and S. S. Stahl, *J. Am. Chem. Soc.*, 2014, **136**, 12166; (b) J. M. Hoover, B. L. Ryland and S. S. Stahl, *J. Am. Chem. Soc.*, 2013, **135**, 2357; (c) J. M. Hoover, B. L. Ryland and S. S. Stahl, *ACS Catal.*, 2013, **3**, 2599; (d) J. M. Hoover and S. S. Stahl, *J. Am. Chem. Soc.*, 2011, **133**, 16901.
4. Control experiments show that hydrogen peroxide does not accumulate under the reaction conditions, and KI-starch test strips did not indicate the presence of an oxidizing agent in solution.
5. Evaporation of solvent from the reaction mixture *via* a hotplate significantly increases the risks of the experiment. The operation should be performed in a fume hood under the supervision of an instructor. Students should add a boiling stick to the mixture prior to commencement of heating. A griddle-type hotplate is recommended so that multiple evaporations can be performed and monitored simultaneously.

12.1.2. Chemoselective Oxidation of 1,2-Tetradecanediol

Siedlecka Renata* and Skarżewski Jacek

Faculty of Chemistry, Organic Chemistry Department, Wroclaw University of Technology, Wyb. Wyspianskiego 27, 50-370 Wroclaw, Poland. *E-mail: renata.siedlecka@pwr.edu.pl

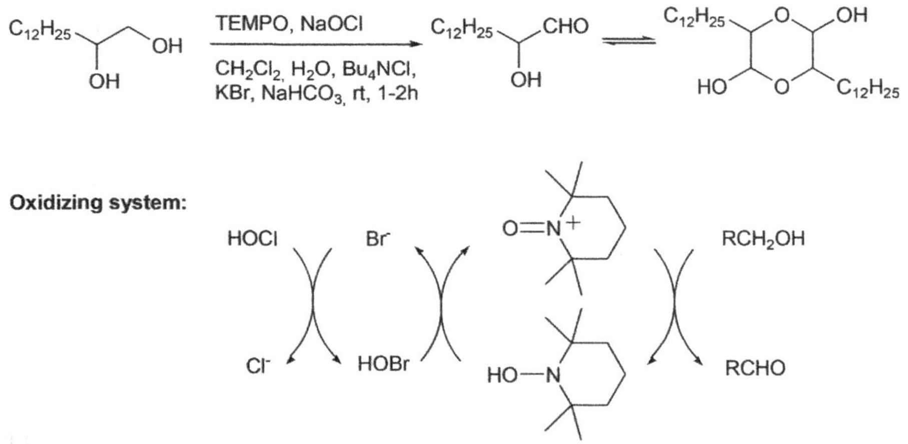

Number of sessions (duration of each session)	Hazard level	Difficulty level	Level of study
2 (3 h each)	Moderate	Low	Intermediate

Class names Vicinal diols, α-hydroxyaldehydes, hemiacetals

Concepts involved This experiment gives a method of selective oxidation of primary alcohols to aldehydes in the presence of secondary hydroxyl group. α-Hydroxyaldehyde formed in this reaction spontaneously dimerizes to hemiacetal

Chemicals needed 1,2-Tetradecanediol, TEMPO (2,2,6,6-tetramethylpiperidine-1-oxyl), sodium hypochlorite (commercial 1.95 M soln.), potassium bromide, tetrabutylammonium chloride, dichloromethane, sodium bicarbonate (saturated soln.), brine, sodium sulfate (anhydrous), chloroform, ethyl acetate, silica gel on TLC Alu foils (optional), deuterated chloroform

Equipment and experimental techniques involved Usual laboratory glassware, magnetic stirrer, liquid–liquid extraction, Rotavapor, melting point, thin-layer chromatography (TLC), NMR

Keywords Chemoselective oxidation, crystallization, extraction, hemiacetal, ^{1}H NMR, mp

Background

Oxidation belongs to the oldest and the most important ways of transformations in organic as well as inorganic chemistry and biochemistry.[1–3] Numerous oxidation methods have been used on the laboratory scale and in industry. Oxidation can be defined in different ways, depending on the field in which it is applied. In organic chemistry, we rarely calculate the actual oxidation

Comprehensive Organic Chemistry Experiments for the Laboratory Classroom
Edited by Carlos A. M. Afonso, Nuno R. Candeias, Dulce Pereira Simão, Alexandre F. Trindade, Jaime A. S. Coelho, Bin Tan, and Robert Franzén.

Published by the Royal Society of Chemistry, www.rsc.org

state of carbon. We more often use the mnemonic: *oxidation* is loss of C–H bond or gain of C–X bonds (Scheme 12.1.2.1).

Oxidation is important due to the transformation of one functional group into another, which leads to the preparation of new reactive compounds. The most important oxidative transformations involve oxidation of alcohols to carbonyl compounds, namely, aldehydes, ketones, carboxylic acids or esters – products desired and useful in organic synthesis.[4,5] In the literature anyone can find many procedures of oxidation, but not all of them are selective and applicable for a specific purpose. Sodium hypochlorite as the oxidant has a small weight of the oxygen equivalent and it is cheap compared to other oxidants. It has been used directly[6] and also in catalytic procedures.[7–9] In the described procedure, sodium hypochlorite is used together with the nitroxyl radical (TEMPO) as the catalyst.

TEMPO is employed in organic synthesis as a catalyst for oxidation,[10,11] in particular of primary alcohols to aldehydes.[12] The actual oxidant is the *N*-oxoammonium salt. In a catalytic cycle with sodium hypochlorite as the stoichiometric oxidant, hypochlorous acid generates the *N*-oxoammonium salt from TEMPO (Scheme 12.1.2.2). TEMPO oxidations also exhibit chemoselectivity, being inert towards secondary alcohols, but the reagent used can convert aldehydes into carboxylic acids. That is why the two-phase conditions are recommended.

The aim of this experiment is to demonstrate the unusual selectivity of the used catalytic system. Moreover, primary–secondary diols are selectively oxidized to hydroxyaldehydes, which are valuable substrates for further transformations (Scheme 12.1.2.3).[13]

Application of 1,2-diol as the substrate (*e.g.* 1,2-tetradecanediol) gives α-hydroxyaldehyde, which dimerizes under the reaction conditions and the cyclic (six-membered ring) hemiacetal is isolated as the product. The addition of an alcohol nucleophile to an aldehyde or ketone is

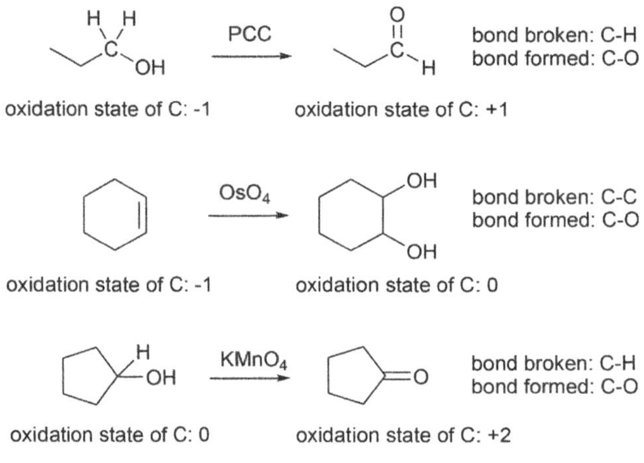

Scheme 12.1.2.1 Examples of oxidation processes.

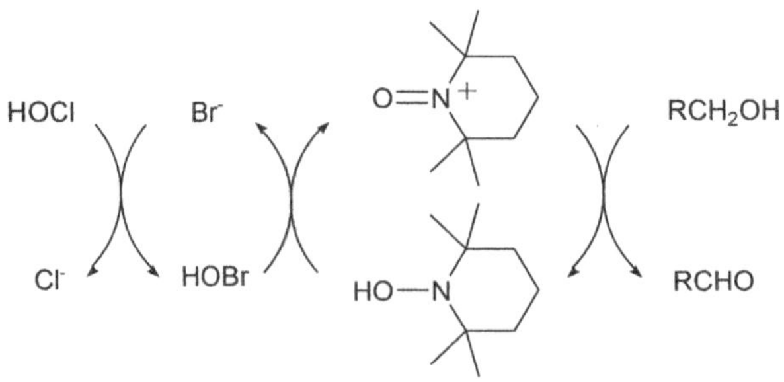

Scheme 12.1.2.2 NaOCl/TEMPO catalytic system.

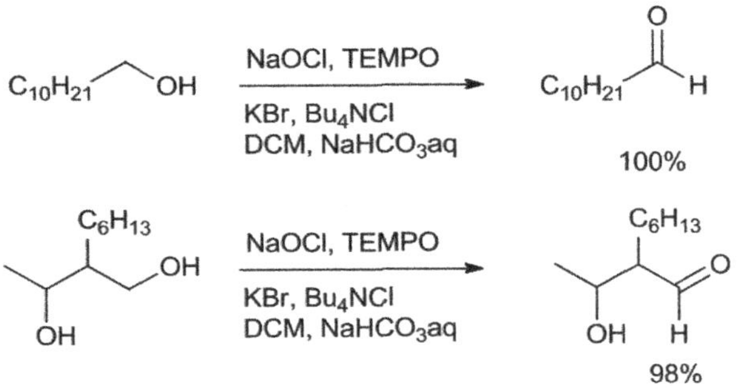

Scheme 12.1.2.3 Selective oxidation with NaOCl/TEMPO catalytic system.

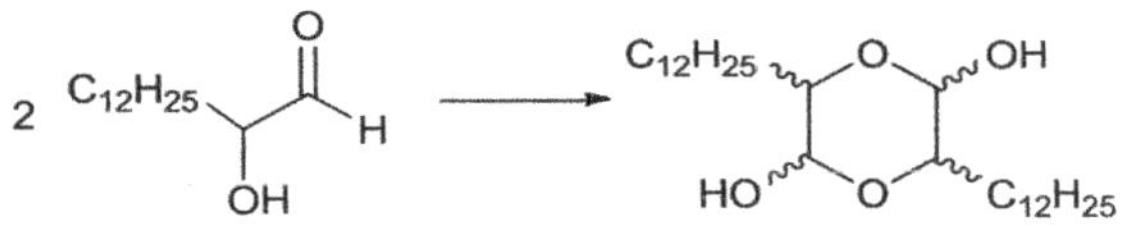

Scheme 12.1.2.4 Formation of cyclic hemiacetal from α-hydroxytetradecanal.

a well-known nucleophilic addition reaction in biological chemistry. When an alcohol adds to an aldehyde, the result is called a *hemiacetal*.[14,15] Molecules that have both a hydroxyl and a carbonyl groups can undergo an inter- or intramolecular reaction to form a cyclic hemiacetal (Scheme 12.1.2.4). The hydroxyl group at the α position has sufficient nucleophilic character to attack the carbonyl group of a second molecule without the need for an acid catalyst.

Additional Safety

The procedure does not require harsh reaction conditions and special precautions. The reaction can be performed on the bench. Personal safety equipment, such protective gloves, protective clothing, and safety goggles should be used.

NaOCl – skin corrosion. Causes severe skin burns and eye damage. However, only solutions containing more than 40% sodium hypochlorite by weight are considered hazardous oxidizers. Solutions less than 40% are classified as a moderate oxidizing hazard. Very toxic to aquatic organisms.

TEMPO – skin corrosion possible. Avoid dust formation. Ensure adequate ventilation.

$CHCl_3$ – accidental splashing into the eyes has caused irritation. Harmful if swallowed or inhaled. May cause drowsiness or dizziness. Suspected of causing cancer. Avoid inhalation of vapor. Ensure adequate ventilation.

Experimental Procedure

Laboratory Session 1 (3 h): Oxidation of 1,2-Tetradecanediol

1. Equip a 250 mL round-bottomed flask with a dropping funnel and magnetic stirrer.
2. Place 2.3 g of 1,2-tertradecanediol (10 mmol), 50 mL of dichloromethane, 0.015 g of TEMPO (1 mol%), 0.119 g of potassium bromide (10 mol%), 0.139 g of tetrabutylammonium chloride (5 mol%) and 15 mL of saturated solution of sodium bicarbonate into the flask, and charge the dropping funnel with 10 mL of a saturated solution of sodium bicarbonate, 20 mL of brine and 6.75 mL of sodium hypochlorite (1.95 M solution, 13 mmol).
3. At room temperature add dropwise this solution to the well-stirred mixture in the flask.
4. Addition should be completed within 30 min, but stirring should be continued for an additional 1.5 h.

5. In the meantime, TLC analysis can be made to control the reaction progress (optional).[16] The reaction is complete when the substrate is no longer observed in the reaction mixture.

Laboratory Session 2 (3 h): Product Isolation and the Yield Determination

1. After an appropriate time (about 1.5–2 h), pour the mixture into a separatory funnel. Separate the lower organic layer and extract the aqueous layer with chloroform or with dichloromethane (5 × 50 mL).
2. Wash the combined organic extracts with water (2 × 100 mL) and then with brine (100 mL), and dry over anhydrous sodium sulfate.
3. After evaporation of solvent the white solid crude product should be obtained.
4. Crystallize the product from ethyl acetate (about 120 mL should be used for this scale; *warning*: the product is poorly soluble in most solvents).
5. Measure the melting point of the obtained product. The product can be characterized also by NMR. To prepare the sample dissolve 5–6 mg of the product in 0.6 mL of deuterated chloroform.

Waste Disposal

The neutralized aqueous solution may be diluted and discarded in the sink. NaOCl – sodium thiosulfate is an effective chlorine neutralizer. Rinsing with a 5 mg L^{-1} solution, followed by washing with soap and water, quickly removes chlorine odor from the hands. Transfer organic residues and solutions not suitable for regeneration to the waste container.

Results Interpretation and Additional Questions

1. Calculate the reaction yield, remembering the dimerization of α-hydroxyaldehyde.
2. Compare the melting point of your product and the one reported in the literature.[13] Explain the wide range of the melting point?
3. Indicate other substrates that are suitable for this reaction.
4. What would be the product if the 1,4-pentanediol would be subjected to the oxidation under the same conditions?
5. Why are two-phase conditions recommended for such an oxidation?

References

1. M. Hudlicky, *Oxidation in Organic Chemistry*, American Chemical Society, Washington, 1990.
2. W. Kroutil, H. Mang, K. Edegger and K. Faber, *Adv. Synth. Catal.*, 2004, **346**, 125.
3. R. A. Sheldon, I. W. C. E. Arends and A. Dijksman, *Catal. Today*, 2000, **57**, 157.
4. M. J. Schultz and M. S. Sigman, *Tetrahedron*, 2006, **62**, 8227.
5. J. Muzart, *Tetrahedron*, 2003, **59**, 5789.
6. G. A. Mirafzal and A. M. Lozeva, *Tetrahedron Lett.*, 1998, **39**, 7263.
7. A. Weickmann and K. P. Zeller, *Methoden der Organische Chemie, Houben-Weyl*, Georg Thieme Verlag, Stuttgart, 1981, vol. 4/1a, p. 494.
8. J. Skarżewski and R. Siedlecka, *Org. Prep. Proced. Int.*, 1992, **24**, 623.
9. R. A. Sheldon, *Bull. Soc. Chim. Belg.*, 1985, **94**, 651.
10. P. L. Anelli, C. Biffi, F. Montanari and S. Quici, *J. Org. Chem.*, 1987, **52**, 2559.
11. Recent reviews on the use of TEMPO: R. Ciriminna and M. Pagliaro, *Org. Process Res. Dev.*, 2010, **14**, 245; T. Vogler and A. Studer, *Synthesis*, 2008, 1979.
12. One typical reaction example is the oxidation of (*S*)-(-)-2-methyl-1-butanol to (*S*)-(+)-2-methylbutanal: P. L. Anelli, F. Montanari and S. Quici, *Org. Synth., Coll.*, 1990, **69** and **8**, 212 and 367.
13. R. Siedlecka, J. Skarżewski and J. Młochowski, *Tetrahedron Lett.*, 1990, **31**, 2177.
14. P. Lehenaff, *Bull. Soc. Chim. Fr.*, 1968, **11**, 4687.
15. S. E. Bode, M. Wolberg and M. Muller, *Synthesis*, 2006, 557.
16. Theoretical basis of thin layer chromatography: B. Spangenberg, C. F. Poole and C. Weins, *Quantitative Thin-Layer Chromatography A Practical Survey*, Springer-Verlag, Berlin Heidelberg, 2011, p. 13.

12.1.3. Catalyzed Oxidation of Naphthalene to 1,4-Naphthoquinone

Jacek Skarżewski

Department of Organic Chemistry, Wroclaw University of Technology, Wyb. Wyspianskiego 27, 50-370 Wroclaw, Poland. *E-mail: jacek.skarzewski@pwr.edu.pl

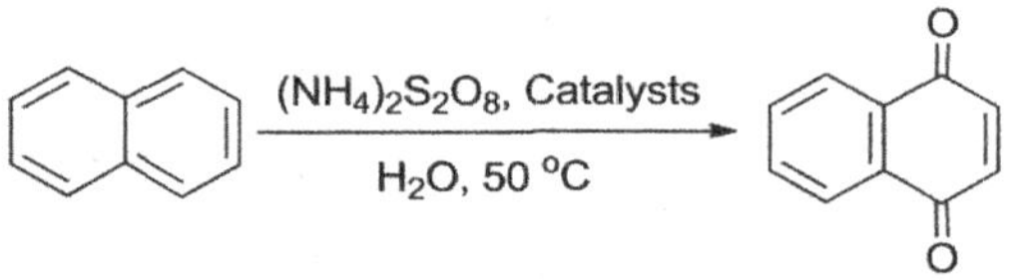

Catalysts: $(NH_4)_2Ce(SO_4)_2$, $AgNO_3$, $C_{12}H_{25}OSO_3Na$

Number of sessions (duration of each session)	Hazard level	Difficulty level	Level of study
2 (4 h + 2 h)	Moderate	Low	Intermediate

Class names Aromatic hydrocarbons, quinones

Concepts involved This experiment gives a method for the biphasic oxidation of polycyclic aromatic hydrocarbons to the respective quinones in an aqueous solution of ammonium persulfate used as a stoichiometric oxidant, in the sub-stoichiometric presence of cerium ammonium sulfate, silver nitrate, and sodium dodecyl sulfate as the reaction promoters. The student should learn the basics of green chemistry (reaction in water, effluents work-up). The chemistry involves a single-electron oxidation, formation of radical-cations, and the catalytic oxidation cascade

Chemicals needed Naphthalene, ammonium persulfate, cerium ammonium sulfate, silver nitrate, sulfuric acid, sodium dodecyl sulfate, ethyl acetate, sodium sulfate, methanol

Equipment and experimental techniques involved Magnetic stirrer with a hot plate, Erlenmeyer flask, separation funnel, rotator evaporator, analytical thin-layer chromatography (SiO_2 precoated foil), melting point apparatus, sintered-glass filter, filtering flask, ^{1}H NMR spectrometer

Keywords Evaporation, liquid–liquid extraction, mp determination, NMR, oxidation, quinone, recrystallization, thin-layer chromatography

Background

1,4-Naphthoquinones are represented by many natural products, including simple 2-methyl-derivatives (menadione, vitamin K_3), and important anticancer drugs, such as streptonigrin, actinomycins, mitomycins. On the other hand, quinones find many industrial applications as polymer modifiers and synthetic intermediates.[1] They are good dienophiles in the Diels–Alder reaction as well as acceptors for various nucleophiles in the Michael reaction.[2] The most straightforward synthetic approach to these useful compounds is based on the selective oxidation of the respective hydrocarbons. The controlled oxidation of aromatic hydrocarbons is still a challenging task. Fortunately, here partially oxidized species (*i.e.* phenols) are much easier to oxidize than the corresponding precursor molecules and the final quinones are essentially resistant

Comprehensive Organic Chemistry Experiments for the Laboratory Classroom
Edited by Carlos A. M. Afonso, Nuno R. Candeias, Dulce Pereira Simão, Alexandre F. Trindade, Jaime A. S. Coelho, Bin Tan, and Robert Franzén.

Published by the Royal Society of Chemistry, www.rsc.org

towards further oxidation. Nevertheless, there is still a problem of finding a strong enough and ecologically acceptable oxidant to replace traditionally used CrO_3.[3] Ceric sulfate as a single-electron oxidant offers an electrochemical potential (*ca.* 1.44 V) high enough to remove one electron from the polycyclic aromatic hydrocarbon. The formed radical-cation reacts with water, loses the next electron and is transformed into phenol. This intermediate is easily further oxidized to the final quinone.[4] However, the whole process requires as much as six oxidant equivalents. From an economical point of view (also atom economy!) it is an inefficient method. To circumvent this obstacle a catalytic cycle has been invented.[5] It allows for *in situ* reoxidation of the used ceric ions (reduced to Ce^{3+}) by cheap and environmentally benign persulfate. To make this reaction faster silver ions are applied as a mediatory between the two-electron oxidant ($S_2O_8^{2-}$, persulfate) and the one-electron reductant (Ce^{3+}). The catalytic cascade resembles that observed in the Wacker–Tsuji oxidation.[6] Moreover, the ionic redox reaction goes smoothly in water, so the hydrophobic aromatic substrate has to be solubilized in this medium. In fact, it is emulsified by the addition of sodium dodecyl sulfate, an anionic surfactant and the oxidation is carried out under phase-transfer catalytic conditions.[7]

Additional Safety

All the necessary care should be taken when handling concentrated sulfuric acid. In agreement with regular good laboratory practices, one should avoid grinding solid naphthalene together with ammonium persulfate (strong oxidant!). Just simple, gentle mixing of both solid reagents is quite safe. The mixture of naphthalene and the oxidant should be added under a fume hood. In addition, for the recrystallization from methanol use a fume hood or/and a reflux condenser. Avoid direct contact of the product with your skin (use nitrile gloves).

Experimental Procedure

Laboratory Session 1 (4 h): Oxidation of Naphthalene

1. On a magnetic hot plate place a water bath and set its temperature at 50 °C.
2. Prepare a 350 mL Erlenmeyer flask with a magnetic stirring bar and place in it 250 mL of distilled water, sulfuric acid (4.7 mL, d 1.84), dihydrated cerium ammonium sulfate (15.8 g, 25 mmol), silver nitrate (1.27 g, 7.7 mmol), and sodium dodecyl sulfate (72 mg, 0.25 mmol) and stir until all the components are dissolved (5 min).
3. Mix gently with a plastic spoon *separately* ground solid naphthalene (5.3 g, 41.6 mmol) and ammonium persulfate (35.0 g, 150 mmol).
4. Place the flask in the bath, stirring the reaction mixture, maintaining a constant temperature of 50 °C and over a 3 h period add small portions of the prepared solid mixture of naphthalene and ammonium persulfate.
5. Remove the flask from the bath and allow it to cool until room temperature.
6. Withdraw a small sample (*ca.* 2 mL), place it in a test tube and extract with the same amount of ethyl acetate. Separate the organic layer (use a Pasteur pipette) to a second test tube and dry it by passing an air-stream.
7. Analyze the test tube content by TLC (precoated silica gel 60 plate, eluent: *n*-hexane–ethyl acetate (3:1 v/v), UV-lamp, naphthalene $R_f = 0.60$ and 1,4-naphthoquinone $R_f = 0.41$).
8. Place the sample $CDCl_3$ solution into an NMR tube and record its ^1H NMR spectrum. Ascribe signals to naphthalene and 1,4-naphthoquinone and compare with the literature. From the integration of signals for the substrate and the product calculate the reaction progress. Based on that additional reaction time is needed.
9. Keep the reaction mixture stirred at room temperature until the next session (overnight).

Laboratory Session 2 (2 h): Product Isolation and Yield Determination

1. Place the reaction mixture in a separatory funnel and extract it three times with 100 mL portions of ethyl acetate. Wash the combined extracts with saturated NaCl solution, briefly dry it with Na_2SO_4 and place in a pre-weighed round-bottom flask.
2. Evaporate the solvent under reduced pressure, leaving the solid crude product.
3. Weigh the flask with the crude product and note its weight.
4. Dissolve the crude product in a minimal amount of methanol (*ca.* 20 mL) at *ca.* 65 °C (pre-heated water bath).
5. Cool the obtained clear solution to room temperature, then leave it in a fridge (+5 °C) for 15 min.
6. Filter off the separated yellow crystals of 1,4-naphthoquinone (sinter-glass funnel) and dry it by passing an air-stream.
7. Weigh the obtained product and calculate the reaction yield.
8. Determine the melting point and compare it with the literature value.
9. In a NMR tube dissolve a 10 mg sample in $CDCl_3$ and acquire the ^1H NMR spectrum.

Waste Disposal

The spent organic solvents can be distilled and reused. The aqueous solution of catalysts can be used for further oxidation. However, before the next run, a fresh portion of sodium dodecyl sulfate (72 mg, 0.25 mmol) should be added. After all the experiments are finished, the spent aqueous catalytic solution is treated with sodium hydroxide and the mixture is left overnight. The precipitates formed are filtered off for separate disposal and the alkaline filtrate can be diluted and poured down the sink.

Results Interpretation and Additional Questions

1. Interpret the ^1H NMR spectra obtained and compare the obtained melting point with the literature value.
2. Calculate the reaction yields (i) after 3 h (based on the ^1H NMR) and (ii) final yield based on the amount of the recrystallized product.
3. Write out the mechanism of naphthalene oxidation with cerium ammonium sulfate in water involving a single electron transfer.
4. Write the catalytic reaction cascade and calculate the turn-over number (TON) for the Ce^{4+} ions.
5. Explain the role of sodium dodecyl sulfate in the catalytic system taking into account its concentration.

References

1. *Ullman's Encyclopedia of Chemical Industry*, Wiley-VCH Verlag GmbH, Weinheim, 2003, vol. 22, p. 471.
2. N. N. Kutyrev, *Tetrahedron*, 1991, **47**, 8043.
3. (a) E. A. Braude and J. S. Fawcett, *Org. Synth.*, 1953, **33**, 50; (b) *Org. Synth. Coll.* 4, 1963 and 698; (c) Organic syntheses, Working with hazardous chemicals, http://www.orgsyn.org/content/pdfs/procedures/cv4p0698.pdf; (d) B. S. Furniss, A. J. Hannaford, P. W. G. Smith and A. R. Tatchell, *Vogel's Text-Book of Practical Organic Chemistry*, Longman, Harlow, 5th edn, 1989, p. 1022.
4. M. V. Bhatt and M. Periasamy, *J. Chem. Soc., Perkin Trans. 2*, 1993, 1811.
5. J. Skarżewski, *Tetrahedron*, 1984, **40**, 4997.
6. See: http://www.organic-chemistry.org/namedreactions/wacker-tsuji-oxidation.shtm.
7. See: http://www.organic-chemistry.org/Highlights/2007/01February.shtm.

12.1.4. Alcohol Oxidation: Menthone Preparation by Menthol Oxidation Using Pyridinium Chlorochromate Immobilized in Silica Gel

Rafael F. A. Gomes* and Carlos A. M. Afonso*

iMed.UL, Faculty of Pharmacy, University of Lisbon, Av. Prof. Gama Pinto, 1649-003 Lisboa, Portugal
*E-mail: rafael.gomes@campus.ul.pt, carlosafonso@ff.ulisboa.pt

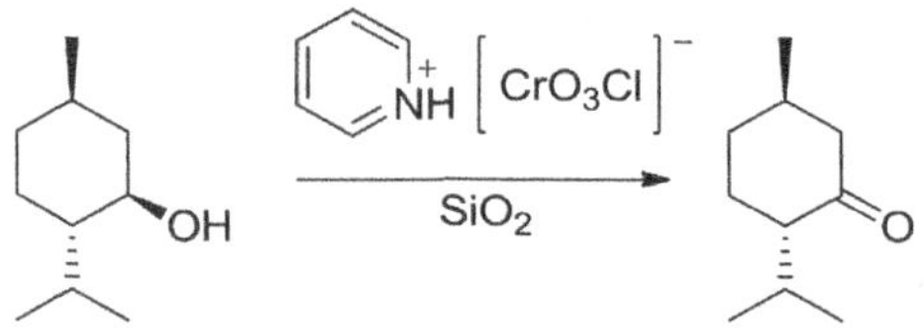

Number of sessions (duration of each session)	Hazard level	Difficulty level	Level of study
1 (3 h)	Moderate	Medium	Intermediate
Class names Alcohols, ketones			
Concepts involved This experiment gives a method for oxidation of a secondary alcohol into a ketone through a metal ion reduction (chromium(VI)) and the immobilization of a reagent (pyridinium chlorochromate) in a solid support			
Chemicals needed (−)-Menthol, pyridinium chlorochromate, silica, dichloromethane, diethyl ether, phosphomolybdenic acid solution in ethanol			
Equipment and experimental techniques involved Pestle and mortar, round-bottom flask, magnetic stirrer, TLC plates, Büchner funnel; reagent immobilization on a solid support, thin-layer chromatography (TLC) to follow a reaction, silica filtration, polarimetry, infrared (IR) and nuclear magnetic resonance (NMR)			
Keywords Alcohol, chromatography, immobilization, ketone, NMR, oxidation			

Background

Oxidation of alcohols is a highly used method in organic chemistry. Several total synthesis of natural compounds, for example, need to use oxidation methods to obtain the final compound. For example, Kang *et al.* described the total synthesis of (−)-crinipellin A where they used both the Swern oxidation and Dess–Martin periodinane to complete the synthesis.[1] Many oxidation methods are known, such as the Swern,[2,3] Oppenhauer,[3,4] Dess–Martin periodinane,[3,5] TEMPO[6] and many chromium based reagents, such as the Jones reagent (CrO_3) (Table 12.1.4.1).[5,7]

Pyridinium chlorochromate (PCC)[5] is frequently used for the conversion of primary and secondary alcohols into aldehydes and ketones, respectively. Unlike the Jones reagent, PCC does not

Comprehensive Organic Chemistry Experiments for the Laboratory Classroom
Edited by Carlos A. M. Afonso, Nuno R. Candeias, Dulce Pereira Simão, Alexandre F. Trindade, Jaime A. S. Coelho, Bin Tan, and Robert Franzén.

Published by the Royal Society of Chemistry, www.rsc.org

Table 12.1.4.1 Common oxidation methods.

Oxidation	Reagents	Key Intermediate	Product (from primary alcohol)
Swern	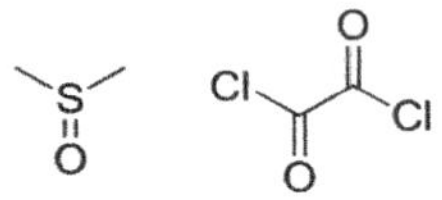	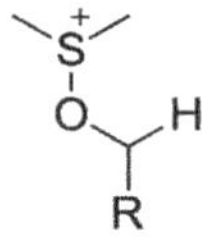	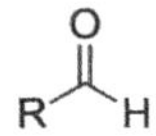
TEMPO	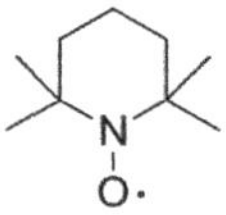	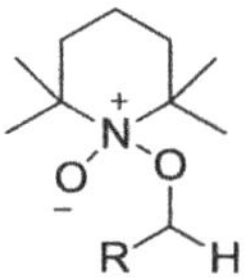	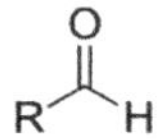
Dess–Martin	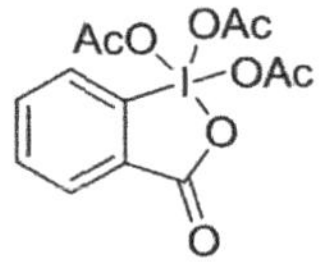	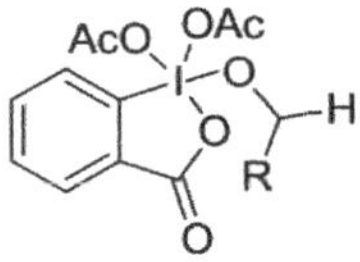	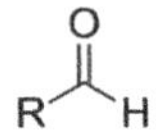
Jones	CrO_3	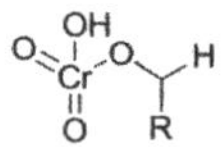	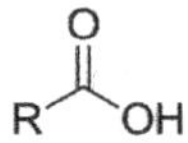

oxidize the aldehyde into its corresponding carboxylic acid. Despite this, the reagent presents some purification difficulties due to the formation of polymeric side products of the reduced chromium species. The purification uses water extraction, which presents an environmental issue of chromium contaminated aqueous residues. One way to reduce this problem is to use PCC previously immobilized on silica.[5,8,9]

Even though the water contamination is reduced, a toxic reagent such as chromium is still used and in stoichiometric quantities, and the silica is still contaminated with the catalyst, which allows the discussion of greener alternatives.[10]

Additional Safety

The student ought to be extremely careful with the oxidant (pyridinium chlorochromate) and the immobilization should be done in the fume hood.

Pyridinium chlorochromate is a strong oxidizer. Contact with other material may cause a fire. Contact with PCC may cause cancer by inhalation, sensitization by skin contact and eye, skin, and respiratory tract irritation. It is also toxic to aquatic organisms and may cause long-term adverse effects in the aquatic environment.

Phosphomolybdenic acid is corrosive and an oxidizer. It can cause eye and skin burns, severe respiratory and digestive tract irritation with possible burns. Contact with other material may cause fire.

Silica if inhaled may cause respiratory tract irritation.

Dichloromethane is very hazardous in case of eye contact (irritant), of ingestion, inhalation and of skin contact (irritant, permeator). Inflammation of the eye is characterized by redness, watering, and itching. It has been proven carcinogenic by OSHA.

Diethyl ether is hazardous in case of skin contact (irritant), eye contact (irritant), ingestion and inhalation. It is slightly hazardous in case of skin contact (permeator). Highly flammable.

Menthol may cause irritation to eyes, skin, and mucous membranes.

Experimental Procedure

Oxidant Preparation

1. Add pyridinium chlorochromate (PCC) (4.85 g, 22.5 mmol) to silica gel (4.85 g, 35–70 mesh) and mix the two solids using a pestle and mortar until the mixture is homogeneous and orange.

(−)-Menthol Oxidation

2. To the previously prepared solid in a 100 mL round-bottom flask equipped with a magnetic stirrer add dichloromethane (50 mL).
3. To this suspension add (−)-menthol (2.34 g, 15 mmol) and follow the reaction by TLC (detection with a phosphomolybdenic acid solution in ethanol) using hexane : diethyl ether (1 : 1) as eluent.
4. After the reaction has ended (less than 90 min) add diethyl ether (120 mL) to the reaction mixture.
5. Filter through a Büchner funnel containing silica gel (2.5–3 cm, 35–70 mesh) and wash the solid with diethyl ether (20 mL).
6. Evaporate the solvent under vacuum, and analyse by TLC. If there is still starting material, purify by column chromatography using the eluent chosen for the TLC.
7. Weigh the product, measure the optical rotation ($[\alpha] = -29.6°$ in ethanol) and analyse by IR and ^{1}H NMR spectroscopy, comparing the data with that of a pure sample.

Optical Rotation Measurement

8. Weigh 10 mg of the product into a volumetric flask.
9. Add ethanol to reach a total volume of 10 mL.
10. Stopper the volumetric flask and mix it well.
11. Measure the optical rotation using a clean and dry polarimeter cell and record the observed rotation in your laboratory notebook.

Waste Disposal

After evaporation of the filtrate, the solvent (diethyl ether and dichloromethane) should be poured into the halogenated solvents disposal containers. The silica should be handled with care (since it contains chromium species) and poured into the solid residues disposal container.

Results Interpretation and Additional Questions

1. Describe the mechanism involved in the oxidation of an alcohol with a chromium catalyst.
2. Why is the immobilization of PCC used?
3. How did you follow the reaction by TLC?
4. Describe other methods used to oxidize alcohols into ketones?
5. If a primary alcohol was submitted to these reaction conditions, which functional group would you expect in the product? In addition, if you used the Jones reagent would the product be the same? Justify your answer.
6. Analyse the IR and ^{1}H NMR spectra.
7. Comment on the observed optical rotation and determine the absolute configuration.

References

1. T. Kang, S. B. Song, W. Y. Kim, B. G. Kim and H. Y. Lee, *J. Am. Chem. Soc.*, 2014, **136**, 10274.
2. Y. Q. Liu and J. C. Vederas, *J. Org. Chem.*, 1996, **61**, 7856.
3. J. B. Arterburn, *Tetrahedron*, 2001, **57**, 9765.
4. C. R. Graves, B. S. Zeng and S. T. Nguyen, *J. Am. Chem. Soc.*, 2006, **128**, 12596.
5. M. B. Smith, *March's Advanced Organic Chemistry: Reactions, Mechanisms and Structure*, John Wiley & Sons, 7th edn, 2013, 1442.
6. L. De Luca, G. Giacomelli and A. Porcheddu, *Org. Lett.*, 2001, **3**, 3041.
7. M. H. Ali and C. J. Wiggin, *Synth. Commun.*, 2001, **31**, 3383.
8. F. A. Luzzio, R. W. Fitch, W. J. Moore and K. J. Mudd, *J. Chem. Educ.*, 1999, **76**, 974.
9. (a) A. R. Tatchell, B. S. Furnis, A. J. Hannaford, and P. W. G. Smith, *Vogel's Textbook of Practical Organic Chemistry*, 5th edn, Longman, 1989, pp. 607–610; (b) Y.-S. Cheng, W.-L. Liu and S.-H. Chen, *Synthesis*, 1980, 223; (c) G. Piancatelli, A. Scettri and D' Auria, *Synthesis*, 1982, 245.
10. R. A. Sheldon, *Catal. Today*, 2015, **247**, 4.

12.1.5. Heterocyclic Target Synthesis – Three-Step Syntheses from Benzaldehyde

Iain A. Smellie*, Nigel P. Botting, Brian A. Chalmers, Andrew D. Harper, and Iain L. J. Patterson

School of Chemistry, University of St Andrews, North Haugh, St Andrews, Fife, KY16 9ST, UK
*E-mail: ias10@st-andrews.ac.uk

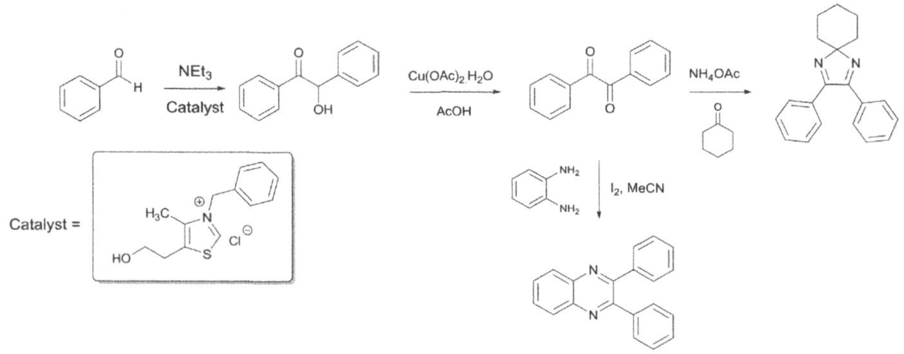

Number of sessions (duration of each session)	Hazard level	Difficulty level	Level of study
4 (3 h each) or 3 (3 h each) if only one heterocycle is prepared	Moderate	Medium	Intermediate

Class names Aldehydes, amines, α-hydroxyketones, ketones, spiro-imidazole quinoxalines

Concepts involved This experiment is designed to demonstrate multi-step heterocyclic synthesis and could be used to introduce key concepts in synthesis planning. The experimental section describes a two-step conversion of benzaldehyde into benzil, followed by two potential heterocycle syntheses. In the first step a biomimetic synthesis of benzoin is accomplished. In the second step the benzoin intermediate undergoes oxidation by cupric acetate to afford benzil. The final stage of the sequence can either be (1) reaction of benzil with 1,2-diaminobenzene to afford the corresponding quinoxaline or (2) reaction of benzil with cyclohexanone and ammonium acetate to form the corresponding spiro-imidazole (2,3-diphenyl-1,4-diazaspiro[4.5]deca-1,3-diene)

Chemicals needed Benzaldehyde, 3-benzyl-5-(2-hydroxyethyl)-4-methylthiazolium chloride, deuterochloroform, triethylamine, ethanol, hydrochloric acid, cupric acetate monohydrate, acetic acid, ammonium acetate, cyclohexanone, 1,2-diaminobenzene, Iodine, acetonitrile, magnesium sulfate, sodium thiosulfate

Equipment and experimental techniques involved Heating mantle or hotplate, glassware for reflux assembly; magnetic stirring, liquid–liquid extraction, vacuum filtration, recrystallization

Keywords Heterocycle synthesis, liquid–liquid extraction, multi-step synthesis, NMR, recrystallisation, vacuum filtration

Comprehensive Organic Chemistry Experiments for the Laboratory Classroom
Edited by Carlos A. M. Afonso, Nuno R. Candeias, Dulce Pereira Simão, Alexandre F. Trindade, Jaime A. S. Coelho, Bin Tan, and Robert Franzén.

Published by the Royal Society of Chemistry, www.rsc.org

Background

The ability to efficiently design and conduct multi-step synthetic sequences is a key skill that students must acquire before working in an organic chemistry research laboratory. Advanced organic chemistry teaching laboratory courses therefore often contain example multi-step syntheses, a popular example being the preparation of 2,3-diphenylquinoxaline (**1**) from benzaldehyde.[1] The quinoxaline synthesis reported here is a modification of the classic teaching laboratory route; in addition, an option to prepare spiro-imidazole **2** (2,3-diphenyl-1,4-diazaspiro[4.5]deca-1,3-diene)[2] is presented. This compound has found application as a precursor for enantiomerically pure 1,2-diaryl-1,2-diaminoethanes.[2,3] Both heterocycles serve as convenient heterocyclic synthetic targets in a teaching laboratory setting and are representative of well-established routes to quinoxalines[4] and imidazoles[5] (see Scheme 12.1.5.1 for a retrosynthetic overview). Overall the sequence employed in this series of experiments involves benzoin condensation of benzaldehyde, oxidation of benzoin to benzil and final assembly of the target heterocycle.

In addition to teaching students how to design and successfully perform multiple step reaction sequences, the examples presented illustrate the synthesis of biologically relevant classes of heterocycle.[4,6] The initial reaction step is typically a thiazolium catalysed benzoin condensation of benzaldehyde; this reaction is of particular interest since it provides a link to the concept of biomimetic synthesis (the chemical principles in this case are related to the function of vitamin B_1).[7]

Benzaldehyde undergoes a cyanide catalysed benzoin condensation in good yield;[1,8] however, this poses a significant toxicity problem. Safer alternatives to cyanide salts are fortunately available, thiamine[1] and 3-benzyl-5-(2-hydroxyethyl)-4-methylthiazolium chloride[9,10] being convenient catalysts for benzoin reactions. In the experiment described, 3-benzyl-5-(2-hydroxyethyl)-4-methylthiazolium chloride is employed as the catalyst.

Oxidation of benzoin to benzil is a well-known reaction that can be accomplished using a wide range of reagents including alumina supported manganese dioxide,[11] copper sulfate/pyridine,[12] oxammonium salts[13,14] and *o*-iodoxybenzoic acid[15] (IBX). Nitric acid is perhaps the most widely employed oxidant for benzoin oxidation in a teaching laboratory setting.[1] In the protocol described, the conditions of Depreux *et al.*[16] have been employed. In this case the reaction is relatively fast and does not require the use of corrosive nitric acid.

The traditional method for preparation of 2,3-diphenylquinoxaline involves heating benzil and 1,2-diaminobenzene and allowing the starting materials to form a melt.[1] In the procedure described, the protocol of Yao *et al.*[17] has been adapted for teaching laboratory use. The latter method does not require heating and the 1,2-diaminobenzene starting material does not require prior purification. A reliable large scale synthesis of 2,3-diphenyl-1,4-diazaspiro[4.5]deca-1,3-diene has been reported by Corey and Pikul.[2] The reaction involves heating a mixture of benzil,

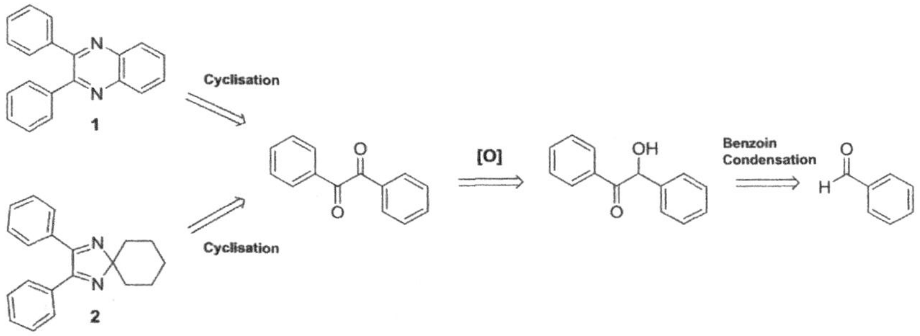

Scheme 12.1.5.1 Retrosynthetic overview of the preparation of quinoxaline **1** and spiro-imidazole **2** from benzaldehyde.

cyclohexanone and ammonium acetate in glacial acetic acid. In the experiment reported herein, the original procedure has been adapted for teaching laboratory work.

^{1}H NMR spectra of the products from each step are instructive and in conjunction with melting point analysis allow the products at every stage to be unambiguously identified.

Additional Safety

All operations should be performed in a fume hood with appropriate protective clothing (laboratory coat, safety goggles and gloves). 1,2-Diaminobenzene is harmful by inhalation, ingestion or skin contact), 1,2-diaminobenzene (like all other aromatic amines) is a suspected carcinogen. Iodine (vapour in particular) is toxic and can cause burns on inhalation or prolonged skin contact). Deuterochloroform is potentially carcinogenic. Glacial acetic acid and triethylamine are flammable and corrosive; exposure can be *via* inhalation, ingestion or skin contact. Cyclohexanone is flammable and harmful upon inhalation, ingestion or skin contact. Benzaldehyde could be harmful if ingested. Sodium thiosulfate should be treated as an irritant and corrosive, especially to the eye. Sodium thiosulfate can release sulfur and sulfur dioxide (corrosive and toxic gas) if exposed to acids. Acetonitrile and ethanol are highly flammable solvents. 3-Benzyl-5-(2-hydroxyethyl)-4-methylthiazolium chloride should be treated as an irritant.

Experimental Procedure

Laboratory Session 1: Synthesis of Benzoin (3 h)

1. Transfer 6.5 g of benzaldehyde, 24 mL of ethanol, 0.86 g of 3-benzyl-5-(2-hydroxyethyl)-4-methylthiazolium chloride and 2.5 mL of triethylamine to a 100 mL round-bottom flask.
2. Equip the flask with a reflux condenser, calcium chloride drying tube and magnetic stirrer and heat the stirred mixture to reflux for 1 h.
3. After cooling to room temperature, pour the reaction mixture onto crushed ice (*ca.* 60 g).
4. Once the ice has melted add *ca.* 18 mL of 2 M HCl in small portions with stirring, until the reaction mixture reaches pH 1.
5. Collect the benzoin, as a yellow precipitate, by filtration and wash it with ice-cold water. Purify the benzoin by recrystallisation from aqueous ethanol, obtaining off-white, crystalline material.
6. Filter off the resulting yellow solid under suction and wash the filter cake with ice-cold water (2 × 15 mL).

Purification of benzoin: recrystallize the crude product from a minimal volume of aqueous ethanol

Weigh the product and determine the % yield. Measure the melting point and record a ^{1}H NMR ($CDCl_3$) spectrum of your product.

Laboratory Session 2: Synthesis of Benzil (3 h)

1. Transfer 4 g of benzoin, 30 mL of glacial acetic acid, 10 mL of water and 7.5 g of cupric acetate monohydrate to a 100 mL round-bottom flask.
2. Equip the flask with a reflux condenser and magnetic stirrer and heat the stirred mixture to reflux for 15 min. Using a pre-heated funnel (plugged with a small quantity of glass wool), remove the resulting cuprous oxide by filtering the hot reaction mixture.
3. After cooling to room temperature, filter off the resulting yellow-green solid under suction and wash the filter cake with ice-cold water (2 × 15 mL).

Purification of benzil: recrystallize the crude product from a minimal volume of aqueous ethanol.

Weigh the product and determine the % yield. Measure the melting point and record a ^1H NMR ($CDCl_3$) spectrum of your product.

Laboratory Session 3 (Option 1): Synthesis of 2,3-Diphenylquinoxaline (3 h)

1. Transfer 0.75 g of benzil, 5 mL of acetonitrile and 0.47 g of 1,2-diaminobenzene to a 25 mL round-bottom flask.
2. Transfer 0.1 g of iodine and 3 mL of acetonitrile to a small flask or sample vial, swirl the solution until the iodine dissolves.
3. Equip the round-bottom flask with a magnetic stirrer and add the iodine solution dropwise over 5 min to the stirred reaction mixture. Allow the reaction mixture to stir for 15 min after completion of the addition of iodine.
4. Add 10 mL of ethyl acetate and transfer the reaction mixture to a separating funnel. Rinse the reaction flask further with ethyl acetate (2×15 mL).
5. Wash the organic layer with 10% sodium thiosulfate (2×50 mL), water (2×50 mL) and dry over anhydrous magnesium sulfate.
6. Filter the ethyl acetate solution through a small cotton wool plug directly into a clean and dry round-bottomed flask. Wash the cotton wool with ethyl acetate and press dry with a glass rod.
7. Remove the solvent using a rotary evaporator to afford crude product.

Purification of 2,3-diphenylquinoxaline: recrystallize the crude product from a minimal volume of ethanol.

Weigh the product and determine the % yield. Measure the melting point and record a ^1H NMR ($CDCl_3$) spectrum of your product.

Laboratory Session 3 (Option 2): Synthesis of 2,3-Diphenyl-1,4-diazaspiro[4.5]deca-1,3-diene (3 h)

1. Transfer 2 g of benzil, 12 mL of glacial acetic acid, 1 mL of cyclohexanone and 7.5 g of ammonium acetate to a 50 mL round-bottom flask.
2. Equip the flask with a reflux condenser and magnetic stirrer and heat the stirred mixture to reflux for 1 h.
3. After partially cooling, pour the reaction mixture into 60 mL of vigorously stirred water in a 250 mL beaker. Cover the beaker and store overnight.
4. Filter off the resulting green precipitate under suction.

Purification of 2,3-diphenyl-1,4-diazaspiro[4.5]deca-1,3-diene: recrystallize the crude product from a minimal volume of aqueous ethanol.

Weigh the product and determine the % yield. Measure the melting point and record a ^1H NMR ($CDCl_3$) spectrum of your product.

Waste Disposal

Waste solutions containing iodine should be treated with a solution of sodium metabisulfite or sodium thiosulfate. Aqueous washings from the extraction steps should be diluted with water and disposed of down a sink. All organic solvent waste should be retained and transferred to an appropriate container for disposal. Cupric oxide waste should be retained and transferred to an appropriate container for disposal.

Results Interpretation and Additional Questions

1. Interpret the ^{1}H NMR spectra you obtained from all of the reaction steps. Use the NMR spectra and melting points to confirm the structure of your compounds unambiguously.
2. Provide a curly arrow mechanism for the benzoin reaction you carried out in step **1**.
3. Suggest what function the copper acetate has in the oxidation of benzoin to benzil.

References

1. L. F. Fieser and K. L. Williamson, *Organic Experiments*, D. C. Heath and Company, Lexington MA, 7th edn, 1992, ch. 46, pp. 54 and 55.
2. S. Pikul and E. J. Corey, *Org. Synth.*, 1993, **71**, 22.
3. E. J. Corey, D.-H. Lee and S. Sarshar, *Tetrahedron: Asymmetry*, 1995, **6**, 3.
4. A. E. A. Porter, *Comprehensive Heterocyclic Chemistry I*, ed. A. R. Katritzky and C. W. Rees, Pergamon, Oxford, 1984, ch. 2.14.
5. M. R. Grimmet, *Comprehensive Heterocyclic Chemistry I*, ed. A. R. Katritzky and C. W. Rees, Pergamon, Oxford, 1984, ch. 4.08.
6. C. A. Matuszak and A. J. Matuszak, *J. Chem. Educ.*, 1976, **53**, 280.
7. R. Breslow, *J. Am. Chem. Soc.*, 1958, **80**, 3719.
8. R. Adams and C. S. Marvel, *Org. Synth.*, 1921, **1**, 33.
9. H. Stetter and H. Kuhlmann, *Org. Synth.*, 1984, **62**, 170.
10. A. G. Neo, J. Delgado and C. F. Marcos, *Chem. Educ.*, 2006, **11**, 327.
11. R. D. Crouch, M. S. Holden and J. S. Burger, *J. Chem. Educ.*, 2001, **78**, 951.
12. H. T. Clarke and E. E. Dreger, *Org. Synth.*, 1926, **6**, 6.
13. T. S. Straub, *J. Chem. Educ.*, 1991, **68**, 1048.
14. P. L. Anelli, S. Banfi, F. Montanari and S. Quici, *J. Org. Chem.*, 1989, **54**, 2970.
15. J. D. Moore and N. S. Finney, *Org. Lett.*, 2002, **4**, 3001.
16. P. Depreux, G. Bethegnies and A. Marcincal-Lefebre, *J. Chem. Educ.*, 1988, **65**, 553.
17. S. V. More, M. N. V. Sastry, C.-C. Wang and C.-F. Yao, *Tetrahedron Lett.*, 2005, **46**, 6345.

12.1.6. Oxidation of Activated Phenols

Luísa M. Ferreira* and Paula S. Branco

LAQV, REQUIMTE, Departamento de Química, Faculdade de Ciências e Tecnologia, Universidade Nova de Lisboa, Campus da Caparica, Portugal. *E-mail: lpf@fct.unl.pt

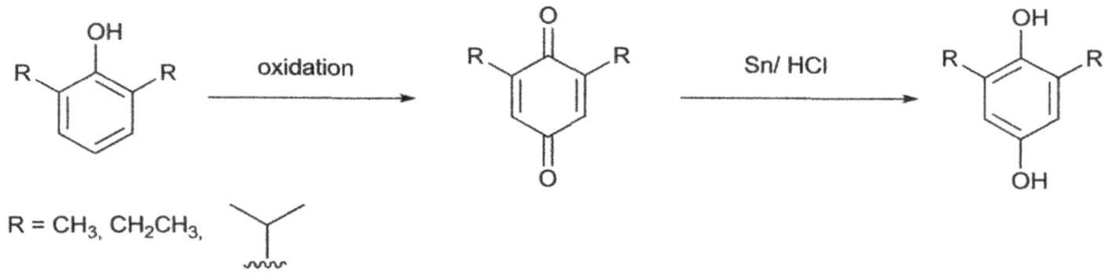

oxidants - O_2/CoSalen, PbO_2, CAN

Number of sessions (duration of each session)	Hazard level	Difficulty level	Level of study
3 (4 h each)	High	High	Intermediate

Class names Quinones, phenols

Concepts involved This experiment gives a method for phenol preparation on an activated aromatic ring by an oxidative process. Such method is an alternative to the traditional method of phenol synthesis by the hydrolysis of diazonium salts. The presented experiment is biomimetic of biologic transformations like those promoted by tyrosinase

Chemicals needed 2,6-Diisopropylphenol, acetone, ceric ammonium nitrate (CAN), *N,N'*-bis(salicylidene)ethylenediaminocobalt(II)perchloric acid, (Co(Salen)), acetic acid, methylene chloride, acetone, ethyl acetate, hexane, silica for column chromatography and pre-coated foil analytical thin-layer chromatography, tin, gaseous oxygen, ethanol, sodium sulfate, hydrochloric acid, brine, pentane

Equipment and experimental techniques involved Reflux, liquid–liquid extraction, evaporation under reduced pressure, monitoring by analytical thin-layer chromatography (SiO_2 pre-coated foil), column chromatography, reaction under controlled atmosphere, melting point apparatus, FTIR spectrometer, NMR spectrometer

Keywords Aromatic hydroxylation, chromatography, NMR, oxidation of activated phenols, quinone, reduction by tin

Background

Propofol (2,6-diisopropylphenol) is a short-acting hypnotic agent used for inducing and maintaining anaesthesia. One of the metabolic pathways on the propofol elimination is the *para* hydroxylation of propofol to produce 4-hydroxypropofol (2,6-diisopropyl-4-hydroxyphenol).[1]

Selective hydroxylation of aromatic compounds is among the most challenging of chemical reactions in synthetic organic chemistry. This oxidation has gained increasing attention in recent years, particularly because phenol and hydroxylated arenes are structural scaffolds frequently found in important pharmaceuticals, agrochemicals, polymers, and biologically active

Comprehensive Organic Chemistry Experiments for the Laboratory Classroom
Edited by Carlos A. M. Afonso, Nuno R. Candeias, Dulce Pereira Simão, Alexandre F. Trindade, Jaime A. S. Coelho, Bin Tan, and Robert Franzén.

Published by the Royal Society of Chemistry, www.rsc.org

natural compounds. A wide variety of oxidative and non-oxidative classical methods have been developed over the years to synthesize phenol and derivatives in high yields, including nucleophilic aromatic substitutions of activated aryl halides, copper-promoted conversions of arene diazonium salts, benzyne methodologies, and anodic oxidation of arenes, among others.[2]

Hydrolysis of diazonium salts (a Sandmeyer-type reaction)[3] is traditionally described as the most important laboratory method for phenol synthesis.[4] To apply the method sacrificial nitrogen atoms must be introduced in the aromatic ring, which are converted into diazonium leaving groups that are replaced by the desired hydroxyl group. The use of reagents for the introduction and conversion of sacrificial atoms results in a time consuming and low atom economy method. A synthesis of 2,6-diisopropyl-4-hydroxyphenol by this method has been published.[1]

Here an oxidation/reduction methodology is used by two independent processes for conversion of the phenol into a quinone, which is later reduced to the diphenol by metal (tin) in acidic media.

The introduction of the oxygen atom is achieved by the use of a catalytic method (oxygen in presence of a CoSalen catalyst) and an excess of a stoichiometric oxidant (ceric ammonium nitrate, CAN). Cerium(IV) compounds represent the most notable oxidants among lanthanide reagents. In particular, CAN has been utilized extensively for a variety of oxidative transformations. The chemistry of Ce(IV) oxidations of organic molecules is dominated by radical chemistry. The fate of these reactive intermediates determines the nature of the organic oxidation products isolated.[5] The mechanisms involved and the influence of the reaction conditions are discussed in the literature based on the properties of cation radicals and radicals resulting from quantum mechanical calculations.[6] The formation of coloured dimers can be prevented using acidic conditions, thereby reducing the attack on the newly produced quinone by the remaining propofol.[7]

When the catalytic system is used, the CoSalen catalyst adds to O_2 to form a catalytically active Co-superoxo adduct and the superoxo adduct abstracts a phenolic hydrogen from the substrate resulting in a phenoxy radical, which combines with a second equivalent of the superoxo adduct. The final step results in the formation of the quinone oxidation product.[8] The method has the advantages of a catalytic method with a low production of metallic residues and has as its main drawback the inherent danger of using a high pressure source of molecular oxygen.

Additional Safety

It is essential to work in a fume hood in any step of the experiment. The procedure involves corrosive acids and toxic metal salts. Regular laboratory good practices should always be taken into account, especially the use of gloves and safety glasses when handling tin, CAN, organic reagents, separator funnel and vacuum apparatus. As tin is not compatible with oxidation reagents, store it in tightly closed containers away from moisture. Silica gel dust is irritating to the respiratory tract and may cause irritation to the skin and eyes, so precautions should be taken. The use of oxygen requires the complete absence of flames in the laboratory and additional care to prevent oxygen enrichment by ensuring that equipment is leak-tight and in good working order. Oxygen gas cylinders should not be taken into confined spaces; the gas can be fed in by using hoses. The hoses should be removed from the confined space when work is finished or suspended.

Experimental Procedure

Session 1 (4 h)

Oxidation by Ceric Ammonium Nitrate (CAN)

1. Add a stirrer magnetic bar, 70% perchloric acid (2.5 mL), acetone (15 mL) and ceric ammonium nitrate(IV) (1.38 g, 5 mmol) in a 50 mL round-bottom flask and add propofol (2,6-diisopropylphenol) (0.2 mL, 1.1 mmol).

2. After magnetic stirring for about 30 min at room temperature perform a TLC plate analysis on silica gel using a slightly polar solvent (*e.g.* hexane : ethyl acetate, 99.5 : 0.5). Use a solution of propofol in acetone (20 µL per 1.5 mL) as standard to check the total consumption of the starting material.
3. Put the reaction mixture into a separatory funnel and extract with ethyl acetate (3 × 20 mL).
4. Wash the resulting organic phase with water until there is a complete elimination of acid (wash water pH > 6), dry the organic phase over anhydrous sodium sulfate, filter the drying agent by gravity to a weighed round-bottomed flask and evaporate the solvent to dryness in a rotary evaporator.
5. Weigh the round-bottom flask with the dry residue and determine the residue mass.
6. Purification of the products is carried out by column chromatography, with hexane/ethyl acetate (9.5 : 0.5) as eluent using 70× more silica than the residue mass.
7. Evaporate the collected fractions with the desired compound to dryness in a weighed round-bottom flask and reserve for the next session.

Oxidation by Air Oxygen in the Presence of CoSalen Catalyst

When the oxidation by CAN is being stirred start this reaction in parallel.

1. To obtain an oxygen saturated solvent, place 5 mL of dimethylformamide (DMF) in a 25 round-bottom flask with a stirrer magnetic bar and pass gaseous oxygen through for 10 min.
2. Add propofol (0.2 mL, 1.1 mmol) and the Co(Salen) catalyst (3 mg). Add a stirrer magnetic bar and perform the reaction at room temperature with magnetic stirring under an oxygen atmosphere (provided by take-off from a balloon filled partly with pure oxygen) for one week while monitoring by TLC.

Session 2 (4 h)

Oxidation by Air Oxygen in the Presence of CoSalen Catalyst

1. The oxidation by air is terminated by addition of the mixture to water (50 mL) and the resulting solution is extracted with ethyl acetate (3 × 20 mL) in a separatory funnel.
2. Wash the resulting organic phase with water (1 × 50 mL), dry it with anhydrous sodium sulfate, filter and dry under vacuum.
3. Purify the products by column chromatography, with hexane/ethyl acetate (9.5 : 0.5) as eluent. The collected fractions with the desired compound are evaporated to dryness in a weighed round-bottom flask. The resulting yellow oil is weighed and compared by TLC with the isolated main product of the oxidation by CAN.
4. Weigh the obtained product and calculate the reaction yield.

Oxidation by Ceric Ammonium Nitrate (CAN)

1. Weigh the resulting yellow oil, calculate the reaction yield, and analyse the product using IR and NMR spectroscopies.
2. The infrared analysis should be done using a drop of the obtained residue between a pair of NaCl discs (film sample).
3. The magnetic resonance analysis should be done using deuterochloroform as solvent and using the prescribed procedures of your laboratory facilities.

Session 3 (4 h)

Reduction of the Quinone by Sn/HCl

The 2,6-diisopropyl-1,4-quinone synthesized in the previous sessions is transformed in this reduction reaction.

1. Introduce the quinone (0.19 g, 1 mmol) in a round-bottom flask (50 mL) with boiling chips.
2. Weigh tin (0.20 g) in a beaker and add 37% hydrochloric acid (35 mL) and ethanol (2 mL). Transfer the beaker contents into the round-bottom flask, heat and stir the reaction under reflux in an oil bath ($T = 150$ °C) for about 2 h.
3. Monitor the reaction by TLC by transferring a few drops of the acid solution into a sample tube with ethyl ether (0.5 mL), stir the mixture and use the supernatant organic phase on the TLC plate. Use hexane/ethyl acetate (9 : 1) as eluent.
4. After the total consumption of the starting material, stop the heating, remove magnetic bar and the boiling chips and drop the mixture into water (10 mL).
5. Extract the aqueous phase with ethyl acetate (3 × 15 mL). Wash the organic phase with brine (NaCl saturated solution) (2 × 15 mL). Dry the resulting organic phase over anhydrous sodium sulfate, filter and dry under vacuum in a weighed round-bottom flask.
6. The resulting product can have an oily appearance and exhibit a yellow/orange coloration. Wash it with pentane and the product precipitates.
7. Weigh the dried white crystals, calculate the reaction yield, and analyse the product using IR and NMR spectroscopies.
8. The infrared analysis should be done using the preparation of a KBr pellet (solid sample).
9. The magnetic resonance analysis should be done using deuterochloroform as solvent and using the prescribed procedures of your laboratory facilities.

Waste Disposal

All organic solvents have to be poured into special containers for subsequent disposal in the manner of treatment of chemical waste. The silica and other solid residues, like metallic tin residues, should be discharged into a special container for later collection and destruction of chemical waste.

Results Interpretation and Additional Questions

1. Interpret the ^{1}H NMR spectra obtained.
2. On oxidation by CAN some coloured side products are formed. Present a rationalization for this.
3. Write out the mechanism of propofol oxidation with CAN.
4. Can the quinone compound react with a nucleophile? If yes, present the mechanism and the structure for the reaction between quinone and water.
5. Present the advantages, discussed in terms of sustainability, of the three described methods for the synthesis of 4-hydroxyphenols: nitration/reduction/diazotization *versus* oxidation/reduction (considering the two oxidation methods proposed).

References

1. J. Helfenbein, C. Lartigue, E. Noirault, E. Azim, J. Legailliard, M. J. Galmier and J. C. Madelmont, *J. Med. Chem.*, 2002, **45**, 5806.
2. D. A. Alonso, C. M. Pastor and M. Yus, *Chem.-Eur. J.*, 2010, **16**, 5274.

3. K. Dimroth, A. Berndt, H. Perst and C. Reichardt, *Org. Synth.*, 1973, **5**, 1130.
4. D. S. Wulfman, *Diazonium and Diazo Groups*, ed. S. Patai, John Wiley & Sons, Inc., New York, 1978, p. 296.
5. G. Molander, *Chem. Rev.*, 1992, **92**, 29.
6. S. Domagala, V. Steglinska and J. Dziegiec, *Monatsh. Chem.*, 1998, **129**, 761.
7. K. Omura, *Synthesis*, 1998, 1145.
8. T. Elder, J. J. Bozell and D. Cedeno, *Phys. Chem. Chem. Phys.*, 2013, **15**, 7328.

12.2. Oxidative Cyclizations and Dehydrogenation

12.2.1. Synthesis of 2-(5-Phenylthien-2′-yl) benzothiazole

Maria Manuela Marques Raposo*, Susana Paula Graça da Costa, and Rosa Maria Ferreira Batista

Department of Chemistry, University of Minho, Campus de Gualtar, 4710-057 Braga, Portugal
*E-mail: mfox@quimica.uminho.pt

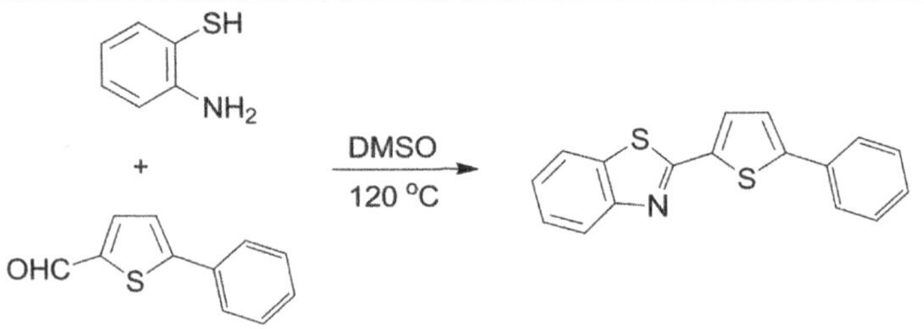

Number of sessions (duration of each session)	Hazard level	Difficulty level	Level of study
2 (3 h each)	Moderate	Low	Intermediate

Class names Amines, thiols, aldehydes, benzothiazoles

Concepts involved Synthesis of a 1,3-benzothiazole derivative by oxidative cyclization between 2-aminothiophenol and an aldehyde; characterization by ^1H and ^{13}C NMR spectroscopy; liquid–liquid extraction; rotary evaporation; thin-layer and column chromatography on silica gel

Chemicals needed 2-Aminobenzenethiol, 5-phenylthiophene-2-carbaldehyde, dimethyl sulfoxide (DMSO), ethyl acetate, dichloromethane, deuterochloroform, petroleum ether (40–60 °C), magnesium sulfate (anhydrous), silica gel (230–400 mesh), silica gel thin-layer chromatography (TLC) plates with fluorescent indicator

Equipment and experimental techniques involved Column chromatography on silica gel, ^1H and ^{13}C nuclear magnetic resonance spectroscopy (NMR), melting point; apparatus for reaction with temperature (120 °C) and magnetic stirring, thin-layer chromatography (TLC), liquid–liquid extraction, gravity filtration, rotary evaporation

Keywords Benzothiazole, chromatography, intramolecular oxidative cyclization, NMR, thiophene

Background

Benzothiazole derivatives have been widely used in various areas due to their interesting optical properties, for example as fluorescent brightening agents for textiles, plastics, inks and paints, as tuneable dye lasers, as fluorescent dopants in the fabrication of organic light emitting devices (OLEDs), as solvatochromic probes, and in the preparation of materials with good

Comprehensive Organic Chemistry Experiments for the Laboratory Classroom
Edited by Carlos A. M. Afonso, Nuno R. Candeias, Dulce Pereira Simão, Alexandre F. Trindade, Jaime A. S. Coelho, Bin Tan, and Robert Franzén.

Published by the Royal Society of Chemistry, www.rsc.org

photochromic and nonlinear optical properties.[1,2] In addition, a large number of therapeutic agents and fluorescent labels in medical analysis are based on the benzothiazole nucleus, which has been shown to possess significant biological activities such as anti-inflammatory, antimicrobial, anti-diabetic, anti-cancer and anti-HIV activity.[3]

Thiophene and its derivatives also exhibit interesting biological activity and important electroluminescent properties with wide application in polymer science, which led to their application in energy transfer systems in optical and electronic devices, as well as sensors and fluorescent markers.[4] Therefore, the combination of a thiophene moiety with a benzothiazole nucleus should result in a compound exhibiting enhanced fluorescence properties.

Benzothiazoles are usually prepared from a substituted aromatic aldehyde, carboxylic acid, acyl chloride or nitrile by condensation with *o*-aminobenzenethiol and through other methods,[3] such as the Jacobsen cyclization or the Hoffman method; however, most of these synthetic procedures include disadvantages, such as aggressive reaction conditions (strongly acidic and high temperatures), lengthy procedures, or expensive catalysts that may be environmentally damaging. In this work the reaction is performed through condensation of a formyl derivative with *o*-aminobenzenethiol, in DMSO at 120 °C.[2a–d]

Additional Safety

The experimental work, in two sessions, should be performed in a well-ventilated fume hood with proper safety equipment such as protective clothing, gloves, and safety goggles. 5-Phenylthiophene-2-carbaldehyde and DMSO may cause skin, eye and respiratory tract irritation. 2-Aminothiophenol may cause skin burns and eye damage and has a very intense odour. Dichloromethane is irritating to eyes and skin, toxic by inhalation and the vapours may cause dizziness. Petroleum ether (40–60 °C) is highly flammable, irritating to skin and the vapours may cause dizziness. Ethyl acetate may cause eye irritation. Silica gel may cause respiratory irritation, do not inhale the dust.

Experimental Procedure

Laboratory Session 1 (3 h): Preparation of 2-(5′-Phenylthien-2′-yl) benzothiazole

1. Prepare a 25 mL round-bottomed flask with a magnetic stir bar, and add 100 mg (0.5 mmol, 1 equiv.) of 5-phenylthiophene-2-carbaldehyde, 66 mg (0.53 mmol, 1 equiv.) of 2-aminobenzenethiol and 5 mL of DMSO.
2. Adapt a water condenser into the round-bottomed flask. Stir the reaction rapidly for 1 h at 120 °C in an oil bath over a magnetic stirring plate.
3. Confirm that the reaction is complete by TLC using dichloromethane/petroleum ether (1 : 1) as eluent and compare with the starting aldehyde (see ESI).
4. Allow to cool to room temperature and add 10 mL of water. Transfer to a separating funnel.
5. Extract with ethyl acetate (2 × 15 mL). Combine the organic extracts, dry with anhydrous magnesium sulfate, filter the drying agent using gravity filtration and evaporate the solvent under reduced pressure in the rotary evaporator. Keep the flask with the crude in a vacuum excicator until the next session.

Laboratory Session 2 (3 h): Purification of the Product

1. Purify the previously obtained crude by column chromatography on silica gel (column height and diameter: 40 cm and 2.5 cm; silica weight: 35 g; silica height: ≈20 cm; column packing solvent: dichloromethane/petroleum ether (1 : 3)). Dissolve the evaporated crude residue in

the minimum volume of packing solvent, apply to the top of the packed column and begin eluting with dichloromethane/petroleum ether (1:3) (≈30 mL). Continue the elution with dichloromethane/petroleum ether (1:2) (≈50 mL) and dichloromethane/petroleum ether (40–60 °C) (1:1) (≈30 mL). Collect numbered fractions of about 15 mL each.

2. Check the various fractions collected by TLC using as eluent a mixture of dichloromethane/petroleum ether (1:1) and combine the fractions containing the product.
3. Evaporate the solvent to dryness under reduced pressure on a rotary evaporator.
4. Weigh the yellow solid and check the purity by ^{1}H and ^{13}C NMR spectroscopy and melting point. Compare the results with literature data[2a] as well as with the spectra provided in the ESI to confirm the compound purity.

Waste Disposal

Petroleum ether (40–60 °C), DMSO and ethyl acetate should be collected in a non-chlorinated waste solvent container. Dichloromethane should be collected in a chlorinated waste solvent container Used silica gel must be stored in a closed container in the hood for subsequent disposal. The aqueous solutions should be neutralized and disposed of in the sink.

Results, Interpretation and Additional Questions

1. Interpret the ^{1}H and ^{13}C NMR spectra obtained and compare with the literature data.
2. Discuss the relevance of the use of DMSO as the reaction solvent.
3. Write out the reaction mechanism for the formation of the benzothiazole.
4. Suggest another synthetic route to obtain the title benzothiazole using other reagents.
5. Explain what the expected effect would be on the yield of formation of the corresponding benzothiazole if 5-nitro-2-aminobenzenethiol was used.

References

1. (a) B. M. Krasovitski and B. M. Bolotin, *Organic Luminescent Materials*, VCH, Weinheim, 1988; (b) Y. Tao, C. Yang and J. Qin, *Chem. Soc. Rev.*, 2011, **40**, 2943.
2. (a) S. P. G. Costa, J. A. Ferreira, G. Kirsch, A. M. F. Oliveira-Campos, *J. Chem. Res.*, 1997, (S) 314–315; (M) 2001; (b) R. M. F. Batista, S. P. G. Costa and M. M. M. Raposo, *Tetrahedron Lett.*, 2004, **45**, 2825; (c) S. P. G. Costa, R. M. F. Batista, P. Cardoso, M. Belsley and M. M. M. Raposo, *Eur. J. Org. Chem.*, 2006, 3938; (d) R. M. F. Batista, S. P. G. Costa, E. L. Malheiro, M. Belsley and M. M. M. Raposo, *Tetrahedron*, 2007, **63**, 4258; (e) J. Pina, S. S. de Melo, H. D. Burrows, R. M. F. Batista, S. P. G. Costa and M. M. M. Raposo, *J. Phys. Chem. A*, 2007, **111**, 8574; (f) J. Pina, S. S. de Melo, R. M. F. Batista, S. P. G. Costa and M. M. M. Raposo, *Phys. Chem. Chem. Phys.*, 2010, **12**, 9719; (g) M. M. M. Raposo, M. C. R. Castro, A. M. C. Fonseca, M. F. S. Cardoso, M. Belsley, L. M. Carvalho and P. Coelho, *Dyes Pigm.*, 2011, **91**, 62; (h) J. Garcia-Amorós, M. C. R. Castro, P. Coelho, M. M. M. Raposo and D. Velasco, *Chem. Commun.*, 2013, **49**, 11427; (i) J. Garcia-Amorós, M. Reig, M. C. R. Castro, A. Cuadrado, M. M. M. Raposo and D. Velasco, *Chem. Commun.*, 2014, **50**, 6704; (j) R. M. F. Baptista, D. Isakov, M. M. M. Raposo, M. Belsley, I. Bdikin, A. Kholkin, S. P. G. Costa and E. de Matos Gomes, *J. Nanopart. Res.*, 2014, **16**, 2502.
3. (a) B. Chandrasekhar, *J. Sulfur Chem.*, 2014, **35**, 538; (b) M. Singh and S. K. Singh, *Anti-Cancer Agents Med. Chem.*, 2014, **14**, 127; (c) S. Noel, S. Cadet, E. Gras and C. Hureau, *Chem. Soc. Rev.*, 2013, **42**, 7747; (d) M. Henary, S. Paranjpe and E. A. Owens, *Heterocycl. Commun.*, 2013, **19**, 89; (e) V. Facchinetti, R. R. Reis, C. R. B. Gomes and T. R. A. Vasconcelos, *Mini-Rev. Org. Chem.*, 2012, **9**, 44.
4. (a) G. Barbarella, M. Melucci and G. Sotgiu, *Adv. Mater.*, 2005, **17**, 1581; (b) M. M. M. Raposo, B. García-Acosta, T. Ábalos, P. Calero, R. Martínez-Manez, J. V. Ros-Lis and J. Soto, *J. Org. Chem.*, 2010, **75**, 2922; (c) L. E. Santos-Figueroa, M. E. Moragues, M. M. M. Raposo, R. M. F. Baptista, R. C. M. Ferreira, S. P. G. Costa, F. Sancenón, R. Martínez-Máñez, J. V. Ros-Lis and J. Soto, *Tetrahedron*, 2012, **68**, 7179.

12.2.2. Synthesis of *N*-*tert*-butyloxycarbonyl-[2-(thien-2′-yl)benzoxazol-5-yl]-L-alanine methyl ester

Susana P. G. Costa*, M. Manuela M. Raposo, Cátia I. C. Esteves and R. Cristina M. Ferreira

Department of Chemistry, University of Minho, Campus de Gualtar, 4710-057 Braga, Portugal
*E-mail: spc@quimica.uminho.pt

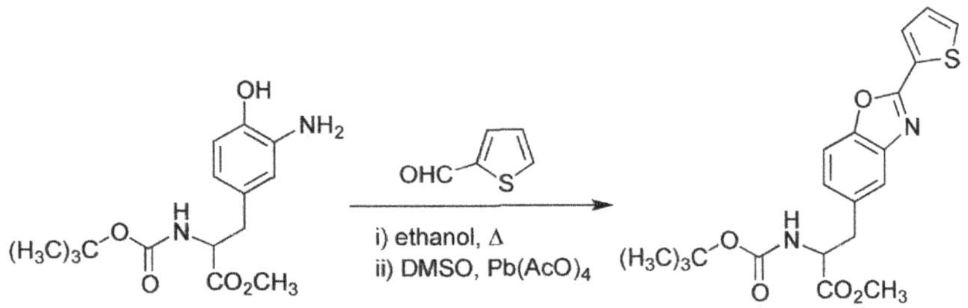

Number of sessions (duration of each session)	Hazard level	Difficulty level	Level of study
3 (4 h + 2 h + 4 h)	High	Low	Intermediate

Class names Aldehydes, amines, imines, alcohols, benzoxazoles

Concepts involved Preparation of a fluorescent heterocyclic amino acid by oxidative cyclization of aminotyrosine with an aldehyde, yielding a benzoxazole; filtration under vacuum; rotary evaporation; liquid–liquid extraction; separation by thin-layer and column chromatography; characterization by ^{1}H and ^{13}C NMR and IR spectroscopy

Chemicals needed *N*-*tert*-Butyloxycarbonyl-3-amino-L-tyrosine methyl ester, thiophene-2-carboxaldehyde, lead tetraacetate, magnesium sulfate (anhydrous), sodium sulfate (anhydrous), ethyl acetate, absolute ethanol, dimethyl sulfoxide, silica gel (230–400 mesh), silica gel thin-layer chromatography (TLC) plates with fluorescent indicator

Equipment and experimental techniques involved Heating mantle for heating at reflux, melting point apparatus, vacuum filtration, rotary evaporation, liquid–liquid extraction, thin-layer chromatography and column chromatography on silica gel

Keywords Amino acid, (benz)oxazole, chromatography, nuclear magnetic resonance, oxidative cyclization, thiophene

Background

Benzoxazoles exhibit excellent optical properties (broad spectral windows, high molar absorptivity coefficients and high fluorescence quantum yields) and recently have been described as excellent fluorescent probes in pharmaceutical analysis and as fluorescent sensors for ions.[1] In addition, benzoxazoles display biological activity such as antibacterial, antimicrobial, antifungal, anti-inflammatory and analgesic activity.[2]

Comprehensive Organic Chemistry Experiments for the Laboratory Classroom
Edited by Carlos A. M. Afonso, Nuno R. Candeias, Dulce Pereira Simão, Alexandre F. Trindade, Jaime A. S. Coelho, Bin Tan, and Robert Franzén.

Published by the Royal Society of Chemistry, www.rsc.org

Two methods are generally used to synthesize substituted benzoxazoles. One is the condensation of carboxylic acids with 2-aminophenols by dehydration, and the other method is the oxidative cyclization of phenolic derivatives and imines, in the presence of an oxidant.[3,4]

Thiophene and its derivatives exhibit interesting biological activity and important electroluminescent properties with wide application in polymer science, which led to their application in energy transfer systems in optical and electronic devices, such as sensors and fluorescent markers.[5,6]

Structural modification at the side chain of natural amino acids may lead to innovative building blocks for the construction of functional peptides, since their incorporation into peptidic chains may result in the appearance of the amino acid properties in the resulting peptide. Properties such as increased UV/Vis absorption or fluorescence allow for the use as energy donors in conformational studies of peptides by means of fluorescence or to be used as fluorescence markers. As a result, there is an interest in the synthesis of unnatural amino acids with specially tailored optical properties.[7]

The aim of this work is the synthesis of a fluorescent heterocyclic amino acid, namely a benzoxazolyl-alanine, using simple experimental techniques and commercially available reagents. The starting aminotyrosine derivative and thiophene-2-carboxaldehyde will be combined in an oxidative cyclization to the oxazole in the presence of an oxidizing agent. The determination of the structure of the compound synthesized will be performed by ^{1}H and ^{13}C NMR spectroscopy. The experiment involves 3 sessions including the preparation of the intermediate *N-tert*-butyloxycarbonyl-3-[(thien-2′-ylmethylene)amino]-L-tyrosine methyl ester and the preparation and purification of *N-tert*-butyloxycarbonyl[2-(thien-2′-yl)benzoxazol-5-yl]-L-alanine methyl ester.

Additional Safety

All sessions should be done in a well-ventilated fume hood, with proper safety equipment, such as protective clothing, gloves and safety goggles. Thiophene-2-carboxaldehyde may cause skin and eye irritation. Lead tetraacetate is very toxic to the aquatic environment and toxic to humans, especially to the reproductive system. May cause skin, eye and respiratory tract irritation. Ethanol is highly flammable in liquid and vapor form and may cause eye and respiratory irritation, methanol is toxic in contact with skin, if swallowed and inhaled, ethyl acetate may cause eye irritation; and dichloromethane is irritating to skin, eyes and respiratory tract. The preparation of the silica gel column and the separation of the components of the mixture should be carried out in a well-ventilated fume hood due to the use of volatile solvents such as dichloromethane, and methanol, and silica gel may cause respiratory irritation, do not inhale the dust.

Experimental Procedure

Laboratory Session 1 (4 h): Synthesis of *N-tert*-butyloxycarbonyl-3-[(thien-2′-ylmethylene)amino]-L-tyrosine methyl ester

1. Prepare a 25 mL round-bottomed flask with boiling chips, and add 310 mg (1 mmol, 1 equiv.) of *N-tert*-butyloxycarbonyl-3-amino-L-tyrosine methyl ester, absolute ethanol (5 mL), 95 μL of thiophene-2-carboxaldehyde (1 mmol, 1 equiv.) and anhydrous sodium sulfate (0.3 g).
2. Insert a water condenser into the round-bottomed flask and heat the solution at reflux in a heating mantle for 3.5 h.
3. Allow the flask to cool and store, properly stopped, in a vacuum desiccator until the next session.

Laboratory Session 2 (2 h): Synthesis of *N-tert*-butyloxycarbonyl[2-(thien-2′-yl)benzoxazol-5-yl]-L-alanine methyl ester

1. Filter the previous reaction mixture under vacuum, through a sintered glass disc funnel, wash the filtered solid in the funnel with absolute ethanol (5 mL), and evaporate the solvent to dryness. The crude residue obtained is used directly in the next reaction without further purification. Record the ^{1}H and ^{13}C NMR spectra in $CDCl_3$ and an IR spectrum in a KBr disc of the crude.
2. Dissolve the crude residue obtained in dimethyl sulfoxide (DMSO) (5 mL), add 0.660 g (1.5 mmol, 1.5 equiv.) of lead tetraacetate (caution in handling) and a magnetic stirrer. Stop the flask properly and stir the mixture for 3 days at room temperature.

Laboratory Session 3 (4 h): Purification of *N-tert*-butyloxycarbonyl[2-(thien-2′-yl)benzoxazol-5-yl]-L-alanine methyl ester

1. To the previous mixture add water (15 mL), transfer to a separating funnel and extract with ethyl acetate (3 × 10 mL).
2. Separate and combine the organic layers and dry with anhydrous magnesium sulfate. Filter and evaporate the solvent to dryness thus obtaining a solid residue.
3. Check the composition of the crude residue by TLC by comparing with the aldehyde reagent (eluent: dichloromethane/methanol (100 : 1)).
4. Purify the solid by column chromatography on silica gel (column height and diameter: 40 cm and 2.5 cm; silica weight: 35 g; silica height: ≈20 cm; column packing solvent: dichloromethane). Dissolve the solid in the minimum volume of dichloromethane and use as eluent dichloromethane/methanol (100 : 1). Check by TLC each collected fraction, combine the fractions containing the product and evaporate to dryness in a rotary evaporator.
5. Determine the mass of the product and calculate the yield. Record ^{1}H and ^{13}C NMR spectra in $CDCl_3$ to confirm purity, a IR spectrum in a KBr disc, determine the melting point and compare with literature data.[4]

Waste Disposal

Ethanol and ethyl acetate can be recovered after evaporation in a rotary evaporator and later distilled, if possible. If not, solvents should be collected in a non-chlorinated solvent waste recipient. Used silica gel should be stored in a closed container in the hood. The aqueous solution containing lead tetraacetate should be stored in a lead waste container for proper destruction.

Results Interpretation and Additional Questions

1. Calculate the product yield.
2. Write out the reaction mechanism for the preparation in session 1.
3. Why should the crude product from session 1 be kept from water and moisture?
4. In session 2, does DMSO work only as a solvent or has it any other role in the reaction?
5. Write the reaction mechanism for the intramolecular oxidative cyclization in session 2.
6. Interpret the ^{1}H and ^{13}C NMR spectral data obtained for the intermediate imine and product, with the corresponding assignment of signals.
7. Relate the ^{1}H NMR chemical shift of the aromatic protons in the intermediate and the product's structure and compare this with the data of the reagent (previously obtained in Experiment 3.1.14) in terms of electronic effects.

References

1. (a) M. Taki, J. L. Wolford and T. V. O'Halloran, *J. Am. Chem. Soc.*, 2003, **126**, 712–713; (b) M. Shamsipur, T. Poursaberi, A. R. Karami, M. Hosseini, A. Momeni, N. Alizadeh, M. Yousefi and M. R. Ganjali, *Anal. Chim. Acta*, 2004, **501**, 55–60; (c) M. Milewska, A. Skwierawska, K. Guzow, D. Szmigiel and W. Wiczk, *Inorg. Chem. Commun.*, 2005, **8**, 947–950; (d) X.-B. Zang, J. Peng, C.-L. He, G.-L. Shen and R.-Q. Yu, *Anal. Chim. Acta*, 2006, **567**, 189–195.
2. (a) J. Vinsova, K. Cermakova, A. Tomeckova, M. Ceckova, J. Jampilek, P. Cermak, J. Kunes, M. Dolezal and F. Staud, *Bioorg. Med. Chem.*, 2006, **14**, 5850–5865; (b) P. K. Jauhari, A. Bhavani, S. Varalwar, K. Singhal and P. Raj, *Med. Chem. Res.*, 2008, **17**, 412–424; (c) M. L. McKee and S. M. Kerwin, *Bioorg. Med. Chem.*, 2008, **16**, 1775–1783.
3. W.-H. Chen and Y. Pang, *Tetrahedron Lett.*, 2009, **50**, 6680–6683.
4. S. P. G. Costa, E. Oliveira, C. Lodeiro and M. M. M. Raposo, *Sensors*, 2007, **7**, 2096–2114.
5. G. Barbarella, M. Melucci and G. Sotgiu, *Adv. Mater.*, 2005, **17**, 1581–1593.
6. (a) S. P. G. Costa, E. Oliveira, C. Lodeiro and M. M. M. Raposo, *Tetrahedron Lett.*, 2008, **49**, 5258–5261; (b) S. P. G. Costa, R. M. F. Batista and M. M. M. Raposo, *Tetrahedron*, 2008, **64**, 9733–9737; (c) C. I. C. Esteves, M. M. M. Raposo and S. P. G. Costa, *Tetrahedron*, 2010, **66**, 7479–7486; (d) E. Oliveira, D. Genovese, R. Juris, N. Zaccheroni, J. L. Capelo, M. M. M. Raposo, S. P. G. Costa, L. Prodi and C. Lodeiro, *Inorg. Chem.*, 2011, **50**, 8834–8849.
7. (a) C. I. C. Esteves, A. M. F. Silva, M. M. M. Raposo and S. P. G. Costa, *Tetrahedron*, 2009, **65**, 9373–9377; (b) C. I. C. Esteves, M. M. M. Raposo and S. P. G. Costa, *Amino Acids*, 2011, **40**, 1065–1075; (c) R. M. F. Batista, R. C. M. Ferreira, M. M. M. Raposo and S. P. G. Costa, *Tetrahedron*, 2012, **68**, 7322–7330.

12.2.3. Synthesis of Flavones (2-Aryl-4*H*-chromen-4-ones)

Artur M. S. Silva*, Augusto C. Tomé, Diana C. G. A. Pinto, Fernando Domingues, Graça M. Oliveira Rocha, José A. S. Cavaleiro, M. Graça P. M. S. Neves, M. Amparo F. Faustino, and Mário M. Q. Simões

Chemistry Department & QOPNA, University of Aveiro, Campus de Santiago, 3810-193 Aveiro, Portugal.
*E-mail: artur.silva@ua.pt

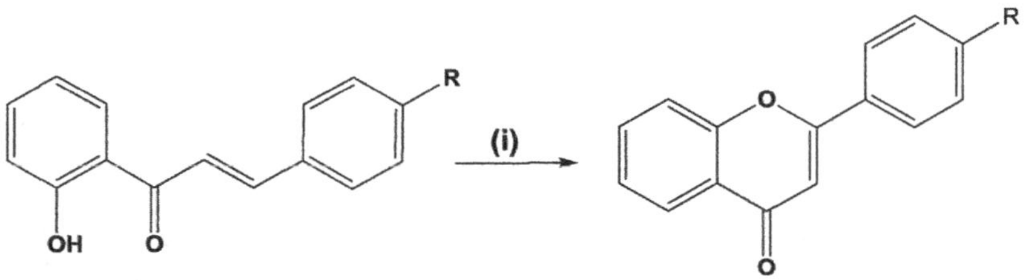

R = H, OCH_3, NO_2
(i) DMSO/I_2 at reflux

Number of sessions (duration of each session)	Hazard level	Difficulty level	Level of study
1 (3 h)	Moderate	Medium	Intermediate
Class names Alcohols, alkenes, unsaturated ketones			
Concepts involved This experiment involves the synthesis of flavones by the cyclodehydrogenation of 2′-hydroxychalcones. The cyclisation is performed in refluxing dimethyl sulfoxide (DMSO) in the presence of a catalytic amount of iodine. Furthermore, purification techniques such as recrystallization and flash chromatography can be learnt			
Chemicals needed (*E*)-1-(2-Hydroxyphenyl)-3-phenylprop-2-en-1-ones (2′-hydroxychalcones), iodine, sodium chloride, sodium sulfate, silica gel, DMSO, dichloromethane, ethanol, deuterated chloroform ($CDCl_3$) with TMS			
Equipment and experimental techniques involved Stirring/heating plate, NMR, melting point apparatus			
Keywords Cyclodehydrogenation reaction, NMR, oxidation, recrystallization and filtration			

Background

Flavones, 2-aryl-4*H*-chromen-4-ones, are a class of natural compounds presenting a 4*H*-1-benzopyran moiety and belonging to the important, and widely distributed in the plant kingdom, group of oxygen heterocyclic secondary metabolites known as the flavonoids. Flavones are known for their important functions in nature, such as protection of plants from UV radiation, attracting insects for pollination and participation in interactions with soil microbes, but also for their valuable biological properties for human health, such as antioxidant and antitumor.[1]

Comprehensive Organic Chemistry Experiments for the Laboratory Classroom
Edited by Carlos A. M. Afonso, Nuno R. Candeias, Dulce Pereira Simão, Alexandre F. Trindade, Jaime A. S. Coelho, Bin Tan, and Robert Franzén.

Published by the Royal Society of Chemistry, www.rsc.org

There are several methodologies towards the synthesis of flavones[2] but the cyclodehydrogenation reaction of 2′-hydroxychalcones[3] is one of the most efficient and easy to perform. The cyclization is performed in refluxing DMSO in the presence of a catalytic amount of iodine, illustrates an intramolecular cyclization and also the oxidation concept in organic chemistry. The reaction mechanism hypotheses involve several undergraduate organic chemistry concepts. For instance, electrophilic addition of halogens to double bonds, elimination reactions, specifically dehydrohalogenation, but also can be used to introduce enol and enolate chemistry. The α-hydrogen acidity, the base- and acid-catalysed enolization and the α-halogenation of ketones can also be studied using this experiment.[4] Furthermore the experimental procedure allows the development of several experimental skills such as reflux heating, reaction control by TLC, filtration under vacuum, melting point determination and characterization by NMR analysis.

Additional Safety

This work requires regular laboratory good practices and personal safety equipment (safety glasses, laboratory coat and nitrile gloves). The synthetic and extraction procedures and chromatographic purifications should be done in a fume hood working properly and with the glass material in perfect condition. Extra care should be taken when handling the heating plate due to the high reaction temperature. If the flash chromatography is used a dust mask should be used to avoid inhalation of silica gel dust.

Experimental Procedure

Laboratory Session: 3 h

This protocol describes the synthesis of 2-phenyl-4*H*-chromen-4-one. Other derivatives can be obtained with an equivalent protocol.

1. In a fume hood heat an oil bath or a sand bath in a stirring plate and keep the temperature around 190–200 °C.
2. In a 100 mL round-bottom flask equipped with a magnetic stirring bar, prepare a solution of 2′-hydroxychalcone (10 mmol) in DMSO (10 mL).
3. Add a catalytic amount of iodine (0.1 mmol) and set up the reaction refluxing system onto the clamp stand.
4. Insert the round-bottom flask of the reaction system into the heated oil or sand bath and stir the reaction for 30 min, maintaining the oil bath at 190–200 °C.
5. After the refluxing period, remove the flask from the oil or sand bath and allow it to cool.
6. Pour the reaction mixture into ice (20–25 g) and water (20 mL), add a spatula of sodium thiosulfate (~0.5 g) and remove the solid by filtration or extract the mixture with dichloromethane (steps 7–9).
7. Extract the mixture with dichloromethane (3 × 20 mL) and collect the organic layer.
8. Wash the dichloromethane solution with 2 × 10 mL of water, followed by 10 mL of saturated aqueous NaCl.
9. Dry the organic layer over Na_2SO_4, then filter to remove the drying agent and remove the solvent by evaporation.
10. The obtained solid can be crystallized in ethanol (5–10 mL). If the experiment is planned for two sessions purification by column chromatography can be performed in the second session (steps 11–16).
11. Check the solvent system by TLC. An appropriate solvent system for flash chromatography is one that produces an R_f of approximately 0.4 on TLC. For flavones, try 40–60% dichloromethane/hexane. When the appropriate solvent is decided prepare 500 mL of it.

12. Obtain a 20 mm diameter flash chromatography column and use a pen to mark 15 cm. Fill the column with silica gel (in the fume hood), tapping gently to pack it down. Carefully add 1 cm of sand to the top of the column of silica gel.
13. Clamp the column securely, place a clean 250 mL flask under the column, open the stopcock, and carefully fill the column from the top with your eluent. Attach a piece of tubing equipped with a column adapter to an airline and start a slow flow of air. Hold the column adapter on the top of the column and push the solvent rapidly through the silica gel. When the solvent level is approximately 1 cm from the sand at the head of the column, remove the air. If the column is not completely wetted, refill the column with eluent and repeat the above procedure. Once the column is completely wetted with solvent, lower the eluent level in the column until it is level with the top of the sand. Recombine the solvent you collected during the wetting of the column with any unused eluent.
14. Dissolve the crude product in 2–3 mL of the eluent. Use a Pasteur pipet to add this mixture to the top of the flash column. Then briefly place the air back on the column and push this mixture onto the top of the silica gel, taking care not to let the eluent level dip below the level of the silica gel. Rinse the container that contained the crude product with another 2–3 mL of eluent and load this onto the column as you did above. Repeat this procedure several times until there is little or no colour left in the solution.
15. Then carefully fill the column with eluent and use the air to force the eluent through the column at a slow rate; collect 10 mL fractions. (You should have previously obtained about 20 clean test tubes for this purpose. It is not necessary to stop the air flow between fractions. With a little practice the test tubes can be changed without spilling any eluent.) When the eluent level is 1 cm from the top of the sand, remove the air and add more eluent. Continue this process until you have used all your eluent or obtained 20 fractions. At no time should the eluent level be allowed to drop below the level of the silica gel. Do not discard the column until you are sure that you have isolated your product.
16. Check each of the fractions by TLC using whatever eluent you used for your column. Combine the fractions containing your product and remove the solvent by rotary evaporation.
17. The purified compound is carefully transferred to a labelled watch glass pre-weighed and placed in the oven at a temperature of 70 °C for 15 min.
18. Weigh the product obtained and calculate the reaction yield. Measure the melting point, check the compound purity by TLC and prepare the NMR tube (step 19).
19. Weigh approximately 10 mg (0.02 g) of your sample into a clean, dry vial. Add approximately 0.5–0.6 mL of deuterated chloroform ($CDCl_3$) containing a small amount of tetramethylsilane (TMS) and swirl gently until your sample dissolves. Transfer your sample to the NMR tube using a clean, dry Pasteur pipet. You should have approximately 3 cm of solvent in the tube. Finally, put a cap on the NMR tube.

Waste Disposal

All liquids should be collected in a recipient containing ice and sodium hypochlorite and afterwards sent to the proper waste collector recipient for posterior destruction. Solvents used in chromatographic purifications or extractions should be collected for further distillation and reutilization.

Results Interpretation and Additional Questions

1. Calculate the percent yield of the reaction.
2. Write out the chemical equations involved in the oxidation/reduction transformations.
3. Discuss the possible reaction mechanisms with the help of your instructor.

4. Using the obtained TLC chromatogram calculate the flavone R_f and comment on its purity.
5. Interpret the ^{1}H and ^{13}C NMR spectra obtained, and compare the obtained melting point with the literature.

References

1. A. K. Verma and R. Pratap, *Nat. Prod. Rep.*, 2010, **27**, 1571–1593.
2. D. C. G. A. Pinto and A. M. S. Silva, *Curr. Org. Synth.*, 2012, **4**, 561–572.
3. A. M. S. Silva, H. R. Tavares, A. I. N. R. A. Barros and J. A. S. Cavaleiro, *Spectroscopy Lett.*, 1997, **30**, 1655–1667.
4. F. A. Carey, *Organic Chemistry*, McGraw-Hill, 6th edn, 2006.

12.2.4. Hantzsch Synthesis of 3,5-Diethoxycarbonyl-2,6-dimethylpyridine

Iain A. Smellie*, Nigel P. Botting, Brian A. Chalmers, Iain L. J. Patterson, and Catherine M. Schofield

School of Chemistry, University of St Andrews, North Haugh, St Andrews, Fife, KY16 9ST, UK
*E-mail: ias10@st-andrews.ac.uk

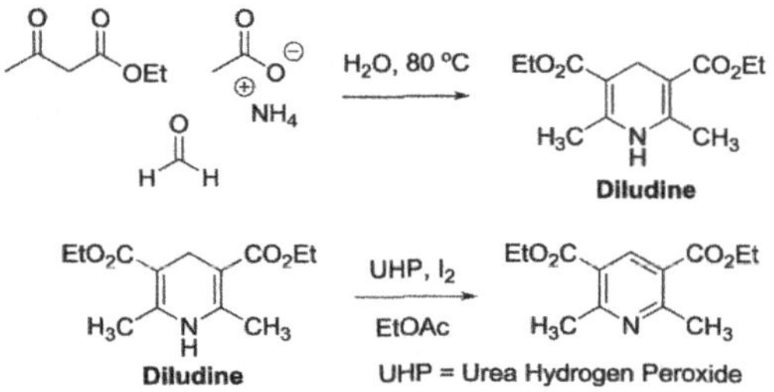

Number of sessions (duration of each session)	Hazard level	Difficulty level	Level of study
2 (3 h each)	Moderate	Medium	Intermediate

Class names Aldehydes, β-keto esters, 1,4-dihydropyridines, pyridines

Concepts involved This experiment involves a two-step pyridine synthesis. In the first step a solvent-free synthesis of a 1,4-dihydropyridine is accomplished. In the second step the intermediate undergoes oxidation by urea hydrogen peroxide to the corresponding pyridine

Chemicals needed Ammonium acetate, deuterochloroform, ethyl acetate, ethyl acetoacetate, formaldehyde (38% aqueous solution), iodine, sodium metabisulfite, sodium sulfate, urea hydrogen peroxide

Equipment and experimental techniques involved Heating mantle or hotplate, glassware for reflux assembly, magnetic stirring, liquid–liquid extraction, vacuum filtration, recrystallization, thin-layer chromatography, NMR spectroscopy, IR spectroscopy

Keywords Heterocycle synthesis, liquid–liquid extraction, multi-component reaction, NMR, oxidation, recrystallization

Background

The pyridine ring is a recurring motif in naturally occurring compounds and also in pharmaceuticals (representative and familiar examples include nicotine, pyridoxal phosphate and isoniazid and nefidipine[1]). Pyridines can be prepared by a variety of routes.[2] One of the best known methods is the "Hantzsch" synthesis,[2–4] a two-step multi-component cyclization/oxidation process. The first part of the reaction is believed to involve a Knoevenagel reaction between a β-keto ester (**1**) and an aldehyde to afford an α,β-unsaturated intermediate (**2**). Simultaneous reaction of an amine (or ammonia) with the β-keto ester is believed to take place to afford enamine **3** which can combine with **2** to afford dihydropyridine **4**. 1,4-Dihydropyridine

Comprehensive Organic Chemistry Experiments for the Laboratory Classroom
Edited by Carlos A. M. Afonso, Nuno R. Candeias, Dulce Pereira Simão, Alexandre F. Trindade, Jaime A. S. Coelho, Bin Tan, and Robert Franzén.

Published by the Royal Society of Chemistry, www.rsc.org

4 can then subsequently be oxidized under relatively mild conditions to afford the corresponding pyridine (Scheme 12.2.4.1).[4–6]

1,4-Dihydropyridines are also found in Nature and comprise an important family of calcium channel blocking drugs that are used to treat cardiovascular disease.[7] Representative examples of 1,4-dihydropyridines include nifedipine and the reduced form of the co-factor nicotinamide adenine dinucleotide (NADH)).[1,2] 1,4-Dihydropyridines have also found application as reducing agents in reductive amination reactions of carbonyl compounds[8,9] and reduction of α,β-unsaturated compounds[10] (see Scheme 12.2.4.2 for an example). This field of research has been recently comprehensively reviewed.[11]

The aim of the experiment is to prepare diethyl 2,6-dimethyl-3,5-pyridinedicarboxylate *via* a Hantzsch sequence. The procedure reported by Dicks *et al.*[7] is employed to prepare diludine in step one (diethyl 2,6-dimethyl-1,4-dihydro-3,5-pyridinedicarboxylate). In the second step, the 1,4-dihydropyridine intermediate is treated (according to the procedure of Vinkovic *et al.*[5]) with urea hydrogen peroxide and iodine to afford the corresponding pyridine. Notably, although there are many reagents available to accomplish this transformation,[5,6] the stated reaction conditions have been found to be mild and convenient for teaching laboratory use as part of an advanced organic chemistry course (typical class size 60–70 students). The ^1H NMR and IR spectra of the intermediate and product are instructive and clearly demonstrate that cyclization and subsequent aromatization has occurred.

Additional Safety

All operations should be performed in a fume hood with appropriate protective clothing (laboratory coat, safety goggles and gloves). Aqueous formaldehyde is toxic by inhalation, ingestion or skin contact. Aqueous formaldehyde and deuterochloroform should be treated as potentially carcinogenic. Ethyl acetate is a volatile and flammable solvent. Urea hydrogen peroxide is an oxidant, contact with skin and eyes must be avoided. Iodine (vapour in particular) is toxic and can

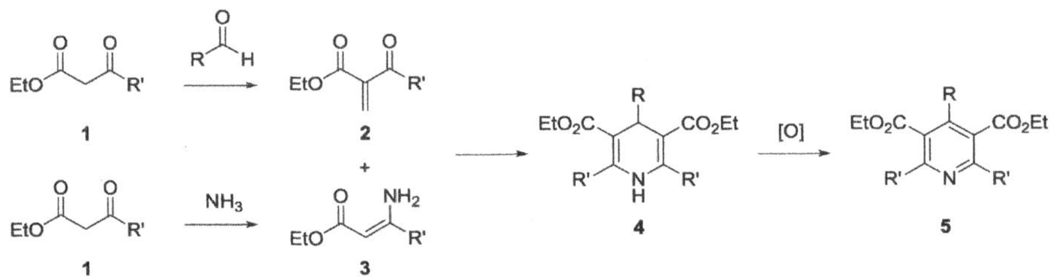

Scheme 12.2.4.1 Outline of the key steps of the Hantzsch synthesis of pyridines.

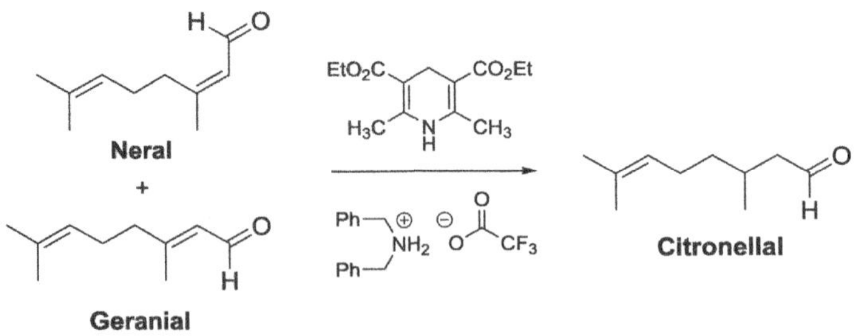

Scheme 12.2.4.2 Reduction of the components of citral to give citronellal.

cause burns on inhalation or prolonged skin contact. Sodium metabisulfite should be treated as an irritant and corrosive, especially to the eye. Sodium metabisulfite can release corrosive and toxic sulfur dioxide gas if exposed to acids.

Experimental Procedure

Laboratory Session 1: Synthesis of Diludine (3 h)

1. Transfer 2.97 g of ethyl acetoacetate, 1.29 g of ammonium acetate and 0.85 mL of formaldehyde (38% aqueous solution) to a 50 mL round-bottomed flask. Add a magnetic stirrer bar and start stirring the solution.
2. Heat the reaction to 80 °C under reflux for 10–15 min. A yellow solid should accumulate in the flask as the reaction proceeds.
3. Allow the flask to cool and add distilled water (*ca.* 10 mL). Break up the solid with a glass rod until a fine suspension is obtained. Filter off the crude product under suction and wash the filter cake with distilled water (*ca.* 20 mL).

Purification Diludine

4. Recrystallize the crude product from a minimal volume of ethanol.
5. Weigh the product and determine the % yield. Measure the melting point and record a ^{1}H NMR ($CDCl_3$) spectrum of your product.

Laboratory Session 2: Oxidation of Diludine (3 h)

1. Transfer 0.78 g of diludine and 15 mL of ethyl acetate to a 100 mL conical flask. Add a magnetic stirrer bar and start stirring the suspension.
2. Add 0.57 g of UHP (urea hydrogen peroxide) in small portions, followed by 0.30 g of iodine.
3. Allow the reaction mixture to continue stirring for 1.5 h at room temperature.
4. Using a disposable pipette, transfer a sample to a small vial for TLC analysis (dilute with ethyl acetate). Confirm the reaction has reached completion (TLC conditions – ethyl acetate : petroleum ether (40–60 °C) (1 : 4) and visualize under a UV lamp).
5. Dilute the reaction mixture with water (*ca.* 20 mL) and add sodium metabisulfite in small portions with gentle swirling until the strong iodine colouration is no longer visible in both phases (they should become completely colourless or very pale yellow).
6. Separate off the organic layer and extract the aqueous layer with ethyl acetate (2 × 15 mL). Dry the combined organic extracts over sodium sulfate and concentrate under reduced pressure.

Purification of Diethyl 2,6-Dimethyl-3,5-pyridinedicarboxylate

7. Recrystallize the crude product from a minimal volume of aqueous ethanol.
8. Weigh the product and determine the % yield. Measure the melting point and record a ^{1}H NMR ($CDCl_3$) spectrum of your product.

Waste Disposal

Waste solutions of iodine should be treated with a solution of sodium metabisulfite or sodium thiosulfate. Aqueous washings from the extraction steps should be diluted with water and disposed of down a sink. All organic solvent waste should be retained and transferred to an appropriate container for disposal.

Results Interpretation and Additional Questions

1. Interpret the ^{1}H NMR spectra you obtained from both reaction steps. Use the NMR spectra and melting points to confirm the structure of your compounds unambiguously and identify diagnostic peaks that demonstrate that oxidation of the 1,4-dihydropyridine intermediate was successfully achieved.
2. Provide a curly arrow mechanism for the reaction you carried out in step 1.
3. In this experiment oxidation of the dihydropyridine ring was achieved using UHP (urea hydrogen peroxide) in the presence of iodine. You should suggest a plausible arrow-pushing mechanism for this process. (Hint – hydrogen peroxide and iodine can react to form hypoiodous acid *in situ*.)

References

1. F. S. Yates, in *Comprehensive Heterocyclic Chemistry I*, ed. A. R. Katritzky and C. W. Rees, Pergamon, Oxford, 1984, ch. 2.09, pp. 511–523.
2. G. Jones, in *Comprehensive Heterocyclic Chemistry I*, ed. A. R. Katritzky and C. W. Rees, Pergamon, Oxford, 1984, ch. 2.08, pp. 482–484.
3. A. Saini, S. Kumar and J. S. Sandhu, *J. Sci. Ind. Res.*, 2008, **67**, 95.
4. B. E. Norcross, G. Clement and M. Weinstein, *J. Chem. Educ.*, 1969, **46**, 694.
5. M. Filipan-Litvic, M. Litvic and V. Vinkovic, *Tetrahedron*, 2008, **64**, 5649.
6. X. Liao, W. Lin, J. Lu and C. Wang, *Tetrahedron Lett.*, 2010, **51**, 3859.
7. L. L. W. Cheung, S. A. Styler and A. P. Dicks, *J. Chem. Educ.*, 2010, **87**, 628.
8. T. Itoh, K. Nagata, A. Kurihara, M. Miyazaki and A. Ohsawa, *Tetrahedron Lett.*, 2002, **43**, 3105.
9. T. Itoh, K. Nagata, M. Miyazaki, H. Ishikawa, A. Kurihara and A. Ohsawa, *Tetrahedron*, 2004, **60**, 6649.
10. A. D. Cunningham, E. Y. Ham and D. A. Vosburg, *J. Chem. Educ.*, 2011, **88**, 322.
11. C. Zheng and S. You, *Chem. Soc. Rev.*, 2012, **41**, 2498.

12.2.5. An "Out-of-the-Box" Example in Heterocyclic Chemistry: Synthesis of 3,6-Bis-(3,5-dimethyl-1*H*-pyrazol-1-yl)-1,2,4,5-tetrazine

Tiago J. L. Silva*[a,b], Paulo J. G. Mendes[b], Ana Isabel Tomaz[a], and M. Helena Garcia[a]

[a]Centro de Química Estrutural, Faculdade de Ciências da, Universidade de Lisboa, Campo Grande, 1049-1016 Campo Grande, Lisboa, Portugal; [b]Departamento de Química, Escola de Ciências e Tecnologia, Centro de Química de Évora, Instituto de Investigação e Formação Avançada, Universidade de Évora, Rua Romão Ramalho 59, 7000-671 Évora, Portugal. *E-mail: tjsilva@ciencias.ulisboa.pt

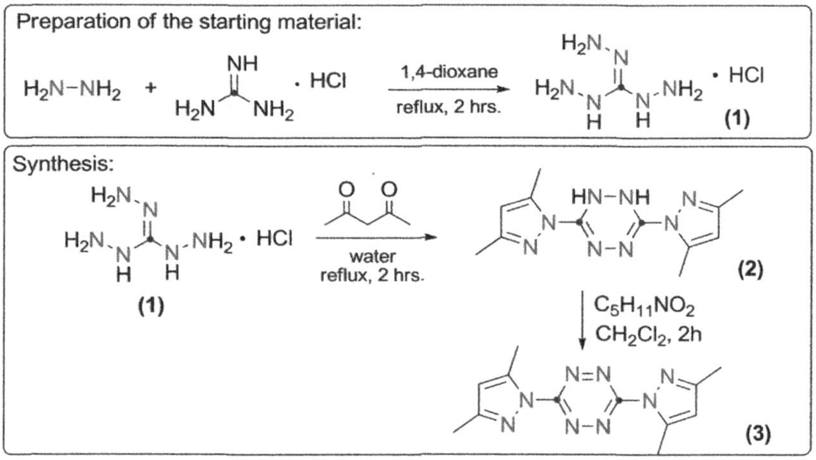

Number of sessions (duration of each session)	Hazard level	Difficulty level	Level of study
3 (3 h + 6 h + 6 h)	High	Medium	Intermediate

Class names Amines, 2,4-diketones, imines, nitrites, tetrazines

Concepts involved This procedure is simple and provides the students with an "out-of-the-box" example of how to synthesize heterocyclic aromatic compounds bearing more than three nitrogen atoms that does not follow the usual Pinner method. The experiment involves reaction routes widely used in heterocyclic ring synthesis: C–heteroatom bond formation through C=O nucleophilic addition to 2,4-dicarbonyl compounds, intramolecular cyclization and an oxidation reaction

Chemicals needed Guanidine hydrochloride, hydrazine hydrate (80% solution), 1,4-dioxane, acetylacetone (2,4-pentanedione), isoamyl nitrite, dichloromethane, anhydrous magnesium sulfate, silica gel, deuterated chloroform

Equipment and experimental techniques involved Vacuum filtration apparatus, liquid–liquid extraction, rotatory evaporation apparatus, addition funnel, chromatography column, laboratory weighing equipment, heating plate with magnetic stirring, reflux apparatus

Keywords Column chromatography, FTIR spectroscopy, heterocyclic ring formation, NMR spectroscopy, oxidation, tetrazines, vacuum filtration

Comprehensive Organic Chemistry Experiments for the Laboratory Classroom
Edited by Carlos A. M. Afonso, Nuno R. Candeias, Dulce Pereira Simão, Alexandre F. Trindade, Jaime A. S. Coelho, Bin Tan, and Robert Franzén.

Published by the Royal Society of Chemistry, www.rsc.org

Background

1,2,4,5-Tetrazines (generally referred to as *s*-tetrazines) are heterocyclic molecules, which consist of a six-membered aromatic ring with four nitrogen atoms (Figure 12.2.5.1). The 3,6-dihydro-*s*-tetrazine is unstable, and hence *s*-tetrazines are more commonly found as 3,6-disubstituted compounds. The nature of these substituents (which can include alkyl, alkynyl, aryl, heteroaryl, alkoxyl and alkylmercapto groups) has a great influence on the stability of the compounds: 3,6-dialkyl-*s*-tetrazines are rather unstable (unless their substituents have electron-rich groups, *e.g.* halogens), while aromatic and heterocyclic aromatic 3,6-diaryl-*s*-tetrazines are the most stable and can be stored for long periods.[1,2]

The most distinct features of tetrazines are their high electron affinity, resulting in an electrochemical reversible reduction potential and their low lying π^* orbitals, which in turn results in an n–π^* transition in the visible region making tetrazines deeply coloured molecules.[1,4,5] These two distinct features are the basis of their unique electrochemical and photochemical properties, making them very attractive for a wide range of applications. 3,6-Dissubsituted tetrazines can be included in functional molecular systems for the development of sensors and optical switches, electroactive polymers or solar cells.[5–7] Tetrazine derivatives, namely dipyridyl-*s*-tetrazines and related molecules, have also been frequently used as bridging ligands in metallo-organic compounds.[3] Their *N*-hydro (di, tetra, hexa) derivatives exhibit biological activity and some have been tested for their anti-tumour activity.[8]

The first report of the *s*-tetrazine ring synthesis was by Pinner, in 1893. This method involves the reaction of hydrazine with a nitrile to afford the dihydrotetrazine, followed by the oxidation of the latter to the fully aromatic tetrazine (Scheme 12.2.5.1). A drawback of this procedure is that it is only effective for aromatic tetrazines where the nitrile is not sterically hindered, but is not suitable for aliphatic ones. In fact, the *s*-tetrazine ring is always synthesized by oxidation of its 1,2- or 1,4-dihydro counterpart. The dihydro counterparts can, in their turn, be obtained by several methods (five main synthetic routes are known).[1,2]

Coburn introduced an easy procedure for the synthesis of the 3,6-bis(3,5-dimethyl-1*H*-pyrazol-1-yl)-*s*-tetrazine with fairly high yield and no tedious purification during the process.[9] This new procedure avoided the harsh conditions of Pinner's method by reacting 1,3-diaminoguanidine monohydrochloride with 2,4-pentadione to obtain the dihydro-*s*-tetrazine skeleton. Oxidation could then be performed with nitrous gas. Audebert later improved the oxidation step by replacing the toxic nitrous gas with sodium nitrite and acetic acid.[1]

This experiment describes a simple procedure for the synthesis of 3,6-bis(3,5-dimethyl-1*H*-pyrazol-1-yl)-*s*-tetrazine in moderate yield with a straightforward work-up. It is based on the

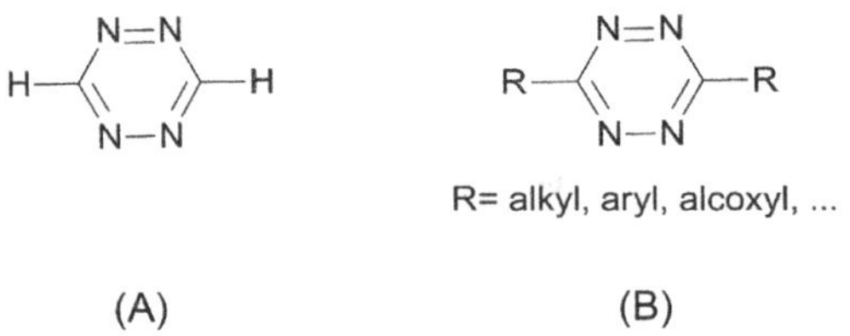

Figure 12.2.5.1 Structure of (A) 3,6-dihydro-*s*-tetrazine and (B) 3,6-disubstituted-*s*-tetrazine.

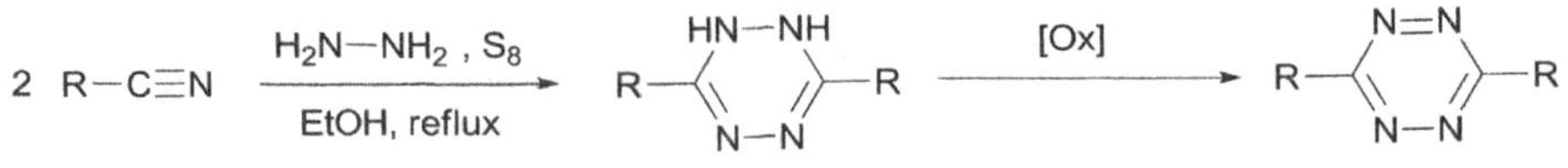

Scheme 12.2.5.1 Pinner's method for the synthesis of *s*-tetrazines. [Ox] refers to an oxidizing agent (*e.g.* sodium nitrite/acetic acid mixture, isoamyl nitrite).

route described by Coburn using isoamyl nitrite as a more convenient and mild oxidizing agent as compared to the sodium nitrite/acetic acid system. The oxidation step with isoamyl nitrite is faster and simplifies the product work-up since it avoids the use of acetic acid and the need to wash the product prior to its purification by chromatography, thus affording the 3,6-bis(3,5-dimethyl-1*H*-pyrazol-1-yl)-*s*-tetrazine in higher yield.

Additional Safety

In all laboratory sessions the use of basic safety protection equipment (protective clothing, gloves and safety goggles) is mandatory. Even though tetrazine derivatives are known for their explosive properties, none of the reactants or products used in this procedure is reported as being explosive. Nevertheless all laboratory sessions should be done in a fume hood. Hydrazine hydrate is highly toxic reagent and should be handled with care. 1,4-Dioxane and dichloromethane are flammable solvents, harmful if swallowed, inhaled, or in contact with the skin and eyes. Thus, care should be taken when handling them.

Experimental Procedure

Laboratory Session 1 (3 h) – Synthesis of 1,3-Diaminoguanidine Monohydrochloride (1)

1. Prepare a 250 mL round-bottomed flask with a big magnetic stir bar (1.5 cm) and add 9.5 g (0.1 mol) of guanidine hydrochloride. Add 50 mL of 1,4-dioxane and stir vigorously to obtain a well-dispersed suspension.
2. Insert an addition funnel charged with 22 mL of an 80% hydrazine hydrate solution (0.5 mol, 5 equiv.) on top of the flask. Add the hydrazine solution dropwise (approx. 1 drop per second) to the suspension prepared in step 1. Two phases will be observed.
3. Remove the addition funnel and replace it with a water condenser. Clamp the flask in an oil bath and heat to reflux with stirring. Maintain refluxing temperature for 1.5 h. A white solid will form.
4. Remove the flask from the oil bath and allow it to cool to room temperature. (The process can be accelerated by placing the flask in a water bath after it has cooled down for a while.)
5. Filter off the white solid by vacuum filtration using a sintered glass filter (porosity 2 or 3), and wash the solid twice with 10 mL of 1,4-dioxane. Transfer the white solid to a 100 mL round-bottomed flask (tared) and dry the product in a vacuum pump until constant mass is obtained.

Laboratory Session 2 (6 h) – Synthesis of 3,6-Bis(3,5-dimethyl-1*H*-pyrazol-1-yl)-1,2-dihydro-1,2,4,5-tetrazine (2)

1. Prepare a 500 mL two-neck round-bottomed flask with a big magnetic stir bar (1.5 cm) and add 12 g (0.085 mol) of triamineguanidine monohydrochloride (**1**). Add 140 mL of water and stir vigorously until all the white solid is dissolved.
2. Insert an addition funnel charged with 17.5 mL of acetylacetone (0.17 mol, 2 equiv.) on the central neck of the flask, a thermometer in one of the side-necks and a cap in the third neck.
3. Place the flask in an ice-cold water bath and allow the temperature inside the flask to drop below 15 °C. Start the addition of the acetylacetone dropwise. If the temperature rises to 20 °C *stop the addition immediately*, and allow the reaction to cool down to 15 °C before continuing the addition.

4. After the addition of acetylacetone is complete, remove the cold water bath and allow stirring to continue for a further 30 min, keeping the reaction mixture at approximately 20 °C.
5. Place the flask in an oil bath and heat the reaction mixture at 70 °C for 4 h.
6. Remove the flask from the oil bath, and allow it to cool down to room temperature. (This process can be accelerated by placing the flask in a water bath after it has cooled down for a while.)
7. Filter off the orange solid by vacuum filtration using a Büchner filter and wash the solid three times with 30 mL of water. Transfer the solid to a 200 mL Erlenmeyer flask and dissolve it in 100 mL of dichloromethane.
8. Dry the dichloromethane solution with anhydrous magnesium sulfate.
9. Filter the suspension and collect the filtrate to a 200 mL round-bottomed flask (tared). Evaporate the solvent in a rotatory evaporator until dryness. Weigh the flask.
10. Store the orange solid product for the next reaction under nitrogen in the fridge until the next laboratory session.

Laboratory Session 3 (6 h) – Synthesis of 3,6-Bis(3,5-dimethyl-1*H*-pyrazol-1-yl)-1,2,4,5-tetrazine (3)

1. Prepare a 250 mL one-neck round-bottomed flask with a big magnetic stir bar (1.5 cm) and add 7 g (0.026 mol) of 3,6-bis(3,5-dimethyl-1*H*-pyrazol-1-yl)-1,2-dihydro-1,2,4,5-tetrazine (**2**).
2. Dissolve the solid in 100 mL of dichloromethane under vigorous stirring until all the yellow solid is completely dissolved.
3. To the previous solution add 3.5 mL of isoamyl nitrite (0.026 mol, 1 equiv.) all at once but carefully, and stir the solution at room temperature *in a well-ventilated fume hood* for 4 h, making sure the flask is not fully closed with a cap. The colour of the reaction mixture will change to red.
4. Transfer the obtained dark red solution to a separatory funnel, and wash the organic phase three times with water (30 mL).
5. Collect the organic phase into a 250 mL Erlenmeyer flask, and dry it with anhydrous magnesium sulfate. Filter the solution and collect the filtrate to a 250 mL one-neck round-bottomed flask. Evaporate the solvent in a rotatory evaporator.
6. Fill in a chromatography column (3 cm diameter × 20 cm height) with a slurry of silica gel in dichloromethane. To the flask containing the product, add 1–2 mL of dichloromethane. Carefully load the chromatography column with this solution.
7. Chromatograph the mixture using dichloromethane. Collect the red coloured band to a 200 mL flask (previously tared).
8. Evaporate the solvent in the rotatory evaporator. Weigh the flask.

Waste Disposal

Hydrazine solution is disposed of by dilution with water to produce a lower than 40% solution. After neutralization with dilute sulfuric acid, the resulting solution can be drained into the sewer with abundant water. Non-chlorinated organic solvents should be collected in a non-chlorinated waste collector recipient, and dichloromethane should be collected in a recipient for chlorinated compounds. Silica gel and drying agents (magnesium sulfate) should be collected in an appropriate solid waste collector recipient for posterior destruction.

Results Interpretation and Additional Questions

1. Calculate the yield of each reaction.
2. Draw the mechanism for the formation of the 1,2-dihydro-*s*-terazine.

3. In reaction 2, the temperature of the reaction cannot exceed 25 °C. Why is such tight temperature control needed in the reaction mixture during the addition of the diketone?
4. Prepare a NMR sample by dissolving 20 mg of the final product in 0.5 mL of $CDCl_3$. Acquire and interpret the NMR spectra (^1H and ^{13}C). Compare the results with data found in the literature.[9]

References

1. G. Clavier and P. Audebert, *Chem. Rev.*, 2010, **110**, 3299.
2. B. Stanovnik, U. Grošelj and J. Svete, *Comprehensive Heterocyclic Chemistry Volume III*, ed. A. T. Katritzky, C. A. Ramsden, E. F. V. Scriven and R. J. K. Taylor, Elsevier Science, Philadelphia, 2008, ch. 9, pp. 641–714.
3. (a) W. Kaim, *Coord. Chem. Rev.*, 2002, **230**, 127; (b) S. Patra, B. Sarkar, S. Ghumaan, J. Fiedler, W. Kaim and G. K. Lahiri, *Inorg. Chem.*, 2004, **43**, 6108.
4. P. Audebert, S. Sadki, F. Miomandre and G. Clavier, *Electrochem. Commun.*, 2004, **6**, 144.
5. P. Audebert, F. Miomandre, G. Clavier, M. C. Vernières, S. Badré and R. Méallet-Renault, *Chem.–Eur. J.*, 2005, **11**, 5667.
6. K. Parimal, E. H. Witlicki and A. H. Flood, *Angew. Chem., Int. Ed.*, 2010, **49**, 4628.
7. Z. Li, J. Ding, N. Song, J. Lu and Y. Tao, *J. Am. Chem. Soc.*, 2010, **132**, 13160.
8. G.-W. Rao and W.-X. Hu, *Bioorg. Med. Chem. Lett.*, 2006, **16**, 3702–3705.
9. (a) M. D. Coburn, G. A. Buntain, B. W. Harris, M. A. Hiskey, K.-Y. Lee and D. G. Ott, *J. Heterocycl. Chem.*, 1991, **28**, 2049; (b) M. D. Coburn and D. G. Ott, *J. Heterocycl. Chem.*, 1991, **27**, 1941.

12.3. Epoxidation

12.3.1. Catalytic Epoxidation of *cis*-Cyclooctene with MTO(VII) and Pyrazole

Elisabete P. Carreiro* and Anthony J. Burke

Centro de Química de Évora, Institute for Research and Advanced Studies, University of Evora, Chemistry Department, School of Science and Technology, Rua Romão Ramalho, no. 59, 7000-671 Évora, Portugal
*E-mail: betepc@uevora.pt

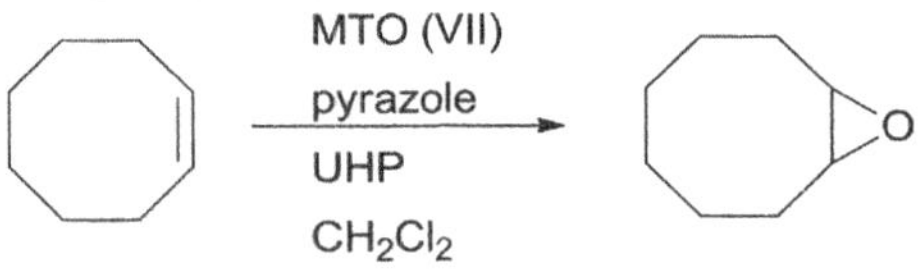

Number of sessions (duration of each session)	Hazard level	Difficulty level	Level of study
1 (3 h)	Moderate	Low	Intermediate

Class names Olefins, epoxides

Concepts involved This experiment involves the selective epoxidation of olefins using MTO(VII) as catalyst, and pyrazole as a ligand in the presence of urea–hydrogen peroxide (UHP) as oxidant in an anhydrous system

Chemicals needed *cis*-Cyclooctene (95%), methyltrioxorhenium(VII) (contains 71–76% Re), pyrazole, urea–hydrogen peroxide, dichloromethane, 1,3,5-trimethylbenzene, distilled water, manganese dioxide, $CDCl_3$

Equipment and experimental techniques involved Reaction flask, separatory funnel, magnetic agitator; NMR spectrometer

Keywords Catalytic epoxidation, evaporation under reduced pressure, MTO(VII), NMR spectrum, olefins, pyrazole

Background

The epoxide is a very important function, which in many cases is present in various biologically active compounds,[1] and for this reason the epoxidation process has much significance in organic synthesis. Methyltrioxorhenium(VII) (MTO), when combined with H_2O_2, acts as an efficient catalyst for a wide range of olefins. This procedure was first developed by Herrmann's group in 1991.[2] Other useful peroxo complexes of early transition metals have been used, such as Ti(IV), Mo(VI), W(VI), *etc.* Sharpless and co-workers showed the application of pyridine as ligand in the catalytic epoxidation with MTO(VII), in order to improve the selectivity and accelerate the reaction.[3] Pyrazole,[4] 3-cyanopyridine[5] and halide substituted Schiff-bases[6] gave very good yields. These heterocyclic aromatic amines can coordinate with the MTO(VII) catalyst to form a complex that is more stable towards

Comprehensive Organic Chemistry Experiments for the Laboratory Classroom
Edited by Carlos A. M. Afonso, Nuno R. Candeias, Dulce Pereira Simão, Alexandre F. Trindade, Jaime A. S. Coelho, Bin Tan, and Robert Franzén.

Published by the Royal Society of Chemistry, www.rsc.org

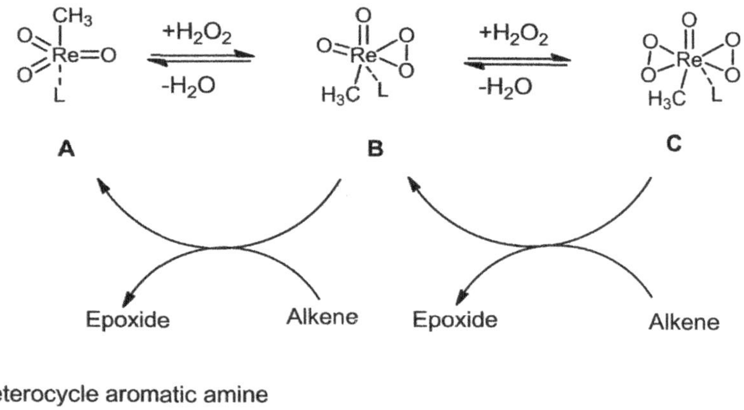

Scheme 12.3.1.1 Proposed mechanism for the catalytic epoxidation of alkenes with MTO(VII) and H_2O_2.[9] MTO(VII)–ligand complex (A) and equilibrium formation of the η2-peroxo species B and C.

decomposition, also reducing the acidity of the catalyst, lessening the degree of epoxide hydrolysis. This ligand also accelerates both the H_2O_2 binding process and the epoxidation process. Aqueous hydrogen peroxide has been mostly used, but sensitive epoxides undergo hydrolysis. Adam and Mitchell[7] utilized an anhydrous source of H_2O_2, namely urea–H_2O_2 (UHP), reducing the occurrence of this unwanted side reaction; however, the reaction time increased. Carreiro *et al.*[8] developed a new method using MTO(VII) with pyrazole as ligand and UHP as oxidant; the reaction conversions were high and unwanted hydrolysis in the case of sensitive epoxides was reduced. The general mechanism for this reaction is shown in Scheme 12.3.1.1. In this experiment *cis*-cyclooctene epoxide is prepared from *cis*-cyclooctene using MTO/UHP in the presence and in the absence of pyrazole.

Additional Safety

The reaction should be conducted in a fume hood. UHP is corrosive.

Experimental Procedure

Laboratory Session (3 h): Catalytic Epoxidation of *cis*-Cyclooctene

Experiments A and B are conducted simultaneously.

Experiment A: Catalytic Epoxidation of cis-*Cyclooctene with MTO(VII)/Pyrazole and UHP*

1. In the fume hood, prepare a 25 mL round-bottomed flask with a magnetic stir bar, and add 3 mL of CH_2Cl_2, 4.8 mg (0.014 mmol) MTO(VII), 22 mg (0.32 mmol) pyrazole and 512 mg UHP (2 equiv.) and stir the mixture at room temperature until the reaction becomes yellow.
2. After a few minutes, add 313 mg (2.7 mmol) of *cis*-cyclooctene and stir the reaction for over 1 h.

Experiment B: Catalytic Epoxidation of cis-*Cyclooctene with MTO(VII) and UHP in the Absence of Pyrazole*

1. In the fume hood, prepare a 25 mL round-bottomed flask with a magnetic stir bar, and add 3 mL of CH_2Cl_2, 4.8 mg (0.014 mmol) MTO(VII), and 512 mg UHP (2 equiv.) and stir the mixture at room temperature until the reaction becomes yellow.
2. After, a few minutes add 313 mg (2.7 mmol) of *cis*-cyclooctene and stir the reaction for over 1 h.

Isolation of the Product: Same Procedure for Both Experiments

3. Reaction work-up: add 5 mL of CH_2Cl_2, 5 mL H_2O and a small quantity of MnO_2 to the mixture and stir until there is no further oxygen evolution (approx. 10 min).
4. Transfer the mixture to a separatory funnel and perform a biphasic extraction with CH_2Cl_2 (3 × 5 mL).
5. Dry the combined organic phases with anhydrous $MgSO_4$, filter to a 50 mL (tared) round-bottomed flask and evaporate the solvents on a rotator evaporator under reduced pressure.
6. Weigh the obtained products (m_1) and calculate the yield.
7. Record the ^1H NMR spectra to confirm the purity of the compounds. Compare the spectra with those in the literature data.

Yield Determination

8. 50 mg of the product mixture (m_2) and 15 mg of the internal standard (IS) 1,3,5-trimethylbenzene (m_3) are dissolved in 0.50 mL of $CDCl_3$ and added to an NMR tube. It is important to note the exact amount of both product (m_2) and internal standard (m_3)
9. Acquire the ^1H NMR spectrum and calculate the reaction yield.

Waste Disposal

The aqueous solution can transferred to an appropriate recipient for aqueous waste, for later destruction. Dichloromethane should be transferred to a chlorinated waste recipient, for later destruction.

Results Interpretation and Additional Questions

1. Interpret the obtained ^1H NMR spectra and compare the spectra obtained for both experiments.
2. Calculate the product yield [by NMR (internal standard (IS) – 1,3,5-trimethylbenzene)]:

$$\text{mol} = \frac{\text{area epoxide NMR peak } (2.92\,\text{ppm})}{\left[\text{area IS NMR peak } (6.820\,\text{ppm}) \times 0.67\right]} \times \frac{\text{IS mass } (m_3)}{\text{IS MW}}$$

$$\text{Epoxide total mass} = \text{mol epoxide} \times \text{epoxide MW} \times \frac{m_1}{m_2}$$

3. Calculate the conversion of both reactions, compare and discuss the differences between experiments A and B, particularly in the context of reaction rate and selectivity:

$$\text{Conversion} = \frac{\text{area epoxide NMR peak } (2.890\,\text{ppm})}{\left[\text{area epoxide NMR peak } (2.890\,\text{ppm}) + \text{area olefin NMR peak } (5.616\,\text{ppm})\right]} \times 100$$

4. Rationalize the epoxidation mechanism.
5. Discuss the importance of the pyrazole ligand and the UHP on the reaction process.

References

1. E. G. Lewars, in *Comprehensive Heterocyclic Chemistry*, ed. A. R. Katritzky and C. W. Rees, Pergamon Press, 1997, vol. 7, ch. 5.05, pp. 96–129, and references cited therein.
2. W. A. Herrmann, R. W. Fischer and D. W. Marz, *Angew. Chem., Int. Ed. Engl.*, 1991, **30**, 1638.
3. J. Rudolph, K. L. Reddy, J. P. Chiang and K. B. Sharpless, *J. Am. Chem. Soc.*, 1997, **119**, 6189.
4. W. A. Herrmann, R. M. Kratzer, H. Ding, W. R. Thiel and H. Glas, *J. Organomet. Chem.*, 1998, **555**, 293.
5. H. Adolfsson, A. Converso and K. B. Sharpless, *Tetrahedron Lett.*, 1999, **40**, 3991.
6. P. Altman, M. Cokoja and F. E. Kühn, *J. Organomet. Chem.*, 2012, **701**, 51.
7. W. Adam and C. M. Mitchell, *Angew. Chem., Int. Ed. Engl.*, 1996, **35**, 533.
8. E. P. Carreiro, A. J. Burke, M. J. M. Curto and A. J. R. Teixeira, *J. Mol. Catal. A: Chem.*, 2004, **217**, 69.
9. S. M. Nabavizadeh, *Dalton Trans.*, 2005, 1644, and references cited therein.

12.3.2. Stereoselective Epoxidation of Cholesterol by *m*-Chloroperoxybenzoic Acid

Maria de Jesus Perry*, Ana Paula Francisco, Ana S. Ressurreição and Francisca Lopes*

Instituto de Investigação do Medicamento (iMed. ULisboa), Faculdade de Farmácia da Universidade de Lisboa, Av. Prof. Gama Pinto, 1649-003 Lisboa, Portugal. *E-mail: mjprocha@ff.ulisboa.pt, fclopes@ff.ulisboa.pt

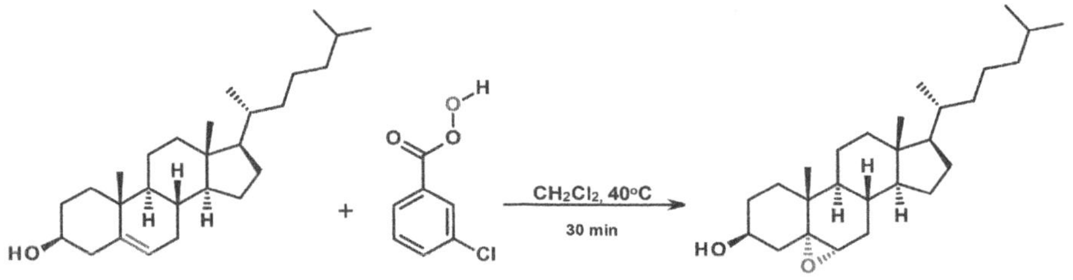

Δ^5-Cholesterol   *m*-CPBA   5α,6α-Epoxycholestan-3β-ol

Number of sessions (duration of each session)	Hazard level	Difficulty level	Level of study
1 (3 h)	Moderate	Medium	Intermediate
Class names Cycloalkene, epoxides, organic peroxyacids, carboxylic acids			
Concepts involved Epoxidation, cholesterol oxides, stereoselective reactions			
Chemicals needed Cholesterol, *meta*-chloroperoxybenzoic acid (*m*-CPBA), dichloromethane, aluminium oxide (90 active neutral), sulfuric acid, acetic acid, ethanol, ethyl acetate, acetone, deuterochloroform (for NMR)			
Equipment and experimental techniques involved Reflux, column chromatography, thin-layer chromatography (TLC), multi-solvent recrystallization, NMR, heating and stirring plate, rotatory evaporator, melting point apparatus			
Keywords Cholesterol epoxide, column chromatography, melting point, NMR, peroxyacids			

Background

Steroids is the generic name for a broad group of compounds having in common a basic carbon framework consisting of four fused rings – three cyclohexane rings and one cyclopentane ring – labeled and joined together as shown in Figure 12.3.2.1. This carbon framework is denominated as the cyclopentane–perhydro-phenanthrene system.

Steroid compounds are widely distributed in nature and are challenging substrates for the synthesis of a wide variety of important biologically active molecules. In this context, the oxidation of several positions of the steroid core is of great importance. Epoxides, which are an archetypal oxidation product, are versatile intermediates in organic syntheses. In fact, as epoxides are three-membered cyclic ethers they have specific reactivity patterns owing to the highly polarized oxygen–carbon bonds in addition to a highly strained ring. Further, several natural and synthetic β- and α-epoxysteroids have relevant biological activities, which accounts for the interest in their stereoselective synthesis over the years.[1,2]

Comprehensive Organic Chemistry Experiments for the Laboratory Classroom
Edited by Carlos A. M. Afonso, Nuno R. Candeias, Dulce Pereira Simão, Alexandre F. Trindade, Jaime A. S. Coelho, Bin Tan, and Robert Franzén.

Published by the Royal Society of Chemistry, www.rsc.org

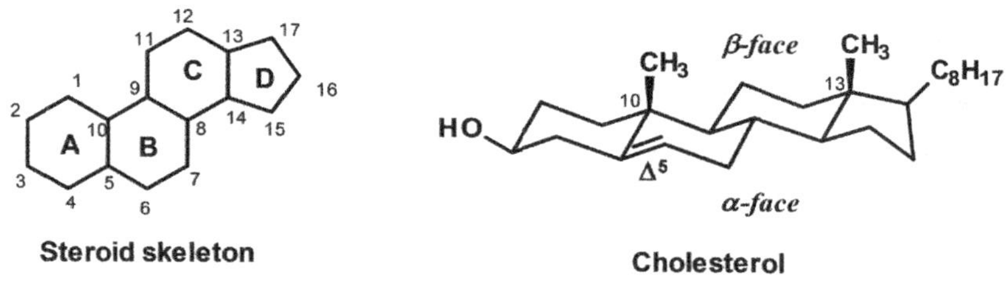

Figure 12.3.2.1 Cyclopentane–perhydro-phenanthrene system and cholesterol structure.

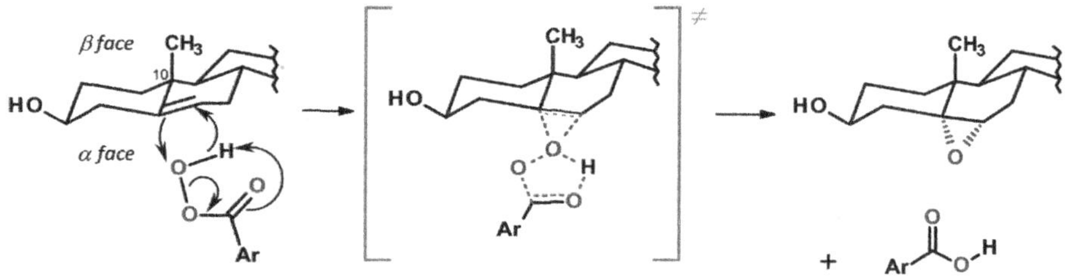

Scheme 12.3.2.1 2D representation of mechanistic and stereochemistry highlights in the epoxidation of Δ^5-cholesterol.[6]

It is well established that the stereochemistry of epoxidation reactions on steroidal substrates is modulated by the shielding of the β-face on the steroid nucleus, due to the two axial methyl groups at C10 and C13 (Figure 12.3.2.1, cholesterol), and therefore the α-epoxide is invariably the major reaction product. Steroidal α-epoxides have been traditionally obtained by use of peroxyacids as oxidizing agents (Scheme 12.3.2.1).[3]

In the present work we will proceed to the epoxidation of cholesterol (Figure 12.3.2.1), specifically the double bond between C5 and C6 (Δ^5), in order to explore the stereoselectivity of the reaction with an organic peroxyacid in stoichiometric amounts. The epoxidation reaction is analysed by thin-layer chromatography (TLC).

The final product is characterized by NMR spectroscopy, since this technique is particularly good for visualization and quantification of the stereoisomers.

Additional Safety

The experimental procedure should be done in a fume hood, with proper safety equipment, such as protective clothes, rubber gloves and safety goggles. *m*-Chloroperoxybenzoic acid (*m*-CPBA) should be stored in a refrigerator in tightly closed containers. It is a flammable solid and contact with heat or oxidizable material should be avoided. Furthermore, it irritates the mucous membranes and respiratory tract, eyes and skin. Skin contact with *m*-CPBA results in burns and blisters.

Sulfuric acid spray reagent (stain solution) – this reagent, used for general detection of steroids, is composed by 10% w/v of sulfuric acid and 5% w/v of acetic acid in ethanol. It is highly flammable, toxic by inhalation, and in contact with skin causes severe burns.

Dichloromethane (CH_2Cl_2) causes irritation of the skin and respiratory system, may cause chemical burns, and it is a possible carcinogen.

Regarding alumina (A_2O_3), it is not expected to present a significant hazard under anticipated conditions of normal use. Do not breathe the dust.

Deuterochloroform ($CDCl_3$) causes irritation of the skin and respiratory system, may cause chemical burns, and it is a possible carcinogen.

Experimental Procedure

Laboratory Session 1 (3 h): Stereo Epoxidation of Cholesterol

1. Preparation of the Reagent Solutions

In a round-bottomed flask equipped with a magnetic stirrer dissolve 1 g of cholesterol (2.6 mmol) with 4 mL of CH_2Cl_2. If necessary slightly warm up the solution in a water-bath to completely dissolve the cholesterol. In an Erlenmeyer flask prepare a solution of 3-chloroperoxybenzoic acid (*m*-CPBA) by dissolving 3.5 mmol in 4 mL of CH_2Cl_2. Note: pay attention to the purity grade of *m*-chloroperoxybenzoic acid, since commercial reagents contain up to 15% of benzoic acid.

2. Monitoring of the Reaction by TLC

Before mixing the reagents prepare the TLC plate to follow the progress of the reaction. The students will collect samples of the reaction mixtures at times zero, 15 and 30 min of reaction. Standards of cholesterol and *m*-CPBA should be used to compare the R_f values and analyse the results. Use ethyl acetate as eluent.

3. Mixing the Reagents

Add dropwise the peroxyacid solution over the cholesterol solution. As soon as the first drop of the oxidizing agent is added, stir and remove a sample, for TLC analysis, with a Pasteur pipette (time zero). Continue to add the solution of peroxyacid with gentle shaking and, finally, assemble a reflux heating system, allowing the reaction to be complete after 30 min at 40 °C.

Elute the TLC plates and analyze the result, sequentially using two viewing methods: (a) UV lamp (254 nm), to view the 3-chlorobenzoic acid. Tick the purple spots with a pencil. (b) Revelation with sulfuric acid-spray reagent, to view the cholesterol and the epoxy-cholesterol spots. In this procedure the TLC plate is sprayed with the ethanolic acetic/sulfuric acid staining solution, allowed to dry for 5 min in the air and then heated at 120 °C on a hot surface (or with a heatgun), until the color developed has reached its maximum. Apart from their different R_fs, cholesterol is revealed as a carmine stain; epoxy-cholesterol is revealed as a spot of green color.[5]

4. Product Separation

To isolate and purify the epoxy-cholesterol, the reaction mixture is eluted in an alumina chromatographic column, since the excess of acid will stay strongly adsorbed by the alumina. Take advantage of the reflux time (30 min) to prepare the chromatographic column by slurry method with 15 g of aluminium oxide (Al_2O_3) and CH_2Cl_2. Once the reaction mixture is cooled, pour it into the chromatography column and elute with 200 mL of CH_2Cl_2. Do not allow the column to dry and collect the solvent in a 250 mL round-bottomed flask (previously weighed), and evaporate the solvent in the rotary evaporator.

5. Epoxy-Cholesterol, Final Purification

Dissolve the crude epoxy-cholesterol in 7.5 mL of hot acetone and transfer the solution to a 25 mL Erlenmeyer flask, add 1 mL of water to this solution, heat the mixture to re-dissolve the precipitate and allow the solution to cool down to room temperature. Cool the mixture on ice and collect the crystals formed in a Hirsch funnel. Wash the crystals with 1 mL of cold acetone, and dry using a vacuum. Determine the weight, the melting point and calculate the yield of the reaction.[5a,b]

6. Analyse the Product by ^1H and ^{13}C NMR, using $CDCl_3$ as the Solvent

$$\% \ \alpha \text{ isomer} = [\alpha \text{ signal integration value}/\alpha + \beta \text{ signals integration values}] \times 100 \qquad (12.3.2.1)$$

Waste Disposal

Non-halogenated organic solvents must be collected in a proper collector recipient and halogenated organic solvents must be collected in a halogenated collector recipient for further destruction. The alumina used must be placed in a container and disposed of during the regular hazardous waste pickups.

Results Interpretation and Additional Questions

1. Draw the mechanism for the epoxidation reaction of cholesterol.
2. Discuss the possible stereochemical effects that can deliver a stereoselective epoxidation reaction.
3. What side products are generated during this reaction?
4. Interpret the ^1H and ^{13}C NMR spectra obtained, paying special attention to the possible presence of two stereoisomers. Note: to calculate the % of the two stereoisomers present in the final product, use eqn (12.3.2.1) (experimental procedure, point 6).
5. Indicate other methods for epoxidation or other enantioselective methods.

References

1. J. A. R. Salvador, S. M. Silvestre and V. M. Moreira, *Curr. Org. Chem.*, 2008, **12**, 492.
2. J. A. R. Salvador, S. M. Silvestre and V. M. Moreira, *Curr. Org. Chem.*, 2012, **16**, 1243.
3. J. F. S. Carvalho, M. M. C. Silva, J. N. Moreira, S. Simões and M. L. Sá e Melo, *J. Med. Chem.*, 2009, **52**, 4007.
4. E. Stahl, *Thin Layer Chromatography: A Laboratory Handbook*, Springer-Verlag, 2nd edn, Berlin, 1969, p. 900.
5. (a) L. F. Fieser and M. Fieser, *Steroids*, Reinhold Publishing Corporation, 1959, New York, pp. 194 and 197; (b) http://steraloids.com/cholest-basesd/cholestan-5-6-epoxy-3-ol.html, mp 141–141.5 °C.
6. J. Clayden, N. Greeves, S. Warren and P. Wothers, *Organic Chemistry*, Oxford University Press, 1st edn, 2001, p. 506.

12.3.3. Green Oxidation of Organic Compounds Using Metalloporphyrins

Rose A. Clark, Anne E. Stock, and Edward P. Zovinka*

Department of Chemistry, Saint Francis University, Loretto, PA 15940, USA. *E-mail: ezovinka@francis.edu

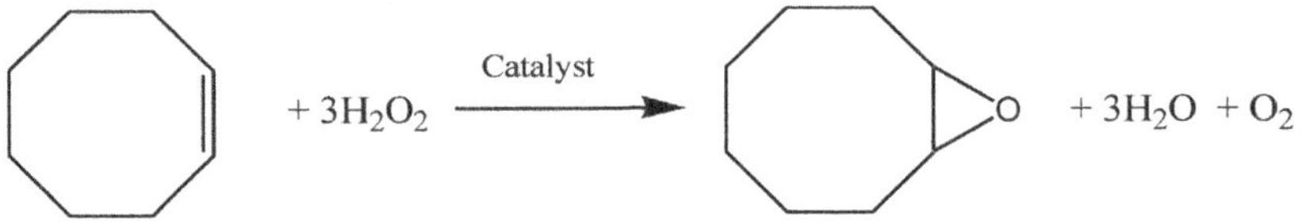

Number of sessions (duration of each session)	Hazard level	Difficulty level	Level of study
1 (4 h)	Medium	Medium	Intermediate
Class names Alkenes, alkanes, epoxides, alcohols			
Concepts involved The experiment presents the oxidation of organic substrates (alkenes and alkanes) using two different metalloporphyrin catalysts			
Chemicals needed Cyclohexane, cyclohexanol, *cis*-cyclooctene, cyclooctene oxide, decane, 30% H_2O_2, methylene chloride, methanol, chloro[5,10,15,20-tetraphenylporphyrinate]iron(III), chloro[5,10,15,20-tetrakis-(pentafluorophenyl)porphyrinate]iron(III); all porphyrins were purchased from Sigma-Aldrich			
Equipment and experimental techniques involved Gas chromatograph/mass spectrometer (GC-MS) or gas chromatograph (GC), gas tight syringes, calibrated micropipets			
Keywords Catalysis, gas chromatography, green chemistry, mass spectrometry, synthesis			

Background

Hydrocarbon feedstocks, the source of many commercial products, need to be functionalized to allow the construction of specialized chemicals. The chemical industry and the consumer industry as a whole have an economic stake in the oxidation of organic substrates to produce their specialized chemical products. Developing green methods to produce the commercial product will benefit the companies as well as the environment. Traditional chemical methodology has focused solely on the desired product with little regard to the other reaction by-products. For example, iodosylbenzene has been used for years to functionalize organic substrates.[1] However, during the redox process, iodosylbenzene breaks down into more hazardous components, such as carcinogenic benzene.

To "green up" the process, and make the synthesis more economically profitable, industrial chemists and researchers are increasingly interested in catalytic reactions. Catalysis reduces the energy needs of a reaction by lowering the reaction activation energy.

A catalytic reaction is highlighted in this laboratory experiment: hydrogen peroxide is used as a co-oxidant, producing an iron(IV) ferryl porphyrin π cation radical from the starting metalloporphyrin without the formation of harmful materials.[2] The oxidized porphyrin then oxidizes the organic compound.[2] The green chemistry advantage of using H_2O_2 is that the desired functionalized organic compound is obtained with only water and oxygen as by-products (Scheme 12.3.3.1).[3]

Comprehensive Organic Chemistry Experiments for the Laboratory Classroom
Edited by Carlos A. M. Afonso, Nuno R. Candeias, Dulce Pereira Simão, Alexandre F. Trindade, Jaime A. S. Coelho, Bin Tan, and Robert Franzén.

Published by the Royal Society of Chemistry, www.rsc.org

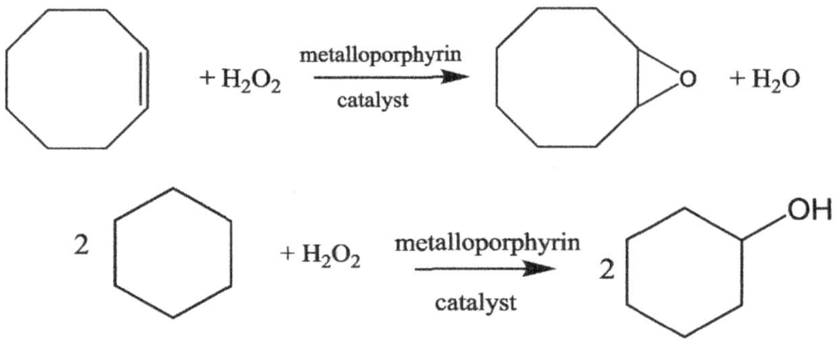

Scheme 12.3.3.1 Organic substrate oxidation using hydrogen peroxide and a metalloporphyrin catalyst.

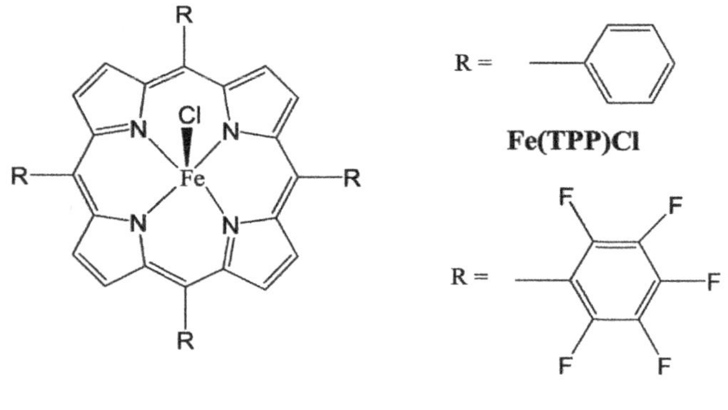

Figure 12.3.3.1 Structure of the metalloporphyrin catalysts, Fe(TPP)Cl and Fe(TPPF$_5$)Cl.

The oxidation reaction will be monitored using a gas chromatograph/mass spectrometer (GC/MS) to detect the functionalized organic compounds. Specifically, cyclooctene or cyclohexane will be oxidized by H_2O_2 with the reaction catalyzed by an iron(III) porphyrin, Figure 12.3.3.1, and the yield of the oxidation product obtained during the reaction will be calculated. Literature reports have shown that the simpler and less hazardous hydrogen peroxide produces nearly equal yields to those of reactions completed using the less desirable iodosylbenzene as an oxidant.[4] Two different catalyst systems are compared and the reaction catalyzed by Fe(TPPF$_5$)Cl yields much better results in comparison to the Fe(TPP)Cl catalyzed reaction. Since the simple (and readily synthesized) Fe(TPP)Cl metalloporphyrin catalyst is not as robust, it degrades during the oxidation reactions, leading to minimal oxidation product.[5] However, the use of chloro[tetrakis(pentafluorophenyl)porphyrinate]iron(III), Fe(TPPF$_5$)Cl, produces a higher yield of oxidized organic product.[6] The amount of oxidized organic product will vary substantially depending on the reactivity of the functional groups in the molecule. The challenge of functionalizing alkanes is clearly demonstrated during the experiment as the cyclohexane oxidation produces very little product using either catalyst.

Additional Safety

Be very cautious working with 30% H_2O_2. It is corrosive to the skin and other body parts. Methanol is toxic to eyes and may be toxic to the kidneys, liver, brain, peripheral nervous system, upper respiratory tract, skin, central nervous system (CNS), and optic nerve. Dichloromethane presents a possible cancer hazard while cyclooctene and cyclohexane are irritants.

Experimental Procedure

Laboratory Session (4 hours)

1. Two catalysts, Fe(TPP)Cl and Fe($TPPF_5$)Cl, will be investigated for their catalytic potential.[7] Prepare solutions for the assigned metalloporphyrin catalyst and assigned substrate. All solutions are prepared using a 3:1 methanol:dichloromethane solvent (see below). The metalloporphyrin concentration is 1×10^{-3} M in the same solvent system. Laboratory data is compiled and shared so comparisons between reactions can be made (Scheme 12.3.3.1).

Gas Chromatography Standards

2. An internal standard calibration curve for the product, cyclooctene oxide (or cyclohexanol), is created to determine the concentration of product produced in the substrate oxidation reactions.
3. Prepare five standard solutions (5–10 mL volumes) of product with concentrations of 8×10^{-3}, 4×10^{-3}, 6×10^{-4}, 2×10^{-4} and 1×10^{-5} M from a ~4×10^{-2} M stock solution.
4. An aliquot of each solution is analyzed using a GC/MS using the same conditions as the reaction (see below). The reaction product standards are diluted to volume using a stock solvent (100 mL of a 3:1 mixture of methanol:dichloromethane with 5 mL of 59×10^{-3} M decane solution added as an internal standard). The ratio of peak area (or peak height) for the cyclooctene oxide/decane peaks or cyclohexanol/decane peaks is to be plotted against the concentration ratio of cyclooctene oxide/decane or the concentration ratio of cyclohexanol/decane. The decane internal standard helps alleviate errors in concentration due to small changes in injection volume.

Reaction Procedure

- Solvents: make 100 mL of 3:1 methanol:dichloromethane with 5 mL of 59×10^{-3} M (1.18×10^{-4} mol) decane as internal GC standard.
- GC conditions:
  - Choose the appropriate GC/MS instrument conditions based on the compound assigned (Table 12.3.3.1).

Control Experiments

  - Control 1: a mixture of substrate, solvent with internal standard, and the metalloporphyrin (no peroxide). Use an additional 92 μL of solvent to keep the volume constant (4000 μL).
  - Control 2: cyclooctene or cyclohexane mixed with hydrogen peroxide (no metalloporphyrin). Use an additional 80 μL of solvent to keep the volume constant (4000 μL).
  - Before starting the reaction, inject an aliquot of the reaction solution before addition of the peroxide (control 1) and after 45 min of reaction inject an aliquot of the solution that contains the substrate and peroxide without the metalloporphyrin (control 2).
- Reaction (4000 μL total volume):
  add to a small vial (needs to hold ~5 mL);
  - Solvent: 3048 μL of methanol:dichloromethane (3:1) (with internal standard).
  - Catalyst: 80 μL of metalloporphyrin (1×10^{-3} M) (8×10^{-8} mol).
  - Substrate: 780 μL of cyclohexane (0.0072 mol) or cyclooctene (0.006 mol).
- Chill the reaction and control vial in an ice bath for 5 min.
- Best results are obtained when:
  - maintaining ice bath conditions during the reaction.
  - reducing the amount of light exposure (wrap vials in foil).
  - maintaining the slow addition of hydrogen peroxide to the reaction mixture.

Table 12.3.3.1 GC-MS reaction conditions.

Compound	Starting temp (°C)	Hold time (min)	Ramp (°C min^{-1})	Ramp temp (°C)	Hold time (min)
Cyclohexane	40	1.00	20	120	1.00
Cyclooctene	50	1.50	15	310	1.00

- Add 92 μL (in *small portions*) of 30% hydrogen peroxide while the vials are in the ice bath over a period of 45 min. Injections were made at 5 min intervals.
- Inject an aliquot of the reaction solution into the GC/MS shortly after adding the final aliquot of hydrogen peroxide, total reaction time 45 min (also inject an aliquot of control 2).
    - Use the same conditions for the GC/MS based on the substrate (Table 12.3.3.1).

Waste Disposal

All waste materials need to be disposed of in properly labeled containers. Separate the halogenated waste from the non-halogenated waste.

Results Interpretation and Additional Questions

1. Generate calibration curves by plotting the ratio of organic substrate/decane signal *versus* concentration ratio of organic substrate/decane.
2. After running the reaction, determine the concentration of the organic substrate in the sample by using the organic substrate/decane signal to determine the concentration ratio from the calibration curve. Once you have the concentration ratio, use the known decane concentration to solve for the concentration of the organic substrate.
3. Questions to consider:

    Why is it important for chemists to use green chemistry methods?
    Why are the substrate oxidation results so poor when the Fe(TPP)Cl catalyst is used?
    Why is it more difficult to functionalize the alkane substrate in comparison to the alkene?
    Are the catalysts under study good examples (or not!) of Green Chemistry?
    Examine the solvent system, do they fit Green Chemistry Principles?
    Why is dichloromethane part of the solvent system in the procedure?

Acknowledgement

The authors thank the Saint Francis University CHEM 304 classes that have completed the experiment in the course of their studies. The experiment protocol is adapted from: *J. Chem. Educ.*, 2012, **89**, 271–275. Copyright (2012) American Chemical Society.

References

1. J. T. Groves, T. E. Nemo and R. S. Myers, *J. Am. Chem. Soc.*, 1979, **101**, 1032.
2. W. Nam, Y. M. Goh, Y. J. Lee, M. H. Lim and C. Kim, *Inorg. Chem.*, 1999, **38**, 3238.
3. W. Nam, S.-Y. Oh, Y. J. Sun, J. Kim, W.-K. Kim, S. K. Woo and W. Shin, *J. Org. Chem.*, 2003, **68**, 7903.
4. I. D. Cunningham, T. N. Danks, J. N. Hay, I. Hamerton, S. Gunathilagan and C. Janczak, *J. Mol. Catal. A: Chem.*, 2002, **185**, 25b.
5. D. Dolphin, T. G. Traylor and L. Y. Xie, *Acc. Chem. Res.*, 1997, **30**, 251.
6. B. S. Lane and K. Burgess, *Chem. Rev.*, 2003, **103**(7), 2457.
7. R. A. Clark, A. E. Stock and E. P. Zovinka, *J. Chem. Educ.*, 2012, **89**, 271.

12.3.4. Catalytic Epoxidation of Carbamazepine

Artur M. S. Silva, Augusto C. Tomé, Diana C. G. A. Pinto, Fernando M. J. Domingues, Graça M. S. R. O. Rocha, José A. S. Cavaleiro, Maria da Graça P. M. S. Neves, Maria do Amparo F. Faustino, and Mário M. Q. Simões*

Chemistry Department, University of Aveiro, Campus de Santiago, 3810-193 Aveiro, Portugal
*E-mail: msimoes@ua.pt

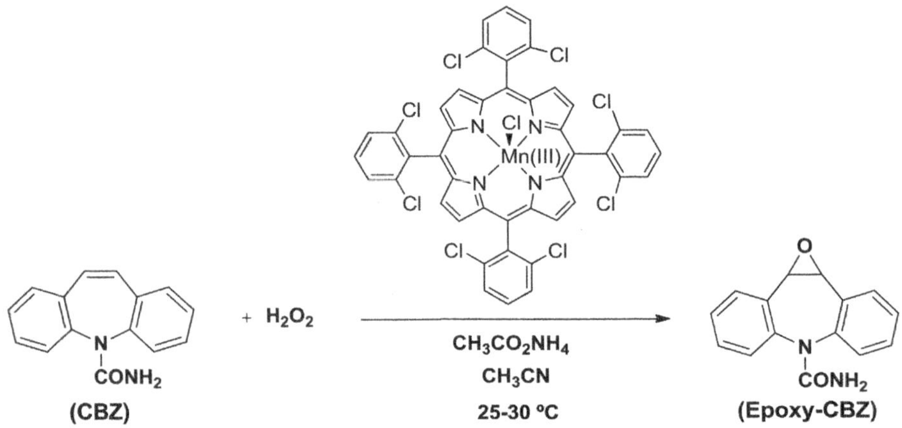

Number of sessions (duration of each session)	Hazard level	Difficulty level	Level of study
1 (4 h)	Low	Medium	Intermediate

Class names Alkenes, epoxides

Concepts involved This experiment involves the use of a manganese(III) porphyrin to mimic the oxidative metabolism of carbamazepine. The *in vitro* oxidation of carbamazepine results in the formation of the corresponding 10,11-epoxide, the main *in vivo* metabolite. Ammonium acetate is used as co-catalyst and aqueous hydrogen peroxide as oxidant

Chemicals needed Carbamazepine, acetonitrile, ammonium acetate, hydrogen peroxide (30% w/w), dichloromethane, anhydrous sodium sulfate, deuterated chloroform, manganese(III) 5,10,15,20-tetrakis(2,6-dichlorophenyl)porphyrin chloride

Equipment and experimental techniques involved Liquid–liquid extraction, stirring hot plate, water bath, TLC, reverse-phase HPLC, ^{1}H NMR spectroscopy

Keywords Catalysis, epoxidation, ^{1}H NMR, HPLC, liquid–liquid extraction, TLC

Background

Epoxides are very easy to prepare *via* reaction of an alkene with a peroxyacid and the chemistry of epoxides is routinely discussed in organic chemistry manuals.[1] In recent years, introduction of the epoxidation reaction in the undergraduate organic chemistry laboratories has increased.[2]

Comprehensive Organic Chemistry Experiments for the Laboratory Classroom
Edited by Carlos A. M. Afonso, Nuno R. Candeias, Dulce Pereira Simão, Alexandre F. Trindade, Jaime A. S. Coelho, Bin Tan, and Robert Franzén.

Published by the Royal Society of Chemistry, www.rsc.org

However, it seems that metalloporphyrins as catalysts and hydrogen peroxide as oxidant have not yet been explored in epoxidation reactions for undergraduate teaching laboratories. The developments achieved recently for epoxidation catalysed by metalloporphyrins make this an appealing system to perform oxidative transformations in such laboratories.[3]

In this work, which is planned for one session of 4 h, a non-enzymatic system will be used to mimic the oxidative metabolism of carbamazepine (CBZ), one of the most prescribed drugs in the treatment of epilepsy.[4]

The metabolism of drugs and other xenobiotics is often initiated by oxidation reactions catalysed by cytochrome P450 (CYP) enzymes.[5] Since the isolation of the biological entities can become complex and time-consuming, synthetic models have been developed for mimicking the function of CYP in living organisms. Actually, there are several synthetic models that can mimic, *in vitro*, the role of CYP, including metalloporphyrins.[5–8]

This is an extremely rewarding subject, because it introduces students of an undergraduate organic chemistry laboratory to the synthesis and characterization of a porphyrin and its manganese complex (please refer to the laboratory experiments entitled "Synthesis of 5,10,15,20-tetrakis(2,6-dichlorophenyl)porphyrin" (5.2.4) and "Synthesis of the manganese(III) complex of 5,10,15,20-tetrakis(2,6-dichlorophenyl)porphyrin" (17.6)), allowing the use of chromatographic and spectroscopic techniques, and the assessment of the performance of a catalyst, namely based on percent conversion, selectivity and turnover number (TON). For that purpose, students will use the previously prepared manganese(III) 5,10,15,20-tetrakis(2,6-dichlorophenyl)porphyrin chloride [Mn(TDCPP)Cl] (Figure 12.3.4.1), which is an extensively studied catalyst in oxidation processes, mimicking a wide range of reactions catalysed by CYP enzymes.[9]

The oxidation reaction of carbamazepine (CBZ) using Mn(TDCPP)Cl results in the formation of the corresponding epoxide (Epoxi-CBZ), known as its main *in vivo* metabolite, and involves chromatographic techniques such as thin-layer chromatography (TLC) and high-performance liquid chromatography (HPLC), allowing students to follow up the reaction and to observe the formation of the epoxide along with the disappearance of the substrate. The epoxide will be subsequently characterized by ^{1}H NMR spectroscopy. In this reaction, dilute aqueous hydrogen peroxide, a safe, easily available and environmentally friendly oxidant is used.[10,11]

The objective of this laboratory experiment is the synthesis of the carbamazepine 10,11-epoxide (Epoxi-CBZ) by the reaction of carbamazepine (CBZ) with hydrogen peroxide catalysed by the manganese(III) complex of 5,10,15,20-tetrakis(2,6-dichlorophenyl)porphyrin in the presence of ammonium acetate (as co-catalyst), in acetonitrile at 25–30 °C. The reaction can be monitored by TLC and HPLC. The structure of Epoxi-CBZ is confirmed by ^{1}H NMR spectroscopy by comparison with the ^{1}H NMR spectrum of the carbamazepine (CBZ).

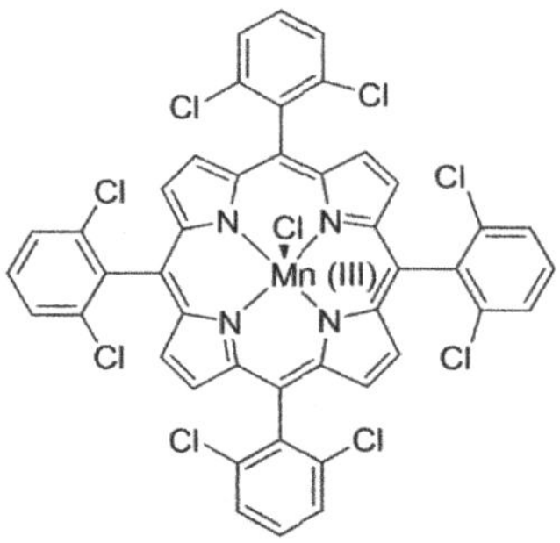

Mn(TDCPP)Cl

Figure 12.3.4.1 Chemical structure of [Mn(TDCPP)Cl].

Additional Safety

All chemicals used in this work should be handled with appropriate care. All of them can be considered toxic and irritant and inhalation or any contact with skin and eyes should be avoided. Laboratory coat, safety glasses and adequate gloves are compulsory and all the experimental work should be done in a well-ventilated fume hood.

Aqueous 30% hydrogen peroxide is an oxidant that can be toxic by ingestion, corrosive and may be harmful if inhaled and causes skin and eye burns. Acetonitrile is a flammable liquid, an irritant, can cause serious eye damage and is harmful by ingestion, inhalation and skin absorption. Dichloromethane is considered to be a possible carcinogen.

Experimental Procedure

Catalytic Epoxidation of Carbamazepine

1. Prepare the catalyst stock solution by dissolving about 10 mg of the manganese(III) complex of 5,10,15,20-tetrakis(2,6-dichlorophenyl)porphyrin in a 10.0 mL volumetric flask, with a small amount of acetonitrile (if necessary, use ultrasound to help dissolve the catalyst), and then fill it up to the mark.
2. Prepare the oxidant solution by diluting 1.0 mL of H_2O_2 30% (w/w) with CH_3CN in a 5.0 mL volumetric flask.
3. In a pear-shaped flask of 10 or 25 mL, add 0.1 mmol of carbamazepine (23.6 mg), the required volume of the catalyst stock solution (6.67×10^{-4} mmol for a substrate/catalyst molar ratio of 150), 15 mg of ammonium acetate (the co-catalyst) and acetonitrile for a final total volume of 2 mL. Keep the reaction mixture under stirring at 25–30 °C and protected from light.
4. Add 25 μL (0.05 mmol) of the oxidant diluted solution at regular intervals of 15 min (Note 1).
5. Follow the reaction by TLC (silica gel plastic sheets with F_{254} indicator) using ethyl acetate as eluent and by reverse-phase HPLC (Note 2).

Product Isolation

6. Add water (2 mL) to the reaction mixture and extract twice with 2 mL of dichloromethane. Filter the organic phase through a glass funnel with cotton and sodium sulfate into a round bottom flask (25 or 50 mL). Evaporate the solvent in a rotary evaporator (Note 3).

Structure Determination

7. Confirm the structure of the epoxide product by ^{1}H NMR using the entire residue (Note 4). Place the residue in the NMR tube followed by the addition of 0.5 mL of $CDCl_3$ (containing tetramethylsilane as the internal standard).
8. Acquire the ^{1}H NMR spectrum.

Note 1. Take a 10 μL aliquot from the reaction mixture before each oxidant addition (including time zero) and dilute the aliquot with 2 mL of acetonitrile for the chromatographic analyses.

Note 2. The best results are obtained with a substrate/catalyst molar ratio of 150, which guarantees a total conversion of the substrate after two hours. Suggestion: different groups of students may test other substrate/catalyst molar ratios, such as 300 (with 3.33×10^{-4} mmol of catalyst) or 600 (with 1.67×10^{-4} mmol of catalyst).

Note 3. Since the product resulting from the oxidation of carbamazepine is an epoxide, it is advisable to use the water bath at room temperature to evaporate the solvent in order to prevent the epoxide ring opening.

Note 4. The carbamazepine ^{1}H NMR spectrum should also be obtained for comparison with the carbamazepine 10,11-epoxide ^{1}H NMR spectrum.

Waste Disposal

Ethyl acetate, the hydrogen peroxide solution in acetonitrile, and the solution of the manganese(III) complex of 5,10,15,20-tetrakis(2,6-dichlorophenyl)porphyrin may be collected in appropriate recipients for non-chlorinated solvents, for posterior destruction/recycling. Dichloromethane may be collected in a recipient for chlorinated solvents, for posterior destruction/recycling.

Results Interpretation and Additional Questions

1. Interpret the ^{1}H NMR spectrum obtained for carbamazepine (CBZ).
2. Interpret the ^{1}H NMR spectrum obtained for carbamazepine 10,11-epoxide (Epoxi-CBZ).
3. Determine the product yield.
4. When hydrogen peroxide is chosen as oxidant for catalytic processes, its dismutation reaction must be taken into account. Relate this practice with the excess of hydrogen peroxide normally used.

References

1. A. F. Carey, *Organic Chemistry*, McGraw-Hill, 7th edn, New York, 2008, ch. 6.
2. (a) L. M. Bradley, J. W. Springer, G. M. Delate and A. Goodman, *J. Chem. Educ.*, 1997, **74**, 1336; (b) J. Hanson, *J. Chem. Educ.*, 2001, **78**, 1266; (c) W. C. Broshears, J. J. Esteb, J. Richter and A. M. Wilson, *J. Chem. Educ.*, 2004, **81**, 1018; (d) K. K. W. Mak, Y. M. Lai and Y. H. Siu, *J. Chem. Educ.*, 2006, **83**, 1058.
3. (a) D. Mansuy and P. Battioni, *The Porphyrin Handbook*, ed. K. Kadish, K. M. Smith and R. Guilard, Academic Press, New York, 2000, vol. 4; (b) J. T. Groves, K. Shalyaev and J. Lee, in *The Porphyrin Handbook*, ed. K. Kadish, K. M. Smith and R. Guilard, Academic Press, New York, 2000, vol. 4; (c) B. Meunier, A. Robert, G. Pratviel, and J. Bernadou, in *The Porphyrin Handbook*, ed. K. Kadish, K. M. Smith and R. Guilard, Academic Press, New York, 2000, vol. 4.
4. (a) I. Bernus, W. D. Hooper, R. G. Dickinson and M. Eadie, *Epilepsy Res.*, 1995, **21**, 65; (b) I. Bernus, R. G. Dickinson, W. D. Hooper and M. Eadie, *Epilepsy Res.*, 1996, **24**, 163; (c) A. L. Faria, T. C. O. Mac Leod and M. D. Assis, *Catal. Today*, 2008, **133**, 863.
5. W. Lohmann and U. Karst, *Anal. Bioanal. Chem.*, 2008, **391**, 79.
6. J. Bernadou and B. Meunier, *Adv. Synth. Catal.*, 2004, **346**, 171.
7. D. Mansuy, *C. R. Chim.*, 2007, **10**, 392.
8. J. Hartmann, P. Bartels, U. Mau, M. Witter, W. van Tümpling, J. Hofmann and E. Nietzschmann, *Chemosphere*, 2008, **70**, 453.
9. (a) R. A. Sheldon, *Metalloporphyrins in Catalytic Oxidations*, Marcel Dekker, New York, 1994; (b) C. M. B. Neves, M. M. Q. Simões, I. C. M. S. Santos, F. M. J. Domingues, M. G. P. M. S. Neves, F. A. A. Paz, A. M. S. Silva and J. A. S. Cavaleiro, *Tetrahedron Lett.*, 2011, **52**, 2898; (c) M. M. Q. Simões, R. De Paula, M. G. P. M. S. Neves and J. A. S. Cavaleiro, *J. Porphyrins Phthalocyanines*, 2009, **13**, 589; (d) S. L. H. Rebelo, M. M. Pereira, M. M. Q. Simões, M. G. P. M. S. Neves and J. A. S. Cavaleiro, *J. Catal.*, 2005, **234**, 76.
10. (a) H. Adolfsson, in *Modern Oxidation Methods*, ed. J. E. Bäckvall, Wiley-VCH Verlag GmbH, Weinheim, 2004, p. 21; (b) J. Brinksma, J. W. de Boer, R. Hage, and B. L. Feringa, in *Modern Oxidation Methods*, ed. J. E. Bäckvall, Wiley-VCH Verlag GmbH, Weinheim, 2004, p. 295; (c) C. W. Jones, *Applications of Hydrogen Peroxide and Derivatives*, The Royal Society of Chemistry, Cambridge, 1999.
11. R. O. Ragsdale, J. C. Vanderhooft and A. P. Zipp, *J. Chem. Educ.*, 1998, **75**, 215.

12.3.5. Regioselective Epoxidation of Geraniol by $VO(acac)_2$ Immobilised in Polystyrene

Ana I. Vicente* and Carlos A. M. Afonso

iMed.ULisboa, Faculdade de Farmácia da Universidade de Lisboa, Portugal
*E-mail: carlosafonso@ff.ulisboa.pt, ana_vicente@ff.ulisboa.pt

Session 1

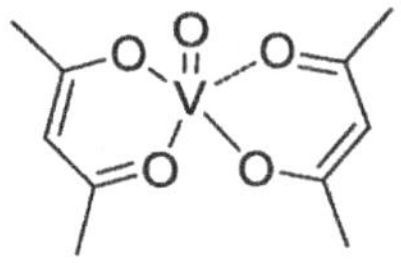

$VO(acac)_2$
Immobilised in polystyrene

Session 2

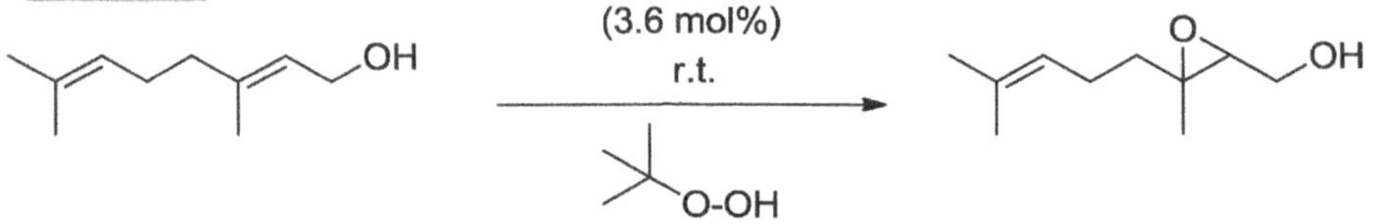

Number of sessions (duration of each session)	Hazard level	Difficulty level	Level of study
2 (4 h each)	Moderate	Medium	Intermediate

Class names Allylic alcohols, epoxides, hydroperoxides, olefins

Concepts involved Demonstration of the technique used to immobilise the catalyst in polystyrene by microencapsulation; mechanism of the regioselective epoxidation of allylic alcohols by *tert*-butyl hydroperoxide and catalysed by $VO(acac)_2$

Chemicals needed Polystyrene (M_w = 280 000), cyclohexane, $VO(acac)_2$, hexane, geraniol, *tert*-butyl hydroperoxide (solution 5–6 M in decane), chloroform-*d*, phosphomolybdic acid

Equipment and experimental techniques involved Precipitation, NMR, decantation, stirring hotplate, vacuum drying, TLC plates

Keywords Allylic alcohols, catalyst, decantation, epoxidation, NMR

Background

The epoxides are versatile functional groups because they allow the synthesis of several organic structural blocks, taking advantage of the epoxide opening by a range of nucleophiles. There are several methods for the epoxidation of olefins.[1–3] Specifically, we are focused on the epoxidation of allylic alcohols, for which there are mainly two synthetic methods reported in the literature, one that uses $VO(acac)_2$ as catalyst and the other the Sharpless asymmetric epoxidation ($Ti(O\text{-}^iPr)_4$ and diethyl tartrate).[4] To illustrate a regioselective epoxidation the reaction was performed with $VO(acac)_2$ on the olefin vicinal to an alcohol in the presence of an olefin, as the

Comprehensive Organic Chemistry Experiments for the Laboratory Classroom
Edited by Carlos A. M. Afonso, Nuno R. Candeias, Dulce Pereira Simão, Alexandre F. Trindade, Jaime A. S. Coelho, Bin Tan, and Robert Franzén.

Published by the Royal Society of Chemistry, www.rsc.org

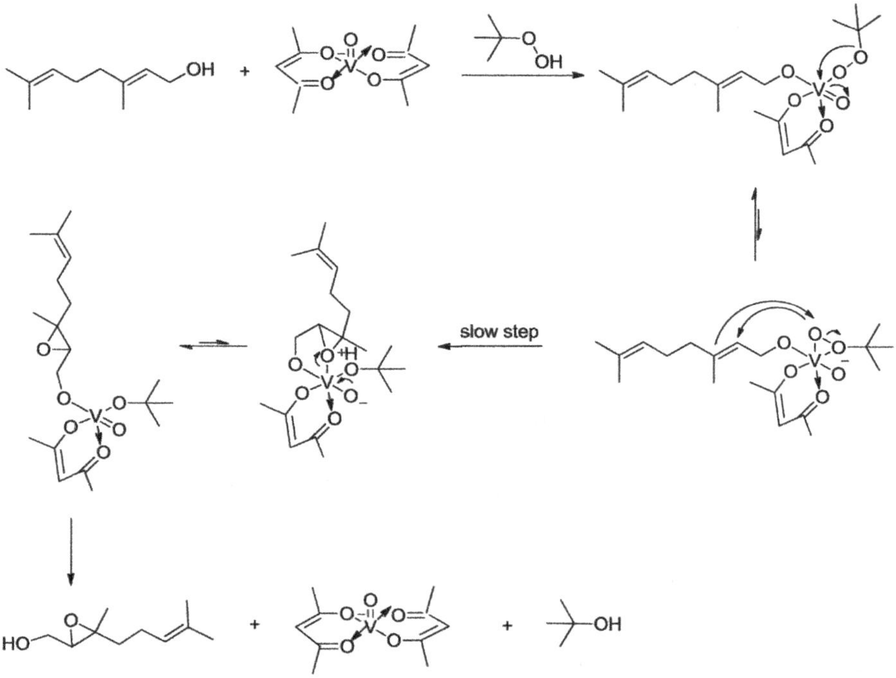

Scheme 12.3.5.1 Proposed mechanism for the vanadium-catalysed epoxidations.[5]

mechanism shows (Scheme 12.3.5.1). The determinant step of the epoxidation is thought to be the oxygen-transfer step and this is also the step in which the stereoselectivity is determined.[5] Supported catalysts allow an easy separation from the product mixture at the end of a reaction and also promote a rapid purification of the product mixture, allowing reuse of the supported complex in subsequent reactions. One of the disadvantages is the lower association of the catalytic active metals in their higher oxidation state compared with a more reduced one, which could lead to leaching of the metal complex from the support. There are several processes of immobilisation, such as those that involve the formation of one (or more) chemical bonding of the catalyst (many times adding a spacer) to the support or by physical immobilisation. The microencapsulation technique has been applied for the immobilisation of several organometallic catalysts and efficiently used in several reactions. This technique consists in the physical involvement of the catalyst by a polymer, through dissolution of the catalyst and the polymer in a solvent, followed by the evaporation of the solvent. This experiment is focused on the immobilisation of the catalyst $VO(acac)_2$ in polystyrene by microencapsulation and its use in the epoxidation of geraniol (as an example of an allylic alcohol). It also shows that in some cases the microencapsulation increases the efficiency of the catalytic system, allowing us to perform the reaction under softer conditions compared to the not immobilised catalyst.

Additional Safety

It is necessary to wear protective gloves, eye protection and face protection to handle the reagents. Cyclohexane and hexane are easily inflammable, harmful, irritant and dangerous for the environment. Geraniol provokes skin irritation and serious damage to the eyes. $VO(acac)_2$ is harmful by ingestion, provokes skin, ocular and respiratory irritation. Additional attention is needed in handling *tert*-butyl hydroperoxide as it is hazardous in case of skin contact (irritant), eye contact (irritant), ingestion, or inhalation and most importantly because it can be explosive.

Experimental Procedure (Ref. 6)

Laboratory Session 1 (4 h): Preparation of $VO(acac)_2$ Immobilised in Polystyrene

1. To a round-bottom flask of 100 mL equipped with a magnetic stirrer add polystyrene (1.0 g and M_w = 280 000) and cyclohexane (50 mL) and heat to 70 °C with stirring until complete solubilization.
2. Add slowly $VO(acac)_2$ (0.2 g) and leave stirring for 1 h at 70 °C.
3. Remove the oil bath and cool the solution with stirring until it reaches room temperature and posteriorly with an ice bath to promote the formation of a dispersed solid in the cyclohexane.
4. Remove the solid by decantation.
5. Add hexane to the solid (30 mL) and stir the solution for 1 h.
6. Decant or filter (depending on whether you obtained a pasty precipitate or a solid) and wash several times with hexane (5 times; 5 × 10 mL) and leave to dry under vacuum (preferentially <20 mmHg) overnight (gradually a precipitate is formed). Alternatively, the solution can be left inside a desiccator under vacuum for several days.

Laboratory Session 2 (4 h): Epoxidation of the Geraniol

1. In a 100 mL round-bottom flask equipped with a magnetic stirrer at room temperature add immobilised $VO(acac)_2$ (0.3 g, approx. 3.6 mol%[6]), geraniol (0.77 g, 5 mmol), hexane (20 mL) and slowly add (with the help of a pipette) *tert*-butyl hydroperoxide (1.2 mL, approximately 1.3 equiv., 5–6 M solution in hexane; *be careful handling it*) and stir the solution until total consumption of the geraniol (2–3 h, detection by TLC using hexane diethyl ether (1:2) and a phosphomolybdic acid solution in ethanol stain).
2. Filter the reaction mixture and evaporate the solvent.
3. Analyse the obtained oil by TLC and ^1H NMR (dissolve 5–10 mg of compound in 400 μL of $CDCl_3$).

Waste Disposal

$VO(acac)_2$ should be disposed of in the solid residues container and *tert*-butyl hydroperoxide can be destroyed by adding sodium metabisulfite solution and stirring the mixture at room temperature.

Results Interpretation and Additional Questions

1. Verify by ^1H NMR and TLC if you obtained the desired product and check the purity.
2. Determine and comment on the turnover number (TON), atom efficiency and the E-factor for the epoxidation step.[7]
3. Complete the mechanism presented with all the intermediates.
4. Do you think allyl alcohol is more reactive than geraniol?

References

1. S. A. Hauser, M. Cokoja and F. E. Kühn, *Catal. Sci. Technol.*, 2013, **3**, 552.
2. K. A. Joergensen, *Chem. Rev.*, 1989, **3**, 431.
3. H. Adolfsson, Transition Metal-Catalyzed Epoxidation of Alkenes, in *Modern Oxidation Methods*, J.-E. Bäckvall, Wiley-VCH Verlag GmbH, 2nd edn, Weinheim, Germany, 2010.
4. M. B. Smith, *Organic Synthesis*, McGraw-Hill, 2nd edn, 2002, vol. 226.
5. K. B. Sharpless and T. R. Verhoeven, *Aldrichimica Acta*, 1979, **12**(4), 63.
6. A. Lattanzi and N. E. Leadbeater, *Org. Lett.*, 2002, **4**, 1519.
7. R. A. Sheldon, *Pure Appl. Chem.*, 2000, **72**(7), 1233.

12.3.6. Organocatalysed *trans*-Dihydroxylation of Olefins

Andreia A. Rosatella* and Carlos A. M. Afonso*

iMed.ULisboa, Faculdade de Farmácia da Universidade de Lisboa, Av. Prof. Gama Pinto, 1649-003 Lisboa, Portugal. *E-mail: rosatella@ff.ulisboa.pt, carlosafonso@ff.ulisboa.pt

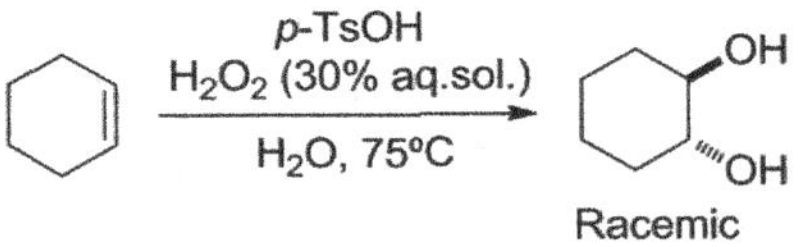

Number of sessions (duration of each session)	Hazard level	Difficulty level	Level of study
2 (5 h + 2 h)	Moderate	Medium	Intermediate

Class names Olefins, alcohols

Concepts involved The experiment gives a method for cyclohexene oxidation to *trans*-1,2-cyclohexanediol using *p*-toluenesulfonic acid (*p*-TsOH) as promoter and hydrogen peroxide as oxidant in a biphasic system

Chemicals needed Cyclohexene, hydrogen peroxide (30 wt%), *p*-toluenesulfonic acid, chloroacetic acid, sodium bicarbonate, sodium sulfite, deuterium oxide

Equipment and experimental techniques involved Apparatus for reaction with temperature (50 °C) and magnetic stirring, liquid–liquid extraction apparatus, rotary evaporator, melting point apparatus, NMR

Keywords Alkene dihydroxylation, diol, liquid–liquid extraction, NMR, oxidation

Background

1,2-Diols play a role as building blocks for the synthesis of pharmaceuticals, agrochemicals, pheromones, or liquid crystals,[1] but are also important as chiral auxiliaries or ligands for asymmetric synthesis.[2] There are several methodologies to synthesize 1,2-diols, such as *via* ketone reduction, epoxide ring-opening, dihydroxylation, and pinacol coupling reactions[2,3] (Scheme 12.3.6.1).

Olefin dihydroxylation is a straightforward method for the synthesis of 1,2-diols that can provide either *syn* or *anti* isomer (Scheme 12.3.6.1). Several methods for the *anti* dihydroxylation of alkenes have been reported,[4–6] but only a few use non-organic solvents as reaction media. Adkins and Roebuck[7] established in 1948 the use of formic acid as solvent for the *trans*-dihydroxylation of cyclohexene, with an aqueous solution of hydrogen peroxide as oxidant. This method uses a large excess of formic acid, and because the reaction is extremely exothermic the olefin was added slowly. Recently, the same reaction has been reported in a microreactor,[8] which eliminates the slow addition of the olefin, but needs a second step to neutralize the acidic solvent, to hydrolyze the formed formate and to isolate the product. Sato *et al.*[9] reported the dihydroxylation of olefins in an aqueous medium with H_2O_2 as oxidant and Nafion® as a catalyst. However, Nafion® is a quite expensive sulfonic resin and the reaction works only above room temperature (70 °C). A different method for the dihydroxylation of olefins uses Oxone® as oxidant, which is a commercially available triple salt ($2KHSO_5{\cdot}KHSO_4{\cdot}K_2SO_4$) synthesized from H_2O_2 and H_2SO_4,

Comprehensive Organic Chemistry Experiments for the Laboratory Classroom
Edited by Carlos A. M. Afonso, Nuno R. Candeias, Dulce Pereira Simão, Alexandre F. Trindade, Jaime A. S. Coelho, Bin Tan, and Robert Franzén.

Published by the Royal Society of Chemistry, www.rsc.org

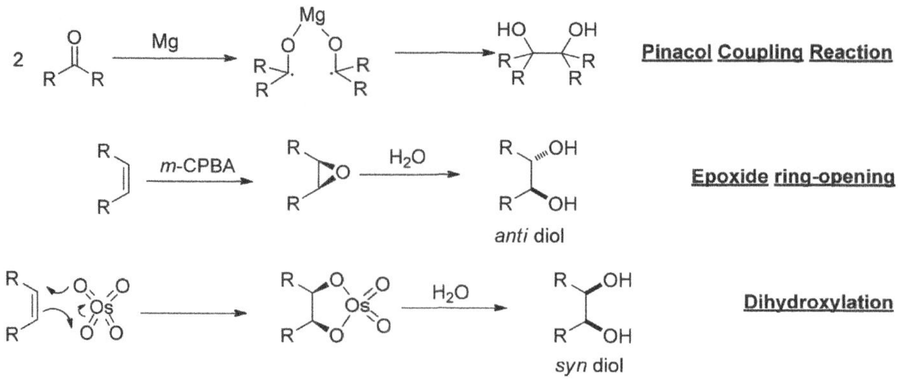

Scheme 12.3.6.1 Methodologies for the synthesis of 1,2-diols.

followed by neutralization with K_2SO_3.[10] The dihydroxylation of the olefin can be accomplished in water at room temperature, but for some substrates an increase of temperature is needed, resulting in a faster decomposition of Oxone®.

The objective of this work is the synthesis of 1,2-cyclohexanediol without the use of organic solvents or metal catalyst. For that, *p*-toluenesulfonic acid (*p*-TsOH) is used as reaction promoter, and an aqueous solution of hydrogen peroxide is used as oxidant agent and also as reaction solvent. Initially, it is a biphasic reaction where the initial substrate is insoluble in the aqueous reaction medium; however, as the product is formed it will be dissolved into the water phase and, consequently, it is possible to observe the substrate disappearance throughout the experiment. This experiment is adjustable since the reaction time can be reduced, with a consequent temperature increase (Methods A and B). In addition, the product isolation can be performed by sublimation, without the use of organic solvents (ESI).

Additional Safety

All sessions should be done in a fume hood, with proper safety equipment, such as protective clothing, gloves and safety goggles. 30% hydrogen peroxide is a corrosive oxidant that, in contact with combustibles, may cause fire. *p*-Toluenesulfonic acid is a corrosive acid and may cause severe skin and eye burns. Cyclohexene is a flammable reagent, is harmful if swallowed, or in contact with the skin.

Experimental Procedure

Laboratory Session 1 (5 h): Dihydroxylation of Cyclohexene (Method A or B)

Method A

1. In a fume hood, heat an oil bath at 75 °C on a magnetic hot plate.
2. To an 50 mL round-bottomed flask (tared) with a magnetic stir bar, add 1.9 g (10 mmol, 1 equiv.) of *p*-toluenesulfonic acid monohydrate, and 2.67 mL of H_2O_2 (30 wt%, 2 equiv.) at room temperature and stir until all the *p*-toluenesulfonic acid is dissolved (5 min).
3. After complete dissolution, add 1.0 mL (10 mmol, 1 equiv.) of cyclohexene. Two phases should be observed.
4. Insert a water condenser into the round-bottomed flask and clamp the flask in the preheated oil bath. Stir the reaction rapidly for 4 h, maintaining a constant temperature of 75 °C.
5. Remove the flask from the oil bath, allowing it to cool to room temperature. Keep the flask at 0 °C until the next session.

Method B

1. In a fume hood, heat an oil bath at 50 °C on a magnetic hot plate.
2. To an 50 mL round-bottomed flask (tared) with a magnetic stir bar, add 1.9 g (10 mmol, 1 equiv.) of *p*-toluenesulfonic acid monohydrate, and 2.67 mL of H_2O_2 (30 wt%, 2 equiv.) at room temperature and stir until all the *p*-toluenesulfonic acid is dissolved (5 min).
3. After complete dissolution add 1.0 mL (10 mmol, 1 equiv.) of cyclohexene. Two phases should be visible.
4. Insert a water condenser into the round-bottomed flask and clamp the flask in the preheated oil bath. Stir the reaction rapidly for 21 h, maintaining a constant temperature of 50 °C.
5. Remove the flask from the oil bath and allow it to cool to room temperature. Keep the flask at 0 °C until the next session.

Laboratory Session 2 (2 h): Yield Determination and Product Isolation

Yield Determination

6. After the reaction reaches completion, weigh the round-bottom flask containing the reaction mixture and determine the final weight (m_1).
7. Remove 200 mg of the reaction mixture (m_2) to an NMR tube followed by 10 mg of standard chloroacetic acid (m_3) – it is of pivotal importance to know the exact amount of sample (m_2) and internal standard (m_3).
8. Add 200 μL of D_2O to the NMR tube.
9. Acquire the ^{1}H NMR spectrum and calculate the reaction yield.

Product Isolation

10. With stirring, slowly add (dead slow) 1.2 g of $NaHCO_3$ to the 50 mL round-bottom flask in order to neutralize the solution.
11. In a water bath at room temperature, add slowly 1.0 g of Na_2SO_3 to the round-bottomed flask to reduce excess H_2O_2. A solid may be formed; add 5 mL of distilled water, if necessary.
12. Transfer the solution into a separatory funnel and perform the extraction with 3 × 25 mL of diethyl ether. Combine the organic layers and dry with anhydrous $MgSO_4$, filter and evaporate the solvent.
13. Weigh the final product and calculate the yield. Measure the melting point, and record ^{1}H and ^{13}C NMR spectra to confirm the compound purity. Compare the results with literature data.

Waste Disposal

The neutralized aqueous solution may be diluted and discarded in the sink. Diethyl ether may be reused or collected in a non-chlorinated waste collector recipient, for posterior destruction.

Results Interpretation and Additional Questions

1. Interpret the ^{1}H NMR spectra obtained, and compare the obtained melting point with the literature data.
2. Calculate the reaction conversion by NMR (internal standard (IS) – chloroacetic acid):

$$\text{mol} = \frac{\text{diol NMR peak}(2.9\text{ ppm})}{\left[\text{IS NMR peak}(3.8\text{ ppm}) \times 0.67\right]} \times \frac{\text{IS mass}(m_3)}{\text{IS MW}}$$

$$\text{Diol total mass} = \text{mol diol} \times \text{diol MW} \times \frac{\text{total sample}(m_1)}{\text{NMR sample}(m_2)}$$

3. Calculate the yield, atom economy,[11] and *E*-factor,[12] for the overall process. Compare your atom economy and *E*-factor with those reported in the previously published experiment.[9–12]
4. Rationalize the involved mechanism of epoxidation and diol formation considering the epoxide ring-opening by water or by *p*-toluenesulfonic acid.
5. Discuss the importance of the promoter *p*-toluenesulfonic acid in the overall process.

Acknowledgement

Reprinted (adapted) with permission from *J. Chem. Educ.*, 2011, **88** (7), 1002–1003. Copyright (2011) American Chemical Society.

References

1. P. Caron, R. W. Dugger, S. G. Ruggeri, J. A. Ragan and D. H. B. Ripin, *Chem. Rev.*, 2006, **106**, 2943–2989.
2. K. C. Bhowmick and N. N. Joshi, *Tetrahedron: Asymmetry*, 2006, **17**, 1901–1929.
3. L. Emmanuel, T. M. A. Shaikh and A. Sudalai, *Org. Lett.*, 2005, **7**, 5071–5074.
4. P. Dupau, R. Epple, A. A. Thomas, V. V. Fokin and K. B. Sharpless, *Adv. Synth. Catal.*, 2002, **344**, 421–433.
5. B. T. Burlingham and J. C. Rettig, *J. Chem. Educ.*, 2008, **85**, 959–961.
6. B. Plietker, M. Niggemann and A. Pollrich, *Org. Biomol. Chem.*, 2004, **2**, 1116–1124.
7. A. Roebuck and H. Adkins, *Org. Synth.*, 1948, **28**, 35–37.
8. A. Hartung, M. A. Keane and A. Kraft, *J. Org. Chem.*, 2007, **72**, 10235–10238.
9. Y. Usui, K. Sato and M. Tanaka, *Angew. Chem., Int. Ed.*, 2003, **42**, 5623–5625.
10. W. M. Zhu and W. T. Ford, *J. Org. Chem.*, 1991, **56**, 7022–7026.
11. B. M. Trost, *Science*, 1991, **254**, 1471–1477.
12. R. A. Sheldon, *Green Chem.*, 2007, **9**, 1273–1283.

12.4. Other Oxidations

12.4.1. Synthesis of 6-Nitrosaccharin

Dulce Simão* and Ana C. Cerdeira

Department of Chemistry, Instituto Superior Técnico, University of Lisbon, Centro de Química Estrutural, Av. Rovisco Pais, 1, 1049-001 Lisbon, Portugal. *E-mail: dulcesimao@tecnico.ulisboa.pt

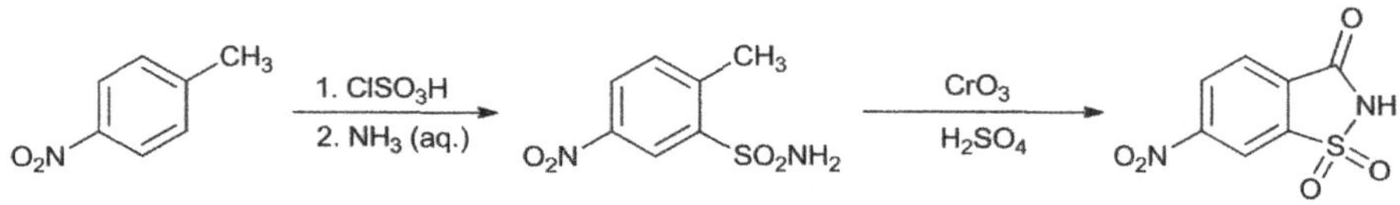

Number of sessions (duration of each session)	Hazard level	Difficulty level	Level of study
2 (4 h each)	High	Medium	Intermediate

Class names Aromatic nitro compounds, sulfonyl chlorides, amines, sulfonamides, carboxylic acids

Concepts involved Two-step reaction sequence to obtain an artificial sweetener illustrating an aromatic electrophilic substitution to produce a sulfonyl chloride followed by reaction with ammonia to origin a sulfonamide; in the second step a benzylic oxidation proceeds with spontaneous cyclization

Chemicals needed 4-Nitrotoluene, chlorosulfuric acid, diethyl ether, aqueous ammonia, methanol, concentrated sulfuric acid, chromium(VI) oxide, deuterated acetone

Equipment and experimental techniques involved Liquid–liquid extraction, ^{1}H NMR, apparatus for reflux with steam bath, magnetic stirring, vacuum filtration and recrystallization; temperature-controlled reactions

Keywords Aromatic electrophilic substitution, cyclization, ^{1}H NMR, liquid–liquid extraction, oxidation reaction

Background

Sucrose is a natural sugar with excellent taste qualities but with the disadvantage of having high energy content. As a consequence, non-nutritious artificial sweeteners, which are often much sweeter than sucrose, were developed and have found wide applications within the food industry. For example, saccharin, discovered in 1879, is 300 times sweeter than sucrose while cyclamate, developed in the 1950s, is 30–50 times sweeter. However, several studies performed in the late 1960s linked both sweeteners to malignant tumors in rats. This generated a controversy around these products, leading to an eventual ban on cyclamate by the US Food and Drug Administration (US FDA) (saccharin is still legal in the USA, but is banned in certain countries). This

Comprehensive Organic Chemistry Experiments for the Laboratory Classroom
Edited by Carlos A. M. Afonso, Nuno R. Candeias, Dulce Pereira Simão, Alexandre F. Trindade, Jaime A. S. Coelho, Bin Tan, and Robert Franzén.

Published by the Royal Society of Chemistry, www.rsc.org

controversy led to the development of new sweeteners such as Aspartame® and Acesulpham®.[1] In this experiment, the 6-nitro derivative from saccharin is prepared[2,3] from 4-nitrotoluene, and illustrates several important aspects of aromatic compounds.[4a] The first step involves an aromatic electrophilic substitution with the chlorosulfonation of 4-nitrotoluene to give 4-nitrotoluene-2-sulfonylchloride. Chlorosulfonic acid is a very strong acid and protonates itself to produce $ClSO_2^+$, which is an electrophile.[4b] So, in this reaction no Lewis acid is required. The reaction of the 4-nitrotoluene-2-sulfonylchloride with aqueous ammonia produces the corresponding sulfonamide by nucleophilic substitution at the sulfur atom.[5] In the second step, the toluene methyl group is oxidized with chromium(VI) oxide in sulfuric acid to form *ortho*-sulfonamidobenzoic acid, which spontaneously cyclizes into the final product with dehydratation.[3,4a] It was found in literature that 6-nitrosaccharin has an initial sweet taste followed by a bitter taste.[3]

Additional Safety

This experiment must be conducted inside a fume hood, since chlorosulfuric acid, aqueous ammonia and sulfuric acid are corrosive. Chromium(VI) oxide is a poisonous, irritant, corrosive solid and has been found to cause cancer. It is very hazardous in case of skin contact. Care should also be taken while handling diethyl ether (flammable). Do not taste the product obtained. Even though the product can be used in food industry, the presence of residual amounts of chromium by this procedure can make your sample highly toxic.

Experimental Procedure

Session 1 – Preparation of 4-Nitrotoluene-2-Sulfonamide

1. To a 50 mL round-bottom flask add 4-nitrotoluene (6.85 g, 50 mmol) and then carefully add chlorosulfonic acid (12 mL, 21 g, 0.18 mol).
2. Fit the flask with a reflux condenser and heat on a water-bath for 30 min with magnetic stirring.
3. Cool the flask on an ice bath and slowly pour the reaction mixture into a 600 mL beaker containing crushed ice (about 100 g) and stir the reaction mixture vigorously with a glass rod.
4. Transfer the mixture to a separatory funnel and extract with diethyl ether (2 × 40 mL).
5. Transfer the combined ether extracts to a 250 mL Erlenmeyer flask containing a magnetic bar, rapidly stir the solution and then slowly add a solution of aqueous ammonia (20 mL).
6. Continue to stir until a light brown precipitate appears. Filter by vacuum and wash the precipitate thoroughly first with cold diethyl ether (20 mL) and then with cold water (40 mL).
7. Allow the solid dry on the funnel with vacuum for a few minutes.
8. Recrystallize from hot water. Record the yield and melting point. Obtain the IR spectrum.

Session 2 – Preparation of 6-Nitrosaccharin

1. Fit a 100 mL three-neck round-bottomed flask with a reflux condenser and a thermometer.
2. Add concentrated sulfuric acid (12 mL) and then add dry 4-nitrotoluene-2-sulfonamide (2.15 g, 10 mmol).
3. Heat this mixture to 65 °C on a water-bath with gentle magnetic stirring.
4. Once it reaches 65 °C add chromium(VI) oxide (3 g, 30 mmol) at such rate that the temperature is maintained between 65 and 70 °C. This addition should take between 15 and 30 min, during which the reaction mixture becomes green and viscous.
5. When the addition is complete stir the mixture for an additional 10 min at 65–70 °C, and then cool it in an ice bath.

6. Pour the reaction mixture into a 250 mL Erlenmeyer flask containing cold water (50 mL) and stir for a few minutes until a solid precipitates.
7. Filter off the solid with a vacuum and wash thoroughly with cold water.
8. Dry the product in the funnel with vacuum for a few minutes and recrystallize from water. Record the yield and melting point, and obtain both IR and ^{1}H NMR spectra.

Waste Disposal

Acid aqueous filtrate solutions containing residual amounts of chromium can be disposed of in a container specifically designated for chromium waste;[6] nevertheless, hexavalent chromium can be reduced to the less toxic trivalent state and precipitated as an oxide. Chromium solutions of lower oxidation states can be disposed of in metal solution or alkali containers.[7,8] Diethyl ether used in washings can be stored in a flask for later purification.

Results Interpretation and Additional Questions

1. Determine the reaction yields and the overall yield.
2. Compare the melting points with those from the literature and evaluate the purity of the obtained compounds.
3. Interpret both IR and ^{1}H NMR spectra.
4. Write out the reaction mechanisms of the reactions involved.
5. Starting from toluene instead of 4-nitrotoluene, what product (or products) would you expect to obtain in the first step?

References

1. M. R. Weihrauch and V. Dieh, *Ann. Oncol.*, 2004, **15**, 1460.
2. L. M. Harwood, C. J. Moody and J. M. Percy, *Experimental Organic Chemistry*, Blackwell Science, 2nd edn, 1999, pp. 633–635.
3. N. C. Rose and S. Rome, *J. Chem. Educ.*, 1970, **47**, 649.
4. J. Clayden, N. Greeves, S. Warren and P. Wothers, *Organic Chemistry*, Oxford University Press, Oxford, 1st edn, 2001(a) p. 644; (b) p. 563.
5. S. Ege, *Organic Chemistry*, D. C. Heath and Company, 2nd edn, 1989, p. 810.
6. *Procedures for Disposal of Chemical Wastes*, October 2007, http://www.ab.ust.hk/hseo/sm06/Chem_Wastes_Disposal_Procedure.pdf, accessed November 2015.
7. F. Thompson, *Disposal of Chromium Laboratory Wastes*, June 1988, http://infohouse.p2ric.org/ref/19/18012.pdf, accessed November 2015.
8. *Procedures for the Laboratory-Scale Treatment of Surplus and Waste Chemicals*, https://www.unige.ch/sciences/chiorg/matile/12-ProceduresforLabTreatmentofWasteChemicals.pdf, accessed November 2015.

13. Reductions

13.1. Reduction of a Ketone Using Sodium Borohydride. Control of a Reaction by TLC

Luis Constantino*[a,b], Catarina Dias[a], and Emília Valente[a]

[a]Faculty of Pharmacy, Universidade de Lisboa Av. Professor Gama Pinto, 1649-003 Lisbon, Portugal; [b]Research Institute for Medicines (iMed.ULisboa), Faculty of Pharmacy, Universidade de Lisboa, Av. Professor Gama Pinto, 1649-003 Lisbon, Portugal. *E-mail: constant@ff.ulisboa.pt

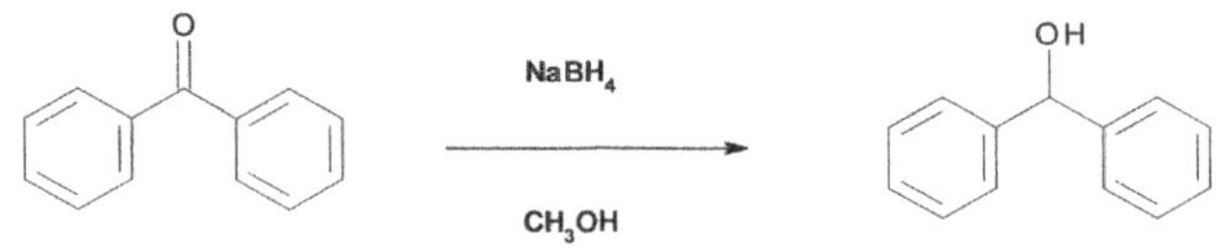

Number of sessions (duration of each session)	Hazard level	Difficulty level	Level of study
1 (4 h)	Moderate	Low	Introductory
Class names Ketones, alcohols			
Concepts involved This experiment uses the reduction of benzophenone in presence of sodium borohydride as a starting point to introduce the subject of reaction control by TLC (thin-layer chromatography). It also includes the recrystallization of the product from a volatile organic solvent. The resulting solid is tested for purity by a melting point determination. IR (infrared spectroscopy) will be performed on the starting material and product to identify the relevant bands in the spectra			
Chemicals needed Diphenyl ketone (benzophenone, diphenylmethanone), diphenylmethanol (benzhydrol, diphenylcarbinol), sodium borohydride, methanol, dichloromethane, petroleum ether (40–60 °C), water			
Equipment and experimental techniques involved *Equipment*: TLC plates with F_{254} fluorescent indicator, capillaries for TLC, elution camera, UV lamp (254 nm), vacuum source, melting point apparatus, capillaries for melting point determination, Erlenmeyer flasks, small funnel, vacuum flask, Büchner funnel, graduated cylinders, ice bath. *Experimental techniques*: addition of reactants, stirring, recrystallization, vacuum filtration, TLC, IR, melting point			
Keywords Benzophenone, crystallization, IR, reduction, sodium borohydride, TLC			

Background

The reduction of ketones and aldehydes to alcohols is a very important reaction both in nature and in the organic laboratory.

Although various methods are available in the laboratory, the most frequently employed is the use of complex metal hydride reagents such as lithium aluminium hydride ($LiAlH_4$) and sodium borohydride ($NaBH_4$). $LiAlH_4$ is more powerful, reducing most carbonyl containing compounds including aldehydes, ketones, esters, and amides. $NaBH_4$ is more selective and only reacts with aldehydes and ketones.

Comprehensive Organic Chemistry Experiments for the Laboratory Classroom
Edited by Carlos A. M. Afonso, Nuno R. Candeias, Dulce Pereira Simão, Alexandre F. Trindade, Jaime A. S. Coelho, Bin Tan, and Robert Franzén.

Published by the Royal Society of Chemistry, www.rsc.org

Both reducing agents are a source of H^-, which are transferred from the boron or aluminium to the carbonyl group. Since the reduction requires two hydrogen atoms, the second comes from the solvent (when using $NaBH_4$) or from added acid in the work-up (when using $LiAlH_4$). Since $LiAlH_4$ is very reactive, it produces rapidly H_2 in protic solvents and, therefore, the reaction is performed in a polar aprotic solvent such as diethyl ether or THF. Although being formally a reduction, the reaction is also a nucleophilic addition to a carbonyl group, one of the most important reaction types in organic chemistry.

The reduction of ketones using sodium borohydride, initially studied by Garrett and Lyttle,[1] can be performed in the presence of a protic solvent, *e.g.* methanol or water. This allows the formation of the corresponding alcohol without the need to include a further reaction step to decompose a tetraalkoxyborate intermediate that is formed under other conditions (*e.g.* reductions in the absence of a proton source).[2–4] The mechanism of the reaction in methanol can be understood as follows: initially H^- is added to the carbonyl carbon; this forms the C–H bond, and breaks the C=O π bond (Scheme 13.1.1). A proton from the solvent is then added to the oxygen, forming the alcohol product. After the alcohol is produced, the remaining methoxide ($-OCH_3$) reacts like a Lewis base with BH_3, to form a borate. This process continues until all the hydrogens from the borohydride are used. A molecule of sodium borohydride is thus able to transfer, in principle, four hydride ions. However, in practice an excess of sodium borohydride (50–100%) is used as the reagent also reacts with the protic solvent to form hydrogen.

The mechanism for $LiAlH_4$ is very similar (Scheme 13.1.2). Due to its reactivity, $LiAlH_4$ reductions are carried out in aprotic solvents (in the absence of a proton source). After transferring the hydride ion to the carbonyl group the remaining neutral AlH_3 binds to the negative oxygen atom, forming an alkyl aluminate, which can react three additional times to produce a tetra-alkyl aluminate. This complex is then hydrolysed in a separate step to yield the final alcohol.

A similar reaction is also performed in nature with NADH or NADPH. Biological reductions of aldehydes and ketones are catalysed by alcohol dehydrogenases that use nicotinamide adenine dinucleotide as cofactor. Other reductases use NADPH, a similar cofactor. Both compounds may

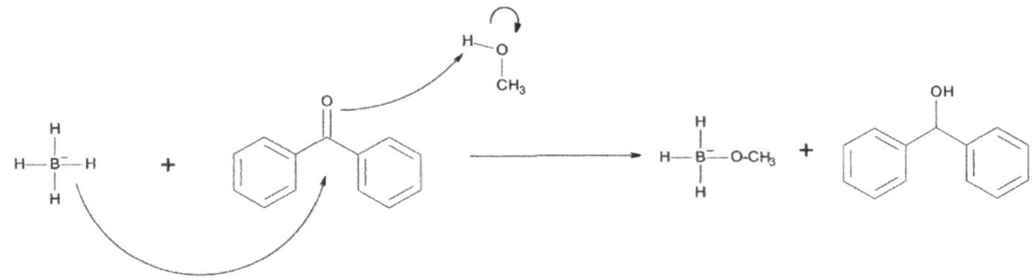

Scheme 13.1.1 Reaction of benzophenone with borohydride in the presence of methanol. This process is exemplified for the transfer of the first hydride; however, it can continue until all the hydride ions from $NaBH_4$ have been consumed.

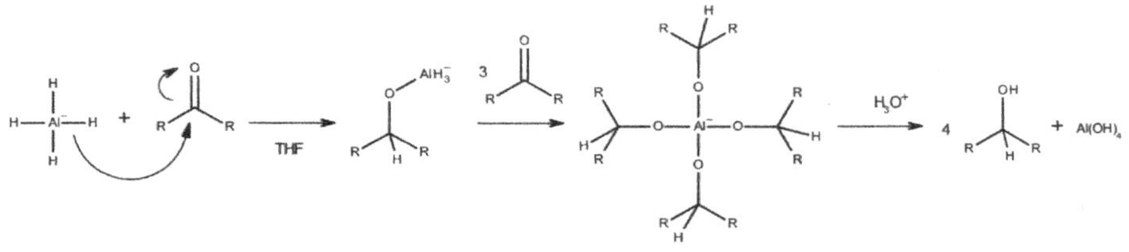

Scheme 13.1.2 Reduction of a ketone using lithium aluminium hydride. The reaction is carried out initially in an aprotic solvent, forming a tetra-alkyl aluminate. Water or acid is then added to hydrolyse the aluminate and form the alcohol.

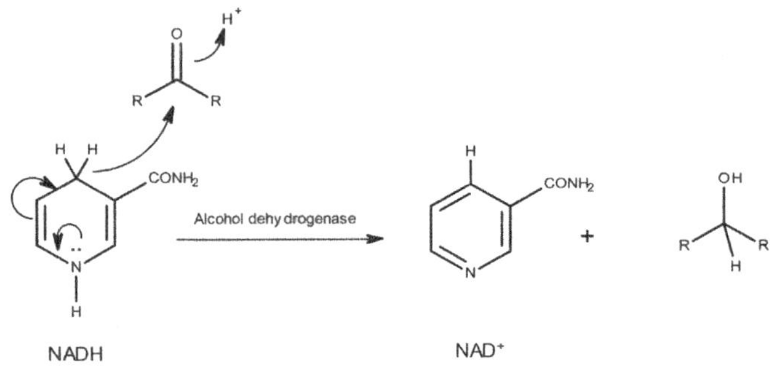

Scheme 13.1.3 Reduction of a ketone performed by an alcohol dehydrogenase. The enzyme uses the cofactor NADH, which can be considered the reducing agent in a manner similar to $NaBH_4$ or $LiAlH_4$. The enzyme often also catalyses the oxidation of the alcohol in the presence of excess NAD^+.

be considered as the reducing agent capable of supplying hydride in a similar manner to $LiAlH_4$ or $NaBH_4$.

During the reducing sequence alcohol dehydrogenase transfers a hydride from NADH to the carbonyl carbon and a proton from water is used to complete the process (Scheme 13.1.3). The cofactor becomes oxidized to NAD^+.

One important difference between chemical reductions using main group hydrides and enzymes is stereospecificity. Chemical reductions of simple ketones will involve hydride addition from either face of the planar ketone and if a new chiral centre is created this normally leads to a racemic alcohol product (with exceptions). The enzymatic reaction is usually much more stereospecific, many times producing a single enantiomer.[5]

In this protocol, the reaction is performed in an ice bath to reduce its rate in order to follow it easily by TLC. Monitoring is performed at 0, 5, 10 and 15 min or every 5 min until the reaction is complete. After the first 5 min the reaction is taken out of the ice bath and is left at room temperature. The results are obtained rapidly because TLC plates are developed as the reaction takes place and the spots are easily visualized under UV light. The end of the reaction is easily monitored by the disappearance of substrate. The students can thus decide by themselves when the reaction is finished. Finally, the product is precipitated and crystallized. The recrystallization is performed using petroleum ether (40–60 °C). This solvent is appropriate to increase the skills of students that only have executed a recrystallization using a less volatile solvent. In fact, since this solvent is volatile and evaporates during the recrystallization, students must be aware of that and compensate for the loss of solvent and must not waste time during the dissolution of the product. After purification the product is tested for purity by melting point determination. IR can also be carried out in order to identify the characteristic bands and the spectrum can be compared with the IR spectrum of the benzophenone.

The limiting reagent should be identified and the yield calculated and discussed.

Additional Safety

Sodium borohydride should be stored in a desiccator and the reaction should be performed preferably in a fume hood as the sodium borohydride decomposes in methanol, giving rise to hydrogen. The use of an ice bath is not mandatory although is slows the reaction and makes it easier to follow by TLC. The recrystallization is performed with a volatile solvent and should also be performed in the fume hood. A water bath should be used as a source of heat. Heating should not be performed directly over a hot plate. The flash point of petroleum ether (40–60 °C) is *ca.* −20 °C and its auto-ignition temperature is *ca.* 290 °C.

Experimental Procedure

Preparation of the Initial Reagent Solution

1. In a 50 mL Erlenmeyer flask dissolve 750 mg of benzophenone in 10 mL of MeOH. Cool the solution in an ice bath and keep it there while proceeding with the validation of the TLC control procedure (see below).

Validation of the TLC Procedure

2. On a silica-gel plate apply one spot of the benzophenone standard, one spot of the diphenylmethanol standard and one spot of the benzophenone solution you prepared in the previous step. Use a TLC capillary to apply the spots. Control the intensity of the spots under UV light (254 nm) to judge the appropriate amount applied in each spot before eluting. You should see clearly three spots in the plate. Elute the TLC plate in a chamber containing dichloromethane–petroleum ether (4:1) as eluent. After the solvent migrates to the desired solvent front height, remove the plate, let it dry, and mark the spots under the UV light with a pencil.
3. You should see clearly the spots of the standard with different R_f and the spot of the benzophenone solution with the same R_f as the corresponding standard. If this TLC was successful you can proceed with the reaction. If not you should solve any problem before proceeding.

Synthesis

4. After you successfully realize the validation of the TLC system, weigh 0.2 g of sodium borohydride.†
5. Keeping the Erlenmeyer flask in the ice bath, add the sodium borohydride slowly to the benzophenone solution. Swirl.
6. Immediately take a sample of the reaction mixture with a capillary and apply it to a TLC plate. On the same plate apply also a spot of each standard as in the previous run of the TLC. Develop the plate under the same conditions as previously and mark the spots.
7. At 5 min repeat the procedure. Take the solution out of the ice batch and leave it at room temperature.
8. At 10 min repeat the TLC control. Repeat also at 15 min and then every 5 min thereafter until the reaction is complete.
9. After you have two TLC controls in which you cannot see the benzophenone spot in the reaction mixture consider the reaction to be complete.
10. If the reaction is not complete continue with the control every 5 min. Make sure you added the correct amount of sodium borohydride and that the reaction mixture was swirled appropriately.

Separation of the Product from the Reaction Mixture

11. Add 20 mL of pre-chilled water to the reaction mixture and swirl.
12. Recover the resultant solid by Büchner filtration and leave the crystals on the Büchner funnel with the vacuum on for 3 min in order to dry the product partially.
13. With the vacuum off carefully transfer the solid cake out of the Büchner funnel along with the filter paper onto a larger piece of filter paper. With a spatula scrape the damp filter cake from the filter paper (but do not scrape any paper fibres into your product). Repeatedly press the crystals between dry sheets of filter paper, and change sheets until the solid is no longer damp as indicated by no transferal of moisture onto the filter paper.

---

†Sodium borohydride should be kept in a desiccator. After weighing the required amount, return the rest to the same desiccator.

Purification by Recrystallization from Petroleum Ether

14. As the solvent used in the recrystallization (petroleum ether) is not miscible with water the solid must not be too damp. If necessary the solid can be partially dried in a drying oven for 5–10 min. Make sure the temperature is set bellow the melting point of the product (65–67 °C).
15. Transfer the dried solid to an Erlenmeyer flask.
16. Heat approximately 25 mL of petroleum ether in an Erlenmeyer flask in a water bath (pay attention to the boiling point of the petroleum ether (40–60 °C)).
17. Dissolve the solid in the minimum amount of boiling petroleum ether.
18. Note if you need to perform any decolouration of the solution and/or hot gravity filtration, add 20% excess of solvent (as the petroleum ether is very volatile and evaporates during the procedure) and then perform the operation(s).
19. Let the solution cool gradually in order to obtain the crystals of the product.
20. When the Erlenmeyer flask is close to room temperature cool it further in an ice bath to complete the crystallization procedure.
21. If no crystals appear but the solution is supersaturated, you may induce the formation of crystals by scratching the surface of the flask close to the solution level using a glass rod.
22. Separate the crystals by Büchner filtration.
23. Wash the product with ice cold petroleum ether.
24. Dry the crystals to constant weight.
25. Measure the melting point of the product.
26. Obtain the IR spectra of the product and of the starting material (ask your instructor for help).
27. Determine the yield of the synthesis.

Waste Disposal

The filtrate containing sodium borohydride is placed in a container labelled “sodium borohydride residues”. Sodium borohydride residues should be eliminated after the class by dilution with water and treatment with acetic acid. The filtrate of the recrystallization and the mixture of eluents used in TLC will be discharged in the container for non-halogenated solvent residues.

Results Interpretation and Additional Questions

1. Identify the IR bands corresponding to the ketone in the starting material and the alcohol in the product.
2. Discuss the results and the purity of the product (use IR and mp data).
3. How many moles of benzophenone can be reduced with one mole of sodium borohydride?
4. Could you perform the same reaction with lithium aluminium hydride as the reducing agent? What would you do differently in such a case?
5. If sodium borohydride was added to benzoic acid what product would you expect? What if the reducing reagent was lithium aluminium hydride?
6. Besides the evident care regarding inhalation of a volatile solvent, what precautions should you take in performing a recrystallization with a solvent with a low boiling point such as petroleum ether?

References

1. E. R. Garrett and D. A. Lyttle, *J. Am. Chem. Soc.*, 1953, **75**, 6051–6052.
2. D. C. Wigfield and F. W. Gowland, *Tetrahedron Lett.*, 1976, **17**, 3373–3376.
3. D. C. Wigfield, *Tetrahedron*, 1979, **35**, 449–462.
4. M. M. Kayser, S. Eliev and O. Eisenstein, *Tetrahedron Lett.*, 1983, **24**, 1015–1018.
5. T. Matsuda, R. Yamanaka and K. Nakamura, *Tetrahedron: Asymmetry*, 2009, **20**, 513–557.

13.2. Regioselective Catalytic Transfer Hydrogenation of Citral

Jaime A. S. Coelho* and Carlos A. M. Afonso*

iMed.UL, Faculdade de Farmácia da Universidade de Lisboa, Av. Prof. Gama Pinto, 1649-003 Lisboa, Portugal
*E-mail: jaimeacoelho@ff.ul.pt, carlosafonso@ff.ulisboa.pt

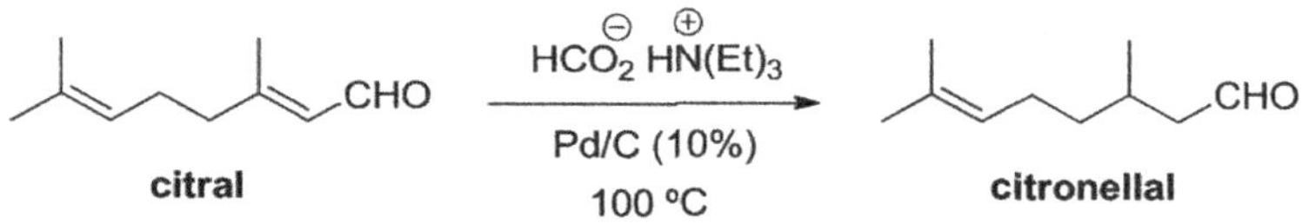

Number of sessions (duration of each session)	Hazard level	Difficulty level	Level of study
2 (1 h + 2 h)	Moderate	Medium	Intermediate

Class names Alkenes, aldehydes, enals

Concepts involved This experiment involves the regioselective Pd/C-catalysed hydrogenation of alkenes using triethylammonium salt as the source of hydrogen instead of the commonly used hydrogen gas. The regioselectivity of the reaction is evaluated by IR and NMR spectroscopy

Chemicals needed Pd/C (10%), citral, triethylamine, formic acid, distilled water, conc. HCl, diethyl ether, sat. aq. sol. $NaHCO_3$, anhydrous Mg_2SO_4, deuterated chloroform ($CDCl_3$)

Equipment and experimental techniques involved Hotplate stirrer (100 °C), condenser, equipment for inert atmosphere, extraction funnel, universal pH indicator paper, rotary evaporator, liquid–liquid extraction

Keywords IR, liquid–liquid extraction, NMR, palladium on carbon, regioselectivity, transfer hydrogenation

Background

Catalytic hydrogenation, including the asymmetric version, and hydrogenolysis are very important transformations for both industrial and laboratorial processes. There are many efficient catalytic systems for the regio- and enantio-selective reduction of several functional groups, such as alkenes, alkynes, aromatic compounds, imines, alkyl halides and carbonyl-containing molecules.[1]

Hydrogenation reactions on both laboratory and industrial scales are usually performed by employing gaseous hydrogen. The implementation of this type of hydrogenation reaction in academic environments, namely in teaching laboratory, is difficult due to the inherent risks of handling hydrogen gas (combustion occurs when hydrogen and oxygen are in contact with the catalyst). Although there are didactic protocols for the use of hydrogen in the teaching classes,[2] there are several alternatives to the handling of hydrogen gas, namely the *in situ* preparation of hydrogen (*e.g.*, using $NaBH_4$[3] or electrochemical methods[4]) or *via* hydrogen transfer (also known as transfer hydrogenation), using hydrogen sources such as formic acid, ammonium formate, cyclohexene, cyclohexadiene, hydrazine, phosphinic acid, sodium hypophosphite,[5] and more recently *via* asymmetric organocatalysis.[6]

Comprehensive Organic Chemistry Experiments for the Laboratory Classroom
Edited by Carlos A. M. Afonso, Nuno R. Candeias, Dulce Pereira Simão, Alexandre F. Trindade, Jaime A. S. Coelho, Bin Tan, and Robert Franzén.

Published by the Royal Society of Chemistry, www.rsc.org

In this experiment the regioselective transfer hydrogenation of citral will be studied using triethylammonium formate as the source of hydrogen and Pd/C (10%) as catalyst. This methodology is based on the work reported by Heck and co-workers,[7] and has been used for the reduction of a diversity of substrates, such as cooking oils,[8] alkynes, nitro compounds, in the hydrogenolysis of *O*-, *N*-allyl and benzyl derivatives, and in the deoxygenation of sulfates and triflates.[9]

The selected substrate (citral) is a readily available monoterpenoid and allows the preparation of citronellal, which is an important compound used in its chiral form in the perfume industry and also presents insect repellent properties.[10] Furthermore, citral hydrogenation is a model reaction for the study of selective reductions since the molecule offers different reactive sites: isolated and conjugated C=C bond, and a carbonyl group.[11]

Additional Safety

The experiment should be conducted in a fume hood and all students should be equipped with protective clothing, gloves and safety glasses. Special attention should be taken regarding the high reflux temperature reaction performed overnight. The reaction mixture must be cooled to room temperature before opening the system to perform the TLC analysis or work-up. Triethylamine and formic acid are hazardous in case of eye or skin contact (irritant) and inhalation or ingestion. Concentrated HCl is very hazardous in case of skin contact (corrosive, irritant, permeator), of eye contact (irritant, corrosive), of ingestion, and of inhalation (lung sensitizer).

Experimental Procedure

Laboratory Session 1 (1 h): Transfer Hydrogenation of Citral

1. Prepare the reaction apparatus consisting of a round-bottom flask (50 mL) and a condenser connected on top to a bubbler containing liquid paraffin or silicone oil.
2. Add Pd/C (10%, 0.21 g, 0.2 mmol), citral (3.04 g, 3.4 mL, 20 mmol) and triethylamine (4 mL, 29 mmol) to the round-bottom flask.
3. Use a water bath to keep the reaction temperature at room temperature and add, dropwise, formic acid (0.83 mL, 22 mmol) to the reaction mixture.
4. Leave the reaction under vigorous stirring at 100 °C overnight (15–24 h).

Laboratory Session 2 (2 h): Reaction Work-Up

5. Cool down the reaction mixture.
6. Check the reaction for completion by TLC analysis (eluent: hexane/ethyl acetate (9 : 1)). For that, perform a work-up of an aliquot of the reaction mixture in a small vial by acidifying the sample with conc. HCl, adding diethyl ether and shaking. The organic phase is spotted onto the thin-layer plate using a capillary.
7. Add distilled water (25 mL) to the reaction mixture followed by conc. HCl until pH 1.
8. Extract the product with diethyl ether (2 × 25 mL) and wash the collected organic phases with saturated aqueous solution of $NaHCO_3$.
9. Dry the organic phase over $MgSO_4$, filter, evaporate the solvent in the rotary evaporator, and weigh the residue.
10. Dissolve 5–10 mg of the final residue in deuterated chloroform and transfer it to an NMR tube for analysis.

Waste Disposal

The acid aqueous solutions should be diluted, neutralized and collected in proper containers. The Et_2O and $MgSO_4$ should be collected in proper residue containers.

Results Interpretation and Additional Questions

1. Analyse the regioselectivity of the reaction by IR and ^{1}H NMR.
2. Why is there two aldehyde signals in the ^{1}H NMR of the starting material citral?
3. Explain the mechanism of the transformation, namely the role of the catalyst and the regioselectivity of the transfer hydrogenation.
4. Suggest other possible products for citral hydrogenation.

References

1. (a) M. B. Smith, *Organic Synthesis*, McGraw-Hill Higher Education, 2nd edn, 2002, p. 306; (b) R. C. Larock, *Comprehensive Organic Transformations*, Wiley-VCH Verlag GmbH, Weinheim, 2nd edn, 1999; (c) I. Ojima, *Catalytic Asymmetric Synthesis*, Wiley-VCH Verlag GmbH, Weinheim, 2000.
2. J. A. Landgrebe, *J. Chem. Educ.*, 1995, **72**, A220.
3. B. Plummer, *J. Chem. Educ.*, 1989, **66**, 518.
4. D. M. A. F. Navarro and M. Navarro, *J. Chem. Educ.*, 2004, **81**, 1350.
5. B. K. Banik, K. J. Barakat, D. R. Wagle, M. S. Maghar and A. K. Bose, *J. Org. Chem.*, 1999, **64**, 5746, and reference therein.
6. J. W. Yang, M. T. H. Fonseca and B. List, *Angew. Chem., Int. Ed.*, 2004, **43**, 6660.
7. N. A. Cortese and R. F. Heck, *J. Org. Chem.*, 1978, **43**, 3985.
8. M. A. Tike and V. V. Mahajani, *Chem. Eng. J.*, 2006, **123**, 31, and reference therein.
9. A. O. King and I. Shinkai, Palladium–triethylamine–formic acid in *Encyclopedia of Reagents for Organic Synthesis*, John Wiley & Sons, 2001.
10. (a) *The Merck Index*, ed. S. Budavari, A. Smith, M. J. O'Neil, P. E. Heckelman and J. F. Kinneary, Merck & Co, 12th edn, 1996; (b) J.-K. Kim, C.-S. Kang, J.-K. Lee, Y.-R. Kim, H.-Y. Han and H. K. Yun, *Entomol. Res.*, 2005, **35**, 117.
11. A. Stolle, T. Gallert, C. Schmöger and B. Ondruschka, *RSC Adv.*, 2013, **3**, 2112.

13.3. Regioselective 1,2-Reduction of an α,β-Unsaturated Ketone. A Green Experiment

M. Manuela A. Pereira

REQUIMTE, Chemistry Department, Faculdade de Ciências e Tecnologia, Universidade Nova de Lisboa, 2827-516 Caparica, Portugal. *E-mail: manuela.pereira@fct.unl.pt

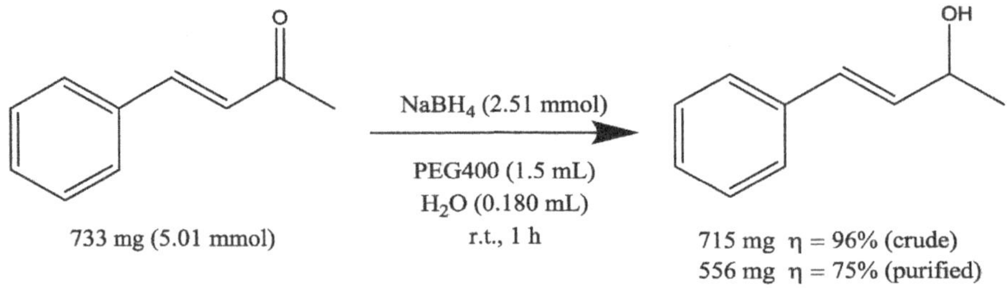

Number of sessions (duration of each session)	Hazard level	Difficulty level	Level of study
2 (3 h each)	Moderate	Medium	Intermediate

Class names α,β-Unsaturated ketones, allylic alcohols

Concepts involved This experiment involves a regioselective 1,2-reduction of an α,β-unsaturated ketone in PEG400/water, a safer low toxic friendly environment medium that possesses little toxicity to human health and the environment (one of the twelve principles of green chemistry)[1]

Chemicals needed (*E*)-4-Phenylbut-3-en-2-one, sodium borohydride, poly(ethylene glycol) 400 (PEG400), diethyl ether, hexane, ethyl acetate, anhydrous sodium sulfate, TLC plates (silica gel 60 F_{254}), silica gel (60 Å; 70–230 mesh ASTM), deuterated chloroform

Equipment and experimental techniques involved Magnetic stirrer, liquid–liquid extraction, thin-layer chromatography, column chromatography purification, IR equipment, NMR equipment

Keywords Chromatography, IR spectroscopy, NMR spectroscopy, 1,2-reduction, 1,4-reduction, regioselectivity

Background

Green chemistry, also called sustainable chemistry, is the design of chemical products and processes that reduce or eliminate the use and generation of hazardous substances.[1] In 1998 Paul Anastas and John C. Warner developed the 12 principles of green chemistry:[2] prevention of waste, atom economy to maximize the incorporation of all materials used in the process into the final product, less hazardous chemical syntheses, designing safer chemicals, safer solvents and auxiliaries, design for energy efficiency, use of renewable feedstocks, reduce derivatives, use catalytic reagents (as selective as possible), design for degradation products and by products, real-time analysis for pollution prevention and inherently safer chemistry for accident prevention.

Comprehensive Organic Chemistry Experiments for the Laboratory Classroom
Edited by Carlos A. M. Afonso, Nuno R. Candeias, Dulce Pereira Simão, Alexandre F. Trindade, Jaime A. S. Coelho, Bin Tan, and Robert Franzén.

Published by the Royal Society of Chemistry, www.rsc.org

Allylic alcohols represent an important and highly versatile class of chiral building blocks for organic synthesis[3] with special applications in the syntheses of complex molecules such as natural products and potential therapeutic agents.[3]

Metal hydrides are valuable reagents in modern organic chemistry. Sodium borohydride[4–7] and lithium aluminum hydride[6,7] have been used widely in organic synthesis.

Sodium borohydride is a mild, inexpensive and invaluable reagent for applications in a wide range of reduction processes. It is the reagent of choice for the reduction of aldehydes and ketones to alcohols[8] and imines[9] or iminium salts[10,11] to amines with protic solvents. The carboxylic acids, esters, amides and nitriles are more resistant towards $NaBH_4$, but the union of sodium borohydride and carboxylic acids has yielded an amazingly versatile and efficient set of reducing reagents.[5]

Reduction of ketones and aldehydes using sodium borohydride in alcohol or water solvent is a general and efficient procedure to prepare alcohol. The mechanism by which the sodium hydride effects reduction involves activation of the carbonyl group by coordination with sodium cation and nucleophilic transfer of hydride to the carbonyl group.[12] Hydroxylic solvents also participate in the reaction, and as the reduction proceeds and hydride is transferred, the Lewis acid character of boron becomes a factor.[13] Reduction with $NaBH_4$ is usually done in aqueous or alcoholic solution and the alkoxyboranes formed as intermediates are rapidly solvolyzed.

An important case of chemical selectivity arises in the reduction of α,β-unsaturated carbonyl compounds. Reduction can occur at the carbonyl group, giving either a saturated ketone at the double bond or an allylic alcohol.[12] These alternative reaction modes are called 1,4- and 1,2-reduction, respectively. If hydride is added at the carbonyl, the allylic alcohol is usually not susceptible to further reduction. If a hydride is added at the β-position, the initial product is an enolate. In protic solvents this leads to the ketone, which can be reduced to the saturated alcohol. The extent of reduction to saturated alcohol depends on the reagent used and in this respect $NaBH_4$ is superior to $LiAlH_4$.

Various methods of enhancement of reactivity and selectivity of $NaBH_4$ using additives for applications in organic synthesis have been described on literature.[4,14] The $NiCl_2$–$NaBH_4$ system was reported to afford dihydrochalcones.[4] The use of copper or cobalt chloride in place of $NiCl_2$ is not as efficient for this application. Use of $NaBH_4$ in combination with cerium chloride (*Luche reagent*) results in clean 1,2-reduction.[15,16]

Since 1998, several reactions using new solvents[17–19] and reaction media,[20] supercritical carbon dioxide,[21] bio-based solvents,[22] nano-catalyst[23] and immobilized enzymes[24] have been explored.

In recent years poly(ethylene glycol) (PEG) aqueous solutions have been widely used in many different kinds of reaction systems.[25–28] PEG is water soluble, and various moderate and high molecular weight PEGs are also solid at ambient temperature and as such are easy to transport. Various molecular weights are approved for use by the Food and Drug Administration of the United States for medical use, and find applications in various commercial products, such as in skin creams, toothpaste and some soft drinks, as well as therapeutic applications, such as drug delivery.[29] All these characteristics make PEG a green solvent of choice.[30]

Their low toxicity, low volatility, and biodegradability represent important environmentally benign characteristics, which are particularly attractive when combined with the relatively low cost as a bulk commodity chemical.[27] Aqueous solutions of PEG are biocompatible and are utilized in tissue culture media and for organ preservation. PEG may be recovered from aqueous solution by extraction with a suitable solvent or by direct distillation of water or solvent.

Most reductions were carried out in organic solvents like ethanol, methanol or tetrahydrofuran. The objective of this work is the regioselective reduction of an α,β-unsaturated ketone, (*E*)-4-phenylbut-3-en-2-one, to the corresponding allylic alcohol, (*E*)-4-phenylbut-3-en-2-ol, in PEG400[27] and water, a low toxic and friendly environment medium. This green medium is an alternative to the usual use of volatile and toxic organic solvents. Lower molecular weight PEG400 can be used as solvent with or without addition of water.[27]

Additional Safety

All sessions should be done in a fume hood, with proper safety equipment, such as protective clothing, gloves (for splash protection: nitrile rubber with minimum layer thickness of 0.11 mm) and safety glasses.

Sodium borohydride on contact with water releases flammable gases that may ignite spontaneously. It is toxic if swallowed and in contact with skin and causes severe skin burns and eye damage. For fire extinction use dry sand, dry chemical or alcohol resistant foam. Avoid breathing vapors or gas. Ensure adequate ventilation.

Do not breathe dust when using silica gel 60 in column purification.

Experimental Procedure

Laboratory Session (3 h): Reduction of (*E*)-4-Phenylbut-3-en-2-one

1. In a fume hood prepare a 50 mL round-bottomed flask with a magnetic stir bar, and add 733 mg (5.01 mmol) of (*E*)-4-phenylbut-3-en-2-one, water (0.18 mL, 0.1 mmol) and 0.5 mL of poly(ethylene glycol) 400 (PEG400) at room temperature. Heat the reaction flask slowly with a hair dryer to solubilize the unsaturated ketone (melting point = 36 °C).
2. After complete dissolution add, slowly, sodium borohydride (95 mg, 2.51 mmol).
3. Stir the reaction mixture for 1 h at room temperature, until complete consumption of the starting material, (*E*)-4-phenylbut-3-en-2-one. (TLC on silica gel 60; eluent: hexane/ethyl acetate (8 : 2 v/v).)

Product Isolation

4. Add carefully to the reaction mixture 10 mL of water.
5. Transfer the mixture into a separatory funnel and perform the extraction with 3 × 10 mL of diethyl ether. Wash the combined organic layers with 10 mL of brine, dry with anhydrous Na_2SO_4, filter to a round flask (weigh the empty round flask) and evaporate the solvent in a Rotavapor.
6. Dry the crude mixture over P_2O_5 and under vacuum in a desiccator.

Laboratory Session 2 (3 h): Yield Determination and Product Purification.

Crude Yield Determination

1. Weigh the crude yellow oil and calculate the crude yield.

Product Purification and Characterization

2. Weigh 70 g of silica gel 60 (silica gel 60 (0.063–0.200 mm), ref. 7734, Merck) and suspend it in a hexane/ethyl acetate mixture (8 : 2 v/v).
3. Place the mixture in a vertical glass column and compact the silica gel. Add washed sand to get a layer of 1 cm over the silica gel. Remove the eluent to 5 mm from the top of the silica gel column.
4. Place the mixture to be separated by chromatography inside the top of the silica gel column and pass the eluent through silica gel by gravity. Collect samples (5 mL) in sample tubes.
5. Analyze the collected samples by thin-layer chromatography (TLC silica gel 60 F_{254} ref. 5554 Merck) and add together the samples with the desired compound in a weighed round flask Evaporate the solvent to get an oil.
6. Dry in a desiccator under vacuum and in the presence of phosphorus pentoxide.

7. Weigh the purified compound and calculate the yield. Measure the melting point and record the IR spectrum.
8. Take the ^{1}H and ^{13}C NMR spectra to confirm the compound structure and the purity. Record the IR and ^{1}H and ^{13}C NMR spectra of the starting material, (*E*)-4-phenylbut-3-en-2-one. Compare them with literature data.

Waste Disposal

The aqueous solutions may be diluted and discarded in the sink. Diethyl ether, ethyl acetate and hexane may be reused or collected in a waste collector recipient for subsequent destruction.

The silica gel from column chromatography may be collected in a waste collector recipient for subsequent destruction.

Results Interpretation and Additional Questions

1. Interpret the IR spectrum of (*E*)-4-phenylbut-3-en-2-one and the IR spectrum of the product. Identify the C=O stretching band in the IR spectrum of (*E*)-4-phenylbut-3-en-2-one and confirm its disappearance in the allylic alcohol spectrum and appearance of the new OH stretching band. Identify the bands corresponding to the unsaturated carbon bond and to the aromatic ring.
2. Interpret the ^{1}H NMR spectra and identify the allylic alcohol, (*E*)-4-phenylbut-3-en-2-ol, and the saturated alcohol, 4-phenylbutan-2-ol. Calculate the percentage of the side product in the final mixture, after product purification.
3. Rationalize the mechanism involved in the reduction of the α,β-unsaturated ketone, (*E*)-4-phenylbut-3-en-2-one, to the allylic alcohol, (*E*)-4-phenylbut-3-en-2-ol, and explain the presence of the saturated alcohol, 4-phenylbutan-2-ol, as side product in the purified product.
4. Discuss the advantages of using PEG400/water and sodium borohydride, an environment friendly solvent system, instead of a volatile organic solvent like diethyl ether or tetrahydrofuran, in ketone reduction.

References

1. P. T. Anastas, *Aldrichimica Acta*, 2015, **48**, 3.
2. P. T. Anastas and T. C. Williamson, Frontiers in green chemistry, in *Green Chemistry: Frontiers in Benign Chemical Syntheses and Processes*, ed. P. T. Anastas and T. C. Williamson, Oxford University Press, USA, 1998, pp. 1–26.
3. A. Lumbroso, M. L. Cooke and B. Breit, *Angew. Chem., Int. Ed.*, 2013, **52**, 1890.
4. M. Periasamy and P. Thirumalaikumar, *J. Organomet. Chem.*, 2000, **609**, 137.
5. G. W. Gribble, *Chem. Soc. Rev.*, 1998, **27**, 395.
6. A. M. Smith and R. Whyman, *Chem. Rev.*, 2014, **114**, 5477.
7. P. Daverio and M. Zanda, *Tetrahedron: Asymmetry*, 2001, **12**, 2225.
8. H. C. Brown and S. Krishnamurthy, *Tetrahedron*, 1979, **35**, 567.
9. G. W. Gribble and C. F. Nutaitis, *Org. Prep. Proced. Int.*, 1985, **17**, 317.
10. L. Guerrier, J. Royer, D. S. Grierson and H. P. Husson, *J. Am. Chem. Soc.*, 1983, **105**, 7754.
11. J. L. Marco, J. Royer and H. P. Husson, *Synth. Commun.*, 1987, **17**, 669.
12. F. A. Carey and R. J. Sundberg, *Advanced Organic Chemistry. Part B: Reactions and Synthesis*, Springer, New York, 2007.
13. F. A. Carey and R. J. Sundberg, *Advanced Organic Chemistry. Part A: Structure and Mechanisms*, Springer, 5th edn, 2007.
14. A. L. Gemal and J. L. Luche, *J. Am. Chem. Soc.*, 1981, **103**, 5454.
15. J. L. Luche, *J. Am. Chem. Soc.*, 1978, **100**, 2226.
16. J. L. Luche, L. Rodriguezhahn and P. Crabbe, *J. Chem. Soc., Chem. Commun.*, 1978, 601.
17. M.-O. Simon and C.-J. Li, *Chem. Soc. Rev.*, 2012, **41**, 1415.
18. Q. Zhang, K. D. O. Vigier, S. Royer and F. Jerome, *Chem. Soc. Rev.*, 2012, **41**, 7108.

19. Y. Gu and F. Jerome, *Green Chem.*, 2010, **12**, 1127.
20. L. Alvarez de Cienfuegos, R. Robles, D. Miguel, J. Justicia and J. M. Cuerva, *ChemSusChem*, 2011, **4**, 1035.
21. X. Han and M. Poliakoff, *Chem. Soc. Rev.*, 2012, **41**, 1428.
22. Y. Gu and F. Jerome, *Chem. Soc. Rev.*, 2013, **42**, 9550.
23. V. Polshettiwar and R. S. Varma, *Green Chem.*, 2010, **12**, 743.
24. R. A. Sheldon and S. van Pelt, *Chem. Soc. Rev.*, 2013, **42**, 6223.
25. M. M. Cecchini, C. Charnay, F. De Angelis, F. Lamaty, J. Martinez and E. Colacino, *ChemSusChem*, 2014, **7**, 45.
26. C. K. Z. Andrade and L. M. Alves, *Curr. Org. Chem.*, 2005, **9**, 195.
27. J. Chen, S. K. Spear, J. G. Huddleston and R. D. Rogers, *Green Chem.*, 2005, **7**, 64.
28. A. G. West, C. Barner-Kowollik and S. Perrier, *Polymer*, 2010, **51**, 3836.
29. F. M. Veronese and G. Pasut, *Drug Discovery Today*, 2005, **10**, 1451.
30. S. Perrier, H. Gemici and S. Li, *Chem. Commun.*, 2004, 604.

13.4. Reduction of Diphenyl Sulfoxide Catalyzed by the Dioxo-Molybdenum Complex $MoO_2Cl_2(H_2O)_2$

Ana C. Fernandes* and Tiago A. Fernandes

Centro de Química Estrutural, Instituto Superior Técnico, Universidade de Lisboa, Av. Rovisco Pais, 1, 1049-001 Lisboa, Portugal. *E-mail: anacristinafernandes@tecnico.ulisboa.pt

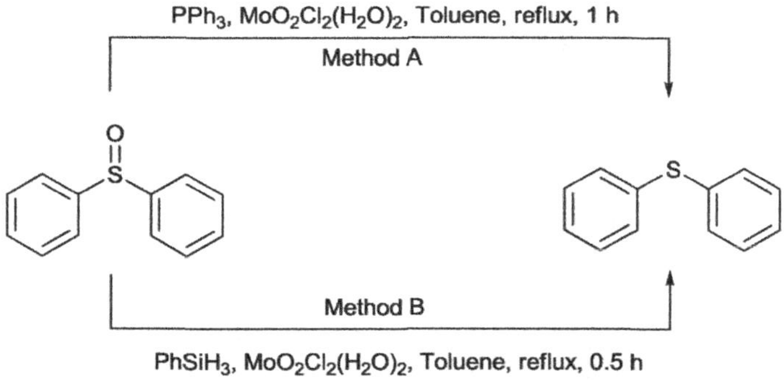

Number of sessions (duration of each session)	Hazard level	Difficulty level	Level of study
2 (3 h each)	Moderate	Medium	Intermediate

Class names Sulfoxides, sulfides

Concepts involved This experiment involves two alternatives methods for the reduction (deoxygenation) of sulfoxides using a dioxo-molybdenum complex as catalyst and triphenylphosphine or phenylsilane as reductants. The evolution of the reaction is followed by thin-layer chromatography, the product is purified by column chromatography and characterization of the structures of the sulfoxide and the sulfide is achieved by infrared and NMR spectroscopy

Chemicals needed Diphenyl sulfoxide, triphenylphosphine, phenylsilane, toluene, dichloromethane, *n*-hexane, ethyl acetate, silica gel (70–230 mesh column chromatography grade), deuterated chloroform, potassium bromide, anhydrous calcium chloride (granular), diethyl ether solution of $MoO_2Cl_2(H_2O)_2$

Equipment and experimental techniques involved *Apparatus for reaction*: a two neck round-bottomed flask, magnetic stir bar, clamp, water condenser, drying tube, oil bath, magnetic stirring apparatus with temperature (110 °C), rotary evaporator, 1 mL pipette, 30 mL graduated cylinder; *apparatus for thin-layer chromatography (TLC)*: TLC plates, developing container for TLC or a beaker with a watch glass, ultraviolet lamp; *apparatus for purification*: column chromatography with a column ~1.5 cm in diameter and ~50 cm high, Erlenmeyer flask, cotton, sand, liquid funnel, powder funnel, 50 mL round-bottomed flask, 30 assay tubes of 20 mL, capillaries or microcapillaries; *apparatus for characterization*: infrared spectrometer and NMR spectrometer

Keywords Column chromatography, infrared spectroscopy, NMR spectroscopy, oxo-molybdenum complexes, reduction of sulfoxides, thin-layer chromatography

Comprehensive Organic Chemistry Experiments for the Laboratory Classroom
Edited by Carlos A. M. Afonso, Nuno R. Candeias, Dulce Pereira Simão, Alexandre F. Trindade, Jaime A. S. Coelho, Bin Tan, and Robert Franzén.

Published by the Royal Society of Chemistry, www.rsc.org

Background

The reduction (or deoxygenation) of sulfoxides to the corresponding sulfides is an important organic and biological reaction, frequently employed in natural products and pharmaceutical synthesis, which require mild conditions, selectivity, and functional group tolerance.[1]

Among the variety of metal complexes that have been used as catalysts for the deoxygenation of organic compounds, oxo-molybdenum complexes have attracted considerable attention because a monooxo- or a dioxo-molybdenum unit is found in the active sites of several molybdoenzymes such as sulfite oxidase, DMSO reductase, xanthine oxidase and nitrate reductase (Figure 13.4.1).[2–4]

These enzymes are known to catalyze various two-electron redox reactions involving a net exchange of an oxygen atom between the Mo-enzymes and the substrate. For example, the deoxygenation of biological important sulfoxides such as dimethyl sulfoxide (DMSO), methionine sulfoxide and biotin-*S*-oxide are catalyzed by the molybdoenzymes DMSO reductase, methionine sulfoxide reductase and biotin-*S*-oxide reductase, respectively (Scheme 13.4.1). In this perspective, the study of deoxygenation reactions using oxo-molybdenum complexes as catalysts has led to better insight in understanding the structure and functions of various enzymes.

Several methods for the deoxygenation of sulfoxides catalyzed by oxo-molybdenum complexes have been developed using different reducing agents such as phosphorus compounds,[5] silanes,[6] boranes,[7] hydrogen[8] and alcohols.[9]

The objective of this work is the reduction (deoxygenation) of diphenyl sulfoxide to the corresponding diphenyl sulfide using triphenylphosphine or phenylsilane as reducing agents and the dioxo-molybdenum complex $MoO_2Cl_2(H_2O)_2$ as catalyst. The easy and inexpensive preparation of the diethyl ether solution of $MoO_2Cl_2(H_2O)_2$, by extraction from a hydrochloric acid solution of Na_2MoO_4 with diethyl ether,[10] and its air stability make this oxo-complex a useful and practical catalyst. This reaction intends to mimic the reduction of a sulfoxide by a molybdoenzyme.

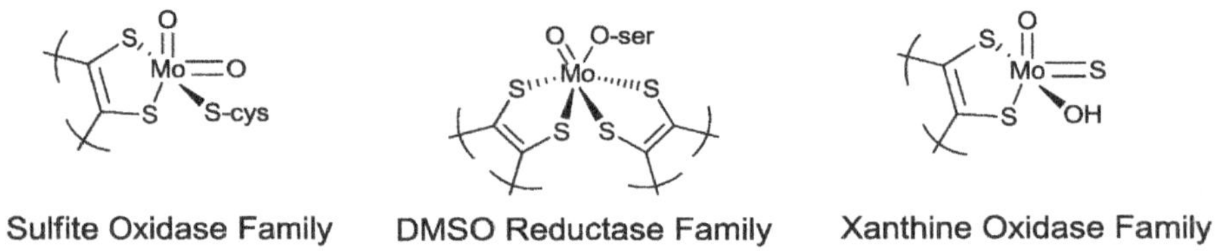

Figure 13.4.1 Active sites of molybdoenzymes.

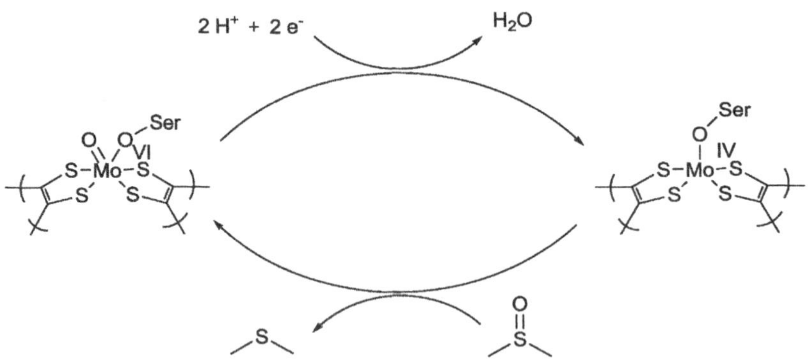

Scheme 13.4.1 Proposed mechanism of DMSO reductase.

Additional Safety

When using the oil bath at 110 °C be careful with the hot surface of the magnetic stirring apparatus. Diphenyl sulfoxide is not classified as dangerous. Triphenylphosphine is harmful if swallowed; it may cause an allergic skin reaction and may cause damage to organs (nervous system) through prolonged or repeated exposure if inhaled. Phenylsilane is harmful if swallowed, inhaled or in contact with eyes and skin. It is highly flammable and reacts violently with water. Toluene, *n*-hexane, ethyl acetate and dichloromethane are highly flammable solvents and are harmful if swallowed or inhaled. They can cause skin irritation, eye irritation, drowsiness or dizziness. $MoO_2Cl_2(H_2O)_2$ is corrosive and may cause severe skin and eye burns.

Experimental Procedure

Laboratory Session 1 (3 h): Reduction of Diphenyl Sulfoxide

Note: this experiment requires the previous preparation of the diethyl ether solution of $MoO_2Cl_2(H_2O)_2$ and the instructor should provide the students the exact concentration of this solution.

Method A Reduction of Diphenyl Sulfoxide with the System $Ph_3P/MoO_2Cl_2(H_2O)_2$

1. In a 50 mL two-neck round-bottomed flask with a magnetic stir bar place 0.202 g (1.0 mmol) of diphenyl sulfoxide.
2. Dissolve the sulfoxide with 5 mL of toluene and add 10 mol% of the diethyl ether solution of $MoO_2Cl_2(H_2O)_2$ previously prepared. Ask your instructor for the exact concentration of this solution.
3. Add 0.314 g (1.20 mmol) of triphenylphosphine at room temperature.
4. Place the round-bottomed flask in the preheated oil bath (110 °C) and clamp the flask around the neck. Insert a water condenser into the round-bottomed flask, and set up the drying tube with anhydrous calcium chloride granular on the top of the condenser.
5. Stir the reaction mixture for 1 h at reflux temperature. Monitor the course of the reaction by thin-layer chromatography (TLC) at approximately 20 min intervals using the mixture of ethyl acetate : *n*-hexane (1 : 3) as eluent (see the instructions below).
6. The reaction is complete when the spot corresponding to the diphenyl sulfoxide is not visible in the TLC. Remove the reaction flask from the oil bath, allowing it to cool before removing the condenser.
7. Transfer the reaction mixture to a 50 mL round-bottomed flask and remove the solvent under reduced pressure using a rotary evaporator. Keep the flask at 0–4 °C until next session.

Method B Reduction of Diphenyl Sulfoxide with the System $PhSiH_3/MoO_2Cl_2(H_2O)_2$

1. In a 50 mL two-neck round-bottomed flask with a magnetic stir bar place 0.202 g (1.0 mmol) of diphenyl sulfoxide.
2. Dissolve the sulfoxide with 5 mL of toluene and add 10 mol% of the diethyl ether solution of $MoO_2Cl_2(H_2O)_2$ previously prepared. Ask your instructor for the exact concentration of this solution.
3. Add 0.15 mL (1.20 mmol) of phenylsilane at room temperature.
4. Place the round-bottomed flask in the preheated oil bath (110 °C) and clamp the flask around the neck. Insert a water condenser into the round-bottomed flask and set up the drying tube with anhydrous calcium chloride granular on the top of the condenser.

5. Stir the reaction mixture for 30 min at reflux temperature. Monitor the course of the reaction by thin-layer chromatography (TLC) at approximately 10 min intervals using ethyl acetate : *n*-hexane (1 : 3) as eluent.
6. The reaction is complete when the spot corresponding to the diphenyl sulfoxide is not visible in the TLC. Remove the reaction flask from the oil bath, allowing it to cool before removing the condenser.
7. Transfer the reaction mixture to a 50 mL round-bottomed flask and remove the solvent under reduced pressure using a rotary evaporator. Keep the flask at 0–4 °C until next session.

Monitor the Course of the Reaction by Thin-Layer Chromatography (TLC)

1. *Preparation of the developing container*: to develop the TLC a specially designed container or a beaker with a watch glass on the top can be used. Add ethyl acetate : *n*-hexane (1 : 3) into the container to a depth of 0.5 cm and cover it.
2. *Preparation of the TLC plates*: cut approximately six TLC plates with 6 cm × 4 cm dimensions. Make sure you handle the plates carefully, without disturbing the coating of adsorbent or getting them dirty. Measure 1.0 cm from the bottom of the plate and with a pencil gently draw a line across, making the origin line, where you will spot the plate. To monitor the course of the reaction by method A, under the line lightly mark three spots corresponding to diphenyl sulfoxide, triphenylphosphine and to the reaction mixture, with enough space between the spots so that they will not run together. For the reaction carried out by method B mark two spots, one for diphenyl sulfoxide and the other for the reaction mixture. Identify each spot with numbers or names. On the top of the plates identify them with the corresponding reaction time.
3. Place your samples and the reaction mixture on the spots with the help of capillaries or microcapillaries. In the case of your reagents, dissolve about 5 mg in 1 mL of ethyl acetate. Make sure you use a clean capillary each time. Immerse the capillary tip into the solution and then gently touch, with the tip, the proper spot location on the TLC plate.
4. Place the prepared TLC plate in the developing container, cover it and leave it undisturbed. Allow the solvent to rise up the plate until about 1.0 cm from the top. Remove the TLC plate from the developing container using a pair of tweezers and immediately mark the solvent front with a pencil before the solvent begins to evaporate. Let it dry for 2–3 min.
5. Visualize the results of your chromatogram under ultraviolet (UV) radiation (254 nm) and circle the dark spots with a pencil.

Laboratory Session 2 (3 h): Purification of the Product by Column Chromatography

1. To the round-bottomed flask containing your product add 5 mL of dichloromethane and 1.0 g of silica gel. Evaporate the solvent under reduced pressure using a rotary evaporator until you obtain a dry powder.
2. To prepare the column chromatography, add a small piece of cotton as a plug, and push it to the narrow end of the column with a glass rod.
3. Clamp the column in a vertical position and place an Erlenmeyer flask underneath, in order to collect the *n*-hexane that you will be adding to the column. Weigh 20 g of silica gel in a beaker and add enough *n*-hexane to make a thin slurry. Transfer the silica slurry to the column using a powder funnel.
4. Fill the column 4–5 times with *n*-hexane to pack it well. Make sure that no part of the column dries out. The eluted *n*-hexane can be reused for this purpose.

5. After the silica is packed, add a 2–3 mm layer of sand above the silica by letting it settle through the *n*-hexane. Allow the *n*-hexane level to almost reach the top of the layer of sand.
6. Transfer your product to the column using a powder funnel, being careful not to disturb the surface of the stationary phase.
7. For method A use *n*-hexane as eluent and collect the eluent in 20 mL assay tubes. Collect the first four fractions with 20 mL and the other 16 fractions with 10 mL. For method B use ethyl acetate : *n*-hexane (1 : 3) as eluent and collect 10 fractions with 10 mL.
8. The column must *never* be allowed to dry out once it is made and the solvent level should *never* be allowed to go below the upper level of the stationary phase.
9. Prepare several TLC plates with five applications points each and monitor all the collected fractions in order to find the fractions containing the product.
10. Develop the TLC plates with the *n*-hexane (method A) or with ethyl acetate : *n*-hexane (1 : 3) (method B). Record the results in your laboratory note book.
11. Recover your purified product by evaporating the pure fractions in a tared round-bottomed flask under reduced pressure using a rotary evaporator. Weigh your purified product.
12. Record the Infrared spectra of the sulfoxide and the sulfide and the ^{1}H and ^{13}C NMR spectra of the product in $CDCl_3$.

Waste Disposal

Pour the dichloromethane from the evaporation of the reaction mixture into a chlorinated waste collector recipient. The toluene, the eluent from the developing container and the eluent from the column chromatography may be collected in a non-chlorinated waste collector. Place the TLC plates and the silica gel from the column in the container for inorganic solid wastes.

Results Interpretation and Additional Questions

1. Determine the mass of the product and calculate the percent yield.
2. Calculate the R_f values of the sulfoxide and the product using ethyl acetate : *n*-hexane (1 : 3) as eluent.
3. Identify the main difference between the infrared spectra of diphenyl sulfoxide and the product.
4. Analyse the purity of the product based on the ^{1}H and ^{13}C NMR spectra, under the supervision of your instructor.
5. Describe the role of triphenylphosphine or phenylsilane in the reaction.

References

1. M. Madesclaire, *Tetrahedron*, 1988, **44**, 6537.
2. J. H. Enemark, J. J. A. Cooney, J.-J. Wang and R. H. Holm, *Chem. Rev.*, 2004, **104**, 1175.
3. C. Kisker, H. Schindelin and D. C. Rees, *Annu. Rev. Biochem.*, 1997, **66**, 233.
4. R. Hille, *Chem. Rev.*, 1996, **96**, 2757.
5. R. Sanz, J. Escribano, R. Aguado, M. R. Pedrosa and F. J. Arnáiz, *Synthesis*, 2004, **10**, 1629.
6. A. C. Fernandes and C. C. Romão, *Tetrahedron*, 2006, **62**, 9650.
7. A. C. Fernandes and C. C. Romão, *Tetrahedron Lett.*, 2007, **48**, 9176.
8. P. M. Reis, P. J. Costa, C. C. Romão, J. A. Fernandes, M. J. Calhorda and B. Royo, *Dalton Trans.*, 2008, 1727.
9. N. García, P. García-García, M. A. Fernández-Rodríguez, R. Rubio, M. R. Pedrosa, F. J. Arnáiz and R. Sanz, *Adv. Synth. Catal.*, 2012, **354**, 321.
10. F. J. Arnáiz, R. Aguado, M. R. Pedrosa and A. De Cian, *Inorg. Chim. Acta*, 2003, **347**, 33.

13.5. Synthesis of Allylic Esters by Reduction of Fructone Followed by Wittig Olefination

Thomas A. Logothetis

University of Southampton, Chemistry, Highfield, Southampton, Hampshire, SO17 1BJ, UK
*E-mail: thomas.logothetis@southampton.ac.uk

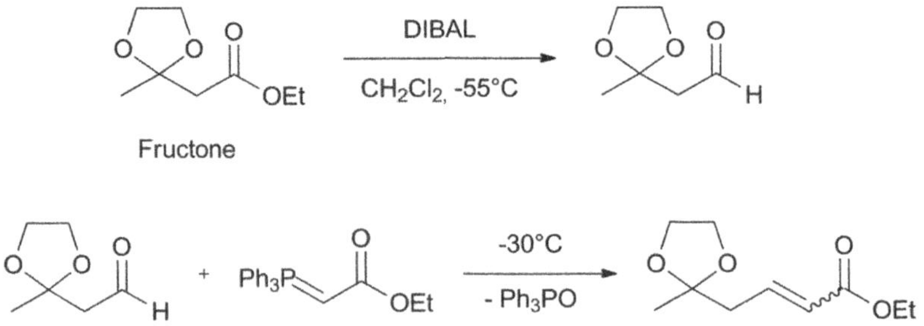

Number of sessions (duration of each session)	Hazard level	Difficulty level	Level of study
2 (4 h + 3 h)	Moderate	High	Intermediate

Class names Alkylic esters, aldehydes, conjugated/allylic esters, stabilised phosphoranylidenes

Concepts involved This experiment is conducted under a protective, inert gas atmosphere and involves the selective reduction of an ester to an aldehyde using DIBAL at reduced temperatures, followed by *in situ* coupling with a stabilised Wittig reagent to afford a conjugated ester as mixture of stereoisomers. The (*E*/*Z*)-isomers are then separated by flash column chromatography and analysed by NMR and IR

Chemicals needed Fructone, anhydrous dichloromethane, diisobutylaluminium hydride (DIBAL solution in hexanes, 1 M), methanol, ethyl triphenylphosphoranylidene acetate, sodium potassium tartrate (Rochelle's salt), sodium chloride, diethyl ether, light petroleum ether (40–60 °C), silica for chromatography (40–63 μm), dry ice, acetone, anhydrous magnesium sulfate, deuterated chloroform

Equipment and experimental techniques involved Handling of pyrophoric reagents under inert gas atmosphere, cooling with a dry ice–acetone bath, flash chromatography, TLC analysis, rotary evaporation, three-neck round bottom flask (100 mL), septum, nitrogen inlet/bubbler, Dewar (or alternative cooling bath), disposable needles and syringes (5–10 mL), separating funnel, conical flasks, beakers, test tubes in rack, silica TLC plates and chamber, UV lamp, permanganate dip, bellows, glass rod, tweezers, glass column (ø 2 cm × ~50 cm), evaporation flask

Keywords DIBAL reduction, (*E*/*Z*)-isomerism, (flash) column chromatography, fructone, NMR spectroscopy, Wittig olefination

Background

Fructone (**2**) can act as building block in syntheses of more complex molecules and natural products which involve a derivatisation of the ester moiety, *e.g.* reduction to an aldehyde (**4**),[1] or complete reduction to an alcohol (**3**),[2] which might be followed by oxidation to an aldehyde (**4**).[3] These transformations necessitate the protection of the more reactive keto-carbonyl moiety of

Comprehensive Organic Chemistry Experiments for the Laboratory Classroom
Edited by Carlos A. M. Afonso, Nuno R. Candeias, Dulce Pereira Simão, Alexandre F. Trindade, Jaime A. S. Coelho, Bin Tan, and Robert Franzén.

Published by the Royal Society of Chemistry, www.rsc.org

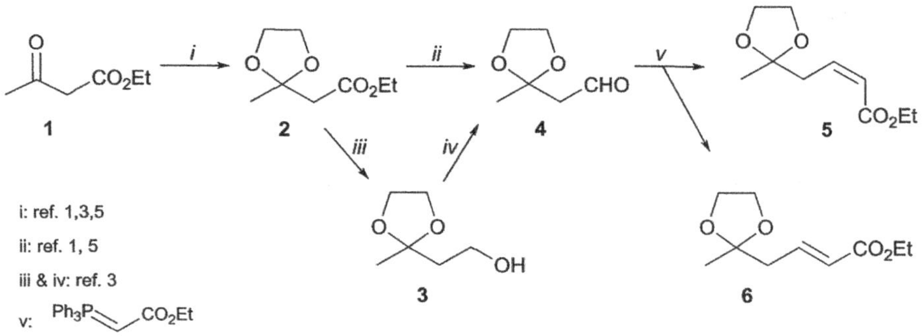

Scheme 13.5.1 Synthesis of allylic esters.

the readily available ethyl acetoacetate (**1**). One possible protection is the acetalisation of the carbonyl moiety, more specifically the formation of a 1,3-dioxolane ring[4] using ethane-1,2-diol, which is another versatile reagent that is typically used as coolant in antifreeze formulations (see Experiment 4.1.2.1 ("Synthesis of Fructone") for the sequence **1** → **2**).

A possible way to achieve chemoselectivity is the proper choice of the reducing agent: whereas lithium alanate would reduce fructone (**2**) to the corresponding alcohol (**3**), the synthesis of the reactive aldehyde **4** involves the use of a complex hydride, namely diisobutylaluminium hydride (DIBAL), at low temperatures. In this way the reduction stops at the aldehyde **4** rather than the corresponding alcohol **3** (Scheme 13.5.1).

This technique was elegantly employed in the synthesis of a radiolabelled tracer starting with the DIBAL reduction of fructone in dichloromethane at −78 °C.[5] The corresponding acid of fructone was also used as a starting material for the synthesis of a conjugated allenic ester *via* a Wittig reaction with a ketene.[6] Ketenes and aldehydes often act as intermediates in syntheses due to their inherent reactivity. One very useful reaction is named after the Nobel laureate Georg Wittig.[7]

In the Wittig reaction the aldehyde **4** is an electrophile that readily reacts with phosphorylide or phosphoranylidene nucleophiles. The Wittig reaction and its Horner–Wadsworth–Emmons variant afford a double bond typically with high stereoselectivity.[8] The Wittig reaction is also scalable and has been used widely in industry,[9] *e.g.* successfully to synthesise vitamin A (BASF Hoffmann-La Roche). Recent research has been directed to finding environmentally sustainable ways to conduct this important reaction.[10]

This practical combines the reduction of the alkylic ester fructone (**2**) utilising DIBAL to synthesise the aldehyde **4**, with the Wittig reaction, and exploits NMR (coupling constants) to determine the nature of the produced allylic ester(s) and the overall selectivity (in respect of the formation of **5** *versus* **6**) of this two-step process.

Additional Safety

All procedures should be performed in a well-ventilated fume cupboard, with proper safety equipment such as protective clothing, safety spectacles and gloves. DIBAL is a pyrophoric reagent and its solutions in hexane are highly flammable. It must be handled under an inert gas atmosphere. Syringes and needles should be rinsed with methanol followed by water before disposal. Sand should be kept ready for dealing with spillages and as extinguishing medium. All organic compounds and solvents bar dichloromethane are flammable and harmful/irritants. Diethyl ether is extremely flammable and may form explosive peroxides. Dichloromethane is a suspected carcinogen. Methanol is toxic by inhalation, in contact with skin and when ingested. Silica should be weighed in a fume cupboard and inhalation of dust must be avoided. Dry ice poses a risk of frost burn. Gloves should be used when handling it or the cooling bath.

Experimental Procedure

Session 1: Synthesis and Work-Up (4 h)

Reaction

1. In a fume cupboard set up a dry apparatus consisting of a three-neck round bottom flask, internal thermometer, magnetic stirrer bar, septum and pressure-equalised inert gas inlet. Assemble this apparatus with dry components: all glassware apart from the thermometer should be oven dried and the apparatus set up while this equipment is still hot. Then let it cool down to room temperature under inert gas atmosphere.
2. Charge the apparatus with anhydrous dichloromethane (15 mL) using a syringe with needle and a reservoir of inert gas to keep the stock solution anhydrous.
3. Add fructone (5.74 mmol, 1.00 g) to the stirred solvent.
4. Cool the solution to −55 °C using a dry ice–acetone bath in a Dewar.
5. Using a syringe with needle and a reservoir of inert gas to keep the stock solution anhydrous, slowly add a solution of DIBAL (1 M in hexanes, 7.5 mL, 7.5 mmol, 1.3 equiv.) at such a rate that the internal temperature does not rise above −55 °C.
6. Stir at this temperature for 10 min, then add methanol (3 mL) to quench excess DIBAL.
7. Remove the cooling bath and allow the mixture to warm to 0 °C.
8. Re-cool the reaction mixture to −30 °C and add ethyl triphenylphosphoranylidene acetate (0.95 equiv., 5.45 mmol, 1.90 g).
9. Remove the cooling bath and stir the mixture until it warms to room temperature.

Workup

10. Pour the reaction mixture into a solution of Rochelle's salt (10% in water, 60 mL). Add dichloromethane (60 mL) and stir vigorously for 10 min.
11. Separate the phases and extract the aqueous phase with dichloromethane (20 mL).
12. Combine the organic phases and wash with brine (50 mL).
13. Dry with magnesium sulfate and, after filtration, concentrate *in vacuo.*
14. Triturate the viscous oil with light petroleum ether (10 mL) before decanting the supernatant. Repeat this process and keep the combined liquid phases. Discard the obtained white solid residue.

Session 2: Purification and Isolation of Stereoisomers by Column Chromatography (3 h)

Flash Chromatography

1. Prepare the eluent (~1000 mL) by mixing diethyl ether (3 parts) with light petroleum ether (7 parts). This is used for TLC and flash chromatography.
2. Weigh silica (100 g) in a fume cupboard and slurry it into the eluent (350 mL).
3. Transfer the slurry into a glass chromatography column and let the silica settle before running out the surplus eluent.
4. Transfer the product-containing petroleum ether solution to the column and run all liquid into the silica without it drying out at the top.
5. Carefully fill the column with eluent and connect the bellows.
6. Elute under pressure and collect fractions (±10 mL, 50–60) in test tubes.
7. Analyse all fractions using TLC analysis under UV light and oxidative conditions.
8. Combine those fractions that contain a pure isomer – separately for each stereoisomer – into a pre-weighed evaporation flask; discard those that contain more than one spot/substance.
9. Rotary evaporate *in vacuo.*

Analysis

10. Determine the yield for each stereoisomer.
11. Acquire IR spectra for each stereoisomer.
12. Acquire ^{1}H-, ^{13}C- and DEPT-135 NMR spectra for each stereoisomer.

Waste Disposal

The neutralised aqueous solutions – including dissolved $MgSO_4$ – may be diluted and discarded in the sink. Dichloromethane may be recycled or collected as chlorinated waste. All other organic solvents may be recycled or, if any by-products are dissolved in it, collected as non-chlorinated waste.

Results Interpretation and Additional Questions

1. The precursor to fructone is ethyl acetoacetate. In fructone the keto-moiety is transformed into a dioxolane. Why was it necessary to protect ethyl acetoacetate before carrying out the reduction reaction? Which products would form without protection?
2. Rationalise why Rochelle salt was used in the work-up and how it worked chemically. Could a mineral acid or base or citrate buffer be used instead? Explain your reasoning.
3. Formulate the mechanism for the DIBAL reduction and the Wittig reaction!
4. Interpret the NMR spectra of both stereoisomers and use the coupling constants of the allyl moiety to determine the geometry around the C=C double bond for each isomer. Explain your reasoning!
5. Assign the C–H, C=O and C=C stretching absorptions in the IR spectra of starting material and both products. These three substances all contain an ester moiety. Explain the observed differences using arguments based on resonance stabilisation and bond order!

References

1. D. G. Velazquez and R. Luque, *Tetrahedron Lett.*, 2011, **52**, 7004.
2. S. Raghavan and S. C. Joseph, *Tetrahedron Lett.*, 2003, **44**, 8237.
3. M. Nogawa, S. Sugawara, R. Lizuka, M. Shimojo, H. Ohta, M. Hatanaka and K. Matsumoto, *Tetrahedron*, 2006, **62**, 12071.
4. (a) T. W. Greene, P. G. M. Wuts, *Protective Groups in Organic Synthesis*, John Wiley and Sons, Inc., New York, 3rd edn, 1999, ch. 4, pp. 307–308 & 312–322; (b) T. W. Greene, P. G. M. Wuts, *Green's Protective Groups in Organic Synthesis*, John Wiley & Sons, Inc., Hoboken, 4th edn, 2007, ch. 4, pp. 448–449 & 454–466.
5. A. C. Jordan, L. C. Axford, J. R. Harding, Y. O'Connell, T. J. Simpson and C. L. Willis, *J. Labelled Compd. Radiopharm.*, 2007, **50**, 338.
6. S. Rougnon-Glasson, C. Tratrat, J.-L. Canet, P. Chalard and Y. Troin, *Tetrahedron: Asymmetry*, 2004, **15**, 1561.
7. G. Wittig, *Angew. Chem.*, 1980, **92**, 671.
8. J. Clayden, N. Greeves and S. Warren, *Organic Chemistry*, Oxford University Press, Oxford, 2nd edn, 2012, ch. 27, pp. 689–693.
9. H. Pommer, *Angew. Chem., Int. Ed.*, 1977, **16**, 423.
10. (a) C. J. O'Brien, J. L. Tellez, Z. S. Nixon, L. J. Kang, A. L. Carter, S. R. Kunkel, K. C. Przeworski and G. A. Chass, *Angew. Chem., Int. Ed.*, 2009, **48**, 6836; (b) H. A. van Kalkeren, A. L. Blom, F. P. J. T. Rutjesa and M. A. J. Huijbregts, *Green Chem.*, 2013, **15**, 1255.

13.6. Preparation of a Thia-Tetraaza Macrocyclic Compound Through a Dual-Step Synthesis

Judite Costa*, João Franco Machado, and M. Fátima Cabral*

Research Institute for Medicines (iMed.ULisboa), Faculty of Pharmacy, Universidade de Lisboa, Av. Professor Gama Pinto, 1649-003 Lisbon, Portugal. *E-mail: jcosta@ff.ul.pt; fcabral@ff.ul.pt

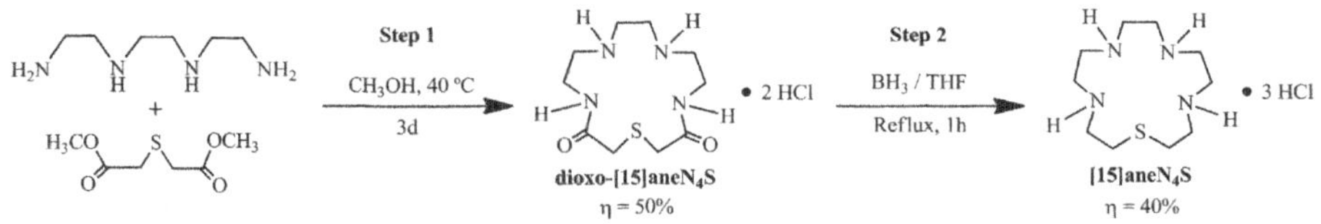

Number of sessions (duration of each session)	Hazard level	Difficulty level	Level of study
2 (4 h each)	Moderate	High	Intermediate

Class names Esters; amides; amines

Concepts involved This experiment consists of a two step synthesis in which the students may develop various skills in terms of laboratory techniques, spectroscopic analysis and team work, giving useful practical experience in the preparation of macrocyclic compounds, which are frequently used for a range of biomedical applications. The synthesis of 1-thia-4,7,10,13-tetraazacyclopentadecane ([15]aneN$_4$S) involves the condensation of triethylenetetramine with dimethyl thiodiglycolate in dry methanol, through two addition–elimination reactions, followed by a reduction final step performed with a large excess of borane in dry tetrahydrofuran. After purification using column chromatography, this compound, as well its cyclic diamide precursor 1-thia-4,7,10,13-tetraazacyclopentadecane-3,14-dione (dioxo-[15]aneN$_4$S), will be properly characterized by ^{1}H NMR and IR spectroscopies

Chemicals needed Triethylenetetramine hydrate, dimethyl thiodiglycolate, dry methanol, dichloromethane, aluminium oxide 90 active neutral, hydrochloric acid, potassium hydroxide, borane tetrahydrofuran complex solution, dry tetrahydrofuran

Equipment and experimental techniques involved Apparatus for reaction with temperature and magnetic stirring, under nitrogen; apparatus for column chromatography; apparatus for filtration; apparatus for ^{1}H NMR and IR spectroscopies

Keywords Addition-elimination reaction, amide reduction, column chromatography, cyclization, IR, NMR, thia-tetraaza macrocycle

Background

A macrocyclic compound is a cyclic molecule with three or more potential donor atoms (*e.g.* N, S or O) in a ring of at least nine atoms.[1] Macrocyclic chemistry has expanded phenomenally since the 1960s and still remains a focus of scientific attention.[2–4] In fact, this class of compounds continues to be actively sought and developed due to a myriad of potential applications, namely in medicine as radiopharmaceuticals, MRI contrast agents, SOD mimetics and chelating agents for the removal of metal ions from the organism.[5,6]

Comprehensive Organic Chemistry Experiments for the Laboratory Classroom
Edited by Carlos A. M. Afonso, Nuno R. Candeias, Dulce Pereira Simão, Alexandre F. Trindade, Jaime A. S. Coelho, Bin Tan, and Robert Franzén.

Published by the Royal Society of Chemistry, www.rsc.org

The objective of this work is to synthetize the macrocycle 1-thia-4,7,10,13-tetraazacyclopentadecane ([15]aneN_4S) and to properly characterize it by ^{1}H NMR and IR spectroscopies, as well as its cyclic diamide precursor 1-thia-4,7,10,13-tetraazacyclopentadecane-3,14-dione (dioxo-[15]aneN_4S).[7] The overall experiment is designed to be carried out conveniently by a group of two students in two laboratory sessions of four hours each.

To prepare the macrocycle [15]aneN_4S, firstly it is necessary to obtain the dioxo-[15]aneN_4S macrocyclic precursor, by adding the dimethyl ester of thiodiglycolic acid (previously synthetized by the instructor according to a reported procedure[8]) to triethylenetetramine, in dry methanol. The progress of this reaction can be checked by using TLC. In this step, two addition–elimination reactions occur – a first intermolecular one, between the starting reagents, which will lead to the second reaction, intramolecular, wherein the ring is formed.[8] The purification of dioxo-[15]aneN_4S is performed by column chromatography, wherein the pure compound is obtained in the third eluate by using a mixture of dichloromethane/methanol (4:1) as eluent. The expected reaction yield for this first step is about 50%. Dioxo-[15]aneN_4S is then reduced to the desired product by using a large excess of borane in dry tetrahydrofuran. The choice of this reduction agent is preferable to the common ones, such as lithium aluminium hydride, once the tendency for C–N bond cleavage to yield an alcohol is completely absent.[9] [15]aneN_4S is then purified by column chromatography, wherein the pure compound is obtained in the third eluate by using methanol as eluent. The expected reaction yield for this second step is about 40%.

Both dioxo-[15]aneN_4S and [15]aneN_4S should be properly characterized by ^{1}H NMR and IR spectroscopies. The first technique will allow the students to verify if the synthetized compound corresponds to the desired pure product for each step of the reaction, while by analysing the IR spectra students may confirm if the reduction step was appropriately executed (there is an absence of the C=O characteristic band in the reduced product).

Additional Safety

All experimental procedures should be carried out in a well-ventilated fume hood, with proper safety equipment, such as protective clothing, disposable vinyl or latex gloves and chemical-resistant safety goggles. Borane reacts violently and exothermically with water, moist air, acids or alcohols, liberating flammable vapours that can ignite or cause severe damage to the eyes. Tetrahydrofuran is a highly inflammable reagent and is harmful if swallowed or in contact with the skin or the eyes. Hydrochloric acid, potassium hydroxide and triethylenetetramine are corrosive agents that can cause severe burns when in contact with the skin or the eyes and when swallowed. Methanol is a highly volatile reagent, the vapours of which can ignite or cause blindness. Dichloromethane is hazardous in case of eye/skin contact, of inhalation or of ingestion.

Experimental Procedure

Previous to the First Laboratory Session: Preparation of Step 1 Reaction

1. In a fume hood, heat an oil bath to maintain a temperature of 40 °C on a magnetic hot plate.
2. Equip a double-necked round-bottomed flask (50 mL) with a magnetic stirrer bar and a water-cooled reflux condenser, with a nitrogen inlet above the condenser.
3. Add, under nitrogen, to the double-necked round-bottomed flask 0.21 g (1.44 mmol) of triethylenetetramine hydrate and 0.20 g (1.12 mmol) of dimethyl thiodiglycolate, followed by dry methanol (15 mL). Close the free joint with a take-off.

(*Note*: dimethyl thiodiglycolate should be previously synthetized by the instructor according to the procedure reported in the literature;[8] methanol should be previously dried by the instructor according to the procedure reported in the literature.[10])

4. Stir the solution under nitrogen at 40 °C for 3 days.

First Laboratory Session (4 h)

(a) Isolation and Characterization of Dioxo-[15]aneN_4S

1. Remove the solvent from the reaction mixture by evaporation under reduced pressure, using a rotary evaporator, and dissolve the obtained residue in the minimum amount of dichloromethane.

(*Note*: it may be necessary to add 2/3 drops of methanol in order to completely dissolve the residue.)

2. Prepare a column (20 cm × 1.0 cm) for chromatography with aluminium oxide (≈25 g) slurred in dichloromethane (≈40 mL). Load the solution containing the crude mixture into the column and elute it with the following solvents: [1] ≈ 30 mL of dichloromethane; [2] ≈ 30 mL of a mixture of dichloromethane/methanol (9:1); [3] ≈ 60 mL of a mixture of dichloromethane/methanol (4:1); [4] ≈ 30 mL of methanol.
3. Remove the solvent from each eluate collected in step 2 by evaporation under reduced pressure, using a rotatory evaporator, and dissolve each residue in methanol.
4. For each solution prepared in step 3 add, slowly and under stirring, a solution of hydrochloric acid in methanol (0.1 M) until pH ≈ 4.
5. Evaporate the solutions obtained in step 4 in a rotary evaporator, until they reach a constant mass.

(*Note*: do not forget to, beforehand, tare the flasks used.)

6. For each residue obtained in step 5, retain a small sample for ^{1}H NMR spectroscopy (using D_2O as solvent).
7. After analysing the ^{1}H NMR spectra, collect a small sample from the residue that contains the pure product (dioxo-[15]aneN_4S) and acquire the corresponding IR spectrum (using KBr pellet).

(b) Preparation of Step 2 Reaction

1. In a fume hood, prepare an oil bath on a magnetic hot plate.
2. Equip a double-necked round-bottomed flask (50 mL; previously dried in an electric oven for 1 h at 110 °C) with a magnetic stirrer bar and a water-cooled reflux condenser, with a nitrogen inlet above the condenser.
3. Weigh the dioxo-[15]aneN_4S obtained in the previous synthesis.

(*Note*: it is of crucial importance to know the exact weight of this compound.)

4. Add, under nitrogen, to the double-necked round-bottomed flask the dioxo-[15]aneN_4S and 10 mL of dry tetrahydrofuran.

(*Note*: tetrahydrofuran should be previously dried by the instructor according to the procedure reported in the literature.[10])

5. Equip the double-necked round-bottomed flask with a sleeve septum stopper in the free joint.
6. Heat the mixture to reflux, under nitrogen, with magnetic stirring.
7. After 10 min, add slowly by syringe (previously dried in an electric oven for 1 h at 110 °C), *via* the septum inlet, a solution of borane tetrahydrofuran complex (1 M, 10 mL).

Replace the septum with a glass stopper and heat and stir the mixture to reflux under nitrogen for 1 h.

8. Stir the mixture at room temperature overnight, under nitrogen.

Second Laboratory Session (4 h): Isolation and Characterization of [15]aneN_4S

1. Cool the reaction mixture in an ice-salt bath and very slowly add dry methanol (5 mL) by pipette down the condenser, to quench the excess borane. After the effervescence has ceased, add a solution of hydrochloric acid in methanol (3 M, 5 mL) and stir the mixture for 10 min.

(*Note*: a vigorous evolution of hydrogen occurs. Take care to add the first few drops of dry methanol particularly slowly.)

2. Add, slowly and under stirring, a solution of potassium hydroxide in methanol (1 M) until pH ≈ 9. Remove the white inorganic precipitate by filtration.
3. Evaporate the obtained filtrate, using a rotary evaporator, and dissolve it in the minimum amount of dichloromethane.
4. Prepare a column (20 cm × 1.0 cm) for chromatography with aluminium oxide (≈25 g) slurred in dichloromethane (≈40 mL). Load the solution containing the crude mixture onto the column and elute it with the following solvents: [1] ≈ 30 mL of dichloromethane; [2] ≈ 30 mL of a mixture of dichloromethane/methanol (4:1); [3] ≈ 60 mL of methanol; [4] ≈ 30 mL of a mixture of methanol/hydrochloric acid (1 M) (9:1).
5. Remove the solvent from each eluate collected by evaporation under reduced pressure, using a rotatory evaporator, and dissolve each residue in methanol.
6. For each solution prepared in step 5 add, slowly and under stirring, a solution of hydrochloric acid in methanol (0.1 M) until pH ≤ 4.
7. Evaporate the solutions obtained in step 6 on a rotary evaporator, until they reach a constant mass.

(*Note*: do not forget to, beforehand, tare the flasks used.)

8. For each residue obtained in step 7, retain a small sample for ^{1}H NMR spectroscopy (using D_2O as solvent).
9. After analysing the ^{1}H NMR spectra, collect a small sample from the residue that contains the pure product ([15]aneN_4S) and acquire the corresponding IR spectrum (using a KBr pellet).

Waste Disposal

The methanolic solutions must be collected in a non-chlorinated waste collector recipient, for posterior destruction, whereas the solutions that contain dichloromethane must be collected in a chlorinated waste collector recipient. Aluminium oxide must be collected in an appropriate recipient for solid waste.

Results Interpretation and Additional Questions

1. Briefly indicate the aims of the experiment.
2. Analyse the ^{1}H NMR spectra obtained in both laboratory sessions and identify the fractions that contain the pure compounds dioxo-[15]aneN_4S and [15]aneN_4S.
3. Interpret the IR spectra obtained for dioxo-[15]aneN_4S and for [15]aneN_4S, and compare them with the data provided in the ESI.

4. Calculate the percent yield for each step of the reaction and for the overall process.

(*Note*: dioxo-[15]aneN_4S is obtained as a dihydrochloride salt with a molecular weight of 333.3 g mol^{-1}; [15]aneN_4S is obtained as a trihydrochloride salt with a molecular weight of 341.8 g mol^{-1}.)

5. Suggest reaction mechanisms for steps 1 and 2.
6. Write a short conclusion.

References

1. E. C. Constable, *Coordination Chemistry of Macrocyclic Compounds*, Oxford Chemistry Primers, Oxford University Press, New York, 1999, vol. 72, p. 1.
2. G. A. Melson, *Coordination Chemistry of Macrocyclic Compounds*, ed. G. A. Melson, Plenum Press, New York, 1979, p. 2.
3. L. F. Lindoy, *The Chemistry of Macrocyclic Ligand Complexes*, Cambridge University Press, Cambridge, 1989, p. 1.
4. S. Kim, L. F. Lindoy and S. S. Lee, *Coord. Chem. Rev.*, 2014, **280**, 176.
5. B. P. Burke and S. J. Archibald, *Annu. Rep. Prog. Chem., Sect. A: Inorg. Chem.*, 2013, **109**, 232.
6. R. E. Mewis and S. J. Archibald, *Coord. Chem. Rev.*, 2010, **254**, 1686.
7. N. Torres, S. Gonçalves, A. S. Fernandes, J. Franco Machado, M. J. Villa Brito, N. G. Oliveira, M. Castro, J. Costa and M. F. Cabral, *Molecules*, 2014, **19**, 550.
8. P. Vollhardt and N. Schore, *Organic Chemistry: Structure and Function*, W. F. Freeman and Co., New York, 6th edn, 2011, p. 939.
9. H. C. Brown and P. Heim, *J. Org. Chem.*, 1973, **38**(5), 912.
10. A. I. Vogel, B. S. Furniss, A. J. Hannaford, P. W. G. Smith and A. R. Tatchell, *Vogel's Textbook of Practical Organic Chemistry*, Longman Scientific & Technical, England, 5th edn, 1989, p. 165.

13.7. Asymmetric Reduction of Acetophenone with Borane Catalyzed by *B*-Methoxy-oxazaborolidine

Marek P. Krzemiński

Department of Organic Chemistry, Faculty of Chemistry, Nicolaus Copernicus, University in Toruń, Poland
*E-mail: mkrzem@chem.umk.pl

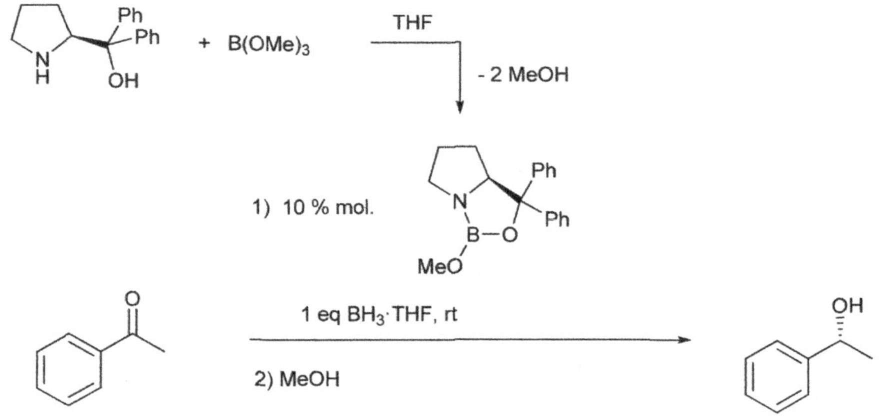

Number of sessions (duration of each session)	Hazard level	Difficulty level	Level of study
2 (4 h each)	Moderate	High	Advanced

Class names Ketones, alcohols, amino alcohols

Concepts involved This experiment involves the catalytic asymmetric reduction of acetophenone with borane catalyzed with *in situ* generated *B*-methoxy-oxazaborolidine. Students will become familiar with highly enantioselective reduction of ketone using CBS catalysis. They will also learn to measure the optical rotation on a polarimeter

Chemicals needed Acetophenone, (*S*)-1,1-diphenylprolinol (synonyms: α,α-diphenyl-L-prolinol, (*S*)-2-(diphenylhydroxymethyl)pyrrolidine), trimethylborate, borane–THF solution or borane–dimethyl-sulfide adduct, tetrahydrofuran, methanol, hexane, ethyl acetate

Equipment and experimental techniques involved Round-bottom flask (25 mL), gas-inlet adapter, oil bubbler, magnetic stirrer, polarimeter, HPLC with Chiralcel OD-H column, work under an inert atmosphere

Keywords Asymmetric reduction, borane, column chromatography, evaporation under low pressure, NMR, oxazaborolidines

Background

The asymmetric reduction of prochiral ketones with borane–oxazaborolidines is a simple and efficient reaction widely used in organic synthesis to obtain chiral secondary alcohols. This reaction was discovered by Itsuno[1] but the development and the recognition of oxazaborolidines

Comprehensive Organic Chemistry Experiments for the Laboratory Classroom
Edited by Carlos A. M. Afonso, Nuno R. Candeias, Dulce Pereira Simão, Alexandre F. Trindade, Jaime A. S. Coelho, Bin Tan, and Robert Franzén.

Published by the Royal Society of Chemistry, www.rsc.org

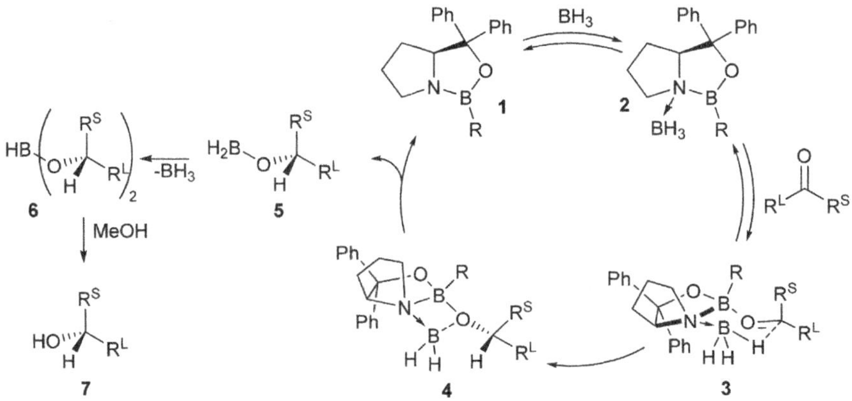

Scheme 13.7.1 Mechanism of the ketone reduction with borane catalyzed by oxazaborolidines (R^L – larger R substituent, R^S – smaller R substituent).[3a,6]

belongs to Corey, Bakshi and Shibata,[2] therefore this reaction is usually called Corey–Bakshi–Shibata (CBS) reduction. Excellent reviews describing the synthesis and utilization of oxazaborolidines are available.[3]

Oxazaborolidines can be obtained from 1,2-amino alcohols in reactions with borane, boronic acids or boroxins. Since the work of Masui and Shioiri,[4] which simplified the reduction protocol by the *in situ* generation of *B*-methoxy-oxazaborolidine, and Ponzo and Kaufmann,[5] which extended the methodology to other *B*-alkoxy-oxazaborolidines, trialkylborates are also used for the preparation of oxazaborolidines.

The proposed mechanism of the oxazaborolidine catalyzed reduction of prochiral ketones with borane is shown on Scheme 13.7.1.[3a,6] Oxazaborolidine **1**, obtained from (*S*)-1,1-diphenylprolinol, forms a complex **2** with borane, and in the next step a carbonyl group of the ketone coordinates to the endocyclic boron atom. The transfer of hydride ion takes place through the six-membered transition state **3**, in which the ketone is arranged to minimize steric interactions of the larger substituent (R^L). The intermediate **4** eliminates monoalkoxyborane **5** and the catalyst is regenerated. The monoalkoxyborane **5** undergoes fast disproportionation to borane and dialkoxyborane **6**, which is hydrolyzed to the product alcohol **7**.

The well-known mechanism of reaction allows the reduction of aryl alkyl ketones as well as various functionalized ketones, including α-halo-, α-hydroxy- and α-azido ketones, α,β-enones and keto esters, producing secondary alcohols with high enantioselectivities and predictable configurations.

Due to the efficiency of oxazaborolidines in reductions of various types of ketones with borane, these catalysts have been used as key step reagents in the syntheses of numerous compounds. Very interesting applications are the synthesis of biologically active natural compounds or fragments thereof, such as epothilones,[7] octalactin,[8] bisacumol[9] or cylindricine C.[10] Several β-adrenergic agonists used in the therapy of asthma, bronchitis and congestive heart failure, *e.g.* formoterol,[11] 2-fluoroepinephrine[12] or octopamine,[13] were also synthesized using this methodology.

Additional Safety

All sessions should be done with protective clothing, gloves and safety goggles. Borane–tetrahydrofuran solution reacts violently with water and is flammable as liquid and vapor. Methanol is a highly flammable liquid, is toxic and is readily absorbed by ingestion and inhalation.

Experimental Procedure

Laboratory Session 1 (4 h): Asymmetric Reduction of Acetophenone

1. Prepare a 25 mL round-bottom flask with a magnetic stirring bar, and add 25.5 mg (0.1 mmol, 0.05 equiv.) of (*S*)-(−)-α,α-diphenyl-2-pyrrolidinemethanol.
2. Insert a gas-inlet adapter capped with a rubber septa, and connected to an oil bubbler, into the round-bottom flask and gently flush the flask with an inert gas (nitrogen or argon) for at least 10 min.
3. Add 1 mL of tetrahydrofuran, and 12.5 μL (0.11 mmol, 0.055 equiv.) of trimethylborate at room temperature, and stir this solution for 30 min.
4. Add 1 mL of tetrahydrofuran again, and 2 mL (2 mmol, 1 equiv.) of 1 M borane–THF solution. Next, add slowly (at least over a 10 min period) a solution of 240 mg (234 μL, 2 mmol, 1 equiv.) of acetophenone in 3 mL of tetrahydrofuran and stir the reaction mixture for 30 min.

1-Phenylethanol Isolation

Method 1

5. Add carefully 10 mL of water and transfer the reaction mixture into a separatory funnel and perform an extraction with 3 × 10 mL of diethyl ether.
6. Wash the combined organic extracts thoroughly with 10 mL of 1 M hydrochloric acid followed by 10 mL of brine. Dry the organic layer with anhydrous $MgSO_4$, filter and evaporate the solvents under vacuum on a rotary evaporator.

Method 2

7. Add carefully 1 mL of methanol to the reaction mixture, stir the solution for 1 h and evaporate volatiles under vacuum on a rotary evaporator.
8. Alternatively, dissolve the oily crude product (from point 7) in 5 mL of diethyl ether and add a solution of hydrogen chloride in diethyl ether or dioxane (0.15 mmol, 0.075 equiv.).
9. Filter off the amino alcohol hydrochloride using a Büchner funnel and evaporate solvents under vacuum.

Laboratory Session 2 (4 h): Product Isolation and Purification

10. Purify the oily crude product (from point 6 or 7 or 9) using column chromatography on silica gel with hexane/ethyl acetate (80 : 20) as an eluent.
11. Collect the proper fractions (TLC analysis) and evaporate solvents.
12. Weigh the isolated product and calculate the yield. Measure the rotation, and record ^{1}H and ^{13}C NMR spectra to confirm the compound purity. Compare the results with literature data.

Waste Disposal

The aqueous solution may be diluted, neutralized and discarded in the sink. Organic solvents from a rotary evaporator should be collected in a non-chlorinated waste container.

Results Interpretation and Additional Questions

1. Calculate the percent yield of the isolated 1-phenylethanol.
2. Measure the optical rotation of 1-phenylethanol and calculate the specific rotation using the following formula:

$[\alpha]_D^{temp.} = \frac{\alpha}{l \times c}$	α – the measured angle of rotation of polarized light plane; temp. – measurement temperature; *D* – the wavelength of yellow sodium light line, 589 nm; *l* – the length of polarimetric tube the in dm; *c* – the concentration of 1-phenylethanol in g per mL
The specific rotation is written as: $[\alpha]_D^{temp.} = +\text{ or } - \ldots\ldots(c\ldots\ldots,\ CHCl_3)$	Here, *c* is the concentration of 1-phenylethanol in g per 100 mL

3. Determine the absolute configuration and enantiomeric purity of the obtained 1-phenylethanol, comparing the sign and the value of optical rotation with the literature data.

Lit.[14] for (*S*)-1-phenylethanol (>99% ee): $[\alpha]_D^{25}$ -55.1 (c 1.63, $CHCl_3$).

4. Examine an enantiomeric excess of the obtained 1-phenylethanol using HPLC analysis on a chiral column (Chiralcel OD-H) and compare with the results obtained from the optical rotation measurement.
5. Using the mechanistic model of the CBS reduction predict the (*R*/*S*) configuration of the product alcohol obtained from the reduction of 2-chloroacetophenone catalyzed with oxazaborolidine prepared from (1*S*,2*R*)-1-amino-2-indanol.

References

1. (a) A. Hirao, S. Itsuno, S. Nakahama and N. Yamazaki, *J. Chem. Soc., Chem. Commun.*, 1981, 315–317; (b) S. Itsuno, M. Nakano, K. Miyazaki, H. Masuda, K. Ito, A. Hirao and S. Nakahama, *J. Chem. Soc., Perkin Trans. 1*, 1985, 2039–2044.
2. (a) E. J. Corey, R. K. Bakshi and S. Shibata, *J. Am. Chem. Soc.*, 1987, **109**, 5551–5553; (b) E. J. Corey, R. K. Bakshi, S. Shibata, C. P. Chen and V. K. Singh, *J. Am. Chem. Soc.*, 1987, **109**, 7925–7926.
3. (a) E. J. Corey and C. J. Hellal, *Angew. Chem., Int. Ed.*, 1998, **37**, 1986–2012; (b) B. T. Cho, *Tetrahedron*, 2006, **62**, 7621–7643.
4. M. Masui and T. Shioiri, *Synlett*, 1997, 273–274.
5. V. L. Ponzo and T. S. Kaufmann, *Synlett*, 2002, 1128–1130.
6. D. J. Mathre, I. Shinkai, M. Zaidlewicz, M. Krzemiński, K. Łączkowski, A. Patti and S. Pedotti, Tetrahydro-1-methyl-3,3-diphenyl-1*H*,3*H*-pyrrolo[1,2-*c*][1,3,2]oxazaborole, *e-EROS Encyclopedia of Reagents for Organic Synthesis*, 2013.
7. E. A. Reiff, S. K. Nair, B. S. N. Reddy, J. Inagaki, J. T. Henri, J. F. Greiner and I. G. Georg, *Tetrahedron Lett.*, 2004, **45**, 5845–5847.
8. J. Bach and J. Garcia, *Tetrahedron Lett.*, 1998, **39**, 6761–6764.
9. A. Li, G. Yue, Y. Li, X. Pan and T.-K. Yang, *Tetrahedron: Asymmetry*, 2003, **14**, 75–78.
10. H. Abe, S. Aoyagi and C. Kibayashi, *J. Am. Chem. Soc.*, 2005, **127**, 1473–1480.
11. H. S. Wilkinson, G. J. Tanoury, S. A. Wald and C. H. Senanayake, *Org. Process Res. Dev.*, 2002, **6**, 146–148.
12. S.-f. Lu, B. Herbert, G. Haufe, K. W. Laue, W. L. Padgett, O. Oshunleti, J. W. Daly and K. L. Kirk, *J. Med. Chem.*, 2000, **43**, 1611–1619.
13. B. T. Cho, S. K. Kang and S. H. Shin, *Tetrahedron: Asymmetry*, 2002, **13**, 1209–1217.
14. K. Nakamura and T. Matsuda, *J. Org. Chem.*, 1998, **63**, 8957–8964.

13.8. Synthesis of 4,5-Bis(benzoylthio)-1,3-dithiole-2-thione

Dulce Simão* and Ana C. Cerdeira

Department of Chemistry, Instituto Superior Técnico, University of Lisbon, Centro de Química Estrutural, Av. Rovisco Pais, 1, 1049-001 Lisbon, Portugal. *E-mail: dulcesimao@tecnico.ulisboa.pt

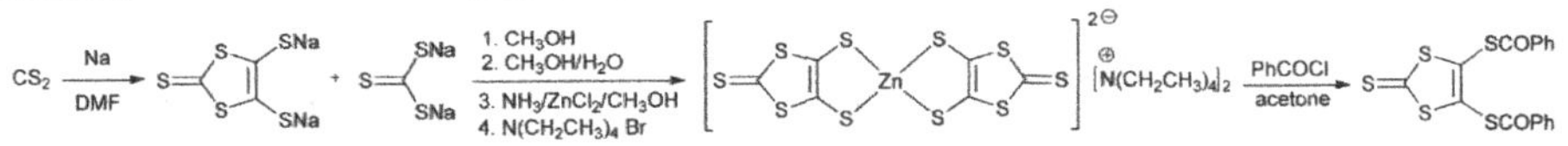

Number of sessions (duration of each session)	Hazard level	Difficulty level	Level of study
6 (6 h + 6 h + 3 h + 6 h + 3 h + 3 h)	High	Medium	Advanced
Class names 1,3-Dithiole-2-thiones, thioesters, acyl chlorides		**Concepts involved** Two-step reaction involving a reduction with sodium, product isolation through metal complexation and thioester formation	
Chemicals needed Sodium, carbon disulfide, dimethylformamide, methanol, distilled water, ammonia (30%), zinc chloride, tetraethylammonium bromide, isopropyl alcohol, diethyl ether, dry acetone, benzoyl chloride, chloroform, activated charcoal			
Equipment and experimental techniques involved Crystallization, IR, mechanical and magnetic stirring, apparatus for dry conditions reaction with inert atmosphere and temperature control, vacuum filtration, rotary evaporator, recrystallization, melting point			
Keywords Crystallization, IR, metal complexation, reduction, thioester, thione			

Background

4,5-Bis(benzoylthio)-1,3-dithiole-2-thione is a key intermediate in the synthesis of bis(ethylenedithio)tetrathiafulvalene (BEDT-TTF or "ET") (Figure 13.8.1) that forms conducting and superconducting charge transfer salts known as "organic metals".[1–3]

The synthesis of 4,5-bis(benzoylthio)-1,3-dithiole-2-thione[4–6] begins with the reduction of carbon disulfide with sodium metal, to produce sodium 1,3-dithiole-2-thione-4,5-dithiolate (dmit) and sodium trithiocarbonate in equimolar amounts.[7,8] Only the dithiolate is able to chelate with transition metals to form square-planar species. This complexation with zinc separates the trithiocarbonate from the thione to give a quaternary ammonium zincate salt and provides a more stable material for further reaction with benzoyl chloride through nucleophilic substitution at the carbonyl group.[9] This reaction is characterized by a nucleophilic addition–elimination mechanism[10] initiated by sulfur atoms to carbonyl carbon atom. In the second step the tetrahedral intermediate formed eliminates the living group, Cl^-, regenerating the carbonyl group to afford 4,5-bis(benzoylthio)-1,3-dithiole-2-thione. This synthesis provides easy and inexpensive access to the 4,5-dithio-1,3-dithiole-2-thione system.

Comprehensive Organic Chemistry Experiments for the Laboratory Classroom
Edited by Carlos A. M. Afonso, Nuno R. Candeias, Dulce Pereira Simão, Alexandre F. Trindade, Jaime A. S. Coelho, Bin Tan, and Robert Franzén.

Published by the Royal Society of Chemistry, www.rsc.org

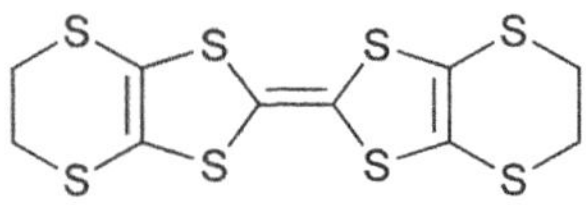

Figure 13.8.1 Bis(ethylenedithio-fulvalene), known as BEDT-TTF or "ET".

Additional Safety

This experiment requires the use of a fume hood due to carbon disulfide (highly flammable and toxic) and benzoyl chloride (corrosive and irritant). Sodium metal is a corrosive and potentially flammable solid. It reacts violently with water, and in case of fire, use dry chemical powder. Other flammable solvents are used (ethyl ether, acetone and methanol) and should be carefully handled. The mechanical stirrer should be manipulated with suitable precautions due to its rotational parts.

Experimental Procedure

Preparation of Tetraethylammonium Bis(1,3-dithiole-2-thione-4,5-dithiol) Zincate

Session 1 (6 h)

All glassware used in this step must be dried in the oven.

1. Remove the glassware and immediately start to assemble the reaction apparatus while the material is still hot.
2. Fit a 1 L four-neck round-bottomed flask with a mechanical stirrer, reflux condenser, pressure equalizing dropping funnel, gas outlet and inlet tubes (connected to nitrogen or argon).
3. Under inert atmosphere, add sodium metal (2.88 g, 0.13 mol) *via* a solid addition funnel and place the apparatus in an ice-water bath.
4. Add CS_2 (22.5 mL, 0.38 mol) through the dropping funnel.
5. Add dropwise DMF (25 mL) with stirring over 4 h, while keeping the temperature of the reaction mixture low with ice-water cooling. After the addition of several mL of DMF a red color appears.
6. When the addition is complete, remove the ice-water bath.
7. Keep the mixture under stirring overnight at room temperature. During this period, the color should turn from red to a very dark shade of violet-red.

Session 2 (6 h)

1. Visually inspect the flask for residual sodium metal.
2. Cool it with an ice bath (0–5 °C) and slowly add methanol (7 mL) with stirring for 1 h.
3. A mixture of methanol (50 mL) and distilled water (63 mL) is degassed under vacuum and is rapidly added to the reaction mixture through the dropping funnel.
4. In a 500 mL Erlenmeyer flask stir NH_3 (30%) (63 mL) with $ZnCl_2$ (2.5 g, 0.019 mol) and add methanol (63 mL).
5. Add this solution to the reaction mixture through the dropping funnel.
6. Prepare a solution of tetraethylammonium bromide (6.7 g, 0.032 mol) in distilled water (32 mL) and add it through the dropping funnel with vigorous stirring for at least 4 h.
7. After addition, allow the mixture to stir overnight, during which time the product precipitates as a red solid.

Session 3 (3 h)

1. Filter under vacuum and wash immediately with distilled water (63 mL), then with 50 mL portions of isopropyl alcohol until the filtrate is colorless.
2. Wash with diethyl ether (25 mL), collect the product (red powder) and dry under vacuum.
3. Record the yield, melting point and IR spectrum.

Preparation of 4,5-Bis(benzoylthio)-1,3-dithiole-2-thione

Session 4 (6 h)

1. Fit a 1 L two- or three-neck round-bottomed flask with a reflux condenser, $CaCl_2$ tube, pressure-equalizing dropping funnel and magnetic stirring.
2. Add dry acetone (200 mL) and tetraethylammonium bis(1,3-dithiole-2-thione-4,5-dithiol) zincate (8 g, 0.011 mol).
3. With stirring add slowly benzoyl chloride (20 mL, 24.2 g, 0.173 mol) through the dropping funnel for 4 h.
4. When the addition is completed, leave the mixture under stirring overnight.

Session 5 (3 h)

1. Collect the resulting yellow-brown solid under vacuum and wash it with water (250 mL) and acetone (150 mL).
2. In a 500 mL round-bottomed flask dissolve the crude material in chloroform (175 mL), add activated charcoal (0.25 g) and reflux it for 10 min.
3. Filter this mixture while it is hot and wash the filter cake with more hot chloroform (25 mL) to dissolve any residual product.
4. Concentrate this solution to 75 mL with a rotary evaporator.
5. Add methanol (25 mL) dropwise with stirring and leave this solution in the refrigerator overnight.

Session 6 (3 h)

1. Filter the product under vacuum and collect the resulting crystalline precipitate.
2. Record the yield, melting point and IR spectrum.

Waste Disposal

All toxic materials should be stored for later elimination. Residual sodium must be carefully destroyed with ethanol. Aqueous solutions are diluted and collected in a waste collector recipient for aqueous solutions. Chloroform and acetone may be stored to further purification. All other organic solvents are collected in non-chlorinated waste collector.

Results Interpretation and Additional Questions

1. Determine the reaction yields.
2. Assess the purity of products by comparing the melting points with the ones reported in the literature.
3. Assign the characteristic absorption bands in the IR spectrum.
4. Write out the mechanism of thiol ester formation.

References

1. M. Mizuno, A. F. Garito and M. P. Cava, *J. Chem. Soc., Chem. Commun.*, 1978, 18.
2. D. O. Cowan, J. A. Fortkort and R. M. Metzger, *Low-Dimensional Systems and Molecular Electronics*, Plenum Press, 1991, vol. 248, p. 1.
3. V. Z. Kresin and W. A. Little, *Organic Superconductivity*, Plenum Press, 1990, p. 295.
4. G. Steimecke, H.-J. Sieler, R. Kimse and E. Hoyer, *Phosphorus and Sulfur*, 1979, vol. 7, p. 49.
5. T. K. Hansen, J. Becher, T. Jergensen, K. S. Varma, H. Khedekar and M. Cava, *Org. Synth., Coll.*, 1996, **73**, 270.
6. C. A. M. Afonso, D. P. Simão, L. P. Ferreira, M. S. Serra and M. M. M. Raposo, *100 Experiências de Química Orgânica*, IST Press, 2011, p. 189.
7. G. G. Abashev and E. V. Shklyaeva, *Chemistry of Heterocyclic Compounds*, 2006, vol. 42, p. 423.
8. J. C. Lodmell, W. C. Anderson, M. F. Hurley and J. Q. Chambers, *Anal. Chim. Acta*, 1981, **129**, 49.
9. J. March, *Advanced Organic Chemistry*, John Wiley & Sons, 4th edn, 1992, p. 409.
10. J. Clayden, N. Greeves, S. Warren and P. Wothers, *Organic Chemistry*, Oxford University Press, 1st edn, 2001, p. 281.

13.9. Organocatalytic Asymmetric Reduction of (*E*)-*N*,1-Diphenyl-1-propanimine to *N*-(1-Phenylpropyl)aniline with Trichlorosilane

Pedro Barrulas[†]* and Anthony Burke*

Departamento de Química e Centro de Química de Évora, Universidade de Évora, Rua Romão Ramalho, 59, 7000 Évora, Portugal. *E-mail: pbarrulas@uevora.pt, ajb@uevora.pt

Two Step Synthesis of (*S*)-*N*-(1-phenylpropyl)aniline: Hydrosilylation of (1*E*)-N,1-diphenyl-1-propanimine

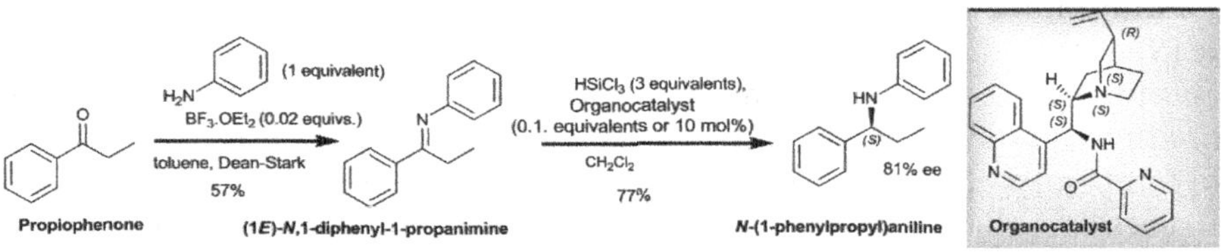

Number of sessions (duration of each session)	Hazard level	Difficulty level	Level of study
2 (4 h each)	Moderate	Medium	Advanced

Class names Ketones, ketimines, amines

Concepts involved Nucleophilic addition to an sp^2 imine carbon, geometric isomers, enantioselectivity, organocatalysis, hydrosilylation reaction

Chemicals needed Aniline and propiophenone to make the ketimine, boron–trifluoride etherate, trichlorosilane, dichloromethane, organocatalyst (8*S*,9*S*)-9-picolinamide(9-desoxy)-*epi*-cinchonidine, saturated solution of $NaHCO_3$, anhydrous $MgSO_4$

Equipment and experimental techniques involved *Session 1*: Dean–Stark apparatus, microdistillation kit, ^1H and ^{13}C NMR spectroscopy. *Session 2*: apparatus for magnetic stirring, ice-bath, liquid–liquid extraction (separatory funnel), column chromatography with silica gel, ^1H and ^{13}C NMR spectroscopy, HPLC and chiral column (Daicel Chiracel OD-H or Chiracel IB)

Keywords Column chromatography, enantiomers, ^1H and ^{13}C NMR spectroscopy, high-performance liquid chromatography (HPLC) using a chiral column, microdistillation, organocatalyst

Background

Organocatalysis is a subdivision of catalysis which relies on the use of wholly organic molecules as catalysts. These catalysts are highly useful and have already made significant contributions in the production and discovery of new pharmaceuticals, agrochemicals and other high added value compounds.[1] The discovery of the Hajos–Parrish–Eder–Sauer–Wiechert reaction was the starting point.[2] But only relatively recently has there been a renaissance of this field, when Barbas III, Lerner, List and MacMillan very successfully applied simple organocatalysts

[†]Current address: Laboratório HERCULES, Universidade de Évora, Largo Marquês de Marialva, N° 8, 7000-809 Évora, Portugal.

Comprehensive Organic Chemistry Experiments for the Laboratory Classroom
Edited by Carlos A. M. Afonso, Nuno R. Candeias, Dulce Pereira Simão, Alexandre F. Trindade, Jaime A. S. Coelho, Bin Tan, and Robert Franzén.

Published by the Royal Society of Chemistry, www.rsc.org

(*e.g.* L-proline (and derivatives) and imidazolidinones) in several asymmetric reactions.[3] In general the catalysts are optically pure and the reactions are highly enantioselective.

The hydrosilylation reaction is a useful reaction for reducing imines to amines and is a widely employed method for accessing enantiomerically enriched amines, many of which are shown to be biologically active. Some 45% of drug molecules on the market contain a chiral amine unit, well-known examples include (among others) rasagiline, sertraline, indatraline, cetirizine and cromakilm.[4]

The catalytic asymmetric hydrosilylation reaction of ketimines is extremely useful from an industrial perspective, as it affords chiral products using typically very cheap reducing agents.[5]

In this experimental procedure the simple ketimine (as a mixture of (1*E*)-(major) and (1*Z*)-1-diphenyl-1-propanimine (minor isomer)), which will be prepared in the first session, is reduced to the corresponding chiral amine *N*-(1-fenilpropil)aniline with an enantioselectivity of 81% ee in favour of the (*S*)-enantiomer[6] using $HSiCl_3$ and the cinchona alkaloid organocatalyst (+)-*N*-(9-deoxy-*epi*-cinchonin-9-yl)picolinamide. Imines are versatile substrates for a plethora of asymmetric transformations,[7] which includes hydrosilylation[8] and arylation.[9] The reaction takes only 30 min at room temperature.[6,7] Regarding the handling of the $HSiCl_3$, and boron trifluoride etherate in the first session, the hazard level is considered medium. Due to the formation of geometrical isomers of 1-diphenyl-1-propanimine in the first session and the mechanism of activation of trichlorosilane by the organocatalyst in the second session, this practical experiment would be considered to be of an advanced chemistry level.

The final amine product is purified by liquid chromatography on silica gel and the enantiomeric excess determined using HPLC with a chiral column.

Overall it is expected that the students will learn a good deal from these laboratory sessions, as they are very broad reaching incorporating synthesis, with chemical analysis, purification techniques, stereochemistry and mechanisms.

Additional Safety

This practical class should be conducted in a fume hood, with a laboratory coat, disposable nitrile gloves and safety goggles or glasses.

Boron trifluoride etherate, which is used in the first session, is a toxic liquid that causes damage to organs through prolonged or repeated exposure if inhaled. It is a flammable liquid and vapor. This reagent reacts violently with water.

Trichlorosilane, which is used in the key reaction in the second session, is a flammable, toxic and corrosive reagent that can cause skin and eye burns and can irritate the respiratory system in case of inhalation. This reagent reacts violently with water to liberate toxic gases.

Experimental Procedure

Laboratory Session 1 (4 h) – Synthesis of (1*E*)-*N*,1-Diphenyl-1-propanimine (Ref. 10)

1. Propiophenone (4.87 g, 36.33 mmol) and aniline (3.38 g, 36.33 mmol) are placed in a 250 mL round-bottom flask, with toluene (150 mL) and boron trifluoride etherate (90 μL, 0.73 mmol).
2. The flask is placed on a suitable fitting on a heating mantle and attached to a Dean–Stark assembly maintained under a nitrogen atmosphere. Bring the reaction to reflux, and monitor the water removal.
3. After 3 h, allow the reaction to cool to room temperature. Remove the solvent on a rotatory evaporator, but during the process transfer to a smaller flask, ideally of 25 mL volume.
4. Set up a microdistillation assembly and distil the product. Collect the first fraction, which will be unreacted propiophenone (50–60 °C @ 1 mmHg) and then the *title compound* as an

aromatic sweet smelling oil at (120–130 °C @ 1 mmHg). This compound solidifies at room temperature.

5. Weigh the isolated product and calculate the reaction yield. Obtain both ^{1}H and ^{13}C NMR spectra.
6. Using ^{1}H NMR calculate the ratio of geometrical isomers present.

Laboratory Session 2 (4 h) – Asymmetric Hydrosilylation of (1*E*)-*N*,1-Diphenyl-1-propanimine to *N*-(1-Phenylpropyl)aniline

1. In a 10 mL round-bottomed flask equipped with a magnetic stir bar, add the (1*E*)-*N*,1-diphenyl-1-propanimine (0.07 g, 0.33 mmol), the organocatalyst[11] (0.013 g, 0.033 mmol, 10 mol%) and CH_2Cl_2 (1 mL) (as the organocatalyst is both air and water stable this can be performed in a stoppered flask using an a non-dry solvent).
2. After complete dissolution of the reagents, put the reaction mixture on an ice-bath for 15 min.
3. When the reaction mixture has cooled, add *via* a syringe dropwise 1.5 equivalents of $HSiCl_3$ (0.134 g, 0.99 mmol). During the addition, you may probably notice the formation of a yellowish or orange coloration.
4. When the addition is finished, let the reaction mixture stir at room temperature for another 30 min to 1 h.
5. After this time, quench the reaction by adding *via* syringe, dropwise, 2 mL of a saturated solution of $NaHCO_3$. *Caution!* During this addition the mixture heats up, with the liberation of small amounts of hydrogen gas, and will produce an effervescent effect (both the hydrides and the excess $HSiCl_3$ will be neutralized).
6. Transfer the final mixture to a separating funnel and extract the organic phase (containing the desired product) with CH_2Cl_2 (3 × 10 mL) and combine all the organic phases.
7. Remove the remaining water from the organic layer by adding anhydrous $MgSO_4$, filtering, and evaporating the solvent to dryness on a rotatory evaporator.
8. Purify the resulting residue using silica gel column chromatography (70–200 μm) using CH_2Cl_2 as eluent.
9. Collect the desired aliquots from the column chromatography. Combine them and evaporate the solvent to dryness using a rotatory evaporator.
10. Weigh the isolated product and calculate the reaction yield. Obtain both ^{1}H and ^{13}C NMR spectra.
11. Separate the enantiomers of the chiral amine and record the respective chromatogram using HPLC with the following conditions:

- Chiral column: Daicel Chirapak OD-H
- Mobile phase: hexane/isopropanol (80 : 20)
- Flow: 1 mL min^{-1}
- λ = 254 nm
- t_r = 5.0 min (*S*)-enantiomer, t_r = 5.6 min (*R*)-enantiomer

Or:

- Chiral column: Chiracel IB
- Mobile phase: hexane/isopropanol (99 : 1)
- Flow: 0.8 mL min^{-1}
- λ = 254 nm
- t_r = 7.9 min (*S*)-enantiomer, t_r = 8.5 min (*R*)-enantiomer

Waste Disposal

The aqueous residue of the reaction, neutralized with a saturated solution of $NaHCO_3$, should be collected in a proper waste recipient for future destruction.

The dichloromethane can be distilled and reused, or alternatively the chlorinated solvent can be stored in a waste recipient for chlorinated residues for future destruction.

Interpretation of the Results and Questions

1. Analyse both the ^1H and ^{13}C NMR spectra of both products and assign the different signals of the molecule.
2. Calculate the overall yield for the two steps from propiophenone.
3. Calculate the enantiomeric excess using the HPLC chromatographic data:[12]

$$\%\text{ee of product} = \frac{\text{area enantiomer }(S) - \text{ area enantiomer }(R) \times 100}{\text{area enantiomer }(S) + \text{area enantiomer }(R)}$$

4. Discuss the reaction mechanism and present a model to explain the obtained reaction enantioselectivity.
5. How might the enantioselectivity be improved?
6. Since 2000, the number of publications on organocatalysis has increased exponentially year after year. In your opinion, what are the main benefits of using this approach for asymmetric catalysis?

References

1. For key reviews/papers: (a) J. Alemán and S. Cabrera, *Chem. Soc. Rev.*, 2013, **42**, 774; (b) M. J. Gaunt, C. C. C. Johansson, A. McNally and N. T. Vo, *Drug Discovery Today*, 2007, **12**, 8; (c) T. A. Davis and J. N. Johnson, *Chem. Sci.*, 2011, **2**, 1076; (d) G. Coulthard, W. Erb and V. K. Aggarwal, *Nature*, 2012, **489**, 278.
2. (a) U. Eder, G. Sauer and R. Weichert, *Angew. Chem., Int. Ed.*, 1971, **10**, 496; (b) Z. G. Hajos and D. R. Parrish, *J. Org. Chem.*, 1974, **39**, 1615.
3. (a) B. List, R. A. Lerner and C. F. Barbas III, *J. Am. Chem. Soc.*, 2000, **122**, 2395; (b) K. A. Ahrendt, C. J. Borths and D. W. C. MacMillan, *J. Am. Chem. Soc.*, 2000, **122**, 4243.
4. S. K. Ritter, *Chem. Eng. News*, 2013, **29**, 34.
5. (a) T. C. Nugent and M. El-Shazly, *Adv. Synth. Catal.*, 2010, **352**, 753; (b) A. K. Ghose, V. N. Viswanadhan and J. J. Wendoloski, *J. Comb. Chem.*, 1999, **1**, 55.
6. P. C. Barrulas, A. Genoni, M. Benaglia and A. J. Burke, *Eur. J. Org. Chem.*, 2014, 7339.
7. R. Bloch, *Chem. Rev.*, 1998, **98**, 7339.
8. S. Guizzetti and M. Benaglia, *Eur. J. Org. Chem.*, 2010, 5529.
9. C. S. Marques and A. J. Burke, *ChemCatChem*, 2011, **4**, 635.
10. J. S. M. Samec and J. I. Bäckvall, *Chem.–Eur. J.*, 2002, **8**, 2955.
11. This organocatalyst, (+)-*N*-(9-deoxy-*epi*-cinchonin-9-yl)picolinamide, is available from both TCI Europe N.V. and Chiratecnics Lda (http://www.chiratecnics.com).
12. M. E. Andersson, D. Aslam, A. Clarke and H. Roeraade, *J. Chromatogr. A*, 2003, **1005**, 83.

14. Rearrangements

14.1. Benzilic Acid Rearrangement

Paulo Coelho* and Céu Sousa

Centro de Química – Vila Real, Universidade de Trás-os-Montes e Alto Douro, 5001-801 Vila Real, Portugal
*E-mail: pcoelho@utad.pt

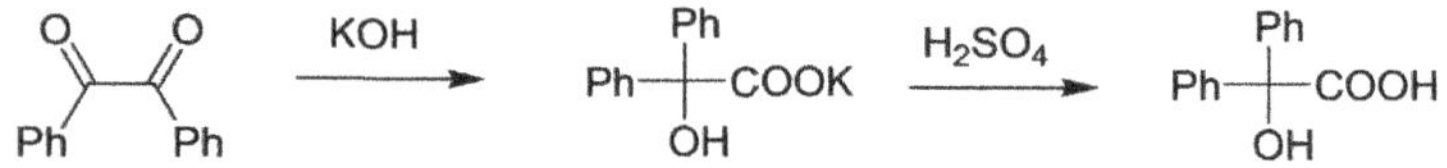

Number of sessions (duration of each session)	Hazard level	Difficulty level	Level of study
1 (2 h)	Low	Low	Intermediate

Class names Ketones, α-hydroxyacids

Concepts involved Some α-hydroxyacids can be easily prepared by a rearrangement of 1,2-diarylethane-1,2-diketones in strong basic media. The reaction involves the addition of one hydroxyl ion and the 1,2-migration of one aryl group. The final carboxylic acid is obtained after acid hydrolysis and can be recrystallized from water

Chemicals needed 1,2-Diphenylethane-1,2-dione, KOH, H_2SO_4, ethanol

Equipment and experimental techniques involved Heating mantle, apparatus for reaction under reflux, filtration

Keywords Filtration, NMR, α-diketones, α-hydroxyacids, migration, rearrangement

Background

In common organic reactions there is a change in the functional group but the skeleton of the molecule remains intact. However, sometimes, migration of part of the molecule occurs leading to a quite different molecule. These rearrangements can be inter- or intramolecular. In the first case part of one molecule or reagent is transferred to another molecule. In the second a group is transferred from one carbon within a molecule to another carbon and forms a constitutional isomer (Figure 14.1.1).

One of the best examples of this phenomenon is the pinacol rearrangement whereby treatment of 2,3-dimethyl-1,2-ethanediol with acid leads to the formation of 3,3-dimethylbutan-2-one, involving the 1,2 shift of a methyl group between two adjacent atoms (Scheme 14.1.1). Movements over longer distances are also possible and are observed in several pericyclic reactions in which multiple bonds are broken and formed along with a shift of a group, usually a hydrogen.[1]

Another very common rearrangement occurs when aromatic 1,2-diketones are treated with strong base in ethanol under reflux. This reaction is called the *benzilic acid rearrangement* and was first reported by Justus Liebig in 1838.[2] This reaction is representative of a 1,2-rearrangement in which one group migrates to the adjacent carbon. Usually the reactive species in these intramolecular transformations is a carbocation but this reaction is unusual because it involves an anionic intermediate. The long established reaction mechanism involves the attack of the

Comprehensive Organic Chemistry Experiments for the Laboratory Classroom
Edited by Carlos A. M. Afonso, Nuno R. Candeias, Dulce Pereira Simão, Alexandre F. Trindade, Jaime A. S. Coelho, Bin Tan, and Robert Franzén.

Published by the Royal Society of Chemistry, www.rsc.org

hydroxyl anion to the ketone and the 1,2-migration of one phenyl group with formation of an α-hydroxyacid after acid treatment (Scheme 14.1.2).[3,4]

This rearrangement is not exclusive to aromatic ketones but requires that no acidic hydrogens are present α to the carbonyl. Starting from unsymmetrical diketones, the same product will be obtained whichever group migrates; however, it has been determined that phenyl groups carrying electron attracting substituents migrate to a greater extent.[5] The reaction can be extended to cyclic diketones and in this case a nice ring contraction can be observed (Scheme 14.1.3).[6] This idea has been successfully applied to steroids, with high stereospecificity, generating derivatives that display biological activity.[7]

The objective of this experiment is the synthesis of 2-hydroxy-2-phenylbenzilic acid from benzil, a simple aromatic diketone (Scheme 14.1.4). The reaction will afford an insoluble salt that

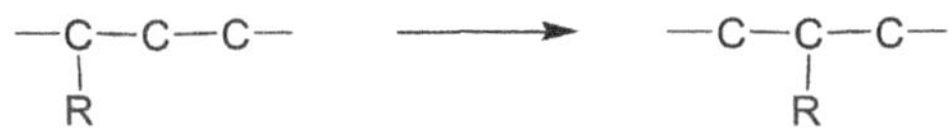

Figure 14.1.1 Intramolecular rearrangement to form a constitutional isomer.

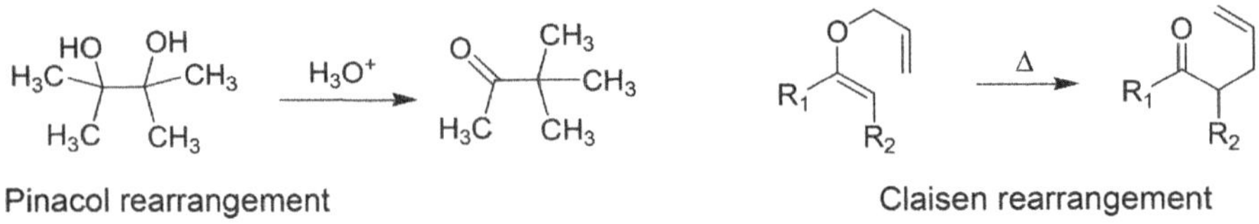

Scheme 14.1.1 Pinacol and Claisen rearrangements.

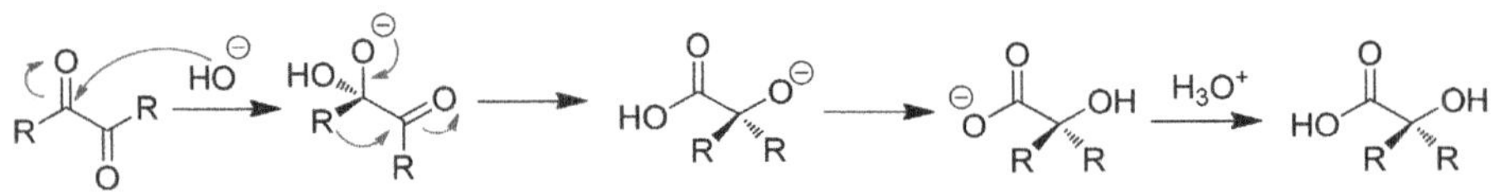

Scheme 14.1.2 Benzilic acid rearrangement.

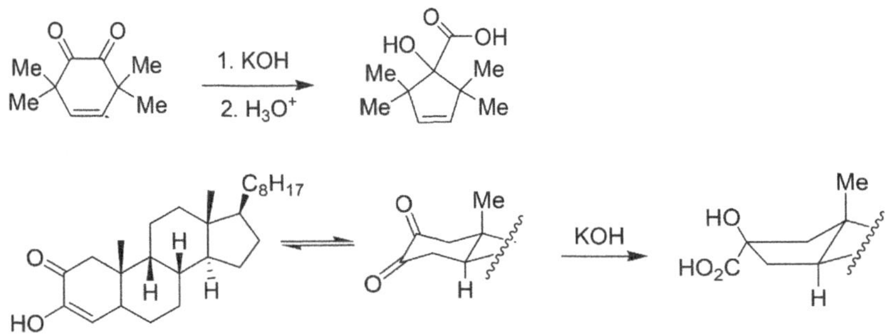

Scheme 14.1.3 Ring contraction *via* benzilic acid rearrangement.

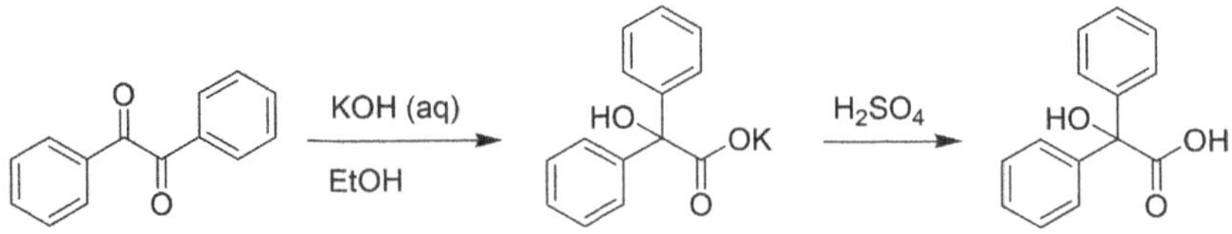

Scheme 14.1.4 Synthesis of 2-hydroxy-2-phenylbenzilic acid from benzil.

precipitates from the reaction mixture and will then be converted into the carboxylic acid after treatment with H_2SO_4.[8,9] The final compound may be recrystallized from water.

Additional Safety

This session should be done in a fume hood. KOH and H_2SO_4 are corrosive and should be handled with care.

Experimental Procedure

1. Add benzil (1.0 g, 4.76 mmol), water (1.5 mL), ethanol (3.0 mL) and KOH (1.0 g, 17.8 mmol) to a round-bottom flask with one or two boiling chips and attach a reflux condenser.
2. Attach the water tubing to the condenser, ensuring that water goes in the bottom and out the top. Turn on the water so that you have a slow but steady stream of water passing through the condenser.
3. Place a heating mantle below the flask and heat the mixture under reflux for 20 min. The solution becomes violet once the reflux starts.
4. Remove the condenser and cool the flask in an ice bath for 5–10 min. Scratch the flask with a spatula to promote the precipitation of the carboxylate salt.
5. Assemble a vacuum filtration apparatus using a Büchner funnel. Wet the filter paper with a small amount of cold ethanol, filter the crystals and wash them with a little cold ethanol (~5 mL).
6. Dissolve the crystals in 50 mL of water and add a magnetic stirrer. Place the solution under a stirring plate and add slowly H_2SO_4 (2 M, approximatively 3 mL) until pH 2–3.
7. Wet a filter paper with a small amount of cold water, filter the crystals in a Büchner funnel and wash them with cold water (~10 mL).
8. Recrystallize the product from boiling water.
9. Collect your product and allow it to dry overnight in an oven (60 °C). Weigh the product.
10. Measure the melting point of the product and evaluate its purity.
11. Record the IR spectra of the product.

Waste Disposal

The aqueous solution present in the Büchner funnel can be diluted and discarded in the sink.

Results Interpretation and Additional Questions

1. Determine the limiting reagent and calculate the product yield.
2. Interpret the IR spectra of the product. Why is the band at 3400 cm^{-1} sharp instead of broad?
3. Why does this reaction not work with hexane-3,4-dione?

References

1. J. March, *Advanced Organic chemistry*, 4th edn, John Wiley & Sons, 1992, p. 1051.
2. J. Liebig, *Ann. Chem.*, 1838, **25**, 1.
3. C. K. Ingold, *Annu. Rep. Prog. Chem.*, 1928, **25**, 124.
4. J. Hine and H. W. Haworth, *J. Am. Chem. Soc.*, 1958, **80**(9), 2274.
5. S. Selman and J. F. Eastham, *Q. Rev., Chem. Soc.*, 1960, **14**, 221.
6. B. P. Mundy, M. G. Ellerd and F. G. Favaloro, *Name reactions and reagents in organic synthesis*, 2nd edn, 2005, John Wiley & Sons, p. 86.
7. H. R. Nace and M. Inaba, *J. Org. Chem.*, 1962, **27**(11), 4024.
8. A. R. Tatchell, B. S. Furnis, A. J. Hannaford and P. W. G. Smith, *Vogel's textbook of practical organic chemistry*. Longman Scientific & technical, 5th edn, 1989, p. 1041.
9. D. L. Pavia, G. S. Kriz, G. M. Lampman and R. G. Engel, *A Microscale Approach to Organic Laboratory Techniques*, Brooks/Cole Laboratory Series for Organic Chemistry, 5th edn, 2012, p. 301.

14.2. Preparation of Phenyl Acetate and Its Conversion into 4-Hydroxyacetophenone

Eimíle Sheehy and Paul Evans*

Center for Synthesis and Chemical Biology, School of Chemistry, University College Dublin, Dublin 4, Ireland
*E-mail: Paul.evans@ucd.ie

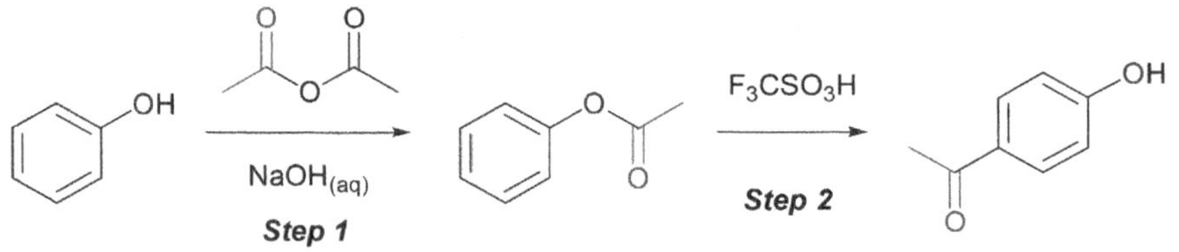

Number of sessions (duration of each session)	Hazard level	Difficulty level	Level of study
2 (3 h + 3 h)	Moderate	Medium	Intermediate

Class names Alcohols, anhydrides, esters, ketones, aromatic rings

Concepts involved This two-step experiment involves, initially, a nucleophilic acyl substitution reaction to prepare phenyl acetate under basic conditions from phenol and acetic anhydride. In the second step, phenyl acetate is subjected to a Fries rearrangement, mediated by the strong Brønsted acid trifluoromethanesulfonic acid

Chemicals needed Phenol, acetic anhydride, sodium hydroxide (4 mol dm^{-3}), dichloromethane, sodium bicarbonate, magnesium (or sodium) sulfate, trifluoromethane sulfonic acid, deuteriochloroform (optional)

Equipment and experimental techniques involved Apparatus for magnetic stirring, liquid–liquid extraction, rotary evaporation, distillation apparatus, nuclear magnetic resonance spectrometer (optional), infrared spectrometer (optional)

Keywords Distillation, electrophilic aromatic substitution, ^{1}H NMR and IR spectroscopy, isomerisation, liquid–liquid extraction, nucleophilic acyl substitution, recrystallisation

Background

This experiment marries concepts taught in carbonyl chemistry and aromatic chemistry. The sequential nature of the experiments, designed to occupy two separate 3 hour time slots, also introduces the idea of multi-step synthetic sequences and an added benefit is that the reaction products can also be used as an aid to the teaching of spectroscopic methods. The experiment could be run sequentially, whereby students use the product from step 1 to perform step 2, or they could be used as individual parts. In the first step, under basic reaction conditions phenol undergoes rapid conversion into the ester phenyl acetate, using acetic anhydride.[1] The nucleophilic acyl substitution involved in this step is fundamental in areas of organic and bioorganic chemistry and is of historic significance, applied, for instance, in the synthesis of aspirin,

Comprehensive Organic Chemistry Experiments for the Laboratory Classroom
Edited by Carlos A. M. Afonso, Nuno R. Candeias, Dulce Pereira Simão, Alexandre F. Trindade, Jaime A. S. Coelho, Bin Tan, and Robert Franzén.

Published by the Royal Society of Chemistry, www.rsc.org

diacetylmorphine (heroin) *etc.* The reaction product is sequentially purified by an aqueous extraction followed by a preparative distillation, which can be performed at atmospheric pressure, or under moderate reduced pressure.

In the second session, a portion of the thus produced phenyl acetate is isomerized in a Brønsted acid-promoted Fries rearrangement, in order to form 4-hydroxyacetophenone.[1] Substituted aromatic ketones of the type produced in this experiment are useful intermediates[2] and 4-hydroxyacetophenone itself has been isolated on numerous occasions from natural sources.[3] Mechanistically the Fries rearrangement is not fully understood and using these conditions the reaction is regioselective, producing high yields of the 4-substituted isomer. Notably, trifluoromethanesulfonic acid is classed as a "superacid" and is approximately 1000 times more acidic than concentrated sulfuric acid.[4]

Additional Safety

All sessions should be conducted in a fume hood. Safety equipment: protective clothing, gloves and safety glasses/goggles should be used. Phenol, sodium hydroxide and particularly trifluoromethanesulfonic acid are corrosive and contact with skin and eyes causes burns (severe). Dichloromethane should be treated with care due to its classification as an agent that may cause cancer.

Experimental Procedure

Step 1: Preparation of Phenyl Acetate (3 h)

1. Add phenol (1.20 g) to a 25 mL round-bottom flask (RBF).
2. With a Pasteur pipette add a solution of 4 mol dm^{-3} aqueous NaOH (5 mL). Swirl to effect dissolution, whereby phenol ionizes forming the water soluble sodium phenoxide.
3. Add a magnetic stirrer bar and ice-chilled water (8 mL).
4. Stir the mixture and at room temperature introduce acetic anhydride (1.5 mL) in a dropwise fashion using Pasteur pipette. Following this addition stir for 5 min.
5. Transfer the reaction mixture to a 100 mL separating funnel.
6. Wash RBF with dichloromethane (10 mL) and add contents to a separating funnel. Shake (venting occasionally) to thoroughly mix phases, then allow phases to separate.
7. Transfer the organic extract to a clean 100 mL conical flask and re-extract the aqueous layer with dichloromethane (10 mL). In the separating funnel wash the combined organic extracts with saturated sodium bicarbonate solution (20 mL).
8. Dry the resultant organic layer with anhydrous magnesium (or sodium) sulfate. Filter into a clean dry 50 mL RBF and evaporate under reduced pressure to provide the target compound as a colourless liquid.
9. Further purify by distillation, either at atmospheric pressure or under moderate vacuum (15–30 mmHg).
10. Calculate a yield, reporting the observed boiling point (range).

Step 2: Fries Rearrangement: Conversion of Phenyl Acetate into 4-Hydroxyacetophenone (3 h)

1. In a clean, dry 5–25 mL RBF, equipped with a magnetic stirrer bar, introduce trifluoromethanesulfonic acid (*care*) (0.5 mL) and cool to 0 °C.
2. To this add phenyl acetate (50 μL) and stir for 0.5 h.
3. Using a Pasteur pipette gradually add (*care*) the contents of the RBF to a mixture of ice-water (approx. 25 mL).

4. Allow the ice to melt and extract in a 100 mL separating funnel with dichloromethane (15 mL).
5. Transfer the organic layer to a clean 100 mL conical flask and re-extract the aqueous layer with dichloromethane (15 mL).
6. Wash the combined organic extracts with saturated sodium chloride solution (25 mL).
7. Dry the organic layer with anhydrous magnesium (or sodium) sulfate. Filter and evaporate under reduced pressure to provide the target compound as a colourless solid.
8. Further purification may be achieved by recrystallization from toluene.
9. Calculate the yield and record the melting point. Optional additional product characterization and comparison: thin-layer chromatography; proton and carbon NMR; infrared spectroscopy.

Waste Disposal

Halogenated organic waste should be separately stored and may be recycled by distillation. Aqueous waste resulting from step 1 should be neutralized, diluted and discarded in the sink. Aqueous acidic waste from step 2 should be neutralized and due to the presence of trifluoromethane sulfonate salts disposed of by a licensed professional contractor.

Results Interpretation and Additional Questions

- Compare recorded boiling point and melting points with literature figures.
- Provide a mechanism for the formation of phenyl acetate in step 1 and describe how the different purification protocols serve to remove by-products and impurities.
- Suggest a mechanism for the conversion of phenyl acetate into 4-hydroxyacetophenone in step 2.
- Provide structures for the possible alternative isomers and suggest a reason(s) why the observed product might be favoured.
- Interpret ^{1}H NMR spectra obtained, explaining the chemical shifts and coupling patterns.
- Rationalize key peaks in the respective IR spectra.

Acknowledgement

Taken with permission from: R. Murashige, Y. Hayashi, S. Ohmori, A. Torii, Y. Aizu, Y. Muto, Y. Murai, Y. Oda and M. Hashimoto, *Tetrahedron*, 2011, **67**, 641.

References

1. R. Murashige, Y. Hayashi, S. Ohmori, A. Torii, Y. Aizu, Y. Muto, Y. Murai, Y. Oda and M. Hashimoto, *Tetrahedron*, 2011, **67**, 641.
2. For key references demonstrating the usefulness of products obtained from the Fries rearrangement see: (a) G. Leclerc, J. C. Bizec, N. Bieth and J. Schwartz, *J. Med. Chem.*, 1980, **23**, 738; (b) R. Martin, *Org. Prep. Proced. Int.*, 1992, **24**, 369.
3. Natural product isolation, see for example: L.-J. Zhu, Y.-L. Hou, X.-Y. Shen, X.-D. Pan, X. Zhang and X.-S. Yao, *Fitoterapia*, 2013, **88**, 44.
4. G. A. Olah, G. K. S. Prakash, J. Sommer and A. Molner, *Superacid Chemistry*, John Wiley & Sons, 2009; For a review concerning the preparation and use of trifluoromethanesulfonic acid see: R. D. Howells and J. D. McCown, *Chem. Rev.*, 1977, **77**, 69.

14.3. Multistep Synthesis of Dilantin

Paula C. Castilho* and Pedro Ideia

CQM, Centro de Competências de Ciências Exactas e das Engenharias, Universidade da Madeira, Campus Universitário da Penteada, 9020-105 Funchal, Portugal. *E-mail: castilho@uma.pt

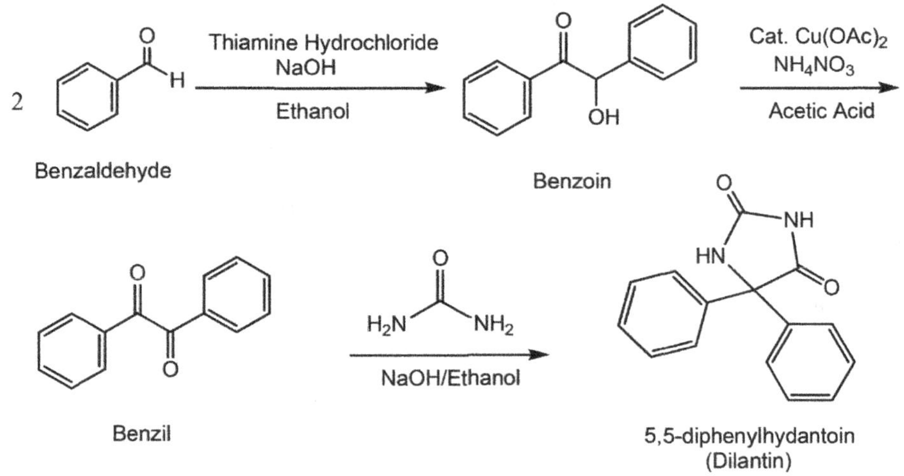

Number of sessions (duration of each session)	Hazard level	Difficulty level	Level of study
5 (3 h + 3 h + 3 h + 4 h + 1 h)	Moderate	Medium	Intermediate

Class names Carbonyl groups, alcohols, amino groups, amides, aldehydes, ketones

Concepts involved This three-step process leads to the total synthesis of the antiseizure drug Dilantin® (phenytoin, 5,5-diphenylhydantoin or 5,5-diphenyl-2,4-imidazolidinedione). Benzoin synthesis from benzaldehyde is a typical example of a benzoinic condensation currently performed using vitamin B1 as catalyst, replacing toxic cyanide ions. The oxidation to benzil can be performed with different oxidizing agents, leading to different yields, allowing for considerations of hazards *versus* yields of products, time of reaction, energy consumption and waste generation. The third step is a condensation reaction between benzil and urea, followed by benzilic rearrangement, leading to the desired drug

Chemicals needed Benzaldehyde, thiamine hydrochloride, copper(II) acetate, ammonium nitrate, glacial acetic acid, urea, sodium hydroxide, ethanol, concentrated hydrochloric acid

Equipment and experimental techniques involved Stirring plate, heating mantle, apparatus for vacuum filtration and reflux, IR equipment, NMR equipment (if available)

Keywords Condensation, crystallization, green catalysts, IR analysis, oxidation, synthesis of pharmaceuticals

Comprehensive Organic Chemistry Experiments for the Laboratory Classroom
Edited by Carlos A. M. Afonso, Nuno R. Candeias, Dulce Pereira Simão, Alexandre F. Trindade, Jaime A. S. Coelho, Bin Tan, and Robert Franzén.

Published by the Royal Society of Chemistry, www.rsc.org

Background

This experiment consists of the synthesis of 5,5-diphenylhydantoin from benzaldehyde in three steps.

The sodium salt of 5,5-diphenylhydantoin is an anticonvulsant drug sold under a variety of trade names including Dilantin, Dihydan, Dilabid, and Zantropil. The drug was originally developed as an antiepileptic and acts as an anticonvulsant to stabilize grand mal seizures. The preparation of Dilantin® using this route was first published in Germany in 1908.[1] The anticonvulsant and antiepileptic properties of this compound were not discovered unit 1938.

Epilepsy is a chronic neurological disorder characterized by recurrent seizures caused by sudden, excessive electrical activity in the brain. Epilepsy affects approximately 1% of the world's population, with the highest rates occurring in children and adolescents. Dilantin (a Parke-Davis product) is one of the most frequently prescribed antiepileptic drugs (AED), although several new AEDs have been approved by the US Food and Drug Administration (US FDA) in recent years. Most of the new AEDs on the market enhance the activity of γ-aminobutyric acid (GABA), the primary inhibitory neurotransmitter in the brain.[2]

This multistep synthesis is performed in several chemistry laboratory programmes using various catalysts; the present proposal presents the best possible combination of environmental friendly catalysts and reaction yields.

Part 1. Synthesis of Benzoin from Benzaldehyde

Benzaldehyde does not possess alpha hydrogens and therefore does not undergo a self-aldol condensation. In the presence of a catalyst, however, benzaldehyde undergoes a unique self-condensation reaction, called the benzoin condensation. The product is the α-hydroxyketone called benzoin. The reaction proceeds *via* sequential nucleophilic additions to the carbonyl groups of the two benzaldehyde molecules. Discovered in 1832, this reaction was typically carried out using a catalytic amount of a cyanide salt, but it was later discovered by Breslow in 1958 that the conjugated bases of thiazolium salts were also effective catalysts. The key structural feature of these salts is the thiazole ring, which is easily deprotonated to yield a carbanion. This nucleophile allows the two benzaldehyde molecules to react in a rational manner.

Thiamine (vitamin B1) also contains a thiazole ring. Indeed, thiamine is an important coenzyme in many important biochemical reactions, which were finally understood and categorized as analogues of the benzoin condensation.[3]

Part 2. Synthesis of Benzil from Benzoin

In the second step of this multi-step synthesis the α-diketone, benzil, is prepared through the oxidation of the α-hydroxyketone, benzoin. This oxidation may be easily performed using mild oxidizing reagents such as Fehling's solution (alkaline cupric tartrate), palladium(II) acetate in dimethyl sulfoxide under an oxygen atmosphere, copper sulfate in pyridine, manganese(II) Schiff base and nitric acid.[4,5]

In the present experiment, this oxidation is performed with a solution of ammonium nitrate in the presence of a catalytic amount of copper(II) acetate.[6]

The conversion of benzoin into benzil is a catalytic oxidation reaction. The benzoin is oxidized to benzil by the copper(II) ions, which are reduced to copper(I) ions in the process.[6] Upon oxidizing benzoin, the Cu(II) is reduced to Cu(I). At this stage it is no longer an effective oxidant, and must be re-oxidized to Cu(II) so that it can oxidize another molecule of benzoin. Thus another oxidant is present, ammonium nitrate, to carry out the oxidation of Cu(I) to Cu(II). The ammonium nitrate, in turn, is reduced to ammonium nitrite, which decomposes to produce nitrogen, water and some possible intermediate oxides. Cupric salts are mild oxidizing agents, which do not attack the diketone product, and in the absence of Cu(II) the ammonium nitrate will not oxidize benzoin at a significant rate.

The benzoin radical cation loses a proton to an acetate ion (AcO^-), forming acetic acid (AcOH) and a resonance stabilized radical, depicted by structures B and C in Scheme SM 14.3.2 in the ESI. Another redox cycle between Cu^{2+} and the radical takes place, forming a second Cu^+ ion and cation D, which loses a proton to another acetate ion to form benzil. The Cu^+ ions formed in these redox cycles are reoxidized to Cu^{2+} by ammonium nitrate, which is present in excess as shown in Scheme SM 14.3.1 in the ESI. This reaction forms ammonium nitrite (NH_4NO_2), which decomposes to nitrogen and water under the reaction conditions.

Part 3. Synthesis of Dilantin from Benzil

In the final step of the linear synthesis, benzil is condensed with urea, followed by a 1,2-phenyl migration. Finally, dehydration under base catalysis yields our target molecule, Dilantin®. An interesting feature of this reaction is the structural rearrangement in which both phenyl groups end up on the same carbon atom.

Additional Safety

Sodium hydroxide is highly caustic and can cause chemical burns. The dissolution of solid sodium hydroxide in water is a highly exothermic reaction. Sodium hydroxide is corrosive to several metals, like aluminum, which reacts with the alkali to produce flammable hydrogen gas on contact. Safety precautions must be taken during its handling.

Glacial (anhydrous) acetic acid is corrosive to skin and must be handled with appropriate care, since it can cause skin burns, permanent eye damage, and irritation to the mucous membranes. Latex gloves offer no protection, so nitrile gloves should be worn instead.

Because Dilantin® has significant biological activity it is very important that protective gloves and safety goggles be worn when handling the product of this reaction!

Experimental Procedure

Session 1. Synthesis of Benzoin from Benzaldehyde (3 h)

1. Dissolve 1.2 g of thiamine hydrochloride in 3.3 mL of distilled water in a 50 mL Erlenmeyer flask.
2. Add 10 mL of 95% ethanol and cool the solution by swirling the flask in an ice-water bath.
3. Meanwhile, place about 3.3 mL of 2 M sodium hydroxide solution in a small Erlenmeyer flask. Cool this solution in the ice bath as well.
4. Slowly, add the cold NaOH solution to the thiamine solution. This should take about 10 min.
5. Measure 6.7 mL of benzaldehyde and add it to the yellow reaction mixture. Cork the flask and wrap Parafilm around the cork. Allow the mixture to sit at room temperature for one week. (Be sure to label your flask.)

Session 2. Purification and Characterization of Benzoin Prepared in the Previous Session (3 h)

1. Calculate the amount of hot ethanol you will need to recrystallize your crude benzoin. The solubility of benzoin in hot ethanol is ≈ 0.05 g mL^{-1}.

2. The solid benzoin should have separated from the solution.† Cool the mixture in an ice-water bath, then collect your product by vacuum filtration, wash the crude product with several portions of ice cold water.
3. Weigh the crude product and then recrystallize it from 95% ethanol. The solubility of benzoin in boiling 95% ethanol is about 12–14 g per 100 mL.
4. Weigh the product, calculate the percentage yield, and determine its melting point and IR and NMR (if available) spectra.

Session 3. Synthesis of Benzil from Benzoin (3 h)

1. Weigh out 1.0 g of benzoin and place it into a dry 25 mL round-bottom flask.
2. Carefully add 3.0 mL of glacial acetic acid to the flask.
3. Next, add 0.47 g of ammonium nitrate and 1.0 mL of a 2% (wt/wt) aqueous solution of $Cu(OAc)_2 \cdot H_2O$.
4. Place a boiling stone in the round-bottom flask and attach a water-jacketed condenser. Keep a slow, steady flow of water circulating through the condenser at all times while heating. Warm the reaction to gentle reflux using the heating mantle. Maintain a steady reflux for 60 min. As the reaction progresses, gas evolution will begin. This can be followed by attaching some rubber tubing to the top of the condenser, connecting the other extremity to a Pasteur pipette and introducing the tip in water.
5. After refluxing the reaction for 60 min, remove the round-bottom flask from the heating mantle and allow it to cool for 5 min.
6. Pour the reaction mixture into a 30 mL beaker containing 5–7 mL of ice water. Stir the resulting mixture with a glass rod for 5 min. Try to break up the hard, yellow solid that precipitates.
7. Isolate the yellow product using vacuum filtration through a Büchner funnel. Wash the benzil product with small aliquots of ice water. Dry the product with aspiration through the Büchner funnel for 5–10 min.
8. Place the benzil product in a clean beaker (covered and labelled) and allow it to dry until the next laboratory period.

Session 4. Synthesis of Dilantin from Benzil (4 h)

1. Obtain a mass, percent yield, and a melting point for your benzil product at the beginning of this laboratory period. Generally, the crude benzil is pure enough to be used in the next step. If the melting point range of your product is >4 °C, benzil should be recrystallized from 75% aqueous ethanol in order to ensure a successful preparation of Dilantin®.
2. Weigh out 600 mg of benzil into a dry, tared 25 mL round-bottom flask. Add 8 mL of 95% ethanol and swirl the round bottom flask until the benzil dissolves.
3. Then, add 0.36 g of urea and 1.6 mL of a 30% (wt/wt) aqueous solution of NaOH.
4. Place a boiling stone into the flask, and attach a water-jacketed condenser. Keep a slow, steady flow of water circulating through the condenser at all times while heating. Warm the reaction to a gentle reflux using a heating mantle. Maintain reflux for at least 1 h.
5. After refluxing for an hour, remove the reaction flask from the heating mantle, and allow the reaction mixture to cool to room temperature.

---

†If no solid product has separated from your solution, but a yellow oil is present, gently heat the solution with stirring until the mixture is homogenous. Remove the mixture from the heat and, with frequent (almost constant as the temperature approaches room temperature) scratching of the inside of the flask with a glass rod, allow the mixture to cool slowly to room temperature, then in an ice-water bath. Collect and wash the crystals as described above.

6. Now, add 12 mL of distilled water to the round-bottom flask. If solids are present, remove these solids by vacuum filtration through a Büchner funnel. *Save the filtrate!!!* Dispose of the solids in the recovered organic solids container.
7. Pour the filtrate into a 30 or 50 mL beaker, and slowly add concentrated HCl dropwise until the pH of the filtrate is 2–3. Cool the resulting acidic solution in an ice-water bath until all the product has precipitated. Collect the product by vacuum filtration through a Büchner funnel. Wash the product with small aliquots of ice water. Dry the Dilantin® product with aspiration through the Büchner funnel for 5–10 min.
8. Recrystallize the crude Dilantin® from ethanol. Transfer to a clean beaker (covered and labelled) and store it for the next (final) period.

Session 5. Characterization of Dilantin® (1 h)

1. Obtain the mass, percent yield, and a melting point for your 5,5-diphenylhydantoin product.
2. Run an IR spectrum (and NMR, if available).

Waste Disposal

The filtrates resulting from the isolation of the crude product and recrystallization of the product should be placed in an aqueous products recovery bottle. Dilantin® product should be placed in a specific Dilantin® recovery bottle.

Results Interpretation and Additional Questions

1. Compare the yields for the individual reactions of the multiple synthesis with literature data.
2. Calculate the percentage yield of the overall process.
3. Draw the complete mechanism of each reaction.
4. Compare the IR data for benzaldehyde, benzoin, benzyl and phenytoin and interpret the most significant variations.

References

1. H. Biltz, *Chem. Ber.*, 1908, **41**(1), 1379.
2. M. O. Cunningham, A. Dhillon, S. J. Wood and R. S. G. Jones, *Neuroscience*, 1999, **95**(2), 343.
3. K. P. C. Vollhardt and N. E. Schore, *Organic Chemistry*, WH Freeman, New York, 4th edn, 2003, pp. 1038–1044.
4. M. C. Hashjin, R. Ciyabi, M. Baharloui, N. Taherizadeh and G. Hosseini, *Res. Chem. Intermed.*, 2014, **40**, 2811 (and references therein).
5. J. Safari, Z. Zarnegar and F. Rahimi, *J. Chem.*, 2013, **2013**, 1.
6. P. Depreux, G. Bethegnies and A. Marcincal-Lefebvre, *J. Chem. Educ.*, 1988, **65**(6), 553.

14.4. Preparation of Isoborneol Through the Wagner–Meerwein Rearrangement Reaction of (1*R*)-(+)-Camphene

Marek P. Krzemiński

Department of Organic Chemistry, Faculty of Chemistry, Nicolaus Copernicus University in Toruń, Poland
*E-mail: mkrzem@chem.umk.pl

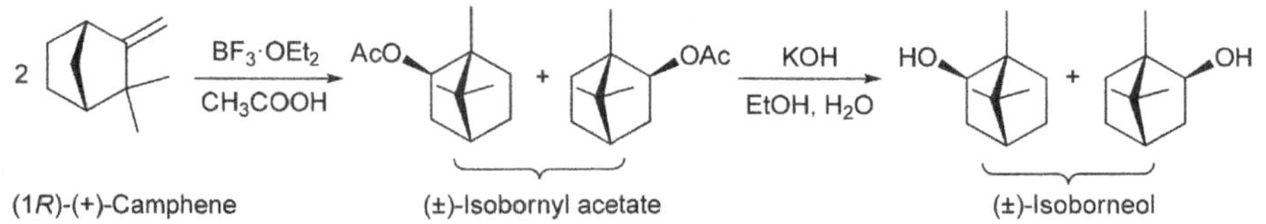

Number of sessions (duration of each session)	Hazard level	Difficulty level	Level of study
2 (4 h each)	Moderate	Medium	Intermediate

Class names Olefins, esters, alcohols

Concepts involved The main objective of this experiment is to familiarize students with the rearrangement reaction and the concept of non-classical carbocations. Another objective is to understand the racemisation process and stereochemistry in bicyclic systems. The second step of this experiment involving hydrolysis of an ester reminds the student of the mechanism of the basic hydrolysis of an ester. Students will also learn sublimation as the method for isolation and purification of the product

Chemicals needed (1*R*)-(+)-Camphene, acetic acid, boron trifluoride etherate, sodium carbonate, potassium hydroxide, petroleum ether (bp 40–60 °C) or hexane, ethanol, deuterated chloroform

Equipment and experimental techniques involved One-neck round-bottom flask (50 mL), reflux condenser, magnetic stirrer with heating plate, oil bath, separatory funnel, rotary evaporator, sublimation apparatus, polarimeter, NMR and/or IR apparatus, gas chromatograph

Keywords Bicyclic monoterpenes, NMR, non-classical carbocation, sublimation, Wagner–Meerwein rearrangement

Background

The Wagner–Meerwein rearrangement is one of the most studied reactions in organic chemistry. When this reaction was originally observed it covered rearrangements in bicyclo[2.2.1] systems, but currently it is assumed that this type of rearrangement includes [1,2]-shifts of hydrogen atoms or alkyl groups in carbocations. The reaction of α-pinene with hydrogen chloride to form bornyl chloride was described by Wagner and Brickner (Scheme 14.4.1).[1] In this paper, Wagner suggested that the acid-catalyzed dehydration of borneol into camphene engaged a molecular rearrangement. Several years later, Meerwein with co-workers presented the behavior of camphene hydrochloride (Scheme 14.4.1) and postulated the formation of a carbocation with structural changes occurring around it.[2]

The historical background of the research leading to the discoveries of Wagner and Meerwein is described by Birladeanu.[3]

Comprehensive Organic Chemistry Experiments for the Laboratory Classroom
Edited by Carlos A. M. Afonso, Nuno R. Candeias, Dulce Pereira Simão, Alexandre F. Trindade, Jaime A. S. Coelho, Bin Tan, and Robert Franzén.

Published by the Royal Society of Chemistry, www.rsc.org

The general mechanism of the Wagner–Meerwein rearrangement involves the formation of the carbocation in the first stage (Scheme 14.4.2). The initial carbocation can be obtained, for example, by protonation of olefins with protic acid or Lewis acid (LA), dehydration of alcohols or solvolysis of halides or alkyl sulfonates. This carbocation undergoes rearrangement to a thermodynamically more stable carbocation by [1,2]-alkyl, -aryl, or -hydride shift. Ring-expansion of strained cyclopropanes and cyclobutanes to more stable cyclopentane and cyclohexane rings is also observed. The last most stable carbocation reacts with a nucleophile present in the reaction mixture or can lose a proton to give an alkene (Scheme 14.4.2).

The proposal of carbocation as intermediate, introduced by Meerwein, was a milestone towards understanding the mechanisms of ionic reactions. These statements initiated studies that lasted several decades, with the involvement of the greatest chemists.

For many years, there was a discussion about the nature of carbocations and the possible existence of nonclassical carbocations. Brown proposed a mechanism involving classical carbocations, in which the positive charge is localized on one carbon atom.[4]

Currently, the mechanism shown in Scheme 14.4.3 is accepted with the participation of non-classical carbocations A and B with the delocalized positive charge.[5] The basis of this interpretation is spectroscopic studies of carbocations at low temperatures.[6]

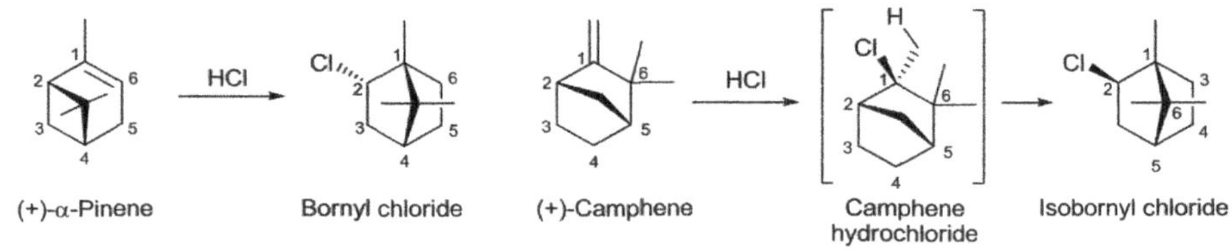

Scheme 14.4.1 Rearrangements of (+)-α-pinene and (+)-camphene in the presence of HCl.

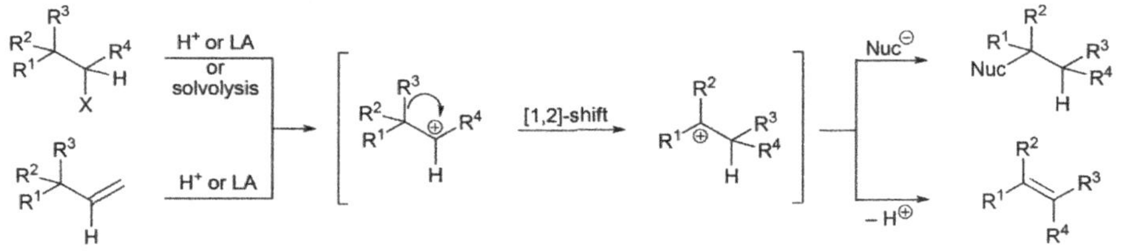

Scheme 14.4.2 General mechanism of the Wagner–Meerwein rearrangement.

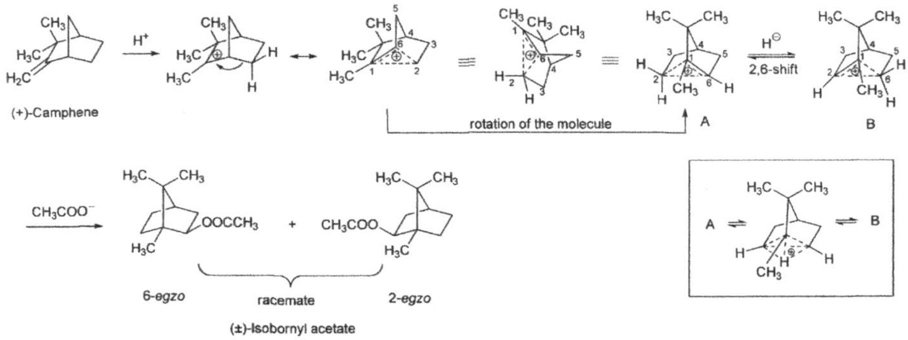

Scheme 14.4.3 Mechanism of the Wagner–Meerwein rearrangement with non-classical carbocations A and B.

Recently, camphene was converted into isobornyl acetate in the presence of homogeneous or heterogeneous acidic catalysts in acetic acid/water solutions (0–10 vol% of water).[7]

Additional Safety

Glacial acetic acid may cause severe irritation to the respiratory system and eye damage, therefore all work should be done wearing protective clothing, gloves and safety goggles or glasses. Boron trifluoride etherate is a fuming liquid and it is a flammable, corrosive, moisture-sensitive liquid. Extreme caution should be applied when handling $BF_3 \cdot Et_2O$. The addition of reagents to the reaction flask should be done under the fume hood, but the reaction can also be carried out outside the fume hood.

Experimental Procedure

Laboratory Session 1 (4 h): Preparation of Isobornyl Acetate

1. Prepare a 50 mL round-bottom flask with a magnetic stirring bar, and add 3.41 g (25 mmol, 1 equiv.) of (1*R*)-(+)-camphene, and 7.2 mL of glacial acetic acid (125 mmol, 5 equiv.) at room temperature and stir until all the (+)-camphene is dissolved (1–3 min).
2. Add 0.5 mL of boron trifluoride etherate.
3. Attach a water condenser into the round-bottom flask and immerse the flask in an oil bath. Stir the reaction mixture for 1 h, maintaining the temperature of the bath in the range of 55–60 °C.
4. Remove the flask from the oil bath and cool the flask in a water bath to room temperature.

Isobornyl Acetate Isolation

5. Transfer the reaction mixture into a separatory funnel. Add 20 mL of water and perform the extraction with 3 × 15 mL of hexane (or petroleum ether).
6. Wash the combined extracts with 10 mL of water, 10 mL of 10% aqueous Na_2CO_3, and 2 × 10 mL of water. Dry the combined organic layers with anhydrous $MgSO_4$, filter and evaporate the solvent on a rotary evaporator.
7. Weigh isobornyl acetate and calculate the yield of the crude product. Record IR spectra to confirm the formation of an ester and, if possible, a GC chromatogram to confirm the compound purity. Keep the ester until the next session.

Laboratory Session 2 (4 h): Hydrolysis of Isobornyl Acetate

8. Prepare a 100 mL round-bottom flask with a magnetic stirring bar, and add the obtained isobornyl acetate, 22 mL of ethanol, 5 mL of water and 3.00 g (53 mmol) of KOH.
9. Attach a water condenser to the round-bottom flask. Stir and heat the reaction mixture at reflux for 1.5 h. Cool the flask in a water bath to room temperature.

Product Isolation

10. Transfer the reaction mixture into a separatory funnel. Add 50 mL of water and perform the extraction with 3 × 20 mL of hexane (or petroleum ether). Wash the combined organic layers with 2 × 20 mL of water, and dry the organic solution with anhydrous $MgSO_4$. Filter the organic solution to a tared round-bottom flask and evaporate the solvent under vacuum on a rotary evaporator (it is crucial to completely evaporate the solvents).
11. Weigh the flask with a crude solid product and calculate the yield.
12. Attach a sublimation head to the flask, connect a vacuum source (water pump) and cooling water to the condenser. Release the pressure in the apparatus and heat the flask with the heating mantle.

13. Collect the purified/sublimed product, weigh and calculate the yield. Measure the melting point in a double end sealed capillary tube. Record ^{1}H and ^{13}C NMR spectra of the product dissolved in deuterated chloroform and GC chromatograms to confirm the compound purity.
14. Weigh, on an analytical balance with an accuracy of at least 1 mg, approximately 0.4 g of camphene into a volumetric flask with a capacity of 10 mL. Fill the volumetric flask with ethanol and measure the optical rotation. Repeat the same operations with the obtained isoborneol.

Waste Disposal

The neutralized aqueous solution may be discarded in the sink. Organic solvents from a rotary evaporator should be collected in a non-chlorinated waste container.

Results Interpretation and Additional Questions

1. Calculate the percent yield for each step, for the crude isobornyl acetate and the isolated isoborneol. Calculate also overall yield of isoborneol from camphene.
2. Calculate the specific rotation of camphene using the following formula and the enantiomeric purity comparing the sign and the value of specific rotation with the literature data. Literature[8] for (1*S*)-camphene (>99% ee): $[\alpha]_D^{21}$ – 119.1 (*c* 2.33, C_6H_6). Literature[9] for (1*S*)-camphene (58.8% ee): $[\alpha]_D^{21}$ – 48.6 (*c* 4, EtOH).

$$[\alpha]_D^{temp.} = \frac{\alpha}{l \times c}$$

α – the measured angle of rotation of polarized light plane
temp. – measurement temperature
D – the wavelength of yellow sodium light line, 589 nm
l – the length of the polarimetric tube in decimeters
c – the concentration of (+)-camphene in g mL^{-1}

The specific rotation is written as: $[\alpha]_D^{temp.}$ = +……(*c*……, EtOH)

Here, *c* is the concentration of (+)-camphene in g per 100 mL

---

3. Calculate the specific rotation of sublimed isoborneol according to the method described in the second point. Estimate the enantiomeric purity of isoborneol, comparing the sign and the value of the specific rotation with the literature data. Literature[10] for (1*R*)-isoborneol (>99% ee): $[\alpha]_D^{20}$ – 34.3 (EtOH).
4. Calculate the amount of boron trifluoride etherate in mmoles and equivalents and explain its role in the reaction.
5. Write out the complete mechanism of alkaline hydrolysis of isobornyl acetate.

References

1. G. Wagner and W. Brickner, *Chem. Ber.*, 1899, **32**, 2302–2325.
2. H. Meerwein, K. van Emster and J. Joussen, *Chem. Ber.*, 1922, **55B**, 2500–2528.
3. L. Birladeanu, *J. Chem. Educ.*, 2000, **77**, 858–863.
4. H. C. Brown and P. v. R. Schleyer, *The Non-Classical Ion Problem*, Plenum Press, New York, 1977.
5. (a) G. A. Olah, *Acc. Chem. Res.*, 1976, **9**, 41–52; (b) M. D. Saltzman and C. L. Wilson, *J. Chem. Educ.*, 1980, **57**, 289–290; (c) G. A. Olah, *J. Org. Chem.*, 2001, **66**, 5943–5957.
6. P. C. Myhre, G. G. Webb and C. S. Yannoni, *J. Am. Chem. Soc.*, 1990, **112**, 8991–8992.
7. (a) K. A. da Silva, I. V. Kozhevnikov and E. V. Gusevskaya, *J. Mol. Catal. A: Chem.*, 2003, **192**, 129–134; (b) A. L. P. de Meireles, K. A. da Silva Rocha, I. V. Kozhevnikov and E. V. Gusevskaya, *Appl. Catal., A*, 2011, **409–410**, 82–86.
8. G. W. Hana and H. Koch, *Chem. Ber.*, 1978, **111**, 2527–2532.
9. H.-R. Buser and M. D. Muller, *J. Agric. Food Chem.*, 1994, **42**, 393–400.
10. *The Merck Index*, ed. M. J. O'Neil, P. E. Heckelman, C. B. Koch and K. J. Roman, Merck & Co., Inc., Whitehouse Station, NJ, USA, 14th edn, 2006, p. 890.

14.5. Umbelliferone: A Natural Scaffold Suitable for the Synthesis of *ortho*-Acetylhydroxycoumarins *via* Fries Rearrangement Reaction

Saleta Vazquez-Rodriguez*[a], Maria João Matos[a], Eugenio Uriarte[b], Lourdes Santana[b] and Fernanda Borges[a]

[a]CIQUP/Department of Chemistry and Biochemistry, Faculty of Sciences, University of Porto, Portugal; [b]Department of Organic Chemistry, Faculty of Pharmacy, University of Santiago de Compostela, Spain
*E-mail: svre77@gmail.com

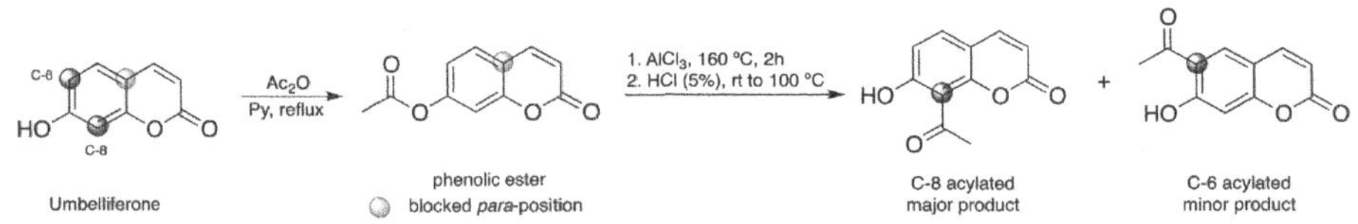

1 2 3

Number of sessions (duration of each session)	Hazard level	Difficulty level	Level of study
3 (4 h each)	Moderate	High	Advanced

Class names Phenols, esters, ketones

Concepts involved This experiment describes the preparation of *ortho*-acetylumbelliferones *via* the Fries rearrangement reaction in a two-step synthesis. The first step is an esterification reaction of the umbelliferone to produce the corresponding phenolic ester. Treatment of the phenolic ester with a Lewis acid gives the Fries rearrangement of the acyl group to the *ortho* position (C6 and C8)

Chemicals needed Umbelliferone (7-hydroxycoumarin), acetic anhydride, pyridine, water, methanol, anhydrous aluminium chloride, argon, hydrochloric acid (5% aqueous solution), ethyl acetate, brine, sodium sulfate anhydrous, silica, hexane, acetone, deuterium dimethyl sulfoxide (DMSO-d_6)

Equipment and experimental techniques involved One- and two-neck round-bottom flasks, drying tube, apparatus for reactions with temperature and magnetic stirring, rotatory evaporator, filtration, recrystallization, liquid–liquid extraction, flash column chromatography, ^{1}H NMR

Keywords Acylation, column chromatography, esterification, evaporation under reduced pressure, liquid–liquid extraction, NMR

Background

Structural modifications of natural products have become a powerful tool in drug discovery. The introduction of acyl groups is a key transformation that allows ulterior reactions, amplifying the versatility of a particular natural product to be further modified.

Phenolic esters of aliphatic and aromatic carboxylic acids, when treated with a Lewis acid as catalyst, undergo a rearrangement reaction to yield *ortho*- and *para*-acylphenols. This reaction

Comprehensive Organic Chemistry Experiments for the Laboratory Classroom
Edited by Carlos A. M. Afonso, Nuno R. Candeias, Dulce Pereira Simão, Alexandre F. Trindade, Jaime A. S. Coelho, Bin Tan, and Robert Franzén.

Published by the Royal Society of Chemistry, www.rsc.org

is commonly known as the Fries rearrangement reaction and is an important method for the synthesis of hydroxyaryl ketones.[1,2] The Fries rearrangement proceeds *via* ionic intermediates but the exact mechanistic pathway (whether it is inter- or intramolecular) is still under debate and is not rigorously known.[3–5] There are many reports in the literature that present evidence to support either of the pathways, but it appears that the exact route and the actual reactivity depends on substrate structure and reaction conditions.

Acylation of the natural product umbelliferone represents a good example of the Fries rearrangement and the *ortho* regioselectivity in organic reactions. Traven and collaborators have studied the Fries rearrangement in a series of acyloxycoumarins using crossed experiments with different acceptor aromatic compounds (including hydroxycoumarins) and have shown the intramolecular character of the process.[6] However, when Sha and collaborators carried out the Fries rearrangement of 7-acetoxycoumarin at 140 °C, only the 8-acetyl-7-hydroxycoumarin was isolated. In contrast, when the reaction takes place at 160 °C, they also isolated the 6-acetyl-7-hydroxycoumarin as a minor product.[7] Obviously, the *para*-substitution does not take place since the pyrone ring of the coumarin blocks this position. The preference of the C8 position over C6 in the 7-acetoxycoumarin derivatives for the Fries rearrangement has been attributed to the aromaticity of the pyrone ring of the coumarin.[8]

The Fries rearrangement reaction has many synthetic applications, especially in drug discovery as a very useful reaction to prepare new pharmacologically active compounds such as protein kinase inhibitors,[9] antitumor agents,[10] or adenosine receptor ligands.[11]

The objective of this work is the synthesis and separation of *ortho*-acetyl-7-hydroxycoumarins in a two-step process. The procedure will allow students to become familiar with the basic concepts and techniques of organic chemistry and compound identification.

Additional Safety

All sessions must be carried out in a fume hood, with proper safety equipment, such as protective clothing, gloves and safety glasses. Acetic anhydride, anhydrous aluminum chloride and hydrochloric acid are highly corrosive and may cause severe skin and eye burns; Ethyl acetate and pyridine are highly flammable and may cause skin and eye irritation. Methanol is highly flammable, may cause irritation of eyes, skin and mucous membranes and is toxic by inhalation, in contact with skin and if swallowed. Hexane is toxic if inhaled, flammable and toxic to aquatic life.

Experimental Procedure

Laboratory Session 1 (4 h): Preparation of 7-Acetoxycoumarin (1)

1. In a fume hood use a heating mantle to maintain a temperature of 115 °C (pyridine reflux temperature) on a magnetic hot plate.
2. Add 2.43 g (15 mmol) of umbelliferone and 40 mL of pyridine in a 250 mL round-bottomed flask with a magnetic stirred bar. Next, add 2.8 mL (30 mmol) of acetic anhydride.
3. Insert a water condenser into the round-bottomed flask and clamp the flask in the preheated oil bath. Stir the reaction under reflux until totally dissolved (for 2–3 h, approximately).
4. Leave the reaction mixture at room temperature and pour into ice-water (200 mL). Neutralize the reaction (100 mL of 5% HCl solution) and filter the precipitate by vacuum filtration.
5. Recrystallize the solid obtained in hot methanol (30 mL). The precipitate appears after 1 h at room temperature.
6. Filter the final product 7-acetoxycoumarin (**1**) by vacuum filtration, and collect the white needles. The expected yield of this reaction is around 95–98%.

Laboratory Session 2 (4 h): Preparation of *ortho*-acetyl-7-hydroxycoumarins 2 and 3

1. In a fume hood heat an oil bath to 160 °C on a magnetic hot plate.
2. In a mortar, mix 2.65 g (13 mmol) of compound **1** and 6.40 g (48 mmol) of anhydrous aluminum chloride.
3. Add the mixture to a 50 mL two-neck round-bottomed flask with a calcium chloride drying tube under an argon atmosphere (fill a balloon with argon; see Figure 2 in the ESI).
4. Stir the mixture at 160 °C for 2 h and then cool it down to room temperature.
5. Add 30 mL of a 5% HCl solution and stir the reaction for 1 h at room temperature.
6. Heat the reaction in a steam bath for another 0.5 h.
7. Extract with EtOAc (3 × 30 mL), and wash with brine the combined organic layers.
8. Dry the combined organic layers over anhydrous sodium sulfate and filter the solution.
9. Evaporate the solvent under vacuum. A yellow solid is obtained.

Laboratory Session 3 (4 h): Separation of Compounds 2 and 3

1. Run a TLC of the yellow solid obtained in session 2 on hexane : acetone (5 : 1 v/v mixture) (two times) and observe the presence of different spots.
2. Prepare the yellow solid sample by the dry method to be loaded in the flash column. First dissolve the sample to be analyzed in the minimum amount of solvent and add about 2.5 g of silica. Evaporate the solvent in a rotatory evaporator and stop only when a dry powder remains.
3. Purify the residue by flash chromatograph using hexane and acetone (6 : 1, v/v, 2 L) as eluent (1.5 h). Prepare the flash chromatography column (60 cm high, 29/32 cm wide). Make a slurry of the silica (80 g) by adding 250 mL of your eluent and mix it thoroughly by swirling vigorously to remove all the air from the silica. Using a powder funnel, carefully and slowly pour the slurry into the column, and compact the column using compressed air.
4. After loading the silica, the dry material on silica must now be prepared and loaded on the column.
5. Add a protective layer of sand to the top of the silica. This should be relatively level and about 2 cm thick.
6. Carefully fill the column with your eluent. At first, add the solvent *via* Pasteur pipet and keep adding the solvent by pipet until a few centimeters of solvent are above the column. Now slowly add the solvent through a powder funnel, letting it first run down the side of the column.
7. Run the column, applying compressed air, and collect fractions in test tubes (40 mL).
8. Monitor the progress of the column by TLC of the different fractions.
9. Once you have determined that all of the compounds you are interested in have eluted from the column, you are ready to wrap everything up. Start to combine fractions that contain only one compound.
10. Remove the solvent to dryness in a rotatory evaporator for each combined fractions.
11. Weigh the products and calculate the yield of the Fries rearrangement reaction and the overall yield. Measure the melting point, and record ^{1}H NMR spectra to confirm the compound purity and structure. Compare your data with the literature data.

Waste Disposal

All substances to be disposed of must be put into leak-proof containers that are clearly labeled with the identity or categorization of the contents, any known hazards, and some indication of boiling point range. The neutralized aqueous solution may be both diluted and discarded in the sink or can be categorized as "aqueous solution waste". The waste organic solvents are categorized as "non-chlorinated" waste. Solvent residues after vacuum evaporation are disposed of in the non-chlorinated organic solvents waste. Please make sure they are properly separated to be further treated by qualified staff.

Results Interpretation and Additional Questions

1. Identify each hydrogen atom in the ^{1}H NMR spectra of compounds **2** and **3** and compare them with the ^{1}H NMR of umbelliferone and compound **1**. Explain the different hydrogen displacement for compounds **2** and **3**.
2. Is there any evidence of internal hydrogen bonding in compounds **2** and/or **3**?
3. Calculate the yield for the last step and the overall yield of the reaction for each compound. Why is the C8 acylated compound the major isomer of the reaction?
4. Why do you use the drying tube with calcium chloride and the argon atmosphere? What are the white fumes observed during the Fries rearrangement reaction?
5. Discuss the reaction mechanism of the Fries rearrangement.

References

1. K. Fries and G. Finck, *Ber. Dtsch. Chem. Ges.*, 1908, **41**, 4271.
2. A. Gerecs in *Friedel–Crafts and Related Reactions*, ed. G. A. Olah, John Wiley & Sons Inc., London, 1964, vol. III, part 1, p. 449.
3. M. J. S. Dewar and L. S. Hart, *Tetrahedron*, 1970, **26**, 973.
4. Y. Ogata and H. Tabuchi, *Tetrahedron*, 1964, **20**, 1661.
5. A. Warshawsky, R. Kalir and A. Patchornik, *J. Am. Chem. Soc.*, 1978, **100**, 4544.
6. D. V. Kravchenko, T. A. Chibisova and V. F. Traven, *Russ. J. Org. Chem.*, 1999, **35**, 924.
7. D. N. Sha and N. M. Sha, *J. Org. Chem.*, 1954, **19**, 1681.
8. A. Rusell and J. R. Frye, *Organic Synthesis*, John Wiley & Sons, New York, 1955, collect. vol. III, p. 281.
9. M. Aschi, M. Attina and F. Cacace, *J. Am. Chem. Soc.*, 1995, **117**, 12832.
10. B. Janssens, P. Catry, R. Claessens, G. Baron and P. A. Jacobs, *Stud. Surf. Sci. Catal.*, 1997, **105B**, 1211.
11. S. Vazquez-Rodriguez, M. J. Matos, L. Santana, *et al.*, *J. Pharm. Pharmacol.*, 2013, **65**, 697.

14.6. Preparation of 4,5-Functionalized Cyclopentenone *via* Cyclization of a Stenhouse Adduct

Jaime A. S. Coelho* and Carlos A. M. Afonso

iMed.UL, Faculdade de Farmácia da Universidade de Lisboa, Av. Prof. Gama Pinto, 1649-003, Lisboa, Portugal
*E-mail: jaimeacoelho@ff.ulisboa.pt

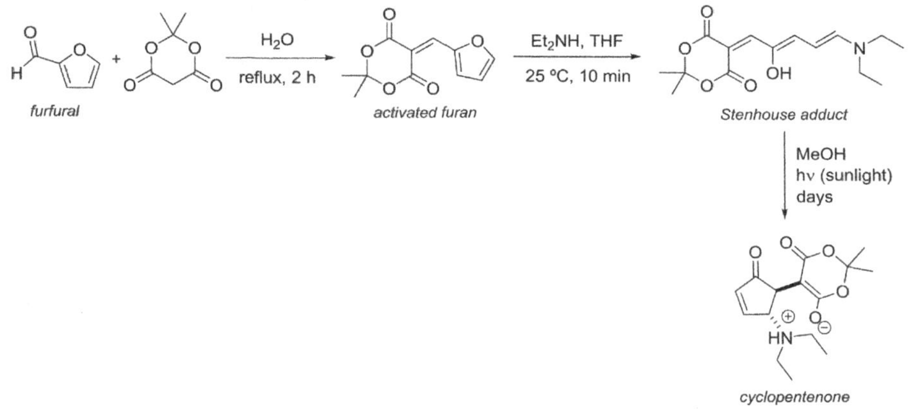

Number of sessions (duration of each session)	Hazard level	Difficulty level	Level of study
3 (4 h + 2 h + 1 h)	Moderate	Medium	Advanced

Class names Conjugated carbonyls, conjugated enamines, aldehydes, amines, esters, furans

Concepts involved This experiment involves the synthesis of cyclopentenones *via* sunlight-promoted cyclization of the corresponding Stenhouse adduct obtained from the reaction of activated furan and diethylamine. The complete process involves a Knoevenagel condensation reaction between furfural and Meldrum's acid followed by 1,8 addition of diethylamine to the activated furan to form the Stenhouse adduct that finally undergoes cyclization to yield the desired cyclopent-2-enone derivative. All the steps use simple work-up protocols with moderate yields and the structure of all the products is confirmed by NMR spectroscopy

Chemicals needed Furfural, Meldrum's acid (2,2-dimethyl-1,3-dioxane-4,6-dione), water, diethyl ether, diethyl amine, tetrahydrofuran, methanol, NaCl saturated aqueous solution, anhydrous magnesium sulfate, deuterated chloroform, deuterated water

Equipment and experimental techniques involved Hotplate stirrer (75 °C), 25 and 50 mL round bottom flasks, 1 mL pipettes, separatory funnel, rotary evaporator, liquid–liquid extraction, NMR

Keywords Cyclopentenone, furfural, liquid–liquid extraction, NMR, precipitation, Stenhouse adduct

Comprehensive Organic Chemistry Experiments for the Laboratory Classroom
Edited by Carlos A. M. Afonso, Nuno R. Candeias, Dulce Pereira Simão, Alexandre F. Trindade, Jaime A. S. Coelho, Bin Tan, and Robert Franzén.

Published by the Royal Society of Chemistry, www.rsc.org

Background

Cyclopentenones are important structural motifs of natural products and pharmaceuticals.[1] Due to their rich reactivity with nucleophiles (at position C3) and with electrophiles (at C5) they are also widely used as intermediates for the synthesis of several active biological compounds, such as carbocyclic nucleosides and prostaglandins. While simple cyclopentenones are industrially prepared by Claisen condensation of hexenedioic acids or their corresponding diesters,[2] functionalized cyclopentenones are typically obtained by Nazarov cyclization or by Pauson–Khand reaction (Scheme 14.6.1).[3]

A very interesting methodology for the synthesis of cyclopentenones relies on the use of biorenewable resources such as furfural or furanol as the source of the five carbon and the oxygen atoms.[4] In this context, it is known that furfurals react with secondary amines in the presence of Lewis acids to yield the corresponding cyclopentenones.[4,5] The mechanism involves a cascade reaction consisting of the *in situ* formation of an activated furan (conjugated iminium-ion) that undergoes furan-ring opening by conjugated amine addition to form the Stenhouse adduct intermediate, which finally cyclizes *via* a Nazarov-like rearrangement.[5] Very recently, the use of activated furans as starting materials allowed the isolation of the Stenhouse adduct intermediates and consequent cyclization promoted by visible light in protic solvents in the absence of Lewis acid catalysis.[6] The activated furan consists of a furan bearing an electron-withdrawing group, and thus is more activated to the nucleophilic addition of an amine. In this experiment the activated furan is easily prepared in one step *via* Knoevenagel condensation of furfural and Meldrum's acid using the protocol reported by Bigi and co-workers.[7] The desired Stenhouse adduct is then formed by reaction of activated furan with diethylamine following the conditions reported by Alaniz and co-workers.[6] Finally, the desired cyclopentenone is obtained by sunlight-promoted cyclization of the previously isolated Stenhouse salt using a modification of the protocol reported by Alaniz and co-workers.[6] All three products are solids and are readily isolated by precipitation from the reaction mixture without the need for further purification by column chromatography.

Additional Safety

The experiment should be conducted in a fume hood and all students should be equipped with protective clothing, gloves and safety glasses. Diethylamine is extremely hazardous in case of ingestion and very hazardous in case of eye and skin contact. Diethyl ether is volatile, flammable and hazardous in case of skin and eye contact or ingestion and inhalation. Cyclopentenones are biological active compounds and even though the hazards of this specific molecule are not

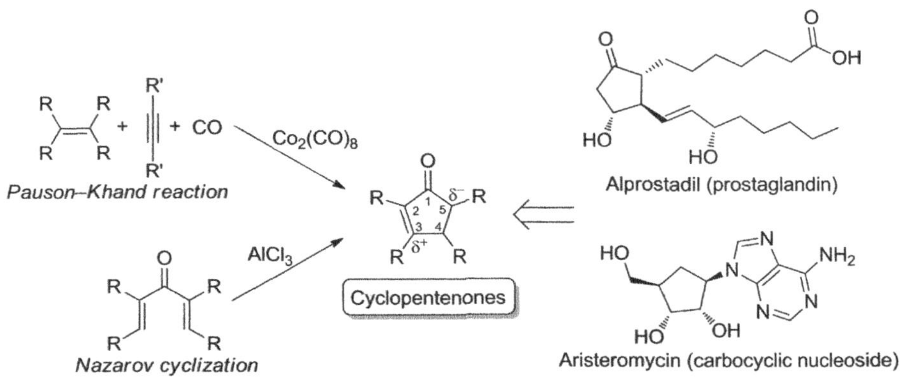

Scheme 14.6.1 Common strategies for the preparation of functionalized cyclopentenones and examples of target molecules synthesized from these rich synthons.

known contact should be avoided. Methanol is toxic in contact with the skin, by inhalation or if swallowed, and is irritant to eyes.

Experimental Procedure

Laboratory Session 1 (4 h): Preparation of Activated Furan

1. Suspend Meldrum's acid (2,2-dimethyl-1,3-dioxane-4,6-dione; 3.3 mmol, 474 mg) in distilled water (9 mL) in a round-bottom flask (50 mL).
2. Add 2-furaldehyde (3 mmol, 0.25 mL) to the suspension.
3. Connect a condenser to the round-bottom flask and stir the resulting mixture at 100 °C using a water bath.
4. Check the reaction completion by TLC (eluent: hexane/ethyl acetate (3 : 1)). Note: for the TLC an aliquot of the reaction mixture should be extracted with ethyl acetate in a small vial and the organic phase spotted on the TLC plate using a capillary.
5. After full conversion of furfural (*ca.* 1.5 h) cool the reaction mixture to room temperature, add dichloromethane (30 mL) and stir the resulting mixture for 10 min.
6. Transfer the solution to a separatory funnel and separate the phases.
7. Wash sequentially the organic phase with a saturated aqueous solution of NaCl (10 mL) and water (10 mL).
8. Dry the organic layer over $MgSO_4$, filter and remove the solvent in the rotary evaporator.
9. Dissolve 10 mg of the compound in deuterated chloroform and acquire ^1H and ^{13}C NMR spectra.

Laboratory Session 2 (2 h): Formation of Stenhouse Adduct

1. Dissolve the activated furan obtained in session 1 (1.35 mmol, 300 mg) in tetrahydrofuran (3 mL) in a 25 mL round-bottom flask.
2. Add diethylamine (1.35 mmol, 0.2 mL) and stir the resulting mixture at room temperature (25 °C) for 10 min followed by cooling at 0 °C for 20 min.
3. Filter the reaction mixture (using a filter paper and gravity) to collect the precipitated solid, and wash it with cold diethyl ether (3 × 30 mL).
4. Allow the solid to dry for a few minutes and then transfer it to a proper recipient.
5. Dissolve 10 mg of the compound in deuterated chloroform and acquire ^1H and ^{13}C NMR spectra.
6. Suspend 150 mg of the prepared Stenhouse adduct in methanol (10 mL) in a 50 mL round-bottom flask and expose the solution to sunlight for several days (consult your instructor for the exact number of days).

Laboratory Session 3 (1 h): Formation of Cyclopentenone

1. Remove the solution from the reaction mixture using a Pasteur pipette and wash the remaining solid with cold diethyl ether (3 × 20 mL) using the Pasteur pipette to remove the diethyl ether from the washings.
2. Dry the solid in the rotary evaporator.
3. Dissolve 10 mg of the compound in deuterated water and acquire ^1H and ^{13}C NMR spectra.

Waste Disposal

Aqueous washings from the extraction step should be diluted with water and disposed of down the sink. All organic solvents waste should be retained and transferred to an appropriate container for disposal.

Results Interpretation and Additional Questions

1. Determine and analyse the reaction yields.
2. Draw out the mechanism of all the transformations.
3. Interpret the ^{1}H NMR spectra obtained and comment on the purity of all the products obtained.
4. The ^{1}H NMR spectrum of activated furan typically shows up to 10% unreacted furfural. Why is the presence of unreacted starting material not a problem going forward in the synthesis?
5. What would be the product of the reaction of furfural and two equivalents of secondary amine under Lewis acid catalysis instead of using the activated furan?
6. Explain the observed colour of the different isolated compounds based on their structure.

References

1. (a) V. B. Kurteva and C. A. M. Afonso, *Chem. Rev.*, 2009, **109**, 6809; (b) M. Iqbal, P. Duffy, P. Evans, G. Cloughley, B. Allan, A. Lledo, X. Verdaguer and A. Riera, *Org. Biomol. Chem.*, 2008, **6**, 4649.
2. S. Liang, A. Haunert and S. Huber-Dirr, Process for preparing cyclopentenone, US EP1422212, 2004.
3. (a) *Name Reactions for Carbocyclic Ring Formations*, ed. J. J. Li, John Wiley & Sons, Inc., 2010; (b) J. J. Li, *Name Reactions: A Collection of Detailed Reaction Mechanisms*, Springer, 2003.
4. G. Piancatelli, M. Dauria and F. Donofrio, *Synthesis*, 1994, 867.
5. S. W. Li and R. A. Batey, *Chem. Commun.*, 2007, 3759.
6. (a) S. Helmy, F. A. Leibfarth, S. Oh, J. E. Poelma, C. J. Hawker and J. R. de Alaniz, *J. Am. Chem. Soc.*, 2014, **136**, 8169; (b) S. Helmy, S. Oh, F. A. Leibfarth, C. J. Hawker and J. R. de Alaniz, *J. Org. Chem.*, 2014, **79**, 11316.
7. F. Bigi, S. Carloni, L. Ferrari, R. Maggi, A. Mazzacani and G. Sartori, *Tetrahedron Lett.*, 2001, **42**, 5203.

14.7. Synthesis of a Chiral Salen. Examples of the Schmidt Rearrangement and Ultrasound Activation

Maria Elisa da Silva Serra and Dina Maria Bairrada Murtinho

Department of Chemistry, University of Coimbra, Coimbra, Portugal
*E-mail: melisa@ci.uc.pt, dmurtinho@ci.uc.pt

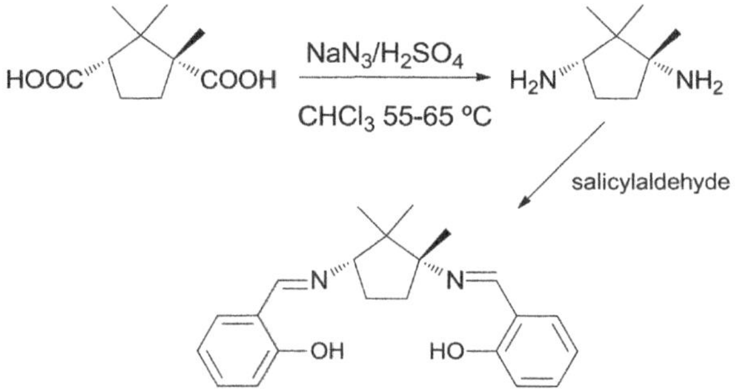

Number of sessions (duration of each session)	Hazard level	Difficulty level	Level of study
3 (2 h + 3 h + 3 h)	High	High	Advanced

Class names Carboxylic acids, amines, imines

Concepts involved This experiment involves the transformation of a carboxylic acid into an amine through the Schmidt reaction using sodium azide and sulfuric acid. The resulting product, a diamine, is subsequently converted into a salen compound by condensation with salicylaldehyde

Chemicals needed Camphoric acid, sodium azide, sulfuric acid (conc.), chloroform, sodium hydroxide, ethanol, silica gel, salicylaldehyde, ethyl acetate, dichloromethane, anhydrous sodium sulfate, deuterochloroform

Equipment and experimental techniques involved Apparatus for reaction with temperature (55–65 °C) and magnetic stirring, two-neck round-bottom flask, liquid–liquid extraction apparatus, ultrasound bath, NMR

Keywords Liquid–liquid extraction, NMR, salen, Schmidt rearrangement, ultrasound

Background

Salen is a general term for diimines that result from the condensation of salicylaldehydes with diamines. The term derives from the condensation product of 1,2-ethylenediamine (en) with salicylaldehyde (sal).[1,2] Salen compounds can easily form stable complexes with a great variety of transition metals and the resulting complexes are widely used in catalytic processes.[3] Of special interest are complexes resulting from chiral salens.

Camphoric acid is a chiral dicarboxylic acid that is obtained from the oxidation of camphor, a chiral natural compound. Due to its abundance, this compound is frequently used as a chiral building block for various purposes.

Comprehensive Organic Chemistry Experiments for the Laboratory Classroom
Edited by Carlos A. M. Afonso, Nuno R. Candeias, Dulce Pereira Simão, Alexandre F. Trindade, Jaime A. S. Coelho, Bin Tan, and Robert Franzén.

Published by the Royal Society of Chemistry, www.rsc.org

Through the Schmidt reaction,[4] camphoric acid is converted into the corresponding diamine. In this reaction, hydrazoic acid adds to carboxylic acids, aldehydes or ketones as well as alcohols and olefins. The most common reaction is between carboxylic acids and sodium azide in the presence of sulfuric acid. This method allows for the synthesis of amines in a one-step reaction with good yields, particularly when the acid is aliphatic. The proposed mechanism for this reaction involves the formation of an acylium ion, with the elimination of water, followed by the attack of hydrazoic acid to the acylium ion formed. Finally, a rearrangement takes place, with simultaneous elimination of N_2. This rearrangement occurs with retention of configuration and thus, in the case of a chiral amine, it has the same absolute configuration as the starting acid.

The diamine, (1*R*,3*S*)-1,3,3-trimethylcyclopentane-1,3-diamine, is prepared according to a previously described procedure,[5] with the introduction of some modifications that simplify execution. Subsequently, the diamine is reacted with two equivalents of salicylaldehyde to give the corresponding salen. This reaction is carried out under ultrasound irradiation using silica gel.[6] This is an alternative to the classical conditions (solvent reflux during several hours), which allows a greener, more efficient reaction in shorter reaction times.

Additional Safety

All sessions should be carried out in a well-ventilated fume hood, using protective clothing, gloves and safety goggles. Concentrated sulfuric acid is corrosive and may cause severe skin burns and eye damage.

Sodium azide, in contact with acid, liberates very toxic gases. Hydrazoic acid is a volatile, toxic and explosive liquid at room temperature and pressure. However, diluted solutions of HN_3 can be safely handled with due precautions. Chloroform is harmful if swallowed, causes skin irritation and is suspected of causing cancer. Avoid prolonged or repeated exposure. Dichloromethane is suspected of causing cancer. Silica gel should be handled with care since the fine particles can be easily inhaled.

Experimental Procedure

Laboratory Session 1 (2 h): The Schmidt Reaction

Warning: hydrazoic acid is a volatile, toxic and explosive liquid at room temperature and pressure. Special care should be taken in the addition of sodium azide.

1. In a fume hood heat an oil bath to a temperature of 55–65 °C on a magnetic hot plate.
2. Prepare a 50 mL two-neck round-bottom flask with a magnetic stirring bar and condenser, and at room temperature, add 1.02 g (5 mmol) of (1*R*,3*S*)-camphoric acid, 3 mL concentrated sulfuric acid and 10 mL chloroform. Start the reaction at 55–65 °C with stirring.
3. Add sodium azide (1 g, 14.3 mmol) in small portions at intervals over 20 min in order to maintain reduced gas evolution.
4. Stir the reaction mixture at the same temperature until gas evolution ceases, usually overnight (18 h).

Laboratory Session 2 (3 h): Diamine Isolation

1. Pour the reaction mixture into a water/ice mixture (10 mL each). With stirring, slowly add solid NaOH pellets in small portions until the pH reaches 14 (add further small portions of ice so that the solution does not overheat).
2. Transfer the solution into a separatory funnel. Extract the product with chloroform (4–5 times 10 mL). Extract the combined organic portions with water and dry over anhydrous Na_2SO_4.

3. Filter the drying agent and evaporate the solvent in a rotary evaporator. Optionally, traces of water may be removed from the resulting diamine by adding toluene and evaporating (azeotropic distillation). The resulting pale yellow oil is used directly in the salen synthesis.
4. Weigh the product and calculate the yield.
5. Place 0.5 g of silica gel in an oven at 200 °C for at least 2 h, for activation.

Laboratory Session 3 (3 h): Synthesis of the Salen Compound

1. Weigh the diamine (1.5 mmol, 0.213 g) into a 25 mL Erlenmeyer flask and add 10 mL of absolute ethanol. To the mixture, add the aldehyde (3 mmol, 0.33 mL) and pre-activated silica (0.5 g).
2. Place the reaction mixture in an ultrasound bath for approximately 15–20 min. Add 15–20 mL of dichloromethane and filter off the silica. Wash the solid with another portion (10 mL) of dichloromethane.
3. Evaporate the solvents. Recrystallize the yellow solid in hot ethyl acetate (smallest quantity required to dissolve the solid). Filter the product.
4. Weigh the salen and calculate the yield. Measure the melting point and record ^{1}H and ^{13}C NMR spectra. Compare your data with the literature data.

Waste Disposal

The neutralized aqueous solutions are collected for subsequent destruction. Chloroform and dichloromethane must be collected in a chlorinated waste collector recipient, and ethyl acetate must be collected in a non-chlorinated waste collector recipient, all for subsequent destruction.

Results, Interpretation and Additional Questions

1. Discuss the mechanism involved in the Schmidt reaction.
2. Calculate the overall yield of the two-step reaction synthesis.
3. Discuss the use of ultrasound radiation and silica gel in the synthesis of the salen compound.
4. Interpret the ^{1}H and ^{13}C NMR spectra of the salen compound and compare the obtained melting point with that reported in the literature.

References

1. K. C. Gupta and A. K. Sutar, *Coord. Chem. Rev.*, 2008, **252**, 1420.
2. A. Zulauf, M. Mellah, X. Hong and E. Schulz, *Dalton Trans.*, 2010, **39**, 6911.
3. P. G. Cozzi, *Chem. Soc. Rev.*, 2004, **33**, 410.
4. R. Bruckner, *Advanced Organic Chemistry: Reaction Mechanisms*, Academic Press, San Diego, 2002, ch. 11, p. 465.
5. Z.-H. Yang, L.-X. Wang, Z.-H. Zhou, Q.-L. Zhou and C.-C. Tang, *Tetrahedron: Asymmetry*, 2001, **12**, 1579.
6. M. E. S. Serra, D. Murtinho, A. Goth, A. M. d'A. R. Gonsalves, P. E. Abreu and A. A. C. C. Pais, *Chirality*, 2010, **22**, 425.

15. Biotransformations

15.1. Bioreduction of *N*-Oxide Moiety

María Laura Lavaggi[a], Mercedes González*[a], and Hugo Cerecetto*[b]

[a]Laboratorio de Química Orgánica, Instituto de Química Biológica, Facultad de Ciencias, Universidad de la República, Iguá 4225, Montevideo 11400, Uruguay; [b]Área de Radiofarmacia, Centro de Investigaciones Nucleares, Facultad de Ciencias, Universidad de la República, Mataojo 2055, Montevideo 11400, Uruguay
*E-mail: megonzal@fq.edu.uy, hcerecetto@cin.edu.uy

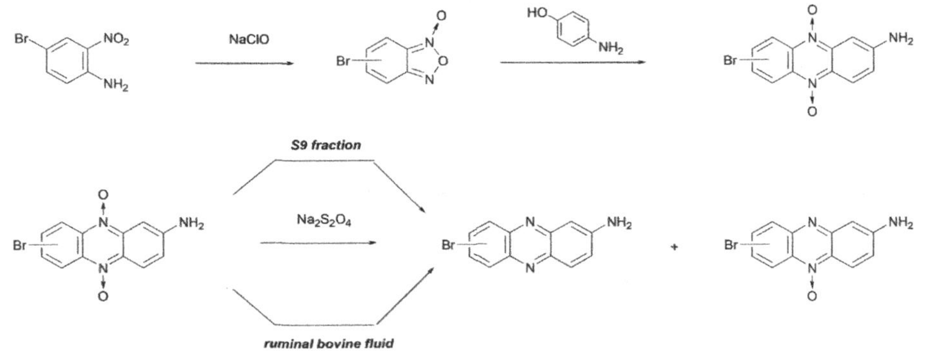

Number of sessions (duration of each session)	Hazard level	Difficulty level	Level of study
5 (2 h + 1 h + 3 h + 4 h + 3 h)	Moderate	High	Advanced

Class names Amino, nitro, *N*-oxide

Concepts involved This experiment involves two methods for *N*-oxide bioreduction, in the phenazine 5,10-dioxide system, to the corresponding amine using different bio-reagents, the S9 fraction and ruminal bovine fluid. Since the monitoring of the biotransformation requires standards of the product of the metabolisms, the first sessions of the experiment involve synthetic procedures to prepare them. Type of reactions in the experiments: oxidation, reduction, biotransformation

Chemicals needed 4-Bromo-2-nitroaniline, sodium hydroxide, ethanol, sodium hypochlorite solution (1–3% active chlorine, solution for domestic household use), sodium, methanol, tetrahydrofuran, *p*-aminophenol, diethyl ether, hydrochloric acid (conc.), sodium dithionite, aqueous saturated solution of sodium carbonate, anhydrous magnesium sulfate, silica gel (230–400 mesh), ethyl acetate, *n*-hexane, hexadeuterated dimethyl sulfoxide or hexadeuterated acetone or tetradeuterated methanol, dimethyl sulfoxide, acetone, phosphate buffer (0.1 M, 1.5 mM EDTA, pH 7.4), aqueous solution of magnesium chloride (1.3 mM in phosphate buffer), nicotinamide adenine dinucleotide phosphate ($NADP^+$) (40 μM in phosphate buffer), glucose-6-phosphate (3.5 mM in phosphate buffer), glucose-6-phosphate dehydrogenase (0.5 U mL^{-1} in phosphate buffer), aqueous solution of potassium chloride (0.9%, w/v), Tris–sucrose buffer (0.05–0.25 M, pH 7.4), McDougall's buffer (sodium bicarbonate, 9.8 g L^{-1}, sodium phosphate, 2.44 g L^{-1}, potassium chloride, 0.57 g L^{-1}, sodium chloride, 0.47 g L^{-1}, magnesium sulfate heptahydrate, 0.12 g L^{-1}, calcium chloride dehydrate, 0.16 g L^{-1})

Comprehensive Organic Chemistry Experiments for the Laboratory Classroom
Edited by Carlos A. M. Afonso, Nuno R. Candeias, Dulce Pereira Simão, Alexandre F. Trindade, Jaime A. S. Coelho, Bin Tan, and Robert Franzén.

Published by the Royal Society of Chemistry, www.rsc.org

Equipment and experimental techniques involved Apparatus for reaction at room temperature/low temperature, magnetic stirring and inert atmosphere, silica gel coated TLC plates, apparatus for vacuum filtration, rotatory evaporator, liquid–liquid extraction apparatus, column for chromatography, chromatography chamber, tubes for NMR spectroscopy, NMR equipment
Keywords Bio-processes, column chromatography, evaporation under reduced pressure, NMR, *N*-oxide, TLC

Background

The reduction, deoxygenation, of *N*-oxide containing aromatic heterocycles to the corresponding heterocycles is an important step in the preparation of new compounds in, for example, the field of medicinal chemistry,[1,2] toxicology,[3,4] environmental sciences,[5] and new materials.[6–8] Several methods have been developed to perform these reductions; however, they often suffer from serious disadvantages, such as incompatibility with other functional groups, low yields, harsh reaction conditions, difficult work-up procedures and environmentally unfriendly methodologies.[9,10]

Specifically, the deoxygenation of pyridine/pyrazine *N*-oxides to the parent heterocycle has been achieved by a large variety of reagents and conditions, such as:[11–15] (i) hydrogenation in the presence of catalysers, *e.g.* Pd; (ii) use of metals in acid milieu, *e.g.* Fe–acetic acid, Zn–NH_4Cl; (iii) use of hydrogen producers, *e.g.* $HCOONH_4$; (iv) use of trivalent phosphorus compounds, *e.g.*, PCl_3, PPh_3, $P(OEt)_3$; (v) use of aqueous sodium dithionite; (vi) using electrochemical procedures.

On the other hand, the deoxygenation of (benzo)furoxan to the corresponding parent heterocycle, (benzo)furazan, has been described only with a small number of reagents. Of them could be mentioned:[16–20] (i) use of trivalent phosphorus compounds, *e.g.* PPh_3; (ii) use of sodium borohydride; (iii) use of Zn in NH_4Cl aqueous solution. The poor aromaticity of the benzofuroxan system impacts, under some reduction conditions, on the product of ring-opening yielding an aromatic system. This is the case of the reaction of benzofuroxans with metals in acid milieu, *e.g.* Cu–HCl,[21] or in the presence of ferrous salts or thiophenol to produce *o*-nitroanilines.[22]

The use of biotransformation, the chemical transformation made by organisms or part of them on a compound, for deoxygenation of *N*-oxides has been scarcely studied. Some examples have described the biotransformation of benzofuroxans in different biosystems, *e.g.* by oxyhemoglobin or by both cytosolic and microsomal fractions from rat liver,[23,24] producing, among others, *o*-nitroanilines or *o*-benzoquinone dioxime, respectively. Similarly, the cytosolic fraction from rat liver was able to deoxygenate phenazine 5,10-dioxide in an O_2- or N_2-purged system.[25] Additionally, the anaerobic microbial community in the bovine rumen was also able to deoxygenate phenazine 5,10-dioxide to produce the corresponding phenazine and to reduce benzofuroxan to yield the corresponding *o*-nitroaniline.[26]

Different procedures to prepare phenazine 5,10-dioxide have been described;[27] in this case the Beirut reaction will be used. This procedure is a heterocycle expansion process, with a suitable benzofuroxan as starting material, where the nucleophile attack results, after atomic rearrangements, in a new diazine system. A further elimination of H_2O aromatizes the new heterocycle.

Because, at room temperature, the substituted-benzofuroxans exhibit isomerization or, as is well known, ring chain tautomerism equilibrium[28] the preparation of them produces a mixture of inseparable positional isomers. For that reason, when the Beirut process uses a substituted-benzofuroxan the final products of reaction are a mixture of phenazine 5,10-dioxide positional isomers, which in general are non-separable by ordinary chromatographic techniques.

Additional Safety

All sessions should be done in a fume hood, with proper safety equipment, such as protective clothing, gloves and safety goggles. 4-Bromo-2-nitroaniline is harmful if swallowed, or in contact with the skin. Sodium hydroxide is a corrosive base and may cause severe skin and eye burns. Sodium hypochlorite solution for domestic household use is a corrosive oxidant that may cause skin and eye allergy. Metallic sodium reacts violently with water. Ethanol, tetrahydrofuran, methanol, diethyl ether, ethyl acetate, *n*-hexane, and acetone are flammable solvents. *n*-Hexane is neurotoxic.

Experimental Procedure

Laboratory Session 1 (2 h): Preparation of 5(6)-Bromobenzofuroxan (Ref. 29)

1. In a fume hood prepare a 250 mL flask with a magnetic stir bar, and add 3.67 g (17 mmol) of 4-bromo-2-nitroaniline, 0.68 g (17 mmol) of sodium hydroxide, and 50 mL of ethanol at room temperature and stir for 5 min.
2. Cool at 0 °C, using an ice-water bath, and add dropwise 60 mL (30 mmol) of sodium hypochlorite solution (1–3%) over 10 min with good stirring.
3. After complete addition check by TLC the advance of the reaction. Stationary phase: SiO_2; mobile phase: *n*-hexane : ethyl acetate (9 : 1); R_f 4-bromo-2-nitroaniline ≈ 0.4 (yellow brilliant spot), R_f 5(6)-bromobenzofuroxan ≈ 0.6 (brown spot).
4. If the reaction is not complete (disappearance of 4-bromo-2-nitroaniline) repeat steps 2 and 3.
5. Remove the flask from the ice-water bath, and allow it to warm to room temperature.
6. Collect the brown-yellow precipitate on a Büchner funnel and wash with 50 mL of water.
7. Dry the solid *in vacuo* on SiO_2 for 24 h.

Laboratory Session 2 (1 h): Preparation of 2-Amino-7(8)-bromophenazine 5,10-Dioxide (Ref. 30)

1. After the drying process, weigh the crude solid and determine the yield.
2. In a fume hood prepare a 50 mL two-necked round-bottomed flask with a magnetic stir bar and a system of inert atmosphere (nitrogen).
3. Prepare a suspension of sodium methoxide, with good stirring and under continuous nitrogen stream, as follows: add 44.9 mg of metallic sodium (1.95 mmol) to 12.5 mL of anhydrous methanol at −5 °C, using an ice-calcium chloride bath.
4. Then add dropwise a mixture of 2.14 g of 5(6)-bromobenzofuroxan (10.0 mmol), 1.09 g of *p*-aminophenol (10.0 mmol), 2.5 mL of anhydrous methanol and 15.0 mL of anhydrous THF, at −5 °C over a period of 15 min with good stirring.
5. Close the necks of the flask with adequate caps and leave at −20 °C for 24 h.

Laboratory Session 3 (3 h): Preparation of 2-Amino-7(8)-bromophenazine and 2-Amino-7(8)-bromophenazine 5-Oxide (Ref. 11)

1. Collect the brown-red precipitate on a Büchner funnel and wash with 2 mL of *n*-hexane, 2 mL of THF, 2 mL of diethyl ether, and dry in air.
2. Weigh the crude solid and determine the yield.
3. Place 15 mg of the solid in an NMR tube and add 500 μL of DMSO-d_6, or acetone-d_6, or CD_3OD.
4. Acquire the ^1H NMR spectrum.

5. In a fume hood prepare a 100 mL round-bottomed flask with a magnetic stir bar and a water condenser.
6. Add 0.76 g of 2-amino-7(8)-bromophenazine 5,10-dioxide (2.5 mmol), 1.74 g of sodium dithionite (10 mmol), 0.5 mL of concentrated hydrochloride acid, and 50 mL of methanol.
7. Insert the water condenser into the round-bottomed flask and clamp the flask in an oil bath. Heat the mixture at reflux for 2 h.
8. Remove the flask from the oil bath and allow it to cool to room temperature.
9. Stir the mixture at room temperature for 24 h.

Laboratory Session 4 (4 h): Isolation of 2-Amino-7(8)-bromophenazine and 2-Amino-7(8)-bromophenazine 5-Oxide

1. Evaporate the solvent *in vacuo*.
2. Treat the residue with 10 mL of an aqueous saturated solution of sodium carbonate and 20 mL of ethyl acetate.
3. Transfer the mixture of aqueous: organic solutions into a separatory funnel. Separate the organic layer and perform two extractions of aqueous solution with 20 mL of ethyl acetate. Remove the water from the combined organic layers with anhydrous $MgSO_4$, filter, and evaporate the solvent *in vacuo*.
4. Apply chromatography to the residue on SiO_2, using a gradient of mobile phase as follows: *n*-hexane : ethyl acetate (8 : 2) to *n*-hexane : ethyl acetate (5 : 5). Check the progress of the separation by TLC. Stationary phase: SiO_2; mobile phase: *n*-hexane : ethyl acetate (5: 5); R_f 2-amino-7(8)-bromophenazine 5,10-dioxide ≈ 0.1 (red spot), R_f 2-amino-7(8)-bromophenazine ≈ 0.5 (orange-yellow spot), R_f 2-amino-7(8)-bromophenazine 5-oxide ≈ 0.3 (orange spot).
5. Collect the fractions with the products of interest ($R_f \approx 0.5$ (orange-yellow spot) and $R_f \approx 0.3$ (orange spot), SiO_2, *n*-hexane : ethyl acetate (5 : 5)) in a tared round-bottomed flask. Evaporate the solvent *in vacuo*.
6. After the evaporation, weigh the round-bottomed flasks containing the products and determine the yields.
7. Place 15 mg of 2-amino-7(8)-bromophenazine in an NMR tube and add 500 μL of DMSO-d_6, or acetone-d_6, or CD_3OD.
8. Acquire the ^1H NMR spectrum.

Laboratory Session 5 (3 h): Bioreduction of *N*-Oxide Moiety (Ref. 25 and 26)

1. In a fume hood prepare a thermostat set to 37 °C and introduce two 10 mL flasks (flasks (A) and (B)). In flask (B) introduce an inert atmosphere (nitrogen).
2. Prepare standards solutions as follows: 10 mg mL^{-1} of 2-amino-7(8)-bromophenazine 5,10-dioxide, 2-amino-7(8)-bromophenazine, and 2-amino-7(8)-bromophenazine 5-oxide in acetone.

Biotransformations Promoted by S9 Fraction

3. In flask (A), at 37 °C, add 1 mL of phosphate buffer (0.1 M, 1.5 mM EDTA, pH 7.4), 10 μL of magnesium chloride (1.3 mM in phosphate buffer), 10 μL of nicotinamide adenine dinucleotide phosphate (40 mM in phosphate buffer), 10 μL of glucose-6-phosphate (3.5 mM in phosphate buffer), and 10 μL of glucose-6-phosphate dehydrogenase (0.5 U mL^{-1} in phosphate buffer).
4. Add 10 μL of 2-amino-7(8)-bromophenazine 5,10-dioxide (40 mM in dimethyl sulfoxide).
5. Add a volume of S9-fraction[31] equivalent to 1 mg of protein per mL to the final volume of the mixture.

6. Incubate the mixture for 60 min at 37 °C.
7. Quench the biotransformation by adding 500 μL of methanol.
8. Extract with ethyl acetate (3 × 500 μL) and evaporate the organic layer to dryness *in vacuo*.
9. Re-suspend the residue in 20 μL ethyl acetate and check the advance of the biotransformation using TLC (SiO_2, *n*-hexane : ethyl acetate (5 : 5), visualizing with and without UV light, 254 nm). Compare the reaction mixture with the prepared standards (2-amino-7(8)-bromophenazine 5,10-dioxide, 2-amino-7(8)-bromophenazine, and 2-amino-7(8)-bromophenazine 5-oxide).

Biotransformations Promoted by Ruminal Bovine Fluid

10. In flask (B), at 37 °C and with an inert atmosphere, add 1 mL of rumen fluid recently collected, and 4 mL of McDougall's buffer (sodium bicarbonate, 9.8 g L^{-1}; sodium phosphate, 2.44 g L^{-1}; potassium chloride, 0.57 g L^{-1}; sodium chloride, 0.47 g L^{-1}; magnesium sulfate heptahydrate, 0.12 g L^{-1}; calcium chloride dehydrate, 0.16 g L^{-1}).
11. Add 10 μL of 2-amino-7(8)-bromophenazine 5,10-dioxide (40 mM in dimethyl sulfoxide).
12. Incubate the mixture for 60 min at 37 °C and gassing with nitrogen.
13. Quench the biotransformation by adding 500 μL of methanol.
14. Extract with ethyl acetate (3 × 500 μL) and evaporate the organic layer to dryness *in vacuo*.
15. Re-suspend the residue in 20 μL ethyl acetate and check the progress of the biotransformation using TLC (SiO_2, *n*-hexane : ethyl acetate (5 : 5), visualizing with and without UV light, 254 nm). Compare the reaction mixture with the prepared standards (2-amino-7(8)-bromophenazine 5,10-dioxide, 2-amino-7(8)-bromophenazine, and 2-amino-7(8)-bromophenazine 5-oxide).

Waste Disposal

The neutralized aqueous solution should be diluted and discarded in the sink. Solvents may be collected in a non-chlorinated waste collector recipient, for posterior destruction.

Results Interpretation and Additional Questions

1. Calculate the yields of the products (5(6)-bromobenzofuroxan, 2-amino-7(8)-bromophenazine 5,10-dioxide, 2-amino-7(8)-bromophenazine, and 2-amino-7(8)-bromophenazine 5-oxide).
2. Rationalize the mechanisms involved in the formation of benzofuroxan and 2-aminophenazine 5,10-dioxide.
3. Interpret and compare the ^1H NMR spectra of 2-amino-7(8)-bromophenazine 5,10-dioxide and 2-amino-7(8)-bromophenazine.
4. Discuss the results of the transformations promoted by biological systems.

References

1. L. Wang, Y.-Y. Zhang, L. Wang, F.-y. Liu, L.-L. Cao, J. Yang, C. Qiao and Y. Ye, *Eur. J. Med. Chem.*, 2014, **80**, 535.
2. U. Kessler, D. Castagnolo, M. Pagano, D. Deodato, M. Bernardini, B. Pilger, C. Ranadheera and M. Botta, *Bioorg. Med. Chem. Lett.*, 2013, **23**, 5575.
3. A. Ray, C. Rentas, G. A. Caldwell and K. A. Caldwell, *Neurosci. Lett.*, 2015, **584**, 23.
4. C. F. McGuigan and X.-F. Li, *Toxicol. In Vitro*, 2014, **28**, 607.
5. W. Zhou, L. Lou, L. Zhu, Z. Li and L. Zhu, *J. Environ. Sci.*, 2012, **24**, 1217.
6. L. Yang, X. Li, Y. Qu, W. Qu, X. Zhang, Y. Hang, H. Ågren and J. Hua, *Sens. Actuators, B*, 2014, **203**, 833.
7. Q. Fan, Y. Liu, M. Xiao, H. Tan, Y. Wang, W. Su, D. Yu, R. Yang and W. Zhu, *Org. Electron.*, 2014, **15**, 3375.

8. K. Imai, S. Uzu, S. Kanda and W. R. G. Baeyens, *Anal. Chim. Acta*, 1994, **290**, 3.
9. P. M. Reis and B. Royo, *Tetrahedron Lett.*, 2009, **50**, 949.
10. M. Toganoh, K. Fujino, S. Ikeda and H. Furuta, *Tetrahedron Lett.*, 2008, **49**, 1488.
11. M. J. Haddadin, G. E. Zahr, T. N. Rawdah, N. C. Chelhot and C. H. Issidorides, *Tetrahedron*, 1974, **30**, 659.
12. Y. Aoyagi, T. Abe and A. Ohta, *Synthesis*, 1997, **8**, 891.
13. R. Balicki and G. Maciejewski, *Synth. Commun.*, 2002, **32**, 1681.
14. S. Chandrasekhar, Ch. R. Reddy, R. J. Rao and J. M. Rao, *Synlett*, 2002, **2**, 349.
15. A. R. Katritzky, C. A. Ramsden, J. A. Joulea and V. V. Zhdankin, Reactivity of six-membered rings, in *Handbook of heterocyclic chemistry*, Elsevier, Oxford, 3rd edn, 2010, pp. 242–382.
16. E. Howard and W. F. Olszewski, *J. Am. Chem. Soc.*, 1959, **81**, 1483.
17. J. H. Boyer and S. E. Ellzey, *J. Am. Chem. Soc.*, 1960, **82**, 2525.
18. J. H. Boyer and S. E. Ellzey, *J. Org. Chem.*, 1961, **26**, 4684.
19. M. Boiani, H. Cerecetto, M. González, M. Risso, C. Olea-Azar, O. E. Piro, E. E. Castellano, A. López de Ceráin, O. Ezpeleta and A. Monge-Vega, *Eur. J. Med. Chem.*, 2001, **36**, 771.
20. D. Castro, L. Boiani, D. Benitez, P. Hernández, A. Merlino, C. Gil, C. Olea-Azar, M. González, H. Cerecetto and W. Porcal, *Eur. J. Med. Chem.*, 2009, **44**, 5055.
21. A. R. Katritzky, C. A. Ramsden, J. A. Joulea and V. V. Zhdankin, Reactivity of five-membered rings with two or more heteroatoms, in *Handbook of heterocyclic chemistry*, Elsevier, Oxford, 3rd edn, 2010, pp. 473–602.
22. A. M. Gasco, C. Medana and A. Gasco, *Synth. Commun.*, 1994, **24**, 2707.
23. C. Medana, S. Visentin, G. Grosa, R. Fruttero and A. Gasco, *Farmaco*, 2001, **56**, 799.
24. G. Grosa, U. Galli, B. Rolando, R. Fruttero, G. Gervasio and A. Gasco, *Xenobiotica*, 2004, **34**, 345.
25. M. L. Lavaggi, M. Cabrera, M. González and H. Cerecetto, *Chem. Res. Toxicol.*, 2008, **21**, 1900.
26. V. Cerecetto, F. Diaz-Viraqué, I. Irazoqui, A. Rodríguez, C. Cajarville, J. L. Repetto, M. L. Lavaggi, M. González and H. Cerecetto, *Rev. Virtual Quim.*, 2013, **5**, 1134.
27. M. González, H. Cerecetto and A. Monge, *Top. Heterocycl. Chem.*, 2007, **11**, 179.
28. H. Cerecetto and M. González, *Top. Heterocycl. Chem.*, 2007, **10**, 265.
29. With modifications from: F. B. Mallory, *Org. Synth., Coll.*, 1963, **4**, 74.
30. H. Cerecetto, M. González, M. L. Lavaggi and W. Porcal, *J. Braz. Chem. Soc.*, 2005, **16**, 1290.
31. From ThermoFisher Scientific: https://www.thermofisher.com/uy/en/home/industrial/pharma-biopharma/drug-discovery-development/microsomes-s9-fractions-cytosol.html.

16. Polymerization Reactions

16.1. Preparation of Nylon 6,6 by Interfacial Polymerization

João P. Telo

Centro de Química Estrutural, Instituto Superior Técnico, Universidade de Lisboa, Av. Rovisco Pais, 1649-003 Lisboa, Portugal
*E-mail: jptelo@tecnico.ulisboa.pt

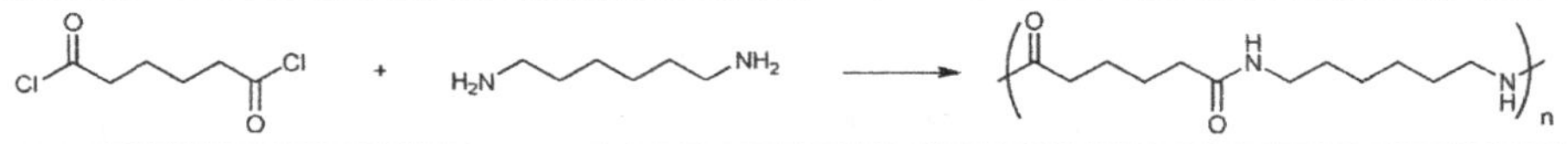

Number of sessions (duration of each session)	Hazard level	Difficulty level	Level of study
1 (2 h)	Moderate	Low	Introductory
Class names Carboxylic acid derivatives, amides, polymers			
Concepts involved Nucleophilic substitution at the carbonyl; polymerization			
Chemicals needed Adipoyl chloride, 1,6-diaminohexane, dichloromethane, sodium carbonate			
Equipment and experimental techniques involved One 250 mL beaker			
Keywords Adipic acid derivatives, interfacial polymerization, nylon 6,6, polymers			

Background

Nylon is a generic designation of a series of synthetic condensation polymers of the polyamide type. They were first introduced by the DuPont Company in 1939 as a textile to produce mainly women's stockings, but soon the whole production was used to produce parachutes to the USA war effort, replacing Asian silk for this purpose at a time when the producing areas of this natural fiber were occupied by enemy forces.

Nylon 6,6 is synthesized industrially by condensation of hexamethylenediamine and adipic acid, each containing six carbon atoms, which give nylon 6,6 its name. The two reactants are combined with water in a reactor, which produces nylon salt. The nylon salt is then heated to evaporate the water and continue the polymerization process, yielding molten nylon 6,6. In 2010 worldwide production was 2 million tons.[1] Nylon 6,6 is a copolymer, *i.e.*, it is made from two different monomers. Other nylons are homopolymers, made from only one monomer. One example is nylon 6, made industrially from the ring-opening polymerization of ε-caprolactam.[2] Most of the world's production of nylons is either nylon 6 or nylon 6,6.

In the experiment described here nylon 6,6 is produced by interfacial polymerization of hexamethylenediamine and adipoyl chloride, producing a collapsed strip of polymer.[3] The experiment is a very popular demonstration for elementary students known as "the nylon rope

Comprehensive Organic Chemistry Experiments for the Laboratory Classroom
Edited by Carlos A. M. Afonso, Nuno R. Candeias, Dulce Pereira Simão, Alexandre F. Trindade, Jaime A. S. Coelho, Bin Tan, and Robert Franzén.

Published by the Royal Society of Chemistry, www.rsc.org

trick". Adipoyl chloride is dissolved in an organic solvent immiscible in water (dichloromethane in this case) and the diamine in the aqueous phase, producing the polymer in the interface of the two solvents. As in most condensation polymers, the reaction mechanism involves a reaction between a dicarboxylic acid derivative and a diamine.

Additional Safety

Acid chlorides are toxic and corrosive and the use of gloves and safety goggles is highly recommended. Dichloromethane is only slightly toxic and may be replaced by the less toxic 1,1,1-trichloroethane. The experiment should be done in a fume cupboard.

Experimental Procedure

Laboratory Session 1 (2 h): Preparation of Nylon 6,6

1. Dissolve 3 mL of adipoyl chloride in 100 mL of dichloromethane in a 250 mL beaker.
2. Add carefully over this without stirring a solution of 4.4 g of hexamethylenediamine and 5 g of anhydrous sodium carbonate in 50 mL of water.
3. Take hold of the film at the interface with tweezers and pull the film continuously and evenly. A large beaker can be used to roll the nylon thread while pulling.
4. Wash the polymer with running water and allow it to dry in the fume cupboard.

Waste Disposal

Shake the two phases to mix the reactants. Remove the lump of nylon with tweezers, squeeze it to the beaker to remove most of the solvents, rinse well with water, and dispose of it in a solid waste container. Separate the lower dichloromethane phase and dispose of it in the "chlorinated solvents" waste container. The aqueous solution should be diluted and discharged in the sink.

Results Interpretation and Additional Questions

1. Write out the reaction mechanism.
2. Nylon 6,6 is a synthetic polyamide. Which natural polymers are also polyamides?
3. Why are nylons and polyamides in general so strong?

References

1. PCI Nylon, 2011, Polyamide 66, http://pcinylon.com/index.php/markets-covered/polyamide-66.
2. PCI Nylon, 2011, Polyamide 6, http://pcinylon.com/index.php/markets-covered/polyamide-6.
3. P. W. Morgan and S. L. Kwolek, *J. Chem. Educ.*, 1959, **36**, 182.

16.2. Synthesis of Copolymer from Acrylamide (AA), 2-Acrylamido-2-methylpropane-sulfonic Acid (AMPS) and *N,N*-Methyleno-bis-acrylamide (BA)

Fátima Coelho*[a] and Dulce Simão[b]

[a]Department of Chemistry, Instituto Superior Técnico, University of Lisbon, CERENA-Centro de Recursos Naturais e Ambiente, Av. Rovisco Pais, 1, 1049-001 Lisboa, Portugal; [b]Department of Chemistry, Instituto Superior Técnico, University of Lisbon, Centro de Química Estrutural, Av. Rovisco Pais, 1, 1049-001 Lisboa, Portugal. *E-mail: fatimacoelho@tecnico.ulisboa.pt

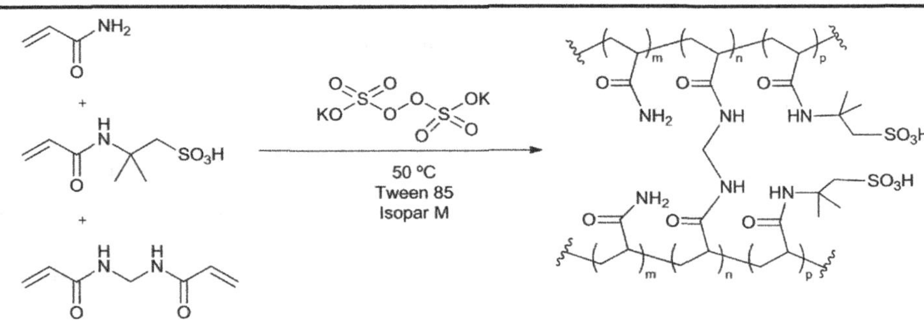

Number of sessions (duration of each session)	Hazard level	Difficulty level	Level of study
1 (4 h)	Low	Low	Intermediate

Class names Alkene, persulfate

Concepts involved Radical polymerization, crosslinker, swelling, superabsorbent polymers, kinetic behaviour, polymerization techniques

Chemicals needed Potassium persulfate, acrylamide (AA), 2 acrylamido-2-methylpropane-sulfonic acid (AMPS), and the crosslinker *N,N*-methyleno-bis-acrylamide (BA); Tween 85 (polyoxyethylene sorbitan trioleate), Isopar M (isoparaffinic hydrocarbon solvents), sodium hydroxide

Equipment and experimental techniques involved Precipitation, FT-IR spectroscopy, mechanical stirring, reflux condenser, thermometer, dropping funnel and a gas inlet tube; free radical polymerization by solution and inverse emulsion

Keywords Crosslinker, FT-IR spectroscopy, polymerization techniques precipitation, radical polymerization, superabsorbent polymers, swelling

Background

Polymers have been used in various industries since the nineteenth century and their properties have been successfully exploited to produce new materials with many applications. A new class of polymeric materials has been labelled "superabsorbent" as they can absorb considerable amounts of fluid without dissolving. This important property is the fundamental reason for the

Comprehensive Organic Chemistry Experiments for the Laboratory Classroom
Edited by Carlos A. M. Afonso, Nuno R. Candeias, Dulce Pereira Simão, Alexandre F. Trindade, Jaime A. S. Coelho, Bin Tan, and Robert Franzén.

Published by the Royal Society of Chemistry, www.rsc.org

attention they have attracted from researchers.[1,2] They are applied in a wide range of fields such as personal care products, agriculture, biomaterials and wastewater treatment.[3,4] This swelling ability depends on the structure of the polymer network and is the result of the monomers used and the production process followed. In fact, the selection of a set of experimental parameters for the synthesis process results in a different absorption capacity.[4] A superabsorbent polymer must have an ionic character, *i.e.*, contain hydrophilic groups such as –OH, –COOH, $-CONH_2$ and $-SO_3H$, which are highly compatible with water and play an important role in the overall effectiveness of the swelling process. Free-radical polymerization by solution or inverse emulsion is the method frequently used for the preparation of superabsorbents. The radical polymerization is a chain reaction that involves the following steps: initiation, propagation and termination. Chain transfer reactions or inhibition may also be involved. The initiation step begins when a molecule of an initiator, *e.g.* potassium persulfate $K_2(SO_4)_2$, decomposes while heat is applied, into two free radicals (Scheme 16.2.1).

This radical with an unpaired electron is extremely reactive and in the presence of the monomer with carbon–carbon double bonds the free radical takes an electron from the double bonds of the monomer, and leaves this molecule with an unpaired electron and a new be active center appears at the end of the chain as shown, for example, in Scheme 16.2.2 with AMPS.

The propagation step then follows and consists of successive unpaired electron molecules adding other monomer molecules, each addition creating a new reactive center (Scheme 16.2.3).

Propagation continues until the growing chain radical is deactivated by a termination process. The termination step occurs with the combination of a polymer chain radical with the initiator radical or another macro radical (Scheme 16.2.4).

Additional Safety

Acrylamide has been shown to be a skin irritant, particularly when used in its solid form. A solution of the monomer is therefore recommended. Although the risks are significantly reduced, the use of protective gloves is still recommended and that the polymerization be carried out in

$$S_2O_8^{2\ominus} \xrightarrow{\Delta} 2SO_4^{\bullet\ominus}$$

Scheme 16.2.1 Initiation step.

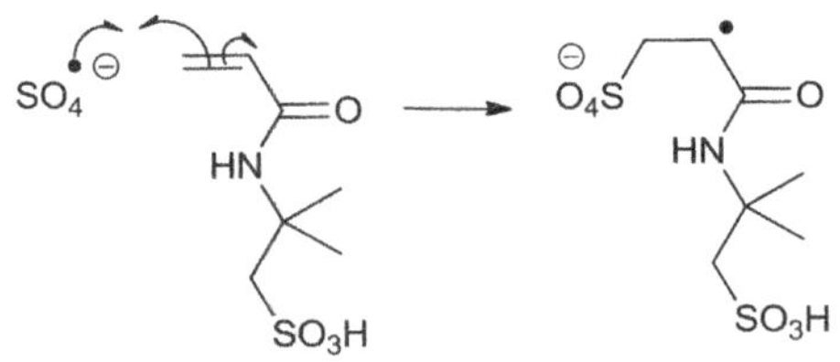

Scheme 16.2.2 Initiation reaction of AMPS with production of initiation radical.

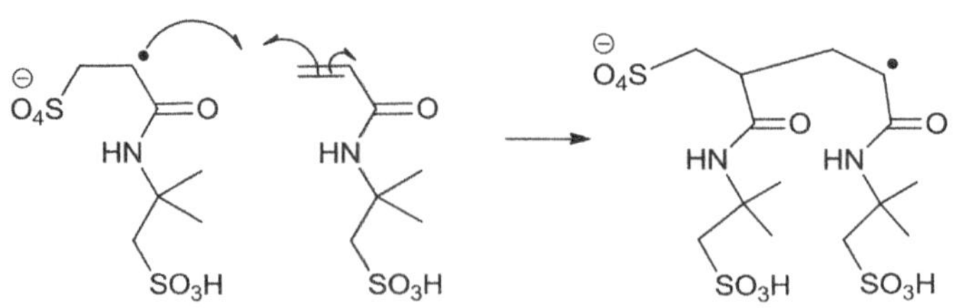

Scheme 16.2.3 Propagation reaction of the polymerization.

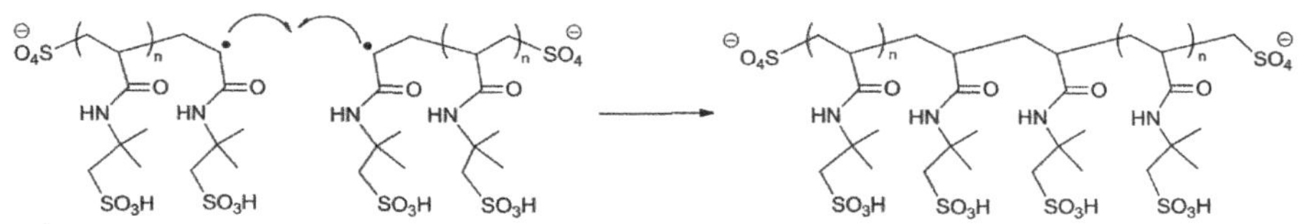

Scheme 16.2.4 Termination reaction by combination of free radicals.

a fume hood. The experiment should be carried out by students provided that they are aware of the risks involved in handling chemicals.

Experimental Procedure

Preparation of the Copolymer AA/AMPS/BA by Free Radical Solution Polymerization Technique

1. In a ventilated fume hood dissolve 9.30 g of AMPS and 0.09 g of BA in 150 mL of distilled water, add 14.70 g of monomer AA (50 wt% solution in water), and adjust the monomeric solution's pH to 10 by slow addition and stirring of aqueous sodium hydroxide 5%.
2. Put this mix in the glass reactor, insert the mechanical stirring apparatus into the centre of the glass reactor, attach a water-cooled condenser, a thermometer, a pressure equalizing dropping funnel and a nitrogen line.
3. Prepare a solution of potassium persulfate ($K_2S_2O_8$) by dissolving 0.0136 g in a volumetric flask (50 mL) and fill the pressure equalizing dropping funnel with 5 mL of this solution.
4. Degas this mixture with a slow nitrogen feed for about 30 min, with constant stirring at 260 rpm, to eliminate the dissolved oxygen.
5. Heat the monomer solution in an oil bath at 50 °C.
6. When the mixture reaches the desired temperature, add the initiator solution of potassium persulfate, dropwise, while stirring.
7. After 1 h 30 min of reaction, with nitrogen flow, under stirring and heat, withdraw the reactor from the oil bath, and cool it to room temperature.
8. Put 25% of the reaction mixture in a 50 mL beaker and add acetone (20 mL), mix well. The copolymer precipitates in the beaker.
9. Repeat this process three more times, until all copolymer is recovered.
10. Finally dry the copolymer in an oven at 60 °C for 10 h. The dried product can be ground to the desired particle size.

Preparation of the Copolymer AA/AMPS/BA by Free Radical Inverse Emulsion Polymerization Technique

In the inverse emulsion polymerization technique, the reaction is carried out in the same apparatus as mentioned above.

1. Fill the glass reactor with the organic phase, 200 g Isopar M (from Exxon Mobil Corp., Fairfax, VA, USA), as organic solvent, and 1.6 g Tween 85 (from Sigma-Aldrich, Lisbon, Portugal) as emulsifier with continuous stirring.
2. Neutralize a mix of the monomers, 2.33 g AMPS, 3.74 g AA (50 wt% solution in water) and 0.021 g BA, with 10 mL of sodium hydroxide (5 wt% in water).
3. Add the neutralized monomer mix, dropwise, into the organic phase (at room temperature) with stirring at 260 rpm and bubbling nitrogen gas for about 30 min.
4. After that, raise the temperature to 50 °C and initiate the polymerization adding by 5 mL of potassium persulfate solution.

5. Allow the reaction to proceed for 1 h 30 min at bath temperature, under a nitrogen atmosphere and with stirring.
6. Finally withdraw the reactor from the oil bath, and cool it to room temperature.
7. Put 25% of the reaction mixture in a 50 mL beaker and add acetone (20 mL), mix well. The copolymer precipitates in the beaker.
8. Repeat this process three more times, until all copolymer is recovered.
9. Finally dry the copolymer in an oven at 60 °C for 10 h. The dried product can be ground to the desired particle size.
10. Prepare a sample of the dry copolymer for FT-IR spectroscopy (KBr). Scan the FT-IR spectra in the range 3500–500 cm^{-1}.

Swelling Measurements

Prepare a sample and determine the swelling of the polymers. The conventional method by which to measure the absorption of the polymer is the tea bag method. Place a certain amount, *e.g.* 0.01 g, of dried superabsorbent into a tea bag, immerse in a beaker with 100 mL of deionized water until equilibrium swelling is reached. Weigh the tea bag (removing the excess of solution) at successive time instants after immersion in water until the maximum value is reached, *i.e.*, the equilibrium value. From these weight measurements, the fraction swelling, $s(t)$, is calculated by eqn (16.2.1):

$$s(t) = \frac{m(t) - m(0)}{m(\infty) - m(0)} = \frac{m(t) - m_{\mathrm{dry}}}{m_{\mathrm{eq.}} - m_{\mathrm{dry}}} \tag{16.2.1}$$

where t = time, $m(t)$ = mass of the swollen superabsorbent (polymer and water) at t. The value of $m(\infty)$ is the final or equilibrium mass, or $m_{\mathrm{eq.}}$, and $m(0)$, or m_{dry}, is the initial mass of (dry) superabsorbent. In addition, from eqn (16.2.2) the equilibrium swelling, S, is calculated:

$$S = [m(\infty) - m_{\mathrm{dry}}]/m_{\mathrm{dry}} \tag{16.2.2}$$

Waste Disposal

In accordance with the rules regarding waste disposal, there are protocols that must be obeyed. In this experiment organic solvent is used and this compound should hence be placed in an appropriate container.

Results Interpretation and Additional Questions

1. Identify the absorption bands in the polymer's FT-IR spectra.
2. Comment on the disappearance of the absorption band corresponding to the C=C group in the FT-IR spectra of the synthesized copolymers.
3. Determine the equilibrium swelling of the polymers and comment on the different values obtained according to the polymerization method used.

References

1. L. Brannon-Peppas and R. S. Harland, *Absorbent polymer technology*, Elsevier, New York, 1990.
2. M. A. Molina, C. R. Rivarola and C. A. Barbero, *Polymer*, 2012, **53**(2), 445–453.
3. F. Masuda, *Chem. Econ. Eng. Rev.*, 1983, **15**, 19–22.
4. F. Rosa, J. Bordado and M. Casquilho, *Polymer*, 2002, **43**(1), 63–70.

16.3. Multi-Step Synthesis of Nylon from Cyclohexene

Marisa G. Weaver and Morgan J. Gainer*

University of California at Santa Barbara, Department of Chemistry & Biochemistry, 9510, Santa Barbara, California, 93106-9510, USA. *E-mail: gainer@chem.ucsb.edu

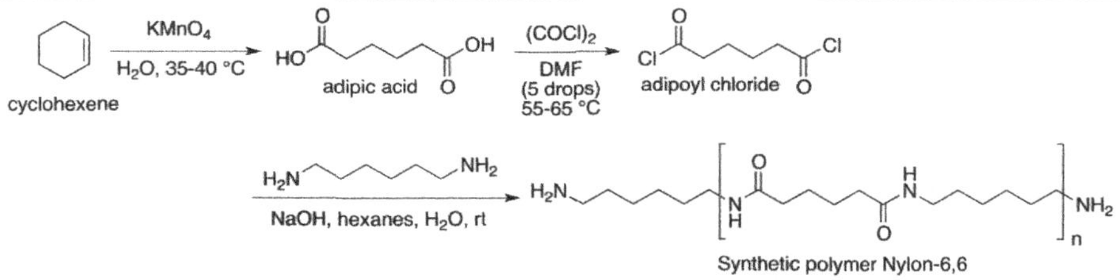

Number of sessions (duration of each session)	Hazard level	Difficulty level	Level of study
2 (4 h each)	Moderate	Medium	Intermediate

Class names Olefins, carboxylic acid, acid chloride, amine, polyamide

Concepts involved This experiment involves a three-step reaction sequence to transform cyclohexene into a nylon-6,6 polymer using oxidative cleavage, acid chloride formation and step growth polymerization

Chemicals needed Cyclohexene, potassium permanganate, concentrated hydrochloric acid, oxalyl chloride, dimethylformamide, methanol, hexamethylenediamine, sodium hydroxide, hexanes, water, phenolphthalein (optional)

Equipment and experimental techniques involved Ice bath, heating set-up for 35–45 °C and 55–65 °C, large mouth Büchner funnel with Erlenmeyer vacuum flask, round bottom flask fitted with an air condenser and a drying tube, copper wire hooks

Keywords Acid chloride formation, IR, oxidation, oxidative cleavage, polymerization, recrystallization

Background

Chemical transformations in organic chemistry provide the foundation for new materials to be made and applied to new products. Nylon-6,6 is one of these materials and was discovered in the late 1930s in a United States DuPont facility by Wallace Caruthers' research group. Since then, many types of durable nylon have been prepared and used in products that require tolerance of mechanical and thermal stress. One method of polyamide formation is *step growth polymerization*. Step growth polymerization chains can grow by addition of monomers (one unit), dimers (two monomer units) or oligomers (multiple monomer units). This is possible because the monomers are bifunctional and have a reactive functional group at each end of the polymer unit. There are many different types of nylon and their names allow one to identify what the monomers are fairly easily. If a nylon (polyamide) is only made from one monomer (amino acid), the name nylon will only be followed by one number (*e.g.* if the monomer contains six

Comprehensive Organic Chemistry Experiments for the Laboratory Classroom
Edited by Carlos A. M. Afonso, Nuno R. Candeias, Dulce Pereira Simão, Alexandre F. Trindade, Jaime A. S. Coelho, Bin Tan, and Robert Franzén.

Published by the Royal Society of Chemistry, www.rsc.org

carbons: "nylon 6"). Those nylons made from two types of monomer systems have two numbers following and denote how many carbons are in the diamine and in the diacid (or diacid chloride) respectively (*e.g.* nylon-6,6).

In this multi-step experiment three common chemical transformations from organic chemistry are used to form nylon-6,6.[1] The process will begin by performing an oxidative cleavage of the commercial reagent cyclohexene to form a linear carbon chain that has a carboxylic acid on each end of the chain. Carboxylic acids are very poor electrophiles and are fairly unreactive towards polymerization, so they must be transformed into more reactive acid chlorides prior to polymerization. The new species is a much more reactive electrophile that allows the polymerization reaction to take place at a lower temperature. The polymerization of hexamethylenediamine with adipoyl chloride is done at room temperature and low concentration using a *step growth unstirred interfacial polymerization*. The formed polyamide is continuously pulled away at the solution interface until one of the monomers is consumed. As a result of diffusion there is no need to induce mixing or ensure correct stoichiometry in this experiment. The mechanisms associated with this reaction sequence are illustrated in Scheme 16.3.1.

The oxidative cleavage by potassium permanganate begins with cycloaddition with the cyclohexene olefin followed by a cycloreversion to afford the linear adipaldehyde (Scheme 16.3.1(I)). The terminal aldehydes are then able to form their corresponding hydrates (under aqueous, sometimes slightly acidic conditions) to allow another permanganate oxidation and form the corresponding carboxylic acids (adipic acid) (Scheme 16.3.1(II)). Adipic acid is converted into the more reactive adipoyl chloride by the Vilsmeier reagent (Scheme 16.3.1(III)). It is formed from a catalytic amount of dimethylformamide (DMF) solvent and oxalyl chloride. The carboxylic acid then undergoes a series of steps to form the acid chloride and restore the DMF catalyst (Scheme 16.3.1(IV)). The hexamethylenediamine nucleophile attacks the more reactive acid chloride to form a tetrahedral intermediate (Scheme 16.3.1(V)). The collapse of this intermediate releases chloride and forms the new amide bond.

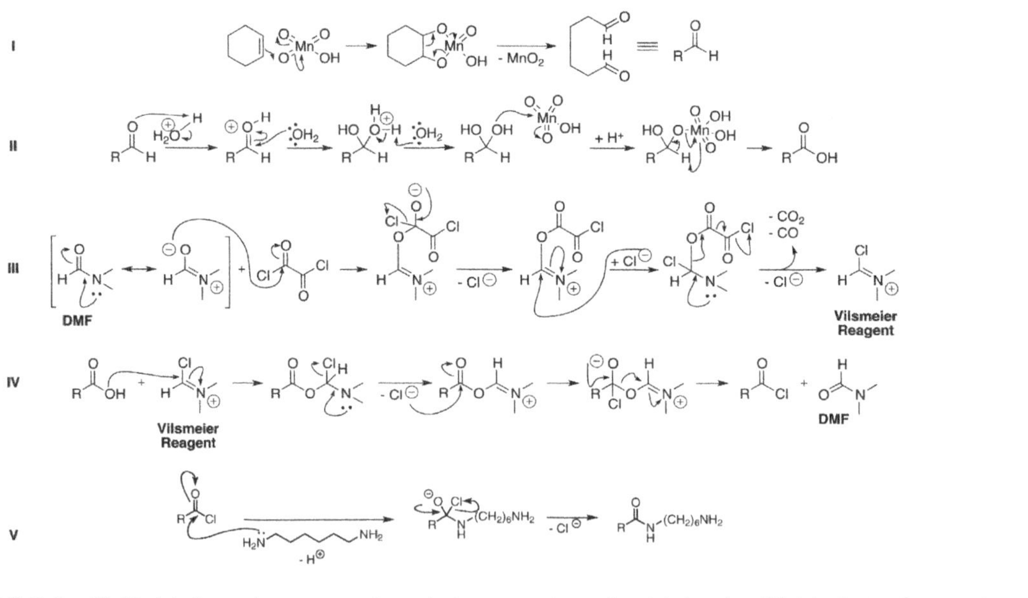

Scheme 16.3.1 (I) Oxidative cleavage of cyclohexene to adipaldehyde. (II) Hydrate formation and oxidation to adipic acid. (III) Vilsmeier reagent formation. (IV) Acid chloride formation. (V) Single polymerization step.

Additional Safety

All reactions should be performed in a fume hood, with proper personal protective equipment such as laboratory coats, gloves and goggles. The oxidation produces a large amount of MnO_2 waste and should be handled carefully as outlined in the procedure. To form the acid chloride, oxalyl chloride is used. Since it is fairly reactive both students and instructor should be more alert when this step is being performed. Glassware from this step should be rinsed with isopropanol, methanol, and then water. Avoid skin contact with any of the chemicals. Hydrochloric acid and sodium hydroxide are both caustic.

Experimental Procedure

Laboratory Session 1 (4 h): Oxidative Cleavage of Cyclohexene (Ref. 2)

1. Prepare an ice bath that is easily accessible for your Erlenmeyer flask (Scheme 16.3.2).
2. Pour 10 mL of H_2O into a 250 mL beaker and mark the meniscus using a permanent marker. Empty the contents and put the beaker aside.
3. Begin heating a hot plate to 40 °C. Monitor the hotplate temperature with a thermometer in a water bath (continuously stirred for even temperature) or a heating block.
4. To a 250 mL Erlenmeyer flask, add 50 mL of water, 2.0 mL of cyclohexene and 8.4 g of potassium permanganate. Stopper the flask *loosely* and begin heating with swirling for 20 min. Monitor the temperature inside the Erlenmeyer flask and ensure it stays between 35 and 45 °C using the hot plate and ice bath.
5. Check the reaction progress by swirling a glass rod in your reaction mixture then touching it to a piece of filter paper. If permanganate is still present, the spot will appear brown (manganese dioxide, oxidant consumed) with a purple ring (unreacted $KMnO_4$) around it. If there is still permanganate present, continue heating and swirling for a further 5 min, then check again.
6. If no progress occurs, add 2–3 mL of methanol then heat (35–45 °C) and check the reaction progress. Repeat steps 5 and 6 until all of the residual purple color from the permanganate disappears.
7. Vacuum filter your reaction mixture through a large Büchner funnel lined with a coffee filter and a properly sized piece of filter paper into a clean filter flask (the coffee filter will contain the MnO_2).
8. Heat 25 mL of 1% sodium hydroxide solution on a hot plate (40–50 °C). Rinse the reaction flask with 10 mL and pour the solution through the filter flask. Repeat again with the remainder of the liquid.
9. Transfer the filtrate and washings into your pre-marked 250 mL beaker. Add a boiling chip or a boiling stick and boil off the solvent until the solution reaches the 10 mL line (light honey syrup color).
10. Cool the solution in your flask on the bench then in your ice-water bath.

Step 1

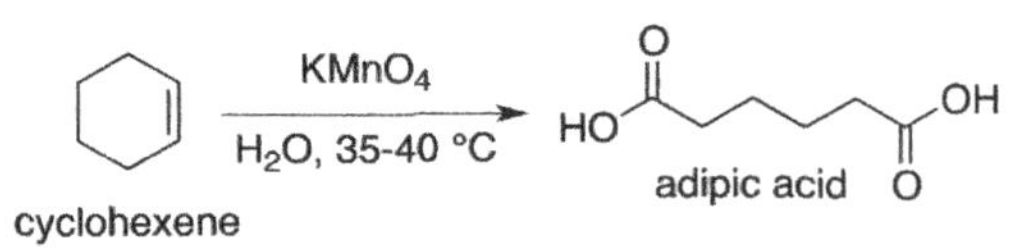

Scheme 16.3.2 Oxidative cleavage of cyclohexene to adipic acid using potassium permanganate.

11. While still cooling, acidify the solution to pH 1 (use pH paper to monitor) by adding concentrated hydrochloric acid dropwise while swirling. Then add an additional 3 mL of concentrated HCl.
12. Allow for crystallization of your product in the ice bath, then collect the solid white adipic acid product using a clean Büchner funnel vacuum filtration set-up and allow it to dry under vacuum.
13. Weigh your crude product and obtain a melting point. If the melting point is not acceptable, perform a recrystallization from <5 mL of water.
14. Collect the pure product and store it safely open to air to dry until the next laboratory period. During the next laboratory session, use the dry product to measure the mass, calculate the percent crude and pure yields and obtain an IR spectrum.

Laboratory Session 2 (4 h): Formation of Adipoyl Chloride (ref. 3) and Polymerization (ref. 4)

Step 2

1. Gently flame-dry a 25 mL round-bottom flask and allow it to cool on a paper towel (Scheme 16.3.3).
2. Prepare a drying tube and start heating a water bath on your hot plate (55–65 °C). Monitor the temperature using a thermometer (ensure that the thermometer does not contact the side of the bath).
3. Weigh your isolated pure adipic acid (step 1) and add stock supply as needed for a starting mass of 1.0 g.
4. In the fume hood, add the adipic acid and a stir bar to the dry 25 mL round-bottom flask, then clamp it so it is stationary.
5. Slowly add oxalyl chloride (2.0 equiv.) followed by 5–10 drops of dimethylformamide (DMF).
6. Immediately attach the air condenser and a calcium chloride drying tube. Vigorous bubbling or fuming generally occurs shortly after DMF addition. Immediately heat this reaction mixture in the hood in the prepared water bath (55–65 °C). When it begins to foam, remove it from the heat, but continue stirring. Return to heat upon stabilization and repeat this process for 5 min total time. After 5 min, proceed even if an un-dissolved white solid or a yellow color appears.
7. Remove the reaction flask from the heat and cool the reaction mixture to room temperature.
8. The flask will mostly contain an oil, but there may be a solid residue. The crude adipoyl chloride product is the oil and is quite soluble in hexanes. The crude adipoyl chloride will be isolated from this reaction flask as a solution in hexanes (step 3.3) and will be used to form the nylon-6,6 polyamide. *Only* add what is soluble in the hexanes solvent as instructed for the polymerization.
9. Calculate the theoretical yield of adipoyl chloride formed (hint: use mmol of adipic acid).

Step 2

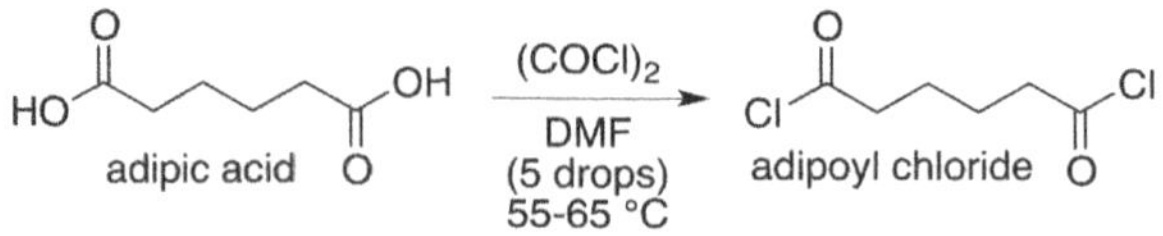

Scheme 16.3.3 Formation of adipoyl chloride from adipic acid using oxalyl chloride.

Step 3

1. Add 10 mL of a 5% hexamethylenediamine/5% NaOH solution (aqueous) to a 50 mL beaker (Scheme 16.3.4).
2. Measure the pH of this solution and ensure that it is greater than pH 7. If it is more acidic, add 20% sodium hydroxide solution until the pH measures >7 (add phenolphthalein for fuschia colored nylon).
3. Dissolve the adipoyl chloride in 20 mL of hexanes (or cyclohexane) to make a 5% solution. Slowly and steadily pour *all* of this solution on the side of the beaker to form *two* phases. Do not agitate.
4. Form a hook on the end of a 15 cm piece of copper wire. Gently scrape the sides of the beaker and slowly move the acquired mass to the center of the solution interface.
5. Slowly and steadily begin pulling the film from the beaker interface to create a long rope of nylon polyamide until no more polymer forms. Rinse the polymer product with water over a waste beaker.
6. Use paper towels to thoroughly dry the nylon product. Obtain the mass of the dried polymer, determine a percent yield of the polymer formed (two steps) and measure the length of your longest nylon strand.
7. Perform a stretch test. Cut a 25 cm piece of nylon from your strand and slowly pull your nylon along the side of a yard stick until it breaks. At the breaking point, record how far the strand stretched.

Waste Disposal

Waste water, hydrochloric acid and sodium hydroxide solutions should be collected in neutral, acidic and basic aqueous waste containers, respectively. Manganese dioxide produced in the oxidation step and nylon polymer should be disposed of in a solid waste container. Organic chemicals, compounds and solvents should be neutralized and disposed of in an organic waste container.

Results Interpretation and Additional Questions

Step 1: Oxidation

1. Why is methanol added if the reaction contains unreacted $KMnO_4$? (Think about the mechanism.)
2. What formed from the reduction of $KMnO_4$?
3. Draw a balanced chemical equation to illustrate the purpose of using concentrated HCl for the reaction work-up.

Step 2: Acid Chloride Formation

1. What product do you anticipate can form if oxalyl chloride and adipic acid sit at room temperature?
2. How could the adipoyl chloride be purified?

Step 3

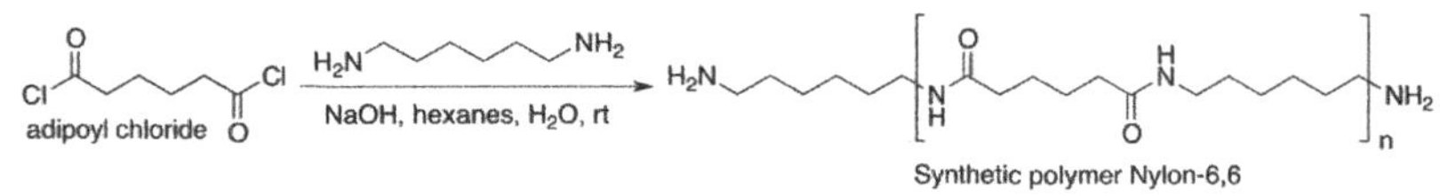

Scheme 16.3.4 Formation of nylon-6,6 synthetic polyamide.

Step 3: Polymerization

1. Why must the polymerization step have a basic solution? (Consider how it would affect your yield.)

Acknowledgements

We would like to express our gratitude to Lisa L. Stamper, the UCSB Undergraduate Lab and Storeroom Manager.

References

1. P. Y. Bruice, *Organic Chemistry*, ed. A. Jawaorski, Prentice Hall, New Jersey, 7th edn, 2013, vol. 27, pp. 1253–1257.
2. (a) R. J. Lancashire, *Preparation of adipic acid from cyclohexene*, University of West Indies, Department of Chemistry (and references therein), http://wwwchem.uwimona.edu.jm/lab_manuals/c10expt12.html, accessed 28 January 2014; (b) J. E. Hutchison and S. M. Reed, *J. Chem. Educ.*, 2000, **77**, 1627.
3. (a) R. M. Roberts, J. C. Gilbert, L. B. Rodewald and A. S. Wingrove, *Modern Experimental Organic Chemistry*, ed. R. M. Roberts, Saunders, Fort Worth, 4th edn, 1985, p. 717; (b) G. Sherer and H.-H. Limbach, *J. Am. Chem. Soc.*, 1994, **116**, 1230–1239 (experimental adaptations by Marisa Weaver, January 2014).
4. (a) D. L. Pavia, G. M. Lampman, G. S. Kriz and R. G. Engel, *Introduction to Organic Laboratory Techniques: A Microscale Approach*, ed. M. Finch and K. M. Mastroianni, Saunders, Fort Worth, 5th edn, 1995, ch. 46B, p. 409; (b) C. E. Carraher Jr, *J. Chem. Educ.*, 1987, **64**, 39.

16.4. Copolymerisation of Styrene and Methyl Methacrylate: An Introduction to Radical Polymerisation and Monomer Reactivity Ratios

Valeria Arrighi and Arno Kraft*

Heriot-Watt University, Institute of Chemical Sciences, School of Engineering & Physical Sciences, Riccarton, Edinburgh EH14 4AS, UK. *E-mail: A.Kraft@hw.ac.uk

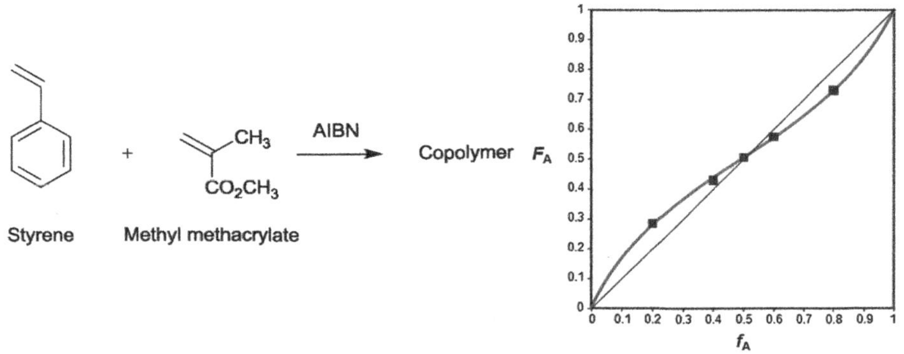

Number of sessions (duration of each session)	Hazard level	Difficulty level	Level of study
1 (4 h)	Moderate	Medium	Advanced

Class names Alkene, ester, radicals

Concepts involved Radical polymerisation, copolymerisation, copolymer reactivity ratios

Chemicals needed Styrene, methyl methacrylate, azobis(isobutyronitrile), methanol, chloroform, deuterated chloroform

Equipment and experimental techniques involved Reactions under inert atmosphere, thermostated bath (60 °C), filtration, preparation of NMR sample

Keywords Copolymer reactivity ratios, copolymerisation, IR, monomers, NMR, precipitation, radical polymerisation

Background

Copolymers of alkene monomers (*e.g.* styrene and methyl methacrylate) can be readily prepared by free-radical polymerisation.[1,2] Typically, a free-radical initiator such as 2,2′-azobis(isobutyronitrile) (AIBN) is added to a mixture of the two monomers (called the feed), and the solution is then heated under nitrogen (Scheme 16.4.1). The copolymers are purified by repeated dissolution/precipitation, isolated by filtration and dried to constant weight over several days.

Most homopolymers that are made from alkene monomers (*e.g.* polyethylene, polypropylene, polystyrene) are generally used on their own except for small amounts of additives (such as a

Comprehensive Organic Chemistry Experiments for the Laboratory Classroom
Edited by Carlos A. M. Afonso, Nuno R. Candeias, Dulce Pereira Simão, Alexandre F. Trindade, Jaime A. S. Coelho, Bin Tan, and Robert Franzén.

Published by the Royal Society of Chemistry, www.rsc.org

stabiliser, pigment, filler or plasticiser) which improve their service life and mechanical properties. Whereas homopolymers such as polystyrene and poly(methyl methacrylate) are immiscible with each other, a copolymer of styrene and methyl methacrylate is able to form a one-phase system with either of the homopolymers, making copolymers the ideal choice as compatibiliser for polymer blends to increase polymer–polymer miscibility. A copolymer often also shows superior properties to the homopolymer. For example, polystyrene tends to be quite brittle, while a styrene–acrylonitrile (SAN) copolymer with *ca.* 25 mol% of acrylonitrile is much tougher.

When two alkene monomers are copolymerised, which comonomer adds next in the propagation step depends on the reactivity of both the monomer and the terminal unit of the active growing chain (Scheme 16.4.2). This so-called terminal model has been used successfully to correctly predict copolymer composition and sequence distribution for many, albeit not all, copolymerisations.[3,4] In the terminal model, four kinetically distinct propagation reactions are possible for a binary copolymerisation, where k_{ij} $(i, j = \text{A, B})$ is the rate constant for the addition of a monomer M_i to a propagating chain end $\sim M_j^{\bullet}$ (the dot represents a radical at the end of a propagating chain). Addition of either the same (homo-propagation) or the other monomer (cross-propagation) can occur.

The reactivity ratios $r_A = k_{AA}/k_{AB}$ and $r_B = k_{BB}/k_{BA}$ indicate the likelihood of homo-propagation compared to cross-propagation. If the reactivity ratio value for monomer A, r_A, is greater than 1, then a growing chain end $\sim M_A^{\bullet}$ is more reactive towards monomer A; if r_A is less than 1, then a growing chain end $\sim M_A^{\bullet}$ will preferentially react with monomer B. Similarly, if the reactivity ratio value for monomer B, r_B, is greater than 1, then a growing chain end $\sim M_B^{\bullet}$ is more likely to add another monomer B, whereas if r_B is less than 1 addition of monomer A will be preferred. Copolymer reactivity ratios for most standard monomer pairs can be found in the literature.[5] From these values it is possible to construct copolymer composition diagrams and then select appropriate feed values to prepare copolymers of different composition. However, sometimes it can become necessary to determine the copolymer reactivity ratios for a different temperature, a different solvent or a new pair of monomers. For this, a copolymerisation is carried out where care is taken to keep conversion low (usually below 10%). The copolymer is then isolated and purified before its composition is determined by FT-IR, ^{1}H NMR or elemental analysis. All these

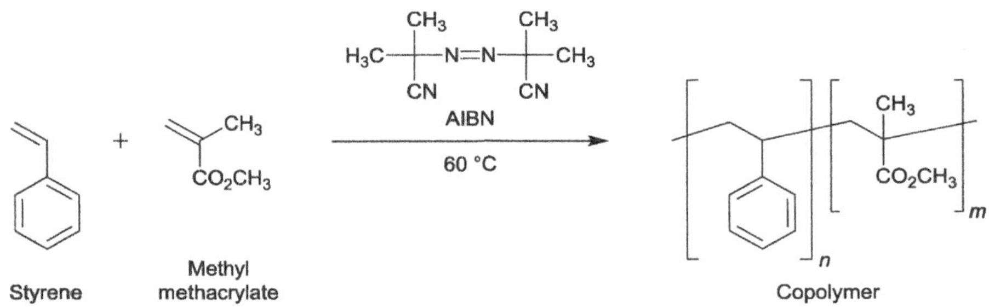

Scheme 16.4.1 Copolymerisation of styrene and methyl methacrylate in the presence of AIBN as free-radical initiator.

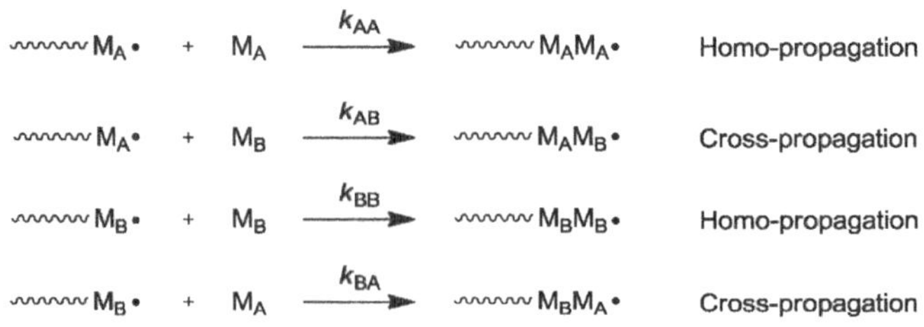

Scheme 16.4.2 The four kinetic steps in the terminal model of a binary copolymerisation.

techniques are equally suitable, although FT-IR or ^{1}H NMR spectroscopy are more convenient. A calibration procedure is needed to determine copolymer compositions from FTIR data.[6] Since ^{1}H NMR spectra are usually recorded to check the degree of purity of copolymer samples, the copolymer composition can be obtained from an integration of the ^{1}H NMR spectra.

The mole fraction of monomer A in the copolymer (F_A) will usually differ from the mole fraction of A in the monomer feed (f_A) and a composition diagram of copolymer composition (F_A) *versus* feed ratio (f_A) can be constructed. The reactivity ratios are then obtained from a non-linear least-squares curve fit, eqn (16.4.1):

$$F_A = \frac{r_A f_A^{\,2} + f_A(1 - f_A)}{r_A f_A^{\,2} + 2f_A(1 - f_A) + r_B(1 - f_A)^2} \tag{16.4.1}$$

Eqn (16.4.1) is the theoretical equation that relates the copolymer composition F_A with the monomer feed ratio f_A and the two reactivity ratios r_A and r_B.

Additional Safety

The preparation should be carried out in a fume hood, with proper safety equipment, such as protective clothing, gloves and safety goggles. Styrene and methyl methacrylate are flammable, irritating to eyes, respiratory system and skin, and may cause sensitisation by skin contact. Styrene is also known to cause damage to organs through prolonged or repeated exposure, particularly by inhalation, and is suspected of damaging the unborn child. Azobis(isobutyronitrile) is harmful by inhalation and if swallowed; owing to its harmful effect on aquatic organisms its solutions should be disposed in the non-halogenated waste container. Methanol is highly flammable and toxic by inhalation, in contact with skin and if swallowed can cause irreversible damage to the eyes. Chloroform causes skin irritation, is harmful if swallowed and suspected of causing cancer, and may cause damage to organs through prolonged or repeated exposure.

Experimental Procedure

Laboratory Session

Monomer Purification (1 h)
Free-radical polymerisations, copolymer isolation, analysis (3 h)

Monomer Purification: Removal of Inhibitor

1. Use a fume cupboard for this experiment. Most vinyl monomers contain an inhibitor, usually a phenol, which needs to be removed by washing with dilute base. In a large separating funnel wash 120 mL of styrene with 3 × 50 mL aliquots of 0.1 M NaOH, one aliquot (*ca.* 50 mL) of 0.1 M HCl, and finally several times with water until the washings are neutral to litmus paper.
2. Dry the styrene over anhydrous $MgSO_4$, then filter through a large fluted paper.
3. Repeat the monomer purification procedure to remove the inhibitor from methyl methacrylate.

Free-Radical Polymerisations

4. Free-radical polymerisations of styrene/methyl methacrylate mixtures are to be carried out in polymerisation tubes. Weigh out 0.16 g of AIBN (1.0 mmol) in a weighing boat, then transfer into a labelled polymerisation tube.

Table 16.4.1 Suggested compositions for a series of monomer mixtures ranging from 10% to 90% styrene monomer and methyl methacrylate.

Sample	Styrene (mL)	MMA (mL)	Styrene[a] (mmol)	MMA[a] (mmol)	f_S
1	9.0	1.0	78.6	9.3	0.89[b]
2	8.0	2.0	69.8	18.6	0.79
3	7.0	3.0	61.1	27.9	0.69
4	6.0	4.0	52.4	37.2	0.58
5	5.0	5.0	43.6	46.4	0.48
6	4.0	6.0	34.9	55.7	0.39
7	3.0	7.0	26.2	65.0	0.29
8	2.0	8.0	17.5	74.3	0.19
9	1.0	9.0	8.7	83.6	0.09

[a]Calculated using a density of 0.909 g cm^{-3} and a molar mass of 104.15 g mol^{-1} for styrene, and a density of 0.93 g cm^{-3} and a molar mass of 100.12 g mol^{-1} for methyl methacrylate.
[b]This mole fraction is calculated as follows: $f_S = 78.6/(78.6 + 9.3) = 0.89$.

5. Using a 10 mL measuring cylinder and reading to ±0.1 mL accuracy, prepare a series of monomer mixtures ranging from 10% to 90% styrene monomer and methyl methacrylate (Table 16.4.1). Add each monomer mixture to a polymerisation tube already containing the AIBN.
6. Close the polymerisation tube with a Suba seal stopper and flush with nitrogen gas. For this, pierce the seal with a long needle connected to your nitrogen supply, and a second needle to allow the air to escape. Flush nitrogen through the monomer mixture for about 2–3 min and gently agitate until all the AIBN has dissolved. Then remove both needles at the same time.
7. Carry out the copolymerisations at 60 °C in a thermostatic bath for 40 min.[7]

Copolymer Isolation

8. Check that the viscosity of your copolymer solutions has increased noticeably during the 40 min polymerisation time; this indicates a successful polymerisation. Remove the Suba seals and pour the solutions into methanol (150 mL).
9. Allow the copolymer to settle, decant most of the supernatant and transfer into a 50 mL beaker. Decant the remaining supernatant. Re-dissolve the sample in *ca.* 10 mL of chloroform and reprecipitate into methanol (150 mL).
10. Transfer the precipitated sample to a pre-weighed sample bottle. Leave uncovered and allow to dry in an oven at 80 °C.

Analysis

11. Weigh each polymer sample and determine the yield.[8]
12. Record a ^{1}H NMR spectrum of each polymer sample (*ca.* 20–30 mg) in deuterated chloroform to confirm the absence of any residual monomer.
13. Record an FT-IR spectrum of each copolymer sample.

Waste Disposal

Residues of monomer mixtures should be collected in the non-halogenated waste container, for disposal by an external contractor. Methanol/chloroform mixtures from the re-precipitation, as well as all NMR samples in $CDCl_3$, should be disposed of in a halogenated waste collector recipient, for posterior destruction.

Results Interpretation and Additional Questions

1. Commercial monomers generally contain a stabiliser. Which stabilisers are used for styrene and methyl methacrylate? Why was a stabiliser present in the monomer? How otherwise might you store a monomer?
2. The decomposition rate constant, k_d, of AIBN at 60 °C is $8.5 \pm 10^{-6}\ s^{-1}$. Calculate the half-life, $t_{1/2}$, of the initiator.
3. Determine the copolymer composition from the ^{1}H NMR and/or IR data, then plot a copolymer composition diagram. Determine the reactivity ratios for the copolymerisation of styrene and methyl methacrylate.

References

1. J. M. G. Cowie and V. Arrighi, *Polymers: Chemistry and Physics of Modern Materials*, CRC Press, Boca Raton, 3rd edn, 2008, ch. 5, pp. 121–127.
2. S. Zhu and D. R. Paul, *Polymer*, 2003, **44**, 3009.
3. M. L. Coote and T. P. Davis, *Prog. Polym. Sci.*, 1999, **24**, 1217.
4. A. Gallardo, M. R. Aguilar, G. A. Abraham and J. S. Román, *J. Chem. Educ.*, 2004, **81**, 1210.
5. J. Brandrup, E. H. Immergut, E. A. Grulke, A. Abe and D. R. Bloch, *Polymer Handbook*, John Wiley & Sons, Inc., New York, 4th edn, 2003.
6. M. K. Bellamy, *J. Chem. Educ.*, 2010, **87**, 1399.
7. Polymerisation times vary depending on the combination of monomers. More reactive monomers require less time, less reactive monomers should be given more time to polymerise. Reaction times should be adjusted in order to reach a *ca.* 10% conversion of monomer to polymer.
8. When determining reactivity ratios it is important that the conversion ("yield" of copolymer) is kept low and does not exceed 10%. A low conversion of <5% is ideal provided the copolymerisation has yielded enough sample for analysing the purity and composition of the copolymer.

16.5. Polymerization of ε-Caprolactone Using a Ruthenium(II) Mixed Metallocene Catalyst

Andreia Valente*[a], Tiago J. L. Silva[a,b], Paulo J. G. Mendes[b], Ana Isabel Tomaz[a], and M. Helena Garcia[a]

[a]Centro de Química Estrutural, Faculdade de Ciências da Universidade de Lisboa, Lisboa, Portugal; [b]Departamento de Química, Escola de Ciências e Tecnologia, Centro de Química de Évora, Instituto de Investigação e Formação Avançada, Universidade de Évora, Évora, Portugal. *E-mail: amvalente@fc.ul.pt

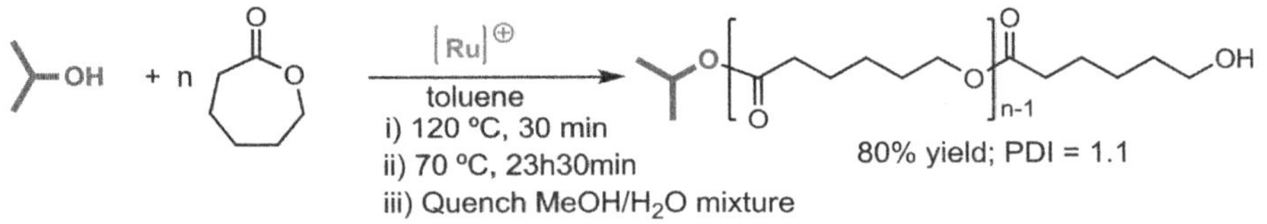

Synthesis of the Ru(II) catalyst:

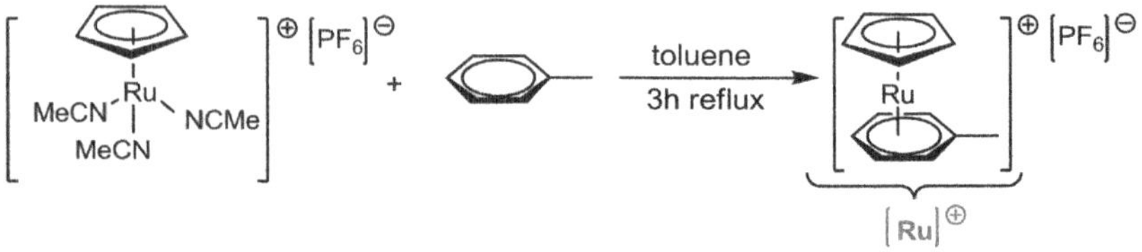

Number of sessions (duration of each session)	Hazard level	Difficulty level	Level of study
2 (4 h + 2 h)	Moderate	High	Advanced
Class names Alcohol, ester			
Concepts involved This experiment involves a method for the synthesis of a ruthenium(II) mixed metallocene compound and its use as catalyst in the polymerization of ε-caprolactone. This procedure is based on the ring-opening of the ε-caprolactone *via* an activated monomer mechanism (using isopropyl alcohol as initiator and the ruthenium(II) as catalyst)			
Chemicals needed ε-Caprolactone, isopropyl alcohol, toluene, *n*-hexane, tris(acetonitrile)cyclopentadienylruthenium(II) hexafluorophosphate, deuterated acetone, deuterated benzene, methanol, dichloromethane, sodium metal, magnesium metal and calcium hydroxide as drying agents			
Equipment and experimental techniques involved Apparatus for reaction with temperature (40–150 °C) and magnetic stirring, oil and sand baths, Schlenk type flasks, vacuum/nitrogen line, NMR tube equipped with a Young valve, desiccator, round-bottom flasks			
Keywords Liquid–solid separation, NMR, polycaprolactone, ring-opening polymerization, ruthenium organometallic catalyst			

Comprehensive Organic Chemistry Experiments for the Laboratory Classroom
Edited by Carlos A. M. Afonso, Nuno R. Candeias, Dulce Pereira Simão, Alexandre F. Trindade, Jaime A. S. Coelho, Bin Tan, and Robert Franzén.

Published by the Royal Society of Chemistry, www.rsc.org

Background

Aliphatic polyesters, prepared by ring opening polymerization (ROP) of lactones, are versatile polymers with good mechanical properties, hydrolyzability, biocompatibility and biodegradability that make them quite suitable for medical and environmental applications. ROP can proceed *via* a cationic, anionic, coordination–insertion or activated monomer mechanism, depending on the type of initiator. In the cationic "activated monomer (AM) mechanism",[1,2] a Lewis acid will act as the catalyst that firstly activates the monomer upon coordination to its carbonyl oxygen; then a protic nucleophilic species, usually alcohols, will initiate the polymerization. This kind of mechanism has attracted a lot of attention due to several aspects: (i) the occurrence of back-biting is considerably reduced since there is no active species at the chain end (contrarily to conventional cationic mechanisms); (ii) the possibility of one step chain-end functionalization; (iii) molecular weight control; and (iv) catalyst economy. These features are achievable since the protic species, which initiate the polymerization by attacking the electrophilic carbon atom of the carbonyl group (ring-opening occurs *via* acyl-oxygen bond cleavage), will also act as a chain transfer agent (CTA). The CTA initially introduced and the hydroxyl end-capped macromolecules H-polymer-OH already generated during the polymerization will all play the role of external nucleophile. All these hydroxyl end-capped molecules are involved in exchange and/or transfer reaction equilibria as "active", *i.e.* coordinated to the Lewis acidic centre, and dormant, *i.e.* Lewis acidic-free. Ideally, each molecule of ROH will originate a new polymer chain enabling the growth of several polymer chains per metal, as described in the pioneering work of Inoue *et al.*[3] As in coordination–insertion mechanisms, if these transfer reactions occur rapidly and reversibly we could have an immortal polymerization.

The objective of this work is the ROP of ε-caprolactone *via* the activated monomer mechanism (using isopropyl alcohol as an initiator and a ruthenium(II) mixed metallocene catalyst).[4] This polymerization is quantitative and occurs in a controlled manner, yielding a product with a dispersity (*Đ*) in the range 1.1–1.3. The ruthenium catalyst is exceptionally stable under the experimental conditions and can be recovered at the end of the reaction. Since the polymerization is slow enough, this experiment is suitable for illustrating a mechanistic study by ^{1}H NMR spectroscopy. NMR spectra clearly show that the η^6–η^4 ring slippage of the coordinated arene ligand at the Ru(II) centre is occurring.[4] The ε-caprolactone is activated upon coordination of the carbonyl oxygen to the catalyst, enabling the attack by the alcohol initiator (*e.g.* isopropyl alcohol) to the electrophilic carbon atom of the carbonyl group. This will enable the ring-opening of the ε-caprolactone and the formation of one growing polymer chain with a reactive hydroxyl end group. This mechanism is repeated until the reaction is quenched by addition of a methanol/water mixture.

Additional Safety

All operations should be done in a fume hood, with proper safety equipment, such as protective clothing, gloves and safety goggles. Toluene, dichloromethane, isopropyl alcohol, *n*-hexane, methanol and magnesium metal are flammable reagents, harmful if swallowed, inhaled, or in contact with the skin and eyes. Sodium metal is very hazardous in case of skin contact, of eye contact and it is extremely flammable in the presence of moisture. Calcium hydroxide is very corrosive in case of eye and skin contact.

Experimental Procedure

All the experiments should be carried out under a nitrogen atmosphere using standard Schlenk techniques. Solvents should be previously dried using standard methods.[5] ε-Caprolactone should be previously dried over calcium hydride and distilled under reduced pressure before use. Isopropyl alcohol should be previously dried with sodium metal for 2 h at room temperature, distilled and further refluxed over magnesium and distilled before use.

Laboratory Session 1 (4 h): Catalyst Synthesis

1. In a fume hood add, in a Schlenk type flask equipped with a magnetic stirrer, add 200 mg (0.46 mmol) of tris(acetonitrile)cyclopentadienylruthenium(II) hexafluorophosphate (CAS: 80049-61-2) to ≈30 mL of dry toluene.
2. Insert a water condenser equipped with a vacuum adapter with a stopcock into the Schlenk flask and allow the suspension to reflux for *ca.* 3 h in an oil bath. The formation of a white toluene insoluble product can be observed during the experiment time.
3. Remove the flask from the oil bath and allow it to cool down to room temperature.
4. Decant the remaining solution and wash the white product with *n*-hexane (3 × 15 mL).
5. Dry the product under vacuum for 5 min and take *ca.* 3 mg to carry out the ^{1}H and ^{31}P NMR in acetone-d_6.
6. Dry the remaining white product overnight to afford a pure compound in ≈90% yield.

Laboratory Session 2 (2 h): Polymerization Work-Up and Mechanistic Study

Previous Work of Laboratory Session 2

Since the polymerization is a time-consuming reaction, students should initiate the reaction 24 h before laboratory session 2. This prior work takes ~40–60 min, following steps 7 and 8.

In addition, one of the groups (or the laboratory technician) should also prepare a parallel experiment in a NMR tube for the mechanistic study (see step 4, Session 2).

7. In a small Schlenk type flask equipped with a magnetic stirrer and purged with dry nitrogen, add, under nitrogen atmosphere, in the following order: 0.5 mL of toluene, 18.2 mg of ruthenium catalyst (45.1 μmol), 0.5 mL of ε-caprolactone (4.5 mmol) and 17.3 μL of isopropyl alcohol (0.23 mmol). Seal the reactor with a rubber septum and place it in a sand bath that was previously set at 120 °C.
8. After 30 min, change the temperature of the sand bath to 70 °C and allow the mixture to react for additional 23 h 30 min.

Laboratory Session 2

1. After 24 h the polymerization reaction is completed and can be taken off the sand bath. The reactor can now be opened to the air. Add some drops of a methanol/water mixture (1 : 1) to quench the reaction using a Pasteur pipette (carefully since the reaction mixture is still hot).
2. Recover the polymer by dissolving it with *ca.* 1 mL of CH_2Cl_2 and add this mixture dropwise to a 100 mL round-bottomed flask (tared) with a methanol/water mixture (~50 mL) and a magnetic stirrer.
3. Let the mixture stir for *ca.* 15 min and then allow the polymer to fully precipitate. Once this has happened decant the solvent and dry the polymer under vacuum until constant weight in a desiccator (meanwhile take off *ca.* 5 mg for ^{1}H NMR characterization).

Mechanistic Studies

4. The laboratory technician or one of the groups can prepare a NMR tube, equipped with a Young valve, to monitor the reaction for the mechanistic study: 0.5 mL of C_6D_6, 1.82 mg of ruthenium catalyst (4.51 μmol), 0.05 mL of ε-caprolactone (0.45 mmol) and 1.73 μL of

isopropyl alcohol (22.63 μmol). Run the ^{1}H NMR spectrum of this solution ($t = 0$). Then, place this tube in the same sand bath (120 °C) as used for the polymerization reactions. Record ^{1}H NMR spectra at $t \approx 15$ and 24 h. Then quench the reaction with an excess of methanol and obtain the final ^{1}H NMR.

Waste Disposal

Non-chlorinated, chlorinated and metal waste solutions should be collected in adequate recipients, for posterior waste disposal management.

Results Interpretation and Additional Questions

1. Interpret and integrate the ^{1}H and ^{31}P NMR spectra of the ruthenium catalyst synthesized and compare with those from the literature.[4]
2. Discuss, in terms of NMR chemical shifts, the η^6–η^4 slippage in the coordination mode of the toluene ring in the catalyst during the polymerization procedure.
3. Calculate the yield of the polymerization.
4. Interpret and integrate the ^{1}H NMR spectrum of the polymer obtained, clearly showing the protons assigned to the main polymer chain and those from the end groups, and compare them to the spectrum from the literature.[4]
5. Calculate the approximate polymer molecular weight by NMR:

$$M(\text{polycaprolactone}) = \frac{\textbf{CH}\text{ integration of the main chain}}{\textbf{CH}\text{ integration of the end group}} \times M(\text{caprolactone})$$

6. Calculate the theoretical molecular weight considering the yield and one growing polymer chain per isopropyl alcohol molecule:

$$M_{\text{theoretical}}(\text{polycaprolactone}) = \frac{n_{\text{caprolactone}}}{n_{\text{isopropyl alcohol}}} \times M(\text{caprolactone})$$

7. Discuss the values obtained in questions 5 and 6 in terms of initiation efficiency, bearing in mind that the total efficiency should lead to five polymer chains per Ru considering the alcohol/catalyst ratio:

$$\text{Initiation efficiency} = \left(\frac{M_{\text{n}}\text{theoretical}}{M_{\text{n}}\text{NMR}}\right) \times 100$$

References

1. S. Penczek, M. Cypryk, A. Duda, P. Kubisa and S. Słomkowski, *Prog. Polym. Sci.*, 2007, **32**, 247–282.
2. N. Ajellal, J. F. Carpentier, C. Guillaume, S. M. Guillaume, M. Helou, V. Poirier, Y. Sarazin and A. Trifonov, *Dalton Trans.*, 2010, **39**, 8363–8376.
3. M. Endo, T. Aida and S. Inoue, *Macromolecules*, 1987, **20**, 2982–2988.
4. A. Valente, P. Zinck, A. Mortreux, M. Bria, M. Visseaux, P. J. G. Mendes, T. J. L. Silva and M. H. Garcia, *J. Mol. Catal. A: Chem.*, 2011, **346**, 102–110.
5. D. D. Perrin, W. L. F. Amarego and D. R. Perrin, *Purification of Laboratory Chemicals*, Pergamon, New York, 2nd edn, 1980.

17. Other Transformations

17.1. The Cannizzaro Reaction: Synthesis of *p*-Chlorobenzyl Alcohol and *p*-Chlorobenzoic Acid

Marja Asp-Lehtinen

Department of Chemistry and Bioengineering, Tampere University of Technology, P.O. Box 541, 33101 Tampere, Finland. *E-mail: Marja.asp-lehtinen@tut.fi

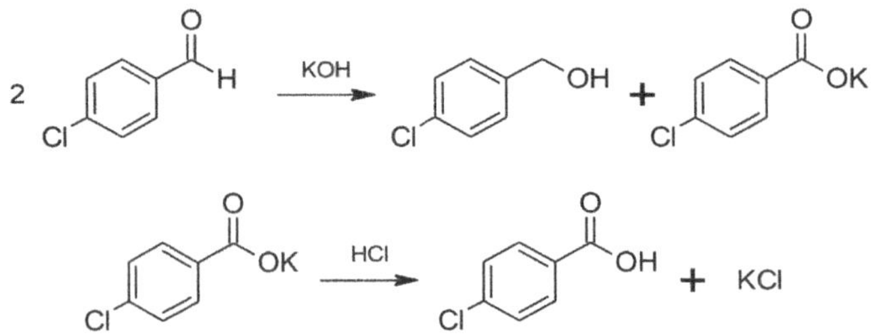

Number of sessions (duration of each session)	Hazard level	Difficulty level	Level of study
1 (7 h)	Moderate	Medium	Introductory
Class names Aldehydes, alcohols, carboxylic acids			
Concepts involved This experiment illustrates the simultaneous oxidation and reduction of an aromatic aldehyde (disproportionation) to form the corresponding benzoic acid and benzyl alcohol			
Chemicals needed *p*-Chlorobenzaldehyde, potassium hydroxide (11 M), methanol, dichloromethane, sodium bicarbonate (saturated solution), sodium sulfate (solid), acetone, hexane, hydrochloric acid (5 M)			
Equipment and experimental techniques involved Apparatus for reaction with temperature (55–60 °C, electrical heat source or oil bath), liquid–liquid extraction apparatus, rotary evaporator; all the products are purified by recrystallization, recrystallization, IR, NMR, melting point, vacuum filtration			
Keywords Disproportionation reaction, IR, liquid–liquid extraction, NMR, recrystallization			

Background

The Cannizzaro reaction is sometimes useful for the preparation of substituted benzoic acid and (or) benzyl alcohols, provided that the starting aldehyde is available.[1] *p*-Chlorobenzaldehyde, for instance, yields *p*-chlorobenzyl alcohol and *p*-chlorobenzoic acid when heated with aqueous KOH. The Cannizzaro reaction takes place by nucleophilic addition of OH^- to an aldehyde to give a tetrahedral intermediate, which expels hydride ion as a leaving group and is thereby oxidized. A second aldehyde molecule accepts the hydride ion in another nucleophilic addition step and is thereby reduced. Aldehydes with α-hydrogen atoms do not undergo this reaction because in the presence of base these α-hydrogen atoms are deprotonated. The resulting enolate generally leads to an aldol reaction.[2–6]

Comprehensive Organic Chemistry Experiments for the Laboratory Classroom
Edited by Carlos A. M. Afonso, Nuno R. Candeias, Dulce Pereira Simão, Alexandre F. Trindade, Jaime A. S. Coelho, Bin Tan, and Robert Franzén.

Published by the Royal Society of Chemistry, www.rsc.org

This experiment illustrates the simultaneous oxidation and reduction of an aromatic aldehyde to form the corresponding benzoic acid and benzyl alcohol. The reaction work-up consists of the extraction of alcohol to an organic solvent while the carboxylic acid is left in the aqueous layer in the carboxylate anion form. Acidification of the water layer allows the acid to precipitate. Both products are obtained as solids and identification and determination of purity can be made after recrystallization by measuring melting points, IR and NMR spectra.

Additional Safety

All sessions should be done in a fume hood, with proper safety equipment, such as protective clothing, gloves and safety goggles. *p*-Chlorobenzaldehyde is harmful if swallowed and it is irritating to eyes, skin and respiratory system. 11 M potassium hydroxide is very hazardous in case of skin and eye contact (irritant, corrosive). Methanol is highly flammable. It is toxic by inhalation, in contact with skin and if swallowed. It can cause blindness. Dichloromethane causes respiratory tract irritation, skin and eye irritation, blood damage, central nervous system depression and cancer hazard (in humans). Acetone is highly flammable. It irritates skin and eyes. It is harmful by inhalation. Vapours may cause drowsiness and dizziness. Hexane is flammable. It causes skin and eye irritation. It causes respiratory tract irritation. Exposure produces central nervous system depression. Vapours may cause dizziness or suffocation. 5 M hydrochloric acid irritates skin and eyes. *p*-Chlorobenzoic acid is irritating to eyes, skin and respiratory system, and may cause central nervous system depression. *p*-Chlorobenzyl alcohol may cause eye and skin irritation. The toxicological properties of this material have not been fully investigated.

Experimental Procedure

Laboratory Session 1 (7 h): Oxidation and Reduction of *p*-Chlorobenzaldehyde, Isolation and Purification of Products (*p*-Chlorobenzyl Alcohol and *p*-Chlorobenzoic Acid)

Reaction and Products Isolation

1. Put 5 mL of 11 M KOH solution in the 100 mL distillation flask.
2. Dissolve *p*-chlorobenzaldehyde (2.00 g, 14.2 mmol) in methanol (6 mL) and add it to the distillation flask containing the potassium hydroxide solution. Add boiling chips (3–4 pieces).
3. Put a very thin layer of grease on the lower part of a condenser (this prevents the condenser and the distillation flask getting stuck together). Attach a condenser to the top of the flask and connect the condenser to the water faucet. Start the water running (remember: inlet at the bottom and outlet at the top of the condenser), and run water through the condenser.
4. Put a suitable heat source under the flask and adjust the heat so that reflux is visible.
5. Reflux the reaction mixture for 1 h.
6. When the reaction time is up, turn off the heat, let the setup cool and dismantle it.
7. Transfer the reaction mixture to the separation funnel, rinsing with water (25 mL).
8. Extract the resulting solution with two 15 mL portions of dichloromethane.
9. After separation of the layers collect the organic layer (layer A) and water layer (layer B) to an Erlenmeyer flask for further handling.

Purification of p-*Chlorobenzyl Alcohol (Layer A)*

1. Wash the combined dichloromethane extracts with saturated sodium bicarbonate solution (10 mL) followed by one 10 mL portion of distilled water (when separating layers you must think carefully which layer is the upper one and which one is the lower one).
2. Dry the dichloromethane layer over anhydrous sodium sulfate for 10 min.

3. After drying the solution evaporate dichloromethane in the rotary evaporator.
4. Purify the crude alcohol by recrystallizing it using a solution of acetone in hexane (1 : 9).
5. Collect the product under reduced pressure using a Büchner funnel.
6. Air-dry the product on a porous clay plate or on a filter paper.

Purification of p-*Chlorobenzoic Acid (Layer B)*

1. Dilute the alkaline phase (layer B) by adding 25 mL of water.
2. Acidify the water layer by adding hydrochloric acid (5 M) until the pH of the solution is under 1 (test using pH paper).
3. Let the solution cool down on the ice bath for 5 min.
4. Collect the product under reduced pressure using a Büchner funnel.
5. Purify the crude acid by recrystallizing it in methanol.
6. Air-dry the product on a porous clay plate or on a filter paper.
7. Weigh both products and calculate the percent yield:

$$\text{Percentage yield} = \frac{\text{mass of actual yield}}{\text{mass of theoretical yield}} \times 100\%$$

8. Determine the melting points of your products.
9. Obtain IR and NMR spectra if possible.

Waste Disposal

The neutralized aqueous solution may be diluted and discarded in the sink. Acetone and hexane may be collected in a non-chlorinated waste collector recipient and dichloromethane may be collected in a chlorinated waste collector recipient, for posterior destruction.

Results Interpretation and Additional Questions

1. Write down the reaction mechanism for the performed Cannizzaro reaction. Explain what occurs at each step of that reaction.
2. Recall how you separated the compounds from the reaction mixture. What is the basis of that separation?
3. What would be the reaction rate order for the following compounds in the Cannizzaro reaction: benzaldehyde, *p*-nitrobenzaldehyde, *p*-chlorobenzaldehyde and *p*-methoxybenzaldehyde. Justify your answer.
4. Compare your melting point values with those found in the literature.
5. Try to find characteristic bands from the IR spectrum for each product.

References

1. J. D. Roberts, R. Stewart and M. C. Caserio, *Organic Chemistry, methane to macromolecules*, W. A. Benjamin, Inc., Canada, 5th edn, 1971, vol. 293, p. 650.
2. J. McMurry, *Organic Chemistry*, Thomson Learning, Inc., USA, 7th edn, 2008, pp. 724–725.
3. J. Clayden, N. Greeves, S. Warren and P. Wothers, *Organic Chemistry*, Oxford University Press, Oxford, 2nd edn, 2008, pp. 1081–1084.
4. D. Mayo, R. Pike and P. Trumper, *Microscale Organic Laboratory*, John Wiley & Sons, Inc., New York, 4th edn, 2000, pp. 155–158.
5. C. G. Swain, A. L. Powell, W. A. Sheppard and C. R. Morgan, *J. Am. Chem. Soc.*, 1979, **101**(13), 3576.
6. A. R. Tatchell, B. S. Furnis, A. J. Hannaford and P. W. G. Smith, *Vogel's A Textbook of Practical Organic Chemistry*, Longman Scientific and Technical, Harlow, 5th edn, 1989, pp. 1028–1031.

17.2. The Kemp Elimination in Water: A Laboratory Experiment for Introductory Organic Chemistry

Caitlin Williamson, Kenneth E. Maly, and Stephen L. MacNeil*

Department of Chemistry and Biochemistry, Wilfrid Laurier University, 75 University Ave W., Waterloo, Ontario, Canada N2L 3C5. *E-mail: smacneil@wlu.ca

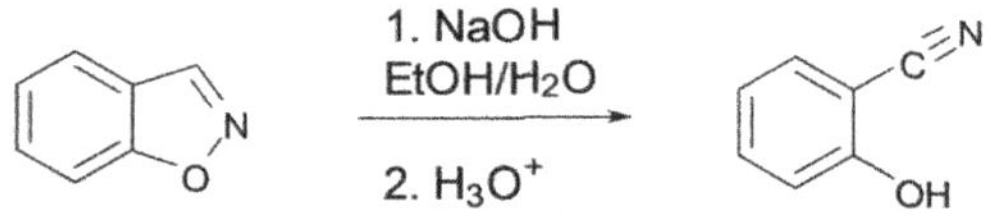

Number of sessions (duration of each session)	Hazard level	Difficulty level	Level of study
1 (3 h)	Moderate	Medium	Introductory
Class names Benzisoxazoles, nitriles, phenols			
Concepts involved This experiment involves the use of the Kemp elimination as an experiment for an introductory organic chemistry course. The reaction occurs quickly under mild conditions in ethanol–water and employs several laboratory techniques that are important for organic chemistry, such as liquid–liquid extraction, filtration, use of a rotary evaporator, and melting point determination. Furthermore, the experiment can be used to reinforce students' understanding of E2 reactions and organic spectroscopy and to introduce students to green chemistry metrics			
Chemicals needed Benzisoxazole, sodium hydroxide (2 M), ethanol, distilled water, concentrated HCl, ethyl acetate, magnesium sulfate			
Equipment and experimental techniques involved Erlenmeyer flasks with magnetic stirring for reaction, separatory funnel for liquid–liquid extraction, funnel and filter paper for gravity filtration, pH paper, disposable Pasteur pipettes, round bottom flask, rotary evaporator for product isolation, melting point apparatus for product characterization			
Keywords Elimination reaction, extraction, green chemistry, Kemp elimination, melting point			

Background

You have probably learned about elimination reactions involving alkyl halides leading to the formation of alkenes. A typical example is shown in Scheme 17.2.1. Depending on the alkyl halide and reaction conditions employed, these elimination reactions may proceed by one of three mechanisms: E1, E2 or E1cb. The latter is relevant to a number of biological elimination reactions. You will learn in this experiment that the elimination of alkyl halides to yield alkenes is only one of many elimination reactions encountered in organic chemistry. In various elimination reactions, the leaving group is not a halide and the product is not an alkene.

The Kemp elimination reaction is an example of a concerted elimination proceeding through an E2 mechanism. While mechanistically resembling an E2 elimination, the Kemp elimination has distinctive features that set it apart from typical elimination reactions. The elimination is carried out on a heterocyclic benzisoxazole and the product of the elimination is a nitrile rather

Comprehensive Organic Chemistry Experiments for the Laboratory Classroom
Edited by Carlos A. M. Afonso, Nuno R. Candeias, Dulce Pereira Simão, Alexandre F. Trindade, Jaime A. S. Coelho, Bin Tan, and Robert Franzén.

Published by the Royal Society of Chemistry, www.rsc.org

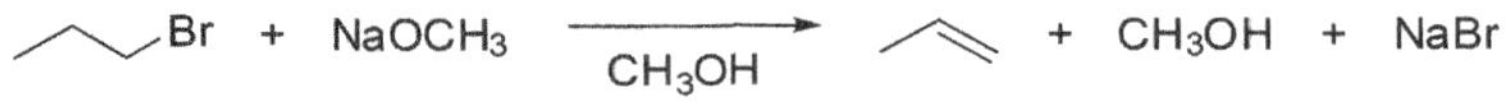

Scheme 17.2.1 Formation of an alkene by elimination.

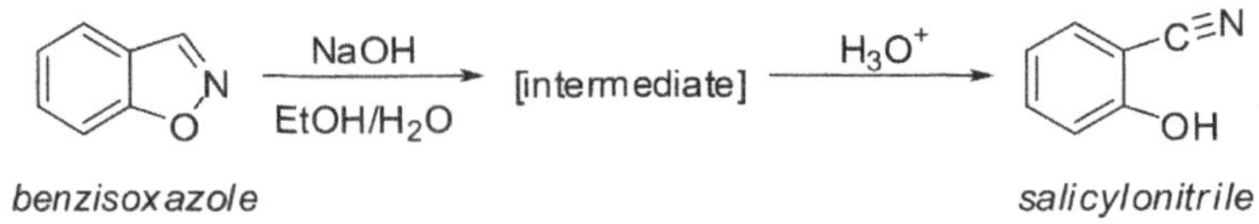

Scheme 17.2.2 The Kemp elimination of benzisoxazole.

than an alkene (Scheme 17.2.2). Furthermore, the elimination is essentially intramolecular, resulting in a ring-opening and the generation of a phenoxide, unlike most E2 eliminations, where a leaving group (halide, OH, or tosylate) is present.

The Kemp elimination reaction is over 40 years old[1] but recently it has attracted a great deal of attention. It is a well-defined model reaction for studying the catalysis of proton transfer from carbon. Abstraction of a proton from carbon is a demanding reaction that is the rate-determining step in several enzyme-catalyzed reactions. For these reasons, the Kemp elimination has been employed for the development of artificial enzymes – relatively small, organic molecules developed to recreate the active site of an enzyme – for proton abstractions from carbon.[2]

In this experiment, you will carry out the Kemp elimination of benzisoxazole in ethanol/water using NaOH as the base. The product of the reaction is salicylonitrile (also known as *ortho*-cyanophenol or 2-hydroxybenzonitrile). The learning objectives for this experiment include:

- Revisiting and reinforcing concepts related to elimination reactions;
- Carrying out, with confidence and efficiency, a synthetic protocol including reaction set up, product isolation and characterization by melting point;
- Performing, with confidence and efficiency, the following laboratory techniques: (i) massing of samples with a top-loading balance; (ii) liquid–liquid extraction; (iii) gravity filtration; (iv) solvent evaporation using a rotary evaporator; (v) melting point determination;
- Applying the 12 Principles of Green Chemistry and green chemistry metrics (see Section 1b of the ESI for an introduction to both). This is an important consideration for all chemists wanting to evaluate the environmental impact of chemical reactions and reduce the use of hazardous and potentially toxic chemicals and solvents.

Additional Safety

Benzisoxazole is a combustible liquid. Benzisoxazole and salicylonitrile may be irritating to eyes, respiratory tract, and skin. Consequently, protective personal wear including laboratory coat, safety goggles, and gloves should be worn, and the handling of benzisoxazole and salicylonitrile should be carried out in a fume hood. Ethanol and ethyl acetate are flammable liquids. Concentrated HCl is corrosive and will burn the skin and eyes. Aqueous (2 M) NaOH is corrosive and an irritant.

Experimental Procedure

Laboratory Session 1 (3 h)

Setting Up the Experiment

1. In a fume hood, weigh out approximately 0.25 g benzisoxazole into a 125 mL Erlenmeyer flask using a top-loading balance. [*Caution*: combustible liquid, no open flames; may be irritating to eyes, respiratory tract and skin.]

2. To the Erlenmeyer flask, add 1.5 mL ethanol [*Caution*: flammable liquid, no open flames], 1.5 mL distilled water and 7.5 mL 2 M NaOH.
3. Add a stir bar to the Erlenmeyer flask and use a magnetic stir plate to stir the reaction at room temperature for 10 min.

Reaction Work-Up

1. Using a disposable Pasteur pipette, add concentrated HCl [*Caution*: corrosive, wear gloves] to the reaction mixture in the fume hood, ten drops at a time until it reaches pH 1. Make sure to swirl the flask and check the pH using pH paper after each addition.
2. Pour the acidified solution into a 125 mL separatory funnel in the fume hood.
3. Perform extractions of the aqueous layer with three 5 mL portions of ethyl acetate, each time rinsing the Erlenmeyer flask with the ethyl acetate before adding to the separatory funnel and collecting the ethyl acetate layer in a clean, dry 125 mL Erlenmeyer flask. [*Caution*: pressure build-up; vent frequently.]
4. Add enough $MgSO_4$ to the ethyl acetate extract to cover the bottom of the Erlenmeyer flask and ensure the $MgSO_4$ is free flowing. If clumping occurs, add more $MgSO_4$ in small quantities. Allow to dry for 5–10 min.

Product Isolation

1. Remove the $MgSO_4$ from your ethyl acetate extract by gravity filtration in the fume hood, collecting the filtrate in a clean, dry and pre-weighed 50 mL round-bottom flask.
2. Use a rotary evaporator to remove the ethyl acetate. Once the ethyl acetate has evaporated, your product should be visible in the round-bottom flask as a white to yellow powdery solid.

Waste Disposal

Dispose of all waste in appropriate aqueous and organic waste containers.

Results Interpretation and Additional Questions

1. Green chemistry is defined as the "design of chemical products and processes to reduce or eliminate the use and generation of hazardous substances".[3] The criteria implemented for development of sustainable chemical processes are summarized by the Twelve Principles of Green Chemistry.[4] With reference to these principles, what was green about this experiment? What was not green? How could the experiment be made greener?
2. Calculate the percentage yield, atom economy, reaction mass efficiency and E-factor for the reaction. For simplification, 2 M NaOH can be assumed to have a density of 1.0. Ignore the mass of concentrated HCl and $MgSO_4$ used since these were not determined accurately.
3. Calculate the cost per gram of product for the reaction using current prices of benzisoxazole, ethanol and ethyl acetate from a reputable vendor. *Note: for this calculation, the cost of $MgSO_4$, HCl, NaOH and water may be ignored (all are very cheap!).* Find the price of the product you synthesized from a reputable vendor, *e.g.*, Sigma-Aldrich Chemical Company, and use this information to determine if your synthesis was cost-effective?[5]
4. Weigh the flask containing your product, calculate the mass of product isolated and determine its melting point. Comment on the success (or lack thereof) of the experiment with reference to the percentage yield and the reported and observed melting point of your product.

5. Draw an arrow-pushing mechanism for this transformation. With reference to this mechanism, determine which compound for each of the following pairs would react faster in a Kemp elimination reaction and explain why.

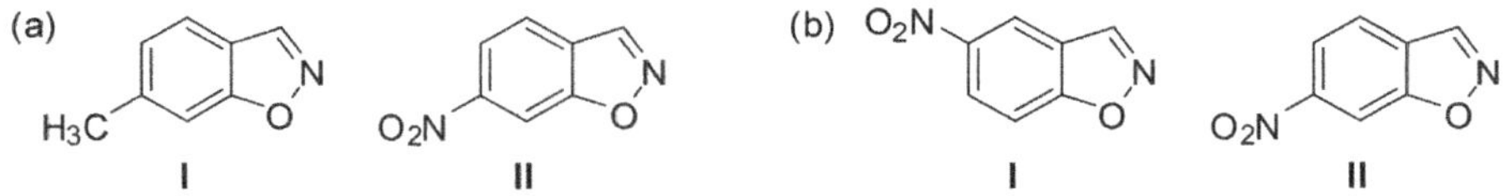

6. The following is the IR spectrum for salicylonitrile.[6] For bands a–d, indicate the functional group and vibration (stretch *versus* bend) responsible.

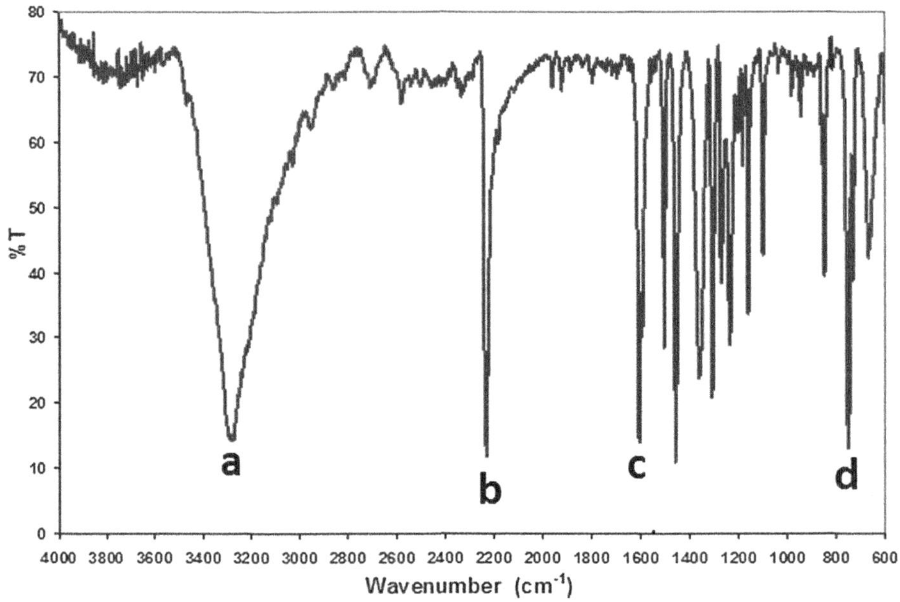

References

1. M. L. Casey, D. S. Kemp, K. G. Paul and D. D. Cox, *J. Org. Chem.*, 1973, **38**, 2294–2301.
2. See M. Sparta and A. N. Alexandrova, *Mol. Simul.*, 2011, **37**, 557–571 and references cited therein.
3. (a) P. T. Anastas and J. C. Warner, in *Green Chemistry: Theory and Practice*, Oxford University Press, New York, 1998; (b) I. Horvath and P. T. Anastas, *Chem. Rev.*, 2007, **107**, 2167; (c) P. T. Anastas and T. C. Williamson, in *Green Chemistry: Designing Chemistry for the Environment*, American Chemical Series Books, Washington, DC, 1996, pp. 1–20.
4. P. Anastas and N. Eghbali, *Chem. Soc. Rev.*, 2010, **39**, 301–312.
5. This cost analysis is necessarily oversimplified in that it ignores costs associated with energy, manpower, over-heads, equipment and waste disposal.
6. AIST, Spectral Database for Organic Compounds, SDBS Web: http://riodb01.ibase.aist.go.jp/sdbs/, (National Institute of Advanced Industrial Science and Technology, 19 June 2013).

17.3. Synthesis of Veratronitrile

Dulce Simão* and Ana C. Cerdeira

Department of Chemistry, Instituto Superior Técnico, University of Lisbon, Centro de Química Estrutural, Av. Rovisco Pais, 1, 1049-001 Lisbon, Portugal. *E-mail: dulcesimao@tecnico.ulisboa.pt

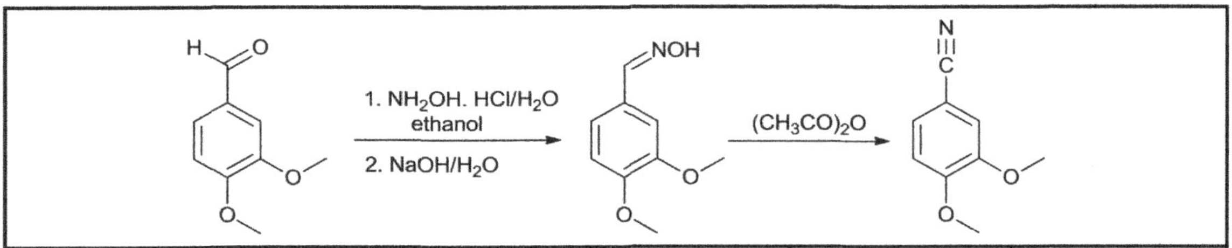

Number of sessions (duration of each session)	Hazard level	Difficulty level	Level of study
3 (4 h + 2 h + 4 h)	Moderate	Medium	Intermediate
Class names Aldehydes, oximes, anhydrides, nitriles			
Concepts involved This experiment gives the two-step reaction of veratronitrile, involving aldoxime formation followed by dehydration to nitrile			
Chemicals needed Veratraldehyde, ethanol, hydroxylamine hydrochloride, sodium hydroxide, carbon dioxide cylinder (g), acetic anhydride, calcium chloride, $CDCl_3$			
Equipment and experimental techniques involved Crystallization, IR, reflux apparatus, magnetic stirring, Bunsen burner and heating mantle, vacuum filtration, ^1H NMR, melting point			
Keywords Crystallization, hydroxylamines, IR, nitrile, oxime formation			

Background

Aldehydes and ketones have the ability to react through condensation with ammonia and primary amines to form water and imine products with the general formula RCH=NR. An oxime (RCH=NOH) is formed when hydroxylamine is used as the ammonia derivative and it is more stable than imines due to the presence of the OH electronegative group attached to a nitrogen atom.[1] Hydroxylamine is not stable, and so the hydroxylamine hydrochloride salt is used in the presence of a base to remove a proton from the nitrogen atom and liberate the reagent.[2] When hydroxylamine attacks the aldehyde a tetrahedral intermediate is generated and further dehydration gives the aldoxime.[3–5] This reaction must be carried out under ideal pH conditions, between 4 and 5. At pH < 4, most of the NH_2OH molecules are converted into the conjugate NH_3OH^+ ions, which cannot attack the carbonyl group. The reaction rate also decreases when the pH > 4.[6a] The dehydration of aldoximes is usually made with acetic anhydride with initial acetylation of the oximino group followed by the elimination of acetic acid to obtain a nitrile (C≡N).[6b,7] Other conditions to perform this conversion in one step[6c] can be found in the literature. In this experiment an aromatic nitrile known as veratronitrile (3,4-dimethoxybenzonitrile) is prepared, starting from veratraldehyde (3,4-dimethoxybenzaldehyde).[7–9] This aromatic aldehyde is a vanillin

Comprehensive Organic Chemistry Experiments for the Laboratory Classroom
Edited by Carlos A. M. Afonso, Nuno R. Candeias, Dulce Pereira Simão, Alexandre F. Trindade, Jaime A. S. Coelho, Bin Tan, and Robert Franzén.

Published by the Royal Society of Chemistry, www.rsc.org

derivative with a characteristic sweet vanilla odor, widely used in the manufacture of fragrances in the cosmetic and perfume industry.[10] Both veratraldehyde and veratronitrile have been used as chemical intermediates in the production of pharmaceuticals and other fine chemicals.[11]

Additional Safety

Acetic anhydride is highly corrosive and should be handled carefully. Hydroxylamine hydrochloride is a corrosive and poisonous solid and is suspected to be mutagenic. The usual caution must be observed when handling ethanol (flammable) and aqueous sodium hydroxide (corrosive).

Experimental Procedure

Session 1: Preparation of Veratraldoxime (4 h)

1. On a 250 mL two- or three-neck round-bottom flask fit a reflux condenser and a pressure-equalizing dropping funnel.
2. Place veratraldehyde (8.3 g, 0.05 mol) in the flask and add warm ethanol (30–40 °C) (20 mL).
3. In a beaker, prepare a solution of hydroxylamine hydrochloride (4.2 g, 0.06 mol) in warm water (30–40 °C) (5 mL).
4. Add the hydroxylamine hydrochloride solution dropwise to the veratraldehyde solution, with magnetic stirring.
5. Add dropwise a solution of sodium hydroxide (3 g) in water (4 mL).
6. Allow the mixture to stand for 2.5 h.
7. Add crushed ice (*ca.* 25 g) and bubble CO_2 gas from a cylinder with a glass Pasteur pipette to saturate the solution. Aldoxime separates as an oil. Allow the mixture to stand in a refrigerator overnight.

Session 2 (2 h)

1. Collect the product by vacuum filtration, wash thoroughly with cold water and dry.
2. Store the oxime under vacuum. No further purification is needed and it can be used in the next step.
3. Record melting point, yield and IR and ^{1}H NMR spectra.

Session 3: Preparation of Veratronitrile (4 h)

1. To a 50 mL round-bottom flask fitted with a reflux condenser and a $CaCl_2$ tube, add veratraldoxime (5.0 g, 0.03 mol) and redistilled acetic anhydride (6 mL, 6.5 g, 0.06 mol).
2. Carefully warm the mixture with a water bath until a vigorous reaction starts.
3. When the reaction subsides reflux gently for 20 min.
4. Carefully and with stirring pour the reaction mixture into a beaker containing cold water (15 mL) and a magnetic bar.
5. Stir and cool in an ice bath.
6. Collect the almost colorless crystals by vacuum filtration and dry.
7. Record melting point, yield and IR and ^{1}H NMR spectra.

Waste Disposal

Filtrate containing hydroxyl amine derivatives should be poured into a container reserved for hazardous aqueous solutions. Acid filtrate should be neutralized, diluted and discarded in the sink.

Results Interpretation and Additional Questions

1. Determine the yields of reaction products.
2. Compare the melting points of the obtained products with the those reported in the literature and evaluate the purity of your products.
3. Interpret the IR and NMR spectra.
4. Write out all reaction mechanisms.
5. Explain why oximes are more stable than imines, showing electron delocalization.

References

1. J. Clayden, N. Greeves, S. Warren and P. Wothers, *Organic Chemistry*, Oxford University Press, Oxford, 2001, p. 351.
2. G. D. Meakins, *Functional Groups: Characteristics and Interconversions*, Oxford Chemistry Primers, Oxford University Press, 1996, p. 63.
3. S. Ege, *Organic Chemistry*, D. C. Heath and Company, 2nd edn, 1989, p. 504.
4. J. Kalia and R. T. Raines, *Angew. Chem., Int. Ed.*, 2008, **47**, 7523.
5. *The Chemistry of Hydroxylamines, Oximes and Hydroxamic Acids*, ed. Z. Rappoport and J. F. Liebman, John Wiley & Sons, 2009, vol. 164, p. 818.
6. J. March, *Advanced Organic Chemistry*, John Wiley & Sons, 4th edn, 1992(a) p. 906; (b) p. 1038; (c) p. 907.
7. A. R. Tatchell, B. S. Furnis, A. J. Hannaford, and P. W. G. Smith, *Vogel's Textbook of Practical Organic Chemistry*, Longman Scientific and Technical, Harlow, 5th edn, 1989, p. 1084.
8. J. S. Buck and W. S. Ide, *Org. Synth. Coll.*, 1943, 2, 622.
9. C. A. M. Afonso, D. P. Simão, L. P. Ferreira, M. S. Serra and M. M. M. Raposo, *100 Experiências de Química Orgânica*, IST Press, 2011, p. 211.
10. The Good Scents Company, Veratraldehyde, http://www.thegoodscentscompany.com/data/rw1030991.html, accessed October 2015.
11. R. J. Chudgar and J. Oakes, *Ullmann's Encyclopedia of Industrial Chemistry*, D. C. Heath and Company, 2nd edn, 1988, vol. A11, p. 200.

17.4. Synthesis of Levulinic Acid from Sucrose

João P. Telo

Centro de Química Estrutural, Instituto Superior Técnico, Universidade de Lisboa, Av. Rovisco Pais, 1049-001 Lisboa, Portugal. *E-mail: jptelo@tecnico.ulisboa.pt

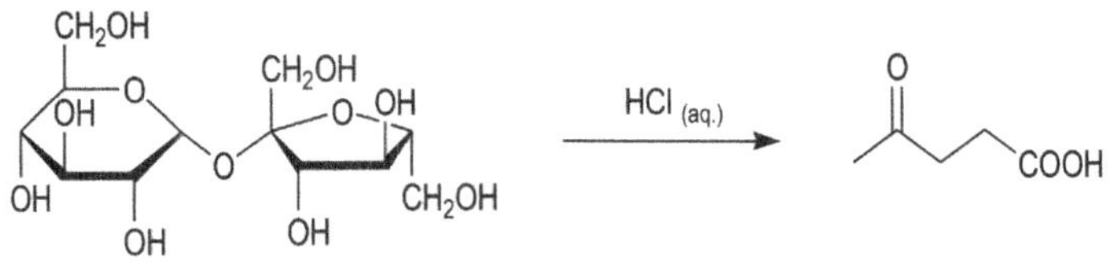

Number of sessions (duration of each session)	Hazard level	Difficulty level	Level of study
2 (2 h + 4 h)	Moderate	Medium	Intermediate

Class names Carboxylic acids, ketals, alcohols, ketones

Concepts involved Levulinic acid, one of the components of caramel, is prepared by acid degradation of commercial sucrose. The product is purified by a sequence of filtration, liquid–liquid extraction and distillation under reduced pressure

Chemicals needed Sucrose (table sugar), HCl (37%), ethyl ether, anhydrous magnesium sulfate

Equipment and experimental techniques involved Apparatus for reaction under reflux, liquid–liquid extraction, vacuum distillation, melting point apparatus, Abbé refractometer

Keywords Fractional distillation, melting point, NMR spectroscopy, refraction index, saccharides

Background

In the last 100 years petroleum and natural gas have been the major sources of the enormous demand of chemical compounds needed for industry. However, growing concerns not only about the environmental impact of petroleum extraction but mostly about the diminishing of natural reserves have revealed that the search for alternative, sustainable and more environmentally friendly sources of chemicals is of critical importance. The renewable biomass resources can be used not only to produce biodiesel but also as a source of chemicals.[1–3]

Levulinic acid is a ketoacid found in caramel and is used in the food industry. It is also used as a precursor of pharmaceutical compounds, polymers, plasticizers and biofuels.[4,5] Levulinic acid is added to tobacco to increase the binding of nicotine to neural receptors.[6] It can be prepared by acid degradation of cellulose, starch, sucrose or other carbohydrates. The mechanism involves the isomerization of glucose into fructose and subsequent dehydration to 4-(hydroxymethyl) furfural,[7] which then hydrolyzes into levulinic acid and formic acid. The complete mechanism can be found in the literature.[8,9] Although this method of preparation has been known for more than a century,[10] it has not been ascribed much importance due to the low yields of the synthesis. Recently, new processes have been optimized to obtain high yields of levulinic acid (around 60%) from recycled paper and organic residues from cellulose and agriculture industries.[11]

Comprehensive Organic Chemistry Experiments for the Laboratory Classroom
Edited by Carlos A. M. Afonso, Nuno R. Candeias, Dulce Pereira Simão, Alexandre F. Trindade, Jaime A. S. Coelho, Bin Tan, and Robert Franzén.

Published by the Royal Society of Chemistry, www.rsc.org

The method described here uses sucrose (table sugar) as reactant, which is a disaccharide of glucose and fructose.[12]

Additional Safety

Hydrochloric acid is corrosive and should be handled with care. The reaction should be done in a fume cupboard due to the possible release of gaseous HCl. Although after the 24 h reflux the amount of hydrochloric acid in the reaction mixture is much smaller, we suggest the use of a water pump coupled to the rotary evaporator, or that the exhaustion of the electric vacuum pump should be directed to a fume cupboard.

Experimental Procedure

Laboratory Session 1 (2 h): Preparation of the Levulinic Acid from Sucrose

1. Add to a 250 mL single-neck round-bottomed flask 50 g of table sugar (sucrose), 100 mL of water and 25 mL of concentrated HCl (37%).
2. Reflux the mixture in a heating mantle for 24 h, during which time some carbonization takes place.

Laboratory Session 2 (4 h): Preparation of the Levulinic Acid from Sucrose

1. Allow the solution to cool, filter the black solid in a Büchner funnel in vacuum and wash the solid in the funnel with 60 mL of water.
2. Concentrate the combined aqueous filtrate in a rotary evaporator until a viscous residue remains in the flask.
3. Extract the viscous residue several times with diethyl ether (4 × 40 mL on average). Make sure the viscous residue is well shaken with the ether.
4. Dry the ether with anhydrous magnesium sulfate (or calcium chloride) and distil in a rotary evaporater.
5. Fractionate the residue under reduced pressure, collecting the fraction between 135 and 145 °C (10 mmHg).
6. Pure levulinic acid is then obtained by redistilling the viscous liquid and collecting the fraction between 137 and 139 °C (10 mmHg), which crystallizes on standing if pure enough.
7. Measure the melting point of the solid. Melt a small amount of solid and measure the refractive index.
8. Obtain the NMR spectrum (^{1}H and ^{13}C) of levulinic acid in $CDCl_3$.

Waste Disposal

The distilled diethyl ether should be stored for future reutilization. The water from the reaction mixture distilled in the rotary evaporator should be diluted and disposed of in the sink.

Results Interpretation and Additional Questions

1. Draw out the reaction mechanism.
2. What is the IUPAC name of levulinic acid?
3. Analyze the NMR spectra of levulinic acid.
4. Calculate the yield of the conversion and assess the purity of levulinic acid by comparing the melting point and refractive index with literature values.

References

1. P. Gallezot, *Chem. Soc. Rev.*, 2012, **41**, 1538.
2. J. B. Binder and R. T. Raines, *J. Am. Chem. Soc.*, 2009, **131**, 1979.
3. S. P. Simeonov and C. A. M. Afonso, *J. Chem. Educ.*, 2013, **90**, 1373.
4. J. J. Bozell, *et al.*, *Resour. Consent. Recy.*, 2000, **28**, 228.
5. G. Metzker and A. C. B. Burtoloso, *Chem. Commun.*, 2015, **51**, 14199.
6. L. Keithly, G. F. Wayne, D. M. Cullen and G. N. Connolly, *Nicotine Tob. Res.*, 2005, **7**, 761.
7. G. Yong, Y. Zhang and J. Y. Ying, *Angew. Chem. Int. Ed.*, 2008, **47**, 9345.
8. J. Horvat, B. Klaic, B. Metelko and V. Sunjic, *Tetrahedron Lett.*, 1985, **26**, 2111.
9. M. J. Antal, W. S. Mok and G. N. Richards, *Carbohydr. Res.*, 1990, **199**, 91.
10. A. Freiherrn, V. Grote and B. Tollens, *Justus Liebigs Ann. Chem.*, 1875, **175**, 181.
11. S. Fitzpatrick, US Pat. 5608 105, 1997.
12. B. F. McKenzie, *Org. Synth. Coll. I*, 1941, 335.

17.5. Synthesis of 1-(4-Bromophenyl)-1*H*-pyrrole by the Clauson-Kaas Reaction

Maria Manuela Marques Raposo*, Maria Cidália Rodrigues Castro, and Sara Sofia Marques Fernandes

University of Minho, Department of Chemistry, Campus de Gualtar, 4710-057 Braga, Portugal
*E-mail: mfox@quimica.uminho.pt

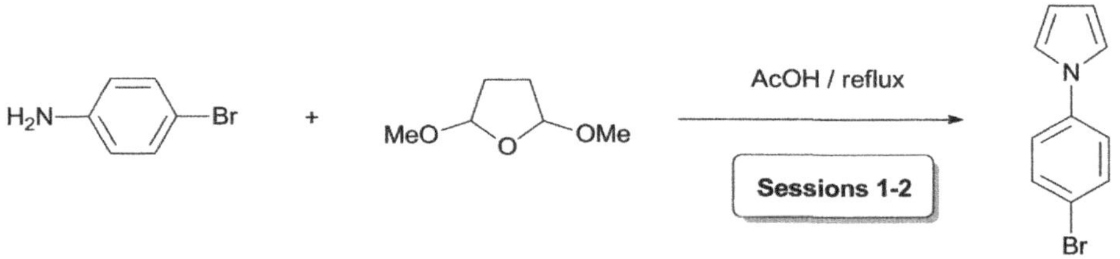

Number of sessions (duration of each session)	Hazard level	Difficulty level	Level of study
2 (4 h + 2 h)	Moderate	Medium	Intermediate
Class names Amines, acetals, aryl bromides, pyrroles			
Concepts involved Synthesis of a pyrrole derivative substituted in position 1; synthesis of the pyrrole nucleus through Clauson-Kaas synthesis; characterization by ^{1}H NMR, spectroscopy			
Chemicals needed 4-Bromoaniline, 2,5-dimethoxytetrahydrofuran, magnesium sulfate (anhydrous), glacial acetic acid, dichloromethane, hexane, deuterated chloroform, and silica gel thin-layer chromatography (TLC) plates with fluorescent indicator			
Equipment and experimental techniques involved Reflux apparatus with magnetic stirring, liquid–liquid extraction, gravity and vacuum filtration, evaporation under vacuum using a rotary evaporator, melting point			
Keywords Clauson-Kaas pyrrole synthesis, liquid–liquid extraction, NMR, pyrrole			

Background

Pyrrole is one of the most important heterocycles; it was first isolated in 1857 from the products of bone pyrolysis. It was identified as biologically relevant when recognized as a structural fragment of heme and chlorophyll. Presently, pyrrole derivatives are of great importance in organic chemistry not just because they are present in nature or find wide application as subunits of natural products and drug molecules but also due to their growing relevance in materials science and supramolecular chemistry with a wide range of applications including semiconducting and photochromic materials, chemosensors for anions and cations of biological, medicinal and environmental relevance, nonlinear optical materials, organic light emitting diodes (OLEDs) and dye sensitized organic solar cells (DSSCs).[1–3] As a result, various strategies have been developed for the synthesis of diverse substituted pyrroles (*e.g.* Knorr, Paal–Knorr, conjugate addition

Comprehensive Organic Chemistry Experiments for the Laboratory Classroom
Edited by Carlos A. M. Afonso, Nuno R. Candeias, Dulce Pereira Simão, Alexandre F. Trindade, Jaime A. S. Coelho, Bin Tan, and Robert Franzén.

Published by the Royal Society of Chemistry, www.rsc.org

reactions, 1,3-dipolar cycloaddition reaction, reductive coupling, aza-Wittig reaction, classical Hantzsch procedure, Barton–Zard cross-coupling reactions, annulations reactions, transition metal-mediated cyclization, multi-components reactions). However, their synthesis still remains challenging in heterocyclic chemistry because they are easily degradable, which makes isolation and purification more difficult.[4] Among the reactions, the Paal–Knorr reaction and Clauson-Kaas reaction are the typical and commonly used protocols to produce *N*-substituted pyrroles. The preparation of substituted pyrroles by the condensation of 1,4-dicarbonyl compounds with ammonia or primary amines is known as the Paal–Knorr pyrrole synthesis. In the Clauson-Kaas pyrrole synthesis instead of a 1,4-dicarbonyl derivatives the 2,5-dimethoxytetrahydrofuran is used as four-carbon source. This reaction converts the primary amino group into a pyrrole group and involves the acid-catalyzed cyclization between an aromatic (or primary aliphatic) amine and 2,5-dimethoxytetrahydrofuran (Scheme 17.5.1).[4,5]

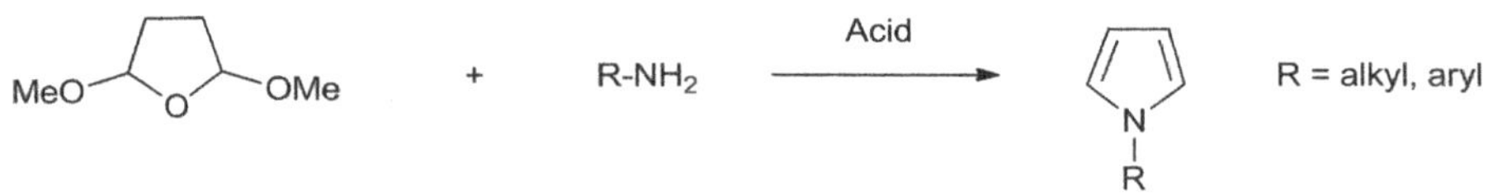

Scheme 17.5.1 Clauson-Kaas pyrrole synthesis.

Although this reaction provides an important and useful way to synthesize *N*-substituted pyrroles, the need of high temperature and acidic conditions can lead to a decomposition of the product, low yields and difficulties with the product purification. To overcome these problems with the classical Clauson-Kaas reaction, several modifications/improvements have been made such as using two-phase systems, initial aqueous hydrolysis of 2,5-dimethoxytetrahydrofuran, and several catalysts have also been used such as glacial acetic acid, phosphorus pentoxide, montmorillonite K-10, cerium(III) chloride, cupric chloride, iron(III) chloride, bismuth nitrate and scandium triflate, MgI_2 etherate, iodine, and nano-organocatalysts. This reaction can also be performed using microwave irradiation.[6,7]

The aim of the present experiment is the preparation, isolation and purification of 1-(4-bromophenyl)-1*H*-pyrrole, using simple experimental techniques and commercially available reagents. The preparation of this compound involves the synthesis of the pyrrole ring through the Clauson-Kaas synthesis. The determination of the structure of the compound prepared will be also performed through ^{1}H NMR spectroscopy.

Additional Safety

The experimental work should be performed in a well-ventilated fume hood with proper safety equipment such as protective clothing, gloves, and safety goggles.

4-Bromoaniline, 2,5-dimethoxytetrahydrofuran, and dichloromethane are toxic reagents and may cause skin, eye and respiratory tract irritation. Glacial acetic acid is a corrosive acid and may cause severe skin and eye burns, and is flammable in liquid and vapor form. Complete information about the safety, toxicity and disposal for all reagents can be found, free of charge, at: http://www.msds.com/index.asp?open=/protected_public/loginsuccessful.asp.

Experimental Procedure

Laboratory Session 1 (4 h): Synthesis of 1-(4-Bromophenyl)-1*H*-pyrrole

1. In a fume hood prepare an apparatus for stirring solutions at high temperatures consisting of a 50 mL round-bottomed flask with a stirrer bar, a water condenser and a magnetic stirring plate and an oil bath.

2. Add to the flask 4-bromoaniline (200 mg, 1.2 mmol, 1 equiv.), 2,5-dimethoxytetrahydrofuran (0.18 mL, 1.4 mmol, 1.2 equiv.) and 4 mL of glacial acetic acid.
3. Insert a water condenser into the round-bottomed flask and clamp the flask in the pre-heated oil bath. Stir the reaction for 2 h, maintaining a constant temperature of 118 °C.
4. Cool the reaction mixture to room temperature and check the evolution of the reaction by TLC comparing the contents of the reaction mixture with the reagents (eluent: dichloromethane/hexane (1 : 1)).
5. Add 20 mL of water to the reaction mixture.
6. Transfer the obtained mixture into a separatory funnel and extract with dichloromethane (2 × 50 mL). Separate and dry the combined organic layers with anhydrous $MgSO_4$, filter the drying agent and evaporate the solvent under reduced pressure with a rotary evaporator.

Laboratory Session 2 (2 h): Purification of 1-(4-Bromophenyl)-1*H*-pyrrole

1. Dissolve the crude product obtained in session 1 in a minimum amount of dichloromethane (approx. 20 mL).
2. Add hexane until obtaining precipitation of the 1-(4-bromophenyl)-1*H*-pyrrole as a beige solid.
3. Filter off the solid that precipitates.
4. Wash the filtered solid with cold hexane.
5. Weigh the product and calculate the yield. Check the purity by ^{1}H NMR spectroscopy and melting point determination. Compare with the literature data[3f] as well as with the spectroscopic data presented in the ESI.

Waste Disposal

Hexane should be collected in a non-chlorinated waste collector recipient whereas dichloromethane should be collected in a chlorinated waste collector recipient. The aqueous solutions should be neutralized and eliminated in the sink.

Results Interpretation and Additional Questions

1. Interpret the acquired ^{1}H NMR spectrum, and compare the obtained melting point with the literature data.[3f]
2. Propose a mechanism for the Clauson-Kaas reaction.
3. Suggests others methods of syntheses for 1-substituted pyrroles.
4. Explain the role of the acetic acid in the Clauson-Kaas reaction.
5. Explain why the use of stronger mineral acids such as HCl and H_2SO_4 should be avoided in the Clauson-Kaas synthesis.

References

1. A. H. Jackson, M. Artico, H. J. Anderson, C. E. Loader, A. Gossauer, P. Nesvadba and N. Dennis, *The Chemistry of Heterocyclic Compounds*, ed. R. A. Jones, John Wiley & Sons, Inc., New York, 1990, vol. 48, Part 1, p. 295.
2. (a) R. E. Niziursky-Mann and M. P. Cava, *Heterocycles*, 1992, **34**, 2003; (b) L. I. Belen'kii, *Heterocycles*, 1994, **37**, 2029.
3. (a) M. M. M. Raposo, Recent developments in the chemistry of 2-thienylpyrroles: synthesis, reactivity and applications, in *Targets in Heterocyclic Systems: Chemistry and Properties*, ed. O. A. Attanasi and D. Spinelli, Italian Society of Chemistry, 2007, vol. 11, pp. 122–154; (b) M. M. M. Raposo, A. M. R. C. Sousa, G. Kirsch, P. Cardoso, M. Belsey, E. Matos Gomes and A. M. C. Fonseca, *Org. Lett.*, 2006, **8**, 3681; (c) J. Pina, J. S. Seixas de Melo, R. M. F. Batista, S. P. G. Costa and M. M. M. Raposo, *Phys. Chem. Chem. Phys.*, 2010, **12**, 9719; (d) D. Collado, J. Casado, S. R. González, J. T. L. Navarrete, R. Suau, E. Perez-Inestrosa, T.

M. Pappenfus and M. M. M. Raposo, *Chem.-Eur. J.*, 2011, **17**, 498; (e) M. M. M. Raposo, M. C. R. Castro, A. M. C. Fonseca, M. F. S. Cardoso, M. Belsley, L. M. Carvalho and P. Coelho, *Dyes Pigm.*, 2011, **91**, 62; (f) M. C. R. Castro, M. Belsley, A. M. C. Fonseca and M. M. M. Raposo, *Tetrahedron*, 2012, **68**, 8147; (g) J. Garcia-Amorós, M. C. R. Castro, P. Coelho, M. M. M. Raposo and D. Velasco, *Chem. Comm.*, 2013, **49**, 11427; (h) J. Garcia-Amorós, M. Reig, M. C. R. Castro, A. Cuadrado, M. M. M. Raposo and D. Velasco, *Chem. Comm.*, 2014, **50**, 6704; (i) R. M. F. Batista, S. P. G. Costa and M. M. M. Raposo, *Dyes Pigm.*, 2014, **102**, 293.

4. (a) V. Estevez, M. Villacampa and J. C. Menendez, *Chem. Soc. Rev.*, 2010, **39**, 4402; (b) V. F. Ferreira, M. C. B. V. de Souza, A. C. Cunha, L. O. R. Pereira and M. L. G. Ferreira, *Org. Prep. Proced. Int.*, 2001, **33**, 411.
5. N. E. N. Clauson-Kaas, *Acta Chem. Scand.*, 1953, **6**, 867.
6. (a) B. S. Gourlay, P. P. Molesworth, J. H. Ryan and J. A. Smith, *Tetrahedron Lett.*, 2006, **47**, 799; (b) K. C. Miles, S. M. Mays, B. K. Southerland, T. J. Auvil and D. M. Ketcha, *ARKIVOC*, 2009, **14**, 181.
7. L. Kurti and B. Czako, *Strategic Applications of Named Reactions in Organic Synthesis*, Elsevier, San Diego, USA, 2005, p. 328.

17.6. Synthesis of the Manganese(III) Complex of 5,10,15,20-Tetrakis(2,6-dichlorophenyl)porphyrin

Artur M. S. Silva, Augusto C. Tomé, Diana C. G. A. Pinto, Fernando M. J. Domingues, Graça M. S. R. O. Rocha, José A. S. Cavaleiro, Maria da Graça P. M. S. Neves*, Maria do Amparo F. Faustino, and Mário M. Q. Simões

Chemistry Department, University of Aveiro, Campus de Santiago, 3810-193 Aveiro, Portugal
*E-mail: gneves@ua.pt

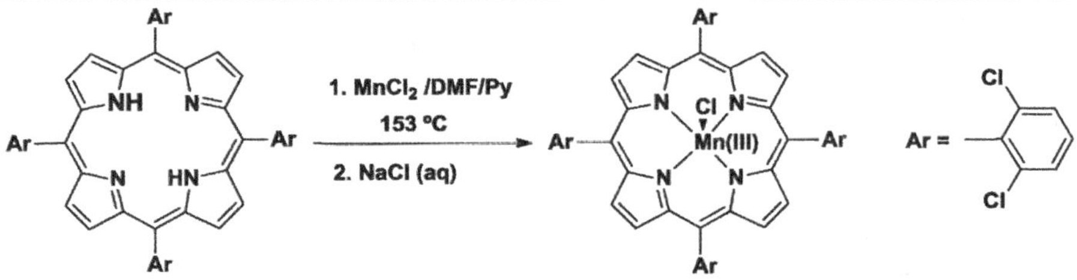

1 2

Number of sessions (duration of each session)	Hazard level	Difficulty level	Level of study
2 (3 h each)	Moderate	Medium	Intermediate

Class names Aromatic heterocycles

Concepts involved Coordination chemistry, the manganese(III) complex of 5,10,15,20-tetrakis(2,6-dichlorophenyl)porphyrin is synthesised by the complexation reaction between 5,10,15,20-tetrakis(2,6-dichlorophenyl)porphyrin and manganese(II) chloride in dimethylformamide (DMF) and pyridine at 153 °C

Chemicals needed *N*,*N*-Dimethylformamide (DMF), pyridine, manganese(II) chloride, anhydrous sodium sulfate, dichloromethane, hexane, toluene, sodium chloride, 5,10,15,20-tetrakis(2,6-dichlorophenyl)porphyrin

Equipment and experimental techniques involved Liquid–liquid extraction, vacuum filtration, stirring hot plate, oil bath, rotary evaporator, thin-layer chromatography, UV-vis spectrophotometry, reaction under nitrogen

Keywords Crystallization, heterocycles, manganese(III) complex, thin-layer chromatography, UV-vis spectrophotometry

Background

The ability of porphyrinic macrocycles to coordinate metals is one of the most widely studied areas in porphyrin chemistry. Metalloporphyrins are important natural compounds involved in electron transfer, respiration and photosynthesis.[1] Some metalloporphyrins can be used as biomimetic models in catalysis, since metalloporphyrins are known to catalyse, *in vitro*, a broad

Comprehensive Organic Chemistry Experiments for the Laboratory Classroom
Edited by Carlos A. M. Afonso, Nuno R. Candeias, Dulce Pereira Simão, Alexandre F. Trindade, Jaime A. S. Coelho, Bin Tan, and Robert Franzén.

Published by the Royal Society of Chemistry, www.rsc.org

range of cytochrome P450 mediated reactions. In the last few decades, the use of efficient and robust metalloporphyrins as catalysts in oxidative systems, particularly Mn(III) complexes, has been the subject of great interest.[2]

The formation of metalloporphyrins is due to the ability of the free-base porphyrins to complex almost any known metal ion with the pyrrole nitrogen atoms through the formation of *N*–metal σ-bonds.[3,4] This area of porphyrin chemistry is well developed and includes not only the chelating of metal centres but also the investigation of metal–metal bonding and organometallic systems. Porphyrins are known to coordinate with most of the metallic and pseudometallic elements, and the synthesis of numerous different organometallic porphyrin derivatives containing one or more metal–carbon bonds is possible.[5]

The objective of this work is the synthesis of the manganese(III) complex of 5,10,15,20-tetrakis(2,6-dichlorophenyl)porphyrin (**2**), which is known to be an excellent model to mimic the action of P450 enzyme, by the reaction of the 5,10,15,20-tetrakis(2,6-dichlorophenyl)porphyrin with manganese(II) chloride in dimethylformamide (DMF) and pyridine at 153 °C. Before coordination, the UV-visible absorption spectra of the free-base porphyrins (with hydrogen atoms linked to two of the inner nitrogen atoms) consist of two distinct regions due to π–π* transitions. In the near-UV region (370–430 nm) there is an intense absorption known as the Soret band, which has typical extinction coefficients of approximately 1×10^5 M^{-1} cm^{-1}. The Soret band is characteristic of the 18π electron system. In the visible region (500–700 nm) several absorptions are present, known as the Q bands, which have extinction coefficients of approximately 1×10^4 M^{-1} cm^{-1}. The free-base porphyrins usually have a four-banded visible absorption spectrum. The relative intensity of these Q bands is dependent on the substituents on the aromatic group. The UV-vis absorption spectra of metalloporphyrins consist also of two distinct regions: the Soret band in the near-UV region and, in general, two bands in the visible region.

Consequently, the reaction can be easily monitored by UV-vis and also by thin-layer chromatography (TLC). At the end of the reaction, porphyrin (**2**) is crystallized in a mixture of dichloromethane and hexane.

Additional Safety

All chemicals used in this work should be handled with appropriate care. All of them can be considered toxic and irritant and inhalation or any contact with skin and eyes should be avoided. Laboratory coat, safety glasses and adequate gloves are compulsory and all the experimental work should be done in a well-ventilated fume hood.

Dimethylformamide is irritant, toxic if inhaled, harmful by skin absorption and exposure may result in foetal death. Pyridine is flammable, irritant, and harmful by ingestion and by skin absorption and may affect fertility. Dichloromethane is considered as a possible carcinogen.

Experimental Procedure

First Laboratory Session: 3 h

Synthesis of the Manganese(III) Complex of 5,10,15,20-Tetrakis(2,6-dichlorophenyl) porphyrin

1. Into a 25 mL round-bottom flask, equipped with a reflux condenser and a magnetic stir bar, weigh 50 mg (0.1 mmol) of porphyrin (**1**) and dissolve it in 5.0 mL of DMF (Note 1). Add 0.5 mL of pyridine and 10 equiv. (198 mg, 1 mmol) of manganese(II) chloride ($MnCl_2 \cdot 4H_2O$).
2. In a well-ventilated fume hood, reflux the solution in the dark, under a nitrogen atmosphere.

3. Monitor the progress of the reaction by UV-vis (CH_2Cl_2 as solvent) and by TLC (CH_2Cl_2 as eluent). The reaction normally takes 2 h to complete.
4. At the end of the reaction, turn off the heating and let the reaction mixture under stirring for about 30 min, in the open air and protected from light.

Second Laboratory Session: 3 h

Product Isolation

1. Remove the solvent in the rotary evaporator and dissolve the residue in 10–15 mL of dichloromethane (Note 2).
2. Wash the organic phase 2–3 times with water (50 mL) and finally with an aqueous saturated sodium chloride solution (50 mL), in a separating funnel.
3. Filter the organic phase through a glass funnel with cotton and anhydrous sodium sulfate into a round-bottom flask (100 mL), remove the solvent in the rotary evaporator and then crystallize the residue in hexane, after dissolution in the minimal amount of dichloromethane.
4. The crystals formed should then be filtered under vacuum, with a Hirsch funnel (small Büchner funnel) with filter paper, washed several times with hexane and placed in an oven (60 °C) to dry for about 1 h. The yield, based on the porphyrin, should be over 90%.

Note 1. Porphyrin (**1**) is commercially available but can be synthesized by a condensation reaction between pyrrole with 2,6-dichlorobenzaldehyde, as described in experiment 5.2.4.

Note 2. As DMF has a very high boiling point (153 °C), addition of a small amount of toluene will facilitate the solvent removal.

Waste Disposal

The mixture containing DMF and pyridine may be collected in a separate recipient, for posterior treatment/destruction. Dichloromethane may be collected in a recipient for chlorinated solvents, for posterior destruction/recovery. Hexane may be collected in a recipient for non-chlorinated solvents, for posterior destruction/recovery.

Results Interpretation and Additional Questions

1. Determine the final product yield.
2. Explain why the structure of this complex cannot be confirmed by ^1H NMR spectrum.
3. Justify the role of pyridine in this synthesis.
4. Identify the type of electronic transitions responsible for the Soret and the Q bands.
5. Explain why a simplification in the Q bands pattern is observed after coordination of the porphyrin macrocycle with the metal.

References

1. R. Guilard and K. M. Kadish, *Chem. Rev.*, 1988, **88**, 1121.
2. M. M. Q. Simões, C. M. B. Neves, S. M. G. Pires, M. G. P. M. S. Neves and J. A. S. Cavaleiro, *Pure Appl. Chem.*, 2013, **85**, 1671.
3. J. W. Buchler, in *The Porphyrins*, ed. D. Dolphin, 1978, vol. 1, p. 389.
4. J. W. Buchler, in *Porphyrins and Metalloporphyrins*, ed. K. M. Smith, 1975, p. 157.
5. M. O. Senge, *Angew. Chem., Int. Ed. Engl.*, 1996, **35**, 1923.

17.7. Synthesis of 1,3-Dithienylbenzo[*c*]thiophene

Tiago J. L. Silva*[a,b], Paulo J. G. Mendes[a], Ana Isabel Tomaz[b], and M. Helena Garcia[b]

[a]Departamento de Química, Escola de Ciências e Tecnologia, Centro de Química de Évora, Instituto de Investigação e Formação Avançada, Universidade de Évora, Rua Romão Ramalho 59, 7000-671 Évora, Portugal; [b]Centro de Química Estrutural, Faculdade de Ciências da Universidade de Lisboa, Portugal
*E-mail: tjsilva@fc.ul.pt

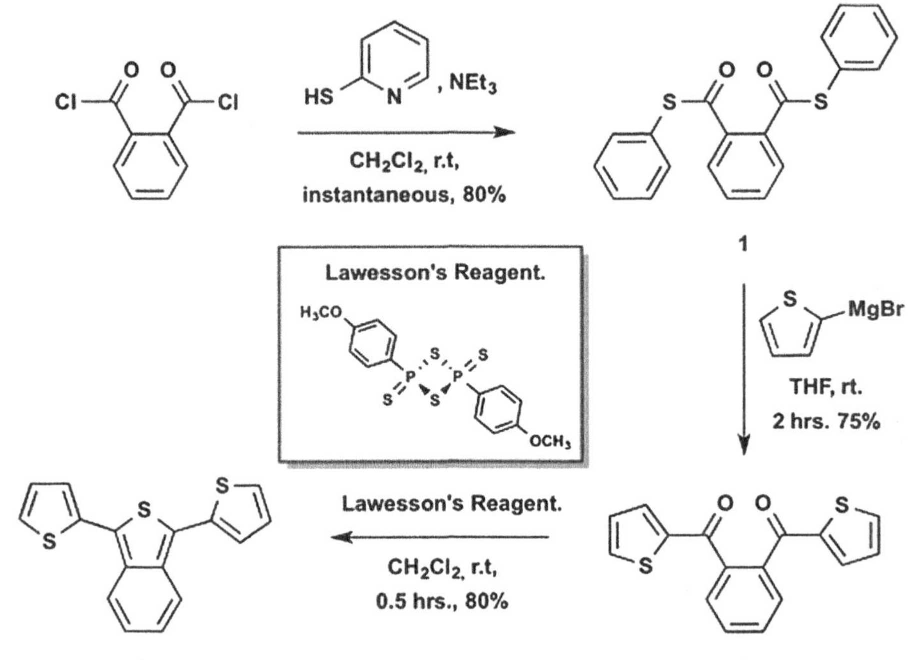

Number of sessions	Hazard level	Difficulty level	Level of study
3 (4 h + 6 h + 6 h)	High	High	Intermediate

Class names Acyl chlorides, ketones, thiophenes, benzo[*c*]thiophenes, thioesters

Concepts involved This experiment provides students with an example of an *S*-heterocyclization reaction (thionation) involving Lawesson's reagent, in a challenging three-step convergent synthetic procedure. Hence, students face three different types of reaction: a thioesterification in the first step, providing a good example of acyl leaving group reactivity; a Grignard reaction in the formation of a *o*-diketone, showing a C–C bond formation reaction; and finally the *S*-heterocyclization of the previous product to afford the product 1,3-dithienylbenzo[*c*]thiophene

Chemicals needed 2-Mercaptopyridine, phthaloyl dichloride, triethylamine, 2-bromothiophene, 2,4-bis (4-methoxyphenyl)-2,4-dithioxo-1,3,2,4-dithiadiphosphetane (Lawesson's reagent), iodine, tetrahydrofuran, dichloromethane, chloroform, ethanol, *n*-hexane, hydrochloric acid, sodium hydroxide, sodium bicarbonate, anhydrous sodium sulfate

Comprehensive Organic Chemistry Experiments for the Laboratory Classroom
Edited by Carlos A. M. Afonso, Nuno R. Candeias, Dulce Pereira Simão, Alexandre F. Trindade, Jaime A. S. Coelho, Bin Tan, and Robert Franzén.

Published by the Royal Society of Chemistry, www.rsc.org

Equipment and experimental techniques involved Liquid–liquid extraction, column chromatography, NMR spectroscopy, Schlenk techniques, rotatory evaporation apparatus, heating plate with magnetic stirring, reflux apparatus
Keywords Column chromatography, fused rings, Grignard synthesis, heterocyclic chemistry, NMR, thiophene

Background

Thiophenes and oligo-, poly- and fused-thiophene derivatives are among the most studied heterocyclic compounds since they exhibit remarkable chemical and physical properties. Common applications of thiophene derivatives are in optical and electronic devices (solar cells, lasers, OLEDs and OFETs), electrochemical and biosensors, functional polymers, and medicinal drugs.[1]

Fused-ring thiophenes, such as benzo[*c*]thiophene (isothianaphthene) derivatives have attractive electric and optical properties and can be used as monomers to build low-bandgap polymers. Benzo[*c*]thiophene compounds are relatively unstable and their synthesis is often laborious. The stability can be improved by the presence of aromatic heterocycles in 1 and 3 positions. For example, 1,3-dithienylbenzo[*c*]thiophene has been studied due to its interesting electro-optical properties.[2,3]

The best way to synthesize 1,3-diarylisothianaphthene derivatives is through the cyclization of *ortho*-diketones with a thionation reagent such as P_4S_{10} or Lawesson's reagent, a procedure known as the Paal–Knorr synthesis (Scheme 17.7.1).[4]

Thionations are long-known important reactions in organic chemistry and can be performed using several reagents, with phosphorous pentasulfide being the most common. However, harsh conditions like high temperatures and a large excess of this thionation reagent are required. The formation of highly toxic H_2S is also a major drawback. At the beginning of the 1980s, Lawesson and co-workers developed an alternative that gradually substituted phosphorous pentasulfide, and became the most used thionation reagent for the formation of five and six-membered aromatic rings. The Lawesson's reagent is a dimeric anhydride of a thiophosphinic acid and is used for the thionation of ketones, esters and amides. The advantages of this reagent over all other available thionation reagents are the higher yields obtained, their easier manipulation and storage, and an easier work-up of reactions.[5]

The crucial step in the synthesis of 1,3-dithienylbenzo[*c*]thiophene is the preparation of the initial *o*-diketone. It is well known that both Diels–Alder[6] and Friedel–Crafts[7] type reactions are possible ways to synthesize these compounds, but present several drawbacks that prevent them being used with a wide range of functional groups. A straightforward way to obtain *ortho*-diketones is the use of Grignard-type reactions on *ortho*-dicarboxylic arenes.[8] The Grignard reaction provides a very straightforward and versatile method for the formation of C–C bonds. This reaction is, in fact, an organometallic reaction in which an alkyl or aryl-magnesium reagent (called a Grignard reagent) reacts *via* an addition reaction to a carbonyl group. Grignard reagents

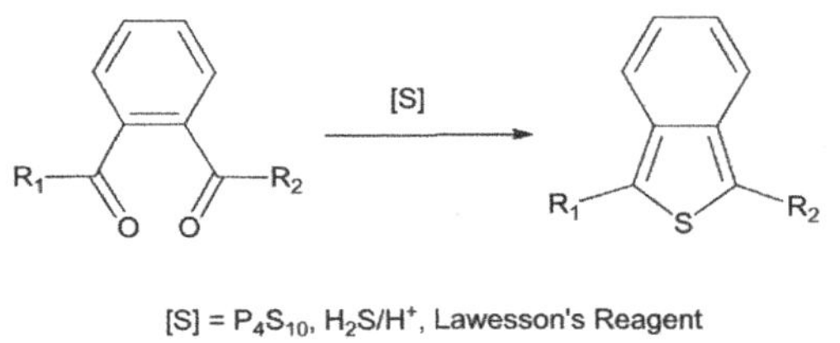

Scheme 17.7.1 Paal–Knorr synthesis for fused-thiophene derivatives.

can be easily obtained by reaction of an alkyl or aryl halide with metallic magnesium in anhydrous ethers such as diethyl ether or tetrahydrofuran (THF), affording an organomagnesium compound of general formula R–MgX (R = alkyl or aryl; X = halogen). They can act as strong nucleophiles or strong bases. The formation of these reagents is not fully understood, but the general accepted mechanism involves radical species.[9]

This experiment proposes the synthesis of 1,3-dithienobenzo[*c*]thiophene by ring closure of an *o*-diketone. The procedure involves three high yielding steps with minimum purification procedures.[10,11] In the first step a diacyl dichloride is transformed into a 1,2-di[*S*-(2-pyridinyl)]benzothioate. The following step is a Grignard reaction that leads to the formation of the 1,2-dithienoylbenzene that can be cyclized to the final product using Lawesson's reagent.

Additional Safety

All sessions involving reflux conditions and Lawesson's reagent should be done in a fume hood, with proper safety equipment, such as protective clothing, gloves and safety goggles. Hydrochloric acid may cause severe skin, lung and eye burns. Tetrahydrofuran and *n*-hexane are flammable solvents, harmful if swallowed, inhaled, or in contact with the skin and eyes. Dichloromethane and chloroform are known for being potential carcinogens and should be handled with care. All laboratory sessions should be done in a fume hood.

Experimental Procedure

Laboratory Session 1 (4 h) – Synthesis of *S*,*S*-Di(pyridin-2-yl)benzene-1,2-bis(carbothioate) (1)

1. Prepare a 250 mL round-bottomed flask with a big magnetic stir bar (1.5 cm) and add 1.15 g (10 mmol) of 2-mercaptopyridine. Add 20 mL of freshly dried THF[12] followed by 2.5 mL (18 mmol, 1.8 equiv.) of dried triethylamine. Place the flask in an ice-cold water bath and stir for 15 min.
2. In the meantime, prepare 100 mL of a 2% (v/v) hydrochloric acid solution in water.
3. In a 50 mL Erlenmeyer flask dilute 1.1 mL (4 mmol, 0.4 equiv.) of phthaloyl dichloride in 20 mL of freshly dried THF.
4. Add the solution of the phthaloyl dichloride to the solution prepared in the first step and *immediately* quench it with the 100 mL hydrochloric acid solution.
5. Transfer the mixture to a separatory funnel, and extract three times with 30 mL of dichloromethane. Wash the combined organic phases twice with 30 mL of a 5% sodium hydroxide solution, 30 mL of a saturated solution of sodium bicarbonate and, finally, 30 mL of water.
6. Dry the organic phases with anhydrous sodium sulfate and collect the filtrate to a 100 mL flask (tared). Evaporate the solvent in a rotatory evaporator.
7. Dissolve the resulting solid in the minimum amount of dichloromethane (approx. 15 mL), and slowly add twice the volume of diethyl ether in order to obtain two phases.
8. Place the flask in an ice bath and wait until full precipitation is obtained (~30 min). Slowly discard the liquid in to a beaker and dry the solid in a vacuum pump for a few minutes.
9. The white solid product *should be kept under a nitrogen atmosphere* until needed.

Laboratory Session 2 (6 h) – Synthesis of 1,2-Phenylenebis(thiophen-2-yl-methanone) (2)

1. Prepare a Schlenk flask with a magnetic stir bar and add 0.164 g of activated magnesium turnings (7 mmol). Purge with nitrogen gas. Add 50 mL of freshly dried and degassed THF[12] and a small crystal of iodine. Stir the solution until all the iodine is dissolved.

2. To the prepared suspension, add 0.61 mL (6 mmol, 1 equiv.) of 2-bromothiophene dropwise, and seal the Schlenk flask. Stir at room temperature for 30 min until the exothermic reaction is finished.
3. Place the Schlenk flask in an oil bath, adapt a condenser and heat the solution to a gentle reflux for 1.5 h (*Solution A*).
4. In the meantime, prepare a second Schlenk with a magnetic stir bar and add 1.09 g (3 mmol, 0.5 equiv.) of *S*,*S*-di(pyridin-2-yl)benzene-1,2-bis(carbothioate) (**1**) prepared in the previous laboratory session. Degas the flask and add 25 mL of freshly distilled and degassed THF. Stir until all the solid is dissolved and place the flask in an ice-cold water bath (*Solution B*).
5. Cool *Solution A* to room temperature and place it in an ice cold water bath for 10 min. Add *Solution A* to *Solution B* dropwise *via* cannula under nitrogen atmosphere, and stir the resulting mixture for an additional 30 min.
6. Quench the reaction by adding the mixture to a 100 mL of 5% hydrochloric acid water solution. Transfer the resulting mixture into a separatory funnel and extract the product three times with 30 mL of dichloromethane.
7. Wash the combined organic phases twice with (i) 50 mL of a 5% sodium hydroxide solution, (ii) 50 mL of a saturated sodium bicarbonate solution and, finally, (iii) 50 mL of water.
8. Dry the organic phases with anhydrous sodium sulfate and filter to a 100 mL flask (tared). Evaporate the solvent in the rotatory evaporator.
9. Dissolve the resulting solid in the minimum amount of dichloromethane (approx. 10 mL), and slowly add twice the volume of hexane in order to obtain two phases.
10. Place the flask in a water-ice bath and wait until full precipitation is obtained (~30 min). Slowly discard the liquid into a beaker and dry the solid in the vacuum pump for a few minutes.

Laboratory Session 3 (6 h) – Synthesis of 1,3-Dithienylbenzo[*c*]thiophene (3)

1. Prepare a 100 mL round-bottomed flask with a big magnetic stir bar and add 0.34 g (1.14 mmol) of 1,2-phenylene-bis(thiophen-2-yl-methanone (**2**)). Add 50 mL of dried dichloromethane. Stir at room temperature until the whole solid is dissolved.
2. To the previous solution, add two portions of 0.648 g of Lawesson's reagent (3.2 mmol, 2.8 equiv.) with a 1 h interval between each addition.
3. Stir the resulting solution for 3 h at room temperature. Meanwhile, prepare the chromatography column for step 8.
4. After completion of the reaction, evaporate the solvent in a rotatory evaporator.
5. To the brown slush obtained, add 20 mL of ethanol and heat the resulting solution for 30 min. Evaporate the solvent in a rotatory evaporator.
6. Dissolve the resulting slurry in 35 mL of dichloromethane and transfer the solution to a separatory funnel. Wash the organic phase twice with 40 mL of a 10% sodium hydroxide solution, 40 mL of a saturated sodium bicarbonate solution and, finally, several times (3–5 times) with 40 mL of water (each time).
7. Dry the organic phase with anhydrous sodium sulfate and filter it to a 250 mL round-bottomed flask. Evaporate the solvent in the rotatory evaporator.
8. Fill a chromatography column (3 cm diameter × 30 cm height) with slurry of silica gel in hexane. Chromatograph the solid obtained in the previous step using hexane, and collect the first yellow band in a 250 mL round-bottomed flask (tared).
9. Evaporate the solvent in the rotatory evaporator.
10. Collect the solid and prepare an NMR tube to record the ^{1}H and ^{13}C NMR spectra of the final purified product.

Waste Disposal

Hydrochloric acid and sodium hydroxide base solutions should be neutralized, diluted and discarded in the sink. Any remaining magnesium should be destroyed by slow addition of the metal into a well-stirred 5% hydrochloric acid solution. This operation must be performed very carefully in a well-ventilated fume hood due to the formation of hydrogen gas. Non-chlorinated organic solvents should be collected in a non-chlorinated waste collector recipient, and dichloromethane and chloroform should be collected in a recipient for chlorinated compounds. All glassware and solvents used in the thionation reaction should be treated with a 5% hypochlorite solution before being washed and, in the case of the solvents, discarded in a proper container. Silica gel and all drying agents (sodium sulfate) should be collected in an appropriate solid waste collector recipient for posterior destruction.

Results Interpretation and Additional Questions

1. Calculate the yield of each reaction and the global yield.
2. The product 1,2-phenylenebis(thiophen-2-yl-methanone) cannot be synthesized directly from phthaloyl dichloride, thus a previous thioester compound has been synthesized (Reaction 1). Explain why. (*Hint*: check the relative reactivity of both carbonyl groups obtained after the addition of the first equiv. of the Grignard reagent.)
3. Draw the mechanism of the Grignard reaction, including the formation of the Grignard reagent.
4. Draw the mechanism of the *S*-heterocyclization using the Lawesson's reagent.
5. Record and analyse the ^{1}H and ^{13}C NMR spectra of the final product. Make the correct attributions for all the protons and carbons and compare the results obtained with those in the literature.[10,11]

References

1. A. Mishra, C.-Q. Ma, J. L. Segura and P. Bäuerle, *Handbook of Thiophene-based Materials: Applications in Organic Electronics and Photonics*, ed. I. F. Perepichka and D. F. Perepichka, John Wiley & Sons Ltd., Chichester, 2009, ch. 1, pp. 1–155.
2. T. Yamamoto, M. Usui, H. Ootsuka, T. Iijima, H. Fukumoto, Y. Sakai, S. Aramaki, H. M. Yamamoto, T. Yagi and H. Tajima, *Macromol. Chem. Phys.*, 2010, **211**, 2138.
3. A. K. Mohanakrishnan, N. S. Kumar and P. Amaladass, *Tetrahedron Lett.*, 2008, **49**, 4792.
4. J. Joule and K. Mills, *Heterocyclic Chemistry*, ed. J. Joule and K. Mills, John Wiley & Sons Ltd., Chichester, 5th edn, 2010, ch. 22, pp. 447–454.
5. T. Ozturk, E. Ertas and O. Mert, *Chem. Rev.*, 2007, **107**, 5210.
6. W. Volz and J. Voss, *J. Prakt. Chem.*, 1991, **333**, 889.
7. P. Bauerle, G. Gotz, P. Emerle and H. Port, *Adv. Mater.*, 1992, **4**, 564.
8. D. R. Shridar, M. Jogibhukta, R. P. Shantan and V. K. Handa, *Synthesis*, 1982, 1061.
9. G. S. Silverman and P. E. Rakita, *Handbook of Grignard Reagents*, ed. G. Silverman and P. E. Rakita, Marcel Dekker, New York, 1996.
10. T. J. L. Silva, P. J. Mendes, M. H. Garcia, M. P. Robalo, J. P. Prates Ramalho, A. J. Palace Carvalho, M. Büchert, C. Wittenburg and J. Heck, *Eur. J. Inorg. Chem.*, 2013, **20**, 3506.
11. R. H. L. Kiebooms, P. J. A. Adriaensens, D. J. M. Vanderzande and J. M. J. V. Gelan, *J. Org. Chem.*, 1997, **62**, 1473.
12. D. D. Perrin, W. L. F. Amarego and D. R. Perrin, *Purification of Laboratory Chemicals*, Pergamon, New York, 2nd edn, 1980.

17.8. Synthesis of 1-Propyl-2-(thiophen-2-yl)-1*H*-pyrrole

M. Manuela M. Raposo

Department of Chemistry, University of Minho, Campus de Gualtar, 4710-057 Braga, Portugal
*E-mail: mfox@quimica.uminho.pt

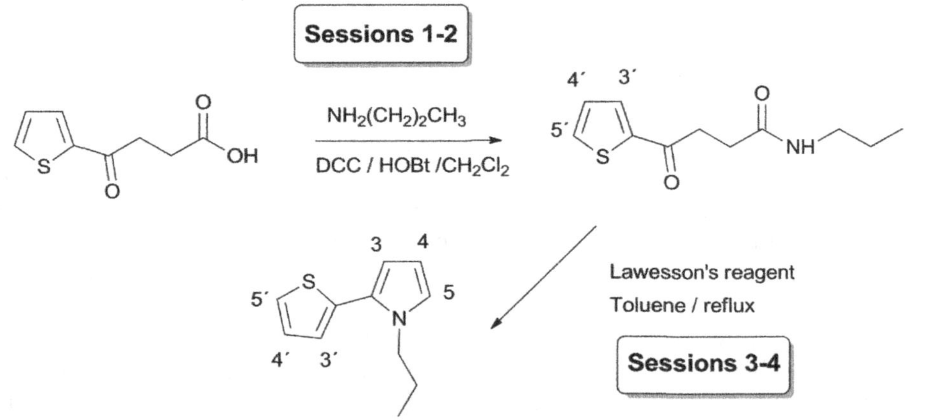

Number of sessions (duration)	Hazard level	Difficulty level	Level of study
5 (4 h + 3 h + 4 h + 2 h+ 4 h)	High	Medium	Advanced

Class names Carboxylic acids, amides, thiophenes, pyrroles

Concepts involved Reactivity of secondary amides with Lawesson's reagent; synthesis of a thienylpyrrole derivative involving several synthetic transformations, namely (i) synthesis of a secondary γ-ketoamide by direct amidation of the 4-oxo-(2′-thienyl) butanoic acid in the presence of DCC/HOBt coupling reagents and (ii) thionation of carbonyl groups using Lawesson's reagent followed by cyclization of the 4-thioxo derivative yielding the thienylpyrrole heterocyclic system; structure elucidation of synthesized compounds by the spectroscopic techniques ^{1}H NMR, ^{13}C NMR, IR and UV

Chemicals needed 4-Oxo-4-(thiophen-2-yl)butanoic acid (synthesized in Experiment 3.1.16 or commercially available), dichloromethane, methanol, 1,3-dicyclohexylcarbodiimide (DCC), 1-hydroxybenzotriazole (HOBt), propylamine, acetone, acetone-d_6, ethanol, toluene, citric acid solution (5%), diethyl ether, magnesium sulfate (anhydrous), Lawesson's reagent (LR), chloroform, deuterated chloroform, petroleum ether (40–60 °C), silica gel for column chromatography (60 Å; 230–400 mesh) and thin-layer chromatography (TLC) plates with a fluorescence indicator

Equipment and experimental techniques involved Apparatus for heating at reflux, rotary evaporator, gravity and vacuum filtration, recrystallization, thin-layer chromatography (TLC), column chromatography, melting point

Keywords Amide, carbonyl thionation, chromatography, ^{1}H NMR, intramolecular cyclization, pyrrole

Comprehensive Organic Chemistry Experiments for the Laboratory Classroom
Edited by Carlos A. M. Afonso, Nuno R. Candeias, Dulce Pereira Simão, Alexandre F. Trindade, Jaime A. S. Coelho, Bin Tan, and Robert Franzén.

Published by the Royal Society of Chemistry, www.rsc.org

Background

Thiophene and pyrrole moieties play important roles in synthetic and medicinal chemistry, being present in a large number of natural products and biologically active compounds.[1] Among five-membered heterocycles, pyrroles possess the highest reactivity and most diverse chemical properties. It is not accidental that it is pyrrole that Nature has "chosen" for constructing the vital pigments chlorophylls and haemoglobin, as well as other vitally important supramolecular structures (various porphyrins, chlorins, prodigiosins, vitamin B_{12}, *etc.*).

For this reason, amongst all five-membered aromatic heterocycles, molecules containing the pyrrole nucleus have attracted the greatest attention of researchers and have been studied in the most detail. Especially, 2-aryl- and 2-heteroaryl-substituted pyrroles are of great interest to the pharmaceutical industry. They are used, for instance, as precursors in the synthesis of chemotherapeutics.[2]

Furthermore, synthetic pyrrole-containing derivatives (*e.g.* thienylpyrroles and 2,5-dithienylpyrroles) and π-conjugated oligo- and polypyrrolic systems are of growing relevance in materials science, supramolecular chemistry, and nanotechnology. For example, they have found applications in anion binding and cation coordination,[3] conducting organic polymers,[4] liquid crystals as well as materials for several optical applications (photochromic, nonlinear optics, organic light emitting diodes (OLEDs) and for dye sensitizer solar cells (DSSC)).[5]

The wide array of interesting properties has inspired the development of several procedures for the preparation of differently substituted pyrroles. Methods of synthesis include the classical Knorr, Paal–Knorr and Hantzsch strategies, transition-metal-catalysed couplings, 1,3-dipolar cycloadditions procedures and multicomponent protocols.[6] On the other hand, methods for the construction of 2-(2′-thienyl)pyrroles remain limited. Often, the yields are low and the regioselectivity is only modest. Therefore, the development of new methods for the synthesis of these heterocycles is an important and challenging objective.

A common approach to the synthesis of both thiophene and pyrrole groups involves the use of 1,4-dicarbonyl compounds. In the thiophene case, the 1,4-dicarbonyl compound is reacted with a source of sulfur, usually H_2S and HCl, phosphorus(V) sulfide or Lawesson's reagent (LR).[7] Similarly, pyrroles have traditionally been prepared *via* the condensation of 1,4-dicarbonyl compounds with ammonia or primary amines through the Paal–Knorr synthesis. However, the preparation of the initial 1,4-diketones and/or amines with thienyl substituents remains a challenge: their synthesis consists of several steps which do not always result in high yields, a fact that in many circumstances limits the applicability of this method.[6]

A new synthetic methodology for the preparation of functionalized pyrroles bearing a thienyl group at position 2 and alkyl or aryl groups at position 1 has been reported. This strategy for the synthesis of 1-(alkyl)aryl-2-thienyl-substituted pyrroles is based on the conversion of 4-oxo-(2-thienyl)butanoic acid into the alkyl and aryl-4-(2′-thienyl)-4-oxobutanamides by direct amidation with several commercial alkyl- and arylamines through DCC/HOBt mediated reaction,[9] followed by construction of the pyrrole heterocycle by thionation of the keto groups in the presence of Lawesson's reagent[7] and cyclization through a nucleophilic intramolecular attack.[8c,10] The reaction of secondary thienyl γ-keto amides with Lawesson's reagent can produce 2-thienylpyrroles, or a mixture of thienylpyrroles and bithiophenes. On the other hand, tertiary amides produce exclusively 5-*N*,*N*-dialkylamino-2,2′-bithiophenes[10,11] (see mechanisms in the ESI).

The goal of this experiment is the synthesis of a thienylpyrrole derivative through five sessions, using as precursor a secondary keto amide[8,12] – session 1: synthesis of 4-oxo-*N*-propyl-4-(thiophen-2-yl)butanamide; session 2: extraction of 4-oxo-*N*-propyl-4-(thiophen-2-yl)butanamide; session 3: purification of 4-oxo-*N*-propyl-4-(thiophen-2-yl)butanamide by recrystallization; session 4: preparation of 1-propyl-2-(thiophen-2-yl)-1*H*-pyrrole and session 5: purification of 1-propyl-2-(thiophen-2-yl)-1*H*-pyrrole using column chromatography.

This thienylpyrrole will be used in Experiment 5.2.14 as precursor for the synthesis of thienylpyrrole derivatives selectively functionalized on the pyrrole or on the thiophene heterocycles.

Additional Safety

All sessions should be done in a fume hood, with proper safety equipment, such as protective clothing, gloves and safety goggles. DCC (1,3-dicyclohexylcarbodiimide) is a potent allergen and sensitizer reagent (oral, dermal, inhalation) that may cause skin rashes and eye damage. HOBt (1-hydroxybenzotriazole) is a toxic reagent (oral, dermal, inhalation) that may cause skin and eye irritation. Propylamine is a corrosive and flammable reagent that may cause skin burns and eye damage. Citric acid may cause eye irritation. Lawesson's reagent is a flammable solid, with a strong odour, which in contact with water liberates H_2S toxic gas, therefore special attention should be taken in the manipulation of this reagent. Methanol, ethanol, chloroform, dichloromethane, diethyl ether, petroleum ether (40–60 °C), acetone and toluene are flammable reagents in both liquid and vapour form and may cause skin, eye and respiratory tract irritation. Additionally, diethyl ether may form explosive peroxides and the vapour may cause dizziness. Repeated exposure to acetone may cause skin dryness and drowsiness or dizziness. The preparation of the silica gel column and the separation of the components of the mixture (session 4) should be run in a well-ventilated fume hood in order to avoid the contact with the silica gel as well as the inhalation of volatile solvents such as chloroform, diethyl ether and petroleum ether. After being used the silica gel must be stored in a closed container in the fume hood for subsequent disposal. Complete information about the safety, toxicity and disposal for all reagents can be found, free of charge, at: http://www.msds.com/index.asp?open=/protected_public/loginsuccessful.asp.

Experimental Procedure

Laboratory Session 1 (4 h): Synthesis of 4-Oxo-*N*-propyl-4-(thiophen-2-yl)butanamide

1. In a fume hood, prepare an apparatus consisting of a 500 mL round-bottomed flask with a magnetic stirrer and an ice bath.
2. Add the 4-oxo-4-(thiophen-2-yl)butanoic acid (4 g, 21.6 mmol) and dichloromethane (120 mL), and stir the mixture until complete dissolution of 4-oxo-4-(thiophen-2-yl)butanoic acid occurs at 0 °C.
3. Add DCC (6 g, 28.4 mmol) and HOBt (4.4 g, 28.4 mmol) to the flask and keep stirring the mixture at room temperature. After 30 min add propylamine (1.84 g, 2.13 mL, 21.6 mmol) and keep stirring the mixture at room temperature for 24 h.

Laboratory Session 2 (3 h): Extraction of 4-Oxo-*N*-propyl-4-(thiophen-2-yl)butanamide

1. In a fume hood, filter with a vacuum the dicyclohexylurea (DCU) formed as a by-product.
2. Transfer the organic solution into a separatory funnel and wash with (3 × 100 mL) of citric acid solution (5%), remove the water from the combined organic layers with anhydrous $MgSO_4$, filter the drying agent and evaporate the solvent under reduced pressure with a rotary evaporator.
3. Dissolve the residue obtained in 50 mL of acetone and leave the solution in a refrigerator at −5 °C until the precipitation of the DCU occurs (approximately 30 min). Usually this procedure should be repeated three times, but it depends on the scale of the reaction. Filter the dicyclohexylurea from the organic solution and check the purity of 4-oxo-*N*-propyl-4-(thiophen-2-yl)butanamide by TLC using a mixture of chloroform/methanol (9 : 1) as eluent. Repeat this operation until the precipitation of DCU stops and check the purity of the amide by TLC.
4. Evaporate the solvent in a 250 mL round-bottomed flask, under reduced pressure with a rotary evaporator.

Laboratory Session 3 (4 h): Purification of 4-Oxo-*N*-propyl-4-(thiophen-2-yl)butanamide by Recrystallization

1. In a fume hood, prepare an apparatus for heating at reflux consisting of a heating mantle, a 250 mL round-bottomed flask that contains the reaction mixture, a reflux condenser and boiling chips.
2. Add 100 mL of ethanol to the flask containing the residue and heat the mixture at reflux until the residue is completely dissolved. If necessary add more ethanol to the mixture.
3. Filter the boiling chips and cool down the ethanolic solution at room temperature to promote precipitation of the product.
4. In a fume hood, filter with a vacuum the 4-oxo-*N*-propyl-4-(thiophen-2-yl)butanamide that precipitates as a colourless solid and wash the filtered solid with cold ethanol.
5. Dry the solid in an oven at 75 °C until constant weight. Weigh the product obtained, determine the melting point and calculate the yield. Record ^1H NMR in $CDCl_3$ and IR spectra and compare them with literature data[8b] to confirm the compound's purity.

Laboratory Session 4 (2 h): Synthesis of 1-Propyl-2-(thiophen-2-yl)-1*H*-pyrrole

1. In a fume hood, prepare an apparatus for heating at reflux consisting of a heating mantle, a 250 mL round-bottomed flask, a reflux condenser, boiling chips and a trap for H_2S.
2. Add 4-oxo-*N*-propyl-4-(thiophen-2-yl)butanamide (3 g, 13.3 mmol), Lawesson's reagent (5.4 g, 13.3 mmol) and 75 mL of toluene to the flask and heat the mixture at reflux for 30 min. Check the contents of the reaction mixture by thin-layer chromatography (TLC) by comparing with the amide precursor using a mixture of petroleum ether (40–60 °C)–diethyl ether (5 : 5) as eluent.
3. Cool the reaction mixture at room temperature, filter the Lawesson's reagent that did not react and evaporate the solvent under reduced pressure with a rotary evaporator.

Laboratory Session 5 (4 h): Purification of 1-Propyl-2-(thiophen-2-yl)-1*H*-pyrrole

1. Purify the residue obtained in session 4 by gravimetry column chromatography with silica gel (60 Å; 200–300 mesh): column height and diameter: 60 and 5 cm, respectively; silica weight: 150 g; silica height: ≈45 cm; column packing solvent: petroleum ether (40–60 °C). Dissolve the residue in the minimum necessary amount of chloroform and start the elution with a mixture of petroleum ether (40–60 °C)–diethyl ether (9.5 : 0.5) and continue with a mixture of petroleum ether (40–60 °C)–diethyl ether (9 : 1). Check by TLC the contents of each collected fraction (eluent: petroleum ether (40–60 °C)–diethyl ether (5 : 5)).
2. Combine the fractions containing the thienylpyrrole product (R_f = 0.51) and evaporate to dryness in a rotary evaporator.
3. Weigh the product obtained as an orange oil and calculate the yield. Record a ^1H NMR spectrum in $CDCl_3$ and an IR spectrum in liquid film and compare them with the literature data,[8b] as well as with spectroscopic data presented in ESI, to confirm the compound's purity.

Waste Disposal

The solvents used in extractions and in column chromatography are recovered after evaporation in a rotary evaporator and should be collected in chlorinated or non-chlorinated waste collector recipients, for subsequent destruction. Used silica gel must be stored in a closed container in the fume hood for subsequent disposal. The drying agents should be disposed of in a proper container after isolation from the organic solutions. Special attention should be paid to the treatment and handling of reaction mixtures in sessions 1 and 4 which contain DCU and H_2S as by-products. DCU should be stored in a closed container in the fume hood for subsequent disposal. Hydrogen sulfide is a toxic colourless gas that is known by its unpleasant "rotten eggs" smell. This gas can be trap by bubbling it through a solution that reacts with the SH group. Examples of solutions used are for this propose are a basic solution of NaOH, a chlorine bleach solution or a potassium permanganate solution in water.

Results Interpretation and Additional Questions

1. Write down the reaction mechanism for the synthesis of the amide through the direct amidation of the precursor carboxylic acid in the presence of DCC/HOBt coupling agents.
2. Write down the structure of the Lawesson's reagent and propose a mechanism for the synthesis of the thienylpyrrole derivative.
3. Suggests an alternate method for the synthesis of 1-propyl-2-(thiophen-2-yl)-1*H*-pyrrole.
4. Interpret the ^{1}H and ^{13}C NMR and IR spectra obtained for all compounds[8b] and compare them with the spectra presented in the ESI.
5. Describe the major differences in the ^{1}H NMR and in the IR spectra that lead to the unambiguous identification of the amide precursor and the thienylpyrrole derivative.

References

1. (a) R. J. Sunderberg, in *Comprehensive Heterocyclic Chemistry*, ed. A. R. Katritzky and C. W. Rees, Pergamond, New York, 1984, vol. 4, p. 313; (b) F. Bohlmann and C. Zdero, in *The Chemistry of Heterocyclic Compounds, Thiophene and its Derivatives*, ed. S. Gronowitz, John Wiley & Sons Inc., New York, 1985, vol. 44; p. 261; (c) A. Fürstner, *Angew. Chem., Int. Ed.*, 2003, **42**, 3582.
2. F. Bellina and R. Rossi, *Tetrahedron*, 2006, **62**, 7213.
3. (a) J. L. Sessler, S. Camiolo and P. A. Gale, *Coord. Chem. Rev.*, 2003, **240**, 17; (b) P. A. Gale, *Acc. Chem. Res.*, 2006, **39**, 465.
4. C. Pozo-Gonzalo, J. A. Pomposo, J. A. Alduncin, M. Salsamendi, A. I. Mikhaleva, L. B. Krivdin and B. A. Trofimov, *Electrochim. Acta*, 2007, **52**, 4784.
5. (a) C. Brosshard, K. Sutter, P. Petre, J. Hulliger, M. Florsheimer, M. Kaatz and P. Gunter, *Organic non-linear optical materials*, Gordon and Breach Science Publishers, Amsterdam, 1995; (b) M. M. M. Raposo, in *Encyclopedia of Colour Science and Technology, Materials: General 1, "Colorant, non-linear optical"*, ed. R. Luo, Springer, New York, 2014, vol. 184–1, pp. 1–14, ISBN 978-3-642-27851-8.
6. (a) G. P. Bean, Pyrroles, in *The Chemistry of Heterocyclic Compounds*, ed. R. A. Jones, John Wiley & Sons, New York, 1990, vol. 48, part 1, p. 105; (b) T. L. Gilchrist, *J. Chem. Soc., Perkin Trans. 1*, 1999, **1**, 2849; (c) V. F. Ferreira, M. C. B. V. de Souza, A. C. Cunha, L. O. R. Pereira and M. L. G. Ferreira, *Org. Prep. Proced. Int.*, 2001, **33**, 411.
7. T. Ozturk, E. Ertas and O. Mert, *Chem. Rev.*, 2007, **107**, 5210.
8. (a) M. M. M. Raposo, A. M. B. A. Sampaio and G. Kirsch, *Synthesis*, 2005, **2**, 199; (b) M. M. M. Raposo, A. M. B. A. Sampaio, A. M. C. Fonseca and G. Kirsch, *Tetrahedron*, 2006, **62**, 3493; (c) M. M. M. Raposo, in *Targets in Heterocyclic Systems: Chemistry and Properties, "Recent developments in the chemistry of 2-thienylpyrroles: synthesis, reactivity and applications"*, ed. O. A. Attanasi and D. Spinelli, Italian Society of Chemistry, 2007, vol. 11, pp. 122–154.
9. (a) C. A. G. N. Montalbetti and V. Falque, *Tetrahedron*, 2005, **61**, 10852; (b) M. Bodansky, *Peptide Chemistry: A Practical Textbook*, Springer, New York, 1988; (c) *Handbook of Reagents for Organic Synthesis: Activating Agents and Protecting Groups*, ed. A. J. Pearson and W. R. Roush, John Wiley and Sons, Chichester, 2000, p. 133; (d) M. Bodansky, *J. Protein Chem.*, 1989, **8**, 461; (e) *Handbook of Reagents for Organic Synthesis: Activating Agents and Protecting Groups*, ed. A. J. Pearson and W. R. Roush, John Wiley and Sons, Chichester, 2000, p. 220.
10. T. Nisho, *Helv. Chim. Acta*, 1998, **81**, 1207.
11. M. M. M. Raposo and G. Kirsch, *Heterocycles*, 2001, **55**, 1487.
12. Information about the synthesis of amides using DCC/HOBt coupling agents (mechanism, advantages, *etc.*) can be found in Experiment 3.1.16.

17.9. Rhodium Carbene C–H Insertion in Water and Catalyst Reuse

Nuno R. Candeias*[a], Pedro M. P. Gois[b], and Carlos A. M. Afonso[b]

[a]Tampere University of Technology, Department of Chemistry and Bioengineering, Korkeakoulunkatu 8, Tampere, FI-33101, Finland; [b]University of Lisbon, Research Institute for Medicines and Pharmaceutical Sciences (iMed.UL), Faculty of Pharmacy, Av. Prof. Gama Pinto, 1649-003 Lisbon, Portugal
*E-mail: nuno.rafaelcandeias@tut.fi, carlosafonso@ff.ul.pt

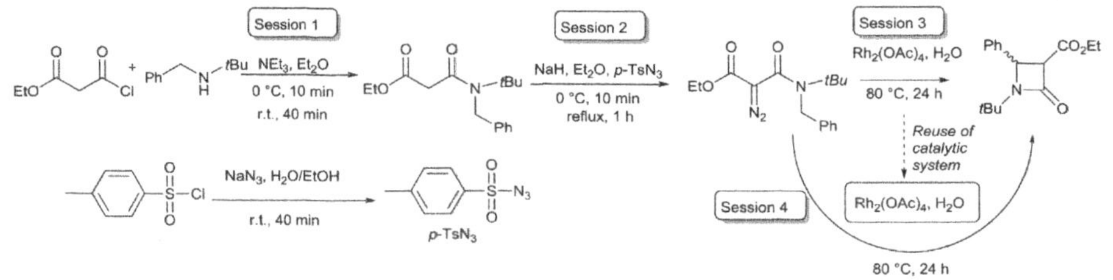

Number of sessions (duration of each session)	Hazard level	Difficulty level	Level of study
4 (4 h each)	High	Medium	Advanced

Class names Acyl chlorides, amines, amides, α-diazoamides, lactams, sulfonyl azides, sulfonyl chlorides

Concepts involved Multi-step synthesis covering an organometallic catalytic process, and reuse of the catalytic system; formation, manipulation and reactivity of sulfonyl azides; amide formation *via* acyl chloride coupling under mild conditions; transfer of diazo group to α-carbonyl position; formation and intramolecular insertion of dirhodium metallocarbene intermediate; synthetic transformations in water; reuse of a catalytic system by selective partition of the catalyst and product in a water–organic solvent bilayer system; formation and interconversion of diastereomers; NMR structural analysis (^{1}H, ^{13}C and two-dimensional techniques – optional)

Chemicals needed Sodium azide, distilled water, *p*-toluenesulfonyl chloride, diethyl ether, hexane, ethanol, triethylamine, *N*-benzyl-*tert*-butylamine, ethyl malonyl chloride, hydrochloric acid (10%), sodium bicarbonate saturated solution, sodium hydride (suspension in mineral oil), sodium chloride saturated solution, dirhodium tetraacetate, TLC plates, silica gel, neutral alumina, deuterated chloroform

Equipment and experimental techniques involved Reactions under anhydrous conditions and in water; isolation by extraction; purification by recrystallization and column chromatography; reuse of catalytic system; rotary evaporator; ^{1}H, ^{13}C and two-dimensional NMR spectral characterization

Keywords C–H insertion, chemistry on-water, α-diazoacetamides, extraction, metallocarbene, NMR

Background

The growing demand for new medicines, materials and general products presents modern chemistry with the challenge of designing new processes to prepare whatever may be required by society.

Comprehensive Organic Chemistry Experiments for the Laboratory Classroom
Edited by Carlos A. M. Afonso, Nuno R. Candeias, Dulce Pereira Simão, Alexandre F. Trindade, Jaime A. S. Coelho, Bin Tan, and Robert Franzén.

Published by the Royal Society of Chemistry, www.rsc.org

Research and development have been carried out in several areas of chemistry in order to achieve more efficient methodologies. Among these, catalysis, in particular organometallic catalysis, organocatalysis and biocatalysis, is certainly a very important topic in modern chemistry mainly due to the unique features for organic transformations such as high reactivity, regio- and stereo-selectivities, and milder conditions and in many cases new transformations are achieved that are still unknown using traditional non-catalysed methodologies. Despite the potential of metal-catalysed reactions, these transformations present some disadvantages that preclude their general utilization, in particular for industrial applications. The metals used are quite often toxic and generally expensive. One way to circumvent such drawbacks is to establish methodologies that allow catalyst reutilization.[1] This experiment covers the synthesis of a β-lactam, an important biological unit, in a sequence of several synthetic steps and involving an intramolecular C–H insertion reaction of a dirhodium metallocarbene as a key-step.[2] As this step is performed using water as solvent,[3] it is possible to reuse the catalytic system by the preferential partition of the catalyst into water and the product into diethyl ether in the reaction work-up. Carbonyl compounds substituted at the α-position with a diazo group have been widely used in C–C bond forming processes *via* C–H activation. Such activation is usually achieved by insertion of a metallocarbene generated from the diazo moiety and a metal complex into a C–H bond.[4] On the other hand, metallocarbenes tend to react with SH, NH and OH functional groups and their combination with water or protic solvents is limited. However, similar to previous reports, the herein described transformation is performed in water,[5] aiming to stimulate an attempt to rationalize the specificity of the transformation considering the hydrophobic effect[6] and on-water[7] reactions.

The starting material needed for the intramolecular C–H insertion is an α-diazo carbonyl compound, prepared by diazo transfer reaction with *p*-toluenesulfonyl azide (Regitz diazo transfer). α-Diazo carbonyl compounds are valuable intermediates in organic synthesis since they can be interconverted into a panoply of functional groups. The principal methods for preparation of such compounds are the above-mentioned diazo transfer reaction with organic azides, or the acylation of diazo alkanes with acyl chlorides or anhydrides.[8] Other methods for introduction of a diazo group at the α-carbonyl position are: diazotization of α-acceptor-substituted primary aliphatic amines; dehydrogenation of hydrazones; base treatment of sulfonyl hydrazones; alkaline cleavage of *N*-alkyl-*N*-nitroso sulfonamides, carboxamides, ureas and urethanes; and electrophilic substitution at diazomethyl compounds.[9] Concerning the diazo transfer strategy, several diazo transfer reagents have been developed mainly because of the potential hazard of *p*-toluenesulfonyl azide and the purification problems in removing the *p*-tosylamide side product. Some of the diazo transfer agents are shown in Figure 17.9.1.[10]

Additional Safety

The use of a fume hood for this experiment is highly advisable and the manipulation of *p*-toluenesulfonyl chloride, sodium hydride and *p*-toluenesulfonyl azide prepared in first session should be performed only by students with appropriate skills. Due to the intrinsic explosive nature of azides and diazo compounds, they should not be heated over 80 °C and the scale used in the experiment should not be higher than 2–3 g. Some minute amounts of highly toxic HN_3

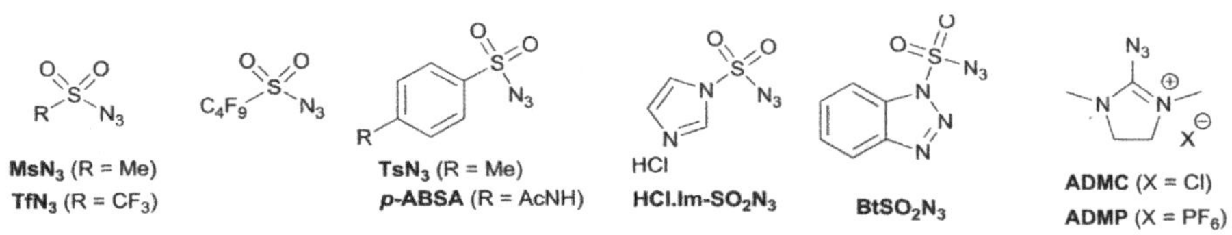

Figure 17.9.1 Diazo-transfer reagents.

can be present in the aqueous mixture in the first step. Diazo compounds can be highly reactive when in contact with some metals due to the entropically favoured liberation of molecular nitrogen and some toxicity intrinsic to the diazo group has been reported. As one of the steps includes heating a mixture of ethyl ether and sodium hydride at reflux temperature, the fume hood sash should be closed and the system properly ventilated.

Experimental Procedure

Session 1 (4 h)

Step I – Preparation of p-*Toluenesulfonyl azide*

Caution: *para*-Toluenesulfonyl azide and diazo compounds, like so many other azides, are well known to be explosive. Take special care in manipulating these compounds.

1. Rapidly add a solution of 2.48 g of sodium azide (38.1 mmol) in 6 mL water to a solution of 6.0 g of *p*-toluenesulfonyl chloride (31.4 mmol) in 30 mL of ethanol.
2. Stir the mixture at room temperature for 40 min, followed by addition of 120 mL of water.
3. After 5–10 min, extract the mixture with diethyl ether (3 × 90 mL), dry over anhydrous sodium sulfate, filter and remove the solvent under reduced pressure.
4. *p*-Toluenesulfonyl azide should be obtained as a liquid emulsion that crystallizes after overnight in the fridge. (R_f = 0.28, silica, diethyl ether/hexane (1 : 9).)

Step II – Synthesis of N-*Benzyl*-N-tert-*butyl-α-ethoxycarbonylacetamide*

1. Add 10 mL of dry diethyl ether to a previously flame dried 25 mL two-necked round-bottom flask equipped with a calcium chloride tube and a septum in an ice bath, followed by triethylamine (371 mg, 3.7 mmol, 0.5 mL) and *N*-benzyl-*tert*-butylamine (400 mg, 2.45 mmol, 0.45 mL).
2. Carefully add (dropwise) ethyl malonyl chloride (479 mg, 3.18 mmol, 0.4 mL) to the initial mixture at 0 °C under vigorous stirring.
3. After 10 min stirring at 0 °C, allow the mixture to reach room temperature and leave it stirring for 40 min.
4. After 30 min at room temperature, a strong spot of product should be visible in TLC (R_f = 0.64, silica, ethyl acetate : hexane (1 : 1)).
5. Filter the mixture over a Büchner funnel and wash the solid with approximately 100 mL of ethyl ether.
6. Transfer the filtrate into a separation funnel and wash it with approximately 40 mL of an aqueous HCl solution (10%), and 40 mL of a saturated solution of $NaHCO_3$.
7. Dry the organic layer over anhydrous sodium sulfate, filter and concentrate it under reduced pressure.
8. Filter the obtained residue through a sintered glass funnel covered with silica-gel (about 5 g of silica) using 80 mL of a 4 : 1 mixture of diethyl ether : hexane as eluent.
9. Remove the solvent under reduced pressure to obtain a yellow oil.

Session 2 (4 h)

Step III – Preparation of N-*Benzyl*-N-tert-*butyl-α-ethoxycarbonyl-α-diazoacetamide*

Caution: *para*-toluenesulfonyl azide and diazo compounds, like so many other azides, are well known to be explosive. Take special care in manipulating these compounds.

1. In a flame dried 50 mL two-neck round-bottom flask equipped with a calcium chloride tube in an ice bath and a septum, prepare a suspension of a 55% sodium hydride mineral oil suspension (118 mg, 2.7 mmol) in dry diethyl ether (8.5 mL) and *p*-toluenesulfonyl azide (357 mg, 1.81 mmol).
2. Slowly add a solution of *N*-benzyl-*N*-*tert*-butyl-α-ethoxycarbonylacetamide (502 mg, 1.81 mmol) in dry diethyl ether (2.5 mL) at 0 °C.
3. After 10 min stirring at 0 °C adapt a condenser to the flask, allow the reaction mixture to reach room temperature and then heat it at reflux for 1 h.
4. Confirm the consumption of *p*-toluenesulfonyl azide by TLC.
5. After reaction completion, slowly add water (5 mL, gas release should be expected) followed by diethyl ether (5 mL).
6. Transfer the reaction mixture to a separatory funnel and extract the aqueous phase with diethyl ether (3 × 10 mL).
7. Wash the combined organic phases with a saturated aqueous solution of NaCl (*ca.* 40 mL), dry over Na_2SO_4 and filter.
8. Evaporate the solvent in the rotary evaporator to obtain a yellow solid (R_f = 0.67, silica, ethyl acetate/hexane (1:4)).
9. Time allowing, the diazo compound can be further purified by recrystallization in a minimum amount of boiling diethyl ether and adding hexane as anti-solvent.

Session 3 (4 h)

Step IV – Cyclization of N*-Benzyl-*N*-*tert*-butyl-α-ethoxycarbonyl-α-diazoacetamide in Water and Reuse of the Catalytic System*

1. Add a solution of dirhodium(II) tetraacetate (2.2 mg, 4.9 μmol) in distilled water (5 mL) to a 25 mL round-bottom flask containing the diazo compound (150 mg, 0.49 mmol).
2. Adapt a condenser to the flask and heat the mixture at 80 °C for 24 h (there is no need to have water passing through the condenser).
3. After reaction completion, as judged by TLC, extract the product with diethyl ether (3 × 5 mL) using a Pasteur pipette, and evaporate the solvent under reduced pressure.
4. The aqueous layer should be kept to reuse the catalytic system.
5. Divide the dry residue into two unequal parts, using one portion to prepare a NMR sample (minimal amount for crude NMR spectra) and dissolve the other portion in a minimal amount of ethyl acetate/hexane (1:4).
6. Pass the dissolved portion through a short neutral alumina column (about 5 g of alumina), using 20 mL of ethyl acetate/hexane (1:4) as eluent.
7. Evaporate the solvent to obtain the purified *trans*-γ-lactam. (R_f = 0.58, neutral alumina, ethyl acetate/hexane (1:4)).
8. Reuse of the catalytic system: add more substrate (150 mg, 0.49 mmol) to the aqueous phase containing the dirhodium(II) tetraacetate, and heat the reaction mixture at 80 °C for 24 h.
9. Determine the reaction yield considering the obtained pure lactam and the amount taken for the NMR sample.

Session 4 (4 h)

Step V – Reuse of the Catalytic System and Determination of Reactions Diastereoselectivity

1. Repeat the above-mentioned procedure for product isolation extracting it with diethyl ether and prepare NMR samples of a small amount of the reaction crude mixture and the purified β-lactam.

Table 17.9.1 Data for catalyst reuse.

Run	Reaction diastereoselectivity (*cis*/*trans* ratio of β-lactam in the reaction crude mixture)	Isolated yield (%)	*Cis*/*trans* ratio of isolated β-lactam
1			
2			

2. Determine the diastereoselectivity of the two runs of the reaction by ^{1}H NMR in $CDCl_3$ of the reaction mixture, through integration of the signals at 4.90 and 4.82 ppm for the major (*cis*) and minor (*trans*) diastereomers, respectively. Determine the *cis*/*trans* ratio of the purified sample by ^{1}H NMR in $CDCl_3$ and compare the obtained spectra with the reported spectra for *cis* diastereomer.[11] Fill in Table 17.9.1.

Waste Disposal

The aqueous extracts should not be discarded in the sink. They should be collected for posterior destruction or the traces of sodium azide carefully destroyed with nitrous acid solution (be careful with nitrogen liberation). The extracts and solvents recovered should be collected in the proper containers for posterior destruction.

Results Interpretation and Additional Questions

1. Propose a reaction mechanism for the formation of the acetamide and interpretation of the work-up procedure.
2. Propose a reaction mechanism for the formation of the α-diazo-acetamide and discuss other reagents that can be used to introduce the diazo functional group at the α-carbonyl position.
3. Enumerate the benefits of the use of water as solvent for organic reactions, specifically in catalytic reactions.
4. Suggest plausible reasoning for the different ratios of β-lactam obtained before and after purification with neutral alumina.

Acknowledgement

Reprinted (adapted) from *Quim. Nova*, 2007, 30, 1768, with permission from ©Sociedade Brasileira de Química.

References

1. A. F. Trindade, P. M. P. Gois and C. A. M. Afonso, *Chem. Rev.*, 2009, **109**, 418.
2. (a) C. Zheng and S.-L. You, *RSC Adv.*, 2014, **4**, 6173; (b) M. O. Doyle, R. Duffy, M. Ratnikov and L. Zhou, *Chem. Rev.*, 2009, **110**, 704.
3. (a) C. J. Li and L. Chen, *Chem. Soc. Rev.*, 2006, **35**, 68; (b) A. Chanda and V. V. Fokin, *Chem. Rev.*, 2009, **109**, 725; (c) C. I. Herrerias, X. Yao, Z. Li and C. J. Li, *Chem. Rev.*, 2007, **107**, 2546.
4. N. R. Candeias, C. A. M. Afonso and P. M. P. Gois, *Org. Biomol. Chem.*, 2012, **10**, 3357.
5. N. R. Candeias, P. M. P. Gois and C. A. M. Afonso, *Quim. Nova*, 2007, **30**, 1768.
6. (a) R. Breslow, *Acc. Chem. Res.*, 2004, **37**, 471; (b) W. Blokzijl and J. B. F. N. Engberts, *Angew. Chem., Int. Ed.*, 2003, **32**, 1545.
7. S. Narayan, J. Muldoon, M. G. Finn, V. V. Fokin, H. C. Kolb and K. B. Sharpless, *Angew. Chem., Int. Ed.*, 2005, **44**, 3275.
8. T. Ye and M. A. McKervey, *Chem. Rev.*, 1994, **94**, 1091.
9. G. Maas, *Angew. Chem., Int. Ed.*, 2009, **48**, 8186.
10. (a) M. Presset, D. Mailhol, Y. Coquerel and J. Rodriguez, *Synthesis*, 2011, 2549; (b) M. Kitamura, S. Kato, M. Yano, N. Tashiro, Y. Shiratake, M. Sando and T. Okauchi, *Org. Biomol. Chem.*, 2014, **12**, 4397.
11. M. K. W. Choi, W. Y. Yu and C. M. Che, *Org. Lett.*, 2005, **7**, 1081.

17.10. Determination of the C–C Bond Strength of Substituted Cyclopropanes and Cyclobutanes using Bomb Calorimetry

Steffanie H. Liskey, Alan R. Esker, and J. M. Tanko*

Department of Chemistry, Virginia Tech, 1040 Drillfield Drive, Blacksburg, VA 24061, USA
*E-mail: jtanko@vt.edu

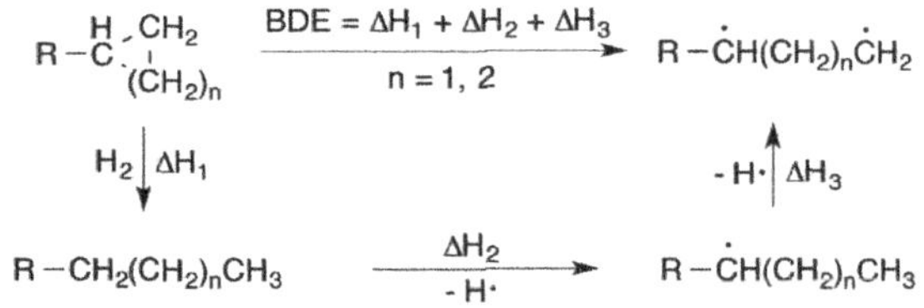

Number of sessions (duration of each session)	Hazard level	Difficulty level	Level of study
2 (3 h each)	Moderate	Medium	Advanced

Class names Strained hydrocarbons, cycloalkanes

Concepts involved The relief of ring strain results in a lower than normal C–C bond dissociation energy (BDE) in cyclopropanes and cyclobutanes. Using a thermochemical cycle, the BDE of various strained hydrocarbons are measured with a bomb calorimetry. Molecular orbital theory and theoretical calculations are suggested to complement the experiment and extend its scope

Chemicals needed Benzoic acid (65-85-0): for calibration of the calorimeters; available from Parr Instrument Company as a 1 gram pellet, and from the National Institute of Standards & Technology (NIST SRM 350b). One (or more if desired) of the following pairs of compounds, all available from Aldrich Chemical Company: (a) cyclopropylbenzene (CA: 873-49-4) and *n*-propylbenzene (CA: 103-65-1), (b) cyclopropyl phenyl ketone (CA: 3481-02-5) and butyrophenone (CA: 495-40-9), (c) cyclobutyl phenyl ketone (CA: 5407-98-7) and valerophenone (CA: 1009-14-9)

Equipment and experimental techniques involved Bomb calorimeter, software for molecular orbital calculations (*e.g.*, Spartan 14, optional)

Keywords Bond strength, calorimetry, molecular modelling, molecular properties and structure, thermochemistry, thermodynamics

Background

The cyclopropyl group is inherently strained for two reasons: (1) bond angle strain (a consequence of placing sp^3 hybridized CH_2s in a three-membered ring where the bond angles are forced to be 60° *versus* 109°) and (2) torsional strain (all the C–H and C–C bonds are eclipsed). As a consequence, much of the chemistry associated with cyclopropanes involves reactions that rupture the three-membered ring, thereby alleviating the strain.[1] The strain energy of cyclopropane is estimated to be 115 kJ mol^{-1}.[2]

Comprehensive Organic Chemistry Experiments for the Laboratory Classroom
Edited by Carlos A. M. Afonso, Nuno R. Candeias, Dulce Pereira Simão, Alexandre F. Trindade, Jaime A. S. Coelho, Bin Tan, and Robert Franzén.

Published by the Royal Society of Chemistry, www.rsc.org

As might be expected, because of ring strain, the carbon–carbon bonds of cyclopropane are weaker than "normal" C–C bonds. Thus, while the C–C bond in ethane has a bond dissociation energy (BDE) of 376 kJ mol^{-1},[3,4] the BDE for the C–C bond of cyclopropane is only 260 kJ mol^{-1};[3,4] the difference is nearly equal to the strain energy of cyclopropane.

In the case of a cyclopropane substituted with an unsaturated group, the C–C bond strength is expected to be even lower because of resonance stabilization. For instance, the C–C bond in ethylbenzene (317 kJ mol^{-1}, reaction (17.10.1))[4] is weaker than the C–C bond of ethane for this very reason, and, on this basis, it could be argued that benzylic stabilization is worth about 376 − 317 = 59 kJ mol^{-1}. If the effect is additive, the C1–C2 bond in cyclopropylbenzene should have a BDE around 260 − 59 = 201 kJ mol^{-1}.

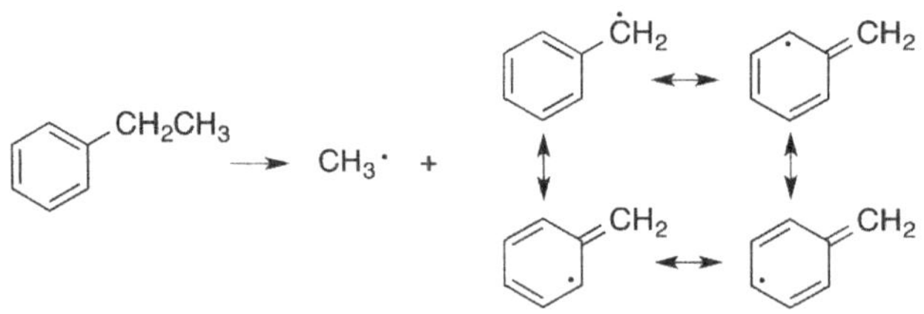

(17.10.1)

The bond angle strain per CH_2 group is expected to be less for cyclobutane (compared to cyclopropane) as a consequence of a C–C–C bond angle of 90° (*versus* 60° for cyclopropane). In addition, cyclobutane adopts a "puckered" conformation that reduces torsional strain. However, there is an additional, strained CH_2 group in cyclobutane, and the net effect is that the strain energy of cyclobutane (111 kJ mol^{-1}) is nearly identical to that of cyclopropane;[1] the C–C bond strength of cyclobutane is reported to be 262 kJ mol^{-1}.[5]

Additional Safety

The organic chemicals used in these experiments pose minor health hazards and are flammable. Students should use safety goggles and gloves when handling these materials, and work in a fume hood or in an area with good ventilation. To avoid any danger of explosion, students must not over pressurize the calorimeter with oxygen, or use too much sample for combustion (<1 g). It is also important to ensure that the valve on the bomb is tightly sealed to prevent leakage upon ignition.

Experimental Procedure

The C1–C2 BDE of either cyclopropylbenzene (**1**), cyclopropyl phenyl ketone (**2**), or cyclobutyl phenyl ketone (**3**) are obtained experimentally by determining the heats of combustion of one of these compounds and the corresponding open-chain analog (*n*-propylbenzene, butyrophenone, and valerophenone, respectively).

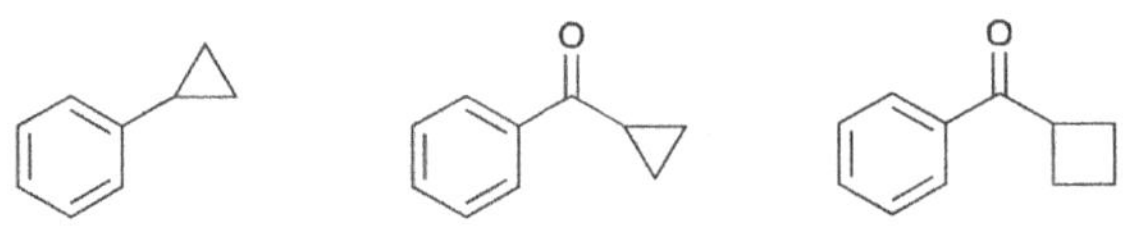

1 2 3

The C–C bond dissociation energy can be estimated using the hypothetical hydrogenation reaction depicted in eqn (17.10.2). For this reaction, the standard reaction enthalpy $\Delta_r\bar{H}^\circ$ can be formulated in two ways: eqn (17.10.3) and (17.10.4). Through bomb calorimetry, students determine the heats of combustion of the pertinent compounds, and convert these values into enthalpies of formation (see below). $\Delta_f\bar{H}^o$ (H_2, g) = 0 (element in its standard state).

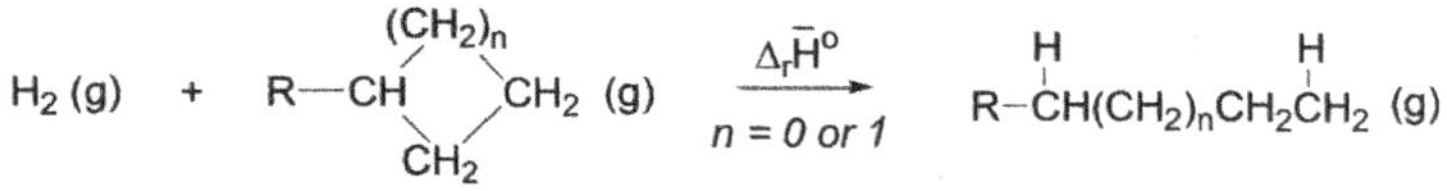

(17.10.2)

$$\Delta_r\bar{H}^o = \Sigma\Delta_f\bar{H}^o(\text{products}) - \Sigma\Delta_f\bar{H}^o(\text{reactants}) \quad (17.10.3)$$

$$\Delta_r\bar{H}^o = \Sigma\text{BDE}(\text{bonds broken}) - \Sigma\text{BDE}(\text{bonds formed})$$
$$= \text{BDE(C–C)} + \text{BDE(H–H)} - \text{BDE(1° C–H)} - \text{BDE(CR–H)} \quad (17.10.4)$$

Of course, the derived heats of formation are for the pure liquids. Application of eqn (17.10.3) and (17.10.4) requires the heats of formation of the compounds in the gaseous state because bond strengths are usually reported in the gas phase. In this treatment, it is assumed that that the enthalpy of vaporization of the two compounds involved in the thermochemical cycle (cycloalkane and corresponding ring-opened form) are the same, and can thus be neglected. This assumption is reasonable because their molecular weights and molecular functionality are very similar. (Cyclopropyl- and *n*-propylbenzene bear this assumption out; their respective enthalpies of vaporization are 50 and 46 kJ mol^{-1}.[6,7]) Pertinent BDEs for appropriate model compounds are summarized in Table 17.10.1.

For these experiments, a bomb calorimeter such as the Parr 1341 calorimeter is required. Procedures for operating the calorimeter and for calculating the heat of combustion are provided by the manufacturer,[8] and are also described in popular laboratory texts.[9,10] The derived enthalpy of combustion is converted into an enthalpy of formation:

$$C_aH_bO_c\,(l) + (a + b/4 - c/2)O_2\,(g) = a CO_2\,(g) + (b/2)H_2O\,(l) \quad (17.10.5)$$

Table 17.10.1 Bond dissociation energies (BDEs) for model compounds pertinent to this experiment.

Compound	BDE (kJ mol^{-1})
H–H	435[a]
CH_3–H	439[a]
RCH_2CH_2–H	423[a]
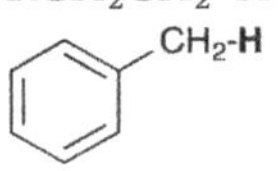	368[a]
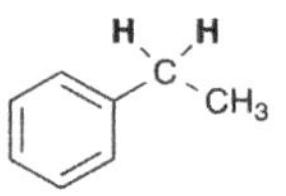	360[a]
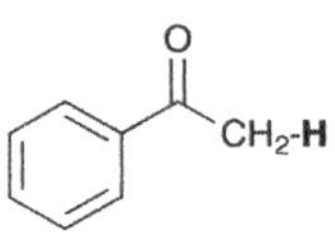	400[a]
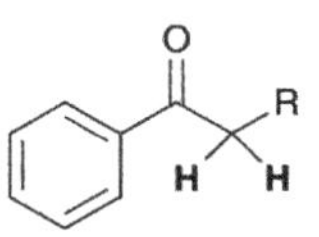	380[b]

[a]Ref. 3.
[b]Ref. 4 and 5.

General Directions

1. A total of three different compounds will be combusted during this experiment: benzoic acid (to determine the heat capacity of the calorimeter), and one of the following three pairs of compounds with the ultimate goal of determining their heats of formation: (a) phenylcyclopropane/*n*-propylbenzene, (b) cyclopropyl phenyl ketone/butyrophenone, or (c) cyclobutyl phenyl ketone/valerophenone.
2. All experiments are to be performed in *duplicate*; to finish in the allotted time, students will need to do three runs per laboratory period. Ideally, one calibration experiment with benzoic acid should be performed in each of the two laboratory sessions to ensure day-to-day reproducibility.
3. Equilibrate the sample 5 min before beginning the run.
4. Take temperature *versus* time measurements (1 min intervals) *starting five minutes prior to ignition*. For two minutes after ignition, take temperature/time measurements every 15 s or so.
5. Continue temperature *versus* time measurements until the *difference* between readings is constant for 5 min, or for a total of 20 min after ignition, whichever is sooner.
6. Calculate ΔT using the procedure in the text and/or operation manual for the calorimeter.
7. The heat capacity of the calorimeter is determined by measuring ΔT for the combustion of a known quantity of benzoic acid (plus ignition wire). Heats of combustion of other substances are determined from ΔT and the heat capacity of the calorimeter (determined above) (Table 17.10.2).

Molecular Orbital Theory (Optional)

To complement this experiment, students can examine how well molecular orbital calculations predict C–C bond strengths. Using heats of formation obtained by semi-empirical methods such as AM1 (Austin Model 1),[11] $\Delta_r\bar{H}^\circ$ for hydrogenation of the substituted-cyclopropanes and cyclobutanes are calculated on the basis of eqn (17.10.3) and (17.10.4). For the systems of interest, these calculations can be performed at the semi-empirical level in a very short period of time using a personal computer with a >1 GHz processor. Higher level calculations such as density functional theory are also practical as the calculations do not involve any odd-electron species.[12–15]

Results Interpretation and Additional Questions

1. Calculate the heat capacity of the calorimeter based upon the combustion of benzoic acid.
2. Calculate the heat of combustion for the compounds studied.
3. Calculate the heat of formation for each of the compounds studied.

Table 17.10.2 Physical properties of compounds used in this experiment.

Compound	MW (g mol^{-1})	Density^a (g mL^{-1})
C_6H_5-*c*-C_3H_5	118.18	0.940
$C_6H_5CH_2CH_2CH_3$	120.20	0.862
C_6H_5CO-*c*-C_3H_5	146.2	1.058
C_6H_5CO-*n*-C_3H_7	148.2	1.021
C_6H_5CO-*c*-C_4H_7	160.2	1.05
C_6H_5CO-*n*-C_4H_9	162.2	0.988

^{a}Values at 25 °C.

4. Calculate the C–C bond dissociation energy based upon eqn (17.10.2)–(17.10.4). Pertinent BDEs for appropriate model compounds are summarized in Table 17.10.1.
5. Calculate the C–C bond dissociation energy using heats of formation derived from molecular orbital theory (optional). How do the calculated values compare to experiment? Are the observed trends reasonable? Explain.

References

1. J. F. Liebman and A. Greenberg, *Strained organic molecules (molecular structure & energetics)*, John Wiley & Sons, Inc., New York, 1986.
2. J. D. Cox and G. Pilcher, *Thermochemistry of organic and organometallic compounds*, Academic Press, London, 1970.
3. J. A. Berson, L. D. Pedersen and B. K. Carpenter, *J. Am. Chem. Soc.*, **1976**(98), 122. See also Darwent, B. B. Bond Dissociation Energies in Simple Molecules, Nat. Stand. Ref. Data Ser., Nat. Bur. Stand. (U.S), 31, Jan. 1970.
4. J. A. Kerr, Strengths of chemical bonds, in *CRC handbook of chemistry and physics*, ed. Lide, D. R., CRC Press, Boca Raton, FL, 84th edn, 2003, pp. 9–52-9-75.
5. S. W. Benson, *Thermochemical kinetics*, John Wiley & Sons, Inc., New York, 1968.
6. *NIST Chemistry WebBook*, http://webbook.nist.gov/chemistry/.
7. W. Acree, Jr and J. S. Chickos, *J. Phys. Chem. Ref. Data*, 2010, **39**, 043101.
8. P. I. Company, *Operating instructions for the 1341 oxygen bomb calorimeter*, Parr Instrument Company, Moline, Illinois USA, 1994.
9. C. W. Garland, J. W. Nibler and D. P. Shoemaker, *Experiments in physical chemistry*, McGraw Hill, 7th edn, New York, 2003.
10. A. M. Halpern, *Experimental physical chemistry: A laboratory textbook*, Prentice Hall, 2nd edn, New York, 1997.
11. M. J. S. Dewar, E. G. Zoebisch, E. F. Healy and J. J. P. Stewart, AM1: A new general purpose quantum mechanical molecular model, *J. Am. Chem. Soc.*, 1985, **107**, 3902–3909.
12. A. D. Becke, Density-functional thermochemistry. III. The role of exact exchange, *J. Chem. Phys.*, 1993, **98**(7), 5648–5651.
13. P. C. Hariharan and J. A. Pople, *Chem. Phys. Lett.*, 1972, **66**, 217.
14. W. J. Hehre, L. Radom, P. v. R. Schleyer and J. A. Pople, *Ab initio molecular orbital theory*, John Wiley & Sons, Inc., New York, 1986.
15. C. Lee, W. Yang and R. G. Parr, Development of Colle-Salvetti correlation-energy formula into a functional of electron density, *Phys. Rev. B: Condens. Matter Mater. Phys.*, 1988, **37**(2), 785–789.

18. Chiral Resolutions

18.1. Resolution of a Chiral Amine and Recovery of Unwanted Enantiomer by Racemization: Towards a Greener Industrial Process

Pedro P. Santos and Pedro F. Pinheiro

CQE, Department of Chemical Engineering, IST, University of Lisbon, Av. Rovisco Pais, 1649-003 Lisbon, Portugal. *E-mail: pepesantos@tecnico.ulisboa.pt, pedro.pinheiro@tecnico.ulisboa.pt

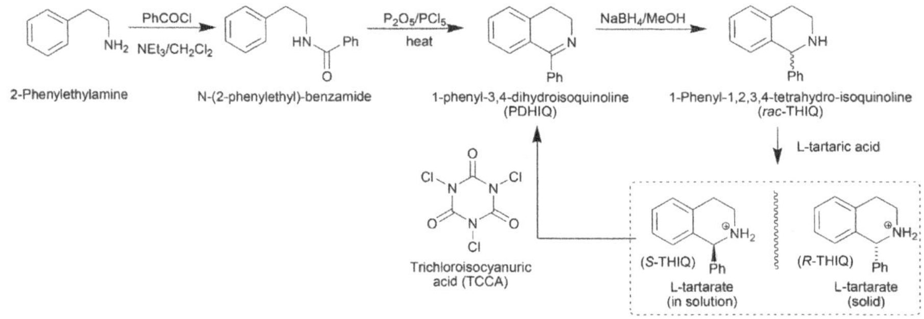

Number of sessions (duration of each session)	Hazard level	Difficulty level	Level of study
5 (3 h + 4 h + 4 h + 4 h + 4 h)	Moderate	Low	Intermediate

Class names Acyl chloride, aromatic ring, amide, amine, carboxylic acid, imine

Concepts involved Green chemistry, classical resolution, isomerization

Chemicals needed *For the synthesis of racemic 1-phenyl-1,2,3,4-tetrahydroisoquinoline*: 2-phenylethanamine, trimethylamine, benzoyl chloride, phosphorous pentoxide, phosphorous oxychloride, sodium borohydride; sodium hydroxide, anhydrous magnesium sulfate, methylene chloride, xylene, methanol, water. *For the resolution and racemization of 1-phenyl-1,2,3,4-tetrahydroisoquinoline*: sodium borohydride, L-(+)-tartaric acid and/or D-(−)-tartaric acid, sodium hydroxide, trichloroisocyanuric acid, methanol, methylene chloride, anhydrous magnesium sulfate, isopropanol, ethyl acetate, water

Equipment and experimental techniques involved Apparatus for reaction with temperature (up to 135 °C) and magnetic stirring, liquid–liquid extraction apparatus, vacuum filtering apparatus, rotary evaporator; silica TLC plates, well ventilated hood, for intermediate and final product analysis: NMR and IV spectroscopy apparatus; polarimeter

Keywords Classical resolution, green chemistry, isomerization, racemization, recovery of unwanted enantiomers

Comprehensive Organic Chemistry Experiments for the Laboratory Classroom
Edited by Carlos A. M. Afonso, Nuno R. Candeias, Dulce Pereira Simão, Alexandre F. Trindade, Jaime A. S. Coelho, Bin Tan, and Robert Franzén.

Published by the Royal Society of Chemistry, www.rsc.org

Background

Most of the current pharmaceutical drugs are chiral and one pure enantiomer is usually preferred over the racemate in many marketed dosage forms.[1–4] It is well known that the opposite enantiomer of a chiral drug often differs significantly in its pharmacological, toxicological, pharmacodynamic and pharmacokinetic properties. With rare exceptions new chiral drugs are now developed in single-enantiomer form, and new racemic drugs are highly unlikely to appear. Three general strategies are frequently applied to produce a chiral enantiopure active pharmaceutical ingredient (API):[1,2]

(a) Using a chiral enantiopure starting material and incorporating it in the API structure (*chiral pool* strategy).
(b) Using an *asymmetric synthesis*. This situation has been regarded as the ultimate production process for enantiopure compounds, even at large scales. In theory, the selective production of the active enantiomer can increase yields, throughput and also decrease wastes. However, processes are generally less robust and more difficult to run at industrial scales. The catalyst used in the asymmetric reaction is sometimes not available in bulk and is frequently very expensive, which has a big impact on its application in large production scale batches.
(c) Using a *resolution* of a racemic mixture. The chiral drug is often synthesized in the racemic form and it is frequently costly to resolve the racemic mixture into the pure enantiomers. Most often the separation of isomers is achieved by a diastereomeric salt formation using an acid–base reaction and the chemists take advantage of the different solubility of the two salts to promote isomer separation by crystallization and filtration (Scheme 18.1.1). With this kind of technology, the resolution at the end of the synthetic route will generate a huge amount of waste as at least one molecule of the unwanted isomer per one of the drug is discarded during the production. As an example, more than 500 tons of Clopidogrel© were produced during 2013 (market value approximately US$10.8 billion). If a resolution is applied at the last synthetic step, 500 tons of wasted material is produced.

However, a classical resolution coupled with a recycling process for the unwanted enantiomer has proven to be frequently the most appropriate process for large scale production. The recovery process of unwanted isomers is often done by racemization of α-carbon carbonyl atoms under basic conditions. An example for the possible recovery of the (*R*) isomer of Clopidogrel® is to perform the racemization by heating a sodium methoxide solution (Scheme 18.1.1).

The recycling of chiral amines is not frequent. In the present experiment, a process to recover the unwanted enantiomer of a chiral amine by an oxidation/reduction process is proposed (*via* an imine intermediate).

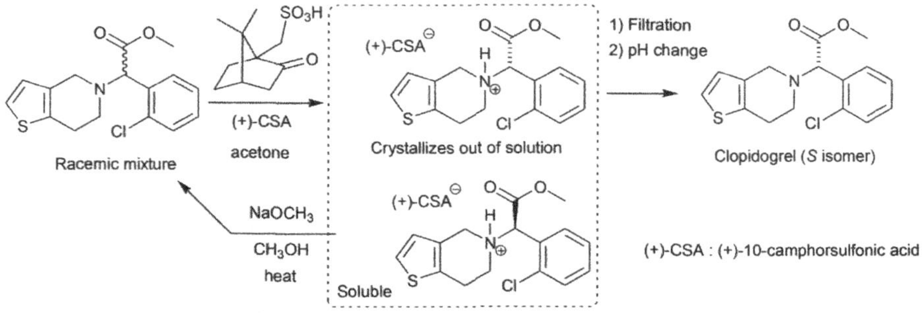

Scheme 18.1.1 Obtaining Clopidogrel® by a classical resolution process and racemization of the (*R*) isomer.

1-Phenyl-1,2,3,4-tetrahydroisoquinoline is a chiral key substructure of several bioactive compounds, *e.g.* Solifenacin. Racemic 1-phenyl-1,2,3,4-tetrahydroisoquinoline is prepared by Bischler–Napieralski cyclization[5–7] of *N*-(2-phenylethyl)-benzamide[8] and reduction of the resulting imine (1-phenyl-3,4-dihydroisoquinoline). Classical resolution of the tetrahydroisoquinoline using tartaric acid in isopropanol/water mixture can efficiently separate the enantiomers.[9] The isomer to be recycled is N-chlorinated with trichloroisocyanuric acid (TCCA), giving a *N*-chloroamine intermediate that eliminates spontaneously HCl to the imine intermediate.[10] This experiment illustrates the power of this strategy to eliminate wastes and to recycle an unwanted enantiomer and develop a greener resolution process with much less chemical wastes.

Additional Safety

Hydrochloric acid is released in sessions A1 and A2. A gas trap may be connected to the reaction mixture or the reaction performed in a ventilated hood. Phosphorous pentoxide and phosphorous oxychloride react vigorously with water.

Experimental Procedure

Laboratory Sessions A1 to A3: Preparation of Racemic Phenyl-1,2,3,4-tetrahydroisoquinoline

Session A1 (3 h) Preparation of N-*(2-Phenylethyl)benzamide*[8] *(Amide Formation)*

1. In a three-neck 250 mL round-bottom flask, equipped with a 25 mL addition funnel, a reflux condenser, a thermometer and a magnetic stirrer, place 10 g (83 mmol) of 2-phenylethanamine and 150 mL of methylene chloride and 12 mL (8.7 g, 86 mmol) of triethylamine.
2. Charge the addition funnel with 10 mL of benzoyl chloride (12.1 g, 86 mmol).
3. Heat the contents of the round-bottom flask to *ca.* 20–30 °C in a water bath. Add the benzoyl chloride in small portions, with vigorous stirring, throughout 30 min. Allow the reaction to proceed for 30 min to 1 h, at 20–30 °C, after the addition of benzoyl chloride.
4. Transfer the contents of the round-bottom flask to a 500 mL separation funnel using up to 50 mL of methylene chloride to wash the round-bottom flask. Add 150 mL of water to the separation funnel and shake vigorously. Drain the organic phase and discard the aqueous phase.
5. Wash the organic phase with 100 mL of 10% HCl solution and again twice with 100 mL of water. Note: check the pH of the last aqueous phase, which should be neutral. If acidic, wash the organic phase again with water.
6. Dry the organic phase over anhydrous magnesium sulfate and evaporate the solvent in the rotary evaporator, yielding the desired product as white crystals (18.3 g, 81 mmol, 97.5% yield). The product can be recrystallized from ethyl acetate. ^{1}H NMR ($CDCl_3$ δ (ppm) 2.94 (2H, t), 3.72 (2H, m), 6.24 (1H, br), 7.23–7.49 (m, 8H), 7.60–7.69 (m, 2H). FT-IR (KBr disk): ν (cm^{-1}) 3343 (N–H stretch), 1639 (C=O stretch), 1545 (N–H bend).

Session A2 (4 h) Synthesis of 1-Phenyl-3,4-dihydroisoquinoline[8] *(Bischler–Napieralski Cyclization)*

1. Place 1 g (4.4 mmol) of *N*-(2-phenylethyl)benzamide in a 250 mL round-bottom flask equipped with a large magnetic stirrer.
2. Dissolve the product in 5 mL of xylene and, with vigorous stirring, add 1.5 g (5.3 mmol) of phosphorous pentoxide followed by 4 mL (6.6 g, 43 mmol) of phosphorous oxychloride.
3. Adapt a reflux condenser connected to a gas trap, and heat the reaction mixture to 140–150 °C. Note: HCl will be released during this reaction. Maintain the reflux until gas release is no longer observable (*ca.* 3 h).

4. Allow the mixture to cool down to room temperature and drain the liquid layer.
5. In the fume hood, carefully dissolve the solid residue in 10% NaOH and adjust the pH of the resulting solution to 9.
6. Extract the aqueous solution with methylene chloride (4 × 25 mL).
7. Combine the organic phases and dry over magnesium sulfate. Evaporate the solvent in the rotary evaporator, yielding the desired product as a yellow oil (0.81 g, 3.9 mmol, 89% yield). ^{1}H NMR ($CDCl_3$): δ (ppm) 2.77 (2H, t), 3.83 (2H, t), 7.24–7.60 (m, 9H). FT-IR (KBr disk): ν (cm^{-1}) 3058, 3026 (sp^3 C–H stretch), 1607 (C=N stretch), 1564 (C=C aromatic stretch).

Session A3 (4 h) Synthesis of Racemic 1-Phenyl-1,2,3,4-tetrahydroisoquinoline (Ref. 10) (Imine Reduction)

1. Transfer 1 g (4.8 mmol) of 1-phenyl-3,4-dihydroisoquinoline to a 100 mL round-bottom flask. Add 50 mL of methanol and stir to dissolve the oil.
2. Add 0.27 g (7.2 mmol) of sodium borohydride in small portions, over 10–15 min. After the addition of sodium borohydride maintain stirring for 30 min at room temperature.
3. Remove the solvent by evaporation in the rotary evaporator.
4. Add 10 mL of water and 10 mL of methylene chloride to the solid residue, stirring vigorously for complete dissolution.
5. Transfer the mixture to a separation funnel, collect and dry the organic layer over magnesium sulfate.
6. Evaporate the solvent in the rotary evaporator, yielding the desired product as a white to yellow solid (0.82 g, 3.9 mmol, 82% yield). ^{1}H NMR ($CDCl_3$): δ (ppm) 1.86 (1H, s), 2.89 (1H, m), 3.08 (m, 2H), 3.27 (m, 1H), 5.11 (s, 1H), 6.76 (d, 1H), 7.04–7.29 (m, 8H). FT-IR (KBr disk): ν (cm^{-1}) 3254 (N–H stretch), 3026, 3019 (sp^3 C–H stretch).

Laboratory Sessions B1 and B2: Resolution of 1-Phenyl-1,2,3,4-tetrahydroisoquinoline and Racemization of Unwanted Isomer

Session B1 (4 h) Enantiomeric Resolution of Racemic 1-Phenyl-1,2,3,4-tetrahydroisoquinoline[9]

1. Dissolve 5 g (24 mmol) of the racemic mixture in 35 mL of isopropanol and 15 mL of water in a 100 mL round-bottom flask, heating the mixture to 60 °C in a water bath.
2. Maintain the yellow solution, stirred, at 60 °C and add 3.6 g (24 mmol) of L-(+)-tartaric acid (note: using D-(−)-tartaric acid will crystalize the opposite isomer).
3. Keep the mixture at 60 °C for 5 min to allow the acid to dissolve completely.
4. Transfer the round-bottom flask to an ice/water bath. Allow the mixture to cool down for 30 min with continuous stirring.
5. Filter the mixture and wash the white solid ((*R*)-1-phenyl-1,2,3,4-tetrahydroisoquinoline tartrate; 3.7 g, 10 mmol, 85% yield) with 2–5 mL of ethyl acetate.

Isolation of the Free Base of (*R*)-1-Phenyl-1,2,3,4-tetrahydroisoquinoline

6. Transfer the solid to a 100 mL round-bottom flask containing 25 mL of distilled water and 25 mL of methylene chloride.
7. Adjust the pH of the aqueous layer to 9, adding small portions of a 10% NaOH solution, mixing thoroughly and allowing the phases to separate between each pH measurement and base addition.

8. After reaching the desired pH value, transfer the mixture to a separation funnel, drain the organic layer and extract the aqueous layer with 2 × 25 mL of methylene chloride.
9. Combine and dry the organic phases over anhydrous magnesium sulfate before removing the solvent by evaporation, yielding (*R*)-1-phenyl-1,2,3,4-tetrahydroisoquinoline (2.1 g. 10 mmol, 98% yield).
10. Measure the optical rotation of a 1 g per 100 mL solution in $CHCl_3$.

The enantiomeric excess is routinely higher than 85%. (*R*)-1-Phenyl-1,2,3,4-tetrahydroisoquinoline: optical rotatory power −10.9° (1.1 g per 100 mL, chloroform; wavelength: 589.3 nm; T = 25 °C).[11]

Isolation of the Free Base of (*S*)-1-Phenyl-1,2,3,4-tetrahydroisoquinoline (from Filtrate)

11. Collect the liquid filtrate (obtained at the end of the reaction with tartaric acid) and reduce the volume to approximately half in the rotary evaporator.
12. Add 25 mL of water and adjust the pH of the resulting solution to 9 with 10% NaOH solution.
13. Extract the aqueous solution with 2 × 25 mL methylene chloride.
14. Combine and dry the organic phases over magnesium sulfate.
15. Remove the solvent in the rotary evaporator, yielding (*S*)-1-phenyl-1,2,3,4-tetrahydroisoquinoline as a white to yellow oily solid (2.18 g, 10 mmol, 87% yield).
16. Measure the optical rotation of a 1 g per 100 mL solution in $CHCl_3$.

The enantiomeric excess is routinely higher than 85%. (*S*)-1-Phenyl-1,2,3,4-tetrahydroisoquinoline optical: rotatory power +13.5° (1.15 g per 100 mL, chloroform; wavelength: 589.3 nm; T = 25 °C).[12]

Session B2 (4 h) Oxidation of (*S*)-1-Phenyl-1,2,3,4-tetrahydroisoquinoline to 1-Phenyl-3,4-dihydroisoquinoline (Ref. 10)

Previous Day

1. Dissolve 1 g (4.8 mmol) of (*S*)-1-phenyl-1,2,3,4-tetrahydroisoquinoline in 10 mL of methylene chloride in a 25 mL round-bottom flask.
2. Cool the solution to *ca.* 5 °C in an ice/water bath.
3. Add 0.41 g of trichloroisocyanuric acid (TCCA, 1.8 mmol) and stir overnight, allowing the mixture and the water bath to heat up to room temperature.

Laboratory Session

4. While stirring, add 10 mL of water and adjust the pH to 8–9 with 10% NaOH.
5. Transfer the mixture to a separation funnel and drain the organic phase.
6. Back-extract the organic phase with 10 mL of water.
7. Dry the organic phase with magnesium sulfate and evaporate the solvent, yielding the desired product as yellow oil (0.88 g, 4.2 mmol, 88% yield).

For the reduction of 1-phenyl-3,4-dihydroisoquinoline to 1-phenyl-1,2,3,4-tetrahydroisoquinoline follow procedure of session A3 above.

Waste Disposal

The aqueous phase must be neutralized before disposal. Mother liquors from the product filtration must be placed in an organic solvent container for disposal.

Results Interpretation and Additional Questions

1. Determine the reaction yields.
2. Calculate the enantiomeric excess of (*R*)- and (*S*)-1-phenyl-1,2,3,4-tetrahydroisoquinoline obtained during session B1.
3. What is the structural relation between the two tartaric acid salts formed during session B1? Is this fact important or not to the separation process by crystallization?
4. Why can the sequence *N*-chlorination, imine formation and reduction be considered as a recycling process for the unwanted enantiomer of 1-phenyl-1,2,3,4-tetrahydroisoquinoline?
5. Write out the mechanisms for the chlorination end elimination reactions to form the 1-phenyl-3,4-dihydroisoquinoline from 1-phenyl-1,2,3,4-tetrahydroisoquinoline.
6. Write out the mechanism of the sodium borohydride reduction.

References

1. *Chiral Drugs: chemistry and biological action*, ed. G. Q. Lin, Q. D. You and J. F. Cheng, John Wiley & Sons, 2011.
2. *Chirality in Industry: The Commercial Manufacture and Applications of Optically Active Compounds*, A. N. Collins, G. Sheldrake and J. Crosby, John Wiley & Sons, 1995.
3. *Chirality in Industry II: Developments in the Commercial Manufacture and Applications of Optically Active Compounds*, ed. A. N. Collins, G. Sheldrake and J. Crosby, John Wiley & Sons, 1997.
4. L. A. Nguyen, H. He and C. Pham-Huy, *Int J Biomed Sci.*, 2006, **2**(2), 85–100.
5. W. M. Whaley and T. R. Govindachari, *Org React*, 1951, **6**, 74.
6. M. D. Rozwadowska, *Heterocycles*, 1994, **39**, 903.
7. B. A. Lorsbach and M. J. Kurth, *Chem Rev*, 1999, **99**, 1549.
8. B. Liang, C. Jiang, Z. Chen, X. Zhang, X. Shi and Y. Cao, *J. Mater. Chem.*, 2006, **16**, 1281.
9. T. Nidam; N. Perlman, PCT application WO2008019055, 2008.
10. C. Bolchi, M. Pallavicini, L. Fumagalli, V. Straniero and E. Valoti, *Org. Process Res. Dev.*, 2013, **17**, 432.
11. R. Pedrosa, *J. Org. Chem.*, 2001, **66**(1), 243.
12. D. Barbier, *J. Org. Chem.*, 1998, **63**(6), 1767.

18.2. Synthesis of Racemic Phenylalanine Methyl Ester and its Kinetic Resolution Catalysed by α-Chymotrypsin

Luca Banfi* and Renata Riva

Department of Chemistry and Industrial Chemistry, University of Genova, via Dodecaneso, 31 16146 Genova, Italy. *E-mail: banfi@chimica.unige.it

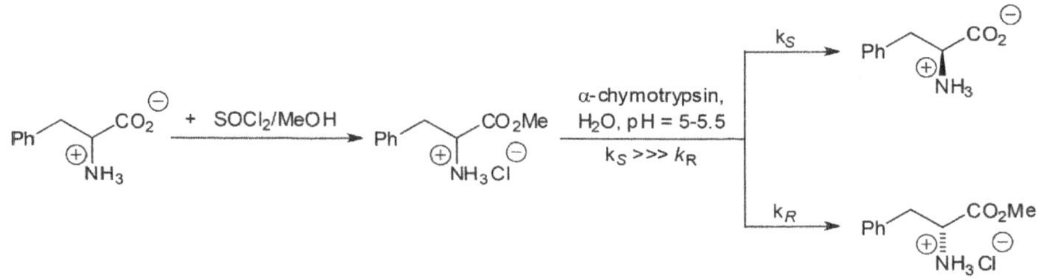

Number of sessions (duration of each session)	Hazard level	Difficulty level	Level of study
2 (4 h each)	Moderate	Medium	Intermediate

Class names Carboxylic acids, alcohols, esters, amines

Concepts involved The main goal of the experiment is to resolve a racemic mixture of an amino acid into its enantiomers, using biocatalysis. This experience will allow the student to learn the concept of kinetic resolution of a racemic mixture, by using a chiral catalyst (in this case an enzyme). The student will learn how to use polarimetry to recognize an enantiomer and to establish the enantiomeric purity. A specific method for the synthesis of esters of amino acids will be used. The student will learn important concepts about the dependence of acid–base equilibria on pH

Chemicals needed D,L-Phenylalanine, thionyl chloride, α-chymotrypsin from bovine pancreas, 1 M HCl solution in diethyl ether, sodium hydroxide, dry Na_2SO_4, NaCl, soda lime, diethyl ether, ethyl acetate, methanol, 96% ethanol

Equipment and experimental techniques involved Hot plate stirrer, Claisen distillation apparatus, desiccator, vacuum pumps, pH meter, burette, liquid–liquid extraction apparatus, rotatory evaporator, polarimeter

Keywords Amino acids, biocatalysis, esterification, kinetic resolution, polarimetry

Background

The obtainment of chiral organic substances in pure enantiomeric form is an important challenge, especially for biological applications.[1] For example, it is well known that enantiomers can have different pharmacokinetics and pharmacodynamics properties.[2] For this reason, although in the past several chiral drugs were used as the racemate, nowadays licensing agencies hardly approve new drugs unless they are marketed in enantiomerically pure form. At the same time, even old drugs that were initially sold in racemic forms are often later launched as pure enantiomers (chiral switching) that are more potent and suffer from less unwanted side effects.

Comprehensive Organic Chemistry Experiments for the Laboratory Classroom
Edited by Carlos A. M. Afonso, Nuno R. Candeias, Dulce Pereira Simão, Alexandre F. Trindade, Jaime A. S. Coelho, Bin Tan, and Robert Franzén.

Published by the Royal Society of Chemistry, www.rsc.org

There are three general methodologies to obtain enantiomerically pure chiral compounds:[3] (a) transformation of chiral building blocks derived from natural sources; (b) asymmetric synthesis; (c) racemic synthesis, followed by separation of the two enantiomers, usually referred to as "resolution". Two enantiomers have identical physical properties in an achiral medium and this makes their separation rather difficult, since common techniques such as crystallization, distillation, extraction, and standard chromatography are not able to distinguish them, unless chiral modifications are implemented. Although chromatography on a chiral stationary phase has witnessed great improvements in recent years,[4,5] often being able to separate two enantiomers, the high cost of this technique makes its application on a large scale troublesome, and therefore in industry it is usually limited to the purification of small amounts of target enantiomers at early development stages.

A more affordable methodology is represented by "classic resolution",[6] where the racemate is converted into a couple of diastereomers (often two diastereomeric salts) by interaction with a chiral, enantiomerically pure, resolving agent. Then separation is made possible by crystallization, since two diastereomers have different physical properties, including solubility. Classical resolution is still the method of choice in the manufacture of active pharmaceutical ingredients (APIs).[7] However, this methodology requires the use of a stoichiometric resolving agent.

A useful alternative is represented by kinetic resolution. In this approach two enantiomers react at different rates under the influence of a chiral catalyst. Therefore, both the substrate and the product are enantiomerically enriched: the product in the faster reacting enantiomer, and the substrate in the slower reacting one. In ideal cases, the rate difference is so high that one of the enantiomers is completely converted into the product, whereas the other one remains totally at the substrate stage. Separation is even easier than for classical resolution, since here we have to separate two completely different substances, endowed with different functional groups. But the greatest advantage, compared to classical resolution, is the use of only a catalytic amount of the chiral resolving agent.

Among the various chiral catalysts, enzymes are particular useful for kinetic resolutions, for their efficiency and selectivity.[8,9] Biocatalytic processes are environmentally friendly, especially compared to those that use heavy-metal catalysis, and require mild reaction conditions (*e.g.* dry conditions and inert atmosphere are not needed, and the reactions can be typically performed at room temperature). Therefore, biocatalysis is increasingly utilized in the production of APIs.[10,11] In particular, hydrolytic enzymes are widely used in organic synthesis, thanks to their broad substrate specificities, to the high number of commercially available varieties, and since they do not need cofactors.

The objective of this experiment is to show an example of biocatalytic kinetic resolution that allows us to isolate the (D) methyl ester of α-amino acid phenylalanine in very high ee. The serine protease α-chymotrypsin will be used towards this goal.[12] The kinetics of methyl ester hydrolysis will be followed by a simple titration technique. Thus the students have the opportunity to experience a real time reaction rate estimation. Since acid–base equilibria play an important role in both the hydrolysis reaction and in the subsequent separation of the unreacted methyl ester from the amino acid, this experience represents a useful tutorial for these important concepts. The first session of the work involves the preparation of the methyl ester from racemic phenylalanine,[13,14] using an interesting method, peculiar to α-amino acids. Finally, the students will use polarimetry to check the optical purity of the obtained product.

Additional Safety

The first session of the work involves the use of thionyl chloride. This reagent reacts violently with water producing toxic gases (HCl and SO_2). Therefore, it is compulsory to use it in a fume hood, with proper safety equipment, such as gloves. All the glassware containing traces of this compound must be treated with aqueous NaOH in order to destroy all $SOCl_2$ before taking it out of the hood. The ethereal HCl solution tends to liberate HCl gas and must be used under the hood with the same precautions. All organic solvents (methanol, diethyl ether, ethyl acetate) must be used under a fume hood and far away from flames, being inflammable. Safety goggles must be worn for both sessions all the time.

Experimental Procedure

Laboratory Session 1 (4 h): Synthesis of Racemic Phenylalanine Methyl Ester Hydrochloride

1. In a fume hood prepare a one-neck 100 mL flat-bottomed flask (29/32 ground joint) equipped with a magnetic stir bar. Place on it (if necessary through a 29/32–14/19 adapter) a Claisen adapter. Put a dropping funnel on one of the two necks of the Claisen adapter.
2. Prepare two bent drying tubes filled with soda lime (with indicator) and place them on the other neck of the Claisen adapter and on the top of the dropping funnel.
3. Prepare a salt-ice bath (its temperature should be between −15 and −18 °C).
4. Pour in the flask 35 mL of absolute methanol, close the whole apparatus and insert the flask in the cooling bath.
5. Pour 3.06 mL (42.0 mmol) of thionyl chloride into the dropping funnel and drop it slowly into the flask, during 10 min, being careful to ensure that the bath temperature remains below −7 °C.
6. Now remove the dropping funnel and the Claisen adapter (and the 29/32–14/19 adapter if used) and add rapidly, through a solid addition funnel, 5.00 g of D,L-phenylalanine (30.3 mmol), previously weighed in a test tube. Then, as soon as possible, to avoid contact with moisture, place a reflux condenser on the flask and place over it one of the soda lime drying tubes previously employed. Connect the condenser to the water faucet.
7. Remove the cooling bath and heat at reflux for 2 h with a heating stir plate. If needed, replace the drying tube when it becomes exhausted, as shown by the indicator.
8. Now, after cooling to room temperature, replace the reflux condenser with a Claisen distiller apparatus ending with a single vacuum adapter dripping into a small (50 mL) round-bottomed flask. Connect the vacuum adapter to a linear drying tube filled with soda lime. Alternatively, distillation can be performed under reduced pressure, connecting a membrane vacuum pump (dual or single stage) to the vacuum hose, and inserting, between the apparatus and the pump, the linear drying tube.
9. Distil most of the solvent, taking care not to bring the mixture to dryness (leaving about 2–5 mL of solvent).
10. Take up the very thick oily residue with 2–5 mL of methanol and grind well with a spatula, inducing crystallization.
11. Add 15 mL of diethyl ether and cool in ice: an abundant precipitate of D,L-phenylalanine methyl ester hydrochloride, as a white solid, is observed.
12. Collect the solid by suction in a Büchner funnel or a sintered glass funnel (using a membrane vacuum pump), and wash it with diethyl ether.
13. Transfer the solid into a tared Petri dish and dry in a desiccator under reduced pressure (2 mbar) overnight.
14. During the next session, determine the melting point (around 160 °C).

Laboratory Session 2 (4 h): Enzymatic Resolution of Racemic Phenylalanine Methyl Ester and Determination of Optical Purity

The reaction itself can be carried out on the bench, but the subsequent extraction and hydrochloride precipitation must be carried out under a well-ventilated fume hood. It is advised to work in pairs.

1. Weigh D,L-phenylalanine methyl ester hydrochloride (2.00 g, 9.27 mmol) in a plastic 100 mL beaker equipped with a magnetic stir bar, and dissolve it in deionized water (20 mL).
2. Insert a low cost portable pH meter into the beaker. Measure the pH and adjust it to about 5.15 by addition, under stirring, of volumetric 0.5 N NaOH from a burette.
3. Weigh α-chymotrypsin from bovine pancreas (type II, >40 units per mg) (60 mg) in a small test tube and dissolve it in deionized water (2–3 mL).
4. Rapidly add the enzyme solution to the beaker containing the methyl ester and mark exactly the start time (or start a chronometer).
5. One of the students must add continuously 0.5 N NaOH from the burette in order to keep the pH exactly at the starting value. Meanwhile, the other student takes note, at given time intervals, of the NaOH volume added from the start of the reaction (*i.e.* after 2, 4, 6 min and so on; towards the end longer time intervals may be used). A graph may be generated (manually or through a computer software) plotting the time on the abscissa and the overall NaOH volume on the ordinate.
6. When the pH no longer decreases and remains constant, further stir the solution for 10 min.
7. Now bring the pH to 9.5 by further addition of NaOH from the burette, and add NaCl (15 g) into the beaker.
8. Extract three times the aqueous solution with 25 mL of ethyl acetate each. This extraction is probably made difficult by emulsion formation. Addition of some NaCl crystals can help in breaking the emulsion.
9. Dry the united organic phases (Na_2SO_4), filter and evaporate to dryness in a tared flask, using a rotatory evaporator. Further strip the oily product at 1 mbar (rotary single stage pump), then weigh it and calculate the number of mmol obtained.
10. Take up the resulting oil in diethyl ether (10 mL) and treat it with a 1.0 M solution of HCl in diethyl ether (1.05 equiv.). A sudden precipitation of a white solid is observed. Collect it by suction, wash it with diethyl ether, transfer to a tared Petri dish and dry under vacuum (1–2 mbar) for 30 min.
11. Determine the melting point of this hydrochloride.
12. Weigh precisely about 400 mg of this hydrochloride in a volumetric flask (20 mL, class A) and dissolve it in 96% ethanol.
13. Measure the optical activity of this solution with an automatic polarimeter (typically with a 1 dm long cell) and calculate the specific optical rotation.

Waste Disposal

Session 1. Both the distilled methanol and the mother liquors from crystallization must be neutralized by addition of small quantities of solid $NaHCO_3$ and then (if necessary after filtration of inorganic salts) disposed of in non-halogenated solvents cans.

Session 2. The aqueous phases from the extraction must be left in a basin under the hood for 24 h to let the traces of ethyl acetate evaporate and then be disposed of with other aqueous solutions. The ethereal mother liquors are treated with some anhydrous K_2CO_3, filtered, and disposed of in the non-halogenated solvent can. The ethyl acetate collected in the rotary evaporator is disposed of in non-halogenated solvent cans.

Results Interpretation and Additional Questions

1. Why is the phenylalanine methyl ester hydrochloride less soluble in diethyl ether than in methanol?
2. What are the products of the reaction of soda lime (which contains CaO and NaOH) with SO_2 and HCl?

3. Why is the amount of NaOH needed to maintain a constant pH equal to (in mmol) the amount of methyl ester hydrolysed?
4. Calculate the overall mmol of NaOH added from the start of the enzymatic reaction to when the pH remains constant. Compare them to the mmol of starting material and explain why they do not match.
5. Calculate the percent yield for both reactions.
6. Calculate the optical purity of the recovered methyl ester hydrochloride and, from it, the relative molar ratio of the D- and L-enantiomers.

References

1. E. J. Ariëns, *Trends Pharm. Sci.*, 1986, 7, 200.
2. A. R. Ribeiro, A. S. Maia, Q. B. Cass and M. E. Tiritan, *J. Chromatogr. B: Biomed. Sci. Appl.*, 2014, **968**, 8.
3. H. Lorenz and A. Seidel-Morgenstern, *Angew. Chem., Int. Ed.*, 2014, **53**, 1218.
4. H. Y. Aboul-Enein and A. Imran, *Chiral Separations by Liquid Chromatography and Related Technologies*, Marcel Dekker, New York, 2003.
5. G. Subramanian, *Chiral Separation Techniques: a Practical Approach*, Wiley-VCH Verlag GmbH, 2007.
6. A. Collet, *Angew. Chem., Int. Ed.*, 1998, **37**, 3239–3241.
7. H. Murakami, in *Novel Optical Resolution Technologies*, ed. K. Sakai, N. Hirayama and R. Tamura, Springer, Berlin Heidelberg, 2007, vol. 269, pp. 273–299.
8. K. Faber, *Biotransformations in Organic Chemistry*, Springer-Verlag, Berlin, 2000.
9. B. G. Davis and D. Boyer, *Nat. Prod. Rep.*, 2001, **18**, 618.
10. V. Gotor, *Org. Proc. Res. Dev.*, 2002, **6**, 420.
11. A. Schmid, J. S. Dordick, B. Hauer, A. Kiener, M. Wubbolts and B. Witholt, *Nature*, 2001, **409**, 258.
12. J.-M. Ricca and D. H. G. Crout, *J. Chem. Soc., Perkin Trans. 1*, 1993, 1225.
13. S. Davies, G. H. J. Sanganee and P. Szolcsanyi, *Tetrahedron*, 1999, **55**, 3337.
14. F. Matsuda, T. Matsumoto, M. Ohsaki, Y. Ito and S. Terasjima, *Bull. Chem. Soc. Jpn.*, 1992, **65**, 360.

18.3. Lipase Catalyzed Kinetic Resolution of Racemic 1-Phenylethanol

Otto Långvik, Tiina Saloranta*, and Reko Leino

Johan Gadolin Process Chemistry Centre, Laboratory of Organic Chemistry, Åbo Akademi University, Piispankatu 8, FI-20500 Turku, Finland. *E-mail: tiina.saloranta@abo.fi

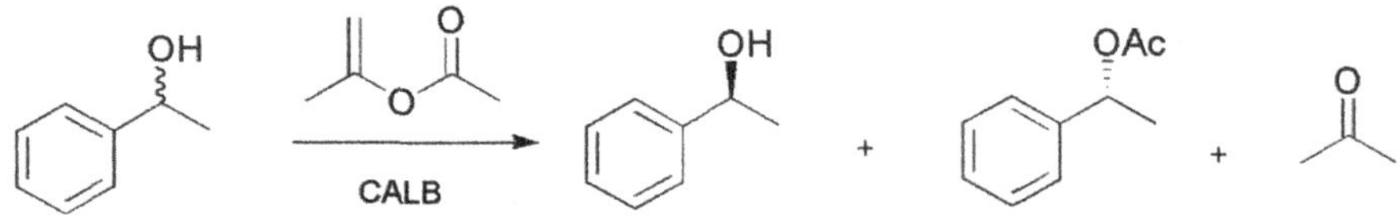

rac-1-Phenylethanol (*S*)-1-Phenylethanol (*R*)-1-Phenylethyl acetate Acetone

Number of sessions (duration of each session)	Hazard level	Difficulty level	Level of study
2 (2 h + 4 h)	Moderate	Medium	Intermediate

Class names Alcohols, enols and esters

Concepts involved The experiment demonstrates the enzymatic kinetic resolution of *rac*-1-phenylethanol utilizing *Candida antarctica* lipase B (CALB) as the catalyst and isopropenyl acetate as acetyl donor

Chemicals needed *rac*-1-Phenylethanol, isopropenyl acetate, *Candida antarctica* lipase B (Novozym 435), methyl *tert*-butyl ether (MTBE, dried over 4 Å mol sieves), (*S*)-1-phenylethanol, acetic anhydride, pyridine, 4-dimethylaminopyridine (DMAP)

Equipment and experimental techniques involved Vials for small scale reactions, micropipettes, syringe filters, chiral gas chromatography instrumentation

Keywords Chiral gas chromatography, enantiomeric excess, enzyme catalyst, keto–enol tautomerism, kinetic resolution

Background

The synthesis of enantiopure compounds is one of the key tasks in current synthetic organic chemistry. Several production methods for such compounds that were originally developed in academia are today applied in the fine chemical industry.[1] In general, the chirality of the synthesis product stems either from the chiral starting material (chiral pool)[2] or is created during the synthesis (stereoselective synthesis).[3] The third approach is based on racemate resolution where the synthesis route itself is not enantioselective but the racemic product is purified in a separate step, called resolution, to yield pure enantiomers.[4] Such resolution is kinetic when performed by utilizing the different reaction rates of the enantiomers in a certain reaction. Enzymatic kinetic resolution makes use of an enzyme catalyzed reaction that is selective for one enantiomer, optimally leaving the other enantiomer unreacted.[5]

Enzymatic kinetic resolution of alcohols employing a selective acylation reaction is illustrated in Scheme 18.3.1. Crucial in enzymatic kinetic resolution is that the rate constant for the reacting

Comprehensive Organic Chemistry Experiments for the Laboratory Classroom
Edited by Carlos A. M. Afonso, Nuno R. Candeias, Dulce Pereira Simão, Alexandre F. Trindade, Jaime A. S. Coelho, Bin Tan, and Robert Franzén.

Published by the Royal Society of Chemistry, www.rsc.org

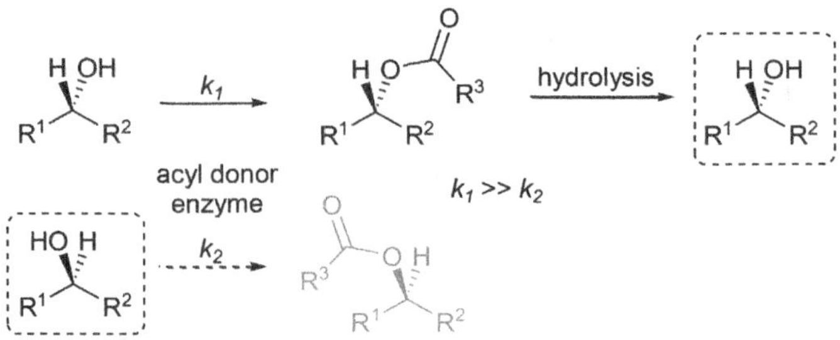

Scheme 18.3.1 Illustration of enzymatic kinetic resolution of a chiral alcohol ($R^1 \neq R^2$) followed by hydrolysis of the reacting enantiomer.

enantiomer (k_1) is significantly higher than the rate constant for the other enantiomer (k_2). This approach produces optimally, at 50% conversion, both alcohol enantiomers in enantiopure form, *i.e.* the more reactive enantiomer after enzyme catalyzed acylation followed by hydrolysis and the less reactive enantiomer remaining unreacted in the enzyme catalyzed acylation.

The objective of this work is to study the *Candida antarctica* lipase B (CALB) catalyzed acetylation of *rac*-1-phenylethanol using isopropenyl acetate as the acetyl donor. The progress of the reaction is monitored by chiral gas chromatography (GC). As the stationary phase, *i.e.* the GC column, is coated with chiral material, such as β-dextrin, the enantiomers will interact differently with the stationary phase and, therefore, the separation of the enantiomers is obtained. The enzyme catalyzed esterification is selective towards one of the 1-phenylethanol enantiomers and the enantiopurity of the esterified product and the starting material are calculated directly from the areas of the corresponding peaks in the chromatograms and reported as enantiomeric excess (ee) (eqn 18.3.1):

$$\mathrm{ee} = \frac{[(R)] - [(S)]}{[(R)] + [(S)]} \qquad (18.3.1)$$

The enzymatic enantioselectivity that describes the enzyme's ability to differentiate between the two enantiomers is measured with the enantiomeric ratio (E-value). The E-value is calculated from the enantiomeric excesses of the substrate ($\mathrm{ee_S}$) or product ($\mathrm{ee_P}$) and the degree of conversion (c) (eqn 18.3.2):

$$E = \frac{\ln\left[(1-c)(1-\mathrm{ee_S})\right]}{\ln\left[(1-c)(1+\mathrm{ee_S})\right]} = \frac{\ln\left[1-c(1+\mathrm{ee_P})\right]}{\ln\left[1-c(1-\mathrm{ee_P})\right]} \qquad (18.3.2)$$

Further, the degree of conversion of the enzyme catalyzed reaction can be calculated from $\mathrm{ee_S}$ and $\mathrm{ee_P}$ (eqn 18.3.3):[6]

$$c = \frac{\mathrm{ee_S}}{(\mathrm{ee_S} + \mathrm{ee_P})} \qquad (18.3.3)$$

The equations presented here are valid under Michaelis–Menten steady state conditions.[7] The derivation of the kinetic model is, however, somewhat beyond the scope of this laboratory work. Nevertheless, theoretically, an E-value > 1000 will provide both the resolution product and the unreacted substrate enantiomer with excellent enantiopurity, $\mathrm{ee_P} > 99\%$ and $\mathrm{ee_S} > 99\%$, at 50% conversion. However, with high E-values even small experimental errors in the determination of ee-values and conversions will distort the E-value considerably. Thus, in the literature, excellent E-values are usually defined as $E > 200$ or $E > 100$.

Additional safety

All experiments should be carried out in a fume hood if possible. Safety goggles, laboratory coat and protective gloves must be used during the work. Note that the chemicals are harmful or toxic (DMAP) if swallowed, cause skin irritation, severe skin burns (acetic anhydride, Ac_2O) or are fatal in contact with skin (DMAP), may cause respiratory irritation or are toxic if inhaled (Ac_2O) and cause serious eye irritation or damage. Although the volumes of liquids handled are small the risk of getting splashes is significant and therefore practical and reliable operation is essential.

Experimental Procedure

Laboratory Session 1 (2 h): Analysis of Three Reference Samples

1. Prepare samples for GC analysis of *rac*-1-phenylethanol and (*S*)-1-phenylethanol. The reference samples should have concentrations of <0.5 mg mL^{-1}. The samples can be prepared, for example, by adding 1–2 drops (approximately 20 mg) of the reference compound in a vial using the Pasteur pipette and diluting it with MTBE. From this solution, 2–4 drops are taken into a GC vial and diluted again using MTBE.
2. Prepare a *rac*-1-phenylethyl acetate sample by derivatizing the diluted solution of *rac*-1-phenylethanol with 1 drop of acetic anhydride and 1 drop of pyridine containing <1% DMAP. Close the cap properly and shake the vial. The derivatization is completed after approximately 10–15 min.

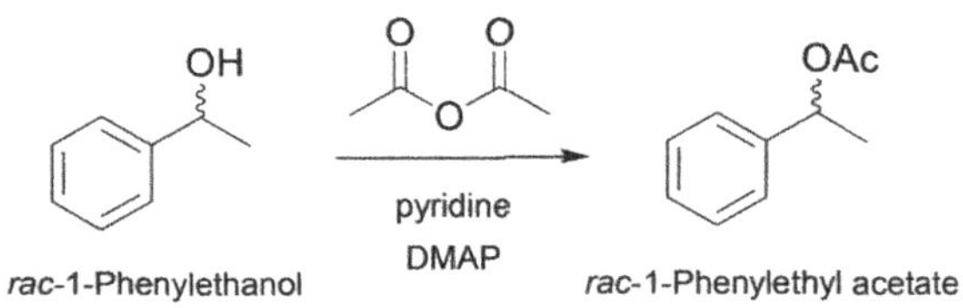

3. Analyze the samples (*rac*-1-phenylethanol, (*S*)-1-phenylethanol and *rac*-1-phenylethyl acetate) with a GC equipped with a chiral column.

Laboratory Session 2 (4 h): Enzyme Catalyzed Kinetic Resolution of *rac*-1-Phenylethanol

1. Calculate the amounts of reagents used in the experiment by filling the gaps in Table 18.3.1.
2. Prepare a 4 mL vial with a small magnetic stirring bar and cap. Adjust the boss head and the clamp in the fume hood so that the vial stays on the magnetic stirrer plate during the reaction.
3. Weigh 10 mg of enzyme beads directly into the vial (±1–2 mg accuracy is acceptable). Use an analytical scale and avoid static electricity if the laboratory climate is dry, preferably by using an anti-static instrument (Zerostat or similar).

Table 18.3.1

	n (mmol)	M (g mol^{-1})	m (mg)	δ (g mL^{-1})	V (μL)
rac-1-Phenylethanol	1	122.16		1.01	
Isopropenyl acetate	2	100.12		0.91	

4. Add 1.7 mL of MTBE using a plastic syringe and the calculated volume of *rac*-1-phenylethanol using a micropipette to the vial. Start the reaction by adding the calculated volume of isopropenyl acetate to the mixture using the micropipette (be sure to note the actual starting time with a proper clock). Close the cap properly and fasten the vial with the clamp. Stir the reaction mixture gently at room temperature. Be sure that the heating, if available, is turned off when using the magnetic stirrer plate.
5. Take 0.15–0.2 mL samples from the reaction mixture after 30, 60 and 180 min using a 1 mL syringe. After taking the sample, attach the syringe filter and filter the sample to an empty 4 mL vial and dilute it with MTBE. From the diluted solution, transfer 2–3 drops to a GC vial and dilute again with MTBE. Analyze the samples by GC equipped with a chiral column.

Waste Disposal

The solid catalyst beads from the reaction mixture can be filtered off and deposited in the solid waste. The organic solutions are collected in the non-halogenated waste container or collected in a designated flask.

Results Interpretation and Additional Questions

1. Draw molecular structures (showing the stereochemistry unequivocally) for all four compounds involved in the resolution: (*R*)-1-phenylethanol, (*S*)-1-phenylethanol, (*R*)-1-phenylethyl acetate and (*S*)-1-phenylethyl acetate.
2. Identify and list the retention times, corresponding to the peaks in GC chromatograms, for (*R*)-1-phenylethanol, (*S*)-1-phenylethanol, (*R*)-1-phenylethyl acetate and (*S*)-1-phenylethyl acetate. The three reference chromatograms are needed to identify the peaks of the starting material and product enantiomers.
3. Calculate the ee-values for the starting material and for the product at 30, 60 and 180 min. Use eqn (18.3.1). Note 1: change the place of [(*R*)] and [(*S*)] if latter is present in a higher concentration. Note 2: kinetic resolution does not provide ee-value = 100%. The highest ee-values are defined as >99.9%.
4. Calculate the enantiomeric ratio (*E*-value) and conversion (c) for each sample using eqn (18.3.2) and (18.3.3), respectively. Note that for eqn (18.3.2) $0 < c < 1$ and $0 < \text{ee} < 1$.
5. Explain the principle of chiral GC analysis and how the separation of enantiomers is accomplished.
6. Explain the formation of acetone as a by-product (see reaction scheme) using a known intramolecular transformation in organic chemistry.

References

1. M. Breuer, K. Ditrich, T. Habicher, B. Hauer, M. Kesseler, R. Stürmer and T. Zelinski, *Angew. Chem., Int. Ed.*, 2004, **43**, 788.
2. (a) G. Casiraghi, F. Zanardi, G. Rassu and P. Spanu, *Chem. Rev.*, 1995, **95**, 1677; (b) *Handbook of Chiral Chemicals*, ed. D. Ager, CRC Press Taylor & Francis Group, New York, 2nd edn, 2005.
3. *Stereoselective Synthesis of Drugs and Natural Products*, ed. V. Andrushko and N. Andrushko, John Wiley & Sons Inc., Hoboken, New Jersey, 2013.
4. E. Fogassy, M. Nógrádi, D. Kozma, G. Egri, E. Pálovics and V. Kiss, *Org. Biomol. Chem.*, 2006, **4**, 3011.
5. (a) B. Kagan and J. C. Fiaud, *Top. Stereochem.*, 1988, **18**, 249; (b) Y. F. Wang, J. J. Lalonde, M. Momongan, D. E. Bergbreiter and C. H. Wong, *J. Am. Chem. Soc.*, 1988, **110**, 7200.
6. A. J. J. Straathof and J. A. Jongejan, *Enzyme Microb. Technol.*, 1997, **21**, 559.
7. (a) C.-S. Chen, Y. Fujimoto, G. Girdaukas and C. J. Sih, *J. Am. Chem. Soc.*, 1982, **104**, 1294; (b) C. S. Chen and C. J. Sih, *Angew. Chem., Int. Ed. Engl.*, 1989, **28**, 695.

18.4 Enzymatic Kinetic Resolution and Separation of *sec*-Alcohols Methodology Based on Fatty Esters

Carlos M. Monteiro*[a], Nuno M. T. Lourenço*[b], and Carlos A. M. Afonso*[a]

[a]iMed. ULisboa, Faculdade de Farmácia da Universidade de Lisboa, Av. Prof. Gama Pinto, 1649-003 Lisboa, Portugal; [b]Department of Bioengineering, IBB – Institute for Biotechnology and Bioengineering, Centre for Biological and Chemical Engineering, Instituto Superior Técnico, 1049-001 Lisboa, Portugal
*E-mail: carlosmmonteiro@ff.ulisboa.pt, nmtl@ist.utl.pt, carlosafonso@ff.ulisboa.pt

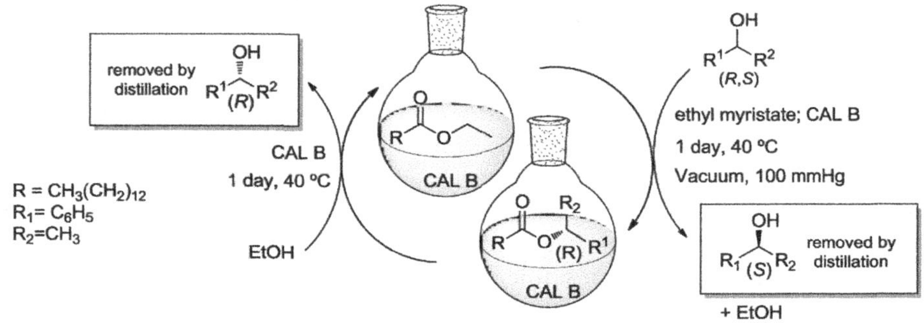

Number of sessions (duration of each session)	Hazard level	Difficulty level	Level of study
3 (1 h + 3 h + 3 h)	Moderate	Medium	Intermediate

Class names Esters, secondary alcohols

Concepts involved Herein is introduced the concept of biocatalysis, using the same enzyme (biocatalyst) in two different reactions, and also (a) concept of concentration of reagents and product in reaction equilibrium (Le Châtelier's principle) and, at the same time, the advantage of this methodology in the resolution of enantiomers of racemic secondary alcohols; (b) application of sustainable methods for the separation of secondary alcohols; (c) physical and spectroscopic characterisation by IR and NMR; (d) chromatographic analysis by GC

Chemicals needed 1-Phenylethanol, myristic acid, ethyl myristate, ethanol, hexane, sodium bicarbonate, magnesium sulfate anhydrous, sulfuric acid, deuterated chloroform, *Candida antarctica* lipase B (CAL B) (Novozym 435®)

Equipment and experimental techniques involved GC, NMR, apparatus for reactions under vacuum, high temperature (100 °C) and magnetic stirring, filtration under vacuum; product purification by distillation under vacuum

Keywords Distillation, chiral GC, enzymatic kinetic resolution, fatty esters, NMR, secondary alcohols

Background of the Experiment

Enzymatic kinetic resolution (EKR) of racemic alcohols is an important and practical methodology in attaining enantiomerically enriched precursors of active drugs.[1,2] The use of enzymes, such as lipases, presents outstanding advantages, given that they catalyse reactions under mild

Comprehensive Organic Chemistry Experiments for the Laboratory Classroom
Edited by Carlos A. M. Afonso, Nuno R. Candeias, Dulce Pereira Simão, Alexandre F. Trindade, Jaime A. S. Coelho, Bin Tan, and Robert Franzén.

Published by the Royal Society of Chemistry, www.rsc.org

conditions with high enantio- or diastereoselectivity, high efficiency and with the possibility of using different reaction media, such as organic solvents.[3] The enzymatic kinetic resolution mediated by lipases is founded on the transesterification, esterification or hydrolysis reactions. From the mechanistic point of view, it is based on formation of an acyl-enzyme intermediate that may be subjected to a nucleophilic attack by different nucleophiles, namely alcohols, thiols or amines. The EKR depends on the difference between the reaction rates of two enantiomers mediated by a biocatalyst in order to afford an enantioenriched sample of the less reactive enantiomer. This fact is explained by the different activation energies of each enantiomer in the presence of a biocatalyst, *i.e.* the biocatalyst reduces the energy of activation of only one enantiomer. Moreover, the EKR is remarkably efficient and numerous successful examples with conversion close to 50% can be found. However, there are still some limitations, economically and environmentally speaking, namely the need for a large amount of acylating agent and the final separation step. To improve the separation of alcohol from the enzymatic product, some elegant separation methods have been developed in order to replace the chromatography techniques, in particular using fluorinated solvents,[4] ionic liquids,[5] membrane or PEG.[6] Since enzymatic esterification and transesterification are generally reversible processes, it is very important to achieve transformation under irreversible conditions *i.e.* it is crucial to shift the equilibrium towards the formation of the ester.

Several efforts have been made and described in the literature to eliminate water or volatile alcohol during the reaction, such as evaporation under vacuum[7] or an azeotropic distillation with benzene or toluene.[8] Another possible way to overcome the reversibility issue would be to design and select different acylating agents. The use of enol esters as acyl donor, namely vinyl esters and isopropenyl esters, is the most popular approach to circumvent the reversibility problem. During the transesterification the leaving group that is an enol instantly tautomerizes to the keto form and consequently no nucleophile remains and the reaction is irreversible.[9,10] Even so, there are still some drawbacks, for instance the formation of reactive side products as, for example, aldehydes and ketones which may provoke some enzymatic deactivation.[11] Against those limitations, other acylating agents were developed with less restrictions, such as succinic anhydride,[12,13] carbonates[14] and fatty esters.[15,16,17]

The possible usage of fatty esters as acylating agents in the enzymatic resolution of racemic secondary alcohols allows the transesterification to be performed without organic solvents, due to these esters being stable liquid compounds with good properties as a solvent.[15,16,17] Moreover, the reversibility of the enzymatic reaction can be avoided by performing the reaction under vacuum. The low vapour pressure of the acylating agent allows an efficient removal of volatile alcohol formed during the transesterification.[15,16,17]

The main aim of this experiment is to show undergraduate students that enzymes can react in both directions under irreversible conditions, depending on those reaction conditions. Using the same enzyme (catalyst) to perform two different reactions, students will understand the effect of reactants and products concentration on chemical equilibrium (Le Châtelier's principle) and at the same time learn the advantage of this methodology in the resolution of secondary alcohols.

Additional Safety

1-Phenylethanol and 1-phenylethyl myristate are toxic, irritant, and may be carcinogenic. Any inhalation, eye/skin contact or ingestion should be avoided. Safety glasses, gloves and good ventilation are required. The ethyl myristate is slightly hazardous in case of skin contact (irritant), eye contact (irritant), ingestion and inhalation. It may also be combustible at high temperatures. Ethanol may be absorbed through the skin. It may cause irritation to eyes, mucous membranes and upper respiratory tract. It also may cause eye damage, nausea, dizziness, headache and central nervous system depression. The vacuum distillation should be carried out with all necessary precautions.

Experimental Procedure

Laboratory Session 1 (1 h + 24 h of Enzymatic Reaction) – Enzymatic Transesterification Reaction

General procedure for enzymatic kinetic resolution and separation of secondary-alcohols:

1. Add CAL B (Novozym 435®; 500 mg) and racemic 1-phenylethanol (2.68 g, 21.94 mmol) to magnetically stirred ethyl myristate (5.63 g, 21.94 mmol) in a plastic test tube (50 mL) inside a glass trap in a vacuum pump system (approx. 1 h is required to set-up the reaction).
2. Stir the reaction mixture for 24 h under reduced pressure (100 mmHg adjusted by a metal needle valve) at 40 °C in a thermostatic water bath as shown in Figure SM 1 (ESI).
3. After this time, keep the reaction mixture in the freezer for the next session.

Laboratory Session 2 (3 h + 24 h of Enzymatic Reaction) – (*S*)-1-Phenylethanol Distillation

1. Filter the above reaction mixture and wash the enzyme with hexane (3 × 15 mL).
2. Evaporate the solvent and distil the reaction mixture under reduced pressure (1 mmHg, 60 °C) to give the (*S*)-1-phenylethanol (Figure 3 in the ESI).
3. Dry the enzyme under reduced pressure (20 mmHg) for 2 h (the activity of the isolated enzyme is similar to that of the commercial sample).
4. After distillation, transfer the recovered enzyme and the collected reaction medium containing the other enantiomer as an ester and the ethyl myristate to a plastic test tube (50 mL) and add absolute ethanol (2.5 equiv.) (requires approx. 15 min to set-up the reaction).
5. Stir the mixture for 24 h at 40 °C in a thermostatic water bath. After this time, keep the reaction mixture in the freezer for the next session.

Laboratory Session 3 (3 h) – (*R*)-1-Phenylethanol Distillation

1. Filter the above reaction mixture and wash the enzyme with hexane (3 × 15 mL).
2. Evaporate the solvent.
3. Distil the reaction mixture under reduced pressure (1 mmHg, 60 °C) to give the (*R*)-1-phenylethanol.
4. Analyse all the compounds obtained by distillation by GC and NMR.

Waste Disposal

The aqueous solutions can be neutralized, diluted and discarded in the sink. The solvents obtained may be put in a non-chlorinated waste collector recipient, for posterior waste disposal.

Results Interpretation and Additional Questions

1. Interpret the spectra and verify the purity of the products obtained by ^{1}H and ^{13}C NMR. Compare them with the literature data.
2. Rationalize the involved mechanism of the esterification and transesterification.
3. Propose other acylating agents in order to perform this reaction. Explain why other acylating agents are irreversible.
4. Discuss the expected effect of the reaction temperature on the ee and yield.
5. Search the literature for other methodologies for the enzymatic resolution of secondary alcohols.

Acknowledgement

Reprinted (adapted) with permission from *J. Chem. Educ.*, 2010, **87**, 423-425. Copyright (2010) American Chemical Society.

References

1. R. N. Patel, *Adv. Synth. Catal.*, 2001, **343**, 527–546.
2. A. Ghanem, *Tetrahedron*, 2007, **63**, 1721–1754.
3. R. Z. Kazandjian and A. M. Klibanov, *J. Am. Chem. Soc.*, 1985, **107**, 5448–5450.
4. Z. Y. Luo, S. M. Swaleh, F. Theil and D. P. Curran, *Org. Lett.*, 2002, **4**, 2585–2587.
5. N. M. T. Lourenco and C. A. M. Afonso, *Angew. Chem., Int. Ed.*, 2007, **46**, 8178–8181.
6. M. Nogawa, M. Shimojo, K. Matsumoto, M. Okudomi, Y. Nemoto and H. Ohta, *Tetrahedron*, 2006, **62**, 7300–7306.
7. J. A. Bosley, J. Casey, A. R. Macrae and G. MyCock, US Patent 56587691, Unichem Chemie BV, 1997.
8. L. Gubicza, A. Kabiri-Badr, E. Keoves and K. Belafi-Bako, *J. Biotechnol.*, 2000, **84**, 193–196.
9. H. B. Kagan and J. C. Fiaud, Kinetic resolution, in *Topics in Stereochemistry*, ed. E. L. Eliet and S. H. Wein, John Wiley & Sons, Inc, 2007, vol. 18, pp. 249–330.
10. J. M. Keith, J. F. Larrow and E. N. Jacobsen, *Adv. Synth. Catal.*, 2001, **343**, 5–26.
11. U. Hanefeld, *Org. Biomol. Chem.*, 2003, **1**, 2405–2415.
12. A. L. Gutman, D. Brenner and A. Boltanski, *Tetrahedron: Asymmetry*, 1993, **4**, 839–844.
13. R. Vaidyanathan, L. Hesmondhalgh and S. H. Hu, *Org. Process Res. Dev.*, 2007, **11**, 903–906.
14. M. Pozo, R. Pulido and V. Gotor, *Tetrahedron*, 1992, **48**, 6477–6484.
15. S. Negishi, S. E. Shirasaw, J. Y. Susuki and M. Yukie, EP0709465, 1996.
16. C. M. Monteiro, N. M. T. Lourenço and C. A. M. Afonso, *J. Chem. Educ.*, 2010, **87**, 423–425.
17. C. M. Monteiro, N. M. T. Lourenço and C. A. M. Afonso, *Tetrahedron: Asymmetry*, 2010, **21**, 952–956.

18.5. Enzymatic Kinetic Resolution and Preparative Separation of Secondary Alcohols

Nuno M. T. Lourenço*[a], Carlos M. Monteiro*[b], and Carlos A. M. Afonso*[b]

[a]Instituto Superior Técnico, Universidade de Lisboa, Av. Rovisco Pais, 1, 1049-001 Lisboa, Portugal;
[b]Faculdade de Farmácia, Universidade de Lisboa, Av. Gama Pinto, 1649-003 Lisboa, Portugal
*E-mail: nmtl@ist.ulisboa.pt, carlosmmonteiro@ff.ul.pt, carlosafonso@ff.ul.pt

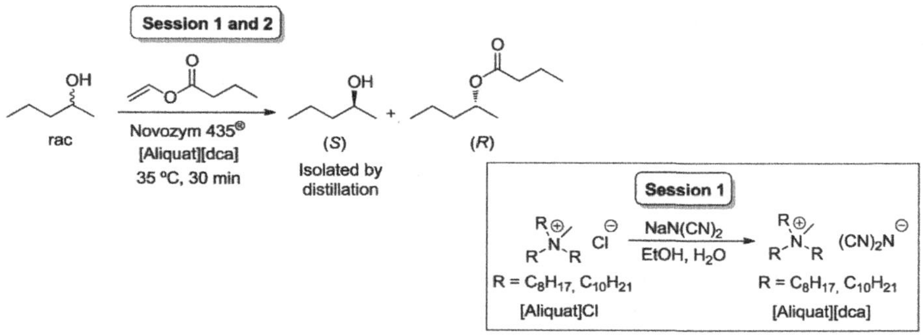

Number of sessions (duration of each session)	Hazard level	Difficulty level	Level of study
2 (4 h each)	Moderate	Medium	Intermediate
Class names Quaternary ammonium salt, alcohol, ester			
Concepts involved Properties of ionic liquids and their preparation by metathesis reactions, kinetic resolution, enzymatic enantioselective transesterification with vinyl alkanoates, enol ether, aldehyde, tautomerism, stereochemistry, optical rotation, specific rotation			
Chemicals needed Aliquat 336®; ethanol, sodium dicyanamide, distilled water, *rac*-2-pentanol, vinyl butyrate, Novozym 435®, diethyl ether, deuterated chloroform, phosphomolybdic acid stain			
Equipment and experimental techniques involved Magnetic stirring hot plate, polarimeter, filtration under vacuum, extraction and distillation under vacuum, infrared spectrometer, NMR spectrometer			
Keywords *Candida antarctica* lipase B, enzymatic kinetic resolution (EKR), liquid–liquid extraction, NMR spectroscopy, polarimetry, vacuum distillation			

Background

Chiral compounds play an important role in the synthesis of drugs, food additives, pesticides and fragrances.[1] Even though, in several cases, only one enantiomer is needed, there are other cases where both enantiomers are important and have distinct applications. Several commercial drugs still occur as racemic mixtures; however, over recent decades, enantiomerically pure compounds have arisen as precursors of important drugs. Consequently, the hydroxyl function group is extremely important and we can find several robust methodologies to attain enantiomerically pure alcohols, namely asymmetric reduction of ketones mediated by chiral catalysts,

Comprehensive Organic Chemistry Experiments for the Laboratory Classroom
Edited by Carlos A. M. Afonso, Nuno R. Candeias, Dulce Pereira Simão, Alexandre F. Trindade, Jaime A. S. Coelho, Bin Tan, and Robert Franzén.

Published by the Royal Society of Chemistry, www.rsc.org

biocatalysts or even both (dynamic kinetic resolution (DKR)). These methodologies are ideal in cases where the goal is to obtain only one enantiomer. However, when both enantiomers are useful, the resolution of racemic mixtures is a key alternative. In this case, chromatographic resolution (*e.g.* bed chromatography) and membrane technology are an alternative, although they are economically unattractive when the resolution is up scaled.[2] The separation of enantiomers from a racemic mixture based on differences induced by diastereomeric interactions combined with crystallization or chromatographic isolation proved to be an effective method in achieving chiral compounds. Another alternative is the resolution by enantioselective derivatisation of the hydroxyl group (kinetic resolution) and separation of the enantiomers, *e.g.* (*R*)-enantiomer as an alcohol and (*S*)-enantiomer as an ester.[3]

The kinetic resolution is based on the difference between the reaction rates of two enantiomers in the presence of a chiral catalyst, resulting in an enantioenriched sample of the less reactive enantiomer. This difference is due to the different activation energies. The chiral enantiomerically pure catalyst reduces the energy of activation of the enantiomers. Though, in order to achieve efficiency of this approach, it is crucial that only one enantiomer reacts, unfortunately the literature does not provide many chemical systems of kinetic resolution. Nevertheless, the enzymatic resolution is extremely efficient, and several successful examples with conversion close to 50% can be found. Due to the difference in concentration of the two enantiomers, only one of them is recognized by the enzyme and the biotransformation ends when the reactive enantiomer is fully consumed (conversion of 50%). The enzymatic kinetic resolution mediated by lipases is based on the reaction of transesterification, esterification or hydrolysis. Mechanistically, it is founded on the formation of an acyl-enzyme intermediate that may be subjected to a nucleophilic attack by different nucleophiles, such as by alcohols, amines and thiols. These transesterifications reactions, *i.e.*, formation of a new ester from a substrate ester, have been widely explored in the preparation of enantiomeric pure alcohols, primary, secondary or even tertiary alcohols. Furthermore, as the formation of esters is an equilibrium reaction, it is very important to shift the equilibrium towards the formation of the ester. The use of enol esters as acyl donor is the most popular approach to overcome the reversibility problem. During the transesterification the leaving group that is an enol instantly tautomerizes to the keto form and consequently no nucleophile remains and the reaction is irreversible.[4,5] Herein is disclosed the enzymatic kinetic resolution (EKR) of the model substrate (*R*/*S*)-2-pentanol with vinyl butyrate as acyl donor mediated by the commercially available *Candida antarctica* lipase B (CAL B (Novozym 435®)).

Ionic liquids (ILs) are liquids that are mainly salts with a melting point below the boiling point of water (100 °C). Normally these salts contain nitrogen-, sulfur- or phosphorous-organic cations with alkyl chains, combined with an anion that may be organic or inorganic. The most common cations are 1,3-dialkyl-imidazolium, tetra-alkylammonium and phosphonium, *N*-alkylpyridinium and hexa-alkylguanidinium. Regarding the anion, the most common ones are BF_4^-, PF_6^-, $(CF_3SO_2)_2N^-$, $CF_3SO_3^-$, $CF_3CO_2^-$ and $(CN)_2N^-$.[6–9] After the discovery of air stable IL during the 1980s, a huge development occurred in this area, especially with the discovery of new IL that are liquid at room temperature. In addition to their usage as reaction medium there is the possibility of an efficient reuse of the catalytic system in organic, inorganic and organometallic transformations, polymers and material chemistry, biotransformations, solar cells, sensors, nanotechnology, lubricants and separation technology.[10] This broad interest is driven by their peculiar properties, namely the high thermal stability, conductivity and electrochemical range, the minor volatility and flammability and the ability to overcome the low solubility of water in organic solvents, only by combination of different cations and anions. However, it presents some drawbacks, namely the production cost, toxicity and biodegradability. In biocatalysis, it was observed that some ILs confer a higher stability to the enzyme than traditional organic solvents, supercritical fluids or water.[11,12]

To assess the efficiency and sustainability of a chemical process, parameters such as the E-factor and atom efficiency can be measured. The E-factor is measured by the ratio of the mass of waste per unit of product. Regarding the atom economy, it is a calculation of how much of the reactants remain in the final product, *i.e.* the comparison of the molecular weight of final product and the sum of the molecular weights of all reactants.

Herein, is presented a procedure to prepare a low cost IL, [Aliquat][dca], by changing the chloride anion from the commercially available salt Aliquat 336® (2 : 1 mixture of tri-*n*-octyl- and tri-*n*-decyl-methylammonium chloride) with dicyanamide cation [dca]. This IL will be used as reaction medium for the EKR of (*R*/*S*)-2-pentanol by enantioselective acylation. One of the enantiomers is removed from the reaction medium by simple distillation, taking advantage of the lack of volatility of ILs under the reaction conditions.

Additional Safety

The Aliquat 336® and [Aliquat][dca] and sodium dicyanamide are toxic and corrosive. They may cause severe respiratory and digestive tract irritation with possible burns, harmful if swallowed and cause eye and skin burns. Any inhalation, eye/skin contact or ingestion should be avoided. Safety glasses, gloves and good ventilation are required. The vinyl butyrate and 2-pentanol are toxic in case of skin contact (irritant), eye contact (irritant), ingestion and inhalation. They may also be combustible at high temperatures. 2-Pentyl butyrate is toxic, may be absorbed through the skin. It may cause irritation to eyes, mucous membranes and upper respiratory tract. Safety glasses, gloves and good ventilation are required.

Experimental Procedure

Laboratory Session 1 (4 h)

Preparation of Ionic Liquid [Aliquat][dca] (Ref. 13–15)

1. In a fume hood, to a 250 mL round-bottomed flask equipped with a magnetic stir bar, add Aliquat 336® (10.0 g, 23.2 mmol, viscous liquid, 2 : 1 mixture of alkyl chains C_8 and C_{10}), ethanol (45 mL), and a solution of sodium dicyanamide ($NaN(CN)_2$) (5.2 g, 57.9 mmol) in distilled water (15 mL).
2. Stir the reaction mixture at room temperature for 30 min.
3. After this time, remove the magnetic stirrer and evaporate the solvents under vacuum in a rotary evaporator (formation of two phases).
4. Transfer the mixture to a separating funnel (100 mL) and wash the ionic phase with water (3 × 20 mL).
5. Evaporate the ionic phase under reduced pressure in a rotary evaporator (if necessary heat it up to 60 °C) to afford tri-*n*-octylmethylammonium dicyanamide, [Aliquat][dca].
6. Keep a small fraction (<500 mg) for spectral characterization (IR, ^{1}H NMR).
7. Use the remaining IL as reaction medium in the enzymatic resolution.

Enzymatic Resolution

1. In a fume hood, to the round-bottomed flask containing the IL, equipped with a magnetic stir bar, add (*R*/*S*)-2-pentanol (10 g, 113.4 mmol), vinyl butyrate (7.9 mL, 0.55 equiv.) and Novozym 435® (3.0 g).
2. Stir the reaction mixture at room temperature for 60 min.
3. After this time, in order to isolate the enzyme, filter the reaction mixture through a fritted funnel under vacuum to a 100 mL round-bottomed flask.

4. Distil the reaction mixture under reduced pressure (<20 mmHg, bp 118–119 °C/760 mmHg).
5. Proceed to the spectral characterization (IR, ^{1}H NMR) of the distilled compound.
6. Keep the distillation round-bottomed flask in the freezer for the following laboratory session.

Laboratory Session 2 (4 h) – (*R*)-(−)-2-Pentyl Butyrate Isolation

1. Carry out a TLC of the mixture in the round-bottomed flask from the previous session (detection with ethanolic phosphomolybdic acid solution; eluent: ethyl acetate : hexane (20 : 80); 2-pentyl butyrate – R_f = 0.85; vinyl butyrate – R_f = 0.70; 2-pentanol – R_f = 0.23).
2. Add diethyl ether (25 mL) and keep it under vigorous stirring for 10 min.
3. Remove the diethyl ether by decantation. Repeat this extraction procedure with diethyl ether (2 × 25 mL).
4. Evaporate diethyl ether from the combined extractions.
5. Weigh the product obtained ((*R*)-(−)-2-pentyl butyrate).
6. Proceed to the spectral characterization (IR, ^{1}H NMR) of the obtained product. In addition, using a polarimeter, measure the angle of rotation of the obtained ester and alcohol ((*R*)-(−)-2-pentyl butyrate, $[\alpha]_D^{25}$ = −16° (pure liquid); (*S*)-(+)-2-pentanol, $[\alpha]_D^{25}$ = +13° (pure liquid)).

Waste Disposal

The solvents obtained may be put in a non-chlorinated waste collector recipient, for posterior waste disposal.

Results Interpretation and Additional Questions

1. Determine the yield of the alcohol and the ester. Verify the purity of the products obtained and determine the enantiomeric excess (ee).
2. Determine and comment on the atom efficiency and the E-factor for each reaction step.
3. What is the advantage of the use of vinyl butyrate instead of butanoic acid?
4. Can the recovered IL be reused?

References

1. A. N. Collins, in *Chirality in Industry: The Commercial Manufacture and Application of Optically Active Compounds*, ed. G. N. Sheldrake and J. Crosby, John Wiley & Sons, 1992.
2. C. A. M. Afonso and J. G. Crespo, *Angew. Chem., Int. Ed.*, 2004, **43**, 5293–5295.
3. T. Kano, K. Sasaki and K. Maruoka, *Org. Lett.*, 2005, **7**, 1347–1349.
4. H. B. Kagan and J. C. Fiaud, in *Kinetic resolution*, ed. E. L. Eliet and S. H. Wein, John Wiley & Sons, Inc, 2007, vol. 18, Topics in Stereochemistry, pp. 249–330.
5. J. M. Keith, J. F. Larrow and E. N. Jacobsen, *Adv. Synth. Catal.*, 2001, **343**, 5–26.
6. T. Welton, *Chem. Rev.*, 1999, **99**, 2071–2084.
7. P. Wasserscheid and W. Keim, *Angew. Chem., Int. Ed.*, 2000, **39**, 3772–3789.
8. C. M. Gordon, *Appl. Catal., A*, 2001, **222**, 101–117.
9. J. Dupont, R. F. Souza and P. A. Z. Suarez, *Chem. Rev.*, 2002, **102**, 3667–3692.
10. A. R. Hajipour and F. Rafiee, *Org. Prep. Proced. Int.*, 2015, **47**, 1–60.
11. F. V. Rantwijk and R. A. Sheldon, *Chem. Rev.*, 2007, **107**, 2757–2785.
12. S. Garcia, N. M. T. Lourenço, D. Lousa, A. F. Sequeira, P. Mimoso, J. M. S. Cabral, C. A. M. Afonso and S. Barreiros, *Green Chem.*, 2004, 466–470.
13. P. S. Kulkarni, L. C. Branco, J. G. Crespo, M. C. Nunes, A. Raymundo and C. A. M. Afonso, *Chem.–Eur. J.*, 2007, **13**, 8478–8488.
14. N. M. T. Lourenço, PhD Thesis, Faculdade de Ciências e Tecnologia da Universidade Nova de Lisboa, 2007.
15. N. M. T. Lourenço, S. Barreiros and C. A. M. Afonso, *Green Chem.*, 2007, **9**, 734–736.

18.6. Catalyzed Resolution and Simultaneous Selective Crystallization

Jaan Parve[a], Lauri Vares[a], Indrek Reile[b], Tõnis Pehk[b], Ly Villo*[c], and Omar Parve[c]

[a]Institute of Technology, University of Tartu, Nooruse 1, 50411 Tartu, Estonia; [b]National Institute of Chemical Physics and Biophysics, Akadeemia tee 23, 12618 Tallinn, Estonia; [c]Department of Chemistry, Tallinn University of Technology, Ehitajate tee 5, 19086 Tallinn, Estonia. *E-mail: ly.villo@ttu.ee

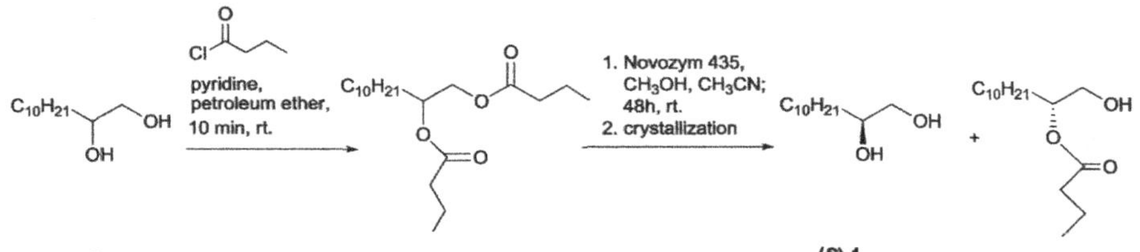

1 er = 50 : 50 — 2 — (*S*)-1 er = 99.9 : 0.1 — 3

Number of sessions (duration of each session)	Hazard level	Difficulty level	Level of study
2 (4 h each)	Moderate	Medium	Intermediate

Class names Alcohols, esters, acid chloride

Concepts involved This experiment involves a prompt separation of (*S*)-1,2-dodecanediol from a racemic mixture of enantiomers, *via* lipase-catalyzed methanolysis of crude chemically prepared diester, followed by *in situ* crystallization of the pure enantiomer from the reaction mixture

Chemicals needed 1,2-Dodecanediol (racemic), butyryl chloride, pyridine, petroleum ether (40–60 °C) (PE), methanol, H_2O, $NaHCO_3$, NaCl, $MgSO_4$, acetonitrile, Novozym® 435 (*Candida antarctica* lipase B immobilized on a solid carrier), ethyl acetate (EtOAc), thin-layer chromatography (TLC) visualization reagent ($EtOH/H_2SO_4$/anisaldehyde 90/10/3)

Equipment and experimental techniques involved Apparatuses for (1) reaction with magnetic stirring, (2) liquid–liquid extraction, and (3) for filtering through paper filter; rotary evaporator, precoated silica gel 60 F_{254} plates for TLC; a removable paper bag (*e.g.* an empty "teabag" or a bag made from filter paper) filled with Novozym® 435; crystallization; apparatuses for vacuum filtering through glass filter and for determination of melting point, polarimeter

Keywords Crystallization, (*S*)-1,2-dodecanediol, kinetic resolution, lipase-catalyzed methanolysis, specific rotation

Background

Single pure enantiomers of 1,2-dodecanediol (**1**) are valuable chiral building blocks for several synthetic applications in the field of biologically active compounds.[1] Pure stereoisomers of alkanediols and alkanetetrols of this type can be prepared by chemical hydrolytic kinetic resolution of terminal epoxides using Jacobsen's catalyst.[2–4] Methods of asymmetric synthesis, such as asymmetric dihydroxylation of olefins[5] or diboration–oxidation of olefins and alkynes[6] can be used as well.

Comprehensive Organic Chemistry Experiments for the Laboratory Classroom
Edited by Carlos A. M. Afonso, Nuno R. Candeias, Dulce Pereira Simão, Alexandre F. Trindade, Jaime A. S. Coelho, Bin Tan, and Robert Franzén.

Published by the Royal Society of Chemistry, www.rsc.org

Starting from racemic 1,2-alkanediols, lipase-catalytic kinetic resolution based on stereoselective acyl transfer reactions has found wide application.[7–12]

In general, enzymes are known to work best by catalyzing reactions of their natural substrates and in their natural environment – water. However, several lipases tolerate artificial substrates and are catalytically very active also in organic solvents.[13] The enzyme preparation Novozym® 435 used in this protocol includes the lipase B from the microorganism *Candida antarctica* immobilized on the solid carrier levatite.

The idea behind the enzyme-catalyzed kinetic resolution of enantiomers is that the enzyme distinguishes between enantiomers because of their different 3D structure. The reaction catalyzed by the enzyme occurs much faster with one enantiomer than the other – and, thus, it is possible to stop the reaction at a suitable point of conversion to afford a kinetically resolved enantiomeric product with the desired enantiomeric ratio (er) value. In the current experiment, spontaneous *in situ* crystallization of (*S*)-1,2-dodecanediol (produced by the lipase-catalyzed methanolysis of butyric diester) from the reaction mixture is stereo- and chemoselective – providing the product in high chemical and stereochemical purity.[12]

Thus, the goal of the current experiment is to demonstrate a prompt, simple method, *e.g.* without using column chromatography, for separation of the crystalline (*S*)-enantiomer of 1,2-dodecanediol from a racemic mixture that allows us to gain the product with very high enantiomeric purity (er > 99.8/0.2).

Additional Safety

All steps should be carried out in a fume hood using personal protective equipment (protective clothing, gloves, and safety goggles). Butyryl chloride is flammable, corrosive, reacts violently with water, and may cause skin and eye burns. Pyridine is flammable and toxic. Petroleum ether is flammable. Methanol is both flammable and toxic.

Experimental Procedure

Laboratory Session 1 (4 h)

Chemical Acylation of 1,2-Dodecane Diol

1. Prepare a 100 mL round-bottomed flask with a magnetic stirring bar, then add 1 g (5 mmol, 1 equiv.) of racemic 1,2-dodecanediol (**1**). After that, add 4 mL of pyridine, with gentle stirring and warming until the diol has dissolved.
2. Add 12 mL of PE and stir until the solution is homogeneous.
3. Add dropwise 1.3 mL (12 mmol, 2.4 equiv.) of butyryl chloride while ensuring effective stirring of the forming heterogeneous solution. Let the reaction stir for 15 min.
4. Check the completion of the reaction by performing TLC, comparing the reaction mixture with the starting 1,2-dodecanediol (**1**). Eluent for TLC: mixture of EtOAc (3 mL) and PE (5 mL). Visualization of TLC stains with anisaldehyde reagent on heating.
5. If the reaction has completed, finish the synthesis by adding methanol (0.5 mL, 12.3 mmol) with the aim of eliminating excess butyryl chloride. Stir for an additional 10 min.
6. Then add 40 mL of PE into the flask followed by 15 mL of 10% $NaHCO_3/H_2O$ solution. The latter should be added carefully due to possible intense gas evolution.
7. Transfer the reaction mixture into a separatory funnel (250 mL), add an additional portion of PE (40 mL), shake the mixture and separate the water layer.
8. Wash the organic layer once with aqueous $NaHCO_3$ solution and twice with brine (2 × 15 mL). Dry over $MgSO_4$ for 15 min.
9. Filter the drying agent from the solution and evaporate the solvents. The obtained crude product of diester **2** is used in the following enzymatic reaction without purification.

Enzymatic Separation of Enantiomers via *Methanolysis Complemented by* In situ *Crystallization*

10. Dissolve the crude diester **2** in acetonitrile (9.6 mL), add methanol (0.4 mL), introduce Novozym 435® (0.5 g) packed in a removable paper bag and a small stirring bar into the reaction flask. Stir on a magnetic stirrer for at least 48 h.

Laboratory Session 2 (4 h)

1. Check the conversion rate of the reaction by performing TLC, comparing the reaction mixture with 1,2-dodecanediol (**1**) and diester **2**. Eluent for TLC: mixture of EtOAc (3 mL) and PE (5 mL). Visualization of TLC stains with anisaldehyde reagent. By TLC, the reaction mixture should contain only trace amounts of the starting diester along with a mixture of comparable amounts of 1,2-dodecanediol (**1**) and its 2-monoester (**3**). The crystalline product should have become observable with the naked eye.

Isolation and Characterization of the Product

2. Stop the reaction by removing the enzyme bag (the bag should be carefully washed with 15 mL of EtOAc in a separate vessel and the washing added later to the crude product/mother liquor from which sequential crystallization of an additional portion of highly pure (*S*)-diol can be performed; see the next procedure).
3. Now filter out the crystalline product using a glass filter. Approximately 100 mg of the product crystals are gained (yield: 10%). The residual crude product/mother liquor (with added washing of the enzyme bag) can be recrystallized from chloroform (dissolve in 2 mL on heating, then stir briefly on an ice bath and filter the product) in order to gain an additional portion of the pure (*S*)-diol and to increase the yield up to 29%.[12]
4. The crystalline product should be dried in a vacuum desiccator for 15 min in order to separate solvent residues. The product should be weighed, the yield calculated, and the homogeneity of the product determined by TLC analysis.
5. Measure the optical rotation of the obtained product in EtOH and calculate the specific rotation.
6. Measure the melting point of the obtained pure enantiomer as well as of the racemic 1,2-dodecanediol.

Waste Disposal

The aqueous solution from liquid–liquid extraction should be collected into a special container of waste water. The solvents evaporated and condensed in a rotary evaporator should be collected and disposed of in a non-chlorinated waste container.

Results Interpretation and Additional Questions

1. Compare the specific rotation value determined for the separated enantiomer with the literature data and, based on this data, calculate the enantiomeric excess (ee) as well as the enantiomeric ratio (er), of the product.
2. Compare the melting point of the separated (*S*)-enantiomer with that of the racemic sample as well as with corresponding literature data; explain the results.[12]
3. Calculate the yield of crystalline (*S*)-1,2-dodecanediol obtained from the racemic material.
4. Explain the mechanism of the lipase-catalyzed sequential methanolysis used for resolution of the enantiomers.

References

1. C. A. Challenger, *Chiral Intermediates*, John Wiley & Sons, London, 2001.
2. M. Tokunaga, J. F. Larrow, F. Kakiuchi and E. N. Jacobsen, *Science*, 1997, **277**, 936.
3. S. Chow and W. Kitching, *Chem. Commun.*, 2001, 1040.
4. J. Bredihhina, P. Villo, K. Andersons, L. Toom and L. Vares, *J. Org. Chem.*, 2013, **78**, 2379.
5. H. C. Kolb, M. S. Van Nieuwenhze and K. B. Sharpless, *Chem. Rev.*, 1994, **94**, 2483.
6. Y. Lee, H. Jang and A. H. Hoveyda, *J. Am. Chem. Soc.*, 2009, **131**, 18234.
7. Y. Shimada, H. Sato, S. Minowa and K. Matsumoto, *Synlett*, 2008, **3**, 367.
8. M.-J. Kim and Y. K. Choi, *J. Org. Chem.*, 1992, **57**, 1605.
9. L. Poppe, L. Novák, M. Kajtár-Peredy and C. Szántay, *Tetrahedron: Asymmetry*, 1993, **4**, 2211.
10. E. Santaniello, S. Casati, P. Ciuffreda and L. Gamberoni, *Tetrahedron: Asymmetry*, 2005, **16**, 1705.
11. P. Virsu, A. Liljeblad, A. Kanerva and L. T. Kanerva, *Tetrahedron: Asymmetry*, 2001, **12**, 2447.
12. O. Parve, I. Reile, J. Parve, S. Kasvandik, M. Kudrjashova, S. Tamp, A. Metsala, L. Villo, T. Pehk, J. Jarvet and L. Vares, *J. Org. Chem.*, 2013, **78**, 12795.
13. I. Vallikivi, Ü. Lille, A. Lookene, A. Metsala, P. Sikk, V. Tõugu, H. Vija, L. Villo and O. Parve, *J. Mol. Catal. B: Enzym.*, 2003, **22**, 279.

Subject Index

Illustrations and figures are in **bold**.

absolute configuration 154–7
acetals 100, 103–6, 188–93, 241–3, 247–51, 257–60, 359–62
acetaminophen *see* paracctamol
acetanilide 80, 472–6
acetic anhydride 129–32, 142–6, **160**, 169–71, 172–4, 472–6, 818–20
acetone 342–4, 345–7, 363–5
acetonides 100
acetophenones 331–4, 335–8, 339–41, 420–3, 803–6
acetyl glycine 403–6
acetylation 80–3, 112–16, 129–32, 142–6, 188–93
acid anhydrides 443–7
acid–base equilibria 914–18
acid–base extraction 298–300
acid–base reactions 1–5, 909
acid–base titrations 64–7
acid–base washing 588–91
acid chlorides 359–62, 853–8
acid hydrolysis 312–14
acid resins 139–41
acids, Pearson's HSAB principle 68–71
acrylamide 849–52
acrylates 408–10
acyl chlorides 133–5, 136–8, 147–8, 165–8, 327–30
acyl halides 261–3
acylanilines 456–61
C-acylation 107–10, 219–22
acylation 169–71, 175–7, 194–7, 359–62, 493–5, 500–3, 504–7, 685–9, 830–3, 931–4
N-acylhydrazones 277–81
4-acylpyrazol-5-ones 367
acylpyrazolones 107–10
N-acylthiazolidines 194–7
Addison–Reedijk degree of trigonality (τ) 652
adenine 92–4
adipic acid derivatives 847–8
adsorption 9–12, 20
aesculetin 376, **377**
β-alanine 370–2, 380–6
alcohols *see also* phenols
  acylation 169–71
  aldehyde synthesis 704–8, 709–12
  alkenes synthesis 600–2
  alkyl chloride synthesis 76–9
  amide synthesis 380–6
  dehydration 608–13
  enantiomer determination 154–7
  ester synthesis 818–20
  ether synthesis 241–3, 312–14
  Friedel–Crafts alkylation 477–9
  Grignard reactions 306–11, 316–18
  ketone synthesis 716–19
  protecting groups 241–3
  synthesis 252–6, 322–6, 327–30, 439–42, 776–80, 784–7, 803–6, 826–9, 868–70
alcoholysis 175–7, 500–3
aldehydes *see also* carbonyls
  alcohol synthesis 868–70
  aldol condensation 272–6, 331–4, 335–8, 339–41, 345–7, 348–51, 352–4, 420–3, 559–63
  aldol reaction 363–5
  alkene synthesis 617–19, 620–2, 623–6, 627–30, 631–4, 635–9, 640–5
  alkylation 120–4
  amide synthesis 380–6, 387–90
  amine synthesis 264–8
  amino acid synthesis 435–8
  α-arylation 112–16
  benzoin synthesis 821–5

aldehydes (*continued*)
benzothiazole synthesis 730–2
benzoxazole synthesis 733–6
carboxylic acids synthesis 370–2, 376–9, 868–70
cyclic acetal synthesis **189**
Diels–Alder reaction 298–300
dihydropyran heterocycles 428–31
dimedone identification 216, **217**
Doebner–Miller reaction 301–5
Grignard reactions 316–18
hydrazone formation 272–6, 277–81, 282–4
imidazole synthesis 269–71
isoindole synthesis 194–7
macrocyclic synthesis 482–4
nitriles synthesis 875–7
oxazolone synthesis 391–4, 403–6
Petasis borono–Mannich reaction 439–42
porphyrins synthesis 485–9, 490–2
synthesis 103–6, 514–16, 554–8, 704–8, 709–12
aldol condensation 272–6, 331–4, 335–8, 339–41, 345–7, 348–51, 352–4, 376–9, 420–3, 559–63, 685–9
aldol reaction 331–4, 335–8, 339–41, 342–4, 345–7, 348–51, 355–8, 359–62, 363–5
alkaloids 26–9, 38–41
alkanes, epoxidation 758–61
alkanoyl halides 517–21
alkenes
Diels–Alder reaction 656–9, 664–7
dihydroxylation 769–72
electrophilic addition 223–5, 226–8, 229–30, 231–3, 234–6, 237–40, 241–3, 244–6, 247–51, 252–6
epoxidation 758–61
Horner–Wadsworth–Emmons reaction 640–5
metathesis 646–9, 650–5
radical addition 701–3
reduction 781–3
synthesis by elimination 600–2, 603–7, 608–13, 614–16
Wittig reactions 617–19, 620–2, 623–6, 627–30, 631–4, 635–9
alkyl bromides 306–11
alkyl cyanides 172
alkyl halides 56–60, 61–3, 64–7, 68–71, 72–5, 76–9, 92–4, 95–8, 117–19, 223–5, 603–7, 635–9, 871–2
alkyl iodides 301–5
alkyl lithium 415–18
alkylation 120–4
N-alkylation 72–5, 92–4, 95–8
O-alkylation 87–91
alkylsilanes 564–7
alkynes 564–7, 588–91, 668–70
allenes 635–9
allyl bromide 87–91, 316–18
allyl zinc 316–18
allylation 125–8
allylic alcohols 766–8, 784–7
allylic alkylation 650–5
alpha-cyano acetamides 387–90
amidation 206–11
amides
hydrolysis 592–4
maleimides synthesis 162–4
nitrile synthesis 172–4
reduction 798–802
synthesis 46–50, 129–32, 133–5, 147–8, 151–3, 159–61, 165–8, 182–6, 206–11, 380–6, 387–90, 443–7
amine protection/deprotection 592–4
amines *see also* anilines
amide synthesis 133–5, 159–61, 165–8, 387–90, 472–6
amino acid synthesis 435–8
aza-Diels–Alder reaction 298–300
Boc-protection 468–71
enaminone synthesis 178–80
ester synthesis 169–71
imidazolin-5-one synthesis 403–6
imine synthesis 277–81
isoindole synthesis 194–7
Knorr pyrrole synthesis 373–5
Petasis borono–Mannich reaction 439–42
protection/deprotection 462–4
recycling of isomers 908–13
synthesis 95–8, 264–8, 811–14, 838–40
amino acids 175–7, 183, 199, 298–300, 327–30, 345–7, 435–8, 733–6, 914–18
aminoalcohols 198–201, 327–30
2-aminobenzamides 84–6
aminocatalysis 112–16
4-aminophenol 129–32

aminotyrosine derivatives 733–6
AMPS (2 acrylamido-2-methylpropane-sulfonic acid) 849–52
anhydrides 125–8, 151–3, 159–61, 162–4, 169–71, 172–4, 212–15, 462–4, 500–3
1,5-anhydro-2-deoxy-1-enitols 247–51
anilines 151–3, 159–61, 206–11, 301–5, 456–61, 462–4, 472–6
anomeric effect 247–51
anthocyanins 10, 19, **20**
anthracenes 617–18
anticholinergic agents 38–41
antioxidant assays 380–6, 559–63
antipyrin 366, **367**
APT-^{13}C NMR (attached proton test) spectroscopy 264–8
arenediazonium salts 448–50
arenes 454–5, 504–7
aroyl chlorides 107–10
aroyl pyrazolones 107–10
aryl bromides 554–8, 881–4
aryl halides 312–14, 451–3, 564–7, 577–80, 584–7
aryl nitros 454–5
aryl sulfonic acids 448–50
arylamines 448–50
α-arylation 112–16
arylboronic acids 411–14, 559–63
N-arylmaleamic acids 159–61
N-arylmaleimides 162–4
arylthiophene-2-carboxaldehydes 554–8
asymmetric synthesis 198–201, 294–7, 345–7, 415–18, 803–6, 909
ate-complexes 117–19
atom economy (AE) 285–9, 317, 388, 436, 726, 929
atropine 38–41
aurones 352–4
aza-Diels–Alder reactions 298–300
azeotropic mixtures 198–201, 257–60
azides 668–70
azo dyes 448–50, 526–31
azoalkenes 671–5

BA (*N*,*N*-methyleno-bis-acrylamide) 849–52
Baeyer–Drewsen process 363–5
Baker–Venkataraman rearrangement 261–3, 359–62
ball mills 290–3
Barbier reaction 316–18
bases, Pearson's HSAB principle 68–71
Baylis–Hillman reaction 408–10
Beckmann rearrangement 76, **77**
Beer–Lambert expression 232
Beirut reaction 842
benzalacetophenones *see* chalcones
benzaldehydes 316–18, 331–4, 335–8, 339–41, 342–4, 348–51, 352–4, 420–3, 720–4, 821–5
benzalphthalide 212–15
benzenes 477–9, 480–1, 527, 568–72
benzil 720–4
benzilic acid rearrangement 815–17
benzisoxazoles 871–4
benzodithioles 120–4
benzodithiolylium tetrafluoroborate 120–4
benzofuran-3(2*H*)-one 352–4
benzofuroxans 841–5
benzoic acid 1–5, 237–40
benzoin condensation 415–18, 720–4, 821–5
benzonitriles 627–30
benzophenone 312–14, 697–700, 776–80
benzopinacolone 697–700
benzopyrans 206–11, 399–402 *see also* chromones
benzoquinone 664–7
benzothiazoles 730–2
benzo[*c*]thiophenes 888–92
benzotriazole 539–41
benzoxazoles 733–6
benzoyl chloride 147–8, 261–3
benzoyl glycine 403–6
benzoylation reactions 147–8
2-benzoylbenzoic acid 198–201
N-benzoylglycine 147–8
o-benzoyloxyacetophenone 261–3
4-benzyl-3-phenylmorpholin-2-ol 439–42
benzyl bromides 568–72
benzyl halides 620–2, 627–30, 631–4
benzylidene acetals 100, 188–93
2-benzylidenebenzofuran- 3-(2*H*)-ones *see* aurones
bicyclic monoterpenes 826–9
bicyclopropyls 660–3
bimolecular elimination 603–7, 609
biocatalysis 914–18, 923–6
biodiesel 202–5, 878
biomimetic synthesis 720–4, 725–9
bioreduction 841–5
bisdimedones 216, **217**
black pepper 26–9

Boc-protection 468–71
bomb calorimetry 903–7
BOP (benzotriazol-1-yl-oxy-tris-(dimethylamino)phosphonium hexafluorophosphate) 209
boranes 252–6, 803–6
boronates 117–19
boronic acids 411–14, 439–42, 554–8, 559–63, 568–72, 573–6, 577–80
bromine addition 223–5, 226–8, 229–30, 451–3, 468–71, 472–6, 493–5, 504–7, 603–7, 694–6
bromoalkanes 694–6
p-bromoaniline 472–6
bromoarenes 504–7
bromobenzenes 568–72
bromobenzofuroxan 841–5
trans-2-bromocyclohexanol 231–3
4-bromomethylbenzoic acid 694–6
2-bromophenol 514–16
3-bromosalicylaldehyde 514–16
N-bromosuccinimide (NBS) 468–71, 694–6
bromotoluene 568–72
bromovanillin 559–63
Brønsted acids 237–40, 247–51, 818–20
Bürgi–Dunitz angle of attack 319–20
2-butanone 306–11
n-butyl acrylate 573–6
butyl chloride 56–60, 64–7

C–H insertion 898–902
caffeic acid 380–6
caffeine 30–4
Cahn–Ingold–Prelog (CIP) system 46–50, 87–91
calorimetry 903–7
camphene 826–9
camphoric acid 838–40
Cannizzaro reaction 868–70
ε-caprolactone 864–7
carbamates 125–8, 175–7, 468–71, 493–5, 595–9
carbamazepine 762–5
carbenes 660–3
carbenium ions 120–4
carbocations 608–13
carbohydrate chemistry 188–93, 247–51
3-carbonylchromomes 359–62
carbonyls *see also* aldehydes; ketones
  1,2-addition 257–60, 261–3, 264–8, 269–71, 272–6, 277–81, 282–4, 366–9, 370–2, 373–5, 376–9, 380–6, 387–90, 391–4, 396–8, 399–402, 403–6, 408–10, 411–14, 415–18
  1,4-addition 285–9, 290–3, 294–7, 298–300, 301–5, 420–3, 424–7
  aldol reaction 331–4, 335–8, 339–41, 342–4, 359–62
  Barbier reaction 316–18
  Grignard addition 306–11, 312–14, 319–21, 322–6, 327–30
  reductions 784–7
  thionation 508–12, 888–92, 893–7
carboxamides 206–11, 443–7
carboxylic acid derivatives 159–61, 162–4, 172–4
carboxylic acids
  amidation 182–6
  amide synthesis 165–8
  amine synthesis 838–40
  amino acid synthesis 435–8
  anhydride synthesis 172–4
  bromine addition 603–7
  carboxamides synthesis 443–7
  Claisen condensation 517–21
  coumarins synthesis 493–5
  Fischer esterification 237–40
  isoindole synthesis 194–7
  ketone synthesis 581–3
  oxazolone synthesis 391–4
  polymerization reactions 847–8
  as reducing agents 784–7
  synthesis 206–11, 370–2, 376–9, 380–6, 399–402, 868–70, 878–80
carnosol 43–5
β-carotene 10, **19**
carotenoids 10, 19
(−)-carvone 6, 244–6
cascade reactions *see* domino reactions
Casiraghi reaction 515
catalysts *see also* enzymes; transition metal catalysts
  acids 84–6, 100, 139, 237–40, 241–3, 244–6, 247–51, 257–60, 352–4, 443–7, 477–9, 480–1, 517–21, 614–16, 725–9
  bases 147–8, 261–3, 408–10, 420–3
  CBS reduction 803–6
  DMAP 169–71
  heterogeneous 522–5

homobenzotetramisole (HBTM) 154–7
organocatalysis 112–16, 120–4, 294–7, 298–300, 345–7, 370–2, 415–18, 758–61, 762–5, 811–14, 898–902
phase-transfer 87–91, 95–8, 301–5, 561, 623–6, 660–3
solid acids 139
vitamins 821–5
catchins **32**
CEC method (competing enantioselective conversion) 154–7
cellulose **32**
ceric ammonium nitrate (CAN) 725–9
ceric ions 713–15
chalcones 229–30, 294–7, 331–4, 335–8, 339–41, 420–3
chavicine **27**
cheletropic reactions 656–9, 660–3
chemiluminescence 504–7, 617–19
chemoselectivity 107–10, 709–12
chiral auxiliaries 125–8
chiral inductors 198–201
chiral resolutions 908–13, 914–18, 919–22, 923–6, 927–30, 931–4
chirality 6–8, 14–17, 46–50, 87–91, 112–16, 125–8, 345–7, 415–18, 577–80
p-chlorobenzoic acid 868–70
p-chlorobenzyl alcohol 868–70
chlorogenic acid **32**
4-[5-(4-chlorophenyl)-1*H*-pyrazol-3-yl] piperidine 355–8
chlorophylls 10, 18–22, 671, **672**
cholesterol 142–6, 754–7
chromanone-2-carboxamide 443–7
chromatography
column 9–12, 35–7, 43–5, 95–8, 142–6, 247–51, 277–81, 335–8, 435–8, 789–93
flash column 112–16, 125–8, 322–6, 380–6, 577–80, 623–6, 794–7
gas chromatography 68–71, 522–5, 610, 758–61, 919–22, 923–6
HPLC 46–50, 51–5, 811–14
reverse-phase methods 51–5
thin-layer 1–5, 18–22, 26–9, 43–5, 46–50, 51–5, 95–8, 129–32, 142–6, 247–51, 335–8, 435–8, 456–61, 690–3, 776–80, 789–93
chromene-3-carboxamide 387–90
chromone-3-carboxylic acid 443–7
chromones 206–11, 261, 359–62, 399–402
α-chymotrypsin 914–18
cinchonidine **295**
cinchonine 294–7
cinnamaldehyde 23–5
cinnamic acid 223–5, 237–40, 603–7
cinnamon 23–5
citral 781–3
civetone 496, **497**
Claisen condensation 108, 206–11, 216–18, 517–21, 835
Claisen rearrangement **816**
Claisen–Schmidt aldol reaction 339–41, 342–4, 352–4, 420–3
Clauson-Kaas pyrrole synthesis 881–4
click chemistry 668–70
coffee 14–15, 30–4
cold finger sublimation 30–4
colorimetry 598
column chromatography 9–12, 35–7, 43–5, 95–8, 142–6, 247–51, 277–81, 335–8, 435–8, 789–93
flash 112–16, 125–8, 206–11, 322–6, 380–6, 577–80, 623–6, 794–7
competing enantioselective conversion (CEC) method 154–7
condensation reactions *see also* cyclocondensation
aldol condensation 272–6, 331–4, 335–8, 339–41, 345–7, 348–51, 352–4, 376–9, 420–3, 559–63, 685–9
amines 103–6, 264–8
benzilic acid rearrangement 821–5
benzoin condensation 415–18, 720–4
Claisen condensation 108, 206–11, 216–18, 517–21, 835
Hantzsch synthesis 432–4
Knoevenagel condensation 370, 376–9, 380–6, 387–90, 399–402, 561, 676–80, 834–7
Pechmann condensation 377, 493–5
Petasis borono–Mannich reaction 439–42
phenol–formaldehyde 482–4
porphyrins synthesis 485–9, 490–2
Verley–Doebner condensation 370–2, 382
conformational analysis 231–3
conjugate addition 294–7, 420–3, 424–7, 640–5, 685–9
π-conjugated systems 399–402

coordination chemistry 885–7
copolymers 847, 849–52, 859–63
copper 550, 564–7, 588–91, 668–70
copper bromide 704–8
Corey–Bakshi–Shibata (CBS) reduction 803–6
correlation spectroscopy (^{1}H–^{1}H COSY) 264–8
CoSalen catalyst 725–9
coumarins 376–9, 493–5, 584–7, 830–3
counterion effects 68–71
coupling agents 182–6, 206–11
Cram's rule 319–21
crosslinkers 849–52
cuprates 424–7
cyalumes 505, 618
cyanides 172
cyanidin-3-glycoside 10
cyanoacetamides 387–90
cyanuric chloride 76–9
cyclative-cleavage 595–9
cyclic sulfites 690–3
cyclization reactions 84–6, 261–3, 282–4, 403–6, 508–12, 517–21, 730–2, 733–6, 773–5, 798–802, 834–7
cycloaddition 664–7, 668–70, 681–4, 690–3
cycloalkanes 903–7
cyclocondensation 198–201, 269–71, 285–9, 290–3, 432–4, 532–4
cyclodehydration reactions 162–4
cyclodehydrogenations 737–40
trans-cyclohexane-1,2-diol 234–6
cyclohexanol 231–3, 234–6
cyclohexene 600–2, 656–9, 769–72, 853–8
N-cyclohexyl-*N*-methyloctanamide 165–8
cis-cyclooctene 750–3
cyclopentadiene 664–7
cyclopentadienone 690–3
cyclopentenones 834–7
cyclopropanation 660–3
cysteine 194–7

DCC (dicyclohexyl carbodiimide) 182–6, 893–7
deadly nightshade 38–41
Dean–Stark apparatus 99–101, 198–201, 257–60, 811–14
decarboxylation 370–2, 443–7, 581–3
dehydration 237–40, 517–21, 600–2, 608–13, 614–16
dehydrogenations 737–40
dehydroiodination 701–3
2-deoxy-glycoside 247–51
dialkylation 477–9
1,3-diamides 133–5
diamines 838–40
1,3-diaryl-2,3-dibromopropan-1-ones 229–30
2,6-diarylpiperidin-4-ones 298–300
1,3-diarylprop-2-en-1-ones *see* chalcones
diastereoisomers 46–50, 223–5, 660–3
diastereoselectivity 194–7, 198–201, 319–21, 898–902
diazenes 526–31
diazo coupling 448–50, 526–31
α-diazoacetamides 898–902
diazonium salts 448–50, 539–41, 668–70, 726
DIBAL (diisobutylaluminium hydride) 794–7
dibenzalacetone 272–6
dibenzylideneacetone 342–4
2,3-dibromo-3-phenylpropionic acid 603–7
α,β-dibromochalcones 229–30
dibromocinnamic acid 603–7
1,3-dicarbonyls 370–2
α-dicarboxylic acids 532–4
4,5-dichloro-1,2-dicyanobenzene 172–4
o-dichloroquinoxalines 532–4
4,5-dicyanobenzenedithiol 542–4
N,*N*′-dicyclohexyl-*N*,*N*′-dimethyl-propan-1,3-diamide 133–5
Diels–Alder reaction 546, 656–9, 664–7, 671–5, 690–3
dienes 656–9, 660–3, 664–7, 690–3
dienophiles 656–9
diesters 133–5, 216–18
diethyl 2,3-*O*-isopropylidene-L-tartrate 99–101
diethyl 3,5-dimethyl-1*H*-pyrrole-2,4-dicarboxylate 373–5
diethyl malonate 376–9
diethylamine 834–7
dihydro-4*H*-pyran 241–3
dihydropyran heterocycles 428–31
dihydropyrazoles 285–9
1,4-dihydropyridines 432–4, 741–4
dihydroxylation 769–72
diimines 838–40
(−)-diisopinocampheylborane 252–6
α-diketones 815–17
1,2-diketones 269–71
1,3-diketones 216–18, 261–3
dilantin 821–5
diludine 741–4
dimedone 216–18
dimers 269–71, 690–3

2,2′-dinitrobiphenyl 550–3
2,4-dinitrophenylhydrazine 272–6
diol protection 99–101
diols 234–6, 769–72
dioxolanes 257–60
diphenyl sulfoxide 789–93
5,5-diphenylhydantoin 821–5
2,3-diphenylindenone 212–15
(*S*)-diphenyl(pyrrolidin-2-yl)methanol 327–30
1,3-dipolar cycloaddition 681–4
dipyrromethanes 671–5
distillation
fractional 6–8, 600–2
fundamentals 1–5, 306–11
microdistillation 811–14
steam 6–8, 23–5, 244–6, 312–14
vacuum 125–8
trans,trans-distyrylbenzene 631–4
diterpenes 43–5
1,3-dithienylbenzo[*c*]thiophene 888–92
1,3-dithiole-2-thiones 807–9
dithiols 542–4
o-dithioquinoxalines 532–4
DMAP (4-(dimethylamino) pyridine) 169–71
DNA 92
Döbner rearrangement 561
(*S*)-1,2-dodecanediol 931–4
Doebner–Miller reaction 301–5
domino reactions 421, 440, 650–5
DPPH (2,2-diphenyl-1-picrylhydrazyl) 380–6
Dragendorff's reagent 38–41
dulcin 592–4
dyeing 363–5, 504–7

E-factor *see* environmental factor
E/*Z* isomers 627–30
E1 elimination 234–6, 603–7, 609
E2 elimination 603–7, 609, 871–4
edaravone 366–9
electrophilic aromatic substitution
amine protection/deprotection 462–4
Boc-protection 468–71
carbonyl thionation 508–12
diazo coupling 448–50, 526–31
formylation 514–16, 526–31
Friedel–Crafts acetylation 496–9
Friedel–Crafts acylation 500–3, 504–7, 522–5
Friedel–Crafts alkylation 477–9, 480–1, 496–9
halogenation 451–3, 468–71, 472–6, 493–5
intramolecular cyclization 517–21
macrocyclic synthesis 482–4, 485–9, 490–2
nitration 456–61, 462–4, 465–7, 496–9
photochromic pyridines 454–5
Vilsmeier–Haack reaction 508–12, 514, 526–31
elimination reactions 600–2, 603–7, 608–13, 614–16, 871–4
enamines 112–16, 120–4, 359–62, 396–8, 432–4
enaminones 178–80, 290–3
enantiomers 6–8, 14–17, 46–50, 125–8, 811–14, 908–13, 914–18, 919–22, 923–6, 927–30, 931–4
enantioselectivity 112–16, 154–7, 294–7, 415–18, 919–22
enol ethers 219–22, 241–3, 247–51
enolate acylation 685–9
enols 112–16, 216–18, 261–3, 363–5, 517–21
enones 216–18, 219–22, 294–7, 335–8, 342–4, 420–3
environmental factor (E-factor) 285–9, 317, 391–4, 616, 929
enzymes 112–13, 777–8, 790, 914–18, 919–22, 923–6, 927–30, 931–4
eosin 504–7
epoxides 234–6, 750–3, 754–5, 758–61, 762–5, 766–8
equilibria 257–60
equilibrium constants 61–3
Erlenmeyer–Plöchl reaction 391–4
essential oils 6–8
esters
alkaline hydrolysis 182–6, 206–11
amide synthesis 387–90
carboxamides synthesis 443–7
coumarins synthesis 493–5
flavone synthesis 261–3
Grignard addition 322–6, 327–30
Grignard reactions 306–11
hydrolysis 826–9
ketone synthesis 257–60
Knorr pyrrole synthesis 373–5
Matteson homologation 117–19
reductions 794–7
synthesis 68–71, 136–8, 139–41, 142–6, 169–71, 175–7, 202–5, 237–40, 376–9, 500–3, 818–20, 830–3, 914–18
Zemplén deprotection 247–51

ethane-1,2-diol 257–60
ethers 80–3, 87–91, 312–14, 477–9, 592–4
ethyl acetate 306–11
ethyl acetoacetate 257–60, 366–9, 373–5, 420–3
ethyl acrylate 408–10
ethyl benzoate 312–14
ethyl cyanoacetate 387–90
ethyl glyoxylate 411–14
ethyl mandelate 411–14
ethylenediamine 264–8
ethylmagnesium bromide 306–11
evaporative ionization mass spectrometry (EIMS) 339–41
extraction
    acid–base 298–300
    liquid–liquid 1–5, 6–8, 30–4, 38–41, 46–50, 68–71, 84–6
    liquid–solid 14–17
    solid adsorption 9–12
    solid–liquid 18–22, 26–9, 30–4, 35–7
    Soxhlet 23–5
    ultrasound 38–41, 43–5
extrusion, gas-phase 690–3

fatty esters 923–6
febrifugine 85
Felkin–Anh model 319–21
Ferrier rearrangement 247–51
ferulic acid 559–63
filtration 1–5, 14–17
Fischer esterification 68, 136, 139, 237–8
flash column chromatography 112–16, 125–8, 206–11, 380–6, 577–80, 623–6, 794–7
flash vacuum pyrolysis (FVP) 690–3
flavones 261–3, 737–40
flavonoids 10, 19, **20** *see also* anthocyanins
fluorescein 504–7
fluorescence 504–7, 617–19, 733–6
formamides 76, 172–4
formylation 514–16, 526–31
formylchromone 399–402
fractional distillation 6–8, 600–2
free radical formation 269–71
Friedel–Crafts acetylation 496–9
Friedel–Crafts acylation 500–3, 504–7, 522–5
Friedel–Crafts alkylation 477–9, 496–9
Fries rearrangement 108, 818–20, 830–3
fructone 257–60
fructones 794–7
fructose 614–16
furan 527, 834–7
Fu's catalysts 169, **170**

Gabriel synthesis 95–8
D-galactopyranose **190**
gallic acid **32**
gas chromatography 68–71, 522–5, 610, 758–61, 919–22, 923–6
gas-phase extrusion 690–3
gas phase reagents 237–40
geraniol 76–9, 169–71, 766–8
Gilman reagent 424–5
α-D-glucopyranoside 189
glucose 614
D-glucuronic acid 282–4
glycals 247–51
glycine 147–8
glycosides 247–51
glyoxal 439–42
glyoxylic acid 439–42
Graebe–Ullmann reaction 539–41
green chemistry
    acid resins 139–41
    aerobic oxidations 704–8, 725–9
    boronic acids 412
    heterogeneous catalysis 522–5
    ionic liquids(ILs) 72–5, 219–22, 285–9
    mechanochemical techniques 290–3, 331–4, 339–41, 550–3
    metrics 285–9
    microwave irradiation (MW) 285–9, 335–8, 348–51, 359–62, 443–7, 581–3
    organocatalysis 428–31, 758–61, 762–5
    principles 9–12, 550–3, 873
    Prins reaction 317
    solvent-free conditions 178–80, 285–9, 331–4, 387–90, 391–4, 490–2, 550–3
    solvent selection 370–2, 784–7
    vitamin catalysts 821–5
Grignard reactions 212–15, 306–11, 312–14, 515, 888–92
grinding methods 290–3, 331–4, 339–41, 550–3

halides 316–18
haloalkanes 694–6
halogenation 451–3
halohydrins 231–3
halosuccinimides 468–71, 694–6

Hammett equation 391–4
Hansch constant 54–5
Hantzsch synthesis 432–4, 741–4
Heck reaction 555, 573–6, 584–7
heme 671, **672**
hemiacetals 709–12
Henry-type condensation 348–51
hetero-Diels–Alder reaction 671–5, 676–80
heterocycles
  carbonylchromones synthesis 359–62
  Claisen condensation 517–21
  coumarins synthesis 493–5
  Hantzsch synthesis 432–4
  imidazolin-5-ones synthesis 403–6
  keto ester synthesis 500–3
  Knorr pyrazole synthesis 366–9
  porphyrins synthesis 485–9
  quinazolinones synthesis 84–6
  synthesis 539–41, 720–4, 741–4
N-heterocyclic carbene (NHC) catalysts 415–18
heterogeneous catalysis 522–5
hexamethylenediamine 847–8
hexamethylmelamine (HMM) 535–8
hexyl caffeate 380–6
N-hexylcarboxamide 380–6
hippuric acid 147–8, 391–4
HMBC (heteronuclear multiple bond correlation) spectroscopy 264–8
HOBt (1-hydroxybenzotriazole) 182–6, 893–7
homobenzotetramisole (HBTM) 154–7
homocuprate 424–5
homologation 117–19
homopolymers 847, 859–60
Horner–Wadsworth–Emmons reaction 640–5, 795
HPLC 46–50, 51–5, 345–7, 811–14
HSQC (heteronuclear single quantum coherence) spectroscopy 264–8
Hughes–Ingold mechanism 92–4
Huisgen cycloaddition 668–70
hydantoins 595–9
hydrazides 277–81, 282–4
hydrazines 272–6, 285–9, 290–3, 355–8, 366–9, 671–5
hydrazones 272–6, 277–81, 282–4, 355–8, 671–5
hydroboration–oxidation 252–6
hydrogen peroxide 758–61, 762–5, 769–72
hydrogenation 87–91, 781–3
hydrolysis 64–7, 312–14, 376, 592–4, 826–9
hydrosilylation reactions 811–14
α-hydroxy ketones 415–18
o-hydroxyacetophenone 261–3, 359–60
2-hydroxyacetophenone 399–402
4-hydroxyacetophenone 818–20
α-hydroxyacids 815–17
hydroxyaldehydes 709–12
hydroxybenzaldehydes 376–9
hydroxychalcones 737–40
hydroxycinnamic acid derivatives 380–6, 559–63
7-hydroxycoumarin 830–3
α-hydroxyesters 411–14
hydroxylamines 875–7
hydroxylation 725–9
5-hydroxymethylfurfural (HMF) 614–16
hygrine **39**
hyoscyamine 38–41

ibuprofen 46–50
imidazoles 269–71, 720–4
imidazolin-5-ones 403–6
imides 95–8, 162–4, 172–4
imines 194–7, 264–8, 272–6, 277–81, 581–3, 733–6, 745–9, 811–14, 908–13
iminium ions 439–42
2-imino-chromene 387–90
immobilization 716–19, 766–8
indanones 212–15
indenones 212–15
indigo 363–5
indoloquinolines 517–21
indoxyls 363–5
inert atmosphere techniques 117–19, 424–7, 542–4, 564–7, 646–9, 794–7, 859–63
infra-red spectroscopy 23–5, 30–4, 229–30, 231–3, 399–402, 610
interfacial polymerization 847–8, 853–8
intramolecular cyclization 282–4, 508–12, 517–21, 730–2
iodination 451–3
1-iodo-2-nitrobenzene 550–3
iodoalkanes 701–3
6-iodoquinaldine 301–5
iodosulfonylation 701–3
ionic liquids (ILs) 72–5, 219–22, 285–9, 927–30
isoamyl acetate 139–41
isoamyl nitrite 745–9

isoborneol 826–9
isobutylene 237–40
isochavicine **27**
isocyanates 592–4, 595–9
isocyanides 435–8, 443–7
isoindoles 194–7
isomannide 87–91
isomerization 27, 627–30, 631–4, 794–7, 818–20, 908–13
isoniazid 277–81
isonitriles 76, **77**
(−)-isopinocampheol 252–6
isopiperine **27**
isopropanol 697–700
isopropylidene 189, **190**
isoxazolines 681–4

Kaiser test 598
Kemp elimination 871–4
ketals 99–101, 241–3, 322–6
ketenes 635–9
ketimines 811–14
keto acids 198–201
γ-keto-amides 182–6
β-keto esters 420–3, 432–4
γ-keto esters 500–3
ketoacids 878–80
ketones *see also* carbonyls
- acylation 107–10
- aldol condensation 272–6, 331–4, 335–8, 339–41, 345–7, 420–3
- alkene synthesis 617–19, 635–9
- benzilic acid rearrangement 815–17, 821–5
- conjugate addition 424–7
- cyclic acetal synthesis **189**
- cyclocondensation 285–9
- Diels–Alder reaction 298–300, 664–7
- dihydropyran heterocycles 428–31
- Grignard reactions 306–11
- Knorr pyrrole synthesis 373–5
- Michael addition 294–7
- pyrazoles synthesis 355–8
- reduction 776–80, 784–7, 803–6, 811–14
- squaraine dye synthesis 396–8
- synthesis 178–80, 206–11, 212–15, 244–6, 257–60, 261–3, 496–9, 522–5, 581–3, 697–700, 716–19, 737–40
- thionation 888–92

kinetic resolution 914–18, 919–22, 923–6, 927–30, 931–4
kinetics 64–7, 603–7
Knoevenagel condensation 370, 376–9, 380–6, 387–90, 399–402, 561, 676–80, 834–7
Knorr pyrazole synthesis 356, 366–9
Knorr pyrrole synthesis 373–5
Kostanecki–Robinson synthesis 261
Ktetradecanediol 709–12

lactams 198–201, 415–18, 898–902
lactones 864–7
Lamiaceae family 43–5
latentiation 151–2
Lawesson's reagent 508–12, 888–92, 893–7
Le Châtelier's principle 62, 923–6
leaves, pigment extraction 9–12, 18–22
levulinic acid 878–80
Lewis acids 403–6, 412, 443–7, 477–9, 480–1, 543, 830–3
Lewis bases 408–10, 415–18
lichen 14–17
(+)-limonene 6–8, 244–6
lipases 919–22, 923–6, 927–30, 931–4
lipophilicity 51–5, 380–6, 559–63
liquid–liquid extraction 1–5, 6–8, 30–4, 38–41, 46–50, 68–71, 84–6
liquid–solid extraction 14–17
lithiation 508–12, 640–5
lithium reagents 424–5
littorine **39**
lophine 269–71
low temperature techniques 117–19, 424–7
lupanine 35–7
lutein 10, **19**

macrocyclic synthesis 482–4, 485–9, 490–2, 798–802
Makosza's method 660–3
maleamic acid 159–61
maleic anhydride 159–61
maleimides 162–4
malonamide 133–5
malonate 650–5
malonic acid 370–2
malononitrile 428–31
Markovnikov's rule 604
mass spectrometry, evaporative ionization (EIMS) 339–41

Matteson homologation 117–19
mechanochemical techniques 290–3, 331–4, 339–41
Meisenheimer complexes 542
Meldrum's acid 380–6, 650–5, 834–7
melting points 223–5
menthone 716–19
metal carboxylates 68–71
metal complexation 807–9
metal–halogen exchange 568–72
metallocarbenes 898–902
metalloporphyrin catalysts 758–61, 762–5, 885–7
metathesis 646–9, 650–5
3-methoxalylchromone 359–62
trans-4-methoxycinnamic acid 370–2
4-methoxymethylbenzoic acid 694–6
2-methoxytetrahydropyran 241–3
3-methyl-1-phenyl-pyrazol-5-one 107–10
methyl methacrylate 859–63
methyl orange 448–50
methyl triphenylmethyl ether 312–14
C-methyl[4]resorcinarene 482–4
methylcyclohexanols 608–13
methylxanthines 31 *see also* caffeine; theobromine
Michael addition 178–80, 216–18, 294–7, 355–8, **377**
Michaelis–Menten steady state 920
microdistillation 811–14
microencapsulation 766–8
microscale chemistry 202–5
microwave irradiation (MW) 285–9, 335–8, 348–51, 359–62, 443–7, 517–21, 535–8, 559–63, 581–3, 595–9
migration 815–17
mixtures, separation 1–5
Mizoroki–Heck reaction 573
molecular modelling 903–7
molecular orbital theory 664–7
molecular sieves 103–6
monosaccharides 614–16
monoterpenes 826–9
Morita–Baylis–Hillman reaction 408–10
MTO (methyltrioxorhenium(VII)) 750–3
multicomponent reactions 428–31, 432–4, 435–8, 439–42, 443–7
muscone 496, **497**
musk ketone 496–9
naphthalene 1–5, 465–7, 713–15
1,4-naphthoquinones 713–15
Nazarov cyclization 835
Negishi reaction 555
neoxanthin **19**
nifedipine 432–4
NIR dyes 396–8
nitration 456–61, 462–4, 465–7, 496–9
nitrile oxides 681–4
nitriles 172–4, 387–90, 428–31, 545–9, 581–3, 627–30, 685–9, 871–4, 875–7
nitrites 745–9
2-nitro-4-methylaniline 462–4
nitroalkanes 348–51
nitroanilines 456–61, 668–70
nitroarenes 496–9
2-nitrobenzaldehyde 363–5
4-nitrobenzaldehyde 408–10, 704–8
4-nitrobenzoyl chloride 136–8
4-nitrobenzyl alcohol 704–8
nitrogen heterocycles 298–300
nitronaphthalene 465–7
6-nitrosaccharin 773–5
nitrosation 373–5
nitrosos 244–6
nitrostilbenes 623–6
nitrostyrene 348–51
NMR spectroscopy 23–5, 30–4, 35–7, 43–5, 72–5, 229–30, 264–8, 610–12
non-steroidal anti-inflammatory drugs (NSAIDs) 46–50
NSAIDs (non-steroidal anti-inflammatory drugs) 46–50
nucleophilic acyl substitution 818–20
nucleophilic addition–elimination 129–32, 133–5, 151–3, 165–8, 206–11, 212–15, 237–40, 396–8 *see also* Grignard reactions
nucleophilic aromatic substitution 532–4, 535–8, 539–41, 542–4, 545–9
nucleophilic substitution 92–4, 95–8, 137, 159–61, 175–7, 182–6, 694–6
nucleophilic unimolecular substitution *see* S_N1 reactions
nucleophilicity 380–6
nylon 847–8, 853–8

octanoic acid 165–8
n-octylamine 95–8
OH-protection 107–10

olefins 223–5, 252–6, 573–6, 608–13, 646–9, 650–5, 750–3, 769–72, 826–9
oligothiophenes 508–12
one-pot reactions 428–31, 595–9, 650–5, 701–3
optical rotation 6–8, 14–17, 194–7, 244–6, 294–7
orange oil 6–8
organoboranes 252–6
organocatalysis 112–16, 120–4, 294–7, 298–300, 345–7, 370–2, 428–31, 758–61, 811–14, 898–902
organocuprates 424–7
organolithium reagents 117–19
organomagnesium compounds 322–6
see also Grignard reactions
organometallics 306–11, 312–14, 411–14, 646–9
organosilicon reagents 564–7
ornithine **39**
orthogonality 175–7
oxalate diesters 617–19
oxalic acid 237–40, 244–6
oxazaborolidines 803–6
oxazolidinones 125–8
oxazolones 391–4, 403–6
oxidations
aldehyde synthesis 704–8, 709–12
benzaldehyde 720–4
cyclization reactions 730–2, 733–6, 773–5
dehydrogenations 737–40
epoxidation 750–3, 754–7, 758–61, 762–5, 766–8, 769–72
Hantzsch synthesis 741–4
ketone synthesis 716–19
phenol synthesis 725–9
quinone synthesis 713–15
tetrazine synthesis 745–9
oxidative cleavage 853–8
N-oxide bioreduction 841–5
oximes 244–6, 373–5, 681–4, 875–7
oxo-molybdenum complexes 789–93
oxonitriles 685–9
oxycarbenium ion 247–51
ozonolysis 685–9

Paal-Knorr pyrrole synthesis 882, 894
palladium 87–91, 554–8, 559–63, 564–7, 568–72, 573–6, 577–80, 581–3, 584–7, 588–91, 650–5, 781–3
paracetamol 80–3, 129–32, 592–4
partition coefficients 51–5, 380–6
Pauson–Khand reaction 835
Pearson's HSAB principle 68–71
Pechmann condensation 377, 493–5
PEG-400 (poly(ethylene glycol)-400) 370–2, 784–7
pentyl chloride 61–3
pericyclic reactions
click chemistry 668–70
cyclopropanation 660–3
Diels–Alder reaction 656–9, 664–7, 671–5, 676–80, 690–3
1,3-dipolar cycloaddition 681–4
oxonitriles 685–9
Perkin reaction 376, **377**
peroxides 694–6
peroxyacids 754–7
peroxyoxalates 617–19
Petasis borono–Mannich reaction 439–42
phase-transfer catalysis 87–91, 95–8, 301–5, 561, 623–6, 660–3
phenacetin 80–3, 592–4
phenol–formaldehyde condensation 482–4
phenolic acids 380–6
phenols
amide synthesis 387–90
amine synthesis 264–8
coumarins synthesis 376–9, 493–5, 830–3
ester synthesis 261–3
formylation 514–16
Friedel–Crafts acylation 504–7
Kemp elimination 871–4
macrocyclic synthesis 482–4
synthesis 725–9
phenyl acetate 818–20
phenyl halides 415–18
phenylacetic acid 212–15
phenylalanine methyl ester 914–18
(*S*)-phenylalaninol 198–201
phenylboronic acid 411–14, 439–42, 559–63
2-phenylchromen-4-one 261–3
o-phenylenediamines 532–4
1-phenylethanol 76–9, 919–22
5-phenylferulic acid 559–63
phenylglycine 439–42
phenylhydrazines 366–9
phenylmagnesium bromide 212–15, 312–14, 322–6

N-phenylmaleimide 656–9
2-phenyloxazol-5(4*H*)-ones 391–4
2-phenylpropanal 319–21
phenylsilane 789–93
5-phenylvanillin 559–63
phosphines 620–2
phosphonium salts 95–8, 617–19, 620–2, 631–4
phosphoranes 631–2, 635–9
photochemical cycloaddition 664–7
photochromism 269–71, 454–5
photodynamic therapy (PDT) 301–5
photoreactions 697–700
phthalic anhydrides **160**, 172–4, 212–15, 504–7
physical properties 1–5
piezochromism 269–71
pinacol rearrangement 697–700, 815, **816**
(+)-α-pinene 252–6
piperettine **27**
piperidin-4-ones 298–300
piperidines **27**, 355–8, 468–71
piperine 26–9
plant oils 202–5
plant pigments 9–12, 18–22
polarimetry 15, 16
polarity inversion 415–18
polycaprolactone 864–7
polymerization reactions 847–8, 849–52, 853–8, 859–63, 864–7
polyphosphoric acid 517–21
porphyrins 18, 485–9, 490–2, 671, 762–5, 885–7 *see also* chlorophylls
potato starch 9–12
Prins reaction 317
privileged structures 207
prodrugs 151–3
(*S*)-proline 327–30
L-proline 298–300, 345–7
propyl acetate 306–11
protecting groups 175–7, 188–93, 241–3, 327–30, 468–71, 472–6, 595–9
protection/deprotection 564–7
psoralen 376, **377**
purification *see also* extraction
  acid–base washing 588–91
  column chromatography 9–12, 35–7, 43–5, 142–6
  distillation 6–8, 23–5, 59, 306–11
  recrystallization 1–5, 16, 26–9, 30–4, 43–5
  sublimation 30–4
putrescine **39**
PyBOP (benzotriazol- 1-yl-oxy-tris-pyrrolidino-phosphonium hexafluorophosphate) 206–11, 380–6
pyran derivatives 677
pyrano[2,3-*d*]pyrimidine 676–80
pyrans 428–31
pyrazole 72–5
pyrazoles 290–3, 355–8, 366–9, 468–71, 545–9, 750–3
pyrazolones 107–10, 366–9
pyridazines 546
pyridines 226–8, 261–3, 370–1, 380–6, 432–4, 454–5, 545–9, 741–4, 885–7
pyridinium chlorochromate 716–19
pyrocarbonates 468–71
pyrroles 373–5, 485–9, 490–2, 526–31, 671–5, 881–4, 893–7
pyrrolidin-2-ones 640–5

quantum yield 698
quinazolinones 84–6
quinidine **295**
quinine **295**
quinolines 301–5
quinolizidine alkaloids (QAs) 35–7
quinones 713–15, 725–9
2,3-quinoxalinedithiol 532–4
quinoxalines 532–4, 720–4

racemic mixtures 46–50, 826–9, 908–13, 914–18, 919–22, 923–6, 927–30, 931–4
radical polymerization 849–52, 859–63
radical reactions 694–6, 697–700, 701–3
reaction mass efficiency (RME) 285–9
reaction optimization 61–3
rearrangements
  benzilic acid rearrangement 815–17
  cyclization reactions 834–7
  dilantin synthesis 821–5
  Fries rearrangement 108, 818–20, 830–3
  Schmidt reaction 838–40
  Wagner–Meerwein rearrangement 826–9
recrystallization 1–5, 14–17, 26–9, 30–4, 43–5, 129–32
redox-neutral reactions 103–6

reduction/oxidation 363–5
reductions
alkenes 781–3
esters 794–7
ketones 776–80, 784–7, 803–6, 811–14
macrocyclic synthesis 798–802
sulfoxides 789–93
thioesters 807–9
reflux 27, 306–11
refraction index 6–8, 244–6, 878–80
regioselectivity
Boc-protection 468–71
cyclic acetals 188–93
formylation 514–16
Heck reaction 573–6, 584–7
kinetic studies 603–4
Knorr pyrazole synthesis 367
nitration 456–61
oxidations 766–8
radical reactions 701–3
reductions 781–3, 784–7
S_N2 reactions 92–4
resorcinol 482–4, 504–7
rhodium based catalysts 411–14, 898–902
ring-closing metathesis (RCM) 646–9, 650–5
ring-opening polymerization 864–7
RNA 92, 676–7
Robinson annulation 216–18, 420–3
ruthenium 646–9, 650–5, 864–7
Ruthenium catalysts 412

saccharides 878–80
saccharin 773–5
sage 43–5
H_2salen 264–8
salens 838–40
salicylaldehyde 103–6, 264–8, 376–9, 387–90, 514–16, 838–40
salicylonitrile 871–4
salt, kitchen 9–12
Salvia officinalis 43–5
saponification 182–6
Schiff bases 264–8
Schlenk equilibrium 323
Schlenk techniques 566, 580, 588–91
Schlosser modification 628
Schmidt reaction 838–40
Schotten–Baumann conditions 147–8, 165–8
scopolamine **39**
separation *see also* chromatography
fundamentals 1–5
racemic mixtures 46–50
slow addition 306–11
S_N1 reactions 56–60, 61–3, 64–7, 68–71, 120–4, 312–14
S_N2 reactions 57–8, 64–5, 68–71, 76–9, 80–3, 87–91, 92–4, 95–8, 234–6, 620–2, 631–4, 635–9
sodium borohydride 776–80, 784–7
sodium hypochlorite 709–12
Solanaceae family 38–41
solid–liquid extraction 18–22, 26–9, 30–4, 35–7
solid-phase synthesis 595–9
Sonogashira cross-coupling 564–7, 588–91
Soxhlet extraction 23–5
sp carbon addition 592–4, 595–9
sp^2 carbon addition (non-aryl)
1,2-addition to carbonyl compounds 257–60, 261–3, 264–8, 269–71, 272–6, 277–81, 282–4, 366–9, 370–2, 373–5, 376–9, 380–6, 387–90, 391–4, 396–8, 399–402, 403–6, 408–10, 411–14, 415–18
1,4-addition to α,β-unsaturated carbonyl compounds 285–9, 290–3, 294–7, 298–300, 301–5, 420–3, 424–7
aldol reaction 331–4, 335–8, 339–41, 342–4, 345–7, 348–51, 352–4, 355–8, 359–62, 363–5
alkene electrophilic additions 223–5, 226–8, 229–30, 231–3, 234–6, 237–40, 241–3, 244–6, 247–51, 252–6
Barbier reaction 316–18
Grignard addition to carbonyl compounds 306–11, 312–14, 319–21, 322–6, 327–30
multicomponent reactions 428–31, 432–4, 435–8, 439–42, 443–7
sp^2 carbon modification (non-aryl)
acylation 147–8, 151–3, 169–71, 194–7
amino acid derivatives 175–7
anhydride aminolysis 159–61
competing enantioselective conversion (CEC) method 154–7
coupling agents 182–6, 206–11
cyanides 172–4
cyclic acetals 188–93
cyclocondensation 198–201
cyclodehydration reactions 162–4

esterification 136–8, 139–41, 142–6, 202–5
Michael addition 178–80, 216–18
nitriles 172–4
nucleophilic addition–elimination 129–32, 133–5, 165–8, 206–11, 212–15, 396–8
sp^3 carbon modification
C-acylation 107–10
N-alkylation 72–5
allylation 125–8
α-arylation 112–16
condensation reactions 103–6
counterion effects 68–71
cyclisation reactions 84–6
homologation 117–19
S_N1 reactions 56–60, 61–3, 64–7, 120–4
S_N2 reactions 76–9, 80–2, 87–91, 92–4, 95–8
specific rotation 194–7, 931–4
spectroscopy
infra-red 23–5, 30–4, 229–30, 399–402, 610
NMR 23–5, 30–4, 35–7, 43–5, 72–5, 229–30, 231–3, 264–8, 610–12
UV-vis 212–15, 399–402, 885–7
spinach 18–22
spiro-imidazoles 720–4
spirooxindole 681–4
squaraine dyes 396–8
squarylium cyanine dye 301–5
Staudinger reaction 96
steam distillation 6–8, 23–5, 244–6, 312–14
Stenhouse adducts 834–7
step growth polymerization 853–8
stereochemistry assignment 277–81
stereoisomers 6–8, 194–7
stereospecificity 226–8, 247–51, 623–6, 754–7, 778
steroids 754–5
trans-stilbene 226–8
Stille coupling 578
styrene 701–3, 859–63
styryls 584–7
sublimation 30–4, 252–6, 826–9
substitutions
S_N1 reactions 56–60, 61–3, 68–71
S_N2 reactions 57–8, 64–5, 68–71, 76–9, 80–3, 87–91, 92–4, 95–8
succinic anhydride **160**, 500–3
sucrose 878–80
sugars
dehydration 614–16
reducing 282–4
regioselectivity 188–93
surfactants 247–51
uracil derivatives 676–80
sulfathiazole 151–3
sulfides 294–7, 789–93
sulfinates 701–3
3-sulfolene 656–9
sulfonamides 51–5, 151–3, 773–5
sulfones 701–3
sulfonyl azides 898–902
sulfonyl chlorides 773–5, 898–902
sulfonyl iodides 701–3
sulfoxides 789–93
superabsorbent polymers 849–52
sustainable chemistry *see* green chemistry
Suzuki cross-coupling 568–72, 577–80, 584–7
Suzuki–Miyaura cross-coupling 554–8, 559–63
swelling measurements 852
Swern oxidation 76, 77

TADDOL 100, 322–6
tannins **32**
tautomerism 269–71, 366–9, 443–7, 454–5, 532–4, 919–22, 927–30
TBTU 46–50
tea 14, 30–4
temperature regulation 117–19
TEMPO 704–8, 709–12
terpenes 244–6
tert-butanol 237–40
tetradentate ligands 264–8
1,2,3,4-tetrahydroisoquinoline 103–6
tetrahydropyranyl ether 241–3
tetraphenyl porphyrin (TPP) 490–2
tetrazines 545–9, 745–9
TFAA (trifluoroacetic anhydride) 219–22
theobromine 31, 32
theophylline 31
thermochemistry 903–7
thermochromism 269–71
thia-tetraaza macrocycle 798–802
thiazolidine 194–7
thienylpyrrole derivatives 526–31

thin-layer chromatography 1–5, 18–22, 26–9, 37, 43–5, 46–50, 51–5, 95–8, 129–32, 142–6, 247–51, 335–8, 435–8, 456–61, 690–3, 776–80, 789–93
thioesters 807–9, 888–92
thioethers 542–4
thiols 194–7, 294–7, 542–4, 730–2
thiones 807–9
thionyl chloride 165–8, 172–4, 175–7
thiophenes 182–6, 435–8, 500–3, 508–12, 522–5, 526–31, 554–8, 640–5, 730–2, 733–6, 893–7
thiophenol 294–7
tin reduction 725–9
toluene derivatives 568–72
p-toluenesulfonic acid (*p*-TsOH/pTSA) 241–3, 257–60, 290–3, 769–72
p-toluenesulfonyl iodide 701–3
p-toluidine 1–5
p-tolylboronic acid 573–6
tosylhydrazide 282–4
β-tosylstyrene 701–3
TPHB (triphenylphosphane hydrobromide) 247–51
transesterification 202–5
transfer hydrogenation 781–3
transition metal catalysts
  alkene hydrogenation 781–3
  compared to other catalysts 112–13, 550
  epoxidation 750–3, 766–8
  Heck reaction 573–6, 584–7
  ketone synthesis 581–3
  metathesis 646–9, 650–5
  oxidations 704–8
  phenylation 411–14
  polymerization reactions 864–7
  reductions 789–93
  Sonogashira cross-coupling 564–7, 588–91
  Suzuki cross-coupling 568–72, 577–80, 584–7
  Suzuki–Miyaura cross-coupling 554–8, 559–63
  Ullmann coupling 550–3
transketalisation 99–101
trialkylboranes 252–6
trialkylborates 803–6
triazines 535–8
triazoles 539–41, 668–70
triazolium salts 415–18
triethylamine 370–2
trifluoroacetic anhydride (TFAA) 219–22
trifluoromethanesulfonic acid 818–20
triglycerides 202–5
trigonality 652
trimethylsilyl group 564–7
triphenylmethanol 312–14
triphenylphosphine 789–93
tropine **39**
tropinone **39**
tyrosine 175–7

Ugi reaction 435–8
Ullmann coupling 550–3
ultrasound extraction 38–41, 43–5, 838–40
ultrasound methodologies 539–41
umbelliferone 830–3
umpolung 415–18
unimolecular elimination 600–2, 603–7, 609
unitary operations, fundamentals 1–5
uracil derivatives 676–80
uranine 504
ureas 428–31, 592–4, 595–9, 821–5
usnic acid 14–17
UV-vis spectroscopy 212–15, 399–402, 885–7

vacuum distillation 125–8
vacuum filtration 14–17
vacuum fractional distillation 6–8
vanillin 451–3, 559–63
Vedejs's catalysts 169, **170**
veratronitrile 875–7
Verley–Doebner condensation 370–2, 382
vicinal diols 709–12
Vilsmeier–Haack reaction 399–402, 508–12, 514, 526–31
4-vinylbenzoic acid 620–2
vinylic halides 588–91
violaxanthin **19**
viscosity 202–5
Vitali–Morin reaction 38–41
vitamin catalysts 821–5
$VO(acac)_2$ 766–8

Wagner–Meerwein rearrangement 826–9
warfarin 376, **377**
Williamson ether synthesis 80–3, 592–4
Wittig reactions 617–19, 620–2, 623–6, 627–30, 631–4, 635–9, 681–4, 794–7
Wurtz-type coupling 323

xanthenes 504–7
xanthophylls 12, 19
m-xylene 496–9

ylides 617–19, 620–2, 623–6, 627–30, 631–4, 635–9, 640–5

Zaitsev's rule 608–13
Zemplén deprotection 247–51
zeolites 522–5
zinc, Barbier reaction 316–18
zinc tetraphenyl porphyrin 490–2